Cornelia Craft

D1611312

# *Clinical Avian Medicine and Surgery*

## including aviculture

GREG J. HARRISON, DVM

All Animal Clinic and Bird Hospital, Lake Worth, Florida;
Adjunct Professor, Department of Biological Sciences,
Florida Atlantic University, Boca Raton, Florida;
President, Research Institute for Avian Medicine, Nutrition,
and Reproduction, Inc., Lake Worth, Florida

LINDA R. HARRISON, BS

Education Coordinator and Newsletter Editor,
Association of Avian Veterinarians,
Lake Worth, Florida;
Secretary/Treasurer and Seminar Coordinator,
Research Institute for Avian Medicine, Nutrition,
and Reproduction, Inc., Lake Worth, Florida

W. B. SAUNDERS COMPANY

*Philadelphia, London, Toronto, Mexico City, Rio de Janeiro, Sydney, Tokyo, Hong Kong*

W. B. Saunders Company: West Washington Square
Philadelphia, PA 19105

**Library of Congress Cataloging-in-Publication Data**

Harrison, Greg J.

Clinical avian medicine and surgery.

1. Cage-birds—Diseases. 2. Cage-birds—Surgery. I. Harrison, Linda. II. Title. [DNLM: 1. Bird Diseases. 2. Birds—surgery. SF 994.2.A1 H319c]

SF994.2.A1H37 1986 636.5′0896 85–31759

ISBN 0–7216–1241–5

Thanks to Tarzan, model for cover drawing and Amazon Parrot-in-residence at *My Best Friend's Pet Shop*, West Lafayette, Indiana. Drawing by Algernon R. Allen.

Acquisition Editors: Ray Kersey, Robert Reinhardt, and Darlene Pedersen
Manuscript Editor: Donna Walker
Designer: Lorraine Kilmer
Production Manager: Frank Polizzano
Illustration Coordinator: Kenneth Green
Page Layout Artist: Joan Sinclair
Indexer: Nancy Guenther

Clinical Avian Medicine and Surgery ISBN 0-7216-1241-5

© 1986 by W. B. Saunders Company. Copyright under the Uniform Copyright Convention. Simultaneously published in Canada. All rights reserved. This book is protected by copyright. No part of it may be reproduced, stored in a retrieval system, or transmitted in any form or by any means, electronic, mechanical, photocopying, recording or otherwise, without written permission from the publisher. Made in the United States of America. Press of W. B. Saunders Company. Library of Congress catalog card number 85-31759.

Last digit is the print number: 9 8 7 6 5 4 3

# Contributors

H. JOHN BARNES, DVM, PhD

Professor of Avian Medicine and Head, Avian Medicine Section, Department of Food Animal and Equine Medicine, School of Veterinary Medicine, North Carolina State University, Raleigh, North Carolina.

*Parasites*

TERRY W. CAMPBELL, MS, DVM

Instructor and Resident in Clinical Pathology, College of Veterinary Medicine, Kansas State University, Manhattan, Kansas.

*Clinical Chemistries; Cytology; Mycotic Diseases; Neoplasia*

DANIEL P. CAREY, DVM

Manager, Veterinary Services Laboratories, Purina Mills, Inc., St. Louis; Bird Clinic, St. Charles, Missouri.

*Zoonotic Diseases*

CARL H. CLARK, DVM, PhD

Professor, Department of Physiology and Pharmacology, Auburn University, Auburn, Alabama.

*Pharmacology of Antibiotics*

SUSAN L. CLUBB, DVM

Staff Veterinarian, Pet Farm, Inc., Miami, Florida

*Therapeutics; Sex Determination Techniques; Diseases of Imported Birds (App. 2)*

CHRISTINE DAVIS, BS

Veterinary guest lecturer and writer for avian publications.

*Captive Behavior and Its Modification*

F. JOSHUA DEIN, VMD, MS

Department of Pathology, National Zoological Park, Smithsonian Institution, Washington, DC.

*Hematology*

JAMES E. DOYLE, MD, Diplomate, American College of General Surgeons and American College of Plastic Surgeons

Clinical Assistant Professor of Surgery, University of Texas Medical School, Galveston, Texas. Private Practitioner, Maxillary and Cosmetic Surgery, Arlington, Texas.

*Introduction to Microsurgery*

LYNNE A. DREWES, MT (ASCP)

Chief Microbiologist, Aviculture Institute, Newhall, California.

*Clinical Microbiology*

RICHARD H. EVANS, DVM, MS

Veterinary Services Department, Purina Mills, Inc., St. Louis, Missouri.

*Zoonotic Diseases*

KEVEN FLAMMER, DVM

Assistant Professor, Department of Companion Animal and Special Species Medicine, and Staff, Exotic and Wild Bird Medicine Service, School of Veterinary Medicine, North Carolina State University, Raleigh, North Carolina.

*Choosing A Bird; Clinical Microbiology; Aviculture Management; Pediatric Medicine; Survey Weights of Common Aviculture Birds (App. 4); Average Breeding Characteristics of Some Common Psittacine Species (App. 5); Sample Weight Gains of Selected Hand-Raised Psittacines (App. 6)*

ALAN M. FUDGE, DVM

Staff Veterinarian, Avian Medical Center of Sacramento and California Avian Laboratory, Citrus Heights, California.

*Preliminary Evaluation of a Case; Differential Diagnoses Based on Clinical Signs; Diagnoses to Consider from Hematologic Results (App. 6)*

HELGA GERLACH, DR MED VET HABIL

Faculty Member, Institute for Avian Diseases, Ludwig-Maximilians University, Munich, West Germany.

*Virology; Viral Diseases; Bacterial Diseases; Mollicutes; Chlamydia*

JAMES E. GRIMES, PhD

Associate Professor of Veterinary Microbiology and Parasitology, College of Veterinary Medicine, Texas A & M University, College Station, Texas.

*Serology*

GREG J. HARRISON, DVM

Adjunct Professor, Department of Biological Sciences, Florida Atlantic University, Boca Raton; Director, All Animal Clinical and Bird Hospital, Lake Worth; President, Research Institute for Avian Medicine, Nutrition, and Reproduction, Inc., Lake Worth, Florida.

*Choosing A Bird; Husbandry Practices; Captive Behavior and Its Modification; Clinical Anatomy; Management Procedures; Preliminary Evaluation of a Case; Differential Diagnoses Based on Clinical Signs; Clinical Chemistries; Endoscopy; Biopsy Techniques; Miscellaneous Diagnostic Tests; What To Do Until A Diagnosis Is Made; Symptomatic Therapy and Emergency Medicine; Nutritional Diseases; Toxicology; Disorders of the Integument; Miscellaneous Diseases; Evaluation and Support of the Surgical Patient; Anesthesiology; Surgical Instrumentation and Special Techniques; Selected Surgical Procedures; Reproductive Medicine*

LINDA R. HARRISON, BS

Education Coordinator and Newsletter Editor, Association of Avian Veterinarians, Lake Worth; Secretary/Treasurer and Seminar Coordinator, Research Institute for Avian Medicine, Nutrition, and Reproduction, Inc., Lake Worth, Florida.

*Choosing A Bird; Management Procedures; Preliminary Evaluation of a Case; Nutritional Diseases*

LORRAINE G. KARPINSKI, VMD, Diplomate, American College of Veterinary Ophthalmologists

Adjunct Instructor, University of Miami School of Medicine; Veterinary Ophthalmologist, South Dade Animal Hospital and Bascom Palmer Eye Institute, Miami, Florida.

*Differential Diagnoses Based on Clinical Signs; Ophthalmology; Symptomatic Therapy and Emergency Medicine*

GEORGE V. KOLLIAS, JR., DVM, PhD, Diplomate, American College of Zoological Medicine

Associate Professor, Department of Special Clinical Sciences, College of Veterinary Medicine, University of Florida, Gainesville; Attending Staff Clinician, Wildlife and Zoological Medicine, Veterinary Medical Teaching Hospital, University of Florida, Gainesville, Florida.

*Biopsy Techniques; Relationships of Avian Immune Structure and Function to Infectious Diseases*

ALBERT H. LEWANDOWSKI, DVM

Adjunct Professor of Small Animal Medicine and Pathology, College of Veterinary Medicine, Michigan State University; Zoological Veterinarian, Chief of Staff, Veterinary Services, Detroit Zoological Park, Royal Oak, and Belle Isle Zoo and Belle Isle Aquarium, Detroit, Michigan.

*Clinical Chemistries*

CLINTON D. LOTHROP, JR., DVM. PhD

Assistant Professor, Department of Environmental Practice, College of Veterinary Medicine, University of Tennessee, Knoxville, Tennessee.

*Miscellaneous Diagnostic Tests; Miscellaneous Diseases*

LINDA J. LOWENSTINE, DVM, PhD, Diplomate, American College of Veterinary Pathologists

Assistant Professor of Veterinary Pathology, School of Veterinary Medicine, University of California, Davis; Pathologist, Veterinary Medical Teaching Hospital, University of California, Davis; Collaborative Researcher and Pathology Consultant, California Primate Research Center, Davis, California.

*Necropsy Procedures*

RONALD LYMAN, DVM, Diplomate, American College of Veterinary Internal Medicine

Private practice, Fort Pierce, Florida.

*Neurologic Examination; Neurologic Disorders*

JOHN S. McKIBBEN, DVM, PhD

Clinician, Sylvan Lake Veterinary Hospital, Rome City, Indiana.

*Clinical Anatomy*

MICHAEL S. MILLER, MS, VMD, Diplomate, American Board of Veterinary Practitioners

Vice President of Clinical Affairs, Cardiopet Division of Animed, Inc., Brooklyn; Staff Consultant, Electrocardiography, Roslyn; Staff Clinician, A & A Veterinary Hospital, Inc., Franklin Square, New York.

*Electrocardiography*

PATRICK T. REDIG, DVM, PhD

Associate Professor, College of Veterinary Medicine, University of Minnesota; Clinical and Field Programs Director, Raptor Research and Rehabilitation Program, St. Paul, Minnesota.

*Evaluation and Nonsurgical Management of Fractures; Basic Orthopedic Surgical Techniques*

ROBERT K. RINGER, PhD

Professor, Departments of Animal Science and Physiology, Michigan State University, East Lansing, Michigan.

*Selected Physiology for the Avian Practitioner*

WALTER J. ROSSKOPF, JR., DVM

Research Associate, California State University, Dominquez Hills; Practicing Veterinarian, Animal Medical Center of Lawndale, Hawthorne, and Avian and Exotic Animal Hospital of Orange County, Fountain Valley, California.

*Differential Diagnoses Based on Clinical Signs; Symptomatic Therapy and Emergency Medicine*

RHONDA K. SAYLE, AHT

Avian Technician, All Animal Clinic and Bird Hospital, Lake Worth; Member, Board of Directors, Research Institute for Avian Medicine, Nutrition, and Reproduction, Inc., Lake Worth, Florida.

*Evaluation of Droppings*

DAVID L. SCHULTZ, BVSc

Private Practice, Hawthorn Veterinary Clinic, Millswood, South Australia.

*Miscellaneous Diseases*

RONALD R. SPINK, DVM

Private Practice, Kindness Animal Hospital, Cape Coral, Florida.

*Aerosol Therapy*

TAMMY UTTERIDGE, BVSc, Dip Vet Path (Syd)

Veterinary Clinical Pathologist, Adelaide, South Australia.

*Miscellaneous Diseases*

MICHAEL T. WALSH, DVM

Staff Veterinarian, Sea World of Florida, Inc., Orlando, Florida.

*Radiology*

RICHARD WOERPEL, DVM

Practicing Veterinarian, Animal Medical Center of Lawndale, Hawthorne, and Avian and Exotic Animal Hospital of Orange County, Fountain Valley, California.

*Differential Diagnoses Based on Clinical Signs; Symptomatic Therapy and Emergency Medicine*

# Acknowledgments

We are very proud and appreciative of the authors who were willing to sacrifice their personal time in order to participate in this project. It is out of their commitment to making a difference in the quality of clinical medicine available for the avian pet that they contributed their knowledge and expertise.

In particular, we are grateful to Terry Campbell, who was willing to assume responsibility for writing and reviewing additional chapters beyond his original agreement. Other authors who graciously reviewed and/or offered editorial comments were George Kollias, Helga Gerlach, Robert Ringer, and Albert Lewandowski. Additional reviews and editorial comments were provided by Julian Baumel, Bruce Glick, Howard Evans, David Graham, and Connie Thurlow.

A special acknowledgment is deserved by John McKibben and Algernon Allen for their contribution of many hours beyond what was anticipated in the preparation of the original Amazon anatomic illustrations; by Helga Gerlach, who made major contributions to this work in a language other than her own; and by H. John Barnes, who completed his chapter in spite of serious health problems at the time.

We are appreciative of the artistry provided by Jody Sayle, Carol Duguid, James McKibben, and Rick Volkmar and the photographic support from Lorraine Karpinski, Ellman International Manufacturing, Inc., Scott McDonald, Aviculture Institute, R. Wolf Medical Instruments Corp., Medical Diagnostics, William Satterfield, Zoological Society of San Diego, Veterinary Learning Systems, Inc., Richard Evans, H. John Barnes, Ross Perry, David Pass, George Kollias, A. Bickford, Chris Murphy, L. Munger, Walter Rosskopf, Angell Memorial Hospital, University of Pennsylvania, Association of Avian Veterinarians, and others from whom we borrowed insights and information.

In the day-to-day progression of "the book" we want to thank our daughters, Dana and Tanya, who were very tolerant of their intense parents; Rhonda Sayle, Greg's "right arm" technician, for her devotion and dedication to excellence; Donna Walker of W. B. Saunders Company, whose patience, commitment to detail, and good-natured support were far "above and beyond the call of duty"; and especially Becky Grubbs, whose office support was contributory to the completion of this project.

GREG J. HARRISON
LINDA R. HARRISON

In addition to those who directly participated in this book, I want to publicly express my appreciation to some other individuals who have made valuable contributions to my knowledge of caged bird medicine.

Drs. Ted Lafeber, Murray Fowler, Robert Altman, George Gee, Mitchell Bush, David Wildt, Steve Seager, Berne Levine, Alan Abrey, David Russell; in particular, David Graham, who has always persisted in his expert pursuit of pathologic causes of clinical cases; and Jack Hanley, who, before his retirement, provided pathology service "beyond the call of duty."

Aviculturists: Bob Wall, Mac Sweat, Dick Vaughn, Ramon Noegel, Tom Ireland, Ron LeClair, Laura McMahon, Doug and Claudine Trabert, John Stoodley, and Jim and Beth Wiley.

Art Clexton of Richard Wolf Medical Instruments Corporation and Jon Garito of Ellman International Manufacturing, Inc., who continue to make their instruments available to practitioners for the development of avian surgical skills.

Drs. Susan Clubb, Dan Wasmund, Kaci Christensen, and Branson Ritchie, who came to my practice to learn and in fact taught me.

Microbiologist Patricia Avery Wainright for opening up the possibility to me that few of our primary diseases are bacterial in nature.

And my father, Harry Harrison, who inspired me through his own interest in wild birds.

GREG J. HARRISON

# *Foreword*

Pet bird medicine has become an increasingly important facet of companion animal medicine. Ten or 15 years ago only a handful of pioneers devoted significant practice time to birds. Now there are many full-time avian practices, some with three or more clinicians and full laboratory support. Hundreds of practitioners devote a major percentage of their time to birds. This publication meets a long-felt need of practitioners for comprehensive clinical information.

The authors of this book are actively involved in avian medicine. They are clinicians, diagnosticians, radiologists, pathologists, and microbiologists. All have been frequent speakers at local, regional, and national meetings. That some of their views may be controversial is expected and healthy. Avian medicine is not an exact science; nevertheless, within the covers of this book are to be found the foundations upon which clinicians may build their own experiences and practices.

The clinical emphasis will be readily apparent to the reader. Case management is presented in the same manner as the practitioner would face the problem. First, equipment, supplies, and special diagnostic material necessary to approach an avian case are described. The chapter on differential diagnoses based on clinical signs is unique in calling the reader's attention to diseases of organ systems associated with a given sign. The extensive tables in this chapter will be a tremendous help to both the novice and the experienced avian practitioner. Signs are evaluated and followed by special diagnostic tests such as hematology, serum chemistries, radiology, cytology, and many more. Practical clinical tips are included, and details are given for the most sophisticated state-of-the-art procedures that modern medicine can supply.

Recommended therapy follows, with artful presentations on general and specific regimens. The writers of these chapters have dealt with thousands of birds and are sharing a wealth of experience.

Conditions managed by surgery are dealt with in a separate section. Not only are the author's techniques described, but ample use of the literature makes this a comprehensive discussion of anesthesia and surgery.

The support chapters for the clinician are likewise included, with anatomy and physiology presented in great detail yet clinically oriented. The work on infectious and parasitic diseases is the most comprehensive, up-to-date material available anywhere in the world.

A unique contribution for a medical text is the inclusion of information dealing with avicultural practices and reproduction. Included is a chapter on sex identification by various means. The tables listing species' secondary sex characteristics have been gleaned from many sources and will serve as a quick reference on this subject.

This volume is destined to become "the book" of pet bird medical practice. Profound thanks are due to busy, active practitioners and scientists who have approached an awesome task with enthusiasm and dedication.

MURRAY E. FOWLER
*Chief of Zoological Medicine*
*School of Veterinary Medicine*
*Davis, California*

# Preface

The remarkable growth and development of pet bird medicine, surgery, and aviculture over the last few years prompted the creation of *Clinical Avian Medicine and Surgery* for use as a reference/textbook for veterinarians and students. As such, it was designed to address the most common concerns of the pet bird practitioner and to introduce the reader to some new philosophies, concepts, and trends in captive bird care. Much of this information is also of value to aviculturists, zoologists, wildlife biologists, avian pathologists, pet shop owners, and ultimately, the pet bird owner.

The Amazon parrot has been used, when applicable, as a model for addressing psittacine species from the context of the individual pet bird. The dynamics of multibird ownership are discussed in the Aviculture section. Avian diseases, which are frequently pansystemic, are described primarily from the etiologic standpoint rather than by physiologic system, and the patient is assessed from the holistic viewpoint of the practitioner. Because of this format, there is some overlap of information.

Documented material on caged bird species which was available at the time of publication is included where appropriate and as background information. In those cases in which information has been extrapolated from poultry data, its applicability to the many exotic species commonly kept as pets can only be assumed.

In those areas where little, if any, scientific investigation has been done, contributing authors with a significant avian caseload have been willing to share their experiential knowledge to provide to the practitioner some direction into the diagnosis and therapy of those avian disorders. While this clinical material is clearly empirical and often anecdotal, it is nevertheless included in the text as a "stepping-off point" to the next level of scientific data. We are well aware that much of this is subject to change, and in fact, new information was constantly evolving during the preparation of the book. Current activity is particularly lively in the areas of nutrition, identification of viral diseases, and clinical treatments. Some of the surgical procedures have never been previously described and are subject to revision.

Anyone who finally sees in print the results of many long hours of work tends to breathe a sigh of relief, yet in a field such as clinical avian medicine and surgery, we have just begun. The editors invite your comments and suggestions for a future edition.

GREG J. HARRISON
LINDA R. HARRISON

# Contents

*Section One*

# THE PET BIRD: GENERAL CONSIDERATIONS

# *Chapter 1*

# CHOOSING A BIRD

LINDA R. HARRISON,
KEVEN FLAMMER,
GREG J. HARRISON

Usually by the time a practitioner sees a pet bird, the owner has already finalized the purchase. However, many new owners become dissatisfied with their choice, and often the birds are given away, donated to a zoo, or neglected. In some cases the bird owners may not have been knowledgeable about the responsibilities involved; in other cases, they may have been unfamiliar with characteristics of the specific species they purchased and may have selected a bird unsuitable for their purposes.

Choosing a pet bird is not unlike the selection of a dog or cat as a pet. In addition to the usual considerations of health, age, source, and price, specific dog breeds tend to exhibit characteristics that lend themselves to a particular life style of the owner, and a small animal practitioner may be called upon for advice concerning these breed tendencies.

Similarly, it would be helpful for the avian clinician to become familiar with the varieties of exotic birds most commonly kept as pets and with the characteristics most likely to be found within a particular genus or species category. This information may also be helpful to the practitioner in providing an appropriate hospital environment.

Other factors that influence the adaptability of available birds to pet and aviculture purposes will be briefly discussed.

## CHOOSING A PET BIRD

The total health of the pet bird depends not only on the physical care, but on the psychological adjustment in a particular household environment. Response of the owner to the bird, acclimation of the bird to its routine, and adaptation of a wild, newly imported bird to a captive environment all influence the owner/bird relationship and ultimately the health of the bird.

A wide range of choices is available in the pet bird market. A potential buyer needs to be aware of exactly what qualities he is looking for in a pet; for example, a "talker" may not enjoy being cuddled, and vice versa. Macaws, lories, and conures are more likely to display the playfulness found in dogs, whereas cockatoos often have a dignified aloofness reminiscent of cat behavior. The buyer should assess the amount of bird "bulk" he is willing to accommodate in terms of cage and living space, and the time and attention required by the bird. Price may also be a consideration and, for a particular species, may vary to some extent depending upon the source.

### *Source*

#### IMPORTED

Imported birds are primarily divided into three groups: wild-caught adults, wild-caught babies being hand fed, and birds that are aviary bred in other countries. Some species of birds are available only through importation and may add new genetic blood lines to domestic collections. Many species are less expensive and more easily obtained from import sources than those that are domestically raised.

The primary disadvantages of this source are the bird's exposure to disease and the stress involved in the import process (see Diseases of Imported Birds as Related to Country of Origin and Species in Appendix 2). Some species have particularly high mortality rates as a result of poor husbandry at the point of capture, overuse of medications, and change in environment. Wild-caught adults tend to be hardier in this respect than wild-caught babies, but some adults may never become totally domesticated or capable of reproducing in captivity.

Wild-caught babies may be imprinted (therefore tame) and are usually less expensive than the domestically raised counterpart; however, these young birds have a high susceptibility to diseases associated with importation because of their inadequately developed immune system at the time of import.

Imported aviary-raised birds are more likely to breed in captivity and less likely to be malnourished, especially if they are from European facilities; however, few species are bred in numbers sufficient to supply the overseas market. These birds are also subjected to quarantine rigors.

Although many wild-caught, imported birds on the pet market are initially frightened and defensive, the birds can often adjust to a captive environment and with the passage of time, an appropriate environment, and proper taming, they can learn to direct their natural social inclinations toward the family members. The qualities found relatively consistently throughout the genus or species become more evident in adapted birds.

#### DOMESTICALLY RAISED

Domestically raised birds are better-adapted to captivity and tend to be reproductively active at an earlier age. A buyer of these birds may have more knowledge about the reputation and cleanliness of the breeding facility, the age of the bird, and the history of the parents. The National Cage and Aviary Bird Improvement Plan (NCABIP) has been proposed to standardize the quality of aviculture facilities and disease prevention programs for domestically raised birds.

If hand-reared, birds are familiar with humans and are more apt to be comfortable and gregarious. However, because they lack fear, some of the negative characteristics of the species are also likely to be exhibited, such as chewing on furniture, screaming for attention, and chasing and biting at people's feet.

#### SMUGGLED BIRDS

Smuggling is influenced by the price of birds and import/export restrictions. Many smuggled birds, particularly Amazons, are commonly brought from Mexico into bordering states and sold for low prices at flea markets, on street corners, or from the backs of vans. This illegal practice is potentially dangerous for the introduction of velogenic viscerotropic Newcastle disease (VVND) into this country.

Illegal species may also be smuggled into the United States through quarantine stations by the use of false-bottom boxes, falsified documents, or altered appearances of the birds, or as unfeathered, unidentifiable babies.

### *Age*

#### YOUNG

Birds acquired at a young age are more likely than adults to be easily tamed and trained as acceptable pets. However, they are similar to puppies in that inappropriate behavior must be corrected at the onset before it becomes established. The behavior will change as the bird matures. Young birds more readily accept a wide variety of foods, which ultimately results in the long-term health of the bird. If the species is well-adapted to mimicking, young birds have the most potential for developing talking ability.

#### ADULT

Buying an older bird is similar to choosing an adult dog from the pound. One does not have knowledge of its previous experiences, but the "puppy" stages are already past and the behavior tends to be more predictable. Adult birds generally are more calm and sedate, and the plumage usually is more attractive. Older birds are more set in their ways and are less likely to become tame if they are wild-caught. They are also less likely to breed in captivity, particularly if they have bred in the wild.

## CHOOSING A BREEDING BIRD

One of the unfortunate trends of thinking is that if a bird is not useful as a pet for some behavioral reason, it can be used as a breeder. From the standpoint of marketable birds in the future, aviculturists need to consider inherited behavioral tendencies and select for breeding those individuals with good parenting ability and pet qualities. The *ideal* breeding bird would be domestically raised (preferably third generation) from parents with good breeding history and adequate gene pool. If the bird under consideration has not already produced offspring, it should be purchased young, relatively tame (not necessarily hand-raised), eating a wide variety of foods, and compatible with its mate (see Chapter 52, Reproductive Medicine).

Aviculture success is facilitated by starting with easily bred, responsive, domestic lines of birds for which there is ample pet demand (e.g., cockatiels) before attempting more complicated genera. Adaptation to the captive climate is a consideration in many species. Specialization of the aviculturist to limited genera tends to increase success.

## GENUS AND SPECIES CHARACTERISTICS

Recognizing that there are individual differences within a genus or species, the following characteristics are found relatively consistently among acclimated members of selected species of birds. The reader is referred to the Appendices for a list of scientific names, average weights, and additional aviculture information. Characteristics of sexual dimorphism are presented in Chapter 51, Sex Determination Techniques.

### Psittaciformes

#### *Budgerigar*

**Comments:** A budgerigar makes an extremely good pet, as it is gregarious and easily tamed, and has entertaining antics. It is inexpensive to buy and maintain, as it requires a small cage and minimal clean-up. A wide variety of colors is available. Budgerigars are easy to breed and easy to hand raise (although hand raising is seldom done commercially because the birds are so inexpensive and easy to tame). They are not loud or destructive and are good talkers. One budgerigar in London was documented to recite up to 10 rhymes and 383 sentences, with a total vocabulary of 531 words.[1] An East German budgerigar recited over 600 words.[2] Unfortunately, these birds are commercially bred for size and color rather than longevity or freedom from tumors.

**Origin:** Australia

**Common Problems:** Malnutrition, tumors (especially lipomas and renal adenocarcinomas); scaly face mites (the susceptibility is believed to be influenced by heredity); fatty liver degeneration, overgrowth of beak and nails, thyroid-responsive conditions, bumblefoot, budgerigar fledgling disease (papovavirus), giardiasis, bacterial infections, gout, diabetes mellitus, leg band injuries, French molt; probable carriers of *Mycoplasma* and *Chlamydia*.

#### *Cockatiels*

**Comments:** One of the best choices for a first pet bird, cockatiels make excellent pets, are inexpensive, and tame with ease (Fig. 1–1). They are long-lived, entertaining, quiet, nondestructive birds that require small living quarters. They have excellent whistling ability, although talking may be limited (males are reported to be the better talkers). Cockatiels are also recommended as the species with which new aviculturists can get experience in preparation for breeding larger species. Cockatiels are prolific, year-round breeders and are sexually mature around one year of age.

**Figure 1–1.** A young domestically raised cockatiel is recommended as a beginning bird, whether for pet or breeding purposes.

**Origin:** Australia

**Common Problems:** Chlamydiosis, conjunctivitis (primarily of suspected mycoplasmal etiology), obesity, feather picking, giardiasis, intestinal ascarids, chronic egg laying, bacterial infections, "diabetes mellitus–like" condition, lead poisoning. Breeding birds are prone to egg binding and malnutrition. *Candida, Giardia, Chlamydia*, and enteric bacteria commonly cause nestling mortality without causing problems in adults. Hand-raised babies are reluctant to wean. An increasing incidence of neurologic problems and genetic baldness has been noted, possibly related to inbreeding. Hybrid strains appear shorter-lived and weaker. South African aviary-bred birds are somewhat "spooky."

### *Parakeets*

**Comments:** The many Australian species classified as parakeets make poor pets unless they are hand-raised, and even then will revert to wild characteristics easily. They are therefore more suitable as aviary birds. Rosellas are especially colorful, popular aviary birds but are somewhat delicate healthwise. Most species are desert-adapted and consume little water; therefore, they are difficult to medicate via the drinking water. All species are active and are best housed in a long (10- to 20-foot) aviary that permits flight. Males are very aggressive, and similar species will fight to the death in the breeding season; therefore, they are usually housed one pair per cage. Cages must be separated by double wire, or the birds will chew each other's toes.

Some aviculturists successfully raise 10 to 12 birds in an exceptionally large aviary if multiple nest boxes are available to reduce the incidence of fighting. As pair bonds form (of the birds' choice), the first few pairs are moved to breeding cages and the rest are sold to the pet market. The increased compatibility is reported to increase breeding success. In general these species are good parents, but some individuals habitually eat their own eggs.

Brotogeris parakeets originate from Central and South America and include the Grey-cheeked, Canary-winged, and Bebe. Grey-cheeked Parakeets, in particular, are naturally tame and affectionate birds but can be very noisy.

**Origin:** This is a broad classification of small psittacine birds that may come from Australia, New Guinea, New Zealand, Asia, Africa, or Central or South America (depending on the species).

**Common Problems:** Intestinal and proventricular worms, pox, respiratory problems in damp or humid environments, bacterial infections; usually not noted for longevity. The young of many species fly immediately from the nest at fledging and may crash into the cage wires; hanging burlap sheets at either end of the cage will help the bird to see the cage walls. Brotogeris parakeets are susceptible to bacterial infections, sarcoptiform mange, chronic bacterial hepatitis, tuberculosis, and chlamydiosis.

### *Lovebirds*

**Comments:** Lovebirds are not destructive, loud, or obnoxious but are poor talkers. Unless they are hand-raised, they are difficult to tame. However, they are relatively easy to hand raise and can become very attached to their owner. They love to hide (in shirt pockets, long hair, etc.).

**Origin:** Africa

**Common Problems:** Chlamydiosis, egg binding (especially in inactive or malnourished birds bred before one year of age), bacterial infections, candidiasis, pox, feather loss and feathering problems of undiagnosed etiology (see Chapter 41, Disorders of the Integument), adenovirus infection, and a number of diseases of unknown etiology (probably viral). Because of the potential for disease problems, lovebirds for pet and breeding stock should be purchased from reputable clean sources. Lovebirds are very territorial and may kill new additions to their cage. They will kill weak birds that act depressed or fly erratically in the cage, including recently fledged young. Lovebirds will cannibalize birds that die for any reason, typically chewing the face and back of the skull; this should not be confused with aggression and the real cause of death missed. They are sensitive to heat and will often refuse to care for young in the nestbox if temperatures exceed 100°F.

### *Conures*

**Comments:** Conures make excellent pets; they are affectionate, playful, gregarious, and entertaining and love to be touched. Conures can be trained to fly back to the owner outside. They crawl around in bags, carry items around, sleep on their backs, roll over, and are very social and strongly body-oriented. However, they are quick-tempered, noisy, destructive chewers, with very limited talking ability and no song. Conures can be bred in a colony but may fight to the death. They are prolific, and the young are easy to hand raise.

**Origin:** Mexico and Central and South America

**Common Problems:** Some species (Patagonian, Painted, Nanday, and White-eyed) are known carriers of Pacheco's parrot disease, and other species are suspected to be carriers, although the Halfmoon Conure is highly susceptible to the herpesvirus. Other problems are nestling papovavirus infection, bleeding and clotting deficiency syndromes, and bacterial infections. Conures are suspected of carrying a number of unknown disease syndromes, probably also of viral etiology.

**SUN, JENDAY, HYBRIDS, GOLD-CAPPED, DUSKY, BLUE-CROWNED, MAROON-BELLIED:** More apt to have pet qualities.

**NANDAY:** Least expensive, least acceptable as pets; most raucous voice; least likely to breed in captivity; may be more of a feather picker.

### *Caique Parrots*

**Comments:** An unusual genus of small, short-bodied parrots, caiques are exceptionally intelligent comics. However, they are stubborn and resist adopting an acceptable captive diet. They are rarely bred in captivity, but when they are, hand-raised birds are delightful pets.

**Origin:** South America

**Common Problems:** May be difficult to adapt to captivity.

### *Pionus Parrots*

**Comments:** Pionus are similar to the smaller Amazon species, although they are quieter and less aggressive. Pionus parrots tend to hyperventilate when they are nervous. They are not good talkers but can mimic sounds such as sneezing and coughing. Although they are difficult to breed, they are easy to hand raise and make good hand-raised pets.

**Origin:** Mexico and Central and South America

**Common Problems:** Very susceptible to stress. Pox causes high mortality rates (especially in recently imported Blue-crowned Pionus); bacterial infections also occur. High-altitude species such as the Plum-crowned Pionus do not do well in warm, humid climates.

### *Amazon Parrots*

**Comments:** Most species make good pets, although Amazons, as a general rule, do not particularly enjoy being petted and cuddled. Their handling tolerance is low, especially if approached by strangers. They are more often chosen for their talking ability, entertaining antics, and outgoing personality. Most species become very aggressive during the breeding season and may attack anyone entering their cage, including a feeder they recognize. They are somewhat difficult to breed but are fairly easy to hand raise.

**Origin:** Mexico and Central and South America

**Common Problems:** Pox (most common in recently imported birds, especially Blue-fronted Amazons), chronic upper respiratory disease, sinusitis, chlamydiosis, bacterial disease, vitamin A deficiency, obesity, VVND in smuggled birds, cloacal papillomatosis, neurologic problems in Red-loreds, and adenocarcinomas in older Amazons.

**DOUBLE YELLOW-HEADED, YELLOW-NAPED, YELLOW-CROWNED, PANAMA:** These four subspecies of *Amazona ochrocephala* exhibit similar characteristics. They have a tendency to establish one-person relationships, even if hand-raised, and may become jealous and vicious toward others, especially as they approach sexual maturity. They are considered to be excellent talkers. Yellow-napeds may be especially hyperactive, have a short temper, and are rarely quiet. Yellow-crowneds may tend to be quieter and better-behaved than the others.

**BLUE-FRONTED:** There are high losses of Blue-fronted Amazons from importation procedures, as they are highly susceptible to all import diseases. Hand-raised babies can be excellent talkers and good pets.

**ORANGE-WINGED:** To the new parrot buyer, an Orange-winged may resemble a Blue-fronted Amazon; however, it is one of the least likely Amazons to talk. It is usually the least expensive Amazon and is appreciated most for its docile, quiet manner. If tame, it can be handled, and it likes to whistle. Orange-wingeds are often passed from owner to new owner owing to disappointment in the subdued personality.

**GREEN-CHEEKED:** Also called Mexican Red-headed, it can be an affectionate pet if hand-raised, and it has some capacity for talking. These birds are very vocal; they like to whistle and scream.

**RED-LORED:** The Red-lored Amazon can be a personable pet for a family. The bird is not at all dignified; it likes to roll over and play.

**MEALY:** This largest of the Amazons may be enjoyed by people looking for a less active parrot, as it usually just sits on a perch. It is docile and nondestructive, and, if tame, is the Amazon most likely to tolerate handling by a wide group of people. The least obnoxious vocally, it rarely talks but occasionally mimics other sounds such as coughing, laughing, and whistling.

### *African Grey Parrots*

**Comments:** If acquired at a young age, African Greys make good pets and can be excellent mimics and talkers with an extensive vocabulary. The *ability* to mimic a particular voice or sound is species-specific, but the actual performance is an *individual* characteristic: the nontalking African Greys probably outnumber the talking ones by 10 to 1. These birds are alert, intelligent, sensitive, and high strung and can be very noisy. A small percentage of wild-caught birds become very tame; however, they seem to be less intimidated by a female trainer.

If wild, they can be the most obnoxious of all birds. Hand-raised birds can develop a close relationship with the owner and are affectionate and sociable. "Growling" can be an expression of affection. It is difficult to start a pair breeding; however, once they start, they are prolific and are fairly easy to hand raise.

**Origin:** Africa

**Common Problems:** Feather picking, respiratory disease, malnutrition, hypocalcemic syndrome, bacterial infections, aspergillosis, reovirus, tapeworms, and dietary disease in older birds.

### *Eclectus Parrots*

**Comments:** Beautiful in appearance but shy, dull, and lethargic in personality, Eclectus have a "spacy" behavior unique to this genus. They can make fair pets if hand-raised; males may make better pets because of their more gentle temperament. They appear to be less intelligent than other parrots, and they are seldom destructive. Eclectus are not long-lived in captivity. They have some talking ability and love to "mumble." Although they can have loud, harsh voices, they do not have prolonged or regular screams. If there are several pair on the premises, they may call to each other. They are difficult to breed. The young are easy to hand feed until weaning, when they may refuse to eat.

**Origin:** Australia, New Guinea, and the South Pacific Islands

**Common Problems:** Infertility common in first few years of egg laying, vitamin A deficiency (oral abscesses), candidiasis, reovirus, feather picking, sudden death from undetermined etiology (especially noted in Florida). The hen is dominant and may severely injure newly introduced males.

### *Macaws*

**Comments:** Macaws usually are memorable companion animals and can vary in size from less than 200 grams (Noble) to 1500 grams (Hyacinth). The most popular macaws are for people who want a lot of responsibility in caring for a pet bird. They are large, intelligent species that require a lot of living room and daily stimulation. Although they have a formidable beak and love to chew and destroy furniture, their bite is usually no worse than that of other parrots. They make good pets, but wild-caught birds may be difficult to tame. All macaws are noisy, particularly vocalizing at dawn and dusk. All become extremely aggressive and protective of their nest box during the breeding season. They are best housed one pair per cage, or the dominant pair may prevent other cage members from eating and perching. They are easy to hand raise but difficult to wean. Individuals have been "house-trained."

**Origin:** Mexico and Central and South America

**Common Problems:** Feather picking, *Capillaria* infection of the proventriculus, intestinal ascarids and tapeworms, *Haemoproteus*, subclinical chlamydiosis, proventricular dilatation (macaw wasting syndrome), behavioral problems, bacterial infections, depigmentation patches on feet caused by a herpesvirus, sunken eyes due to sinusitis, chronic air sacculitis. Reproductive problems include uterine infections, egg binding, and soft-shelled eggs. Species' tendencies include feather cysts in Blue and Golds, oral and cloacal papillomatosis in Green-wingeds, and self-mutilation in Severe and Yellow-collared Macaws.

**BLUE AND GOLD:** Although they appear nervous, hyperactive, and "stand-offish" with strangers, Blue and Golds are more of a total family bird than other macaws. They are mischievous and may be easily spoiled, and they may even run around the house and bite at the owner's toes. They are the least expensive and easiest to breed of large macaws. They are commonly used to hybridize with Scarlets (to produce Catalinas).

**SCARLET:** Scarlets can be untrustworthy as pets and are more inclined to be one-person pets. They may get unexpectedly nasty, rambunctious, or nippy no matter what the owner/bird relationship is. They are several times the cost of Blue and Golds.

**GREEN-WINGED:** The high-strung Green-winged does not do well as a single bird, as it usually cannot entertain itself. It may need another bird in the house for companionship; it may have more of a tendency to feather pick or scream if it is bored. A very intelligent bird, it is on guard constantly during a clinical examination and is hard to fool and capture. Green-wingeds are often aggressive toward strangers.

**HYACINTH:** As the largest of the genus, the Hyacinth is similar in some respects to the Mealy Amazon. It is mellow, accepting of humans, and usually the least noisy of macaws. It is usually not destructive, except to wood. If purchased young, it can be an excellent and affectionate pet. It has limited ability to talk.

**NOBLE, SEVERE, YELLOW-COLLARED:** The smaller macaws are not as destructive as the larger macaws. They are vocal "watch dogs" when threatened, are similar to conures in behavior, and are medium quality talkers.

### *Cockatoos*

**Comments:** In general, cockatoos are easily stressed and are less suited for captivity than other species of psittacine birds. They are reluctant to try new foods and may feather pick out of frustration, boredom, disease, or neglect. Although it is unusual for adult cockatoos to become tame, they seem to be less intimidated by a female trainer. Hand-raised birds are more tolerant of captivity than wild-caught cockatoos. Pet cockatoos, in general, require a large cage and a lot of time and attention to make good pets; screaming may result from lack of attention. Members of the cockatoo family like to be touched and look for physical petting and affection. Some birds have been "house-trained." Of all the psittacine birds, cockatoos make the best child surrogates. They are more difficult to breed and hand raise than other psittacine species. Most cockatoos have abundant "powder down," which produces a fine dust that will cover surrounding areas.

**Origin:** Australia, New Zealand, and the South Pacific Islands

**Common Problems:** Psittacine beak and feather disease syndrome (which appears to be increasing in frequency of diagnosis), feather picking, malnutrition as a result of reluctance to change diet, lipomas, obesity, bacterial infections, herpesvirus "warts" on the feet. Many species, especially Moluccan and Umbrella cockatoos, carry various hemoparasites (e.g., *Haemoproteus* and microfilaria) and intestinal tapeworms. With breeding pairs, the male may attack and kill the female any time during the breeding season. This behavior cannot be predicted and may occur in pairs that have previously produced offspring for years; some pairs may neglect offspring in the nest.

**MOLUCCAN:** Moluccans are a lot of responsibility for an owner; they must be watched all the time. They are destructive and are powerful enough to break out of chain link cages. They are uptight, high-strung, manipulative, potentially nasty, and very intelligent; however, their attention span is short. Although they are the most trainable for tricks such as roller skating, they are intimidated easily when being trained and may be hand-shy. The noisiest cockatoo, Moluccans get noisier as they get more comfortable. Their mimicking voice is high-pitched and mechanical and is the least likely to be mistaken for a human voice. Breeding Moluccans are especially shy and will spend much of their lives in their nestbox. If hand-raised, they can be affectionate, adorable, smart, uninhibited, vivacious, and demanding.

**UMBRELLA:** A better choice for the average pet owner than the Moluccan, an Umbrella Cockatoo is less bird "bulk" and less responsibility and has a lower price. It can be timid, sedate, or gregarious, or it can be a clinger that loves to be cuddled and stroked. Umbrellas are not good talkers and can be noisy and destructive.

**ROSE-BREASTED:** Hand-raised Rose-breasteds are good pets, although this species especially resists a change in diet and is highly susceptible to obesity and lipomas. Adapted to dry climates, Rose-breasteds rarely breed in humid areas in the United States.

**MEDIUM SULPHUR-CRESTED/CITRON-CRESTED:** Both subspecies of *Cacatua sulphurea*, these birds are more expensive than other cockatoos, are very short-lived in captivity, and cannot be recommended as pets. Individuals are more apt to have a personality consistent with others of the same species.

**GOFFINS:** Goffins are the least expensive cockatoos and are less suitable as pets than other species, as they are high-strung and stubborn. Few adapt well to captivity, and they have a short captive life span. There is no reported beak and feather syndrome in Goffins.

**BARE-EYED:** Although it is the least beautiful cockatoo, Bare-eyed cockatoos (Little Corellas) are natural entertainers and clowns. Hand-raised birds may tend to retain a whimpy, baby personality, but they can be unpredictably aggressive and nippy. They are usually long-lived with few disease problems.

### *Lories and Lorikeets*

**Comments:** Lories are brilliantly colored and playfully active birds. As the most consistently animated of all psittacines, they are the least likely to revert to a wild state if not played with. They will chase, retrieve, and roll over; however, lories are noisy and have limited talking ability. A high-carbohydrate, nectar-like diet is required in addition to seeds and fruit. Because they produce copious liquid droppings, they are ill-suited for caging inside a home and should be housed in an aviary. They love to bathe if provided with clean water daily. Lories can be housed in a colony with known members of their own species but may attack and kill newly introduced birds of their own or other species.

**Origin:** Australia and South Pacific Islands

**Common Problems:** Hypothermia, bacterial and candidal enteritis if the liquid diet spoils, liver diseases, reproductive disorders (especially unexpected loss of females during the breeding season, presumably from malnutrition).

## Passeriformes

Passerines as a general rule are more suitable as caged birds than as pet birds. Although there are individual exceptions, especially with imprinted birds, they are usually not touchable or affectionate. They are generally short-lived compared to psittacines. Passerines probably have their own viruses that have not been well addressed.

### *Canaries*

**Comments:** The raising of canaries has become a culture of man-made hybrids. Breeding pairs are genetically selected for a specific characteristic: brilliant color (Red factor), singing ability (Border), fluffy feathers (Frill). They are prolific and can be raised in small cages.

**Common Problems:** Malnutrition, obesity, liver disease, bacterial infections; parasites (respiratory mites, scaly leg mites); baldness in males; lack of singing; cataracts, feather cysts, overgrowth of nails, atoxoplasmosis. In areas of high mosquito population, canaries must be raised indoors to prevent transmission of pox.

### *Australian Finches*

**Comments:** Australian finches are small, colorful birds popular among breeders and pet owners because they are quiet and entertaining to watch. Although they do not talk, they have a melodious song. Finches are somewhat skittish and short-lived. Although some species may require live food for reproduction, most species can be foster-raised by the inexpensive Society finches, thereby increasing the potential production during the breeding season.

**Origin:** Australia

**Common Problems:** High mortality rates from importation, bacterial infections, and parasitic infections (air sac mites, proventricular worms, ascarids, tapeworms) are much more important in finches than in other birds and may be responsible for considerable losses. Pox virus is another common disease agent, and outbreaks may claim a high percentage of the flock. Egg binding and hatching problems may result from an inadequate level of humidity. Malnutrition is common.

### *Mynahs*

**Comments:** Mynahs remain aloof and usually cannot be touched, although they have high-quality mimicry and may vocalize constantly. Because of their very loose, messy droppings, they are usually housed outdoors.

**Common Problems:** Mynahs infrequently live beyond 10 years and commonly develop a liver/heart/respiratory complex with ascites that is usually fatal; iron storage disease, liver cirrhosis, chronic active hepatitis, bacterial infections, and traumatic keratitis during shipping also occur.

## Piciformes

### *Toucans and Toucanettes*

**Comments:** These birds have a quiet "duck-like" sound and are animated and playful, with unparalleled bathing antics. They enjoy having their beaks scratched. The diet may be a problem, as they like to eat mice and smaller birds.

**Common Problems:** Most must be housed as single pairs because they are highly territorial and will kill strange birds, especially ones of the same species. Fighting between mates is also common.

## Others

Other birds such as peafowl, turkeys, geese, chickens, ducks, pheasants, and pigeons can be remarkable in their pet qualities, especially in imprinted birds. Jays and crows have been found to be highly intelligent and can be taught to talk. Falconers develop a special relationship with their birds. The avian practitioner will be presented with individuals from these groups from time to time and should refer to appropriate literature for their care and medical treatment.

# CAPTURE OF ESCAPED BIRDS

Species of captive birds exhibit some tendencies in regard to their attitude toward freedom and the likelihood of being recaptured if they escape from outdoor aviaries or fly out of open doors in the home. Some species (e.g., conures, macaws, some cockatoos, pigeons, and lories) appear to have more of a "homing" instinct and will stay around the area to perch in a nearby tree, or return at a later time, thus facilitating capture attempts. On the other hand, others are more inclined to "take off" once freedom is theirs, and recapture may be more difficult: cockatiels, budgerigars, mynah birds, canaries, finches, Amazons, and Moluccan Cockatoos. Some birds will stay in the area if their mate is still in captivity. The mate can be used as a lure for recapture.

## REFERENCES

1. Williams, M.: "Sparkie" Williams, the "talking" budgerigar. A copy of the typescript of this unpublished book and accompanying sound tape are on deposit in the Cornell Library of Natural Sounds in Ithaca, NY.
2. Wallschlager, D.: Mitt. Zool. Mus. Berlin [57 suppl.] Ann. Ornithol. 5:3–13, 1981.

## RECOMMENDED READINGS

Bates, H., and Busenbark, R.: Parrots and Related Birds. Neptune City, NJ, T.F.H. Publications, Inc., 1969.

Forshaw, J. M., and Cooper, W. T.: Parrots of the World. Garden City, NY, Doubleday & Company, Inc., 1973.

Fowler, M. E. (ed.): Zoo and Wild Animal Medicine. 2nd ed. Philadelphia, W.B. Saunders Co., 1986.

Hofstad, M. S., et al.: Diseases of Poultry. 8th ed. Ames. IA, Iowa State University Press, 1984.

Low, R.: Parrots: Their Care and Breeding. United Kingdom, Blandford Press, 1980.

Roskopf, W. J., et al.: Pet avian disease syndromes. Proceedings of the Annual Meeting of the Association of Avian Veterinarians, Boulder, CO, 1985, pp. 295–317.

**ACKNOWLEDGMENT**

The authors would like to acknowledge John McGuire for his contributions to this chapter.

# *Chapter 2*

# HUSBANDRY PRACTICES

GREG J. HARRISON

Successful aviculturists tend to be more cognizant of appropriate husbandry practices of cage and aviary birds, as they are more aware of how the management of their flock is directly related to production (see Chapter 50, Aviculture Management). It is the pet bird owner who may be less knowledgeable about providing for the physiologic and psychologic needs of his bird(s). Because feeding and care recommendations change as new information becomes available, it is the avian practitioner's responsibility to stay abreast of state-of-the-art husbandry recommendations and to encourage his clients to incorporate these into pet bird preventive medicine programs.

## ENVIRONMENT

Zoos have found that captive maintenance and reproduction are enhanced by simulation of a natural environment. Many breeding birds are housed outdoors in large cages or exercise flights with access to fresh air, sunlight, temperature variations, and rain. Some of these natural elements can be incorporated into pet bird care. In addition, a client who is aware of species' tendencies (see Chapter 1, Choosing a Bird) is more apt to provide the environment and practices that psychologically nourish the bird.

### Quarantine

Pet bird owners may fail to realize that quarantine recommendations following a new bird purchase are not just for aviculturists. A new pet bird should likewise be isolated in a separate facility and thoroughly checked out before introducing it to the existing collection (see Chapter 50, Aviculture Management). Placing the new bird in a separate room for the 30-day time period is inadequate, because many airborne viruses may be transmitted by central air conditioning/heating systems.

### Caging

Confining a pet bird to a cage is necessary at times to prevent escape and injury of the bird and for protection of the owner's home. However, most pet bird species benefit from a degree of freedom and exercise in the home, and, if possible, this should be provided with supervision. Restricting flight by a wing trim may be preferred.

Traditional recommendations suggest that a cage should be large enough to provide full body extension without contact with the confines. In most cases, the cage would need to be wider than it is tall to accommodate stretched wings; however, ample height should be provided for long-tailed birds. This size is adequate for sleeping or traveling but does not provide sufficient living space for birds that are expected to be confined most of the time. Beyond this minimum size, the largest cage that can be accommodated in the home is recommended for these birds.

An appropriate cage must be strong enough to resist bending or dismantling by the bird, made of nontoxic material, and designed for safety and ease of cleaning. Cleaning of metal cages is facilitated by fewer corners, less surface area, and use of a silicone spray designed for food processing equipment. The cage bars should be designed to prevent injury to the bird. Some bars come together in a "V" at some point, which may entrap the bird's extremities. A wide door facilitates moving the bird in and out. Some door latch systems are easily dismantled by larger birds.

Although recent trends in feeding pet birds suggest two limited feeding times rather than ad lib access to food,[5] multiple feeding dish accommodations should be provided. Use of wide dishes rather than deep cups displays the food more attractively and may encourage the bird to eat new foods. Healthy psittacines with normal ambulatory skills can easily approach the food bowl; therefore, it is not necessary in these cases to place food cups directly near the perch. Birds often overeat or chew on food dishes out of boredom. The food and water bowls should be changed from the outside to prevent escape of the bird through the door. Some birds may need well-anchored food and water dishes to prevent overturning of the contents.

## Perches

The primary purpose of a perch is to supply a clean, comfortable place for the bird to stand. The trend is to supply easily replaceable, natural wood branches from nontoxic (e.g., Northern hardwoods, fruit, eucalyptus, Australian pine) and pesticide-free trees, which with their accompanying leaves and fruit provide a source of beak exercise and mental entertainment. A variety of perch sizes, shapes, and diameters has been suggested to provide exercise for the feet, although the health of a bird's feet may depend more on the adequacy of the diet than on perch qualities. Sandpaper, whether on the perch or the floor, has no place in a bird's cage.

As most psittacines are agile climbers, a single well-placed perch may be adequate. Birds tend to prefer the highest perch and often use only one even if more are provided. Two perches, one on each end of the cage, should be provided for some species, such as finches, which prefer flying or jumping to climbing. Perches should be placed to prevent droppings from contaminating the bird's food and water and to prevent the bird's tail from contacting the food, water, or floor of the cage.

Although nontoxic wood chips, chopped corn cobs, kitty litter, or sand is convenient as a cage substrate below the grating, stool appearance cannot be monitored. Newspapers, paper towels, or other plain cage liner paper may be preferred, especially for new, sick, or convalescent birds. Because of the potential for birds to chew on the substrate, cedar, redwood, and pressure-treated pine chips are contraindicated.

A daily cleaning of the cage floor and bowls prevents problems associated with food spoilage and alerts the owner to signs of potential illness.

Mite boxes have no value; in the rare pet bird with external parasites other methods of treatment are more effective.

## Cage Covers

Cage covers may be useful in providing a visual barrier for pet birds. Clinical experience has suggested that frequent use of cage covers in newly imported birds may tend to concentrate disease organisms in the confined area and precipitate a disease status that previously had been circumvented by access to fresh circulating air in the import facility or pet shop.

## Toys

Psittacines are intelligent, active animals that need their psychologic as well as physiologic needs addressed. Toys are useful as mental diversions and tend to encourage physical exercise and beak wear; however, they must be selected with care for the safety and psychologic health of the bird. The number of toys in the cage should be limited at any one time, kept clean, rotated to retain interest, and removed if they appear to stimulate a sexual bond in the bird, which is particularly common in budgerigars.

There is little quality control in the manufacture of commercial toys for birds. The chewing characteristics of birds are often not considered, as some balsa wood or leather chew toys contain clips or metal skewers that can snag the bird's beak. The popular lead-weighted penguins are not harmful for budgerigars but can be lethal if chewed and opened by an Amazon. Small link chain items can interlock with a split leg band and break a leg or hang a bird.

In addition to wood perches, safe items that can be offered to promote natural beak activities in lieu of psychologic feather picking include pine cones, clothes pins, natural fiber rope, egg cartons, and soft white pine.[4] Some food forms can serve as "boredom relievers" or "occupational therapy" in helping to provide sources of activity: whole nuts, berries, buds and leaves, corn on the cob, bones, sprouts, whole fruit, coconut husks, and small seeds that take longer to hull and eat.

## Security

Most birds benefit from the availability of a retreat inside the cage for a sense of privacy. Some birds (e.g., conures and lories) enjoy crawling and sleeping in a small paper bag or box. Although *all* pet birds respond favorably to a consistent routine, some species (e.g., cockatoos) are especially sensitive to new situations, and the owner should strive to keep stress to a minimum (see Chapter 1, Choosing a Bird).

### Exercise

Allowing a bird freedom outside the cage may contribute to an exercise program.[5] Supervised flying may be provided for flighted birds. Unflighted birds may be placed on a hand-held perch that is repeatedly moved up and down for 5 minutes at a time. This forces a wing-flapping, which can contribute to conditioning of the cardiovascular, muscular, and respiratory systems.

### Light and Fresh Air

Pet birds need ample exposure to fresh air and sunlight under supervision. Sunlight passing through household windows is not effective for activating vitamin D production because the ultraviolet rays are filtered out. Indoor breeding facilities often use fluorescent fixtures specifically designed to simulate a complete spectrum of light, required also for indoor plant growth. Manipulation of artificial light with timers to mimic increasing day length in the natural environment has been successful in stimulating reproduction in some species.

Contrary to what has been mentioned in pet bird publications, the author has been unable to find a correlation between excessive light periods and feather picking or other illness, although normal seasonal light periods are recommended for healthy birds.

### Temperature

It is unfortunate that publications continue to propagate the notion that drafts kill birds. The truth is that healthy, well-nourished pet birds can withstand a wide range of temperatures and wind and when adapted to a cooler climate may choose to bathe in 30 to 40° F weather or sleep in the rain in preference to a nest box or shelter. Certainly any comfortable ambient temperature for the owner is adequate for the healthy bird.

In contrast, temperature becomes critical in the sick bird. Supplemental heat is a primary first-aid measure in most sick birds, which may require an ambient temperature of 85 to 90° F depending on body weight, fat, and normal body temperature. A sudden drop in temperature with wind (draft) may further compromise the patient and result in its death. In the author's experience, if an aviary bird is found dead following a cold front in Florida, a necropsy usually reveals a compromising factor, such as severe parasitism, which is believed to be the primary factor in its death. In temperatures above 90° F, most pet birds will show evidence of overheating by holding their wings away from their body and panting; heat stroke and death usually occur at higher temperatures.

### Humidity

Exotic pet birds appear to flourish in a captive climate similar to their native climate, although they apparently have some capacity for adaptation to differences in humidity as well as temperature. Birds from tropical origins may benefit from localized increased humidity in the home, such as going into the shower with the owner or frequent spraying of the feathers with water. Again, one may look to avicultural results to see which exotic species are able to adapt to a climate to the degree that reproduction occurs (see Chapter 52, Reproductive Medicine).

### Disease Prevention

Direct exposure to a sick bird is not the only method of disease transmission to a pet bird. Contaminated seeds, toys, cages, and other products brought into the home from shops housing birds or from flea markets (and used without disinfection) may transmit the organisms. Unfortunately, many pet shops move cages of sick birds to the "back room" where seed supplies are also stored. Although bulk-purchased and repackaged seed may in some cases be fresher than prepackaged boxed food, it should be purchased from a store that sells only bird supplies. Otherwise, boxed products can be wiped off with a disinfectant and the contents transferred to another container at home. Owners should insist that nail grinders or scissors for wing clips be disinfected prior to each use. Boarding a healthy bird somewhere other than a pet shop or veterinary bird hospital is wise.

## FEEDING PRACTICES

In addition to physiologic and psychologic requirements, the diet chosen for the pet bird must allow some consideration to ease of preparation for client compliance. The reader is referred to Chapter 31, Nutritional Diseases, for information on factors that may alter the nutritional needs of pet birds, clinical syndromes associated with extreme dietary deficiencies or excesses, and those clinical signs suspected to be related to marginal diets. Reducing diets for obese birds are presented in Chapter 42, Miscellaneous Diseases.

When the specific nutritional requirements of the many species of birds commonly kept as pets become known, adaptations will no doubt be required for the captive diet. As an example, feral Nanday Conures were recently observed to consume exceptionally large quantities of insects.[8] Insects are normally not offered to this species in captivity. Other similar reports occur in the ornithologic literature regarding feeding habits in the wild of common pet bird species.

## Feeding Schedule

Lafeber[5] has advocated removal of ad lib feeding dishes from the cages of pet bird species and the establishment of specific morning and evening feeding times. This technique simulates the food-gathering times in the wild and provides the opportunity for the bird to develop a normal appetite. In this way, the bird may be more likely to accept new foods, and a positive relationship between the feeder and the bird can evolve, especially if the bird is fed outside the cage. Obesity and other nutritionally related problems are less likely to occur, as food is not used as a boredom reliever. In addition, the owner may more readily note any change in the bird's appetite and attitude.

## Diet Content

Recent work at the University of California[3] evaluated the nutritional value of sunflower seed kernels, the mainstay of many "parrot mixes," and compared them to the working estimates (based on poultry and practical experience) of adult nonbreeding cockatiels. Sunflower kernels were found to meet or exceed the needs for energy, protein, and lipids, but no conclusion was reached regarding whether this food source meets vitamin and mineral requirements. Contrary to some advertising claims, this study could find no evidence that sunflower seeds contain a narcotic substance that causes an addiction in birds.

Some information from the current emphasis on fitness and evidence relating to dietary habits in humans may be extrapolated to a degree for animal diets. With some alterations, one may even analogize to pet birds the "basic four" human food categories (I. Fruits and Vegetables; II. Breads and Cereals; III. Meats and Meat Products; IV. Milk and Milk Products) taught to elementary school children. Using this analogy, clients may be more easily persuaded to evaluate the diet offered to pet birds. The relatively sedentary life style of most pet species would preclude a high-fat, high-calorie diet for fitness.

**I. Fruits and Vegetables.** Not only do fresh vegetables and fruits provide essential minerals and vitamins, but they provide a source of seasonal variety with relatively low caloric consumption as well. With regard to the presumed dietary needs of pet birds, emphasis should be placed on feeding dark leafy green or dark yellow meaty vegetables and fruits that contain high levels of nutrients (e.g., carrots, spinach, broccoli, endive, parsley, winter squash, yams, pumpkin, apricots, citrus).

In the author's years of seeing "typical" pet birds, the pet Amazon that most closely approximated the appearance of a healthy bird in the wild was owned by a vegetarian who fed the bird exclusively from her table and skillfully combined legumes and other vegetable products to provide an apparently balanced diet.

**II. Breads and Cereals.** The seeds normally provided to birds correspond to the cereals category in this analogy. If the B vitamins associated with this food group are contained in the hulls, birds that hull their seeds before eating may not be receiving sufficient amounts. Whole-grain bread or commercial products such as monkey biscuits adequately supply these nutrients. Monkey biscuits also supply other nutrients assumed to be required by pet birds, including vitamin $D_3$, and are relished by Amazons and macaws as a dry supplement.[3] Budgerigars, cockatiels, and other small species may prefer these soaked in water or orange juice.

Of the primate biscuits commercially available, the author particularly recommends ZuPreem, Science Diet, and Holiday brands because the vitamins included in the composition are in the form of coated microchips for the purpose of retarding oxidation and maintaining a guaranteed analysis. These products are also packaged in an air-tight plastic-lined bag, which further protects the quality and freshness of the ingredients. Surplus quantities can be frozen until needed.

Seeds do not have to be eliminated from the diet but should be restricted, primarily because of their high fat and low calcium content. Seeds should be selected according to the species (Fig.

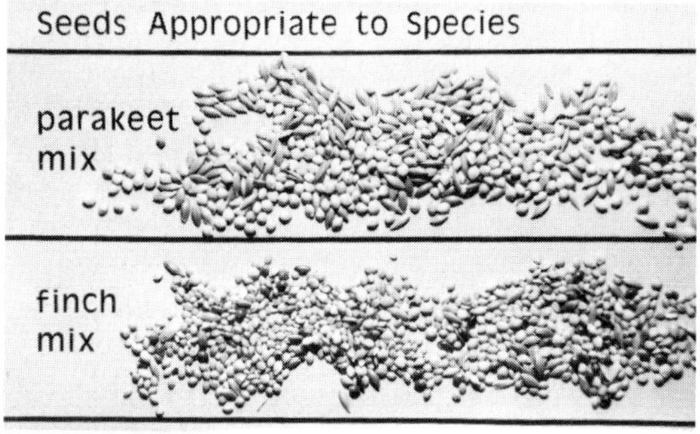


**Figure 2–1.** The seeds offered to birds must be appropriate to the species. For example, finches are not capable of eating the larger seeds packaged as "parakeet mix."

2–1) and tested occasionally for viability by sprouting. Use of ultraviolet light for possible detection of aflatoxin in seeds is a superficial clinical test; however, the results may not be definitive (see Chapter 36, Mycotic Diseases).

**III. MEATS AND MEAT PRODUCTS.** Small amounts of supplemental animal protein (cheddar cheese, hard-cooked eggs, chicken, tuna) or balanced amino acids from vegetable sources (mature legumes) can be offered in an adult maintenance diet. These items are increased in quantity during the breeding and molting seasons. When a hand-feeding formula based on monkey biscuits was tried, a suspected lysine deficiency resulted in yellow pigmented feathers in abnormal areas on Sun Conures, although this has been shown not to be true in cockatiels fed a lysine-deficient diet.[7] (Fig. 2–2).

**IV. MILK AND MILK PRODUCTS.** The analogy to this group of the human basic four is less obvious, although this classification may be considered generally to address a bird's need for calcium. Although the intake of large quantities of milk products (with levels of lactose from 10 to 30 per cent) has been reported to cause diarrhea in birds owing to their lack of the enzyme lactase to digest lactose, this has not been noted by the author. Dairy products such as hard cheeses, cottage cheese, and yogurt are often fed to birds. However, these products cannot be expected to provide totally for the calcium needs of birds, especially for laying hens. The traditional seed diets of birds provide such a badly skewed phosphorus:calcium ratio (from 5.6:1 to 9:1) that provision of calcium to approach the recommended balance of 1:1.5 must be made from almost pure sources of calcium (Fig. 2–3). Calcium carbonate, a superior source of dietary calcium, may be found in oyster shell, mineral block, cuttlebone, and dry wall board (treated or water-proofed dry wall board may be toxic). If cuttlebone is used, the soft side is placed toward the bird; contrary to some reports, however, it is not abrasive enough to be considered a beak conditioner.

The necessity for additional nutritional supplements to a diet similar to that in Table 2–1 is controversial. The primary factor to consider in the selection of a vitamin/mineral preparation for birds is a guaranteed label analysis from a reputable manufacturer. Additionally, one must provide vitamin $D_3$ rather than $D_2$.

Nutritional supplements for birds is another area that has little quality control, and many supplements are extremely overpriced. Many manufacturers combine unprotected vitamins and trace minerals in the same product, which potentiates the oxidation of vitamins. Other antagonistic factors contributing to the oxidative degradation of supplementary vitamins include oxygen, time, moisture, heat, and low pH.[1] These factors, plus the increased potential for bacterial growth in the water dish and the clinical observations suggesting hypovitaminosis

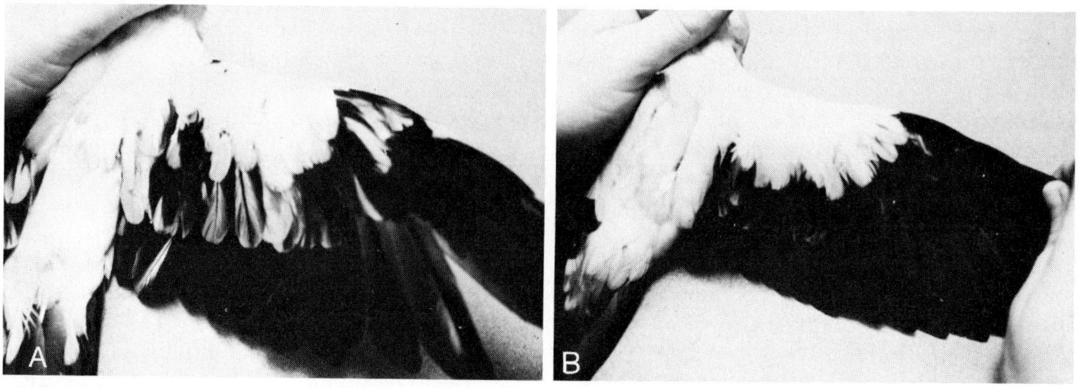


**Figure 2–2.** Yellow pigmentation, believed at the time to be a lysine deficiency in the hand-feeding formula, was observed in the primary and secondary coverts of Sun Conures (*A*). Contrast this to the normal coloration (*B*) that resulted from the addition of cheese to the diet.

**Figure 2–3.** A number of products may be offered to the pet bird for calcium supplementation: *a*, powdered calcium carbonate; *c*, white oyster shell; *d*, mineral/salt block made for rabbits; *e*, aviary size mineral block. An unsuitable supplement *(b)* contains black bits of charcoal (which absorb valuable nutrients) and grey-colored oyster shell that demonstrates a highly radiopaque density suggestive of metal contamination.

in birds offered this form of supplementation, appear to lend support to the author's opinion that the addition of vitamin products to the drinking water is not a good practice. The oil coating of seed hulls for adhesion of vitamins is contraindicated, because the hulls are discarded and the oil serves only to dirty the feathers.

If one tends to accept the recent advertising publicity regarding the nutritional inadequacy of commercially prepared dog foods, it could be assumed that many commercial products and vitamin mixtures being fed to birds, including colored seeds or bits of foodstuff, do not meet the label analysis. Birds on some "complete" seed diets show clinical signs similar to those in chronically malnourished birds (see Chapter 31, Nutritional Diseases). Compressed and sweetened "seed treats" are useless in the husbandry program.

Table 2–1 and Figure 2–4 illustrate a suggested adult maintenance diet that has been used by the author. Table 2–2 offers one method of incorporating additional foodstuff into a pet bird's seed diet. The quantities can be increased for young birds or used as a substitute for seeds.

## Formulated Diets

A number of commercial products have been formulated, primarily from the known nutritional requirements of poultry, and are available in appropriately sized pellets for feeding captive birds. Pellets and crumbles have been specifically developed for cockatiels as a result of

**Table 2–1.** A SUGGESTED DIET FOR MAINTENANCE OF ADULT CAGED BIRDS

| | Amount Per Bird | | | | | |
|---|---|---|---|---|---|---|
| | Canary | Budgerigar | Cockatiel | Conure | Amazon/ African Grey | Macaw/ Cockatoo |
| **Offer Daily** | | | | | | |
| Whole grain bread cubes or monkey biscuit | ¼ Tbs | ½ Tbs | 1 Tbs | 1½ Tbs | 2 Tbs | 4 Tbs |
| Fresh dark green or yellow vegetables | ½ Tbs | 1 Tbs | 2 Tbs | 3 Tbs | 4 Tbs | ½ cup |
| Protein source (cheese, hard-cooked egg, meat, mature legumes) | ¼ size of pea | ½ size of pea | Size of pea | ¼ tsp | ½ tsp | 1 tsp |
| Dry Seeds (two 15-minute periods) | | | | | | |
| Sunflower | 0 | 0 | 1 Tbs | 2 Tbs | 2 Tbs | 4 Tbs |
| Small seeds (canary, niger, poppy, rape, millet, safflower, hemp) | Ad lib | Ad lib | Ad lib | Ad lib | Ad lib | Ad lib |
| **2 to 3 Times Weekly** | | | | | | |
| Fruit (cantaloupe, apricot, apple) | ⅛ tsp | ⅛ tsp | 1 tsp | 1/12 apple | 1/12 apple | 1/6 apple |
| Citrus fruit | 0 | 0 | 0 | 1/12 orange | 1/12 orange | 1/6 orange |
| Fresh corn on the cob | 3 or 4 kernels | ⅛″ piece | ¼″ piece | ½″ piece | ½″ piece | 1″ piece |
| Peanuts | 0 | 0 | 0 | 1 | 2 | 4 |
| **Add Temporarily for New Birds** | | | | | | |
| Vitamin A (from 10,000 IU capsule) | 1 drop/week | 2 drops/week | 3 drops/week | 1 drop/day | 1–2 drops/day | 4 drops/day |
| Yogurt | drop | few drops | ¼ tsp | ¼+ tsp | ½ tsp | 1 tsp |
| **Always Available** | | | | | | |
| Calcium/mineral supplements (cuttlebone, mineral treat block, oyster shell, calcium lactate) | | | | | | |

**Figure 2–4.** Daily rations for captive psittacines should contain fresh wholesome foods suitable for human consumption.

nutritional studies at the University of California. Pellets have been used exclusively in some aviculture facilities; however, they are often supplemented with fresh greens or small amounts of seeds. Converting a bird to a pelleted diet is frequently time-consuming and may be facilitated by mixing the pellets with more familiar food initially (see Chapter 3, Captive Behavior and Its Modification). Table 2–3 lists some sources of formulated diets for pet birds. Diets developed for meat-producing birds (e.g., turkey rations) have an inappropriate protein-to-energy ratio for pet species.[9]

**Table 2–2.** A SUGGESTED DAILY SUPPLEMENT FOR ADULT SEED-EATING BIRDS

**Ingredients**

3 cups whole grain bread cubes or 30 monkey biscuits
½ cup shredded cheddar cheese or 2 hard-cooked eggs, chopped
10 (10-gr) calcium carbonate or lactate tablets
½ cup Clovite vitamin/mineral supplement
Warm water (Flavoring such as maple, anise may be added)

**Directions**

Add enough warm water to bread or monkey biscuits to soften into a dough.
Crush calcium tablets to a powder, mix with Clovite and cheese.
Combine two mixtures.
Roll into balls approximately equal in size to the following measurements:

| Canary | Budgerigar | Cockatiel | Conure | Amazon/ African Grey | Macaw/ Cockatoo |
|---|---|---|---|---|---|
| ¼ Tbs | ½ Tbs | 1 Tbs | 1½ Tbs | 2 Tbs | 4 Tbs |

**To Serve**

Mix daily amount of chopped fresh greens or yellow meaty vegetables into single portion. Remaining balls can be frozen until needed.

**Feeding Schedule**

Remove seeds from cage at night.
Offer seeds for only 15 minutes in the morning.
Remove seeds and replace with ½ portion of supplement.
Repeat for evening meal.
Increase amounts during times of growth, molt, egg-laying, feeding young.
Decrease in summer or if bird is becoming overweight.

## Grit

In the author's opinion, sufficient grit for the efficient functioning of the ventriculus in psittacines is accomplished by providing 10 to 12 appropriately sized grains two to four times a year. Some digestive disturbances and impactions have been associated with overconsumption of ad lib grit. Because grit does not dissolve in the ventriculus, oyster shell and other dissolvable substances do not serve as grit products.

## Seasonal Variations

Avicultural diets tend to incorporate variations during the year. During the heat of the summer following breeding season, the quantity, caloric intake, and protein levels are reduced, with vegetable matter providing the bulk of the intake. Birds tend to need higher concentrations of nutrients, particularly vitamin A, in hot weather because they consume less food.[6] In the fall, a "flush" diet, or sudden increase in the volume and quality of food, signals the anticipation of breeding season. This seasonal change can be incorporated into pet bird care, with the flush diet correlating with molt.

## Feeding Wild Birds

Provisions for wild bird feeding stations should also include supplemental items. A suet/seed mixture is not sufficient, especially for birds that become dependent upon the feeders.

## Special Diet Requirements

Lories, lorikeets, and related species require specialized diets in captivity. A nectar-type formula that contains the assumed nutrients is usually provided. In the author's experience, a satisfactory semiliquid diet is composed of blended monkey biscuits, molasses, fresh apple, and carrots and water. Commercial products are available for nectars. These can be supplemented with spray millet, small seeds, flower heads, fruits, vegetables, meal worms, sunflower seeds, and fresh branches. Because of the high sugar content of the nectars, frequent changing of the solution and cleanliness of the bowls are particularly important to prevent bacterial growth.

## GROOMING

Grooming is not usually necessary for birds that are free of genetic or other disease problems, eat a varied diet, and are kept in a large aviary with access to fresh air, sunlight, rain, and fresh wood for chewing. Indoor pet birds that resist a diet change are prone to require more attention in the care of the beak, nails, and feathers. Owners prefer pets with blunted nails and beaks for comfort in handling the bird. (See Chapter 6, Management Procedures, for beak, nail, and wing trim instructions.)

Although it is a rare practice, it might be ideal for pet birds to be maintained in a natural aviary during solitary hours and then brought into a household cage or stand for contact time with the family.

### Feather Care

Frequently misting or otherwise wetting the feathers appears to be necessary for the overall appearance of most psittacine birds. The application of water also encourages pin feather preening and other grooming activities by the bird. Owners preparing a bird for a show often use cold water misting as often as daily for three to four weeks to "tighten" their bird's feathers, which results in a sleeker, more uniform appearance.

Pin feathers are normally covered with a keratin sheath that the bird grooms to open the feather. These normal keratin flecks may resemble "dandruff" and are occasionally of concern to owners.

Bathing with a detergent product may be necessary for some cases of dirty feathers, especially evident in white cockatoos. In a warm room, a mild solution of Woolite can be gently manipulated into the feathers, allowed to remain for 3 to 5 minutes, and rinsed thoroughly.

**Table 2–3.** SELECTED SOURCES OF FORMULATED DIETS FOR PET BIRDS

| |
|---|
| Lafeber Company<br>R.R. #2<br>Odel, IL 60460 |
| Zeigler Bros., Inc.<br>P.O. Box 95<br>Gardners, PA 17324 |
| The Bird Company<br>P.O. Box 866<br>Poway, CA 92064 |
| Roudybush<br>P.O. Box 331<br>Davis, CA 95617–0331 |
| Nekton USA, Inc. (Lory diet premix)<br>1917 Tyrone Blvd.<br>St. Petersburg, FL 33710 |

Tame birds may be towel- and blow-dried with care. Less tame birds may prefer to dry in the sun.

Oil may be removed from feathers by a similar application of a stronger cleansing product, such as Amway's LOC or Dawn detergent. Light oil may be absorbed and removed by repeated dusting of the feathers with a light powder (e.g., corn starch) and subsequent brushing out with a very soft child's toothbrush.

## REFERENCES


1. Adams, C. R. : Effect of environmental conditions on the stability of vitamins in feeds. *In* Effect of Processing on the Nutritional Value of Feeds. Washington, DC, National Academy of Sciences, 1973.
2. Fowler, M. E.: Metabolic bone disease. *In* Fowler, M. E. (ed.): Zoo and Wild Animal Medicine. 2nd ed. Philadelphia, W. B. Saunders Co., 1986.
3. Grau, C. R.: Exotic Bird Report No. 1, Dept. of Avian Sciences, University of California, Davis, 1983.
4. Harrison, G. J.: Feeding practices for psittacines and passerines. *In* Fowler, M. E. (ed.): Zoo and Wild Animal Medicine. 2nd ed. Philadelphia, W. B. Saunders Co., 1986.
5. Lafeber, T. J.: Material presented at Mid-Atlantic States Avian Veterinary Seminar. Atlantic City, NJ, 1984.
6. Morrison, F. B.: Feeds and Feeding. Clinton, IA, The Morrison Publishing Co., 1959.
7. Roudybush, T. E., and Grau, C. R.: Lysine requirements of cockatiel chicks. Proceedings of the Western Poultry Disease Conference, Davis, CA, 1985, pp. 113–115.
8. Schmidt, R. E., and Matison, M. T.: Insectivory by the Blackhooded parakeet (*Nandayus*) in southern New York. Kingbird 33:13–14, 1983.
9. Schultz, D. L.: Nutrition. *In* Macwhirter, P. (ed.): Bird Health in Cage, Loft and Aviary. Australia. In Press.

# *Chapter 3*

# CAPTIVE BEHAVIOR AND ITS MODIFICATION

GREG J. HARRISON
CHRISTINE DAVIS

Training encompasses any activity that is used to modify behavior. With regard to birds, training is frequently thought of as the teaching of tricks; however, training also includes taming and changing undesirable habits in pets.

Bird owners are usually most concerned with the results of training when they want to tame a wild bird into an affectionate pet. Frequently they know little about the bird and lack the commitment to obtain the appropriate knowledge and devote the required time to the project. Unfortunately, violence in some form—hitting the bird or throwing it to the floor after being bitten—is too often a part of the process. If the owner is ill-prepared to train the bird, alternatives should be sought. An already tame or hand-raised bird may be purchased, or a knowledgeable experienced trainer may be hired initially to modify the bird's behavior.

In the rare instances in which a true "bronco" or untrainable bird is encountered, the bird is destined to be only a display bird. Many behavioral tendencies are inherited, and it is not recommended to breed birds with unpleasant dispositions.

Behavioral problems of tame birds may relate to boredom or lack of discipline. All bird owners need to avail themselves of information on training if they want a bird that remains adaptive to the lifestyle of the family. A knowledge of basic modification principles is essential for maintaining desirable pet qualities in the face of behavioral deviations of the bird associated with age, hormones, or environmental changes.

Creating a situation in which the bird performs tasks (stepping up on a perch, allowing handling) in exchange for reinforcement (affection, attention) provides a purpose for the bird and results in approved behavior. Methods of obedience or "forced" training, such as restraining the bird into submission, may break the bird's will to resist, but the relationship will be based on fear. To have a satisfying, trustful relationship with a pet bird, one must use behavior modification. Behavior modification accomplishes feats not possible with the use of force.

## GENERAL PRINCIPLES OF BEHAVIOR MODIFICATION

B.F. Skinner trained pigeons early in the development of his behavior modification principles. The basic philosophy was to reward and reinforce positive behavior and ignore or avoid negative actions.

According to Karen Pryor, whose book *Don't Shoot the Dog* offers behavioral training guidelines applicable to birds, good training insights do not necessarily lend themselves to verbal expression. She believes many trainers are concerned with the methods and not with the key principles that can be applied to all animals.

### Communication

Pryor believes training is a two-way communication between an instructor and a subject in which something is done *with* the subject, not *to* the subject, to create a change in subsequent behavior. One needs to create a language or some form of communication to convey one's approval of the behavior. In order for communication to evolve, the trainer must know the species, body language, and appropriate reinforcement for the species.

From experience, the authors believe verbal communication with psittacines is a powerful adjunct to training of this species. Irene Pepperberg's work with Alex, the African Grey, suggests that a parrot is able to participate in some form of interspecies communication and is capable of demonstrating a degree of cognitive behavior beyond simple mimicry of human speech patterns. Functional vocalization by Alex was developed relating to a variety of items after it was established that the bird had definite preferences in characteristics of objects: color, size, shape, and texture.

The primary technique utilized in training Alex was what Pepperberg called the model/rival approach. In this procedure, humans demonstrated to the parrot the types of interactive responses desired. One human acted as a trainer of a second human, asking questions, giving praise and reward for correct answers, showing disapproval for incorrect answers. The second human acted both as a model for the bird's responses and as a rival for the trainer's attention. Rather than extrinsic rewards such as food, Alex was rewarded with acquisition of the object (blue clothespin, three-sided rawhide). In this way, Alex was trained not only to mimic the words, but to identify, request, and/or refuse more than 30 objects by means of verbal labels.

These experimental levels of training a psittacine are beyond the average pet owner's concerns, but they suggest that levels of communication and relationship beyond what were previously believed possible can be developed with African Greys and probably other psittacines.

The result of purposeful communication during training is what is known as shaping. Shaping is customized for the individual bird and developed by the trainer with the bird, whereby small tendencies in the right direction are reinforced one step at a time toward the ultimate goal. Shaping includes knowing when to press on, when to let up, how to move on to the next goal, what to do when you run into trouble, and, most importantly, when to quit. Table 3–1 lists the 10 rules for shaping developed by Pryor, which can be applied to working with birds.

**Table 3–1.** TEN RULES FOR SHAPING BEHAVIOR*

1. Raise criteria in increments small enough so that the subject always has a realistic chance for reinforcement.
2. Train one thing at a time; don't try to shape for two criteria simultaneously.
3. Always put the current level of response onto a variable schedule of reinforcement before adding or raising criteria.
4. When introducing a new criterion, temporarily relax the old ones.
5. Stay ahead of your subject. Plan your shaping program completely so that if the subject makes sudden progress you are aware of what to reinforce next.
6. Don't change trainers in midstream; you can have several trainers per trainee but stick to one shaper per behavior.
7. If one shaping procedure is not eliciting progress, find another; there are as many ways to get behavior as there are trainers to think them up.
8. Don't interrupt a training session gratuitously; that constitutes a punishment.
9. If behavior deteriorates, "go back to kindergarten"; quickly review the whole shaping process with a series of easy reinforcements.
10. End each session on a high note, if possible, but in any case quit while you're ahead.

*From Pryor, K.: Don't Shoot the Dog. New York, Simon and Schuster, 1984. Reprinted with permission.

## Reinforcement

A primary factor in behavior modification is reinforcement. Positive reinforcement is anything that, occurring in *conjunction* with an act, tends to increase the probability that the act will occur again. Constant reinforcement is needed in the learning stages but can become random and unpredictable as the skill is improved. Psychologists have found that a variable schedule of reinforcement is far more effective in maintaining a behavior than a constant, predictable schedule of reinforcement.

Reinforcements are relative, not absolute, and depend upon the species and the individual. Examples of positive reinforcements are stroking, scratching, social interaction, toys, food bits, water, attention, and a soft voice. Because of the nature of psittacines, stroking, scratching, and tone of voice are effective reinforcements. It is important that the bird associate the reinforcement with the specific activity desired. One of the authors (Davis) uses a soft, rhythmic speaking voice in conjunction with slow, deliberate movements to train a bird to accept touching, the initial step in her taming program. The most common problem with beginning trainers is reinforcing too late.

Negative reinforcement is something the subject is willing to work to avoid. Such an unpleasant event or stimulus, no matter how mild, can be halted or avoided by the subject's changing its behavior. Negative reinforcement, such as a sharp verbal "No!", can be effective, again if in direct association with the undesirable activity. Other negative reinforcements are blowing, time out, and rebuke. "Out time" has been successfully applied to children as well as animals and serves as social isolation for continued negative behavior. The bird can be put in a dark closet for one minute for the first offense, progressing up to 15 minutes if necessary; a longer time period is apparently ineffective. This may be an effective treatment for a screaming bird; the bird should be moved from "out time" the moment the screaming stops. A separate area not normally occupied by the bird can be utilized for "out time." It is not appropriate to use the bird's cage or other location associated with good behavior for this purpose.

Negative reinforcement must be differentiated from reinforcement of negative behavior. Running in to yell at the bird to stop picking at itself is a "reward" for negative behavior; the bird gets what it wants—attention from the owner—and that method is reinforced as the way to get it.

Punishment is not an effective modifier of behavior. In addition to creating fear, anger, resentment, and mental states not conducive to learning, it is usually administered after (in some cases, long after) the behavior has occurred, when the bird has no way to relate it to the behavior.

Although food is used in some training programs, it is not an effective enforcement if the subject is full, nor is severe food deprivation advised. If modifications of this method are used, a portion of the daily rations can be saved for use during training. Food reinforcement is less commonly used by modern animal trainers.

Some birds, such as young birds with a short attention span, require special attention. Physically impaired, recuperating, or fearful birds require extra patience. One must gradually teach wild birds that have been mistreated to have faith in people again.

## TAMING A BIRD

Prior to taming or other behavior modification procedures, the bird should be checked by a veterinarian to confirm its health, and it should be sufficiently adapted to family life to perform normal preening and eating activities in the presence of family members. It is essential to plan ahead and establish a set routine for training with a predictable location, time of day, person, and perhaps even a training uniform. The bird will soon associate this with the work time and should not be expected to initially perform for reinforcement at times other than those specified for training. Maintaining set times lets the bird know what to expect, and a higher level of cooperation is achieved.

A separate environment such as a particular corner of a room or a bathroom is used to

**Figure 3–1.** Placing the bird on a waist-high T-stand perch conveys dominance of the trainer.

eliminate distractions, so the full attention of the bird is directly upon the trainer. One of the authors (Davis) prefers to use a low T bar stand for initial taming techniques (Fig. 3–1). The trainer's behavior must be firm, predictable, and appropriate and must assert dominance without creating fear in the bird. A bird's wings are clipped for training to assist in conveying to the bird that the owner has control of the situation. A prospective trainer may wish to practice in front of a mirror to develop and project a calm, confident manner and to practice tones of voice: demand or command ("No!"), praise or baby talk, conversation. If unpleasant tasks such as nail grinding have to be performed during the time period that active training is going on, the owner/trainer is advised to don a costume or mask to avoid interruption of the relationship. Alternatively these tasks can be performed by someone else.

### Taming and Handling Procedures

If the bird is made to feel pleasure from the hands at the onset of taming, he will no longer be frightened of the prospect of being touched, and perch or arm training will follow in extremely rapid progression. Figures 3–2 to 3–5 illustrate the steps Davis incorporates into the initial taming sequence. (Methods of capturing a bird are presented in Chapter 7, Preliminary Evaluation of a Case.)

Touching the bird should be intermittent, stopping after each gentle approach for a few seconds, then slowly "nibbling" the feathers

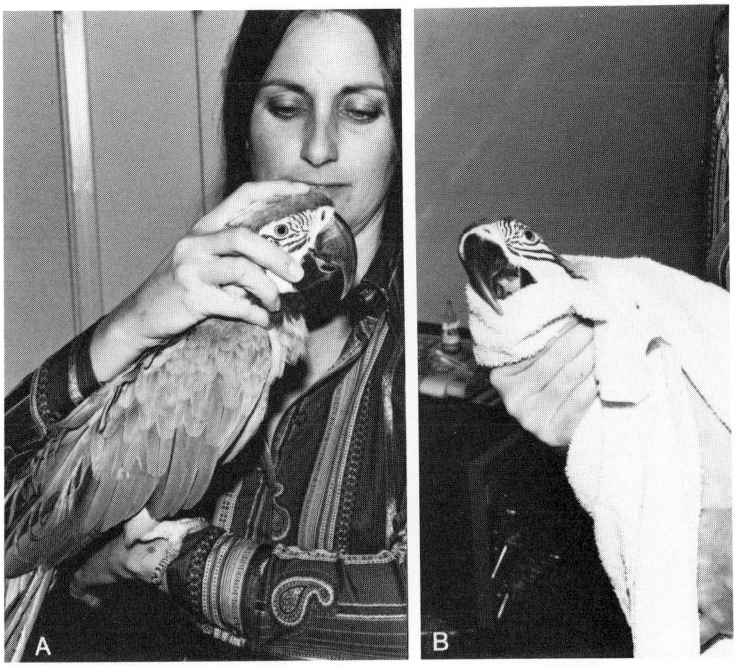


**Figure 3–2.** Davis uses a "helmet" grip *(A)* or, in the case of extremely large birds, a neck grip *(B)* to gain initial control over the bird.

with the thumb and forefinger to imitate the bird's natural preening behavior. When the handler feels secure, he can start at the back and sides of the bird's head. The bird should not be approached, however, over the top of the head, as this is interpreted by the bird as threatening. It is important to stop the handling while the bird is enjoying it (positive note) to encourage the anticipation of future sessions.

When the bird is accepting petting, a stick can be placed under its belly and against its legs to encourage stepping up. This process may have to be repeated until the bird remains on the stick. When the trainer is working from a perch, the bird must be returned to the security of the perch immediately upon successful completion of stepping on the stick. This time. period is extended after the bird gains some familiarity with the procedure.

When the bird accepts being seated on the stick for extended time periods, going to the hand is the next step. As the stick is slanted upward toward the handler's body at a 45-degree angle, the bird will move up to seek the highest point and the hand holding the stick can be carefully maneuvered under the bird.

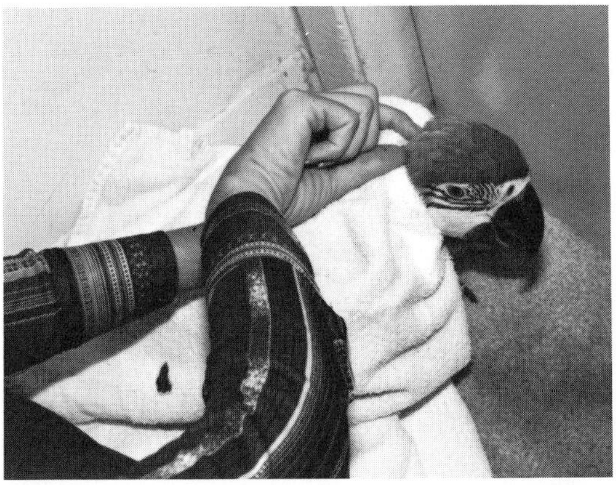

**Figure 3–3.** After the bird is in a normal standing position in the corner, perhaps with the capture towel still resting on its shoulders, the trainer can begin to gently preen its tail feathers (near the oil glands) or the back of the neck.

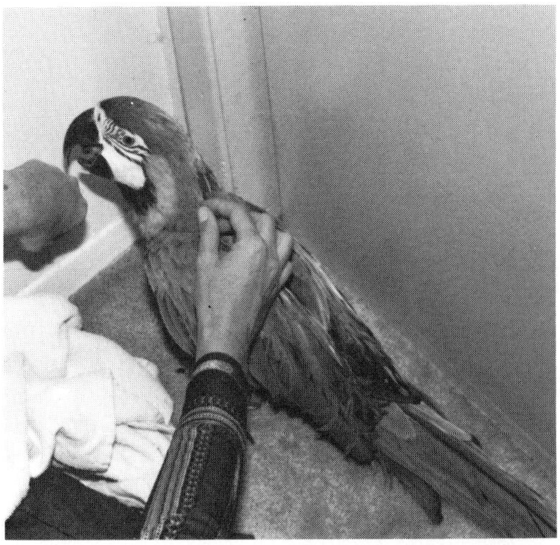

**Figure 3–4.** The back of the tightly fisted free hand may be offered as a decoy when approaching the head. Do not offer fingers or knuckles.

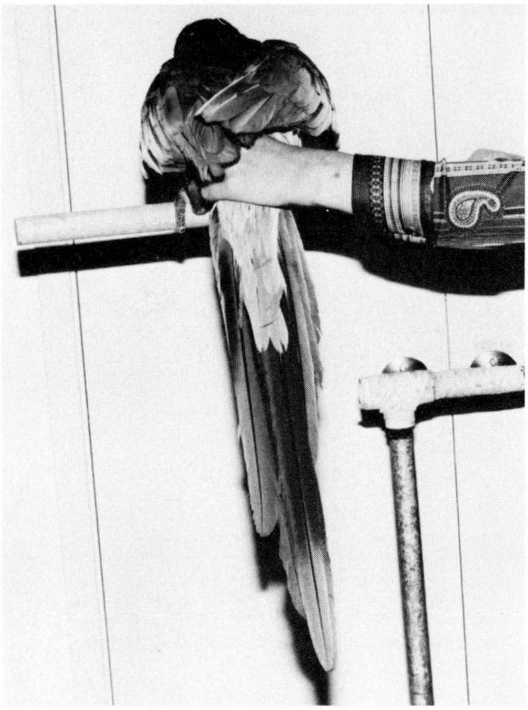

**Figure 3–5.** The trainer may need to gently hold on to the fleshy base of the bird's tail to prevent escape during stick training.

The free hand is used as a decoy to prevent biting. One must resist doing too much too fast. Many small steps can make the training process faster, and the experience is ultimately less frightening for the bird.

## Changing the Diet

Attempts to change the diet are not recommended during times of stress for the birds, such as immediately following a change in environment or in times of illness. An acclimation time in the home is required for a newly purchased bird with regard to diet change as well as other behavior modification attempts.

Altering the diet of the pet psittacine requires persistence and time as well as ingenuity. If the bird consumes its seed diet satisfactorily but will not touch fresh food, pieces of soft fruits and vegetables can be very thinly sliced, shredded, or chopped finely over the dry food mixture in the cup. Even if the bird merely discards the food on its way to the seed, bits will adhere to the beak and be tasted. Altering the feeding schedule from ad lib to specific time periods will encourage the bird to try what is offered because of hunger (see Chapter 2, Husbandry Practices). Limiting the feeding times to two meals of 15 minutes each following a 15-minute training session seems to produce favorable results, if the bird has not been upset by the training.

To incorporate taming with the positive feeding ritual, the bird should be encouraged to come out of its cage for the food bits. One taming suggestion is to offer bits of food from a spoon, gradually moving one's hand closer to the bowl of the spoon until the bird will accept the food from the hand. Perching the bird by the family's dinner table has been successful in some cases to encourage socialization and varied food consumption. Use of another bird as a role model may also introduce a reluctant bird to unaccustomed food items.

Because of the perceived preferences of parrots in regard to taste, shape, size, texture, color, and temperature, some of these factors may be incorporated into items as "bait" for attempted diet change. A preferred "bait" may be mixed with seeds and then mixed with the desired items. Some owners have gone so far as to color different shapes of pound cake to use as bait. Other successful bait items include popcorn, pretzels, and miniature peanut butter and jelly or grilled cheese sandwiches.

The color and taste of vegetables may be accepted more readily if they are offered initially in a pureed form with the drinking water a few days a week. Alternatively, V-8 juice may be diluted and served.

## IMPRINTING

Imprinting, or the identification of an animal with what is normal in its environment, was first described by the European animal behaviorist Konrad Lorenz in his studies of Greylag geese. Because young animals may imprint on other than their own species, this characteristic may be used to advantage in the hand rearing of baby psittacines. Hand-raised and imprinted psittacines identify with humans and readily accept their companionship. Because dietary habits in the wild are based on imprinting (eating what the parents eat), the introduction of a wide variety of foods to the human-imprinted bird avoids many of the difficulties in converting a wild-caught bird to an acceptable captive diet.

The imprinting of hand-raised psittacines to humans was initially believed to interfere with the species' mating preferences. On the contrary, the physiologic influences are apparently more dominant, as imprinted psittacines have been shown to exhibit normal reproductive behavior. These birds in some cases have reached sexual maturity at an earlier age and are superior in parenting abilities, as they are less disturbed by human intrusion. They usually exhibit normal aggression toward outsiders during breeding season.

Some negative aspects of the imprinting of hand-raised birds on humans may be expressed as vices.

## VICES

The particular manifestation of aberrant behavior, such as biting, screaming, feather picking, or nervous mannerisms, may be dependent upon the species, the individual bird's personality, or any external environmental situation that may incite it. Hand-raised psittacines that have been pampered may become so demanding of the owner's time that vices occur as a result of normal separation. Recognition of this possibility and setting a schedule of regular limited play times that the bird can depend on from the very beginning may encourage a more satisfying relationship.

There have been instances in the authors' experience of commercial hand-raising facilities producing nonsocialized animals owing to insufficient individual attention. These birds tend to prefer isolation and require behavior modification to establish pet quality personalities.

### Biting

Regardless of the reason for biting, opportunities for the bird to bite should be avoided in order to discourage the habit from being established.

A bird may bite for several reasons : (1) fear, (2) jealousy (which may be analogous to a perceived threatened relationship by a small child), (3) increased aggression due to sexual maturity, (4) warning behavior, and (5) expression of dominance. Some distinction should be made between biting that is an instinctive reaction to fear and aggressive biting.

To prevent biting during the initial training of a bird to leave the cage and perch on a stick, a removable barrier between the owner's hand and the bird's end of the stick is useful. Because, with few exceptions, birds will seek heights, this barrier protects the trainer's shoulders and upper body as well as his hand as the bird climbs the stick. When the bird is not on the protected stick, it must be allowed only two places: inside its cage or on a perch level with the trainer's waist. Often fear and aggressive behavior subside when the bird is in this subordinate position.

Baby birds, like puppies, need to be trained early not to use their mouths in playing or affection. They must be gently discouraged from undesirable biting of the owner by being offered a wide variety of colors and textures of chewable rewards.

If the bird still becomes "nippy," it must be given a verbal reprimand such as a sharp "No" and may be isolated ("out time") for a period of 5 to 10 minutes. A sudden onset of biting may be related to an external situation and may be corrected by identifying and removing the cause.

### Screaming

It is important to determine when loud vocalization is most evident in order to correct the problem. A number of factors may contribute to increased noise levels in the psittacine:

1. The A.M./P.M. "screamer" is analogous to a crowing rooster; it is a natural response and can be difficult to control. Some suggested methods may be attempted. The food can be removed six to eight hours before the bird becomes noisy and then offered just *before* the onset of the yelling episodes. It is important that the bird not be fed *in response* to loud vocalizations. As birds are creatures of habit, the timing can be fairly reliable. A toy can be offered in lieu of food if the bird is not hungry, or exercise may be provided. Alternately, the bird may be distracted by moving it closer to the owner. The bird must be distracted *before* the verbalization begins so that the reward will not reinforce the negative behavior.

2. Exuberant screaming is usually in response to the owner's returning home. Avoidance of the problem works well in cases such as this. The owner may go directly to the bird upon arrival and provide attention (perhaps a favorite toy that has been withheld during the day) at the onset of the owner's return to prevent prolonged noisiness in the pet and a more enjoyable parrot-owner relationship. If possible, perching the bird near the owner while he or she goes about normal household duties satisfies the bird's need to be near its owner without actually being held.

3. Screaming may be an expression of discontent, such as if a bird has been moved to an

unfamiliar location within the home and becomes fearful. The only solution may be to return the bird to its preferred location and introduce it gradually to the new area.

4. A good "talking" bird will usually be noisier in general than nonverbal birds. Frequently, if other birds are in the household, the calling to each other may reach deafening proportions. The A.M./P.M. times may be natural for verbalization training.

## Psychogenic Water and Food Consumption

Although overconsumption of water may be an inappropriate response for the recently weaned psittacine unsure of eating, it may also be a nervous habit, and the principal cause of anxiety must be determined and controlled. Because this may become habitual, access to water should be limited. Fruits and vegetables with high water content should be substituted to compensate for the desire for increased liquids. A small mammal watering bottle with a metal tube tip can be used to provide access to, but decreased consumption of, water.

Psychogenic eating may result from boredom as well as nervousness, and restricted access to food is advised. This is usually not a problem when specific feeding times are scheduled.

## Other Nervous Behavior

Panting, wheezing, sneezing, coughing, and foot-stomping may all be expressions of fear and insecurity. The environment of a high-strung bird must be controlled to minimize the anxiety-producing situations until appropriate behavior modification techniques have produced the desired results. As mentioned, training should not be instituted until the bird is comfortable in the family.

# MODIFICATION OF SEXUAL BEHAVIOR IN THE SINGLE PET BIRD

Budgerigars and other species domestically bred for the pet market are genetically selected for high production and multiple clutches and have high reproductive interest year round. These normal physiologic drives can be expressed in a number of ways by the single pet bird. Perhaps many of the antics of birds, such as kissing, strutting, and even talking, may be exaggerated or aberrant prenuptial activity.

**Figure 3–6.** Normal courtship activity in budgerigars includes regurgitation, which may be misinterpreted as vomiting in the single pet bird.

## Regurgitation and Masturbation

Regurgitation is the most common form of sexual display observed in pet birds, most commonly seen in male budgerigars (Fig. 3–6). Toys and mirrors in the cage become the object of pair bonding. Some owners consider this activity a form of entertainment, while others find it distressing. It is Harrison's opinion that such activity on a constant year-round basis, coupled with malnutrition, may lead to digestive and/or hormonal disturbances. Although genetic tendency is also a factor, it is interesting to speculate whether this constant prenuptial readiness may relate in some way to the high incidence of kidney and gonadal tumors in single pet budgerigars, as these tumors are seldom seen in paired aviary birds of the same species.

In the wild, an abundance of food and longer daylight hours signal the appropriate time for breeding; thus, reduction of caloric intake and light for the single pet bird may reduce this tendency. One suggestion that has produced favorable clinical response is to limit the total daylight during this time to five or six hours each day for a period of five days. The daylight would not be offered all at once, however. One could place the bird and cage in the dark for four hours, bring it back to the light for an hour so the bird could eat, then back into the dark for another four hours, and so on. At the same time, cheese, high-fat seeds, and other high-calorie foods would be limited.

Sometimes regurgitation may be prevented by removal of the toys or mirrors in the cage or rotation of toys so that the bird does not become attached to a particular one. Regurgitation has also been observed in larger species, especially Amazons and macaws; in these cases it is most often the owner who has become the pair bond partner.

Masturbation has been observed most commonly in budgerigars but sperm has also been noted in cockatiels, cockatoos, Amazons, and macaws. Owners report incidences of the bird mounting various toys in the cage and even the hands, arms, or shoes of the owner. Some birds actually deposit a copulatory fluid, but sperm is usually absent in these samples. The owner should ignore this activity and substitute a positive behavior. Techniques similar to those offered for regurgitation may help in reducing the incidence by reducing the physiologic stimuli.

## Chronic Egg Laying

Many owners are surprised to learn that a single female pet bird can lay eggs. Ovulation in most psittacines has nothing to do with the presence of a male. Chronic egg laying by a single pet bird most often occurs in highly domestic species, such as cockatiels that are well-adjusted to their environment, but can occur in Amazons and macaws as well as other species.

The most significant problems associated with this are the potential for malnutrition in the hen from depletion of body stores of calcium and other nutrients and the complaint by the owners that the bird is not a good pet during this time.

Owners often ask if the eggs should be removed as they are laid or allowed to accumulate. One may advise with larger species such as Amazons or African Greys, which are specifically seasonal in their breeding, to leave the eggs where they are. The incubation patch on the chest "assesses" the number of eggs laid and stops reproductive activity when the appropriate number for that species' clutch size is reached. Once the full clutch size is laid, the body hormones change and the female shifts into broodiness until the time the eggs should be hatched, completing the cycle. Taking away each egg as it is laid may only prolong the process.

If the owners elect to allow for the completion of the reproductive cycle, they will note during the brooding stage that the bird will stay on the nest, reduce food consumption, and possibly increase aggressive behavior. At the end of the normal incubation period for that species, the eggs may be removed and discarded.

Year-round breeders, such as cockatiels, may go back into another reproductive cycle as soon as one is complete. Chronic egg laying in cockatiels can be controlled to some degree by lowering caloric intake, by avoiding stroking and handling by the owner, or by progesterone- or testosterone-derived hormone injections. Effects of the hormone injections are short-lived, and higher doses may be required in subsequent treatments. Some minor stress in the bird's life to break the pattern, such as moving the bird to a different room to make it less secure about its nesting area, may work in some cases. It is Harrison's opinion that hysterectomy is the treatment of choice for chronic egg-laying cockatiels (see Chapter 48, Selected Surgical Procedures).

## Other Sexual Behavior

Amazons and macaws exhibit sexual patterns that are sometimes misinterpreted by the owner as problems. These species have a tendency to crouch low on the perch, raise the wings at the elbows, and flutter the wings in sexual display. Amazons in particular may fan out the tail feathers. Imprinted or very tame birds may pair bond with the owner, choosing a particular individual within the family. The choice was once thought to be sex-related (male bird attracted to female owner, etc.), but this has not been substantiated. These larger species may become more protective of their territory (cage) and exhibit aggression toward other family members during this prenuptial time.

In a household with several species of pet birds, a tendency toward extraspecies attraction may occur. It is not uncommon to note preening and mating behavior between a cockatoo and macaw, for example.

Some birds will demonstrate excessive tearing of newspaper or other cage liners and will stay at the bottom of the cage. Observant owners can sometimes predict imminent egg laying by the bird's desire for stroking. Budgerigars, finches, or canaries may be found nesting in a seed cup, causing the owner to think that the bird is sick or has a broken leg.

Birds going through prenuptial activity become more vocal, especially in the morning and evening, and the appetite will noticeably increase in females up to the point of egg laying. The cuttlebone in the cage will often be devoured; this is believed to relate to changed estrogen levels from follicular development. Immediately preceding egg laying, the female's food consumption will dramatically decrease.

## Other Symptoms of Sexual Frustration

Some cases of feather picking and other self-mutilation may be related to misplaced aggression, whether directed at a cage mate or self. Scruffy feathers may result from inability to complete the normal reproductive cycle, as molting usually occurs following the breeding season in the wild. There is not the opportunity to complete this cycle in the single pet bird.

## RECOMMENDED READINGS


1. Boyd, T. L., and Levis, D. J.: Exposure is a necessary condition for fear-reduction: A reply to DeSilva and Rachman. *In* Rachman, S. (ed.): Behavior Research and Therapy, 21:2, 1983.
2. Davis, C.: The biting parrot. Bird World, Sept., 1983, pp. 8–9, 16–17.
3. Davis, C.: Parrot taming forum. Bird World, 6(4):63–64; 6(5):14, 19–22; 6(6):14, 16, 30, 1984; 7(1):20–24, 26, 28, 1984.
4. Edwards, C. A., et al.: Imitation of an appetitive discriminatory task by pigeons. Bird Behav., 2:57–86, 1980.
5. Foster, J. W.: Socialization as a means of enrichment for mother raised zoo animals. Ann. Proc. Am. Assoc. Zoo Vet., 1978, pp. 81–87.
6. Foster, J. W.: Behavior of captive animals. *In* Fowler, M. E. (ed.): Zoo and Wild Animal Medicine. 2nd ed. Philadelphia, W. B. Saunders Co., 1986.
7. Foster, J. W.: Animal behavior and its application to aviculture. Proceedings of the Veterinary Seminar, American Federation of Aviculture, Las Vegas, NV, 1981.
8. Fox, M. W.: Animal freedom and well-being: Want or need? Appl. Anim. Ethol., 11(3):205–209, 1984.
9. Gibbs, M. E.: Memory and behavior: Birds and their memories. Presidential Address, Australian Society for the Study of Animal Behavior, University of New England, May, 1982, Vol. 4, pp. 93–108.
10. Howell, D. J.: Optimization of behavior: Introduction and overview. Am. Zool., 23(2):257–260, 1983.
11. MacRoberts, M. H., and MacRoberts, B. R.: Animal communication theory: Mentalism versus naturalism. Bird Behav., 2:57–86, 1980.
12. MacRoberts, M. H., and MacRoberts, B. R.: The referent in animal communication. Bird Behav., 1(3):83–92, 1979.
13. Pepperberg, I. M.: Cognition in the African Grey parrot: Preliminary evidence for auditory/vocal comprehension of the class concept. Anim. Learn. Behav., 11(2):179–185, 1983.
14. Pryor, K.: Don't Shoot the Dog. New York, Simon and Schuster, 1984.
15. Rogers, L. J., and McCulloch, H.: Pair-bonding in the Galah *(Cacatua roseicapilla)*. Bird Behav., 3(3):80–92, 1981.
16. Schultz, D. L.: Nutrition. *In* Macwhirter, P. (ed): Bird Health in Cage, Loft, and Aviary. Australia. In press.

*Section Two*

# THE NORMAL BIRD

# Chapter 4

# CLINICAL ANATOMY

## WITH EMPHASIS ON THE AMAZON PARROT

JOHN S. McKIBBEN
GREG J. HARRISON

Although the anatomy of the chicken and budgerigar has been extensively investigated, fewer anatomic studies have been conducted in larger psittacines, and many of those have little clinical significance. In approaching literature and in recording personal observations, the authors have endeavored to include anatomic data that have direct application in clinical situations. Further in-depth descriptions of avian structure may be found in references cited at the end of the chapter.

## THE INTEGUMENT

### Skin

A complete treatise on avian integument has been published by Lucas and Stettenheim.[21] The skin of birds lies upon the subcutis, which is often thick and contains an abundance of fat in older psittacines. Fat may also be dermal, which may predispose certain birds to xanthomatosis.

Except for the meibomian glands of the eyelid, the uropygial gland, and holocrine glands of the external ear canal, the skin of birds is deficient in glands. However, lipoid spheres are elaborated by the epidermis in various parts of the skin.[21] Infections of the meibomian glands and occasional neoplasms of the uropygial gland have been noted.

The uropygial (oil or preen) gland of many birds is bilobed and lies dorsal to the levator muscles of the tail and the pygostyle. Its excretory ducts empty at the oil gland circlet on the oil gland papilla. Uropygial glands are absent in Amazon Parrots. Well-developed uropygial glands have been observed in budgerigars and Green-winged Macaws, whereas they appeared less developed in cockatiels, finches, canaries, cockatoos, lovebirds, African Grey Parrots, Eclectus Parrots, and Blue and Gold Macaws. Uropygial glands have not been observed in the Hyacinth Macaw.

Modifications of the skin of birds may take various forms.[21] In psittacines these include scales on the legs and feet, claws, beak, cere, crest, cheek patches, and incubation patch. Crests, dorsal on the head of birds, can be bony, fleshy, or plumed as in cockatoos and cockatiels. Genus variations exist with these modifications. For example, the cere of the Amazon Parrot has tiny bristle (setae) feathers on it, whereas the cockatoo's cere is completely covered with contour feathers.[8]

Many of these modified structures are often colorful and may vary between males and females (see Chapter 51, Sex Determination Techniques). Cheek patches of some psittacines are able to "blush" or redden when the bird is alarmed. This should not be interpreted as a pathologic condition. This area may also show evidence of bruising, which is initially pink and later changes to a darker color.

Smooth muscles deep to the skin are attached to the walls of feather follicles and are responsible for the feather fluffing often observed in cold or sick birds trying to conserve heat. Plumed crests are raised by dermocutaneous muscles.

The wing skin fold, or propatagium (wing web), is located in the angle between the arm and forearm. This membrane has been used to tattoo psittacines but does not retain the detail achieved by tattoos on the skin over the breast muscle.

The ventral skin of the bird modifies during nesting to form an incubation area (brood patch).[21] This area has fewer feathers and increased vascularity and serves to increase warmth for incubation of eggs. Through a feedback mechanism, the area may help regulate the number of eggs laid within a standardized clutch size for the species.

## Feathers

Feathers are epidermal structures analogous to hairs of mammals which comprise about 10 per cent of the body weight. Feathers serve as insulators for maintaining body temperatures (usually between 102° F and 109° F), as displays during courtship and fighting, as nest material, and as necessities for flight. Feathers arise from tubular invaginations of the skin known as feather follicles. These folicles (Fig. 4–1) are arranged in patterns or rows (feather tracts or ptertylae) that are separated by relatively unfeathered areas (apteria). Vessels, bones, ligaments, and muscles, which may be seen through the frequently transparent skin of the apteria, have occasionally been misidentified as

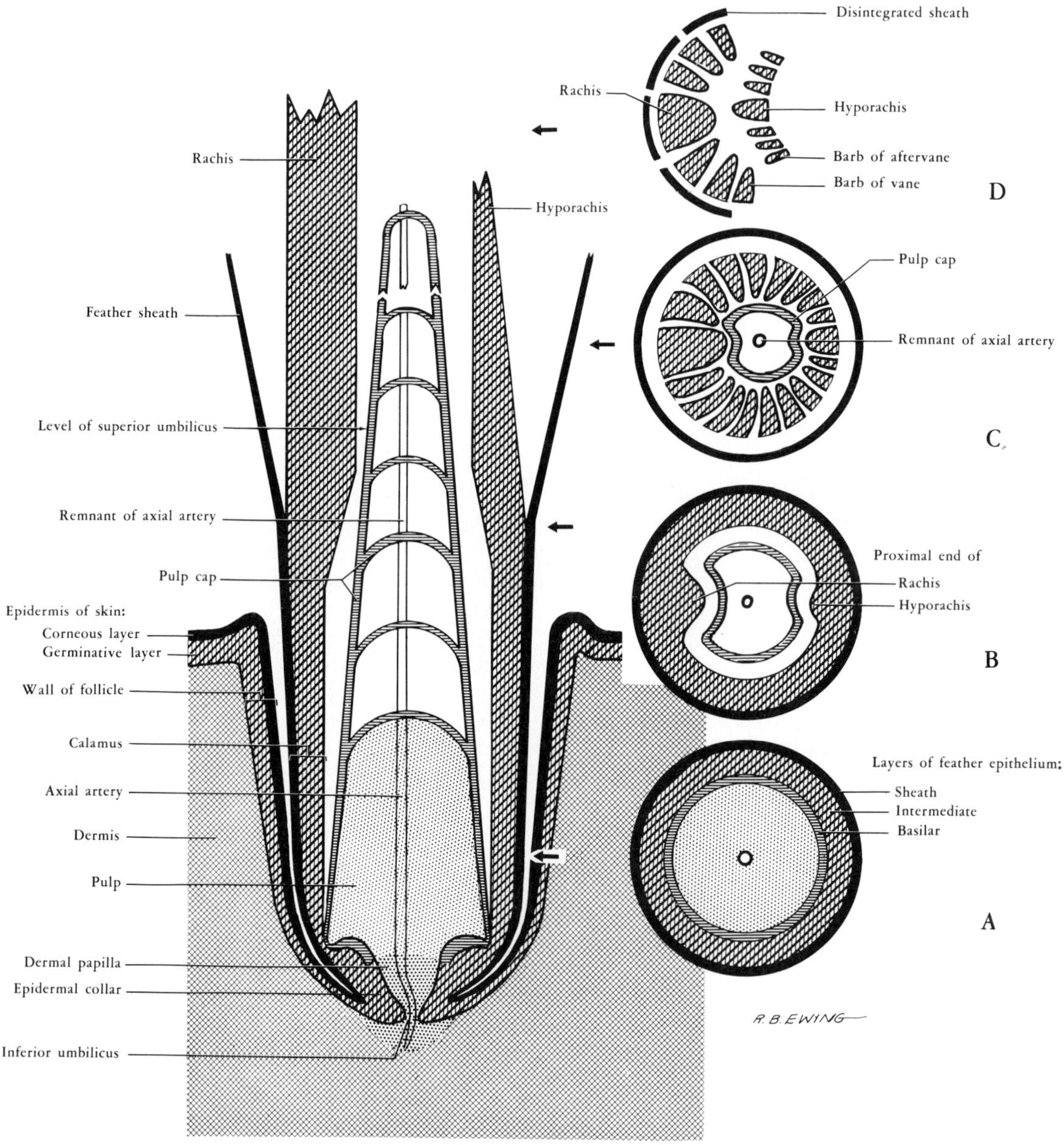


**Figure 4–1.** Layers of the epidermis and the dermis in a growing contour feather. Insets *A* through *D* are cross-sections through the feather at the levels shown by the heavy arrows (From Lucas and Stettenheim, 1972).

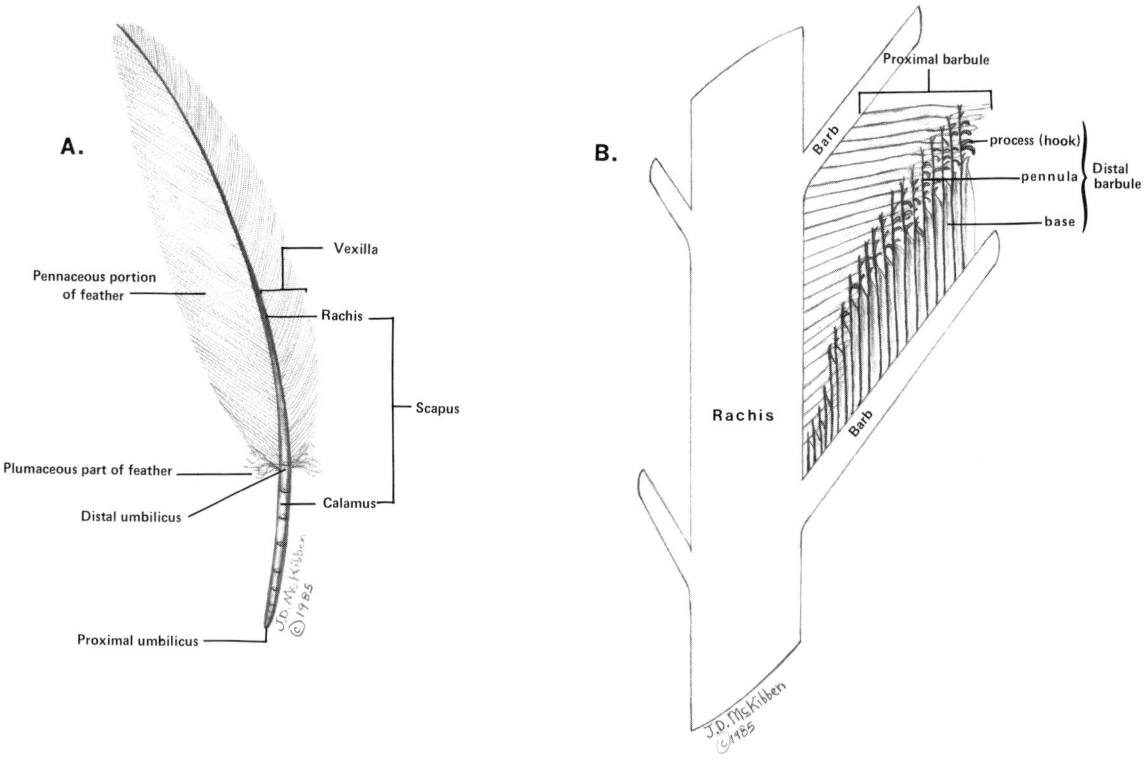


**Figure 4–2.** Contour feather of the Amazon Parrot. *A*, General morphology. *B*, 100 × magnification.

lacerations or "raw" spots when the feathers have been damp, oily, or removed.

*Nomina Anatomica Avium* classifies feathers into contour, plume, and semiplume types.[20]

## Contour Feathers

Contour feathers (Fig. 4–2*A*) have flat, closely knit vanes, although the basal part of the vane may be variably plumulaceous. Remiges (wing flight feathers or quills), rectrices (tail flight feathers), coverts, and general feathers of the body, head, neck, and limbs comprise this group.

The axis of remiges and rectrices is the scapus (shaft or quill). Its hollow bare portion closest to the body is the calamus, whose proximal portion is implanted in the feather follicle. The portion of the scapus where the vexilla or vane expands is the rachis. The vexilla is formed by barbs extending at 45-degree angles from each side of the rachis; barbules arise from each side of the barbs. On contour feathers, barbules interlock with adjacent barbules from other barbs by a system of hooks or processes arising from distal barbules (Fig. 4–2*B*). Preening restores interlocking of the barbules when they become dislodged. Waterproofing is a result of this physical barrier created by interlocking feathers rather than application of an oily substance from the uropygial gland to the feathers. The uropygial gland is not present in all "waterproofed" birds, and surgical removal of the gland apparently does not impair a bird.

Immature or pin feathers have a vascular calamus and rachis, which can be exposed to increased incidence of trauma in pet psittacines as replacement feathers grow subsequent to wing trims. Such damage results in hemorrhage, and treatment requires extraction of the calamus.

Modified feathers resembling hairs that occur on the head of birds are called setae or bristles. Examples are the eyelashes of the Amazon and budgerigar. These eyelashes may not regrow following infection with pox or other viral diseases (see Chapter 19, Ophthalmology).

The feather coat of the Amazon Parrot is illustrated in Figure 4–3*A*. The wing flight feathers (remiges) of the Amazon Parrot may be divided into three groups: (1) *Primary remiges* are numbered from the carpal region outward and are attached proximally to the carpometacarpal and digital areas (Fig. 4–3*B*). The primary feathers are strongly connected to muscle,

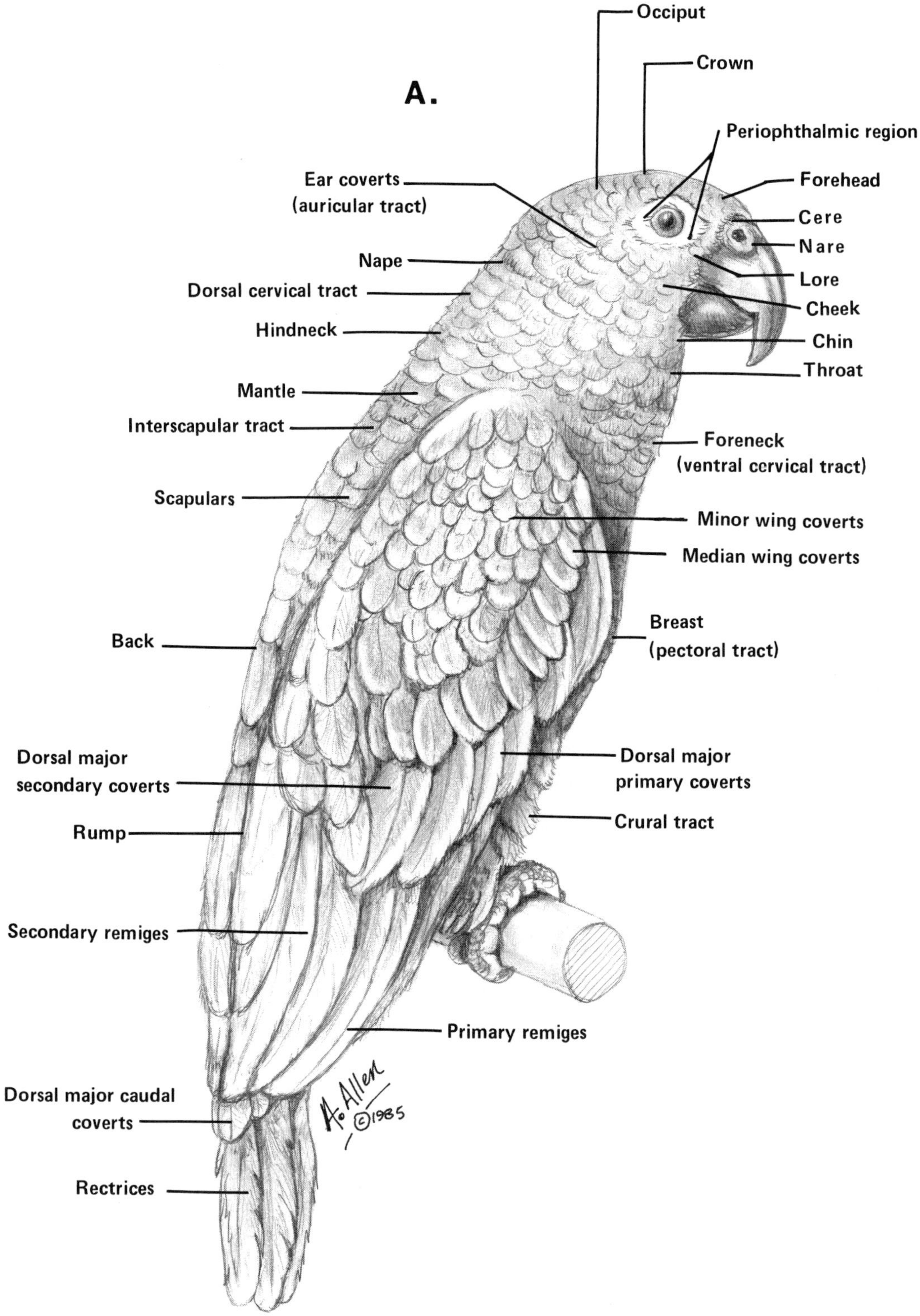


**Figure 4–3.** Plumage of the Amazon Parrot. *A*, Body areas and tracts.

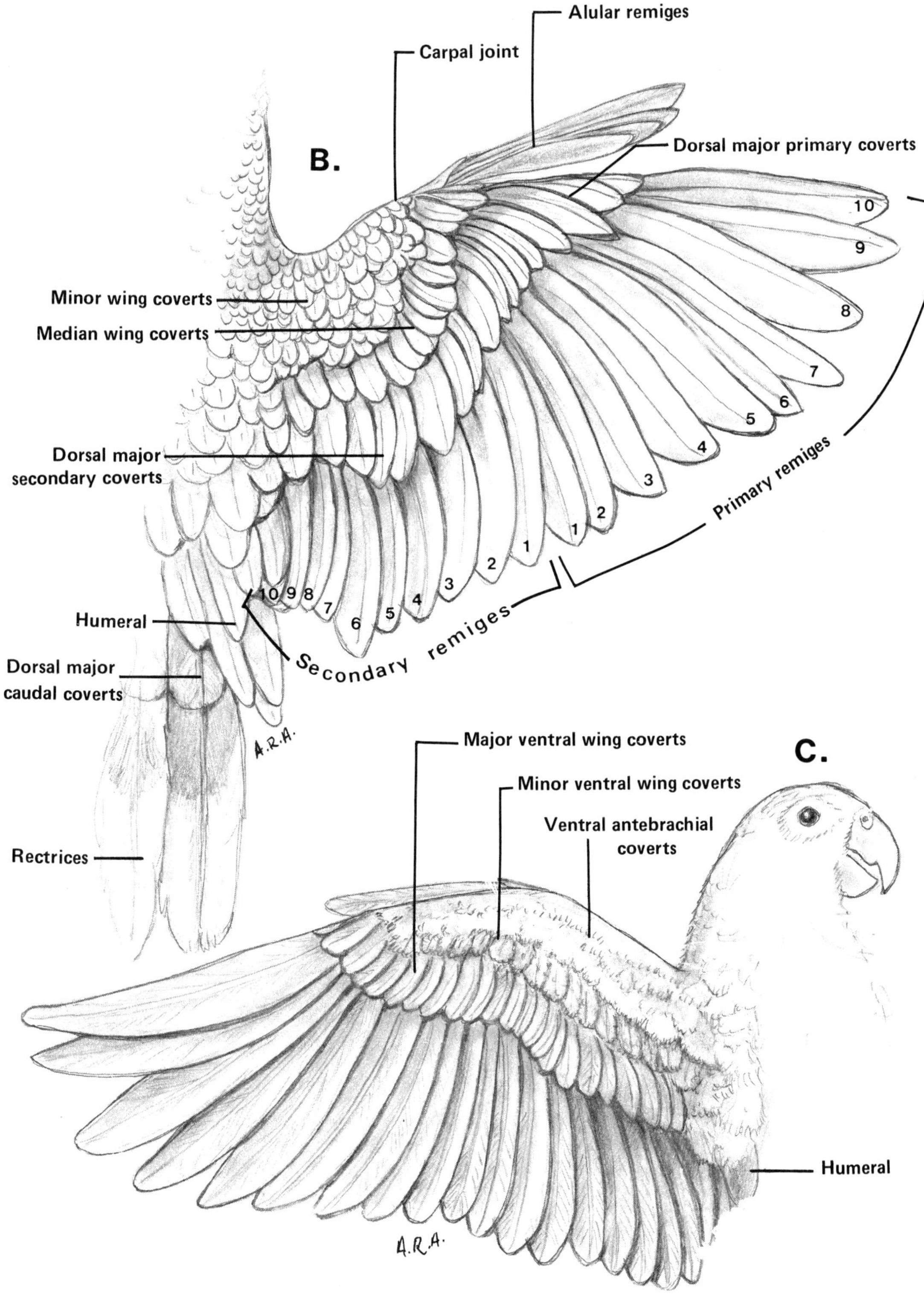


**Figure 4–3** *Continued. B,* Dorsal wing feathers. *C,* Ventral wing feathers.

fascia, and bone of the manus by fibrous tissue and possess much less mobility than the secondaries. This immobility of the primaries makes surgical approach to feather disorders of this area difficult. (2) *Secondary remiges* are numbered from distal to proximal in the antebrachium. (3) *Alular remiges* vary in number from two to seven in birds[20] and attach to the alular digit. Tertiary remiges, extending from the humeral area, are also described.[7]

Although individual variations occur, the number of primary remiges is reported to be species-specific, whereas the number of secondary remiges is more variable within species.[20] Amazon Parrots were found generally to have 10 primary remiges (9 to 11 range), 10 to 13 secondary remiges, and 4 alular remiges. A gap, known as a diastema, is present in the wing in many species of birds just proximal to the fourth secondary remex.[20] This increased space between the fourth and next proximal remex is barely noticeable in Amazon Parrots. Secondary remiges are generally numbered consecutively.

The remainder of the feathers that cover the wing and tail are called coverts[20] or tectrices. The most proximal group of wing feathers (Fig. 4–3) is referred to as shoulder feathers by poultrymen, scapulars in ornithologic terminology, or humeral tract coverts.

Generally 12 large tail quills (rectrices) arise from the uropygium of the Amazon Parrot.

## Plume and Semiplume Feathers

Plume and semiplume feathers are distributed over the body of adult psittacines. Plumes or down feathers have a rachis that is shorter than their longest barb. The fluffy appearance results from slender barbs and filamentous, noninterlocking barbules. These feathers usually underlie contour feathers and are associated with their follicles.

Some birds have modified plumes called powder down feathers (Fig. 4–4). Powder from these feathers is reported to be keratin material that originates from surface cells of barb-forming tissue within the feather germ. As the powder down feathers lose their powder and break open, they look like ordinary feathers. The function of powder is unknown[21]; however, it is speculated to serve as an adhesive agent within the feather, to serve as a water repellant, or to produce bloom. In a clinical case observed by Harrison, a cockatoo that was apparently congenitally deficient in powder down had a dirty, unkempt appearance to the feathers.

**Figure 4–4.** A powder down feather from a Sulphur-crested Cockatoo (*Cacatua sulphurea*).

The amount of powder down produced is species-specific and may vary with the stage of molt.[21] Powder down "patches" occur dorsal to the femoral pterylae and ventral to the pelvic pterylae areas in several cockatoo species (Umbrella, Moluccan, Red-vented, Bare-eyed, Sulfur-crested, Rose-breasted), as well as in cockatiels, African Grey Parrots, and some macaws. White and Moluccan Cockatoos appear to produce the most powder down. The powder may be visible on dark-colored clothing of a handler after direct contact with one of these birds. Powder down patches are apparently absent in Amazon Parrots, budgerigars, lovebirds, Eclectus Parrots, Palm Cockatoos, some conures, and some macaws (Blue and Gold, Green-winged, Scarlet).

A semiplume feather has fluffy (plumulaceous) vanes, but its rachis exceeds the length of its longest barb. Filoplume feathers are the hairlike feathers remaining on the body after plucking.

The color of feathers is influenced by pigments and diffracted light on cells or oil layers. Colors can be altered by hormones, diet, bleaching agents, dirt, surface oils, age, physical damage, disease, and temporarily by water.

Molting, the process of replacing old feathers with new ones, is influenced by season, temperature, nutrition, egg laying, species, and sex. It has been reported[8] that all birds molt at least once a year, many species twice, and a few, three times. Molting cycles generally begin after completion of the breeding season. Amazon Parrots appear to molt continually during the year. Postnuptial molt of wing feathers may extend over several months in psittacines, and new molt cycles may commence before previous cycles are completed.

According to limited personal observations (Harrison) in young Green-winged Macaws and Bare-eyed Cockatoos, the breast and head feathers are the first to be replaced, starting at three to four months of age. The first molt may extend over one and one half to two years, with the wing and tail feathers molted last. Several generations of feathers can be present at one time owing to different molting rates.[21] Some passerines such as canaries generally start molting in the spring and continue throughout the summer.

During the development of feathers and their accompanying vascular supply, the body's metabolic rate increases 30 per cent.[2] This results in increased nutritional demand on the bird and may produce significant stress, increasing the bird's vulnerability to disease-producing organisms.

## SKELETAL SYSTEM

Birds possess both spongy and compact bones, which have both endosteum and periosteum. Avian bones generally have thin cortices, predisposing them to fracture. Because air sacs

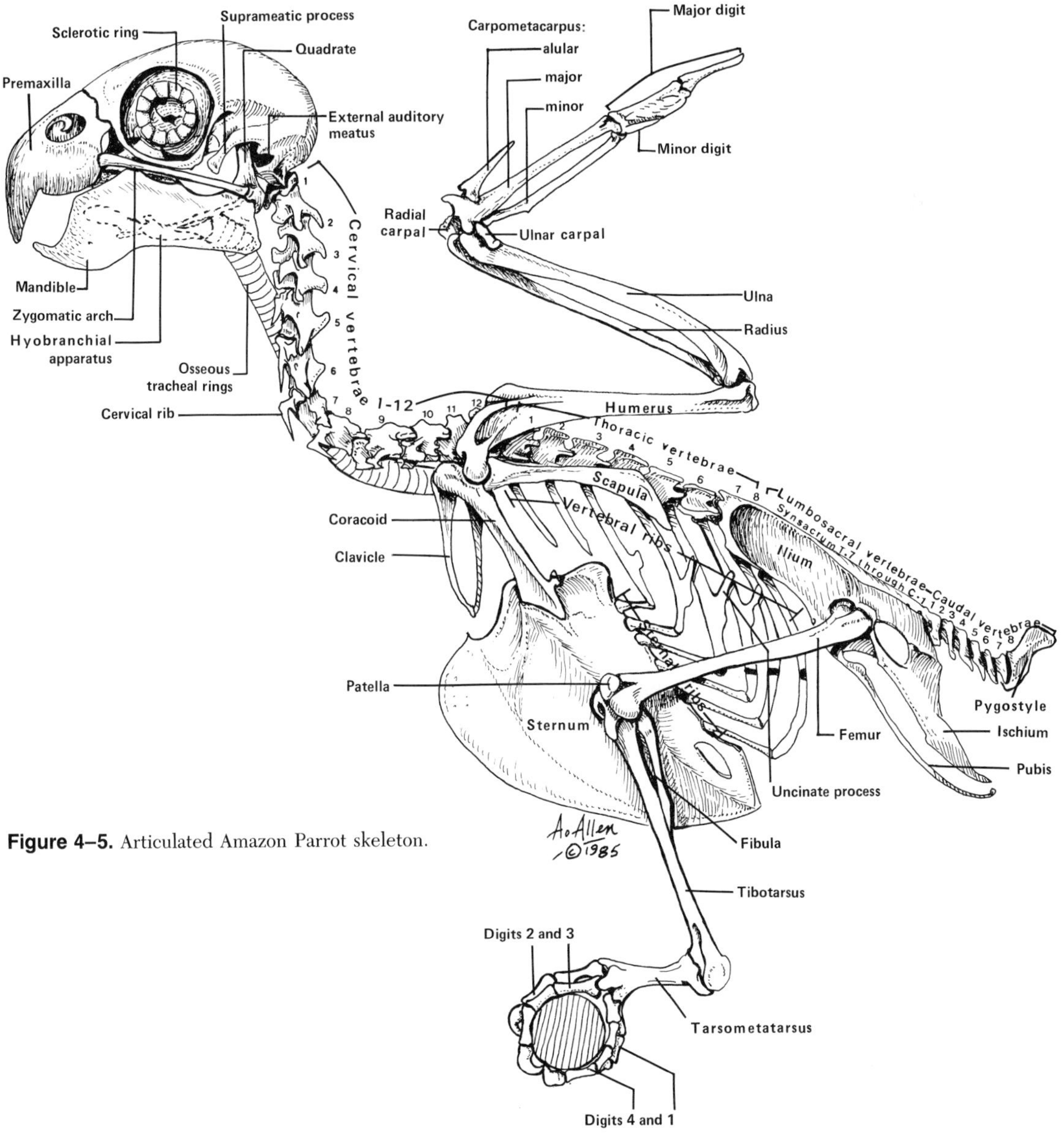


**Figure 4–5.** Articulated Amazon Parrot skeleton.

communicate with some bones such as the humerus, osteomyelitis associated with fractures can also result in infection of the diverticula of the air sacs. Air sac infections such as aspergillosis may also extend into bone through these communications[19] (see Chapter 14, Radiology).

Pneumatized bones in the Blue-fronted Amazon include the ribs, vertebrae, humerus, coracoid, clavicle, sternum, ilium, ischium, pubis, and cranial bones. Bone marrow, where present in bones, is rich in fat. The tibiotarsal bone may be the easiest and most useful site from which to obtain bone marrow specimens for

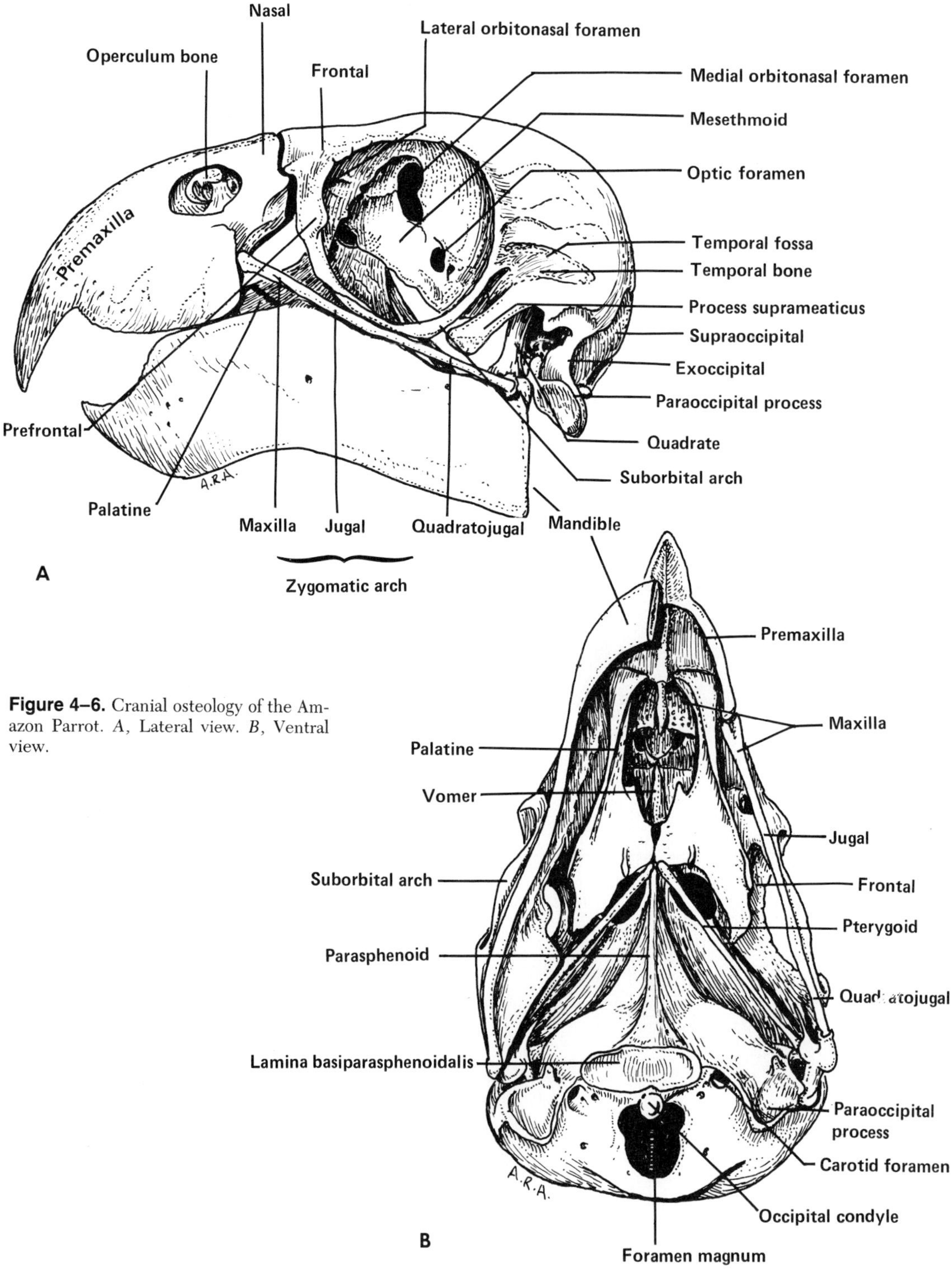


**Figure 4–6.** Cranial osteology of the Amazon Parrot. *A*, Lateral view. *B*, Ventral view.

clinical cytologic evaluation. Evaluation may be more difficult in aged birds because the marrow content decreases as birds mature.

## Axial Skeleton

The axial skeleton is composed of bones of the skull, vertebral column, ribs, and sternum (Fig. 4–5). Developmentally, cranial bones are more numerous in birds than in mammals, but in mature birds, suture lines are not usually observed owing to fusion. Figure 4–6 illustrates cranial osteology of the Amazon Parrot as identified by the senior author.

Unusual bones in the heads of birds are

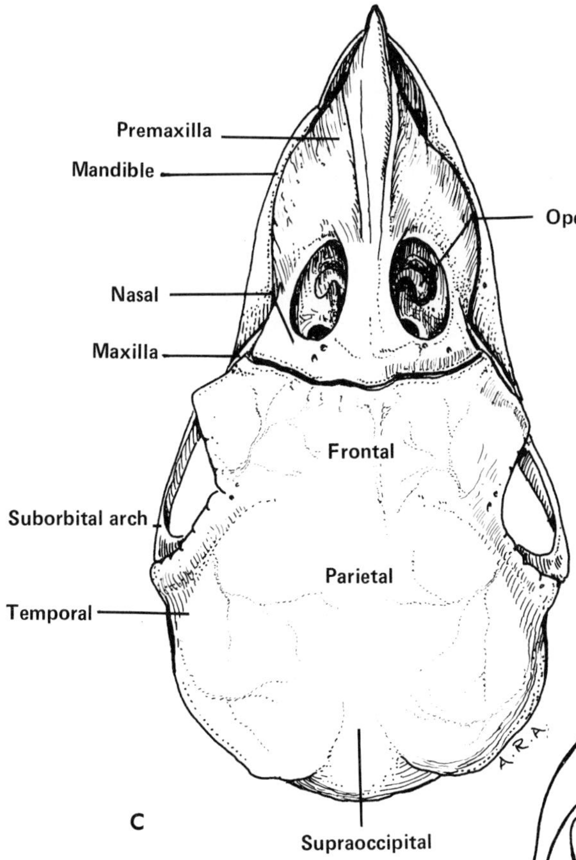


C

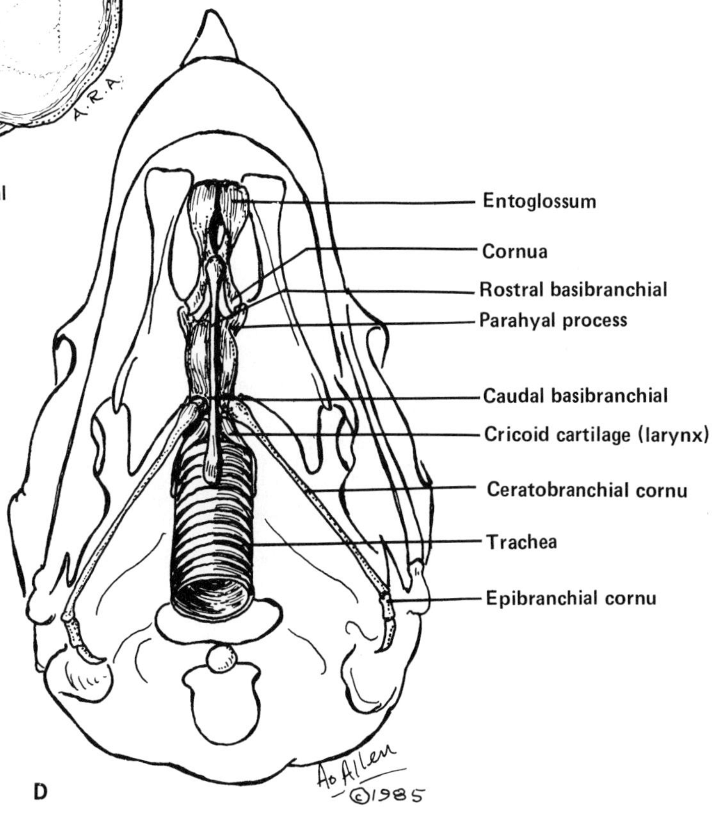


D

**Figure 4–6** *Continued. C,* Dorsal view. *D,* Hyobranchial apparatus.

scleral ossicles, which form a ring in each eye. There are 12 such ossicles within each eye of the Amazon Parrot. These can be observed by transillumination, particularly in albino cockatiels and by radiography. A small C-shaped opercular bone lies within the Amazon's operculum, which inhibits direct observation of the nasal cavity. The hyobrancheal apparatus of the Amazon Parrot extends nearly to the tip of the tongue and should not be confused radiographically with lingual foreign bodies.

The skull articulates with the atlas by a single occipital condyle. Practitioners who are familiar with paired occipital condyles in mammals should not diagnose the normal single occipital condyle of birds as pathologic.

The premaxilla forms the upper jaw and is covered by the upper beak. The mandible forms the lower jaw and is partially covered by the lower beak. The functional anatomy of the avian jaw is described by Bühler.[3]

The vertebral formula of the Amazon Parrot appears to be like that of the budgerigar: C—12, T—8, LS—8, and C—8. Fusion of components makes it difficult to ascertain the number of lumbar, sacral, and caudal components. However, this area should be palpated during a physical examination to determine abnormalities such as curvature or periosteal thickening that may have resulted from trauma, malnourishment, or incubation deformities. These findings may be associated with clinical signs of clumsiness or inability to preen feathers, particularly those of the tail.

An odontoid process is present on the axis (second cervical vertebra) as in mammals. The lumbar, sacral, last two thoracic, and first one or two caudal (coccygeal) vertebrae fuse to form

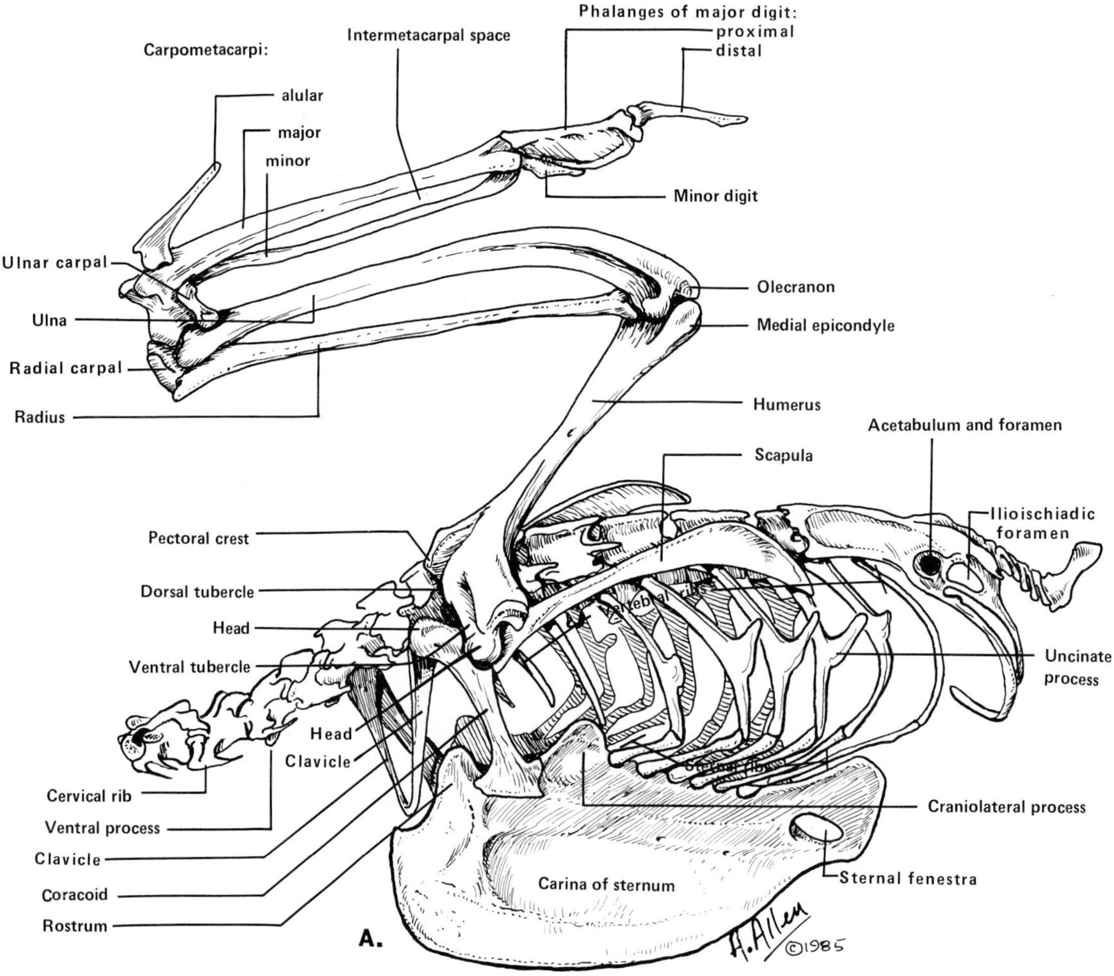


**Figure 4–7.** Osteology of the axial skeleton and wing of the Amazon Parrot (skull and first seven cervical vertebrae are illustrated in Figures 4–5 and 4–6). *A*, Craniolateral view.

the synsacrum. The terminal caudal vertebrae fuse to form the pygostyle. Unlike many birds, Amazon Parrots have vertebrae with concave caudal surfaces in the thoracic region. Synovial joints lie between unfused cervical, thoracic, and caudal vertebrae. In the Amazon Parrot, synovial joints generally occur between the last seven caudal vertebrae.

Ribs are classified as cervical, thoracic, and lumbar. Cervical ribs appear as fused, caudally directed lateral spines on all cervical vertebrae other than C-12. An articulating short cervical rib approximately 5 mm long is embedded in the ventrolateral neck muscles associated with the twelfth cervical vertebra in the Amazon Parrot.

The first two thoracic ribs of Amazon Parrots consist of only vertebral components. Ligaments join these vertebral segments to the third sternal rib. The next five thoracic ribs have osseous vertebral and sternal components with joints between them. The sternal portion of the last thoracic rib may not reach the sternum.

A caudally directed uncinate process arises from each of the second through sixth vertebral ribs. The absence of this process may be used to locate the seventh rib as a landmark for endoscopy, sexing birds, diagnostic procedures such as radiology, and surgical intervention, particularly of the proventriculus (see Chapter 15, Endoscopy; Chapter 51, Sex Determination Techniques; and Chapter 48, Selected Surgical Procedures). A long, slender sternal rib may extend from a long muscular attachment on the

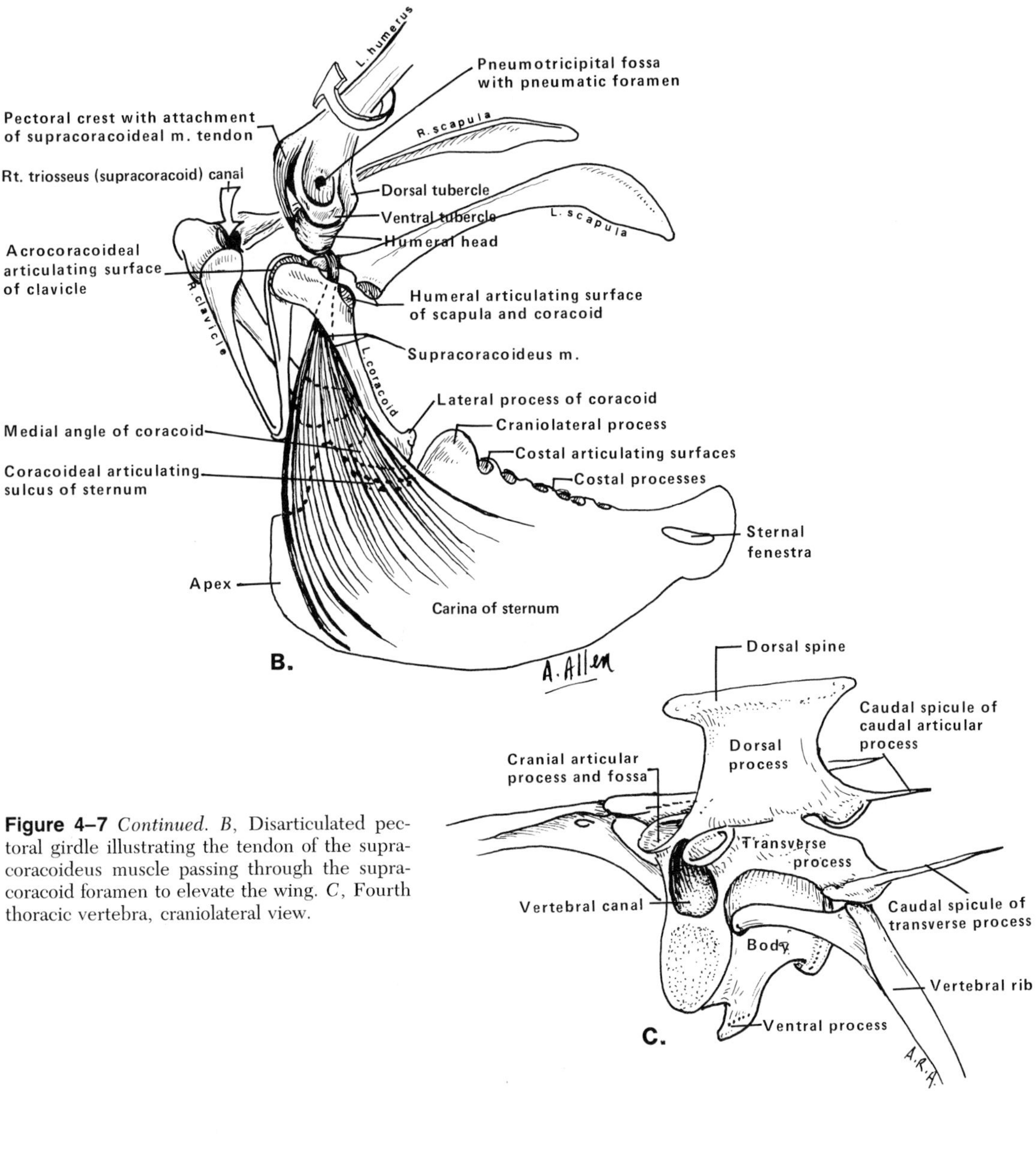


**Figure 4–7** *Continued. B*, Disarticulated pectoral girdle illustrating the tendon of the supracoracoideus muscle passing through the supracoracoid foramen to elevate the wing. *C*, Fourth thoracic vertebra, craniolateral view.

pubis to parallel the eighth thoracic sternal rib ventrally. No lumbar ribs were observed in the Amazon Parrot.

Air passages in the Amazon Parrot extend between the lungs and the vertebral portion of thoracic ribs. The interclavicular air sac communicates with the sternum and sternal ribs.

The sternum is long, with a large midventral carina (keel) serving as the attachment of the origin of the major flight muscles. The craniodorsal projection of the carina is the rostrum. Palpation of the carina is a valuable aspect of the physical examination to determine the relative amount of muscle, degree of straightness, and mineralization. The carina also serves as an anatomic guide for positioning patients for radiography.

Sternal ribs articulate with the costal processes of the sternum via synovial joints. The sternum articulates with the coracoid bone be-

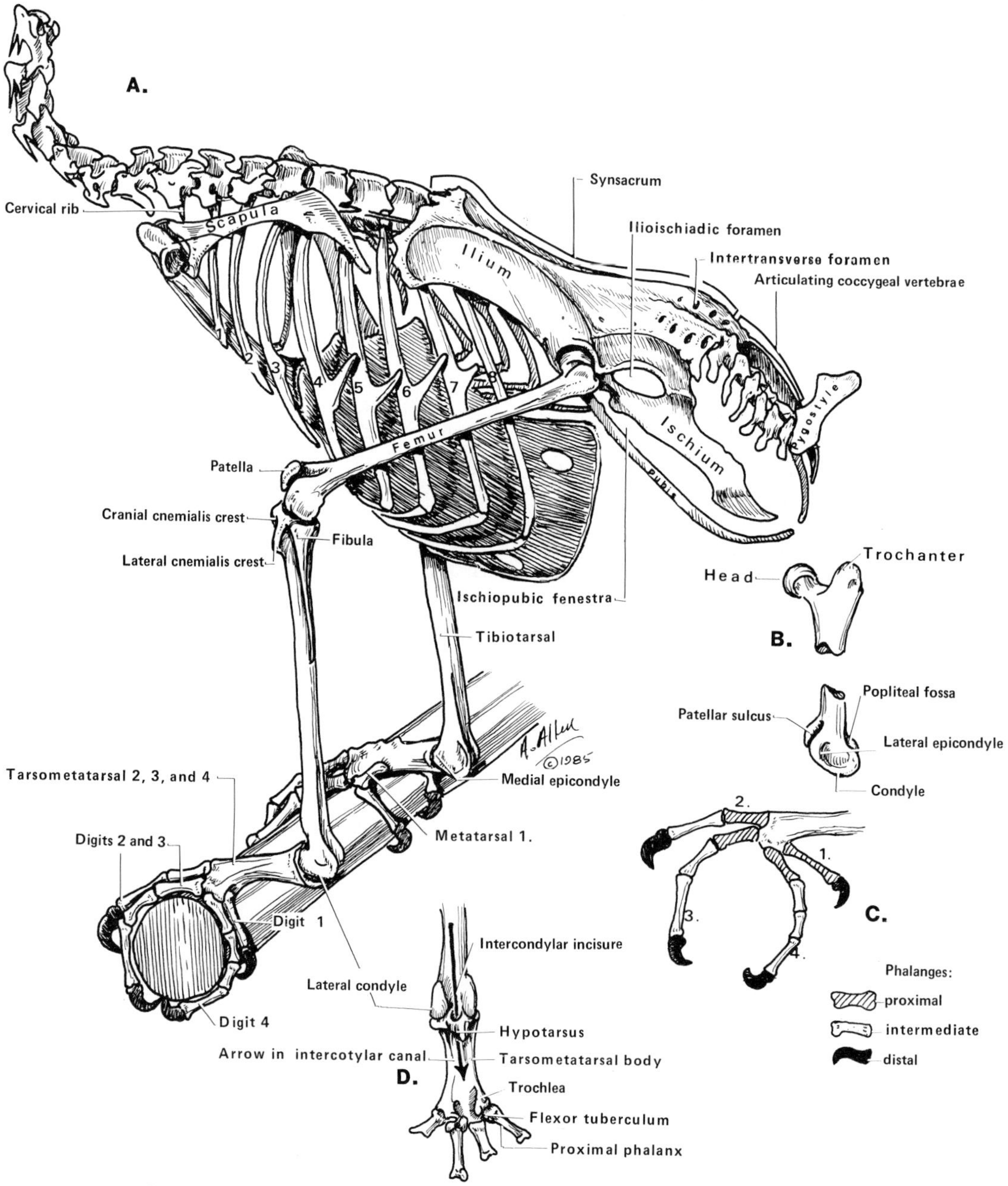


**Figure 4–8.** Osteology of the axial skeleton and leg of the Amazon Parrot (skull and first seven cervical vertebrae are illustrated in Figures 4–5 and 4–6). *A*, Caudolateral view. *B*, Proximal and distal femur. *C*, Digits. *D*, Plantar view of the ankle area.

tween the rostrum and the craniolateral process (Fig. 4–7).

### Appendicular Skeleton

The appendicular skeleton is composed of the bones of the limbs, thoracic girdle, and pelvic girdle (Figs. 4–7 and 4–8). The thoracic girdle consists of the scapula, coracoid, and clavicles, which come together to form the triosseal (supra-coracoid) foramen through which the tendon of the supracoracoideus muscle passes to abduct the wing. In Amazon Parrots the clavicles are thin, complete bones that fuse midventrally to form the furcula. Distal to the scapula the wing is composed of the humerus, ulna, radius, radial and ulnar carpals, carpometacarpus, alular digit, major digit, and minor digit.

The ulna is larger in diameter than the radius in the Amazon Parrot. The carpus consists of two components, the radial carpal and ulnar carpal bones. The carpometacarpal bone represents the distal row of carpal bones fused with the three metacarpals. One phalanx is present in the alular and minor digit and two in the major digit.

The pelvic girdle is equivalent to the hip bone (Fig. 4–8), consisting of the fused ilium, ischium, and pubis. The pubis lies ventral to the ischium, from which it is separated by the ischiopubic fenestra. The pubis is a thin bone and does not form a pelvic symphysis as it does in mammals. Aviculturists sometimes attempt to determine the gender of some birds by evaluating the amount of space between the paired pubic bones. This is not a consistently accurate method of sex determination. However, from a clinical viewpoint, the relationship of the pubis to the caudal aspect of the sternum is significant. This distance increases in cases of abdominal disease, such as ascites, organ enlargement, and tumors.

The ilium is the largest bone of the pelvic girdle. The ilia are located dorsal to the three divisions of the kidneys, similar to their location in mammals. In the Amazon Parrot, the ilia fuse with the synsacrum. Distal to the pelvic girdle the bones of the leg are the femur; patella; tibiotarsus; fibula; fused tarsometatarsals 2, 3, 4; metatarsal 1; and in Amazon Parrots, digits 1, 2, 3, and 4. The tibiotarsus in most birds is the largest bone of the limb. The proximal fibula articulates laterally with the femur; its thin prolongation extends distally about one-third the length of the tibiotarsal bone, to which the distal end is attached.

Some species of birds have two digits and some three, but none has over four digits. In Amazon Parrots, digit 1 has 2 phalanges, digit 2 has 3, digit 3 has 4, and digit 4 has 5. In parrots digits 2 and 3 extend cranially and 1 and 4 caudally. Woodpeckers, toucans, and cockatoos resemble parrots in having two digits extending cranially and two caudally.[7]

## THE MUSCULAR SYSTEM

The muscular system of birds is composed of striated and smooth muscles. Striated skeletal muscles are of both light and dark fibers, as in domestic animals. Unusual occurrences of skeletal muscle in the bird include the ciliary and sphincter pupillae muscles of the eye (see Chapter 19, Ophthalmology). Cardiac striated muscle resembles that of mammals. Figures 4–9 and 4–10 illustrate muscles associated primarily with the appendicular skeleton, which may be important surgically and when approaching fractures for repair.

Skeletal muscles of the bird may be classified as axial or appendicular. Vanden Berge's writings[30, 31] serve as a guide for nomenclature and descriptions in avian myology.

Axial muscles of birds may be grouped to include cutaneous muscles, muscles of the head and hyobranchial apparatus, muscles of the neck and back, tail and cloacal muscles, and trunk muscles.

### Cutaneous Muscles

Cutaneous muscles lie superficially on the head, neck, thorax, and abdomen and do not appear clinically unique in Amazon Parrots.

### Muscles of the Head and Hyobranchial Apparatus

Although intrinsic tongue and hyobranchial muscles are poorly developed in most birds, they are well-developed in Amazon Parrots, and the paraglossus is uniquely large in the tongue.[25] Muscles of the hyobranchial apparatus have been described[30] and are related to tongue movement and swallowing.

### Muscles of the Neck and Back

Back muscles in the bird are poorly developed owing to the immobility of the synsacrum.

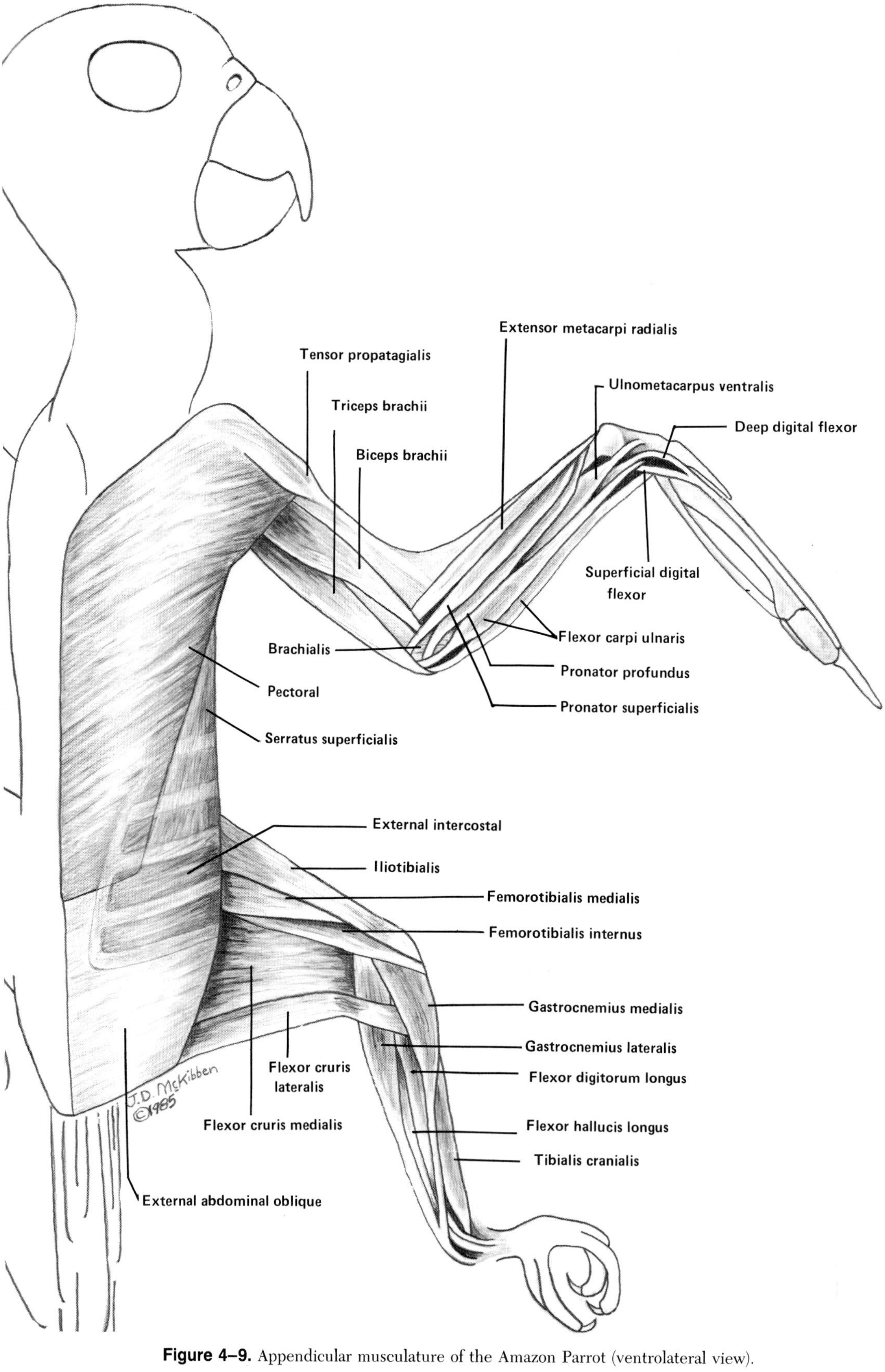


**Figure 4–9.** Appendicular musculature of the Amazon Parrot (ventrolateral view).

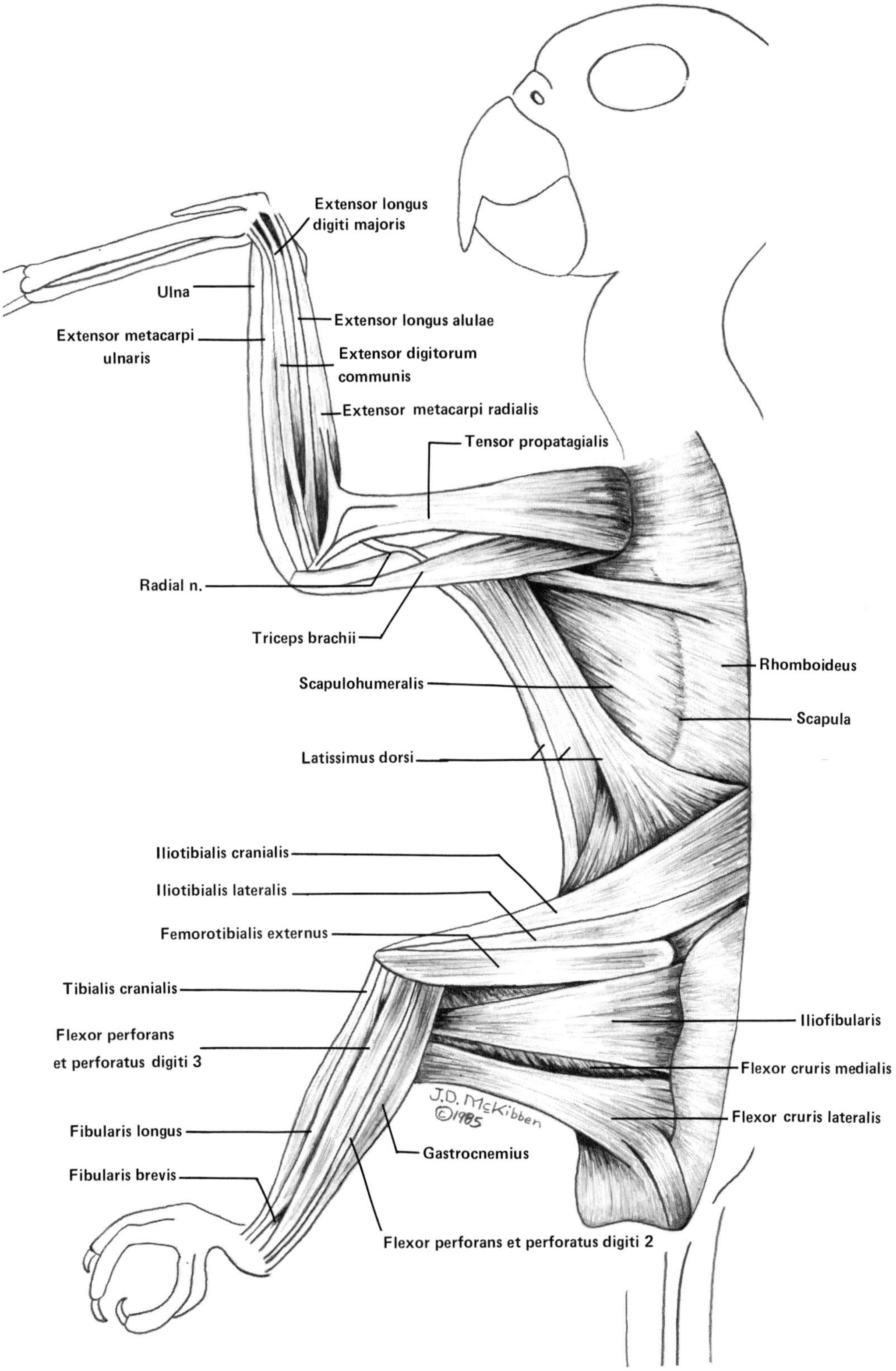


**Figure 4–10.** Appendicular musculature of the Amazon Parrot (dorsolateral view).

The complexus (a dorsolateral cervical muscle) is sometimes referred to as the hatching muscle. It is enlarged in newly hatched psittacines and is especially large in neonatal macaws. It should not be diagnosed as an abnormal structure, as its size diminishes during the first week after hatching.

### Tail and Cloacal Muscles

Since the tail of birds is important in flight as well as in functional and behavioral activities related to mating, egg laying, and defecation, its muscles are generally well-developed.

### Trunk Muscles

Because the oblique and horizontal septa (so-called diaphragm) of the psittacine are not highly muscular, respiration is controlled mainly by the muscles of the thorax and abdomen. The horizontal septum is closely applied to the ventral surface of the lungs. The delicate vertical or oblique septum extends from the midline between the lungs, at the base of the heart, to the sternum and body wall. The heart lies on the abdominal side of the septum in birds, in contrast to the heart-diaphragm relationship in mammals.

Appendicular muscles in birds extend to the wing or leg or within these appendages. Figures 4–9 and 4–10 illustrate these muscles in the Amazon Parrot as dissected by the primary author.

### Wing Muscles

Muscles of the thoracic girdle, brachium, and manus are named and described in the fowl by Vanden Berge.[30, 31] Approximately 20 per cent of a bird's weight is attributed to the largest muscle of the body, the pectoral muscle.[6] The pectoral muscle adducts and depresses the wing in flight. The supracoracoideus muscle lies deep to the pectoral muscle and elevates and abducts the wing in flight (see Fig. 4–7*B*).

The humeral tendon of the triceps brachii muscle is not present in all birds.[31] It is present in Amazon Parrots. The triceps brachii muscle aids in extension of the wing; its tendon is sometimes unilaterally transected to deflight birds, although this technique is seldom employed in pet birds. The 13 muscles of the antebrachium and 10 of the manus are described in detail by Vanden Berge.[30]

### Muscles of the Leg

Muscles within the hindlimb are well-developed, and in some birds ossification of the distal tendons may occur. No tendinous ossifications have been observed in dissections by the primary author in the Amazon Parrot. Muscles of the legs have been described in White-fronted Amazons[1] and other birds.[30, 31]

## RESPIRATORY SYSTEM

The respiratory system of birds is complex. Smith[29] has described the microscopic and submicroscopic structure of the respiratory system in the budgerigar.

Air entering the external nares in psittacines passes into the nasal cavity and then into the oral cavity through the choana, a median slit in the roof of the mouth. The nasal (salt) gland is located dorsomedial to the eye and superficial to the frontal bone in most birds; however, its structure and function have not been studied in Amazon Parrots. An operculum, protruding into the external nasal opening of Amazon Parrots, helps prevent inhalation of foreign objects bilaterally.

Air in the nasal cavity passes around and between the caudal and middle conchae (turbinates) located caudoventral to the operculum (Fig. 4–11). These important bony structures have an extensive vascular supply that should be avoided or manipulated carefully when irrigating or surgically exploring the nasal cavity. The relationship of these structures, and that of the infraorbital sinus, to the porous calvaria and the brain makes them clinically important in rhinitis and sinus infections (Fig. 4–11). Chronic rhinitis in Amazon and African Grey Parrots often results in a caseous necrotic obstruction of the external nares, nasal cavity, and choana; the nasal conchae and nasal septum are often destroyed.

The infraorbital sinus is the only paranasal sinus in the Amazon Parrot. It lies cranial, ventral, and medial to each eyeball (Fig. 4–11). Periorbital swelling may be associated with infection of this sinus. Diverticula extend from this triangular sinus into the upper beak and mandible and communicate with extensive pneumatized sections of the skull. These extensive communications make sinus infections dif-

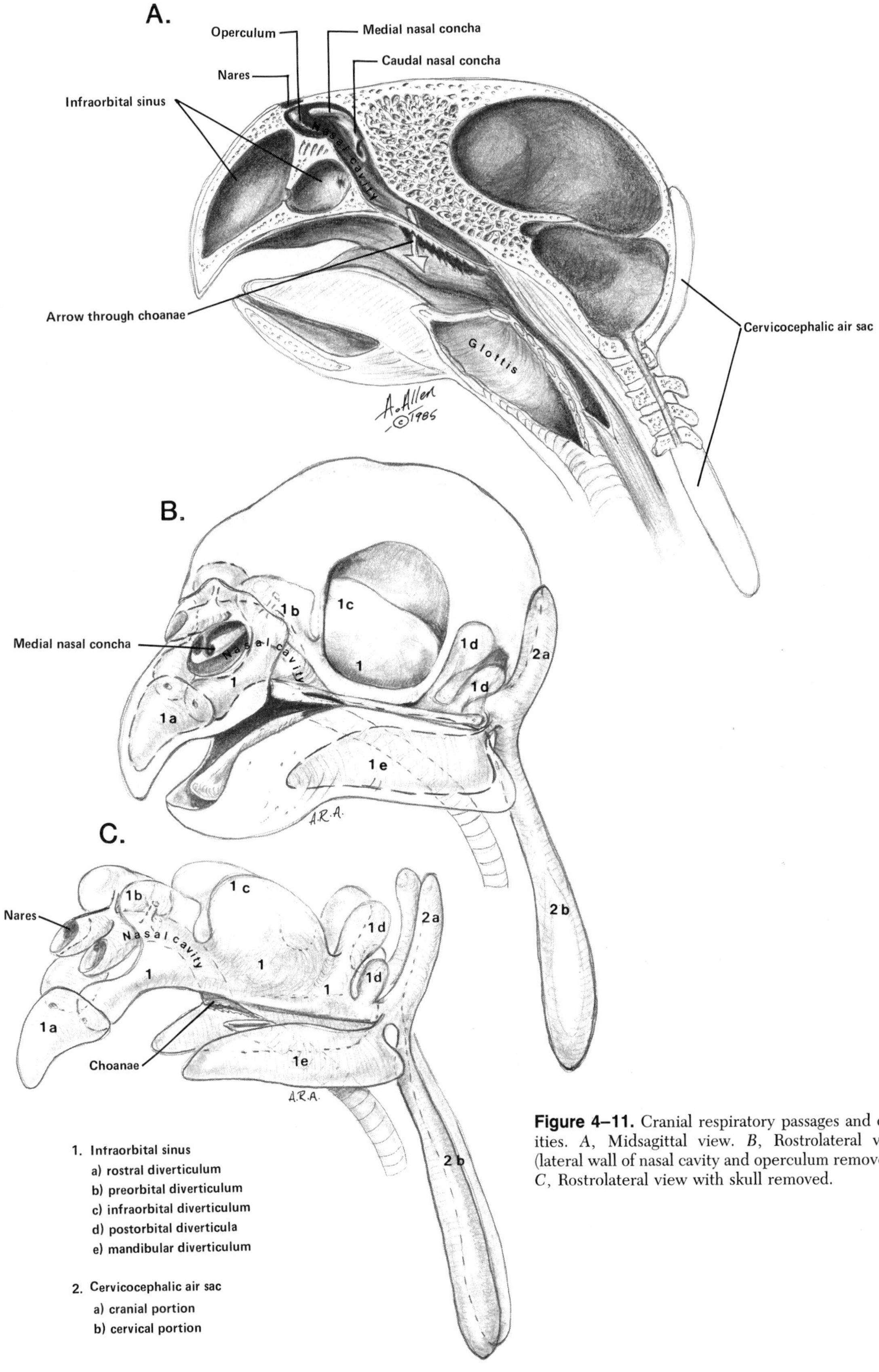


**Figure 4–11.** Cranial respiratory passages and cavities. *A*, Midsagittal view. *B*, Rostrolateral view (lateral wall of nasal cavity and operculum removed). *C*, Rostrolateral view with skull removed.

ficult to treat. A regional anatomic knowledge of this area is essential for surgical drainage or irrigation. Right and left infraorbital sinuses intercommunicate in psittacine birds but not in passerines. Therefore, aspiration of both sides is indicated for cultures in passerines when isolating causative agents of sinusitis.[4]

Olfactory mucosa is similar to that of mammals. Inspired air traversing the nasal cavity enters the oral cavity and pharynx through the choana, a median elongated triangular opening in the roof of the mouth. Air from the pharynx enters the larynx. During nasal breathing the glottis is placed directly under and opposed to the choana. Care should be taken in tube feeding psittacines to avoid entering the larynx, since no epiglottis is present. The coracoid cartilages are the largest laryngeal cartilages, but no vocal cords are present in the larynx.

## Trachea and Syrinx

The trachea conducts air from the larynx to the syrinx. Complete rings lend structural support to the trachea. Although these rings are cartilaginous in many species and in young birds, they are calcified in the adult Amazon Parrot. The trachea of the Amazon Parrot lies primarily on the right side of the neck, ventral and to the left of the esophagus, which also lies to the right of the cervical vertebrae (see Fig. 4–6). Within the thorax, the trachea terminates caudally in the syrinx, or vocal organ.

The syrinx of birds may be tracheal, tracheobronchial, or bronchial. The syrinx of Amazon Parrots is tracheobronchial in type. The last few tracheal rings fuse into a syringeal box, upon which the first pair of bronchial semirings articulate.

Air passing through the syrinx of psittacines vibrates two external and two internal tympaniform membranes to produce sound. The external tympaniform membrane extends between the first and second bronchial semirings. The inner tympaniform membrane covers the medial surface of the tracheobronchial bifurcation and spaces between the semirings. The bronchial muscles extending from the syringeal box to the first bronchial semiring and the bronchotrachial muscle extending from each bronchus to the trachea stretch or relax the tympaniform membranes and alter the size of the syrinx, thereby altering sounds. The narrowness of the trachea and recessive location of the typaniform membranes inhibit surgical approach to this area (e.g., foreign body removal or devocalization procedures) with current technology.

The syringeal muscles and syringeal functions have been described by Gaunt[10] in five species of psittacines. The wall of the syrinx in healthy Amazon Parrots is transparent enough to allow visualization of the heart through its caudal wall using endoscopy. The area around the outside of the syrinx is a frequent site for respiratory infections.

## Lungs

A main bronchus connects the syrinx to each lung. Each lung lies ventral and in intimate contact with the first through eighth thoracic vertebrae and ribs. Dorsally each lung fits laterally into the intercostal spaces, giving them a wavy appearance when observed at necropsy.

Radiographically the lungs resemble a sponge in appearance. This sponginess and pale pink color can be readily observed by endoscopy through the caudal thoracic or abdominal air sacs. Each main bronchus divides into secondary bronchi, which divide into parabronchi (tertiary bronchi) in the nonlobulated lungs. Parabronchi extend between secondary bronchi and anastomose with other parabronchi. The primary, secondary, and tertiary bronchi are conducting pathways. From each parabronchus arise numerous blind air capillaries that are the respiratory exchange sites.

## Air Sacs

Ventral secondary bronchi anastomose with the caudal thoracic air sac. The caudal continuation of the main bronchus is the abdominal air sac. A catheter may be passed via the trachea into the abdominal air sac. An ostium (orifice) on the lungs may be observed from these air sacs during endoscopy.

Birds may have cervicocephalic, pulmonary, pharyngeal, and tracheal air sacs (Figs. 4–11 to 4–13). The cervicocephalic air sac system of many psittacines connects with the infraorbital sinuses at the level of the tympanic area. This air sac system has cervical and cephalic components. In conures, budgerigars, and cockatiels, Walsh[32] describes cephalic air sacs arising from the infraorbital sinus and extending dorsally between the cere and eyes to cap the dorsal skull (Fig. 4–12). This was not noted by the primary author in the Amazon Parrot. A cephalic portion does, however, extend dorsally adjacent to the occipital bone in the Amazon Parrot (Fig. 4–13). The connecting cervical portion of the cervicocephalic air sac in the Amazon

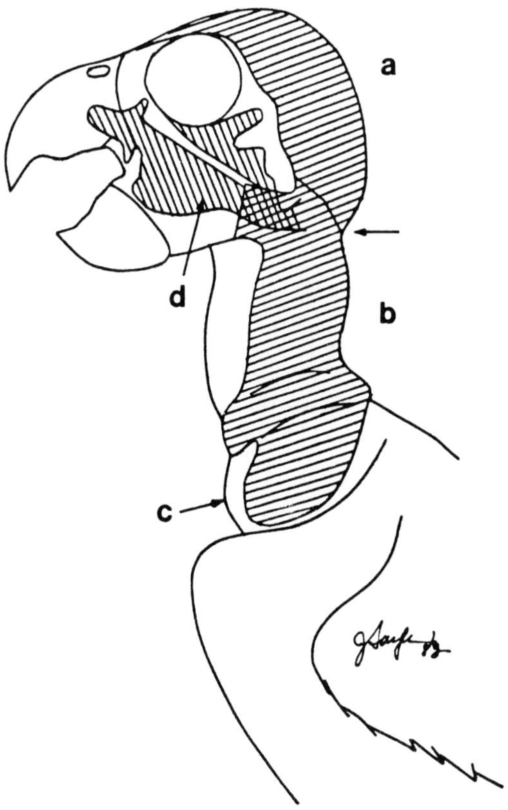


**Figure 4–12.** Relationship of the infraorbital sinus to the cervicocephalic air sacs in the cockatiel. Unlettered arrow indicates the division between the cephalic portion (*a*) and the cervical portion (*b*). The crop is *c* and infraorbital sinus is *d*. The caudal aspect of the infraorbital sinus connects to the cervicocephalic air sacs. (From Walsh, M.: Clinical manifestations of cervicocephalic air sacs of psittacines. The Compendium, 6:785, 1984. Used with permission of Veterinary Learning Systems Co., Inc.)

Parrot extends bilaterally dorsolaterally in the neck from the head to the level of approximately the seventh cervical vertebra. The cervicocephalic and pulmonary air sacs do not communicate.

Most birds have four paired and one unpaired pulmonary air sacs with several diverticula. These air sacs are thin-walled, lack diffuse blood vessels, and do not communicate.

In contrast to the epithelia of the trachea and bronchi, which in budgerigars have been reported to be pseudostratified ciliated columnar epithelia,[29] pulmonary air sacs are lined with a single layer of low cuboidal or simple squamous epithelial cells peripherally and simple columnar ciliated cells near the lungs. This is covered by a thin layer of connective tissue attaching the sacs to adjacent tissues. An outer layer of mesothelium and then a connective tissue layer make up the rest of the air sac wall. In air sacculitis the caudoventral location, avascular nature, and absence of peripheral cilia to remove purulent material within air sacs results in chronic respiratory infections. Resorption of materials is slow, and related infections are difficult to treat.

Psittacines have paired cervical, cranial, caudal thoracic, and abdominal pulmonary air sacs connecting to the lungs. An unpaired clavicular air sac lies dorsal to the crop between the clavicles. Paired extrathoracic diverticula pneumatize the humerus, clavicles, and coracoid bones. The intrathoracic diverticulum lies just caudal to the clavicles between the coracoid bones. It extends into the sternum and the sternal ribs. Extrathoracic diverticula may also connect to vertebral pneumatic spaces. Small cervical air sacs extend cranially, dorsal to the last two or three cervical vertebrae and medial to the extrathoracic diverticula of the clavicular air sac. They connect to pneumatized cervical vertebrae. Thoracic vertebrae and vertebral ribs are pneumatized directly from the lungs. Cranial thoracic air sacs approximate the thoracic wall cranial and lateral to the caudal thoracic air sacs on each side of the pericardium. Abdominal air sacs, which are the largest air sacs, lie on either side of the viscera. During some endoscopy procedures, especially in areas near the proventriculus, sparse blood vessels have been encountered by Harrison between air sac chambers; these must be avoided during surgical procedures.

## DIGESTIVE SYSTEM

The digestive system of the Amazon Parrot consists of the beak (rostrum), oral cavity, tongue, salivary glands, taste buds, pharynx, cervical esophagus, crop, thoracic esophagus, proventriculus, ventriculus, duodenum, jejunum, ileum, rectum, cloaca, pancreas, and liver. Figures 4–14 to 4–18 illustrate the position and structure of these organs.

Birds lack teeth but use their horny beak for prehension and incising. The egg tooth, a dorsal process on the beak, aids hatching in chicks and disappears soon after hatching. The oral and nasal cavities communicate through the choana, a cleft in the palate.

The Amazon Parrot's tongue is blunt and very muscular. Elongated papillae form a brush-like tongue in lory-type parrots. This modification aids in gathering nectar, harvesting pollen, and pressing food into forms suitable for swallowing.

Birds' tongues have few taste buds. Taste buds are located in the palate, floor of the mouth, base of the tongue, and floor of the pharynx[23, 24] and on the laryngeal mound in some birds.

Salivary glands are distributed in the walls of the oral cavity and pharynx. They include maxillary glands, palatine glands, sphenopterygoid glands, glands of the corner of the mouth, cheek glands, mandibular glands, lingual glands, and cricoarytenoid glands.[23] They are mucigenic and important in lubricating food. Clinically these glands and related lymphoid tissue are frequent sites of infection. Pox or hypovitaminosis A lesions may involve these glands, resulting in oral or pharyngeal abscesses. Preanesthetic atropine to control serous oral secretions is not necessary in psittacines (see Chapter 44, Evaluation and Support of the Surgical Patient).

The cervical esophagus passes distally on the right side of the neck to the right of the trachea. Near the thoracic inlet the cervical esophagus enlarges in the Amazon Parrot to form the ingluvies (crop), where food is stored. In its empty state, the crop is teardrop- or U-shaped and lies primarily to the right and ventral to the trachea and cervical vertebrae. In very young birds it appears almost bilobed. The crop is reduced in size in lory-type parrots.

When ingluviotomies are necessary for removal of foreign bodies, the left lateral cranial portion of the sac is approached surgically (see Chapter 48, Selected Surgical Procedures). This site avoids the large right jugular vein and vessels branching from the right common carotid artery which supply the crop. This location also minimizes the pressure from the interior.

Caudal to the crop, the thoracic esophagus enters the thoracic inlet to the right of the trachea and continues caudally coursing from right to left, dorsal to the trachea, syrinx, and heart and ventral to the hilus of the lungs. The esophagus enters the proventriculus (glandular stomach) at the craniodorsal border of the left lobe of the liver (Figs. 4–14 and 4–16). Radiographically on the lateral view, the esophagus appears to enter the proventriculus at a 45-degree angle. Because of the position of the proventriculus near the liver, when it is full, it may radiographically resemble hepatomegaly, especially on the ventrodorsal view (see Chapter 14, Radiology). Regurgitated food used in feeding young birds may originate from the proventriculus in psittacines. In psittacines, a strong muscular sphincter separates the proventriculus from the muscular ventriculus (gizzard or non-glandular stomach), which receives ingested food into its dorsal portion from the proventriculus (Figs. 4–14 and 4–16).

The ventriculus in Amazon Parrots is thick-walled and composed primarily of smooth muscles. Its glandular secretion forms a hard keratinized plate that serves as a rough grinding surface against which grit grinds food. In birds not requiring extensive grinding of food, such as lory-type, the ventriculus is soft-walled and bag-like.

Extensive information on the gross, histologic, and scanning electron microscopic morphology of the digestive tract of 134 psittacine species is reported by Güntert.[11]

The ventral ligament of the ventriculus is continuous with the falciform ligament of the liver. Together with the medial air sac walls, they partially divide the peritoneal cavity into right and left portions. This division of the right and left halves may be further appreciated by superimposing the sternum over the vertebral column. An imaginary line between the acetabula of the pelvis further divides the abdomen into cranial and caudal right and left quadrants (Fig. 4–16). Use of these quadrants may be helpful in anatomic orientation, particularly in radiographically identifying structures.

The cranial right and left quadrants normally contain the heart, liver lobes, lungs, gonads, adrenal glands, caudal thoracic air sacs, and part of the abdominal air sac dorsally. The proventriculus, spleen, and part of the ventriculus additionally lie in the cranial left quadrant. In the female the persistent gonad usually lies in the cranial left quadrant. The duodenum and pancreas lie in the right caudal quadrant. Structures in the caudal right and left quadrants include the intestinal loops, the rectum, the caudal and a portion of the middle divisions of the kidneys, and most of the abdominal air sac. The major portion of the ventriculus extends into the left caudal quadrant.

The valvelike fold called the pyloric valve lies between the ventriculus and the duodenal loop. This valve tends to restrict solid objects from leaving the ventriculus. Ingesta leaves the ventriculus ventrally near the junction of the left cranial and caudal quadrants of the abdomen and courses caudally in the duodenal loop in the right caudal quadrant. Much of the pancreas is enclosed in and associated with the antimesenteric border of the ascending portion of the duodenal loop. The remaining ileal and jejunal loops are suspended by a long mesentery. The posthepatic septum resembles the mammalian omentum, but no true omentum covering the viscera is present in psittacines.

A hepatoduodenal ligament attaches the duodenal loop to the liver cranially. At the apex of

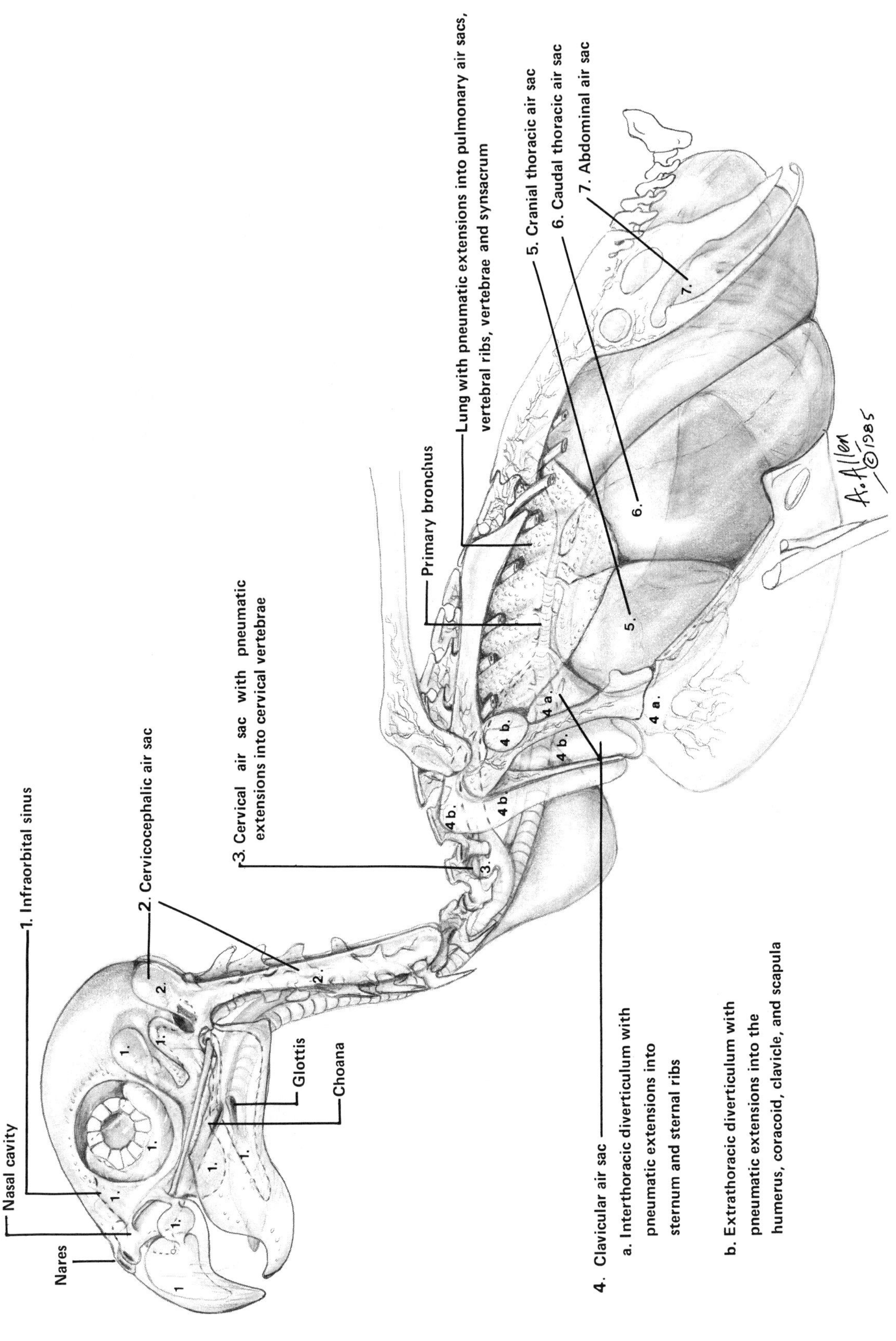


**Figure 4–13.** Cervicocephalic and pulmonary air sacs of the Amazon Parrot.

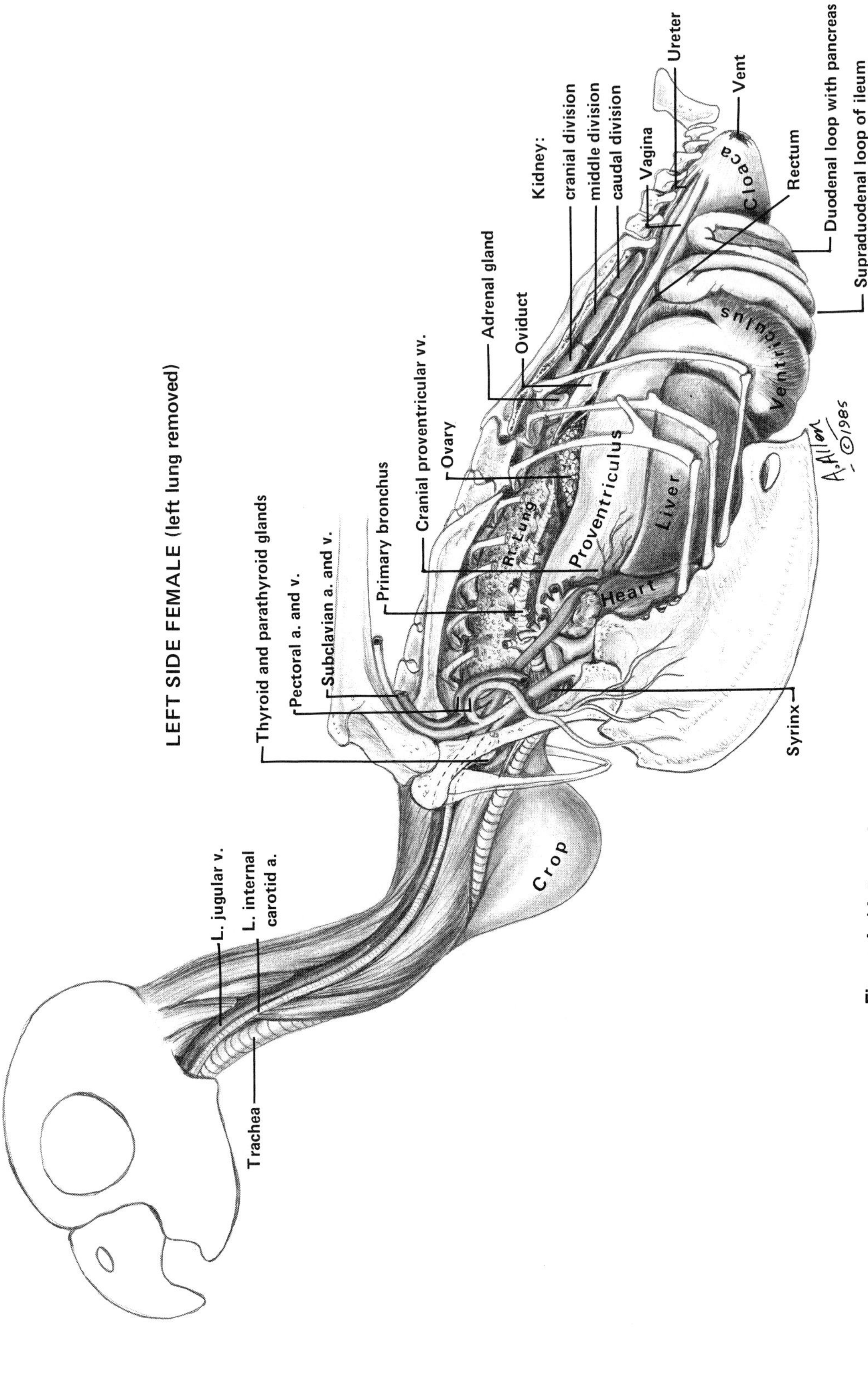


**Figure 4–14.** Internal structures of the immature female Amazon Parrot (left lateral view).

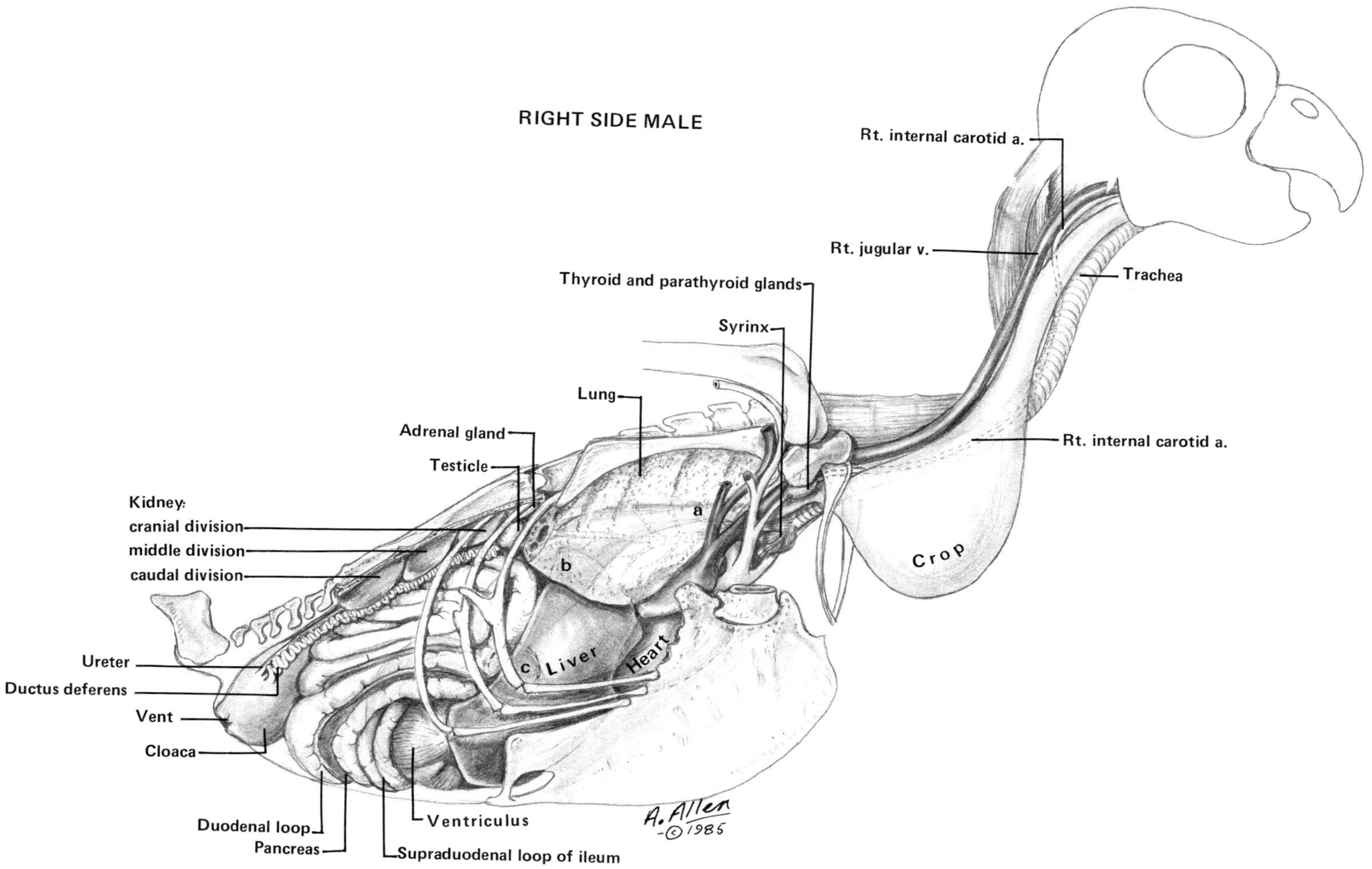


Deep structures shown in shadow: a. thoracic esophagus
b. proventriculus
c. spleen

**Figure 4–15.** Internal structures of the male Amazon Parrot (right lateral view).

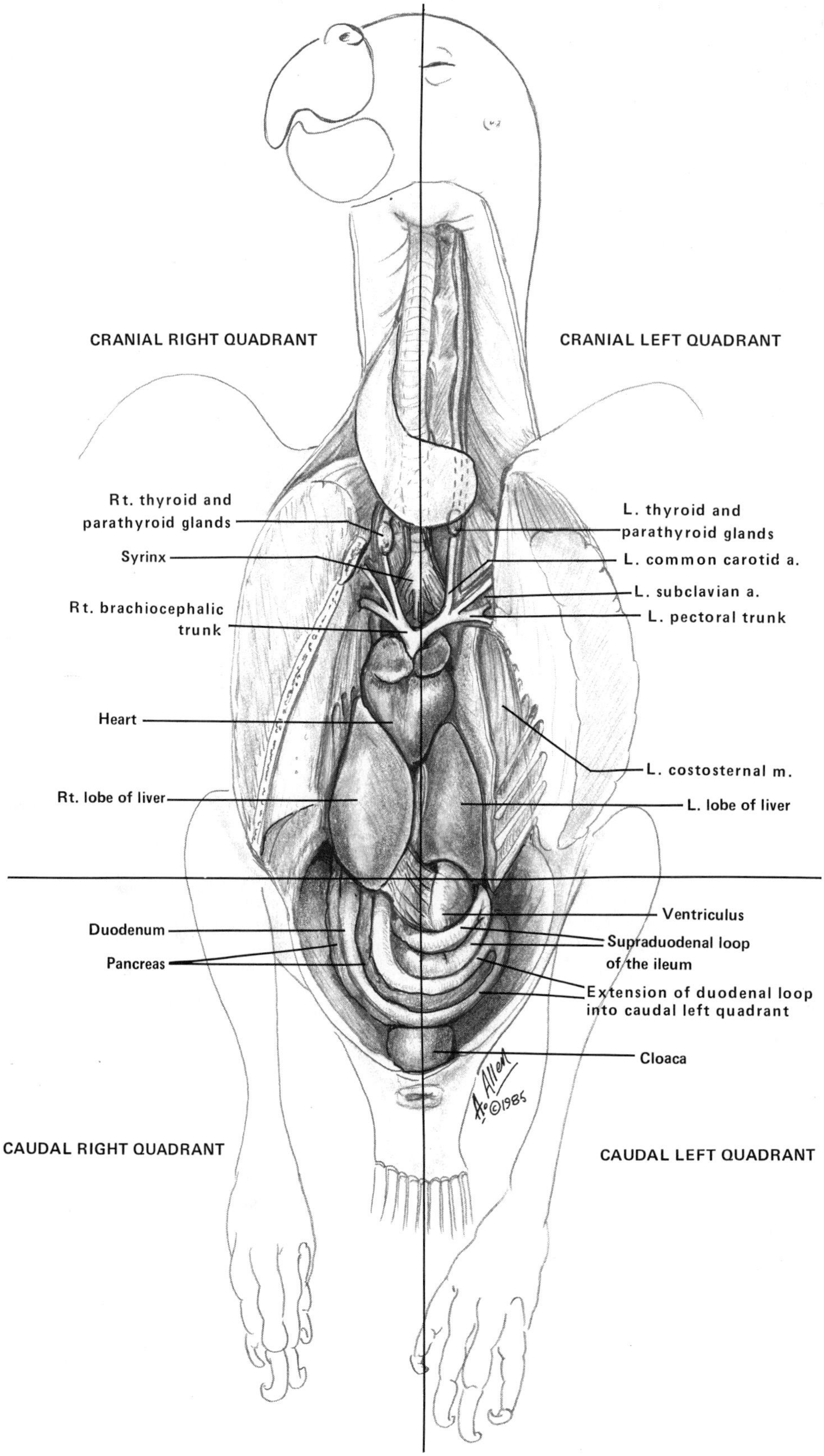


**Figure 4–16.** Internal structures and quadrants of the Amazon Parrot (superficial ventrolateral view).

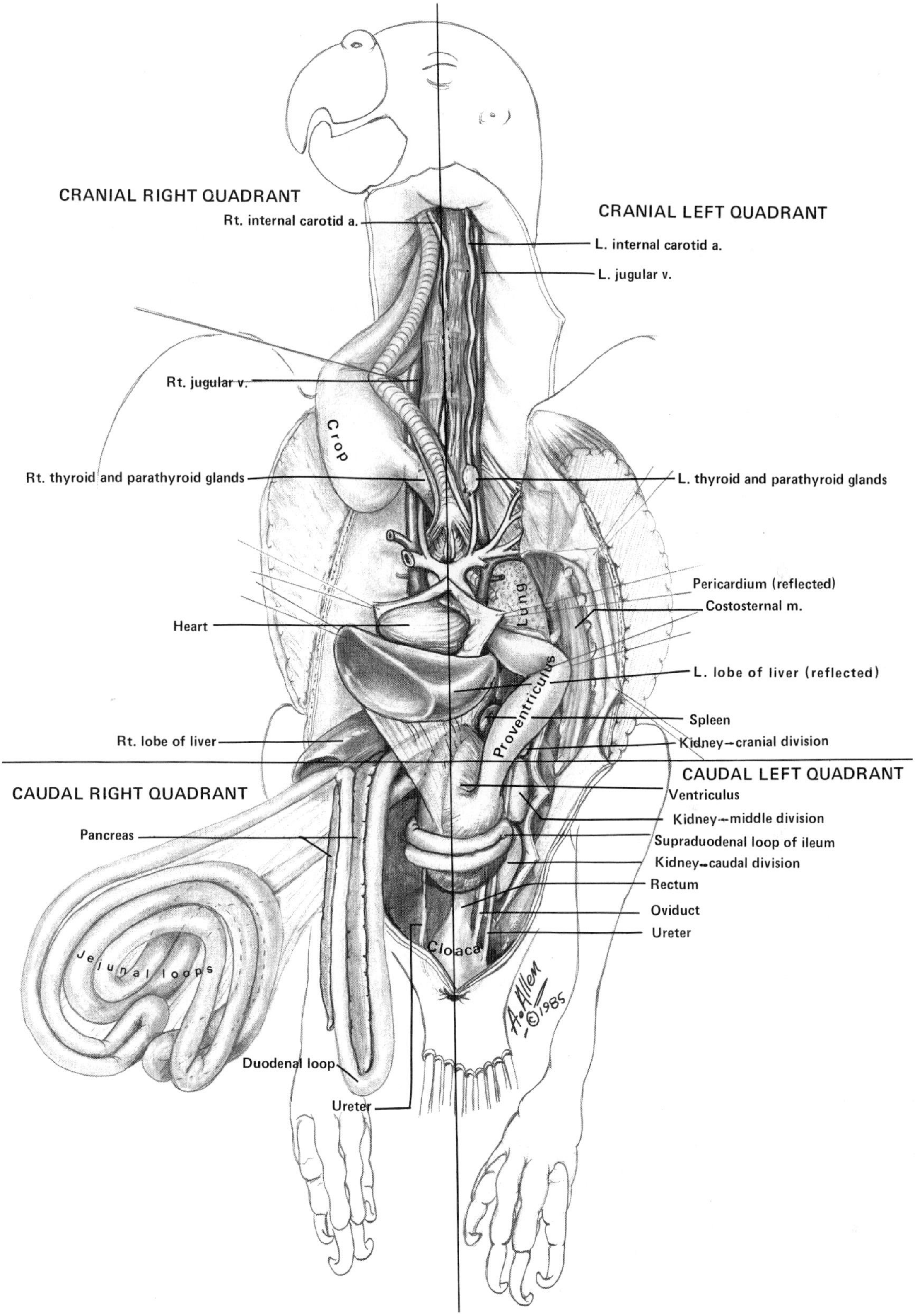


**Figure 4–17.** Internal structures and quadrants of the Amazon Parrot (deep ventrodorsal view with reflection of viscera).

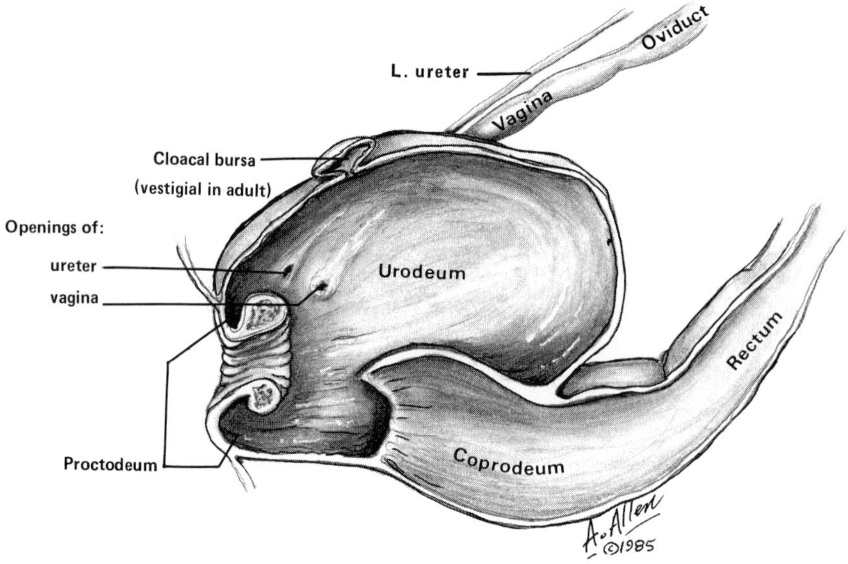


**Figure 4–18.** Cloaca of the female Amazon Parrot (left internal structures).

the axial loop the remnant of the yolk sac (diverticulum vitellinum or Meckel's diverticulum) sometimes persists as a small projection from the antimesenteric surface, especially in the budgerigar. This serves as the arbitrary junction of the jejunum and ileum. The long supraduodenal loop in the Amazon Parrot is the most distal loop of the ileum.

The usual gross or histologic distinctions in mammals between the portions of the small intestine (duodenum, jejunum, and ileum) are not as clear in birds. It is justifiable, however, for descriptive purposes and from the standpoint of naming regional blood vessels (Fig. 4–19) to use the names of the small intestine as duodenum, jejunum, and ileum. Ceca are absent in the Amazon Parrot and all other psittacines.[18]

In macaws, Amazons, and cockatoos, the rectum is much shorter and generally narrower than the intestinal loops of the small intestine. The rectum is the straight portion of the intestinal tract extending between the ileum and the cloaca. One must be cautious in making midventral surgical abdominal incisions to avoid transecting the supraduodenal loop of the ileum, which traverses the caudal quadrants in direct apposition to the ventral abdominal musculature in this area.

The rectum, descending caudally in the left caudal quadrant, is short in the parrot and serves little function other than water absorption. Being of small diameter, it should not be confused surgically with the uterus in juvenile birds.

Terminally the rectum opens with the urogenital system into the cloaca (Fig. 4–18). The midventral portion of the cloaca into which the rectum terminates is the coprodeum. A fold separates this area from the dorsal area, called the urodeum, into which the ureters and oviduct or ductus deferens empty. The coprodeum and urodeum continue over a shallow septum as the proctodeum, which opens externally by way of the vent. Hysterectomy, replacement or amputation of vaginal or rectal prolapses, removal of cloacal papillomas, manipulation or surgical removal of retained eggs, and other cloacal surgery require knowledge of the anatomic location of all cloacal orifices to avoid occlusion of any of these vital openings.

The pouchlike diverticulum on the mid-dorsal wall of the proctodeum is called the cloacal bursa (of Fabricius). It is surrounded by tissue resembling that of the thymus. The cloacal bursa is largest in young birds but atrophies after the bird reaches maturity. The liver and pancreas, very important in digestion, will be discussed below under the Endocrine System.

## UROGENITAL SYSTEM

The kidneys of birds are metanephric in type and are partially divided into three sections, which are accurately termed divisions rather than lobes[15] (Figs. 4–19 and 4–20). The cranial renal division extends between the eighth thoracic rib and cranial synsacrum ventral to the

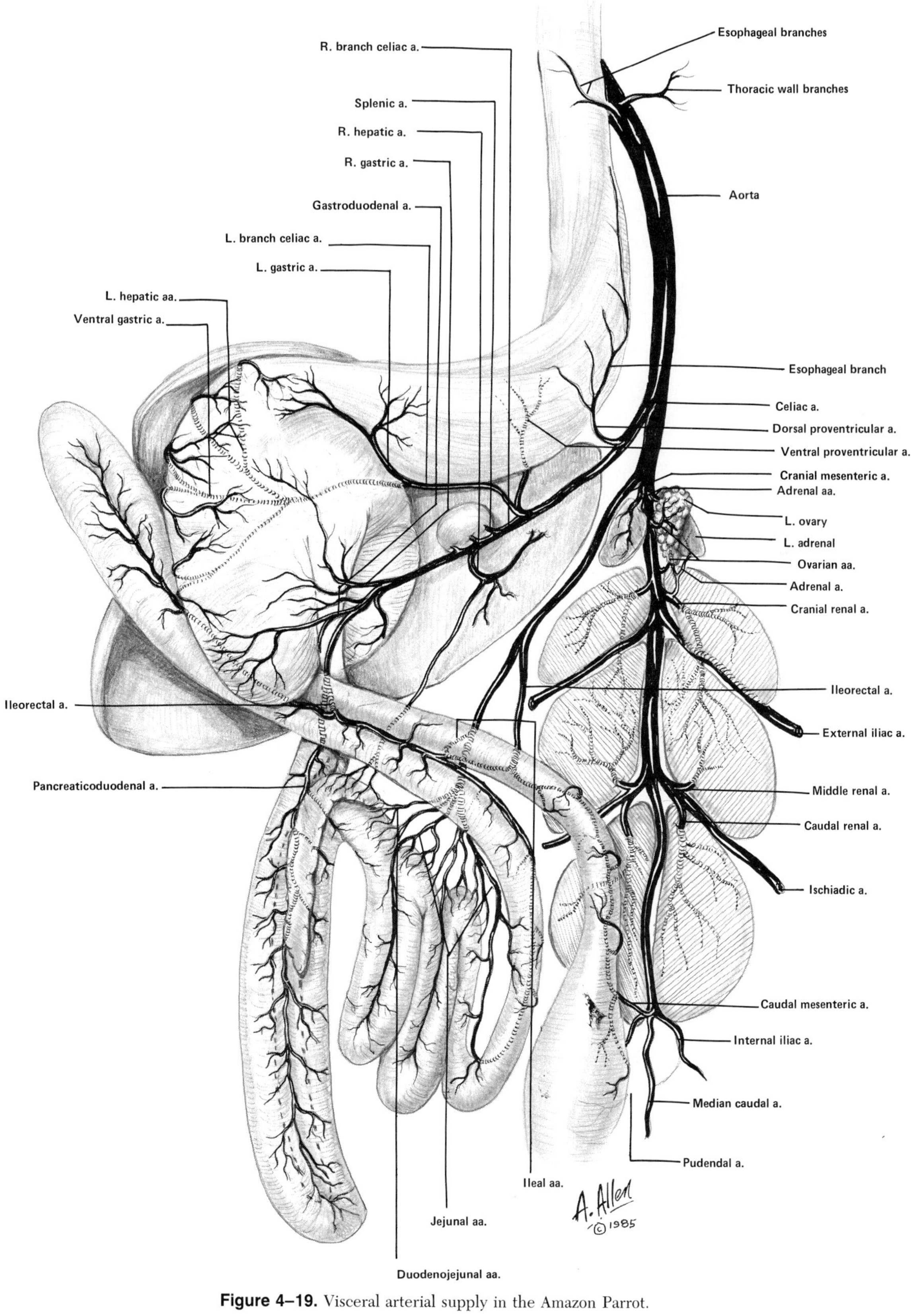


**Figure 4–19.** Visceral arterial supply in the Amazon Parrot.

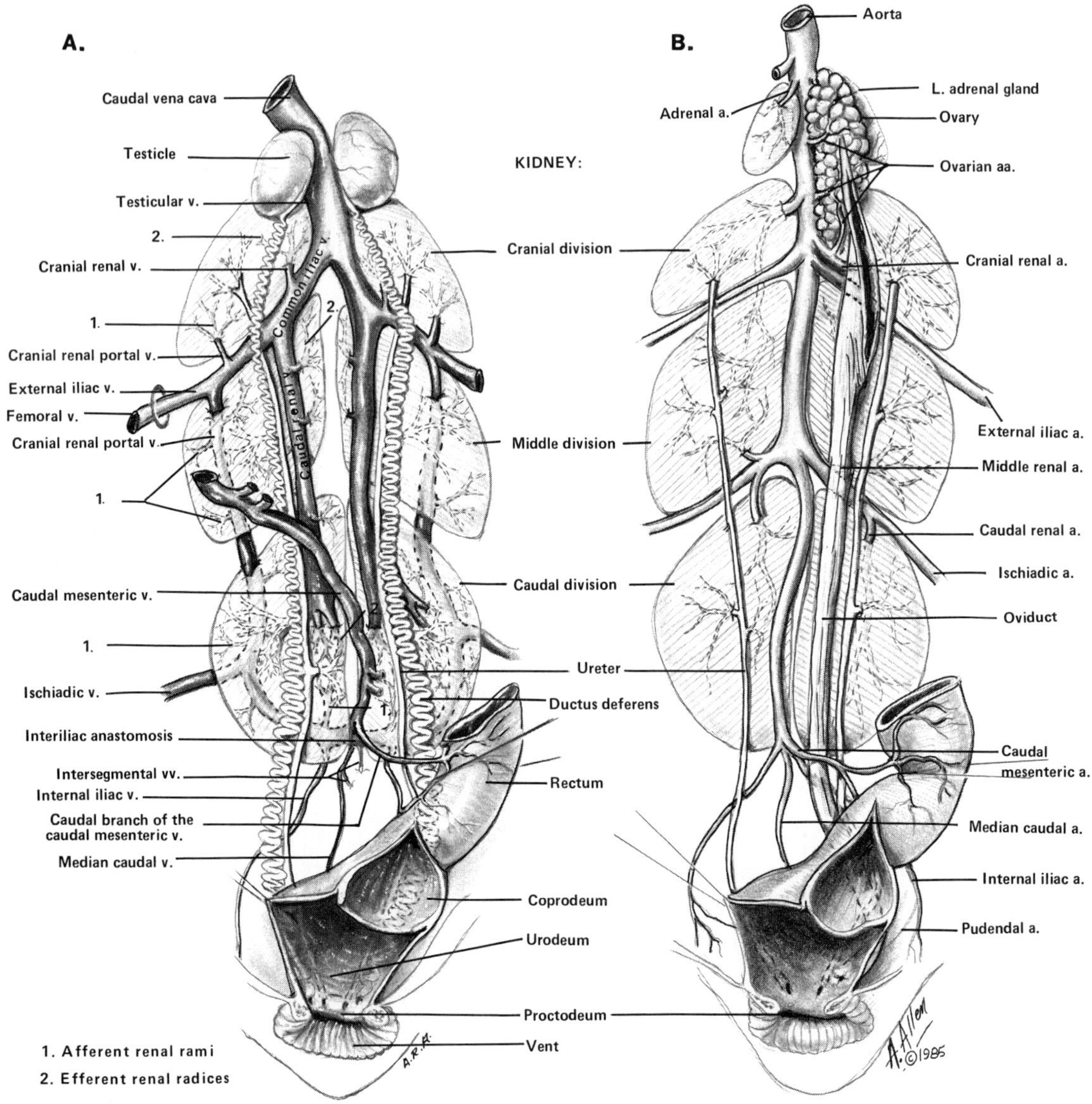


**Figure 4–20.** Urogenital system of the Amazon Parrot. *A*, Male with veins. *B*, Female with arteries.

ilium. The middle renal division is recessed between the lumbar transverse processes ventral to the synsacrum, and the caudal renal division lies in apposition to the sacrum. The glomeruli of birds are smaller and proximal tubules larger than in mammals. A renal pelvis is absent in birds. As in other organs of birds, aggregates of lymphocytes are observed microscopically in the kidney. A slurry uric acid excretion, rather than the liquid allontoin of mammals, accounts for more than 60 per cent of the urinary nitrogen excreted in birds.

As pointed out by Johnson,[14] the avian kidney is deeply recessed into bony depressions or fossae. Large nerve trunks and blood vessels traverse the kidney substance as they pass the synsacral region. The combined effect of fossae, vessels, and nerves secures the kidneys in place rather intricately. Thus it is generally difficult to excise them. The additional blood supply and fibrous adhesions accompanying renal neoplasms generally prevent surgical removal.

The ureters receive urinary ducts from each division, are ventral to each division, and extend caudally to the cloaca. The ureters open dorsal to the opening of the oviduct in the female and medial to the ductus deferens in the male. During endoscopic visualization, semisolid urates can be observed moving in the ureters like cars of a monorail train. The Amazon Parrot, like other psittacines, lacks a urinary bladder and urethra.

### Female Reproductive System

As in mammals, the female genital tract is composed of the ovary, oviduct, uterus, and vagina. In the Amazon Parrot, as in most birds, the right ovary is vestigial while the left one is functional. The unpaired oviduct, uterus, and vagina vary considerably in size and relationship to the ureters during different periods of the breeding cycle. The isthmus becomes very large and tortuous in the reproductively active female, displacing the rest of the abdominal viscera to the right, with the exception of the traversing ileum, which is forced even more tightly against the midventral abdominal wall. These anatomic differences are important in endoscopy for sex determination and evaluation of the reproductive tract in birds.

The ovary is suspended by mesovarium, the oviducts by mesosalpinx, and the uterus by mesometrium. The vagina of birds opens into the urodeum of the cloaca rather than directly to the outside. The vaginal cloacal prominence protrudes into the left dorsolateral urodeum, resembling a volcano in mature hens (Fig. 4–18). The vagina is easily entered for endoscopic viewing or diagnostic culturing in egg-laying birds. If renal biopsy is to be performed in a potential breeding female, the right kidney divisions should be chosen, since the infundibulum of the oviduct fans out over the kidney and must be penetrated in order to biopsy the left renal divisions.

The glandular region of the oviduct (magnum) first receives ova from the ovary through its ostium or infundibulum cranially. The oviduct continues as the isthmus, which terminates in the uterus. Here more fluid albumin and the calcereous shell are added. The short vagina connects the uterus to the left side of the cloaca (Fig. 4–18). Egg formation is described in Chapter 5, Selected Physiology for the Avian Practitioner. The vagina of the cockatiel is quite bulbous. It is the caudal site of ligation for hysterectomies in chronic egg-laying birds or birds with egg peritonitis. In hysterectomies the thin semitransparent oviduct is transected cranially following proper ligation of uterine vessels (see Chapter 48, Selected Surgical Procedures).

### Male Reproductive System

The male psittacine's reproductive tract consists bilaterally of a testis, epididymis, and ductus deferens. The testes are intra-abdominal, lying just cranial to the kidney in immature or non-nuptial males. In some males and most females in breeding condition, the gonads may cover the cranial divisions of the kidneys and in some cases the adrenal glands as well. The testes are intimately associated with the abdominal air sacs, whose function in regulating testicular temperature and spermatogenesis is controversial.[17] The left testis generally is slightly larger than the right, particularly in younger birds.

The testes increase in size greatly during sexually active periods. During this period they should not be confused radiographically with neoplasms. They are generally whiter when active and become darker in the inactive state.[18] Sperm in the testicles pass from the seminiferous tubules of the testicles to the rete testis, efferent ductules, connecting ductules, epididymal duct, and ductus deferens. The epididymis is not grossly visible. The tortuous ductus deferens arises from the medial side of the testis and parallels the ureter caudally to empty into the cloacal promontory as ejaculatory ducts. The prominence of the ductus deferens in mature males is obvious during endoscopic procedures, which is a further aid in identification of the ureters.

Macroscopic organized accessory sex glands are absent in the Amazon Parrot. Seminal fluid is formed by the seminiferous tubules and efferent ducts; a phallus is absent in the Amazon Parrot.

## CARDIOVASCULAR AND HEMOPOIETIC SYSTEMS

The heart of psittacines, although proportionately larger than that of mammals, has four similar chambers: the right atrium, right ventricle, left atrium, and left ventricle. The ventral portion of the heart is cradled by the sternum.

In the Amazon Parrot, the right atrium receives blood from the periphery via divided right and left cranial venae cavae, single caudal vena cava, and vessels directly from the proventriculus (Fig. 4–21). Atrioventricular valves are thin and exhibit poorly defined cusps. The right atrioventricular valve is muscular. Superficial and deep branches arise from both the right and left coronary arteries and supply the myocardium; unlike mammals, deep branches are dominant.[24] From the right atrium blood enters the right ventricle and then the pulmonary artery, passing to the lungs. Pulmonary veins return blood from the lungs to the left

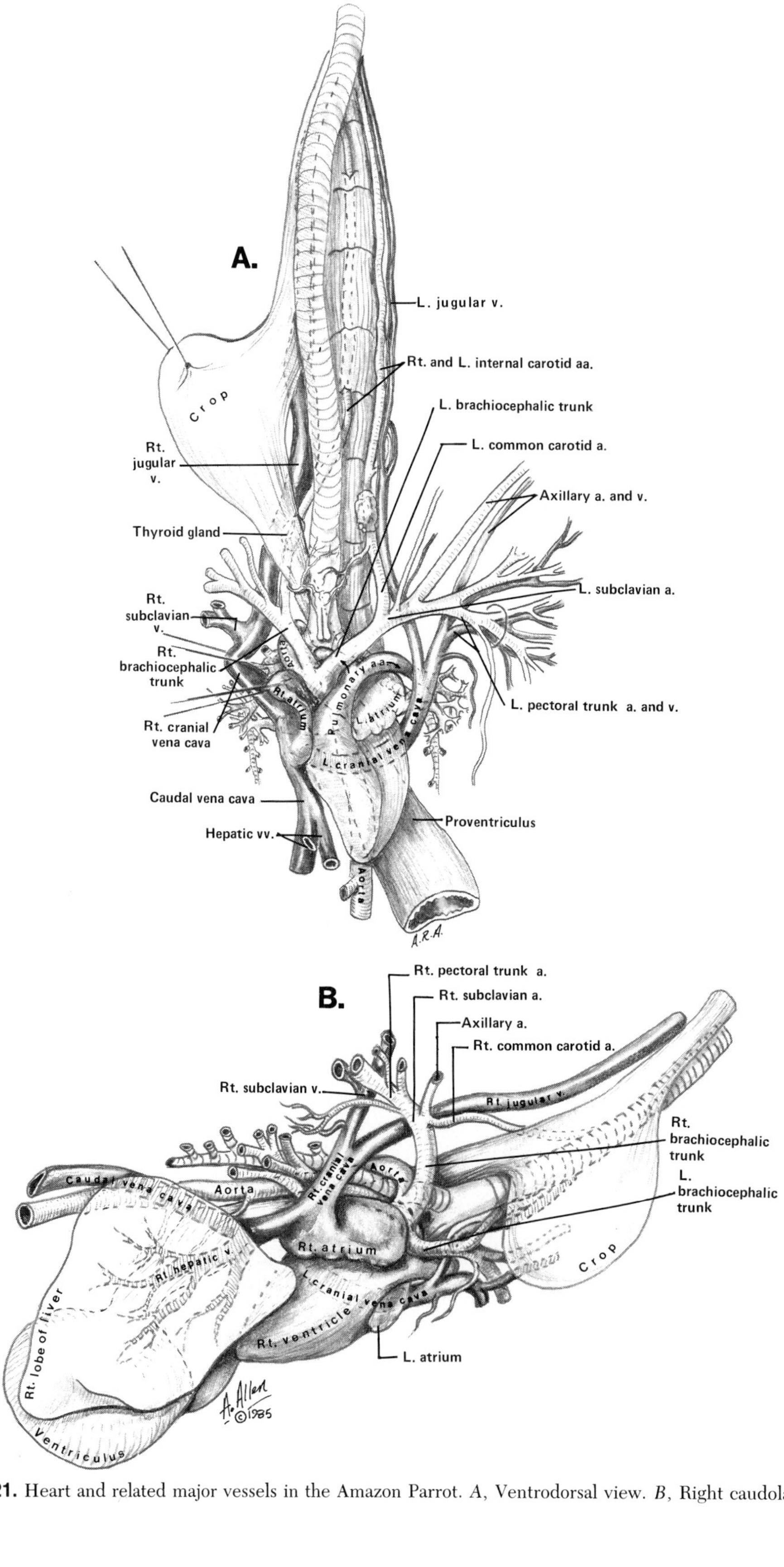


**Figure 4–21.** Heart and related major vessels in the Amazon Parrot. *A*, Ventrodorsal view. *B*, Right caudolateral view.

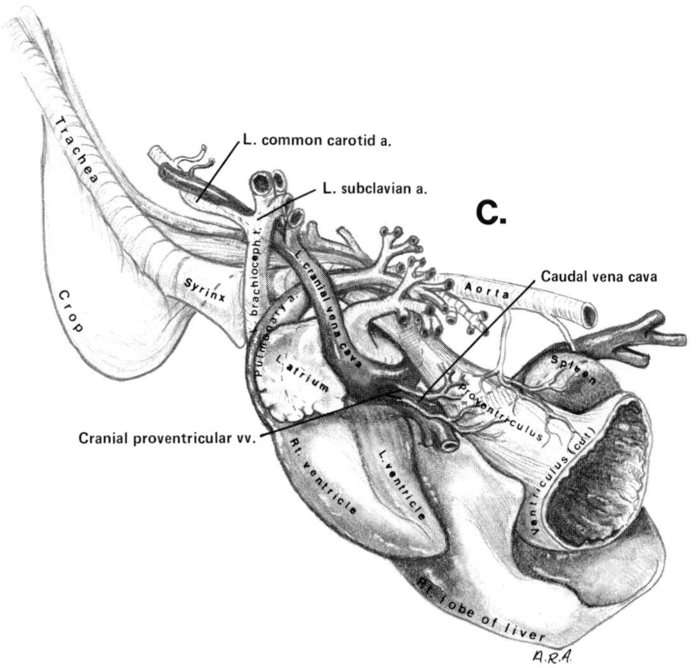


**Figure 4–21** *Continued. C,* Left lateral view.

atrium. It then enters the left ventricle, where it is pumped into the aorta for distribution throughout the body. Arterial supply to the appendages and viscera of the Amazon Parrot as dissected by McKibben is illustrated in Figures 4–19, 4–22, and 4–23.

In Amazon Parrots the left internal carotid artery accompanies the left jugular vein in the neck. The right internal carotid artery in the proximal neck lies deep to the longus colli muscle. It emerges near the head to accompany the right jugular vein to the head.

In birds the internal vertebral venous sinus, vertebral veins, and jugular veins return blood from the head and neck to the cranial venae cavae. The right jugular vein, which can be used for venipuncture, is much larger in caliber than the left in the Amazon Parrot.

Caudal to the cranial mesenteric artery, between the kidneys, the cranial renal arteries arise from the aorta and supply each cranial division of the kidneys. The middle and caudal divisions of the kidney are supplied by middle and caudal renal arteries originating from the ischiadic artery (Fig. 4–19). The cranial division of the kidney is smallest. In cockatiels, the cranial artery is easily lacerated if this division is biopsied. The middle and caudal divisions are recommended for biopsy, as the renal arteries in these areas are better protected by a more bulky parenchyma.

On the left side, branches of the aorta or cranial renal artery supply the ovary, oviduct, and left adrenal gland. Because the ovarian arteries of a cockatiel are difficult to ligate owing to the short length and large diameter of the vessels, the ovary is not removed in a hysterectomy of this species. However, the infundibulum is sparsely vascularized and can be bluntly dissected from its attachments with little hemorrhage.

The right adrenal gland receives arterial branches directly from the aorta or right cranial renal artery. Testicular arteries originate from the cranial renal artery or directly from the aorta. There is no pampiniform plexus in birds. Testicular veins empty into the caudal vena cava. Paired lumbar arteries originate from the aorta between the kidneys and supply the lumbar area.

Near the middle division of each kidney, each external iliac artery originates from the aorta, passes dorsal to the kidneys, and divides into a cranial and a caudal branch (Fig. 4–23). Renal neoplasms impinging on the external iliac arteries may cause ischemia, paralysis, and necrosis of the legs. The cranial branch of the external iliac artery supplies the medial cranial thigh muscles. The caudal branch supplies the abdominal wall and pelvic area and enters the limb as the femoral artery. It anastomoses with the ischiadic artery near the knee. The ischiadic artery arises from the aorta between the middle and caudal divisions of the kidney and supplies most of the hindlimbs (Figs. 4–19 and 4–23). The caudal mesenteric artery originates from the aorta caudal to the origin of the paired ischiadic arteries. It divides into one branch

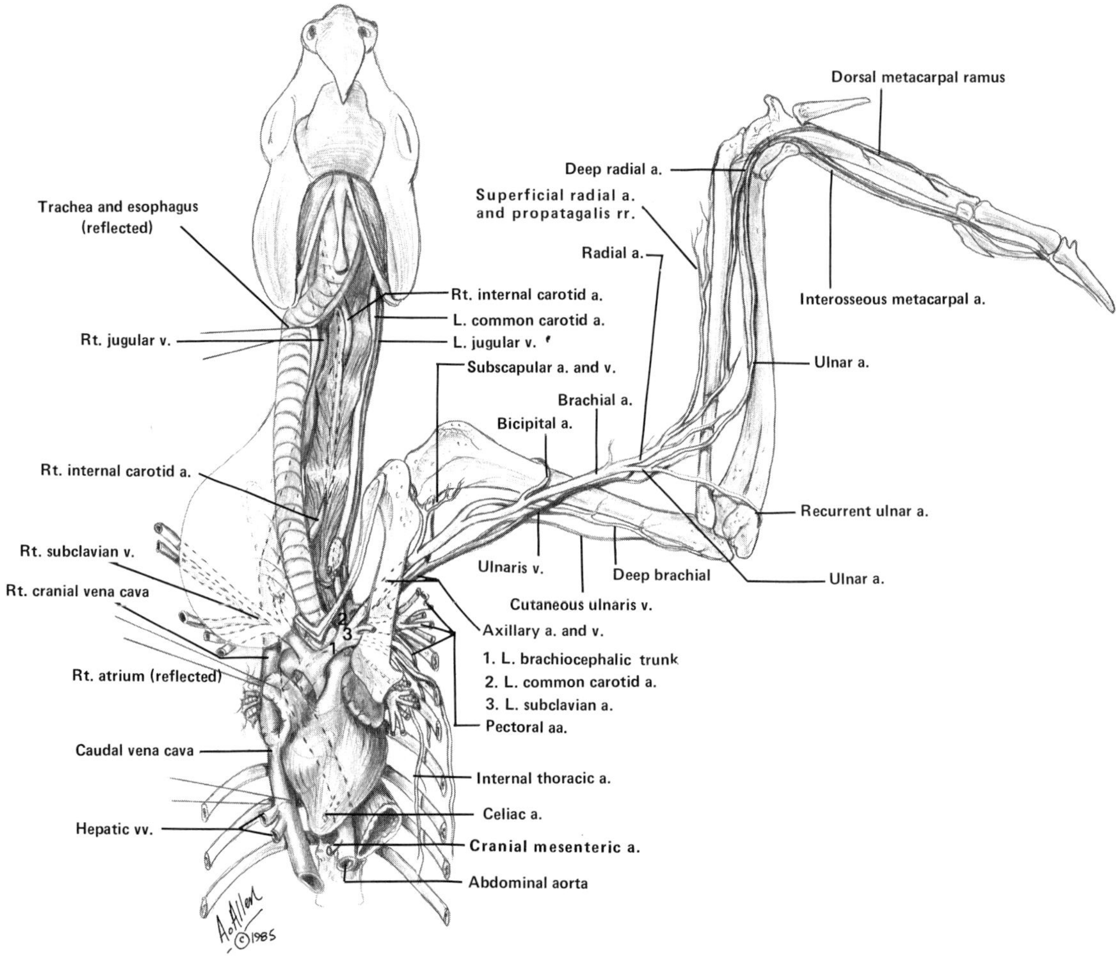


**Figure 4–22.** Arterial supply to the wing of the Amazon Parrot.

accompanying the caudal mesenteric vein to the ileum (intestines) and the other supplying the rectum. The aorta terminates into the right and left internal iliac arteries, which supply the dorsolateral wall of the cloaca and pelvic musculature lateral to the pubis.

Mesenteric and iliac veins drain the abdominal and pelvic regions. The caudal mesenteric (coccygeomesenteric) vein receives blood from the caudal small intestines, rectum, and cloaca. It terminates in both the hepatic and renal portal systems. Usually blood in the caudal mesenteric vein flows toward the kidney, but it may reverse its flow toward the liver.[15] The renal portal shunt is discussed in Chapter 5.

Lymphatic vessels are less numerous in birds than in mammals. Inside the trunk they usually accompany arteries and outside the trunk follow veins.[28] Their names are generally similar to those of the vessel they accompany. Muscular dilatations of lymphatic vessels called lymphatic hearts occur in some birds. Valves are present in lymphatic vessels of birds.

Primary lymphoid tissue includes the thymus and cloacal bursa. These organs are involved in the immune response. The thymus gland is a multilobed paired structure extending along the jugular vein and vagus nerve in the neck. Grossly it appears to regress in the adult Amazon Parrot (see Chapter 24, Relationship of the Avian Immune System to Infectious Diseases).

The secondary or peripheral lymphoid tissues of birds include the spleen, bone marrow, and mural lymphoid nodes (small collections of lymphoid tissue within or close to lymphatic vessels which are not generally grossly visible in psittacines). In addition, considerable lymphoid tissue is present in diffuse or aggregated areas in the wall of the digestive tract from the oral cavity to the end of the intestines. Both solitary and aggregated lymphoid nodules have also been demonstrated in the oculonasal re-

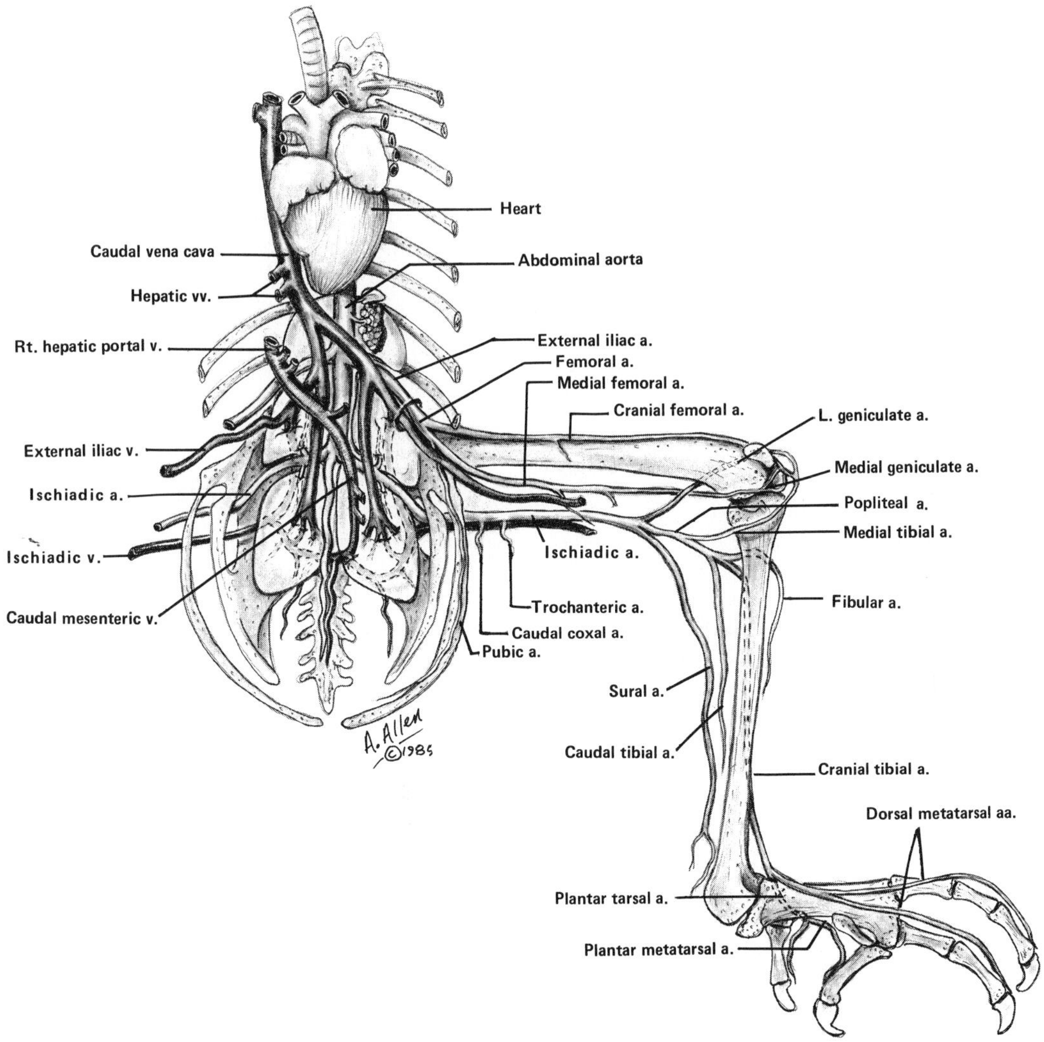


**Figure 4–23.** Arterial supply to the leg of the Amazon Parrot.

gion, liver, pancreas, kidney, lung, endocrine organs, gonads, peripheral nerves, and skin.[30]

The spleen lies cranial to the ventriculus on the right of the ventral border of the proventriculus (see Figs. 4–16 and 4–17). The left lateral approach is used for biopsy; the surgeon proceeds dorsally over the proventriculus and moves the coiled small intestines caudally to expose the spleen. Although rounded in the Amazon Parrot, the appearance of the spleen varies according to the species (see Chapter 23, Necropsy Procedures). Enlargement in disease is a common postmortem and radiographic finding.

## SENSORY ORGANS

Vision is the dominant special sense in birds. The eyes are relatively large, and the muscles controlling the movement of the globe and lids are well-developed. Enucleation and surgical removal of cataracts have been described in recent veterinary literature, emphasizing the necessity for a knowledge of the anatomy of the avian eye and related structures (see Chapter 19, Ophthalmology).

In addition to dorsal and ventral eyelids, the nictitating membrane, or third eyelid, is well-developed in the Amazon Parrot. The gland of the nictitating membrane (harderian) and lacrimal glands are present. The sclerotic ring of 12 bones previously described lends stability to the iris in the Amazon Parrot. Although the musculature of the iris is commonly considered striated, Oliphant et al.[26] describe myoepithelial, smooth muscle, and striated muscle components in the Great Horned Owl. The striated component is the primary pupillary constrictor, and the myoepithelium is the primary pupillary dilator. The annular band of smooth muscle is important in maintaining pupillary size. The yellow color of the iris in some birds is due to fat enclosed in crypts. Age and sex also influence the color of the iris (see Chapter 51, Sex Determination Techniques, and Chapter 7, Preliminary Evaluation of a Case).

A densely pigmented vascular projection called the pecten extends from the retina near the optic disc into the vitreous humor of the eyes (see Chapter 19, Ophthalmology). It may assist in nourishing the retina, serve as a heating element for the eye, a means of detecting small shadows, a regulator of intraocular pressure, or sensor of magnetic fields. A fovea, or depression, in the central area of acute vision of the retina may be absent in some birds, but frontal and central foveae are present in others. These dual foveae permit binocular stereoscopic vision. Cones outnumber rods in birds that are active during the day and rods outnumber cones in nocturnal birds. Diurnal birds probably also have color vision. According to Evans,[7] birds are capable of discriminating color. This was later confirmed by Pepperberg[27] in African Grey Parrots.

Birds have no retinal vessels comparable to those in mammals, but the retina receives its blood supply from the choriocapillaris. Tears drain into the nasal cavity via nasolacrimal ducts, which are formed by dorsal and ventral lacrimal canals in Amazon Parrots.

The psittacine's middle ear has only one bony ossicle, the columella. Sounds pass from the tympanic membrane through the columella to the vestibular window, where they become pressure waves entering the perilymph system of the inner ear. The inner ear has been described by Evans.[7] Pharyngotympanic tubes extend between the middle ears and pharynx, opening by a common slitlike opening into the pharynx.

## ENDOCRINE SYSTEM

Hormones are secreted into the circulatory system. Glands with endocrine functions include the thyroid, parathyroid, ultimobranchial gland, thymus, adrenal, pituitary, and pineal body. The pancreas, testes, ovaries, liver, small intestine, kidneys, uterus, and other organs secrete hormones as well as performing their other functions (see Figs. 4–14 to 4–20).

Bilaterally the reddish-brown thyroid gland of psittacines is located within the thorax. It lies near the syrinx and closely related to the jugular vein and common carotid artery. Enlarged thyroid glands are reported in psittacines.

The parathyroid glands lie on the caudal surface of the thyroid gland or in close proximity to it. The ultimobranchial gland lies caudal to the parathyroid gland associated with the common carotid artery and jugular vein.

Adrenal glands bilaterally lie ventral to the synsacrum just cranial to the kidneys and gonads, immediately cranial to the bifurcation of the caudal vena cava (see Figs. 4–19 and 4–20). Endoscopically, they usually appear oblong, and orange, yellow, or cream-colored. Microscopically, they are not clearly divided into a cortex and medulla.[12]

The pineal gland is a mid-dorsal outgrowth from the roof of the diencephalon, lying between the cerebral hemispheres and the cerebellum. The hypophysis lies within the hypophyseal fossa of the sella turcica portion of the basisphenoid bone ventral to the diencephalon and just caudal to the optic chiasma. The adenohypophysis is divided into a pars distalis and pars tuberalis. The pars intermedia is absent.[12] The neurohypophysis is composed of the infundibulum, median eminence, and neural lobe. These areas are extensively supplied with nerve tracts from the hypothalamus.[12] The gland secretes nine hormones.

The liver is divided into right and left lobes, with the left lobe often smaller than the right (see Figs. 4–14 to 4–16). The liver has little connective tissue. The liver may be approached for organ biopsy procedures by a transabdominal percutaneous needle biopsy technique[16] or in conjunction with laparoscopy with the bird in right lateral recumbency (see Chapter 15, Endoscopy, and Chapter 16, Biopsy Techniques). A knowledge of the anatomic relationships is essential to avoid penetrating other organs, particularly the heart, lungs, and proventriculus.

A gallbladder is absent in Amazon Parrots and Indian Ring-necked Parakeets (*Psittacula krameri*), although it is present in some psittacine species. The presence of a gallbladder appears to be variable even in individuals within a species, as observed by Gadow[9] in cockatiels and other birds.

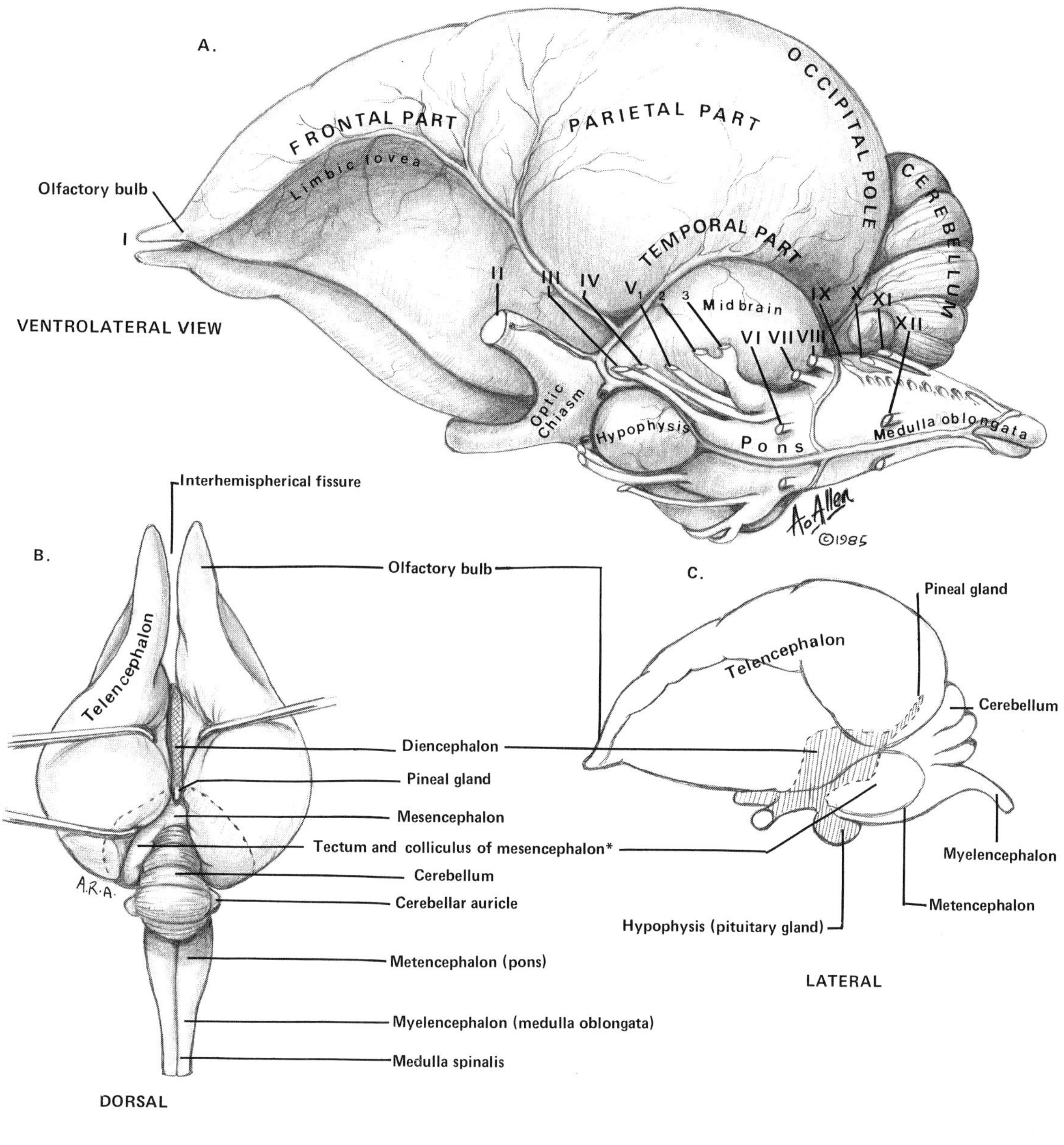


**Figure 4–24.** The brain of the Amazon Parrot. *A*, Left ventrolateral view. *B*, Dorsal view. *C*, Lateral view.

The pancreas in the Amazon Parrot lies enclosed in and adjacent to the duodenal loop. Evans[5] reports a three-lobed pancreas in the budgerigar, as does Jain[13] in the Indian Ring-necked Parakeet, with two ducts entering the proximal medial ascending duodenal loop and one duct on the opposite side. The bile ducts also enter in the area of the first two pancreatic ducts.

## NERVOUS SYSTEM

The clinical significance of the avian central and peripheral nervous systems is discussed in Chapter 20, Neurologic Examination, and Chapter 38, Neurologic Disorders. Figure 4–24 illustrates the brain of the Amazon Parrot.

## REFERENCES

1. Berman, S. L.: The hindlimb musculature of the White-fronted amazon *(Amazona albifrons)*. The Auk, 101:74–92, 1984.
2. Blackmore, F. H.: The effect of temperature, photoperiod and molt on the energy requirements of the house sparrow *(Passer domesticus)*. Comp. Biochem. Physiol., 30:433, 1969.
3. Bühler, P.: Functional anatomy of the avian jaw apparatus. *In* King, A. S., and McLelland, J. (eds.): Form

and Function in Birds. Vol. II. New York, Academic Press, 1981.
4. Campbell, T. W.: Infra-orbital sinus aspiration. Assoc. Avian Vet. Newsletter, 4:38, 1983.
5. Evans, H. E.: Anatomy of the budgerigar. *In* Petrak, M. L. (ed.): Diseases of Cage and Aviary Birds. Philadelphia, Lea and Febiger, 1969.
6. Evans, H. E.: Dissection and Study of the Chicken and Budgerigar. Ithaca, NY, Cornell University, 1972.
7. Evans, H. E.: Anatomy of the budgerigar. *In* Petrak, M. L. (ed.): Diseases of Cage and Aviary Birds. 2nd ed. Philadelphia, Lea and Febiger, 1983.
8. Forshaw, J. M., and Cooper, W. T.: Parrots of the World. New York, Doubleday & Company, Inc., 1973.
9. Gadow, H.: Vogel. *In*: Bronn's Klassen und Ordnungen des Thierreichs, Anat. Theil. Vol. 6. Leipzig, C. F. Wintersche, 1891.
10. Gaunt, A. S., and Gaunt, S. L.: Electromyography of the syringeal muscles in parrots. *In* Abstracts of Presented Posters and Papers, 101st Stated Meeting of American Ornithologists' Union, New York, 1983.
11. Güntert, V. M.: Morphologische Untersuchungen zur adaptiven Radiation des Verdauungstraktes bei Papagien (Psittaci). Zool. Jb. Anat., 106:471–526, 1981.
12. Hodges, R. D.: Endocrine glands. *In* King, A. S., and McLelland, J. (eds.): Form and Function in Birds. Vol. II. New York, Academic Press, 1981.
13. Jain, D. K.: Histomorphology of digestive glands of frugivorous, carnivorous, and omnivorous species of birds. Zool. Rec., 114:119, 1977.
14. Johnson, O. W.: Urinary organs. *In* King, A. S., and McLelland, J. (eds.): Form and Function in Birds. Vol. I. New York, Academic Press, 1979, pp. 185–186.
15. King, A. S.: Systema urogenitale. *In* Baumel, J. J., et al. (eds.): Nomina Anatomica Avium. New York, Academic Press, 1979.
16. Kollias, G. V., Jr.: Liver biopsy techniques in avian clinical practice. Vet. Clin. North Am. [Small Anim. Med.], 14(2):287–298, 1984.
17. Lake, P. E.: The male in reproduction. *In* Bell, D. J., and Freeman, B. M. (eds.): Physiology and Biochemistry of the Domestic Fowl. Vol. 3. New York, Academic Press, 1971.
18. Lake, P. E.: Male genital organs. *In* King, A. S., and McLelland, J. (eds.): Form and Function in Birds. Vol. II. New York, Academic Press, 1981.
19. Lowenstine, L.: Skeletal differences between birds and mammals. Assoc. Avian Vet. Newsletter, 4:49, 1983.
20. Lucas, A. M.: Integumentum commune. *In* Baumel, J. J., et al. (eds.): Nomina Anatomica Avium. New York, Academic Press, 1979.
21. Lucas, A. M., and Stettenheim, P.: Avian Anatomy—Integument. Pts. I and II. Agriculture Handbook 362. Washington, DC, U.S. Government Printing Office, 1972.
22. McKibben, J. S., and Christensen, G. C.: The venous return from the interventricular system of the heart: A comparative study. Am. J. Vet. Res., 25:512–517, 1964.
23. McLelland, J.: System digestorium. *In* Baumel, J. J., et al. (eds.): Nomina Anatomica Avium. New York, Academic Press, 1979.
24. McLelland, J.: Digestive system. *In* King, A. S., and McLelland, J. (eds.): Form and Function in Birds. Vol. I. New York, Academic Press, 1979.
25. Mudge, G. P.: On the myology of the tongue of parrots, with a classification of the order based upon the structure of the tongue. Trans. Zool. Soc. London, *16*:211–278, 1903.
26. Oliphant, L. W., Johnson, M. R., et al.: The musculature and pupillary response of the great horned owl iris. Exp. Eye Res., 37:583–595, 1983.
27. Pepperberg, I. M.: Cognition in the African Grey parrot; preliminary evidence for auditory/vocal/comprehension of the class concept. Ann. Learn. Behav., 11(2):179–185, 1983.
28. Rose, M. E.: Lymphatic system. *In* King, A. S., and McLelland, J. (eds.): Form and Function in Birds. Vol. III. New York, Academic Press, 1981.
29. Smith, J. H., et al.: Microscopic and submicroscopic structure of the budgerigar *(Melopsittacus undulatus). In* Abstracts of Presented Posters and Papers, 101st Stated Meeting of the American Ornithologists' Union, New York, 1983.
30. Vanden Berge, J. C.: Aves myology. *In* Getty, R. (ed.): The Anatomy of the Domestic Animals. Philadelphia, W. B. Saunders Co., 1975.
31. Vanden Berge, J. C.: Myologica. *In* Baumel, J. J., et al. (eds.): Nomina Anatomica Avium. New York, Academic Press, 1979.
32. Walsh, M. T.: Clinical manifestations of cervicocephalic air sacs of psittacines. Comp. Cont. Ed., 9:783–789, 1984.

## ACKNOWLEDGMENTS

I wish to thank Drs. Susan Clubb and Berne Levine for providing specimens for anatomic dissection, injection, and maceration. The sacrifices and support of my wife Marge made completion of this work possible. To her I am particularly grateful. The artistry and dissection assistance contributed by my son, James McKibben, and by Professor Algeron Allen of Purdue University help visualize descriptions much more clearly. The reviews by Dr. Howard E. Evans are greatly appreciated.

JOHN S. McKIBBEN

The reviews and editing by Dr. Julian J. Baumel are greatly appreciated. The anatomy drawings were funded in part by a grant from the Research Institute for Avian Medicine, Nutrition and Reproduction, Inc., and completed by the donation of extensive amounts of time by Professor Allen and Dr. McKibben.

GREG J. HARRISON

*Chapter 5*

# SELECTED PHYSIOLOGY FOR THE AVIAN PRACTITIONER

ROBERT K. RINGER

Our knowledge of the physiologic functioning of such birds as the parrot is very fragmentary; hence we base our understanding on the domesticated species about which extensive information has been gathered in recent years. However, extrapolation to other species is always fraught with considerable chance of error. Excellent detailed reviews on anatomy and physiology of birds are contained within the books listed at the end of this chapter. These reviews, together with selected journal articles, provided much of the information included in this chapter.

## BLOOD AND CIRCULATION

### Blood

Blood volume of most birds ranges from about 6.5 to 10 per cent of the body weight, with packed cell volumes from 28 to 48 per cent. Plasma volume is in the range of 4 to 8 per cent of body weight. There is an inverse relation to the amount of body fat; thus, older birds with somewhat more fat than younger birds have lower blood volumes as a percentage of body weight. Those bird species studied have a greater capacity to withstand hemorrhage without death than do mammals. This capacity is attributed to an ability to mobilize fluid from the tissues to replenish the lost blood at a greater rate than mammals. Birds do not demonstrate irreversible shock following prolonged reduced blood pressure due to hemorrhage.

The specific gravity of whole blood averages 1.04 to 1.06. Viscosity of whole blood is influenced by the total erythrocyte concentration and that of plasma by the concentration of plasma proteins. Factors that alter the concentration of these two components alter the viscosity of the blood.

Avian erythrocytes are somewhat larger than mammalian erythrocytes (10 to 16 $\mu m \times 6$ to 8 $\mu m$), oval in shape, and nucleated. Their concentration in millions per cubic millimeter ranges from 1.9 to 5.9. The life span of avian blood cells is shorter than that of mammalian cells, being approximately 28 to 45 days, depending on the species and how the span was measured. This shorter life span is attributed to the greater metabolic rate and higher body temperature of birds than mammals. Erythrocytes are formed in the bone marrow from erythroblast cells. Erythropoietin, produced by the kidney, stimulates these blast cells to produce erythrocytes. Mammalian erythropoietin has been shown to be inactive in avian species and vice versa. Following the loss of red blood cells, reticulocytes are observed in abundant number in peripherally circulating blood, their release stimulated by erythropoietin.

Birds exhibit at least two types of avian hemoglobin with different electrophoretic mobilities and molecular weights. Hemoglobin concentration in blood averages approximately 9 to 16 gm/dl. The mean corpuscular hemoglobin concentration (MCHC), based on the hematocrit value and the hemoglobin concentration, is more variable in avian blood than in mammalian blood. The MCHC value for avian blood is about 30 to 35 gm/dl.

Some differences in blood constituents between mammals and birds are exemplified by glucose and the plasma proteins. Blood glucose in avian species is about twice that of mammalian blood. The albumin:globulin ratio in birds is generally less than 1.0. How high globulin and low albumin content aid in the maintenance of osmotic pressure and normal blood volume or antibody production, the primary functions of plasma proteins, remains a question.

Avian species do not have megakaryocytes in their blood and therefore lack blood platelets. The thrombocyte, which has its origin in the bone marrow of birds, apparently is involved in blood clotting and is the cell that aggregates following the loss of blood from a blood vessel aiding in the formation of a clot. Thromboplastin released from damaged tissue is very important in initiating clotting of avian blood, because avian plasma lacks or is deficient in several of the coagulating factors common to mammalian blood. Thus in surgery a clean cut by a scalpel will bleed more profusely than a jagged cut by a scissors or a tear.

Thrombocytes in avian blood range in most species from 20,000 to 40,000 per cu mm. They are characterized by their clear cytoplasm, almost round nucleus, and one or more bright eosinophilic-staining granules at the pole of the nucleus.

The leukocytes are composed of three granular, polymorphonuclear cells (heterophil, eosinophil, and basophil) and two nongranular cells with a somewhat round nucleus (lymphocyte and monocyte). These cells are described in Chapter 12, Hematology. The leukocytes form the "buffy coat" of the centrifuged, anticoagulant-treated packed cell volume measurement. The leukocytes contribute from 0.5 to 1.5 per cent of the cell volume of whole blood. In avian blood their concentration is between 12,000 and 40,000 per cu mm. Hormones such as estrogen and ACTH, dietary deficiencies, disease, and environmental stress can alter the leukocyte concentration. The chief function of the leukocytes is protection of the body against diseases.

### Circulation

The heart of avian species on a body weight basis is larger than that of mammalian species (1.4 to 2.0 times). In general, small birds have relatively larger hearts than larger birds as a result of greater metabolic activity in smaller birds per unit of weight. The four-chambered heart is innervated by the sympathetic and parasympathetic nervous systems via the vagus nerve. The parasympathetic tone predominates, as both vagotomy and atropinization increase heart rate. The avian heart is also innervated by the cardioaccelerator nerves arising from the sympathetic chain.

The heart of a bird has a specialized conducting system consisting of the sinoatrial node and the atrioventricular node with the AV bundle. In addition, birds possess additional fibers from the AV node and its branch that course around the right AV valve and aortic root. The SA node is the primary pacemaker, and electrical excitation precedes mechanical contraction. The wave of excitation spreads from the SA node to the AV node and hence by way of the bundle and branches. The electrocardiogram of birds can be recorded by placing electrodes on the junction of the wings and the body or on the wings and the thigh muscles of the leg for the conventional leads. Spatial vectors can be taken by placing electrodes on the chest muscles. The electrocardiogram can be altered by vitamin and mineral deficiencies, drug toxicosis, and certain diseases (see Chapter 21, Electrocardiography).

Heart rates taken under restraint are normally higher than those taken by telemetry. Age, season, and excitement have a major impact on heart rate. Drugs known to alter heart rate in mammals generally also alter avian heart rates. These effects may be direct or reflex responses to a change in blood pressure. This reflex response occurs apparently from activation of mechanoreceptors in the aortic arch and pulmonary artery. Attempts in the chicken to demonstrate a functional baroreceptor response in the carotid sinus region resulted in failure to elicit reflex responses to changes in pressure.

The major blood vessels in birds are innervated by both adrenergic and cholinergic fibers. Adrenergic and cholinergic blockers and stimulators and hormones that are pressor or depressor in mammals have the same response in birds. Sodium phenobarbital and sodium pentobarbital have a marked depressor effect on blood pressure. Because blood pressure steadily decreases during prolonged constraint and anesthesia normally drops blood pressure markedly, lengthy surgical procedures subject the bird to possible cardiovascular failure.

Avian species are prone to develop atherosclerosis. In domesticated species the abdominal aorta develops intimal thickening with plaque formation characteristic of the human lesion. In the aortic arch the media is infiltrated with lipid material that ultimately becomes calcified. Males are subject to a greater incidence than females in the abdominal aortic type, but females are more prone to demonstrate the medial form. Both turkeys and chickens exhibit aortic rupture caused by dissecting aneurysms. In turkeys this rupture has been associated with atherosclerosis and hypertension.

## DIGESTION

### Food and Water Consumption

Appetite in birds is controlled by a center located in the lateral hypothalamus, with a satiety center in the ventromedial hypothalamus. In some birds it has been demonstrated that estrogen stimulates appetite; hence birds increase food consumption at the onset of reproduction and this increase continues until the cessation of the reproductive period.

Water consumption is dependent upon water balance; input must equal output. Water input is derived from that contained in the food consumed plus metabolic water added to the amount of water consumed. Output is equivalent to water in excreta plus respiratory water loss and evaporative water loss through the skin and, in the female, the water content of the egg produced, about 65 per cent. Hence, the nature of a bird's droppings depends on the type of food consumed. Foods with high protein content, which require additional water consumption for excretion of extra uric acid, result in very moist droppings, whereas the consumption of grains low in protein results in dry droppings. Also affecting excreta water content are such factors as dietary concentrations of sodium and potassium salts, fiber level, and the presence of a disease condition within the bird (see Chapter 9, Evaluation of Droppings).

## The Digestive System

The digestive system varies considerably by species as the result of adaptation to varied nutritional regimens. The beak and tongue take on many shapes; the crop, ceca, and gallbladder may be present or totally absent; and food may pass through the system in minutes or hours. The parrot may use its tongue to move food in the mouth into the pharynx, while other birds must throw their heads backward and forward rapidly to move food within the buccal cavity.

The beak is a good example of adaptation to function—it serves as a prehension tool, prepares food for swallowing by crushing, hulling, or tearing, and in the parrot also serves in locomotion. In the prehension of food, the parrot is able to grasp food in one foot and raise it up to the beak. The beak of birds grows continuously, and it is recommended in parrot management to provide something, such as wood or hard food, with which the beak can be worn down naturally. Trimming the beak may cause it to split and heal crookedly; therefore grinding is recommended for shortening the beak (see Chapter 2, Husbandry Practices).

The mouth possesses mucous glands to moisten and lubricate food. The salivary glands also moisten food and are the source of enzyme secretion (amylase) for starch digestion in some birds. In parrots, which have a dry mouth, salivary secretion is apparently low, or little mucous is secreted by the glands.

The tongue is covered with cornified epithelium with papillae. In parrots the tongue is very muscular and has few taste buds, yet the bird can taste and has an excellent tactile sense provided by Herbst corpuscles. Pacinian corpuscles, which aid in the tactile sense of the mouth, are found in the dermis of the beak in many birds. Shelling of seeds requires that the tongue and beak be manipulated so that the husk is broken, removed, and expelled while the kernel is retained and swallowed. The muscular tongue (parrot), together with the hyoid apparatus and its attachment, aids in this procedure, as does the palatal notch.

The esophagus connects the pharynx to the proventriculus and contains mucous glands in its walls to moisten and lubricate food passing through this connecting tube. The esophageal wall is muscular to move the food along. The crop, which is merely an outpouching of the esophagus, serves as a storage organ and meters food to the stomach. In addition, the crop plays a role in the softening of food by water consumed and a site for the start of enzymatic action by either enzymes contained within plant materials or by salivary secretion.

The proventriculus or true stomach of birds has both digestive and mucous glands within its walls. The digestive glands exhibit pepsinogen granules within the cells that disappear at the time of food ingestion. Hydrochloric acid from the proventriculus converts the pepsinogen to pepsin, the enzyme responsible for protein breakdown.

Ingestion of food causes a reflex stimulation of the gastric mucosa initiating secretion of hydrochloric acid and pepsinogen. As the partially digested food passes in the small intestine, the hormone gastrin is released, which in turn stimulates copious secretion of gastric juices by the proventriculus. Two additional hormones from the small intestine alter gastric secretions. Cholecystokinin-pancreozymin stimulates proventricular secretion, and secretin reduces volume produced as well as pH and pepsin activity.

In general, anesthetics decrease secretion of gastric juices. Since there is only one cell (chief cell) that secretes both pepsinogen and HCl, acidity and enzyme activity increase or decrease together in response to drug or hormone administration.

Drugs such as histamine (subcutaneous infusion) increase gastric juice secretion, whereas atropine injection decreases or inhibits secretion. The fact that the gastrointestinal tract is innervated by the vagus nerve and its branches means that drugs altering the vagus will alter intestinal secretions.

The gizzard or muscular stomach, which is well-developed in parrots, is attached to the proventriculus and contains an epithelium which secretes keratinous fluid that hardens and provides an internal protection and a surface against which feedstuffs can be ground. The presence of grit or small stones aids in grinding; however, grit may not be essential for food digestion.

The pylorus is a fold forming a valve between the gizzard and the duodenum. This fold prevents food with a more alkaline pH from reentering the gizzard, where the pH is in the range of 1.5 to 2.5. The primary organ of digestion is the small intestine. The intestine has folds or plicae with villi that increase the surface area for digestion. Also present are glands and circular muscle bands that move the partially digested food along by segmented and peristaltic movement. The bile ducts and pancreatic ducts empty into the distal portion of the duodenal loop. Some species of parrots lack a gallbladder and others possess one. Where a gallbladder is present, the bile is more concentrated, as the gallbladder reabsorbs water from the bile. Bile neutralizes HCl from the proventriculus and plays an important function in the emulsification of lipids. It also aids in the digestion of carbohydrates by its amylase content. The major bile pigment is biliverdin, giving the green color to bile. In most birds chenodeoxycholic acid is the primary bile acid, with cholic and allocholic acid also present.

Ceca are absent in psittacine birds. In species in which they are present, they function as sites for water reabsorption and digestion of a small amount of cellulose (10 per cent). Evacuation of the cecum occurs separately from the coprodeum, and its contents can be identified generally by its composition, appearance, and odor.

Little or no digestion takes place in the rectum of birds, but this portion of the gastrointestinal tract does reabsorb water, as does the cloaca.

Some investigators have suggested that birds when preening obtain vitamin D from the uropygial gland oil that has been irradiated by sunlight. More recent studies indicate that the preen gland contains little, if any, provitamin $D_3$ and is therefore not a source of vitamin $D_3$ for birds.

## Protein Digestion

Proteins consumed by birds are not denatured, and therefore the acid state of the proventriculus and gizzard is essential in the breakdown of the pepsin-sensitive peptide bonds of the native protein molecules. Pepsin initiates proteolysis and forms polypeptides, which are further broken down by intestinal enzymes.

The endopeptidases secreted in the proventriculus and pancreas degrade proteins to small peptides containing from 2 to 6 amino acids and free amino acids. The brush border of the small intestinal villi serves as the digestive-absorptive surface, having the presence of the digestive enzymes and the capacity to absorb small molecules. Intestinal mucosal cells contain cytoplasmic peptidases that hydrolyze these small molecules to free amino acids. These absorbed amino acids are transported by the portal blood supply to the liver, where the majority pass through to the body cells or are stored in muscle cells for future use.

## Digestion of Fats

Lipids form a bile-bile micelle in the duodenum, which solubilizes the material for absorption. Monoglycerides are absorbed intact. In addition, lipases from the pancreas split triglycerides at the ester linkages. The solubilized material is absorbed by the microvillous projections of the intestinal mucosal cells. These microvilli greatly increase the surface area for absorption.

The bile salts are not absorbed in the upper small intestine (only in lower intestine are bile salts absorbed), so they remain in the lumen of the intestine to form new micelles. Within the mucosal cells the absorbed substances are reesterified and form chylomicrons.

The chylomicrons are composed of a protein, cholesterol, and phospholipid structure surrounding the re-esterified triglycerides. It is in this chylomicron form that they circulate in the blood and are absorbed directly into the portal circulatory system to the liver.

## Carbohydrate Digestion

Pancreatic and intestinal enzymes break down starch into simple sugars, which are absorbed primarily in the small intestine. The breakdown of starch is initiated in the crop, where soaking in water softens the granules, and is further enhanced by the grinding action within the gizzard.

## BODY TEMPERATURE

The maintenance of a relatively constant body temperature requires a balance of heat inflow, production, and outflow. The attainment of homeothermy in birds also requires the coordinated development of many related functions, including changes in thermal insulation, evaporative water loss, body weight, surface area–to–volume ratio, behavior, food consumption, basal metabolism, shivering and nonshivering thermogenesis, and thyroid hormone metabolism. Most species of adult birds maintain a core body temperature around 40° C; thus there is less need for evaporative cooling than in animals with lower body temperatures. In birds a greater importance is placed on conduction, convection, and radiation for heat dissipation. High body temperature aids the bird in resisting infection following surgery, because this temperature is unfavorable for bacterial growth. Many young birds, including psittacines and passerines, are poikilothermic for several weeks after hatching until their thermoregulatory ability is fully developed.

Body temperature shows a diurnal variation, with the highest temperature attained during daylight hours and the lowest at night. Exogenous factors, such as repeated handling, may increase body temperature. Endogenous factors may also influence temperature, as evidenced by the relationship of body temperature and thyroid activity.

### Body Temperature Control

There exists a range of environmental temperatures for each species of birds above and below which the bird has to expend energy either to keep warm or to keep cool. This range is known as the range of thermoneutrality. Outside of this range, the bodily work necessary to maintain a relatively constant body temperature causes increased oxygen consumption. The points at which oxygen consumption increase are known as the lower and upper critical temperatures, respectively, for the lower and upper ambient temperatures of the range of thermoneutrality. As the ambient temperature increases, a constant body temperature can be maintained until the rate of heat production exceeds the rate of heat loss; then the body temperature begins to rise. If ambient temperature continues to rise above the upper critical temperature, body temperature rises until the upper lethal temperature is reached, and the bird dies from respiratory and cardiac failure.

The range of thermoneutrality is narrow for young birds and widens as they age and attain greater body size and thermal insulation. The widening is a result of a decline in the lower critical temperature. One of the major means of adjusting to temperatures outside the thermoneutral range is by altering plumage insulation. To maintain a constant body temperature in an environment below the lower critical temperature, heat must be generated at a rate sufficient to compensate for the heat lost to the environment.

During surgery it is essential to maintain a body temperature somewhere between the lower and upper critical temperatures. The lower critical temperature is higher for small birds than for large ones.

### Insulation

Cutaneous blood flow to the feathered regions of the body is less than to the extremities; thus heat exchange across the feathered area is low compared to that in the legs.

Elevation and depression of the body contour feathers for increased or decreased insulation are accomplished by involuntary smooth muscles within the skin under the control of the autonomic nervous system. Observation of feather erection or depression can be a guide to temperature comfort for birds. Fluffing of feathers with raising of wings by healthy birds indicates an excessively hot environment, with the potential to cause dehydration.

### Higher Centers' Control of Thermoregulation

Nerve centers in the anterior hypothalamus control heat loss by regulating both panting and vasodilation, whereas neurons in the posterior hypothalamus control heat maintenance by regulating shivering and heat production.

Sensory receptors in the skin are sensitive to changes in skin temperature and feed this information to higher centers via the spinal cord. In the pigeon, cooling of the spinal cord's thermosensitive structures causes erection of plumage, shivering, and arteriolar vasoconstriction. Heating of these structures causes cessation of shivering, onset of thermal panting, and vasodilation.

### Shivering Thermogenesis

During cold exposure, birds increase heat production by increased contractile activity of skeletal muscle without voluntary movement and external work. In the adult inactive bird, this mechanism is the primary means of generating heat during cold exposure. The development of homeothermy in birds is correlated with the development of shivering ability. Some birds, such as the duck and pheasant, may show shivering during embryonic development, while others exhibit pronounced shivering after hatching.

### Nonshivering Thermogenesis

Some animals increase heat production during cold by mechanisms unrelated to contraction of the skeletal muscles. In many very young mammals, thermogenesis by release of fatty acids from brown adipose tissue is the primary mechanism involved. With aging, shivering replaces nonshivering heat production.

Little evidence for nonshivering thermogenesis has been demonstrated in birds. No brown adipose tissue has yet been found in young birds.

## RESPIRATION

The lungs serve as the gas exchange portion of the respiratory system. Avian lungs are small, based on body weight percentage, compared to mammalian lungs. Attached to the lungs are a variable number of air sacs (depending upon the species). Unlike the mammalian lungs, which expand and contract on inspiration and expiration, the avian lungs are relatively rigid and do not change in volume during the respiratory cycle. Inelastic avian lungs make auscultatory signs of lung pathology difficult or impossible to detect.

### Mechanics of Respiration

The mechanics of the respiratory cycle differ markedly from those of the mammal (Fig. 5–1). Birds have no functional muscular diaphragm as mammals do. On inspiration the intercostal inspiratory muscles pull the ribs laterally and the sternum ventrally and cranially. This action increases the volume of the body cavity (air sacs) and creates a negative pressure with respect to the atmospheric air pressure surrounding the bird. Thus, air rushes into the region of negative pressure and through the lungs. On expiration the intercostal expiratory muscles pull the ribs caudally, raising the sternum and pulling the ribs slightly inward. This muscle constriction increases the internal pressure within the air sacs and forces the air out of the sacs, through the secondary bronchi, parabronchi, bronchi, and trachea. Both inspiration and expiration require energy expenditure.

The bird can be thought of as a bellows arrangement, fixed at the cranial end by the attachment of the sternum to the pectoral girdle and free at the caudal sternum region. In this fashion the abdominal region of the body moves considerably during the respiratory cycle, whereas the cranial region of the body moves only slightly. Therefore, most of the inspired air flows to the caudal group of air sacs on inspiration because it has the greatest volume change; only a small amount of inspired air moves into the cranial group of air sacs because only a small volume change has occurred, although some parabronchial air enters the cranial air sacs on inspiration. This greater flow of air to the caudal sacs explains why the air within these sacs at the end of inspiration more closely resembles ambient air in $O_2$ and $CO_2$ content than does the air in the cranial air sacs, which have $O_2$ and $CO_2$ content almost identical to expired air. This also accounts for the higher incidence of pathology in the caudal as opposed to the cranial air sacs.

The weight of the viscera also aids in the respiratory cycle, since its weight contributes to pulling the sternum down on inspiration but increases the work load on expiration. It should be pointed out that inverting a bird during clinical examination changes the mode of respiration, because the weight of the viscera and the pectoral muscles must now be raised on inspiration. More importantly, since the intercostal muscles must constrict on both phases of the respiratory cycle to move the ribs and sternum, undue constraint of these structures while the bird is being held in the examiner's hand will cause respiratory difficulty and often suffocation from inadequate ventilation of the lungs. Bandaging that restrains the movement of the sternum will interfere with breathing.

The distribution and direction of air flow

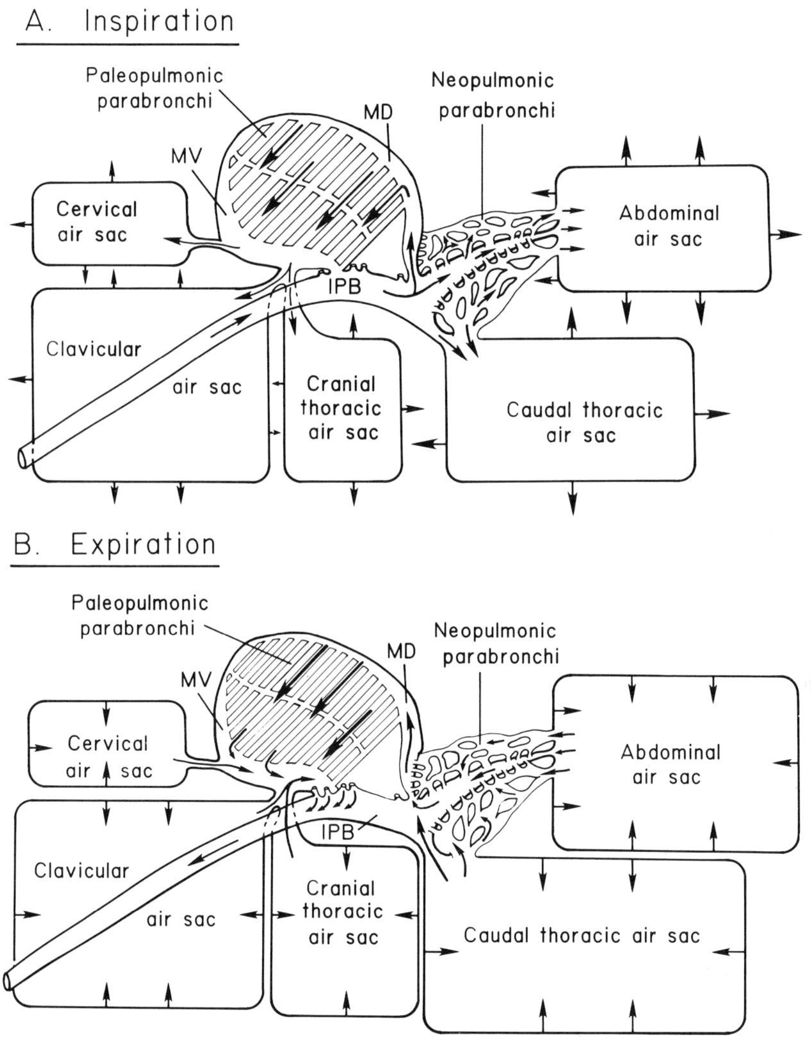


**Figure 5–1.** Structure and gas flow pattern in the avian respiratory system. (Courtesy of Roger Fedde.)

through the avian lung have been the subjects of many studies and much speculation. Recent studies indicate that gas flows through the avian lungs during both inspiration and expiration. All birds possess within their lungs a network of parabronchi; others contain two networks. The first is the paleopulmonic, in which inspiratory and expiratory gases pass unidirectionally, and the second is the neopulmonic, through which gases pass bidirectionally during the respiratory cycle. The paleopulmonic parabronchi constitute the major gas-exchange portion of the lung.

## Gas Exchange and Its Control

Oxygen and carbon dioxide exchange occurs within the parabronchi lobule, in the region of the air capillaries and blood capillaries. Gas that passes through the parabronchi diffuses through the atria and infundibula and into the air capillaries. The air capillaries lie adjacent to and surround the blood capillaries. The air capillaries are 3 to 10 μm in diameter in various species. Mixed venous blood (blood low in $O_2$ and high in $CO_2$) is transported to the lungs by the pulmonary artery. It then courses through smaller arteries into the interparabronchial arteries and into the region of gas exchange by blood capillaries. After perpendicularly traversing the air capillaries, the oxygenated blood courses through venules back to the interparabronchial septum and eventually to the pulmonary vein. This exchange forms a cross-current exchange of the lung gases with the blood and gives a greater efficiency than the mammalian lung arrangement. Moreover, the exchange surface per unit volume of the avian lung is at least ten times that of the human lung.

Oxygen in blood is chemically bound to hemoglobin. Carbon dioxide is carried in blood bound to the proteins, dissolved in the blood, or as the bicarbonate ion in plasma. The arterial pH is generally higher and the $P_{CO_2}$ is generally lower than in most mammals. If the bird is restrained and excited, these differences may be even further exaggerated as the result of hyperventilation.

Many factors influence avian respiration. Anesthetization influences the central nervous system and reduces ventilation, restraint often causes hyperventilation, hyperthermia produces polypnea, and inhalation of noxious gases such as ammonia results in abnormally slow breathing. Severing of both vagi in birds causes marked slowing of the respiratory rate. Low $CO_2$ content of inspired air generally increases minute volume, whereas high $CO_2$ content (6 to 12 per cent) suppresses ventilation. The use of $CO_2$ is an accepted method for euthanasia of birds and is primarily used in the poultry industry.

Humoral control of respiration is by way of the carotid bodies, located near the parathyroid glands. These bodies sense the oxygen tension in the blood coursing to the head region. $CO_2$ chemoreceptors are also located in the lungs and in the brain. Thermoreceptors in the skin and muscles and within the hypothalamus, when excited, cause an increase in respiratory rate and a decrease in tidal volume. This rapid, shallow breathing increases evaporative heat loss but prevents alkalosis.

In surgery it is possible to insert a cannula into the tracheal opening of the larynx, insert a large needle into one of the caudal air sacs, and unidirectionally ventilate the bird. In fact the air flow can be in either direction because the parabronchi of the lungs will be perfused with gas by flow in an inspiratory or expiratory direction.

It should also be pointed out that within the peritoneal cavity the air sacs surround the viscera and attach to the peritoneal wall; therefore, what is termed intraperitoneal injection of drugs in birds is almost certain to be intrapulmonary because it is difficult not to inject into an air sac.

## URINARY SYSTEM

Birds have evolved a urinary system in which the end product of nitrogenous metabolism is uric acid. This type of excretion enables the nitrogenous metabolic wastes to accumulate inside the eggshell during incubation within the allantois without killing the developing embryo. The virtually insoluble uric acid is nontoxic, whereas urea accumulation would kill the embryo. The uric acid excretion also requires less water than the soluble nitrogenous waste, urea. The major osmoregulatory organ is the kidney; however, some birds have a functional extrarenal organ, the salt gland, for the excretion of highly hypertonic saline solutions.

### Avian Kidneys

Avian kidneys are larger in proportion to body weight than are mammalian kidneys (about 1 per cent of body weight) and commonly show three divisions served by several renal arteries. Another difference is that the avian kidney is supplied with afferent venous blood via the functional renal portal system. A common efferent vein drains both the arterial supply and the renal portal supply.

The functional unit of the avian kidney, like that of the mammal, is the nephron. There are two types of nephrons in the avian kidney. Each is composed of a glomerulus and a tubular system. The capillaries of the avian glomerulus are much simpler and larger than the mammalian counterpart. The avian renal medulla contains a countercurrent multiplier system and an associated osmotic gradient. The reptilian type nephron predominates; thus birds' kidneys are less effective than mammalian kidneys in the excretion of electrolytes.

As in the mammalian kidney, the nephrons of the bird possess the capacity for filtration, excretion (secretion), and reabsorption of a substance. The cortical tubules of the nephrons in the bird have the advantage of a renal portal blood supply that is lacking in the mammalian kidney. Tubular secretion and reabsorption are enhanced by this extra blood supply. This renal portal blood supply is regulated by the opening and closing of the renal portal valve located, one on each side, at the junction of the external iliac vein and the renal vein. This muscular valve is richly innervated with adrenergic and cholinergic fibers. *In vitro* studies indicate that acetylcholine causes the valve to close and epinephrine causes it to open. When the valve closes, blood is directed to the kidney tubules in addition to the normal arteriolar blood. Thus, substances absorbed by the intestines, metabolic wastes from the legs and posterior portion of the bird, and those from the oviduct may reach the kidney tubules in the bird before they return to the general circulation. Drugs injected into the leg may also go to the tubules before going into the systemic circulation.

The renal portal system increases the variability obtained in blood sampling; since blood taken from a femoral vein may return to the kidney and be subjected to tubular excretion, it will differ from a brachial vein sample for certain blood constituents.

### Urine Formation and Excretion

Glomerular filtration rate in avian species ranges from approximately 1.7 to 4.6 ml/kg/min. During diuresis in some birds, as little as 6 per cent of the filtered water may be reabsorbed. This figure can be as high as 99 per cent when urine flow is very low.

Water deprivation is important in sick birds because it may affect blood pressure and therefore glomerular filtration, tubular resorption, and cloacal resorption. Sick birds must have water available.

Glucose is normally completely reabsorbed. Sodium chloride and bicarbonate ions are also reabsorbed. Creatinine, uric acid, potassium, histamine, serotonin, acetylcholine, atropine, and epinephrine are among those substances cleared by tubular excretion. The ureters "milk" the semisolid urine along by peristaltic action. This urine empties into the urodeum portion of the cloaca, where the liquid fraction mixes with the fecal material within the cloaca and may pass by retrograde flow (antiperistaltic waves) during dehydration back into the large intestine and into the cecum if one is present. In this process water is reabsorbed and conserved even further.

Cloacal resorption of water is important. Skadhauge (1975) reported the maximal osmotic urine:plasma ratio for a selected group of Western Australian birds. The emu *(Dromaius novaehollandiae)*, crested pigeon *(Ocyphaps lophotes)*, grass parrot *(Neophema bourkii)*, galah *(Cacatua roseicapilla)*, and zebra finch *(Taeniopygia castanotis)* had ratios of 1.36, 1.77, 2.50, 2.58, and 2.78, respectively. Parrots are good concentrators of urine. The galah was also shown to absorb more water per solute molecule from its coprodeum and large intestine than the domestic fowl. Water conservation in the cloaca is important in this species.

Birds increase their water consumption just prior to the onset of egg laying and normally produce excreta with a higher water content at this time.

Some birds have the ability to withstand long periods without drinking water, and others can consume water with considerable salinity. Skadhauge (1975) stated

> The ability of larger birds to drink saline of a certain concentration follows the pattern known from mammals. Except in those species with a salt gland, the osmolality of the saline must be lower than the renal concentrating ability; there must be osmotic room for creation of free water. In some very small birds with a high relative rate of metabolism, and therefore, a high rate of metabolic water production, this pattern of intake of saline changes. Fluids more concentrated than urine can still be drunk but at a rate which is small compared with the total water turnover.

In the chicken, the ability of the kidney to clear blood of creatinine is considerably lower than it is for uric acid or phenol red. Unlike the mammal, the bird possesses only a negligible amount of creatinine in urine relative to creatine.

### Renin-Angiotensin

Histologically birds possess a macula densa, extraglomerular mesangium, and juxtaglomerular cells. These juxtaglomerular cells have been demonstrated to produce renin, but whether in response to changes in blood pressure has not been determined. It is assumed that renin acts upon angiotensin I, an inactive peptide in blood, to form angiotensin II. Angiotensin II is a potent vasoconstrictor and in addition to this activity it also stimulates the adrenal to secrete and release aldosterone. This latter hormone stimulates sodium, chloride, and water reabsorption by the distal convoluted tubules and the collecting ducts. Through this mechanism the kidney exemplifies autoregulation. Studies indicate that the avian antidiuretic hormone arginine vasotocin (AVT) acts as a vasoconstrictor to the afferent arteriole. Glomerular filtration rate is thus influenced by AVT in response to changes in plasma osmolality.

## LYMPHATIC SYSTEM

The avian lymphatic system consists of lymphatic vessels and the lymphatic tissues. The lymphatic vessels transport lymph to the vascular system. This flow is in a return direction only. Numerous valves prevent retrograde flow of lymph. Lymph contains plasma proteins, electrolytes, lipids, hormones, and other substances normally transported by plasma proteins. This extracellular fluid also contains lymphocytes as the primary cell.

The lymphocytes can be divided into two major classes and many subclasses. Those lymphocytes that are thymus-dependent, called T cells, can be distinguished from bursa-dependent cells, B cells, by the surface content of immunoglobulin. B cells are involved in the antibody reaction and contain the immunoglobulins IgA, IgM, and IgG on the surface, whereas T cells are involved in cell-mediated immune responses (see Chapter 24, Relationship of the Avian Immune Structures and Functions to Infectious Diseases).

The primary lymphoid tissues are the thymus and cloacal bursa (bursa of Fabricius). In some avian species regression is caused by the sex steroids (chicken) and recrudescence does not occur; however, in others enlargement occurs after regression in response to postnuptial molt or, in some, to anemia. The principal function of the thymus is immunologic.

The primary function of the cloacal bursa is the differentiation of B lymphocytes from bursal stem cells to antibody-producing lymphocytes. The cloacal bursa undergoes regression in size shortly after hatching and remains involuted. Complete involution is attained at about sexual maturity. Regression can be accelerated by the injection of testosterone, estradiol, or the adrenal steroids. Both thymic and bursal involution can be hastened by dietary exposure to such environmental pollutants as polychlorinated biphenyls, dibenzo-p-dioxins, and polybrominated biphenyls.

The remaining lymphoid tissues consist of tissues to which the T cells and B cells migrate. The spleen is the major tissue, but aggregates of lymphoid tissue can be found in many areas, such as the digestive tract, Harderian gland, and bone marrow. These tissues respond to antigenic stimulation by lymphocyte recruitment. In psittacines, diffuse lymphoid follicles are uncommon.

## ENDOCRINE ORGANS

The endocrine organs of avian species are similar to those of other vertebrates, but in certain aspects they are unique, especially when compared to those of the mammal. Examples are the presence of discretely separate glands called ultimobranchial bodies, the fact that the adrenal is not discretely divided into a medulla and cortex, and the positioning of the thyroid glands away from the trachea as separate structures not linked by an isthmus.

The primary source of information on the endocrine system stems from our knowledge of domesticated species, the mallard, the Japanese quail, the White-crowned sparrow, and several finches. The avian hypophysis has been considered to be the master endocrine gland because it elaborates hormones that stimulate secretion in other glands, but control of hypophysial function is initiated in the hypothalamic portion of the brain. The hypothalamic nuclei are connected to the neurohypophysis by the fiber tracts along which the neurosecretions of the nuclei are transported. Neurosecretory material from the supraoptic and paraventricular nuclei is transported via the axons of these tracts down to the median eminence or the neural lobe. Two hormones are contained within these neurosecretions, oxytocin and arginine vasotocin.

Arginine vasotocin (AVT), the avian antidiuretic hormone, increases reabsorption of water in the collecting ducts and tubules. Antidiuresis by AVT injection has been demonstrated in the chicken with both low doses (5 to 30 μg/kg) and high doses (30 to 200 μg/kg). AVT appears to be released from the neurohypophysis in response to increases in plasma osmolality. Extirpation of the neural lobe produces polydipsia and polyurea in chickens and ducks. Oxytocin injection into birds can cause premature expulsion of the developing egg within the oviduct. This, coupled with the fact that in the chicken plasma vasotocin concentration increases 20-fold at the time of oviposition, provides inferential evidence that the posterior pituitary hormones are involved in oviposition.

Neurosecretory materials are also transported from the hypothalamic nuclei by axons that terminate in the median eminence and from there to the hypophysis by portal blood vessels. These neurosecretory materials contain the releasing hormones. Specific releasing hormones increase the plasma concentration by stimulating the release of the hormone by the cells in the anterior pituitary. LHRH of mammalian origin has been shown to be biologically active in birds even though the two neurosecretions may not be structurally identical.

Mammalian FRH is also biologically active in birds. Seven cell types have been identified in the anterior pituitary, and seven hormones are released by these cells. The following have been identified: the gonadotropic hormones (LH, FSH, and prolactin), the thyrotropic hormone (TSH), the adrenocorticotropic hormone (ACTH), growth hormone (GH), and melanotropin (MSH).

## Thyroid Glands

Diet, climatic conditions, and certain drugs have a marked influence on the size of the thyroid glands. The thyroids are about 0.01 per cent of the body weight. Enlargement may be observed in either a hypo- or hyperfunctioning gland. TSH from the anterior pituitary stimulates the thyroid and produces both hypertrophy and hyperplasia concomitant with the secretion and release of the predominant circulating thyroid hormones, thyroxine ($T_4$) and triiodothyronine ($T_3$). Avian species synthesize thyroid hormones like that of mammals in which iodide, obtained from dietary iodine, is concentrated within the thyroid by the so-called iodide trap. Iodide within the thyroid is converted to $I_2$ and then to $I^+$. $I^+$ iodinates the thyroid proteins, forming both monoiodotyrosine and diiodotyrosine. A peroxidase system converts the iodide to iodine, and a second enzyme combines the iodinated tyrosines within thyroglobulin to form $T_3$ (3,5,3′-triiodothyronine) and $T_4$ (3,5,3′,5′-tetraiodothyronine). The iodinated tyrosine forms must be converted to an iodinated thyronine form for release into the bloodstream from the thyroid. In the blood, these hormones are bound to serum albumin and have a short half-life (three to six hours), indicating a very rapid degradation. It has been demonstrated in some birds, as in mammals, that considerable peripheral conversion of $T_4$ may also result in the formation of 3,3′,5′-triiodothyronine, called reverse $T_3$ or $rT_3$. In mammals there is a category of disorders that inhibits $T_3$ neogenesis referred to as "low $T_3$ syndrome" and attributed to reduction in the 5′-monoiodination of $T_4$. Circulating levels of plasma $T_4$ in many birds range between 0.5 and 2.0 μg/dl and for $T_3$ between 200 and 500 ng/dl. A useful measure of thyroid status is the $T_3/T_4$ ratio. It is also important to report free thyroxine and free triiodothyronine.

Thyroid function can be evidenced by goitrogen administration or thyroidectomy. Depressed thyroid activity as a consequence of goitrogen administration is reflected in reduced metabolic rate, increased fat deposition, and, in some cases, growth depression. Thyroidectomy results in growth retardation, feather structure alteration, and reduced gonadal function. Molting, a complex mechanism in birds, is known to be under the influence of several factors, one of which is the thyroid hormones. Feeding or injection of thyroid material stimulates the feather papilla and produces molt in several days.

Hyperthyroidism can be produced by administration of pituitary extract, TSH, $T_4$, or $T_3$ hormones. Hyperthyroidism can adversely affect reproduction and growth.

## Adrenal Glands

The adrenal glands are located anterior and medial to the cephalic lobe of the kidneys and immediately posterior to the lungs. They are triangular or oval in shape. Histologically, the inter-renal and chromaffin tissue are intermingled; therefore, unlike mammals, birds lack a distinct cortex and medulla.

The size varies with species, sex, age, health, and such factors as stress. The adrenals comprise about 0.005 to 0.01 per cent of the body weight. Nutrition can have an effect on adrenal weight, as evidenced by the fact that vitamin A and riboflavin deficiencies in chickens increase adrenal weight. Injection of adrenocorticotropic hormone (ACTH) will also increase adrenal weight.

ACTH stimulates the synthesis of corticosterone but not aldosterone. The fact that intact but not hypophysectomized birds release corticosterone in response to stress indicates that stress stimulates ACTH release from the pituitary.

The major peripheral circulating adrenal steroid is corticosterone. Aldosterone is the second principal corticosteroid of adrenal secretion in birds. The biologic half-life of these steroids in circulation can be measured in terms of minutes.

Corticosterone is important in carbohydrate, lipid, and electrolyte metabolism. Aldosterone plays a role in retention of sodium and water by acting on the kidney tubules. Corticosterone's mechanism of action on carbohydrate metabolism in birds is through gluconeogenesis. Injection of corticosterone causes both liver glycogen stores and protein catabolism to increase. Corticosterone injection produces lipogenesis characterized by increased liver lipids, depot fat, plasma lipid, cholesterol, and phospholipids. Corticosterone plays a role in electrolyte metabolism and is necessary for normal functioning of the nasal gland. Thus, its main function as a mineralocorticoid is in extrarenal salt excretion. Salt loading of ducks results in elevated plasma levels of corticosterone.

Histochemical studies indicate that there are two types of medullary cells, those that are epinephrine-containing and those that are norepinephrine-containing. The proportion of each varies with species. These cells secrete the catecholamines, epinephrine, and norepinephrine. Release of adrenal catecholamines is evoked by neural stimulation. In the pigeon, turkey, and chicken the plasma norepinephrine:epinephrine ratio is less than 1.0.

Epinephrine and norepinephrine are glycogenolytic and hyperglycemic agents in birds, as they are in mammals. Epinephrine is also a potent beta-adrenergic stimulator. In most birds epinephrine and norepinephrine injection causes a significant hyperglycemia. Perfusion with norepinephrine and epinephrine leads to significant elevation in both systolic and diastolic blood pressure. Heart rate is unchanged by norepinephrine and may even be slowed by epinephrine. Birds, lacking brown fat, do not demonstrate a thermogenic response to norepinephrine as mammals do.

### Parathyroid Glands

Two glands are located on each side of the neck immediately posterior to the thyroids. Histologically the parathyroids of birds lack the oxyphil cells seen in mammals and possess cords of chief cells. Hyperplasia as well as hypertrophy occurs in deficiencies of vitamin D or dietary calcium. Hypertrophy, as a consequence of dietary calcium deficiency, occurs more rapidly in birds than in other vertebrates. In the chicken, high dietary calcium has been reported to result in parathyroid atrophy.

Parathyroidectomy, in those birds studied, produces hypocalcemia and tetany. Injection of mammalian parathyroid hormone has hypercalcemic effects. Thus, the main function of the parathyroid glands and its hormone, parathormone, is calcium homeostasis brought about by controlling resorption of metaphyseal and endosteal bone, control of renal tubular excretion of calcium phosphate, and control of the rate of absorption of calcium from the intestine. In domesticated species of birds, parathormone regulates the ionic calcium level of blood, but estrogen elevates the protein-bound calcium level through inducing synthesis by the liver of a calcium-binding protein.

Vitamin $D_3$ is synergistic with parathormone in mobilizing bone calcium and plays a vital role in absorption of calcium from the gut.

### Ultimobranchial Glands

Located posterior to the parathyroid glands, the ultimobranchial glands (bodies) are composed of C cells arranged in cords or clusters of individual cells. These cells secrete the hormone calcitonin. The physiologic role of calcitonin in avian species is questionable, and its effect on calcium metabolism is unresolved. However, the glands secrete calcitonin in response to hypocalcemia, and dietary calcium is the principal factor regulating calcitonin secretion in birds.

### Endocrine Pancreas

The pancreas, suspended within the duodenal loop as a discrete lobular organ, synthesizes and releases into the circulatory system several peptide hormones that originate from the islet tissue scattered throughout the acinar tissue (exocrine portion). The islet tissue is composed of $\alpha_1$, $\alpha_2$, and $\beta$ cells. $\alpha_2$ cells secrete the hormone glucagon and $\beta$ cells secrete insulin. $\alpha_1$ cells are also known as D cells.

While insulin appears to be the predominant hormone regulator of carbohydrate metabolism in mammalian species, in avian species the key regulator appears to be glucagon. Each hormone plays a secondary but important role in its species. The precise role of avian pancreatic polypeptide has not yet been determined. This hormone is elaborated by an $\alpha_1$ type cell (PP-cell) of the pancreas.

The release of pancreatic glucagon is stimulated by cholecystokinin, insulin, and free fatty acids, whereas its release is suppressed by glucose in some avian species. Insulin is released in several avian species by cholecystokinin, glucagon, and, to a lesser degree, glucose. Food deprivation has little influence on insulin concentrations in blood.

Somatostatin, a polypeptide hormone, is produced and released by the D cells of the pancreas. Somatostatin modulates the release of insulin and glucagon from the pancreatic cells. Thus, the pancreas exerts control over its own secretions.

Glucagon decreases hepatic glycogen (glycogenolysis) by cell membrane receptor activation, resulting in hyperglycemia. Insulin induces hypoglycemia by its antigluconeogenic nature. Glucagon plays a major role in the mobilization of glycerol and fatty acids from adipose tissue depots. Insulin causes increased metabolism of glucose by adipose tissue.

## REPRODUCTION

### Female Reproduction and Egg Formation

During embryonic development birds have both a right and a left ovary. During incubation the right ovary ceases development in most species and the left continues to grow. Posthatching injury to or removal of the left ovary results in the development of right ovarian tissues into an ovitestis. In domestic species, natural sex reversal has been observed; however, the production of gametes by the ovitestis is rare.

The ova remain quiescent during immaturity and at sexual maturation are stimulated by environmental cues, particularly the direction of change in the daily photoperiod. Most avian species respond to increasing periods of daylight with follicular development; however, there are exceptions. Reversal of the direction of the stimulatory photoperiod causes depression of follicular growth and development and the cessation of egg production. The normal pathway for the photoperiodic stimulation is through the pupil to the retina and from thence by way of the short optic nerve to specific nuclei in the hypothalamus. These nuclei produce releasing hormones that traverse the neurons from these nuclei into the region of the median eminence. Here the releasing hormones enter the portal circulation connecting the hypothalamic region with the adenohypophysis (anterior pituitary) and are transported to the specific cells of the anterior pituitary that produce the gonadotropins responsible for ovarian functioning. Diseases that result in cataracts of the lens or even surgical enucleation of the eyes normally do not necessarily prohibit ovarian functioning, since extraretinal receptors have been demonstrated in avian species. Light rays penetrate the soft, spongy bone of the head and are capable of stimulating the hypothalamus to release its neurosecretory substances.

Follicular growth is under the control of follicle-stimulating hormone (FSH). Once an ovum starts its growth within a follicle, maturation takes several days. Yolk is deposited as concentric rings representing daily maturation. Ovulation is under the stimulation of luteinizing hormone (LH). The active ovary produces estrogens, progesterone, and testosterone. Following ovulation, no structure homologous to the mammalian corpus luteum develops. In avian species the injection of progesterone at low levels has been shown to cause premature ovulation, presumably through LH release; however, high doses inhibit ovulation. In domestic fowl the ruptured follicle may play a role in oviposition. Rupture of the follicle occurs within the region called the stigma, which is almost totally avascular to the unaided eye.

Extrusion of the second polar body generally occurs shortly before ovulation; thus the sex of the embryo that will result, should the egg be fertilized, is determined prior to ovulation, since the female bird is heterozygous while the male is homozygous. Fertilization takes place in the first segment of the oviduct, the infundibulum, prior to the initiation of albumen secretion. The infundibulum, under the control of the nervous system, engulfs the follicle prior to ovulation and by peristaltic movement propels the ovulated ovum down the oviduct. Should the infundibulum fail to engulf the ovum, either before ovulation from the follicle or following ovulation into the body cavity, because of neural dysfunctioning caused by a disease, obstruction, or drug therapy, an internal layer results and the ovum remains in the body cavity or is ultimately reabsorbed.

The avian oviduct, which consists of five segments and develops normally only on the left side of the body (the rudimentary right oviduct often becoming cystic in older domestic fowl), changes to the magnum section aborally to the infundibulum. The magnum secretes the thick albumen protein around the ovum. The presence of an object pressing against the secretory glands of the magnum causes the secretion of albumen. The smooth muscles of the magnum force the developing egg aborally to the isthmus, where the two shell membranes are secreted loosely around the ovum and albumen. On the surface of the outer shell membrane are knoblike structures called mamillary bodies where calcium carbonate crystallization takes place in the uterus (shell gland). While the egg is in the shell gland and before calcification proceeds too far, water plus vitamins and mineral salts passes through the semipermeable egg membranes into the egg and migrates toward the yolk, the site of lowest water concentration. It is at this stage in the egg's development that the various layers of egg albumen are formed by the migrating fluids from the shell gland. Following complete crystallization, the egg is oviposited by the eversion of the vaginal section.

In domestic birds and those wild birds studied, the ovum spends about 15 minutes in the infundibulum during which fertilization can take place, 3 hours in the magnum, 1½ hours in the isthmus, and 20 to 21 hours in the shell gland and passes through the vagina almost instantly. Thus, egg formation takes between 24 and 26 hours.

At no point are there anatomic oviductal structures that prevent the egg from undergoing reverse peristalsis. Disturbances of the nervous system may cause the egg to reverse its course through the oviduct and be deposited into the body cavity, often resulting in egg peritonitis. Eggs may accumulate in the oviduct for various reasons, causing egg binding. Since the smooth muscles of the oviduct wall are under control of the nervous system, diseases that disrupt the nervous system, physical disturbance of a bird such as frightening, and neural drugs may interrupt normal egg development with resulting soft-shelled eggs, shell-less eggs, or misshapen eggs. Foreign substances in the shell gland generally cause premature expulsion of an egg. The administration of adrenergic or cholinergic drugs influences the oviduct, with adrenergic drugs causing relaxation of the oviduct and cholinergic drugs causing constriction. Acetylcholine readily produces expulsion of isthmus or uterine eggs. Oxytocin and vasotocin yield similar results. Several of the prostaglandins cause oviposition, whereas prostaglandin antagonists cause delayed oviposition.

In order for a sound shell to be produced it is necessary for metabolic $CO_2$ to be converted to $HCO^-_3$ within the shell gland catalyzed by carbonic anhydrase. Drugs such as sulfonamides and acetazolamide inhibit carbonic anhydrase and result in eggshell thinning. During shell formation the precipitation of calcium carbonate is accompanied by the release of hydrogen ions. These ions enter the blood stream, resulting in increased acidity of the blood. Ultimately the hydrogen ions are excreted via the urine, causing it also to become more acidic during shell formation.

Within the femur and tibia in the marrow cavity of female birds is the specialized medullary bone. Medullary bone formation is stimulated by the synergistic action of estrogens and androgens. Both hormones are necessary, which is in contrast to elevated serum calcium concentrations that can be brought about by estrogen alone.

## Male Reproduction and Sperm Storage

The male reproductive system consists of right and left testes, vas epididymis, and vas deferens. The testes remain juvenile after hatching until environmental cues initiate their growth. Light, temperature, and rainfall have been shown to be such cues, light being the predominant one.

As in the female, the photoperiod of natural or artificial light initiates the growth of the testes and the production of semen through hypothalamic production of the releasing hormones controlling FSH and LH production by the anterior pituitary. In general, the red end of the spectrum is more stimulatory to the gonads, especially the testes, than the blue portion of the spectrum. Once a threshold of stimulation is reached, about 5 to 10 lux, additional light has no further effect.

Testicular atrophy takes place spontaneously in some species following an active period even though a stimulatory photoperiod is continued. This photorefractory period ends with the commencement of a nonstimulatory photoperiod.

The testes undergo considerable changes in size during seasonal changes in spermatogenesis. Spermatogenesis starts with the spermatogonia, then the primary spermatocytes, the secondary spermatocytes, spermatids (which attach to the Sertoli cells), and spermatozoa. Semen is normally white and opaque but, as in the fowl and turkey, may be clear, brownish, and watery depending upon the spermatozoal concentration and the amount of transparent fluid added at ejaculation.

The testes produce testosterones from the Leydig cells responsible for plumage color; feather structure, behavior, and vocal changes. Storage and maturation of sperm in most avian species take place in the vas deferens, particularly the caudal section. No capacitation of spermatozoa is necessary for fertilization to take place.

In natural mating, ejaculated semen is deposited on the everted cloaca of the female at the region of the oviductal opening. Motility propels the spermatozoa into the uterovaginal junction (UVJ) and into channels that connect with the sperm glands, tubular glands into the oviductal wall located at the UVJ. Viable spermatozoa enter these sperm glands (sperm nests or host glands) and remain there until released; in some species this will be days to weeks later. These glands extend the life expectancy of spermatozoa within the female oviduct. Released sperm cells are transported to the infundibulum by retrograde ciliary action and there are harbored in crypts for a limited duration or until a subsequent ovulation occurs. Fertilization takes place in the infundibulum.

When artificial insemination is used, it is advisable to insert the semen into the vagina to a depth approximating the site of the uterovaginal junction. The sperm cells then readily enter the uterovaginal glands for storage. Spermatozoa have a long headpiece, with a pointed acrosome, a short midpiece, and a long tail. Avian spermatozoa are smaller than most mammalian spermatozoa.

## RECOMMENDED READINGS


Bell, D. J., and Freeman, B. M. (eds.): Physiology and Biochemistry of the Domestic Fowl. Vols. 1–3. New York, Academic Press, 1971.

Farner, D. S., and King, J. R. (eds.): Avian Biology. Vols. 1–7. New York, Academic Press, 1971–1983.

Freeman, B. M.: Physiology and Biochemistry of the Domestic Fowl. Vols. 4–5. New York, Academic Press, 1984.

Hodges, R. D.: The Histology of the Fowl. New York, Academic Press, 1974.

King, A. S., and McLelland, J. (eds.): Form and Function in Birds. Vols. 1–2. New York, Academic Press, 1979.

Lucas, A. M., and Jamroz, C.: Atlas of Avian Hematology. Washington, DC, U.S. Department of Agriculture, Monograph 25, 1961.

Ookawa, T. (ed.): The Brain and Behavior of the Fowl. Tokyo, Japan Scientific Societies Press, 1983.

Pearson, R.: Avian Brain. New York, Academic Press, 1972.

Piiper, J. (ed.): Respiratory Function in Birds, Adult and Embryonic. Proceedings, Satellite of the 27th International Congress of Physiological Sciences, Paris, 1977. New York, Springer-Verlag, 1978.

Romanoff, A. L.: The Avian Embryo: Structural and Functional Development. New York, Macmillan and Co., 1960.

Skadhauge, E.: Osmoregulation in Birds. New York, Springer-Verlag, 1981.

Skadhauge, E.: Renal and cloacal transport of salt and water. In Peaker, M. (ed.): Avian Physiology. London, Academic Press, 1975, pp. 97–106.

Sturkie, P. D. (ed.): Avian Physiology. New York, Springer-Verlag, 1976.

Welty, J. C.: The Life of Birds. 3rd ed. New York, Saunders College Publishing, 1982.

*Section Three*

# A CLINICAL APPROACH

# Chapter 6

# MANAGEMENT PROCEDURES

GREG J. HARRISON
LINDA R. HARRISON

The decision to incorporate avian patients into a veterinary clinic and hospital requires some adjustments in the management procedures established for small animals. A different approach may have to be developed for preventing spread of disease (in the examination room as well as in the hospital), submitting diagnostic samples to a laboratory, obtaining specialized equipment and supplies, and developing a clientele.

## COMMUNITY RELATIONS

An important aspect in the management of the avian practice is the veterinarian's relationship with aviculturists, pet shops, and small animal veterinarians in the community, because these are the primary sources of referrals. Efforts put forth in educating these groups will ultimately benefit the quality of care that is given pet birds.

In the author's experience, one program that has been successful with pet shops is the establishment of a store-sponsored three-day warranty on the bird. This allows the customer the opportunity to have the bird examined by the veterinarian. Birds with serious illnesses are rejected and recommended for return. Birds requiring minor therapy are treated immediately at the expense of the shop for a reduced fee. If there are no obvious illnesses or positive diagnostic tests, but some minor suspected lesions have been demonstrated by radiography, a successful option has been to encourage the shop to sponsor an extended warranty (up to 1 ½ years) covering this specific finding. If the bird dies or develops a serious illness from causes directly related to the lesions covered by the warranty, the shop will replace the bird. This offers a diplomatic compromise between rejection of the bird as a healthy specimen and ignoring a potential disease. Birds have rarely needed replacement under these conditions, but such "insurance" alleviates the need to reject less than perfect specimens.

## INSTRUCTIONS TO CLIENTS

The initial direct contact with the client is usually made at the time of the telephone call for an appointment. An educated receptionist is helpful in determining the length of time required for the office visit. In general, avian patients are scheduled for longer office visits than dogs and cats because of the client education aspects (see Chapter 7, Preliminary Evaluation of a Case).

Small birds may be transported to the office in their own cage, a paper bag, or a travel box with sufficient ventilation. Overheating must be avoided, especially if the ambient temperature is over 90° F. Larger birds may be accommodated by an airline kennel or similar container. Many tame birds are brought in on their owner's arm; however, this presents a greater possibility of exposure of the bird to disease organisms of other patients.

The client is advised to bring the following with him:

1. A fresh fecal sample. A piece of plastic wrap may be placed under the perch prior to the visit to collect several droppings. The plastic wrap is then folded and placed in a paper cup or plastic bag with enough ice cubes to keep the sample cool during the trip to the office. If a cage is used for transporting the bird, paper towels or other white paper can be used as a cage liner to assist in gross evaluation of the cloacal output (see Chapter 9, Evaluation of Droppings).
2. Sample of any abnormal discharge that has been noted at home and all information relative to the history (see Chapter 7, Preliminary Evaluation of a Case). Some clients may be embarrassed to say that they have previously con-

sulted another veterinarian, but this information is important. Any reports relating to a previous consultation, including radiographs, diagnostic test results, and therapy recommendation, are helpful.

3. A towel for handling the bird, if desired. Paper towels are used in the author's practice for the capture of large birds for physical examination, but some clients may prefer to bring their own cloth towel (See Chapter 7, Preliminary Evaluation of a Case).

### Initial First Aid

Simple first aid measures can be suggested over the phone to be used until the appointed transport to the clinic, although the urgency of the visit should not be underestimated. Owners often tend to delay seeking professional advice until the bird is well into the disease process.

A first aid provision that is warranted for almost all ill birds is supplemental heat. The transport cage can be wrapped with towels and blankets with a hot water bottle enclosed. For the trip, favorite foods may be spread out on the bottom of the cage (perhaps with the perch removed). A sugar/stimulant solution, such as concentrated coffee with sugar, has been beneficial in reviving some noncomatose birds. Many aviculturists have medical backgrounds or nursing experience that may be used to advantage in providing appropriate at-home supportive care prior to and following the diagnosis.

If someone other than the owner transports the patient to the office, some arrangements must be made to discuss the case and possible monetary and diagnostic consequences directly with the owner.

### Flock Assessment

If the client is requesting diagnostic assistance with a flock problem, he is advised to bring in for examination a typical untreated bird in the beginning stages of the disease. Owners should be encouraged to dampen and freeze, rather than discard, a single dead aviary bird for which diagnostic procedures are not being sought so that in the event of further deaths, an example of an early case would be available.

Owners frequently request postmortem procedures on a bird recently found dead at home. They should be advised to immediately soak the carcass in a soapy solution (¼ tsp. liquid detergent/gallon water) to fully wet the feathers, place the damp carcass in a plastic bag, and immediately chill, either by refrigeration or storage with ice in an insulated box or cooler. The body should not be frozen in these cases.

## MANAGING THE AVIAN CLINIC

The reception area is the first area of concern to the avian practitioner. Educational material made available in the reception area may acquaint the client with the bird care philosophy of the clinic and initiate pertinent questions. Sick birds often tend to have disease conditions that may be highly contagious; therefore, special cautions need to be taken in the clinic facilities. Some provision may need to be made to separate bird patients from each other as well as from aggressive dogs and cats.

### Preventing Spread of Disease in the Examination Room

An efficient air handling system similar to a laminar flow system is probably one of the best deterrents to spreading airborne viruses and other disease organisms. Air conditioning vents should deliver fresh air into the room, wash it over the examination table, and remove the air to the outside of the building by exhaust fans at knee level on the opposite side. Contents of the treatment rooms, including nail/beak grinders and tables, should be thoroughly disinfected prior to bringing a healthy bird into the room. A quaternary ammonium solution* is satisfactory as a table wash. It is essential that bird handlers wash and disinfect their hands† between patients. Frequent spraying with a disinfectant‡ may be necessary for clothing of the technical staff as well as for a DustBuster§ or similar floor-cleaning tool.

### Specialized Avian Equipment and Supplies

Treatment of avian patients can take advantage of much of the equipment and supplies that are already available in a small animal practice, such as radiology equipment, anes-

---

*Roccal-D—Winthrop Veterinary, Division of Sterling Drug, Inc., New York, NY 10016

†Alcare Hand Degermer—Vestal Laboratories, Div. Chemed Corp., St. Louis, MO 63110

‡Staphene Disinfectant Spray and Air Sanitizer—Vestal Laboratories, Div. Chemed Corp., St. Louis, MO 63110

§Black and Decker

**Table 6–1.** SUGGESTED SUPPLEMENTAL SUPPLIES FOR USE WITH AVIAN PATIENTS

Incubator, specially designed cage or other temperature-controlled environment—Thermocare, Inc., P.O. Box YY, Incline Village, NV 89450
Metal crop gavage tubes—Small sizes from Corners Limited, 424 Harrison St., Kalamazoo, MI 49007; larger sizes from I.A.W. Research Institute, P.O. Box 6385, Burbank, CA 91510 or Ejay, Inc., P.O. Box 1835, Glenodora, CA 91740
Gram scale—Tanita Electronic Digital Read-out Scale, American Scale Co, Chicago, IL
Mouth speculum—Lafeber Co., Odell, IL 60460
Microliter syringes
TB syringes—Monoject
26-gauge ⅜-inch needles—Monoject
30-gauge ½-inch needles—MPL, Inc., 1820 West Roscoe St., Chicago, IL 60657
Hand-held hobby drill kit for beak and nail trim—Dremel
Diamond dental grinding wheel for drill
Perch stand on examination table; dowel perch for moving birds
Closed aluminum leg bands and band removers—L and M Bird Leg Bands, P.O. Box 2943, San Bernardino, CA 92406
Split bands for identification—Donna G. Corp., 4903 N. Ardsley Dr., Temple City, CA 91780
Bolt cutter for heavy bands
Nebulizer or vaporizer
Plexiglass board for restraint and positioning for radiography or surgery—Henry Schein
Gas (ethylene oxide) sterilization
Ear protectors (EHP-1 Hearing Protector)—Eastern Safety Equipment
Bird collars (Saf T-Shield)—Ejay, Inc., P.O. Box 1835, Glendora, CA 91740
Avian blood donors
Environmental thermometers
Botulism antitoxin
Appropriate food supplies (see Chapter 2, Husbandry Practices, for sources of formulated diets; Chapter 27, What To Do Until A Diagnosis Is Made, for force-feeding formulas)

**First Aid Supplies**

Wire for splints
Hair dryer
Blunt probang
Powdered activated aquarium charcoal
Corn starch to remove oil
Dawn detergent or Amway's LOC
Masking tape
Tape remover (Detachol)—Ferndale Labs., Inc.

**Laboratory Aids**

Gram's staining kit—Difco Laboratories, Detroit, MI 48232
Gram check slide, Q-check System—Scott Laboratories, Inc., Fiskeville, RI 02823
Sterile culture swabs with transport media:
  Culturette—Marion Scientific Corp., Rockford, IL 61101
  Port-a-Cul—BBL Microbiology Systems, Cockeysville, MD 21030
  Culture-Eze—Medical Media Laboratory, Inc., Boring, OR 97009
  Calcium alginate swabs—Inolex Laboratories, Glenwood, IL
  Vipak—Scott Laboratories, Inc., Fiskeville, RI 02823
  *Chlamydia* transport media—J. E. Grimes, Texas A & M University
Latex agglutination test for *Chlamydia*—J. E. Grimes, Texas A & M University
Microtainer Capillary Blood Serum Separator—Becton-Dickinson
Microhematocrit capillary tubes, plain
Microhematocrit capillary tubes, heparinized
Unopette eosinophil test kit (5877)—Becton-Dickinson
Hemoccult—Smith Kline Diagnostics
Diff Quik staining kit
Macchiavello's stain
Iodine stain
Glucose pediatric strips (Chemstrip bG)—Biodynamics, Inc., Indianapolis, IN 46250
pH paper—MCB Reagents, Cincinnati, OH 45212
10% buffered formalin
Dry Ice shipping box

thetic vaporizer, microscope, centrifuge, and autoclave. Table 6–1 suggests some general items that the practitioner may need to obtain specifically for treating birds (Fig. 6–1). Quality control measures beyond those normally used by a small animal clinic may be indicated (e.g., Q-check System Slide for monitoring Gram's stain quality and technique). Additional equipment for avian use is described in other chapters.

## GROOMING PROCEDURES

Restraint for grooming procedures is a necessity for the safety of both the bird and the handler (see Chapter 7, Preliminary Evaluation of a Case). For beak and nail trims, most veterinarians prefer a hand-held drill (in some

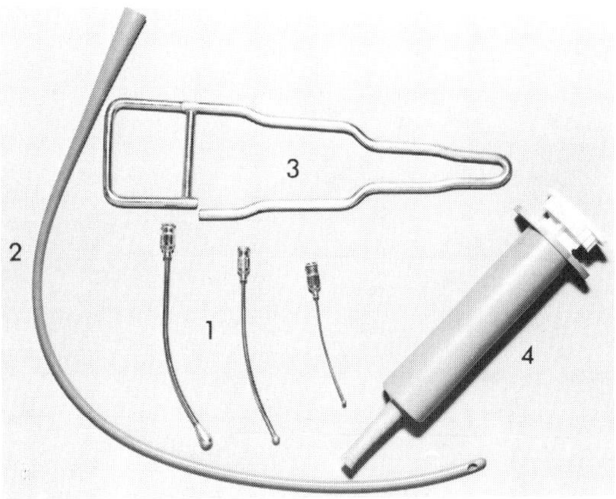


**Figure 6–1.** A variety of utensils is available to assist in force feeding of birds: (1) appropriately-sized metal feeding needles; (2) soft, flexible tubing; (3) mouth speculum; (4) large plastic syringe. (From Harrison, G. J.: Hospitalization of the avian patient. Vet. Clin. North Am. [Small Anim. Pract.], 14:147:164, 1984.)

cases with diamond dental tip) because it provides more control than a nail clipper over the degree of trimming desired. In addition, heat generated by the grinding stone cauterizes small blood vessels during the procedure, making possible a shorter trim. To prevent dust generated by the stone from building up in the nares or eyes, the handler must be aware of the direction the stone is turning (which determines the direction of the dust spin-off) and blow the dust away from the bird.

## Nail Trims

Many owners prefer that their bird's normally sharp nails be blunted regularly to prevent the bird from inflicting minor skin cuts and scratches during handling. This may be especially important in birds the size of conures and larger.

An experienced handler may be able to work alone to perform a nail trim on a bird smaller in size than an Amazon. Larger and highly fractious birds may require a second handler, and in extreme cases, three people may be necessary to carry out the procedure safely.

Manipulation of individual toes is facilitated by rolling the bird on its back while offering a stick for the bird to grasp. It is dangerous to attempt to physically force open the grasp, as fractures or torn ligaments may result.

If the drill can be held in such a way as to transfer the control of the instrument to the muscles of the handler's fingers and hands, rather than to his arms and shoulders (Fig. 6–2), the finger can be use as a fulcrum to increase the accuracy of drill placement and prevent potential damage to surrounding tissue. If the skin is inadvertently abraded, Monsel's solution can be applied.

Suture scissors are preferred for trimming nails of small species. Extremely long nails will often bleed following trimming. A steady grasp of the toe serves as a tourniquet until styptic (e.g., heat, Monsel's solution, silver nitrate) can be applied in a firm grinding manner. Deep cauterization of the vessels is necessary to prevent subsequent bleeding. An abrupt change in nail length may require a period of adjustment for the bird. The perch might be removed from the cage for the trip home to prevent falling accidents.

## Beak Trims

Most psittacines in the wild have a sharply pointed upper beak and may need it blunted in captivity to prevent damage to the owner. A familiarity with the normal length for the species is necessary, as there are some birds that normally have long beaks (e.g., Slender-billed Conure, most lories). The beak of a healthy bird does not require trimming for any other reason.

Through the use of a grinding stone as described above, the upper beak (premaxilla) can be trimmed shorter than with other trimming methods, and sharp edges can be rounded off (Fig. 6–3). The lower beak (mandible) seldom needs trimming, which is fortunate because the tongue may easily be burned with the grinding tool. Thin-layered lateral flaking commonly seen on the beak can be cautiously smoothed with the grinding techniques. A very light application of oil will restore a polish to the beak. An abrupt change in beak length may also require an adjustment period by the bird.

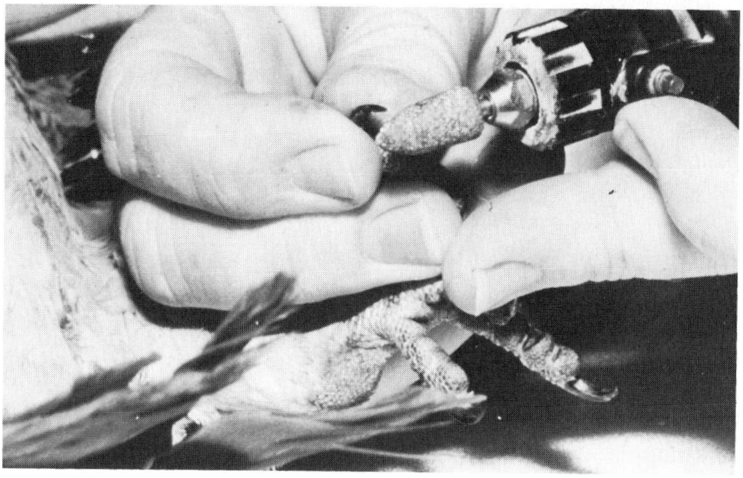

**Figure 6–2.** A grinding tool is maneuvered to provide the greatest degree of control in trimming each individual nail. Restraint is provided by an assistant.

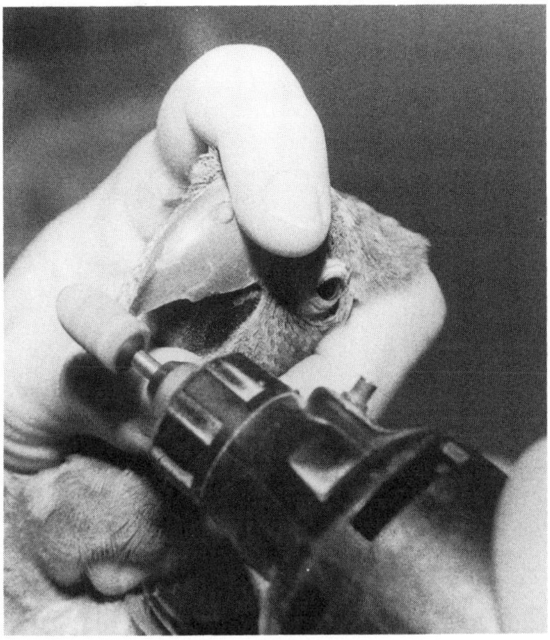

**Figure 6–3.** A beak may be shortened and shaped with a grinding instrument.

Battery-powered cautery units for trimming nails and beaks are commercially available. The author has found the odor produced by this trimming method to be offensive.

## Wing Trims

Wing trims may be required for purposes of preventing escape, confining the bird to a specific area, preventing injury to the bird, or aiding in training and taming. Full-flighted birds, especially large species, are generally less desirable as pets.

The wing-trim procedure in large species will require an assistant for restraint of the bird; however, with smaller species, it may be accomplished by a single person (Fig. 6–4). The style of the wing trim can vary with the weight of the bird, the experience and preference of the veterinarian, and the request of the owner. The overall appearance of a wing trim is improved if each feather is cut individually just below the level of the coverts so the cut end is not visible. Suture scissors are useful for cutting the individual quills, and regular scissors can trim any irregularities.

A single wing trim (Fig. 6–5), in the author's experience, is most appropriate for light-bodied birds with long wings (e.g., cockatiels, conures, budgerigars, Australian parakeets, dwarf macaws, cockatoos). The most effective technique for preventing flight, but also the least attractive, is to trim all the primary and most of the secondary remiges on one wing. This unbalances the bird during flight attempts and results in a softer landing because the bird tends to spin as it falls. One disadvantage of this technique is the lack of supportive feathers for the protection of pin feathers as they emerge. This often results in fractured, bleeding pin feathers. Fledglings of light-bodied species may benefit from the modified single wing trim described below because they tend to make more flying attempts and lack the wing-flapping strength of adult birds.

The modified single wing trim (Fig. 6–6*A*) is more appropriate for obese or heavy-bodied

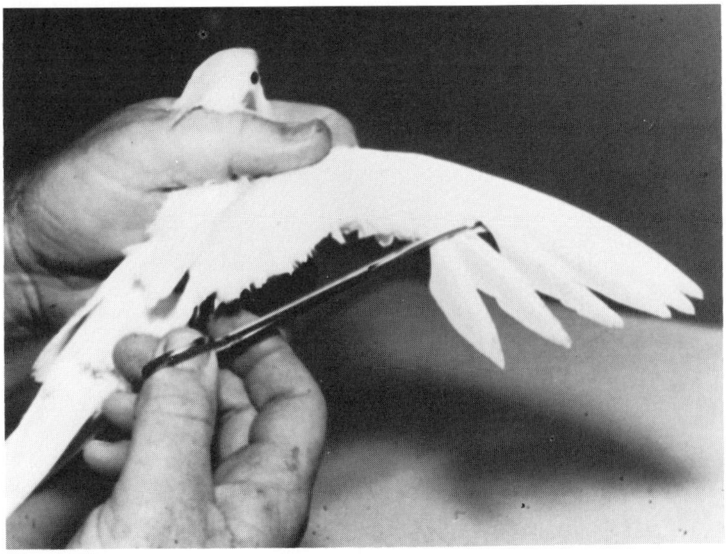

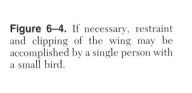

**Figure 6–4.** If necessary, restraint and clipping of the wing may be accomplished by a single person with a small bird.

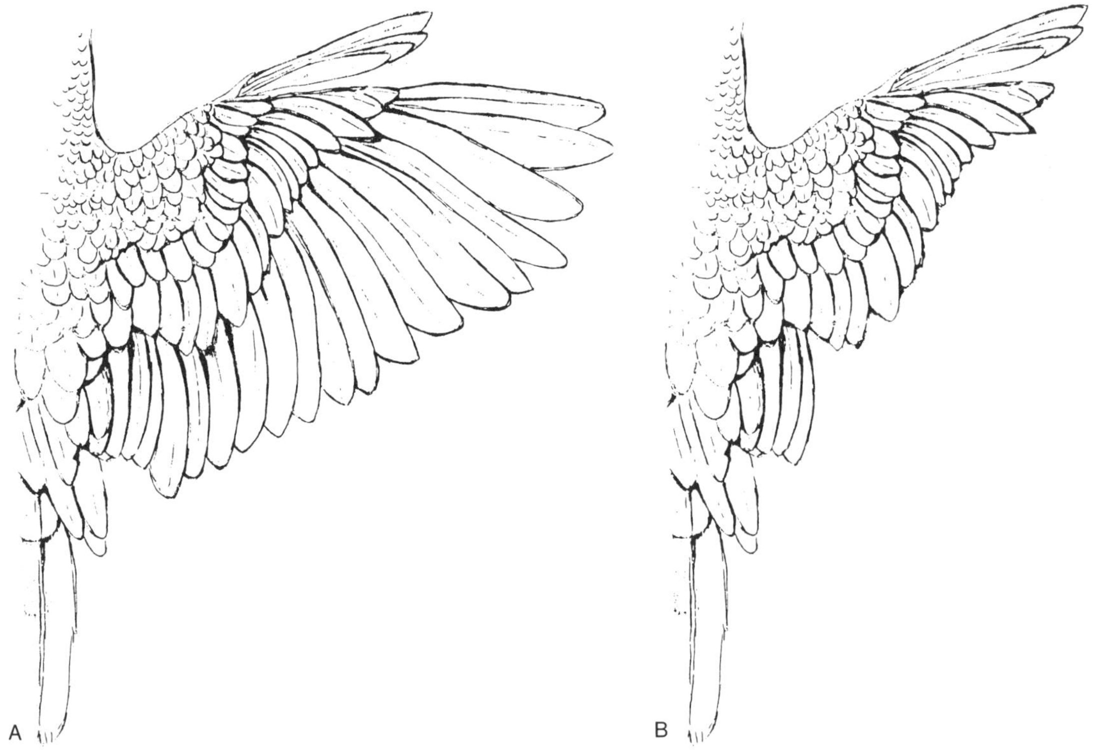

**Figure 6–5.** Flight is significantly restricted when the surface area of an unclipped wing (*A*) is reduced by removing the tips of the primary and most of the secondary remiges in the single wing clip (*B*). (Drawings by A. Allen.)

birds such as African Grey parrots. If all the distal primary feathers are trimmed, the weight of these birds exaggerates the fall to the ground in a flight attempt and may result in sternal trauma. In heavy-bodied birds, two or three distal primary remiges are left on the trimmed wing to provide additional stability. This is one of the most aesthetically pleasing trims when the wings are closed (Fig. 6–6*B*).

For larger cockatoos whose natural wing flap

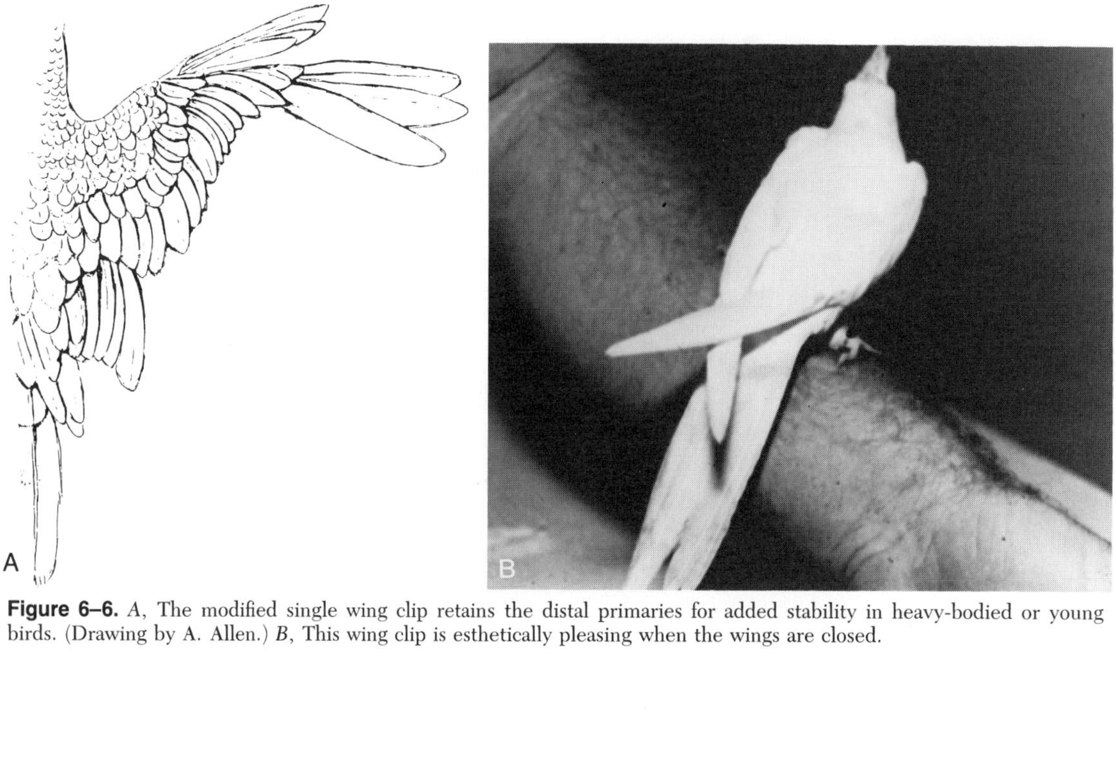

**Figure 6–6.** *A*, The modified single wing clip retains the distal primaries for added stability in heavy-bodied or young birds. (Drawing by A. Allen.) *B*, This wing clip is esthetically pleasing when the wings are closed.

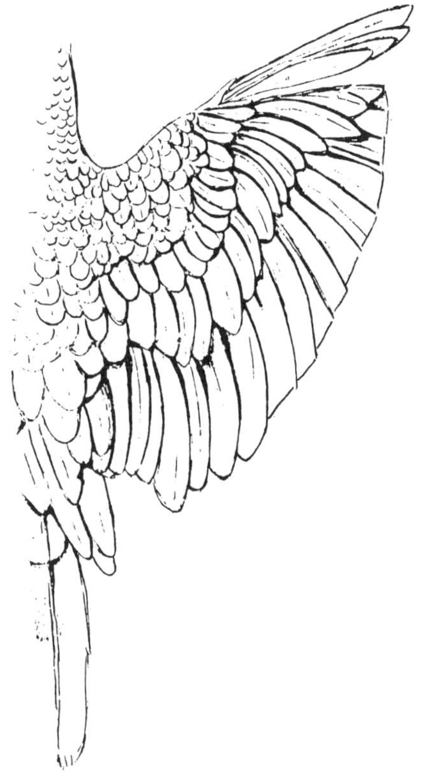

**Figure 6–7.** A partial removal of the feather tips on both wings may be effective for slow-flying species such as cockatoos. (Drawing by A. Allen.)

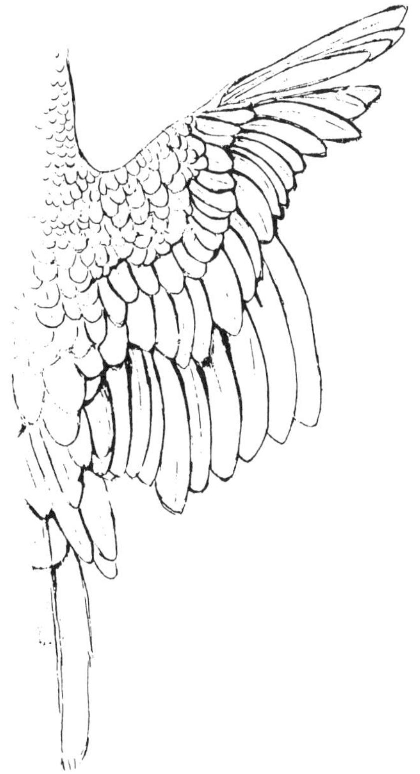

**Figure 6–8.** The clipping of all primaries at the level of the coverts on both wings is the least attractive trim but may be necessary for flight restriction in some species. (Drawing by A. Allen.)

is slower than that of other birds, a successful cosmetic technique is to remove the tips of the primary and the first two secondary remiges on both wings in an arching fashion, with the shortest feathers at the distal end (Fig. 6–7).

A more severe two-wing trim is unsightly and, in the author's experience, is reserved as a last resort procedure for the best fliers. The distal primary remiges are removed under the tips of the coverts of both wings (Fig. 6–8). This procedure also increases the incidence of distal wing trauma. Cockatiels appear to have more flying ability with a two-wing trim than with a single-wing trim.

If a highly vascularized growing pin feather is inadvertently cut during the procedure, the entire feather must be removed from the follicle to prevent further hemorrhage (see Chapter 28, Symptomatic Therapy and Emergency Medicine).

After the wing trim procedure, the veterinarian should test fly the bird in the examination room to determine the degree to which flight has been restricted. Additional feathers may be cut if necessary. One cannot guarantee that a wing trim will totally prevent flight. The author has not found an aesthetically pleasing wing trim that will render a bird totally flightless. Some birds have been able to compensate and fly to a degree in spite of remarkably few intact wing feathers, especially in a strong wind. Trimmed feathers may molt soon after they are cut and be replaced within a short period of time. Owners should be advised to check the trimmed wing periodically and clip newly developing primaries and secondaries as they emerge beyond the coverts (distal to the blood supply), and to exercise caution in taking the bird outdoors.

## Band Removal

Bands are applied to birds for a number of regulatory reasons (see Regulations Involving Birds later in this chapter). For those clients who choose to have these bands removed, a variety of band removers may be necessary to accommodate the sizes, shapes, and tensile strength of avian bands that have been applied. One instrument that is adequate for removing most bands is a modified Olsen-Hager needle holder (Fig. 6–9). Stainless steel quarantine bands may require industrial bolt cutters or two

**Figure 6–9.** An Olsen-Hager needle holder has been modified by grinding it down to beveled points. This effectively spreads and cuts a leg band for removal.

pairs of locking pliers. Large, thick aluminum bands can be removed from an anesthetized bird with a metal saw blade attachment to the drill used for beak and nail trims (with protection to the leg). A commercial source of an aluminum band remover is listed in Table 6–1. A jeweler's ring saw is usually so large that it may cause leg trauma and possibly fractures in small birds. Stability should be provided to both the leg and the band by the handler in all band removals. Special care will need to be exercised in removing bands from swollen or otherwise damaged legs. Anesthesia may be required in these cases.

## SUBMISSION OF DIAGNOSTIC SAMPLES TO A LABORATORY

Specific avian diagnostic procedures are discussed in other chapters. Although some of these may be completed within the avian hospital, the isolation and identification of many disease agents (e.g., *Chlamydia, Mycoplasma, Mycobacterium,* anaerobic bacteria, viruses, fungi, some parasites) require special training, conditions, and supplies for which the services of reference laboratories may be needed. Because a single laboratory facility may not have the equipment or knowledge to perform all of the avian tests required, it is essential to the management of a quality avian practice to develop a rapport with several diagnostic laboratories. Specific laboratories for avian diagnoses are not listed here because the services offered tend to change. Therefore, the practitioner is advised to first contact the veterinary school in his area and the State Diagnostic Laboratory for direction.

The primary factor in submitting diagnostic quality samples is full communication with the laboratory. The practitioner must understand and conform to the submission protocol, clearly identify each sample and the tests required, and include a detailed and legible report on the origin of the sample and the medical and management history of the bird.

The practitioner is advised to have on hand appropriate transport media, shipping boxes, and a method to ship the samples quickly. A source of Dry Ice may need to be located.

In general, the practitioner should:

1. Submit samples from animals that have not been subjected to therapy.
2. Collect specimens aseptically from anatomic sites most likely to contain the suspected pathogens.
3. Select samples from animals in the early or acute stages of the disease, rather than those chronically affected.
4. Submit a meaningful history of the disease process or outbreak.
5. Offer a tentative clinical diagnosis based on observation of the bird(s) and a knowledge of diseases in the area.

### Bacteriology

Samples intended for bacteriologic screening should be kept moist in an appropriate transport medium, refrigerated (not frozen), and sent immediately with sufficient ice to keep them cool during the transport. The speed with which they reach the laboratory determines the value of the results obtained. Hand-delivery (if practical), Express Mail Next Day Service, UPS Next-Day Air, and other courier messenger services are recommended. Shipments should be planned to arrive at the lab early in the week to avoid weekend layovers.

### Chlamydiosis

Postmortem tissues from birds suspected of having chlamydiosis should be refrigerated immediately and sent on ice to the diagnostic laboratory. Isolation of *Chlamydia* from frozen tissues is usually successful only if the tissues are frozen on and shipped with Dry Ice.

Antemortem diagnosis may be made from fecal samples or cloacal swabs submitted in an appropriate transport medium. Brain-heart infusion broth with gentamicin is often used. A special transport medium to preserve *Chlamydia* for isolation is available.*

*J.E. Grimes, Texas A & M University, College Station, TX

### Viruses

Viruses are difficult to isolate from pet bird species. Specimens for viral recovery should be sent frozen with Dry Ice or liquid nitrogen whenever possible. The tissues or carcass should be placed in sealed containers (Zip-Loc or Whirl Pak plastic bags) to avoid contact of the tissues with the carbon dioxide released by the Dry Ice. Other methods of diagnosing viral diseases, such as fluorescent antibody tests and special treatments of organ impression smears, are available through some laboratories. The practitioner is advised to make direct contact with the laboratory for specific submission protocol.

### Other Tests

Preparation of samples for histopathology and toxicology is discussed in Chapter 23, Necropsy Procedures. Prior direct communication with the laboratory is advised for samples intended for mycoplasmal, mycotic, anaerobic bacterial, electron microscopic, or other evaluations.

### Shipping Samples

Under a criminal statute, a shipper of diagnostic samples through the United States Postal Service may be prosecuted if, during shipment, any specimen spillage (from noninfectious as well as infectious agents) damages the mail or equipment or affects personnel. Therefore, a great deal of care should be devoted to preparation of specimens for shipment to the laboratory.

Refrigerant (wet ice or Dry Ice) should constitute 50 per cent of the weight of the contents of the package. Wet ice should be placed in leakproof plastic bags. Conversely, Dry Ice should be packed in containers from which the carbon dioxide can escape after sublimation. Styrofoam boxes, adequately protected from damage by outer coverings of wood or sturdy cardboard, are preferred for shipping refrigerated specimens. Pack with sufficient material around the specimen to absorb any leaking fluid to comply with legal as well as medical responsibilities.

## HOSPITAL ENVIRONMENT FOR THE AVIAN PATIENT

### Housing

An ambient temperature of 85 to 90° F is a prerequisite for hospital care of the sick bird. It would be ideal to have an entire hospital room equipped with a heat source and thermostat so that the individual bird would not have to be removed from this environment for cleaning, feeding, weighing, and treating. Alternatives for the housing of sick birds include stainless steel kennel cages (Fig. 6–10), custom-built isolettes, baby incubators, rabbit laboratory cages, and wire bird cages. Galvin[4] uses three sizes of glass aquariums for the housing of most hospitalized birds: 5 gallon for budgerigar-sized birds; 10 gallon for cockatiels and conures; 20 gallon for Amazons and small to medium cockatoos. He believes that the aquariums with screen-covered tops offer the advantages of being easy to clean and disinfect, preventing seeds and droppings from contaminating the hospital, complete visibility, easy access to the patient, and individual control of the environment.

Wire cages are suitable for birds with injuries or for convalescing birds with noncontagious disorders. If custom cages are built, wire that has been galvanized *after* welding is preferred for its strength, rust-resistance, and ease of cleaning. A light coat of Pam cooking oil or silicone spray on the cages prior to use facilitates cleaning of droppings and food before disinfecting.

**Figure 6–10.** Large hospitalized birds can be adequately housed in stainless steel dog cages if the ambient temperature is controlled. (From Harrison, G. J.: Hospitalization of the avian patient. Vet. Clin. North Am. [Small Anim. Pract.], 14:147:164, 1984.)

If a bird is being hospitalized in its own cage, all sources of grit, oyster shell, lava stone, and similar products are removed to avoid overconsumption of these by the sick bird.

To avoid transmission of contagious diseases, a modified laminar air flow system draws fresh clean air from one side of a room and pulls contaminated air to the outside of the building. Patients suspected of being particularly contagious should be isolated from other hospitalized birds. If boarding is part of the client services, boarding birds should be housed in a totally separate area that has never contained sick birds. A separate area may be provided where convalescent birds that have responded to treatment can be observed briefly under normal environmental temperatures prior to release.

Within the individual hospital environment, perches may or may not be provided. In some cases a sick bird may be too weak to climb down from a perch even if it is hungry; therefore, no perch would be made available. An alternative may be to cut in half large natural perches of nontoxic wood and place the flat side down on the bottom of the cage with the food. Replacement of natural perches for each patient is preferred over recycling because the build-up of disinfectant may irritate the feet.

Heavy ceramic bowls that can be easily disinfected and cannot be easily overturned are preferred as water bowls. A commercial roll of brown wrapping paper is useful for lining cages. Newsprint tends to come off on the feathers of white birds.

Maintaining light in the avian hospital room for 24 hours a day provides the opportunity for birds to eat throughout the night if they have been disturbed by daytime interference. A study involving the use of water medications in pigeons also revealed that blood levels of antibiotics administered in this way were more accurately regulated by consistent intake of water over a 24-hour period when lights were left on.

## STAFF RESPONSIBILITIES

It is useful to have one member of the nursing staff who is not involved in actual treatments to establish a positive relationship with a hospitalized bird by performing only nonthreatening activities: cleaning the cage, bringing the food, talking to the bird. In this way, the bird may become more relaxed when in contact with this person and a more accurate evaluation of the bird's attitude may be observed. Frequent monitoring of the patient is also necessary in regard to food consumption, qualitative and quantitative analysis of droppings, and general condition.

### Weighing the Bird

Hospitalized birds are weighed every day. A gram scale is necessary, and several adaptations are useful for avian patients. A laboratory animal scale with an attached basket; paper bags, boxes, or cones as holding devices on platform scales; or mounted T-stands have been utilized (Fig. 6–11). The weight of any holding device must be subtracted from the total reading. Electronic scales that adjust this automatically are becoming more reasonable in price.

### Preventing Spread of Disease in the Hospital

The staff has significant responsibility for following specific practices within the hospital to avoid contamination from one patient to another and to prevent the spread of diseases.

The normal daily progression of cleaning, feeding, and treating should begin with the healthiest birds (such as boarding birds and blood donors), then the convalescents, and the highly contagious and critically ill birds last. Recent arrivals, for which there is not yet a positive diagnosis, should be regarded as contagious until proven otherwise.

Several disinfectants are used in an avian hospital, as each is specific for certain organisms (see Chapter 50, Aviculture Medicine). Chlorhexidine* is gentle to tissues, is effective against most avian viruses and *Candida* (see Chapter 26, Therapeutics), and can be used in some cases in drinking water, as a mouth swab, or in hand feeding formulas. Chlorhexidine is not potent against chlamydial organisms and certain bacteria.

Quaternary ammonium products are the disinfectants of choice for *Chlamydia*. These are also effective against some bacteria and viruses; however, they must be thoroughly rinsed from equipment before use because ingestion of these products is believed to be nephrotoxic to birds.

A phenol type disinfectant† is recommended for use in U.S.D.A. quarantine stations against

*Nolvasan—Fort Dodge Laboratories, Fort Dodge, IA 50501

†1-Stroke Environ—Ceva Laboratories, Overland Park, KS 66212

**Figure 6–11.** A gram scale is essential for accurate weighing of avian patients. One adaptation for relatively tame birds utilizes an epoxy-covered perch. (From Harrison, G. J.: Hospitalization of the avian patient. Vet. Clin. North Am. [Small Anim. Pract.], 14: 147:164, 1984.)

possible infection with Newcastle disease virus and other organisms. This product is also necessary to include in an avian practice. Because phenol products are harsh on tissues, rubber gloves should be used.

In the author's practice, cages are cleaned and disinfected daily with a quaternary ammonium (quat) solution. When dry, the cages are sprayed with a disinfectant. The nets that are available to capture escaped flighted birds are soaked after use in a quat solution.

The floor of each hospital room is swept as needed, at least once a day, with a broom that has been dipped in the disinfectant and is then mopped with a fresh solution. A vacuum cleaner tends to scatter viruses in the air through the bag; therefore, if it is necessary to use one, the bag is sprayed thoroughly with a disinfectant spray prior to and during use. Each room is totally cleaned before going on to the next room.

Members of the nursing staff wash their hands and apply a foam disinfectant between treatments of individual birds and use a disinfectant spray on their hair and clothes between rooms. If a bird is to be held up against the body for injections or tube feeding, a hospital gown is worn by the technician for a highly contagious bird. A mask should be worn by the handler if chlamydiosis is suspected. If the bird was not highly stressed when admitted, grinding the nails down prior to hospitalization helps protect the handler from being scratched.

Paper towels are used to retrieve hospitalized birds from cages for treatment and are disposed of immediately after use. Disposable paper towels are also good serving dishes in the cage for fresh fruits and vegetables.

Seed and water bowls are thoroughly cleaned and disinfected daily by (1) wiping off heavy debris with a paper towel; (2) soaking in a phenol solution (½ oz/gal water) for 30 minutes; (3) putting through a dishwater cycle with the hottest water available; (4) soaking in fresh phenol solution for another 15 to 20 minutes; (5) rinsing thoroughly, moving to the food preparation room, and air drying.

It is important to remember in working with any disinfectant that it cannot be expected to do its job on microorganisms if there is a substantial amount of organic material left on the surface of the item.

## REGULATIONS INVOLVING BIRDS

The avian veterinarian should be aware of regulations and laws that have been established on both domestic and international levels for the protection of birds.

### State and Federal Protection of Birds

Table 6–2 lists a number of federal agencies that are involved to some degree in the possession and/or movement of birds. Appropriate permits depend upon the species and intended activity.

#### *U.S. Endangered Species Act*

The U.S. Endangered Species Act places restrictions on a wide range of activities involving endangered and threatened animals or plants to help ensure their continued survival. With few exceptions, the act prohibits activities with these protected species unless authorized by a permit from the U.S. Fish and Wildlife Service. Endangered species permits may be issued for scientific research, enhancement of propagation, and survival of the species. Threatened species permits are allowed for the above purposes plus zoological exhibition, education, and special purposes consistent with the act. Applications for permits are directed to the Federal Wildlife Permit Office (Table 6–2) with a $25 application fee. A list of species protected by the act may also be requested from this office. More than 700 species of animals and plants are officially listed as endangered. Depending on the species, other requirements for activity may include import and export documents under CITES, possession permits under the Migratory Bird Treaty Act, and compliance with federal and state laws and laws under the Lacey Act (which supports foreign laws).

**Table 6–2.** FEDERAL OFFICES OF AGENCIES INVOLVED WITH PET BIRDS

| |
|---|
| Federal Wildlife Permit Office<br>U.S. Fish and Wildlife Service<br>1000 North Glebe Road<br>Arlington, VA 22201<br>(703) 235-1903 |
| Chief, Division of Law Enforcement<br>U.S. Fish and Wildlife Service<br>P.O. Box 28006<br>Washington, DC 20005<br>(202) 343-9242 |
| United States Department of Agriculture<br>Animal and Plant Health Inspection Services<br>Hyattsville, MD 20782<br>(301) 436-8172 |
| Commissioner of Customs<br>U.S. Customs Service<br>1301 Constitution Ave., N.W.<br>Washington, DC 20229<br>(202) 566-5286 |

#### *Captive-bred Wildlife*

The Captive-bred Wildlife regulation was established to enhance the propagation and survival of eligible captive-bred wildlife listed under the U.S. Endangered Species Act. As it applies to avians, qualified aviculturists who register with the U.S. Fish and Wildlife Service may buy and sell live endangered or threatened exotic (non-native) birds that were hatched in the United States to other holders of a federal permit for that species. A $25 fee is required for application for a Captive-bred Wildlife permit from the Federal Wildlife Permit Office.

#### *Migratory Bird Treaty Act*

The U.S. Fish and Wildlife Service regulates most of the activity involving migrating birds under the terms of the Migratory Bird Treaty Act (MBTA). The possession of endangered or threatened raptors (other than the Bald Eagle) and other migratory birds requires banding under the terms of a valid permit issued under this Act. An annual report of activity is required.

In the United States, activity involving the Bald Eagle is the most restricted, as this species is protected under the U.S. Endangered Species Act, the Bald Eagle Protection Act, and the Migratory Bird Treaty Act. Other species such as the Peregrine Falcon may be protected under both the Endangered Species Act and the Migratory Bird Treaty Act. Migratory species that are not endangered are protected under the MBTA. There are a few bird species in the United States that are not considered migratory. Exotic species (other than those that are endangered) are not protected by U.S. laws.

No permit is required simply to possess lawfully acquired, properly marked migratory waterfowl or their progeny. "Lawfully acquired" means that the bird was obtained from a person with a permit to dispose of it, the bird was captive-reared, and the bird was properly marked by the time it was six weeks of age. The four ways to acceptably mark these birds are clipping the hind toe on the right foot, wing pinioning, stainless metal band on the leg with identifying number, and tattooing on the web

**Table 6–3.** REGIONAL OFFICES OF U.S. FISH AND WILDLIFE SERVICE DIVISION OF LAW ENFORCEMENT

For information on lists of protected avian species and applicable federal regulations, write to the **Special Agent in Charge, U.S. Fish and Wildlife Service** of the appropriate region:

Region 1 (serving CA, HI, ID, NV, OR, WA): 847 N.W. 19th Avenue, Suite 225, Portland, OR 97232. (503)231–6125

Region 2 (serving AZ, NM, OK, TX): P.O. Box 329, Albuquerque, NM 87103. (505)766–2091

Region 3 (serving IL, IN, IA, MI, MN, MO, OH, WI): P.O. Box 45, Federal Building, Fort Snelling, Twin Cities, MN 55111. (612)725–3530

Region 4 (serving AL, AR, FL, GA, KY, LA, MS, NC, PR, SC, TN): P.O. Box 4839, Richard B. Russell Federal Building, Atlanta, GA. (404)221–5872

Region 5 (serving CT, DC, DE, ME, MD, MA, NH, NJ, NY, PA, RI, VT, VA, WV): P.O. Box 129, New Town Branch, Boston, MA (617)965–2298

Region 6 (serving CO, KS, MT, NE, ND, SD, UT, WY): P.O. Box 25486, Denver Federal Center, Denver, CO 80225. (303)234–4612

Region 7 (serving AK): P.O. Box 4–2597, Anchorage, AK 99503. (907)786–3311

of the foot. A permit *is* required, however, if the possessor of these waterfowl intends to dispose of them either by sale, trade, or donation. Mallards are an exception to this; one can possess and dispose of captive-bred Mallards without a permit. Muscovey Ducks are considered to be domestic species and are not covered under this regulation.

## Medical Treatment of Wild Birds

Avian veterinarians or other people who treat injured wild birds are required to have a Special Purpose Wildlife Permit, which is issued by the U.S. Fish and Wildlife Service (Table 6–2). This permit authorizes the holder to temporarily possess and care for sick and/or injured migratory birds. Any endangered species and Bald or Golden Eagles are to be reported to a Special Agent of the U.S. Fish and Wildlife Service within 48 hours. An annual report of activity under this permit is required.

If a veterinarian without a permit receives an injured endangered species in a non–life-threatening situation, he should call law enforcement personnel of either his state conservation department or the U.S. Fish and Wildlife Service (Table 6–3). They will either recommend transferring the bird to an approved facility or will give authorization to initiate treatment. In a life-threatening situation, the veterinarian should provide emergency medical care prior to contacting law enforcement personnel.

Individual states may have their own regulations governing protected species; therefore, anyone undertaking activities involving endangered or threatened wildlife is expected to have prior contact with his state fish and wildlife agency and to comply with any applicable state laws.

## Interstate Movement of Birds

Birds may be shipped interstate by commercial airline. Flight times and specific freight regulations must be obtained from the individual airline. Federal regulations prohibit shipment of live cargo during periods when the temperature is above 85° F or below 45° F at any point along the route. Summer shipments may be arranged for night flights.

The bird should be shipped in a sturdy wooden box (pressed board is acceptable for some airlines) with ventilation holes covered with sturdy wire mesh (Fig. 6–12). The potential for injury is minimized if the box is just large enough to accommodate the bird: 1 to 2 inches wider than the bird, long enough not to damage the tail feathers, and high enough for

**Figure 6–12.** A custom-made shipping box has separate compartments for three birds, wire-covered ventilation holes, and perches nailed on the bottom. The remains of a millet spray are on the floor.

the bird to stand up straight. Heavy chewing birds may need the box lined with appropriate gauge wire to prevent damage to the box.

A wooden bar, nailed to the floor as a perch, provides the bird with a source of stability. Seeds and fruit may be placed in food cups wired to the side of the box. If the bird does not eat fruit, a small cup of water with a sponge inside to prevent spillage may be used. A recent health certificate must accompany the bird, with clear instructions on the outside of the box as to the recipient's name, address, and phone number. Reservations are required. On some airlines, a single pet may be accommodated in an appropriate box under the passenger seat. Compliance with regulations of other agencies is required.

## Banding

The banding of birds' legs is required for identification under some regulations. Closed and open bands are available in a number of sizes of inside diameter, usually in 1/16 inch increments. Although some sizes may be generally appropriate for a particular species (e.g., 5/16 inch—conure; 7/16 inch—Green-cheeked Amazon; 9/16 inch—Umbrella Cockatoo), it is more important for the band to properly fit the individual bird to prevent damage to the leg. Codes for identification are determined by the purchaser. Only U.S.D.A.-approved quarantine stations may use a three-letter, three-number code: ABC 123.

Closed bands have been proposed as a method of identifying those birds that have been captive-reared because the band must be slipped over the foot of a young bird in the pin feather stage. The caudal digit is pulled back against the tibia and the other three digits are pointed forward for application of the band over the ankle. The band must be large enough to accommodate the adult size of the leg.

Open bands can be applied to birds of any age with an appropriately sized tool similar to a pair of pliers. It is essential that the opposing ends of the band be in close proximity after application and that the band fit comfortably over the ankle. Commercial sources of bands are listed in Table 6–1.

## International Movement of Birds

### *CITES*

The Convention on International Trade in Endangered Species of Wild Fauna and Flora (CITES) consists currently of 88 nations that have established regulations for the import and export of imperiled birds and other wildlife. Appendix I of CITES includes species currently threatened with extinction, and the most stringent controls are directed toward these species. Two permits are required for the movement of birds in Appendix I: one from the importing country and another from the exporting country. Examples of some of the birds in Appendix I are *Aratinga guarouba* (Golden or Queen of Bavaria Conure), *Ara glaucogularis* (Caninde Macaw), *Ara rubrogenys* (Red-fronted Macaw), *Pionopsitta pileata* (Red-capped or Pileated Parrot), and *Rhynchopsitta pachyrhyncha* (Thick-billed Parrot). *Ara macao* (Scarlet Macaw) and *Ara ambigua* (Buffon's Macaw) were moved to Appendix I during the 1985 CITES meeting in Argentina.

Appendix II species are not currently threatened with extinction but may become so unless their trade is regulated. All psittacine birds not in Appendix I with the exception of the *Nymphicus hollandicus* (cockatiel), *Melopsittacus undulatus* (budgerigar), and *Psittacula krameri* (Indian Ring-necked Parakeet or Rose-ringed Parakeet) are in Appendix II. Import permits are not needed for Appendix II species, but a permit from the exporting country is required with few exceptions.

The Federal Wildlife Permit Office (Table 6–2) acts as the U.S. Management Authority for CITES.

### *Commercial Importation*

Exotic birds are imported into the United States for the purpose of resale. To protect the poultry industry from the threat of Newcastle disease virus, the United States Department of Agriculture (U.S.D.A.), Animal and Plant Health Inspection Service (APHIS), administers a mandatory 30-day quarantine period. A list of privately owned and operated commercial stations that perform quarantine procedures is available from the U.S.D.A. (Table 6–2). The Public Health Service is no longer involved in the bird quarantine procedure. Therefore, regulations of the Centers for Disease Control, such as the feeding of chlortetracycline to psittacine birds, are no longer in effect. The U.S.D.A. is continuing to recommend this prophylactic therapy, but it is not enforced. A total of 741,921 birds were released from commercial quarantine stations during the fiscal year 1984.

Smuggling of birds to avoid the quarantine period is illegal. Birds suspected of being smuggled should be reported to any Customs office

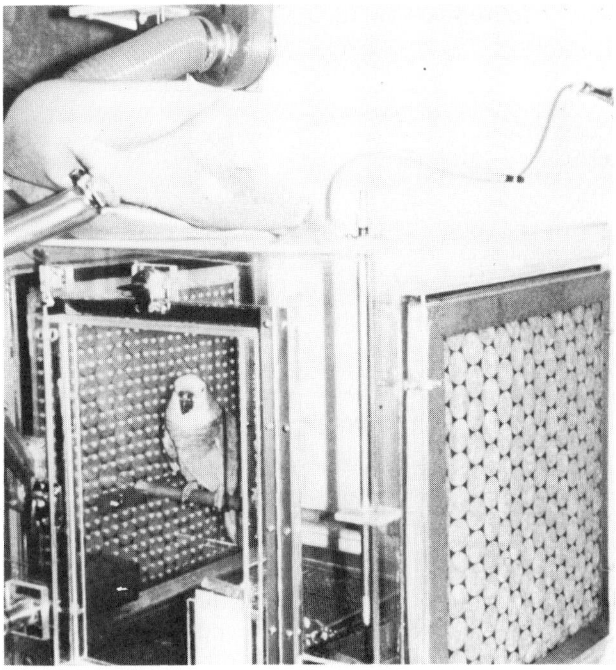

**Figure 6–13.** Birds imported as personally owned pets are quarantined in individual isolettes. (Courtesy of the U.S.D.A.)

or to Keith Hand at APHIS at Hyattsville, MD (301) 436-8065.

### *Pet Bird Quarantine*

Specific regulations have been developed for the importation of personally owned pet birds. A pet bird is any bird, with the exception of poultry, intended for the pleasure of its owner and not for resale. Individuals may reserve space in a U.S.D.A.-operated import facility at one of the eight ports of entry for the mandatory 30-day quarantine period. The previous limit of two birds per person has been lifted; the number is now governed by the amount of space available for a specific quarantine period.

The birds are housed in individually controlled isolation units (Fig. 6–13) and are converted to a pelletized psittacine bird ration containing 1 per cent chlortetracycline during the quarantine period. Young birds requiring hand feeding cannot be accepted. Careful procedure is followed to prevent the spread of disease from one lot of birds to another (Fig. 6–14).

Reservation requests (Form 17-23) are obtained by writing to the Port Veterinarian at the city of arrival (Table 6–4). A deposit of $80 is required with the reservation request. Currently the cost of quarantine services including

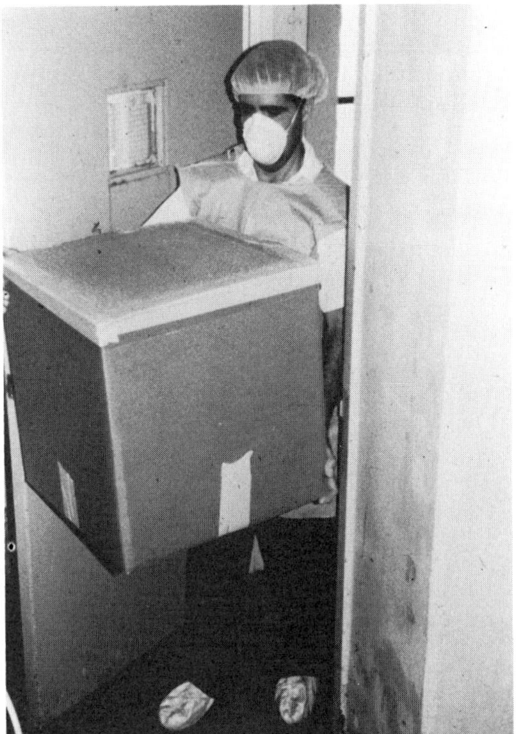

**Figure 6–14.** Special clothing and handling techniques are employed by the U.S.D.A. in pet bird quarantine facilities to reduce the incidence of disease. (Courtesy of the U.S.D.A.)

**Table 6–4.** PORTS OF ENTRY FOR QUARANTINE OF PERSONALLY OWNED PET BIRDS

Write to **Port Veterinarian** at the specific location:

NEW YORK, NY: JFK International Airport, Cargo Bldg. 80, Room 101, Jamaica, NY 11430. (718)917–1727

MIAMI, FL: 8120 N.W. 53rd St., Suite 102, Miami, FL 33166. (305)350–6921

NOGALES, AZ: P.O. Box 1411, Nogales, AZ 85621. (602)287–4717

HIDALGO, TX: P.O. Box 3068, Brownsville, TX 78520. (512)542–7812

EL PASO, TX: 109 N. Oregon, 12th Floor, El Paso, TX 79901. (915)541–7691

HONOLULU, HI: P.O. Box 50001, Honolulu, HI 96850. (808)546–7529

SAN YSIDRO, CA: P.O. Box 126, San Ysidro, CA 92073. (619)428–7332

LOS ANGELES, CA: 5510 West 104th St., Los Angeles, CA 90045. (213)215–2352

necessary tests and examinations is $100 for one bird or $125 per isolation cage if the birds are compatible and can be caged together. Information is available through the Port Veterinarian for local transportation arrangements and for the list of designated brokers for clearance of U.S. Customs Service, U.S. Department of Agriculture, and U.S. Department of the Interior, Fish and Wildlife Service regulations. A health certificate from the country of origin is required. Documents under CITES may be required from the exporting country for protected species.

### *International Travel with Pet Birds*

Pet birds may enter the United States from Canada if they have been in the possession of the owner for 90 days and have been kept separate from other birds. Inspection by an APHIS veterinarian at designated ports of entry is required.

Birds may be taken out of and re-enter the United States through one of the approved ports without quarantine provided the bird has been identified by health certificate and tattoo or leg band number to the port authorities prior to departure and again upon arrival and has been maintained separate from other birds.

Because some regulations are subject to change, the appropriate agency should be contacted prior to any activity.

## REFERENCES


1. Beckley, E. G.: Import Veterinarian/Pet Bird. U.S.D.A, APHIS, Veterinary Services, Miami, FL: Personal communication.
2. Carpenter, J.: Legalities of providing emergency veterinary care to an endangered species. A.A.V. Newsletter, 5(3):77, 1984.
3. Dorrestine, G.: A preliminary report on pharmacokinetics and pharmacotherapy in racing pigeons (*Columba livia*). Proceedings of the International Conference on Avian Medicine, Toronto, Canada, 1984, pp. 9–23.
4. Galvin, C.: Avian practice tips. Proceedings of the American Federation of Aviculture Veterinary Medicine Seminar, San Francisco, CA, 1985.
5. Harrison, L. R., and Herron, A.: Submission of diagnostic samples to a laboratory. Vet. Clin. North Am. [Small Anim. Pract.], 14:165–172, 1984.
6. United States Department of Agriculture: Importing a Pet Bird (Revised). Animal and Plant Health Inspection Service, 1985.
7. U.S.D.A. Documents, Freedom of Information Act, Federal Bldg., Hyattsville, MD 20782.
8. United States Department of the Interior, Fish and Wildlife Service, Washington, DC (FWS-F-006; FWS-F-037; FWS-F-014; CITES; 50 CFR Part 10; FWS-LE-9-83).

# *Chapter 7*

# PRELIMINARY EVALUATION OF A CASE

GREG J. HARRISON,
LINDA R. HARRISON,
ALAN M. FUDGE

Initial contact with client and avian patient during the office visit provides an opportunity to educate the client and form some preliminary opinions on the condition of the bird. This information is obtained through observation, history, and performance of a thorough physical examination. The professional relationship may be completed at the end of these procedures if, for example, a health certificate is requested. If the bird is clinically ill or a subclinical condition is suspected, the practitioner may proceed with appropriate diagnostic procedures and subsequent therapy.

Beginning avian practitioners often underestimate the value of a complete physical and the historic evaluation. A systematic approach to the physical examination will provide the clinician with valuable objective information, assure evaluation of all parts of the external anatomy, and provide the opportunity to be more selective in his diagnostic approach.

## OBSERVATION

A preliminary assessment of the client's knowledge of captive bird care can be made by viewing the cage contents and noting the client/bird relationship. For example, accumulated droppings on the bottom of the cage, dirty food and water dishes, wild bird seed as the only food source, and lack of mineral supplements may suggest a poor human/animal bond. A clean cage with vitamin-colored water, colored seed, and an abundance of plastic toys may suggest a strong bond but lack of knowledge pertaining to proper cage bird husbandry. A bird that is clean, appropriately fed and housed (see Chapter 2, Husbandry Practices), and relatively easy to handle would suggest ownership by one who has made the effort to acquire knowledge of captive birds. Discussion of nutrition and other husbandry factors is most often initiated early in the interview when the client is receptive to suggestions. Technical staff, video presentations, and written material can assist with client education.

The avian patient is first observed whenever possible from a nonthreatening distance during the history/interview, as some indications of ill health may be evident on casual observation. Attitude, body conformation, posture, general appearance, and obvious gross abnormalities may become evident. For calm psittacines it is useful to have the examination room equipped with a perch on a stand, possibly resin-coated, which can be disinfected between uses and can serve as a collection area for fresh droppings (Fig. 7–1). The owner should place the bird on the perch for visual appraisal by the veterinarian prior to handling. In contrast to evaluating the

**Figure 7–1.** A resin-coated perch on a stand in the examination room serves as a neutral zone for the bird, provides a collection site for fresh fecal samples, and can be disinfected between bird patients.

bird on the owner's shoulder, the perch serves as a neutral zone to reduce the defense mechanisms of the bird when it is approached for handling. The assessment can thus be more objective from this viewpoint. In addition, the client's attention is free to be focused on the clinician's questions.

### Attitude

Most birds are somewhat apprehensive outside their own environment. The healthy bird should appear alert and attentive. Extremely tame birds may appear more relaxed and calm. Macaws, Amazons, African Grey Parrots, and conures may exhibit species' characteristics of rowdiness and vocalization. It is abnormal for a bird to exhibit fluffing, head tucking, rhythmic movement of the tail, frequent blinking, general lethargy, or dropping off to sleep in the examination room.

### Body Conformation and Posture

The presence of such abnormalities as swellings or skeletal deformities may be obvious if the practitioner is familiar with the normal body conformation and posture of various species (Fig. 7–2). For example, Amazons and African Grey Parrots tend to be stout; cockatiels are normally more trim. Displacement of contour feathers, most commonly in the abdominal area,

**Figure 7–2.** Gross distortions in body conformation, such as this abdominal swelling in a budgerigar, are obvious from initial observation.

may suggest a tumor or swelling in this area. Drooped wing or body posture may be suggestive of weakness or injury. Restlessness, shifting of body weight, or favoring of one leg may suggest pain or dysfunction from disease or injury. Specifically in budgerigars, gonadal or renal tumors may result in leg disuse (see Chapter 40, Neoplasia). Evaluation of grasping ability is aided by rotating the perch during the observation, or later during the physical examination, by perching the bird on one's finger or arm. Disequilibrium may suggest the possibility of fracture, spinal malformation, ingestion of toxin, or head injury and (see Chapter 20, Neurologic Examination, and Chapter 39, Toxicology).

### Estimation of Age and Sex

Some potential diseases may be suggested by the species or country of origin if the bird is newly imported (see Appendix 2). Knowledge of the age and sex of the bird may assist the clinician in making the differential diagnosis. After fledging, size is usually not a factor in determination of age, as birds reach adult stature relatively rapidly; however, other physical features may be characteristic of an immature bird. The feathers appear relatively dull-colored in many young psittacines. Sun Conures, for example, resemble Jenday Conures as juveniles, obtaining their brilliant yellow plumage at maturity. There are species in which the immature plumage color imitates that of the adults, but the feathers of the tail are usually shorter in the young bird.

Beak color varies with age in some species. When adult birds of a species have dark bills, those of the young are usually light-colored. When the adults have pale beaks, those of juveniles are generally dark or have dark markings at the base of the beak.[1]

Iris color may be an indication of age (Fig. 7–3). In most psittacine species, very young birds have brown or dark irides. In macaws, these fade to grey within the first year, appear white from one to three years, and turn yellow as the bird matures. Iris color in some Amazons changes to red-orange as the bird matures, and in African Greys the irides lighten from brown through grey to white.[2] In Moluccan Cockatoos and most white cockatoo species, the iris color can generally be related to the sex of the bird at maturity (see Chapter 51, Sex Determination Techniques).

Some dichromatic species such as Eclectus Parrots are identical as hatchlings; however,

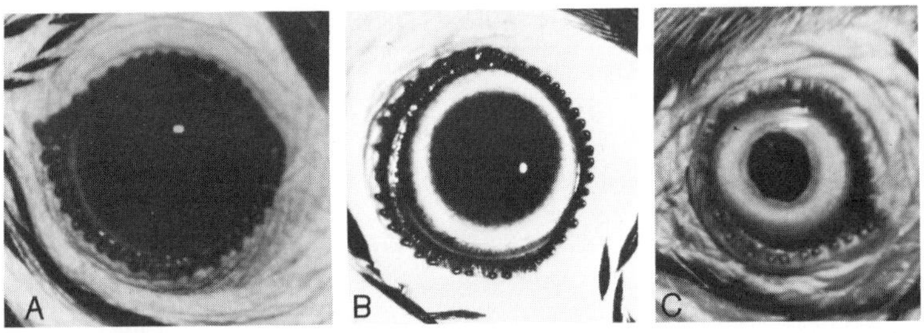


**Figure 7–3.** The dark iris in a very young Blue and Gold Macaw *(A)* will lighten during the first three years *(B)* and turn yellow as the bird matures *(C)*. (Courtesy of Lorraine Karpinski.)

gender is evident at the pin feather stage. Males of some other dichromatic species resemble the female in the immature plumage and obtain gender-related markings at sexual maturity (one to two years of age).

## HISTORY

The history interview must be conducted carefully to elicit pertinent responses from the client. If the bird is now ill, the client may not be aware of early situations that led to the current condition. Basic questions include origin of the bird, length of ownership, housing arrangements, exposure to other birds, normal diet and supplements, owner's evaluation of presenting condition including changes in food or water consumption, droppings, environment, and behavior. Knowledge of previous treatment attempts by the owner or another veterinarian is essential. More specific questions will evolve from the discussion.

Based on the history, it is useful to determine into which of two broad categories the bird falls—"new" or "old"; this distinction gives further direction to the development of a differential diagnosis.

### "New" Bird

The classification of "new" bird includes any bird that has been exposed, directly or indirectly, to other birds or to the possibility of infectious disease agents within the last two years. Recently imported birds, those newly purchased from a pet shop or aviary, or pet birds with any exposure to "new" birds (such as coming in contact with a veterinary hospital or pet shop for boarding or grooming procedures) are considered to be in this category. An example of indirect contact might involve the single pet bird for which seeds or other pet supplies are purchased from locations near bird cages in a pet shop.

"New" bird diseases are among the most difficult to correctly diagnose and treat, as the primary etiology may be undetectable at the time the bird is presented for treatment and the clinician may be encountering secondary or tertiary complications. Table 7–1 enumerates the most commonly occurring disease conditions observed in birds in this classification.

### "Old" Bird

The category of "old" bird is not related to age but to the length of time the bird has been in a stable, noncontaminated environment. Birds that have been maintained under such changeless circumstances with no exposure to other birds for at least two years are more susceptible to problems relating to chronic malnutrition, psychologic disturbances, tumors, and other conditions generally of a noninfectious nature (Fig. 7–4). Problems of neonates from parents classified as "old" birds would also likely be from noninfectious causes. The most common disease conditions associated with "old" birds are included in Table 7–2.

Once the practitioner has made this distinction in the historic background of the patient, other findings become more relevant. Although the history usually entails some discussion of the appetite, this information is not singularly valuable in assessing the condition of the bird. Frequently many very ill birds who have not eaten for hours or days at home may begin food consumption at the veterinarian's office. Conversely, birds that are "going light" in response to disease problems such as giardiasis, chlamydiosis, or diabetes mellitus have been observed to eat constantly.

## RESTRAINT AND HANDLING

One cannot perform a complete physical examination without handling the patient. If the bird is particularly distressed, however, this

**Table 7–1.** COMMON CONDITIONS CLINICALLY ASSOCIATED WITH THE "NEW" BIRD

| Causes | Clinical Manifestations |
|---|---|
| Acute malnutrition | Polyurea, polydipsia |
| Poor quality diet | Weight loss |
| Increased need for vitamin C from stress | Anorexia |
| Malabsorption of nutrients | Diarrhea |
| Trauma | Broken feathers, chewed toes |
| Physical (fright reactions) | Fearfulness, screaming |
| Psychological stress | Feather picking |
| Subjected to rough handling | Poor quality feathers |
| Parasites | Hepatomegaly |
| Hematologic | Splenomegaly |
| External | Air sacculitis |
| Gastrointestinal | Multiple site granulomas |
| Respiratory | Upper respiratory disease |
| Chlamydiosis | Oral abscesses |
| Viral diseases | Periophthalmic disease |
| Herpes (Pacheco's parrot disease) | Beak and feather disease syndrome |
| Pox | High loss of offspring |
| Reovirus | Other reproductive diseases |
| Papovavirus | Infertility |
| Other viruses | Poor hatchability |
| Mycoplasmosis | Increase susceptibility to low virulent organisms |
| Bacterial infections | Sternal ulcers |
| Mycotic infections | Bleeding feathers |
| Candida | Sudden death |
| Aspergillosis | Blood in feces |
| Immunosuppression | Frank hemorrhage |
| Overtreatment with antibiotics, poor hygiene, polluted water, no exposure to sun and rain | Black tarry feces |
| Poisoning | Occult blood |
| Unknown | |

procedure should be avoided or kept to a minimum to prevent the stress of handling from aggravating the symptoms. Although it is extremely rare, the stress of handling an extremely sleepy, weak, fluffing, staggering small bird has been reported to precipitate the sudden death of the bird in the veterinarian's hands. The owners of birds in such condition should be made aware of this potential danger.

**Figure 7–4.** One of the most common conditions associated with the "old" bird is paralysis of the leg from a gonadal or renal tumor.

Practitioners should also be aware that they may experience hearing loss as a result of close contact with noisy psittacines during physical examinations. Ear protection devices are recommended when examining especially vocal patients (see Table 6–1).

## Capture and Restraint

Several general factors should be considered in restraint of birds[2]:

1. Any manipulation that prevents the proper movement of the sternum will interfere with respiration (see Chapter 5, Selected Physiology for the Avian Practitioner).
2. Muscular activity of the bird associated with restraint increases heat production, which, coupled with inability to dissipate heat from the body surface, may produce hyperthermia.
3. Fractures may be induced by rough handling.

**Table 7–2.** COMMON CONDITIONS CLINICALLY ASSOCIATED WITH THE "OLD" BIRD

| Causes | Clinical Manifestations |
|---|---|
| Chronic malnutrition | Overgrowth of beak and nails |
| Nutritional deficiencies | Dry, flaky skin and beak |
| Protein | Loss of skin pattern on plantar surfaces of feet |
| Calcium | Feather problems |
| Vitamins A, B complex | Egg binding |
| Vitamins C, $D_3$, E | Endocrine gland malfunctions |
| Iodine | Increased susceptibility to low virulence organisms |
| Iron | Obesity |
| Salt | Lipomas |
| Trace minerals | Fatty liver degeneration |
| Nutritional excesses | Atherosclerosis |
| Fats | Congestive heart disease |
| Carbohydrates | Neoplasms |
| Phosphorus | Neuropathies |
| Psychological disturbances | Boredom |
| Sexual frustration | Overeating |
| Genetic weakness | Overeating of grit |
| Old age | Feather picking |
| Poisonings/trauma less common than with "new" bird | Behavioral problems |
| | Screaming, destructiveness, biting |
| | Pacing |
| | Regurgitation to toys, mirrors, or owner, resulting in gastrointestinal problems |
| | Aggressive behavior |
| | Short-lived |
| | Reproductive disease |
| | Poor fertility, low clutch size and hatchability, loss of reproducing females, weak young |
| | Polyuria, polydipsia |

4. Twisting the head may restrict air passages.
5. Damage may be inflicted upon a bird by careless netting.

The amount of restraint necessary to expedite a complete hands-on physical examination depends upon the individual bird. Young hand-raised babies respond to slow movements and affectionate scratching behind the neck, so the examination can be effectively performed with little or no restraint.

Capture of a patient that is less tame is best performed by the veterinarian, taking it directly from the bird's cage or carrier. This is facilitated by removing any perches and toys or the entire top of the cage if it is so designed. Small cage doors often do not allow easy access. Capture of birds is sometimes expedited by darkening the room.

A cloth or paper towel, which serves as a visual barrier to corner the bird, is useful in capturing most birds (Fig. 7–5). A crisp, disposable paper towel is preferred. The towel is placed flat up against the palm of the hand with bits of it tucked in between the fingers to hold it in place. One corner of the towel can be accessible for the bird to chew on as the veterinarian's hand is maneuvered behind the bird's head. When the opportunity arises, the bird is grasped from behind the neck with the thumb and forefinger on both sides of the temporomandibular joints. Other birds may respond to gentle stroking with the towel until restraint is applied. Drawing the bird backward out of the cage prevents damage to the wings and simplifies manipulation of the rest of the body.

Some patients who have been former visitors to a clinic, especially Amazons, may soon catch on to the paper towel technique and roll over on their backs. In these cases, one may need to approach the bird with a wooden perch to hold the mandible back while the legs are seized, the bird is stretched out, and the practitioner's hand is positioned behind the neck.

The veterinarian should not attempt to retrieve a bird directly from a client. Some birds are so well-trained by the owner or handler that capture and restraint by a stranger may damage the human/bird bond. The decision to allow an owner to assist with restraint must be made with care, as the veterinarian is ultimately responsible if any injury occurs.

For large birds, the handler should place the thumb and index finger behind the neck to control the beak and bring the bird close to the body, with the other hand gently restraining the wings (Fig. 7–6).

**Figure 7–5.** Paper towels are effective for capturing a bird from its cage and can be discarded following use.

Once the feet are stretched out, the handler may choose to place a finger under the mandible and tilt the head back. If the owner is handling the bird, he should be advised that all birds are capable of biting if frightened; therefore, the restraint is for the owner's protection as well as the practitioner's. In some cases, another person such as the technician may hold the wings and expose the body parts for the examination, leaving the head and feet for the primary holder. It should be pointed out that proper restraint is *minimal* necessary restraint.

A small bird can be restrained in one hand. The neck is placed between the index and middle finger and the head tilted back. The wings are controlled with the ring and little fingers of the handler on one side and the base of the thumb on the other. This frees the thumb, which can then be used to help hold a toenail for grinding, a leg for splinting or band removal, or a wing for trimming. In each case, the instrument can be maneuvered by the other hand. A similar restraint method for small birds can be used for drawing a jugular blood sample or for tube feeding (Fig. 7–7).

If a bird with clipped wings has escaped from one's grasp, it has probably fluttered to the floor. To recapture it, first try the towel technique. If that does not work, attempt to get the bird on a perch and rapidly roll him back toward

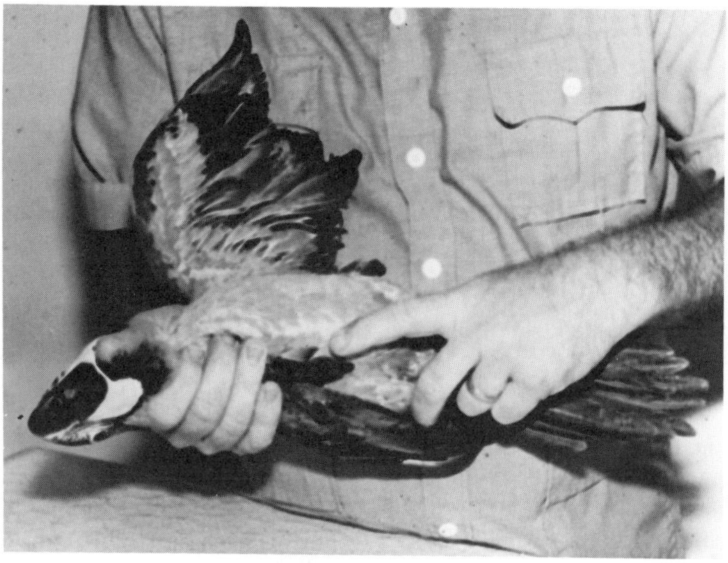

**Figure 7–6.** The mandible of a large bird is controlled with one hand and the feet and tail restrained with the other hand.

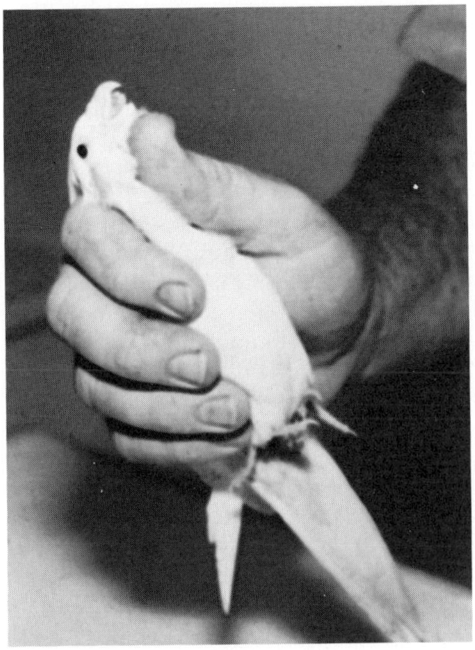

**Figure 7–7.** A small bird may be restrained in one hand. The thumb is placed under the mandible and the wings are held close to the bird's body.

the ground, simultaneously grasping the base of the tail and sliding the perch up to the mandible to pin the head to the ground.

If a cloth towel is to be used for any of these techniques, the owner should be requested to bring it from home. It is usually not practical in a clinic situation to have a supply of sterilized towels on hand for use on each bird.

A sterilized net should be accessible near or in the examination room in case capture of an escaped flighted bird is necessary.

The use of gloves with psittacines should be totally avoided in the average clinical setting, as they lead to rougher handling and serve as fomites of disease transmission, as do reused towels. Properly designed gloves may be necessary, however, in restraint and handling of birds of prey. As control of the talons is of primary importance in these species, these birds should be caught above the talons, with control of the head and wings to follow. Waterfowl and poultry are best restrained by supporting the head and neck while holding the wings up against the body. Control of the beak is necessary for restraint of toucans. A smaller toucan tends to have a more severe bite than one of the larger species such as a Toco. Restraint of penguins, ratites, and other bird species is discussed by Fowler.[2]

### Releasing a Biting Bird

Avian handlers do get bitten occasionally, and often the bird continues to maintain its grip. Several methods have been successful in releasing the bird's bite.

1. Blowing in the bird's face usually startles it into letting go, and one can quickly reposition the restraint or let go completely and start over.
2. Releasing any contact with the bird is sometimes successful in releasing the bite.
3. If nothing else works and the bird is still attached to one's finger or hand, one can try spinning around in a circle.

Trying to pry the beak open manually may fracture or otherwise injure the beak. Use of water seldom works and is potentially dangerous. The best alternative is to prevent the bite by having appropriate control over the beak.

## HANDS-ON PHYSICAL EXAMINATION

As stated before, the patient *must* be handled for a thorough physical examination. Obtaining an accurate weight of the patient is essential. The disposition of the bird and anticipated difficulty in recapturing it from the scale may determine the point at which weighing the bird will be included in the preliminary evaluation of the case.(The reader is referred to Appendix 4 for average juvenile and adult weights of selected species.)

### Mouth

The opportune time for examining the mouth may be immediately following the initial capture while the bird is vocalizing. While holding the head near the temporomandibular joints, smell the oral cavity (as long as chlamydiosis or other zoonotic diseases are not suspected). Most cage birds have a neutral odor. Macaws have a "pine forest" smell, while cockatoos smell slightly musty. A foul-smelling mouth may indicate bacterial pharyngitis/sinusitis or digestive disorders such as those associated with proventricular dilatation.

The beak may be held open with a speculum, and, with the aid of a concentrated light source, the oral cavity can be fully visualized (Fig. 7–8).

Normal oral epithelium is shiny and uniform in coloration. Some cockatoos, Amazons, and macaws may have melanistic pigmentation, while in other species the epithelium may be

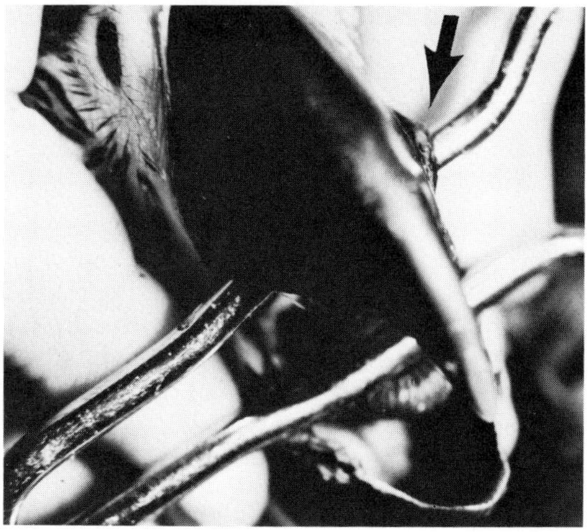

**Figure 7–8.** A mouth speculum must be appropriately sized for the bird. The surface area of this speculum is too narrow for a macaw; damage to the beak has resulted from its use.

pink. In birds suffering from bacterial or fungal diseases of the digestive system, the dorsal epithelium may have a grey-blue devitalized appearance. The presence of white, caseous deposits may suggest inflammatory changes secondary to squamous metaplasia (see Chapter 31, Nutritional Diseases). Oral lesions may be associated with pox, vitamin A deficiency, candidiasis, trichomoniasis, or coliform infections and other disorders.

Many sick birds have a mucous accumulation in the mouth or accumulated food particles under the premaxilla or tongue. Check the surface of the soft palate and the area of the larynx and pharynx for the presence of ulcerations or small abscesses, which may appear only as pin-point swelling. Birds that have recovered from viral diseases may have scar formation around the choana, or the normal papillae on the choanal borders may be missing (Fig. 7–9). Many cases require tranquilization for adequate oral examination. A source of magnification and additional lighting are also helpful.

The normal parrot tongue is smooth-surfaced, symmetrical, thick, and fleshy and has a thick horny layer of epithelium near the tip. Unilateral swellings are often abscesses. Nectar feeders such as lories and lorikeets have a brushlike tip to the tongue covered with elongated papillae that become erect when the bird is eating. Tongue color may vary with the species.

## Beak

A normal beak is smooth and symmetrically colored with a deep sheen. The presence or absence of a notch in the premaxilla is a genus taxonomic characteristic. Beak length, width, and shape also vary according to species. In most species, the beak comes to a sharp tip that clients sometimes prefer to have blunted (see Chapter 6, Management Procedures).

An abnormally fast growing beak, usually observed in older, obese or otherwise malnourished birds, has been clinically associated with chronic liver damage. In the budgerigar, this beak overgrowth is often accompanied by a dark reddish-black discoloration on the anterior margin of the premaxilla and/or a midcaudal reddish hematoma-like growth on the mandible.

Hereditary factors, malnutrition, systemic disease, and traumatic causes have been implicated in cases of twisted beaks, seen most often in fledglings. Most of these conditions are nonresponsive to therapy; therefore, beak trimming may be necessary on a regular basis for aesthetics as well as function. Unless they experience some type of trauma, mature, healthy birds offered a balanced diet and an ample supply of branches or other objects to chew on seldom have beak problems. Contrary to popular opinion, cuttlebone and mineral block will not act as "beak conditioners" (see Chapter 2, Husbandry Practices).

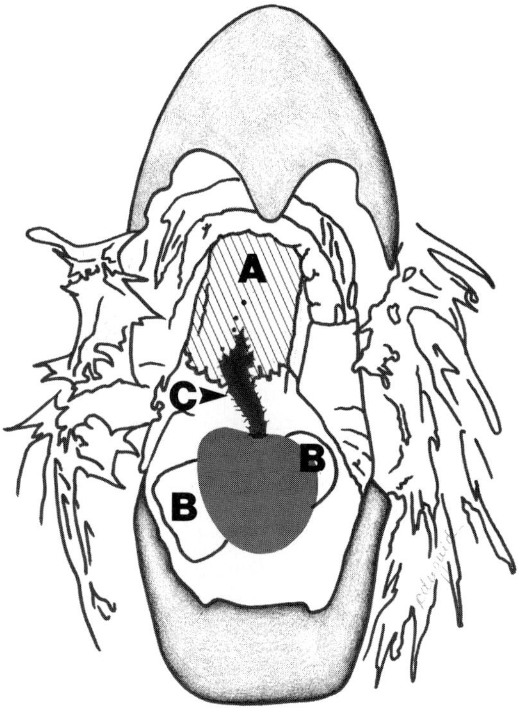


**Figure 7–9.** Common sites for oral abscesses are: *(A)* palatine salivary glands and *(B)* mandibular salivary glands. The choanal slit *(C)* is often tortuous, and the choanal papillae may be swollen or absent.

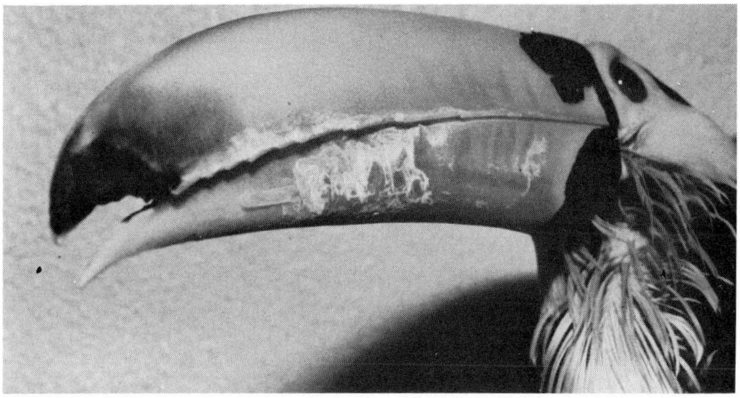

**Figure 7–10.** Flaking layers of beak epithelium were associated with diabetes mellitus in this Toco Toucan.

A flaky or rough appearance to the beak may be associated with malnutrition, lack of proper chewing, or disease conditions. A syndrome described in psittacines includes deterioration of the tip of the premaxilla and palate (see Chapter 41, Disorders of the Integument). Retained horn accumulation has been observed in a Toco Toucan with diabetes mellitus (Fig. 7–10). Examine the beak for fractures or malalignment.

## Cere

The cere should be firm, smooth, neither dry nor wet, and have no accumulations of flakes or debris. The presence or absence of feathers on the cere is a genus variation (see Chapter 4, Clinical Anatomy).

Cere color is often used to identify the sex in budgerigars, but it is age-related and undependable in color mutations (see Chapter 51, Sex Determination Techniques). Brown hypertrophy of the cere in this species may be associated with a breeding condition in a normal older hen or with estrogen-secreting gonadal tumors in males.

## Nares

The nares should be evenly placed in relationship to the cere and be bilaterally symmetrical in size and shape. A change in the diameter of the nares may be the only obvious clinical sign of past or current respiratory infections or nasal tumors.

Healthy birds do not have a discharge from the nares; if a discharge is present, one should note the consistency. The presence of a discharge may be evident only by stains on the feathers of the cere area (Fig. 7–11). The reader is referred to Chapter 10, Clinical Microbiology, and Chapter 17, Cytology, for appropriate sampling and diagnostic techniques. Mild cases of nasal discharge, or rhinorrhea, may be accompanied by diffuse air sacculities or sinusitis with caseous accumulations (see Chapter 14, Radiology). Chronic rhinorrhea may lead to grooves in the beak. Accompanying symptoms of sneezing may suggest impending respiratory disease, foreign body, or occasionally, allergy (see Chapter 15, Endoscopy). Some birds may only be mimicking the sneezing of the owner.

An otoscope or a small diameter endoscope can be used to examine the nares; however, one must exercise caution to avoid causing hemorrhage. Some Amazons will have bristle feathers inside the nares and over the cere.

It is relatively common, although abnormal, for cockatiels to have a flaky, wafer-like plug of epithelium and debris accumulation on the operculum; this debris should be removed (see

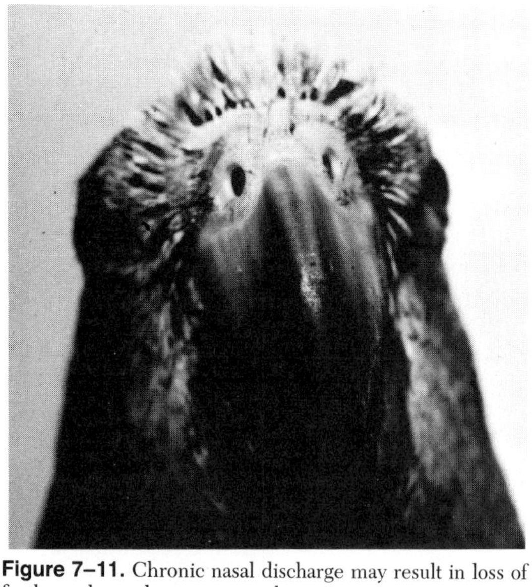

**Figure 7–11.** Chronic nasal discharge may result in loss of feathers above the cere as in this Amazon.

Chapter 28, Symptomatic Therapy and Emergency Medicine). A collection of debris in the nares of African Greys and Amazons is often the precursor of atrophic rhinitis.

## Eyes

Transillumination of the eye enhances the ophthalmic examination. With the practitioner holding the light source at a 45-degree angle from above, or from inside the mouth, the pecten appears as a black, "wormlike body" in the back of the eye if the aqueous humor, lens, and vitreous are clear (see Chapter 19, Ophthalmology). The nictitating membrane in most parrots crosses the cornea in a dorsomedial to ventrolateral direction. The cornea should be clear with a damp, glistening appearance. Visualization of minute ulcers on the eye of a bird is enhanced by fluorescein dye, black light in a darkened room, and/or magnification.

Normal eyelid margins appear symmetrical and flat, with some species variations: those of macaws and conures include papillae (see Fig. 7–3*A*); Amazons have modified feather lashes; eyelid margins of African Grey Parrots are relatively smooth.

Any clinical signs of periophthalmic swelling, epiphora, presence of a discharge around or under the eyelids, conjunctivitis, or loss or matting of feathers around the eye may suggest an infection (Fig. 7–12). Mycoplasmosis in cockatiels, for example, may present in this way. Some birds accumulate a caseous material under the eyelids that should be removed. If a discharge from the eye is clear, one may also consider an allergy or foreign body. A "wet" area on the shoulder or wing may suggest that the bird has been rubbing its eye.

Evidence of problems with the eyes and surrounding tissues may indicate past disease, such as previous upper respiratory infection or pox (see Chapter 32, Viral Diseases). Birds that have recovered from pox may exhibit facial scabs and may have eyelid scars or paralysis of the eyelids, epiphora, and photophobia due to chronic corneal lesions. These conditions may

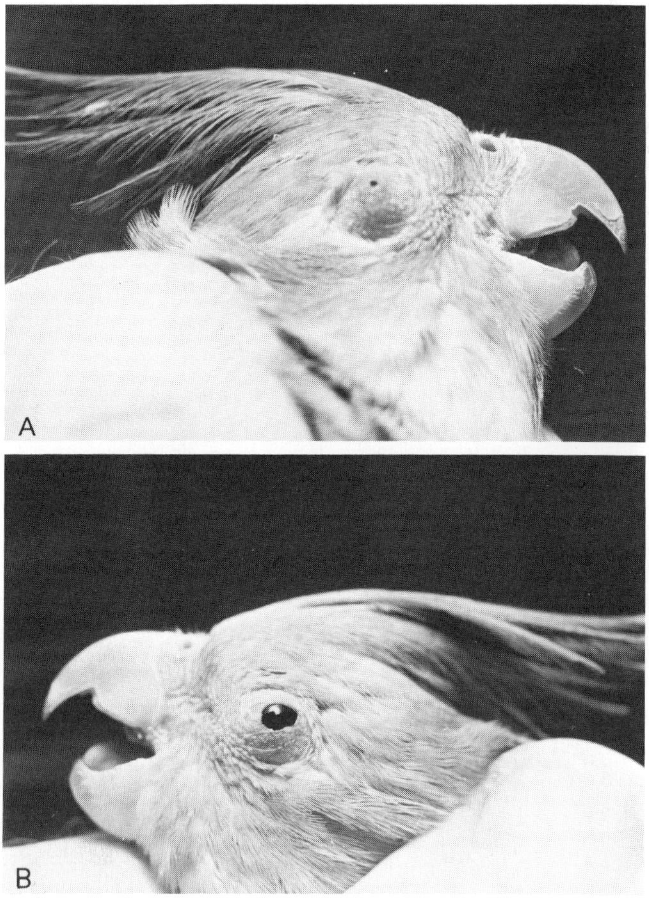


**Figure 7–12.** Some eye conditions, such as this progressive bilateral closing of the aperture of the eyelids, may need a specialized ophthalmologic examination. The left eye *(B)* eventually closed completely, as did the right eye *(A)*.

not respond to conventional therapy. Chronic pox cases require a thorough eye examination (see Chapter 19, Ophthalmology).

Check to see if the bird can follow a moving object with its eyes without the help of sound, touch, or air movement (see Chapter 20, Neurologic Examination). Cataracts give a cloudy appearance to the lens and are relatively common in aged psittacines.

## Ears

Otitis externa does occur but is relatively uncommon. Epithelial debris is often noted in the ears of birds suspected to be malnourished. An endoscope is valuable for this examination.

## Trachea

In small, light-skinned birds, the trachea can be transilluminated and the presence of tracheal mites and/or foreign bodies may be detected (see Chapter 15, Endoscopy). The feathers are wet with water and separated into the tracts for better visualization (Fig. 7–13).

## Wings

Wing injuries are relatively common (see Chapter 30, Evaluation and Nonsurgical Management of Fractures). Many birds may appear normal; however, they may be unable to fly

**Figure 7–13.** Mites or foreign bodies may be detected in the trachea of small, thin-skinned birds with the use of transillumination.

and are useless for purposes of display in some collections. Twisted feathers on the wings may result from genetic, nutritional, or traumatic causes. Large areas of feather loss may result from self-mutilation secondary to dermatitis, suggesting the possibility of more serious internal conditions (see Chapter 41, Disorders of the Integument).

## Turgidity of Wing Vein

A plump and relatively turgid wing vein has generally been found to correlate clinically with normal serum protein and PCV. We have associated flat veins with hypoproteinemia, anemia, and dehydration. While checking this vein, note the skin in the wing web, as dermatitis is common in this area.

## Skin Elasticity

As in mammals, skin elasticity can be used to assess the relative degree of hydration in birds. The skin of the neck and abdomen provides effective sites for this assessment.

## Feet and Legs

The surfaces of normal feet and legs should be shiny, with a uniform scaled pattern not unlike that of reptile skin (Fig. 7–14). Flaking, cracking, peeling, and signs of inflammation are abnormal. Malnutrition, possibly complicated by trauma, is clinically suspected in birds with smooth, worn down, or ulcerated surfaces of the feet. Pododermatitis, with or without ulceration, is not uncommon in psittacines yet is an abnormal finding (see Chapter 41, Disorders of the Integument). A necrotic condition of unknown etiology that requires prolonged therapy has been observed in Amazons (see Chapter 41, Disorders of the Integument).

Although normal parrot nails are relatively long and sharp, excessive length may be responsible for leg or foot injuries resulting from entanglement in cage bars or carpets. Overgrowth of the nails often accompanies beak pathology associated with metabolic/nutritional disorders. This condition is especially noted in lories, finches, budgerigars, canaries, and Amazons. In certain breeds, such as frill canaries, long nails are the accepted norm.

Missing toes or nails may be used as a means of identification. Although this may decrease the value of the bird from an aesthetic view-

**Figure 7–14.** The pattern of scales on the feet of normal healthy Amazons is not unlike reptilian skin.

point, it will usually not affect breeding potential unless perching is impaired.

## Feathers

Notice if body contour feathers appear homogeneous, adhere tightly together, and have a bright sheen. One exception would be the Eclectus Parrot, which characteristically has loose, hairlike feathers. Hold the wing and tail feathers up to the light to observe for signs of mites, color breaks, structural damage such as "stress" lines, fractures, or holes in vanes (Fig. 7–15). Feather lice and mites may be found on newly imported birds; however, external parasites on pet birds are rarely seen. The exception is *Knemidokoptes* mites, which are relatively common (see Chapter 37, Parasites). Abnormal feathers can be pulled for Gram's stain, culture, or microscopic examination (see Chapter 10, Clinical Microbiology, and Chapter 41, Disorders of the Integument).

Birds in continual heavy molt may require dietary supplementation or special lighting conditions. The presence of a high percentage of pin feathers may suggest a potential occult disease.

Feather picking may be caused by sexual frustration, boredom, change in owner/bird routine, disease problems, or dietary deficiency. Damaged or soiled feathers may be due to inappropriate caging or perch placement (see Chapter 2, Husbandry Practices), trauma from handling, damage from falling with severe wing trims, acrobatics, feather picking, "bullying" by cage mates, external parasites, or malnutrition. Damaged feathers may be removed.

In cockatoos, it is especially important to examine the powder down feathers of the rump

**Figure 7–15.** Signs of wear are evident on feathers that have been retained past their normal molt time.

and thigh area and the feathers of the crest for early symptoms of psittacine beak and feather disease syndrome (see Chapter 41, Disorders of the Integument).

A powdery white substance (powder down) that may come off the feathers of white cockatoos, African Grey Parrots, and cockatiels onto the practitioner's hands is normal (see Chapter 4, Clinical Anatomy).

## Palpation of the Body

Palpation should begin with the intermandibular space. Examine for swelling; then pass down the neck, palpating the trachea and esophagus. Evaluate the crop for contents. A completely empty crop is rare in a healthy bird if food has been available. Sick birds sometimes have fluid collection in the crop. The crop may be insufflated and examined with transillumination (see Chapter 15, Endoscopy) or sampled for other diagnostic tests (see Chapter 10, Clinical Microbiology, and Chapter 17, Cytology).

Palpate the entire body, including the joints, for presence of tumors, old fractures, body fat, and other malformations. The sternum should be straight and slightly elevated in comparison to the pectoral muscles. The ratio of muscle to bone and fat is an excellent general indicator of physical condition. Extremely emaciated birds are poor risks for some specific diagnostic and treatment procedures. Obesity may be accompanied by other problems; these birds should be placed on a reducing diet (see Obesity in Chapter 42, Miscellaneous Diseases). In small, lean birds with translucent skin, the feathers can be wet and parted to examine the integument, sternum, muscle conformation, abdominal organs, and appendages.

Normally, one finds a slight abdominal depression while palpating; in the normal bird, the ventral borders of the liver are barely evident. More caudally, the gizzard is usually discernible as a landmark.

Birds of normal weight with a protruding abdomen may be suffering from egg binding, egg peritonitis, tumors, obesity, or other abnormalities (see Abdominal Swelling in Chapter 8, Differential Diagnosis Based on Clinical Signs). Palpation of an enlarged abdomen must be done with caution. In birds with ascites, a condition that is relatively common in mynah birds, it is possible to rupture the abdominal membranes with aggressive palpation, forcing the fluid into the air sacs and causing immediate death.

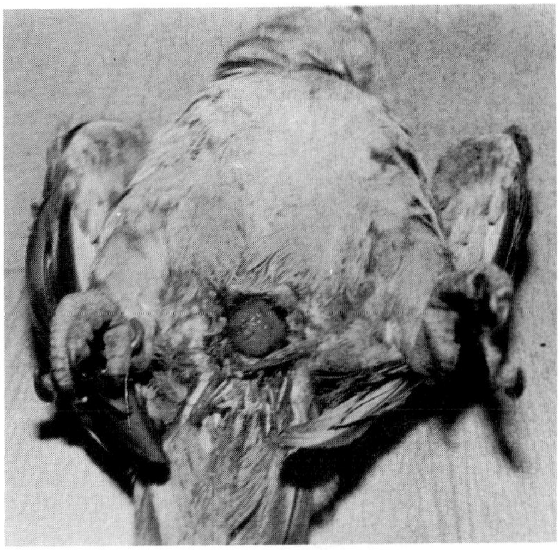

**Figure 7–16.** A protrusion from the cloaca may be other than a cloacal prolapse. Papillomas should be considered and are confirmed by histologic evaluation of the tissue.

## Vent

Soiled vent feathers, resulting from loose droppings, possibly associated with gastrointestinal disease, should be further investigated. Check for evidence of protruding tissue from the vent (Fig. 7–16). Hypertrophy of the mucosal lining, a condition that is commonly misdiagnosed as a prolapsed cloaca, may result from papillomas or other tumors (see Chapter 40, Neoplasia). A uterine prolapse from egg binding and other causes presents with a similar clinical appearance (see Chapter 28, Symptomatic Therapy and Emergency Medicine).

## Respiratory Evaluation

Auscultation of sinuses, trachea, thoracic air sacs, and abdominal air sacs is facilitated by the use of a pediatric stethoscope. Normally, the gentle rushing of air is all that will be heard. The immobile lungs do not produce rales or other pathologic sounds as in mammals. Audible sounds on inspiration or expiration appear to correlate clinically with upper respiratory and lower respiratory pathology, respectively. Abnormal sounds of clicking, rattling, wheezing, squeaking, and honking are often clinically associated with a number of disorders, including parasites, liver problems, air sacculitis, malnutrition, and endocrine gland dysfunction.

Resting birds have a closed-mouth respiration that is barely noticeable. Although some psittacine species, such as the Palm Cockatoo, are incapable of completely closing their beaks, open-mouth breathing is usually the result of nervousness, hyperthermia, compensation for a plugged nostril, anemia, or lung or air sac pathology.

Often the handling for the physical examination is enough to initiate a stress reaction of increased respiration. If not, grasp the feet and allow the bird to flap its wings for a few seconds; place the bird back in the cage and note the amount of time required for it to stop panting. If not hyperthermic, normal birds recover resting respiration rate within one to two minutes (i.e., resiratory recovery time). Lung, air sac, abdominal, or cardiovascular problems are suggested if this time is extended. At the end of the exercise, the practitioner can place his ear close to the bird's dorsal thoracic area and mouth in an attempt to detect abnormal sounds.

## REFERENCES

1. Forshaw, J. M., and Cooper, W. T.: Parrots of the World. Garden City, NY, Doubleday & Company, Inc., 1973.
2. Fowler, M. E.: Restraint and Handling of Wild and Domestic Animals. Ames, IA, Iowa State University Press, 1978, pp. 262–285.
3. Karpinski, L. G., and Clubb, S. L.: Clinical aspects of ophthalmology in caged birds. Proceedings of the Annual Meeting of the Association of Avian Veterinarians, San Diego, CA, 1983.

# Chapter 8

# DIFFERENTIAL DIAGNOSES BASED ON CLINICAL SIGNS

DIAGNOSIS

GREG J. HARRISON,
WALTER J. ROSSKOPF, JR.,
RICHARD W. WOERPEL,
ALAN M. FUDGE,
LORRAINE G. KARPINSKI

The ultimate goal of the avian practitioner is rapid therapeutic response of the patient. Numerous diagnostic procedures are available to the clinician to assist in making a diagnosis and selecting the appropriate therapy. The selection of tests to perform depends upon the clinical signs, the condition of the bird, the financial considerations of the owner, and the knowledge and experience of the veterinarian in interpreting the results. The value of a complete and thorough history interview and physical examination cannot be overemphasized, as many diagnoses can be made solely on the basis of this information and thus eliminate the need for further investigation. For example, the clinical signs associated with most cases of psittacine beak and feather disease syndrome are easily recognizable. No further diagnostic tests are warranted if the physical characteristics of this disease have been confirmed unless information on the true internal status of the bird is desired to prolong its life.

However, many avian diseases are pansystemic, and the clinical signs may not directly indicate an etiology. Based on the assessment of the individual bird, the practitioner must carefully select those procedures that will yield the greatest amount of information to rule out or confirm a specific diagnosis.

## PROBLEM-SPECIFIC DEFINED DATA BASE

Table 8–1 illustrates how a problem-specific defined data base[1] may be developed. If an avian patient exhibits multiple clinical signs, the practitioner may consult the corresponding categories and select appropriate procedures from among those suggested for each. For example, an avian patient may present with oral lesions (Data Base I) as well as weight loss with biliverdinuria (Data Base III). Diagnostic tests beyond the history and physical examination could be selected from among those suggested in both I and III. If the bird also exhibited polyuria, polyurates, and polydipsia, further tests might be indicated in Data Base IV.

A Gram's stain of the feces may be valuable to include in the data base for any disease condition, as it can represent a general assessment of the patient (see Chapter 9, Evaluation of Droppings). The incorporation of further tests under the problem-specific defined data base depends upon the results of previous steps and can be changed as the conditions warrant. The experience one develops in having contact with large numbers of avian patients will further guide the diagnostic approach. In some cases response to therapy may become part of the diagnosis.

The differential diagnoses charts on the following pages have been developed from the experience of the authors and generally represent the most common clinical conditions presented to an avian practitioner. Clinical signs are listed under the basic categories of Behavior, Central Nervous and Musculoskeletal Systems, Digestive and Urogenital Systems, Hemorrhage, Integument, Neonates, Ophthalmology, Respiratory System, Swelling, and Necropsy. Specific disease entities are discussed in the respective chapters. General symptomatic therapy measures are presented in Chapter 27, What to Do Until a Diagnosis Is Made; Chapter 26, Therapeutics; and Chapter 28, Symptomatic Therapy and Emergency Medicine.

**Table 8–1. PROBLEM-SPECIFIC DEFINED DATA BASES**

**I. DIGESTIVE** (Clinical signs of oral/crop/proventricular disorders)

- Gram's stain of feces; saline wet mount and flotation for parasites
- Gram's stain of affected areas
  If increase in number of gram-negative bacteria on Gram's stain, proceed to culture and sensitivity
- Other cytologic stains of affected area; pH
  If specific positive results, proceed to appropriate culture
- Complete blood count (CBC: Serum protein, PCV, WBC, differential, RBC morphology, parasite examination, thrombocytes)
- Special parasitology tests

**Add**

IF CROP IS INVOLVED:

- Cytology of crop (wash)
- Transillumination; insufflation
- Endoscopic evaluation
- Exploratory ingluviotomy

IF PROVENTRICULUS OR VENTRICULUS IS INVOLVED:

- Cytology of cloaca
- Radiograph (possible contrast)
- Blood toxicology (lead)
- Possible exploratory surgical intervention

**II. DIGESTIVE** (Clinical signs of diarrhea)

- Fecal evaluation: Gram's stain; saline wet mount and flotation; pH
- Cloacal cytology; special stains (e.g., trichrome, acid-fast)
- Culture and sensitivity; special cultures
- *Chlamydia* (cytology, culture, serology)
- Radiograph: contrast studies
- CBC
- Basic avian panel (total protein, glucose, uric acid, AST)

**III. LIVER/PANSYSTEMIC** (Clinical signs of "classic sick bird": weight loss, polyuria, polyurates, biliverdinuria, fluffing up, scant stool, poor appetite vomiting, ascites, dyspnea)

- *Proceed as in Data Base II.* (May see hepatomegaly or decreased size and density of liver on radiograph)

**Add**

- LDH, cholesterol, clotting/prothrombin time to basic avian panel for further liver evaluation
- Calcium, phosphorus, electrolytes (sodium, potassium), to basic avian panel for further renal evaluation
- Endoscopy; multiple organ biopsy (cytology, histopathology)
- Exploratory surgery

**IV. UPPER RESPIRATORY** (Clinical signs of open mouth breathing sounds, sneezing, sinus swelling, rhinorrhea, plugs in nares, slow respiratory recovery time, tail bobbing, dyspnea, head shaking, scratching at ears, feather loss around eyes, face and forehead, esp. in Amazons)

- Gram's stain of feces; direct fecal smear for parasites; special parasitology stains
- Cytology of affected area (sinus aspirate, flush; scraping for parasites); special stains
- Radiographs (whole body; sinus views)

**Table 8–1.** PROBLEM-SPECIFIC DEFINED DATA BASES *Continued*

**IV. UPPER RESPIRATORY** *Continued*
- Rhinoscopy (check for foreign bodies)
- Culture and sensitivity (bacteria may be only secondary invaders; special culture procedures may be necessary: chlamydia, virus, mollicutes)
- Biopsy of lesions
- Paracentesis

  Avian hematology panels are optional unless Data Base III symptoms are involved

**V. LOWER RESPIRATORY** (Clinical signs of change in voice, labored respiration, increased respiratory recovery time, coughing, syringeal or upper bronchial symptoms)
- *Proceed as in Data Base IV*

  **Add**
- Radiograph of lungs, air sacs; view with magnification
- Laparoscopy; tracheoscopy

  Cytology of laparoscopic, tracheal samples
- Biopsy of air sacs
- Surgical intervention (air sac granulomas; tracheal foreign body)

**VI. METABOLIC/URINARY/PSYCHOGENIC** (polydipsia, polyurates, polyuria; seldom with accompanying weight loss or "classic sick bird" symptoms)
- Gram's stain of feces
- Urinalysis; cytology
- Radiography
- Blood glucose
- Basic avian panel plus renal evaluation tests (Ca, P, electrolytes, Na, K)
- CBC
- Water deprivation test

**VII. FEATHER PICKING** (psychologic, malnutritional, reproductive frustration, thyroid-responsive disorders)
- Magnification examination of feather structure and surrounding tissues
- Fecal saline wet mount (for *Giardia*) or Trichrome stain
- Response to modifications of diet, behavior, or collar
- Response to parenteral nutritional supplements, tranquilizers, opposite sex hormones, thyroid supplementation, TSH test

**VIII. SELF-MUTILATION FEATHER DISORDERS** (skin lesions accompanying feather loss)
- Gram's stain of feces, saline wet mount of feces
- Gram's and other cytologic stains of affected areas
- Radiograph
- Biopsy of feather follicle/skin; histopathology, appropriate culture
- CBC; basic avian panel
- Endoscopy; multiple organ biopsy; histopathology

DIFFERENTIAL DIAGNOSES BASED ON CLINICAL SIGNS ASSOCIATED WITH BEHAVIOR

| Clinical Signs | Species Affected | History | Possible Causes |
|---|---|---|---|
| Aggressiveness to people | Primarily Amazons macaws others | Sudden onset Associated with breeding season | Normal incomplete cyclic hormonal changes |
| | All | "New" or "old" bird | Fear Dominance |
| Aggressiveness to other birds/mate | Primarily cockatoos lovebirds Amazons | Fighting Feather picking Bullying | Normal hormonal change Misplaced aggression |
| Dropping off to sleep | All | Otherwise normal | Systemic illness Environment too dark Little social stimulation Malnutriton |
| Falling off perch (see Swollen legs, feet, digits, joints) | All | "Old" bird | Loose perch Nails recently trimmed Convulsions Leg paralysis Foot trauma |
| | Cockatiels | | Selenium/vitamin E syndrome |
| Masturbation | All, esp. budgerigars, cockatiels | Single pet | Reproductive frustration |
| On cage bottom | All | Acute | Egg binding |
| | | Egg-laying female | Egg-related peritonitis Septicemia Hypocalcemia (broken leg) Peritonitis from ruptured uterus "Obturator" paralysis Normal egg laying |

DIFFERENTIAL DIAGNOSES BASED ON CLINICAL SIGNS ASSOCIATED WITH BEHAVIOR *Continued*

| Clinical Signs | Species Affected | History | Possible Causes |
|---|---|---|---|
| | | Chronic illness | Infectious disease<br>Neoplasia<br>Metabolic/nutritional<br>Toxic |
| | | Otherwise normal behavior | Foraging on cage bottom<br>Nest preparation<br>Normal for some individuals |
| Shifting weight back and forth | All | | Pododermatitis<br>Parasites<br>Bacterial arthritis (esp. *Staphylococcus, Streptococcus, E. coli)*<br>Mycobacteria |
| | | Trauma | Foot/leg injury |
| | All, esp. cockatoos | | Fear, threat posture |
| Lethargic | All | | Systemic disease<br>Overweight<br>Hypothyroidism<br>Chronic lead poisoning<br>Hypothermia<br>Hypoglycemia<br>Psychologic/depression |
| | All, esp. African Greys | | Hypocalcemia |
| Change in vocalization | All | | Change in environment (loss of mate, jealousy, etc.)<br>Stress<br>Systemic disease<br>Foreign body in trachea, syrinx or bronchi<br>Localized physical changes (syrinx)<br>Deep respiratory problem |
| | Canaries | | Normal in some periods of breeding cycle |

DIFFERENTIAL DIAGNOSES BASED ON CLINICAL SIGNS RELATING TO CENTRAL NERVOUS AND MUSCULOSKELETAL SYSTEMS

| Clinical Signs | Species Affected | History | Possible Causes |
|---|---|---|---|
| Comatose | All | Sudden onset | Shock<br>Trauma<br>Pansystemic disease<br>Toxemia<br>Blood loss |
| Convulsions | All | | Epilepsy<br>Toxins [(lead, zinc, aflatoxins, insecticides), organophosphates, etc.], polytetrafluoroethylene<br>Hepatic encephalopathy<br>Head injury<br>Acute ischemic infarction<br>Hypoglycemia<br>Neoplasia<br>Viral (VVND and other paramyxoviruses; *Herpesvirus*; papovavirus)<br>Terminal behavior<br>Parasite migration<br>Hypocalcemia |
| | | Heavy egg laying | Egg-related peritonitis |
| Droopy wing(s) (see also Anorexia) | All | Physical restraint<br>Warm environmental temp. | Response to stress or hyperthermia |
| | | Acute onset | Pansystemic disease<br>Terminal behavior<br>Fracture<br>Spinal injury<br>Soft tissue sprain, strain |
| | Primarily budgerigars | Gradual onset | Neoplasm |
| Head tilt | All | | Heavy metal (lead) intoxication<br>Infection (inner ear, encephalomyelitis)<br>Encephalopathies<br>Infections<br>Head trauma<br>Neoplasia |
| General incoordination | All | | Bacterial sepsis<br>Toxicities (lead ingestion)<br>Aspergillosis<br>Neoplasia<br>Vitamin deficiencies<br>Terminal behavior |

DIFFERENTIAL DIAGNOSES BASED ON CLINICAL SIGNS RELATING TO CENTRAL NERVOUS AND MUSCULOSKELETAL SYSTEMS *Continued*

| Clinical Signs | Species Affected | History | Possible Causes |
|---|---|---|---|
| | | Acute mortality in flock | VVND, reovirus, other viruses |
| | Lovebirds | "New" bira | Microsporidiosis |
| | Rosellas<br>Cockatiels<br>Budgerigars | Cerebellar signs | Ascarid migration |
| | Neophemas<br>Primarily Amazons, African Greys, cockatoos | Respiratory signs | *Herpesvirus*<br>*Chlamydia* |
| Paralysis (unilateral or bilateral) | All | Trauma | Cervical vertebral fractures or luxations<br>Skull fracture(s) resulting in hematomas<br>Multiple fractures (both wings and legs)<br>Pelvic fracture<br>Dislocations; sprains |
| | | Illness | Neuritis (peripheral nerve)<br>Neuromuscular disorder<br>Encephalitis; encephalomyelitis<br>Cloacal lithiasis<br>Kidney disease<br>Reovirus, papovavirus, Pacheco's virus<br>*Chlamydia*<br>Listeriosis, *Yersinia, Salmonella*<br>Aspergillosis |
| | | Malnutrition | Multiple fractures secondary to metabolic bone disease<br>Vitamin $B_1$ deficiency |
| | Macaws | | "Macaw wasting syndrome" |
| | Cockatiels | | Suspected vitamin E/selenium deficiency |
| | | Egg-laying female | "Obturator" paralysis from difficult delivery<br>Egg binding<br>Broken leg from calcium deficiency<br>Ectopic egg |

*Table continued on following page*

DIFFERENTIAL DIAGNOSES BASED ON CLINICAL SIGNS RELATING TO CENTRAL NERVOUS AND MUSCULOSKELETAL SYSTEMS *Continued*

| Clinical Signs | Species Affected | History | Possible Causes |
|---|---|---|---|
| | Primarily budgerigars | Insidious onset | Renal adenocarcinoma<br>Fibrosarcoma<br>Other tumors, cysts, or space-occupying lesions<br>Hepatic enlargement |
| | | Trauma | Fractures of synsacrum |
| Paralysis (flaccid) | Cockatiels | | *Chylamydia*<br>Selenium/vitamin E syndrome<br>Spinal cord trauma<br>Encephalopathies<br>Botulism and other toxins<br>Septicemia with spinal infection |
| Paralysis of toes | All | | Dislocations; fractures; CNS infection<br>Crushing injuries<br>Emboli of septic response |
| Paralysis of eyelids | All | Trauma<br>Illness | Cranial nerve neuritis<br>Encephalitis<br>Post-pox lesions |
| | All, esp. cockatiels | Chronic epiphora | *Mycoplasma*<br>Selenium/vitamin E syndrome<br>Bacteria<br>Foreign bodies<br>Ingrown feathers of lids |
| Paralysis of beak/mouth | | | Aflatoxins<br>Fracture or other injury<br>Temporomandibular dislocations |
| Paralysis of wing | | | Nerve and vascular damage<br>Ankylosis due to trauma<br>Malunion of fracture<br>Chronic dislocations |
| Paralysis of tail | All | Trauma | Vertebral fractures or luxations (spinal trauma)<br>Malnutrition<br>Improper incubation |
| | All, esp. budgerigars | | Neoplasm of spinal cord |

DIFFERENTIAL DIAGNOSES BASED ON CLINICAL SIGNS RELATING TO CENTRAL NERVOUS AND MUSCULOSKELETAL SYSTEMS *Continued*

| Clinical Signs | Species Affected | History | Possible Causes |
|---|---|---|---|
| Paralysis of cloaca | All | Trauma | Vertebral fractures or luxations<br>Fractures of synsacrum<br>Improper or careless suturing of vent (cloacal prolapse) |
| | | Chronic illness | Cloacal lithiasis<br>Myelitis |
| "Spacey" appearance | All | | Chronic lead poisoning<br>Hepatic encephalopathy<br>Hypoglycemia |
| Torticollis (see Head tilt; General incoordination) | All | | Lead intoxication |
| | | "New" bird | Encephalitis secondary to VVND, other PMV |
| | Amazons | | Acute chlamydiosis |
| Weakness in grip | All | | Localized infirmities<br>Pansystemic disease<br>Malnutrition<br>Selenium/vitamin E syndrome<br>Trauma<br>Nerve damage<br>Fractures<br>Gout<br>Hypocalcemia<br>Arthritis<br>Loose perches<br>Tight leg band<br>Neoplasia |
| Weight loss (see Anorexia) | All | Decreased appetite | Pansystemic diseases<br>Environmental stressors<br>Obstructions |
| | | Normal appetite | Gastrointestinal obstruction<br>Metabolic dysfunction from systemic disease<br>Parasites e.g., *Giardia* |
| | | Heavy egg laying | Nutritional deficiencies |
| | Macaws<br>Cockatoos<br>Other | | "Macaw wasting syndrome" (proventricular dilatation) |

DIFFERENTIAL DIAGNOSES BASED ON CLINICAL SIGNS RELATING TO DIGESTIVE AND UROGENITAL SYSTEMS

| Clinical Signs | Species Affected | History | Possible Causes |
|---|---|---|---|
| Anorexia (see also Failure to crack and consume seeds) | All | | "Sick" bird (autointoxication; pansystemic; multiple etiologies)<br>Liver and kidney disease<br>Oral abscesses<br>Broken jaw; other mouth trauma<br>Paralysis of head muscles<br>Inaccessibility of food<br>Psychologic<br>Blindness<br>Foreign body in mouth |
| | | Egg laying | Egg binding<br>Peritonitis<br>Septicemia<br>Normal brooding |
| Blood in stool (see also Hemorrhage) | All | Black, tarry feces or frank hemorrhage | Bleeding in gastrointestinal system<br>Lead and other toxins<br>Coagulopathies<br>Hepatic disease<br>Vitamin K deficiency<br>Viral diseases (Pacheco's, reovirus) |
| | Conures | | Cloacal papillomas, tumors, cysts<br>Foreign body<br>Ulcer<br>Parasites<br>Impending egg laying<br>Aflatoxicosis<br>Conure bleeding syndrome |
| Blood in urates | | | Lead poisoning<br>Other toxins<br>Nephritis |
| Crop stasis (see also neonates) | All | Force feeding of hand-fed bird | Laceration/rupture/scalding of crop<br>Foreign body<br>Impaction<br>Infectious diseases<br>Paralytic ileus |
| | Macaws, other species | | Proventricular dilataton |
| | Budgerigars | | Thyroid disease |
| Diarrhea (see also Tenesmus) | All | | Dietary change<br>Pansystemic illness<br>Viral diseases (reovirus, adenovirus, paramyxovirus, esp. VVND, *Herpesvirus*)<br>Gastrointestinal obstruction<br>Abdominal hernia<br>Hepatitis<br>Endocrine/metabolic<br>Chronic malnutrition<br>Mycobacteria<br>Pancreatitis<br>Toxins<br>Fecolith, grit impaction |

DIFFERENTIAL DIAGNOSES BASED ON CLINICAL SIGNS RELATING TO DIGESTIVE AND UROGENITAL SYSTEMS *Continued*

| Clinical Signs | Species Affected | History | Possible Causes |
|---|---|---|---|
| | | "New" bird | Overtreatment with antibiotics<br>Mycotic overgrowth<br>Yeast |
| | | With/without respiratory symptoms | *Chlamydia*<br>*Salmonella*, *Pasteurella*, or other bacteria<br>Viral diseases<br>Toxins |
| | | Breeding bird | Impending egg laying<br>Egg binding/peritonitis |
| | All, esp. budgerigars cockatiels | "New" bird | *Giardia*, *Hexamita*<br>Other parasites |
| Concurrent polydipsia/ polyuria/ polyurates diarrhea (see also Polyuria, Polyurates, Diarrhea) | All | | Bacterial diseases, esp. *Salmonella*, *E. coli*<br>Other infectious diseases<br>Reaction to medications<br>Psychogenic<br>Fatty liver and other liver diseases<br>Intoxications<br>Nutritional<br>Diabetes mellitus<br>Starvation<br>Visceral gout<br>Endocrine disorders |
| Failure to crack and consume seeds (see also Swollen tongue; Anorexia) | All | Hungry | Fractured jaw or beak<br>Damage to muscles of beak<br>Candidiasis<br>Trichomoniasis<br>Palatine abscesses<br>Vitamin A deficiency<br>Electrical or other burns |
| | | Change in environment | Psychologic (fear) |
| | Primarily cockatiels | | Selenium/vitamin E responsive from suspected malabsorption due to *Giardia* |
| | Primarily cockatoos | May or may not have affected feathers | Psittacine beak and feather disease syndrome with mouth lesions |

*Table continued on following page*

DIFFERENTIAL DIAGNOSES BASED ON CLINICAL SIGNS RELATING TO DIGESTIVE AND UROGENITAL SYSTEMS *Continued*

| Clinical Signs | Species Affected | History | Possible Causes |
|---|---|---|---|
| Flatulence (see Tenesmus) | All | | Post cloaca-pexy<br>Cloacal papillomas, prolapse<br>Enteritis<br>Stretched cloaca from egg-laying<br>Food allergies<br>"Normal" in some macaws<br>Abnormal bacteria in gut<br>Tapeworms, other parasites |
| Increased swallowing motions and neck stretching | All | | Regurgitation<br>Crop irritation<br>Postnasal drip<br>Foreign bodies<br>Passing seeds from crop to stomach |
| | Budgerigars | | Goiter |
| Passing whole seeds (see Undigested food in droppings) | | | |
| Polydipsia | All | | Liver disease<br>Septicemia<br>Enteritis<br>Kidney disease or damage<br>Toxins (lead, salt)<br>Neoplasia<br>Dehydration<br>Psychogenic<br>Food allergy<br>Diabetes mellitus<br>Pseudodiabetes<br>Increased thirst-producing foods<br>Hyperthermia<br>Calcium imbalance<br>Water deprivation<br>Aminoglycoside therapy<br>Metritis<br>Steroid administration<br>Gout |
| | | Egg-laying female | Egg-related peritonitis |
| | Primarily budgerigars cockatiels | | "Transient" diabetes mellitus |
| | Primarily cockatoos | | Tapeworm infestation<br>Psychogenic water drinking<br>Dietary deficiency |
| | Mynah birds | | Liver disease |
| Polyphagia | All | | Malabsorption<br>Dietary deficiencies<br>Chronic disease<br>Parasites<br>Grit impaction<br>Psychogenic<br>Any disease |

DIFFERENTIAL DIAGNOSES BASED ON CLINICAL SIGNS RELATING TO DIGESTIVE AND UROGENITAL SYSTEMS *Continued*

| Clinical Signs | Species Affected | History | Possible Causes |
|---|---|---|---|
| | | Breeding activity | Normal appetite increase prior to egg laying |
| | | Separation from owner | Starvation<br>Hypoglycemia |
| | All, esp. cockatiels budgerigars | Polydipsia<br>Weight loss<br>Male budgerigar | Diabetes mellitus<br>Pancreatic disorder<br>Proventricular ulceration<br>Renal adenocarcinomas |
| Polyphagia (apparent) | All | | Inappropriate food |
| Polyurates (seldom seen alone) | All | Reduced amount of feces | Excess dietary protein<br>Grit impaction; obstruction<br>Infectious diseases<br>Other causes of anorexia<br>Water deprivation<br>Toxins<br>Normal for raptors; sea birds |
| | Amazons | Change in color of urates<br>Reddish urates | Liver damage<br>Vitamin overdose<br>Lead poisoning<br>Other toxins |
| Polyuria (see Polydipsia) | All | | Viruses<br>Stress<br>Normal for increased vegetables, fruits in diet<br>Food allergies<br>Renal disease [infectious, toxic, neoplastic, metabolic (phosphorus flush from Ca:P imbalance) ]<br>Diabetes mellitus<br>Toxins<br>Neoplasia |
| | | Concurrent drug use | Overdose (esp. aminoglycosides) |
| | | Post I.V. fluids, esp. bolus | Normal |
| | | Scant feces | Infectious diseases affecting the liver<br>*Chlamydia*<br>Trichomoniasis |
| | | Color change: Yellow<br>Pea-green<br>Biliverdinuria | Liver disease<br>Virus [Pacheco's, adenovirus, acute psittacine beak and feather disease syndrome (PBFDS)]<br>*Chlamydia*<br>Vitamins, other pigments<br>*Salmonella*, other bacteria |

*Table continued on following page*

DIFFERENTIAL DIAGNOSES BASED ON CLINICAL SIGNS RELATING TO DIGESTIVE AND UROGENITAL SYSTEMS *Continued*

| Clinical Signs | Species Affected | History | Possible Causes |
|---|---|---|---|
| Regurgitation | All | | Liver disease<br>Foreign bodies<br>Toxicities<br>Reaction to medications<br>Proventricular lesions<br>Neoplasms<br>Enlarged thyroid glands<br>Crop infections<br>Impaction/obstruction (seed hulls, parasites, grit)<br>Food allergy<br>Motion sickness<br>Nervousness<br>Pancreatitis<br>Candidiasis<br>Ascites |
| | | Young birds | Normal weaning behavior |
| | | Breeding | Normal courtship behavior |
| | | Egg-laying | Metritis; peritonitis |
| | | | Egg binding |
| | Primarily macaws | | Proventricular dilatation |
| | Primarily cockatoos | Polydipsia | Tapeworm infestation |
| | | Pica | Kidney disease |
| Regurgitation to mirrors, toys, owners | All, esp. budgerigars Amazons macaws | Single pet | Aberrant sexual behavior |
| Tenesmus (see Diarrhea and Flatulence) | All | | Diarrhea<br>Constipation<br>Infectious diseases<br>Toxins |
| | | Soiled vent feathers<br>No droppings in cage | Cloacal obstruction by pasted feathers |

DIFFERENTIAL DIAGNOSES BASED ON CLINICAL SIGNS RELATING TO DIGESTIVE AND UROGENITAL SYSTEMS *Continued*

| Clinical Signs | Species Affected | History | Possible Causes |
|---|---|---|---|
| | | Egg-laying female | Egg-binding |
| | | Long-term antibiotic therapy | Lack of normal flora |
| | | Treatment for cloacal prolapse<br>Pasty vent<br>Other digestive symptoms | Cloacal papilloma, hypertrophy, or tumor<br>Cloacapexy<br>Fecalith or urolith |
| | | Sudden onset<br>Recent worming | Intestinal obstruction |
| | All, esp. cockatoos | | Tapeworms<br>Other parasites<br>Partial cloacal prolapse |
| Undigested food in droppings (passing whole seeds) | All | | Inflammation, infection, or dysfunction of the digestive tract<br>Liver disease<br>Parasites, e.g., *Ascarids*, *Giardia*<br>Ingestion of oil<br>Pancreatic disease<br>Food allergy<br>Dehydration<br>Grit impaction<br>Overtreatment with antibiotics |
| | Macaws | | Proventricular dilatation |
| | Small species | | Lack of grit |
| Voluminous droppings (see Polyphagia) | All | Egg-laying female | Normal reproductive dilation of cloaca, nerve damage to cloaca |
| | Conures<br>Others | Normal behavior<br>"House-trained" birds | "Morning" droppings held overnight in cloaca |
| | All, esp. budgerigars, cockatiels | | Giardiasis |

DIFFERENTIAL DIAGNOSES BASED ON CLINICAL SIGNS OF HEMORRHAGE

| Clinical Signs | Species Affected | History | Possible Causes |
|---|---|---|---|
| Bleeding from cloaca (see Blood in Stool under Digestive) | All | Audible tenesmus<br>Frank blood in droppings | Cloacitis<br>Aflatoxicosis<br>Infectious enteritis<br>Toxins<br>Hepatitis |
| | | Egg-laying female | Normal<br>Prolapsed cloaca<br>Prolapsed uterus<br>Self-mutilation of prolapse<br>Internal bleeding disorder |
| Bleeding from feather | All | Recent wing trim or molt | Fractured pin (blood) feather |
| Bleeding from nose/mouth | All | | Calcium deficiency<br>Vitamin K deficiency<br>Liver dysfunction<br>Malnutrition |
| | | Trauma | Injured on cage parts<br>Inflicted by cage mate<br>Bleeding of caseous masses from vitamin A deficiency<br>Bacterial infections<br>Parasitic infections |
| | | Upper respiratory infection | Rupture of blood vessels from chronic irritation |
| | Conures | Sudden onset<br>Recovered from previous episode | "Conure bleeding syndrome" |
| | | Sick bird | Neoplasia<br>Reovirus |
| Bleeding from nails/beak | All | Sudden onset | Trauma<br>Incorrect trimming |
| | | Amputated digits | Trauma from cage<br>Inflicted by another bird |
| | Finches, canaries, macaws | | Avascular necrosis<br>Circular swelling of unknown etiology |

DIFFERENTIAL DIAGNOSES BASED ON CLINICAL SIGNS RELATING TO THE INTEGUMENT

| Clinical Signs | Species Affected | History | Possible Causes |
|---|---|---|---|
| **BEAK** | | | |
| Avulsion, fracture | All | | Trauma |
| Brittle, soft, porous | Primarily budgerigars | With white encrustations | PBFDS<br>*Knemidokoptes* mites |
| Malformed | | Overgrown | Liver disease<br>PBFDS<br>Malnutrition |
| | | Malocclusion, other deformities | Congenital<br>Improper incubation<br>Injury<br>Neoplasia<br>*Knemidokoptes* mites |
| Necrosis | Cockatoos<br>Other species | | PBFDS<br>Yeast and bacterial infections of beak |
| **FEATHERS** | | | |
| Bald areas on top of head | Cockatiels<br>Cockatoos | | Normal finding (genetic) |
| | Cockatoos | Prolonged and malformed pin feathers | PBFDS |
| Blood retained in shafts | All | Normal molt | Normal for emerging feathers |
| | Cockatoos<br>Lovebirds<br>Others | Gradual decline in feather quality and quantity | PBFDS |
| Broken/frayed or missing flight/tail feathers | All | | Improper caging<br>Self-mutilation<br>Acrobatic vice for attention<br>Attack by cat/dog/child/other bird |
| | | "Stress lines" | Malnutrition, other stress |
| | | "New" bird | Shipping damage<br>Delayed molt from change in hemisphere<br>Infrequent molt<br>Reduced preening<br>Steroid injections |

*Table continued on following page*

DIFFERENTIAL DIAGNOSES BASED ON CLINICAL SIGNS RELATING TO THE INTEGUMENT *Continued*

| Clinical Signs | Species Affected | History | Possible Causes |
|---|---|---|---|
| Color variation | All | Green to black | Worn feathers<br>Hepatic disease |
| | | Green to white or yellow | Malnutrition<br>Deliberate bleaching and/or color application for smuggling and/or higher prices |
| | | Breeding birds | Mutations |
| | Macaws<br>Mynahs | Red to yellow<br>Black to white | Long-term administration of L-thyroxine |
| | Cockatoos | Gradual browning or blackening | PBFDS<br>Excessive handling<br>Malnutrition<br>Newsprint |
| | Amazons<br>Macaws | Bronzing | Malnutrition<br>Reduced preening<br>Infection or disease |
| Feathers flat against body | All | Panting<br>Wing held away from body | Hyperthermia |
| Fluffed feathers | All | Cool environmental temperature<br>"Illness" | Hypothermia |
| | | | Normal sleep period<br>In molt |
| Disturbances in feather formation | Cockatoos<br>Lovebirds<br>Others | Gradual decline in feather quality and quantity | PBFDS |
| | All | | Survival from acute chlamydiosis |
| | | Excessive pin feathers | Liver disease<br>Malnutrition |
| | All, esp. budgerigars | | Papovavirus (BFD)<br>Adenovirus<br>*Herpesvirus*<br>Budgerigarpox virus |
| Lack of irridescence | All | Water unavailable for bath | Insufficient moisture<br>Chronic disease |
| | | Poor husbandry | Malnutrition<br>Parasites |

DIFFERENTIAL DIAGNOSES BASED ON CLINICAL SIGNS RELATING TO THE INTEGUMENT *Continued*

| Clinical Signs | Species Affected | History | Possible Causes |
|---|---|---|---|
| Feather loss with subsequent replacement | All | | Normal molting |
| Feather loss without subsequent replacement | All | Trauma in past<br>Over cere | Feather follicle damage or inactivity<br>Sinus infections |
| | All, esp. Amazons | Thyroid-responsive | Suspected hypothyroidism |
| | Canaries<br>Finches | Normal skin<br>Balding | Idiopathic baldness |
| | Cockatoos<br>Lovebirds<br>Other species | Gradual decline in feather quality and quantity | PBFDS |
| | All | Ocular problems | Chronic epiphora<br>Topical ointments<br>Sinus infections (*Mycoplasma*)<br>Mycobacteria<br>Other systemic diseases |
| | Primarily budgerigars<br>Reported in other species | With white encrustations | *Knemidokoptes* mites |
| Feather picking/pruritus (solitary bird) | All (cockatiels, conures, lovebirds, macaws most often affected) | Chest pattern<br>"New bird"<br>Stress-induced<br>Poor husbandry | Exaggerated normal preening<br>Change in environment<br>Systemic disease<br>Malnutrition<br>Boredom<br>Reproductive frustration<br>Ectoparasites<br>Quill mites<br>Infectious folliculitis<br>Endoparasites |
| | | Sudden onset | Giardiasis<br>Multifocal necrotizing dermatitis<br>Bacterial/mycotic dermatitis |
| | | "Old" bird | Hypothyroidism<br>Attention-getting device<br>Allergy<br>Endocrine imbalances<br>Internal disease<br>Psychologic disturbances |
| Feather picking (with cage mates) (see also Feather picking—solitary bird) | All | Nesting bird<br>Dominated bird | Ready for next clutch<br>Crowding<br>Diseased<br>Male pushing female to breed<br>Subordinate male<br>Incompatibility |
| | | New bird introduced | Territorial response<br>Self trauma to injury site<br>Malnutrition |

*Table continued on following page*

DIFFERENTIAL DIAGNOSES BASED ON CLINICAL SIGNS RELATING TO THE INTEGUMENT *Continued*

| Clinical Signs | Species Affected | History | Possible Causes |
|---|---|---|---|
| Feather picking with self-mutilation (see also Feather picking—solitary bird) | All | Nonresponsive to dietary and behavior modification | Pansystemic disease<br>Hormonal |
| | | Intermittent response to isolation, collar, tranquilizers, or housing with own species plus negative for systemic disease | Psychological |
| Retained feather sheath | All | | Malnutrition |
| | | Failure to preen | Spinal deformity/injury<br>Prolonged application of collar<br>Illness/injury<br>Beak fracture |
| | Cockatoos<br>Others | Beak deterioration | PBFDS |
| | | Baby birds | Lack of experience |
| Soiled feathers on top of head (see Regurgitation) | | | |
| **SKIN** | | | |
| Erythema of facial patches | All | | Insect bites<br>Pox lesions<br>Bruise from trauma<br>Bleeding disorders |
| | Macaws<br>Palm Cockatoos | Excitement<br>Nervousness | Normal flush |
| | Macaws<br>African Greys | Physical restraint | Digital pressure, bruise |
| | | Severe erythema and skin congestion following parenteral aminoglycosides | Overdose of aminoglycosides |

DIFFERENTIAL DIAGNOSES BASED ON CLINICAL SIGNS RELATING TO THE INTEGUMENT *Continued*

| Clinical Signs | Species Affected | History | Possible Causes |
|---|---|---|---|
| Excoriations (see Swollen feet) | All, esp. Amazons | | Self-mutilation<br>Burns or abrasions<br>Possible viral or bacterial etiology |
| Unusual odor to integument | All | | Multifocal moist necrotizing dermatitis |
| Excessively dry scales of feet | All | Poor husbandry<br>Presence of "corns" | Malnutrition<br>Vitamin A deficiency |
| Wartlike growths on legs | Primarily cockatoos, macaws | | *Herpesvirus* |
| Malformed/missing nails | All | | Trauma |
| | | Poor husbandry<br>"Old" bird; overgrown nails | Malnutrition<br>Hepatic disease |
| | | Yellow-white nodules at joints | Articular gout<br>Septic arthritis |
| | Primarily budgerigars, canaries | White proliferative encrustations | *Knemidokoptes* mites |
| | Grey-cheeks | | Sarcoptiform mange |
| Lack of uropygial gland | Amazons | | Normal anatomic finding |
| Enlargement of uropygial gland | Other species | | Infection; abscessation<br>Plugged gland |
| | Primarily budgerigars | | Neoplasia |

DIFFERENTIAL DIAGNOSES BASED ON CLINICAL SIGNS OF HAND-RAISED NEONATES*

| Clinical Signs | Species Affected | History | Possible Causes |
|---|---|---|---|
| Crop stasis | All | | Improper storage of hand-feeding formula<br>Ingestion of nursery substrate, other foreign material<br>Inadequate consistency of formula (total solids)<br>Environmental temperature too low<br>Atony of crop due to overstretching<br>Scalded crop from too hot formula<br>Infection: bacterial, candidal, viral, parasitic |
| | | "Appearance" of crop stasis | Subcutaneous food from crop puncture or burn |
| | | Pendulous crop | Ruptured air sac<br>Atony from overfeeding or feeding too fast<br>Hernia |
| | | Crop filled with air | Congenital<br>Swallowing air<br>Throat or esophagus irritation<br>Gas-producing organisms |
| Dry, reddish look to skin | All | | Dehydration, low humidity<br>Autointoxication (septicemia)<br>Humidity too low |
| White plaques in mouth | | | Vitamin A deficiency<br>Pox<br>Bacterial abscess<br>Parasites (rare) |
| | | Injured jaw or tongue | Food accumulation |
| | All, esp. cockatiels | | Candidiasis |
| Holding one or both eyes closed | All | | Foreign body<br>Disease problem |
| | | Nursery mates | Scratched by another bird |
| Swollen tongue (see also Swellings) | All | Presence of dead sheets of tissue in mouth | Trauma from feeding tube<br>Scalded from too hot formula |

DIFFERENTIAL DIAGNOSES BASED ON CLINICAL SIGNS OF HAND-RAISED NEONATES* *Continued*

| Clinical Signs | Species Affected | History | Possible Causes |
|---|---|---|---|
| Sneezing, coughing, wheezing, poor doer | All | | Inhalation pneumonia<br>Respiratory foreign body<br>Infections (Chlamydia) |
| | | Malnutrition | Systemic diseases<br>Protozoa, other parasites |
| Bilateral paralysis (see also Central Nervous and Musculoskeletal Systems) | All | | Congenital<br>Malnutrition<br>Malabsorption<br>Parasitic<br>Incubation injury<br>*Chlamydia* |
| Regurgitation | All | | Foreign body<br>Infection of crop or proventriculus<br>Liver disease |
| | | Weaning bird | May be normal |
| Polydipsia | | Weaning age | Unwilling to wean |
| Weight loss in growing bird | All | | Insufficient quantity of food being fed |
| | | Change in environment | Stress<br>Excessive handling<br>First sign of infectious disease<br>Parasites<br>Mycotoxicosis |
| Won't eat on own | All | Process of weaning | Psychologically attached to owner<br>Weaning too young or too old<br>Appropriate weaning foods not available or not easily accessible |
| Sudden death following hand feeding | All | | Asphyxiation from formula inhalation |
| | | Weaning birds | Asphyxiation from struggling during feeding or incorrect restraint |
| Lacerations; fractured beaks | All | Mixed ages and species in nursery | Trauma from "feeding" frenzy of cage mates |

*The most common conditions specific for neonates are included. Diseases listed for adult birds should also be considered in the differential diagnosis.

DIFFERENTIAL DIAGNOSES OF DISEASES OF PARENT-RAISED NEONATES IN NEST

| Clinical Signs | Species Affected | History | Possible Causes |
|---|---|---|---|
| Warm, active baby but no food in crop first 6-8 hours | Large psittacines | Liquid in crop | Normal |
| Weak hatchlings | All | | Malnutrition of parents<br>Diseases in parents<br>Older babies defecating on eggs<br>Too hot or cold environment with multiple nest incubation disturbances |
| Crying, weak, cold, no food in crop | All | Possible first-time parents | Abandonment by parents<br>Rejection of an individual<br>Unavailability of soft foods for feeding<br>Aviary disturbances |
| Crop filled with air or water | All | | Parents aren't feeding enough |
| Failure to defecate | All | | Vent pasted shut<br>Dirty nest box<br>Too many babies together<br>Infectious diseases |
| Food not passing out of crop | All | | Environmental temperature too low<br>Disease conditions (e.g., candidiasis, papovavirus)<br>Toxin ingestion |
| | | Inexperienced parents | Foreign bodies (sunflower hulls) fed if not enough soft foods<br>Overfeeding by foster parents |
| Broken leg | All | | Aggressive or nervous parents<br>Stepped on by large clutch mate |

DIFFERENTIAL DIAGNOSES OF DISEASES OF PARENT-RAISED NEONATES IN NEST *Continued*

| Clinical Signs | Species Affected | History | Possible Causes |
|---|---|---|---|
| | | Banded bird | Catch band on object |
| Folding fractures | All | Poor parenteral diet | Metabolic bone disease<br>Slippery nest surface |
| Lumpy, crooked bones | All | | Callus formation due to undetected fractures<br>Malnutrition<br>Malabsorption |
| Missing babies or eggs | All | | Predation by rats or snakes<br>Destroyed by parents |
| Mutilated babies; amputated legs, wings; pulled feathers | All | New parents or foster parents | Nervous, frustrated, or poor parents<br>Aviary disturbances<br>Rats |
| | Esp. budgerigars, cockatiels | | Parents may be ready to start breeding again |
| Unthrifty, failure to grow correctly | All | | Parasites<br>Malnutrition of parents<br>Environment too hot<br>Poor husbandry<br>Chronic low-grade infectious diseases<br>Aflatoxins<br>Papovavirus |

DIFFERENTIAL DIAGNOSES OF NEONATAL DEATHS*

| History | Possible Causes |
|---|---|
| Babies hatched as poor doers; die shortly after birth | Genetic weakness (inbreeding)<br>Chronic viral, bacterial, yeast, chlamydial or parasitic disease in parents<br>Poor husbandry and aviary management<br>Improper incubation conditions<br>Incubator or nest box contamination<br>Humidity too high<br>Malnutrition of parents<br>Inclement weather |
| Babies hatched healthy; die a few days later | Environmental<br>Food or water contamination<br>Oral infection or pathogens passed by parents<br>Neglected by parents<br>Contamination of hand-raising facilities<br>Trauma from parents<br>Inappropriate total solids in formula<br>Competition for food with older nestlings<br>Malnutrition of parents |
| Babies hatched healthy; die 2-3 weeks later; no signs of trauma | Papovavirus infection<br>Other viral, bacterial, mycoplasmal, chlamydial, fungal, yeast, protozoa<br>Intestinal obstruction by parasites<br>Malnutrition of parents<br>Inappropriate hand-feeding formula |
| Die-offs at a later age | Poor husbandry and sanitation<br>Inadequate nutrition of parents or hand-raising formula<br>Flock diseases |
| Dead after first flight | Trauma, concussion from flying into cage |

*See also Necropsy of adult birds.

DIFFERENTIAL DIAGNOSES BASED ON OPHTHALMOLOGIC CLINICAL SIGNS

| Clinical Signs | Species Affected | History | Possible Causes |
|---|---|---|---|
| Blepharitis | All, primarily lovebirds, Amazons | Pox | Crusty lesions of lid margins and choana<br>Infections |
| | Lovebirds | "New " bird | Stress |
| Blindness | All | | Trauma<br>Neoplasia<br>Toxins (lead, etc.)<br>Encephalopathies<br>Cataracts<br>Infection |
| Conjunctivitis | All | | Extension of any intraocular problem<br>Infection (chlamydial, bacterial, viral sinusitis extensions) |
| | Primarily cockatiels, budgerigars | | *Mycoplasma* |
| | All, primarily Amazons | Nonresponsive to antibiotics | Allergy |
| Corneal opacity | All | | Ulcer<br>Uveitis with corneal edema |
| | | Previous ulceration | Subepithelial crystals |
| | Amazons | | Punctate keratitis |
| | Mynah birds | Shipping injury<br>Stress | Superficial keratitis |
| Corneal lacerations | All | | Trauma |
| Epiphora | All | "New" bird | Pox; other viruses<br>*Chlamydia; Mycoplasma, bacteria* |
| | Primarily cockatoos | | Flukes, nematodes |
| Material in anterior chamber | All | | Hyphema from trauma<br>Bleeding disorder<br>Cells and protein from uveitis |
| Nonresponsive pupil | All | | See Blindness (above) |
| | | Uveitis | Synechiae |
| Periophthalmic swelling (see Swelling) | | | |

DIFFERENTIAL DIAGNOSES BASED ON CLINICAL SIGNS RELATING TO RESPIRATORY SYSTEM

| Clinical Signs | Species | History | Possible Causes |
|---|---|---|---|
| Coughing (acute) | All | | Foreign body |
| | | | Trauma |
| | | "New" bird | Acute pox |
| | | | Tracheitis |
| | | | Acute or chronic upper respiratory infection with postnasal drip |
| | Amazons | "New" bird | Amazon-tracheitis virus |
| | Finches, canaries | | Air sac mites |
| | Conures | | "Conure bleeding syndrome" |
| | All, esp. cockatiels, macaws, African Greys | | Abscess or tumor in lungs near syrinx, elsewhere |
| | Budgerigars, mynah birds | | Neoplasia |
| Coughing (chronic) | All | Chronic infection | Bacterial, viral, fungal, chlamydial, parasitic, yeast, mycobacterial |
| | | | Ascites |
| | | | Presence of walled-off foreign material |
| | | All seed diet | Vitamin A deficiency |
| | | "Talking" species | Mimicry of human |
| | Primarily finches, canaries | | Air sac mites |
| Dyspnea (see Swollen abdomen, Coughing) | All | | Infectious diseases |
| | | | Bacterial (esp. *Staph.*, *Strep.*, *E. coli*, *Pasteurella*, *Pseudomonas*) |
| | | | Viral (paramyxovirus, esp. VVND; reovirus) |
| | | | Mycotic (aspergillosis) |
| | | | Foreign body inhalation |
| | | | Liver disease |
| | | | Kidney disease |
| | | | Internal bleeding from coagulopathies, ruptured lung vessel, other injury |
| | | | "Conure bleeding syndrome" |
| | | | Allergy |
| | | | Toxins (inhaled—polytetrafluororethylene; lung edema) |
| | | | Plugged nostrils |

DIFFERENTIAL DIAGNOSES BASED ON CLINICAL SIGNS RELATING TO RESPIRATORY SYSTEM *Continued*

| Clinical Signs | Species | History | Possible Causes |
|---|---|---|---|
| Dyspnea *Continued* | | | Ascites<br>Heart disease<br>Neoplasia<br>Gout |
| | | "New" bird | Upper respiratory infection with pneumonia<br>Air sacculitis<br>Malnutrition |
| | Amazons | | Amazon-tracheitis virus *(Herpesvirus)* |
| | | "New" bird<br>Exposure to conures<br>Sudden death in flock | Pacheco's disease *(Herpesvirus)* |
| | Primarily canaries | | Systemic pox |
| | Primarily budgerigars | | Thyroid disease<br>Budgerigarpox virus |
| | Old World species<br>Neonatal macaws | Sudden death | Sarcosporidiosis (lung edema) |
| Rhinorrhea (runny nose) | All | | Upper respiratory infection (viral, bacterial, esp. *Haemophilus*, chlamydial, *Mycoplasma*, or combination)<br>Systemic disease<br>Allergy<br>Backflush of water after crop flush<br>Foreign body<br>Congenital nasal membrane |
| Sneezing | All | | Respiratory disease<br>Foreign body<br>Toxic disease<br>Allergy |
| "Sunken" eyes | Macaws | Slow onset | Chronic sinusitis ("sunken eye syndrome") |

DIFFERENTIAL DIAGNOSES BASED ON CLINICAL SIGNS RELATING TO SWELLING

| Clinical Signs | Species Affected | History | Possible Causes |
|---|---|---|---|
| Abdominal swelling | All | "New" bird | Hepatomegaly from bacterial, yeast, viral etiologies<br>Fatty infiltration secondary to other etiologies |
| | | "Old" bird | Fatty liver from inactivity and malnutrition |
| | | Poor diet (free choice seeds)<br>Polyuria<br>Loose droppings<br>Panting | Obesity<br>Tumors or cysts<br>Abdominal hernia |
| | | Weight gain | Progesterone reaction |
| | | "New" bird<br>Respiratory and/or gastrointestinal symptoms | Chlamydiosis<br>Reovirus |
| | | "New" bird<br>Chronic diarrhea | Hepatomegaly—<br>granulomatous disease (TB) |
| | | Female in nest<br>Stops laying<br>Stays in nest | Impending egg laying<br>Egg binding<br>Egg peritonitis<br>Metritis<br>Ectopic eggs |
| | All, esp. mynah birds | "Old" bird<br>Emaciation<br>Polyphagia<br>Dyspnea | "Iron storage disease"<br>Hepatomegaly<br>Congestive heart failure<br>Ascites<br>Neoplasia |
| | | Lethargic | Peritonitis from puncture of gastrointestinal tract<br>Constipation<br>Obstruction of gastrointestinal tract |
| | | Newly purchased seed | Aflatoxin hepatitis<br>Bacterial hepatitis |
| Swollen breast muscles | All | Free choice seeds | Obesity |
| | Primarily budgerigars<br>Cockatiels<br>Amazons<br>Cockatoos | | Lipoma (color: yellowish white)<br>Thyroid responsive conditions |
| | | Necrotic skin ulcers | Necrotic-ulcerative lipoma with cholesterol clefts |
| | All | Intramuscular injections<br>Trauma | Hematoma (color: black/brown to green)<br>Spontaneous bleeding-DIC |
| Periophthalmic swelling | All | | Trauma (hematoma)<br>Foreign body<br>Allergy |

DIFFERENTIAL DIAGNOSES BASED ON CLINICAL SIGNS RELATING TO SWELLING *Continued*

| Clinical Signs | Species Affected | History | Possible Causes |
|---|---|---|---|
| | Conures | "New" bird<br>Previous or current respiratory infection | *Mycoplasma*<br>Chlamydiosis<br>Viral rhinitis/sinusitis with secondary bacterial/fungal infection and chronic malnutrition<br>Nasolacrimal abscess<br>Mycobacteria<br>Bacterial infection |
| | | "Old" bird | Bacterial/fungal infection secondary to chronic malnutrition<br>Glaucoma<br>Neoplasia |
| | All, esp. cockatiels budgerigars<br>Rosellas<br>Lovebirds<br>Passerines<br>Amazons<br>Conures | "New" bird<br>Corneal ulcers | Pox (advanced chronic)<br>Other viruses<br>Trauma<br><br>"Conure bleeding syndrome" |
| Swollen cere (see also Rhinorrhea, Sneezing) | All | Sneezing | Chronic rhinitis<br>Foreign body: seed hulls or feathers in nares<br>Trauma<br>Neoplasia<br>Allergy |
| | All, esp. cockatiels | Rubbing cere on cage or with nails; dried epithelium | Chronic malnutrition<br>Giardiasis |
| | Budgerigar males | Cere color changed to dark brown; dried hypertrophied epithelium | Estrogenic tumors |
| | Budgerigar females | Brown color | Normal female hormone production |
| Swollen cere plus enlarged nares with/without caseous plug | All, primarily Amazons and African Greys | Chronic plugged nares and upper respiratory tract | Bacterial/mycotic/mycoplasmal/nutritional rhinitis and unknown factors |
| Swollen cere, face, legs, vent, pygostyle | Primarily budgerigars | Slowly developing crusty lumps | *Knemidokoptes* mites |
| Swollen tongue | All | With swollen oral tissues<br>Weight loss<br>Scant feces<br>Poor diet<br>General lethargy | Hypovitaminosis A<br>Primary or secondary bacterial infection<br>Yeast infection<br>Trichomoniasis<br>Pox<br>Combination of above |

*Table continued on following page*

DIFFERENTIAL DIAGNOSES BASED ON CLINICAL SIGNS RELATING TO SWELLING *Continued*

| Clinical Signs | Species Affected | History | Possible Causes |
|---|---|---|---|
| | | Trouble eating food<br>Appetite depressed<br>Food dribbling out of mouth<br>Pick up food and drop it | Trauma<br>Poisonous plant ingestion<br>Foreign bodies (splinter of wood, seed hull)<br>Bite on tongue from another bird<br>Burn from chewing on electrical cord |
| Subcutaneous swelling | All | Crop area | Herniated crop<br>Fistula<br>Foreign bodies |
| | | Soft swelling on body; previous laparoscopy | "Ruptured" air sac<br>Injury<br>Chronic or previous infection<br>Unknown causes |
| | | Trauma | Bite wound<br>Abscess |
| | | Subcutaneous fluid therapy | Overdose or inappropriate administration |
| | All, primarily Amazons and neonatal cockatiels | Slow or sudden development; around head and neck | Overinflation of cervicocephalic air sac |
| Swellings on skin | All | | Lipomas<br>Neoplasms<br>Abscesses<br>Hematomas<br>Calcium deposits<br>Pox<br>Idiopathic cysts, uric acid deposits in gout cases |
| | All, primarily macaws | Primarily on wings | Feather cysts |
| | Canaries, budgerigars | Hybrids; inbreeding | Feather cysts |
| | All, esp. cockatiels budgerigars cockatoos | Thickened yellow skin | Xanthomatosis |
| | Primarily columbiformes | Pox survivor | Tumor from pox virus |
| Swollen feet | All | Gradual onset | Parasitic cysts<br>Vascular damage |
| | | Sudden onset | Trauma; sprain, strain<br>Fracture |
| | All, esp. Yellow-naped Amazons | Acute necrotizing dermatitis<br>Pruritus<br>Self-mutilation | "Amazon leg necrosis" of unknown etiology |

DIFFERENTIAL DIAGNOSES BASED ON CLINICAL SIGNS RELATING TO SWELLING *Continued*

| Clinical Signs | Species Affected | History | Possible Causes |
|---|---|---|---|
| | All, esp. budgerigars | Slowly developing crusty lumps | *Knemidokoptes* mites |
| | All, esp. raptors | Poor diet esp. vitamin A deficiency<br>Poor perches<br>Trauma to feet | Pododermatitis (bumblefoot) |
| | Macaws<br>Cockatoos | Single or multiple dry raised flaky patches | *Herpesvirus* |
| | All, esp. canaries | Necrotic areas on toes | Avian pox, injury |
| Swollen digits | All, esp. canaries | Outdoor aviary<br>Poor diet | Callus strangulation from scaly leg mite<br>Staphylococcal infection; secondary malnutrition |
| | All, esp. canaries finches | Nesting birds | Constriction from string or hair in nest<br>Some of unknown etiology<br>Ergot or fungus |
| | | Trauma | Bite wound inflicted by another bird |
| Swollen joints | All | Lameness or wing droop | Fracture (with callus)<br>Pododermatitis<br>Degenerative osteoarthritis<br>Dislocation; sprains<br>Hematomas<br>Generalized infections |
| | | Acute deaths in flock | Salmonella |
| | | Chronic enteric problems | Other bacteria; mycobacteria |
| | All, esp. raptors pigeons | Poor diet<br>Disfiguration | "Rickets"<br>Nutritional secondary hyperparathyroidism |
| | All, esp. Amazons | "New" bird<br>Little pain | Filarial worms |
| | All, esp. budgerigars cockatiels | "Old" bird<br>Yellow-white nodules at joints<br>Pain | Gout |
| | | "New" bird<br>Chronic flock problem<br>Accompanying respiratory/ocular symptoms | *Mycoplasma* |

*Table continued on following page*

DIFFERENTIAL DIAGNOSES BASED ON CLINICAL SIGNS RELATING TO SWELLING *Continued*

| Clinical Signs | Species Affected | History | Possible Causes |
|---|---|---|---|
| Swollen legs(s) | All | Around or distal to leg band | Collapsed band<br>Callus forming under band causing constriction<br>Reovirus |
| | All, esp. budgerigars finches | Disuse of limb | Tumor |
| Swollen tissue protruding from cloaca (see Tenesmus) | All | Straining<br>Passing Blood | Prolapsed cloaca<br>Prolapsed rectum<br>Enteritis<br>Cloacal papilloma or tumor<br>Reproductive disease |
| | | Nest behavior | Prolapsed uterus<br>Egg binding |
| Swollen vent | All | Passing voluminous feces<br>Nesting activity<br>Passing blood | Normal impending egg laying |
| | | Straining or comatose in cage or nest | Egg binding |
| | | Slow onset | Tumor, cyst<br>Cloacal lithiasis |

DIFFERENTIAL DIAGNOSES TO CONSIDER AT NECROPSY

| History | Possible Causes of Death |
|---|---|
| Weight loss<br>Sick bird with lesions present | Chronic disease: bacterial, mycobacterial, viral, fungal, chlamydial, heavy parasitism<br>Blockages<br>Malnutrition<br>Starvation<br>Organ damage from previous disease (liver cirrhosis, kidney failure)<br>Combinations of the above |
| Good weight<br>No obvious gross lesions | Acute viral (esp. Pacheco's, VVND), chlamydial, fungal, protozoal, bacterial disease<br>Acute injury (fractured skull, concussion)<br>Blockages<br>Toxicities<br>Cardiovascular accident<br>Aflatoxicosis |
| Good weight<br>Blood loss or bruising | Injury<br>Clotting deficiency secondary to liver dysfunction, malnutrition<br>Viral disease (reovirus, papovavirus)<br>Egg binding<br>Ruptured aneurysm<br>"Conure bleeding syndrome"<br>Pesticide (warfarin) ingestion |
| Weight loss<br>Blood loss or bruising | Chronic infection involving liver, clotting problems<br>Heavy metal poisoning, bleeding from kidneys, vascular hemolysis<br>Conure bleeding syndrome |
| Good weight<br>Respiratory symptoms | Acute inhalation of foreign body<br>Acute pneumonia<br>Ruptured blood vessel in lung<br>Sarcosporidiosis<br>Toxins (smoke, burned Teflon)<br>Acute anemia<br>Conure bleeding syndrome |
| Weight loss<br>Respiratory symptoms | Chronic chlamydiosis, bacterial, viral, fungal infections<br>Heart disease<br>Liver disease with ascites<br>Any disease resulting in internal bleeding<br>Chronic anemia<br>Blood, gastrointestinal or respiratory parasites<br>Conure bleeding syndrome |

*Table continued on following page*

DIFFERENTIAL DIAGNOSES TO CONSIDER AT NECROPSY *Continued*

| History | Possible Causes of Death |
|---|---|
| CNS symptoms<br>Convulsions (see also Differential Diagnoses: Convulsions) | Head injury<br>Acute septicemia<br>Encephalopathy (viral, bacterial, protozoal, mycotic)<br>Lead or other heavy metal poisoning<br>Other toxins (parasiticides, nitrofurazone, organophosphates, autointoxication)<br>Neoplasms<br>Hypocalcemia |
| Good weight<br>Egg laying | Acute egg binding<br>Egg-related peritonitis<br>Acute shock from egg-related injury<br>Acute metritis<br>Ruptured uterus, peritonitis, shock |
| Weight loss<br>Egg laying | Chronic egg binding<br>Egg-related peritonitis<br>Chronic injury<br>Internal infection secondary to dietary deficiency<br>Metritis |
| Digestive symptoms<br>Diarrhea (see also Differential Diagnoses: Diarrhea) | Acute bacterial, yeast, viral, or protozoal infection<br>Pancreatitis<br>Septicemia<br>Toxicity<br>Neoplasms<br>Parasitism<br>Combinations of the above |
| Digestive symptoms<br>Vomiting (see also Differential Diagnoses: Regurgitation) | Blockages<br>Acute bacterial, viral, yeast, or fungal infections<br>Toxicity, parasitism, neoplasia, combinations |
| Polyuria/polydipsia (see also Differential Diagnoses: Polyuria/polydipsia | Kidney disease, toxicities (e.g., heavy metal poisoning, aminoglycoside reaction, steroid overdose, salt poisoning)<br>Diabetes mellitus<br>Calcium imbalance<br>Septicemia<br>Hyperthermia<br>Neoplastic diseases |

## REFERENCES


1. American Animal Hospital Association: Medical Records Manual. South Bend, IN, AAHA, 1978.
2. Davis, R. B., et al.: Recognition of abdominal enlargement in the budgerigar. VM/SAC, Feb., 1981.
3. Harrison G. J.: Guidelines for treatment of neonatal psittacines. Proceedings of the Annual Meeting of the American Association of Zoo Veterinarians, Tampa, FL, 1983.
4. Kollias, G. V.: Polydipsia, polyuria and diarrhea in birds: A diagnostic approach. Association of Avian Veterinarians Newsletter 6: 23–25, from material presented at the American Animal Hospital Association Annual Scientific Seminar, Orlando, FL, 1985.
5. McMillan, M. C., and Petrak, M. L.: The clinical significance of abdominal enlargement in the budgerigar (*Melopsittacus undulatus*). Comp. Cont. Ed., 3:898–904, 1981.

*Section Four*

# DIAGNOSTIC PROCEDURES

# Chapter 9

# EVALUATION OF DROPPINGS

RHONDA K. SAYLE

Evaluation of the patient's droppings is one of the most important aspects of a complete avian examination. Significant information may be obtained regarding the appetite, gastrointestinal function, and renal and hepatic status in the preliminary formulation of a differential diagnosis. The value of gross and microscopic examination of the droppings, as well as subsequent chemical diagnostic procedures, depends upon samples that are fresh, free of contaminants, and preferably in their undisturbed configuration.

Gross examination includes visual evaluation of each of the components: the fecal portion from the intestinal tract and the liquid urine and pasty urates excreted via the kidney. Color, texture, consistency, and volume of each component, plus the frequency of passage and presence of any odor should be noted.

Factors that may alter the gross appearance of droppings include species, age, time of day, diet content, quantity of food and water consumed, stage of reproductive cycle of a female, disease, parasitic infections, and medications. A thorough history of the patient is beneficial to the practitioner in determining which of these contributing factors may be involved.

In response to nervousness, patients presented for office examination may pass a loose, gas bubble–filled dropping or a sample composed primarily of urates with scant fecal material present. Therefore, a freshly collected and properly stored sample may be brought to the office by the client for a more accurate assessment (see Chapter 6, Management Procedures).

## GROSS EVALUATION OF FECAL COMPONENT

### Normal

The color of normal feces is influenced by the species of bird and the diet consumed. The practitioner is advised to become familiar with species' variables. The recent ingestion of foods such as concord grapes, blueberries, beets, or carrots may be obvious in a color change of the fecal material. Exclusive consumption of pellet or biscuit rations produces soft brown droppings.

Early descriptions of "normal" stool for budgerigars (dry black fecal component with a white urate "bull's eye") were established from birds maintained exclusively on dry seeds. Supplementing the diet with foods of high water content such as fruit and vegetable matter produces normal droppings of a softer consistency, which is not to be mistaken for diarrhea. Neonatal psittacines that are hand-fed a semiliquid formula will have soft, semiformed voluminous stools.

Skadhauge[9] groups droppings into three patterns: rod-shaped, snail-shaped, and fluid. The two pet bird species he studied, Zebra finch and galah (Rose-breasted cockatoo), pass rod-shaped droppings. Fluid droppings are a mixture of liquid urine and feces, but such species often pass separate droppings of "pure" urine separate from the fecal material.

The quantity of feces passed is usually directly correlated with the amount of food consumed. Some birds will not defecate during the night when sleeping and produce a large volume of stool in the morning. This phenomenon also occurs with nesting birds. Increased food consumption prior to egg laying results in increased stool formation. Impending or recent passage of an egg may also cause a temporary distention of the cloaca, allowing a large volume of feces and urates to accumulate before voiding occurs.

Normal psittacine feces may have a faint musty odor. A diet consisting of cat food or other fish products may be reflected in the odor of the stool.

### Abnormal

Any abnormal color of the stool may indicate a problem. Frank hemorrhage in the feces has been clinically correlated with coagulopathies, liver pathology, malnutrition, and enteric disease. Blood in the droppings may also originate from the oviduct, kidneys, testicles, or cloacal growths (e.g., papillomas) or be associated with impending egg laying, poisoning, or the presence of foreign bodies. A dark, tarlike stool may also suggest the presence of blood. On occasion, mahogany-colored to light pink feces have been clinically associated with lead poisoning.

Birds presenting with bright "pea-green" loose feces (true diarrhea) often have severe pansystemic disease. Viral, chlamydial, acute bacterial, or toxic etiologies should be considered in the differential diagnosis.

Less common causes of diarrhea are parasitic infections (e.g., giardiasis), ingestion of petroleum-based products, peritonitis, cloacal papillomas or tumors, diabetes, fungal or yeast infection, and other systemic diseases. Loose droppings may paste around the vent and obstruct the cloaca, interfering with subsequent defecation. Abdominal enlargements that alter the angle of the cloaca may also result in "pasted vents."

Cockatiels and other species may pass a dry white "popcorn"-appearing stool. Although this condition has been reported as possibly related to pancreatic or kidney disease, giardiasis, or candidiasis, the etiology is unknown. Necropsy findings from a budgerigar with intermittent droppings characteristic of "popcorn" stool included acute multifocal moderate hepatic necrosis with no other pathology.

A lumpy consistency to the feces suggests poor digestion; therefore, the possibility of parasites, pancreatitis, proventriculitis, ventriculitis, or intestinal disease should be explored. Microscopic examination of fecal smears may further confirm digestive disturbances if a high percentage of undigested food material is present.

Scant cloacal output may be the result of starvation. On occasion, the owner may assume that the bird is eating, but, in fact, the seed dish may contain only hulls, or an aggressive cage mate may be preventing access to the food. A bird may also be regurgitating the majority of food consumed. Scant stools resulting from decreased food consumption may often be noted before any weight loss in sick birds.

If a scant amount of excreta is associated with normal or high food intake, blockage or dilatation of the digestive tract may be suspected. Ventricular obstruction is especially common in budgerigars that have consumed large quantities of grit. The passing of whole, undigested seeds is often associated with this latter condition.

A strong odor to the stool, of undetermined cause, may be clinically related to disease conditions and has been frequently associated with chronic cloacal papilloma-like conditions, especially in Amazons. Characteristically, the droppings adhere to the cloacal epithelium; attempts to remove the material often result in bleeding of the cloaca.

## GROSS EVALUATION OF THE URATES AND URINE

### Normal

Normal excreta from the kidney in most psittacines appear as white or cream-colored semisolids (urates) that have been incorporated with the feces in the cloaca. Normal urine may also be excreted as a clear liquid without urates under conditions of nervousness, polydipsia, or increased water content of the diet, as a normal component in nectar-feeding species such as lories, and with hand-feeding of baby psittacines. Following iatrogenic manipulation subsequent to egg binding, hens may pass a uterine liquid that may be mistaken for urine.

### Abnormal Urine

Polyuria is an abnormal increase in production of the liquid urine. The fecal component may remain stable, be absent, or concurrently change in consistency. Prolonged polyuria, which may or may not be associated with polydipsia, may lead to dehydration.

A thorough history may be one of the major diagnostic aids in determining which of the following factors are contributing to polyuria:

1. **Therapeutic Agents.** Many drugs that are excreted through the kidneys have the potential to cause polyuria. Gentamicin is a well-documented example (see Chapter 25, Pharmacology of Antibiotics). In a majority of cases, polyuria is not the result of renal damage but of increased uric acid (or water) transport. Diuretics such as furosemide (Lasix) must be used with caution in birds, as a profound polyuria may result, especially in lories.

2. **Endocrine-associated Conditions.** Polyuria is associated with diabetes mellitus in birds as it is in mammals. Excessive stress may increase corticosterone levels, which may man-

ifest as polyuria and transient hyperglycemia. The polyuria associated with egg laying is suspected to relate to hypercalcemia. Conversely, insufficient dietary calcium may produce polyuria necessitated by a need for increased urine flow to remove excess phosphate.[4]

**3. Generalized Disease.** As most diseases of birds are pansystemic and may affect the ability of the kidney to reabsorb water and form urates, polyuria is associated with almost all diseases of psittacine birds. Anorexic birds release large amounts of biologic water associated with metabolization of body tissues for energy.

**4. Other.** Polyuria may also result from acute toxic nephritis related to salt poisoning, e.g., overconsumption of salty snack foods. Chronic interstitial nephritis has been diagnosed in several polyuric birds by means of biopsy. Psychogenic polydipsia and subsequent polyuria are most often noted in recently weaned birds.

### Abnormal Urates

Polyurates are increased quantities of the white urate portion of the droppings. Although large volumes of urates are normal in carnivorous birds, such as hawks and pelicans, which consume large amounts of tissue protein, polyurates as a general rule in psittacines are produced only by a negative energy balance resulting from inadequate food consumption. Usually this is accompanied by a decrease in fecal content. This may result in the phenomenon of "going light," a lay expression describing loss of breast muscle mass.

The presence of biliverdin, the major bile pigment of birds, and other elements such as hemoglobin, may alter the color of the urine and/or urates to shades of yellow, green, or brown and may suggest systemic disease, particularly of the liver. Harrison believes the presence of urate pigments is one of the most dependable indicators of liver problems.[5] Overuse of supplemental vitamins, especially carotenes, has also been incriminated with colored urates.

## MICROSCOPIC AND BIOCHEMICAL DIAGNOSTIC PROCEDURES

### Parasite Examination

Although fecal flotation techniques can be utilized in the same manner as for mammals, some avian nematode ova fail to float consistently; therefore, direct fecal smears have been found to be the most effective technique for the detection of parasite ova, protozoa, fungi, bacteria, and inflammatory cells. Trophozoites or cysts of *Giardia* can be found intermittently in fresh fecal direct smears; however, some practitioners report difficulty in finding positives. The California Avian Laboratory recommends that clients collect very fresh fecal samples and place them in polyvinyl alcohol for preservation and submission. The Trichrome stain identifies flagellates. The reader is referred to Chapter 37, Parasites, for information on the most common gastrointestinal parasites encountered in captive birds.

A fecal smear may be dried and evaluated with further differential stains (see Chapter 10, Clinical Microbiology, and Chapter 17, Cytology, for specific diagnostic techniques).

### Gram's Stain

A Gram's stain of the stool sample provides a general overview of the composition of the intestinal flora and can be used routinely for each examination. Although no definitive diagnosis can be made from the results of a Gram's stain, this tool serves a purpose similar to use of a thermometer in mammalian medicine—it can assist in determining the direction of a differential diagnosis. The practitioner is advised to establish his own guidelines for normal reference compositions, and use Gram check slides for quality control (see Table 6–1).

In Harrison's experience in a Florida practice, normal psittacine feces contain approximately 100 to 200 organisms per oil immersion field.[5] Approximately 60 to 80 per cent of the organisms are gram-positive rods (medium to short) and 20 to 40 per cent are gram-positive cocci. One *Candida*-like yeast and one or two gram-negative rods per oil immersion field are accepted as normal. Incomplete gram-positive rods (e.g., *Bacillus* sp.) are also normal.

Abnormal psittacine fecal smears may include an absence of cocci; decreased or very high numbers of gram-positive rods; many filamentous gram-positive bacteria; a large number of gram-negative rods, yeasts, or white blood cells; or a marked amount of undigested food. Any of the above would indicate the necessity for a culture of the fecal material or other diagnostic measures (see Chapter 10, Clinical Microbiology, and Chapter 27, What to Do Until a Diagnosis Is Made).

Low numbers (<30 per oil immersion field) or total absence of bacteria on a Gram's stain of the stool is also abnormal and may be seen in cases associated with overtreatment with antibiotics, marginal levels of nutrition, and some infectious diseases. These results may also occur if, inadvertently, the main component on the smear is the urate portion of the dropping.

## Examination for Blood

The Hemoccult test is a rapid, simple procedure for detection of blood in the stool, although some false-positive results have been associated with its use. The presence of blood in the stool may be confirmed by microscopic examination for evidence of numerous red blood cells.

## Urinalysis

There is some support for evaluation of the urine as a separate component of the cloacal output. Skadhauge[9] divides the rod-shaped droppings into an oral end and an anal end. Most uric acid is found at the anal end of the dropping. The oral end is not covered by uric acid.

If plastic wrap or another clean, smooth, nonporous surface has been used to collect the droppings, the urine (if present) and urates may be aspirated with a needle and syringe. A commercially produced urinalysis "dipstick" (Multistix, Miles Laboratories, Inc.) has been used by a number of avian practitioners. Based on dipstick methods, urinalysis results for psittacines are:

Glucose—negative to trace
pH—6.0-8.0
Blood—negative to trace
Ketones—negative
Bilirubin—negative (Biliverdin does not react with bilirubin on the dipstick.)
Protein—negative to trace

The normal specific gravity of avian liquid urine is 1.005 to 1.020.

Scullion[8] questions the clinical interpretation of these results. He believes that it is impossible to test the fluid portion of the excrement without contamination by the feces and/or urates and that the presence of uric acid interferes with dipstick reactions. Schultz[7] agrees that fecal contamination of the urine precludes microbiologic assessment of urine as a separate entity.

Skadhauge[9] states that in the budgerigar anal urine is largely unaffected by the fate of the fraction of the urine that moves orally. After carefully collecting, separating, and testing the anal part of natural droppings of graniverous birds or the liquid milky white part of the droppings of mostly carnivorous birds for osmolality and solute concentrations, he concluded that they are equivalent to the ureteral urine and that the error introduced by taking these parts of the natural droppings would presumably be small.

## REFERENCES


1. Avian Medicine. American Animal Hospital Association, undated booklet.
2. Campbell, T. W.: Laboratory diagnosis in avian medicine. Proceedings of the Caged Bird Seminar and Wet Lab, Kansas State University, 1983.
3. Clubb, S. L.: Loose droppings. Proceedings of the American Federation of Aviculture Veterinary Seminar, Washington, DC, 1982.
4. Fowler, M. E.: Metabolic bone disease. In Fowler, M. E. (ed.): Zoo and Wild Animal Medicine. Philadelphia, W. B. Saunders Co., 1978.
5. Harrison, G. J.: Pet bird practice. Proceedings of the Texas Veterinary Medical Association Annual Convention, Corpus Christi, Texas, 1983.
6. Petrak, M. L. (ed.): Diseases of Cage and Aviary Birds. 2nd ed. Philadelphia, Lea and Febiger, 1982.
7. Schultz, D. J.: Proceedings No. 55, Refresher Course on Aviary and Caged Birds, The University of Sydney, Australia, 1981.
8. Scullion, F.: Personal communication.
9. Skadhauge, E.: Osmoregulation in Birds. New York, Springer-Verlag, 1981.
10. Woerpel, R. W., and Rosskopf, W. J., Jr.: Clinical experience with avian laboratory diagnostics. Vet. Clin. North Am. [Small Anim. Med.], 14(2):279–281, 1984.

# Chapter 10

# CLINICAL MICROBIOLOGY

LYNNE A. DREWES
KEVEN FLAMMER

Microscopic examination, isolation, identification, and susceptibility testing of microorganisms have generally been performed by a microbiology reference laboratory. The availability of commercially prepared media and identification kits has made it more feasible and economical for practitioners and properly trained laboratory personnel to perform their own work. This enhances the quality control and the speed with which a diagnosis can be made.

It is important to realize, however, that there are limitations to what can be done in the office laboratory and that good judgment must be used in determining when samples should be sent to a reference laboratory. Certain fastidious bacteria require special conditions, equipment, and selective media or reagents with a short shelf life. There are also many disease agents that are highly infectious and require special training and equipment for safe handling. A summary of factors to consider in submission of microbiology samples to an outside laboratory is given in Chapter 6, Management Procedures, and Chapter 23, Necropsy Procedures.

The methodology described in this chapter is specifically designed for practical, economic, in-house identification of the majority of common avian pathogens.

Regardless of the laboratory used, it is the practitioner who must interpret test results to determine which of the isolated microorganisms is responsible for the disease process. Bacterial and mycotic organisms are frequently identified and must be treated if they are producing clinical signs, but they may in fact be secondary to the true cause of the disease. The reader is referred to other chapters (Chapter 32, Viral Diseases; Chapter 33, Bacterial Diseases; Chapter 36, Mycotic Diseases) for complete discussion of the individual microorganisms, the clinical manifestations, and the interrelationship of several factors in producing a clinical disease.

## DIRECT MICROSCOPIC EXAMINATION OF CLINICAL MATERIAL

Microbial cultures are essential for quantifying, isolating, and speciating bacteria, but one cannot underestimate the value of direct microscopic examination of clinical material. Stained and unstained (wet mount) slide preparations provide immediate results with a minimum of equipment, time, materials, and expense. A direct examination may also be the only diagnostic method available when an organism cannot be cultivated. Large numbers of organisms must be present, however, to be seen on direct smears; a negative result does not necessarily mean that they are absent. The reader is referred to Chapter 17, Cytology, for information on staining procedures and interpretation.

Saline wet mounts are useful for detecting bacteria, protozoa, and yeast in fecal material, gut contents, and exudates. Dilute the specimen with a drop of saline, add a coverslip, and examine under low and high dry power. The addition to the microscope of a special condenser and objective for phase contrast microscopy provides a distinct advantage in distinguishing unstained microorganisms from background material. A 10 per cent solution of potassium hydroxide (KOH) is particularly useful as a wet mount preparation for identifying yeast and fungal elements in clinical samples, especially skin scrapings. Warming the KOH mount over a flame will lyse bacteria and keratinized cells, making yeast and fungal elements more visible.

The Gram's stain[a] forms the basis for classifying bacteria and is useful for identifying organisms grown on media as well as in clinical samples. Gram's stained slides are especially important if one does not set up anaerobic cultures or special growth media. Lack of growth on a primary aerobic culture plate with the presence of bacteria on the smear may indicate that a specimen should be sent to a reference laboratory for speciation of a possible anaerobe or fastidious aerobe.

Gram's stains of fresh fecal material are useful for evaluating the bacterial status of the gut (see Chapter 9, Evaluation of Droppings). Caution must be exercised in interpreting fecal Gram's stains, however, as microorganisms observed on the slide may not necessarily correlate with those recovered from cloacal cultures. Gram's stained cloacal and crop specimens that contain yeast or large numbers of gram-negative organisms may suggest an abnormal state, and cultures should be taken to determine the species and antibiotic sensitivities of suspected pathogens.

## COLLECTION AND HANDLING OF SPECIMENS FOR CULTURE

There are relatively few sites that can be cultured in the live bird. In most cases isolates from the cloaca reflect the microflora of the distal bowel, but in the presence of renal infection, a pathogen from the kidney might be shed in the urine. In addition, hens with infections of the oviduct might also shed bacteria into the cloaca. Cloacal cultures must therefore be evaluated in conjunction with clinical signs. Isolates from the crop are best interpreted with caution as potential pathogens may be transient. A pathogen isolated from the crop should not be considered the cause of disease unless it is present in large numbers or cultured in repeated samples.

The microflora of the upper respiratory system is best reflected in cultures from the sinuses or deep inside the choana. Sinus cultures are collected by surgically lancing the skin at the commissure of the beak and inserting a swab into the infraorbital sinus. Deep choanal cultures are obtained by forcing the beak open and inserting a small swab into the most rostral section of the choana (Fig. 10–1). Care must be taken to avoid contaminating the sample by touching the caudal choana or tongue. Cultures taken from the superficial, oral portion of the choana should be interpreted in the same manner as crop specimens, since many transient organisms may be found there. Cultures from the nares poorly reflect the flora present in the upper respiratory system, as a number of pathogenic organisms, not involved in sinus disease, may be repeatedly isolated from this site. Cultures from the glottis will sometimes reveal pathogens present in both the trachea and lungs.

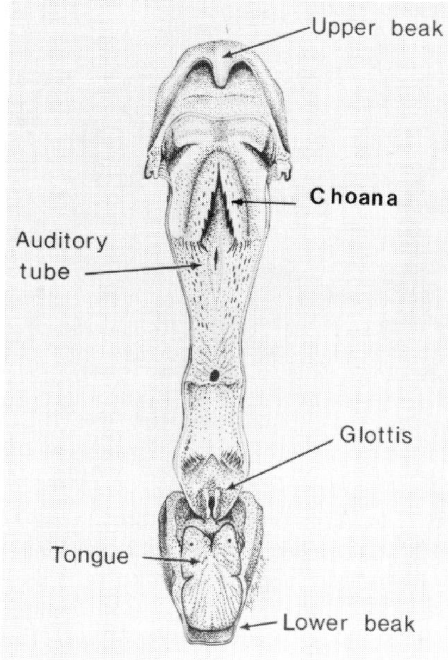


**Figure 10–1.** The choana is a site of culturing for microflora of the upper respiratory system. (Drawing by Beverly Benner from Avery, P. O.: Preliminary bacteriology in psittaforme birds with some procedures in the identification of the bacteria present. Proceedings of the Annual Meeting of the Association of Avian Veterinarians, Atlanta, GA, 1982.)

The lower respiratory system is best evaluated by washing the thoracic or abdominal air sacs with sterile saline and culturing the aspirated fluid (see Chapter 17, Cytology).

Proper collection and handling of a specimen are essential for valid results. Most samples can be collected on sterile, saline-moistened cotton or Dacron swabs. Swabs cannot retain much moisture: therefore, it is essential that they be inoculated onto an agar medium within one hour to prevent drying and subsequent death of organisms present. Samples from small sites, such as the rostral choana, can be collected with calcium alginate swabs,[b] originally designed for human urethral specimens. These small swabs dry quickly and must be plated onto the medium within 30 minutes. Organ and tissue specimens can be inoculated directly into broth and streaked on agar plates or stored temporarily in sterile jars or Petri dishes. Refrigeration of a specimen will help preserve most bacteria and extend the recommended time between collection and plate inoculation by approximately one to two hours.

If a specimen must be sent to a reference laboratory for special studies, or will require a delay of more than one hour before media inoculation, transport jars or media should be used (see Chapter 6, Management Procedures).

## INOCULATION OF MEDIA

A direct culture from the sample is necessary to quantify the amount of growth and relative percentages of organisms present in the sample. Enrichment broth favors the growth of all microorganisms, so that bacteria present in low numbers, which may be missed on the direct plates, can be isolated. A guideline to set-up procedures and the bacteria most likely to be encountered in routinely cultured sites is included in Table 10–1.

Direct agar plates for culturing commonly isolated bacteria should include trypticase soy agar with 5 per cent sheep blood (BAP), eosin methylene blue (EMB), and MacConkey (MAC). Blood agar plates will support the growth of most microorganisms, whereas EMB and MAC agars selectively isolate gram-negative bacteria, which include the majority of avian pathogens. Although EMB plates will grow and isolate all gram-negative bacteria, MAC agar is used because of its ability to distinguish lactose-negative bacteria such as *Salmonella*. MAC agar will not, however, support the growth of certain gram-negative rods, most notably some of the *Pasteurella* species.

Although *Salmonella* will grow on BAP and MAC plates, enrichment media may be necessary to identify carriers. If a carrier state is suspected, inoculate a selenite broth tube and incubate it for 24 hours at 37° C. This broth should then be used to inoculate a brilliant green agar plate, which is examined for pink colonies. Although both of these media specifically favor the growth of *Salmonella*, other organisms such as *Pseudomonas* will also grow, making positive identification procedures necessary.

Plates for fungal isolation should also be inoculated directly from the specimen. Sabouraud-dextrose agar with chloramphenicol is recommended for most samples (see Chapter 36, Mycotic Diseases).

Streak all plates for isolation with a consistent method so that the amount of growth (scant, moderate, or heavy) can be quantified and a relative percentage given to each species present (e.g., 60 per cent *Lactobacillus*, 30 per cent *Micrococcus*, 10 per cent *Streptococcus*) to evaluate its significance.

Thioglycolate broth is suggested as an enrichment broth because of its ability to encourage anaerobic growth. After overnight incubation, a loopful of broth should be subcultured onto MAC (and/or BAP) plates and incubated overnight. Examine the subcultured MAC plate especially for lactose-negative (clear) *Salmonella* colonies. If turbidity is present in the broth and no growth appears on the subplate, anaerobes should be suspected and a Gram's stain of the broth examined for the presence of bacteria.

**MEDIA INOCULATED FOR ISOLATION OF COMMON PATHOGENS**

1. Blood agar (BAP)
2. MacConkey agar (MAC)
3. Eosin methylene blue agar (EMB)
4. Sabouraud-dextrose agar with chloramphenicol (Sab-Dex + Ch)
5. Thioglycolate broth

Fastidious organisms that may be potential pathogens in birds and that require special conditions or nutrients for growth include *Campylobacter*, *Yersinia*, *Mycoplasma*, *Chlamydia*, *Mycobacterium*, and anaerobes. Any clinical specimens suspected of having these bacteria should be sent to a qualified reference laboratory for isolation and identification.

## IDENTIFICATION OF BACTERIA

Identification of bacteria begins with the Gram's stain characteristic (gram-positive or gram-negative) and morphology (rods or cocci) and is completed through the use of biochemical tests. Colony characteristics can also be very useful for initial grouping and presumptive identification before further tests are executed. Key features of the most routinely isolated bacteria on BAP, EMB, and MAC agars are listed in Table 10–2. Macroscopic and microscopic descriptions for most avian pathogens and the normal flora likely to be encountered are provided in the following sections. A comprehensive medical microbiology text[7] should be consulted for more in-depth studies. The more common tests used for identification of avian bacteria are listed at the end of this chapter, but other references[8] should be consulted before these procedures are put into use.

### Aerobic Gram-positive Cocci

There are two major groups of gram-positive cocci: the Micrococcaceae (*Staphylococcus* and *Micrococcus*) and the Streptococcaceae. Of the 13 species of *Staphylococcus*, *Staph. aureus* has the most virulent strains and must be distinguished from the other species and *Micrococcus*. Most species of *Streptococcus* are considered normal flora when isolated from the alimentary tract (see Chapter 33, Bacterial Diseases). Tests and descriptions needed to spe-

*Text continued on page 163*

**Table 10–1.** NORMAL AND PATHOGENIC FLORA EXPECTED FROM ROUTINELY CULTURED SITES AND MEDIA RECOMMENDED FOR THEIR ISOLATION

| Specimen Site | "Normal" Flora | Contaminants (Probably Not Pathogens) | Probable Pathogens | Recommended Media for Isolation |
|---|---|---|---|---|
| Cloaca and crop | *Lactobacillus* sp.<br>*Streptococcus* sp.<br>*Micrococcus/Staph. epidermidis*<br>*Corynebacterium* sp.<br>*Bacillus* sp.<br>*Escherichia coli* in non–seed- or fruit-eating birds may be normal | *Bacillus* sp., fungi<br>*Actinomyces* or *Streptomyces* species<br>Gram-negative rods in low incidence, such as *Enterobacter cloacae* or *Ent. agglomerans*, which may be environmental transients | All gram-negative rods, most commonly *E. coli*, *Enterobacter*, *Klebsiella*, *Citrobacter*, *Proteus*, *Pasteurella*, and *Pseudomonas*<br>*Staph. aureus*, some *Streptococcus* | BAP—blood agar<br>EMB—eosin methylene blue<br>MAC—MacConkey agar<br>Thioglycolate broth |
| | | | Yeast | Sabourand-dextrose agar with chloramphenicol (Sab-Dex + Ch) |
| | | | *Salmonella* (also gram-negative rods) | Selenite broth and brilliant green agar |
| | | | *Listeria*, *Erysipelothrix*, *Mycoplasma*, *Chlamydia*, *Campylobacter*, *Yersinia*, *Vibrio* | Recommend that specimen be sent to a qualified reference laboratory |
| Choana and sinus | *Lactobacillus* sp.<br>*Streptococcus* sp.<br>*Micrococcus/Staph. epidermidis*<br>*Corynebacterium* sp. | Same as above | All gram-negative rods,<br>*Staph. aureus*, some *Streptococcus* sp., yeast, and fungi | BAP, EMB, MAC<br>Thioglycolate broth<br>Sab-Dex + Ch |
| | | | *Haemophilus* | Chocolate agar |
| | | | *Listeria*, *Erysipelothrix*, *Mycoplasma*, *Chlamydia* | Recommend that specimens be sent to reference laboratory |
| Tissue, organs, and body fluids | No growth | Contaminants from the gut such as *Micrococcus/Staph. epidermidis*, *Streptococcus*, or *Lactobacillus* sp.; any organism in an autolyzed specimen | Any organism in significant numbers grown aerobically or anaerobically | BAP, EMB, MAC<br>Chocolate agar<br>Thioglycolate broth<br>Sab-Dex agar with or without chloramphenicol<br>Mycobiotic agar bottles<br>Also do Gram stains, acid fast stains, and Gimenez stains on specimens<br>Possible send-outs to a reference laboratory |

**Table 10–2.** COLONY CHARACTERISTICS OF COMMON BACTERIA ON THREE RECOMMENDED ISOLATION AGARS (24-Hour Incubation)

| Organism | Blood Agar | Eosin Methylene Blue Agar | MacConkey Agar |
|---|---|---|---|
| **Gram-positive Bacteria** | | | |
| *Lactobacillus* | tiny, clear grey, alpha hemolysis | no growth | no growth |
| *Streptococcus* sp. | clear, shiny whitish, alpha, gamma, or beta-hemolysis | no growth | no growth |
| *Micrococcus/Staphylococcus* | white, opaque, large | may grow as white to pink small colonies | |
| *Staph. aureus* | yellowish-white, dull, opaque, usually with beta hemolysis | may grow as white to pink small colonies | |
| *Corynebacterium* | tiny, dry, crinkly, often req. 48 hr. | no growth | no growth |
| *Bacillus* | large, smooth to rough, often crinkly or nebular, dull, grey to yellow | no growth or small pink-yellow crinkly colonies | |
| **Yeast** | white, orange, pink, creamy, dull | may grow pinkish to white but smaller than on blood agar | |
| **Gram-negative Bacteria** | | | |
| Most Enterobacteriaceae | "typical" gram-negative rod; grey-tan with a mucoid to shiny opaqueness, round, convex, smooth; larger and duller than *Strep.* sp.; smaller and shinier than *Staph.* sp. | most with smooth-edged, opaque, pink to purple colonies | most with rounded, opaque, pink colonies |
| *Escherichia coli* | as above, often with beta-hemolysis | dark purple to black centers, often with metallic sheen | brick red |
| *Salmonella* | "typical" gram-negative | clear pink | transparent |
| *Enterobacter agglomerans* | yellowish pigment | opaque pink | pink-yellow pigment |
| *Klebsiella* | extremely mucoid and snotty | mucoid pink | mucoid pink |
| *Proteus* | grey, round with a large clear apron (swarming) | flat clear or with black center, little swarming | |
| *Pseudomonas aeruginosa* | flat irregular edged with greenish sheen and fruity odor, usually with beta-hemolysis | flat clear colonies with irregular edges | |

**Table 10–3.** CHARACTERISTICS AND TESTS TO IDENTIFY AEROBIC GRAM-POSITIVE COCCI*

| Organism | Colony on Blood Agar (24-hr. Incubation) | Gram's Stain Characteristic | Hemolysis† | Catalase | Coagulase | BEA‡ | NaCl‡ | Sorbitol | Growth on MacCorkey Agar |
|---|---|---|---|---|---|---|---|---|---|
| *Micrococcus/ Staphylococcus* | 1-3 mm, opaque, white, butyrous creamy to dry | single, pairs or grape-like clusters | usually gamma | + | usually – | NA | NA | NA | — (occ.) |
| *Staph. aureus* | 1-3 mm, yellowish white, opaque, creamy | as above | usually beta | + | usually + | NA | NA | NA | – (occ.) |
| *Streptococcus* sp. | 0.5-1.0 mm, grey to white, translucent | single, pairs, or chains | alpha<br>beta<br>gamma | – | NA§ | – (most) | – | +/– | – |
| *Enterococcus (Strep. faecium, Strep. durans)* | as above | as above | usually alpha, gamma | – | NA | + | + | – | + |
| *Strep. faecalis* | as above | as above | as above | – | NA | + | + | + | + |
| *Strep. zooepidemicus* | as above | as above | usually beta | – | NA | – | – | + | – |

*Refer to a microbiology text[7] for all test procedures and interpretations.
†Hemolysis: Alpha—partial destruction of red blood cells, characterized by a zone of green discoloration
Beta—total destruction of red blood cells, has a completely clear zone
Gamma—no hemolysis
‡BEA—Bile esculin azide test; NaCl—6% sodium chloride test.
§NA = Not applicable.

ciate potential pathogens from the normal gram-positive cocci are listed in Table 10–3. *Micrococcus* and *Staphylococcus* are catalase-positive, whereas *Streptococcus* is catalase-negative. The pathogenic strains of *Staph. aureus* are coagulase-positive and can therefore be distinguished from other Micrococcaceae. *Streptococcus* species are identified with the bile esculine azide (BEA), 6 per cent sodium chloride (NaCl), and sorbitol tests.

## Aerobic Gram-positive Rods

The most common gram-positive rod isolated from the alimentary tract of psittacine birds is *Lactobacillus*. *Corynebacterium* and *Bacillus* species are also frequently isolated.[3] These organisms, along with the normal gram-positive cocci described above, are considered to be normal flora for psittacine birds and finches and should be the major organisms isolated from a healthy bird. Most speciation of gram-positive rods is done through the use of colony morphology, Gram's stain characteristics, and a few quick tests (Table 10–4).

*Erysipelothrix rhusiopathiae* and *Listeria monocytogenes* will grow on BAP agar but are encouraged by incubation in a candle jar on an enrichment agar such as brain-heart infusion (BHI) medium with 5 per cent serum. Complete species identification procedures have been reviewed elsewhere[5] (see also Chapter 33, Bacterial Diseases).

Filamentous and branching gram-positive rods such as *Nocardia*, *Actinomyces*, and *Streptomyces* may also be encountered. These genera are identified by Gram's stain, acid-fast stain, and colony characteristics (Table 10–5).

## Aerobic Gram-negative Cocci

Gram-negative cocci are rarely isolated from avian sources, and none is considered a disease agent. Certain gram-negative rods, such as *Pasteurella*, can appear slightly coccoid, but *Neisseria* species are the only true aerobic gram-negative cocci of medical importance (although *Moraxella* may be a secondary pathogen in birds). *Neisseria* usually stains as a diplococcus with a coffee-bean shape. Colonies on BAP are yellow to tan, mucoid, and 1 to 2 mm in diameter. They are oxidase- and catalase-positive, indole- and urease-negative, nonmotile and will not grow on MAC agar.

## Aerobic Gram-negative Rods

Gram-negative rods are the most common agents of bacterial disease in birds. When isolated from the alimentary tract of psittacines, which have predominantly gram-positive gut flora, gram-negative rods are considered opportunists or pathogens. *Escherichia coli* cultured from the gut of other species, such as poultry, pheasants, and waterfowl, may be normal. Gram-negative rods isolated from any site other than the gut may be pathogenic. Isolation, identification, and susceptibility testing of these bacteria are therefore important.

Biochemical tests are required for speciating gram-negative rods. Table 10–6 lists a few of the simpler tests and reactions of some of the more common genera. Procedures and interpretation of the tests may be found in a medical microbiology text. Although most of these tests require 24 hours, two quick tests, the oxidase and the indole, can give immediate presumptive identification when used in conjunction with colony morphology (see Table 10–2). *E. coli*, for example, probably the most commonly isolated gram-negative rod, will have a positive indole reaction with characteristic brick red colonies on MAC agar. *Pseudomonas aeruginosa*, another frequent isolate, has a positive oxidase test, and most colonies on BAP have a greenish sheen, beta hemolysis, and a fruity odor.

Positive identification, however, is most accurately obtained with one of the many commercially prepared kits made specifically for this purpose. They are simple to use, have a prolonged shelf life, and are easy to interpret. Brief descriptions of the most common products and their manufacturers are listed in Table 10–7. Two types of kits are available; those for glucose fermenters and those for nonfermenters. Although most gram-negative rods are fermenters (e.g., Enterobacteriaceae species), if there is any doubt, a triple sugar iron (TSI) agar slant should be used to determine fermentative qualities. An acid butt (yellow color on the bottom of the tube) after overnight incubation indicates a fermenter. These organisms can then be set up and speciated on strips such as the API-20E or Rapid E. Nonfermenters will have an alkaline slant and butt (orange with no color change), and can be speciated with the N/F wheel or Rapid NFT strip. Often the fermentative qualities of an organism can be inferred from its presumptive identification, and the appropriate kit can be inoculated without waiting 24 hours for the TSI reaction.

*Text continued on page 167*

**Table 10–4.** CHARACTERISTICS AND TESTS TO IDENTIFY AEROBIC GRAM-POSITIVE RODS*

| Organism | Colony on Blood Agar (24-hr. Incubation) | Gram's Stain Characteristics | Hemolysis† | Catalase | Motility | Oxidase | Spores | $H_2S$ (TSI) |
|---|---|---|---|---|---|---|---|---|
| *Lactobacillus* | <0.5 mm, greyish, clear | long slender rods, some pleomorphic and in chains | alpha | – | – | – | – | – |
| *Corynebacterium* (diphtheroids) | 0.5-1.0 mm, dull, dry, rough, white (often takes 48 hr.) | palisading like Chinese letters | gamma<br>beta | + | – | – | – | – |
| *Bacillus* | >2 mm, square to rough, most crinkly, grey to yellow | large square ends, sometimes spores | alpha<br>beta<br>gamma | + | + (most) | + (some) | + | – |
| *Listeria monocytogenes* | 1.0-2.0 mm, grey, translucent | occurs in pairs, may resemble diplococci | beta | + | + (tumbling) | – | – | – |
| *Erysipelothrix rhusiopathiae* (two species types) | 0.5-1.0 mm, smooth<br>>1 mm, rough | short and in clumps, pleomorphic, beaded, elongated | alpha | – | – | – | – | + |

*Refer to a microbiology text[7] for all test procedures and interpretations.
†Refer to Table 10–3 for explanation of hemolysis reactions.

**Table 10–5.** CHARACTERISTICS AND TESTS TO IDENTIFY BRANCHING GRAM-POSITIVE RODS*

| | Colony Morphology on Blood Agar | Gram's Stain Morphology | Catalase* | Acid-Fast† | Spores* |
|---|---|---|---|---|---|
| *Actinomyces* (anaerobic) | white, rough, nodular to smooth and mucoid; may adhere to agar | short to long rods; branching, pleomorphic filaments | – | – | – |
| *Streptomyces* (aerobic) | tough, glabrous, velvety or most often chalky, white to greyish; often earthy odor | elaborate branching, but nonfragmenting, aerial hyphae form chains of conidia | + | – | + (in chains) |
| *Nocardia* (aerobic) | orange glabrous, heaped and folded; or white to pink, raised and chalky; leathery or crumbly and adheres to agar | filamentous and branching hyphae can fragment into rodlike or coccoid elements | + | partial | – |
| *Dermatophilus* (aerobic) | pinpoint tan colonies with a small zone of beta-hemolysis | filaments of varying thickness; irregular branching hyphae with transverse and longitudinal septa | + | – | + |

*Refer to a microbiology text[7] for catalase and spore staining procedures. See also p. 170.
†See Chapter 17, Cytology, for acid-fast stain procedure.

**Table 10–6.** SELECTED BIOCHEMICAL TESTS FOR COMMON GRAM-NEGATIVE GENERA*

| | TSI Slant | | | | | | Decarboxylase Tests | | | | Quick Tests | | |
|---|---|---|---|---|---|---|---|---|---|---|---|---|---|
| **Organism** | Slant | Butt | Gas | $H_2S$ | **Citrate** | **Urea** | Ornithine | Lysine | Arginine | **Motility** | Indole | Lactose | Oxidase |
| *E. coli* | A or K | A | +/− | − | − | − | +/− | +/− | +/− | +/− | + | + | − |
| *Salmonella* | K | A | + | + | + | − | + | +/− | + | +(most) | − | − | − |
| *Citrobacter* | A or K | A | + | +/− | + | +/− | +/− | − | +/− | + | +/− | +/− | − |
| *Klebsiella* | A or K | A | + | − | + | + | − | + | − | − | +/− | + | − |
| *Enterobacter* sp. | A or K | A | + | − | + | +/− | + | +/− | +/− | + | − | + | − |
| *Ent. agglomerans* | A or K | A | + | − | +/− | +/− | − | − | − | + | +/− | + | − |
| *Shigella* | A or K | A | − | − | − | − | − | − | + | − | +/− | − | − |
| *Edwardsiella* | A or K | A | +/− | +/− | +/− | − | + | + | − | + | +/− | − | − |
| *Proteus* | K | A | + | + | +/− | + | +/− | − | − | + | +/− | − | − |
| *Morganella* | K | A | + | − | − | + | + | − | − | + | + | − | − |
| *Providencia* | K | A | + | − | + | +/− | − | − | − | + | + | − | |
| *Aeromonas* | A or K | A | +/− | − | +/− | − | − | +/− | +/− | + | + | +/− | + |
| *Pasteurella* | A | A | − | − | − | ± | − | − | − | − | +/− | − | + |
| *Pseudomonas* | K | K | − | − | +/− | +/− | | | | + | − | − | +/− |
| *Alcaligenes* | K | K | − | − | + | − | | | | + | − | − | + |
| *Acinetobacter* | K | K | − | − | + | − | | | | − | − | − | − |

*Refer to microbiology texts[7, 8] for test procedures and interpretations. See also p. 170.

**Table 10–7.** COMMERCIALLY AVAILABLE KITS FOR BACTERIAL SPECIATION

| Name | Number of Reactions Tested | Incubation Time | Shelf Life | Approximate Cost* | Manufacturer |
|---|---|---|---|---|---|
| **Fermenters** (Enterobacteriaceae) | | | | | |
| API-20E | 20 | 18 hr | 1 yr | 3.35 | Analytab Products<br>Plainview, NY 11083 |
| Rapid E | 20 | 4 hr | 1 yr | 3.25 | DMS Laboratories<br>Flemington, NJ 08822 |
| Micro-Id | 15 | 18 hr | 1 yr | 3.25 | General Diagnostics<br>Morris Plains, NJ 07950 |
| Enterotube II | 12 | 24 hr | 3–6 mo | 2.50 | Roche Diagnostics<br>Nutley, NJ 07110 |
| **Nonfermenters** | | | | | |
| Uni-N/F Tek Plate | 12 | 48–72 hr | 3 mo | 4.00 | Flow Laboratories<br>Roslyn, NY 11576 |
| Rapid NFT | 20 | 48 hr | 1 yr | 4.80 | DMS Laboratories<br>Flemington, NJ 08822 |
| API-20E | Some nonfermenters will be identified with this strip in 48 hours; see above for details | | | | |
| **Anaerobes** | | | | | |
| Mini-Tek | 20 | 48 hr | 1 yr | 4.00 | BBL Microbiology Systems<br>Cockeysville, MD 21030 |
| API-20A | 20 | 48 hr | 1 yr | 3.50 | Analytab Products<br>Plainview, NY 11083 |
| **Yeast** | | | | | |
| API-20C | 19 | 2–4 days | 1 yr | 3.40 | Analytab Products<br>Plainview, NY 11083 |
| Uni-Yeast Tek | 12 | 1–6 days | 3–6 mo | 4.00 | Flow Laboratories<br>Roslyn, NY 11576 |

*Approximate cost at time of publication; use as a comparative value and consult the manufacturer for actual cost when purchasing.

## Unusual Gram-negative Rods

As mentioned, some organisms require special media or environmental conditions for growth. *Haemophilus* requires either factor X (hemin) or factor V (NAD), both of which are supplied by chocolate agar. *Yersinia,* although fermentative and closely related to the Enterobacteriaceae, are difficult to grow and separate from other gram-negative bacteria without a cold enrichment broth. *Campylobacter* grows best with reduced oxygen and requires special nutrients and vitamins. Although acid-fast stains can be done as a preliminary study on tissues or feces for the presence of mycobacteria, these organisms are highly infectious and difficult to isolate and should be handled only under a biologic safety cabinet. Mycoplasma are also fastidious organisms that need a protein-based medium enriched with serum factors and other special requirements for isolation (see Chapter 34, Mollicutes). *Chlamydia psittaci* can be grown in cell culture or embryonated eggs. All of the above cultures are impractical to perform in most in-house laboratories, and specimens suspected of containing these microorganisms should be sent in appropriate packaging to a reference laboratory.

In addition, isolating and identifying anaerobic bacteria are expensive in both time and money for the average avian practice. However, the majority of normal flora in the avian gut are anaerobic,[12] so it may be important to demonstrate their presence. If an anaerobic infection is suspected or if bacteria are seen on a Gram's stain and there is no growth aerobically, the specimen should be sent in a special anaerobic transport medium to a qualified laboratory. Collection of all specimens into anaerobic vials[d, e] or Culturettes[c] should be done promptly to reduce exposure to air. Samples should be transported by the quickest method possible; ice is not necessary.

Some indications that an anaerobic infection is present are a foul-smelling discharge, gas production, and an infection in proximity to mucosal membranes or associated with the use of aminoglycoside antibiotics.[4] Pale and unevenly stained gram-negative bacilli are likely to be anaerobes. Pleomorphic gram-negative rods with rounded ends and bipolar staining are characteristic of *Bacteroides fragilis*. Those with filamentous forms or tapered ends are likely to be *Fusobacterium*. Thick blunt-ended gram-positive rods are typical of *Clostridium*.

## ANTIMICROBIAL SUSCEPTIBILITY TESTING

The Kirby-Bauer diffusion method for determining antimicrobial susceptibilities is the most reliable, inexpensive, and easily performed test available. Standardized procedures must be followed according to the disk manufacturer's directions in order for valid results to be obtained. Zone of inhibition sizes depend on the concentration and diffusion properties of the antibiotic and the susceptibility of the bacteria. Caution must be exercised when using the manufacturer's recommendations for deciding whether or not a bacterium is sensitive to the antibiotic, as these recommendations are based on the antibiotic concentrations attainable with standard dosage regimens in humans. It may not be possible to achieve the same concentrations in birds, so treatment failure of a "sensitive" organism might be related to the inadequacy of the antibiotic dosage regimen used, rather than to failure of the Kirby-Bauer test.

A few key points are listed here as reminders for accurate results with the Kirby-Bauer procedure:

1. Always pick at least three or four isolated colonies to incubate in the broth. Using multiple colonies reduces the chance that a mutant colony with altered antibiotic sensitivity will be chosen.
2. Be sure that the final turbidity is a 0.5 McFarland standard. A turbid tube can easily be diluted with sterile saline.
3. Blood and chocolate Mueller-Hinton (MH) agar plates are available for gram-positive and fastidious organisms.
4. It is best to allow refrigerated MH plates to warm slightly before inoculating.
5. When reading zones of inhibition, one can ignore the swarming film of *Proteus* species. In many other species the zone surrounding a sulfonamide disk may be ill-defined. This happens because sulfa drugs do not suppress bacterial growth immediately and therefore leave an attenuated growth area. The zone should be measured at the outer rim of "fuzzy inhibition."
6. Isolated colonies within the zone of inhibition may be mutant strains resistant to the drug.
7. Quality control is highly recommended to ensure proper technique and disk viability. Lyophilized disks of many bacteria are available. The control strains of *Staph. aureus*, ATCC 25923; *E. coli*, ATCC 25922; and *Pseu-*

*domonas aeruginosa*, ATCC 27853 are recommended. BBL Microbiology Systems[d] will send sample zone sizes for these control organisms upon request.

8. Although manufacturers recommend that disks be discarded one to two weeks after being opened, the disks (and dispenser) may be kept frozen to prolong the shelf life to approximately one month. Quality control is especially important if this is done. Disks should be thawed slightly prior to use.

## FUNGI

Screening for fungi should be a routine procedure in microbial studies. Birds come in contact with a large number of fungal organisms present in nest box material, seed, and soil, so caution is necessary in determining pathogenicity if a fungus is isolated.

Direct examination of clinical material such as feces, fluids, skin scrapings, and impression smears with a KOH wet mount or Gram's stain can be very important in identifying the presence of mycotic disease before the fungus can be cultured.

Sabouraud-dextrose (Sab-Dex) agar is the best medium for growth and isolation of fungi. The addition of chloramphenicol (Sab-Dex + Ch), however, is recommended for most routine cultures to prevent the growth of bacteria. If dermatophytes, which may cause skin and feather problems, or the systemic diphasic fungi (*Blastomyces dermatitidis, Coccidioides immitis, Histoplasma capsulatum,* and *Sporothrix schenckii*) are suspected, mycobiotic agar should be used. This is Sab-Dex agar with chloramphenicol and cycloheximide added to prevent the growth of bacteria and saprophytic fungi. All Sab-Dex plates with additives must be incubated at room temperature. Incubation at higher temperatures encourages bacterial growth that may override the slower growing fungi.

Sab-Dex and Sab-Dex + Ch plates should be held for one week to look for yeast and the quickly growing fungi (e.g., *Aspergillus*). Mycobiotic agar should be held for one month for slower growing organisms and is best used in bottle form to prevent drying.

Owing to the highly infectious nature of many fungi, laboratories that do not have a biologic safety hood should not work past the point of initial isolation and growth of a fungus. The original bottles or well-taped and bagged plates used for isolation may be mailed to a reference laboratory for identification. Nonpigmented mold colonies that come up after seven days may be highly infectious systemic fungi, such as *Sporothrix* and *Histoplasma,* and are best speciated by a reference laboratory, even when

**Table 10–8.** CHARACTERISTICS OF PATHOGENIC AND COMMONLY ISOLATED FUNGI

| Species | Growth Time | Macroscopic Characteristics on Sabouraud-Dextrose Agars |
|---|---|---|
| **Known Pathogens** | | |
| *Aspergillus* sp. | 1–3 days | first white and then dark green, brown or yellow; velvety to powdery |
| *Sporothrix schenckii* | 3–5 days at room temperature | moist, wrinkled, membranous or radially folded colony; cream to black in color |
| | 7–10 days at 37° C | grey-yellow yeastlike colony |
| *Histoplasma capsulatum* | 2–4 weeks at room temperature | white, dense, cottony colony |
| | 1–2 weeks at 37° C | glabrous, white, moist colony |
| **Frequent Contaminants** | | |
| Possible Pathogens | | |
| *Mucor* sp. | 3–5 days | loose cottony or woolly mycelia that become darkened, grey, or peppery |
| *Rhizopus* sp. | | |
| *Absidia* sp. | | |
| Dermatophytes | 3–7 days | a variety of colony types and sporulation; refer to Mycology manuals |
| *Penicillium* sp. | 3–5 days | first white and then brightly colored green to blue shades; velvety to powdery |
| *Cladosporium* sp. | 6–10 days | greenish-brown surface with velvety greyish nap; reverse pigment is black; folded colony |

a biologic hood is available. When mailing a viable culture, be sure the package is double-wrapped against breakage and marked with warnings.

Most fungal identification can be done with proper evaluation of macroscopic characteristics and the aid of a good pictoral mycology text.[9, 10] The color, texture (e.g., woolly, velvety, hairy), and morphology (e.g., domed, flat, nebular) of the colony are all important for identification, as is the time required for initial growth on a plate (Table 10–8).

The growth of a yeast, either at 37° C or at room temperature on Sab-Dex or blood agar plates, is easily detected. Most colonies are dull, opaque, creamy, rounded domes with a characteristic smell. A saline wet mount examined under high dry power is an easy way to observe the typical oval to rectangular bodies (often budding) of a yeast cell. After a yeast is isolated, a germ tube test should be done to identify *Candida albicans*, the most common clinical isolate. Germ tubes are short germinating hyphae that can be induced to grow when placed in a liquid nutrient environment. Make a light suspension of the yeast in trypticase soy broth (TSB) and incubate at 37° C for four hours. Examine microscopically for characteristic hyphal shoots extending from the larger oval body (blastoconidia). They look like long noses and are distinguished from pseudohyphae and buds by the lack of constriction between the blastoconidia and the hyphae. *Candida albicans* is the only yeast species that makes germ tubes.

If the germ tube test is negative, a yeast isolate can usually be placed into a genus on the basis of morphology on corn meal agar, or the species can be identified with carbon assimilation tests, available in two commercially prepared test kits (see Table 10–7).

There is no standard method to determine susceptibility of yeasts to the antifungal agent nystatin. Nystatin disks[d] are available, but no clinical correlation to inhibition zone sizes has been made.

## FOOD, WATER, AND ENVIRONMENTAL MICROBIOLOGY

Isolation and antibiotic susceptibility testing are the most important jobs when treating an active clinical case, but identification of an organism and its possible source may be necessary to prevent reinfection. Infections may be traced to various sources in the aviary (see Chapter 50, Aviculture Medicine). No standards exist for evaluating the microbial contamination of environmental samples, so the clinician must use his own judgment in deciding the significance of isolates from such things as food, water, and nest material.

The weight or volume of a sample should be determined in order to semiquantify the microorganisms present and accurately interpret results. Small items may be ground with a sterile mortar and pestle. Blenders work well for chopping seeds, nuts, fruits, and nest materials. The blender should be taken apart, cleaned, soaked in a disinfectant, and thoroughly rinsed in sterile water between each specimen. A small amount of sterile saline may be necessary to blend the sample.

Inoculate BAP, EMB, and Sab-Dex + Ch plates directly from the blended material. In addition, inoculate a thioglycolate broth tube and incubate for four hours before subculturing onto BAP and MAC agars. We have found that four-hour incubation of the inoculum in thioglycolate broth will increase the number of organisms present but will still maintain the same ratio of species that was present in the original sample. This enrichment step is sometimes necessary to obtain enough growth to identify organisms present in small numbers. Incubation of the specimen in enrichment medium for a longer period of time is not recommended, as faster-growing bacteria may overwhelm other organisms in the sample. Read and evaluate all plates as previously described. *Enterobacter agglomerans* is a common isolate from sunflower seeds. A variety of *Bacillus* species and opportunistic fungi are frequent isolates from environmental samples and probably are not pathogenic.

The numbers and species of bacteria in water can be determined by filtering the sample and growing the bacteria trapped on the filter. Place a gridded Metricel filter with 0.45 micron pores[f] in a flat-based, autoclavable filter funnel. Pour a 10-ml sample into the funnel and allow the water to drain through by gravity. Cover the funnel with a sterile Petri dish to prevent contamination. Once the water has drained, use sterile forceps to remove the filter and place it onto a blood agar plate. Incubate for 24 to 48 hours, count the number of colonies per square on the filter, and identify the bacterial species by means of the methods previously described. The formula below can be used to quantify the number of bacteria per milliliter of water. If growth is too heavy to count individual colonies, repeat the procedure with a smaller sample.

$$\text{No. of bacteria/ml } H_2O = \frac{\text{No. of colonies/square} \times \text{No. of squares/filter paper}}{\text{No. of ml } H_2O \text{ filtered}}$$

*Pseudomonas*, *Alcaligenes*, and other nonfermenters are common water contaminants. See Bergy's manual[1] to speciate the many organisms likely to be encountered.

## SOURCES OF MICROBIOLOGY PRODUCTS MENTIONED

[a]Gram's stain kit—Difco Laboratories, Detroit, MI 48232

[b]Calcium alginate swabs—Inolex Laboratories, Glenwood, IL 60425

[c]Culturette—Marion Scientific Corp., Rockford, IL 61101

[d]Port-a-Cul—BBL Microbiology Systems, Cockeysville, MD 21030

[e]Culture-Eze—Medical Media Laboratory, Inc., Boring, OR 97009

[f]Metricel filter—Gelman Sciences, Inc., Ann Arbor, MI 48106

## TEST PROCEDURES

**Bile Esculine Azide (BEA).** Used for differentiating *Streptococcus* species. Inoculate the broth with one or two isolated colonies. A turbid black color after overnight incubation is a positive reaction. A grey to yellow color or clearness indicates a negative reaction.

**Catalase Test.** Transfer an isolated colony to a flamed glass slide. (Extreme care must be exercised if the colony is taken from a blood agar plate, as the enzyme catalase is present in red blood cells and can give a false-positive reaction.) Add a drop of 30 per cent hydrogen peroxide to the colony. Strong and immediate production of gas bubbles is a positive reaction. No bubbles or just a few (*Enterococcus* species) is considered negative.

**Coagulase Slide Test.** Used for differentiating *Staph. aureus* from other *Staphylococcus* species. Mix several isolated colonies in a drop of sterile water to form a smooth, milky suspension on a slide. Add one or two drops of reconstituted lyophilized rabbit plasma (BBL) and mix thoroughly. Immediate clumping is a positive test for coagulase, indicating the presence of *S. aureus*. The absence of clumping on the slide is a presumptive negative reaction, and the tube test must be performed for confirmation.

**Coagulase Tube Test.** Emulsify several colonies into 0.5 ml of reconstituted rabbit plasma (BBL) and incubate at 37° C for four hours. If a clot forms, the test is positive; if not, reincubate overnight and examine for coagulation. The four-hour reading is necessary because fibrinolysin production of occasional *S. aureus* strains will lyse the clot within 16 to 18 hours of incubation and thus give a false-negative test.

**Indole Test.** Indole reagent (Kovacs or Ehrlics) is available in dropper bottles (API, BBL) or in convenient capsules (Marion). Place a drop of the indole reagent on a piece of filter paper and gently rub a loopful of bacteria over the reagent. Do not take a colony from plates with dyes (e.g., MAC or EMB), as these can give false-positive results. Development of a pink color within 20 seconds is positive; a delayed reaction or no reaction is negative.

**6% Sodium Chloride (NaCl).** Used to differentiate *Streptococcus* species. Inoculate the broth with one or two isolated colonies and incubate overnight. An organism's ability to tolerate the salt will allow growth and convert the purple color to a turbid yellow. No change is a negative reaction.

**Oxidase.** Oxidase reagent is available in dropper bottles (BBL) or in convenient capsules (Marion) that have an extended shelf life. Place a drop of reagent on a piece of filter paper and gently rub a loopful of bacteria over the reagent. Do not take colonies from plates with dyes (e.g., MAC, EMB); use a platinum loop to prevent false-positive results. The production of a strong blue to purple color within 20 to 30 seconds is a positive reaction.

**Triple Sugar Iron (TSI).** Inoculate by stabbing the loop with an isolated colony through the center of the medium to the bottom of the tube and then streak the surface of the agar slant. Incubate overnight with the tube loosely capped. No change in the orange color is negative (alkaline), while a yellow color is positive (acid). The production of bubbles indicates gas. A black precipitate in the butt (ferrous sulfate) indicates hydrogen sulfide production.

Alkaline slant/Alkaline butt (K/K) = nonfermenter

Alkaline slant/Acid butt (K/A) = glucose fermentation only

Acid slant/Acid butt (A/A) = glucose, sucrose, and/or lactose fermenter

## REFERENCES

1. Buchanan, R. E., and Gibbons, N. E.: Bergy's Manual of Determinative Bacteriology. Baltimore, The Williams and Wilkins Co., 1975.

2. Carter, G. R.: Diagnostic Procedures in Veterinary Microbiology. Springfield, IL, Charles C Thomas, Publisher, 1979.
3. Drewes, L. A., and Flammer, K.: Preliminary data on aerobic microflora of baby psittacine birds. Proceedings of the Jean Delacour/IFCB Symposium on Breeding Birds in Captivity, Universal City, CA, 1983, pp. 73–81.
4. Finegold, S. M., et al.: Cumitech 5: Practical Anaerobic Bacteriology. Washington, DC, American Society of Microbiology, 1977.
5. Hitchner, S. B., et al.: Isolation and Identification of Avian Pathogens. College Station, TX, American Association of Avian Pathologists, 1980.
6. Hofstadt, M. S., et al.: Diseases of Poultry. Ames, IA, Iowa State University Press, 1978.
7. Lennette, E. H., et al.: Manual of Clinical Microbiology. Washington, DC, American Society of Microbiology, 1980.
8. MacFaddin, J. F.: Biochemical Tests for Identification of Medical Bacteria. Baltimore, The Williams and Wilkins Co., 1980.
9. McGinnes, M. R., et al.: Pictoral Handbook of Medically Important Fungi and Aerobic Actinomyces. New York, Praeger Publishers, 1982.
10. Moore, G. S., and Jaciow, D. M.: Mycology for the Clinical Laboratory. Reston, VA, Reston Publishing, 1979.
11. Prescott, J. F., and Munroe, D. L.: *Campylobacter jejuni* enteritis in man and domestic animals. J. Am. Vet. Med. Assoc., 181:1524–1530, 1982.
12. Salanitro, J. P., et al.: Bacteria isolated from duodenum, ileum and cecum of young chicks. Appl. Env. Micro., 35:782–790, 1978.

# *Chapter 11*

# VIROLOGY

HELGA GERLACH

Samples from live birds can be taken from the feces, skin and feathers, pharynx, cloacal swabs, and, depending on the size, from the larynx, trachea, and nasal cavity. Biopsy samples can be obtained during endoscopy or surgery. Samples from dead birds can be taken freely and should include liver, spleen, kidney, lung, air sac, and, according to the circumstances, other tissues such as intestinal tract, brain, and endocrine glands. Special samples of blood, synovial fluid, etc., may be necessary in selected cases. If at all possible, acute and convalescence serum samples should be taken and kept frozen for future use.

Diagnostic procedures for demonstration of virus are usually limited to specialized laboratories, and samples must be sent in special transport media (neutrally buffered salt solution plus protein source with no antibodies) and/or on ice or even Dry Ice. It might be necessary to add antibiotics (water-soluble with no preservatives) in order to prevent bacterial multiplication. Close contact with the diagnostic laboratory about which methods are recommended for each virus group is desirable. Routine methods for virus demonstration in poultry or mammals do not always work without alterations with material from psittaciformes. Therefore, the application of several different methods might be justified. In addition, it is strongly suggested that a variety of "new viruses" exist in this large group of birds and only culture methods are of real value to prove it. Generally, the following principles of virus demonstration are available for material from psittaciformes.

## CULTURAL METHODS

A variety of cell lines is available for viral culture, but for avian samples primary cells are often advantageous. Chicken, duck, or goose is normally the source for fibroblasts, kidney cells, or liver cells. The supplier flock for eggs should be free from all known virus (and *Mycoplasma*) and their antibodies. The latter is of importance for virus culture in embryonated eggs. Since the psittaciformes are taxonomically quite different from chickens, ducks, or geese, adaptation passages (at least three) might be necessary even for a well-known virus group like paramyxovirus. Cell cultures derived from embryonated eggs of the budgerigar *(Melopsittacus undulatus)* or cockatiel *(Nymphicus hollandicus)* are sometimes necessary for successful isolation from original material. The best culture system would be cellular material from the same species as the samples. Normally such eggs are not available.

The identification of isolates is usually done by serologic methods with known antibodies. However, new isolates from psittaciformes might have to be classified taxonomically. Virus isolates also have to be tested experimentally for pathogenicity and relationship with the established disease process.

## ELECTRON MICROSCOPY

Samples designated for this procedure should be fixed at once in special fixatives (preferably 3 to 5 per cent glutaraldehyde) in very small pieces. Positive results can be expected only if the virus is in rather high concentration in the cells. At best, a morphologic genus diagnosis can be made. Such virus demonstrations are sometimes positive, although culture methods show negative results. Histologic demonstration of tissue inclusion bodies may suggest a virus infection, but on these formalin-fixed tissues, proof of a virus infection must be done by electron microscopy.

## SEROLOGIC METHODS

Virus infections, as a rule, cause the production of antibodies within the host organism. Routine methods demonstrate humoral antibodies of either the IgM or the IgG class. A variety of techniques is in use. Positive results indicate that the host and the virus have interacted, but not necessarily that the host is immune. Negative results are less indicative. IgA and cell-mediated reactions are demonstrated by special tests not routinely used. Furthermore, in acute cases, antibody response might not yet have developed. In the psittaciformes, demonstrable humoral antibodies may not be as regularly induced as in poultry. The budgerigar is the best-known example of this phenomenon. Depending on the serologic test and the virus preparation used as antigen, a group-specific or a species-specific diagnosis can be made. The establishment of the relationship of antibodies to the disease process can be difficult, and conclusions drawn from one psittaciforme species cannot simply be applied to other species.

Serologic tests, such as the immunofluorescence, can also be used directly or indirectly for virus demonstration within tissues or feces. Specific controls have to be developed to ensure that the antibodies produced experimentally cause specific reactions and that nonspecific ones are recognized as such. Tissue extracts containing virus antigens may be identified by the ELISA procedure.

The majority of the serologic tests need to be performed in specialized laboratories. The reader is referred to Chapter 18, Serology, for further discussion of this diagnostic method.

# Chapter 12

# HEMATOLOGY

F. JOSHUA DEIN

## INTRODUCTION TO CLINICAL PATHOLOGY

Before utilizing clinical pathology for any species, the clinician must have an understanding of what the test results mean. This is important for the avian clinician, since much of the groundwork defining the avian hematologic and biochemical response to disease has not yet been done. A commonly asked question of the clinical laboratory is whether a test result is "normal." Although this may appear to be a simple question, there are numerous variables, such as species, sex, age, physiologic condition, nutritional status, geographic location, and time of day (or year) when the sample is collected, that may affect the laboratory results. In addition, the term *normal* can refer to the average, most commonly encountered, or most representative value.[55] Because of this confusion over "normal," the terms *reference* and *referent value* are superior when indicating a range of values obtained from diagnostic testing.[24] Ideally, a reference range should be developed from, and utilized for, a group of animals possessing similar characteristics. Hematology and serum chemistry reference ranges could be determined for all apparently healthy Amazon Parrots (*Amazona* spp.) for only Yellow-naped Amazon parrots *(Amazona ochrocephala ochrocephala)*, adult female Yellow-naped Amazon parrots, or only adult female Yellow-naped Amazon parrots during the reproductive season. These different groups may show significant hematologic and/or chemical differences that would necessitate separate reference ranges. Practically, it is difficult to obtain enough samples to determine a proper reference range for smaller, more specific groups, so we must be content using larger ones. Intuitively, it seems that individual species would be the most appropriate groupings; however, the similarity of values among different species may allow combining into larger groups. This can be determined by large-scale sampling over both narrow and broad taxonomic lines.

### Determining Reference Ranges

The determination of an appropriate reference range for a group of animals can be difficult. Generally, a group of animals is sampled, and the mean and standard deviation (SD) are calculated for a specific test or battery of tests. The reference range is established as the mean plus or minus two standard deviations. A few problems are associated with calculating reference ranges in this manner.[44] One is the lack of controls in the selection of the sampled population, such as proper consideration of the variables mentioned above. Are the individuals selected truly healthy or just apparently healthy? Another problem occurs in the calculation of the mean and SD; the assumption is made that there is a normal (Gaussian) distribution of the values obtained when in fact such a distribution does not exist. A further factor is correcting for any analytic bias that is present in the test procedure. For example, any spectrophotometric test is dependent on the clarity of the sample; any degree of opacity (e.g., lipemia) will affect the result. Finally, the use of the mean ±2 SD for the reference range is an arbitrary limit; it is possible that values outside the 2 SD range can be normal for some individuals.[44] Opinions on how many animals to include in the population sampled to establish a reference range vary from 20 to 30 to 300 to 500. Logically, the greater the sampled population, the better the estimate.

The proper development of reference ranges can be a detailed and time-consuming process. It is beyond the scope of this book to describe all the procedures that should be followed. Extensive reviews on the determination of reference ranges are available to the veterinarian.[44, 91, 93]

---

Photos courtesy of Jessie Cohen, National Zoological Park Office of Graphics and Exhibits.

### *Editors' Note*

Examples of reference values for hematology and selected clinical chemistries for several pet bird species as reported by a number of individuals are listed in Appendix 3. Included with information from the United States is a postgraduate dissertation from the University of Munich, West Germany. With the translation assistance of Dr. Helga Gerlach, several points were emphasized:

There are species that consistently have a blood picture that *normally* includes an abundance of a specific cell type in the differential. Three such types were identified: those species with a heterophilic blood picture, those with a lymphocytic blood picture, and those with a normal abundance of eosinophils.

The birds that normally have a high percentage of heterophils in the differential blood picture are Greater Sulfur-crested Cockatoo with approximately 55 per cent; Yellow-crowned Amazon, approximately 54 per cent; African Grey Parrot, approximately 56 per cent; and Rainbow Lory, approximately 49 per cent.

Birds that normally have a predominance of lymphocytes in the differential are the budgerigar with 54 per cent; the cockatiel with 66 per cent; and the Rose-ringed Parakeet (Indian Ring-neck) with 52 per cent.

Of the species studied, two species appear normally to have a balance in the numbers of heterophils and lymphocytes: Blue-fronted Amazon (46 per cent heterophils, 49 per cent lymphocytes) and Fischer's Lovebird (43 per cent heterophils, 49 per cent lymphocytes).

The eosinophilic birds have been found primarily among birds of prey. The common buzzard has 15 per cent and the European kestrel has 24 per cent eosinophils as a normal distribution in the differential.

In appraising representative blood values for birds according to sex, significant differences in hematologic values were found between males and females of the same species.

Age was confirmed to be a further factor affecting the normal values of the blood in a recent study of Brown Pelicans.* Marked differences were found in clinical chemistry values for each of five groups: hatchlings, fledglings, immature and adult captive-raised pelicans, and wild-caught adults. Although direct conclusions cannot be drawn from information provided by study of the Brown Pelican, one would expect similar differences among the ages and captive status of psittacines (see Chapter 13, Clinical Chemistries).

*Wolf, S. H.: Seasonal, age and sex variations in the blood composition of the Brown Pelican (*Pelecanus occidentalis*). Master of Science thesis, University of South Florida, 1984.

### Predictive Value of Diagnostic Tests

The terms *sensitivity* and *specificity* are most often associated with epidemiology but are of great importance in understanding the diagnostic value of tests performed in the clinical laboratory. They indicate the degree of accuracy with which a specific test can predict the presence or absence of disease in a patient.

The sensitivity of a test describes its ability to determine the presence of a disease, that is, the number of true positive results. The specificity describes the test's ability to determine the absence of a disease, or true negative results. It is important that we know the sensitivity and specificity of a test for each disease process so that we can estimate from the laboratory results the chances of a patient's having that disease. The predictive value is calculated from the sensitivity and specificity and is a qualitative evaluation of a test's reliability. Obviously, tests with high predictive value for a particular disease are the most valuable.

However, the clinician should not rely too heavily on any one test result for a positive diagnosis. For example, an increased level of serum aspartate amino transferase (AST, SGOT) may not always detect liver disease, a high WBC can only suggest the presence of an infectious agent, and analysis of serum uric acid will not always indicate renal function. Even batteries of tests (e.g., profiles, screens, SMAC-12 or 20) can be problematic; for each set of 20 tests run on a normal, healthy patient, there is a 64 per cent probability that one test will be abnormal owing to chance alone.[44] The clinical laboratory can help in narrowing the list of differentials, but the clinician must add his or her knowledge of species and physiologic, environmental, and technical variation to determine the final diagnosis.

## HEMATOLOGY

Clinical hematology has long provided some of the most useful indicators of mammalian health. As a connective tissue, blood is in intimate contact with all organs and carries nourishment and information throughout the body. Because of this contact and the blood cells' rapid replacement rate, examination of peripheral blood quickly reveals changes resulting from differing physiologic demands. Veterinarians routinely use blood examination when treating mammals, but it has not been for long a significant part of the diagnostic repertoire of those working with birds. Hematology of domestic species has been studied since the beginning of the century,[90] but it is not commonly used in the evaluation of flock health. The usual practice of sacrificing a few individuals to obtain a definitive postmortem diagnosis has distinct advantages over the speculative nature of antemortem clinical pathology. With the significant rise in the number of pet birds and increased interest in captive breeding, the flock health approach has been replaced by one that places emphasis on the individual. It is important to have effective and reliable methods available for the diagnosis of disease in these patients. Although there are some problems associated with the methods available, hematology can be an important part of the avian clinical work-up. With routine use, any clinician can become comfortable with the techniques. The greater the number of people exploring avian hematology, the greater the amount of information that will be gained. With the development of better testing methods, this accumulated knowledge can be utilized to provide reliable diagnostic insights into disease processes in the avian patient.

This chapter will introduce the basics of avian hematology and discuss the similarities and differences between it and mammalian hematology. It cannot be a complete reference for comparative hematology; it is suggested that the reader refer to larger volumes in veterinary and human hematology for further information.[4, 33, 52, 75] A general laboratory manual of avian hematology techniques and identification is also available.[19]

## Blood Collection

A blood collection technique should be selected on the basis of the size and physical condition of the bird and the amount of sample required. The total blood volume of a bird is 6 to 13 per cent (depending on species) of the body weight.[9, 59] Ten per cent of the blood volume in milliliters (0.6 to 1.3 per cent body weight in grams) can be taken without adverse consequences in healthy birds. This amount should be reduced in sick birds.[34, 35, 63] Usually, 10 per cent of the blood volume is more than is required for diagnostic tests. Most laboratories require 0.2 to 0.3 ml of whole blood for hematology and 0.01 to 0.075 ml serum or plasma for each chemistry. The amount of serum that can be harvested from whole blood depends on packed cell volume and sample handling (see Chapter 13, Clinical Chemistries).

## Venipuncture Equipment

The equipment needed for obtaining blood samples from birds is identical to that used for mammals. A syringe and needle can be used for all but very small vessels. A small volume (1-cc) syringe is best for collecting from small vessels to avoid vessel collapse. The needle and syringe can be coated with a liquid anticoagulant prior to collection by filling the syringe with the anticoagulant of choice and then expelling it back through the needle into the anticoagulant bottle. It is important to expel any excess anticoagulant to avoid dilution of the blood sample.

Using a needle alone is a useful technique for collecting a small sample or for sampling from a small vessel. This technique is not appropriate for high-flow vessels such as the jugular vein. Thin-walled needles (Becton-Dickinson, Terumo) are available and provide a larger bore for blood flow while creating a smaller puncture hole.

## Sampling Sites

The most important criteria for the selection of the sampling site are the species of bird and the amount of sample required. Ease of manipulation of the patient and access to a vessel should be evaluated before a site selection is made. Larger vessels should be used if large or multiple samples are needed.

CAUDAL TIBIAL (MEDIAL METATARSAL VEIN) (Fig. 12–1). This vessel is convenient on all but the smaller species (<75 gm). It runs superficially along the medial side of the tarsometatarsus and is excellent for obtaining samples with a needle and syringe or needle alone. In most species the vein is visible, although in some it can be obscured by wrinkled or pigmented skin. It is usually easiest to cannulate just below the ankle joint. Avoid collecting directly over the joint.

JUGULAR VEIN (Figs. 12–2 and 12–3). The jugular veins are variable in size and location. They are most easily found in the featherless areas adjacent to the cervical vertebrae; the right is often larger than the left. It is best to use a syringe for collection owing to the high volume and pressure of this vessel. The jugular vein is a good routine sampling location and is excellent for large or multiple samples. Although this vessel can be difficult to stabilize, it is possible for one person working alone to bleed smaller birds from this site (Fig. 12–3).

ULNARIS VEIN (Figs. 12–4 and 12–5). The

**Figure 12–1.** Caudal tibial (medial metatarsal vein) (arrow).

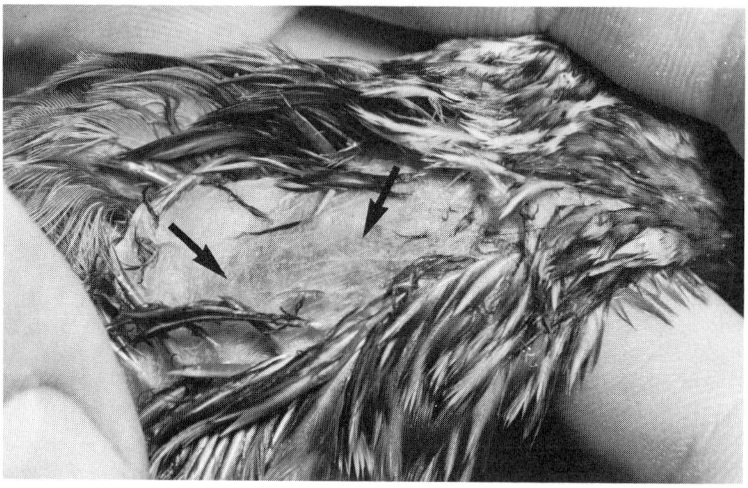

**Figure 12–2.** Jugular vein (arrows).

**Figure 12–3.** One-person collection technique from jugular vein.

**Figure 12–4.** Cutaneous ulnar vein (arrow).

"wing vein" has traditionally been used as a collection site in poultry medicine. It does require firm restraint of the bird in an abnormal position, but there is little difficulty in finding the vessel or collecting from this site. Wetting feathers with alcohol or water, rather than plucking, is the preferred method of preparing the area. A large hematoma often forms at this site but can sometimes be minimized either by directing the needle in a proximal to distal direction or by sliding the overlying skin to one side during sampling and letting the skin slide back over the puncture site after removing the needle.

TOENAIL CLIPPING. The toenail is easily accessible, requires minimal restraint, and presents little risk of large blood loss. The disadvantages of this technique include pain associated with the procedure; the possibility that cell counts may be skewed owing to sample contamination with tissue juices, milking of the vessel, or poor circulation in the toe; the possibility of nail bed damage; and lack of enough blood flow for an adequate sample.[5, 51, 88] It is important that the nail be thoroughly cleaned before clipping, especially to avoid skewed uric acid levels due to urate contamination.[72]

SKIN PRICK (Figs. 12–6 and 12–7). This method can be useful in very small birds. A 25- or 27-gauge needle is used to prick the skin directly over the caudal tibial (medial metatarsal) vein. Blood is then collected directly from

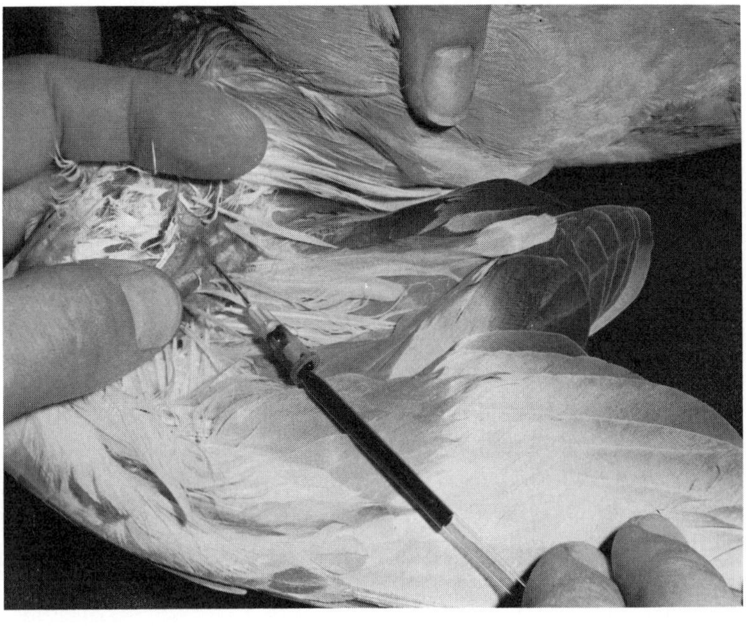

**Figure 12–5.** Collection of blood sample from cutaneous ulnar vein using Caraway capillary tube.

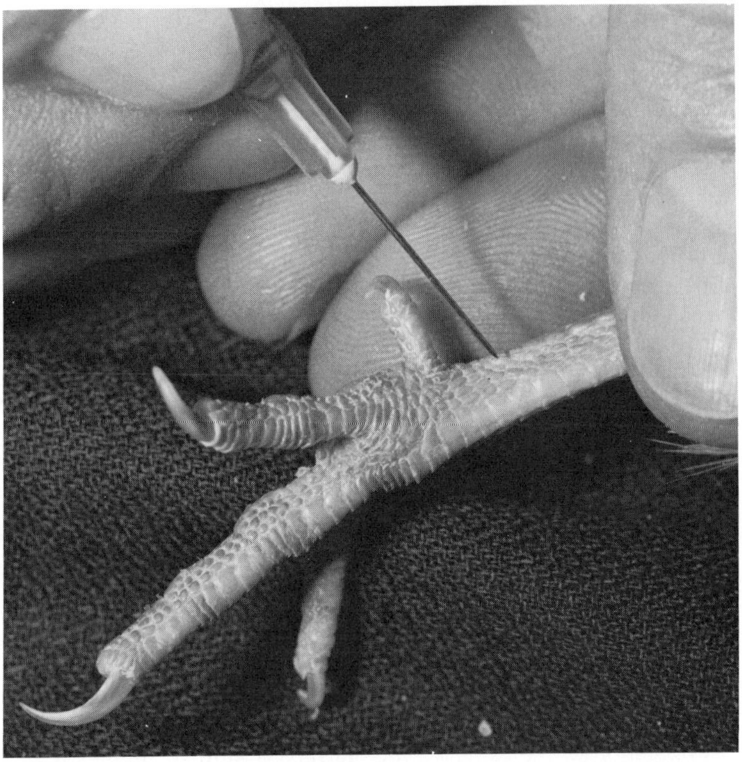

**Figure 12–6.** Location for skin prick over the caudal tibial (medial metatarsal) vein.

the skin. It is essential that the skin be clean and dry before performing this procedure.

**Heart Puncture.** This technique can rapidly provide a large amount of blood but can be dangerous to the bird. The mortality in chickens is estimated at 1 per cent.[34] The bird is placed on its back and a long (1½-inch) needle is inserted in a posterior direction along the ventral margin of the thoracic inlet.*

**Dural Venous Sinus.** A method has re-

*Editor's Note:* This cannot be recommended for pet birds.

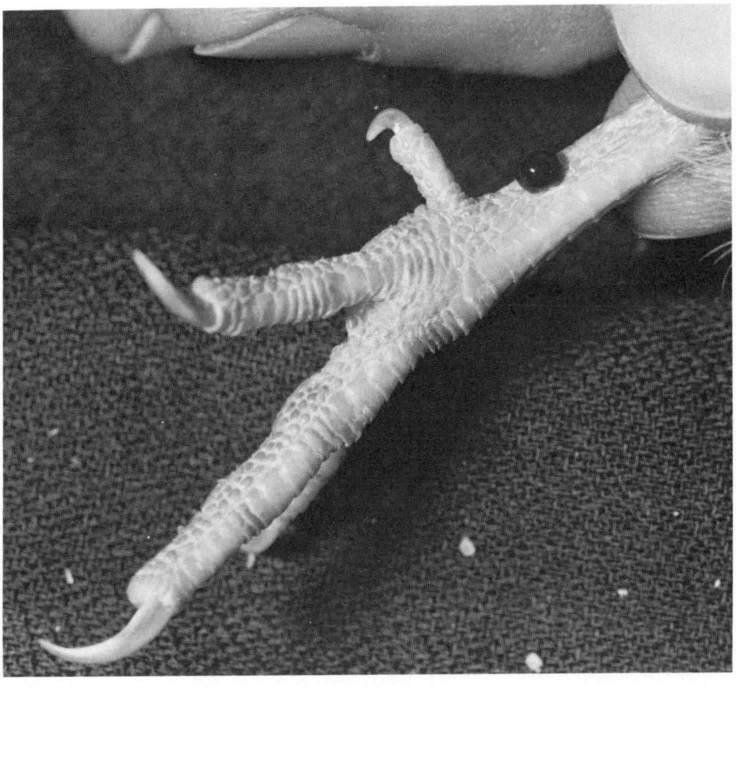

**Figure 12–7.** After skin prick, a drop of blood will form on skin surface which can be collected into capillary tube or container.

cently been described using the venous occipital sinus located at the junction of the base of the skull and the first cervical vertebrae.[89] This method utilizes evacuated blood tubes and permits multiple, large samples. The needle is inserted at a 30- to 40-degree angle to the skull, as the head is held with the beak perpendicular to the neck. When the skin is punctured, the Vacutainer tube is perforated and the needle is pushed a few millimeters deeper. This technique has been described for use only in ducks and geese. Its use in psittacine species may be limited to collection of large volumes of blood prior to euthanasia.

## Sample Containers

Samples are often collected into standard microhematocrit tubes, which have a capacity of 0.07 ml. Other sizes containing from 0.001 to 0.2 ml are also available (Drummond Scientific). Caraway tubes (0.37 ml capacity) and Natelson tubes (0.25 ml) are large-bore capillary tubes that can be purchased from any scientific supply house. Heparin is the most common anticoagulant coating for these tubes, but EDTA-coated capillary tubes can be obtained (Drummond). Samples collected into capillary tubes should be transferred to another container to avoid blood separation in the tubes, making reconstitution of the sample difficult.

Various other blood collection and storage containers are available. W. Sarstedt makes 0.3- and 1.0-ml microcollection devices with a variety of anticoagulant coatings. These tubes can also be used as centrifuge tubes for serum or plasma separation. Owing to the conical design of the devices, small samples may settle in the apex and not mix properly. The Microtainer (Becton-Dickinson) is an excellent, although expensive, sample container/collection device that is available without an anticoagulant, with heparin, or with EDTA. A serum separator variety using a silicon gel simplifies serum collection. Standard mammalian blood tubes (e.g., Vacutainer, Venoject) in sizes from 2 to 15 ml are commonly used. The tubes should be filled to at least half their capacity to avoid sample dilution and excessive anticoagulant concentration; this is most important with the smaller (2-ml) tubes. Glass and plastic test tubes, to which anticoagulant must be added, are manufactured by many different companies and are available in many sizes. Brockway Medical Products makes a 0.8-ml sampler vial that is very useful for small samples.

## Anticoagulants

Any anticoagulants used at the same concentration as in mammalian hematology will usually work with avian blood. However, there is an unusual reaction of the blood of some birds (crows, jays, brush turkeys, hornbills) to EDTA (1 to 2 mg/ml blood). After approximately 15 minutes, the blood becomes dark and viscous. Although the blood does not clot, the cells are disrupted enough to make the sample useless. It is best to use heparin (25 IU/ml blood) for these species.

EDTA is the anticoagulant of choice for hematology; heparin is best for tests requiring plasma. Oxalates, sodium fluoride, and citrate can also be used in special circumstances.

## Sample Handling

**BLOOD SMEARS.** The two-slide push or wedge smear technique[4, 42, 75] that most clinicians use may not be the best method for avian blood. In addition to margination of the granulocytes and monocytes at the edges and tail of the smear, this method seems to be more damaging to avian blood cells. The slide and coverslip method (Fig. 12–8) usually produces smears with fewer "smudge" cells and a more uniform distribution of cells.[2] A variant of this method is to use two square coverslips and pull them apart with a twisting motion.[4, 75] Automatic slide spinners (Corning) make slides with a uniform cell distribution and little cell damage.[68]

**STAINING.** The most crucial aspects of a staining method are consistent results and an equal preservation of all cell types. No one stain is perfect, and some experimentation is required to establish a proper method. Many stain formulations have been proposed for avian blood,[15, 41, 54, 74] and it is useful to have a few different ones available. Some stain formulae and procedures are outlined in Chapter 17, Cytology.

Quick stains (e.g., Diff-Quick) are very convenient but may selectively damage some cell types, making identification difficult and adversely affecting differential counts. Erythrocyte polychromasia is not as apparent with these stains. Wright or Wright/Giemsa stains take longer but produce more consistent results. Buffer/water pH is important with these stains; a solution too acid will favor red colors and one too alkaline will favor blue colors. Slides stained with automatic stainers (Hematek) are usually of very good quality and consistency.

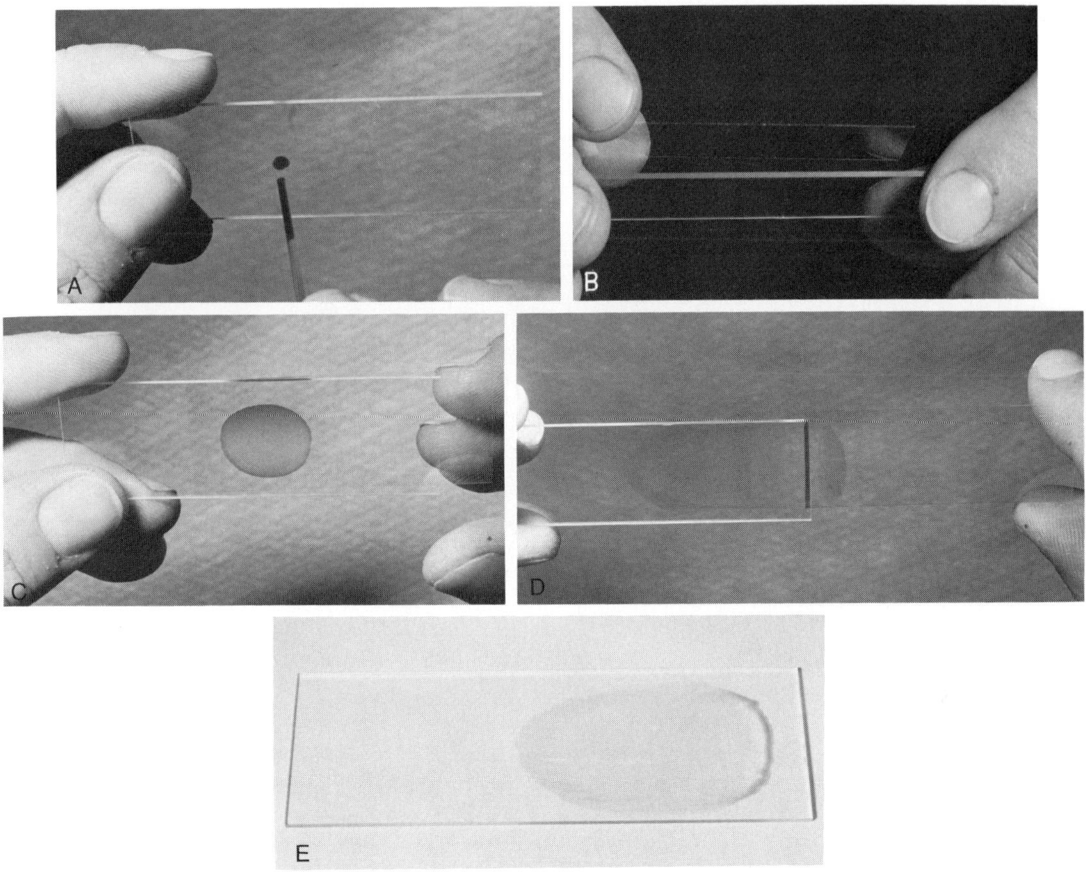


**Figure 12–8.** Slide and coverslip method for preparing blood smears. *A*, A drop of blood is placed on a slide. *B*, A long coverslip (50 to 60 mm) is placed onto the drop of blood. *C*, The blood will quickly spread between slide and coverslip. *D*, The coverslip is then drawn evenly away from the slide. Coverslip must not be lifted, but drawn out parallel to the slide. *E*, The slide is rapidly air dried. Differential leukocyte count is performed by examination of the central area of the smear.

## Cell Identification

Identifying cell types found in the peripheral blood of different avian species can be a frustrating experience. Although there are many similarities to mammalian blood, at first glance the differences may seem overwhelming (Fig. 12–9). One obvious difference is that all avian blood cells are nucleated. Since morphology can vary between birds of the same species or between successive samples from one individual, one must develop criteria that permit consistency from slide to slide. The criteria should not be so loose that classification is meaningless. It is useful to record cell descriptions for each differential for future reference.

**ERYTHROCYTES.** The mature cells (erythrocytes) are oval to elliptic with a similarly shaped nucleus. A good stain will give the cytoplasm a rich orange-yellow color, contrasting with a purple-magenta nucleus containing densely packed chromatin. The immature cells (polychromatophilic erythrocytes, reticulocytes) are rounder, have a blue tinge to the cytoplasm, and have loosely packed nuclear material. Very early cells (basophilic erythroblasts) have a greater nuclear/cytoplasmic (N/C) ratio, a fine-grained magenta nucleus, and dark blue cytoplasm. These cells are usually larger than the mature erythrocytes. In severe anemia, the erythrocyte shape may be distorted. This is due to heightened erythropoiesis that causes the bone marrow to release cells before they have completed development.[26]

**THROMBOCYTES.** These cells are potentially the most confusing owing to their varied appearance. *In vivo* they look like small erythrocytes, with an oval cell and nuclear shape. The cytoplasm is clear or faintly blue. The nucleus is dark with little chromatin visible. Often there will be one to three serotonin-containing magenta granules in the cytoplasm. These granules

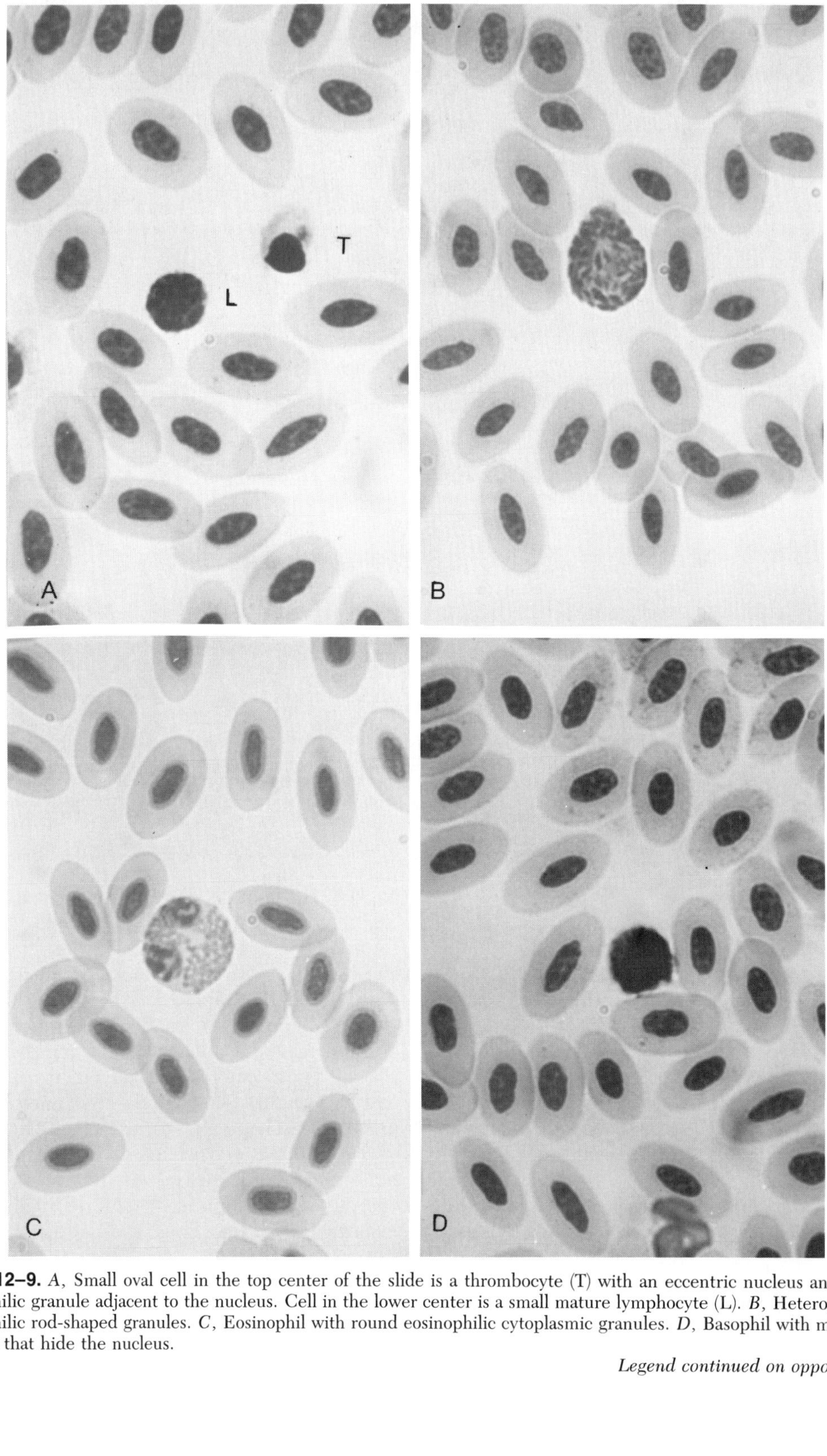


**Figure 12–9.** *A*, Small oval cell in the top center of the slide is a thrombocyte (T) with an eccentric nucleus and a large eosinophilic granule adjacent to the nucleus. Cell in the lower center is a small mature lymphocyte (L). *B*, Heterophil with eosinophilic rod-shaped granules. *C*, Eosinophil with round eosinophilic cytoplasmic granules. *D*, Basophil with many dark granules that hide the nucleus.

*Legend continued on opposite page*

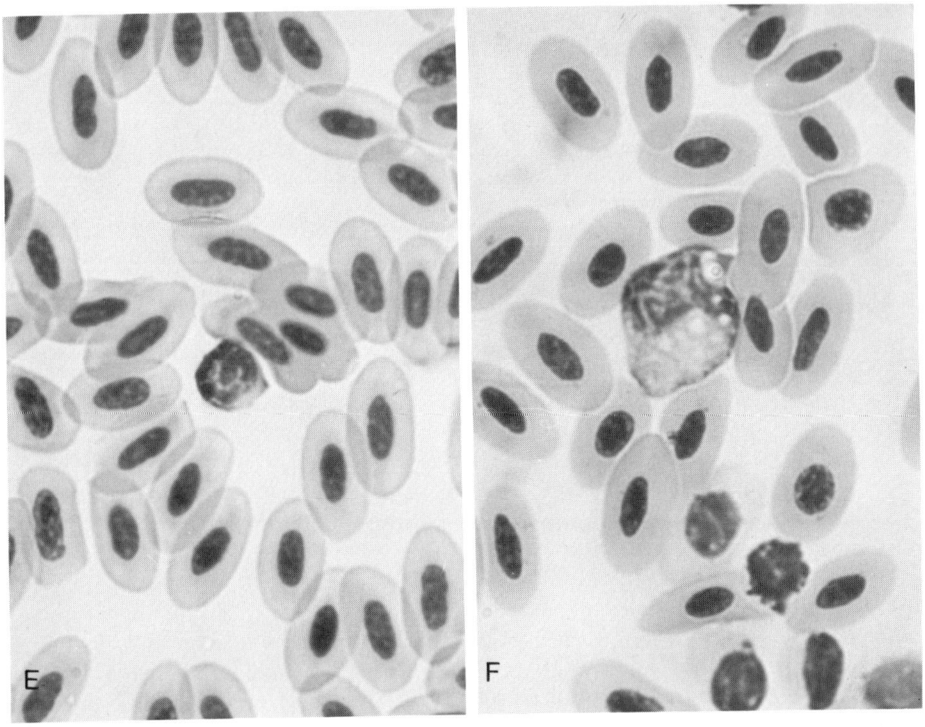


**Figure 12–9** *Continued.* *E*, Small mature lymphocyte. *F*, Monocyte. (Wright's stain, × 1000.) (From Campbell, T. W.: Diagnostic cytology in avian medicine. Vet. Clin. North Am. [Small Anim. Pract.], 14(2):317–344, 1984.)

appear as vacuoles upon degranulation.[41] Thrombocytes will frequently be clumped on a smear, aiding in their identification.

HETEROPHILS/EOSINOPHILS. In most mammals, the major granulocyte types, the neutrophil and eosinophil, are readily distinguished by the color of their cytoplasmic granules. Unfortunately, in birds the heterophil (the neutrophil analogue) and the eosinophil have similar red-orange granules, making their differentiation more difficult. The shape of the heterophil granule is usually described as elongated or rod-shaped and that of the eosinophil granule as round or spherical. Electron microscopic studies show that there are granules of different sizes and shapes in both the heterophil and eosinophil—some rodlike, some oval, and some round.[50, 60] The heterophil granules are subject to water dissolution, and their shape can be dependent on the tonicity of the staining solution.[57] Since the resolution of the light microscope is not always sufficient to distinguish granule shape, other morphologic criteria such as size, nuclear configuration, and staining characteristics should be used to differentiate heterophils and eosinophils. In simple terms, the two cell types look different and a thorough scanning of the slide before initiating the count will usually permit the distinction to be made.

BASOPHIL. These cells are more common in avian blood than in mammalian. The cell's appearance is so similar to its mammalian counterpart that there is little confusion. Basophils in various stages of degranulation are often found.

LYMPHOCYTES/MONOCYTES. Distinguishing between these cells can be difficult. They are not functionally related, but their morphology on stained smears can be very similar. It is easiest to speak of a lymphocyte/monocyte continuum because there are many intermediate forms that are difficult to classify. The same problem exists with mammals; however, the variability among bird species makes differentiation more difficult.

There are two common forms of lymphocytes; one is similar to the characteristic mammalian form, i.e., uniformly round, high N/C ratio, dark blue cytoplasm, and a magenta, eccentrically placed nucleus. The other is a cell with an irregular cell border and light blue cytoplasm, sometimes containing numerous distinct magenta granules. This cell resembles the large granular lymphocyte described in rats and humans. The presence of these granules makes it possible to confuse this lymphocyte with the thrombocyte. The lymphocyte will have more numerous and larger granules than the thrombocyte; the thrombocyte will be smaller and its nucleus will be darker. Dhingra et al. have

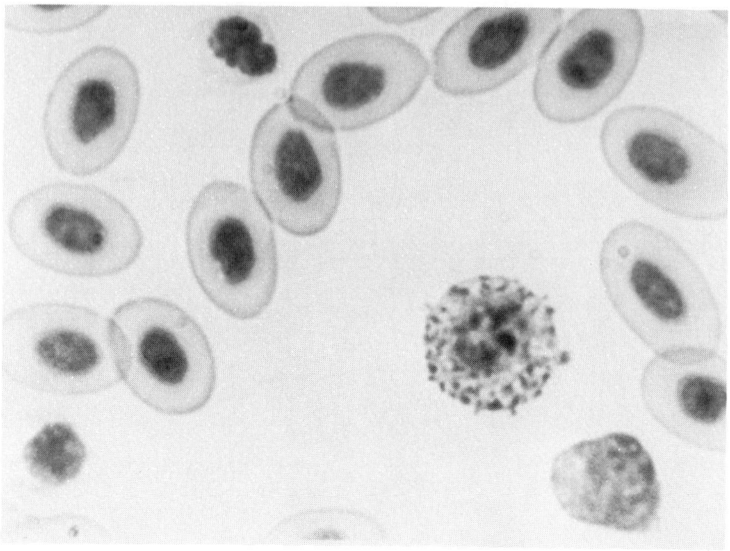

**Figure 12–10.** Basophil, reactive lymphocyte, and thrombocytes in the blood of a chicken with salmonellosis (Wright's stain, × 1000). (Courtesy of Terry W. Campbell.)

described the ultrastructural differences between the cells.[20] Generally, monocytes are larger than lymphocytes. They have more cytoplasm, which is fine-grained, blue, and usually has some degree of vacuolization. The nucleus can be bean-shaped, round, or even bilobed, and is eccentrically placed.

## Cell Counting

The many technical problems associated with counting avian blood cells create variations in the results. It is better to evaluate hematologic results as a comparison or trend, rather than looking at just one count. A rising or falling hematocrit or WBC may be more significant and diagnostically useful than one isolated value. Repeat sampling to assess progress should be a standard procedure.

### *Erythrocytes*

**HEMATOCRIT (HT, PACKED CELL VOLUME, PCV).** The hematocrit (ratio of the volume of erythrocytes to the volume of whole blood) is a good indirect measurement of red cell mass. The microhematocrit method is the most convenient for clinical practice. Seventy-five millimeter microhematocrit capillary tubes (0.07 ml) or tubes as small as 0.001 ml can be used. Filling a 75-mm tube with blood to a height of 20 mm will yield a result as accurate as with a completely filled tube.[38] Centrifugation at 12,800 g for five minutes is necessary to achieve maximum packing of the cells.[14]

The hematocrit is a consistent blood cell parameter, but it can be affected by age, molting, reproductive cycle, and air temperature.[66, 80] There is considerable species variation; most values fall between 37 and 53 per cent. Extensive compilations of hematocrit values for birds are available.[38, 86]

**TOTAL RED BLOOD CELL COUNT (RBC).** The total erythrocyte number is another estimation of red cell mass. It can be determined either with a hemacytometer or an electronic counter.

With the hemacytometer, the RBC can be counted concurrently with the leukocytes (e.g., using Natt and Herrick's solution) or alone using the Unopette system (Test 5850). The counting procedure is identical to that for mammals.[75] Precautions should be taken to assure the greatest accuracy possible when using the hemacytometer.[6]

An electronic particle counter (Coulter Counter, Model $Z_{BI}$) with the lower threshold set at 10 and upper threshold off produces very precise counts (coefficient of variation [CV] $<$ 2 per cent). Electronic counts correlate better with the hematocrit than do hemacytometer counts. Other methods of using electronic counters for RBC have been reported.[32, 81] The RBC of birds has a general range of 2 to 4 million cells/cu mm.[38, 86]

**HEMOGLOBIN (HB).** Measurement of avian Hb is almost identical to the procedure for mammals. The most common method is spectrophotometric analysis of cyanmethemoglobin at 540 nm.[4, 33, 52, 75] The only difference is that avian blood must be centrifuged for 10 minutes at 1000 g following red cell lysis to remove the nuclear and cytoplasmic debris. Reference ranges of 10 to 20 gm/dl have been reported for various species.[38, 86]

**Red Cell Indices.** As for mammals, the mean corpuscular volume (MCV), mean corpuscular hemoglobin (MCH), and mean corpuscular hemoglobin concentration (MCHC) can be calculated from the Ht, RBC, and Hb.

$$\text{MCV (fl)} = \frac{\text{Ht} \times 10}{\text{RBC}}$$

$$\text{MCH (pg)} = \frac{\text{Hb} \times 10}{\text{RBC}}$$

$$\text{MCHC (gm/dl)} = \frac{\text{Hb} \times 100}{\text{Ht}}$$

Reference ranges must be established for each species involved for the indices to be useful. Some values are available and others can be calculated from published data.[38, 86] The indices are useful in describing the cell size and degree of hemoglobinization and in classifying disease states, as shown in Table 12–1.

**Reticulocytes.** When an immature erythrocyte is stained by vital dyes such as new methylene blue or brilliant cresyl blue, the cell develops a bluish stippling or reticular pattern and is called a reticulocyte. The reticulum results from the staining of ribosomal RNA, which is abundant in immature cells. Both the concentration of stain and the staining time affect the amount of reticulum observed. A 1 per cent solution of new methylene blue in normal saline is mixed 1:1 with whole blood in a test tube or covered spotting dish and incubated for 15 minutes; a smear is made and examined.

The maturation stage most often identified as a reticulocyte in conventionally stained preparations is the polychromatophilic erythrocyte, but basophilic erythroblasts and the mature erythrocytes may also contain reticulum. The stippling in avian reticulocyes is much more extensive than in the mammalian reticulocyte, and many more cells will stain. The resultant

**Table 12–1.** MORPHOLOGIC CLASSIFICATION OF ANEMIA

| | MCH Normal | MCH Decreased |
|---|---|---|
| MCV normal | Normocytic Normochromic | Normocytic Hypochromic |
| MCV increased | Macrocytic Normochromic | Macrocytic Hypochromic |
| MCV decreased | Microcytic Normochromic | Microcytic Hypochromic |

**Table 12–2.** POLYCHROMATOPHILIC INDEX

1. Erythrocytes homogenous in size, nuclear shape, and texture. Almost no polychromatophilic cells.
2. Erythrocytes not entirely homogenous. A few polychromatophilic cells present (<10%).
3. Many forms present, from polychromatophilic through mature erythrocytes. Moderate number of polychromatophilic cells (10–20%).
4. Significant number of polychromatophilic cells (20–50%).
5. Large number of polychromatophilic cells (>50%). Significant poikilocytosis.

difficulty in identifying "true" reticulocytes may account for the wide range of values published.[13, 41, 43, 67] One method, which is easy to use, is to count only cells with five or more reticulum clumps as reticulocytes.[11] An alternate method is to identify any cell with a complete ring of reticulum around the nucleus as a reticulocyte.[37, 41] Reference ranges will depend on the method used, but in general, a reticulocyte count of less than 10 per cent is in the normal reference range.

Another method for indicating the incidence of immature erythrocytes is the use of a "polychromatophilic index" as described in Table 12–2. This is a score of the number of distinctly polychromatophilic cells present in a normally stained peripheral blood smear, based on a scale of 1 to 5. This index can be likened to the 0 to +4 polychromasia scale often used in mammalian hematology.[52, 75] At this time, there are no data correlating this index with the number of true reticulocytes in birds, but this system does provide a qualitative, comparative measurement of the erythropoietic response.

### *Leukocytes*

**Total White Blood Cell Count (WBC).** Total WBC is the most difficult parameter to measure in avian blood. Because of the interference of nuclei from lysed RBC and thrombocytes, common electronic counting methods are not applicable. The difficulty in differentiating WBC and RBC nuclei causes a similar problem with hemacytometer counts. Many diluting solutions have been proposed to facilitate WBC counting.[18, 56, 77] Any of these may be used, but some require a reasonable amount of experience for proper cell identification. Others do not permit direct counting of all WBC, and some have limited stability after formulation. Two of the most commonly used solutions are described below.

Natt and Herrick's (N & H) solution (Table 12–3) is a commonly used diluent. The direct

**Table 12–3.** NATT AND HERRICK'S SLUTION[56]

| | |
|---|---|
| NaCl | — 3.88 g |
| $Na_2SO_4$ | — 2.50 g |
| $Na_2HPO_4 \cdot 12\ H_2O$ | — 2.91 g |
| $KH_2PO_4$ | — 0.25 g |
| Formalin (37%) | — 7.50 ml |
| Methyl violet 2B | — 0.10 g |

All ingredients are dissolved in distilled water in the order listed and diluted to a total volume of 1000 ml in a volumetric flask. After standing overnight, the solution is filtered through fine filter paper (Whatman No. 2). The solution has a pH of 7.3.

counting method using N & H solution stains all leukocytes. N & H solution is stable up to two years.[56] A disadvantage is that erythrocytes and thrombocytes, as well as blood parasites, are also stained to varying degrees, which can cause confusion in counting. Species variation in the mononuclear cell morphology may also create a problem. With experience this method can be fairly easy to perform. To use N & H solution, blood is diluted 1:100 in a glass diluting pipette, with graduated pipettes and test tubes, or with Unopette reservoirs filled with N & H solution. (Empty reservoirs are available as a special order from Becton-Dickinson.) Blood and diluent are mixed well, and both chambers of the hemacytometer are charged with the solution. The filled hemacytometer sits covered for five minutes to allow the cells to settle and pick up the dye. At 400 magnification (high dry), all darkly stained cells in all nine large squares of both chambers are counted. To calculate the total WBC, 10 per cent is added to the total count, which is then multiplied by 100.

A prepackaged counting solution is available in the Eosinophil Unopette (B-D, Test No. 5877) which uses the dye phloxine B (Wiseman's solution[94]).[16] The cells are usually easy to identify in the hemacytometer, making this system easy to learn. A disadvantage is that the dye stains only the heterophils and eosinophils; the lymphocytes, monocytes, and basophils must be added indirectly from the differential count. Therefore, a well-prepared blood smear and accurate cell identification are required. The Eosinophil Unopette technique follows the package insert directions. The blood is diluted 1:32 in the reservoir. Both chambers of the hemacytometer are charged and then left for five minutes to allow the cells to settle and pick up the stain. At 100 magnification (low dry), all bright red-orange refractile cells in all nine large squares of both chambers of the hemacytometer are counted. Ten per cent is added to the total cells counted and the result is multiplied by 16. This is the total heterophil and eosinophil count. From the differential leukocyte count on the blood smear (see below), the percentages of heterophils and eosinophils are determined. These values are substituted into the formula below for the total WBC.

$$\text{Total WBC} = \frac{\text{Total Heterophils + Eosinophils}}{\text{\% Heterophils + Eosinophils}} \times 100$$

**DIFFERENTIAL LEUKOCYTE COUNT.** Blood smears are made and stained to ascertain the relative proportions of leukocytes in the sample, especially if the Unopette method is used for total WBC. There can be a considerable variation in the results obtained from similar slides made from the same sample[40, 42]; therefore, it is important to reduce causes of errors created during smear preparation and cell identification.

If slides are prepared by the two-slide push technique, more attention must be paid to the method to obtain accurate results. The battlement method is a way of routinely covering all sections of the slide to maximize accuracy, but other patterns are also recommended.[4, 42] When using the slide and coverslip method to make a smear, the counting pattern is not as important.[2]

Most laboratories count 100 cells for their differential count. Statistically, increasing the number of cells counted does not result in a better estimate of the real population unless at least 400 cells are counted.[42]

### *Thrombocytes*

A total thrombocyte count (TBC) is derived directly using Natt and Herrick's solution (same as for WBC) or indirectly from the peripheral smear. The TBC per 100 leukocytes can be determined when performing the differential count. An alternative estimate of TBC can be made by counting the number of thrombocytes per oil immersion field in a monolayer area of the blood smear.[10] This method is only a reasonable estimate if the packed cell volume is normal and the smear is well-made. Both indirect methods are difficult owing to the tendency of thrombocytes to clump on smears.

## Coagulation

In response to vascular injury, the avian thrombocytes (mammalian platelet analogues) aggregate and form an initial plug.[39, 82] This aggregation is most likely mediated by degranulation of the thrombocyte-specific granules with the release of 5-hydroxytryptamine.[3, 36] The coagulation process from this point is still controversial. There is evidence from a number of species that avian hemostasis is dependent on the activity of the extrinsic clotting system, since there is little intrinsic thromboplastin generated or decreased clotting time after contact with glass.[8, 84] Citing data from chickens with aflatoxicosis, Doerr and Hamilton disagree, claiming that increased partial thromboplastin and recalcification times are evidence of a clotting pathway complementary to the extrinsic system.[21] Obviously, more experimental information is needed to describe the complete avian coagulation picture.

All coagulation tests that are performed in mammals have been done in birds.[1, 21, 52, 84] The procedures are similar with the exception that for prothrombin times, avian thromboplastin, derived by brain acetone dehydration, must be used. Brain extracts for avian species should also be used for partial thromboplastin and activated partial thromboplastin times.[21]

A wide range of values has been reported for coagulation tests in poultry, with results varying with the method used.[1, 3, 8, 21, 39, 84] Until the coagulation process is better understood and testing standardized, normal controls of the same species should be utilized whenever testing is undertaken.

## Interpretation

With some recent exceptions, most experimental data on the correlation of blood cell changes with disease states are derived from work on domestic species. A review of the early literature in this field is provided by Olson[62] and Lucas and Jamroz.[41] With variation in methods and therefore results, the available literature does not fully describe the avian hematopoietic response to disease and toxic agents, even for domestic birds. However, until further definitive work is done on nondomestic species, we must rely on the chicken and quail as our experimental models to provide us with comparative data, keeping in mind that the results may not be directly applicable to other birds.

In addition to the experimentally derived data, there is substantial information reported from clinical practice. These reports can be very important in directing attention to possible clinicohematologic correlations, but they should not be accepted as fact until they have been confirmed in multiple practice situations and/or experimental disease studies.

### *Erythrocytes*

Anemia is a reduction below normal levels of the erythrocyte number and/or hemoglobin concentration per volume of blood.[52] It is the most common erythroid abnormality. As in mammals, anemia itself is not a disease but an indication of a disease process. Once a patient is diagnosed as anemic, the clinician must classify the anemia in order to discover the underlying cause.

In general, all anemias can be divided into two types: regenerative and nonregenerative. This distinction can be made by examination of the peripheral blood for the presence or absence of reticulocytes and polychromatophilic cells. Regenerative anemias are those in which there is active erythropoiesis to replace lost cells; this results in a higher percentage of immature forms in the circulation than is normally present. Causes of regenerative anemia include blood loss and increased erythrocyte destruction. If there are few or no reticulocytes in the peripheral smear, the anemia is nonregenerative. This is due either to erythrocyte maturational defects or to a decrease in bone marrow production of red blood cells.

Anemias from all causes listed above have been reported in birds.[11, 12, 22, 31, 32, 37, 46, 47, 67, 73, 97] Published data on avian anemias can be difficult to interpret because of differing techniques and approaches, but some general features are apparent.

**Regenerative Anemias.** The picture of acute blood loss is similar, whether due to extravascular hemorrhage or intravascular hemolysis. There is a noticeable decline in hematocrit, hemoglobin, and RBC within a few hours. The length of time to maximum decline, as well as to complete recovery of these parameters, varies with the magnitude of the loss and the species involved. Following a hemorrhagic loss of 30 per cent of the initial blood volume, ducklings reach maximum depression of the Ht at 10 hours and return to normal at 72 hours. Cockerels given phenylhydrazine (an experimental hemolytic agent) had their lowest erythroid parameters at 72 hours and returned to normal at 7 to 8 days.[12] The anemia in these studies was initially normocytic, normochromic, but gradually changed to a macrocytic, normochromic or slightly hypochromic picture. In other words, the Ht recovers more quickly than the RBC number because of the large size of the immature cells released into the circulation.

Even though the MCHC may be close to normal, increased numbers of polychromatophilic cells are seen in the peripheral blood, resulting in a higher polychromatophilic index. Ramis and Planas observed a hypochromic, macrocytic anemia in chickens and pigeons in a phenylhydrazine-induced anemia.[64]

Chronic hemorrhage or chronic hemolysis, which develops more slowly, with a smaller loss of red blood cells, results in the same erythroid picture, but not the same magnitude of decline and recovery of erythrocyte values.[11] These patients will show the initial drop in Ht, Hb, and RBC, but it may be slower. During recovery these values may not completely return to baseline until the cause of the anemia is removed.[83]

With some variation depending on the species and disease process, reticulocytes will be seen soon after the drop in other RBC parameters is obvious. Reticulocyte counts greater than 50 per cent are not uncommon in severe anemias.[12]

Severe acute and chronic anemias are characterized by dramatic anisocytosis and poikilocytosis. Since the marrow is releasing all available hemoglobin-carrying cells, the erythroid development is accelerated and immature cells with incomplete microtubule development are released into the peripheral blood.

Parasitism, including both blood and gastrointestinal infections, can result in both acute and chronic regenerative anemias (see Chapter 37, Parasites). Macrocytic, normochromic anemias are common in bacterial infections.[62] Microcytic anemia has been reported in African Grey Parrots (*Psittacus erithacus*) with bacterial infections.[31] Hypochromic anemia has been seen in cranes with *Mycobacterium avium* infection.[32]

Anemias due to toxic substances also have been described. Female chickens fed rapeseed oil for 10 weeks developed a macrocytic, slightly hypochromic anemia.[47] Experimental administration of dimethyl disulfide to chickens and petroleum to gulls and puffins produced anemias in which the erythrocytes contained Heinz bodies, which are products of hemoglobin denaturation.[37, 46]

NONREGENERATIVE ANEMIA. The hallmark of a nonregenerative anemia is the presence of normal to decreased numbers of circulating reticulocytes. This indicates that the bone marrow is not responding to the anemia-caused decrease in oxygen-carrying capacity of the blood. There is significant variation in the reported values of the RBC indices in birds with nonregenerative anemias, perhaps due to species differences.

Classic iron deficiency has been experimentally produced in chickens.[53] In the study, a microcytic, hypochromic anemia readily responded to iron administration. The dietary iron required to maintain the Ht in adult chickens was found to be 35 to 40 ppm in the feed.

Experimental viral infections have resulted in nonregenerative anemias. In these cases, it was a microcytic form that developed slowly after inoculation. The bone marrow was very hypoplastic, suggesting a severe inhibition of hematopoiesis.[97]

Antibiotic administration has also been associated with decreased red cell production. Early work with chloramphenicol showed that erythropoiesis is depressed in ducks given 1000 mg of the drug orally.[67] This was seen as a sharp drop in reticulocytes, with a maximum depression on the third day. Despite an additional dose, reticulocyte counts began to recover from this point. The same study also showed that withholding food, without drug administration, resulted in a complete absence of reticulocytes.

Pesticides have been implicated as a cause of red cell depression. Administration of DDT and a carbamate to quail resulted in a significant decrease in Ht after feeding of the pesticide for 30 or 23 days, respectively. The cause of the resultant normocytic, normochromic anemia is suggested to be a decrease in the normal production of erythropoietin.[22]

### *Leukocytes*

One of the most important aspects of evaluating an animal's leukogram is separating changes due to handling and sampling technique from those caused by a pathologic process. There are also blood changes associated with an animal maintained in a stressful environment over a period of time that must be considered.

Davidson and Flack demonstrated that a single dose of adrenal corticotropic hormone (ACTH) in chickens produced a significant decrease in total white blood cell count within one to two hours.[17] This change was due to a decrease in the number of circulating lymphocytes. Between four and 12 hours, there was an increase in the WBC, primarily due to a heterophilia. The first phase of this response is similar to the stress reaction produced when a bird is captured and a blood sample is taken immediately. The second phase is similar to the stress reaction produced in an animal being held for a period of time before the sample is obtained. The findings of leukocytosis, heterophilia, and lymphopenia following ACTH or corticosteroid administration have been corroborated in many studies in chickens.[58, 78, 96] Gross and Siegel suggest that the heterophil/lymphocyte ratio is a good indicator of stress in chickens.[29] A single survey of the effects of ACTH in nondomestic birds produced varied results. In the common mynah *(Acridotheres tristis)* the response was similar to that in chickens. There were no leukogram changes in the pigeon *(Columba livia)* and a heteropenia and lymphocytosis occurred in the cattle egret *(Bubulcus ibis)*. All samples were collected one hour following injection.[7]

Reproductive hormones experimentally injected into chickens also cause peripheral blood changes. Administration of progesterone produces a heteropenia, eosinopenia, and lymphocytosis.[30] Estrogen causes the opposite effect, a heterophilia and lymphopenia.[61] This has significance when sampling birds during the laying cycle; increased progesterone levels are found prior to ovulation,[79] and estrogen levels three times normal are detected after laying.[76]

Many different avian species respond to a wide range of pathogenic agents with a leukocytosis, commonly a heterophilia. This change has been associated with infections with bacteria,[31, 62, 87] including tuberculosis,[32] fungi,[65] herpesvirus,[27] and parasites,[69] with rapeseed ingestion,[47] and with lead poisoning.[95] It is clear that in many of these processes, the leukocytosis may be present only during one phase, but it is well-documented that in bacterial infections, the leukocytosis/heterophilia is a salient feature. With proper antibiotic therapy, the hematopoietic aberrations disappear and the blood picture returns to the reference range for the species.[31] For this reason, the WBC can be a useful indicator of bacterial disease and an important monitor for effective therapy. As the specificity for blood tests is not high, results should be evaluated in light of other clinical data.

Immature heterophils (bands, stabs) have been described in avian peripheral blood.[41, 43, 47] They have been associated with the heterophilia of bacterial infections, but insufficient data exist to determine if this parameter is a reliable diagnostic indicator.[28, 87]

Leukopenias are frequently associated with viral infections in mammals.[75] This change has only infrequently been described in birds.[62] A case of leukopenia in a Sulfur-crested Cockatoo *(Cacatua sulphurea)* with hepatitis caused by a herpesvirus (Pacheco's parrot disease) has been reported.[70] Experimental infection with the same virus in Quaker Parrots *(Myopsitta monachus)* failed to confirm this finding.[27] Further research and clinical reports will determine if leukopenia is a general feature of avian viral infections.

Eosinophilia in birds has been described in association with parasites.[62] An idiopathic eosinophilia associated with facial edema has also been seen in chickens.[49] Multiple injections of horse serum in chickens can experimentally produce an eosinophilia, but many other agents that produce this change in mammals do not do so in birds.[45, 48]

Monocytosis is frequently described in mammals suffering from chronic infections, especially those involving tissue necrosis and granuloma formation.[52, 75] The author has seen this with *Mycobacterium avium* infections in psittacines and waterfowl. Hawkey et al. report increases in cells with irregular nuclei and vacuolated cytoplasm in cranes that are *M. avium*–positive.[32] These cells were counted as lymphocytes, but because of lymphocyte/monocyte similarities, there is a possibility that this could indicate a monocytosis. Monocytosis has been described in chickens, turkeys, and quail on a zinc-deficient diet.[92]

## REFERENCES


1. Archer, R. K.: Blood coagulation. *In* Bell, D. J., and Freeman, B. M. (eds.): Physiology and Biochemistry of the Domestic Fowl. New York, Academic Press, 1971, pp. 897–911.
2. Beacon, D. N.: Differential blood counts. J. Lab. Clin. Med., 13:366–369, 1928.
3. Belamarich, F. A., and Simoneit, L. W.: Aggregation of duck thrombocytes by 5-hydroxytryptamine. Microvasc. Res., 6:229–234, 1973.
4. Benjamin, M. M.: 1978. Outline of Veterinary Clinical Pathology. Ames, IA, Iowa State University Press, 1978.
5. Berchelmann, M. L., et al.: Comparison of hematologic values from peripheral blood of the ear and venous blood of infant baboons *(Papio cyanocephalus)*. Lab. Anim. Sci., 23:48–52, 1973.
6. Berkson, J., et al.: The error of the estimate of the blood cell count as made with the hemocytometer. Am. J. Physiol., 128:309, 1940.
7. Bhattacharyya, T. K., and Sarkar, A. K.: Avian leucocytic responses induced by stress and corticoid inhibitors. Ind. J. Exp. Biol., 6:26–28, 1968.
8. Bigland, C. H., and Starr, R. M.: Comparison of simple blood coagulation tests in birds. Can. Vet. J., 6:233–236, 1965.
9. Bond, C. F., and Gilbert, R. W.: Comparative study of blood volume in representative aquatic and nonaquatic birds. Am. J. Physiol., 194:519–521, 1958.
10. Campbell, T. W., and Dein, F. J.: Avian hematology:

The basics. Vet. Clin. North Am. [Sm. Anim. Pract.], 14:223–248, 1984.
11. Christie, G.: Hematological and biochemical findings in an anemia induced by the daily bleeding of ten-week-old cockerels. Br. Vet. J., 134:358–365, 1978.
12. Christie, G.: Hematological and biochemical findings in an experimentally produced hemolytic anemia in eight-week-old brown leghorn cockerels. Br. Vet. J., 135:279–285, 1979.
13. Coates, V., and March, B. E.: Reticulocyte counts in the chicken. Poultry Sci., 45:1302–1304, 1966.
14. Cohen, R. R.: Anticoagulation, centrifugation time, and sample replicate number in microhematocrit method for avian blood. Poultry Sci., 46:214, 1967.
15. Cook, F. W.: Staining fixed preparations of chicken blood with combination May-Greenwald(sic)-Wright-phloxine B stain. Avian Dis., 3:272, 1959.
16. Costello, R. T.: A Unopette for eosinophil counts. Am. J. Clin. Pathol., 54:249, 1970.
17. Davidson, T. F., and Flack, I. H.: Changes in the peripheral blood leucocyte populations following an injection of corticotropin in the immature chicken. Res. Vet. Sci., 30:79–82, 1981.
18. DeEds, F.: Normal blood counts in pigeons. J. Lab. Clin. Med., 12:437–438, 1927.
19. Dein, F. J.: Laboratory Manual of Avian Hematology. E. Northport, NY, Association of Avian Veterinarians, 1984.
20. Dhingra, L. D., et al.: Electron microscopy of non-granular leucocytes and thrombocytes of chickens. Am. J. Vet. Res., 30:1837–1842, 1969.
21. Doerr, J. A., and Hamilton, P. B.: New evidence for intrinsic blood coagulation in chickens. Poultry Sci., 60:237–242, 1981.
22. Ernst, R. A., and Ringer, R. K.: The effect of DDT, Zectran, and Zytron on the packed cell volume, total erythrocyte count, and mean corpuscular volume of Japanese quail. Poultry Sci., 47:639–643, 1968.
23. Fourie, F. LeR., and Hattingh, J.: Comparative hematology of some South African birds. Comp. Biochem. Phys., 47A:443–448, 1983.
24. Galen, R. S., and Gambino, S. R.: Beyond Normality: The Predictive Value and Efficiency of Medical Diagnoses. New York, John Wiley & Sons, 1975.
25. Galvin, C.: Acute hemorrhagic syndrome of birds. *In* Kirk, R. W., (ed.): Current Veterinary Therapy VIII. Philadelphia, W. B. Saunders Company, 1983, pp. 617–619.
26. Gazaryan, K. S.: Avian erythropoiesis: Cellular and molecular aspects. Acta Biol. Med. Germ., 35:295–303, 1977.
27. Godwin, J. S., et al.: Effects of Pacheco's parrot disease virus on hematologic and blood chemistry values of Quaker Parrots *(Myopsitta monachus)*. J. Zoo Anim. Med., 13:127–132, 1982.
28. Gross, W. B.: Blood cultures, blood counts, and temperature records in an experimentally produced "airsac disease" and uncomplicated *E. coli* infection of chickens. Poultry Sci., 40:691–700, 1961.
29. Gross, W. B., and Siegel, H. S.: Evaluation of the heterophil/lymphocyte ratio as a measure of stress in chickens. Avian Dis., 27:972–979, 1983.
30. Gupta, S. K., et al.: Leucocytic responses to progesterone in juvenile female ducks. Mikroscopie(Wien), 38:356–358, 1981.
31. Hawkey, C. M., et al.: Haematological findings in healthy and sick African Grey Parrots *(Psittacus erithacus)*. Vet. Rec., 111:580–582, 1983.
32. Hawkey, C. M., et al.: Normal and clinical haematology of captive cranes (Gruiformes). Avian Pathol., 12:73–84, 1983.
33. Henry, J. B. (ed.): Clinical Diagnosis and Management by Laboratory Methods. Philadelphia. W. B. Saunders Company, 1979.
34. Hodges, R. D.: Technical methods. *In* Archer, R. K., and Jeffcott, L. B. (eds.): Comparative Clinical Hematology. Oxford, Blackwell Scientific Publications, 1977, pp. 553–554.
35. Kovach, A. G. B., and Szasz, E.: Survival of pigeon after graded haemorrhage. Acta Phys. Acad. Sci. Hung., 34:301–309, 1968.
36. Kuruma, I., et al.: Ultrastructural observation of 5-hydroxytryptamine–storing granules in the domestic fowl thrombocytes. Z. Zellforschr., 108:268–281, 1970.
37. Leighton, F. P., et al.: Heinz body hemolytic anemia from ingestion of crude oil: Primary toxic effect in marine birds. Science, 220:871–873, 1983.
38. Leonard, J. L.: Clinical laboratory examinations. *In* Petrak, M. L. (ed.): Diseases of Cage and Aviary Birds. Philadelphia, Lea and Febiger, 1982, pp. 269–303.
39. Lewis, J. H., et al.: Comparative hematology: Studies on class aves: Domestic turkey *(Meleagris gallopavo)*. Comp. Biochem. Phys., 62A:735–745, 1979.
40. Lucas, A. J., and Dennington, E. M.: The statistical reliability of differential counts of chicken blood. Poultry Sci., 37:554–549, 1958.
41. Lucas, A. J., and Jamroz, C.: Atlas of Avian Hematology. Washington, DC, USDA Monograph No. 25, 1961.
42. MacGregor, R. G., et al.: The differential leucocyte count. J. Pathol. Bacteriol., 51:337–368, 1940.
43. Magath, T. B., and Higgins, G. M.: The blood of the normal duck. Folia Hematol., 51:230–241, 1935.
44. Martin, H. F., et al.: Normal Values in Clinical Chemistry: A Guide to Statistical Analysis of Laboratory Data. New York, Marcel Dekker, Inc., 1975.
45. Maxwell, M. H.: Attempted induction of an avian eosinophilia using various agents. Res. Vet. Sci., 29:293–297, 1980.
46. Maxwell, M. H.: The production of a "Heinz body" anaemia in the domestic fowl after the ingestion of dimethyl disulphide: A haematological and ultrastructural study. Res. Vet. Sci., 30:233–238, 1981.
47. Maxwell, M. H.: The effect of dietary rapeseed meal on the hematology and thrombocyte ultrastructure of the adult fowl. Avian Pathol., 11:427–440, 1982.
48. Maxwell, M. H., and Burns, R. B.: Experimental eosinophilia in domestic fowls and ducks following horse serum administration. Vet. Res. Comm., 5:369–376, 1982.
49. Maxwell, M. H., et al.: Eosinophilia associated with facial oedema in fowls. Vet. Rec., 105:232–233, 1979.
50. Maxwell, M. H., and Trejo, F.: The ultrastructure of white blood cells and thrombocytes of the domestic fowl. Br. Vet. J., 126:583, 1970.
51. Meites, S., and Levitt, M. J.: Skin-puncture and blood-collecting techniques for infants. Clin. Chem., 25:183–189, 1979.
52. Miele, J. B.: Laboratory Medicine: Hematology. 6th ed. St. Louis, C. V. Mosby Company, 1982.
53. Morck, T. A., and Austic, R. E.: Iron requirements of white leghorn hens. Poultry Sci., 60:1497–1502, 1981.
54. Mukkur, T. K. S., and Bradley, R. E.: Differentiation of avian thrombocytes by use of Giemsa stain. Poultry Sci., 46:1595, 1967.

55. Murphy, E. A.: The normal, and the perils of the sylleptic argument. Perspect. Biol. Med., 15:566–582, 1972.
56. Natt, M. P., and Herrick, C. A.: A new blood diluent for counting the erythrocytes and leucocytes of the chicken. Poultry Sci., 31:735–738, 1952.
57. Natt, M. P., and Herrick, C. A.: Variations in the shape of the rod-like granule of the heterophil leucocyte and its possible significance. Poultry Sci., 33:828–830, 1954.
58. Newcomer, W. S.: Factors which influence acidophilia induced by stresses in the chicken. Am. J. Physiol., 194:251–254, 1958.
59. Newel, G. W., and Shaffner, C. S.: Blood volume determinations in chickens. Poultry Sci., 29:78–87, 1950.
60. Nirmalan, G. P., et al.: Ultrastructural studies on the leucocytes and thrombocytes in the circulating blood of Japanese quail. Poultry Sci., 51:2050–2055, 1972.
61. Nirmalan, G. P., and Robinson, G. A.: Hematology of Japanese quail treated with exogenous stilbestrol dipropionate and testosterone propionate. Poultry Sci., 51:920–925, 1972.
62. Olson, C.: Avian hematology. *In* Biester, H. E., and Swarte, L. H. (eds.): Diseases of Poultry. 5th ed. Ames, IA, Iowa State University Press, 1965, pp. 100–119.
63. Palmer, J., et al.: Haematological changes associated with hemorrhage in the pigeon. Comp. Biochem. Physiol., 63A:587–589, 1979.
64. Ramis, J., and Planas, J.: Hematological parameters and iron metabolism in pigeons and chickens with phenylhydrazine-induced anemia. Avian Dis., 26: 107–117, 1982.
65. Redig, P. T.: Aspergillosis. *In* Kirk, R. W. (ed.): Current Veterinary Therapy VIII. Philadelphia, W. B. Saunders Company, 1983.
66. Rehder, N. B., et al.: Variations in blood PCV of captive American kestrels. Comp. Biochem. Physiol., 72A:105–109, 1982.
67. Rigdon, R. H., et al.: Anemia produced by chloramphenicol in the duck. AMA Arch. Pathol., 58:85–93, 1952.
68. Rogers, C. H.: Blood sample preparation for automated differential systems. Am. J. Med. Technol., 39:435–442, 1973.
69. Rose, M. E., et al.: Peripheral blood leucocyte response to coccidial infection: A comparison of the response in rats and chickens and its correlation with resistance to reinfection. Immunology, 36:71–79, 1979.
70. Rosskopf, W. J., et al.: Chronic endocrine disorder associated with inclusion body hepatitis in a Sulfur-crested Cockatoo. J. Am. Vet. Med. Assoc., 179: 1273–1276, 1981.
71. Rosskopf, W. J., et al.: Pathogenesis, diagnosis, and treatment of adrenal insufficiency in psittacine birds. Calif. Vet., 5:26–29, 1982.
72. Rosskopf, W. J., et al.: Hematologic and blood chemistry values for common pet avian species. V.M./S.A.C., 77:1233–1239, 1982.
73. Rosskopf, W. J., et al.: Trauma-induced anemia in a Patagonian Conure. Calif. Vet., 4:8, 1982.
74. Santamarina, E.: A formalin-Wright stain for avian blood cells. Stain Tech., 39:267, 1964.
75. Schalm, O. W., et al.: Veterinary Hematology. 3rd ed. Philadelphia, Lea and Febiger, 1975, p. 21.
76. Senior, B. E.: Estradiol concentration in the peripheral plasma of the domestic hen. J. Repro. Fert., 41:107–115, 1974.
77. Shaw, A. F. B.: The leucocytes of the pigeon with special reference to a diurnal rhythm. J. Pathol. Bacteriol., 37:411–430, 1933.
78. Siegel, H. S.: Blood cells and chemistry of young chickens during daily ACTH and cortisol administration. Poultry Sci., 47:1811, 1968.
79. Silver, R., et al.: Radioimmunoassay of plasma progesterone during reproductive cycle of male and female ringed doves. Endocrinology, 984:1574–1579, 1974.
80. Smith, E. F., and Bush, M.: Hematologic parameters on various species of strigiformes and falconiformes. J. Wildlife Dis., 14:447–449, 1978.
81. Smith, I. M., and Licence, S. T.: Observations on the use of a semiautomatic system for haematological measurements in birds. Br. Vet. J., 133:585–592, 1977.
82. Stalsberg, H., and Prydz, H.: Studies on chick embryo thrombocytes. II. Function in primary hemostasis. Thromb. Diathes. Haemorrh., 9:291–299, 1963.
83. Stino, F. K. R., and Washburn, K. W.: Responses of chickens with different phenotypes to phenylhydrazine-induced anemia. II. Hematology. Poultry Sci., 49:114–121, 1970.
84. Stopforth, A.: A study of coagulation mechanisms in domestic chickens. J. Comp. Pathol., 80:525–533, 1970.
85. Stoskopf, M. K., et al.: Avian hematology in clinical practice. Mod. Vet. Pract., 64:713–717, 1983.
86. Sturkie, P. D.: Avian Physiology. New York, Springer-Verlag, 1976, pp. 53–75.
87. Tangredi, B. P.: Heterophilia and left shift with fatal diseases in four psittacine birds. J. Zoo Anim. Med., 12:13–16, 1981.
88. Thomas, W. J., and Collins, T. M.: Comparison of venipuncture blood counts with microcapillary measurements in screening for anemia in one-year-old infants. J. Pediatr., 101:32–35, 1982.
89. Vuillaume, A.: A new technique for taking blood samples from ducks and geese. Avian Pathol., 12:389–391, 1983.
90. Warthin, A. S.: Leukemia of the common fowl. J. Infect. Dis., 4:833–835, 1907.
91. Werner, M., and Marsh, W. L.: Normal values: Theoretical and practical aspects. CRC Crit. Rev. Clin. Lab. Sci., 6:81, 1975.
92. Wight, P. A. L., et al.: Monocytosis in experimental zinc deficiency of domestic birds. Avian Pathol., 9:61–66, 1980.
93. Winkel, P., et al.: The normal region—A multivariate problem. Scand. J. Clin. Lab. Invest., 30:339–346, 1972.
94. Wiseman, B. K.: An improved method for obtaining white cell counts in avian blood. Proc. Soc. Exp. Biol. Med., 28:1030–1033, 1931.
95. Woerpel, R. W., and Rosskopf, W. J.: Heavy-metal intoxication in caged birds—I. Comp. Cont. Ed., 4:729–740, 1982.
96. Wolford, J. H., and Ringer, R. K.: Adrenal weight, adrenal ascorbic acid, adrenal cholesterol, and differential leucocyte counts as physiological indicators of "stressor" agents in laying hens. Poultry Sci., 41:1521–1529, 1962.
97. Yuasa, N., et al.: Isolation and some characteristics of an agent inducing anemia in chicks. Avian Dis., 23:366–385, 1978.

# *Chapter 13*

# CLINICAL CHEMISTRIES

ALBERT H. LEWANDOWSKI,
TERRY W. CAMPBELL,
GREG J. HARRISON

Clinical chemistries can be a useful adjunct in the development of a diagnosis and prognosis in avian diseases. However, avian clinical chemistry has yet to receive the same degree of critical study as mammalian clinical chemistry. Normal reference ranges for serum chemistries commonly used in veterinary medicine have been published for use in a variety of avian species. Research has also determined the enzyme activities of various tissues in a number of avian species. The enzyme activities reported show a great variation between the tissues of each species and between species.

It is important to realize that, although the activity of a particular enzyme may be high in a particular tissue or specific for that tissue, if it does not change significantly in the blood when that tissue is damaged, it has little clinical value. Also, a particular enzyme may have a greater activity in one organ (activity per gram of tissue), but a change in serum (or plasma) activity may reflect damage to another tissue with lower enzyme activity because the latter is a larger organ. The large organ with lower enzyme activity may contribute more to the serum enzyme activity than the smaller organ with higher activity. Damage to the smaller organ may not contribute significantly to the serum enzyme concentration, and that enzyme would be a poor test for disease involving that organ.

The reader is referred to Appendix 3 for reference values established by a number of authors. The work by Baron[3] was conducted on birds maintained in a stable environment for at least a year, for which the sex and medical history were known, and for which complete bacteriologic, virologic, and parasitologic evaluations had been performed prior to the study. The macaws evaluated by Raphael[47] had been in a stable environment for a minimum of six months. Reference intervals established by practitioners are much broader, suggesting that "normal" birds presented to veterinarians demonstrate a wide variety of physiologic conditions caused by the variable environmental conditions in which the birds are maintained.

A further difficulty in the use of clinical chemistry values is the extent of the interpretation that must be done by the clinician. No set of tests or normals is a diagnostic panacea, and no normal reference tables are applicable to all birds.

It is often confusing to beginning pet bird veterinarians to submit a serum sample to a laboratory and receive a numerical value for each test listed on the panel. Commercial laboratories performing serum chemistries report values from a predetermined battery of tests that are established for human or domestic mammalian patients. It is the responsibility of the practitioner to determine which of these test results have any significance in making a diagnosis in the avian patient.

Some trends and general principles do apply in the interpretation of avian profiles, although the information obtained is only as good as the sample submitted, the tests selected, and the clinician's ability to correlate test results with the signs that the patient exhibits.

## SPECIMEN HANDLING

Serum is preferred over whole blood or plasma for chemistry analysis to avoid interference by the anticoagulant with the tests being performed. EDTA (ethylene diamine tetraacetic acid) is a sodium or potassium salt that binds calcium to prevent clotting. Heparin is also available as the sodium or potassium salt. To determine electrolytes from such samples would be meaningless. Glucose determinations require either serum or the use of fluoride as the anticoagulant to slow glycolysis in whole blood. Fluoride inhibits enzymes in both sam ples and test reagents. Therefore, fluoride would interfere with plasma enzyme determinations.

Plasma may be analyzed without deleterious effects when lithium heparin is the anticoagulant. Unfortunately, serum enzymes and protein electrophoresis cannot be run on plasma. When an anticoagulant must be used, heparin interferes least with clinical chemistry determinations.

Blood processed within an hour after drawing can be expected to give better quality serum. Allowing a clot to form at room temperature for about 20 minutes, gentle rimming of the clot, and centrifugation at an RCF (relative centrifugal force) of 1000 for 10 minutes works well for standard blood tubes. The supernatant should be promptly removed from the clot and can be refrigerated (4° C) if analysis is performed within three to four hours; otherwise the sample should be frozen (−20° C). Once thawed, several enzymes including SGPT (serum glutamate pyruvate transaminase) and CPK (creatine phosphokinase) become unstable, and analysis must be immediate.

Serum or plasma can be obtained in the highest percentage yield using serum separator devices such as the Microtainer (Becton-Dickinson).

Improper handling of blood will cause numerous analytic inaccuracies, the greatest of which is hemolysis. Hemolysis will increase values of LDH (lactate dehydrogenase), ACP (acid phosphatase), AST (aspartate aminotransferase, formerly serum glutamate oxalacetate transaminase), and potassium.[24, 60] Severe hemolysis will interfere with colorimetric, enzymatic, and chemical reaction–based processes used by laboratories to obtain chemical profiles. Hemolysis can be avoided by using clean, dry equipment and gentle, thoughtful technique. Saturation of the venipuncture site with alcohol without allowing the skin to dry, forcibly expelling blood through the needle, shaking rather than simple inverting the tube to mix the blood with the anticoagulant, excessive rimming of the clot, overcentrifugation, and refrigeration before clotting occurs will contribute to hemolysis. Blood worth drawing is worth handling properly to obtain accurate, reliable results.[18, 21, 24]

## SERUM PROTEIN

One of the easiest tests to perform is the determination of total solids to estimate total protein. Although refractometric methods of estimating total protein values may not accurately correlate with avian serum protein,[42a] a temperature-compensated refractometer has been traditionally used by the practitioner for a prognostic impression, provided that the serum is clear. Hemolysis, lipemia, or cloudiness will give an elevated value. The "normal value," or comparison value, varies with the species of bird being tested but generally falls in the 3- to 5-gm/dl range. Hypoproteinemia may be suspected when the total protein level falls below 2.5 gm/dl.[2, 11, 20, 33] Low total protein can reflect parasitism, chronic disease (especially chronic hepatic and renal diseases), stress, and starvation.[2, 11, 20, 54] Increased values may indicate dehydration, shock, or infection.[20] Extremely high serum protein values (11 to 15 gm/dl) have been seen with chronic lymphoproliferative diseases that resemble leukosis of chickens.

Electrophoresis of avian serum proteins has been experimentally studied to determine sex, investigate taxonomic relationships, and as a tool in immunologic and embryologic research.[12, 33, 43, 44, 56] Electrophoresis has limited application for the general avian practitioner and, except for albumin and globulin, is not routinely performed.

Based on clinical impressions, Galvin[27] reports that albumin is the largest individual protein fraction in avian serum and that the lowering of plasma protein in disease states is usually due to a decrease in the albumin. He believes that in some diseased avian patients the globulin increase can offset the albumin decrease, resulting in a normal protein level. Hypoalbuminemia that fails to respond to treatment is a poor prognostic sign. Birds appear to suffer from hypoalbuminemia with disease more quickly than do mammals. Since serum albumin binds and transports anions, cations, fatty acids, and thyroid hormones, a hypoalbuminemia would affect the blood concentrations of these albumin-transported compounds.

Galvin[27] found that changes in the various components of plasma proteins suggest trends in the avian patient but are not specifically diagnostic, since many disease conditions produce similar changes in this parameter. The most common cause of increased globulins is infection. The alpha globulins include the glycoproteins, haptoglobulins, ceruloplasmin, and alpha-2 macroglobulin.[33] Increases in the alpha globulin fraction can be seen with tissue damage resulting from inflammation, infection, or trauma.[33] Surgery can also cause elevations of alpha globulins.[33] Alpha globulins can be lowered with liver disease, malabsorption, and malnutrition.[27, 33]

The gamma fraction consists primarily of circulating antibodies, and increases usually reflect chronic inflammation or infection.[33] Antibodies can migrate in the beta or gamma range in birds. The IgM migrates mostly in the beta range.[27] Disease states may result in alterations of more than one fraction. The beta and/or gamma globulins are most often elevated with infectious diseases, and alpha globulin is less often elevated. In some cases, gamma, beta, and alpha globulins are elevated.

Age, seasonal changes, and captive status were found to have an effect on serum protein values in birds (such as the Brown Pelican). In comparing the serum protein of fledgling, immature, and adult captive Brown Pelicans, fledglings had significantly higher total protein values than the other groups because of elevated albumin, whereas adults had higher total values than immatures because of elevated globulins.[64] The total serum protein values for the pelicans were elevated from December through May and showed an elevated albumin/globulin (A/G) ratio in adults, while the A/G ratio remained constant in immature birds. This time period corresponded to the breeding season of captive birds and to local winter weather, suggesting that such reserves are important factors in winter survival and reproduction. Wild pelicans also reflected a seasonal increase in total protein, especially in the globulin levels. A change in total protein levels has been reported to parallel change in the packed cell volume (PCV). This was moderately correlated in the pelican, in which the PCV was likewise elevated in adults from December through May. Serum albumin levels may indicate an animal's protein reserves and ability to resist stress, such as infection by malarial parasites. In addition, most avian species undergo a complete molt following breeding. Since feathers make up 20 to 30 per cent of the bird's structural protein, a complete molt may result in a reduced PCV and serum protein. Captive birds may have higher albumin levels than wild birds, owing to the stable diet and limited exercise of captivity.[64]

Serum albumin and globulin concentrations may be affected by egg production. Bobwhite quail demonstrate an increase in serum albumin and decrease in serum globulin during egg production. No change in serum protein is found in egg-laying Canada Geese, whereas chickens show a variable change during egg production. Brown Pelicans show elevated serum albumin and globulins during egg production.[64]

## GLUCOSE

A second commonly assayed chemistry is glucose. A simple, practical means of evaluating blood glucose is with commercially available dipsticks, e.g., Dextrostix (Ames Division, Miles Laboratories, Inc.). A pediatric blood glucose test strip (Chemstrip bG, Bio-Dynamics) has been found useful. Whole blood glucose levels have been found to be no lower than plasma glucose levels and, while not an ideal method, dipsticks are available in almost every practice as a rapid screening test.

Normal birds have a serum glucose between 200 and 500 mg/dl, with stressed birds showing a twofold increase above normal. Glucose levels consistently exceeding 500 mg/dl are considered abnormal. The serum glucose of diabetic birds usually exceeds 750 mg/dl. Glucose values less than 200 mg/dl are evidence of need for immediate attention, and values less than 100 mg/dl have a grave prognosis and often lead to hypoglycemic convulsions. Hypoglycemia may occur with starvation, malnutrition (e.g., hypovitaminosis A), hepatopathies (e.g.,, acute hepatitis, Pacheco's disease, chronic hepatitis), septicemias, endocrinopathies, and laboratory error.[14, 27, 28, 51, 52] Prolonged contact of serum with the clot will result in a 5 per cent per hour loss of serum glucose due to red blood cell metabolism. During starvation, small birds can become hypoglycemic within 24 hours, whereas Amazons have been reported to take two to five days to develop a starvation hypoglycemia.[37]

Hyperglycemia can be ruled out with a negative "urine glucose" using urinalysis reagent strips (e.g., Multistix, Ames Division, Miles Laboratories, Inc.) or tablets (Clinitest, Ames Division, Miles Laboratories, Inc.). The urate or liquid portion of the dropping is collected and tested. Normal birds show a negative to trace urate glucose.

Hyperglycemia may occur during the breeding season, with certain diets, immediately after feeding, and during hyperthermia.[35] Altman et al.[2] recommend repeated glucose tests on all hyperglycemic birds within 24 to 48 hours, because transient hyperglycemias are common and most elevated values tend to drop by the second sample. Transient hyperglycemia has been associated with egg-related peritonitis in cockatiels and other birds.[52] The mechanism may be a temporary pancreatitis secondary to the peritonitis.[52] Birds with diabetes mellitus maintain a high glucose level on the second sampling or show an increase to as high as 1000 to 2000 mg/dl. Diabetes mellitus has been confirmed in budgerigars, cockatiels, Amazons, a Scarlet Macaw, an Umbrella Cockatoo, and a Toco Toucan. Three cases of renal adenocarcinoma in budgerigars were associated with glucose values greater than 1200 mg/dl; however, no gross or histopathologic pancreatic lesions were found.

Blood glucose values may show a normal variation with age, time of day, and state of captivity. Pelicans show an increase in serum glucose from the fledgling stage through the juvenile stages but demonstrate a decrease with maturity.[64] Domestic fowl and Redpolls, both granivorous birds, show significant diurnal variations in serum glucose values. Wild birds tend to have lower serum glucose values than captive birds.

## NONPROTEIN NITROGEN

### Uric Acid

Uric acid is the primary catabolic product of protein, nonprotein nitrogen, and purines in birds. It is excreted by the avian kidney primarily by tubular excretion. Therefore, elevated serum uric acid values are expected in birds with impaired renal function and reduced renal uric acid clearance. Known normal baseline serum uric acid values for the individual avian patient provide comparative values to detect an elevated serum uric acid when the bird suffers from renal disease.

The normal blood uric acid value for most birds ranges between 2 and 15 mg/dl. Values over 20 to 30 mg/dl are considered elevated. Elevated values can be the result of starvation, dehydration, or massive tissue trauma but are usually due to nephrosis, gout, or impaired renal function.[2, 39, 48] Hyperuricemia is frequently reported in birds on the third or fourth day of gentamicin therapy; however, uric acid values seldom exceed 20 mg/dl. Hyperuricemia has been reported in birds during ovulation.[27] Elevated uric acid values result from ureter ligation, gout, and renal disease. Artifactually increased blood uric acid values occur if the toenail clip method is used and the nail is not properly cleaned of urates from the droppings, thus contaminating the sample.

Normal blood uric acid values from a wide variety of birds have been reported. These values can also be affected by age, time of year, and captivity. Higher blood uric acid values in pelicans are seen in fledglings versus nestlings, in wild adults versus captive adults, and between the months of December and May.

### Creatinine

Creatinine is not a major nonprotein nitrogen component in avian blood. It has questionable value in evaluating renal function in birds. The normal value for most birds is less than 0.2 mg/dl (normal values for Blue and Gold Macaws are 0.3 to 0.5 mg/dl). Creatinine will elevate slightly with renal failure (0.5 to 1.5 mg/dl) but is less reliable than uric acid.[27]

One of the reasons creatinine may not give a true assessment of avian renal function is that birds excrete creatine in their urine before it has been converted to creatinine.[5, 18, 61] The measurements obtained as normals for creatinine may well be pseudocreatinines, such as glucose, protein, ascorbic acid, and pyruvic acid, and may not reflect glomerular function. Birds do excrete surplus amino nitrogen as urate, which can be readily measured in serum and remains fairly constant despite fluctuations in either activity or diet.[5, 33, 38]

High levels of serum creatinine have been associated with diets high in animal protein and with failure to clean the diagnostic equipment correctly. Despite the limitations, one author has correlated high values with renal disease, egg-related peritonitis, septicemia (e.g., chlamydiosis), prerenal azotemia (dehydration), renal trauma, and nephrotoxic drugs. Creatinine levels in pelicans were relatively stable, showing no significant variation related to age, sex, season, or captive status.[64]

### Blood Urea Nitrogen (BUN)

Birds are uricotelic and produce uric acid and not urea as the major nitrogenous end product of metabolism. Therefore, blood urea nitrogen is not a useful test of renal function in birds.

## SERUM ENZYMES

### Aspartate Aminotransferase (AST, SAST, SGOT)

Aspartate aminotransferase was formerly known as SGOT (glutamic oxaloacetic transaminase). The test can be performed using serum or plasma. In general caged birds having AST values greater than 230 IU/L are considered abnormal. The most common cause of elevated AST in caged birds is liver disease.[27] Published normals vary according to sex, age, time of year, and breeding activity. Values determined by Baron[3] were significantly lower than those of clinically sampled birds.

The distribution of AST in avian tissues varies among the species. The highest AST activity in chicken, geese, and turkey tissues occurs in the heart, followed by liver and skeletal muscle, with the exception that kidney and brain levels are higher than muscle in turkeys.[7, 8, 33] The distribution in ducks in descending order is skeletal muscle, heart, kidney, brain, and liver.[33] Therefore, AST is not liver-specific. Moderate increases (two to four times normal mean) are often associated with soft tissue injury, whereas liver necrosis results in greater elevations.[17, 33]

Elevated AST values were observed in a suspected allergic reaction in a macaw given chlortetracycline; with severe muscle necrosis from I.M. administration of an I.V. form of doxycycline in a macaw; and with tobramycin and ticarcillin administration suspected of causing liver damage in a Rose-breasted Cockatoo. High AST levels from liver damage due to experimental pesticides and carbon tetrachloride toxicity peaked in 24 to 48 hours and returned to normal in four to six days.[19, 23, 63] Liver damage from histomoniasis in turkeys and excessive rapeseed meal were monitored via AST.[1, 30, 32]

## Alanine Aminotransferase (ALT, SALT, SGPT)

Values can be obtained for ALT from serum or plasma samples. The ALT activity in turkey tissues is greatest in skeletal muscle and low in the liver and heart.[8] The highest ALT activity occurs in the kidney in Mallard Ducks and is higher in the erythrocytes than in serum or plasma.[53] Heart and skeletal muscle, liver, and lung tissues have low ALT activity in chickens and geese.[7]

Although there are some reports of elevated blood ALT values in raptors, chickens, and ducks with hepatic insult and Quaker Parakeets with Pacheco's disease, it is believed that ALT levels do not rise in all cases of hepatic disease and are therefore not a useful diagnostic indicator of liver disease in birds.[7, 9, 27, 28, 41, 53] Raptors show a seasonal variation in ALT values.

## Lactate Dehydrogenase (LDH)

Serum LDH is also a nonspecific enzyme. The highest LDH activity in chickens occurs in skeletal muscle, followed by heart muscle, liver, and lung in descending order.[7] The LDH activity in turkeys is highest in the heart muscle, followed by the liver, skeletal muscle, spleen, and lung.[8] The highest serum LDH activity in ducks occurs with hepatic tissue.[25] The five isoenzymes found in mammalian tissues occur in birds, but with different distributions. Isoenzymes 2, 3, and 4 of LDH are relatively more liver-specific in birds than LDH 1 or 5.[5] Serum LDH increases with hemolysis in some birds and may increase with hepatic disease in psittacine birds.[27] Since LDH is not liver-specific and is highly labile, it is believed to be a poor indicator of liver disease in birds. Rosskopf et al.[51] has found LDH to rise and fall quickly with appropriate treatment in birds with liver disease, whereas AST tends to increase and decrease more slowly. Normal LDH values have seasonal variations.

## Alkaline Phosphatase (AP, ALP)

Serum alkaline phosphatase is another nonspecific enzyme in birds. Chickens show no AP activity in the lung, skeletal muscle, and heart muscle and little activity in the liver.[7] The highest AP activity in turkeys occurs in the testes, followed by the liver.[8] Ducks have high AP activity in the duodenum and kidney.[53] Severe liver disease in raptors has been reported to cause a marked increase in serum AP activity (five to six times normal), whereas osteoblastic activity will cause a lower elevation (two to three times normal).[28] On the other hand, many authors feel serum alkaline phosphatase is a poor test for liver disease but is a better indicator of increased osteoblastic activity in birds (e.g., hyperparathyroidism, fracture repair, impending ovulation, rickets).[2, 28, 33, 37] Values over 40 IU/L are suggestive of secondary hyperparathyroidism.[2] Low serum alkaline phosphatase has been linked to a dietary zinc deficiency.[55]

Baron[3] found seasonal variation in AP test results. Serum AP levels in Quaker Parrots infected with Pacheco's parrot disease virus did not increase in spite of massive liver necrosis.[68] Rosskopf[52] found levels to increase with corticosteroid use. Harrison found that female cockatiels and a pigeon had significantly elevated serum AP levels prior to egg laying.[29]

### Gamma Glutamyltranspeptidase (GGTP)

Gamma glutamyltranspeptidase has its highest tissue activity in the kidney in ducks.[25] Serum GGTP will increase with damage to the hepatobiliary system or pancreas in some birds.[27, 45]

### Creatinine Phosphokinase (CPK, CK)

Creatinine phosphokinase values for most birds normally range from 100 to 200 IU/ml. Serum CK values will increase with neuropathies, lead toxicity, chlamydiosis, and bacterial septicemias.[61] Intramuscular injections usually will not elevate CK values unless the material is highly irritating (e.g., I.V. doxycycline given intramuscularly or oxytetracycline).[27] Graham has seen elevated CK levels in raptors with nutritional myopathy from vitamin E/selenium deficiency (see Chapter 31, Nutritional Diseases).

### Other Serum Enzymes

Several other enzymes that are less commonly assayed warrant further investigation to help diagnose avian liver disease. Plasma GDH (glutamate dehydrogenase) and SDH (sorbitol dehydrogenase) appear to peak between 16 and 24 days following insult and seem to correlate with hepatic cell proliferation.[63] ICDH (isocitrate dehydrogenase) has been useful for the differential diagnosis of liver disease. MDH (malate dehydrogenase) is elevated with hepatic necrosis, and GGT (gamma glutamyl transferase) appears to be a specific and sensitive index of liver damage.[32] A superficial evaluation of 5′-nucleotidase enzyme values in cockatiels indicated that it is not clinically significant. If research in the future shows that these or other enzymes provide better detection of avian diseases, perhaps the development of a multiple enzyme profile tailored specifically to reflect avian physiology will become a reality.

## BILIRUBIN

Bilirubin is not an important test for liver disease in birds. Icterus is rare in birds, probably because biliverdin, a green pigment, is the main bile pigment in these animals. Birds lack the biliverdin reductase enzyme needed to reduce biliverdin to bilirubin.[40, 59] Also, the avian kidney rapidly clears bilirubin from the blood. Bilirubinuria and icterus may be seen occasionally in some birds with severe liver disease. This may be due to nonspecific reduction of biliverdin to bilirubin by bacteria or nonspecific reducing enzymes. It should be emphasized that many birds will normally have a yellow plasma due to carotene pigment and not to bilirubin.

## CHOLESTEROL

Comparison values for serum cholesterol will vary with diet among the different bird families.[15] Carnivorous birds have higher blood cholesterol levels than fruit- or grain-eating birds. Obese birds or birds on a high-fat diet will show an elevated serum cholesterol. Hypercholesterolemia also occurs in birds with hypothyroidism.[42, 52]

Low serum cholesterol values, especially when coupled with a low serum glucose, have been implicated in the development of fatty liver and kidney syndrome.[49] Decreases have also been linked nonspecifically to liver disease.[6] Seasonal fluctuations of blood cholesterol levels appear to be minimal.[37]

## CALCIUM

The calcium measured in most chemical determinations is nondiffusable and bound to serum proteins; therefore, the total serum calcium levels will frequently rise and fall with high or low protein levels. Usually serum calcium levels range from 8 to 12 mg/dl, but ovulating birds have been reported to have values from 20 to 40 mg/dl, possibly due to the increased calcium demand for shell formation.[46, 50] In one limited study,[29] elevated calcium values were not observed in cockatiels and pigeons immediately prior to egg laying.

Increased serum calcium has also been reported with dietary excesses of vitamin $D_3$ and several neoplasms.[6, 33, 52, 60] Extreme hypercalcemia may even cause clotting in the presence of an anticoagulant.[57, 66] High serum calcium values have been noted in clinically normal birds. Primary hyperparathyroidism and pseudohyperparathyroidism are not well-documented in birds.[13]

Serum calcium levels of less than 6.0 mg/dl have been reported to result in tetany, especially in birds that are stressed. However, in a clinical study of pigeons, the average serum calcium level of 29 birds was 4.4 mg/dl, and no evidence of tetany was noted.[29] Hypocalcemic convulsions have been reported in psittacines (e.g., idiopathic African Grey Parrot hypocalcemia syndrome) and raptors.[6, 33, 52] Renal disease can cause low serum calcium due to loss of protein causing a hypoalbuminemia or due to decreased calcium reabsorption. Low body calcium levels are often seen with normal serum calcium due to bone resorption (secondary nutritional hyperparathyroidism).

## PHOSPHORUS

Although the metabolism of calcium and phosphorus is closely linked in the body, the diagnostic value of serum phosphorus in birds is inconsistent. Comparison values of normal birds are usually in the 2 to 6 mg/dl range. Increases are reported with renal disease and nutritional secondary hyperparathyroidism.[6, 52] Dietary excesses or deficiencies of vitamin D have also been known to cause high or low serum phosphorus values.[18, 52, 60] In laying birds the presence of extremely high serum calcium levels has had no effect on inorganic phosphorus levels, which are expected to be decreased.[46] Phospholipids are thought to bind calcium in laying birds. Our understanding of calcium and phosphorus metabolism in mammals may be of limited value when applied to birds. More research and continued case documentation along these lines would be helpful in determining the role of phosphorus in avian diagnostics.

## SODIUM AND POTASSIUM

The roles of sodium and potassium in avian diagnostics also require better documentation. Normal comparison values give serum sodium levels between 130 and 170 mEq/L and serum potassium from 2.5 to 6.0 mEq/L.[4, 6, 46, 49, 50, 52] Abnormalities affecting the adrenal cortical hormones (e.g., aldosterone), the body's state of hydration (e.g., diarrhea or water deprivation), or the salt glands and kidneys of birds would be expected to alter serum sodium and potassium values. Unfortunately, little research has confirmed or disproved these assumptions. Hypernatremia has been seen in salt-loaded, water-fasted birds.

Extrarenal salt excretion occurs in many species of birds by use of a specialized nasal gland (salt or supraorbital gland) that is able to secrete large quantities of sodium as an osmoregulatory response. This mechanism of decreasing the sodium concentration in the serum and urine of birds is influenced by pituitary-adrenal (i.e., aldosterone) responses. Birds without these salt glands remove excess sodium through the kidneys, whereas those birds with salt glands remove the majority of excess sodium by extrarenal salt excretion.[58]

Specimen handling is important when measuring sodium and potassium. Hemolysis or failure to remove serum from the clot will increase serum potassium and decrease serum sodium, invalidating the determination.[60] Erythrocytes contain 23 times as much potassium as plasma, so even slight hemolysis will give erroneous results.[24]

## TEST SELECTION

An avian chemistry profile that can readily be performed should include total protein, glucose, uric acid, and AST. With a limited volume of sample or limited laboratory capabilities, these four tests will provide a rapid general overview. The total protein and glucose can quickly underscore the need for immediate treatment and indicate a prognosis. Uric acid levels help evaluate renal function, and AST provides an indication that liver, heart, or skeletal muscle is affected. More specific testing may be employed if the results of the abbreviated panel are suspicious.

Laboratories offering ultramicroanalysis using 0.1 ml of serum (100 lambda) or less per test are becoming more prevalent as pediatricians and neonatologists require more clinical chemistry information on their patients. The developing technology benefits the avian diagnostician, who is also faced with the constraint of minimal blood volume. A more complete avian panel with ultramicroanalysis would include the four basic tests plus calcium and phosphorus, with information on cholesterol and GGT being highly desirable.

Special studies on muscle damage as seen with exertional, traumatic, or nutritional myopathies can be geared to give high priority to chemistries such as CPK, LDH, and AST, as these appear to be straightforward increases over baseline values by three- to fivefold.[10, 16, 31, 61] Liver-related problems can be better monitored using LDH, cholesterol, AST, GGPT, and a combination of the special chemistries discussed previously, depending on the nature of the problem. Renal dysfunction can be followed with uric acid, total protein, calcium, phosphorus, and the electrolytes sodium and potassium. Serum ALP, calcium, phosphorus, sodium, and potassium can be helpful in the diagnosis and treatment of hormonal abnormalities.

In order for test results to be statistically significant, each laboratory must determine its own set of values for comparison. Ideally, the clinician should compare test results with the normal reference values obtained from the laboratory performing the testing. Given the difficulty of obtaining sufficient numbers of samples for each species and the overwhelming number of species that an avian practitioner may be faced with, tables compiled by researchers and a handful of interested laboratories are frequently the only information available (see Appendix 3). With published values as guidelines and good clinical judgment, progress is slowly being made at upgrading the practice of avian clinical pathology. Continued research and case reports with good documentation from veterinarians in the field will help promote and better define the usefulness of clinical chemistries in the practice of avian medicine.

## REFERENCES


1. Al-khateeb, G. H., and Hansen, M. F.: Plasma glutamic oxalacetic transaminase as related to liver lesions from histomoniasis in turkeys. Av. Dis., 17:269–273, 1973.
2. Altman, R., et al.: Avian clinical pathology evaluation panel. American Association of Zoo Veterinarians Annual Proceedings, 1975, pp. 146–149.
3. Baron, H. W.: Die aktivitatamessung einiger enzyme im blutplasma bzw.—serum verschiedener vogelspezies, Postgraduate dissertation. University of Munich. 1980.
4. Beljan, J. R., et al.: Determination of selected avian blood plasma chemistry values using the Technicon AutoAnalyzer. Poultry Sci., 50:229–232, 1972.
5. Bell, D. J., and Freeman, B. M. (eds.): Physiology and Biochemistry of the Domestic Fowl. New York, Academic Press, 1971.
6. Black, D. G.: Avian clinical pathology. *In* Refresher Course on Aviary and Caged Birds. University of Sydney, 55:224–248, 1981.
7. Bogin, E., and Israeli, B.: Enzyme profile of heart and skeletal muscle, liver and lung of roosters and goose. Zbl. Vet. Med. A., 23:152, 1976.
8. Bogin, E., et al.: Enzyme profile of turkey tissues and serum. Zbl. Vet. Med. A., 23:858, 1976.
9. Bokori, J., and Karsai, F.: Enzyme-diagnostic studies of blood from geese and ducks, healthy and with liver dystrophy. Acta Vet. Acad. Sci. Hung., 19:269, 1969.
10. Brannian, R. E., et al.: Restraint associated myopathy in East African Crowned Cranes. American Association of Zoo Veterinarians Annual Proceedings, 1981, pp. 21–23.
11. Bush, M., and Smith, E. E.: Clinical chemistry and hematology as diagnostic aids in zoological medicine. *In* Montali, R. J., and Migaki, G. (eds.): The Comparative Pathology of Zoo Animals. Washington, DC, Smithsonian Institution Press, 1980, pp. 615–620.
12. Butler, E. J.: Plasma proteins. *In* Bell, D. J., and Freeman, B. M. (eds.): Physiology and Biochemistry of the Domestic Fowl. New York, Academic Press, 1971.
13. Campbell, T. W.: Clinical chemistry. Proceedings of the Caged Bird Seminar and Wet Lab, Kansas State University, 1983, pp. 10–13.
14. Chandra, M., et al.: Renal and biochemical changes produced in broilers by high-protein, high-calcium, urea-containing, and vitamin A–deficient diets. Avian Dis., 28:1, 1983.
15. Conetta, J., et al.: Hematologic and biochemical profiles in normal demoiselle cranes (*Anthropoides virgo*). Lab. Anim. Sci., 24:105–110, 1974.
16. Cornelius, C. E., et al.: Serum and tissue transaminase activities in domestic animals. Cornell Vet., 49:116–126, 1959.
17. Cornelius, C. E., et al.: Plasma aldolase and serum glutamic oxaloacetic transminase activities in inherited muscular dystrophy of domestic chickens. Proc. Soc. Exper. Biol. Med., 101:41, 1959.
18. Davidson, I., and Henry, J. B.: Clinical Diagnosis by Laboratory Methods. Philadelphia, W. B. Saunders Company, 1979.
19. Dieter, M. P.: Further studies on the use of enzyme profiles to monitor residue accumulation in wildlife. Arch. Environ. Contamin. Tox., 3:142–150, 1975.
20. Dolensek, E. P., and Otis, U. S.: Clinical pathology in avian species. Vet. Clin. North Am., 3:159–163, 1973.
21. Dorner, J. L., et al.: Effects of *in vitro* hemolysis on equine serum chemical values. Am. J. Vet. Res., 42:1519–1522, 1981.
22. Ensley, P.: Caged bird medicine and husbandry. Vet. Clin. North Am., 9:499–525, 1979.
23. Fowler, J. S. L.: Chlorinated hydrocarbon toxicity in the fowl and duck. J. Comp. Pathol., 80:465–471, 1970.
24. Frank, J. J., et al.: Effect of *in vitro* hemolysis on chemical values for serum. Clin. Chem., 24:1966–1970, 1978.
25. Franson, J. C.: Enzyme activities in plasma, liver, and kidney of black ducks and mallards. J. Wildlife Dis., 18(4):481, 1981.
26. Fudge, A. M., and McEntee, L.: Avian clinical pathological services for the veterinary practitioner. Citrus Heights, CA, California Avian Lab, 1985.
27. Galvin, C.: Laboratory diagnostic aids in pet bird practice. Proceedings of the American Animal Hospital Association, South Bend, IN, 1980, p. 41.
28. Halliwell, W. H.: Serum chemistry profiles in the health and disease of birds of prey. *In* Cooper, J. E., and Greenwood, A. G. (eds.): Recent Advances in the Study of Raptor Diseases. West Yorkshire, England, Chiron Publications, Ltd., 1981, p. 111.
29. Harrison, G. J., et al.: Clinical comparison of anesthetics in domestic pigeons and cockatiels. Proceedings of the Annual Meeting of the Association of Avian Veterinarians, Boulder, CO, 1985.
30. Hirsch, R. P.: Dynamics of protozoan population density, plasma glutamic oxalacetic transaminase and plasma bilirubin concentrations during histomoniasis in turkeys. Int. J. Parasitol., 9:395–399, 1979.
31. Hollands, K. G., et al.: Plasma creatine kinase as an indicator of degenerative myopathy in live turkeys. Br. Poultry Sci., 21:161–169, 1980.
32. Ibrahim, I., et al.: Plasma enzyme activities indicative

of liver cell damage in laying fowl given a diet containing 20 per cent of rapeseed meal. Res. Vet. Sci., 28:330–335, 1980.
33. Ivins, G. I., et al.: Hematology and serum chemistries in birds of prey. *In* Fowler, M. E. (ed.): Zoo and Wild Animal Medicine. Philadelphia, W. B. Saunders Company, 1978, pp. 286–290.
34. John, T. M., and George, J. C.: Seasonal variation in cholesterol level in the migratory starling, *Sturnus roseus*. Pavo, 5:29–38, 1967.
35. Kumar, S., and Gupta, Y. K.: Seasonal and experimental studies on the blood glucose of *Psittacula krameri*. Pavo, 19:1–10, 1981.
36. Lafeber, T. J.: Respiratory diseases. Vet. Clin. North Am., 3:199–227, 1973.
37. Leonard, J. L.: Clinical laboratory examinations. *In* Petrak, M. L. (ed.): Diseases of Cage and Aviary Birds. Philadelphia, Lea and Febiger, 1982, pp. 291–303.
38. Levine, R., et al.: Concentration and transport of true urate in the plasma of the azotemic chicken. Am. J. Physiol., 151:186–191, 1947.
39. Lewandowski, A. H.: Selected avian biochemistries and their application in clinical diagnosis. American Association of Zoo Veterinarians Annual Proceedings 1982, pp. 5–7.
40. Lin, G. L., et al.: Bile pigments in the chicken. Res. Vet. Sci., 8:280–282, 1967.
41. Lohr, J. E.: Fatty liver and kidney syndrome in New Zealand chickens. N. Z. Vet. J., 23:167, 1975.
42. Lothrop, C. D.: Diseases of the thyroid gland in caged birds. Proceedings of the Association of Avian Veterinarians, Toronto, 1984, p. 85.
42a. Lumeij, J. T. and de Bruijne, J. J.: Evaluation of refractometric methods for determination of total proteins in avian plasma or serum. Avian Pathol. 14:441–444, 1985.
43. Morgan, R. P., et al.: Serum proteins of sandhill cranes and bald eagles. Comp. Biochem. Physiol., 57B:297–298, 1977.
44. Mosher, J. A., et al.: Serum proteins of selected falconiformes and strigiformes. Biochem. System Ecol., 10:373–376, 1982.
45. Pearson, A. W., et al.: Rapeseed meal and liver damage: Effect on plasma enzyme activities in chicks. Vet. Rec. 105:200, 1979.
46. Pitt, M. A., et al.: Enzyme, metabolite and electrolyte levels in the blood of ducks in Israel. Zbl. Vet. Med. A., 27:775–779, 1980.
47. Raphael, B. L.: Hematology and blood chemistry of macaws. American Association of Zoo Veterinarians Annual Proceedings, 1980, pp. 97–98.
48. Rivetz, B., et al.: Biochemical changes in fowl serum during infection with strains of Newcastle disease virus of differing virulence. Res. Vet. Sci., 22:285–291, 1977.
49. Ross, J. G., et al.: Determination of haematology and blood chemistry values in healthy six-week-old broiler hybrids. Avian Pathol., 5:273–281, 1976.
50. Ross, J. G., et al.: Haematological and blood chemistry "comparison values" for clinical pathology in poultry. Vet. Rec., 102:29–31, 1978.
51. Rosskopf, W. J., et al.: Hematologic and blood chemistry values for common pet avian species. V.M./S.A.C., 77:1233–1239, 1982.
52. Rosskopf, W. J., and Woerpel, R. W.: Clinical experience with avian laboratory diagnostics. Vet. Clin. North Am. [Small Anim. Pract.], 14:2, 1984.
53. Rozman, R. S., et al.: Enzyme changes in Mallard ducks fed iron or lead shot. Avian Dis., 18:435, 1974.
54. Schalm, O. W., et al.: Veterinary Hematology. Philadelphia, Lea and Febiger, 1975, pp. 602–630.
55. Schuster, N. H., and Hindmarsh, M.: Plasma alkaline phosphatase as a screening test for low zinc status in broiler hybrid chickens affected with "clubbed down." Aust. Vet. J., 56:499–501, 1980.
56. Sibley, C. G., and Johnsgard, P. A.: Variability in the electrophoretic patterns of avian serum proteins. Condor, 61:85–95, 1959.
57. Small, S. D., et al.: The clotted CBC tube—a sign of severe hypercalcemia. N. Engl. J. Med., 307:11, 1982.
58. Sturkie, P. D.: Avian Physiology. 3rd ed. New York, Springer-Verlag, 1976.
59. Tenhunen, R.: The green color of avian bile: Biochemical explanation. Scand. J. Clin. Lab. Invest. [Suppl. 27], 116:9, 1971.
60. Tietz, N. W. (ed.): Fundamentals of Clinical Chemistry. Philadelphia, W. B. Saunders Company, 1976.
61. Tripp, M. F., and Schmitz, J. A.: Influence of physical exercise on plasma creatine kinase activity in healthy and dystrophic turkeys and sheep. Am. J. Vet. Res., 43:2220–2223, 1982.
62. Vertommen, M., et al.: Infectious stunting and leg weakness in broilers: I. Pathology and biochemical changes in blood plasma. Avian Pathol., 9:133–142, 1980.
63. Westlake, G. E., et al.: The effects of 1,1-di(*p*-chloraphenyl)-2-chloroethylene on plasma enzymes and blood constituents in the Japanese quail. Chem. Biol. Int., 25:197–210, 1979.
64. Wolf, S. H.: Seasonal, age and sex variations in the blood composition of the Brown Pelican *(Pelecanus occidentalis)*. Master of Science thesis, University of South Florida, 1984.
65. Woodard, A. E., et al.: Blood parameters of one-year-old and seven-year-old partridges *(Alectoris chukar)*. Poultry Sci., 62:2492–2496, 1983.
66. Zaremski, J. C., and Smith, E. E.: Evaluation of plasma glucose in anseriformes, falconiformes, gruiformes, and psittaciformes at the National Zoological Park. *In* Montali, R. J., and Migaki, G. (eds.): The Comparative Pathology of Zoo Animals. Washington, DC, Smithsonian Institution Press, 1980, pp. 643–646.
67. Zilva, J. F., and Pannall, P. R.: Clinical Chemistry in Diagnosis and Treatment. Chicago, Year Book Medical Publishers, 1975.
68. Goodwin, J. S., et al.: Effects of Pacheco's parrot disease virus on hematologic and blood chemistry values of Quaker Parrots *(Myopsitta monachus)*. J. Zoo Anim. Med., 13:127–132, 1982.

# Chapter 14

# RADIOLOGY

MICHAEL T. WALSH

Radiography is considered a basic tool for the avian practitioner. It is utilized as a diagnostic aid in ill patients for which there is inadequate data available as well as in apparently healthy individuals for which a full evaluation has been requested. Often the reluctance of the clinician to radiograph a bird is based on the individual practitioner's lack of familiarity with avian restraint, radiographic technique, basic anatomy, or film interpretation.

## RESTRAINT

Three types of restraint are used with the avian patient for radiography. Manual restraint involves maintaining the bird's position with lead-lined gloved hands. This results in increased exposure of the handler to radiation and may be illegal in some states. It should be used only with birds that are not prone to struggle and induce self-trauma. This is often the only method needed for raptors.

Physical restraint, using a plexiglass restraining board, velcro strips, masking tape, or other devices may also suffice for tame or calm individuals. The tolerance of the bird to this procedure should be gauged by its response to initial manual restraint. In many cases, physical restraint may result in excessive patient stress, possible injury from struggling (such as fractured wings), and nondiagnostic films from patient motion or malposition.

Chemical restraint is most often used in combination with positioning techniques to facilitate smooth handling and result in the highest percentage of diagnostic-quality films (see Chapter 45, Anesthesiology). Each patient must be evaluated on an individual basis for chemical restraint, and careful clinical judgment must be used with a critically ill animal. In some cases radiography may need to be delayed until the bird is stabilized and its survival capability increased. In other situations, radiographic evaluation may be the only means to a diagnosis. With full sedation of the patient, masking tape or other light restraint may be used to position the feet, wings, and neck.

## POSITIONING

Malpositioning of the avian patient is often related to a failure to stretch the body out to full extension and is a common cause of nondiagnostic radiographs. It is for this reason that a thoroughly relaxed patient is desirable and chemical restraint is preferred. Two views, the lateral and ventrodorsal (V/D), are taken for all cases. Whole body views will usually supply the most information. For further diagnostic data, specific anatomic sites or different exposure angles may be used.

Interpretation of organ relationships is based on the position of the bird at the time of the radiographic exposure; therefore, consistent placement of the bird and the marker on the cassette is essential for identification during viewing.

It is recommended to radiograph the lateral view first during the maximum degree of sedation, as it tends to be the most difficult to position. For the lateral view (Fig. 14–1*A*), the right side of the bird is always placed against the cassette and the wings are displaced dorsally above the back with the shoulders superimposed. The down wing may be positioned slightly anterior to differentiate right from left. Tape may be applied as proximal as possible on the distal humerus for stabilization of a sedated individual. Thin foam pads may be placed under the lower wing to avoid trauma, and care is taken to avoid excessive dorsal displacement of the upper wing (which may create an oblique angle).

In positioning the legs in the lateral view, the femoral heads should ideally be superim-

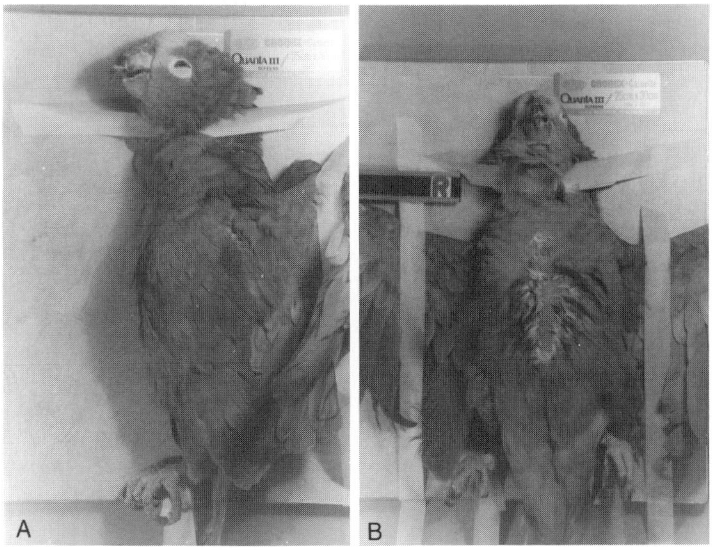


**Figure 14–1.** *A*, Positioning for lateral radiograph. The bird is placed on its right side. The wings are displaced dorsally and the legs pulled caudally and restrained. The head and neck are extended and restrained. The natural shape of the bird's body leads to some rotation. *B*, Positioning for ventrodorsal radiograph. The bird is placed on its back, and the head and neck are gently extended and restrained. The wings are pulled away from the body and restrained at the distal humerus, and the legs are pulled caudally. Wetting the feathers over the sternum assists in alignment with the spine but excessive fluid may interfere with x-ray passage.

posed and the legs pulled caudally as far as possible. To differentiate between the legs when investigating the limbs, the down leg is placed slightly cranial. An individual limb in question should be placed closest to the cassette for best detail. It should be noted that even under heavy sedation the "ideal" position in the lateral view is difficult to attain without padding to finely adjust body positioning.

An anteroposterior view of the wing can be obtained either by moving the cassette to the edge of the table and holding the body off the edge and perpendicular to the table, or by extending the wing over the cassette with the bird held upside down and the cassette elevated from the table to allow room for the hands. Handlers must be extremely careful and observant with this technique.

Orientation to the V/D view (Fig. 14–1*B*) is achieved by placing the identification marker to the bird's right at the time of film exposure. In the V/D view, the sternum should be superimposed over the spine. Failure to do so can change the appearance of the image with even slight deviation (see Soft Tissue Systems, below). To decrease tissue overlap, the head is stretched out and restrained, the feet are pulled caudally and secured around the tarsometatarsal bone, and the wings are extended outward from the body and taped down symmetrically. If the bird is not heavily sedated, applying the tape over the distal humerus, possibly with padding under the elbow, may help avoid trauma. Films should be exposed on inspiration as a standard procedure.

## TECHNIQUE

Radiographic technique is commonly an area of confusion for many individuals. For many of the medium and large parrots a feline technique is often adequate. An effective technique for birds should stress detail and incorporation of the shortest exposure time possible (1/30 second or less). Table 14–1 illustrates examples of techniques that have been used by some authors. No technique is universally applicable. The value of a grid is controversial, although Harrison believes increased detail from the use of a grid (80 lines per inch) with standard small animal radiographic techniques may decrease the number of false interpretations of air sacculitis. Newer, more powerful, and faster radiograph machines in conjunction with ultra detail, rare earth screens, and automatic processing of films have significantly improved avian radiographic results. High detail film is especially helpful when interpreting radiographs of smaller species.

## FILM INTERPRETATION

### Normal Radiographic Anatomy

Viewing of radiographs should be consistent with the positioning of the bird at the time of exposure for appropriate interpretation. When viewing films, the lateral view should be read with the bird's head to the reader's left. The V/D view is read with the bird's right side to

**Table 14–1.** RADIOGRAPHIC TECHNIQUES FOR PSITTACINES*

| Screen† | Film | FFD‡ | KVP | MA | Time | Bird |
|---|---|---|---|---|---|---|
| 1. Ultra detail | Rare earth | 34 | 40–44 | 200 | 1/30 | Budgerigar |
| 2. Par Speed | Par speed | 40 | 70 | 100 | 1/60 | Parrot |
| Ultra detail | Rare earth | | | 200 | 1/30 | Parrot |
| 3. Ultra detail | Cronex 6§ | 40 | 72 | 100 | 1/60 | Macaw |
| 4. Par Speed | Cronex 6 | 40 | 46 | 300 | 1/120 | Parrot |
| 5. Ultra detail‖ | Cronex 7 | 40 | 60 | 50 | 1/5 | Amazon |
| | | | 50 | 50 | 1/5 | Budgerigar |
| | Rare earth¶ | 40 | 76 | 50 | 1/60 | Amazon |
| | | | 65 | 50 | 1/60 | Budgerigar |

*These are suggested guidelines for the practitioner in developing his own effective radiographic techniques. Grids were not used with these techniques.

†*Key to Radiographic Techniques*

1. Robert B. Altman, Franklin Square, NY
2. Marjorie McMillan, Boston, MA
3. Sam Silverman, San Francisco, CA
4. University of Florida, Gainesville, FL
5. Greg J. Harrison, Lake Worth, FL

‡FFD = Focal film distance.

§Dupont Cronex 6, E.I. Dupont DeNemours & Co., Wilmington, DE 19898.

‖Quanta Detail, E.I. Dupont Co.

¶Quanta III, E.I. Dupont Co.

the reader's left. It is essential that the clinician be totally familiar with normal avian anatomy before attempting to interpret the radiographs (see Chapter 4, Clinical Anatomy). Some individuals find it helpful to first evaluate the surrounding bone and external soft tissue and then concentrate on the coelomic cavity. A division of the body cavity into quadrants or layers may be useful when first reading avian radiographs. Several species of Amazon Parrots serve as the primary subjects for the "normal" radiographs illustrated in this chapter.

### *Head, Neck, and Skeletal System*

The skull of the avian species (Fig. 14–2*A* and *B*) is perfused with pneumatic channels that communicate with the sinuses of the head. The reader should be aware of differences in the bones of the skull and the presence of an ossified ocular ring. The cervical vertebrae of the neck are interwoven with the cervical pulmonary air sac and surrounded by the cervical portion of the cervicocephalic air sac system. The caudal portion of the axial skeletal system (Fig. 14–2*C* and *D*) is fused into the synsacrum, which is perfused by the abdominal air sacs. The proximal humerus connects with the interclavicular air sacs, while the femur communicates with the abdominal air sacs. The osteoporotic appearance of the shoulder is normal. The cortices of the long bones are thinner than those in domestic animals. Reproductively active females may have an increase in endosteal and intramedullary density, which is normal prior to egg laying.

### *Soft Tissue Systems*

In the V/D view (Fig. 14–3*A*), it may be helpful for radiographic analysis to mentally divide the peritoneal cavity into quadrants with a vertical imaginary line over the midline and a horizontal one between the acetabula (see Chapter 4, Clinical Anatomy). The crop is often visible on the right side of the vertebrae at the height of the shoulder; however, when it is engorged with food, as is often the case in young birds, it may take on a half-moon shape and extend over to the left of the midline. In the thoracic portion of the body cavity, the heart is centrally located midway in the sternum, with its apex often obscured within the anterior portion of the liver. These two organs form the classic hour-glass shape. In this view the great vessels may appear as circular densities close to the base of the heart; these should not be confused with pathologic densities such as small granulomas. The lungs are located lateral and caudal to the heart.

The lungs are difficult to evaluate on the V/D view because of soft tissue overlap of the cardiac

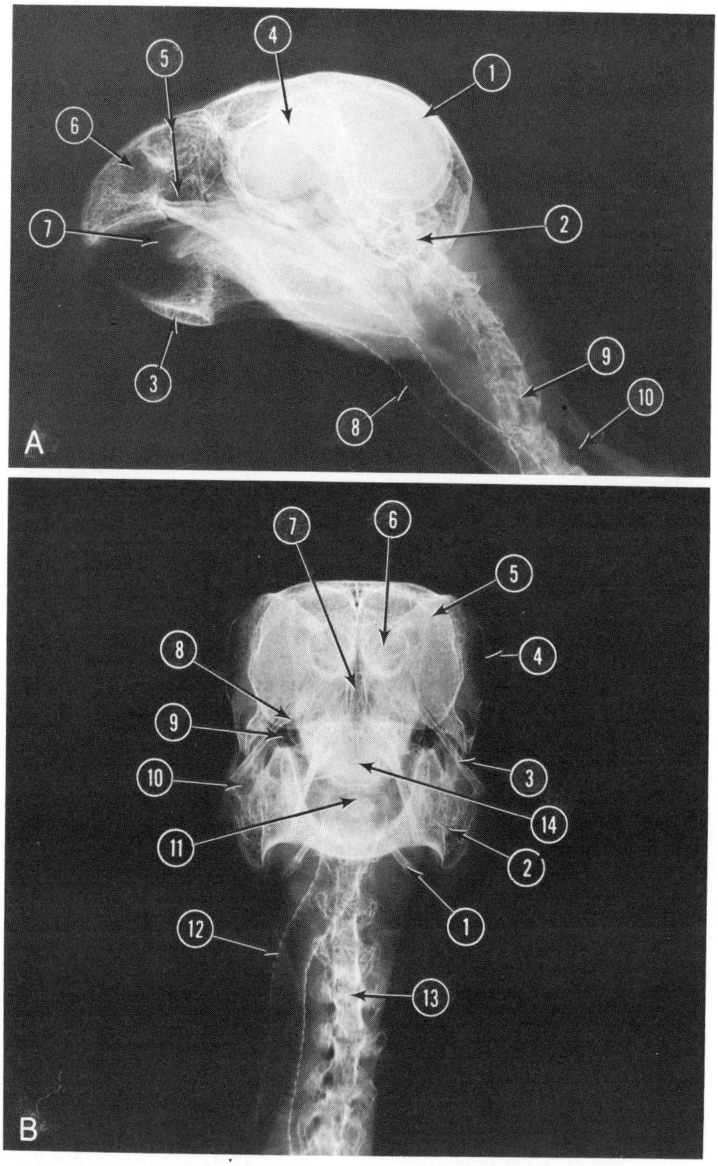


**Figure 14–2.** Normal skeletal structures of the Amazon Parrot. These were taken with a Faxitron unit while the bird was sedated with ketamine. The long exposure time and small focal spot give excellent detail of the hard tissue but may sacrifice soft-tissue detail because of patient movement.

*A*, Lateral view of the skull of a normal Amazon Parrot. 1, Cranial vault. 2, Tympanic area. 3, Mandible. 4, Orbit. 5, Cere (superficial arrow); anterior portion of the paranasal sinus (ventral portion of the arrow). 6, Maxillary sinus. 7, Tongue (contains bones of the hyoid apparatus). 8, Trachea. 9, Cervical vertebrae. 10, Area of cervicocephalic air sac.

*B*, Skyline (cranial to caudal) view of the skull of a normal Amazon Parrot. A ventrodorsal view may provide better delineation of the tympanic area. 1, Hyoid apparatus. 2, Mandible. 3, Zygomatic bone. 4, Orbital ring. 5, Edge of cranial vault. 6, Cere. 7, Nasal septum (nasal cavity). 8, Edge of the tympanic area. 9, Cranial portion of the paranasal sinus. 10, Caudal portion of the paranasal sinus (infraorbital sinus). 11, Glottis. 12, Trachea. 13, Cervical vertebrae. 14, Tongue.

*Illustration continued on opposite page*

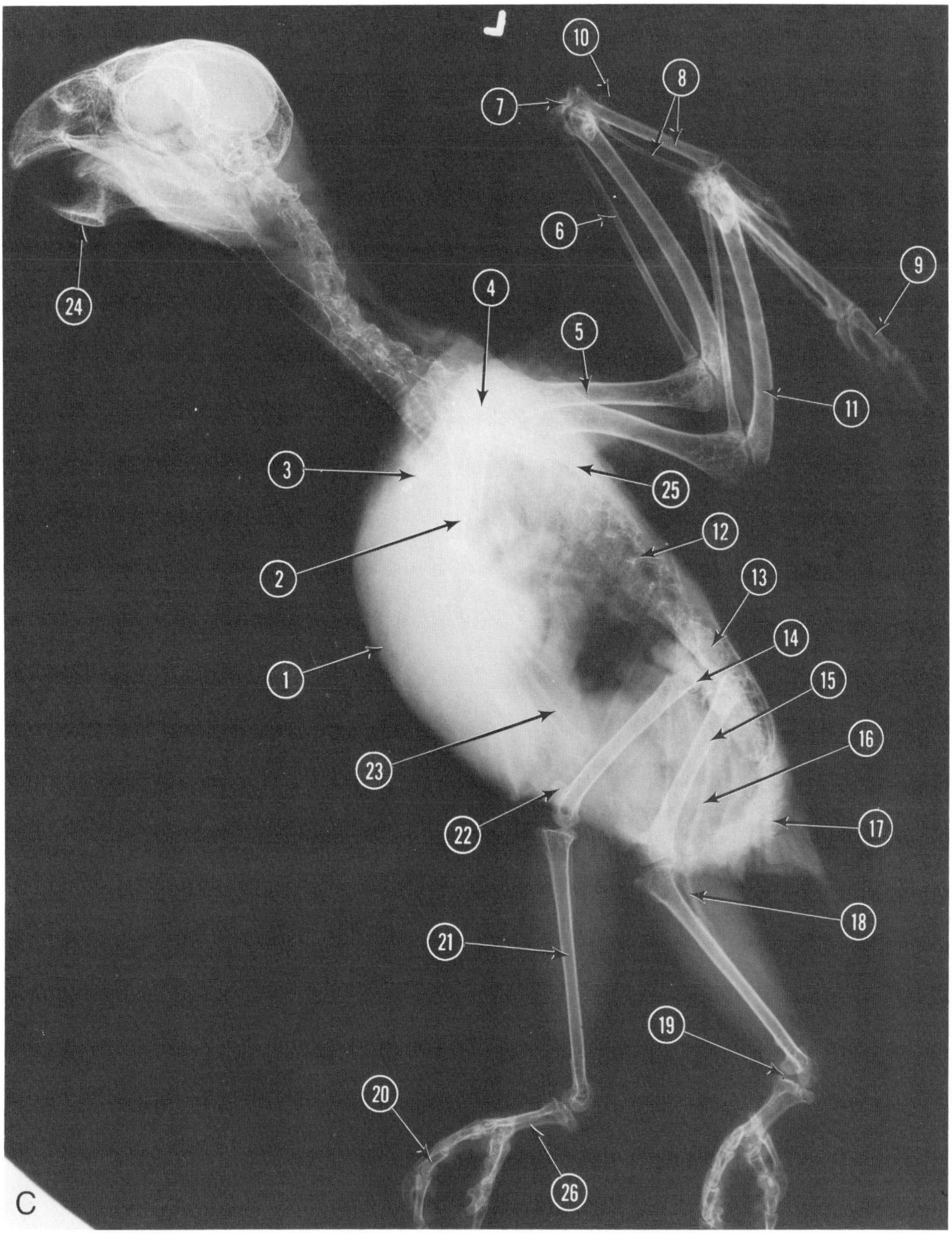


**Figure 14–2.** *Continued. C*, Lateral view of the skeletal structure of an Amazon Parrot. 1, Sternum (keel). 2, Coracoid bone. 3, Clavicle. 4, Shoulder joint. 5, Humerus. 6, Radius. 7, Radial carpal. 8, Metacarpus. 9, Phalanges of major digit. 10, Alular. 11, Ulna. 12, Thoracic vertebrae. 13, Synsacrum. 14, Acetabulum. 15, Femur. 16, Pubic bone. 17, Coccygeal vertebrae. 18, Fibula. 19, Intertarsal (hock or ankle) joint. 20, Phalanges. 21, Tibiotarsus. 22, Patella. 23, Ribs (sternal segment). 24, Mandible. 25, Scapula. 26, Tarsometatarsus.

*Illustration continued on following page*

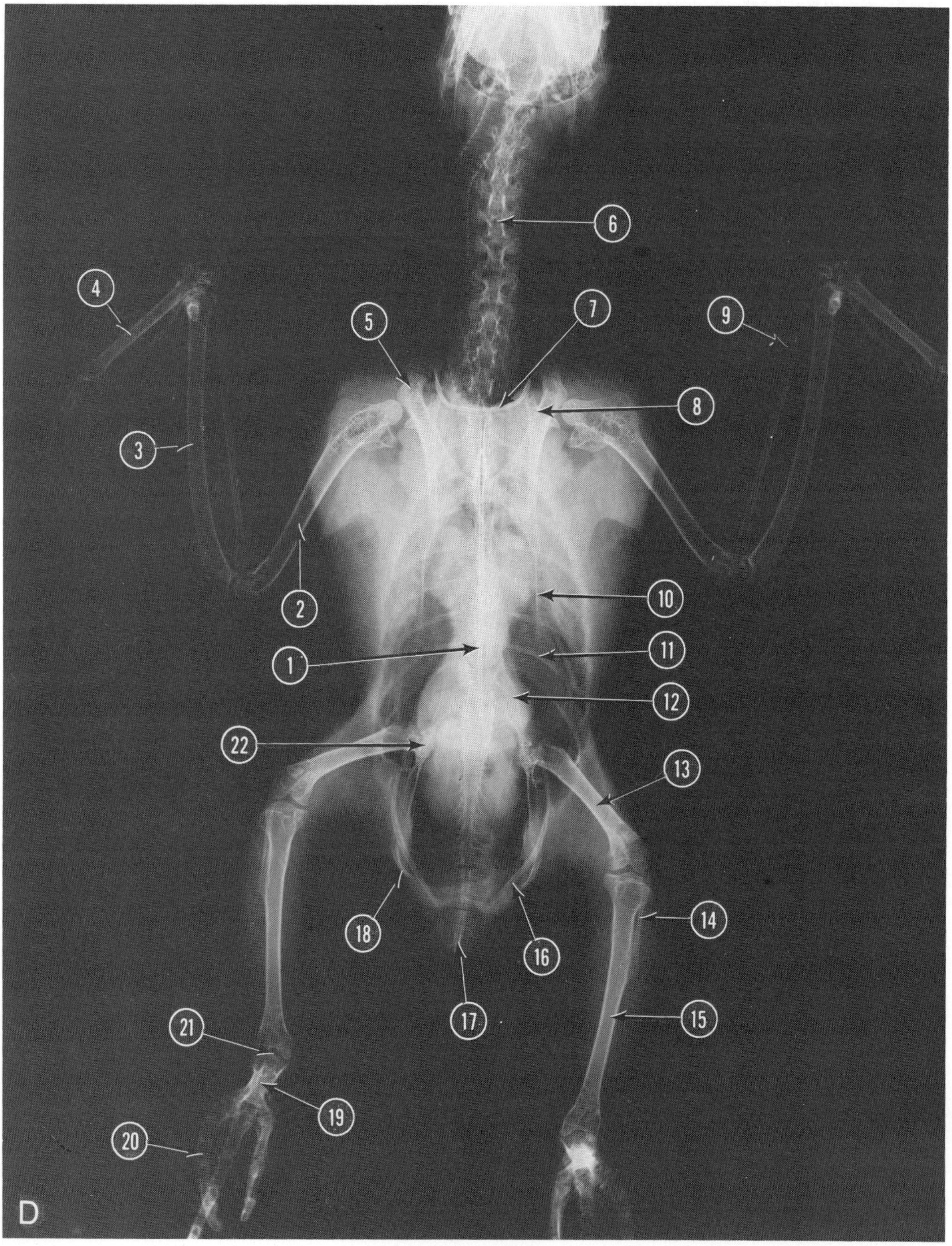


**Figure 14–2.** *Continued. D,* Ventrodorsal view of the skeletal structure of an Amazon Parrot. 1, Sternum. 2, Humerus. 3, Ulna. 4, Metacarpals. 5, Cranial portion of coracoid bone. 6,Cervical vertebrae. 7, Clavicle. 8, Cranial border of the scapula. 9, Radius. 10, Scapula. 11, Ribs. 12, Synsacrum. 13, Femur. 14, Fibula. 15, Tibiotarsal bone. 16, Pubis. 17, Pygostyle. 18, Ischium. 19, Tarsometatarsus. 20, Digit. 21, Intertarsal (hock or ankle) joint. 22, Acetabulum.

silhouette as well as the thoracic air sacs, yet the practitioner can get an overall impression of the bilateral homogeneity of the lung tissue. The cranial portion of the liver is tapered where it enfolds the apex of the heart. The right lobe of the liver is usually larger than the left and may extend to the edge of the sternum. The degree of caudal extension of the normal liver lobe is somewhat variable between species. The left lateral edge of the proventriculus may be seen paralleling the lateral line of the left liver lobe. This may give the appearance of an enlarged liver, especially in poor quality films.

On the V/D view the ventriculus lies caudal to the left liver lobe slightly below a line drawn between the acetabula. To the right and caudal to the ventriculus are the coiled loops of the intestinal tract. The duodenal loop is situated to the right of the ventriculus. When normal in size, the spleen may occasionally be seen on the V/D view between the proventriculus and the ventriculus, near or to the right of the midline. The size and shape of the spleen vary with the species of bird (see Chapter 23, Necropsy Procedures). The anterior section of the kidneys may be visible, especially when the liver is small. The outline of the cloaca may be seen when air is present.

Evaluation of the pulmonary air sac system is difficult until the practitioner can associate radiographic signs with lesions found at endoscopy or necropsy. The air sacs overlie almost the entire thoracic portion of the body cavity so that there may be some misinterpretation of the lesion location. Subtle changes seen on the V/D view may not be visualized on the lateral film because of tissue overlap. Portions of the interclavicular air sac can be seen as radiolucent areas communicating with the humerus in the shoulder area. The thoracic air sacs extend from the lungs almost to the sternum centrally and help to highlight the hour-glass shape of the heart and lungs. The abdominal air sacs are very large but are partially obscured by the digestive tract laterally. Portions of the abdominal air sacs can be seen as finger-like projections extending caudally along the body wall.

In the lateral view (Fig. 14–3*B*), the quadrant system may be confusing, and a three-layer approach to interpretation may be more useful. With the bird's head to the reader's left, the crop (with food present) is seen cranial to the thoracic inlet, ventral or sometimes dorsal to the vertebrae. Portions of the cervicocephalic air sacs are seen as small lucent areas under the skin of the neck area. The ventral layer consists of the heart cranially, followed by the liver, the ventriculus, and the ventral portion of the intestinal loops. This layer is positioned over the sternum (heart and liver) and along the ventral abdominal wall (ventriculus and intestinal tract). The air sacs overlap the dorsal portions of this layer of soft tissue.

The second layer begins cranially with the trachea and continuation of the esophagus from the thoracic inlet. Over the liver the esophagus expands into the proventriculus, which curves slightly ventral to join the ventriculus. The spleen, situated between the proventriculus and the ventriculus, may be observed overlapping the caudal proventriculus or protruding dorsally above it. The air sacs overlap all except the caudal portion of the layer containing the middle portion of the intestinal loops.

The dorsal layer includes the lungs cranially (which have a discrete bronchiolar pattern) and an obvious triangular lucent area dorsal to the proventriculus, composed of a portion of the thoracic and abdominal air sacs. Caudal to the lung the dorsal layer contains the gonads, which may be evident during the breeding season. The oviduct of a reproductively active female may change the normal locations of some soft tissue structures. The kidneys are visualized partially protruding from the synsacrum in the caudal portion of this layer.

## Radiographic Indications of Abnormal Conditions

### *Head and Neck*

Indications for radiographs of the head and neck area of birds differ from those of other species. The complex overlay of structures in the smaller avian patient, extensive pneumatization of the head, and unique respiratory features make lesion detection more difficult until one becomes familiar with these unique features. Films of the head in cases with chronic sinusitis may be helpful in locating areas that are plugged with organized granulomatous debris, especially unilateral abnormalities. Birds with clinical signs of nonresponsive sneezing, chronic rhinorrhea, or swollen periophthalmic tissue are candidates for radiography. Oblique films of the head may be indicated for birds with a head tilt to better define the tympanic area. Trauma to the head may be difficult to detect, but cranial films may be applicable in some cases.

The cervicocephalic air sac system has recently been recognized as having different clin-

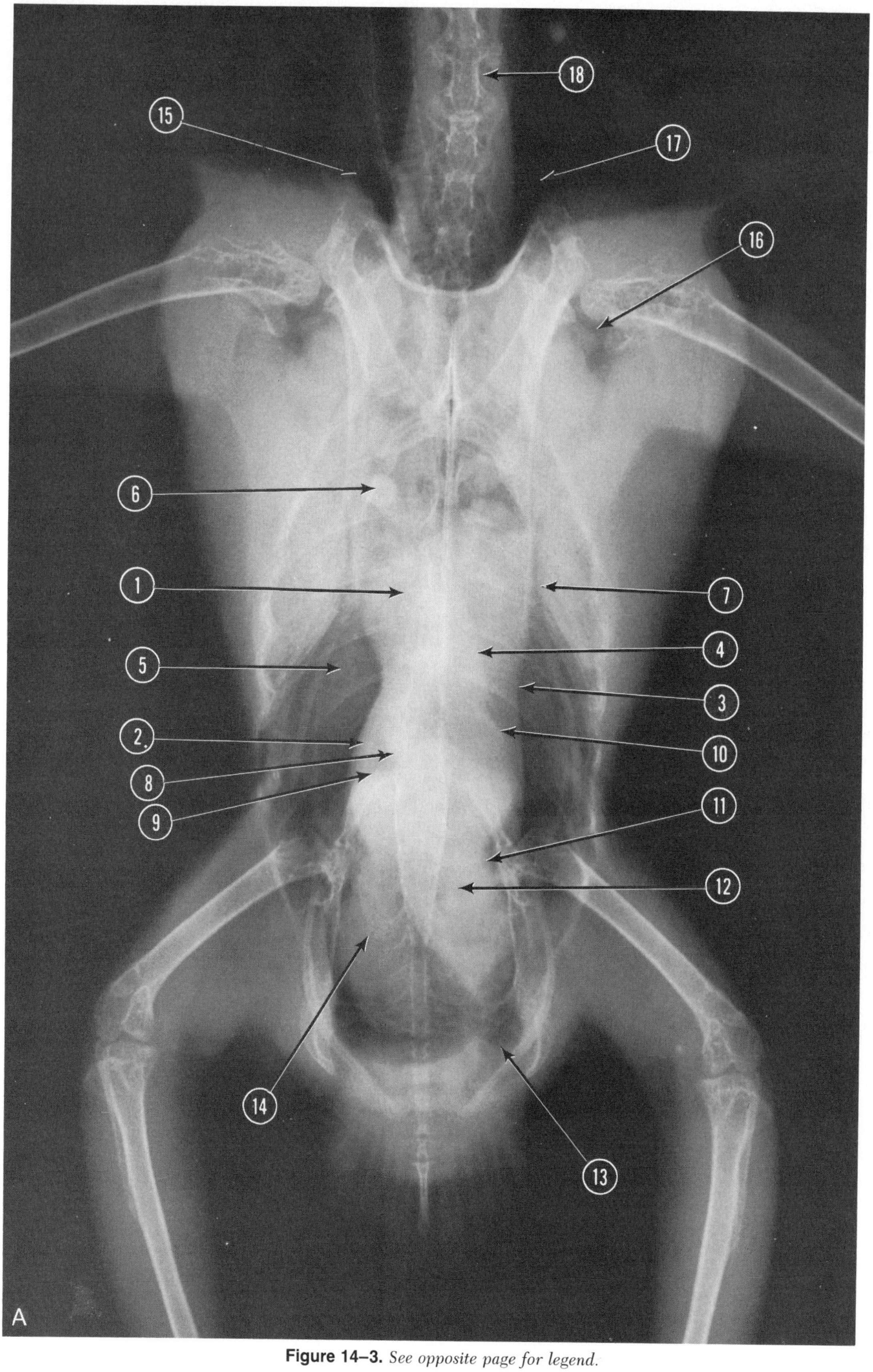


**Figure 14–3.** *See opposite page for legend.*

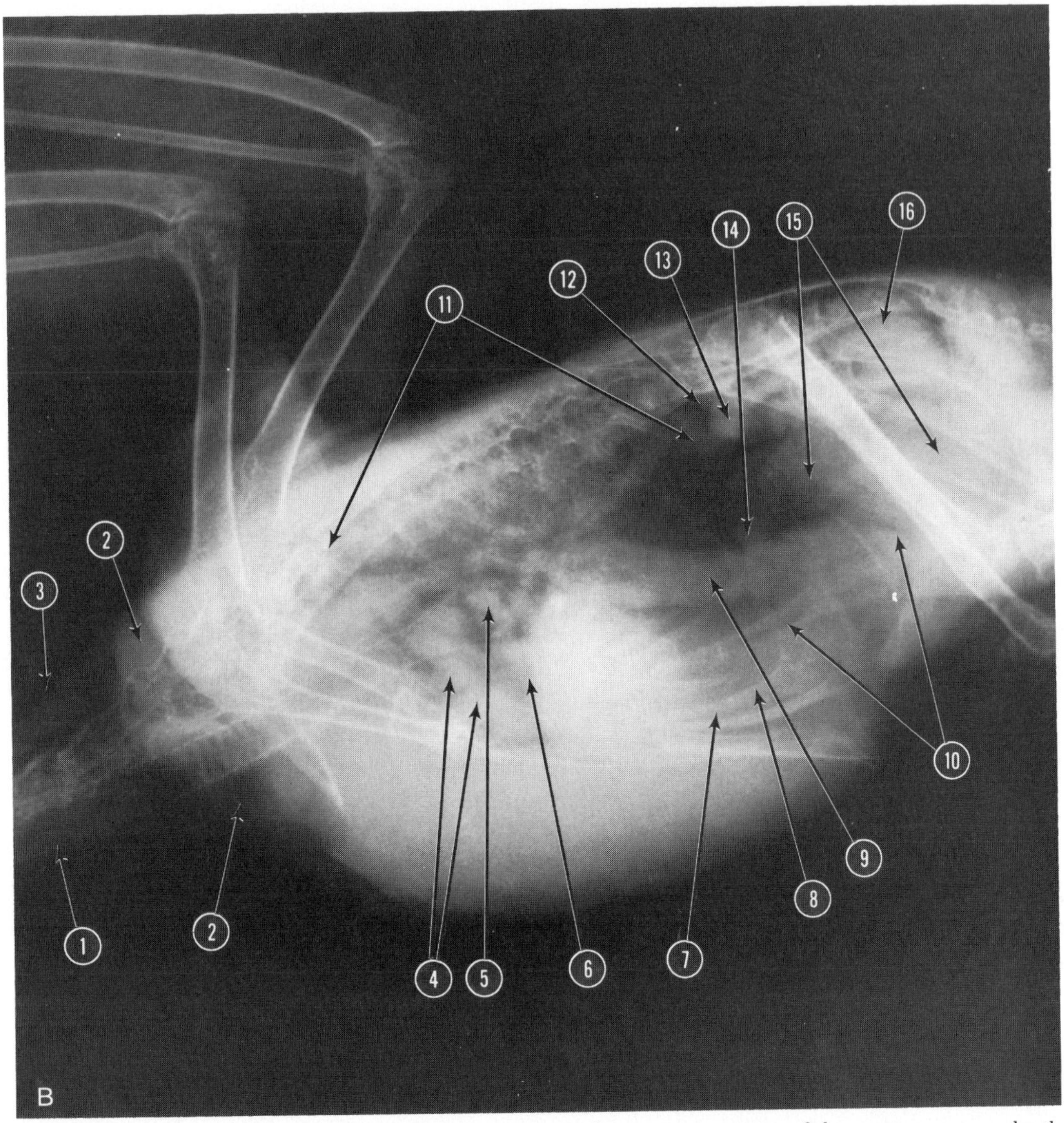


**Figure 14–3.** *A*, Ventrodorsal view of a normal Amazon Parrot. There is some rotation of the specimen as seen by the sternum failing to overlap the vertebral column. Even this amount of rotation can change the appearance of the relationship of the heart (1) and liver (2), with a shift of the normal waist line on the left (3). 4, Thoracic esophagus. 5, Caudal surface of lung. 6, Cardiac vessel. 7, Line formed by soft tissue overlap. 8, Area of spleen. 9, Cranial border of kidney. 10, Lateral border of proventriculus. 11, Approximate caudal border of left caudal thoracic air sac. 12, Ventriculus. 13, Left abdominal air sac caudal border (note smaller than right). 14, Intestinal loops containing pancreas. 15, Crop. 16, Humeral portion of interclavicular air sac. 17, Left cervicocephalic air sac. 18, Cervical vertebrae.

*B*, Lateral view of a normal Amazon Parrot. Note that the acetabula and shoulders are not superimposed. This can lead to misinterpretation if it is extensive. Perfect superimposition is difficult to achieve without supportive measures to prevent rotation of the body. 1, Trachea. 2, Crop. 3, Cervicocephalic air sac area. 4, Major cardiac vessels. 5, Syrinx. 6, Cranial edge of heart. 7, Caudal edge of heart. 8, Liver. 9, Proventriculus. 10, Cranial and caudal border of the ventriculus. 11, Cranial and caudal border of the lung. 12, Cranial border of the kidney. 13, Gonad. 14, Spleen. 15, Coils of the intestine. 16, Caudal segment of the kidney.

ical manifestations that may be detected or confirmed radiographically. Distention of the head and neck area with air may indicate a problem with this system. Distended cervicocephalic air sacs (Fig. 14–4*A*) should be differentiated from subcutaneous emphysema. Although rare, granulomas may form in these air sacs and may be radiographically visible (Fig. 14–4*B*).

### *Skeletal System*

Traumatic fracture detection and interpretation will not be discussed here (see Chapter 30, Evaluation and Nonsurgical Management of Fractures). Healing fractures of the long bones in birds have a greater endosteal callus component than in mammals, so there may be less periosteal component than expected. Septic arthritis occurs most commonly in the intertarsal (hock or ankle) joints.

Osteomyelitis, although considered by some to be rare, does occur in avian species. The pneumatic bones may allow for pathogen access to the respiratory system in addition to hematogenous spread (Fig. 14–5). It may be difficult to differentiate bone infection from bone tumor; however, tumors are uncommon and tend not to cross the joint space. If necessary, a biopsy and culture should be taken to assist in differentiation (see Chapter 10, Clinical Microbiology, and Chapter 16, Biopsy Techniques). Untreated infectious pododermatitis (bumblefoot) may lead to osteomyelitis and septic arthritis of the foot. Chronic cases of pododermatitis should be radiographed at the start of the treatment to gauge the degree of bone involvement. Bone lysis, common in advanced cases, is a poor prognostic sign. In following some cases of fractures that were not compound, it appears that some birds may develop an extensive periostitis without infection.

Metabolic bone disease may exhibit radiographic changes that are age-dependent (see Chapter 31, Nutritional Diseases). Folding or pathologic fractures are a problem in young, rapidly growing birds (Fig. 14–6). These fractures may not show any displacement. Spinal column and sternal deformities are common in malnourished young birds. Hyperostosis (localized increases in bone density in the long bones) may occur in females with ovarian changes.

### *Lower Respiratory Disease*

Diseases of the air sacs and lungs may be viewed radiographically, although pulmonary lesions are more difficult to detect. Clinically, lower air sac diseases may be associated with a nasal discharge in birds, and radiographs are recommended in these cases. Figures 14–7 to 14–10 illustrate common air sac abnormalities. When possible, air sac abnormalities on the V/D view should be confirmed on the lateral view to differentiate them from artifacts and to more accurately locate the lesions. The practitioner may consider attempting to identify a possible etiologic agents (See Chapter 15, Endoscopy; Chapter 16, Biopsy Techniques; Chapter 17, Cytology).

The trachea should be closely examined radiographically for possible obstruction or luminal narrowing in birds with obvious dyspnea and, if indicated, followed by endoscopy for direct visualization.

Lesions of the lung tissue are not as common in birds as in other species and are frequently overlooked on radiographs. Abnormalities of the pulmonary system include focal pulmonary disease (Fig. 14–11), which may be secondary to infection and may appear unilateral.

McMillan[4] reports that air mesobronchograms are a possible finding in the development of pneumonia and may show a loss of the usual reticular pattern. Two views (V/D and lateral) should be taken to more accurately define the lesion. Discrete circular lesions need to be differentiated from end-on major vessels when near the heart and end-on anterior intestinal segments when near the liver. Magnification of the radiograph may assist in the differentiation. Normal segmental rib densities may also cause confusion on the V/D view.

Pneumocoelom may be seen following a loss of normal respiratory integrity (Fig. 14–12), such as with an air sac rupture. The coelomic structures may be more obvious and the heart may be lifted off the sternum, depending on the portion of the respiratory system involved. Subcutaneous emphysema may occur (Fig. 14–13), as seen with a traumatic rupture of the axillary portion of the interclavicular air sac or secondary to endoscopy procedures.

### *Cardiovascular System*

Radiographic evidence of cardiovascular disease is relatively rare. Auscultation may detect possible murmurs; however, these do not appear to be common. It is difficult to judge cardiomegaly and microcardia based on a single film, because the heart may be caught in one stage of the cycle. If size is of concern, additional films or supplemental tests such as con-

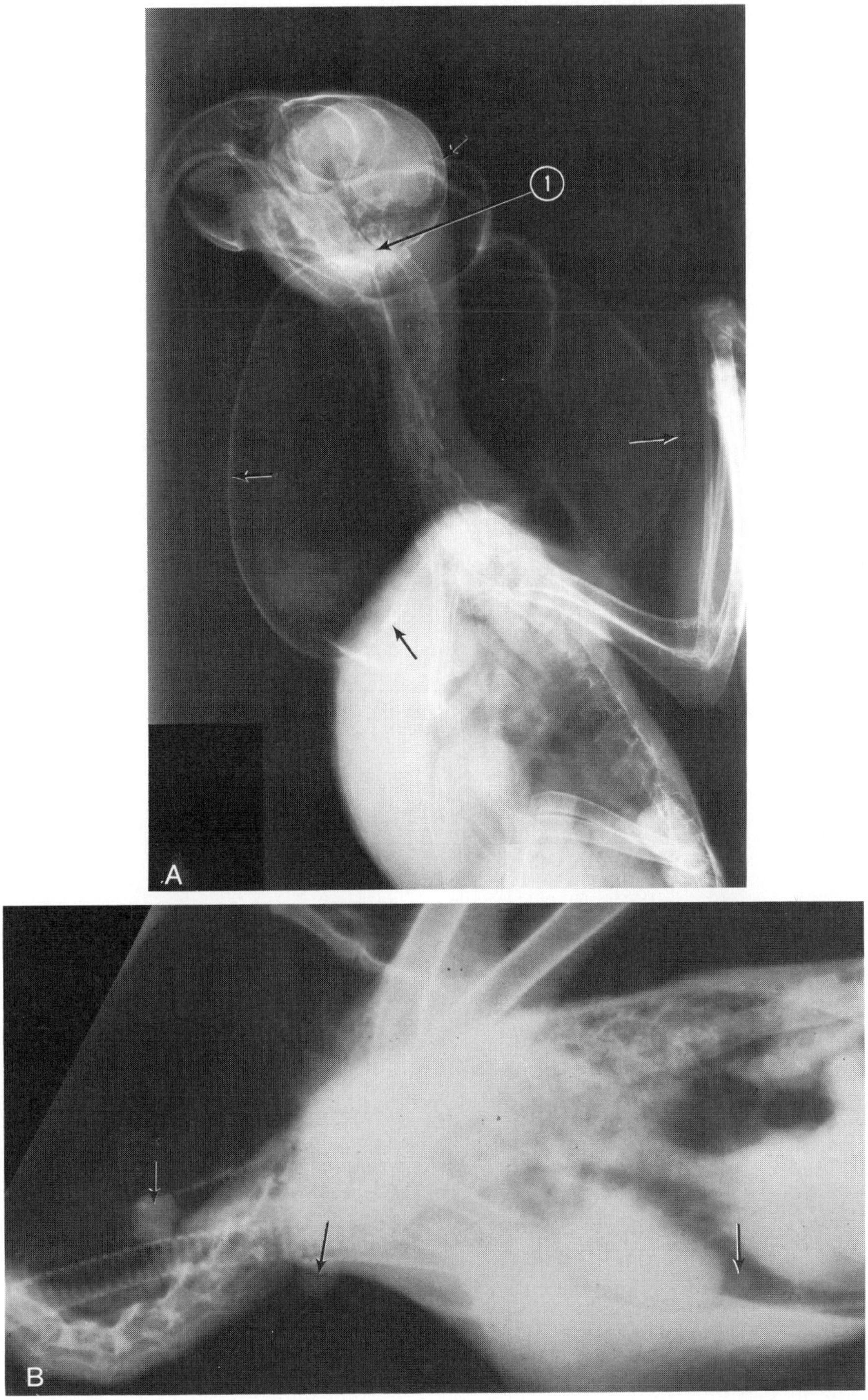


**Figure 14–4.** *A*, Lateral view of the distended cervicocephalic air sacs of an Amazon. Note that the borders (arrows) do not extend past the shoulders ventrally. There is no extension of the cephalic portion over the head. *B*, Tuberculosis granulomas are seen in the cervicocephalic air sacs of a cockatoo (arrows). This bird was extremely wasted at presentation. Note the absence of the normal liver shadow caudal to the heart (arrow). (From The Compendium. Used with permission of Veterinary Learning Systems Co., Inc.)

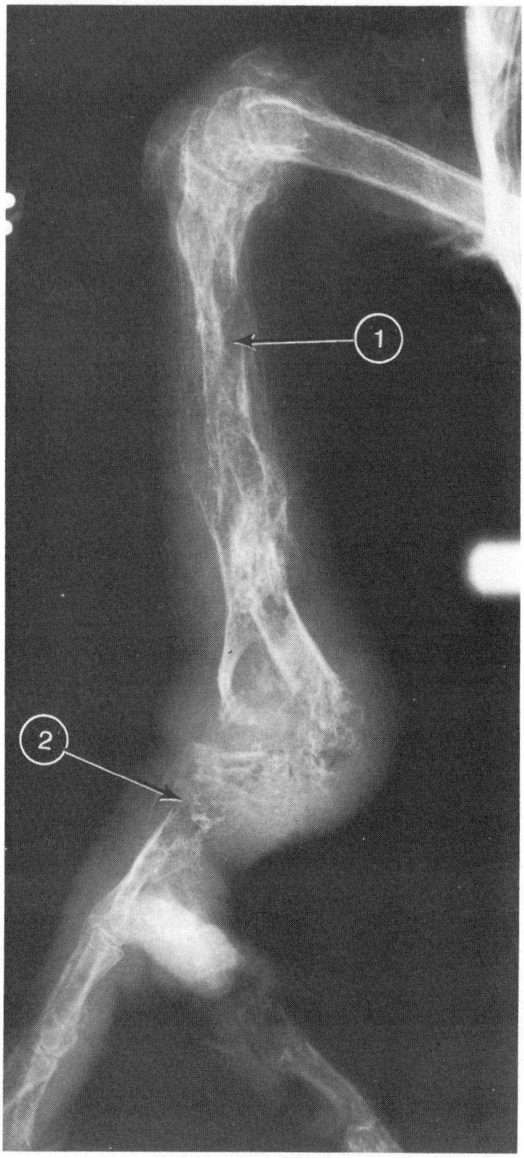


**Figure 14–5.** Osteomyelitis of the tibiotarsal bone (1), which has extended to the femur and tarsometatarsal bone. A pathologic fracture has developed in the tarsometatarsal bone (2).

trast studies, blood work, and ECG's may verify an abnormality (see Chapter 21, Electrocardiography).

The heart shadow in larger psittacines such as Hyacinth Macaws or other avian species (e.g., raptors) may not always show the classic hour-glass shape when viewed with the liver. These differences should not be interpreted as cardiomegaly or a small liver shadow.

Pericarditis may be associated with some disease problems and may be suggested by a change in the cardiac silhouette and subsequent blending with the shadow of the liver, mediastinal tissues, or lungs.

### *Gastrointestinal and Abdominal Disorders*

Gastrointestinal disorders observed on radiographs may involve organ distention (mechanical or adynamic ileus, fluid, gas); the presence of space-occupying lesions (neoplasms, granulomas, retained egg); visceral displacement; loss of normal abdominal detail; or presence of a foreign body.

Diseases of the crop other than foreign body or obstruction are often best approached by direct visualization (endoscopy, insufflation, transillumination) or by crop washing (see Chapter 15, Endoscopy, and Chapter 17, Cytology).

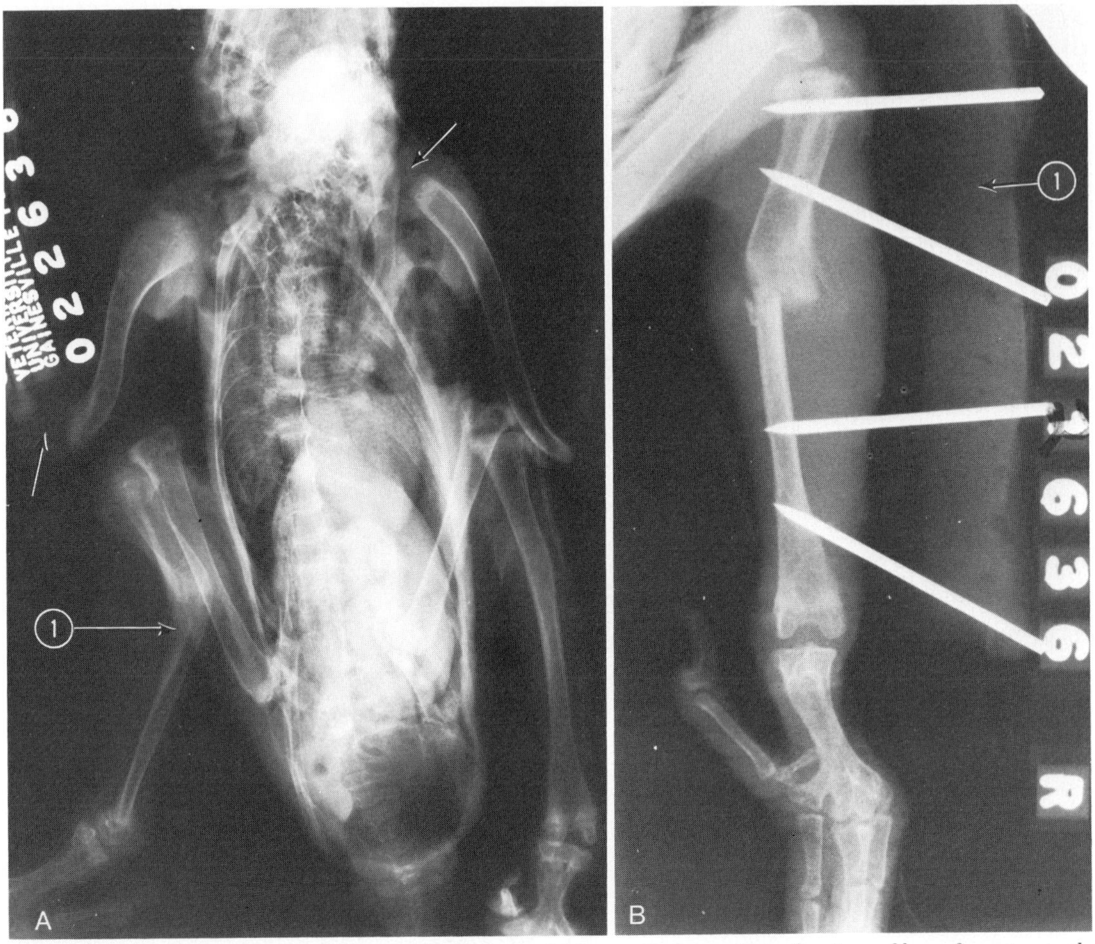


**Figure 14–6.** *A*, Folding fracture (1) developed in a young Moluccan Cockatoo. Note the stage of bone formation in the diaphyseal portions of the long bones (arrows). Positioning was difficult because of skeletal abnormalities in the spine, sternum, neck, and wings. *B*, Postsurgical fixation of the bird shown in *A*, with acrylic splint (1) and Kirschner-Ehmer pins.

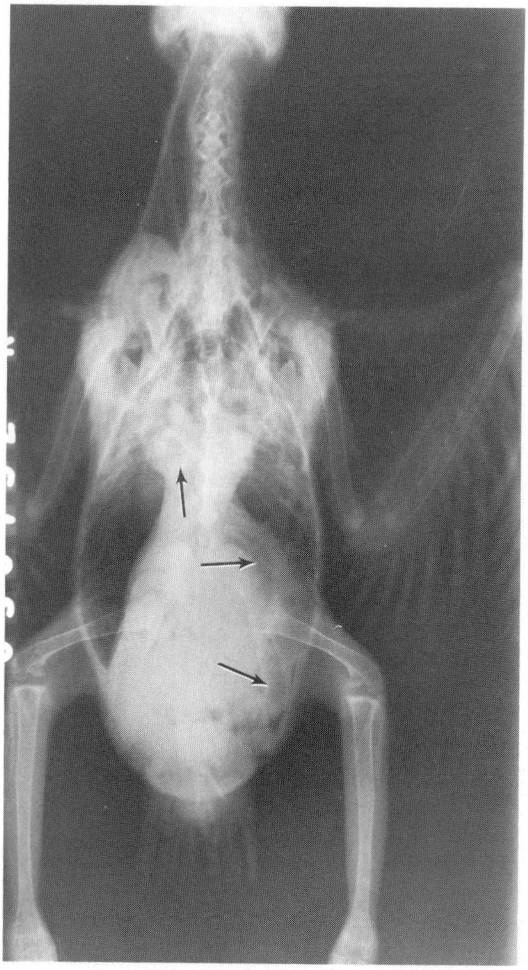

**Figure 14–7.** This young Eclectus Parrot has air sacculitis of the left thoracic and abdominal air sacs (arrows to left of midline), which contains a mixture of air and fluid. Compare it to the right side. There is another abnormality on the bird's right, overlying the heart. Air sac changes may be very subtle, so it is imperative to establish good technique for optimum results in order to detect their appearance. Note that the elbows have not fully developed.

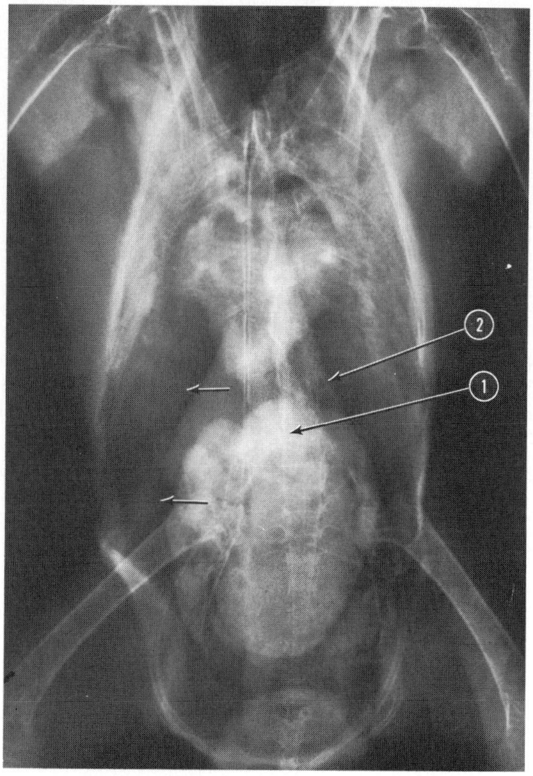


**Figure 14–8.** In this V/D view of a cockatoo, the small arrows indicate feathery-appearing densities on the right side of the coelom. These are consistent with air sac lesions. An enlarged spleen is present (1), and the edge of the proventriculus is evident (2).

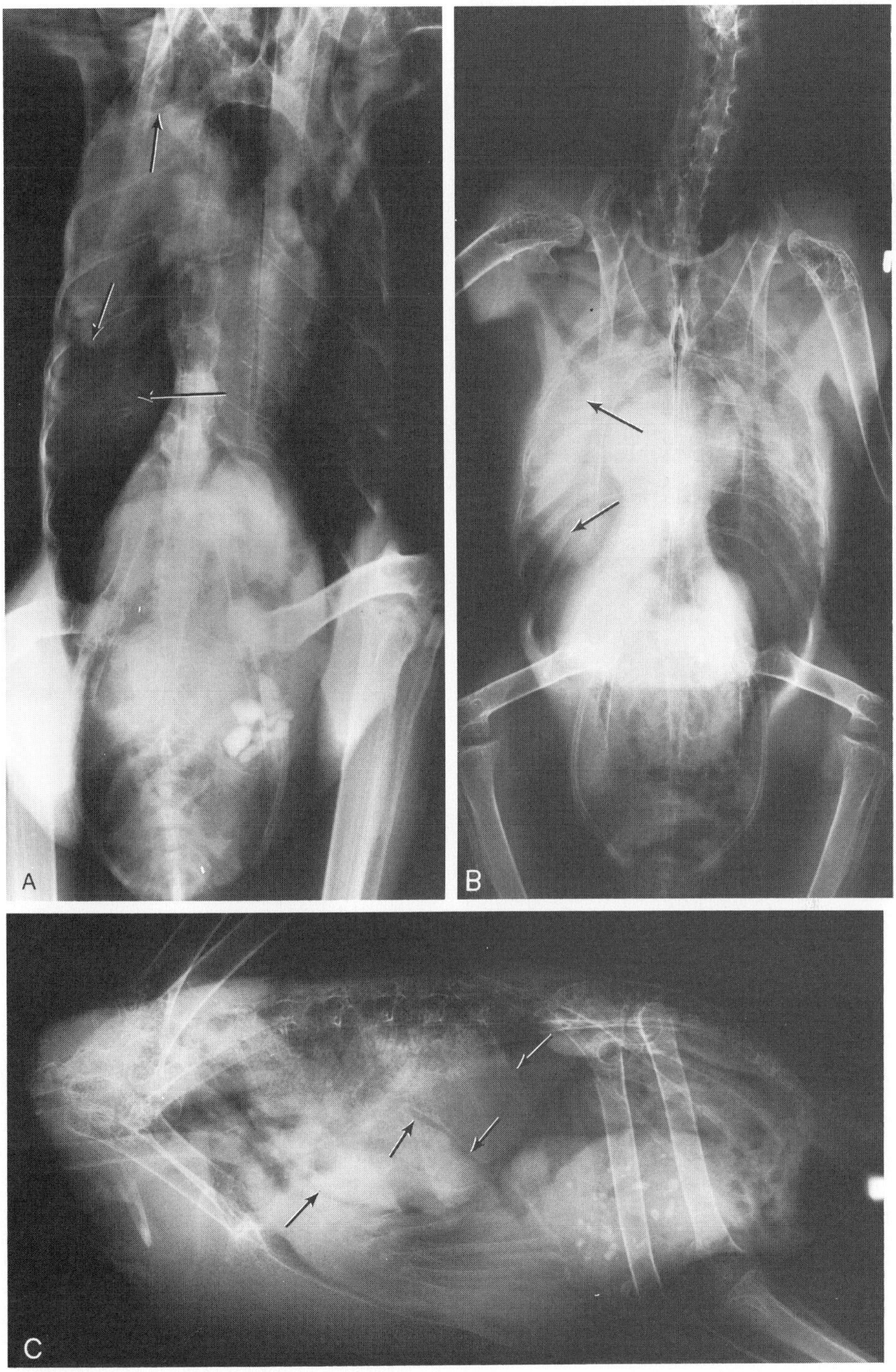


**Figure 14–9.** *A*, A massive air sac granuloma is shown in the V/D view of an African Crowned Crane (arrows). Aspiration and biopsy revealed an acid-fast organism that was compatible with tuberculosis. *B*, V/D view of an organized granulomatous area in an Amazon (arrows). A gram-negative organism was isolated. *C*, A lateral view of the bird in *B*. The lesions are extensive, filling the cranial and caudal thoracic air sacs. The cranial and caudal arrows outline the peripheral borders of the air sacs, while the two closely opposing arrows show the line of division between the two air sacs.

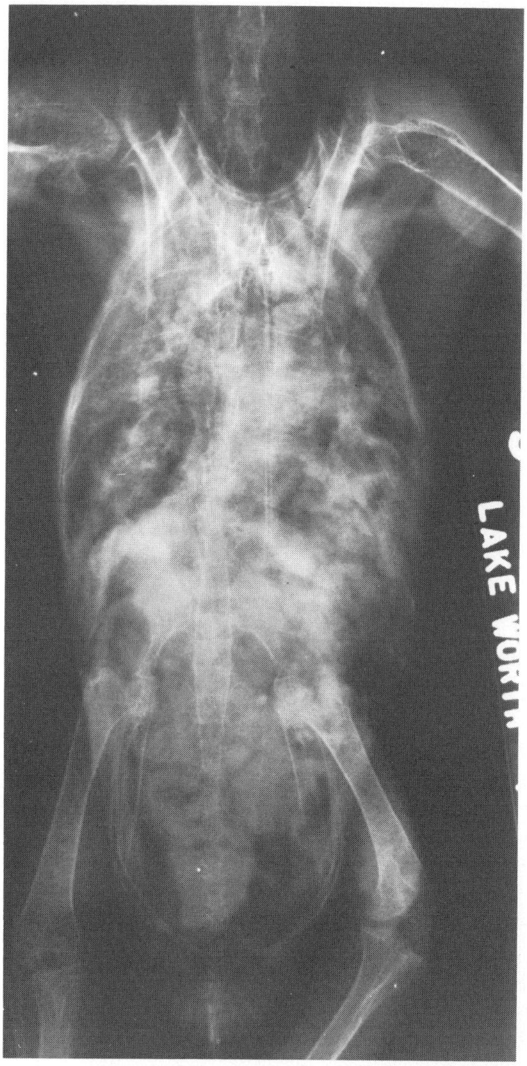


**Figure 14–10.** V/D view of a bird with a diffuse air sacculitis. Air is mixed with radiodense material throughout the air sacs, which gives a mottled appearance to the upper portions of the body. (Courtesy of Greg J. Harrison.)

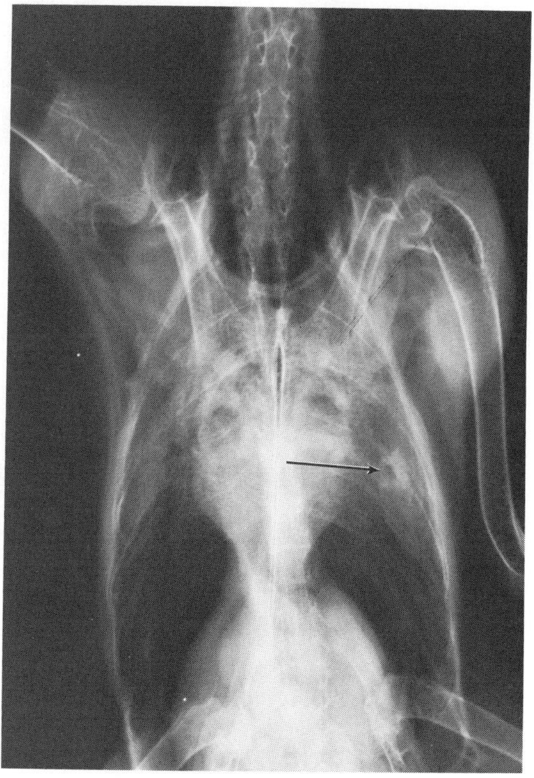

**Figure 14–11.** V/D view of a cockatoo with respiratory difficulty. There appears to be a focal lesion in the area of the left lung. It is possible for an overlying lesion in the air sacs to be confused with a pulmonary lesion.

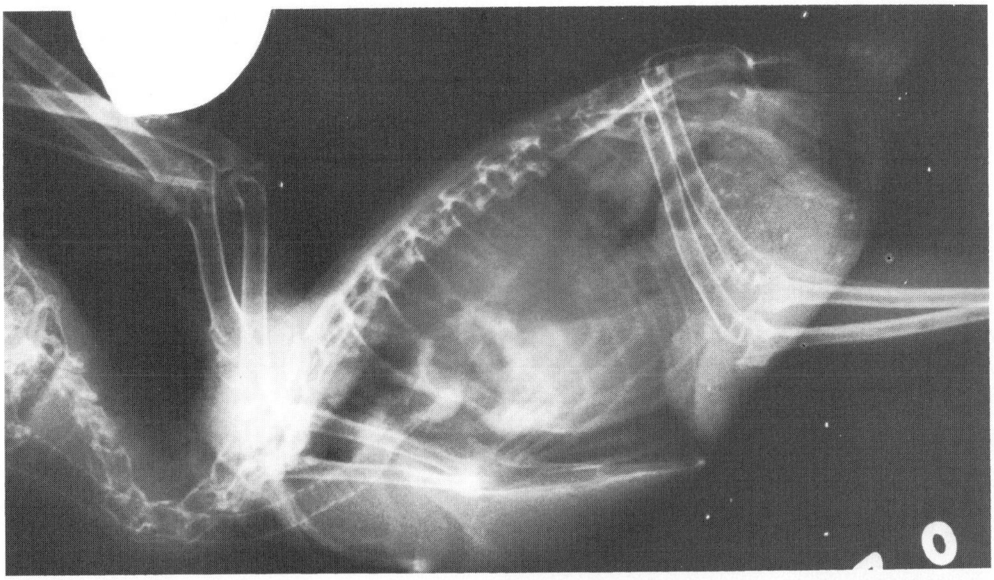

**Figure 14–12.** Cardiac displacement in a mynah bird with dyspnea. Note the elevation of the cardiac shadow and the increased lucency around the syrinx. The heart appears displaced caudally and may be enlarged. The liver shadow appears to be pushed caudally and the usually lucent lung areas appear clouded. Air sac rupture and cardiomyopathy are suspected.

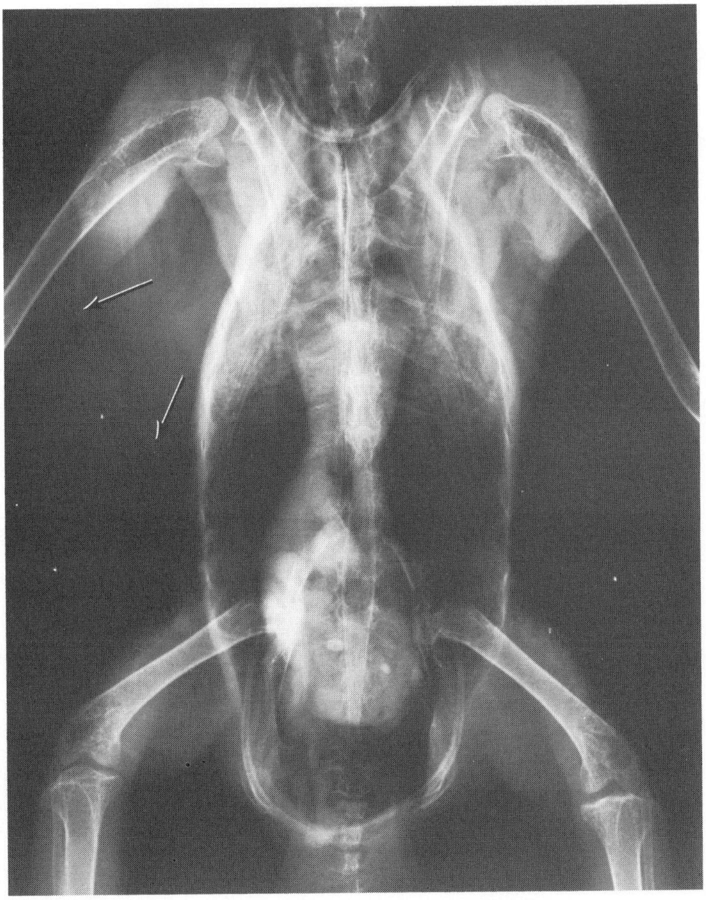

**Figure 14–13.** V/D view of subcutaneous emphysema of the axillary region in a cockatoo that sustained trauma. The proventriculus is displaced to the right of midline.

The proventriculus, like other dynamic organs, may require either more than one film or contrast studies to help diagnose a specific abnormality. Chronic dilatation of the proventriculus may be observed radiographically. This ballooned appearance, which is most prominent in the lateral radiograph, may also be interpreted as proventricular impaction, since it often seems distended with material.

Chronic dilatation in some cases appears to result from a loss of normal intestinal motility. This may possibly be secondary to a primary disease of the proventriculus such as has been reported in macaws and other psittacine species

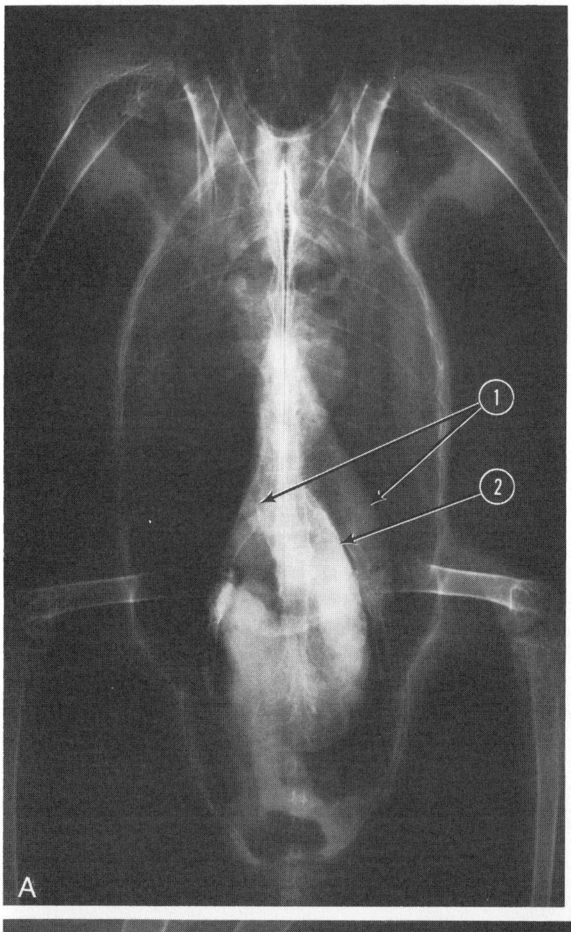


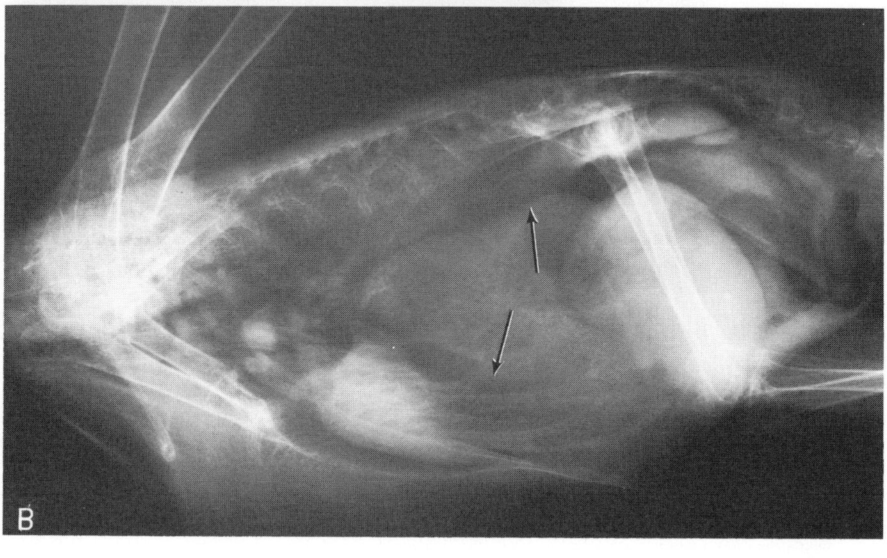


**Figure 14–14.** *A*, V/D view of a regurgitating macaw, illustrating the potential outline of an enlarged proventriculus (1) and the lateral border of the liver (2). This must be confirmed with positive contrast (see Fig. 14–15*A*). *B*, Lateral view of a macaw with proventricular dilatation. The arrows outline the gross distention. The ventriculus is also distended caudally (compare to normals).

(see Chapter 38, Neurologic Disorders). In proventricular dilatation of large psittacines, early clinical signs include regurgitation and weight loss. Radiographs taken at this point may not show true dilatation because the proventriculus may still have some muscle tone. Films should be repeated at a later date when the dilatation may be more pronounced (Fig. 14–14). Contrast studies may help to visualize the proventriculus (Fig. 14–15) and verify the lack of normal motility.

Chronic dilatation may also be related to a generalized adynamic ileus secondary to infection or toxicosis, which may occur in other species (e.g., generalized bacterial enteritis and/or psittacosis). Other disorders of the proventriculus may include foreign bodies (Fig. 14–16), yeast infections, overfeeding of neo-

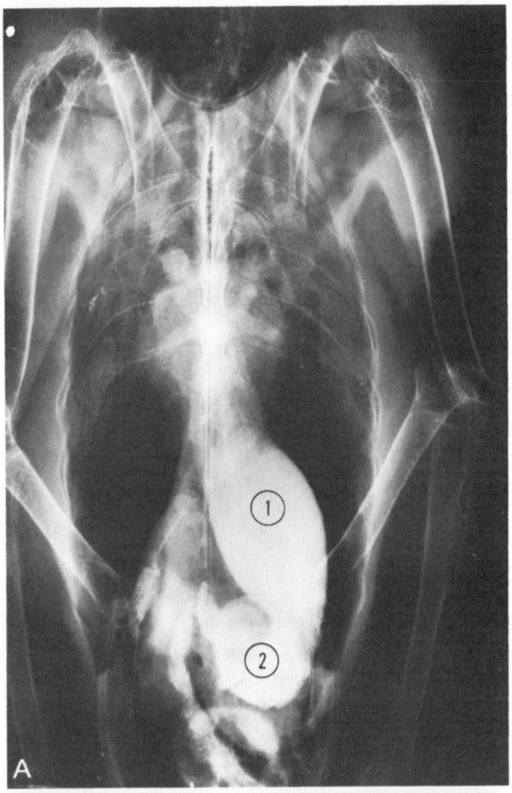


**Figure 14–15.** *A*, V/D view of a barium study in a macaw that is chronically regurgitating. The proventriculus (1) is distended and the ventriculus (2) is shown. Note that the lateral edge of the proventriculus extends past the liver border. Distention may also be found in hand-reared neonates. If in doubt as to the presence of true distention, retake the films some time later to verify. *B*, Lateral view of a barium study in a macaw with proventricular dilatation. The distended proventriculus is shown by (1) and the ventriculus by (2). Compare to Figure 14–35, which illustrates a bird with the same disorder; however, the film was taken prior to the dilatation phase. The "blotchy" pattern is caused by decaying seeds.

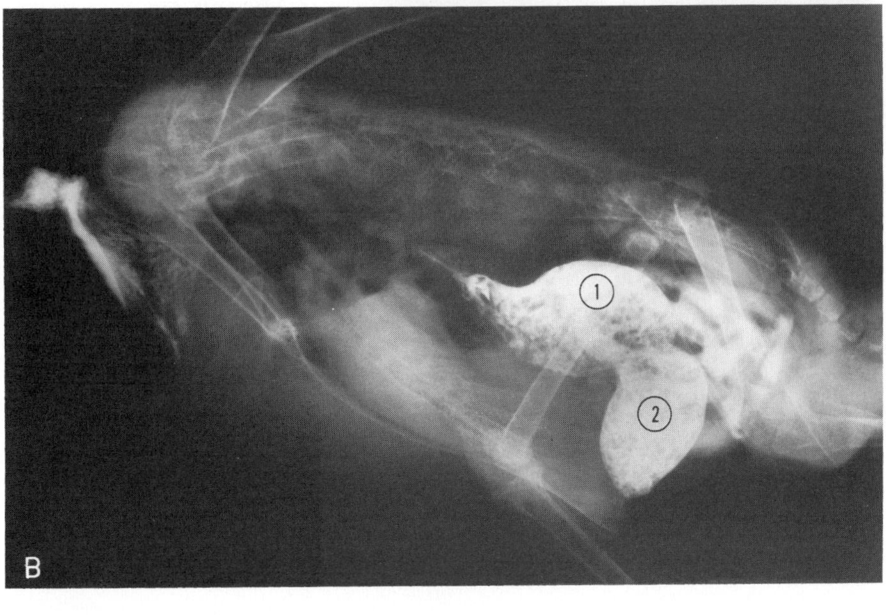

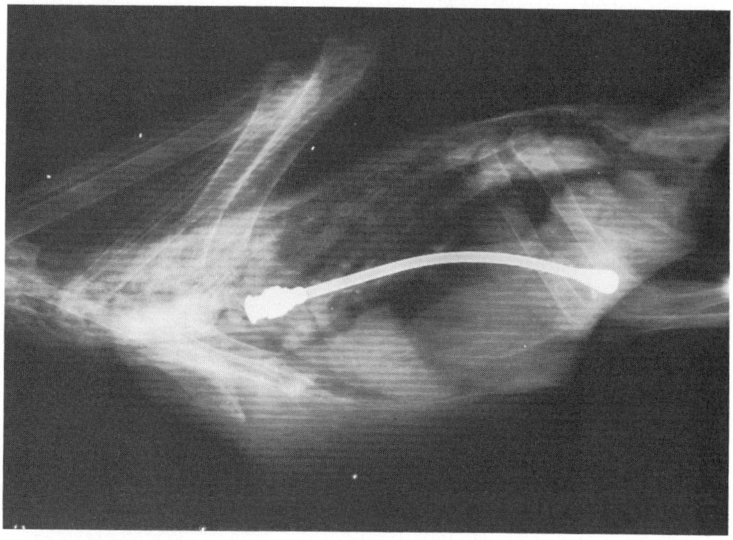

**Figure 14–16.** Lateral view of a macaw with an obvious foreign body. This view illustrates, with the help of the feeding tube, the path of the lower esophagus, proventriculus, and ventriculus. (Courtesy of Greg J. Harrison.)

nates, or tumors (Fig. 14–17). The proventriculus may normally appear larger in neonates than in adults.

The ventriculus (gizzard) is usually the organ most easily identified in birds that receive grit. Common disease conditions involving the gizzard include impaction/obstruction from foreign material or grit, luminal masses, and toxicosis

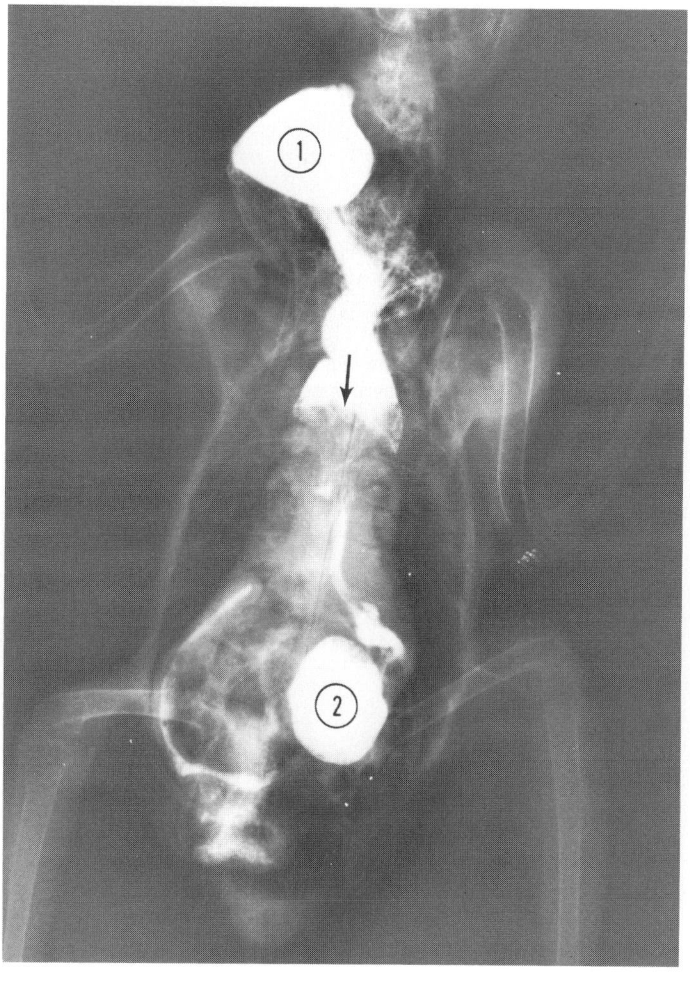


**Figure 14–17.** V/D view of an Amazon with an adenocarcinoma in the proventriculus. This was taken at 10 minutes post–barium ingestion. The crop (1) still retains a great deal of contrast. The arrow shows the cranial extent of the mass. The ventriculus is highlighted (2). The contrast has already passed through a large portion of the intestinal tract. (Courtesy of G. V. Kollias.)

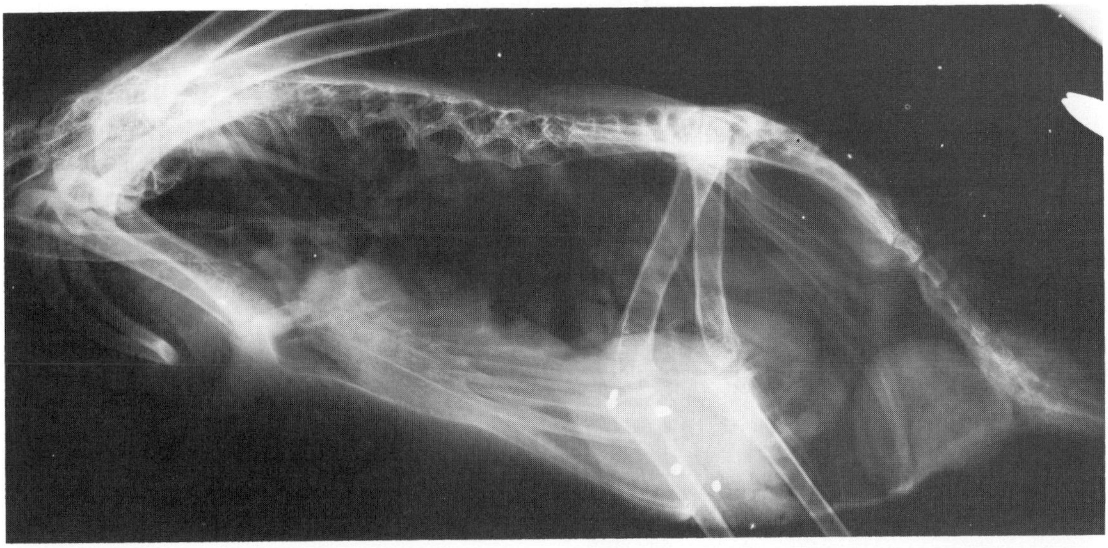

**Figure 14–18.** Lateral view of a Mallard Duck with ingested lead particles. The bird was showing generalized weakness and diarrhea. The absence of metal dense material does not rule out heavy metal intoxication.

from heavy metals (Fig. 14–18). The absence of metal particles on radiographs does not rule out heavy metal toxicosis, nor does the presence of radiodense material confirm the condition, as other materials exhibit a radiographic density approaching heavy metal.

Ventricular displacement is often the most obvious abnormality seen on radiographs and may result from a coelomic mass, organ enlargement, or tumor. The direction of displacement may suggest the source.

The spleen often enlarges in birds that have an active infection. Splenomegaly is more consistently viewed on the lateral view with the spleen situated between the proventriculus and ventriculus. Its image may overlap the proventriculus or may be seen dorsally (Fig. 14–19). On the V/D view the enlarged spleen may be visualized through the liver tissue to the right of the gizzard and slightly anterior (Fig. 14–20). Gross enlargement of the spleen may slightly displace the liver or ventriculus. It must be emphasized that although splenomegaly is commonly seen with psittacosis, it is not pathognomonic for psittacosis, since the spleen may enlarge with other disease processes as well (Fig. 14–21). Following treatment, the spleen can be seen to return to a more normal size (Fig. 14–22).

Hepatomegaly is a commonly diagnosed condition in avian species. Liver enlargement may result from primary liver diseases such as fatty

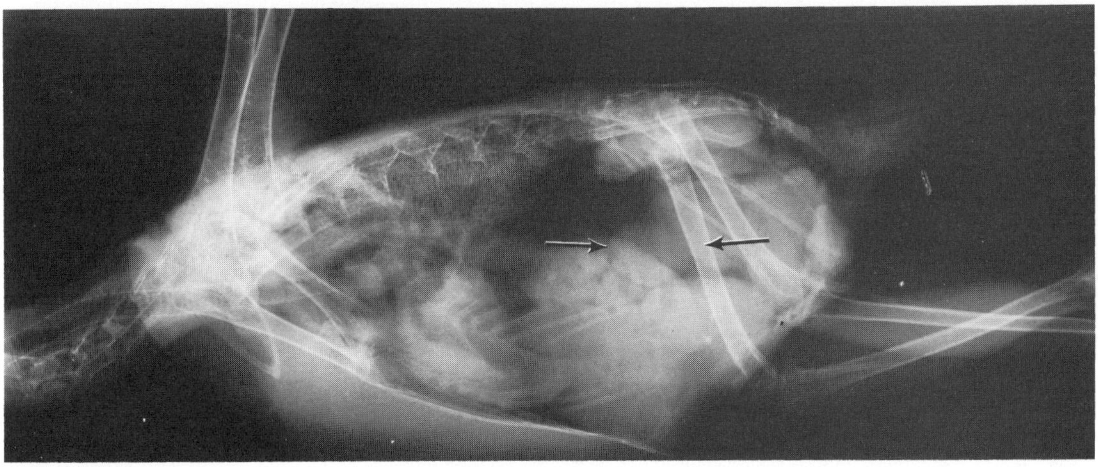

**Figure 14–19.** Lateral view of an Amazon Parrot with splenomegaly related to psittacosis. The opposing arrows outline the spleen. Note its position between the proventriculus and ventriculus.

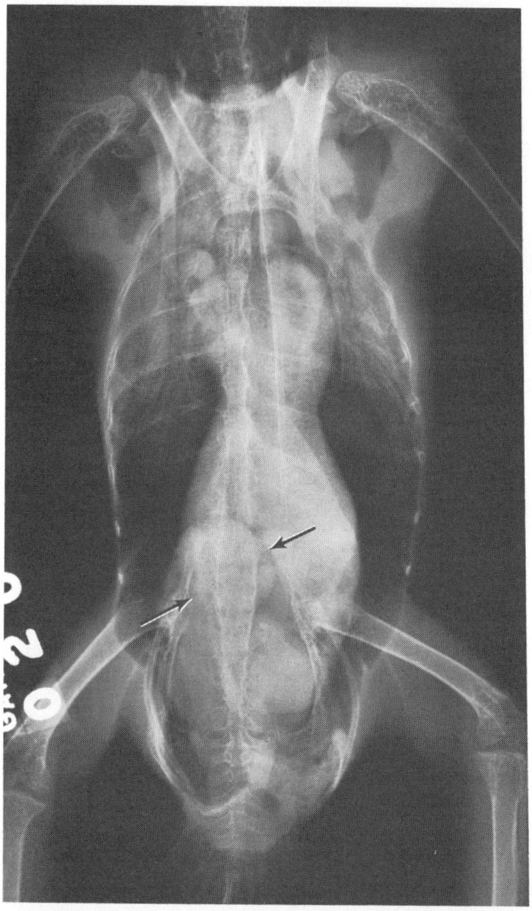

**Figure 14–20.** V/D view of the bird in Figure 14–19. The bird is rotated and the spleen is evident as outlined by the arrows. Its position even without rotation is usually at or to the bird's right of midline, dorsal to a line through the acetabula.

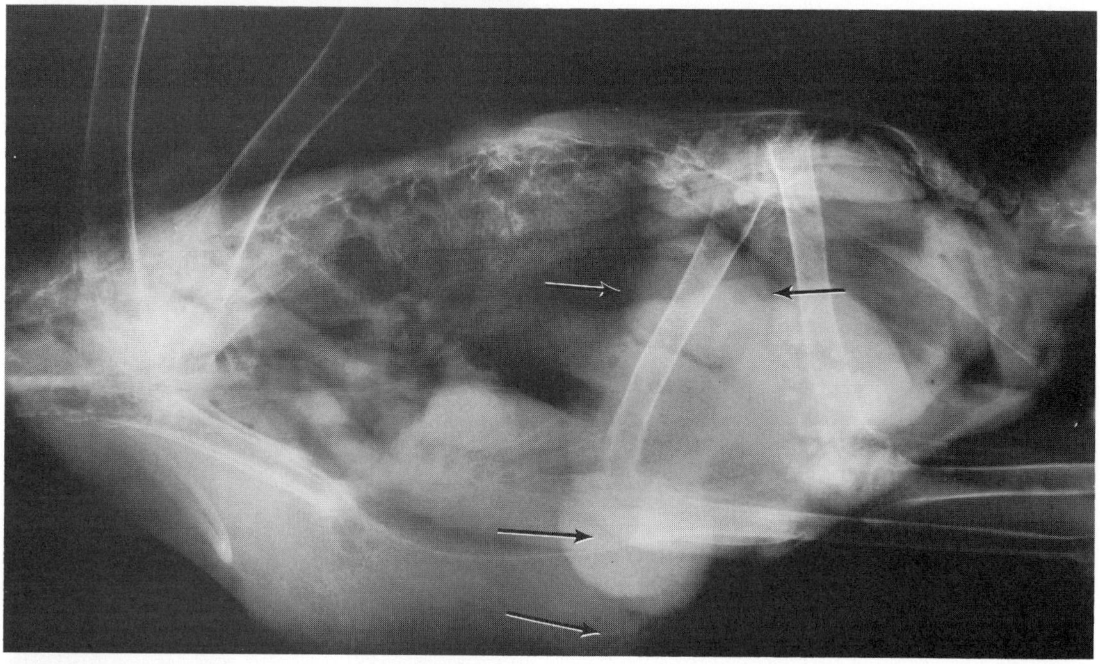

**Figure 14–21.** Splenomegaly in a macaw with granulomas. The opposing arrows outline the massively enlarged spleen. These are multiple subcutaneous granulomas (ventral arrows). This illustrates the fact that an enlarged spleen is not always related to psittacosis.

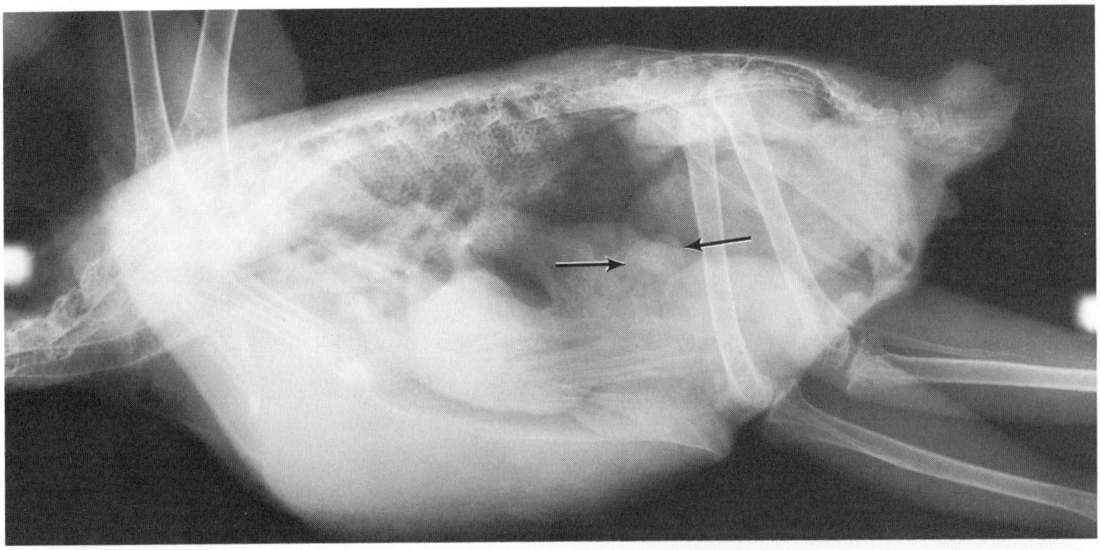

**Figure 14–22.** Post-treatment spleen size reduction of bird shown in Figures 14–19 and 14–20. Measurement ratios of the femur width to the length of the spleen show approximately a 48 per cent reduction in size.

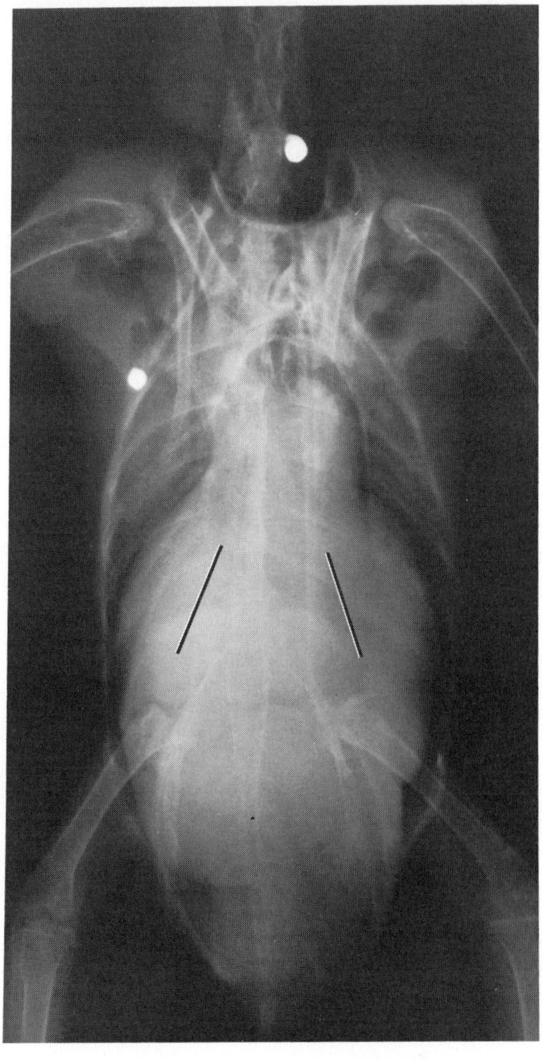

**Figure 14–23.** V/D view of an Amazon with an increased "liver" shadow. This film illustrates the need for the lateral view, since it revealed an image similar to Figure 14–25. The lead pellets were an incidental finding, but this material may be slowly absorbed, creating a chronic intoxication.

infiltration, congestion, hepatitis, and tumors. It should be noted that expiratory films may give the appearance of liver enlargement, so caution must be exercised in diagnosing borderline hepatic enlargement based on one set of films. Apparent liver enlargement may be artifactual from proventricular or other organ displacement. If an overfilled proventriculus is suspected, a second set of films following a fast may be helpful.

On the V/D view hepatomegaly (Fig. 14–23) may be interpreted as a decrease in the lucent air sac space laterally and an open umbrella appearance of the liver shadow. There may be crowding of the heart shadow anteriorly and displacement of the ventriculus caudally On the lateral view there may be elevation of the proventriculus and gizzard, and a decrease in the size of the lucent triangular space between the lung, kidney, and proventriculus. Tumors of the liver (especially the right lobe) may displace the gizzard dorsally (Fig. 14–24A). Large masses from gonadal or renal neoplasms may resemble hepatomegaly on the V/D view.

Ascites and peritonitis, which may be present individually or simultaneously, may be misinterpreted as gross liver enlargement on the V/D view. In some disease processes both ascites and liver enlargement may be present, such as with the mynah bird hepatopathy syndrome

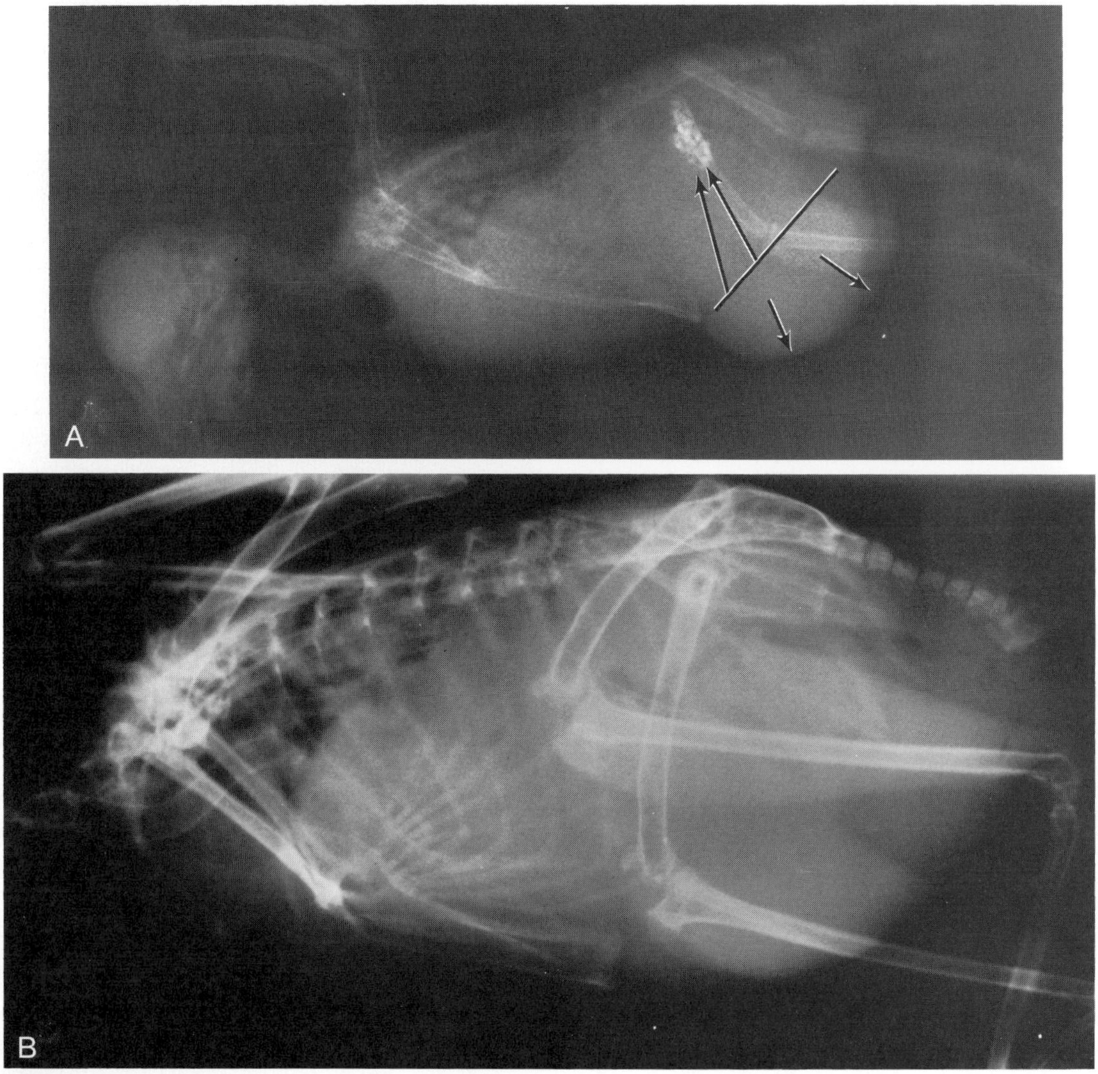


**Figure 14–24.** *A*, Lateral view of a budgerigar with hepatic neoplasm and ascites. The dorsal arrows show the displacement of the ventriculus from the approximate normal abdominal line. The ventral arrows outline the abdominal distention. *B*, Lateral view of a mynah bird with hepatomegaly and ascites. Caution should be exercised in radiographing birds with this syndrome. Handling can be fatal. Note the complete loss of abdominal detail. Positioning was sacrificed for easy handling.

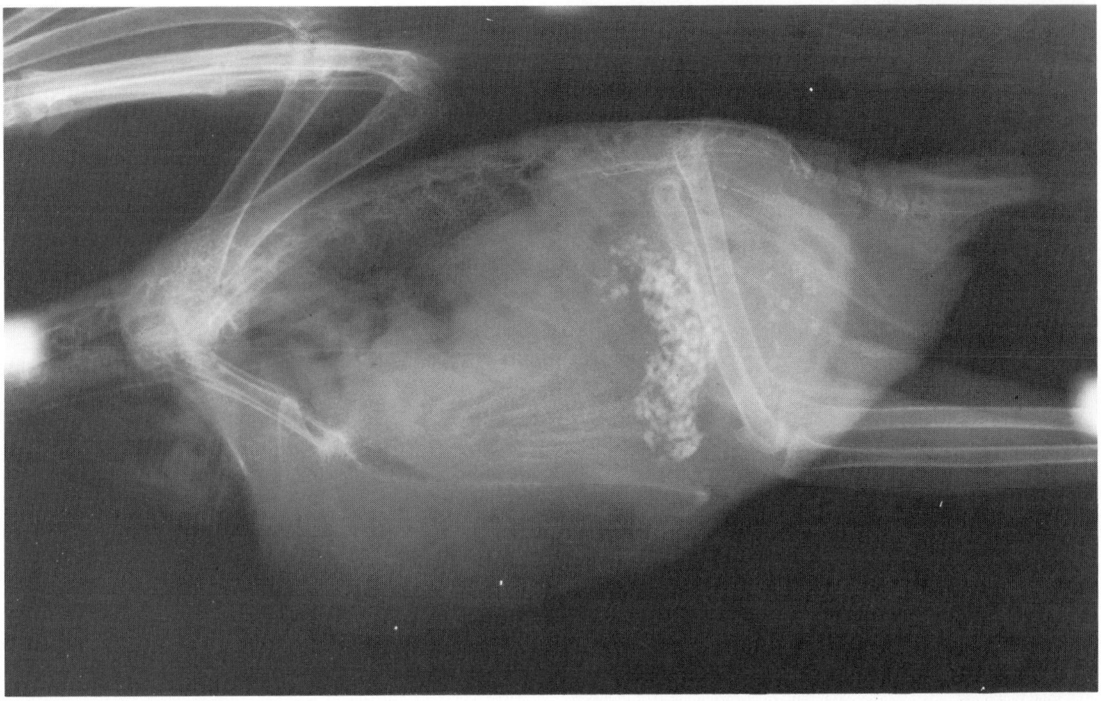

**Figure 14–25.** Generalized peritonitis in an Indian Ringneck Parakeet. Note displacement of the grit-filled proventriculus dorsally, probably related to the cranial displacement of the ventriculus. There is a total obliteration of the lucent triangular space and loss of abdominal detail.

(Fig. 14–24*B*), infectious peritonitis, and egg peritonitis (Fig. 14–25). Although resembling hepatomegaly on the V/D view, the latter conditions show more of an overall loss of organ detail. They are better appreciated on the lateral view. There is often an obliteration of the lucent area dorsal to the proventriculus; although the lung can be viewed, the rest of the field is a general haze with occasional gas shadows from the intestinal tract. Positive contrast may be helpful in outlining the structures of the coelom to see if there is displacement present. Paracentesis is also a diagnostic aid (see Chapter 17, Cytology).

Radiographic signs of bowel disease are somewhat similar to those in other species but are difficult to localize. Air-filled bowel may result from functional atony related to intestinal disease, such as the macaw proventricular syndrome (Fig. 14–26), or secondary to aerophagia, impaction, enteritis, or toxicosis. Fluid-filled bowel may be seen with enteritis (Fig. 14–27) or psychogenic water consumption. The lesions observed radiographically may depend on what stage of a disease the animal is in. Contrast studies may help to determine the presence of an impaction.

Abdominal hernias (Fig. 14–28*A*) are found classically in the female members of the psittacine species, especially the budgerigar, and may be associated with egg laying and weakened abdominal musculature. Common radiographic findings with herniation are loss of the abdominal line integrity, organ displacement, and/or the bulging of viscera into the defect. Herniation may also be related to trauma, an abdominal mass, egg binding, or straining and may be clinically suggestive of malnutrition or endocrine imbalances. Radiographs may help evaluate these possibilities. Contrast studies may be required to determine if viscera are involved in the herniation (Fig. 14–28*B*).

### *Reproductive System*

Egg binding is a fairly common problem in some species of caged birds. Calcified eggs are easy to view radiographically (Fig. 14–29). Larger species frequently show more than one egg, although one is usually soft-shelled. The presence and position of these incompletely developed eggs, which have a leather-like texture, may not be obvious. Occasionally, developing eggs may be released into the abdominal

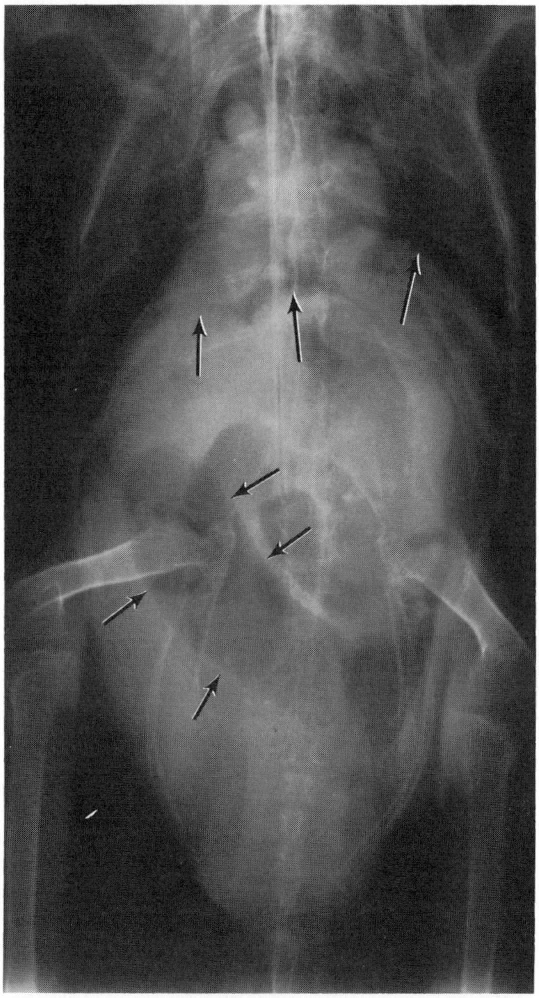

**Figure 14–26.** Ileus of the intestinal tract of a macaw. The cranial arrows show the displacement of the intestinal tract forward. The caudal opposing arrows outline a portion of distended air-filled bowel. While mechanical obstruction is a possibility, it is not as common as a dynamic ileus caused by intestinal disease such as in this bird.

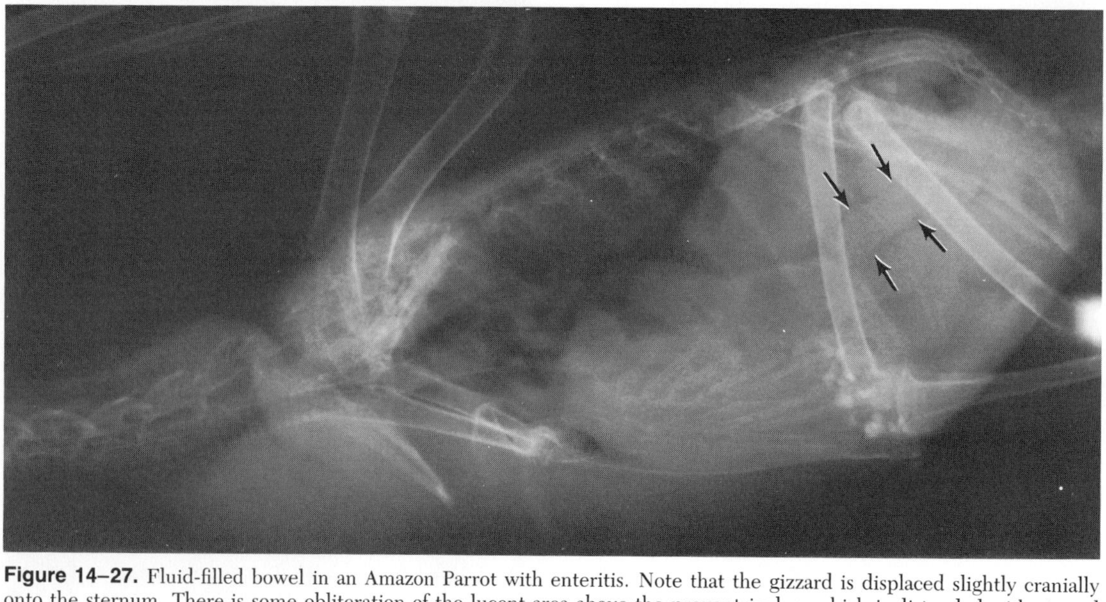

**Figure 14–27.** Fluid-filled bowel in an Amazon Parrot with enteritis. Note that the gizzard is displaced slightly cranially onto the sternum. There is some obliteration of the lucent area above the proventriculus, which is distended with air and fluid. The opposing arrows show the bowel distention.

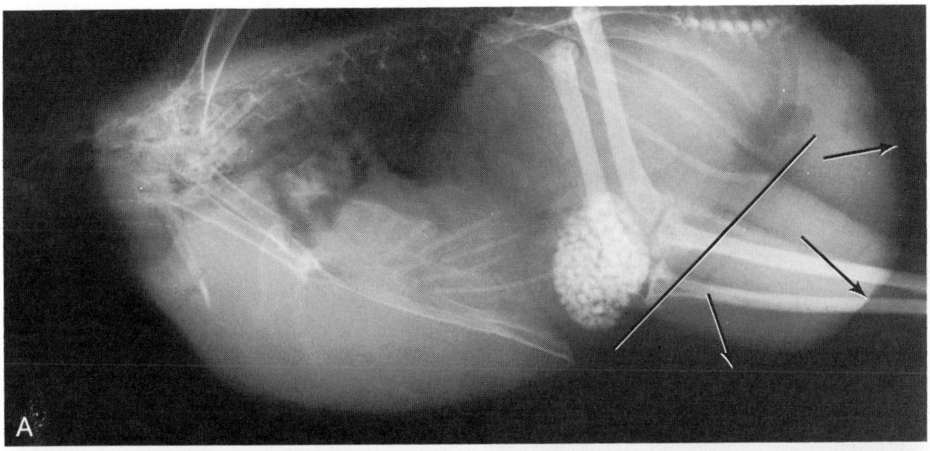


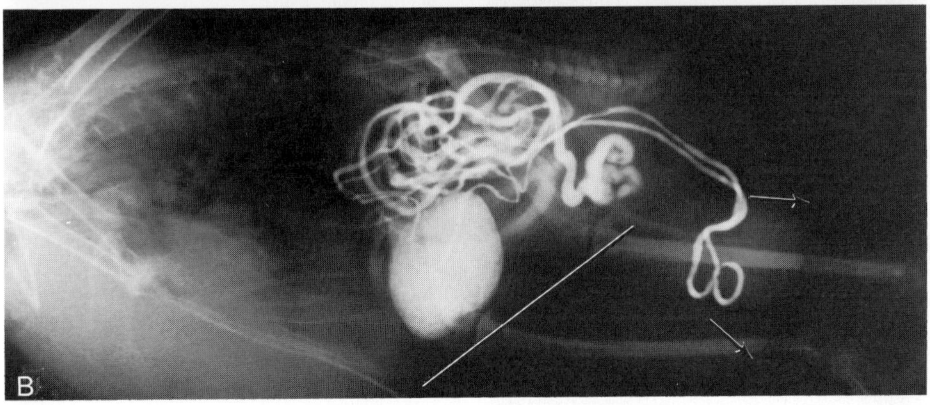


**Figure 14–28.** *A*, Lateral view of an Amazon Parrot with an abdominal hernia. The normal abdominal line is approximately marked. Note the distention marked by the arrows. There is a loss of the "lucent" triangle above the proventriculus. When a female bird is involved, always consider an increase in size of the ovary and oviduct and/or egg peritonitis. *B*, Positive contrast of the gastrointestinal tract of the bird in *A*. The barium is visualized in a portion of the intestinal tract that is present in the abdominal hernia. The approximate normal abdominal line is shown and the distended portion illustrated (arrows).

cavity and may create a wide range of radiographic signs, from peritonitis with a general loss of detail (see Fig. 14–25) to a local mass (Fig. 14–30). If the bird is stable enough, further data on location may be gained from endoscopy, laparotomy, or positive contrast radiography (Fig. 14–31). Some displacement of visceral contents may occur from the enlarged oviduct, which may not be well-defined on the films.

### *Urinary System*

There are few well-documented radiographic manifestations of renal disease. Renal tumors most commonly affect the budgerigar, although they may occur in other species. Clinical signs such as limb paresis or paralysis and abdominal distention may be related to renal neoplasms. Nephritis resulting in swollen kidneys can occasionally be appreciated (Fig. 14–32). Calcification of the kidneys (Fig. 14–33) has been reported (see Chapter 31, Nutritional Diseases). In Harrison's[3] experience, radiographs taken prior to and immediately following intravenous administration of fluids suggest that bolus fluid therapy may cause transient renal enlargement.

## CONTRAST STUDIES

Because visualization of many abnormalities on plain films may be difficult, the use of contrast materials may be helpful in defining the location and size of a lesion. This technique is questionable in cases of sinusitis, although it could be beneficial in locating chronic granulomatous debris. The major difficulty in using contrast material in this location is the extent and distribution of the sinuses (Fig. 14–34). Contrast material may have some potential for use with cervicocephalic air sac disease; how-

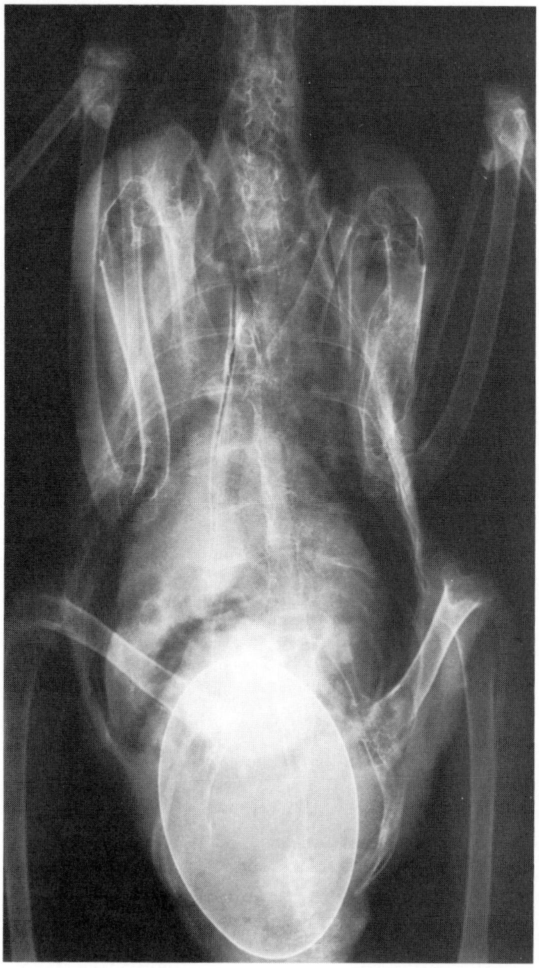

**Figure 14–29.** V/D view of a bird that is eggbound with a calcified egg. Egg binding may also involve noncalcified eggs that are in other areas of the abdomen (see Figs. 14–30 and 14–31). Note the crooked sternum to the right of the vertebral column, an indication of chronically poor calcium metabolism.

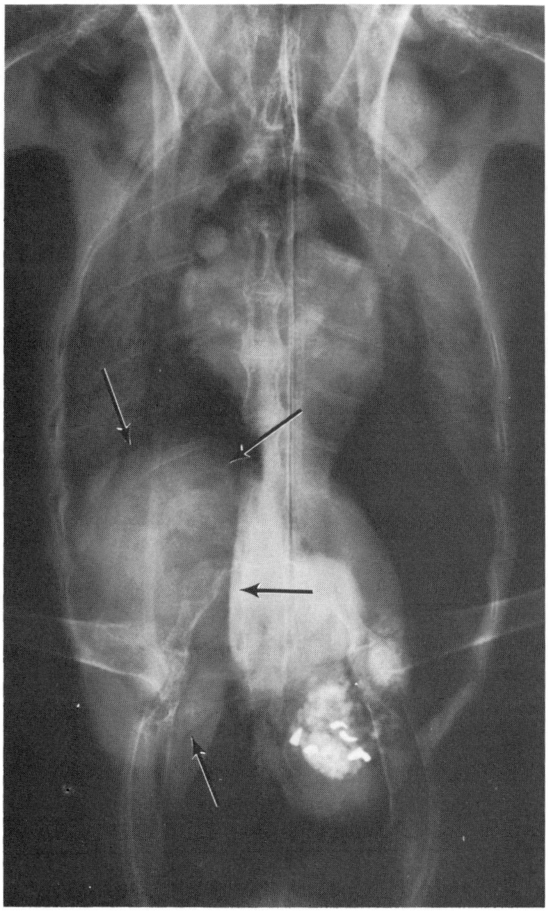

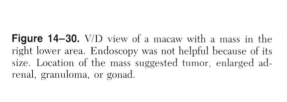

**Figure 14–30.** V/D view of a macaw with a mass in the right lower area. Endoscopy was not helpful because of its size. Location of the mass suggested tumor, enlarged adrenal, granuloma, or gonad.

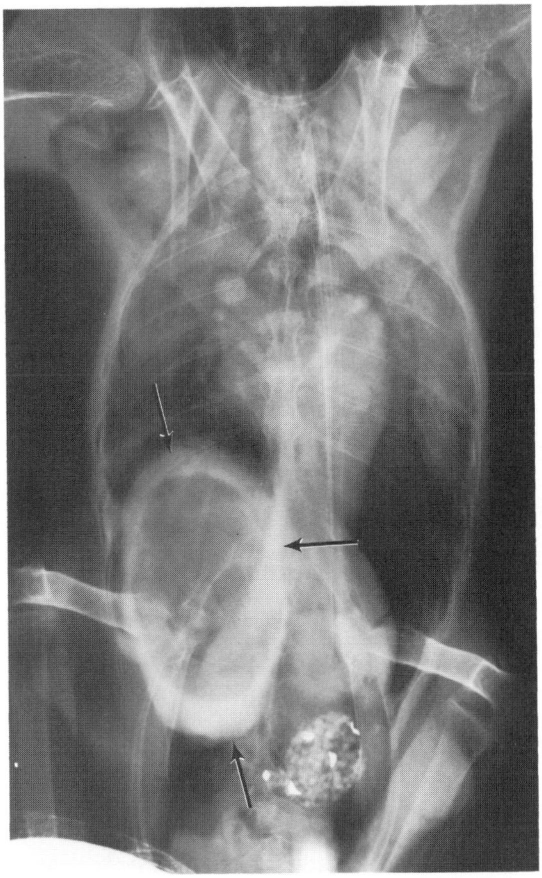

**Figure 14–31.** Utilizing fluoroscopy, a needle (22-gauge) was inserted into the mass visualized in Figure 14–30. Six ml of a transudate were removed and 4 ml of Renografin was injected into the mass. The final diagnosis was a retained abdominal egg. It appears that the contrast material (inside arrows) was placed around the yolk portion.

ever, direct visualization is probably more productive.

Contrast radiography is used most often with the gastrointestinal tract. Indications for intestinal contrast examination (subsequent to plain films) include abdominal enlargement, organ displacement or dysfunction, regurgitation, lack of stool production, persistent diarrhea, or abnormal plain films such as those with loss of abdominal detail. McMillan believes that hemorrhagic enteritis is also an indication.[5]

The most common contrast agents used for gastrointestinal radiography are barium sulfate and Gastrografin (Table 14–2). The main disadvantages of Gastrografin are that (1) it may be irritating to the intestinal tract, (2) it can be absorbed through the intestinal wall so that visualization is lost later in the study, and (3) it is extremely hygroscopic; that is, it will draw fluid into the intestinal tract, which may gravely compromise a dehydrated individual. Because Gastrografin is less likely than barium to result in peritonitis if leaked into the abdomen, its main use has been with patients with a possible intestinal perforation or those in which there is a strong possibility of intestinal surgery following the study.

The material most commonly used for contrast studies is barium sulfate. It is usually used as a 25 to 35 per cent solution (weight/volume). Powdered forms of barium may partially settle out of solution, but premixed solutions have improved this problem. Mucosal detail is also improved, although this is not as evident in small species.

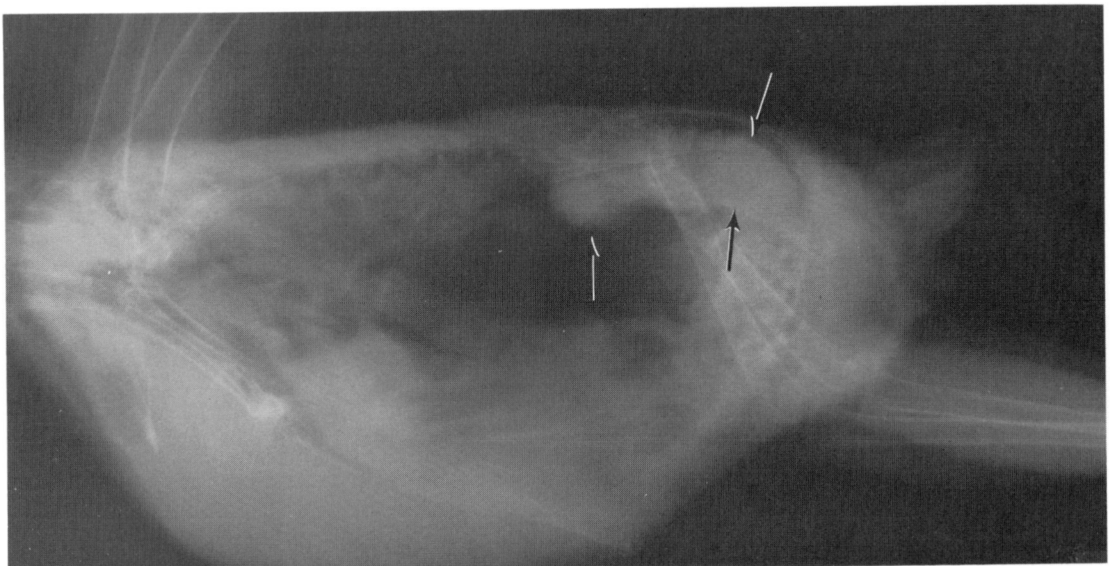

**Figure 14–32.** Possible nephritis in an Amazon Parrot. The size of the kidneys and the tendency for malpositioning make it difficult to accurately diagnose enlargement of the kidneys in birds. Nevertheless, it is important to scan the kidneys routinely for change. Such enlargement can be seen immediately following bolus fluid administration.

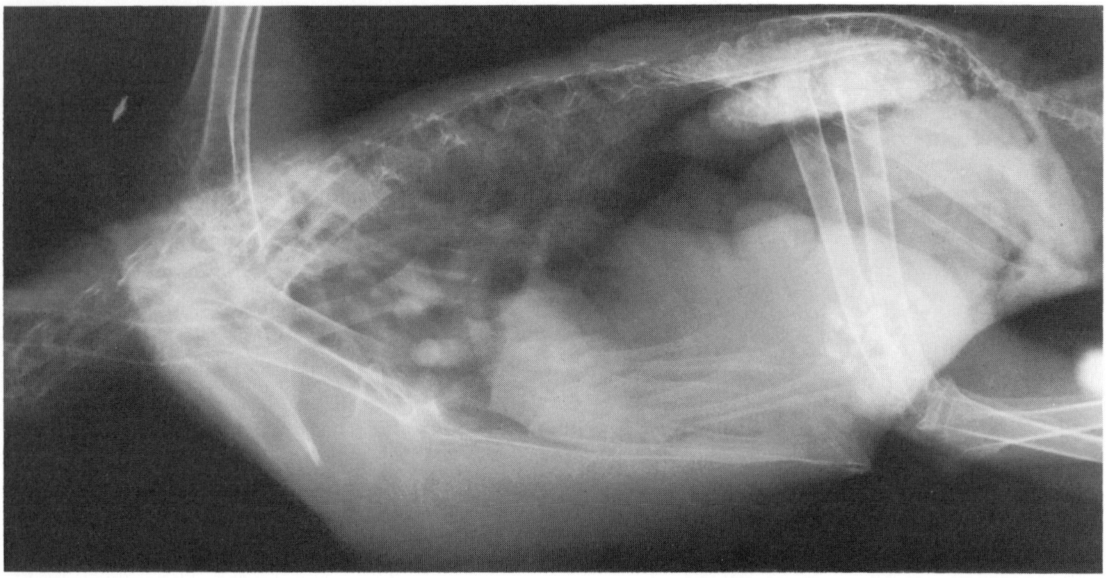

**Figure 14–33.** Lateral view of an African Grey Parrot with a diffuse appearance of calcification of the kidneys. This is often an incidental finding and its significance is unknown. Harrison believes that the condition is abnormal; it commonly occurs in feather pickers.

The use of contrast material is contraindicated in patients that are laterally recumbent or comatose. The patient should be fasted for four hours prior to the study to facilitate visualization of the lumen and decrease the possibility of identifying food particles as lesions; in paralytic diseases, however, food may remain in the gastrointestinal tract for days to weeks. The barium study should always be preceded by an adequate evaluation of the intestinal tract for ingluvitis and enteritis, especially in regurgitating birds. Prior to administration of contrast material, the crop is palpated to determine the presence of mucus and fluid. Removal of some of the fluid facilitates appropriate dosing of barium.

Table 14–3 presents approximate dosages of barium for administration, although the exact amount used depends on the clinical state of the bird. The material is deposited in the crop by gavage. For birds without crops, slower and more deliberate administration is advised.

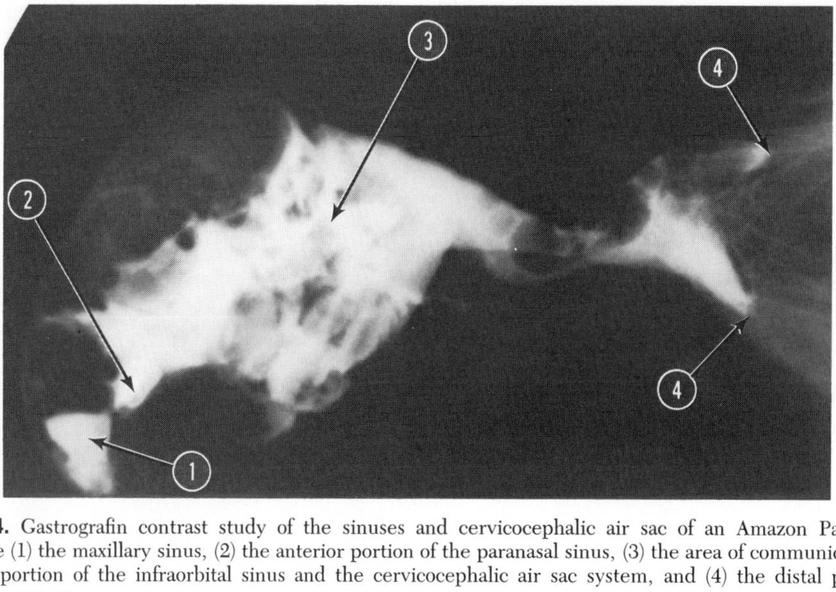


**Figure 14–34.** Gastrografin contrast study of the sinuses and cervicocephalic air sac of an Amazon Parrot. Areas of importance are (1) the maxillary sinus, (2) the anterior portion of the paranasal sinus, (3) the area of communication between the tympanic portion of the infraorbital sinus and the cervicocephalic air sac system, and (4) the distal portions of the cervicocephalic system.

**Table 14–2.** CONTRAST MATERIALS

| Product | Concentration | Vendor |
|---|---|---|
| Hypaque-76 | 370 mg/ml (iodine) | Winthrop Laboratories 11427 Yellow Tail Ct. Jacksonville, FL 32218 |
| Renovist | 370.5 mg/ml (iodine) | E.R. Squibb & Sons, Inc. Princeton, NJ 08540 |
| Renografin-76 | 370 mg/ml (iodine) | E.R. Squibb & Sons, Inc. Princeton, NJ 08540 |
| Novopaque | 60% barium sulfate suspension | Picker International 13900 58th Ct. N.W. Miami Lakes, FL 33166 |
| Colibar | 72% barium sulfate suspension | Veterinary Radiographic Systems 1780 Geronimo Trail Maitland, FL 32751 |
| Gastrografin | 370 mg/ml (iodine) | E.R. Squibb & Sons, Inc. Princeton, NJ 08540 |

Grimm et al.[2] studied the passage time of contrast media through the intestinal tract in several bird species. Preparations of Micropaque (100 per cent weight/volume microdispersion), Microtrast (a contrast paste with 70 per cent barium sulfate), and barium sulfate DAB 7 (100 per cent weight/volume) were used. Because of the consistency, the contrast paste was more difficult to administer to birds; however, better contrasts were achieved with this product in the esophagus. The barium sulfate preparations were comparable in passage time. Table 14–4 illustrates the average times in the species studied. Gastrografin was also evaluated in this study but was not as satisfactory as a contrast medium, probably because of its high osmotic and hygroscopic activity.

The timing of postadministration films may depend on the area of interest; two views should be taken at each interval. The decrease or loss of motility associated with many gastrointestinal diseases may prolong the completion of the study. Any significant abnormality should be reproducible on more than one film. This will help rule out normal peristaltic movement.

**Table 14–3.** BARIUM DOSAGES FOR CONTRAST STUDIES (25 Per Cent Solution)

| | |
|---|---|
| **Passerines** | |
| Finch | 0.2–0.35 ml |
| Canary | 0.25–0.5 ml |
| Mynah | 2.0–4.0 ml |
| **Psittacines** | |
| Budgerigar | 0.5–3.0 ml |
| Cockatiel | |
| Small Conure | 2.0–6.0 ml |
| Small Parrots | |
| Medium Parrots | 5.0–12.0 ml |
| Large Parrots | 10.0–15.0 ml |

Contrast findings may include possible intraluminal masses (Fig. 14–35), changes in lumen size, mucosal thickening, mucosal defects, abnormal intestinal placement, and increased transit time. Ileus (obstruction of the small intestine) may be mechanical in origin, such as from foreign material or an intraluminal mass, or it may be a functional abnormality (adynamic-paralytic). With functional ileus the entire tract may become distended. This may occur secondary to infection (viral, bacterial, fungal, yeast) or toxicosis (heavy metal) or from extraluminal influences such as peritonitis. Paralytic ileus is probably more common than mechanical ileus. Transit time with ileus may exceed 24 hours.

Contrast studies of the cloaca and rectum by enema may be useful but have no common current application. Intravenous pyelograms are also not commonly done but may have potential in the diagnosis of suspected renal tumors or the locating of ureters prior to cloacal or tumor surgery. Figure 14–36 shows the appearance of the ureters in an Amazon.

Pneumocoelography (injection of air into the coelom) may be helpful in delineating abdomi-

**Table 14–4.** AVERAGE PASSAGE TIMES OF BARIUM SULFATE[2]

| Passage Time | Hawks | Buzzards | Budgerigars | Amazons | Canaries | Hens |
|---|---|---|---|---|---|---|
| 5 min | Crop | Crop | Crop | Crop | Crop | Crop |
| 10 min | Crop | Stomach | Crop | Crop | Small Int. | Crop |
| 15 min | Stomach | Stomach | Crop | Crop | Large Int. | Crop |
| 30 min | Stomach | Small Int. | Stomach | Stomach | Cloaca | Stomach |
| 45 min | Small Int. | Small Int. | Stomach | Stomach | Cloaca | Stomach |
| 60 min | Small Int. | Small Int. | Small Int. | Stomach | Cloaca | Small Int. |
| 90 min | Small Int. | Large Int. | Small Int. | Small Int. | Eliminated | Small Int. |
| 120 min | Small Int. | Cloaca | Large Int. | Large Int. | Empty | Small Int. |
| 150 min | Large Int. | Cloaca | Cloaca | Cloaca | — | Large Int. |
| 180 min | Large Int. | Cloaca | — | — | — | Cloaca |
| 210 min | Cloaca | Cloaca | — | — | — | Cloaca |
| 240 min | Cloaca | Eliminated | Empty | — | — | Cloaca |

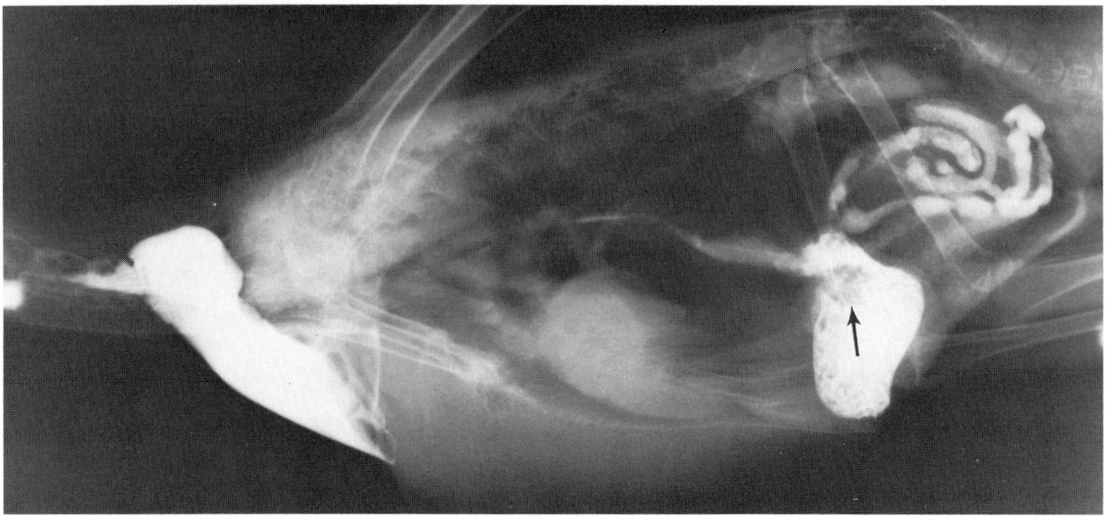

**Figure 14–35.** Barium contrast study of a macaw with regurgitation. Note the comparatively lucent area at arrow. This shadow was a misleading artifact. Originally diagnosed as having a space-occupying lesion, the bird died from a gastroenteropathy of unknown etiology. The barium shadows would eventually change dramatically as the disease progressed (refer to Fig. 14–15).

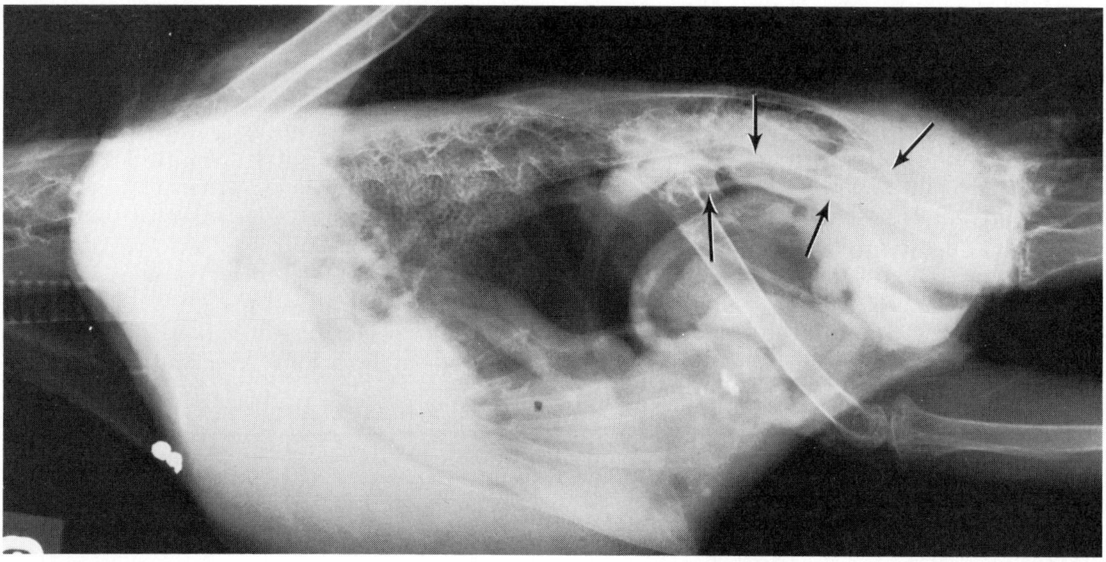

**Figure 14–36.** Intravenous pyelogram in an Amazon. Note the position of the ureters (arrows). This technique is of limited value at present. The V/D view is not shown because it is often difficult to visualize the position of the ureters. The metal dense particles in the crop are a result of the bird's ingesting part of the lead gloves used for restraint. (Courtesy of the University of Pennsylvania.)

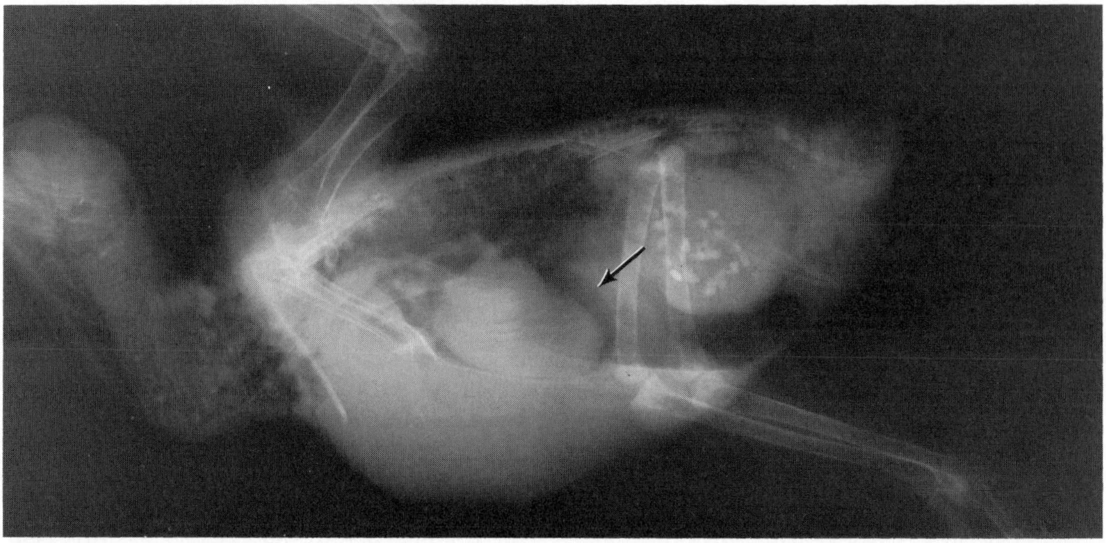

**Figure 14–37.** Air coelogram in a healthy pigeon. A total of 12 ml of air has been injected through a Teflon catheter inserted between the sternum and the liver. Note the separation of the liver from the heart (arrow) and the displacement of the ventriculus. This technique may be useful in distinguishing liver anomalies.

nal masses but is still experimental at this time, as are contrast studies of the lung. The extent of distribution of the air sacs will make accurate placement of air other than on the midline difficult. By means of an indwelling catheter, air can be placed to separate the heart and liver shadows for better visualization of the separate liver lobes. The air tends to pocket rather than distribute so that visualization is limited (Fig. 14–37).

The use of endoscopy in a clinical practice can dramatically reduce the need for multiple and specialized radiographic studies.

## REFERENCES


1. Altman, R. B.: Radiography. Vet. Clin. North Am. [Small Anim. Pract.], 3:165–173, 1973.
2. Grimm, F., et al.: Rontgenkontrastaufnahmen des verdauungskanals-einsatzmoglichkeiten beim vogel (Possible applications of contrast radiography to digestive tract of birds). Proc. Int. Sympos. Erkrankungen Zoo. Brno. Berlin, Akademie-Verlag, 1984.
3. Harrison, G. J.: Personal communication, 1984.
3a. King, A. S., and McLelland, J.: Birds: Their Structure and Function. 2nd ed. London, Baillière Tindall, 1984.
4. McMillan, M. C.: Avian radiology. *In* Petrak, M. L. (ed.): Diseases of Cage and Aviary Birds. 2nd ed. Philadelphia, Lea & Febiger, 1982, pp. 329–360.
5. McMillan, M. C.: Avian gastrointestinal radiography. Comp. Cont. Ed., 5:273–278, 1983.
6. Shively, M. J.: Radiographic anatomy of the barred owl (*Strix varia*). Southwestern Vet., 31:141–150, 1978.
7. Shively, M. J.: Xerographic anatomy of the pigeon (*Columba livia domestica*). Southwestern Vet., 35:101–111, 1982.
8. Silverman, S.: Avian radiographic technique and interpretation. *In* Kirk, R. (ed.): Current Veterinary Therapy VII. Philadelphia, W. B. Saunders Company, 1980, pp. 649–653.
9. Walsh, M. T., and Mays, M. C.: Clinical manifestations of cervicocephalic air sacs of psittacines. Comp. Cont. Ed., 6:783–789, 1984.



## ACKNOWLEDGMENTS

The author would like to thank Dr. Norman Ackerman, the support staff at the University of Florida Radiology Department, and Dr. George Kollias for their help. Thanks also to Dr. D. Martinez, whose financial and emotional support allowed me the time and desire to complete this work.

MICHAEL T. WALSH

# Chapter 15

# ENDOSCOPY

GREG J. HARRISON

Endoscopy is the term applied to visualization of the interior of the body with an instrument consisting of optical fibers designed for high-intensity light transmission and lenses for magnification. Endoscopy is referred to as laparoscopy for surgical examination of the abdominal cavity and as arthroscopy for joint examination. The smallest diameter instruments are most often used in birds. The procedure is well-accepted as a fast, safe diagnostic technique in many species.[1–4, 7, 8, 10–13]

## CLINICAL APPLICATIONS

The endoscope achieved early popularity among avian practitioners as a means to surgically sex monomorphic species of birds. The endoscope's availability in the clinical situation has prompted other uses.

The endoscope and/or accessories may be employed nonsurgically as diagnostic tools during the physical examination. For example, the cable may be used with the light source to enhance evaluation of the eyes and mouth. These same accessories can be used to transilluminate the entire head of small birds for detection of sinus, nares, or beak abnormalities.

With the endoscope attached, body orifices including the choana, nares, ears, and even the cloaca and oviduct can be more completely investigated without surgical invasion. The magnification afforded by the scope can augment ophthalmic examination of the cornea, lids, and conjunctival cul-de-sac.

Some emergency situations may be diagnosed by direct observation with the scope. Acute sneezing, coughing, or shaking the head may indicate a foreign body in the nares or trachea; acute dyspnea in hand-fed babies may be caused by foreign body inhalation. Tracheoscopy is a common procedure even in finches, canaries, and hand-raised birds, using the smallest available endoscope (Fig. 15–1). Retrieval of foreign bodies may be accomplished with the use of a biopsy instrument or retinal ophthalmic forceps in conjunction with the endoscope (Fig. 15–2).

Parasites, tumors, atresia of the nasal passages, and lesions associated with atrophic rhinitis may be observed by endoscopic examination of the nares. The anterior approach must be conducted with caution, as very little penetration is tolerated in the psittacine with normal opercula conchae. The posterior approach requires general anesthesia and insertion of the endoscope into the choanal slit. This latter approach has revealed persistent membranes, hypertrophy, or inflammation of the conchae and other lesions.

When other diagnostic methods have been inconclusive, investigations of the sinuses have been accomplished with the endoscope through the lateral commissures of the mouth or other sites described for sinus flushing.

Transillumination of the crop with the fiber-

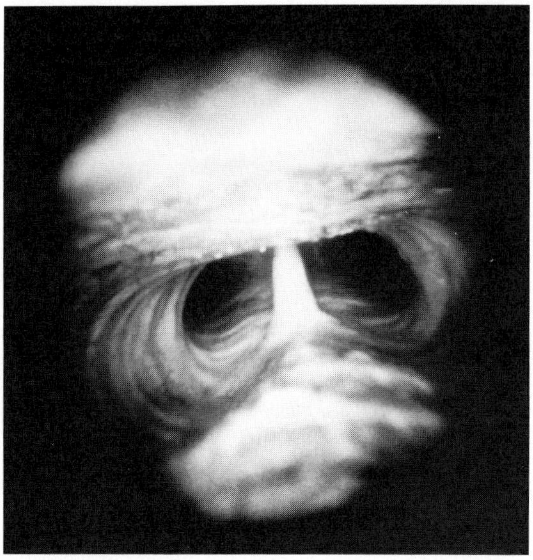

**Figure 15–1.** The syrinx and bifurcation of the trachea into the bronchi are evident in this view through the endoscope. (Courtesy of S. McDonald.)

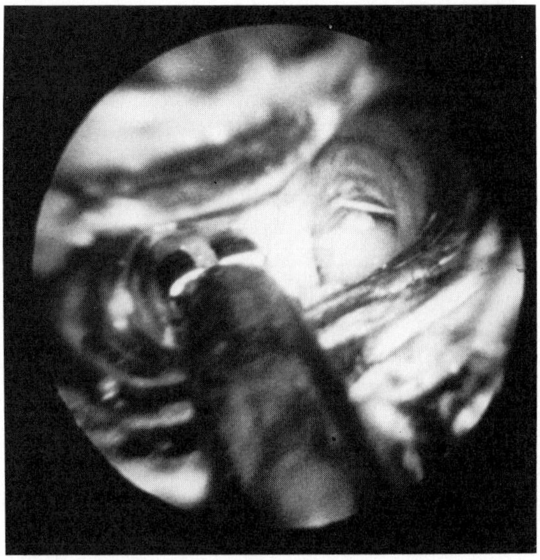

**Figure 15–2.** A foreign body in the trachea can be viewed and removed with biopsy forceps placed alongside the endoscope. (Courtesy of S. McDonald.)

optic light cable is enhanced by insufflation of the crop using a 35-cc syringe with a soft rubber feeding tube attached. Subsequently, the interior of the crop may be directly examined with the endoscope attached to the cable. Light digital pressure may be used around the barrel of the scope to retain the insufflation in the crop. With this method, plaquing associated with candidiasis in hand-fed baby birds with crop stasis or the presence of foreign bodies may be visualized. The upper thoracic esophagus may also be entered by this approach.

Endoscopic examination of the cloaca may be warranted in cases with presenting symptoms of straining, egg binding, bleeding, or tissue protrusion from the cloaca. Insufflation may also be employed in cloacal examinations and digital pressure applied around the scope barrel to retain air for enhanced viewing. After the cloaca is entered, the rectum, vaginal prominence, possible irregularities in the oviduct, or a retained egg may be visualized.

## DIAGNOSTIC LAPAROSCOPY

The decision to use the instrument for diagnostic laparoscopy may be based on inconclusive results of other procedures—history, physical examination, serum chemistry tests, radiology, or cultures. This technique, which offers the opportunity for direct observation of the lesions and/or sample selection, may in fact be the fastest and most efficient means of obtaining a diagnosis.

Birds with presenting symptoms of vomiting, weight loss, metabolic disorders, abdominal swelling, feather picking, unilateral leg paralysis, or chronic failure to thrive and birds nonresponsive to conventional therapy are potential candidates for laparoscopy. Certainly respiratory disorders based on history, physical exam, or radiographic findings would indicate laparoscopy as the diagnostic approach.

Although radiology is often a prerequisite for laparoscopy, it has been the author's experience that birds with signs of upper or general respiratory involvement may or may not show the lesions radiographically (see Chapter 14, Radiology). Conversely, a shadow on a radiograph suggesting air sacculitis may actually turn out to be an artifact, or a presumed radiographic diagnosis of an enlarged liver may in fact be the image of the proventriculus when the organs are viewed laparoscopically.

Evaluation of nonproductive breeding birds may indicate use of the procedure for detection of possible gonadal abnormalities. In addition to gonads, the surgical sexing procedure permits direct observation of adjacent organs and the opportunity to note gross pathologic signs in the air sacs, proventriculus, lungs, liver, kidneys, and adrenal gland. Deliberate manipulations of the scope from this lateral approach facilitate observation of the spleen, aorta, posterior vena cava, mesenteric vessels, intestines, gizzard, ureters, and possibly the fallopian tubes, uterus, or vas deferens in a mature bird.

Evaluation of cultures and biopsies of some obvious lesions has determined the etiologic agent and subsequent appropriate therapy. Amyloidosis of the liver, kidney and spleen, renal gout, avian tuberculosis in the liver and spleen, aspergillosis, air sacculitis, and reproductive diseases have been diagnosed using this approach.[2, 3, 5, 12, 13] In some cases, gross lesions may be absent and the diagnosis may be obtained by detecting histopathologic changes in the organ parenchyma biopsy sample (see Chapter 16, Biopsy Techniques). It is the author's opinion that in appropriate cases, laparoscopy in combination with cultures or biopsies is often preferred to currently available serum chemistry tests as the diagnostic approach. Although abnormal serum chemistry values may suggest tissue damage or dysfunction, they are often nonspecific and lend nothing to determining the etiology.

Endoscopy from ventral and lateral approaches allows direct observation of the avian heart for evidence of cardiomegaly, hydropericardium,[3] and other abnormalities. The absence of a well-developed, muscular diaphragm in birds facilitates observation of the heart. There is no expanding and collapsing of the lungs to contend with, as in mammals.

As an adjunct to major abdominal surgeries, the endoscope can be maneuvered into position and locked in place with a rigid support device* (Fig. 15–3). This then serves as a source of concentrated light and magnification. In this way, one may use the endoscope as a modified operating microscope during a proventriculotomy, multiple organ biopsy, hysterectomy, or abdominal tumor removal, providing the opportunity to develop microsurgical techniques.

Some adaptations of basic examinations may be accomplished with the otoscope[9]; however, visual investigation of internal organs is limited without extensive surgical invasion. The Focuscope† system has intermediate visualization capabilities.

## CONTRAINDICATIONS

Contraindications to laparoscopy are similar to those for any surgical procedure: extremely weak, severely dehydrated, or debilitated birds may succumb to the stress involved. An experienced clinician can often determine the degree of risk involved by evaluating the physical condition of the patient.

*Flexarm—Richard Wolf Medical Instruments Corporation.

†Medical Diagnostics Services, Brandon, FL.

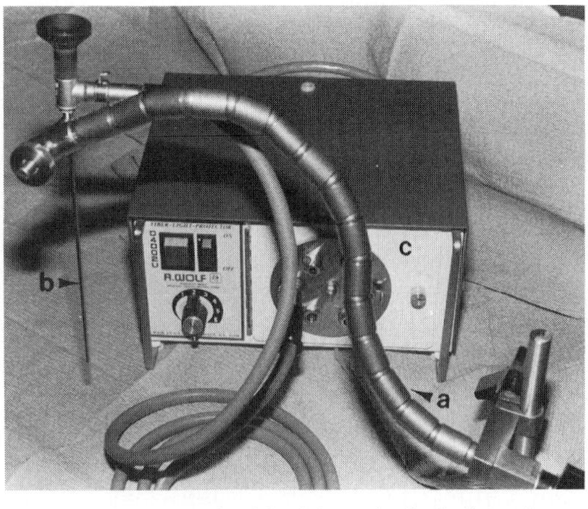


**Figure 15–3.** A rigid cable *(a)* can be locked in place to hold the endoscope *(b)* steady and free the surgeon's hands. The light intensity is controlled by the power source *(c)*. (From Harrison, G. J.: New aspects of avian surgery. Vet. Clin. North Am. [Small Anim. Pract.], 14(2):363–380, 1984.)

In some cases laparoscopy may be worthwhile even in a somewhat debilitated patient if the bird has not responded to any therapy. In these cases, it can become a "last ditch" effort to arrive at a diagnosis to save the bird. As an example, a decision was made by the author to perform a laparoscopy on a severely dyspneic, slightly emaciated Moluccan Cockatoo. Lung and air sac lesions were observed and determined to be induced by *Aspergillus* sp. on culture. Removal of large portions of the lesions with concurrent fungal therapy resulted in recovery of the bird (Fig. 15–4).

Laparoscopy is contraindicated in birds with bleeding disorders, which may become evident at the time of an initial anesthetic injection (if a parenteral agent is utilized) or at the site of feather removal during surgical preparation. Patients with prolonged clotting times may require treatment and conditioning prior to any surgical attempts.

Laparoscopy in birds that have recently ingested food or water should be postponed for a few hours. Abdominal fat in a severely obese bird or imminent egg laying may obstruct any visual capabilities and render the procedure worthless and even risky.

The author sees no reason to put a healthy, single pet bird through surgical sex determination just because the owner "wants to know the sex" if there are no plans to breed the bird (see Chapter 51, Sex Determination Techniques).

No surgery should be performed if the surgeon is insufficiently trained, fatigued, or otherwise psychologically unprepared.

## EQUIPMENT

The basic instrumentation is a light source, a fiberoptic cable, a trocar and cannula set, and the endoscope itself (Fig. 15–5). These are available in a wide selection of optical probe diameters and projection source candlepowers. Fiberbundle light transmitting quality and quantity are prime considerations in selecting the endoscope and light cable and may vary with the manufacturer.

The simplest light source is all that is needed for clinical applications. Photographic or multiscope use would require a more powerful light source.

The flexible fiberoptic cable, which transmits the light from the power source to the scope, is available in a variety of models, although the most basic model is adequate for avian practices. One must be aware of the location of the

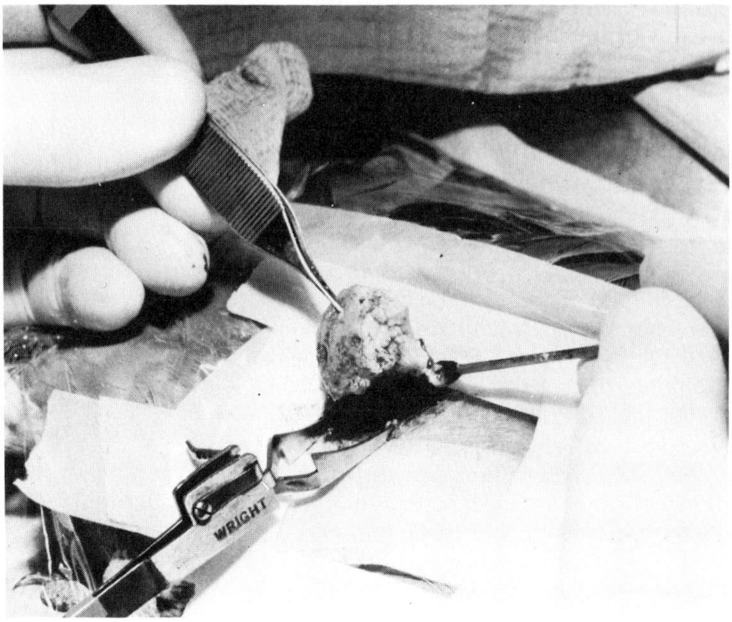


**Figure 15–4.** Laparoscopy confirmed lesions in the air sacs as first suggested by radiograph; the granuloma was subsequently surgically removed.

soft cable in working with unanesthetized birds, as a large psittacine is capable of biting through it.

The trocar and cannula are designed to assist in introducing the endoscope into the surgical site and are recommended by the manufacturers to protect the endoscope from damage during use. These instruments should definitely be used until one gains some experience; however, they may be found cumbersome with small birds (under 250 grams body weight). With experience and care in use, one may not need the added rigidity in avian practice. In addition, repeated accidental bumping of the endoscope on the sides of the cannula during insertion might damage the lens. In some models, the scope is longer than the cannula, which may present a potentially dangerous situation for the beginning endoscopist who may inadvertently expect the cannula to control the depth of penetration. To increase dexterity over the relatively large trocars supplied, small diameter bone pins, both sharp and blunted, can be used to prepare the surgical site in small patients; the endoscope is then inserted alone.

Endoscopes are sized according to the diameter of the operating barrel. For exclusive bird work, one would choose one of the smaller diameter endoscopes (1.9-mm arthroscope* or 1.7-mm Needlescope†). However, most practi-

*Richard Wolf Medical Instruments Corporation.
†Dyonics, Inc.

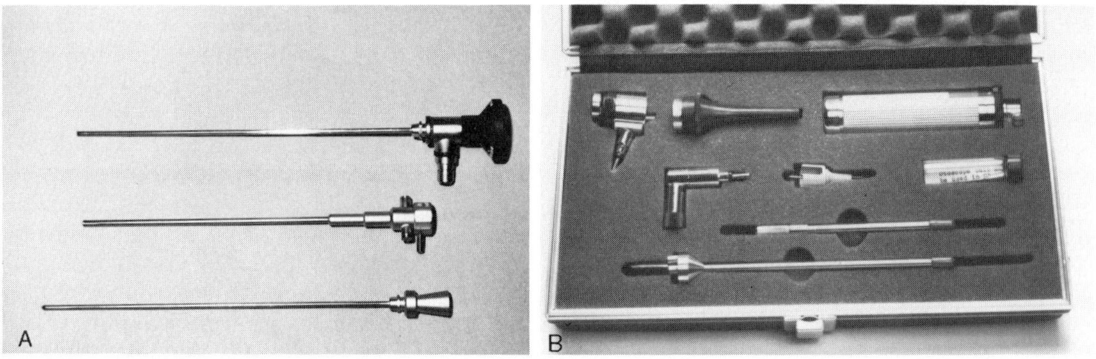


**Figure 15–5.** *A*, Components of an endoscopic system include (from top): 2.7 mm arthroscope, trocar, and blunt obturator. (From Kollias, G. V., Jr.: Liver biopsy techniques in avian clinical practice. Vet. Clin. North Am. [Small Anim. Pract.], 14(2):287–298, 1984.) *B*, A modified industrial optical probe with accessories may serve as a simplified portable endoscopic system. (Courtesy of Medical Diagnostics.)

tioners prefer a versatile model that may be used with small animal species as well.

The author's preference is the Wolf 2.7-mm scope. It is a flexible yet sturdy instrument, has excellent optics, transmits high-quality light, and has not posed problems with delamination of the lens or broken bundles of light fibers. The 100-mm length of the Wolf scope is preferred to the smaller-diameter, shorter-length Needlescope in our practice. This is especially true for use in bronchoscopies, for which the only alternative might be a tracheotomy (a difficult procedure in psittacines due to the complete rings and stiffness of the trachea).

On the other hand, the smaller 1.7-mm diameter Needlescope is useful in very narrow areas, such as for sinus investigations and for tracheoscopies in small birds. Another benefit of this brand is the operating end of the scope, which is slightly bevelled as compared to the blunt Wolf scope. This bevelled point is extremely useful in penetrating air sacs and suspensory ligaments during sexing and biopsy procedures. When a blunt-ended scope is used, a second instrument, such as a trocar or biopsy forceps, is often used to puncture or tear these tissues. This additional step increases the time and precision required for the procedure.

Applications of the endoscope become so numerous in avian practice that two complete sets may be desirable—one for the examination room and one available for surgery.

## INSTRUMENT CARE

The endoscope barrel is delicate and expensive to replace and requires care in handling. Flexing the barrel will fracture the fiberoptic bundles and lead to black dots in the field of view or dislodgement of the lens.

Both ends of the scope need to be well-supported during handling and storage. A rigid padded container, perhaps the original shipping box, will protect the endoscope between uses (Fig. 15–6).

The lens must be cleaned at the end of each procedure with lens paper or a cotton swab dampened in alcohol (Fig. 15–7). The operating and ocular ends must be free of disinfectants and other substances that may blur the field of vision or cut down on light transfer. Attempts to remove films that have already dried may scratch the lens.

Avian endoscopy is performed using sterile instruments and techniques. In the examination room, the instruments may be stored in a solution of chlorhexidine* or 2 per cent glutaraldehyde. Glutaraldehyde is very caustic and may discolor the skin and irritate the surgeon's eyes if not thoroughly rinsed prior to use. For this reason, aqueous chlorhexidine may be preferred. A germicidal spray† may be used on the cable between birds.

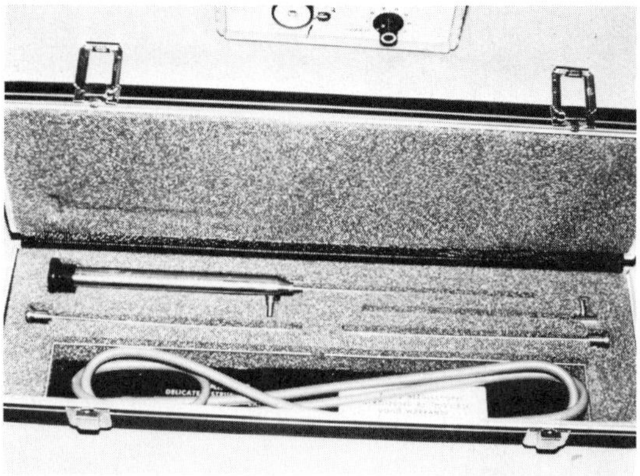

**Figure 15–6.** The delicate endoscopy instruments should be handled and stored with care.

Ethylene oxide may be employed for gas sterilization of the light source and accessories so any adjustments for control of the light can be handled by the surgeon without breaking the sterile field. Some manufacturers approve of autoclaving some of the components; one should consult the owner's manual for specifics.

The disinfectant solution for placement of the scope and auxiliary instruments during surgery should be warm to help prevent fogging of the lens when the scope penetrates the body.[2] If supplementary diagnostic procedures are anticipated, the scope should be rinsed in warmed sterile saline solution prior to entry to prevent the chemical action of the disinfectant from interfering with the culture or histopathologic results. The cannula, when used, should drip dry for a few seconds after rinsing and prior to insertion to prevent bubbling of residual fluid inside the cannula once the air sacs are encountered.

## PREPARING FOR LAPAROSCOPY

In order for laparoscopy to be a safe and effective diagnostic tool, the practitioner must be totally familiar with the instrumentation as well as the normal anatomy of a variety of bird

*Nolvasan—Fort Dodge Laboratories, Fort Dodge, IA, 50501

†Staphene—Vestal Laboratories, St. Louis, MO 63110

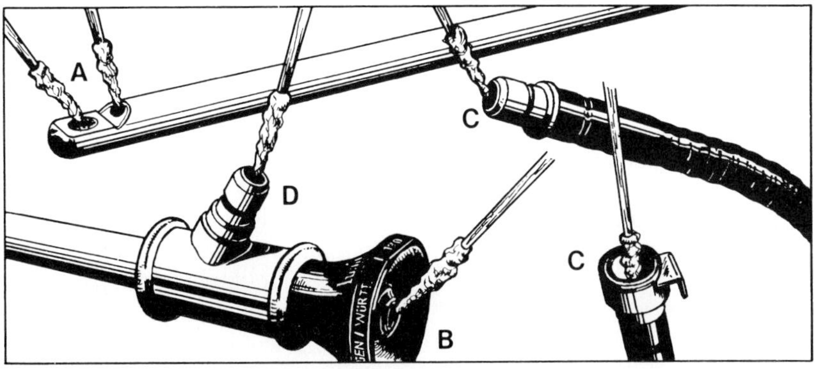


**Figure 15–7.** General cleaning guide for standard fiber illuminated laparoscopes and accessories. 1. Clean thoroughly with soft brush, mild soap, and warm water. 2. Use cleaning rod with cotton for inside of sheaths. 3. Rinse in warm water. 4. All distal objective lenses (A) and ocular windows (B) must be cleaned individually with a cotton swab dipped in alcohol to assure clear viewing. Be careful not to scratch the objective of the scope. 5. All fiber end surfaces on light cables (C) and telescopes (D) must also be cleaned periodically as in Step 4 to guarantee maximum light transmission. (Courtesy of Richard Wolf Medical Instruments Corporation.)

species. Wet labs are available at various times around the country for introduction to the equipment, instruction in positioning of the patient, and practice with basic procedures. Review of anatomic literature and dissection of cadavers will assist in improving performance.

The clinician must gain considerable experience on his own before he attempts the technique with a client's bird. Pigeons are often used as a practice species. Although pigeons appear to have more air sac chambers than psittacines and are ultimately more difficult subjects for laparoscopy, the transition to psittacines is simple once one has developed proficiency in the technique. Pigeons raised specifically for laboratory use are more apt to have clear air sacs than feral or breeding birds and are preferred as beginning practice specimens.

Laparoscopies are performed in the surgery room, and additional supplies should be assembled prior to the procedure:

Appropriate anesthetic agent
Restraint board and Velcro, heavy wire, masking or other tape to position the patient
Sterile transparent surgical drape
Masking tape to hold back feathers adjacent to entry site
No. 11 blade for initial skin incision
Sterile lens paper for clearing the endoscope lens as needed
Warmed disinfectant fluids for storage and cleaning of the scope
Warmed sterile saline to rinse off the disinfectant
Sterile paper towels to dry the scope
Sterile bottles of 10 per cent buffered formalin (alcohol for fixation of tissues in birds suspected of having gout) or culture media for direct deposition of samples
Sterile glass slides with frosted ends for impression smears
Sterile pencil to identify the slides
Sterile 35-cc syringe and tube for insufflation
Circulating water blanket or heating pad for supportive heat
Biopsy forceps (see Chapter 16, Biopsy Techniques)

## PRESURGICAL PROCEDURES

Avian patients should have had a relatively thorough presurgical work-up by the time diagnostic laparoscopy is indicated. Birds intended for surgical sexing without full clinical evaluation should be rejected if any signs of disease are obvious on physical examination (see Chapter 51, Sex Determination Techniques).

Fasting the patient for two to four hours prior to laparoscopy may reduce the risk of suffocation from regurgitation and help standardize the anatomic landmarks and maximize potential spaces during the procedure. If one fails to fast the bird, one may encounter an enlarged proventriculus that interferes with the angle of the approach and obscures the location of other organs.

## ANESTHESIA

Anesthesia is indicated for laparoscopy. Because of the potential dangers inherent in previously popular avian anesthetic agents, some practitioners preferred to perform brief laparoscopic procedures in healthy birds over 200 grams body weight with physical restraint only (see Chapter 51, Sex Determination Techniques). This is no longer necessary, as newer, safer inhalation agents eliminate this danger. Birds smaller than Sun Conures, hyperactive species such as rosellas or lories, clinically ill birds, and birds for which longer procedures are anticipated have always required chemical immobilization (see Chapter 45, Anesthesiology).

## POSITIONING

Beyond knowledge of anatomy and experience with the procedure, a primary factor for repeated success in laparoscopy is positioning. The use of a restraint board with masking tape, heavy wire, or Velcro strips allows adjustments of the bird to obtain true lateral or ventral position, depending on the location of the suspected lesion. The most common position is right lateral recumbency (see Chapter 51, Sex Determination Techniques). The head is secured first; the legs are then stretched firmly caudally and positioned with the appropriate attachment. Unless full anesthesia is achieved, an assistant restrains the wings together at the humerus close to the body to prevent wing fracture. In this lateral approach, the tip of the upper wing should be slightly elevated so the body is not rotated. Respiration is monitored closely and the trachea palpated to prevent impingement.

If familiar landmarks are not obvious when entering the bird, remove the scope and recheck the positioning (Fig. 15–8). The most common cause of malpositioning is failure to stretch the bird out firmly; the condensed viscera tend to obscure the air sacs and inhibit visualization.

When performing tracheal endoscopy, the neck must be fully extended and the feet taped down or held. Unless an air sac is trocarized for supplemental air supply, the procedure must be performed rapidly, especially if the endoscope occludes the trachea.

## SURGICAL PREPARATION

The entry site for most laparoscopies is the same as used for surgical sexing, dorsal to the sternum and anterior to the femoral muscles. The site is palpated by pressing in this area with the index finger and noting the sternal notch, a V-shaped indented fossa. Depending on the species, the entry site has been reported as caudal to the last rib.[2] Anatomic dissection of an Amazon has established the author's point of entry as between the seventh and eighth ribs (Fig. 15–9).

As with the surgical sexing procedure, the feathers are plucked from the site and povidone-iodine* surgical scrub applied and wiped off. To prevent potential hypothermia, use of alcohol or water is avoided or severely limited by this author.

*Betadine—Purdue-Frederick Co., Norwalk, CT 06856

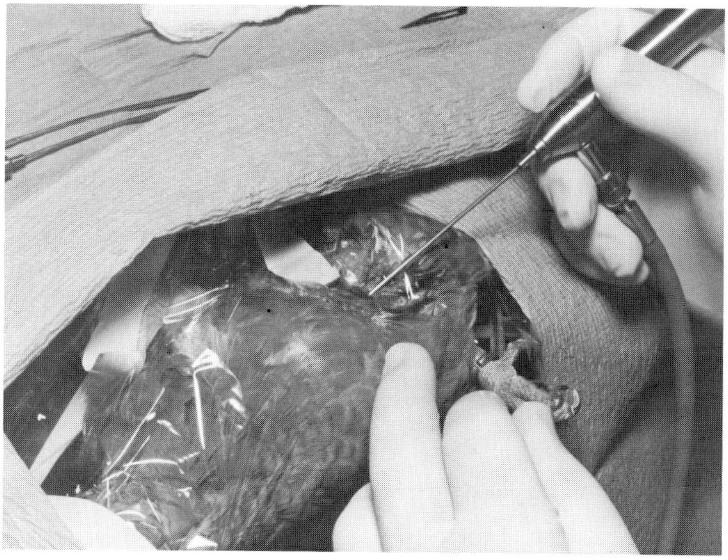

**Figure 15–8.** Disorientation during laparoscopy may be due to malpositioning of the bird. Clear surgical drapes facilitate observation of patient.

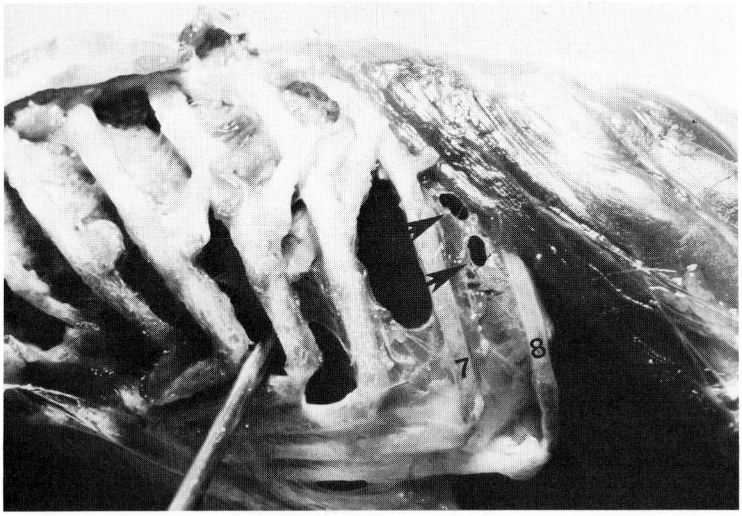


**Figure 15–9.** Anatomic dissection of an Amazon Parrot reveals that the author's usual entry sites for endoscopy (arrows) are located between the seventh and eighth ribs.

## LAPAROSCOPY

Prior to inserting a No. 11 blade under the skin at the described anatomic site for a small incision, a hemostat may be used to momentarily clamp the skin to reduce superficial bleeding. A sharp, pointed bone pin, half the diameter of the scope, is first used to enter the incision site. A twisting motion rather than a pushing motion is used to penetrate the intercostal muscles.

The pin is held like a writing pen with the middle finger acting as a fulcrum to control it and prevent sudden penetration after the tissue is separated. To avoid organ laceration, a slightly larger, smooth, rounded bone pin is used to enlarge this opening by rotating it around the perimeter of the hole and stretching the musculature outward. Prior to insertion, the scope is tested for clarity by bringing a finger up to the viewing field. The light source is usually set at the maximum setting, especially with the small diameter scopes.

Visualization should begin as soon as the scope is introduced into the incision. If one has entered at the appropriate location, the tip of the scope should be in the lateral portion of the abdominal air sac. If the top of the occupied chamber is highly membranous and the dorsum of the proventriculus difficult to observe, one can assume that one is in the caudal thoracic air sac. The dorsal membrane must be penetrated to reach the abdominal air sac (Fig. 15–10).

From the abdominal air sac in psittacines, one can swing the tip in an anterior direction and view the pink, lacy-looking lung. (If the viewing field were the face of a clock, the lung would be located between 8 and 1 o'clock.) The lung can be followed craniomedially to a point where it is joined on the right side by the medial air sac wall. Any lesions in the lungs or air sacs can be noted.

As one proceeds along the medial air sac wall, the scope may need to be retracted somewhat so that it does not come in direct contact with

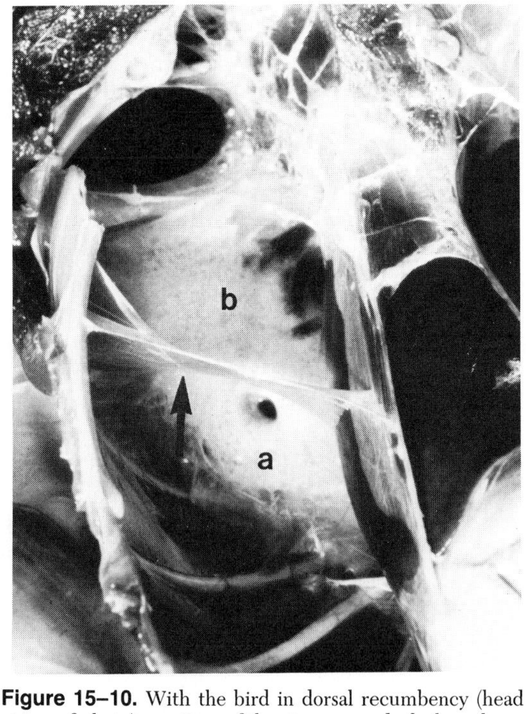


**Figure 15–10.** With the bird in dorsal recumbency (head at top of photo), portions of the sternum and ribs have been removed to reveal the membrane (arrow) separating the caudal thoracic air sac (b) from the abdominal air sac (a). The "hole" in the caudal surface of the lung above (a) is the normal ostium. The dark structure to the right is the liver. (Courtesy of S. McDonald.)

tissues that obstruct vision. The proventriculus, a smooth, long, white organ, is found at the midventral aspect of this air sac.

Referring back to the lung, place the scope against the medial air sac wall in preparation for penetration. One may carefully use the tip of the scope to tear through the air sac or use a small trocar or biopsy punch for a smaller hole to accommodate the scope. Once inside, the endoscopist may orient himself by locating familiar organs; the dorsal dark mahogany kidney may be the easiest to identify. Anterior to this is the gonad (see Chapter 51, Sex Determination Techniques) with the adrenal gland intimately associated with its anterior and dorsal surfaces. In juvenile birds, the three structures may often be seen in the same field if the scope is retracted to enlarge the view. In mature psittacines, view of the adrenal gland may be obscured by the gonad, especially in females.

There may be some deviations from these specific directions when using pigeons as the practice species. The lungs may not be initially visible in a pigeon until an anterior transverse air sac membrane is penetrated. An additional membrane creates an even smaller viewing field in this species. Mature male pigeons may have gonads so large that the endoscopist may fail to recognize them as testicles until the posterior pole is located. The adrenal gland is very difficult to visualize in pigeons.

Under direct laparoscopic observation, appropriate biopsies may be taken at this time (see Chapter 16, Biopsy Techniques) or cultures, using a small Calgiswab* (Fig. 15–11).

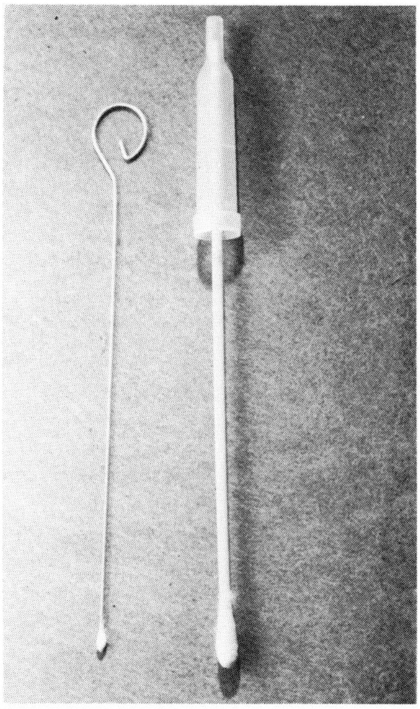

**Figure 15–11.** Sterile cotton swabs may be inserted into the laparoscopy entry site for direct cultures of tissues.

Through the same air sac hole, the proventriculus can be maneuvered slightly ventral with the tip of the endoscope to permit visualization of the spleen—a small, cylindrical, purple to brownish-red tissue in psittacines (elongated in pigeons). Noting one or both poles of the spleen differentiates this organ from the similarly colored, large mesenteric vessel that is also located in this area.

Retracting the scope from the medial air sac hole and proceeding ventral and lateral to the proventriculus, one will locate the liver. The overlying membranes (air sac, peritoneum, liver capsule) may be relatively thick and may have to be penetrated and reflected prior to a biopsy of the liver to ensure a sample of parenchymal tissue.

Intimately associated with the proventriculus and liver is the heart. Although movement of the heart may be evident through the membranes, occasionally one may need to move these membranes laterally toward the lung in order to see the organ itself.

It is useful to develop a pattern in the viewing of the organs, so that if one gets disoriented, the location of the scope can be more easily resolved (e.g., go back to the lung and proceed again).

## COMPLICATIONS OR UNTOWARD RESULTS OF LAPAROSCOPY

Some potential problems resulting from laparoscopy may be prevented by thorough preparation of the surgeon and prudent evaluation of the case involved. Others may not be evident until the procedure is underway. Causes of untoward results may include the following:

1. **Lack of endoscopy experience** or failure to understand and follow the principles of working under magnification (see Chapter 47, Introduction to Microsurgery). As mentioned previously, malpositioning may be a common error.
2. **Lack of knowledge of anatomic differences** among species. Physical "anomalies" may actually be lack of understanding of these differences. Locating specific tissues in pigeons, if

*Spectrum Diagnostics, Inc., Glenwood, IL 60425

one is using guidelines developed for psittacines, may be somewhat more difficult if one is not aware of subtle differences. Penetration of the intercostal space used for psittacine sexing may result in lung lobe or liver trocarization in other species such as waterfowl or herons. On two occasions, the author has noted hermaphrodism, a true anomaly.

3. **Visual interference.** Blurring of the visual field may be the result of blood, cells, or pieces of tissue clinging to the lens. Occasionally one can lightly dab the lens on a familiar organ such as the proventriculus to remove the particles. Otherwise one must remove the scope, clear the lens in the disinfectant solution, rinse, wipe, and re-enter the bird.

Disorientation from the magnification may be a problem at times. The endoscope may be in contact with the air sacs, membranes, or organs. One may be advised to pull the scope slowly back to the original entry site, locate familiar landmarks, and proceed again to the desired location, identifying new reference points. Recheck the position of the bird.

Some tissues, especially abnormal or adolescent gonads, may not be easily identified (see Chapter 51, Sex Determination Techniques).

Not only does excessive fat on the obese bird interfere with visualization of the viscera, but the lens of the endoscope becomes easily contaminated with lipid material. One might attempt elevating the surgical restraint board to a perpendicular plane in order to suspend the viscera and locate the gonads, kidneys, and adrenals. However, endoscopy is best avoided in obese birds.

4. **Trauma to the patient.** Puncture of a blood vessel can be serious. The procedure should be immediately suspended and the bird removed from the anesthetic and placed in a warm, quiet environment for observation. Administration of vitamin K may be indicated. Blood or any other liquid in the air sacs is capable of causing airway obstruction. If the bird is still anesthetized, place its body head-up at a 45-degree angle so that the blood remains in the caudal portion of the air sac.

Tracheal impingement or inhibition of respiration by physical restraint, such as in tightly stretched small birds, may also cause suffocation. In the author's experience, dilation and "glassing over" of the eyes and raising of the nape feathers may be signs of grave danger. Release of the bird and application of respiratory assistance may reverse the process.

Accidentally bumping into the liver or kidney with the scope may produce some bleeding, but if the preoperative parameters are normal (see Chapter 44, Evaluation and Support of the Surgical Patient), this should not be serious.

Puncture of the abdominal viscera, especially the gastrointestinal tract, requires laparotomy and suturing of the wound.

Fracture of wings from improper restraint may be encountered, particularly in inapparent cases of malnutrition.

Subcutaneous emphysema is uncommon and usually insignificant if it does occur unless the air sacs have been excessively damaged. A single aspiration of air normally suffices, rarely requiring a suture. A recurrence may require re-entry and suturing of the musculature.

A report from the Netherlands[6] warns of the danger of penetrating both body cavities (abdominal air sacs), which may lead to cessation of egg laying in domestic hens. It recommends that the surgical trauma of laparoscopy be limited to only one of the abdominal air sacs.

Although abdominal fluids from edema or ascites are aspirated prior to any entry, occasionally one may encounter residual liquids foaming in the air sacs. This is often more of a visual hindrance than a danger to the patient. However, elevating the bird to a 45 degree angle may be suggested.

Anesthetic deaths are rare, especially with the low doses of parenteral drugs or new inhalation agents commonly used for this procedure (see Chapter 45, Anesthesiology). The open drip method of inhalation anesthesia may not be as predictable, and the author strongly advises against this method. Fatal cardiac arrest may occur if a hyperventilating patient is induced with some inhalation agents.

5. **Mechanical failures.** Fracture of the endoscope barrel or fiber bundles or other damage to the equipment obviously interrupts the procedure. Most units have a built-in spare light bulb that should be replaced when used.

## POSTSURGICAL CARE

Air sacs are normally not sutured following a routine laparoscopy. It is interesting to note, however, that spontaneous repair of large air sac holes may not occur. Rents in air sacs from previous laparoscopic examinations (usually 3 to 5 mm in size) have been noted in subsequent procedures by the author months later. Smaller holes appear to heal.

Suturing of the skin incision is usually unnecessary. Food and water may be offered as soon as the birds are ambulatory, and appropriate follow-up treatment, based on the results of the procedures, is instituted.

### *Sources of Equipment Mentioned in Text*

Richard Wolf Medical Instruments Corp.
7046 Lyndon Avenue
Rosemont, IL 60018
(312) 298-3150

Dyonics, Inc.
71 Pine Street
Woburn, MA 01801
(617) 935-5900

## REFERENCES


1. Bottcher, M.: Endoscopy of birds of prey in clinical veterinary practice. *In* Cooper, J. E., and Greenwood, A. G. (eds.): Recent Advances in the Study of Raptor Diseases. West Yorkshire, England, Chiron Publications, Ltd., 1981.
2. Bush, M.: Diagnostic avian laparoscopy. *In* Cooper, J. E., and Greenwood, A. G. (eds.): Recent Advances in the Study of Raptor Diseases. West Yorkshire, England, Chiron Publications, Ltd., 1981.
3. Bush, M.: Laparoscopy in birds and reptiles. *In* Harrison, R. M., and Wildt, D. E. (eds.): Animal Laparoscopy. Baltimore, Williams and Wilkins Company, 1980, pp. 192–193.
4. Bush, M., et al.: Laparoscopy in zoological medicine. J. Am. Vet. Med. Assoc., 9:1081–1087, 1978.
5. Bush, M., et al.: Sexing birds by laparoscopy. *In* International Zoo Yearbook, 18:197–199, 1978.
6. Frankenhuis, M. T., and Kappert, H. J.: Infertility due to surgery on body cavity in female birds—cause and prevention. The Netherlands.
7. Harrison, G. J.: Endoscopic examination of avian gonadal tissue. Vet. Med. Small Anim. Clin., 73:479–484, 1978.
8. Harrison, G. J.: New aspects of avian surgery. Vet. Clin. North Am. [Small Anim. Pract.], 14(2):363–380, 1984.
9. Ingram, K. A.: Laparotomy technique for sex determination of psittacine birds. J. Am. Vet. Med. Assoc., 9:1244–1246, 1978.
10. Jones, B. D., and Roudebush, P.: The use of fiberoptic endoscopy in the diagnosis and treatment of tracheobronchial foreign bodies. J. Am. Anim. Hosp. Assoc., 20:497–504, 1984.
11. McDonald, S. E.: Laparoscopy in birds. *In* Burr, E. (ed.): Companion Bird Medicine. Ames, IA, Iowa State University Press, in press.
12. Satterfield, W. C.: Diagnostic laparoscopy in birds. *In* Kirk, R. (ed.): Current Veterinary Therapy VII. Philadelphia, W. B. Saunders Company, 1980, pp. 659–661.
13. Satterfield, W. C.: Early diagnosis of avian tuberculosis by laparoscopy and liver biopsy. *In* Cooper, J. E., and Greenwood, A. G. (eds.): Recent Advances in the Study of Raptor Diseases. West Yorkshire, England, Chiron Publications, Ltd., 1981.
14. Seager, S. W. J., and Wildt, D. E.: Laparoscopy, A Method of Diagnosis in Small Animal Medicine. St. Louis, MO, Ralston Purina Co., 1977.

# Chapter 16

# BIOPSY TECHNIQUES

GEORGE V. KOLLIAS, JR.
GREG J. HARRISON

Physical and historic findings associated with many diseases of birds are often nonspecific and may vary with etiology as well as with the species involved. Additionally, physical findings may be associated with disease processes in multiple organs and tissues of the body. Although a number of clinical laboratory tests have been recommended as aids in the recognition of liver or other organ dysfunction in avian species, they lack specificity in establishing a definitive etiology.[2, 3] Other conditions may be more occult and elude detection by these methods. The presence of lesions in the air sacs, for example, may not be reflected in a change in the hemogram or serum chemistry profile.

Radiography and laparoscopy are valuable adjuncts to clinical laboratory tests and assist in establishing a differential diagnosis. Biopsy of internal organs and tissues is an important tool for more specific evaluation.

Biopsy techniques, in conjunction with laparoscopy, have been used in mammals for evaluation of the liver, spleen, ovary, and kidney[1] and in birds for diagnosis of liver disease.[3, 4] Using similar techniques, other organs and tissues in birds may be biopsied for the purpose of determining the etiology, duration, or prognosis of a pathologic process. This is particularly important when other diagnostic tools fail to provide such information.

A transabdominal percutaneous needle technique, without the aid of a fiberoptic endoscope, has been described for biopsy of the avian liver.[3] Despite the minimal time required to carry out this procedure and economic considerations relative to the cost of optical equipment, this "blind" technique precludes direct visual observation of the liver and increases the probability of penetrating some other organ (e.g., heart, proventriculus, or gallbladder). In addition, one does not have the benefit of observing morphologic changes or the distribution of lesions. Because of the experience required of the operator and the potential risk to the patient, the remarks below will be confined to techniques that can be utilized in conjunction with a fiberoptic endoscope.

## INDICATIONS

Once considered a diagnostic technique for highly selected cases, biopsies are commonly being obtained as an adjunct to physical examination, history, clinical pathology, and radiology. Biopsy becomes a relatively simple procedure when performed with the aid of a fiberoptic endoscope (Fig. 16–1). Biopsies of the liver, spleen, kidney, lung, and air sacs may be important in the evaluation of such problems as chronic wasting, chronic respiratory syndromes, and therapeutically nonresponsive feather picking. Biopsies need not be confined to organs with visually observable lesions because histopathologic changes may be present before gross morphologic changes are evident.

## CONTRAINDICATIONS

Any abnormality in the patient's hemostatic capabilities is an important consideration. Important prebiopsy parameters to consider are prothrombin and clotting times.[2, 3] Pretreatment with whole blood or vitamin K 12 to 24 hours prior to biopsy may be crucial when bleeding disorders exist. Additional factors used in evaluating hemostatic capacity include packed cell volume and estimation of thrombocyte numbers on a stained blood smear.

Additional contraindications include birds with severe circulatory compromise, severe debilitation, or depression and an inexperienced operator (see Chapter 15, Endoscopy). The lateral approach is contraindicated in patients with ascites. The use of the forceps for kidney

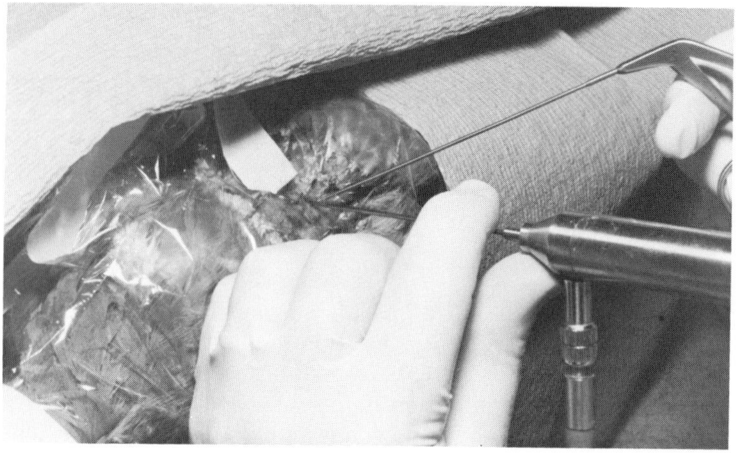

**Figure 16–1.** The biopsy forceps may be inserted into the same entry site as the endoscope. This allows for direct visualization of the forceps during the procedure. (From Harrison, G. J.: New aspects of avian surgery. Vet. Clin. North Am. [Small Anim. Pract.], 14:363–380, 1984.)

biopsies has limited application owing to excessive bleeding following sample acquisition in small (< 200 grams) birds.[2a] A straight punch biopsy instrument* takes a very fine slice from the tissue and may be more suitable for use in these individuals.

## INSTRUMENTATION AND TECHNIQUES

The use of forceps for tissue biopsy can be combined with endoscopic examination of abdominal viscera and other structures. Many practitioners working with avian species have acquired endoscopic equipment for use as diagnostic tools and to determine the sex of monomorphic species. A relatively small additional cost can provide the instrumentation necessary for biopsy procedures. The Flexarm described in Chapter 15, Endoscopy, is useful for maintaining stability of the instrument during these procedures.

Flexible biopsy forceps† (Fig. 16–2) can be employed with an endoscope as small as 2.2 mm. Disinfectants must be rinsed off the forceps prior to entry to prevent alteration of the samples. The forceps is passed alongside the endoscope when visual observation of the organ has been completed. If focal morphologic changes are apparent, the specific site is biopsied. In the majority of cases the disorder is diffuse and involves the entire organ; therefore, a sample is obtained from the visible border.

The tissue is grasped with the cups or capsule of the forceps and the forceps is closed. The operator may need to pierce or remove a small section of organ capsule or air sac membrane to facilitate sampling of the parenchyma. This is particularly important when using the lateral approach where air sacs or peritoneal membranes may cover the organs. Hemorrhage is usually insignificant; however, the sampling site should be observed for signs of bleeding following biopsy. Generally there is an ovoid-shaped tissue sample obtained (Fig. 16–2). If not, a second sample site can be selected. The sample is gently touched to a sterile glass slide to make impressions for cytologic examination and marked for identification (see Chapter 17, Cytology). Half of the sample is then fixed in 10 per cent buffered formalin for histopathologic

*R. Wolf Medical Instruments (No. 821150).
†R. Wolf Medical Instruments (No. 8150.00).

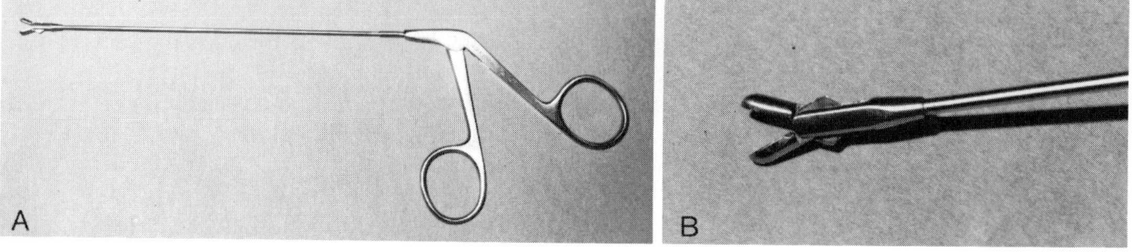


**Figure 16–2.** Flexible biopsy forceps *(a)*; an ovoid-shaped tissue sample is generally obtained owing to the shape of the forceps capsule *(b)*. (From Kollias, G. V., Jr.: Liver biopsy techniques in avian clinical practice. Vet. Clin. North Am. [Small Anim. Pract.], 14:287–298, 1984.)

examination. The second section is handled using sterile technique and is submitted for bacterial, viral, or fungal culture (see Chapter 10, Clinical Microbiology).

Some stains, such as Gram's, Macchiavello's, Diff-Quik (see Chapter 17, Cytology) can be performed by an assistant at the time of the procedure to lend some direction to immediate treatment. Depending on the gross lesions observed and the suspected etiology, one may wish to prepare samples in appropriate fixation media (Fig. 16–3). This is critical when considering such problems as gout, *Mycoplasma*, anaerobic bacteria, or viral infections or when preparing tissues for electron microscopic evaluation. Biopsy samples should be submitted in individual containers to avoid loss of the often tiny specimens.

A rigid optical biopsy forceps that attaches to the endoscope has been described[3] (see Chapter 15, Endoscopy).

Suction biopsy needles of varying gauges and lengths have been used by one of the authors (Kollias) in the ventral approach to the liver (described below). Following insertion of the endoscope and visual orientation, the needle is passed alongside or perpendicular to the scope through another entry site. The depth of penetration into the liver is preset on the needle by a movable locking collar or a protective sleeve that is placed over the needle. A 6-ml syringe containing 1 to 2 ml of sterile saline solution is attached directly to the needle hub. The site to be biopsied is observed, 4 ml of negative pressure are applied to the syringe, and the needle is rapidly thrust into the liver and withdrawn from the abdominal cavity. Evidence of an adequate sample is determined by ejecting 0.5 to 1.0 ml of saline solution from the syringe through the needle onto a sterile gauze sponge or into a vial containing a small quantity of sterile saline solution. The sample is then handled as previously described in the cup biopsy procedure.

## REGIONAL ANATOMY AND APPROACHES

### Left Lateral Coelomic

The lateral approach described in Chapter 15, Endoscopy, has the distinct advantage of viewing and biopsying several organs and tissues from a single entry site. From this approach samples of kidney, spleen, lungs, air sacs, liver, and testes can be obtained. Identification of air sac mites or other internal parasites has been accomplished by removing a sample with the forceps through the lateral endoscopic approach. The reader is referred to Chapter 15, Endoscopy, for information on positioning the patient and manipulation of the instruments.

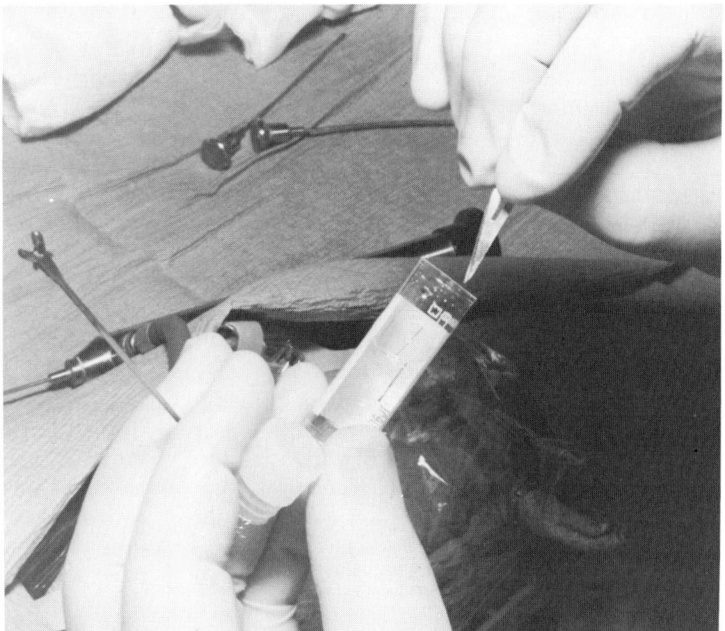

**Figure 16–3.** Containers of appropriate fixation media should be available for biopsy samples.

### Ventral Coelomic

The ventral approach is useful for biopsy and visualization of both liver lobes. The patient is placed in dorsal recumbency with the wings and legs extended laterally and attached to an appropriate restraint device. The feathers covering the area over the distal border of the keel and anterior right and left quadrants of the abdomen are removed or retracted laterally away from their respective tracts. Extensive feather removal is not necessary as long as a sterile surgical field can be maintained. The skin in this area is surgically prepared. After placement of the surgical drape, a 4- to 6-mm skin and muscle incision is made medially 0.5 to 1.0 cm below the apex of the keel. Electrosurgical cutting or blunt dissection of the muscle virtually eliminates hemorrhage. The telescope (and cannula, if used) is inserted on a line parallel to the vertebral column to a depth necessary to pass through the abdominal musculature.

Zealous penetration with the trocar or telescope at this point can result in laceration of the liver or large blood vessels. More seriously, puncture of the heart may occur owing to its location craniad between the right and left lobes of the liver.

The endoscope and forceps are inserted. Often abdominal fat at this site obscures direct visualization of the liver and must be dissected free and moved laterally. From this approach, both lobes of the liver, the heart, the pericardium, the caudal borders of both lungs, and cranial aspects of the abdominal air sacs can be examined. In obese birds, a lateral examination may be necessary, possibly with the bird in an upright position.

### Endotracheal

Granulomas or other lesions involving the trachea, primary bronchi, and anterior portions of the lungs are accessible via the tracheal approach. In addition to visualization and subsequent biopsy of these lesions the forceps may be useful in retrieving foreign bodies from the tracheal lumen. A supplemental air supply may need to be provided by trocarization of the abdominal air sacs during tracheoscopy.

## SPECIFIC ORGAN BIOPSY

The liver, kidney, and spleen are often involved in pansystemic diseases and are optimal tissues from which valuable information for diagnostic purposes can be made. If all three organs are to be biopsied with the flexible biopsy forceps, the kidney should be approached first, then the spleen, and the liver last, to reduce the possibility of hemorrhage obscuring the visual field. Birds with chronic, nonresponsive, or undiagnosed conditions are candidates for such a procedure.

In using laparoscopy as a diagnostic tool, an initial tendency may be to visually examine multiple organs and terminate the examination if no lesions are evident. In the experience of one of the authors (Harrison), several cases have emphasized the necessity of including biopsy of visually normal tissues to obtain a diagnosis. One example included a Hyacinth Macaw that was suffering from chronic weight loss and lack of vitality, with radiographic evidence of air sacculitis. Endoscopic examination revealed the air sacs and other tissues to be grossly normal. Cultures and biopsies were submitted from the spleen, liver, kidney, and air sac. Histologic examination of the tissues revealed no lesions; however, the bird continued to decline. *Mucor* sp. was subsequently isolated from the air sac sample. Ketoconazole therapy was initiated and within days the bird dramatically improved. In this case information needed to initiate therapy was not obtained from laparoscopic or histopathologic evaluation alone, but resulted from culture of the air sac membrane sample.

### Air Sacs

Biopsy or acquisition of fluid or exudate for culture from the air sacs is indicated in cases of chronic respiratory disease as evidenced by radiography. Therapeutically nonresponsive feather picking, especially in areas of the skin overlying the air sacs, is another consideration for endoscopic examination and biopsy.

### Lung

Although lung lesions are often difficult to identify radiographically, they can be apparent through endoscopic examination. Biopsy of the lung is carried out from a lateral approach. Visualization and potential sites for biopsy are restricted to the caudal/medial surfaces of the lung. Through the tracheal approach one may have access to the bronchi and cranial aspect of the lung. Unfortunately, lesions are often associated with the lateral aspects of the lungs which elude visualization from these approaches.

In the experience of one of the authors (Harrison), few respiratory diseases involve only the lung. Usually there is involvement of the air sac(s), nasal cavity turbinates, and infraorbital sinus. Clinical manifestations may include rhinorrhea, sneezing, wheezing, and coughing.

### Testes

One of the authors (Harrison) has had limited experience with biopsy of avian testes. This technique may prove to be a valuable tool in evaluation of male infertility. For example, biopsy samples taken from the testicle of a mature Blue-fronted Amazon, paired with an infertile egg-laying female, were found to be morphologically normal. However, the numbers of spermatozoa present were less than those observed in postmortem samples taken from a male that was actively breeding and producing viable offspring. Biopsy of the testes would be indicated only after extragonadal causes for infertility had been ruled out. Behavioral and husbandry considerations (such as lack of visual barriers from other birds, aggressive female) may result in neuroendocrine dysfunction that results in gonadal atrophy (see Chapter 52, Reproductive Medicine).

## POSTBIOPSY CARE

The authors have experienced few complications following biopsy. However, the patient should remain quiet to minimize complications associated with hemorrhage; observation should continue for 24 to 48 hours. Complications such as bile peritonitis become evident during this time period resulting in paresis, anorexia, and depression. Treatment consists of eliminating the continued source of leakage, removal of free bile from the coelomic cavity, administration of anti-inflammatory agents, and supportive care.

## PROCESSING BIOPSY SAMPLES

The avian practitioner may have little difficulty in obtaining a biopsy sample. However, proper and rapid evaluation of the sample may be a problem. The choice of a facility for histopathologic and cytologic evaluation of samples should depend upon the institution's commitment to avian medicine and pathology. It is often important to develop a personal relationship with the pathologist(s) in order to improve the quality of such diagnostic services.

## REFERENCES


1. Bush, M., et al.: Laparoscopy in zoological medicine. J. Am. Vet. Med. Assoc., 9:1081–1087, 1978.
2. Hardy, R.M.: Hepatic biopsy. *In* Kirk, R.W. (ed.): Current Veterinary Therapy VIII. Philadelphia, W.B. Saunders Company, 1983, pp. 813–817.

2a. Harrison, G. J., et al.: A clinical comparison of anesthetics in domestic pigeons and cockatiels. Proceedings of the Annual Meeting of the Association of Avian Veterinarians, Boulder, CO, 1985.

3. Kollias, G.V.: Liver biopsy techniques in avian clinical practice. Vet. Clin. North Am. [Small Anim. Pract.], 2:287–298, 1984.
4. Satterfield, W.C.: Early diagnosis of avian tuberculosis by laparoscopy and liver biopsy. *In* Cooper, J.E., and Greenwood, A.G. (eds.): Recent Advances in the Study of Raptor Diseases. London, England, 1980, pp. 105–106.

# *Chapter 17*

# CYTOLOGY

TERRY W. CAMPBELL

Cytologic examination of tissues and fluids can provide a rapid antemortem diagnosis or a diagnosis at necropsy. Frequently, the etiologic agent for the disease and a presumptive or definitive diagnosis supported by cytologic evidence can be obtained in the clinical setting. It is important that cytologic specimens be obtained from fresh tissues. Samples should be processed for cytologic examination immediately after collection for best results. Ideally, the samples should be collected immediately after the bird's death when obtained during necropsy. Cellular degeneration can occur rapidly, especially in birds that have been dead for several hours and kept at room temperature.

Cytologic evaluation of tissues and fluids is part of the data base that is easily obtained by the practicing veterinarian. Other data include the clinical history, physical examination, laboratory examination of samples removed from the bird (i.e., complete blood count, blood chemistries, fecal examination, and gross necropsy findings). Histopathologic evaluation of tissues usually requires the assistance of an outside laboratory. Cytologic and histologic evaluations of tissues should complement and not compete with each other. Histopathology often confirms a diagnosis or provides one in cases in which the diagnosis is elusive.

## SPECIAL SAMPLING PROCEDURES

### Crop Aspiration

Indications for the examination of crop content include vomiting, repeated regurgitation, delayed crop emptying, and the presence of other signs suggestive of a crop disorder. A crop aspiration procedure requires a mouth speculum, a sterile soft plastic or rubber tube, and a sterile syringe. The head and neck are extended to straighten the esophagus to avoid puncturing it with the tube. An appropriately sized tube is passed into the mouth, down the esophagus (usually lying on the right side of the neck), and into the crop. A mouth speculum prevents the bird from biting the tube. The tube should pass freely; if any resistance is encountered, the procedure should be stopped and repeated. A tube should never be forced down the esophagus. The crop material is aspirated into a syringe attached to the exposed end of the tube. Excessive vacuum should be avoided to prevent ischemic lesions on the crop mucosa that may result if the tube opening lies against the crop wall. The aspirated material can be submitted for microbiologic culture and microscopic examination.

### Upper Respiratory Sinus Aspiration

A sinus aspirate is obtained from avian patients with sinusitis. This procedure allows sampling of the sinus content with a minimal amount of surface contamination for microbial culture and cytologic evaluation. One method involves inserting a hypodermic needle (i.e., 22-gauge) with syringe attached into the commissure of the mouth and directing the needle vertically to a point midway between the eye and the external naris. The needle should be held parallel to the skin surface and guided under the zygomatic bone (this bone crosses diagonally from the lower corner of the upper beak to the ear in psittacine birds). Peripheral blood contamination of the sample may result if the ocular orbit or surrounding muscle mass is penetrated. Holding the mouth open using an oral speculum will provide a larger lateral surface at the aspiration site when using this procedure.

A second sinus aspiration technique involves the aspiration of the sinus space just below the eye. The needle is inserted just caudal to the commissure of the mouth and directed ventral to the zygomatic arch ending just under the eye. An alternative method is to approach the sinus directly, entering at a perpendicular angle to the side of the head.

During any procedure, the puncturing of the

ocular orbit and globe should be avoided. The head should be held firmly, with proper limb and body restraint.

### Intratracheal Aspiration

Intratracheal aspiration is indicated whenever a disease of the trachea, syrinx, or bronchi is suspected. The procedure is simple but may require a general anesthetic. An appropriately sized, sterile, soft plastic or rubber tube is passed through the glottis into the trachea, ending near the syrinx at the thoracic inlet. Sterile saline solution is infused into the trachea and reaspirated into the sterile syringe. The use of an excessive amount of fluid should be avoided (0.5 to 1.0 ml/kg body weight can be used safely in most cases). A large-bore hypodermic needle inserted into the intraclavicular air sacs will aid respiration during the procedure if the bird is severely dyspneic. Tracheal wash samples usually require a concentrating technique to examine the cells obtained. Some contamination of the sample is expected as the tube is passed through the oral cavity into the glottis. A sterile technique for obtaining uncontaminated samples involves passing the sterile tube through a sterile endotracheal tube previously in place. Tracheal swab samples can be obtained by passing a small sterile cotton swab directly into the trachea.

### Air Sac Sampling

Air sac paracentesis can be useful in the diagnosis of chlamydial, fungal, and bacterial air sac disorders. Samples can be obtained during otoscopic laparotomy, laparoscopy, or exploratory laparotomy procedures. Samples are obtained using a sterile cotton swab rubbed across an air sac lesion or by imprinting an excised lesion. The swab is rolled across the surface of a microscope slide (a sterile slide should be used if the swab is also used for microbiologic culture) to provide the cytologic specimen.

### Abdominocentesis

Abdominal fluid should be collected and evaluated for birds with ascites, peritonitis, hemoperitoneum, or other abdominal fluid accumulations. The site of entry should be prepared as for surgery. A 21- to 25-gauge needle is inserted along the ventral midline immediately distal to the point of the sternum (keel). The needle is directed to the right side of the abdomen to avoid puncturing the ventriculus (gizzard). Peritoneal washings can be attempted in birds with little abdominal fluid accumulations by infusing sterile saline solution into the abdominal cavity and reaspirating. However, one must consider the possibility of washing an abdominal air sac instead of the abdominal cavity. Usually large fluid accumulations in the abdominal cavity will decrease the air sac volume that can be seen radiographically.

## SAMPLING EXPOSED SURFACES

Cytologic samples of exposed lesions in the oral cavity or on the body surface can be obtained by scraping the lesion with a cotton swab or spatula blade. Masses can be aspirated or excised. Conjunctival and corneal samples are obtained by using a sterile moist swab or a metal or plastic spatula to gently scrape the lesion. Local ophthalmic anesthetics may be required to reduce the pain of the procedure.

## SAMPLE PREPARATION

Contact smears are made by imprinting masses or tissue removed during the necropsy procedure. Smears are made by gently touching a glass microscope slide to the freshly cut surface of the tissue. The imprinted surface should be fairly dry and free of blood. A clean, absorbent material (e.g., paper towel) can be used to blot the cut surface of the tissue. Several imprints should be made on each slide. Some tissues exfoliate poorly and produce smears with low cellularity. Exfoliation of cells can be improved by scraping the surface of the tissue with a scalpel blade. The roughened surface can be imprinted or the material remaining on the scalpel blade can be used to make the smear.

Any abnormal accumulation of fluid discovered during the necropsy should be collected. An analysis of the fluid, including cytologic examination, should be performed. The fluid can be obtained by aspiration into a clean, sterile syringe or by a sterile cotton swab. Direct smears of the fluid can be made using the same techniques used for making blood smears. A feathered edge should be created with these types of smears. Highly viscid fluid or fluid containing solid tissue fragments (or clotted material) requires dragging the sample slowly across the slide using a spreader slide. A "squash" preparation can also be made by pressing another glass slide to the smear and spreading the two slides apart. This will result in two smears, one on each glass slide. Fluid collected on a cotton swab can be applied to the slide by gently rolling the swab across the slide surface. Fluids with poor cellularity require concentration techniques such as sediment smears (smears of the cellular sediment after centrifugation) or cytocentrifugation.* Cells can be concentrated at the feathered edge by purposely marginating the cells in the fluid when making a smear. This is done by slowly moving the spreader slide along the surface of the smear slide and quickly lifting the spreader slide just before the end of the spreading technique. This will concentrate many cells at the feathered edge of the smear for examination.

A variety of stains and staining techniques is available for cytologic evaluation. Commonly used general stains include Wright's, Giemsa, new methylene blue, and Diff-Quik† stains. Commonly used specific stains include the Gram's stain for bacteria, acid-fast stain for tubercle bacilli, Giménez and Macchiavello's stains for *Chlamydia*, and Sudan stains for fat. The reader should refer to the Appendix to this chapter for staining methods and interpretations. The choice of stain depends upon the desired staining effect and in part upon the preference of the cytologist. Frequently, more than one stain is used, since each stain may provide different information. Therefore, more than one slide of the sample should be made to provide the opportunity to use more than one stain. Cytologic interpretations described in this text are based primarily on smears stained with a rapid modified Wright-Giemsa stain,† since these types of stains are in common use in veterinary practices.

Stained slides are initially examined using scanning (45×) or low (100× or 200×) magnifications to estimate the cellularity, determine the best location for cytologic examination, and identify tissue structures or large pathogenic agents (e.g., fungal elements). Higher magnifications (e.g., high dry [400×] and oil immersion [1000×]) are used to identify cellular structures, bacteria, and other small objects.

*Cytospin, Shandon Southern Instruments, Inc., Sewickley, PA 15143

†Diff-Quik, Harleco, Gibbstown, NJ 08027

## BASIC CYTOLOGIC INTERPRETATIONS

Cells observed in cytologic preparations are usually derived from one of four major tissue groups: hemic, epithelial-glandular, connective, and nervous.[1] Usually not every cell in a smear can be identified but can be placed into one of the major groups. Occasionally, cells from poorly differentiated neoplasms cannot be classified. The cytologic appearances of many avian cells resemble those described for mammalian cytology. The avian practitioner should consult mammalian as well as avian cytology references as a guide for cytologic interpretations.[2, 11, 12]

Cells obtained from hemic tissue are derived from the peripheral blood, bone marrow, and ectopic hematopoietic sites. Normal peripheral blood contains the definitive cells of hematopoiesis: mature erythrocytes, thrombocytes, heterophils, eosinophils, basophils, lymphocytes, and monocytes. The reader should refer to the avian hematology chapter and other references for the cell descriptions.[4, 10] The immature hemic cells are found in the various hematopoietic sites or peripheral blood showing pathologic changes.

Cells of epithelial origin are usually round to oval (except mature squamous epithelial cells, which tend to be polygonal) with abundant cytoplasm and distinct cytoplasmic borders. The nuclei are typically round or oval with smooth chromatin. Epithelial cells exfoliate easily and often appear in clusters or sheets. Normal epithelial cells are uniform in appearance.

Connective tissue cells exfoliate poorly. Scraping of the tissue is usually required to exfoliate a significant number of cells for cytologic evaluation. Connective tissue cells have a variable amount of cytoplasm and variable nuclear shape. The cytoplasmic margins are often indistinct.

Cells from nervous tissue are seldom seen on cytologic specimens. Stellate cells with a variable number of cytoplasmic projections obtained from samples near the central nervous system are most likely of nervous origin.

## CYTOLOGIC FEATURES OF INFLAMMATION

Cytologic evidence of inflammation includes the presence of heterophils, eosinophils, macrophages, lymphocytes, and plasma cells. These cells actively migrate to the site of inflammation drawn by chemotactic factors derived from the immune system, living organism (e.g., microorganisms), tissue destruction due to nonliving agents (e.g., traumatic, thermal, or chemical), or neoplastic tissue. The backgrounds of smears obtained from inflammatory lesions often show granular precipitation and protein aggregation. Fibrin may be present and can be outlined using new methylene blue stain.[11]

Acute inflammation is associated with a predominance of heterophil granulocytes (usually greater than 70 per cent of the inflammatory cells). Overwhelming bacterial infections or other toxic environments create a degenerative appearance to the heterophils. Mild degenerative changes are indicated by an increase in cytoplasmic basophilia, vacuolization, and loss of granules. Greater degenerative changes in heterophils are indicated by karyolysis, in which the cell nucleus becomes swollen and the chromatin has a poorly defined, smooth, pink, homogeneous appearance. Karyorrhexis (nuclear fragmentation) may be seen in marked degenerative changes. Nuclear pyknosis (shrunken, dense, deeply basophilic nucleus) appears in cells that have undergone a slow, progressive aging in a nontoxic environment. These cells can be seen in non- or mildly inflammatory lesions.

Chronic active inflammation is indicated by a mixture of leukocytes. Approximately half of the cells are nondegenerate heterophils. The remainder of the cells are a variable number of lymphocytes, plasma cells, and macrophages.

Chronic inflammation is indicated by a predominance (greater than 50 per cent of the inflammatory cells) of mononuclear cells (macrophages and lymphocytes).[12] Inflammatory multinucleated giant cells or macrophages forming netlike sheets represent a granulomatous reaction. Frequently, the causative agent (e.g., foreign bodies, fungal elements, tubercle bacilli) can be seen associated with this type of reaction.

Exudative effusions are abnormal accumulations of fluids with inflammatory characteristics. Exudates vary according to the etiology, host reaction, and duration of time. They are characterized by high cellularity (primarily inflammatory cells), a specific gravity greater than 1.020, and a total protein greater than 3.0 gm/dl. Exudative effusions vary in color, turbidity, viscosity, and odor. They frequently clot and may require an anticoagulant (e.g., EDTA) for cytologic evaluation; however, clotting usually has occurred if the bird died other than just prior to necropsy. Septic exudation is characterized by intracellular bacteria and degenerative heterophils. Plasma cells are seen in chronic exudates.

## CYTOLOGIC FEATURES OF HYPERPLASTIC TISSUE

Tissue hyperplasia is a proliferative response to injury.[12, 13] Cells from hyperplastic tissues are uniform in appearance and show signs of immaturity (e.g., cytoplasmic basophilia). Epithelial and fibrous proliferation are examples of hyperplasia that occur in areas of chronic inflammation. Cells from hyperplastic tissue should not be confused with those from neoplastic tissue, which also appear immature.

## CYTOLOGIC FEATURES OF MALIGNANT NEOPLASIA

No single cytologic feature is diagnostic for neoplasia; therefore, the diagnosis should be based on a combination of features. Many cells should be examined. Neoplastic cells in individual smears appear to have been derived from a common source (the cells resemble each other). However, they exhibit a variable amount of pleomorphism and do not have a uniform appearance. This feature aids in differentiating between neoplastic and hyperplastic tissue.

The cell nucleus often provides cytologic evidence of neoplasia. Neoplastic features of cell nuclei include nuclear enlargement, variable nuclear sizes and shapes, nucleolar changes, multinucleation, irregularity of the nuclear chromatin or membrane, abnormal mitoses, and variable nuclear/cytoplasmic (N/C) ratios. Cells with large nuclei (larger than the size of a heterophil, for example) are suspect for neoplasia. Large, swollen nuclei with a pink, homogeneous chromatin due to exposure to hypotonic solutions should not be confused with neoplastic nuclear enlargement. Variations in nuclear size and shape are neoplastic features that are especially significant when cells occur in clumps or sheets. Marked variations in the N/C ratio (or cells with higher than normal N/C ratios) are suggestive of neoplasia. Neoplastic change can also be represented by variations in the shape, size, number, and staining quality of nucleoli. Very large nucleoli are suggestive of neoplasia. Multinucleated giant cell formation (especially uneven nuclear numbers) is suggestive of asynchronous cell division; however, this feature can also be found in nonneoplastic lesions (Fig. 17–1). Cells with abnormal or high percentages of mitoses can represent neoplastic change. Abnormal mitotic figures appear to have more than two poles of nuclear division. Occasionally, cells with cyto-

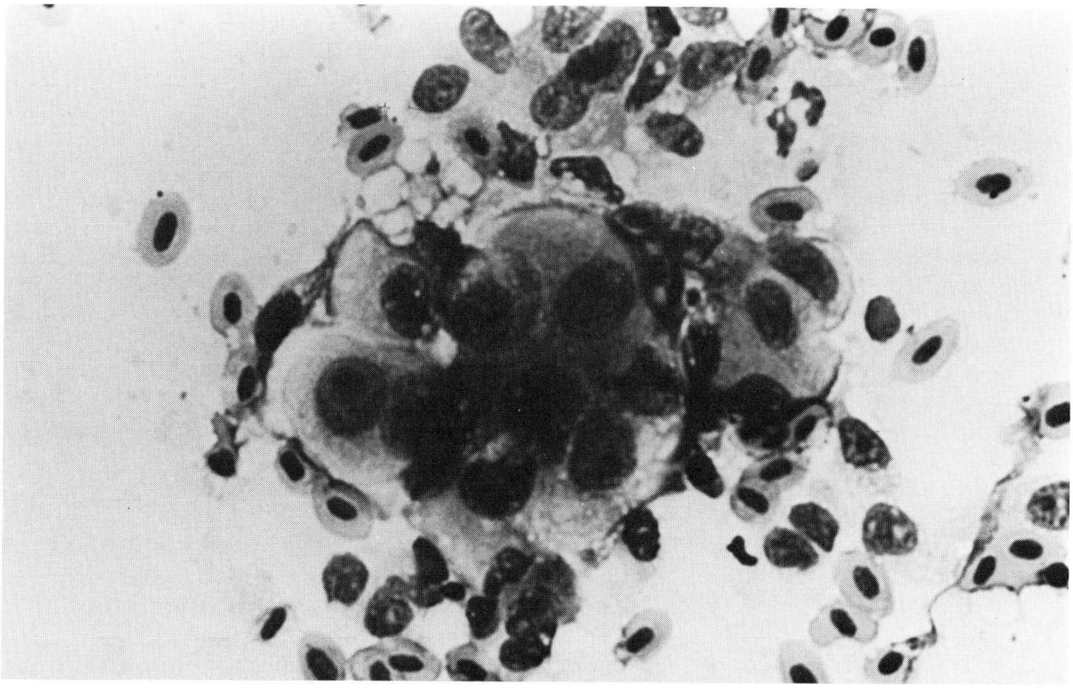

**Figure 17–1.** A multinucleated giant cell found in the abdominal fluid from a seven-year-old female budgerigar (*Melopsittacus undulatus*) with an ovarian cystadenocarcinoma (Diff-Quik, ×1000).

plasmic nuclear fragments result from abnormal cell divisions. Nuclei with irregular coarse chromatin and nuclear membrane margins are clues to neoplasia.

Cytoplasmic features of neoplasia include variations in cytoplasmic borders, basophilia, vacuolation, and inclusions. Indistinct cytoplasmic borders are features of certain types of neoplasms (e.g., sarcomas). Cytoplasmic basophilia represents a high concentration of cytoplasmic RNA due to the rapid metabolic rate commonly seen in neoplastic tissue.[12] Large cytoplasmic vacuoles occur in certain types of neoplastic cells (e.g., adenocarcinomas). Inclusions such as nuclear fragments, pigment granules, and other cells within phagocytic vesicles (cannibalism) are additional clues to neoplasia.

Cells originating from neoplastic epithelial tissue (i.e., carcinomas) tend to exfoliate in clusters or sheets. The cells usually appear round or oval with distinct cytoplasmic borders. Adenocarcinomas are represented by neoplastic-appearing epithelial cells with large cytoplasmic secretory vacuoles and occasionally giant cell formation (Fig. 17–1).

Cells from connective tissue neoplasms (i.e., sarcomas) usually exfoliate poorly and provide smears with low cellularity. The cells tend to be elongated or spindle-shaped and occur singly. Sarcoma cells often have indistinct cytoplasmic borders and large nuclei (often multinucleated) with prominent nucleoli (Fig. 17–2).

Discrete cell neoplasms are derived from individual cells that have no cellular interactions.[12, 13] Cytologically, discrete cell tumors exhibit individual round or oval cells that exfoliate individually. Lymphoid leukosis (lymphosarcoma) is an example of a discrete cell neoplasm that occurs in birds. It is characterized by a marked number of immature lymphocytes either in peripheral blood or solid lymphoid masses (Fig. 17–3). The immature lymphocytes are large, round cells with dark blue cytoplasm. The nucleus has a smooth chromatin pattern as compared to the coarse chromatin of mature lymphocytes. A prominent nucleolus indicates a lymphoblast. Mitotic figures are common with lymphoid leukosis. The background of imprints from lymphoid masses is usually thick and contains many pale blue cytoplasmic fragments.

## CYTOLOGIC FEATURES OF ABDOMINAL EFFUSIONS

Effusions can be classified as pure transudative, modified transudative, nonseptic and septic exudative, malignant exudative, and hemorrhagic.[6] The accumulation of body fluids is

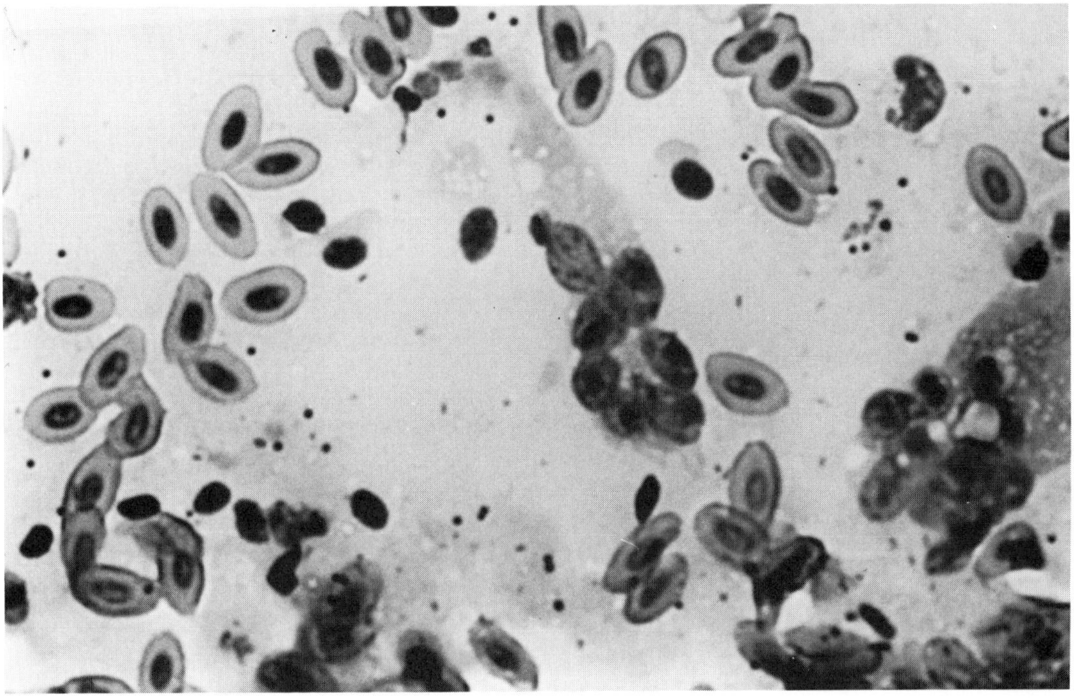

**Figure 17–2.** Multinucleated spindle-shaped cells from the scraping of an oral lesion in a Double Yellow-head Amazon Parrot *(Amazona ochrocephala)* (Diff-Quik, ×1000). Multinucleated fibroblasts with odd number of nuclei are suggestive of a connective tissue neoplasm (e.g., fibrosarcoma).

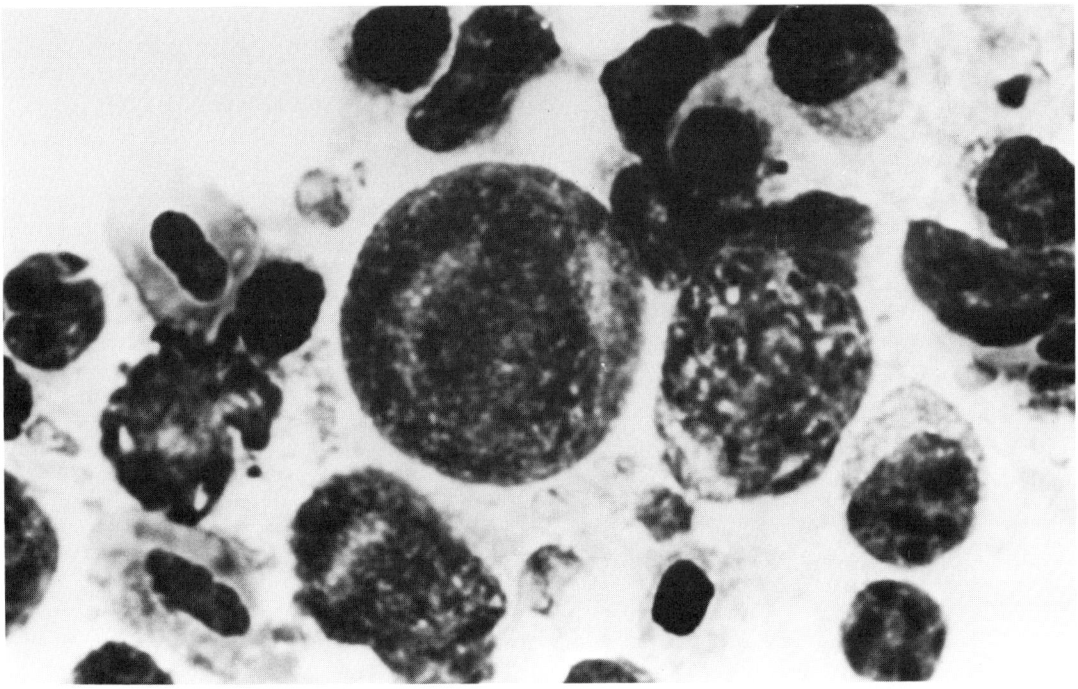

**Figure 17–3.** Immature lymphocytes from an impression smear taken from a periorbital lymphoid mass in an Amazon Parrot (*Amazona* sp.) (Diff-Quick, ×1000). Compare the large lymphoblast with its smooth chromatin to the small, mature lymphocytes with condensed chromatin.

mainly confined to the abdominal cavity in birds. Normally little or no abdominal fluid can be obtained; therefore, any amount of fluid (0.5 ml or greater) that is easily collected is considered abnormal. Normal abdominal fluid has features of transudates. The normal fluid is poorly cellular, with an occasional mesothelial cell and macrophage. Mesothelial cells are round or oval and vary in size. They are derived from serous membranes and exfoliate singly or in clusters. They have abundant basophilic cytoplasm and centrally positioned round nuclei. Occasionally, mesothelial cells have fine eosinophilic villus-like margins.[12] Mesothelial cells become reactive with irritation of the serous membranes and convert from a flat to a cuboidal shape. The reactive mesothelial cells are larger and more basophilic than normal, and occasionally multinucleated forms and mitotic figures are present. Reactive mesothelial cells can transform into phagocytic cells and resemble active histiocytes or macrophages.[6] Macrophages are large cells that are irregularly shaped, have an abundant blue cytoplasm, and often contain phagocytic vacuoles and foreign material (Fig. 17–4).

Transudative fluids accumulate as a result of oncotic pressure changes or other circulatory disturbances. The usual causes of transudative fluid accumulation are hypoproteinemia, cardiac insufficiency, hepatic cirrhosis, and chronic renal disease.[14] Transudates have a clear to pale yellow color, low cellularity, specific gravity less than 1.020, and a total protein of 3.0 gm/dl or less. The cytology reveals predominantly mononuclear cells on a clear background. Transudates undergo modification with hydrostatic pressure changes or chronic irritation to serous membranes. Modification results in an increase in the cellular and protein content of the fluid.

Exudative effusions have been previously described. Malignant effusions have exudative characteristics. Neoplastic cells may be present. Occasionally, hemorrhagic effusions are associated with neoplasms.

Acute hemorrhagic effusions resemble peripheral blood. Thrombocytes indicate active hemorrhage but quickly disappear and are not usually found in fluids obtained at necropsy. Chronic or resolving hemorrhagic effusions reveal varying amounts of erythrophagocytosis. This is demonstrated by macrophage or heterophil phagocytosis of intact erythrocytes, erythrocyte fragments, and iron pigment (Fig. 17–5).

## CYTOLOGY OF THE UPPER ALIMENTARY TRACT (ORAL CAVITY, ESOPHAGUS, CROP)

White to yellow plaques, nodules, and ulcers are frequently found in the upper alimentary

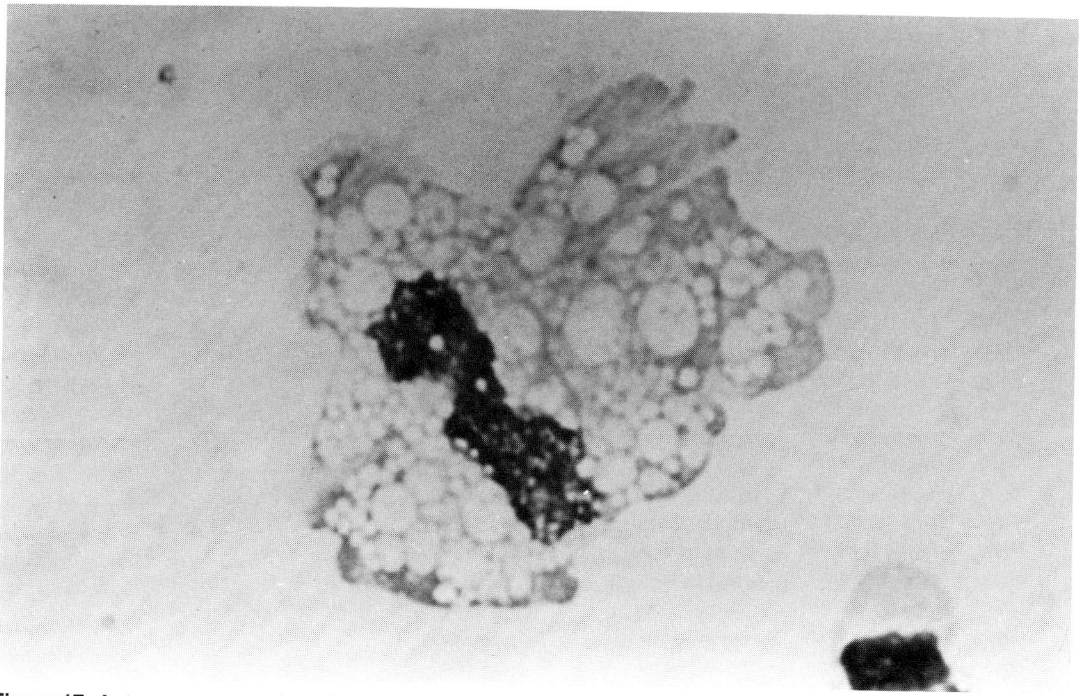

**Figure 17–4.** A reactive macrophage from the abdominal fluid of a male Button Quail *(Excalfactoria chinensis)* (Diff-Quik, ×1000).

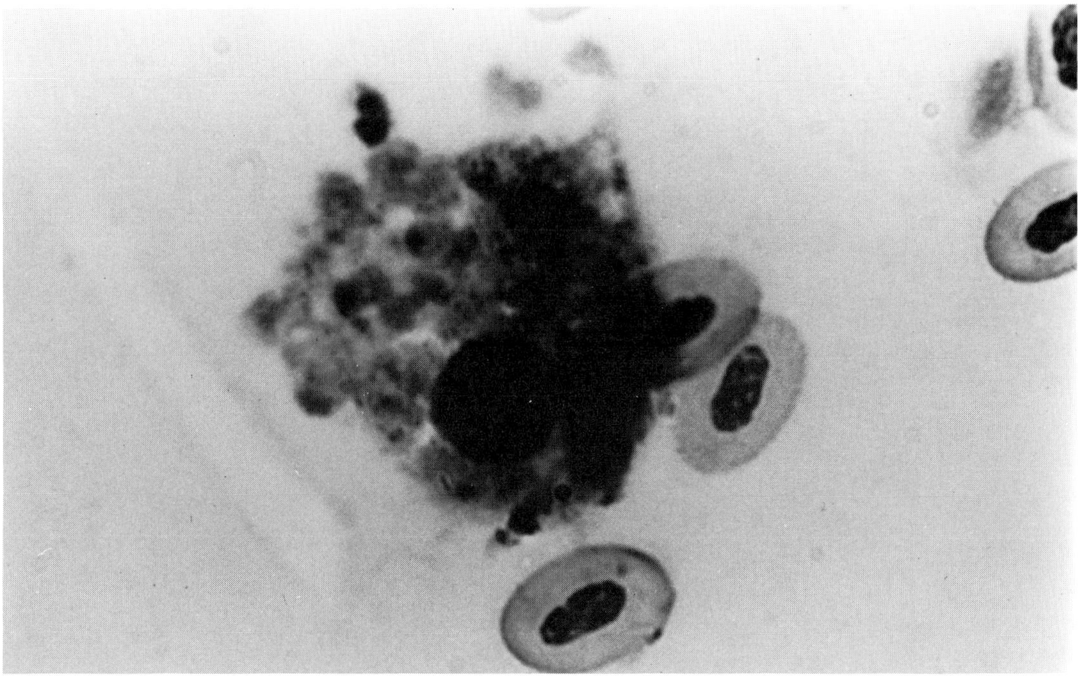

**Figure 17–5.** A macrophage in the abdominal fluid of a five-year-old female budgerigar *(Melopsittacus undulatus)* with a hemoperitoneum (Diff-Quik, ×1000). The intracytoplasmic iron pigment is indicative of erythrophagocytosis.

tract of birds. Cytologic examination of these lesions will aid in the determination of the cause. These lesions can be caused by candidiasis, trichomoniasis, capillariasis, bacterial abscessation, and hypovitaminosis A. These conditions are easily differentiated with microscopic examination.

The avian buccal cavity consists of a cornified beak or bill and an oral cavity lined with cornified squamous epithelium.[1] The esophagus and crop are lined by stratified squamous epithelium. Normal cytology of the upper alimentary tract reveals many squamous epithelial cells with varying degrees of cornification. Young squamous cells are round to oval with abundant basophilic cytoplasm and a centrally positioned vesicular nucleus. The larger mature squamous epithelial cells are polygonal, with occasional angular or folded margins. Mature squamous cells have relatively smaller, condensed nuclei compared to the younger cells. Many extracellular bacteria are present and frequently associate with squamous epithelial cells. The normal bacterial flora of the upper alimentary tract in pet birds shows a predominance of gram-positive bacteria with Gram's staining. *Alysiella filiformis* is a normal bacterial inhabitant of the upper alimentary tract of birds (Fig. 17–6). This organism consists of pairs of cells arranged in unbranched, ribbon-like filaments that are frequently associated with squamous cells.[9] These organisms should not be confused with fungal hyphae. The normal cytology of the upper alimentary tract contains a moderate amount of debris and foreign material (e.g., plant fibers and crystals) (Fig. 17–7). *Candida*-like yeasts are considered normal if found in low numbers (fewer than one per high power field).[3]

Inflammatory lesions exhibit a variable number of squamous epithelial cells and background debris, large numbers of inflammatory cells, and bacteria. Ulcerative lesions may reveal parabasilar cells. These cells are epithelial cells with large nuclei and scant, deeply basophilic cytoplasm. Chronic ulcerative lesions may show fibroblasts due to fibrous proliferation. Fibroblasts are spindle-shaped cells with oval to elongate single nuclei (Fig. 17–8). Lesions due to bacterial pathogens show degenerate heterophils and leukocyte phagocytosis of the bacteria (Fig. 17–9). Lesions caused by candidiasis reveal numerous oval, thin-walled budding yeasts (3 to 6 μ diameter) that stain dark blue with Wright's stain and gram-positive with Gram's stain (Fig. 17–10). There is usually a marked amount of background debris but few inflammatory cells associated with candidiasis unless mucosal ulceration has occurred. Tissue invasion by *Candida* sp. is indicated by the presence of blastospores and pseudohyphae (or

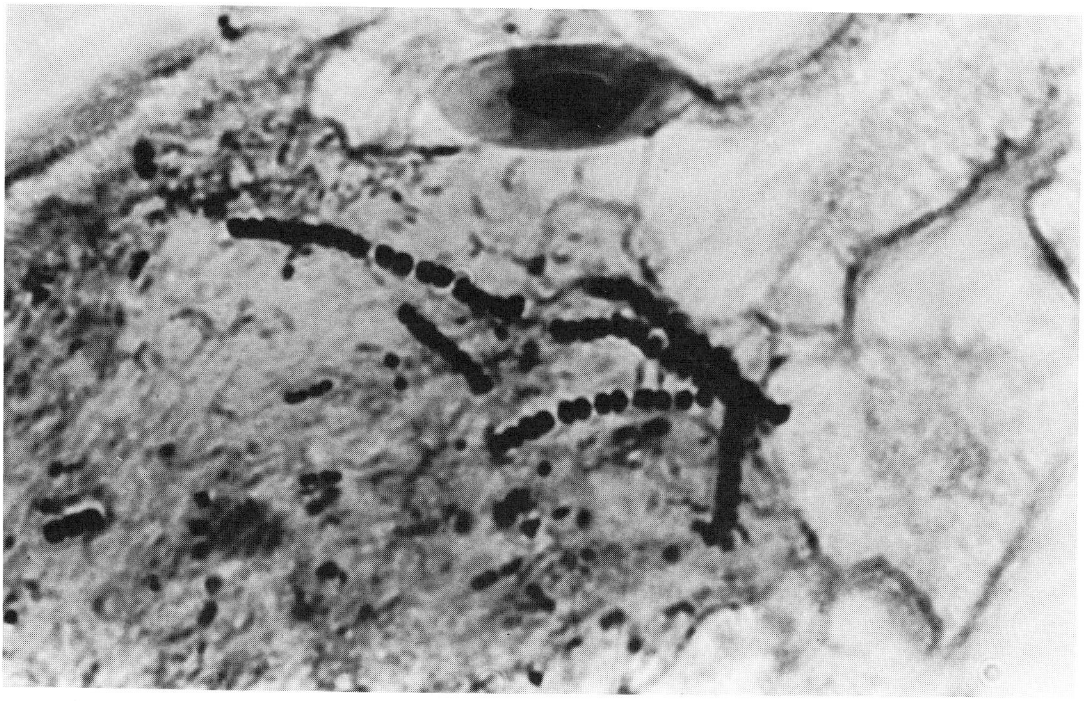

**Figure 17–6.** *Alysiella filiformis* is a large, ribbon-like bacterium often associated with squamous epithelial cells and is considered a normal inhabitant of the avian buccal cavity, esophagus, and crop (Diff-Quik, ×1000). (From Campbell, T. W.: Diagnostic cytology in avian medicine. Vet. Clin. North Am. [Small Anim. Pract.], 14:317–344, 1984.)

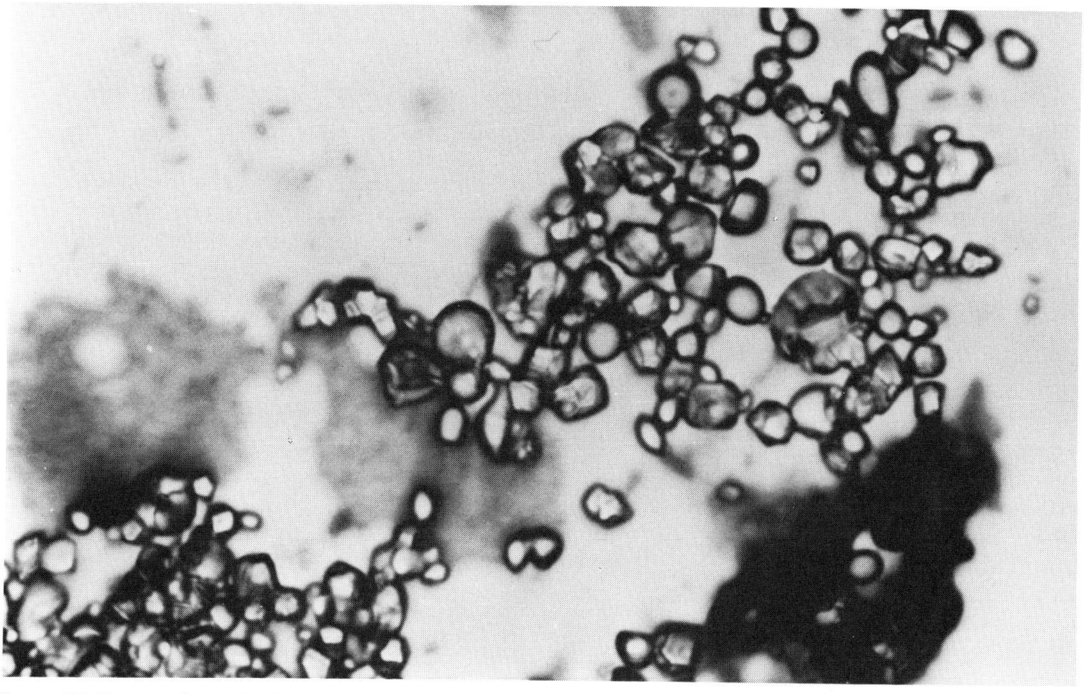

**Figure 17–7.** Normal crop fluid from a Quaker Parakeet *(Myiopsitta monachus)* containing numerous crystals (Diff-Quik, ×1000).

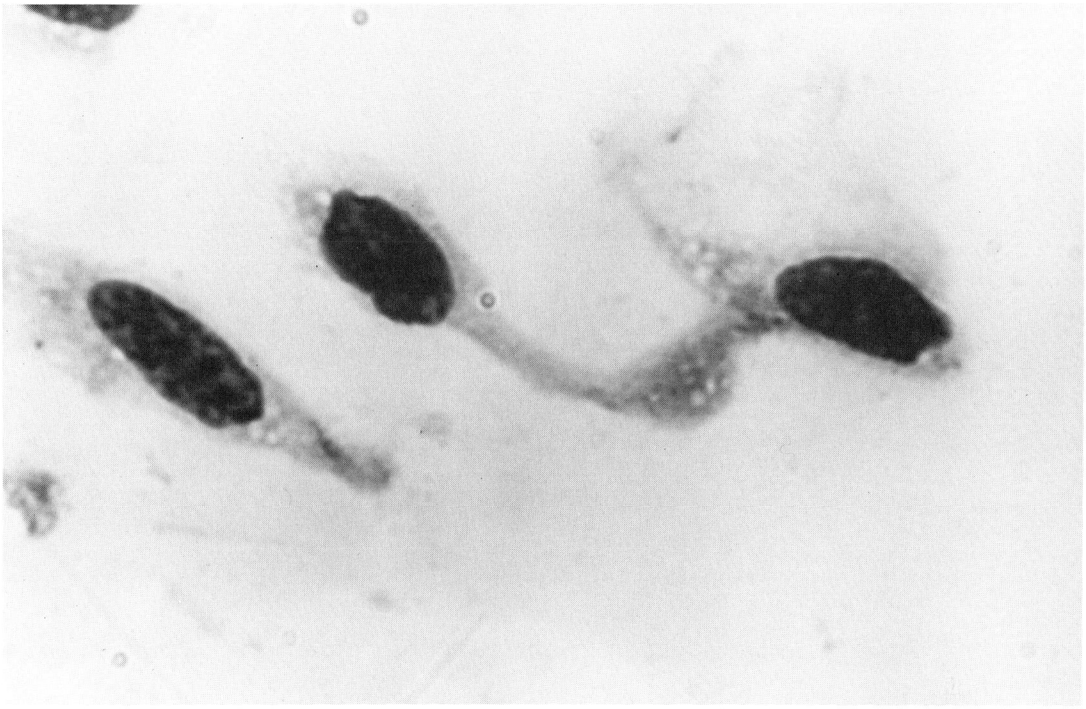

**Figure 17–8.** Normal fibroblasts are spindle-shaped cells that are uniform in appearance (Diff-Quik, ×1000). Note the coarse chromatin and indistinct cytoplasmic borders.

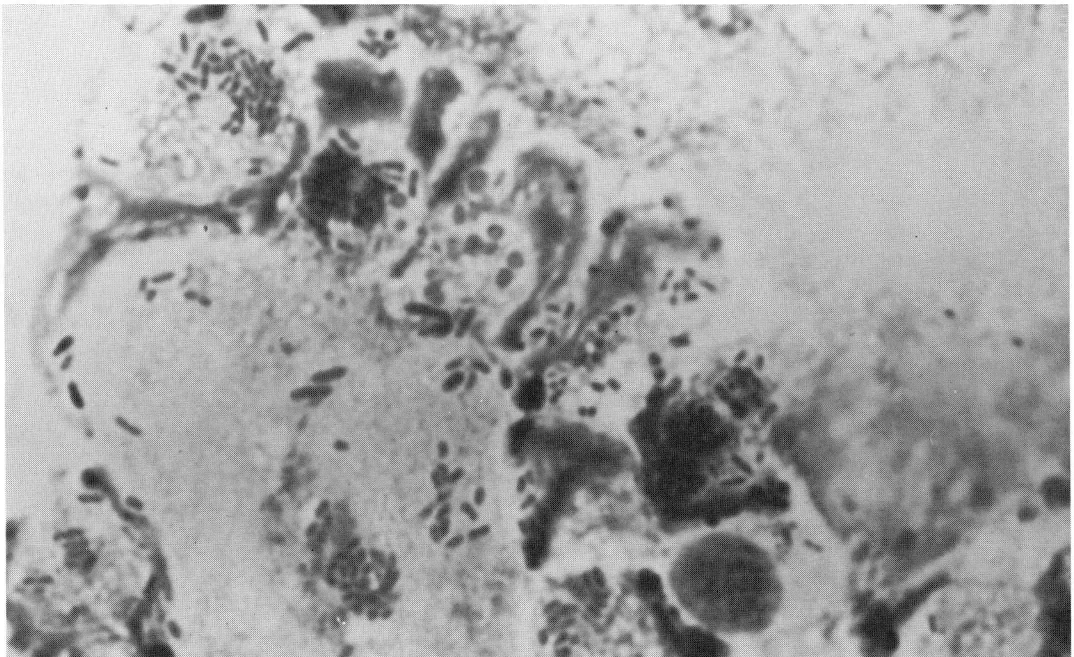

**Figure 17–9.** Crop fluid from a cockatiel *(Nymphicus hollandicus)* with septic ingluvitis. The cytology reveals numerous heterophils and macrophages containing intracellular bacteria (Diff-Quik, ×1000).

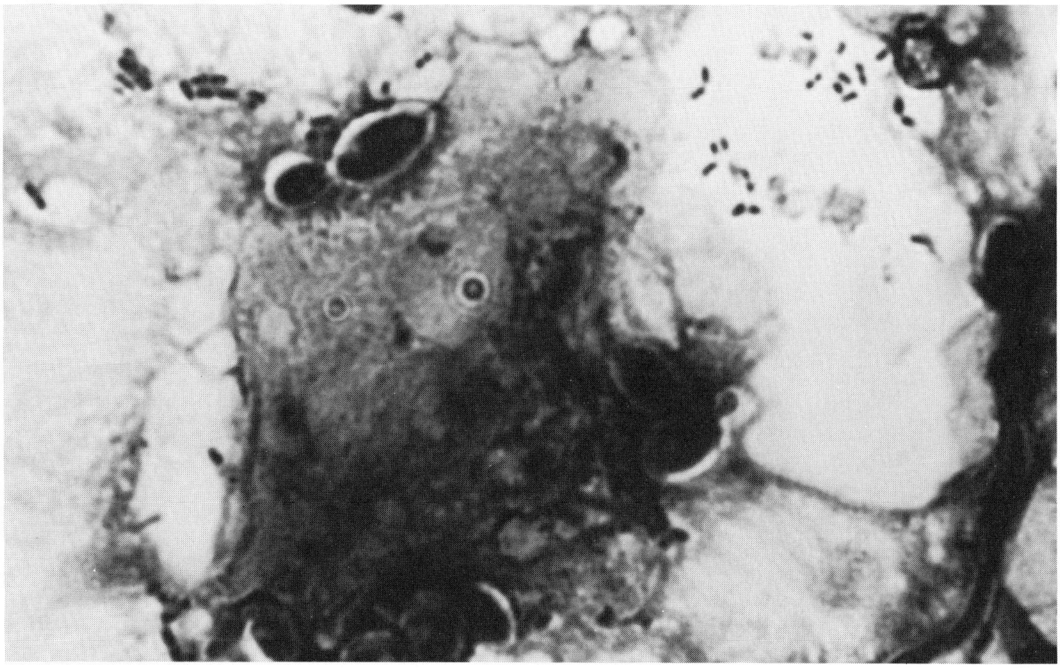

**Figure 17–10.** Crop fluid from a cockatiel (*Nymphicus hollandicus*) with candidiasis. The cytology reveals numerous budding yeasts (Diff-Quik, ×1000).

true mycelia) (Fig. 17–11).[8] Cytologic evidence of trichomoniasis is the presence of piriform flagellate protozoa with an undulating membrane (Fig. 17–12). The trichomonads are best identified by a wet mount preparation. *Capillaria* involvement of the upper alimentary tract can be detected by the presence of the small nematode parasites when viewed in a wet mount preparation under scanning magnification (45×). Female capillariae can be crushed to release the typical double operculated ova. Lesions due to hypovitaminosis A can resemble lesions caused by infectious agents. Cytologic examination usually reveals a marked number of cornified squamous epithelial cells without evidence of inflammation.

## CYTOLOGY OF THE UPPER RESPIRATORY TRACT

The nasal and infraorbital sinuses of birds are lined by stratified squamous epithelium (nonkeratinizing), and the trachea and primary bronchi are lined with pseudostratified ciliated columnar epithelium with goblet cells.[1] The mucosa of the syrinx (located at the junction of the trachea and primary bronchi) consists of either bistratified squamous or columnar epithelium.[1] Smears made from normal sinuses reveal few cells (occasional squamous epithelial cells) and a few extracellular bacteria. Normal tracheal and bronchial cytology is also poorly cellular with an occasional ciliated respiratory epithelial cell and goblet cell. Respiratory epithelial cells exfoliate singly or in clusters of varying sizes and shapes. These cells have an abundant basophilic granular cytoplasm with fine eosinophilic cilia located on the margin opposite the nucleus (Fig. 17–13). Goblet cells exfoliate singly or in clusters and vary in size and shape. They have abundant cytoplasm with eosinophilic granules and have no cilia (Fig. 17–14). The nucleus is large, oval, and eccentrically positioned within the cell. Goblet cells produce mucin. An occasional leukocyte or mesothelial cell may be found in the normal upper respiratory tract cytology.

Inflammatory lesions of the upper respiratory tract often involve exudative fluid accumulation

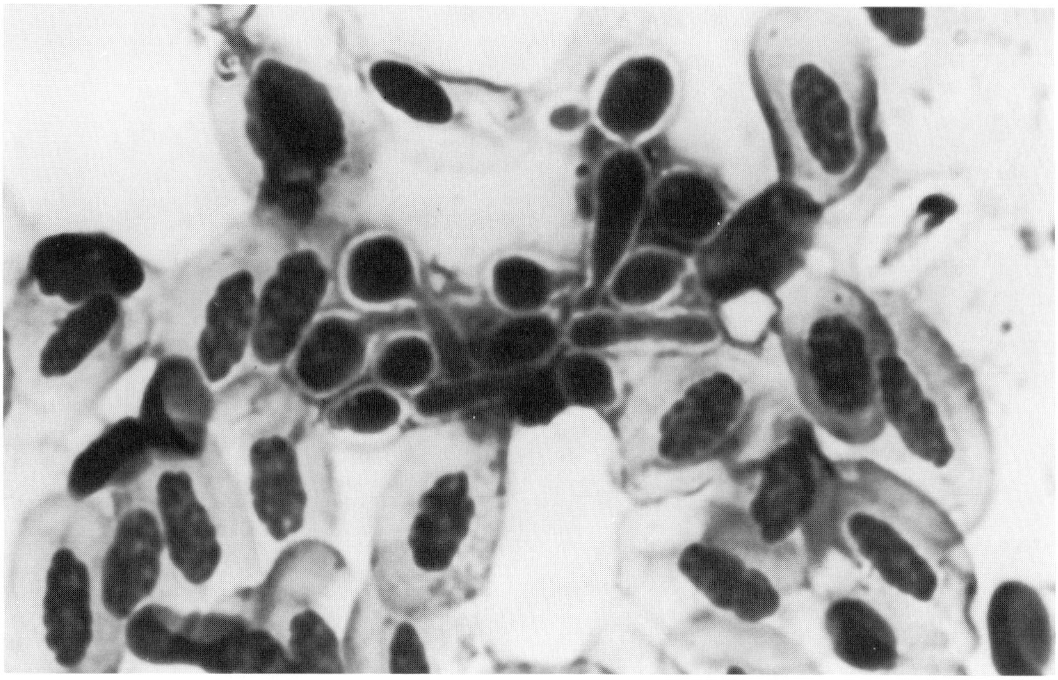

**Figure 17–11.** Impression of the lung from a cockatiel *(Nymphicus hollandicus)* with candidiasis. Note the pseudohyphae and budding yeasts (Diff-Quik, ×1000).

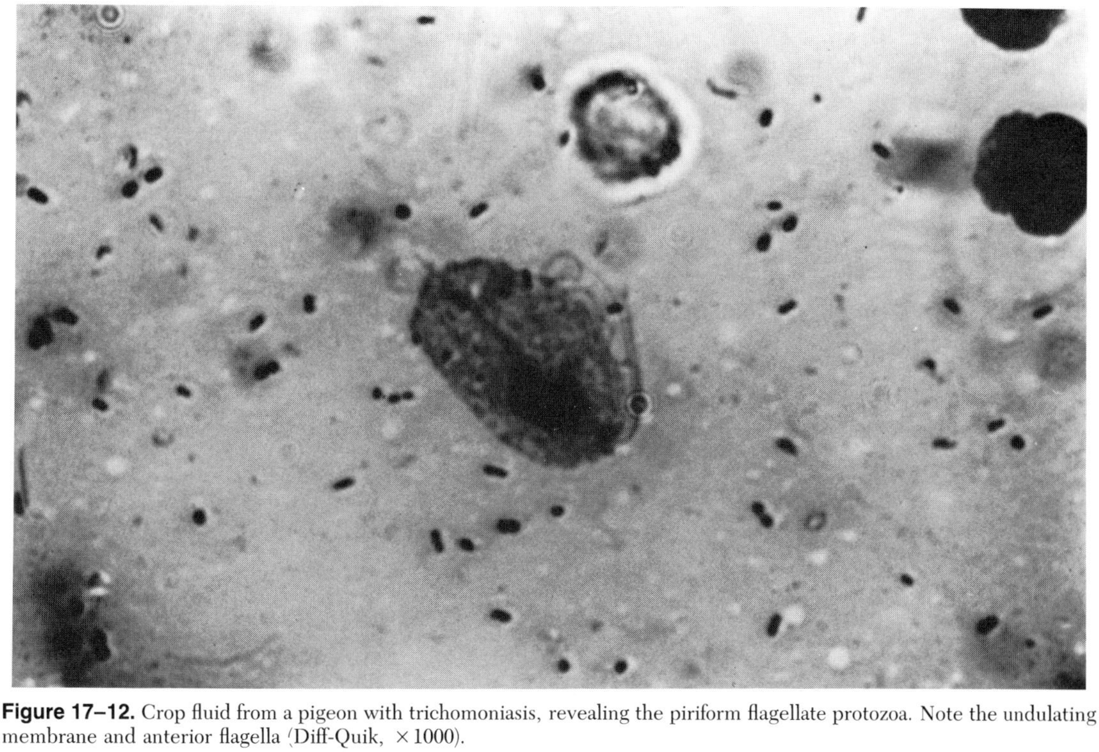

**Figure 17–12.** Crop fluid from a pigeon with trichomoniasis, revealing the piriform flagellate protozoa. Note the undulating membrane and anterior flagella (Diff-Quik, ×1000).

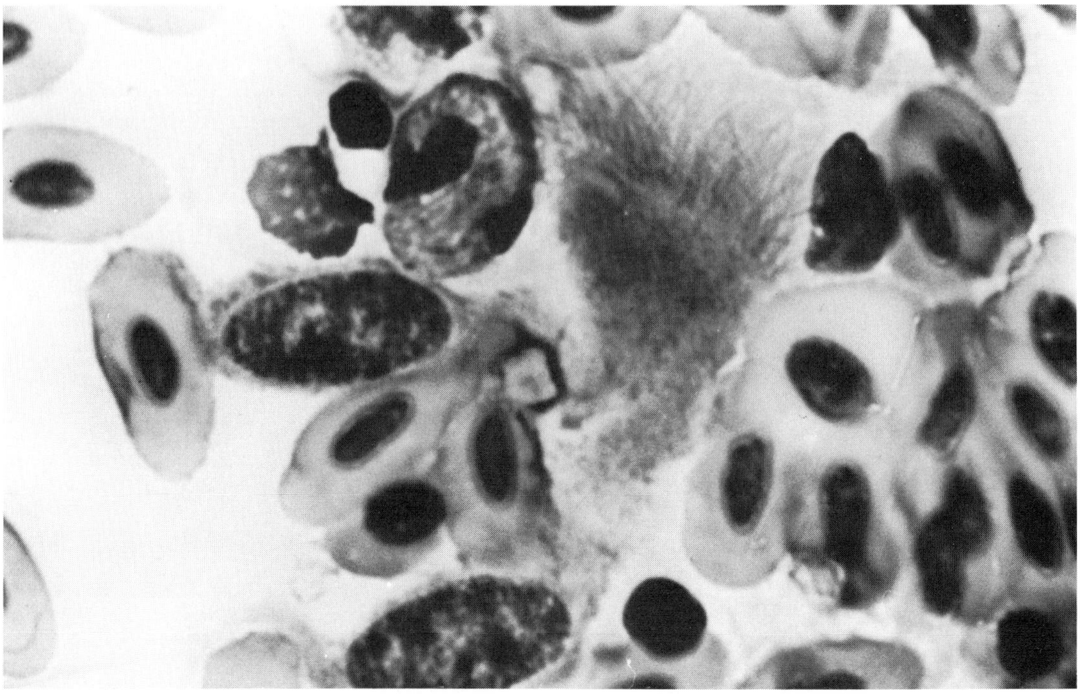

**Figure 17–13.** A respiratory epithelial cell from a turkey *(Meleagris gallopavo)*. Note the eosinophilic cilia on one pole of the cell and the oval nucleus on the opposite pole (Diff-Quik, ×1000).

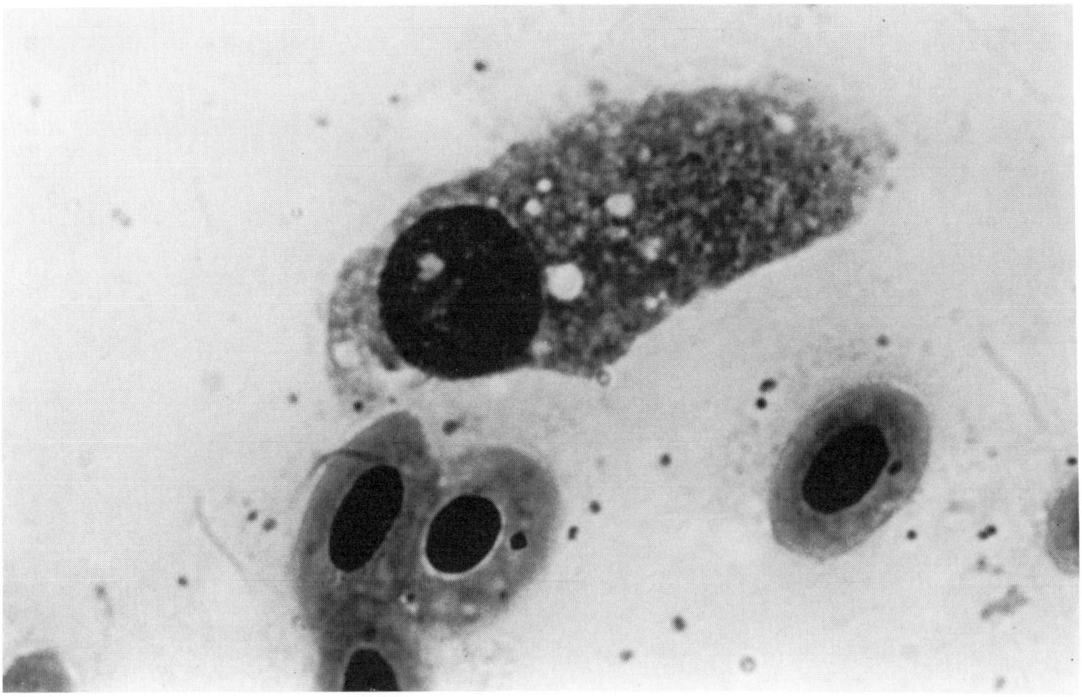

**Figure 17–14.** A respiratory goblet cell from the trachea of a common night hawk *(Chordeiles minor)* (Diff-Quik, ×1000). (From Campbell, T.W.: Diagnostic cytology in avian medicine. Vet. Clin. North Am. [Small Anim. Pract.], 14:317–344, 1984.)

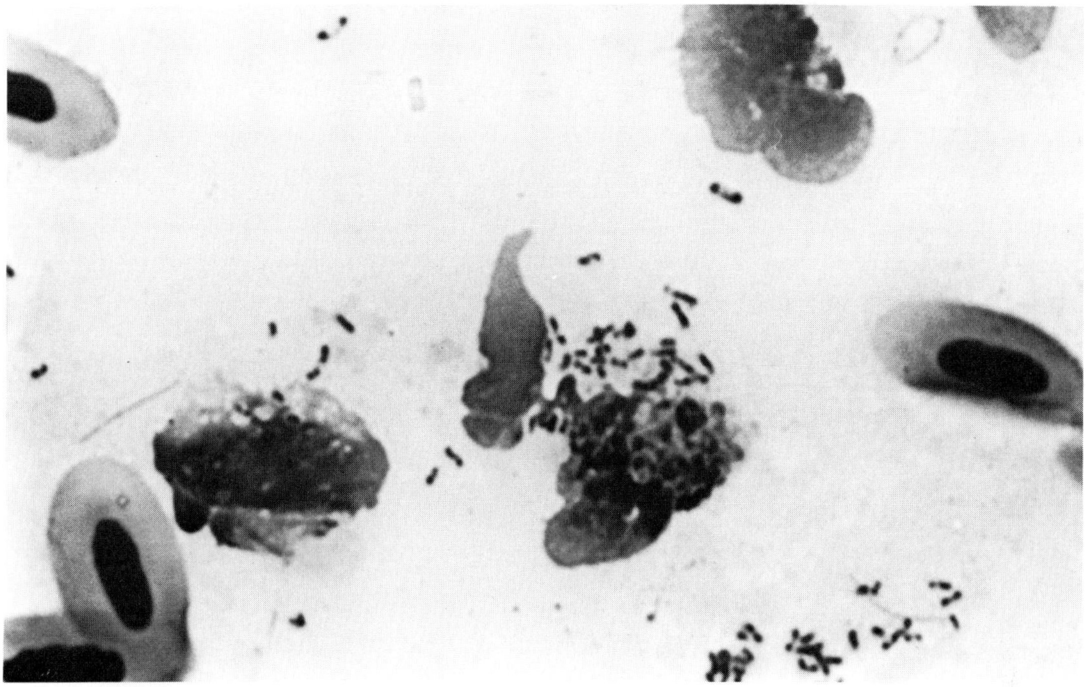

**Figure 17–15.** Septic sinusitis in an Orange-wing Amazon Parrot *(Amazona amazonica)* showing degenerate leukocytes and numerous bacteria (Diff-Quik, ×1000).

(i.e., sinusitis). The cytology reveals a variable number of inflammatory cells, degenerate epithelial cells, and occasionally the causative agent (i.e., bacteria) (Fig. 17–15). Many plasma cells indicate a chronic disorder. Avian chlamydiosis (psittacosis) can be detected by demonstrating the small coccoid intracytoplasmic inclusions within epithelial cells or leukocytes. These are best identified by special stains (Giménez and Macchiavello's stains). Mycotic infections may show a chronic inflammatory or granulomatous response to fungal elements (e.g., spores and hyphae). The fungal elements are best demonstrated with new methylene blue stain rather than alcohol-based stains (e.g., Wright's stain).[11]

## CYTOLOGY OF THE LOWER RESPIRATORY TRACT (LUNGS AND AIR SACS)

Impression cytology of lung and air sac lesions may be useful in the diagnosis of chlamydial, fungal, and bacterial respiratory diseases. The lung air capillaries and air sacs are lined by simple squamous epithelium, and the normal cytology reveals an occasional squamous epithelial cell on a clear background. Owing to the vascular nature of the lungs, it is difficult to obtain imprints without peripheral blood contamination. Inflammatory lesions (i.e., pneumonia and air sacculitis) are indicated by a large number of inflammatory cells and a variable amount of background debris. Phagocytized bacteria indicate a bacterial etiology, whereas the presence of fungal elements confirms mycotic involvement. Bacteria and fungi are frequently secondarily involved with viral and nutritional diseases. *Aspergillus* is a common fungal pathogen of the lower respiratory tract in birds. It is detected cytologically by the presence of branching septate hyphae and occasionally spores and conidiophores (Fig. 17–16). Mycotic nodules can be crushed to expose the fungal elements. Chlamydial air sacculitis can be determined by specialized staining of air sac imprints. The small chlamydial elementary bodies (200 to 300 μm) stain red with Giménez and Macchiavello's stains and purple with Giemsa stain (Fig. 17–17).[7] The spherical elementary bodies are found intracellularly and extracellularly. The larger initial bodies (900 to 1000 μm) stain blue with Macchiavello's and Giemsa stains and occur only intracellularly.[7] Neoplastic lesions involving the lungs can be detected by demonstrating cells with features of neoplasia in the lung imprints.

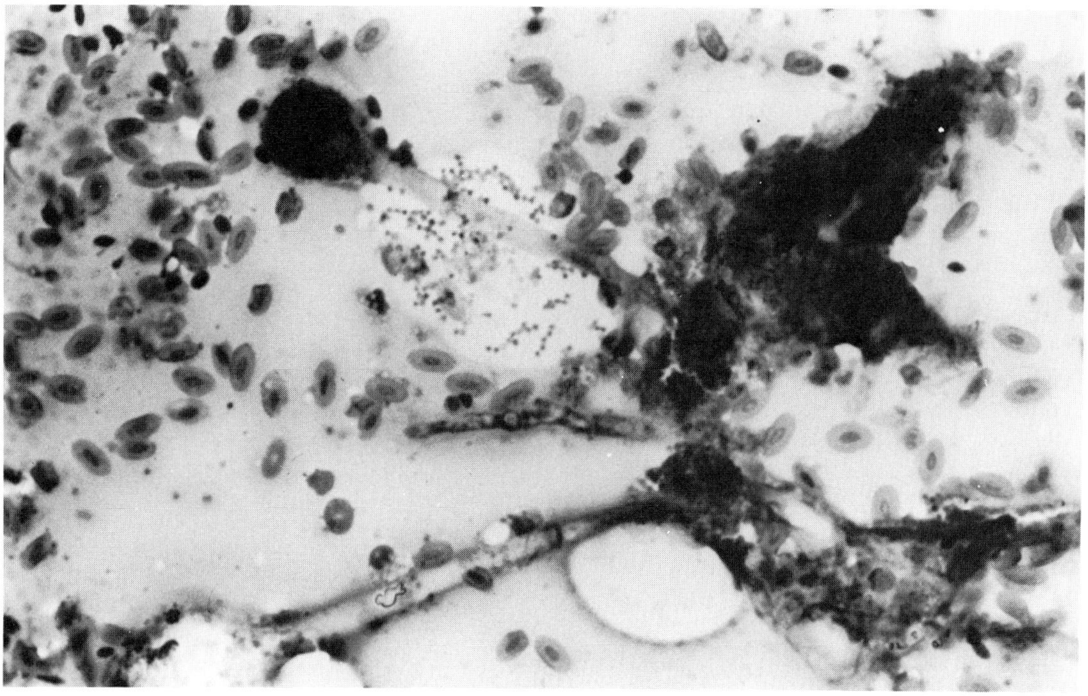

**Figure 17–16.** Impression cytology of the lungs from an Orange-wing Amazon Parrot *(Amazona amazonica)*, revealing branching septate hyphae and conidiophores with many spores (Diff-Quik, ×200). A diagnosis of respiratory aspergillosis was made.

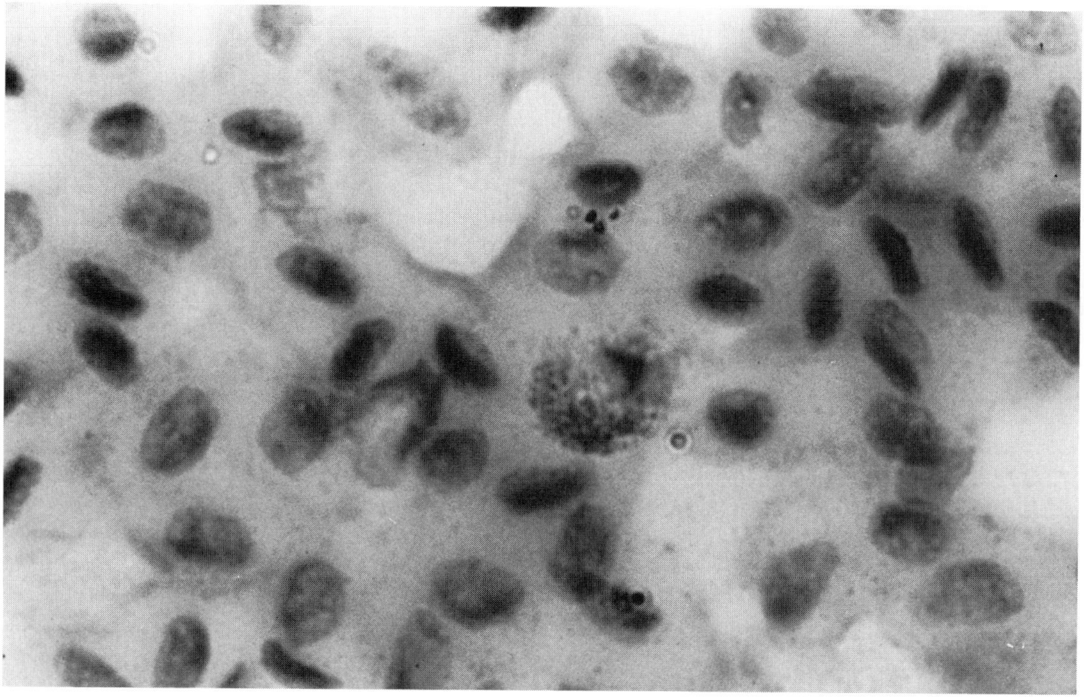

**Figure 17–17.** Impression cytology of the air sacs from an African Grey Parrot *(Psittacus erithacus)* with avian chlamydiosis. The slide stained with Giménez stain revealed the red intracytoplasmic inclusions of *Chlamydia psittaci* (Giménez stain, ×1000).

## CYTOLOGY OF ABNORMAL ABDOMINAL FLUIDS

Abdominal fluid from birds with ascites, peritonitis, hemoperitoneum, or other abnormal fluid accumulations should be examined cytologically. Ascitic fluid often has features of transudates or modified transudates. Birds with peritonitis often have an exudative fluid accumulation. Hemorrhagic effusions are seen in birds that have suffered from a hemoperitoneum. Yolk peritonitis is a common disorder of caged birds and is a result of yolk material free in the abdominal cavity. The resulting effusion is exudative in character but will vary depending on the amount of material in the abdomen, length of duration, and presence of bacteria.[2] The background material is frequently heavy and granular and may contain fat droplets. The fat droplets can be detected using Sudan III stain (see Appendix), giving the droplets a red-orange color (Fig. 17–18). Many macrophages are present with a variable number of heterophils and erythrocytes (Fig. 17–19). If a bacterial peritonitis occurs, many degenerate heterophils will be seen. Chronic yolk effusions are indicated by many lymphocytes and plasma cells.

## CYTOLOGY OF SYNOVIAL FLUID

Normal avian joints contain very little fluid; therefore, any fluid-distended joint should be examined. Normal synovial fluid is poorly cellular and contains an occasional mononuclear cell (primarily macrophages). The background is normally thick and granular. Septic synovial fluids have exudative features with intracellular bacteria. Traumatic arthritis is indicated by a variable number of erythrocytes and leukocytes. A chronic traumatic joint lesion reveals an increased number of macrophages showing erythrophagocytosis. Degenerative joint fluids are highly cellular, with a predominance of mononuclear cells. Articular gout is indicated by a thick white to yellow fluid containing numerous inflammatory cells and urate crystals. The urate crystals can be detected as birefringent needle-like crystals under polarized light. Adult filarial worms or their microfilariae have been found in joint and periarticular fluid (e.g., tendon sheaths) of birds.[5]

## CYTOLOGY OF THE SKIN AND SUBCUTIS

Masses involving the skin should be examined cytologically. Bacterial infections (i.e., ab-

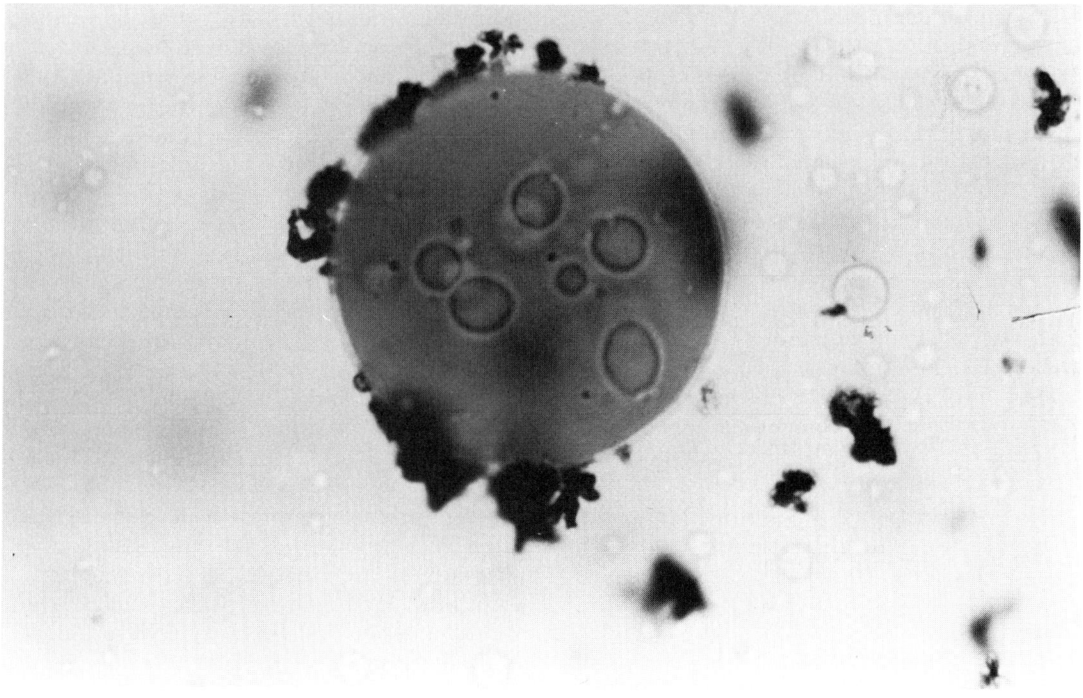

**Figure 17–18.** A large lipid droplet from an impression smear of a lipoma in a budgerigar *(Melopsittacus undulatus)* (Sudan III, ×400). Fat droplets in avian tissues can be identified using fat stains.

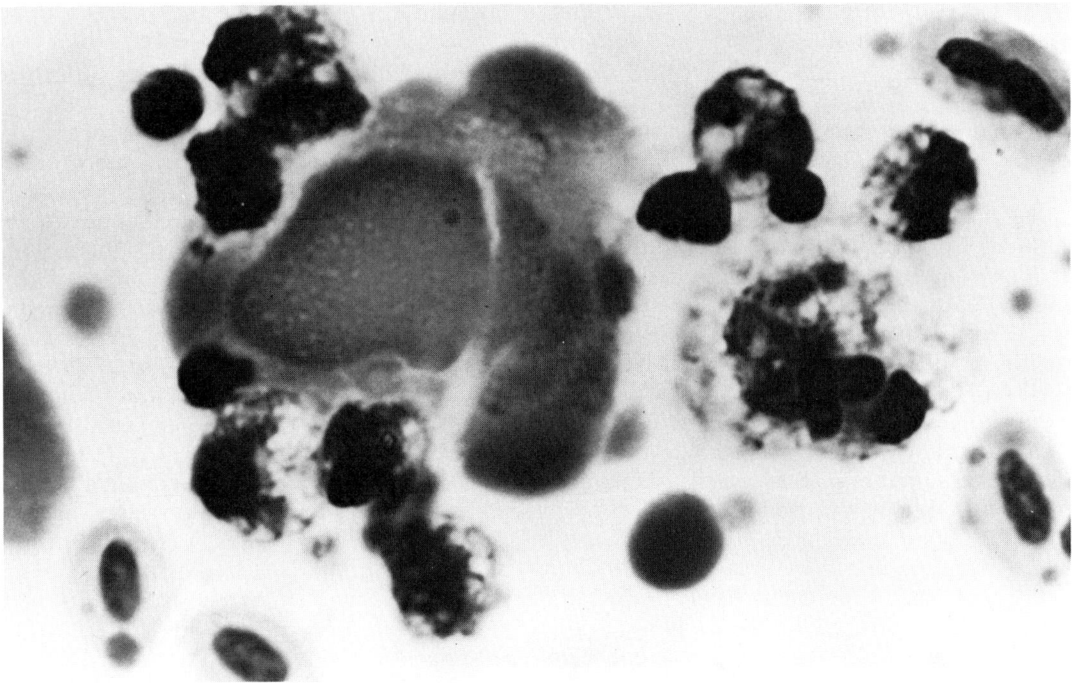

**Figure 17–19.** Yolk peritonitis in a budgerigar *(Melopsittacus undulatus)*. The cytology of the abdominal fluid reveals a large number of leukocytes and erythrocytes and large basophilic protein globules (Wright's stain, ×1000).

scessations) show large numbers of inflammatory cells (primarily heterophils), cell debris, and intracellular bacteria. Chronic abscessations often reveal a marked amount of amorphous background material with only a few degenerate cells present. Active lesions will show a marked inflammatory response at the margins. Fungal involvement may show a granulomatous reaction and reveal fungal elements. Foreign bodies also produce granulomatous reactions. Pox lesions can be detected cytologically by swollen epithelial cells with round intracytoplasmic inclusions (Bollinger bodies) that tend to force the cell nucleus to one side of the cell. Subcutaneous lipomas are indicated by the presence of adipose cells and numerous background fat droplets. The fat droplets can be detected using Sudan stains. The adipose cells are round with foamy cytoplasm that contains large secretory (fat) vacuoles (Fig. 17–20). The nucleus is usually pushed to one margin of the cell.

## CYTOLOGY OF THE LIVER

Liver imprints are usually highly cellular, with a heavy background containing many free nuclei and cell fragments. Normal hepatocytes undergo rapid degeneration, and imprints should be made from fresh liver. Normal hepatocytes are large cells with abundant basophilic, granular cytoplasm which occur in sheets or clusters (Fig. 17–21). Often the cytoplasmic borders between cells occurring in groups are indistinct. The nuclei are round to oval and uniform in appearance among hepatocytes. A prominent nucleolus is usually present. An occasional binucleated hepatocyte or normal mitotic figure can be seen. A variable number of lymphocytes, plasma cells, macrophages, and spindle-shaped stromal cells are present in liver imprints. Macrophages often contain iron pigment from erythrophagocytosis. Lymphoid aggregates are common in avian liver; therefore, imprints may reveal large numbers of mature lymphocytes. A large number of plasma cells may indicate lymphoid reaction to a systemic antigen (Fig. 17–22). Lymphoid neoplasia involving the liver would reveal a large number of immature lymphocytes. Degenerate hepatocytes are suggestive of postmortem autolysis or hepatic disease. Certain hepatotoxins (e.g., aflatoxins) may cause hepatocyte degeneration. Fatty livers are indicated by swollen hepatocytes with lipid material. Inflammatory lesions show numerous inflammatory cells and degenerate hepatocytes. A neoplastic involvement of the liver is revealed by cells with neoplastic characteristics.

Avian tuberculosis (primarily caused by *My-*

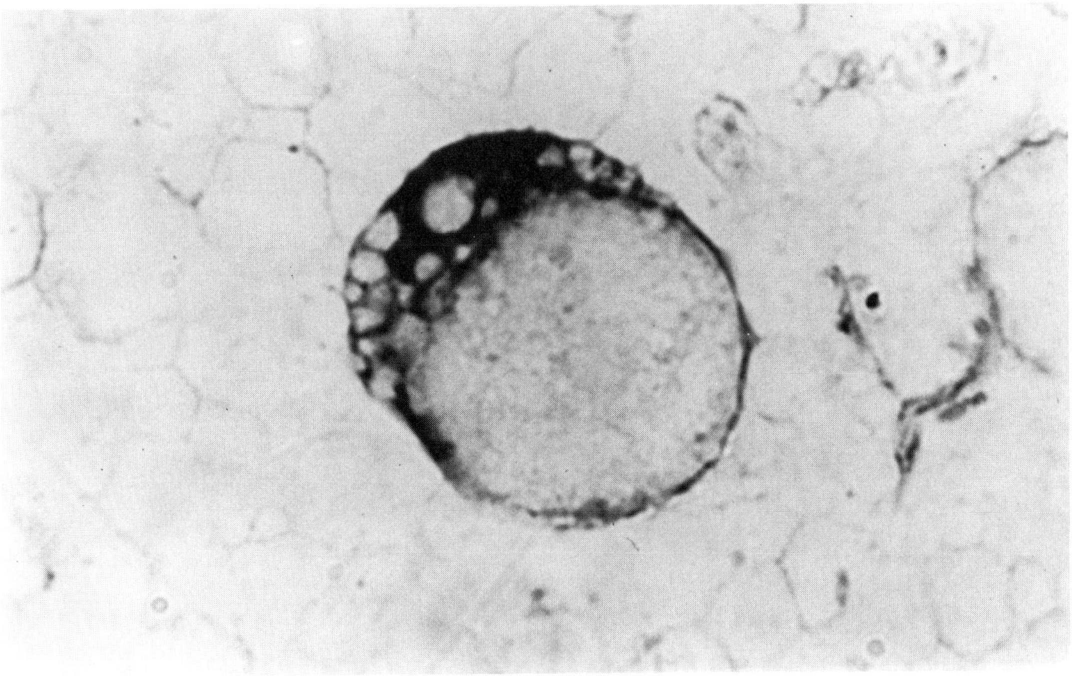

**Figure 17–20.** A large adipose cell from the impression of a lipoma in a budgerigar *(Melopsittacus undulatus)* (Diff-Quik, ×1000).

*cobacterium avium*) affects the liver in most birds but can involve the lungs, spleen, bones, and intestines. Cytology of the lesions may reveal a granulomatous pattern or numerous macrophages and many acid-fast bacilli when stained with Ziehl-Neelsen acid-fast stain (see Appendix at end of chapter).

## CYTOLOGY OF THE SPLEEN

The avian spleen contains hematopoietic and lymphoid tissue. Normal splenic cytology shows a marked amount of peripheral blood, cellular debris, and free nuclei. Occasionally, groups of lymphoid cells in various stages of maturation

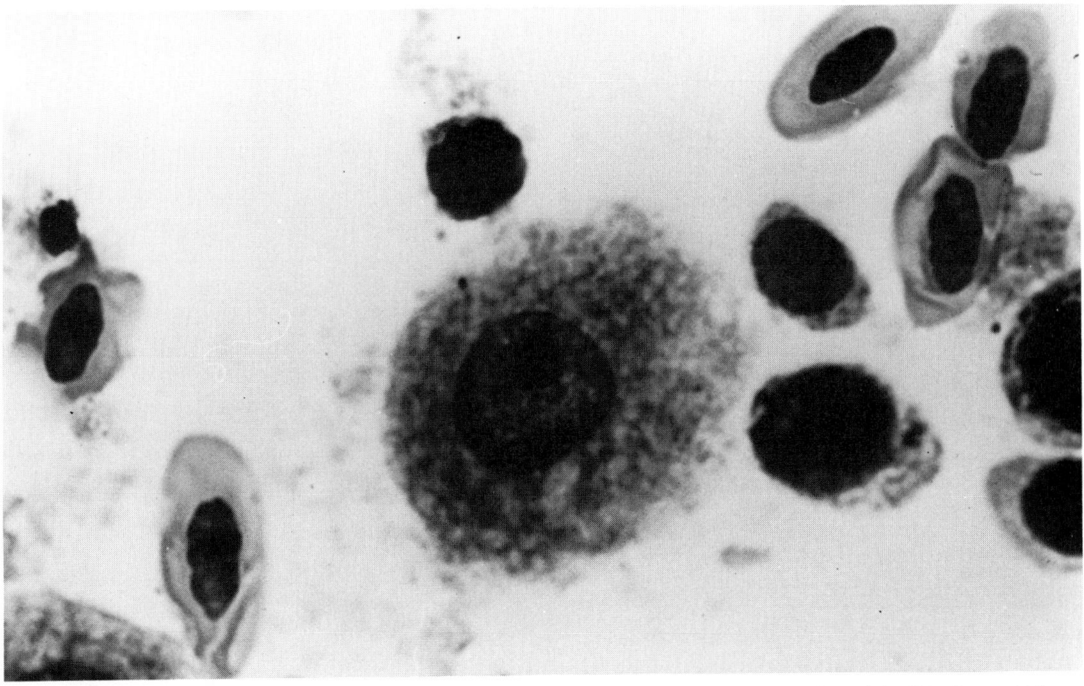

**Figure 17–21.** A normal avian hepatocyte from an African Grey Parrot *(Psittacus erithacus)* (Diff-Quik, ×1000).

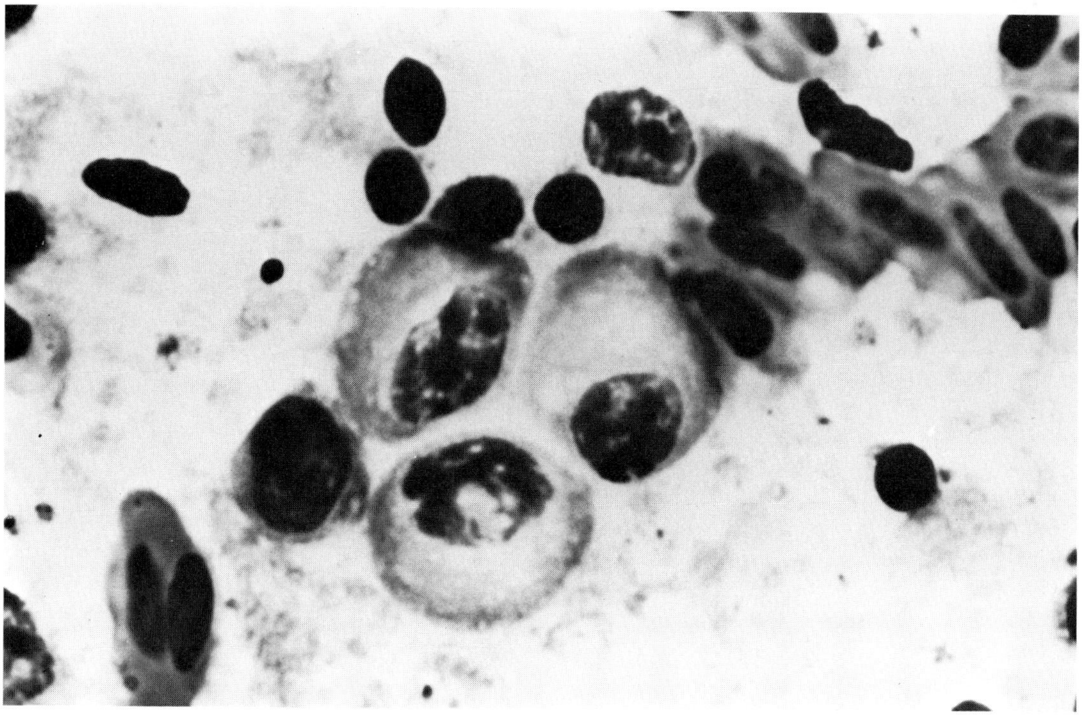

**Figure 17–22.** Plasma cells in an impression smear from the liver of an African Grey Parrot *(Psittacus erithacus)* (Diff-Quick, ×1000).

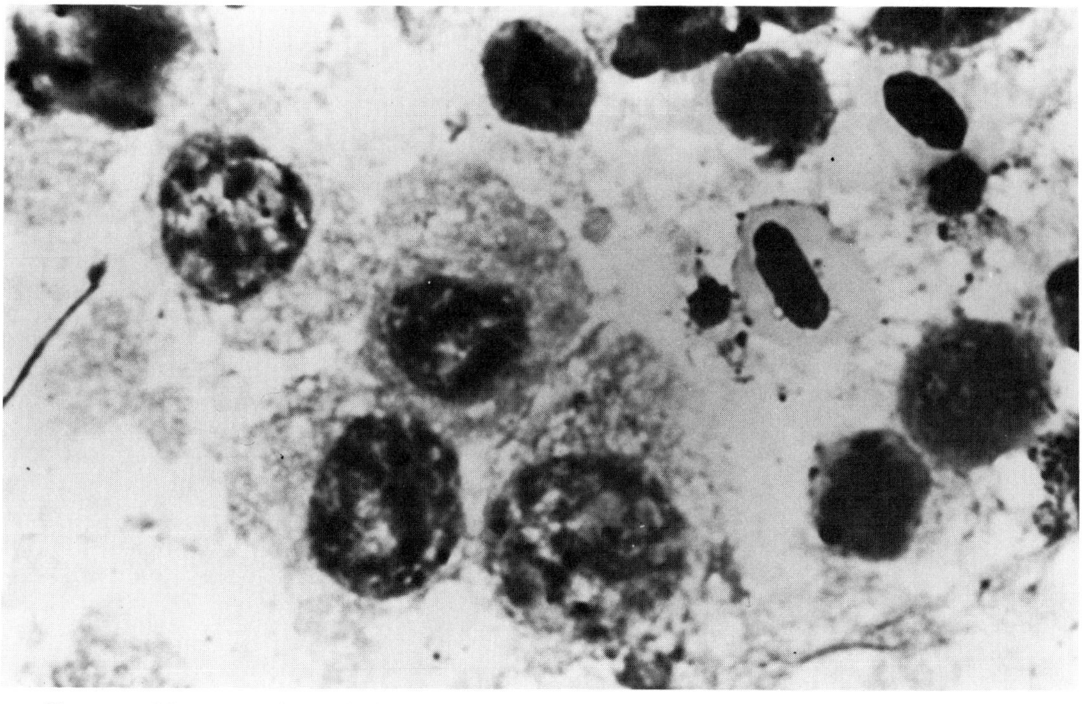

**Figure 17–23.** Normal splenic cells from a Peach-faced Lovebird *(Agapornis roseicollis)* (Diff-Quik, ×1000).

can be seen. Macrophages often show erythrophagocytosis. The splenic stromal cells have a pale blue cytoplasm with an indistinct cytoplasmic border (Fig. 17–23). The nucleus of these cells is eccentrically located within the cells and round to oval with a coarse chromatin pattern. Impression cytology of the spleen is a good source for the detection of chlamydia and tubercle bacilli.

## ARTIFACTS

The veterinarian should be aware of the artifacts that occur in cytologic specimens. Excessive peripheral blood contamination of a specimen will dilute and mask the diagnostic cells, which will make interpretation difficult. The presence of thrombocytes in a smear may indicate peripheral blood contamination of the sample or an active hemorrhage within the lesion. Evidence of erythrophagocytosis indicates that all or part of the erythrocytes in the sample were present in the lesion at the time of sampling and are not contaminants of the sample. Stain precipitate on the smear is a common artifact and should not be confused with bacteria or cellular inclusions. Stain precipitate is variable in size and shape and will be more refractile than bacteria or most cellular inclusions. These precipitates appear to sit on top of the cellular plane on close examination. Ruptured cells and nuclei are found in many cytologic smears. The nucleoprotein strands that are created by streaking the ruptured nuclei across the slide should not be confused with fungal elements. These strands are usually basophilic, vary in thickness, and can often be traced back to a ruptured nucleus. Exogenous materials can also be found in cytologic specimens. Common exogenous contaminants include talc and starch crystals from examination gloves, cotton or plant fibers, pollen, and feather fragments. These should not be confused with any pathologic agents. Exposure of cytology slides to certain chemical fumes (e.g., formalin) often alters the staining of the smears. Exposure to formalin will cause a blue-green discoloration of erythrocytes, prevent staining of leukocyte granules, and create a halo around the cells on a Wright's stained smear. Artifacts commonly occur in smears that are too thick. Thick smears do not allow the cells to properly expand on the slide and may prevent proper staining.

## APPENDIX OF CYTOLOGIC STAINS

### Harleco Diff-Quik Stain

#### *Staining Solutions*

A Diff-Quik stain set contains three solutions.

1. A methanol-based fixative solution is provided.
2. Solution I is a buffered eosin Y solution.
3. Solution II is a buffered solution of methylene blue and azure A dyes.

#### *Staining Procedure*

1. The air-dried slide is dipped into the fixative for five one-second dips and the excess is allowed to drain off.
2. The slide is placed into Solution I for five one-second dips and the excess is drained off.
3. The slide is placed into Solution II for five one-second dips and the excess is drained off.
4. The slide is rinsed in distilled water and allowed to air dry.

#### *Staining Results*

Diff-Quik stain produces results similar to those of Wright's stain (see Wright's Stain). The above procedure can be modified according to the desired staining results or the smear thickness. The overall staining can be intensified by increasing the number of dips into Solutions I and II. Likewise, a paler stain is obtained with fewer dips (a minimum of three dips in each solution is required). Eosinophilic staining is enhanced by increasing the number of dips in Solution I, and basophilic enhancement is achieved by increasing the number of dips in Solution II.

### Wright's Stain

#### *Staining Solutions*

1. Wright's stain solution. The stain is prepared by dissolving 0.2 to 0.3 gm of Wright's stain powder in 100 ml methyl alcohol (absolute, neutral, acetone-free). The solution is allowed to stand for five to seven days before using.
2. Wright's buffer. The buffer is prepared by dissolving 3.80 gm $Na_2HPO_4$ (dibasic) and 5.47 gm $KH_2PO_4$ (monobasic) in 500 ml of

distilled water. The solution is brought to 1000 ml with distilled water and adjusted to about pH 6.5.

### *Staining Procedure*

1. The air-dried smear is flooded with Wright's stain and allowed to stand for one to three minutes.
2. An equal amount of Wright's buffer is added and mixed by gently blowing on the fluid surface until a metallic green sheen appears. The slide is allowed to stand twice as long as the undiluted stain (usually six minutes).
3. The slide is rinsed under running tap water and allowed to air dry.

### *Staining Results*

In properly stained slides, the erythrocytes have a dark purple nucleus and yellowish red cytoplasm. Heterophils have a pale to dark violet nucleus and a colorless cytoplasm with red-orange rod-shaped granules. Eosinophils have a pale to dark violet nucleus, pale blue cytoplasm, and red to red-orange round cytoplasmic granules. Basophils have a purple nucleus and dark purple cytoplasmic granules (may vary with species). Lymphocytes have dark purple nuclei and a pale blue cytoplasm. Thrombocytes have dark purple nuclei and a colorless cytoplasm that often contains red granules.

## Giemsa Stain

### *Staining Solutions*

Giemsa stain solution. Dissolve 0.5 gm of Giemsa stain powder in 33 ml of glycerol by allowing to stand for 1.5 to 2 hours at 55 to 60°C. Add 33 ml of methanol and allow to stand for at least 24 hours. The staining solution is made by adding 1.5 ml of the above solution to 30 ml of distilled water.

### *Staining Procedure*

1. Air-dried smears are fixed in methyl or ethyl alcohol for two to seven minutes.
2. After drying, the slide is immersed in the Giemsa stain for 15 to 40 minutes. (Several test samples with varying staining times may be required to obtain the best stain.)
3. Rinse the slide in tap water and allow to air dry.

### *Staining Results*

Giemsa stain provides results similar to Wright's stain (see Wright's Stain) except that the cell nuclei appear reddish purple instead of violet. Chlamydial elementary bodies (200 to 300 μm) stain purple with Giemsa, whereas the larger initial bodies (900 to 1000 μm) stain blue. *Mycoplasma* (0.2 to 0.5 μm) stains pink or purple.

## New Methylene Blue Stain

### *Staining Solutions*

Dissolve 0.5 gm of new methylene blue in 99.0 ml of 0.85 per cent saline and add 1.0 ml of 40 per cent formalin. The solution should be filtered and stored in a dark container.

### *Staining Procedure*

The smear should be allowed to dry completely on the slide before staining. Place a small drop of the new methylene blue stain in the smear and add a coverslip.

### *Staining Results*

Granulocytes stained with new methylene blue have purple nuclei and pale blue cytoplasm. The erythrocytes have purple nuclei and a distinct cytoplasmic border, since new methylene blue does not stain hemoglobin. Most red blood cells contain a variable amount of blue-staining reticulum, and frequently the cytoplasm stains a pale greenish blue. New methylene blue is water-soluble, thus providing a more distinctive chromatin and nucleolar pattern to cell nuclei than alcohol-based stains, which tend to "smudge" the nuclear structure. New methylene blue also outlines lipid droplets that are dissolved in alcohol-based stains. Heterophil and eosinophil granules are not stained by new methylene blue but can be seen outlined within the cytoplasm. Fibrin is outlined (not stained) in smears using new methylene blue, which cannot be demonstrated using alcohol-based stains.

## Gram's Stain

### *Staining Solutions*

1. Crystal violet stain. A stock solution is prepared by dissolving 2 gm of crystal violet in

20 ml of 95 per cent ethanol and mixing with 80 ml of 1 per cent aqueous ammonium oxalate. The solution can be stored for months.
2. Gram's iodine solution. This solution is prepared by dissolving 1 gm of iodine ($I_2$) and 2 gm of potassium iodide (KI) in 300 ml of distilled water. This solution should be prepared fresh every three weeks.
3. 95 per cent ethyl alcohol.
4. Safranin stain. A solution is prepared by dissolving 2.5 gm of safranin 0 in 100 ml of 95 per cent ethanol. Ten milliliters of this solution is diluted with 90 ml of distilled water to make the staining solution (0.25 per cent safranin solution).

#### *Staining Procedure*

1. Smears made on glass slides are allowed to dry before heat fixation. Heat fixation is performed by gently passing the slide, smear side up, through a low flame from a Bunsen burner.
2. The smear is flooded with crystal violet for 1 minute.
3. Gently rinse in tap water for 1 to 5 seconds.
4. Flood the slide with Gram's iodine solution for 1 minute.
5. Gently wash in tap water for 1 to 5 seconds.
6. Decolorize the smear by applying 95 per cent ethyl alcohol until stain is no longer being eluted from the smear (15 to 30 seconds).
7. Gently rinse in tap water for 1 to 5 seconds.
8. Flood the slide with safranin for 1 to 2 minutes.
9. Wash the slide in tap water for 1 to 5 minutes and allow to air dry.

#### *Staining Results*

The Gram's stain technique is variable owing to the nature of the staining procedure. Reliable staining can be achieved if known gram-positive and gram-negative organisms from young cultures are added to the slide for comparison. A commercially prepared Gram check slide (Scott Laboratories) makes this quality control more practical. Properly stained slides show gram-positive organisms as deep violet and gram-negative organisms as red. Most eukaryotic cells (except for yeasts) appear gram-negative. Smears can vary in thickness on the same slide, which can affect the staining quality of different parts of the smear. Excessive decolorization may occur in very thin areas, causing gram-positive organisms to appear gram-negative. Thicker areas may be poorly decolorized, causing gram-negative organisms to appear gram-positive. Therefore, thinner smears that are even in thickness provide better Gram's staining material than thick or uneven smears.

### Acid-Fast Stain (Ziehl-Neelsen carbol-fuchsin)

#### *Staining Solutions*

1. Ziehl-Neelsen carbol-fuchsin. Three grams of basic fuchsin are dissolved in 100 ml of 90 per cent ethyl alcohol. A 5 per cent phenol solution is prepared by dissolving 5 gm phenol in 100 ml distilled water. The Ziehl-Neelsen carbol-fuchsin is prepared by mixing 10 ml of the alcoholic basic fuchsin with 90 ml of 5 per cent phenol and allowed to stand for 24 hours. The solution is filtered prior to use.
2. Acid alcohol solution. This solution is prepared by mixing 2.0 ml concentrated hydrochloric acid with 98 ml of 95 per cent ethyl alcohol.
3. Methylene blue stain. A saturated stain solution is prepared by adding 1.5 gm of methylene blue to 100 ml of 95 per cent ethyl alcohol. The powder is dissolved into the alcohol by rubbing small amounts of powder into increasing amounts of alcohol. Once the powder is dissolved, 100 ml distilled water, 0.1 ml 10 per cent potassium hydroxide, and 30 ml of the saturated alcoholic solution of methylene blue are added together. Prior to use, the solution is filtered and diluted 1:20 with distilled water.

#### *Staining Procedure*

1. The smear is allowed to dry, then heat fixed (see the Gram's Stain staining procedure).
2. The smear is flooded with the carbol-fuchsin stain and heated on a water bath (gently steaming) for 3 to 5 minutes.
3. The slide is rinsed with tap water.
4. Acid alcohol is used to decolorize the smear until little of the red color remains visible to the unaided eye.
5. The slide is rinsed twice in tap water.
6. The dilute aqueous methylene blue stain is flooded on the smear for 1 minute (depending on the thickness of the smear).
7. The slide is rinsed with tap water and allowed to air dry.

### Staining Results

This stain is used to detect the presence of tubercle bacilli *(Mycobacterium* sp.) in tissues and exudates. These organisms are differentially stained from other bacteria, tissue cells, and tissue debris. The tubercle bacilli are distinctly red, whereas other bacteria, leukocytes, and cellular debris appear blue.

## Modified Giménez Stain

### Staining Solutions

1. Carbol basic fuchsin stock solution. Dissolve 10 gm of basic fuchsin in 100 ml ethyl alcohol. To 100 ml of the alcoholic fuchsin solution add 250 ml of 4 per cent phenol. Next add 650 ml of distilled water and incubate at 37°C for 48 hours. The stock solution can be stored for several months.
2. 0.1 ml phosphate buffer (pH 7.45). The buffer is prepared by mixing 17.6 ml of 0.1 M monobasic sodium phosphate ($NaH_2PO_4 \cdot H_2O$) and 82.4 ml of 0.1 M anhydrous dibasic sodium phosphate ($Na_2HPO_4$).
3. Working carbol basic fuchsin stain. Mix 4 ml of the carbol basic fuchsin stock solution with 10 ml of the 0.1 M phosphate buffer and filter twice before staining slides. Filtration minimizes the amount of red stain precipitate on the smear that may make microscopic interpretations difficult.
4. Malachite green stain. A 0.8 per cent malachite green stain is prepared by dissolving 0.8 gm of malachite green oxalate in 100 ml of distilled water.

### Staining Procedure

1. Impression smears are allowed to dry and are heat fixed over a flame (see Gram's Stain staining procedure).
2. The slide is flooded with the working solution of carbol basic fuchsin and allowed to stand for 1 to 2 minutes.
3. Wash the slide thoroughly in tap water.
4. Flood the slide with malachite green stain for 6 to 9 seconds.
5. Rinse the slide in tap water and recover with malachite green stain for an additional 6 to 9 seconds.
6. Wash the slide with tap water and allow it to air dry.

### Staining Results

The Giménez stain is designed to detect chlamydial organisms in tissues and exudates. The small chlamydial elementary bodies (200 to 300 μm) stain red against a blue-green cellular background. The organisms may be found both intracellularly and extracellularly.

## Macchiavello's Stain

### Staining Solutions

1. Basic fuchsin stain. A 0.25 per cent solution is prepared by dissolving 0.25 gm of basic fuchsin chloride into 100 ml of distilled water. This solution should be prepared fresh with each day of use.
2. Citric acid solution. A 0.5 per cent citric acid solution is prepared by dissolving 0.5 gm citric acid into 100 ml of distilled water. The citric acid solution should be made fresh when mold growth occurs.
3. Methylene blue stain. A 1 per cent methylene blue solution is prepared by dissolving 1.0 gm of methylene blue chloride in 100 ml of distilled water.

### Staining Procedure

1. Air-dried smears are heat fixed by gently passing the slide over a flame (see Gram's Stain staining procedure).
2. The slide is flooded with the basic fuchsin stain and allowed to stain for 5 minutes.
3. The slide is quickly washed in tap water and dipped 1 to 10 times in the citric acid solution (1 to 3 seconds).
4. The slide is rinsed in tap water.
5. The slide is counterstained with the 1 per cent aqueous methylene blue solution for 20 to 30 seconds.
6. The slide is washed in tap water and allowed to air dry.

### Staining Results

In a properly prepared slide, the chlamydial elementary bodies (200 to 300 μm) stain red and the larger initial bodies (900 to 1000 μm) stain blue. Some smears contain nonchlamydial particles that stain red, making the interpretation difficult. Heterophil and eosinophil granules frequently stain red. *Mycoplasma* colonies may resemble *Chlamydia.* Excessive decolorization with citric acid may decolorize the elementary bodies, making them appear blue.

Therefore, the citric acid decolorization step may be omitted or shortened in some smears.

## Sudan III Stain

### *Staining Solutions*

A 0.7 per cent Sudan III solution is prepared by dissolving 0.7 gm Sudan III in 100 ml propylene glycol. The solution is filtered prior to use.

### *Staining Procedure*

The stain can be applied to a wet or dry smear. A small drop of the stain is placed on the smear and a coverslip is added.

### *Staining Results*

Sudan III stains fat droplets or globules a red-orange color. Cell nuclei usually appear blue, and the cytoplasm stains a pale green. Erythrocyte cytoplasm often appears green.

Commercially prepared individual stains, staining solutions, and staining kits using certified biologic stains are available from:

1. American Scientific Products, a division of American Hospital Supply Corporation, 1430 Waukegan Road, McGaw Park, IL 60085
2. Fisher Scientific, a division of Allied Company, 711 Forbes Ave., Pittsburgh, PA 15219

## REFERENCES TO THE STAINING PROCEDURES

Clark, G. (ed.): Staining Procedures. Baltimore, Williams and Wilkins Company, 1981.

Lillie, R.D. (ed.): H. J. Conn's Biological Stains. Baltimore, Williams and Wilkins Company, 1977.

## REFERENCES

1. Banks, W.J.: Histology and Comparative Organology: A Text-atlas. Baltimore, Williams and Wilkins Company, 1974
2. Campbell, T.W.: Diagnostic cytology in avian medicine. Vet. Clin. North Am. [Small Anim. Pract.], 14:317–344, 1984.
3. Campbell, T.W.: Disorders of the avian crop. Compend. Contin. Educ. Pract. Vet., 5:813–824, 1983.
4. Campbell, T.W., and Dein, F.J.: Avian hematology. Vet. Clin. North Am. [Small Anim. Pract.], 14:223–248, 1984.
5. Clubb, S.L., and Cramm, D.: Blood parasites of psittacine birds; a survey of the prevalence of *Haemoproteus*, microfilaria, and trypanosomes. Proceedings of the Annual Meeting of the American Association of Zoo Veterinarians, 1981, pp. 32–37.
6. Duncan, J.R., and Prasse, K.W.: Cytological examination of the skin and subcutis. Vet. Clin. North Am. [Small Anim. Pract.], 6:637, 1976.
7. Gillespie, J.H., and Timoney, J.F.: Hagan and Bruner's Infectious Diseases of Domestic Animals. Ithaca, Cornell University Press, 1981.
8. Jungerman, P.F., and Schwartzman, R.M.: Veterinary Medical Mycology. Philadelphia, Lea and Febiger, 1972.
9. Kaiser, G.E., and Starzyk, M.J.: Ultrastructure and cell division of an oral bacterium resembling *Alysiella filiformis*. Can. J. Microbiol., 19:325–327, 1972.
10. Lucas, A.J., and Jamroy, C.: Atlas of Avian Hematology. Washington, DC, U.S.D.A. Monograph No. 25, 1961.
11. Perman, V., et al.: Cytology of the Dog and Cat. South Bend, IN, American Animal Hospital Association, 1979.
12. Rebar, A.H.: Handbook of Veterinary Cytology. St. Louis, Ralston Purina Company, 1978.
13. Rebar, A.H., and Boon, G.D.: Diagnostic cytology in small animal practice. Proceedings of the 47th Annual Meeting of the American Animal Hospital Association, 1980, p. 131.
14. Takahashi, M.: Color Atlas of Cancer Ctyology. Philadelphia, J.B. Lippincott, 1971.

# *Chapter 18*

# SEROLOGY

JAMES E. GRIMES

Serologic tests often can aid the veterinary practitioner in making a definitive diagnosis. The tests may help to rule out an infection as well as to confirm it. Tests useful in the diagnosis of avian infections include immunodiffusion (or agar gel diffusion), agglutination, hemagglutination inhibition, virus neutralization, enzyme-linked immunosorbent assay, and complement fixation. Each has its own application in specific situations. Some procedures have yet to be thoroughly tested for their applicability to psittacine and other pet bird species.

In order to interpret serologic test results, it must be kept in mind how and when antibodies are elicited. This may vary among avian species and may depend upon the infecting organism. The detection of antibody activity, except by very sensitive methods, generally is preceded by several days of overt illness. Two serum samples, one taken during the acute phase and another collected during convalescence, are often required for retrospective confirmation of an infection. A four-fold rise in titer in the second sample over that of the first is indicative of a recent infection.

The poultry industry makes extensive use of serologic testing of flocks to detect infections such as those caused by *Mycoplasma*, *Salmonella*, viruses causing avian influenza and infectious bursal disease, and various other agents. For a number of reasons, serology perhaps cannot be as readily applied for testing of psittacine birds and other species kept as pets.

For one reason, flocks of psittacine birds, except those used for breeding, are not generally as static in their composition as poultry flocks. The author also surmises that the veterinary practitioner often may not be presented with a patient until after disease has progressed well past the acute stage. Antibody production already may have occurred to the extent that a moderate or even high level of activity is detectable on the first sample. In addition, little or no antibody activity may be detected in cases of carrier birds. If the disease is occurring in a flock, individuals in various stages of disease may be present, and that must be taken into account.

Another consideration is the size of the bird. Some species may not be large enough to allow collection of an adequate sample from an individual. Fortunately, even for larger birds, most serologic tests are done in microamounts, allowing use of smaller samples.

## SPECIMEN COLLECTION AND SHIPPING

Proper sample collection and handling are of paramount importance because test results are directly dependent upon the serum quality. The individual laboratory should be consulted for sample submission protocol. The author prefers blood collected in tubes containing a blood serum separator because maximum recovery of high quality serum can be effected with their use.

The minimum amount of blood needed is 0.5 ml. This should yield enough serum for testing. Because it may be necessary to retest a sample, it is strongly urged that more than the minimum be collected when possible, such as from larger birds. If it is necessary to bleed a very small bird with a small blood volume, it is acceptable to use an anticoagulant and thus test plasma; however, this should not be done unless absolutely necessary. Combining samples collected over a few days from a sick or small bird may provide a sufficient amount.

The sample should be centrifuged to avoid hemolysis, which can sometimes render a sample useless in a test. Samples collected in a serum separator may be submitted intact following centrifugation. If other collection methods are used, the serum is decanted into a sterile tube. Blood and serum should be handled as aseptically as possible because gross bacterial contamination will either render the sample unsuitable for testing or will produce unreliable results.

Because most testing procedures are not

practical for many veterinary practices, it is necessary to ship the sample. Shipment of samples by first class mail in styrofoam boxes, padded envelopes, or other sturdily constructed containers is absolutely necessary to prevent loss or crushing of the sample container due to machine handling of mail. Air or express mail or commercial carriers should be used for shipment over long distances if possible, when cost is not a factor. Securely closed (taped if necessary), pliable plastic tubes are preferable because rigid plastic is easily crushed. Glass tubes are acceptable provided that they are fully protected against breakage.

There is considerable interest of practitioners in the detection of chlamydial antibody activity in pet birds; therefore, the disease of chlamydiosis will be emphasized here. Other diseases for which serologic tests are available will be considered in lesser detail because research on their applicability is lacking.

## CHLAMYDIOSIS

Currently the test method most commonly used for serologic testing for chlamydial antibody activity is direct complement fixation. Although this test has been reported to be unsuitable with sera from budgerigars, young African Greys, cockatiels, canaries, and finches, the author has found the test to be usable with cockatiel serum; however, only low titers are obtained. The interpretation of this is described later.

The individual bird showing signs of chlamydiosis should be tested (see Chapter 35, Chlamydia). Because chlamydiosis may be mimicked by other diseases, the test is sometimes indicated to rule out a chlamydial infection. Serologic evaluation may also be indicated as part of a "new" bird screening; random sampling may be appropriate for screening of groups of clinically normal birds.

### Interpretation of Serologic Test Results

Interpretation of the complement fixation test results presents some problems for the clinician. Ideally, tests are performed on properly spaced "paired" specimens (four weeks apart). In the case of chlamydiosis, however, it has been found that test results on a single sample can be of *some* value to the practitioner. A bird sick from chlamydiosis will often have a "significantly high" titer to chlamydial antigen in the first sample taken because of the nature of the infection. In these cases, the bird has had a systemic infection some time before it is brought for examination, and the antigen stimulation is thus great enough to induce a high level of antibody. In acute infections, a high titer of antibody activity may not be detected until a second sample is tested and a significant rise ($\geq$ four-fold) in antibody titer is demonstrated.

Interpretations of test results are given by the laboratory as an aid to the veterinary practitioner according to the titers. A titer of $<8$ (i.e., no antibody activity detectable at a serum dilution of $\geq 1{:}8$) has been interpreted as "negative"; a titer of 8, "probably negative"; titers of 16 and 32, "possibly infected"; and titers of $\geq$ 64, "probably infected."

Information gained from testing a second sample would be expected to yield more conclusive evidence concerning the disease status of the bird. However, if treatment is initiated, it should be borne in mind that titer changes are not likely to occur.

It should be pointed out that the final diagnosis rests with the veterinary practitioner because serologic test results are only an aid. Laboratory personnel can only *suggest* the possible meaning of results on a single serum tested against chlamydial antigen. The veterinary practitioner who sees that bird must take many things into consideration relative to the bird's condition. Therefore, whenever equivocal serologic results are obtained, it is suggested that a bird be cultured for *Chlamydia*, keeping in mind the limitations of that test.

A latex agglutination (LA) serologic method has been developed, tested, and compared with direct complement fixation (DCF) results on over 3000 sera. Results show that the LA test has a sensitivity of 89.7 per cent and a specificity of 92.8 per cent when compared with DCF results. There was about 1 per cent nonspecific reactivity, rarely above the 1:8 dilution, with control latex. Therefore, it is now possible to make more definite statements in many cases concerning the status of a given bird relative to its being infected with *Chlamydia psittaci.*

The LA serologic method is currently being used in conjunction with DCF testing on serum samples submitted to Texas A & M University. Table 18–1 illustrates the possible interpretations formulated by incorporating results of both tests on a single serum sample.

The LA testing method has also been correlated with isolation attempts. When serologic reactivity was obtained by both DCF and LA on clinical serum specimens, the isolation rate

**Table 18–1.** LABORATORY INTERPRETATION OF SEROLOGIC RESULTS FOR CHLAMYDIOSIS

| Complement Fixation Titer | Latex Agglutination Titer | Code | Interpretation |
|---|---|---|---|
| ≥8* | ≥8* | A* | Currently infected or in early recovery stage. |
| ≥64 | <8 | B | Currently infected if signs so indicate or in recovery stage from recent systemic infection. |
| ≥32 | <8 | C | Currently infected if signs so indicate or in convalescent stage from recent systemic infection. |
| ≥8 or ≥16 | <8 | D | Currently infected if signs so indicate or infected systemically in distant past if signs absent. |
| <8 | <8 | E | Negative for antibody but possibly in acute stage of infection if signs so indicate. A second specimen in three to four weeks is indicated for confirmation. |

*Complement fixation titers usually ≥ 32 and latex agglutination titers usually ≥ 16 in serum of culture-positive birds.

from feces or cloacal swabs submitted with the serum was 41 per cent. The isolation rate was 15 per cent when reactivity was obtained by DCF only. From birds with no demonstrable serologic activity by either DCF or LA, the isolation rate was 4 per cent.

The interpretation of serologic evidence is further complicated by several other factors. Many birds may never produce antibody at detectable levels, even though the bird is known to be infected, because of positive isolation results. Only research over time will prove whether this is due to an individual bird's being anergic, is the result of species' differences, or is because of "strain" differences of infecting chlamydiae. The species of bird must be considered in a serologic interpretation. For example, it has been found that known infected cockatiels may develop a titer no higher than 16.

It is evident from experience that high titers (≥ 64) are stable and decrease slowly over a period of months after infection; thus, a static titer is considered to be indicative of cessation of an active systemic infection. However, the possibility of an intestinal carrier state infection must be considered in these cases. Such an infection could presumably, by more or less continual antigenic stimulation, cause a titer to remain elevated.

The possibility of chlamydial infection occurring without any demonstrable antibody activity also must be kept in mind, although this seems to occur in only 5 to 10 per cent of cases. Subsequent to culture and in cases in which clinical signs of the disease are present, rapid response to appropriate therapy may further imply the presence of the organism.

In spite of the enigmatic nature of chlamydial infections, serologic tests should be relied upon to aid in the diagnosis or ruling out of the disease.

## Considerations for the Future

There are some things that should be considered in refining old or in developing new chlamydial serologic tests. An ELISA (enzyme-linked immunosorbent assay) could be used, but its sensitivity cannot be effective or efficient unless the indirect method is used. Prior to such use it must be established whether or not immunoglobulins of all psittacine species are detectable by a single anti-immunoglobulin conjugate. Otherwise, the test procedure would be cumbersome because of the need for anti-immunoglobulin conjugates to each species or related species.

It would be useful to be able to differentiate antibody activity in a current case from antibody activity resulting from an infection in the distant past. This would involve knowing the differences, if any, of the class(es) of antibody produced during different stages of infection, which may be related to different antigenic components. A test measuring such differences might be too complicated for common usage, although one such procedure was developed for detecting human IgM activity against a particular antigenic component of *Coxiella burnetii*.

There is a need for a simple and direct test that could be easily performed in veterinary practitioners' laboratories. The LA test has possibilities here. However, this still involves the variables that occur with individuals within a biologic system. The need for careful interpretation of results in the proper perspective of various possibilities will remain.

## ASPERGILLOSIS

Immunodiffusion and crossed immunoelectrophoresis have been used with avian serum to detect antibody activity against *Aspergillus*. However, controversy exists concerning the usefulness of this method. A great deal of research needs to be done concerning serologic reactions with this organism. More sensitive and more rapid procedures may prove to be useful. It has been suggested that a method of antigen detection, rather than serology, should be developed and evaluated.

## NEWCASTLE DISEASE VIRUS AND AVIAN INFLUENZA VIRUS INFECTIONS

Testing for antibody activity in psittacine bird serum against these agents may be indicated if birds have been exposed to known infected birds. However, attempted isolation of the virus is the preferred method, and testing for antibody activity probably serves no useful purpose except in poultry. In poultry, the test of choice is hemagglutination inhibition for Newcastle disease virus; it can also be used for testing against specific types of influenza viruses. Immunodiffusion test applies to influenza viruses in general because a group antigen is used in the test.

## PACHECO'S DISEASE *(Herpesvirus)*

Virus neutralization testing of psittacine bird serum might prove useful in detecting carriers, but research needs to be done to prove its usefulness (see Chapter 11, Virology, and Chapter 32, Viral Diseases).

## SALMONELLOSIS

To the author's knowledge there has been no testing of psittacine birds for antibody activity against *Salmonella*. *Salmonella* usually causes severe disease, making its presence known by high morbidity and mortality. The agglutination testing method used is valuable for testing of poultry, however.

## MYCOPLASMOSIS

Although breeder poultry flocks are routinely subjected to mycoplasmosis testing as a control measure, further research is needed before agglutination and hemagglutination inhibition methods are used to test for this disease in respiratory infections of psittacine birds.

In a study of an experimental infection of budgerigars and chickens with *Mycoplasma* isolates from an upper respiratory infection of Amazons, serologic findings were ambiguous and therefore considered unreliable. In another study, a *Mycoplasma*-associated respiratory disease was reported in Severe Macaws. Although the diagnosis was supported by evidence from other measures, none of the sera of the birds involved in the outbreak reacted serologically with *Mycoplasma gallisepticum* or *Mycoplasma synoviae* (see Chapter 34, Mollicutes).

## RECOMMENDED READINGS


Boseman, L.H., et al.: Mycoplasma challenge studies in budgerigars *(Melopsittacus undulatus)* and chickens. Avian Dis., 28:426–434, 1984.

Bryant, N.J.: Laboratory Immunology and Serology. Philadelphia, W.B. Saunders Company, 1978.

Cottral G.E.: Manual of Standardized Methods for Veterinary Microbiology. Ithaca, Cornell University Press, 1978.

Gaskin, J.M., and Jacobson, E.R.: A mycoplasma-associated epornitic in Severe Macaws *(Ara severa several)*. Proceedings of the Annual Meeting of the American Association of Zoo Veterinarians, Denver, CO, 1979, pp. 59–61.

Grimes, J.E., et al.: Complement-fixation and hemagglutination antigens from a chlamydial agent (ornithosis) grown in cell culture. Can. J. Comp. Med., 34:256–260, 1970.

Grimes, J.E., and Page, L.A.: Comparison of direct and modified direct complement-fixation and agar-gel precipitin methods in detecting chlamydial antibody in wild birds. Avian Dis., 22:422–430, 1978.

Grimes, J.E.: Latex agglutination: A rapid serologic diagnostic aid for psittacine chlamydiosis. Proceedings of the Annual Meeting of the Association of Avian Veterinarians, Boulder, CO, 1985.

Grimes, J.E.: Enigmatic psittacine chlamydiosis: Results of serotesting and isolation attempts, 1978 through 1983, and considerations for the future. J. Am. Vet. Med. Assoc., 186:1075–1079, 1985.

Grimes, J.E.: Results of direct complement fixation and isolation attempts for *Chlamydia psittaci* infections on clinical specimens from psittacine birds. Avian Dis., 29:873–877, 1985.

Hitchner, S.B., et al.: Isolation and Identification of Avian Pathogens. College Station. TX. American Association of Avian Pathologists, 1980.

Lennette, E.H., and Schmidt, N.J.: Diagnostic Procedures for Viral and Rickettsial Diseases. 3rd ed. New York, NY, Am Public Health Association, Inc., 1964.

McDonald, S.E., and Bayer, G.V.: Psittacosis in pet birds. Calif. Vet., 4:6–17, 1981.

Napolitano, R.L., et al.: Serodiagnosis of aspergillosis in avian species. Proceedings of the American Association of Zoo Veterinarians, Washington, DC, 1978, pp. 265–267.

Schachter, J., et al.: Psittacosis: The reservoir persists. J. Infect. Dis., 137:44–49, 1978.

Tizard, I.R.: An Introduction to Veterinary Immunology. Philadelphia, W.B. Saunders Company, 1977.

# *Chapter 19*

# OPHTHALMOLOGY

LORRAINE G. KARPINSKI

The eyes and periocular tissues are involved in a number of avian diseases. A preliminary differential diagnosis may be formulated from clinical signs noted in this area; however, specialized diagnostic techniques are required for a complete ophthalmologic evaluation. Some ophthalmologic treatment regimens are also unique to birds.

## DIAGNOSTIC PROCEDURES

An ophthalmologic evaluation should begin with the determination of vision. Pupillary constriction to light is not an accurate indication of vision, and a bird's ability to voluntarily control his pupil size makes this determination even more unreliable. A menace response should be tested bilaterally. Function of the eyelids and integrity of the lid margins should be evaluated. Symmetry, position, and motility of the globes should be noted. Using magnification and a point light source, the external structures should be examined. If indicated, conjunctival culture and sensitivity testing should be done. Conjunctival scraping with immediate fixation and Wright's, Giemsa, Gram's, and/or Macchiavello's staining may indicate the presence and type of any infectious agent and the predominant cell type. The corneal surface should be smooth and shiny and any irregularities should be stained with fluorescein. Using the slit setting on a direct ophthalmoscope, depth and optical clarity of the anterior chamber should be determined. The iris should be flat and thin and have a freely moving pupillary border.

Because the avian iris is resistant to commonly used mydriatics, that is, parasympatholytics and sympathomimetics, examination of the lens and fundus of a bird's eye is difficult. Dilation of the avian pupil may be achieved by several applications over a 15-minute period of a 3 mg/ml solution of dry crystalline *d*-tubocurarine in 0.025 per cent benzalkonium chloride. The solution must be freshly mixed before each session. Commercially prepared aqueous solutions of tubocurarine do not readily pass through the corneal layers.

During dilation, clarity of the lens should be determined. The evaluation of the fundus should reveal an evenly reflective, avascular retina with a pecten, the heavily pigmented, vascular pleated structure extending from the optic disc into the vitreous (Fig. 19–1). Detail of the optic disc is often obscured by the pecten.

## OPHTHALMOLOGIC DISORDERS

Newly imported parrots from South and Central America are often infected with pox (see Appendix 2). Eye lesions associated with parrot pox are seen 10 to 14 days following infection. A mild blepharitis and serous ocular discharge escalate to form dry, crusty scabs along the lid margins. The eyelids swell and are sealed shut (Fig. 19–2). The corneas may ulcerate. The scabs become leathery in texture and subsequently dry and fall off. Clinical illness lasts two to six weeks. Eye lesions are treated daily by washing with dilute baby shampoo followed by application of mercurochrome solution (1 oz. 2 per cent mercurochrome to 4 oz. eye wash); an

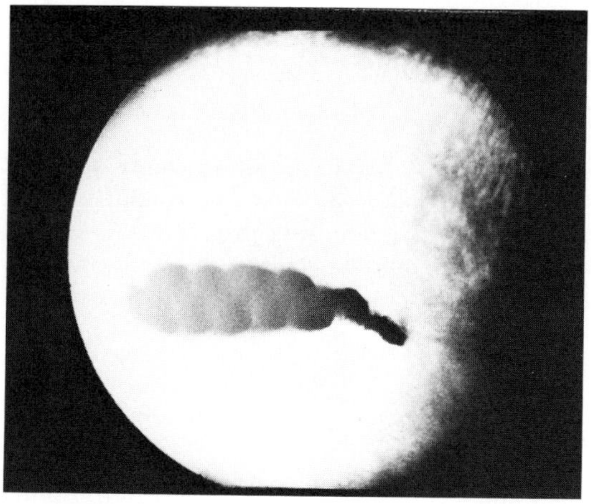

**Figure 19–1.** The pecten of the eye may be visible in an ophthalmologic examination.

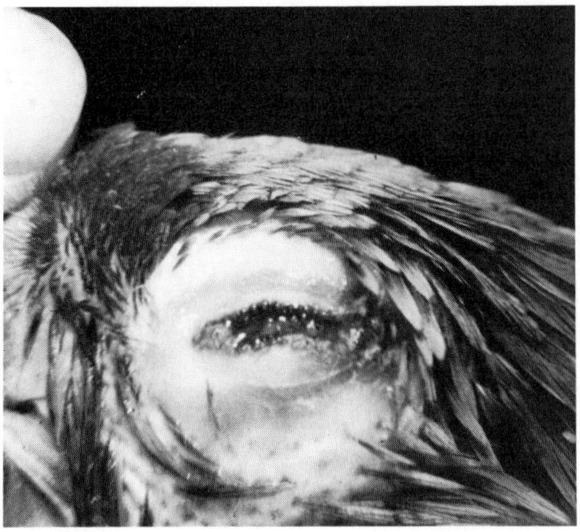

**Figure 19–2.** Eyelids may be sealed shut from serous ocular discharge associated with pox virus.

antibiotic ointment is instilled onto the eye under the scabs. The scabs should not be removed because of the potential for damage to the lids. Systemic supportive care is indicated in parrot pox (see Chapter 32, Viral Diseases). Residual effects include distorted lids, loss of periocular pigmentation and lashes, subepithelial corneal crystals, chronic corneal ulcers, and epiphora (Fig. 19–3). Eye lesions due to avian pox infections may also be seen in canaries, birds of prey, pigeons, and gallinaceous birds.

Lovebird eye disease is a severe and highly fatal disease seen in captive-bred mutation lovebirds, especially Peach-faced. The first signs are generalized depression, blepharitis, and serous ocular discharge. As the disease progresses, the lids become hyperemic and swollen and the ocular discharge becomes more severe (Fig. 19–4*A*). Affected birds are often assaulted by cagemates (Fig. 19–4*B*). Birds do not respond to treatment, and death usually occurs within a few days. The only measure that has been helpful in the management of the disease in groups of birds is placing the birds in a nonstressful environment and using no treatment.

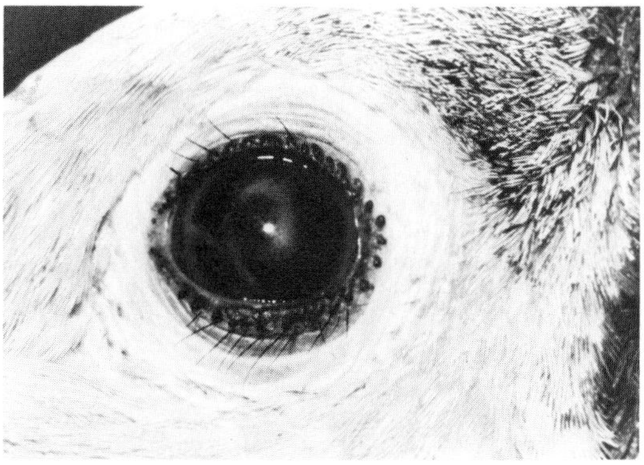

**Figure 19–3.** Corneal irregularities and loss of normal periocular pigmentation may be a residual effect of pox infection.

Three distinct eye problems have been observed in mynahs. Corneal scratches that may occur during shipping re-epithelialize with no treatment. Keratitis, which is seen one week after shipping, regresses in several weeks to several months regardless of treatment. Chronic keratoconjunctivitis manifests itself as a large mass of infiltrated conjunctiva under the lower lid with concurrent superficial keratitis. Surgical removal of the mass may cause resolution of the keratitis.

A transient punctate keratitis may be seen in Amazons imported from Central America. The lesion appears as a slight irregularity of the medial cornea. Some cases will progress to involve the entire cornea with concomitant anterior uveitis. Most cases will regress spontaneously in one to two weeks. Treatment is recommended for cases that develop a mucoid or mucopurulent nasal discharge, deep corneal ulcers, or uveitis.

Cockatiel conjunctivitis affects white and albino mutations most commonly. The suspected etiology is a mycoplasma. The conjunctivae become inflamed, protrude from under the lids, and become infiltrated (Fig. 19–5). This conjunctivitis is often associated with upper and lower respiratory infections. Therapy with tetracycline systemically or with tylosin systemically and topically (Tylan powder mixed 1:10 with water and sprayed in the eyes) is effective. Caseated material that may appear under the lids must be removed. Recurrence is common. This disease may be transmitted to offspring, so affected birds should not be used as breeders.

Chronic conjunctivitis in Amazons may be associated with upper respiratory infections. Conjunctival tissue becomes edematous, inflamed, and infiltrated. Treatment consists of surgical removal of the redundant tissue and application of a topical steroid.

Periorbital abscess may be a sequela to chronic upper respiratory disease. Surgical removal of the caseated material and the use of systemic antibiotics are indicated. Lacrimal sac abscess may follow sinusitis. It is manifested by a firm swelling below the medial canthus. Often caseated material is visible in the lacrimal punctum. Flushing the nasolacrimal system through a feline indwelling intravenous catheter with saline containing an appropriate antibiotic is the recommended treatment.

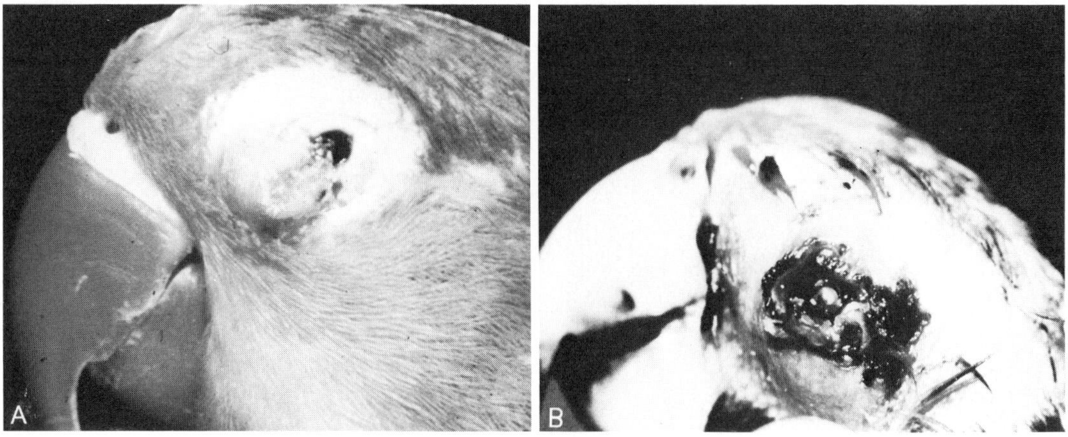

**Figure 19–4.** Lovebird eye disease involves a progressive swelling of the eyelids and ocular discharge *(A)*; affected birds may be attacked by cagemates *(B)*.

A sinusitis causing globe recession and collapse of the tissues over the sinuses is seen in macaws, especially in Scarlet and Green-winged Macaws (Fig. 19–6). A thick mucoid to mucopurulent material can be flushed from the sinuses through the nostrils. Large volumes of saline (10 to 30 ml per nostril) with appropriate antibiotics are used daily for two weeks. A systemic antibiotic, routinely carbenicillin and either gentamicin or amikacin, is also recommended. A similar condition seen in Amazons is not responsive to therapy and may be unrelated.

Cataracts are seen sporadically, especially in older birds, and as a result of previous intraocular inflammation (Fig. 19–7). Norwich and Yorkshire canaries exhibit a genetic cataract, inherited as a recessive trait. When indicated, cataracts may be removed surgically.

Knemidokoptic mites cause scaley face lesions of the cere and periorbital area in budgerigars and other species. A single dose of ivermectin is effective treatment.

*Oxyspirura* is a small, slender nematode that may be found behind the nictitating membrane or in the conjunctival sac and nasolacrimal duct of cockatoos. The eyelids may appear swollen, and the bird may scratch at the eye. Worms may be manually removed. Ivermectin administered systemically is effective. Dead and dying worms must be washed from the conjunctiva.

## SURGICAL TREATMENT

Enucleation in birds is a disfiguring procedure because of the relatively large volume of the globe (Fig. 19–8). A transconjunctival approach is used, and care must be taken to minimize traction and trauma to the optic

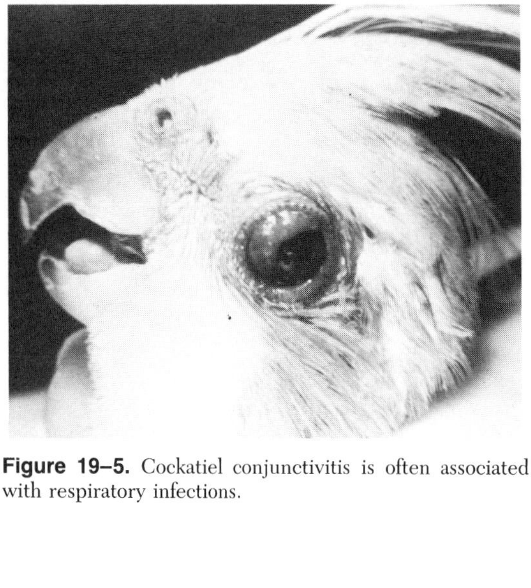

**Figure 19–5.** Cockatiel conjunctivitis is often associated with respiratory infections.

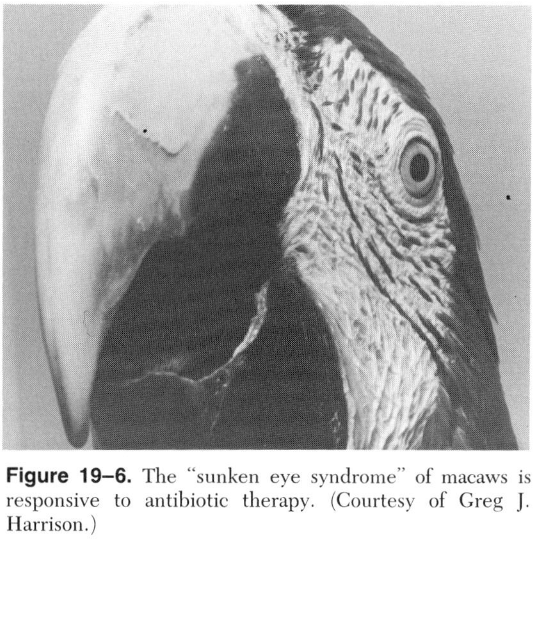

**Figure 19–6.** The "sunken eye syndrome" of macaws is responsive to antibiotic therapy. (Courtesy of Greg J. Harrison.)

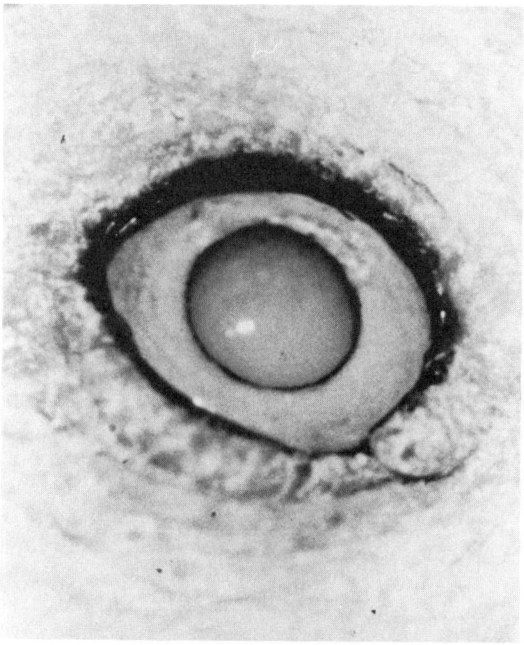

**Figure 19–7.** Cataracts may occur in older birds.

nerve, which is short and close to the optic chiasm. Excessive trauma in the posterior pole of one orbit may affect the optic nerve of the other, resulting in contralateral blindness. After the globe is removed, fine-diameter absorbable sutures should be placed vertically from the dorsal to the ventral orbital rim. This reduces collapse of the overlying tissue and improves the cosmetic result. The eyelid borders should be closely trimmed and sutured.

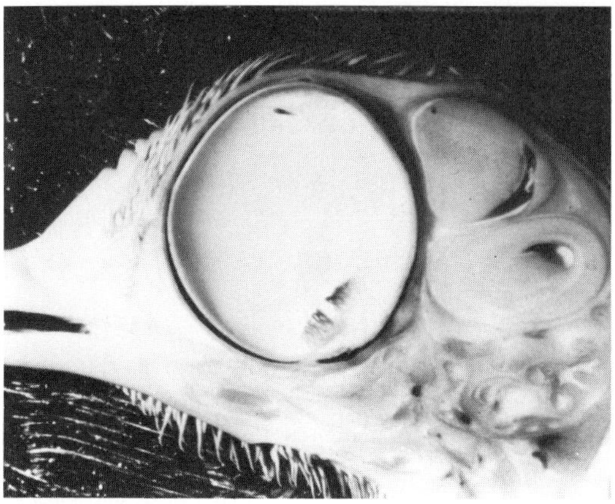

**Figure 19–8.** The enormous size of the globe in comparison to the head is apparent in this cross section of a chick. (Courtesy of Chris Murphy.)

Evisceration of the ocular contents is preferable. The cornea is removed and the intraocular contents are teased free and removed. The conjunctiva is closed over the globe with small-diameter absorbable sutures. The eyelid borders are trimmed close to their margins and sutured.

## RECOMMENDED READINGS

Bellhorn, R.W.: Laboratory animal ophthalmology. *In* Gelatt, K.N. (ed.): Textbook of Veterinary Ophthalmology. Philadelphia, Lea and Febiger, 1981.

Murphy, C.J., et al.: Enucleation in birds of prey. J. Am. Vet. Med. Assoc., 183:1234–1237, 1983.

# *Chapter 20*

# NEUROLOGIC EXAMINATION

RONALD LYMAN

The basis for diagnosis, prognosis, and therapy in neurologic disease lies in the ability to localize lesions. A routine neurologic history and examination are the most useful tools in this process. This chapter discusses a protocol that may be part of a detailed physical examination. Practice on healthy birds will allow confident interpretation of normal versus abnormal response.

The objective is to locate the lesion in one of the following areas: head, cervical spinal cord, thoracolumbar cord, or multifocal (more than one of these areas). A generalized neuromuscular or metabolic lesion must be considered if the examination fails to localize a lesion. True spinal reflexes are difficult to assess objectively in the avian species. Attention must be directed toward strength, muscle tone, or atrophy. Special interest focuses on symmetry of response. Accurate localization of a lesion allows for appropriate diagnostic and therapeutic steps.

In general, all patients with a serious neurologic problem should have preliminary screening blood tests such as WBC, differential count, serum chemistries (see Chapter 12, Hematology, and Chapter 13, Clinical Chemistries) and/or multiple organ biopsies (see Chapter 16, Biopsy Techniques).

Neurologic findings may be grouped according to the areas they represent.

## HEAD SIGNS

Head signs include loss of cerebral, cerebellar, brain stem, or cranial nerve functions. The neurologic history begins with a review of the recent mental status, including degree of consciousness, response, ability to perform normal activity and/or mimic, and personality. The presence of convulsive or syncopal episodes should be noted.

Seizures may manifest themselves in various ways; however, the typical avian seizure consists of a short period of disorientation with head and body ataxia followed by loss of grip on the perch. The bird usually falls to the cage floor and remains rigid or has some form of major motor activity for varying lengths of time from seconds up to a few minutes (Fig. 20–1). Cloacal discharge may occur. The postictal activity is variable. Syncopal episodes are difficult to differentiate from seizures on appearance alone.

An historical account of cranial nerve (C.N.) function can occasionally be obtained. Loss of olfaction (C.N. I) may affect appetite or recognition of food. Vision (C.N. II) impairment is recognized by failure to avoid obstacles in the cage or environment. Loss of balance, head tilt, nystagmus, or deviation of the eyes may implicate the vestibular system (C.N. III, IV, VI, VIII, brain stem, cerebellum). Loss of beak strength in eating or climbing may indicate a motor function in C.N. V. Loss of hearing may come from a C.N. VIII lesion. Dysphagia may represent a problem with Cranial Nerves IX and X. Deviation or atrophy of the tongue results from a C.N. XII lesion. An abnormal gait may result from various lesions, but head incoordination is typically a "head sign" (cerebellar/vestibular).

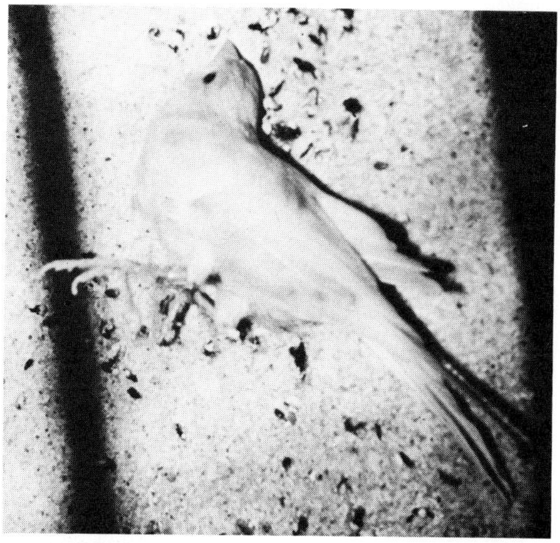

**Figure 20–1.** Characteristic of avian seizures, a canary remains rigid on the cage floor. (Courtesy of Greg J. Harrison.)

The following reflexes/responses are commonly used to test for head signs:

**Menace Response** (evaluates C.N. II, VII, cerebrum, cerebellum, brain stem). Bringing the hand or finger toward each individual eye should result in an eyeblink, pulling away of the head, or aggressive action with the beak. Asymmetry or loss of the response indicates a lesion somewhere in the head.

**Pupillary Light Reflexes** (evaluates C.N. II, III, sympathetic N.S., cerebrum, brain stem). Introduction of light to one pupil should cause a constriction of that pupil (direct) and the opposite pupil (consensual). Absence of these reflexes indicates an ocular or head lesion. More precise localization requires consideration of other signs.

**Eye Movements** (evaluates C.N. III, IV, VI, VIII, brain stem, cerebellum). A normal physiologic nystagmus or eye jerk, called a doll's eye response, results while moving the head from side to side or up and down. Loss of this response occurs on rare occasions with bilateral peripheral C.N. VIII lesions or with severe brain stem lesions accompanied by altered consciousness. The latter lesions have a guarded prognosis.

Spontaneous nystagmus occurs with lesions of the vestibular system. Precise localization depends on combinations of signs such as head tilt and radiographic evaluation.

Strabismus (ocular deviation) is a head sign indicating vestibular or specific cranial nerve dysfunction (III, IV, VI).

**Head Posture**. A head tilt usually indicates a vestibular (C.N. VIII) lesion on the tilted side.

**Head Coordination**. Head tremors or incoordination constitutes a head sign, usually of cerebellar or vestibular origin.

**Pupil Symmetry** (evaluates C.N. III/sympathetic nervous system). In the absence of intraocular inflammation or structural lesions, pupils should be symmetrical. Asymmetry indicates either a C.N. III lesion (dilated, poorly responsive) or a sympathetic lesion. Horner's syndrome, a miotic pupil that will not fully dilate in darkness, is not necessarily a head sign, since cervical interruption of the sympathetic tract may cause this sign.

**Eyeblink** (C.N. V, VII). Touching the lateral canthi should elicit a symmetrical eye blink. Absence or asymmetry of this response is a head sign.

**Facial Sensation** (sensory C.N. V). Touching each side of the face while obstructing vision should elicit similar responses. Absence or asymmetry constitutes a head sign.

**Beak Strength, Swallowing, and Tongue Function.** These have been discussed elsewhere in this chapter.

## WING SIGNS

Asymmetrical wing carriage without fracture, atrophy of the proximal musculature, and decreased wing strength during manipulation constitute "wing signs" of neurologic disease. Inability to fly in previously flighted individuals is also a sign. These signs should be differentiated from generalized weakness caused by systemic or abdominal disease. Whereas reflexes are not reliable to interpret, an electromyographic ex-

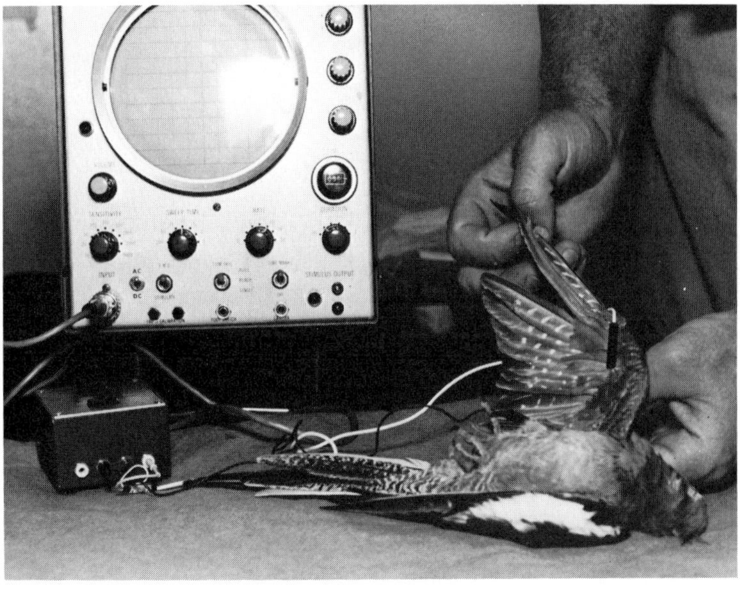

**Figure 20–2.** An electromyogram (EMG) may be useful in determining the location of neuromuscular lesions.

amination (Fig. 20–2) may confirm a lesion somewhere in the motor units of the wing.

## LEG SIGNS

Weak gait, absence or weakness of withdrawal response (toe-pinch), inability to grasp, and proximal muscle atrophy constitute leg signs. Knuckling of the digits or contraction of a limb may occur with severe neurologic impairment. Again, electromyography can confirm a lesion within the motor units of the limbs. A crossed extensor response (extension of the opposite limb during flexion of one limb) probably indicates a lesion *above* the level of the limbs in the spinal cord or head. This occurs as a result of loss of normal inhibitions at the spinal level.

## CLOACAL SIGNS

Absent or weakened cloacal response to a gentle pinch is suggestive of a lesion in the lower spinal cord. Constant dripping of cloacal contents and/or severe soiling of the tail area may occur with lesions anywhere in the head or spinal cord. The practitioner must rule out peripheral sphincter nerve damage.

## SENSORY SIGNS

Complete loss of pain sensation in the conscious avian patient implies a poor prognosis for neurologic recovery. Interpretation of sensation requires a conscious reaction to a painful stimulus. For example, vocalization, attempting to bite, and attempting to escape while a painful

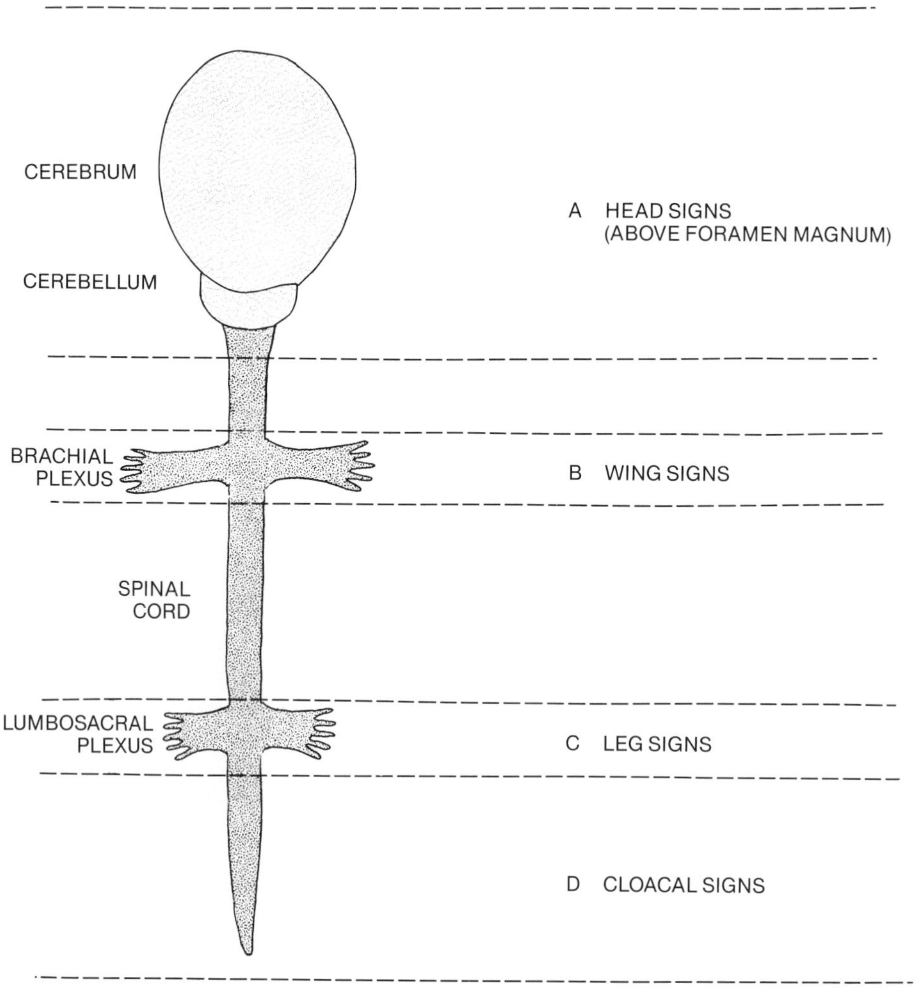


**Figure 20–3.** Localization of neurologic lesions.

**Table 20–1.** LOCALIZATION CHART FOR AVIAN NEUROLOGIC EXAMINATION

| Location of Lesion | Head Signs | Wing Signs | Leg Signs | Cloacal Signs | Useful Diagnostic Aids |
|---|---|---|---|---|---|
| Head lesion | + | 0 | 0 | 0 | I |
| Head lesion, multifocal lesion, neuromuscular, toxic, or metabolic lesion | + | + | + | + | II |
| Thoracolumbar or lower | 0 | 0 | + | + | |
| or | 0 | 0 | + | 0 | III |
| Cervical or multifocal lesion | 0 | + | + | + | IV |

+ = signs present; 0 = signs absent

**Useful Diagnostic Aids**

I. Radiographs, EEG, blood profile, multiple organ biopsies, ECG
II. Radiographs, EEG, blood profile, multiple organ biopsies, $T_3$ and $T_4$ (pre- and post-TSH stimulation), blood lead, ECG
III. Radiographs, EMG
IV. Radiographs, EMG

stimulus is being applied are signs that sensation is intact. On the contrary, withdrawal of a limb may simply indicate that a local reflex arc is intact and does *not* imply that sensation is perceived at higher levels. The prognostic importance of sensory determination occurs most commonly in traumatic spinal lesions. Needle stimulation and forceps closure on the skin represent the most common means of applying stimuli. This part of the examination should be performed last, to prevent other signs from being altered by the painful stimuli.

The findings of the neurologic examination must then be consolidated to determine a general location of the neurologic lesion. Figure 20–3 and Table 20–1 clarify this process.

Neurologic signs may occur from a lesion at the level of the dysfunction or at any level "higher" (closer to and including the cerebrum). If the examination implicates dysfunction at more than one level, the lesion is assumed to be located at the "higher" location, or to be multifocal.

In reference to Figure 20–3, three examples of selected specific neurologic problems illustrate the neuroanatomic localization:

1. Normal A, B, D, abnormal C: The lesion is most likely to be in the cord between the areas of B and C, or directly in the area of C (thoracolumbar).

2. Normal A, abnormal B, C, D: The lesion is likely to be in the cord between A and B, or directly in the area of B (cervical).

3. Abnormal A, B, C, D: The lesion may be in A, or may be a multifocal lesion, metabolic disorder, or toxicity.

The reader is referred to the Recommended Readings at the end of Chapter 38, Neurologic Disorders.

# *Chapter 21*

# ELECTROCARDIOGRAPHY

MICHAEL S. MILLER

The electrocardiogram (ECG) is defined as a graphic record of sequential, electrical depolarization-repolarization patterns of the heart. Each deflection of the electrocardiograph (galvanometer) stylus delineates the summation of transmembrane potentials of the cardiac cells at a particular instant, as recorded by electrodes on the surface of the body.

Electrocardiographic data are documented in a wide variety of avian species (chicken, turkey, duck, seagull, buzzard, quail, and pigeon). Normal values for caged birds such as the parakeet, cockatiel, and parrot have only recently been established.[3a, 11, 12] Poultry research is pursuing use of the electrocardiogram (ECG) for diagnosing specific diseases.[6] For example, round heart disease in turkeys causes an early change in the ventricular mean electrical axis that is diagnostic for the disease.[9] The veterinary literature contains few clinical case studies in which a diagnosis of primary cardiac disease is based on history, clinical examination, radiography, and electrocardiography, with confirmation on necropsy.

## INDICATIONS

Indications for an electrocardiogram in birds are similar to those for all veterinary patients. Birds with listlessness, coughing, dyspnea, or auscultatable murmurs and rhythm disturbances require an ECG as part of their medical data base. Clinical signs such as syncope or seizures may be due to an underlying arrhythmia or conduction disturbance. Pharmacologic therapy may result in a cardiotoxic side effect that can be recognized by an abnormal ECG. An abnormality on the ECG may support a diagnosis of a specific nutritional problem or infectious disease. An abnormal ECG correlated with cardiomegaly on a thoracic radiograph may aid in the diagnosis and prognosis for a patient with primary heart disease.

A complete medical data base on a newly purchased bird may include a baseline ECG. The high purchase price of macaws, parrots, and cockatoos and the stress of quarantine and domestication warrant inclusion of an ECG in the initial evaluation. An ECG may be indicated preoperatively and is useful in monitoring a patient under a general anesthesia.

## TECHNIQUE

Many published reports discuss recording the ECG in the anesthetized bird. If possible, the bird should be awake, since it is recognized that anesthetic agents may alter the ECG. Restraint of the unanesthetized bird may significantly change the heart rate.

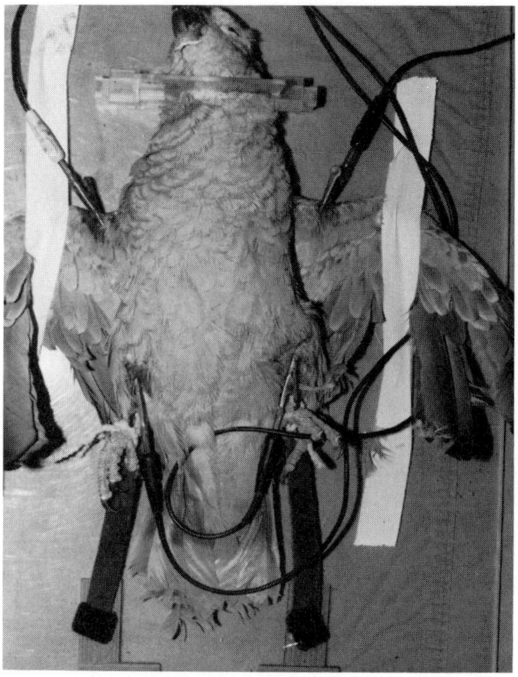

**Figure 21–1.** A parrot is immobilized on a plexiglass restraint board. The alligator clips of the electrocardiograph leads are positioned on the skin at the proximal cranial wing margins and on the skin at the cranial midtibiotarsus.

The bird is restrained by holding the head and body in a ventrodorsal position. The wings are taped to a nonmetal table or restraining board (Fig. 21–1) in a technique similar to that used for radiography, with tape securing the spread wings at the elbow joint. Standard alligator clips or substituted metal needles (for less trauma to the skin) are placed on the proximal cranial wing margins. The alligator clips or needles are placed on the immobilized legs at the cranial midtibiotarsal region. The electrodes are placed in contact with the skin, not the feathers, to minimize electrical artifact. A small quantity of alcohol, saline, or electrode jelly is applied to the sites of electrode placement. A paper speed of 50 mm/second or greater is desirable when recording the avian ECG.

Well-muscled or large birds (macaws and parrots) may show ECG complexes that are too small to permit accurate interpretation of the normal rhythm or recognition of arrhythmias and conduction disturbances. A precordial lead, applying the V electrode to the right or left side of the sternum over the area of the heart, may result in greater amplitudes of all complexes.

## THE NORMAL ELECTROCARDIOGRAM OF THE PARAKEET AND PARROT

The ECG of birds is comparable to that of mammals, with the P, QRS, and T waves being the primary deflections (Fig. 21–2). This similarity is not unexpected, as the avian heart contains specialized pacemaker and conduction tissue (sinus node, AV node, ventricular Purkinje cells) as well as atrial and ventricular myocardial cells.[9] Regarding the QRS complex (ventricular depolarization), the negative S wave is the predominant deflection versus the positive R wave in most mammals, including man. The heart rate is extremely variable depending on the bird species, temperature, health profile, and status (anesthetized versus awake) while recording the ECG.

The heart rate in the anesthetized parakeet varies from 600 to 750 beats per minute (bpm), versus 120 to 780 bpm in the anesthetized parrot,[1, 11, 12] and 496 to 604 bpm in the cockatiel.[3a] Unanesthetized birds also have marked variation in heart rate, although it is usually above 250 bpm. The major difference in the normal ECG of the bird versus the dog, cat, or

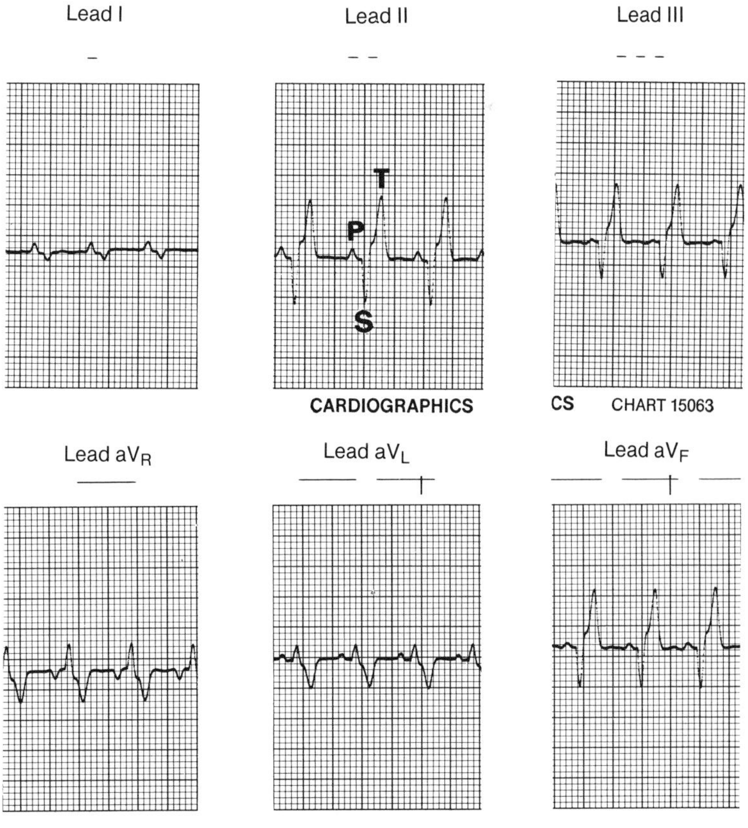


**Figure 21–2.** A normal six-lead ECG obtained from a parrot sedated with ketamine hydrochloride and xylazine recorded at 50 mm/second paper speed. Lead I (-), Lead II (--), Lead III (---), Lead $aV_R$ (-), Lead $aV_L$ (--), and Lead $aV_F$ (---) are indicated at the top of the strip. The heart rate is 300 bpm. The MEA is approximately −100 degrees. The S wave is the major deflection in the QRS complex in Leads II, III, and $aV_F$.

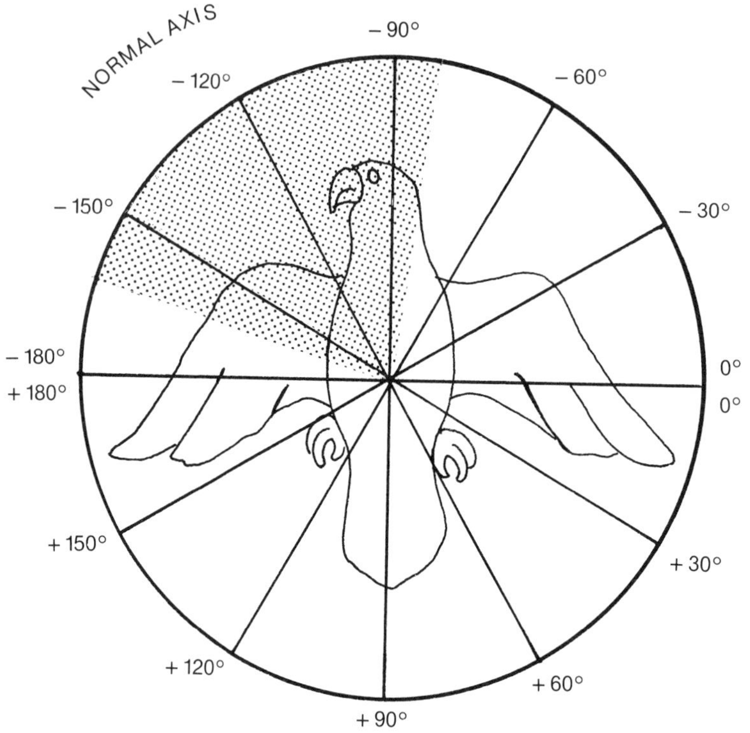


**Figure 21–3.** In the parakeet and parrot the normal MEA in the frontal plane is −83 degrees to −162 degrees. This summation vector is oriented craniad and toward the right side of the body, and is not a result of a hypertrophied right ventricular myocardium. An extensive Purkinje network may explain the mean electrical axis range.

man is a ventricular mean electrical axis (MEA) of −83 to −162 degrees (Fig. 21–3), with the QRS complex showing a negative polarity in leads II, III, and $aV_F$ (see Fig. 21–2).[9]

The P wave is positive in leads I, II, III, and $AV_F$ and may vary in leads $aV_R$ and $AV_L$. Owing to the rapid heart rate that is often present, there may be fusion of the P wave (atrial depolarization) with the preceding T wave (ventricular repolarization) (Fig. 21–4). The P wave may not be recognized even if the chart speed on the recorder is greater than 50 mm/second, suggesting that the atria depolarize before the ventricles are completely repolarized.[9] The P wave and T wave are not always fused at rapid heart rates. The P wave duration is 0.01 to 0.02 second in the parakeet and parrot.

The P-R interval (atrioventricular [AV] conduction time) is 0.01 to 0.04 second in the parakeet versus 0.03 to 0.07 second in the parrot. If there is no R wave in the lead II rhythm strip, the S wave is used for analysis of the P-R interval.

The QRS complex polarity is similar in the parakeet and parrot. In leads II, III, and $AV_F$, a negative deflection (S or RS) is consistently present. There is no Q wave in a bird's ECG in the standard limb leads I, II, or III.[9] If there is a positive and negative deflection during the QRS complex, the positive deflection is always seen first. This pattern is also seen in other bird species (cockatoo, pigeon, and mynah). The QRS complex polarity varies in lead I, and a positive deflection is usually seen in leads $aV_R$ and $aV_L$. The QRS complex duration is 0.01 to 0.03 second in the parakeet and parrot. The amplitudes in lead II vary between −0.2 to −1.1 mV in the parakeet and parrot.

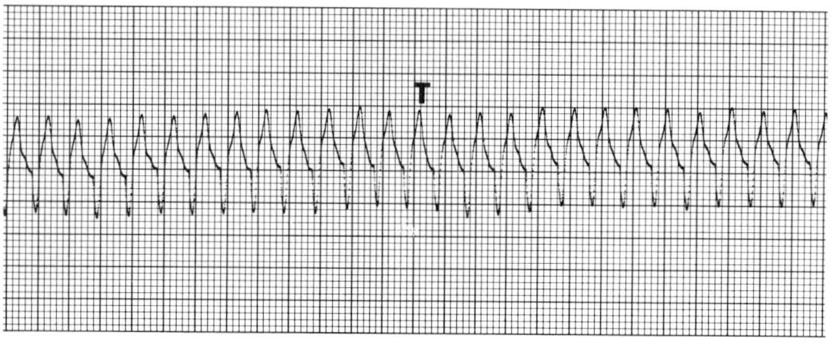


**Figure 21–4.** A Lead II ECG in an awake parrot recorded at 50 mm/second paper speed. The heart rate is 600 bpm. The P wave is not clearly visible but is likely fused with the declining limb of the T wave. The QRS complex is also fused to the following T wave, resulting in an inaccurate measurement of amplitude, duration, and polarity of these complexes.

The T wave and preceding QRS complex often fuse at rapid heart rates (Fig. 21–4), resulting in an inability to accurately measure amplitude, duration, and polarity (positive or negative) of these complexes. Even the veterinarian well-trained in electrocardiography may have difficulty accurately interpreting the avian electrocardiogram at rapid heart rates. Since the QRS complex amplitude in lead I is often small, leads II and III are used to analyze the mean electrical axis. If the QRS complex amplitude is small in lead II, a precordial chest lead is required to successfully evaluate the cardiac rate and rhythm.

The T wave is positive in leads II, III, and $aV_F$ in the parakeet and parrot. The T wave polarity is variable in lead I and negative in leads $aV_R$ and $aV_L$. A combined P and T wave may affect the true morphology of each individual complex.

Sturkie[10] reports that there are sex differences in the ECG in chickens. The mean amplitude of the QRS complex in lead II is significantly greater in males than in females. The administration of estrogen to males over a two- to three-week period depressed the amplitudes of all complexes approximately 35 per cent. There is a greater tendency toward fusion of the T and P waves in females. There are also marked differences in cardiac rate in male versus female chickens and turkeys but not in some species of quail, pigeons, or ducks.[7] It is not known at this time whether sexual ECG variations occur in psittacine birds.

## ELECTROCARDIOGRAPHIC ABNORMALITIES

Electrocardiographic abnormalities in chickens have been associated with nutritional deficiencies, infectious disease, hypertension, toxicity, and primary congestive heart failure. The underlying causes (congenital versus acquired) of congestive heart failure in birds are not well-documented.

Chickens that were experimentally fed polychlorinated biphenyl (PCB) in their ration developed ECG changes such as bradycardia and attenuated voltages of the S waves in leads II and III (may be related to hydropericardium).[5] Other ECG changes noted in less than 50 per cent of the birds included sinoatrial block and P wave and QRS complex amplitude and polarity alterations.

Acute potassium deficiency in the growing chick produces an abnormal ECG in 70 per cent and mortality in 100 per cent of these birds in two to four weeks.[8] The ECG abnormalities include sinus arrhythmia, sinoatrial (sinus) block, incomplete AV block, sinus bradycardia, premature nodal (AV junctional) beats, and ventricular premature contractions (VPC's) and escape beats. The most common abnormality is second-degree AV block. Atropine sulfate abolishes the conduction disturbances, suggesting that this electrolyte abnormality increases parasympathetic (vagal) tone to the sinus and AV nodes. Elevated potassium results in tall (peaked) T waves like those reported in ducks that emerge from a dive.[9]

Acute thiamine deficiency in pigeons and chicks is correlated with major ECG alterations and high mortality.[9] ECG abnormalities include sinus arrhythmia, sinus bradycardia, sinus arrest, and ST segment depression. Chronic thiamine deficiency may result in VPC's and sinus arrhythmia. Vitamin E–deficient birds show sinus arrhythmia, right axis deviation, VPC's, and ST segment elevation. Excess cadmium administered to chickens produces myocardial infarction and hypertrophy and also results in AV block and the reversal of T-wave polarity.

Round heart disease is an infectious problem that results in a change in the mean electrical axis in turkeys (−85 degrees in the normal turkey to +70 degrees in the diseased bird[4]). The T wave is depressed or inverted in selected leads of these birds. Experimental *Escherichia coli* infection in chickens causes myocarditis and pericarditis.[3] ECG changes include accentuated P waves, increased amplitude of the S wave, and attenuation of the R wave in leads I and $aV_F$. Also, there are variations in the mean electrical axis in infected birds.

Arrhythmias and conduction disturbances are also detected in turkeys infected with a pathogenic strain of avian influenza virus (AIV) and chickens with viscerotropic velogenic strain of Newcastle disease virus (NDV).[6] A recent study shows delayed AV conduction in chickens infected with both AIV and NDV.[6] Specific changes in AIV include a prolonged R-S interval, ST segment, T-P interval, and P-R interval. Also, there is an increase in amplitude of the P waves. A fusion of the T and P waves and tented T waves develop in NDV-infected chickens. Arrhythmias that are seen include VPCs and paroxysmal ventricular tachycardia. Occasionally VPCs are seen in normal control chickens. This report concludes that the ECG may be useful in presumptive diagnoses of specific disease states, although this technique requires further investigation.

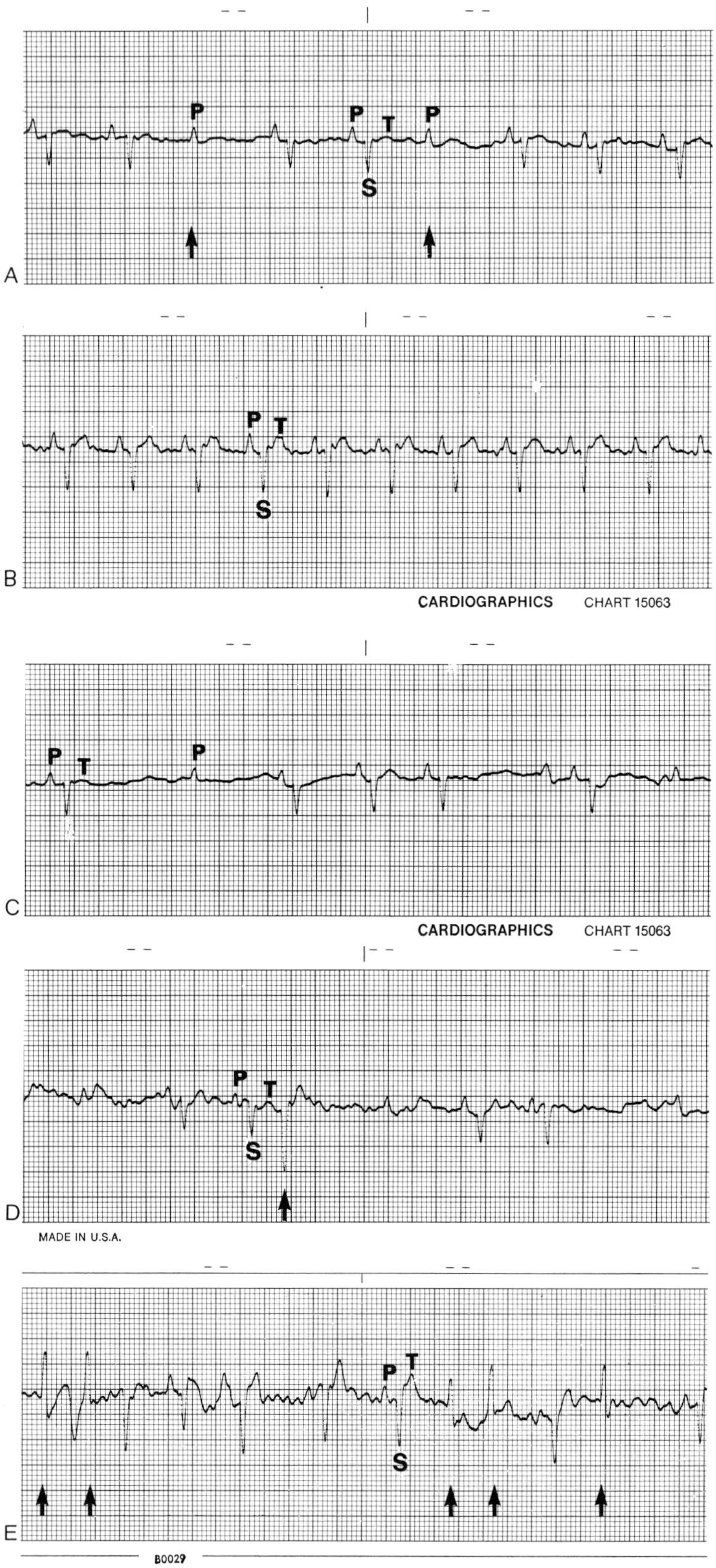


**Figure 21–5.** A Lead II ECG in an awake pigeon recorded at 50 mm/second paper speed. *A*, A Lead II ECG demonstrates second-degree AV block. The cardiac rate is approximately 120 bpm with several P waves (arrows) not followed by QRS-T complexes. *B*, A Lead II ECG recorded five minutes after administration of 0.1 mg atropine sulfate. The cardiac rate increases to 220 bpm and the second-degree AV block is abolished. The MEA is also changed, as evidenced by an alteration in amplitudes of the negative QRS complexes. There is also a change in the T wave morphology. *C, D, E.* Lead II ECG's recorded on successive days following the administration of high oral doses (0.1 mg/pound/day) of digoxin. Sinoatrial conduction is diminished *(C)*, while ectopic beats and/or aberrant conduction develops *(D* and *E*; arrows). Excessive baseline movement also expresses the possibility of artifact. Further clinical and experimental studies of the ECG in birds receiving toxic levels of digitalis are needed to clarify the above abnormalities.

Clinical information regarding ECG abnormalities in caged birds is sparse. Transient arrhythmias develop in the parakeet and parrot during induction of general anesthesia with ketamine hydrochloride (15 to 20 mg/kg) and maintenance with halothane.[12] There were no arrhythmias noted in cockatiels in conjunction with halothane, methoxyflurane, or isoflurane in one study.[3a] Arrhythmias are difficult to interpret owing to the rapid heart rates and include changes in polarity of the QRS complexes and premature beats. The abnormalities disappear during maintenance anesthesia with halothane.

Conduction disturbances including second-degree AV block in an awake pigeon (Fig. 21–5*A*) were abolished by atropine administration (Fig. 21–5*B*). Atropine also results in an increase in the cardiac rate, supporting a decrease in parasympathetic (vagal) tone to the sinus and AV nodes, as in mammals.

Digitalis, a drug commonly prescribed for congestive heart failure, has positive inotropic and electrophysiologic effects on the mammalian heart. Toxic levels of digitalis may cause cardiac arrhythmias and impair sinoatrial and AV conduction in the awake pigeon (Fig. 21–5*C*, *D*, and *E*). An oral dosage of digoxin as high as 0.1 mg per pound per day is required to induce toxic changes on the pigeon ECG. Atropine also increased the cardiac rate in a parrot that was sedated with ketamine hydrochloride and xylazine (Fig. 21–6).

The conduction disturbance, first-degree AV block (prolonged PR interval), occurs simultaneously with a decrease in body temperature and cardiac rate during prolonged inhalation anesthesia with halothane in healthy parrots and pigeons.[1] Progressive hypothermia during maintenance anesthesia results in a decrease in the cardiac rate below 100 beats per minute, with the PR interval increasing to more than 3.5 times the normal duration. The development of second-degree heart block is not observed.

Few clinical reports in the literature show a cardiac arrhythmia that correlates with primary heart disease in a caged bird. A recent report[2] documents atrial fibrillation in an anesthetized Pukeko with a history of respiratory distress and possible syncope. Thoracic radiographs

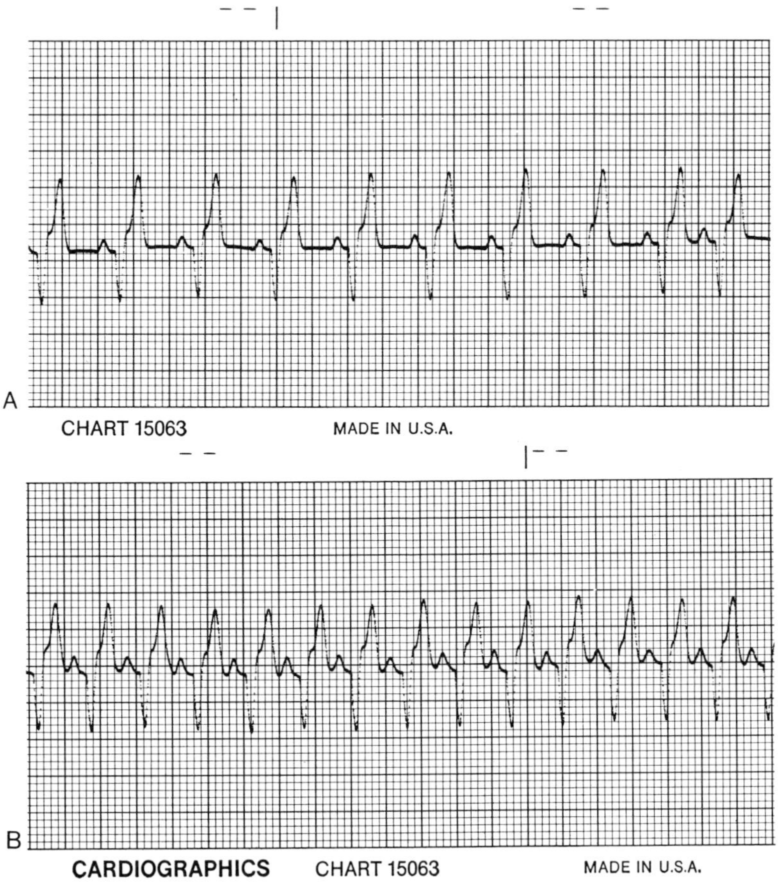


**Figure 21–6.** A Lead II ECG in a parrot sedated with ketamine hydrochloride and xylazine recorded at 50 mm/second. The heart rate prior to atropine sulfate administration is 280 bpm. *A*, Fifteen minutes after intramuscular administration of 0.2 mg atropine, the cardiac rate increases to 400 bpm. *B*, The P wave is slightly increased in amplitude and the sinus arrhythmia evident in *A* is abolished.

demonstrate cardiac enlargement, and postmortem examination reveals mitral valve abnormalities and myocardial pathology.

## CONCLUSION

Clinical cardiology in caged birds is in its infancy. Practitioners are advised to approach patients with potential cardiac disease in a consistent manner. The ECG is accepted as one of the principal methods of examining the cardiac status. All parts of the data base including history, physical examination, radiography, electrocardiography, and echocardiography are needed to definitively diagnose primary cardiac disease.

When cardiac disease is appropriately defined, the practitioner may utilize a therapeutic regimen of the current armamentarium for cardiac failure (diuretics, digitalis, glycosides, bronchodilators, vasodilator therapy, and new inotropic agents) in a rational manner. Favorable or poor clinical response must be correlated with dosages and serum levels of medications in order to determine therapeutic serum levels. Clinical trials comparing individual and combination treatment regimens must be evaluated.

As we gain experience in interpreting the avian ECG, abnormalities correlated with infectious disease, nutritional disease, toxicities, and electrolyte and metabolic derangements will be documented. This information will further aid in the diagnosis, treatment, and prognosis of specific disease conditions.

## REFERENCES

1. Altman, R. B., and Miller, M. S.: Effects of anesthesia on the temperature and electrocardiogram of birds. Proceedings of the Annual Meeting of the American Association of Zoo Veterinarians, Denver, CO, 1979, p. 42–44c.
2. Beehler, B. A., et al.: Mitral valve insufficiency with congestive heart failure in the Pukeko. J. Am. Vet. Med. Assoc., 177:934–937, 1980.
3. Gross, W. B.: Electrocardiographic changes of *Escherichia coli* infected birds. Am. J. Vet. Res., 27:1427–1436, 1966.
3a. Harrison, G. J., et al.: A clinical comparison of anesthetics in domestic pigeons and cockatiels. Proceedings of Annual Meeting of the Association of Avian Veterinarians, Boulder, CO, 1985, pp. 7–22.
4. Hunsaker, W. G., et al.: The effect of round heart disease of the electrocardiogram and heart weight of turkey poults. Poultry. Sci., *50*:1712–1724, 1971.
5. Iturri, S. J., and Ringer, R. K.: The effect of dietary PCB on the electrocardiogram of cockerels. IRCS Med. Sci., 7:254, 1979.
6. Mitchell, B. W., and Brough, M.: Comparison of electrocardiograms of chickens infected with viscerotropic velogenic Newcastle disease virus and virulent avian influenza virus. Am. J. Vet. Res., 43(12):2274–2278, 1982.
7. Ringer, R. K.: Personal communication, 1984.
8. Sturkie, P. D.: Further studies of potassium deficiency on the electrocardiogram of chicken. Poultry Sci., 31:648–650, 1954.
9. Sturkie, P. D.: Heart: Contraction, conduction and electrocardiography. *In* Sturkie, P. D. (ed.): Avian Physiology. New York, Springer-Verlag, 1976, pp. 114–121.
10. Sturkie, P. D.: Role of estrogen on the sex differences of the electrocardiogram of the chicken. Proc. Soc. Exp. Biol. Med., 94:731–733, 1967.
11. Zenoble, R. D., and Graham, D. L.: Electrocardiography of the parakeet, parrot and owl. Proceedings of the Annual Meeting of the American Association of Zoo Veterinarians, 1979, pp. 42–44c.
12. Zenoble, R. D.: Electrocardiography in the parakeet and parrot. Comp. Cont. Educ. Pract. Vet., 3(8): 711–714, 1981.

## RECOMMENDED READING

Tilley, L. P.: Essentials of Canine and Feline Electrocardiography. Interpretation and Management. Philadelphia, Lea & Febiger, 1985.

### ACKNOWLEDGMENT

I wish to express thanks to Amy Miller for her excellent illustration (Figure 21–3).

# Chapter 22

# MISCELLANEOUS DIAGNOSTIC TESTS

CLINTON D. LOTHROP, JR.
GREG J. HARRISON

A number of specialized tests are being investigated as possible diagnostic aids for avian practitioners.

## CHLAMYDIA

A chlamydia identification kit* developed for the diagnosis of *Chlamydia trachomatis* in humans utilizes a peroxidase-antiperoxidases (PAP) method in conjunction with a genus-specific monoclonal antibody. Preliminary work with the test as a rapid, in-house method of identifying *Chlamydia psittaci* in birds suggests a high correlation with culture results.[11] The product was designed to detect chlamydia in McCoy tissue culture cells; however, in birds it was found to identify positive reactions in formalin-fixed, paraffin-embedded tissues, direct staining of impression smears of the spleen, or from choanal and cloacal swabs. Because the test involves several steps and considerable time, it could be conducted on a weekly rather than daily basis.

Research is currently being conducted to develop effective methods to differentiate chlamydial strains.[1] Monoclonal antibodies and DNA restriction endonuclease analysis both have potential for use in differentiating strains of chlamydia. Monoclonal antibodies will be easier, faster, and less expensive to use for diagnostic purposes; however, the DNA restriction endonuclease analysis will provide improved accuracy for epidemiologic studies.

## THYROID FUNCTION TESTING

A number of investigators have evaluated thyroid function testing to confirm or eliminate a suspicion of hypothyroidism or hyperthyroidism. Because birds with thyroid dysplasia or goiter are functionally hypothyroid, an iodine response test is necessary to differentiate this disease from true hypothyroidism.

Normal resting thyroid hormone concentrations have been reported for selected species.[14] It is important not to diagnose thyroid disease on the basis of a single random blood sample. Many factors such as stress, fever, systemic illness, and drug administration can falsely lower the serum thyroxine concentration into the hypothyroid range. However, the capacity of the thyroid gland to respond to thyrotropin (TSH) will not be reduced and the serum thyroxine concentration ($T_4$) will increase in the TSH stimulation test unless the individual is hypothyroid. Also, hypothyroidism should not be ruled out with a normal resting thyroxine concentration but only after a nonresponse is seen in a thyroid function test.

The iodine response test consists of intramuscular injections of 20 per cent sodium iodide solution for three to five days (chronically affected budgerigars receive 0.01 cc daily; acutely affected, critical budgerigars receive 0.02 to 0.03 cc). If there is no clinical response to the iodine and the $T_4$ level remains low following a subsequent TSH test, goiter is ruled out and true hypothyroidism is confirmed.

Hyperthyroidism could be confirmed in a bird with suggestive clinical signs and elevated serum thyroxine or triiodothyronine concentration and/or exaggerated response in the TSH stimulation test.

Most diagnostic laboratories have developed RIA techniques for measuring hormone concentrations commonly seen in human diseases. Because there is considerable variation in hormone concentrations between birds and man, it is inappropriate to submit avian samples for hormone quantification and expect meaningful results unless the diagnostic laboratory has

*Cultureset—Ortho Diagnostic Systems, Inc., Raritan, NJ 08869 (800–631–5807).

**Table 22–1.** SERUM THYROXINE CONCENTRATIONS AFTER TSH STIMULATION IN PSITTACINE BIRDS[a]

| | Thyroxine (ng/ml) | |
|---|---|---|
| | BASELINE | POST-TSH |
| Cockatoo Species[b] (*Cacatua* sp.; n=6) | 13.63 ± 6.53 | 35.10 ± 13.16* |
| Amazon Species[c] (*Amazona* sp.; n=8) | 8.19 ± 6.90 | 27.40 ± 15.93* |
| Scarlet Macaws (*Ara macao;* n=9) | 1.34 ± 0.51 | 6.46 ± 3.10*** |
| Blue and Gold Macaws (*Ara ararauna;* n=8) | 3.41 ± 1.78 | 12.36 ± 6.34*** |
| African Greys (*Psittacus erithacus;* n=6) | 1.42 ± 0.44 | 9.30 ± 2.90*** |
| Conure Species[d] (*Aratinga* sp.; n=5) | 1.76 ± 0.77 | 13.50 ± 7.71* |
| Cockatiels (*Nymphicus hollandicus;* n=3) | 11.83 ± 6.76 | 39.00 ± 5.66** |

[a]Significance (p value) was determined with the Student's T test, comparing baseline and post-TSH stimulated thyroxine concentrations. *p <0.01; **p <0.005; ***p <0.001. Baseline blood samples were collected, 1 IU of TSH administered I.M., and the post-TSH stimulated sample collected 4 to 6 hours later.

[b]Goffin Cockatoo (1); Umbrella Cockatoo (2); Moluccan Cockatoo (2); Greater Sulfur-crested Cockatoo (1).

[c]Blue-fronted Amazon (4); Yellow-naped Amazon (1); Yellow-cheeked Amazon (1); Lilac-crowned Amazon (2).

[d]Mitred Conure (2); Nanday Conure (3).

properly modified procedures specifically to quantitate avian hormones. There are no apparent side effects to TSH administration in the psittacine birds examined.[6, 7]

Thyroxine concentrations ($T_4$) before and after TSH stimulation for selected psittacine birds are summarized in Table 22–1. Although there is considerable variation with species in the absolute thyroxine concentration, in this study there was at least a doubling in serum thyroxine concentration after TSH stimulation in the psittacines tested.[6, 7]

Table 22–2 lists serum triiodothyronine concentrations ($T_3$) before and after TSH stimulation in selected psittacine species. There appears to be a slight increase in serum triiodothyronine concentration after TSH stimulation, except for the macaw species.

The age[15] and sex[13] of the bird and time of the year[3] have been shown to relate to variable

**Table 22–2.** SERUM TRIIODOTHYRONINE CONCENTRATION AFTER TSH STIMULATION IN PSITTACINE BIRDS[a]

| | Triiodothyronine (ng/ml) | |
|---|---|---|
| | BASELINE | POST-TSH |
| Cockatoo Species (*Cacatua* sp.; n=3) | 1.22 ± 0.35 | 1.67 ± 0.50 |
| Amazon Species (*Amazona* sp.; n=2) | 1.26 ± 0.21 | 1.61 ± 0.41 |
| Scarlet Macaws (*Ara macao;* n=9) | 0.84 ± 0.26 | 0.69 ± 0.26 |
| Blue and Gold Macaws (*Ara ararauna;* n=8) | 0.75 ± 0.16 | 0.70 ± 0.26 |
| African Greys (*Psittacus erithacus;* n=4) | 1.31 ± 0.13 | 1.47 ± 0.24 |
| Conure Species (*Aratinga* sp.; n=5) | 0.84 ± 0.22 | 1.15 ± 0.58 |
| Cockatiels (*Nymphicus hollandicus;* n=2) | 1.45 ± 0.35 | 1.67 ± 0.21 |

[a]TSH stimulation testing is described in a footnote to Table 22–1.

thyroxine and triiodothyronine concentrations in other avian species. For example, plasma $T_4$ levels in Canada Geese reached their highest values during the spring postmigratory and breeding periods and their lowest during the fall postmigratory period. Peak $T_3$ levels were found to occur during the spring premigratory period and minimum levels to occur in the fall premigratory phase. No relationship was evident between thyroid activity and molt in this study. Variation in $T_4$ levels was attributed to the effects of variation in ambient temperatures and daylength on thyroid and animal activities.

From a sound medical and scienctific perspective, the TSH response may be the one best test for definitive diagnosis of primary hypothyroidism. However, this requires collecting blood samples both before and after TSH administration and, for accurate results, samples must be mailed to a valid reference laboratory. It is Harrison's opinion that there may be select clinical situations in which the TSH test is not appropriate (e.g., budgerigars). A tentative diagnosis of goiter might be made by clinical response to supplemental iodine. Thyroid-responsive conditions (not necessarily true hypothyroidism) may be suggested by positive response to thyroid supplementation. In these cases, the practitioner is advised to be alert to iatrogenic *hyper*thyroidism.

## ADRENAL FUNCTION TESTING

Confirmation of a clinical suspicion of adrenal disorders can be made by determining glucocorticoid concentrations (corticosterone in birds; cortisol in humans and dogs) before and after provocative stimulation. Quantification of corticosterone requires an assay specific for this hormone. Thus, diagnostic techniques for determining cortisol concentrations should not be used for corticosterone concentration measurements.

ACTH stimulation tests have been performed with both repository ACTH gel (Adrenomone—Burns-Biotec, Omaha, NE 68103) and aqueous ACTH (Cosyntropin—Organon, Inc., West Orange, NJ 07082). Significant stimulation of corticosterone occurs with administration of either form of ACTH in normal birds.

To perform an ACTH stimulation test, a baseline blood sample is collected (50 μl plasma), 16 to 25 units of ACTH are administered intramuscularly, and a second blood sample is collected between one and two hours later (50 μl plasma).

Corticosterone concentrations before and after ACTH stimulation for a variety of popular psittacine birds are summarized in Table 22–3.[5] Although there is some variation in resting corticosterone concentrations, the magnitude of corticosterone increase after ACTH stimulation is similar for most psittacines. Adrenal insufficiency should be suspected when no increase in corticosterone is seen after ACTH administration. Cushing's syndrome may be suspected with elevated corticosterone concentrations before and after ACTH administration, although this disease has not been reported in birds.

## FLUORESCENT ANTIBODY

Fluorescent antibody (FA) tests are being developed for the rapid diagnosis of a number of avian diseases.[2] A tentative diagnosis of Pacheco's disease may be made by demonstration of the inclusion bodies on impression smears of the liver stained with the hematoxylin and eosin technique for frozen sections. FA tests can also be done on impression smears; therefore, virus isolation in eggs or cell culture is not necessary for diagnosis of Pacheco's disease in laboratories equipped for FA tests.

Fluorescent antibody conjugates have been produced for use as a screening test for psittacine reovirus. Generalized parrot papovavirus (including budgerigar fledgling disease), chlamydia, and other organisms may also be identified by this method.

## INDOCYANINE GREEN

Indocyanine green (ICG) (Cardio Green—Hynson, Westcott and Dunning, Baltimore, MD) is under investigation as a possibly useful and more sensitive measure of avian liver function than other serum chemistry tests. ICG is a water-soluble tricarbocyanine, cholephilic dye that is bound to plasma protein after intravenous injection. Because it is eliminated almost exclusively by hepatic parenchymal cells and is excreted via the bile, the plasma dye clearance is a sensitive indicator of hepatobiliary function. ICG has been tested in humans, dogs, cats, rabbits, and ducks[10] and more recently in three species of raptors.[9] The avian studies did not correlate the results with clinical illness or liver biopsies.

**Table 22–3.** PLASMA CORTICOSTERONE CONCENTRATIONS AFTER ACTH STIMULATION IN PSITTACINE BIRDS[a]

| | Corticosterone (ng/ml) | |
|---|---|---|
| | BASELINE | POST-ACTH |
| Cockatoo Species (*Cacatua* sp.; n = 7) | 6.7 ± 2.9 | 29.1 ± 13.4 |
| Scarlet Macaws (*Ara macao;* n = 6) | 3.58 ± 4.0 | 32.3 ± 11.4 |
| Blue and Gold Macaws (*Ara ararauna;* n = 6) | 1.83 ± 1.6 | 25.3 ± 5.5 |
| Green-wing Macaws (*Ara chloroptera;* n = 5) | 3.1 ± 3.07 | 18.7 ± 5.7 |
| Amazon Species (*Amazona* sp.; n = 6) | 2.66 ± 3.16 | 32.1 ± 10.7 |
| Conure Species (*Aratinga* sp.; n = 3) | 2.7 ± 3.1 | 16.5 ± 10.8 |
| Lorikeet Species (*Charmosyna* sp.; n = 2) | 1.72 ± 1.89 | 22.5 ± 3.5 |

[a]Baseline blood samples were collected, 16 units ACTH (Adrenomone) administered I.M., and post-ACTH stimulation samples collected between 1½ and 2 hours later.

## GALACTOSE DEGRADATION CURVE

Galactose was administered to a group of clinically normal doves and subsequently evaluated for postadministration degradation.[12] Because the liver is the site of galactose removal, it is hypothesized that this could be a valuable liver function test. No clinical or histopathologic case correlation was done.

## BIOASSAY

There are occasions when an avian practitioner may not have access to a laboratory equipped to perform specialized tests to identify infectious organisms. The use of sentinel birds is one technique that the clinician may employ to determine if a recently necropsied bird had a contagious disease or if a newly acquired bird is shedding contagious organisms. A closed flock of budgerigars with known medical history may be used as sentinels. Such sentinel birds can be placed next to the new birds or housed separately and force-fed small samples of cloacal and pharyngeal swabs every few days for a minimum of 10 days. The sentinel birds are then closely observed for at least three weeks following the exposure.

Another bioassay test may determine the pathogenicity of bacterial cultures. A 0.05-cc broth sample may be administered subcutaneously to healthy budgerigars. In a number of cases studied by Harrison, although most birds showed transient lethargy for several hours owing to metabolites in the broth, pathology or disease from isolated and inoculated organisms was rarely seen. This appears to further verify that the predisposing factor in the disorder of the cultured patient was probably related to management (stress, hygiene, malnutrition) or individual cases of previous viral diseases or immunosuppression rather than to the bacteria. From these findings it is hypothesized that it is inappropriate to treat an entire flock for a bacterial organism isolated from one or two birds if the true cause is other than bacterial.

Bioassay birds should be euthanized and complete necropsy and histopathologic evaluation performed to detect subtle disease. Some birds are very genetically resistant to certain diseases (e.g., chlamydiosis in budgerigars and pigeons).

## WATER DEPRIVATION TEST

Withholding water for 12 hours in polyuric birds will assist in diagnosing kidney function. If the specific gravity of the urine following the water deprivation test is greater than 1.025, psychogenic polydipsia and nonrenal metabolic diseases other than diabetes insipidus should be expected. Fixed concentrations of 1.008 to 1.012 suggest renal involvement in the disease process. If urine does not become concentrated with deprivation, nephrogenic diabetes insipidus, psychogenic polydipsia, or medullary washout should be considered.[4]

## BONE MARROW ASPIRATES

Bone marrow aspirates may be obtained from the cranial aspect of the tibia in psittacines using 22-gauge spinal needles.[16] The site is aseptically prepared and the needle directed through the patellar ligament into the tibia. To avoid blood contamination, the sample is quickly aspirated and the needle removed. Larger needles may be used with the same technique under anesthesia to obtain bone biopsies.

## CAT SCANS

Computerized axial tomography (CAT) scans have been experimentally used in birds of prey.[8] Extensive decreased brain density in the retrobulbar region and cerebral cortex of a Barred Owl were noted. Evaluation of Hounsfeld units suggested that the lesion represented brain edema and/or hemorrhage. Sophisticated equipment for human diagnostic procedures may in the future have value in avian patients.

## REFERENCES

1. Andersen, A.: The use of monoclonal antibody and DNA restriction endonuclease analysis for typing of *Chlamydia psittaci.* Abstract in Proceedings of the 57th Northeastern Conference on Avian Disease and 6th Mid-Atlantic States Avian Medicine Seminar, Columbia, MD, 1985.
2. Graham, D. L.: Update on Pacheco's disease. AAV Newsletter, 6(2):51, 1985, from material presented at Annual Meeting of the Association of Avian Veterinarians, Boulder, CO, 1985.
3. John, T. M., and George, J. C.: Circulating levels of thyroxine ($T_4$) and triiodothyronine ($T_3$) in the migratory Canada Goose. Physiol. Zool., 51:361–370, 1978.
4. Kollias, G. V.: Polydipsia, polyuria and diarrhea in birds: A diagnostic approach. AAV Newsletter, 6(1):23–25, 1985, from material presented at AAHA Annual Meeting, Orlando, FL, 1985.
5. Lothrop, C. D.: Diagnosis of adrenal diseases in caged birds. In Proceedings of the International Conference on Avian Medicine, sponsored by the Association of Avian Veterinarians, Toronto, Canada, 1984.
6. Lothrop, C. D.: Diseases of the thyroid gland in caged birds. In Proceedings of the International Conference on Avian Medicine, sponsored by the Association of Avian Veterinarians, Toronto, Canada, 1984.
7. Lothrop, C. D., et al.: Thyrotropin stimulation test for evaluation of thyroid function in psittacine birds. J. Am. Vet. Med. Assoc., 186(1):47–48, 1985.
8. McMillan, M. C.: Computerized axial tomography in birds of prey. AAV Newsletter, 6(2):60, 1985.
9. Olsen, G. H., and Holmes, R. A.: Indocyanine green as an indicator of liver function in raptors. Unpublished data.
10. Patton, J. F.: Indocyanine green: A test of hepatic function and a measure of plasma volume in the duck. Comp. Biochem. Physiol., 60:21–24, 1978.
11. Petrak, M. L.: The use of an immunoperoxidase–monoclonal antibody procedure in the diagnosis of avian chlamydiosis. In Proceedings of the Annual Meeting of the Association of Avian Veterinarians, Boulder, CO, 1985, pp. 265–266.
12. Rich, B., and Schultz, D.: Personal communication.
13. Ringer, R. K.: Personal communication.
14. Rosskopf, W. J., et al.: Normal thyroid values for common pet birds. VM/SAC, 77:409–412, 1982.
15. Spiers, D. E., and Ringer, R. K.: Thyroid hormone changes in the Bobwhite *(Colinus virginianus)* after hatching. Gen. Comp. Endocrinol., 53:302–308, 1984.
16. VanDerHeyden, N.: Technique for bone marrow aspiration. AAV Newsletter, 6(2):57, 1985, from material presented at the Annual Meeting of the Association of Avian Veterinarians, Boulder, CO, 1985.

### ACKNOWLEDGMENTS

The contributions to this chapter by Robert D. Zenoble, Robert J. Kemppainen, and James L. Sartin are appreciated.

# Chapter 23

# NECROPSY PROCEDURES

LINDA J. LOWENSTINE

A thorough postmortem examination of birds that die or are euthanized is a necessary adjunct to any good clinical practice. If a practitioner has access to the services of a veterinary pathologist who will perform the gross necropsy, procedures will be established for the submission of the intact carcass. More commonly, the avian practitioner or a technician performs the postmortem examination and submits appropriate tissues to a diagnostic laboratory. The quality of information received from such an examination is directly proportional to the quality and choice of the specimens submitted and the information that accompanies them. The pathologist will make use of clinical history (including hematology and blood chemistry), the gross description, culture results, and other data as well as the histologic appearance of the lesions to make a diagnosis. Absence of any of these or incorrect submission of tissues will hamper this process.

## EQUIPMENT NECESSARY

It is helpful to have a set of instruments designated for postmortem examinations. These should be thoroughly sterilized (i.e., autoclaved) after use in a necropsy. One should not use instruments that are used around living birds. Some pathogens, such as avian pox virus, are extremely resistant to casual disinfection. Instruments sterilized in chemical disinfectant should be rinsed in sterile water prior to use to avoid killing pathogens on contact, thus altering the diagnostic picture.

The instrument pack should include forceps, scalpel handle, necropsy knife, stout scissors and/or poultry shears (for cutting bone), and fine scissors for dissection (Fig. 23–1). Tiny birds such as finches require fine instruments such as iris scissors. For large birds (e.g., ratites or waterfowl), instruments appropriate for small or large mammal necropsy including a vibrating (cast-cutting) saw may be used.

In addition to instruments, one should have on hand 10 per cent buffered formalin or a similar fixative (e.g., Carson's), 100 per cent ethyl alcohol (for fixing specimens suspected of having gout), and appropriate containers.

Other materials for ancillary diagnostic procedures include syringe and needle to obtain samples for serology, hematology, or cytology; clean glass slides for impression smears; sterile swabs or culture tubes with appropriate transport media for bacterial, fungal, chlamydial, or mycoplasmal cultures; and sterile Petri dishes or freezer-proof tubes for submission of tissues for viral isolation. One may choose to have a camera available for documentation of gross lesions.

A standard necropsy report form will assist in recording observations (Fig. 23–2.)

## EUTHANASIA

The method of euthanasia may affect specimens submitted to the pathologist. High doses of barbiturates given intravenously are caustic to tissues and may cause spurious lesions. Crystallization in and on the heart may be mistaken for early gout. Giving such agents slowly to effect is helpful.

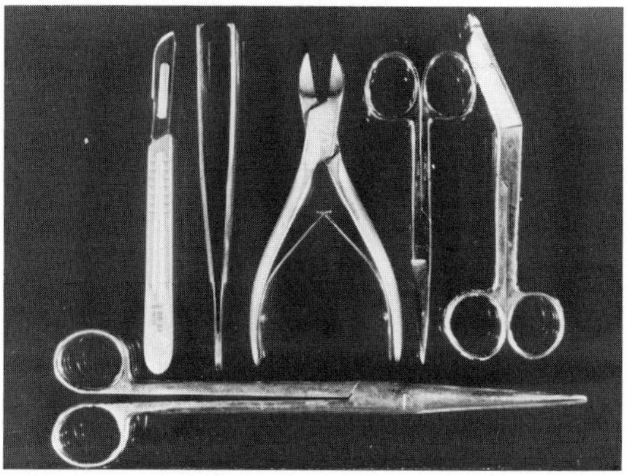

**Figure 23–1.** Instruments suitable for postmortem examination of small to medium-sized birds.

## AVIAN NECROPSY REPORT FORM

| | |
|---|---|
| Clinic No.________________ | Pathology No.________________ |
| Owner________________ | Address________________ |
| Species________________ | Age________ Sex________ |
| Name________________ | Band No.________________ |
| Postmortem interval________________ | General condition________________ |

External Examination

- Integument:
- Oral cavity:

Internal Examination

- Subcutis:
- Musculoskeletal:
- Visceral arrangement:
- Air sacs/pleura/peritoneum:
- Digestive tract
    - Upper:
    - Stomachs:
    - Intestines:
    - Cloaca:
- Liver:
- Pancreas:
- Urinary tract:
- Reproductive tract:
- Respiratory tract
    - Infraorbital sinuses:
    - Lungs:
- Heart and vessels:
- Lymphoid system (thymus, spleen, bursa):
- Endocrine system (thyroids, pituitary, adrenals):
- Special senses (eyes, ears, nares):
- Bone marrow:

Ancillary Diagnostics

- Bacteriology (mycology):
- Virology:
- Toxicology:
- Parasitology:
- Other:

Tissues Saved:

Tissues Submitted for Histopathology:

Case Summary (tenative diagnoses):

**Figure 23–2.** Avian necropsy report form. Use of a postmortem report form prevents many oversights.

The commonly used euthanasia agent, T-61, can be administered intravenously or intramuscularly.*

Hematologic samples may be obtained prior to euthanasia. Large samples may be obtained from the jugular vein or direct heart puncture through the thoracic inlet. The blood may be centrifuged and serum submitted or saved and frozen pending necropsy results. This may be helpful in diagnosis of endocrine disorders or viral infections. Routine hematologic tests may also be performed on this sample.

## THE POSTMORTEM EXAMINATION

There are probably as many ways to dissect a bird as there are pathologists. Several procedures have been published (see Recommended Readings). One should choose a procedure with which one is comfortable and use it consistently. As when performing an exploratory laparotomy, a thorough examination done in a standard manner will preclude overlooking a lesion. The procedure that follows is one example and not the definitive method.

The first step is to perform a thorough external examination of the carcass. Abnormalities that may not have been apparent in the clinical setting may be detected. Palpation followed by direct examination helps to sharpen one's clinical skills. Make note of leg band numbers or other identifying marks at this time.

If the body has been submitted by a client, it should already be wet from the soapy water soak recommended prior to cooling to reduce the insulation qualities of the feathers (see Chapter 6, Management Procedures). A recently dead bird should be wetted prior to necropsy and plucked to allow for better visualization of the skin and to prevent loose feathers from irritating the prosector and contaminating the viscera. Some birds, e.g., waterfowl, are too heavily feathered to permit easy plucking. The feathers in these birds should be wetted and parted to permit incision of the skin. The skin in normal adult birds is opaque owing to dermal fat stores (Fig. 23–3). Transparent skin and a prominent keel bone suggest weight loss. These changes can occur rapidly in small birds because of their high metabolic rate.

The bird is positioned on its back with the wings tucked underneath. The skin of the medial thighs and adductor muscles is cut and the hips are disarticulated by forcing the knees outward. This helps to stabilize the carcass. One may wish to pin the wings and legs of tiny birds to a dissecting board to keep the carcass steady.

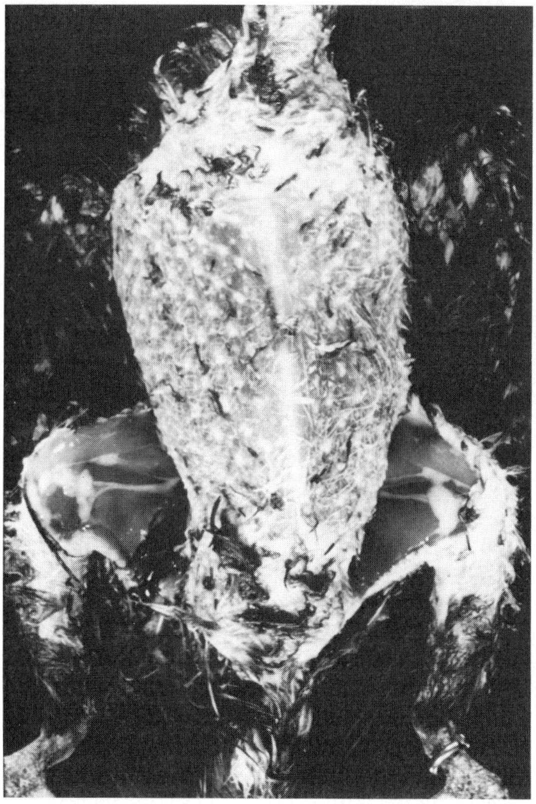

**Figure 23–3.** Plucked parrot demonstrates the opaque skin indicative of adequate fat stores.

An incision is made in the skin along the ventral midline from the mandible to the vent. The skin is reflected to expose the underlying crop, keel, and pectoral muscles. In most normal cage and aviary birds, the pectoral muscles are plump and red-brown. Pallor or pale streaking may be a reflection of necrosis (as in white muscle disease or exertional myopathy) or a cellular infiltrate of inflammation or neoplasia (Fig. 23–4).

Next, incisions are made through the pectoral muscles along the sides and around the posterior border of the sternum through the abdominal muscles (Fig. 23–5). The sternum can be grasped and reflected anterodorsally (Fig. 23–6). At this point cultures of air sac exudates may be taken before they are contaminated by handling (Fig. 23–7). Heavy rongeurs, scissors, or poultry shears are then used to cut through ribs, coracoid bones, and clavicle to remove the sternal plate (Fig. 23–8). A midline incision is

*Editors' Note: An extremely fast and humane euthanasia is achieved by injection of T-61 into the spinal cord area at the base of the skull with the head flexed.

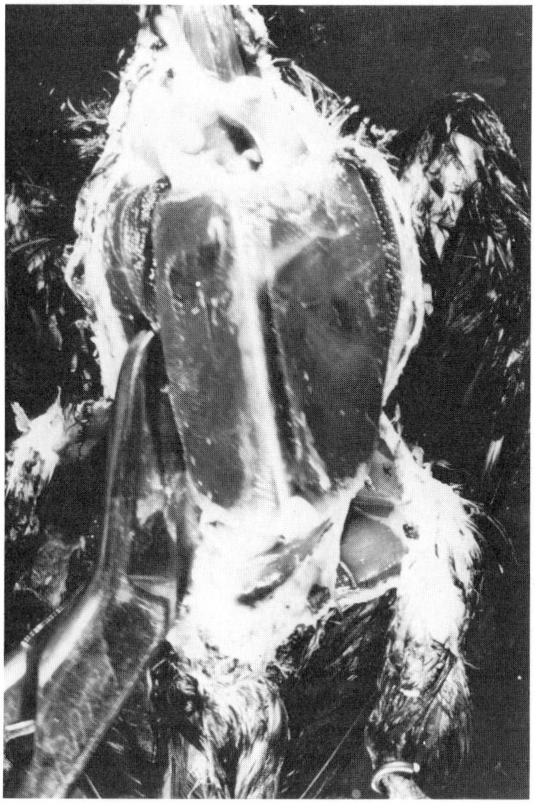

**Figure 23–4.** The skin has been incised and reflected to reveal crop, pectoral muscles, and abdomen.

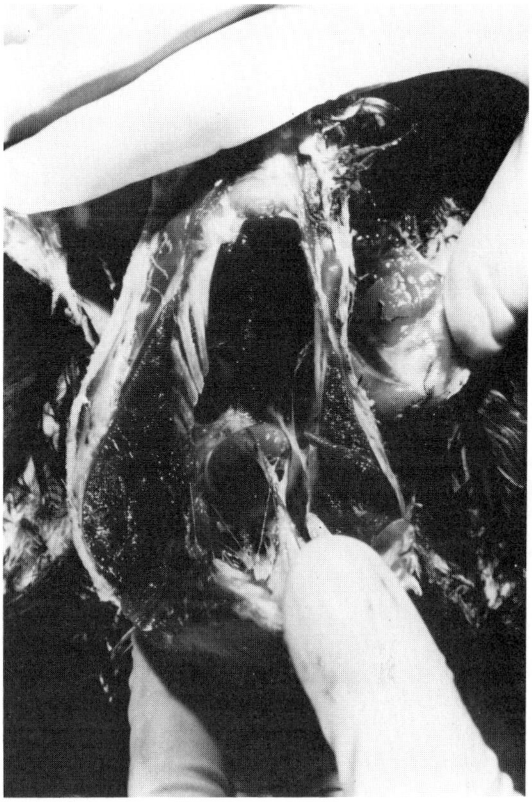

**Figure 23–6.** The sternal plate can be forced upward and forward.

**Figure 23–5.** Incision is made through the pectorals and around the posterior sternum.

made through the thin abdominal muscles. The viscera may then be examined *in situ* (Fig. 23–9).

The air sacs will have been partially ruptured in removing the sternal plate; nevertheless, sufficient portions usually remain for examination. Normal air sacs are transparent to slightly translucent and contain no fluid. Similarly, the pericardial sac should be nearly transparent. Well-fed birds have an abdominal fat pad similar to the falciform ligament in mammals. There may also be a pelvic fat pad around the cloaca. These are often excessive in inactive cage birds (especially budgerigars and galahs) and in waterfowl. Fat is also found within the coronary groove of all species and surrounding the epicardium in some species (e.g., ostrich). A gelatinous appearance to these areas is an indication of serious atrophy of fat secondary to severe inanition.

The liver is easily examined *in situ*. A uniform dark mahogany brown color is consistent among species of birds, whereas the shape and size may vary. The lobes are nearly equal in raptors and galliforms. In psittacines the right lobe is usually larger. In piscivorous birds the liver is

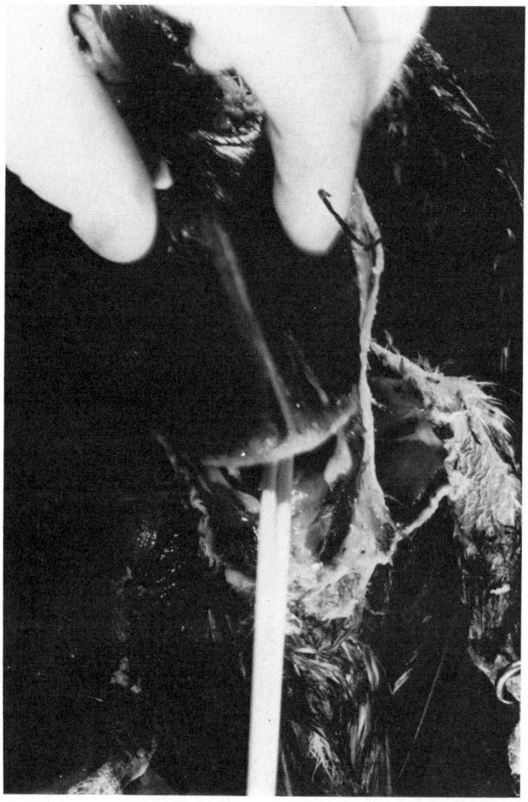

**Figure 23–7.** Cultures of air sacs should be made prior to contamination of the body cavity by subsequent manipulation.

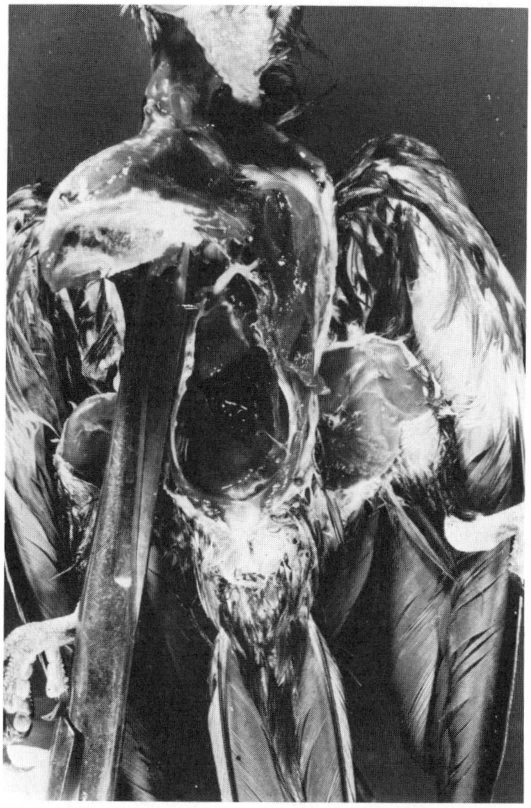

**Figure 23–8.** Stout scissors, rongeurs, or poultry shears may be used to cut coracoids and clavicle.

usually quite large and asymmetrical, with the right lobe extending to the pelvis.

The spleen can be examined by grasping the gizzard with forceps and rotating it toward the right side (Fig. 23–10). This exposes the spleen in the angle between the proventriculus and ventriculus. By gently pushing all viscera aside, one can examine the adrenals, gonads, and kidneys. It is helpful to examine the gonads and adrenals at this point, as they may be overlooked later. Testicles in birds undergo cyclic physiologic atrophy and enlargement. They may be quite large in breeding birds (Fig. 23–11) and should not be mistaken for tumors or lymph nodes (which most birds do not have). Finally, the lungs may be examined *in situ* by gently rotating the heart from side to side. Identify the thyroids above and lateral to the syrinx along the carotid arteries (Fig. 23–12).

Once all the viscera have been examined *in situ*, abnormalities noted; and cultures taken, the viscera should be removed for further examination. To do this, free the tongue by cutting along the base and through the hyoid apparatus. Using combined blunt and sharp dissection, cut through the posterior pharynx and retract the tongue, trachea, esophagus and

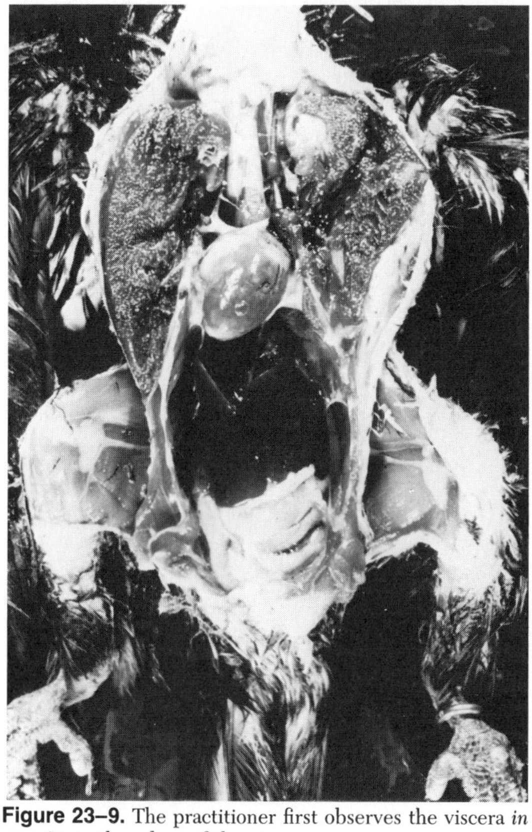

**Figure 23–9.** The practitioner first observes the viscera *in situ*. Note the edges of the air sacs.

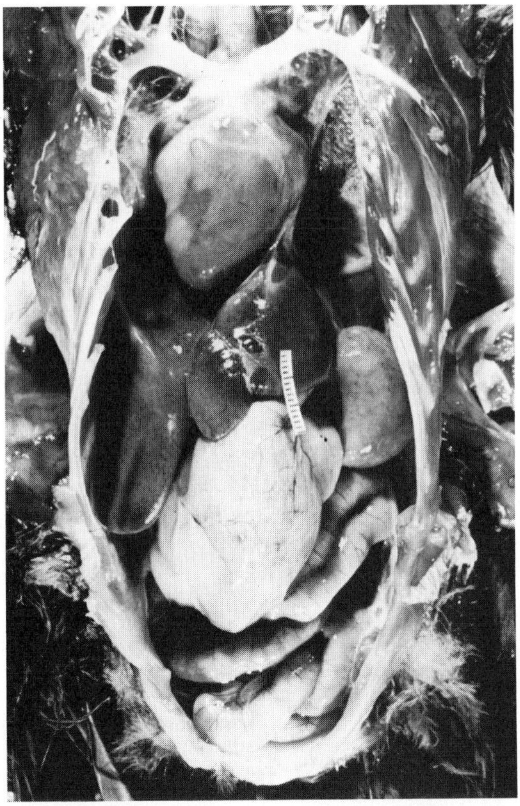

**Figure 23–10.** The greatly enlarged spleen in this macaw with psittacosis has been exposed by pushing and rotating the viscera toward the bird's right side.

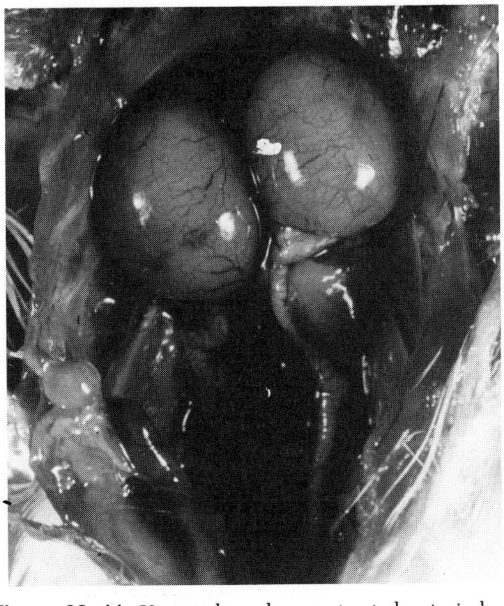

**Figure 23–11.** Viscera have been retracted anteriorly to expose kidneys and gonads in the pelvic fossa. The active testes in this breeding budgerigar are many times larger than in the nonbreeding male. The pale area in the left kidney below the testicle is an early renal adenocarcinoma. (Courtesy of Angell Memorial Animal Hospital.)

vessels with attached thymus, thyroids, and parathyroids.

To free the lungs, apply *gentle* traction to the trachea and esophagus and cut the attachment to the ventral ribs and the backbone at the thoracic inlet. With continued traction, push the lungs away from the ribs with the scalpel handle. This may be a difficult step, as there is no pleural space in birds. However, removal of the lungs is essential, as many pneumonic lesions in birds occur on the lateral and posterior surfaces of the lungs. By applying constant traction one can remove these tissues to the level of the anterior pole of the kidneys (Fig. 23–13). A combination of blunt and sharp dissection will free these anterior divisions.

Leave the anterior viscera and grasp the vent with forceps. With scalpel or scissors, cut around the vent and through the musculature. Retract the vent anteriorly and upward, exploring the bursa and posterior divisions of the kidneys (Fig. 23–14). Loosen these divisions with blunt dissection. The middle divisions from the sciatic nerve may be cut with scissors or scalpel. One should then be able to lift the

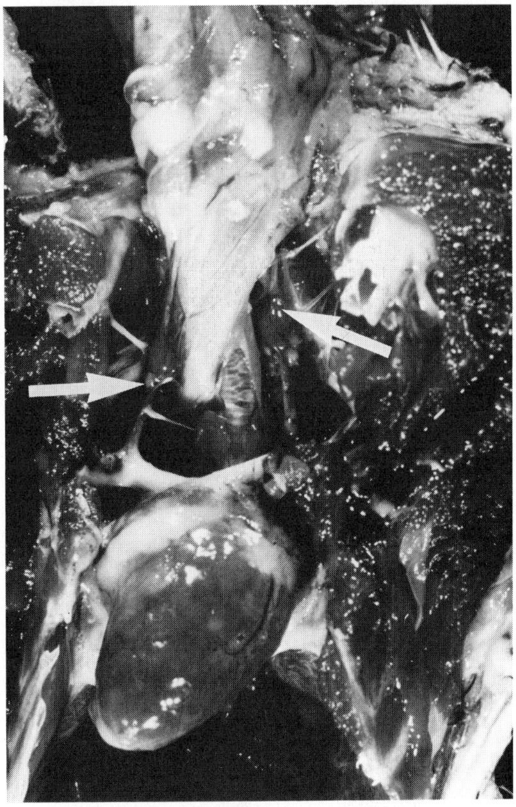

**Figure 23–12.** The thyroids are located at the heart base along the carotid arteries. They should not be confused with the muscles of the syrinx.

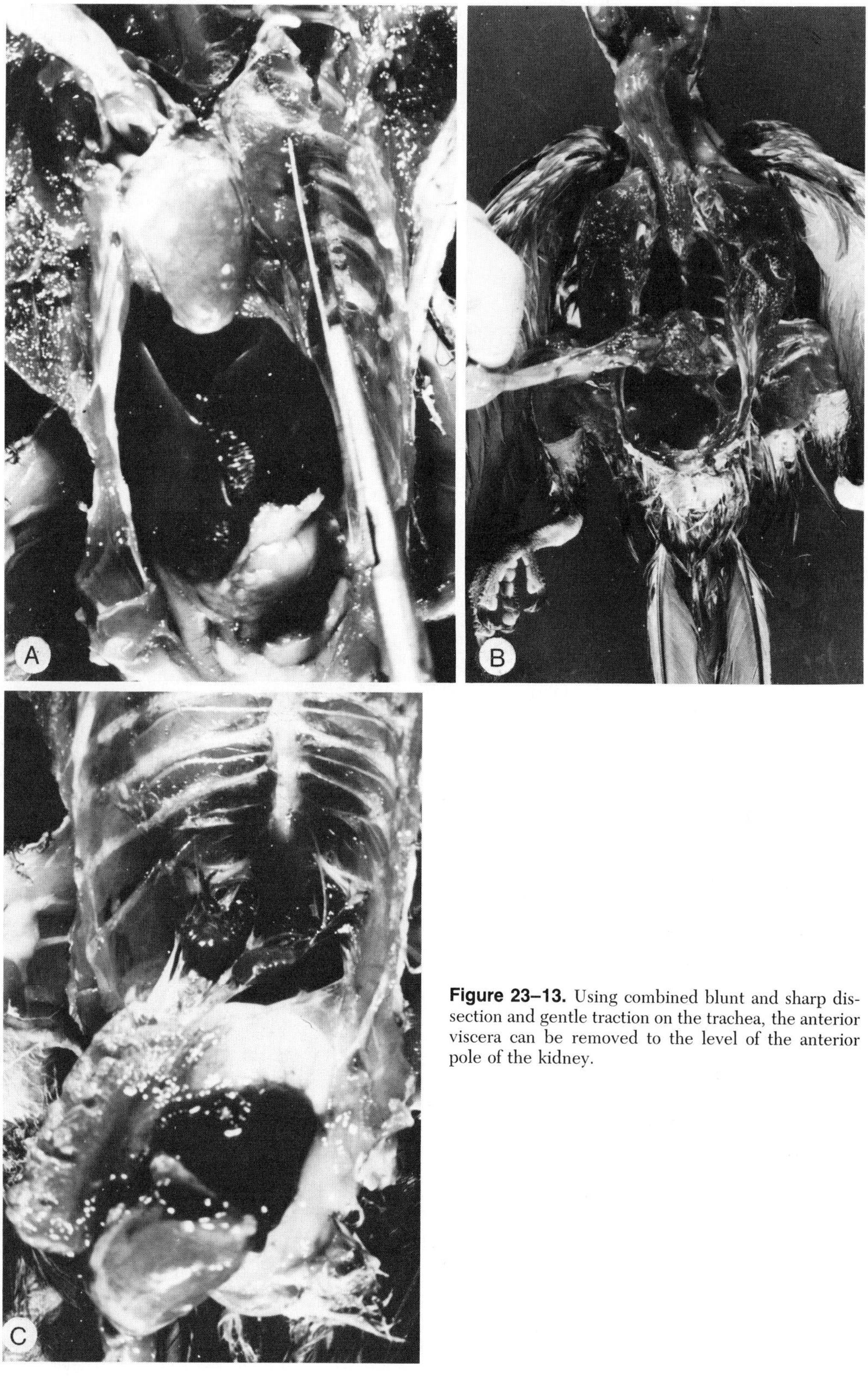


**Figure 23–13.** Using combined blunt and sharp dissection and gentle traction on the trachea, the anterior viscera can be removed to the level of the anterior pole of the kidney.

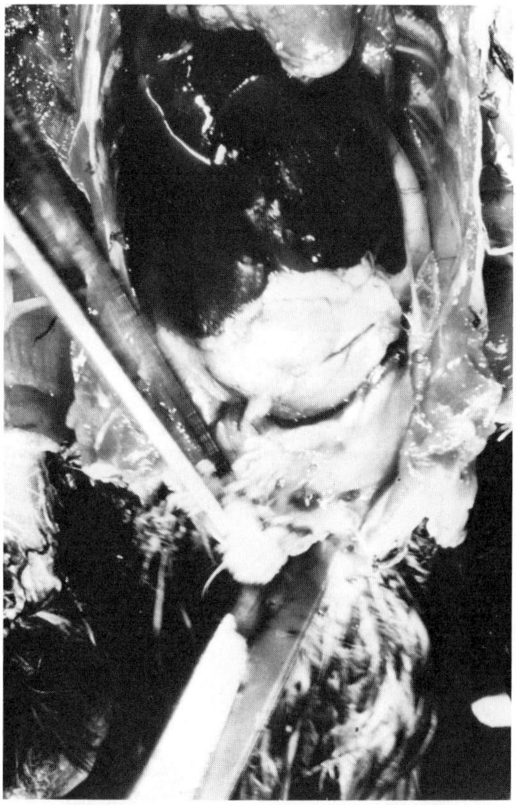

**Figure 23–14.** The posterior viscera may be freed by cutting around the vent.

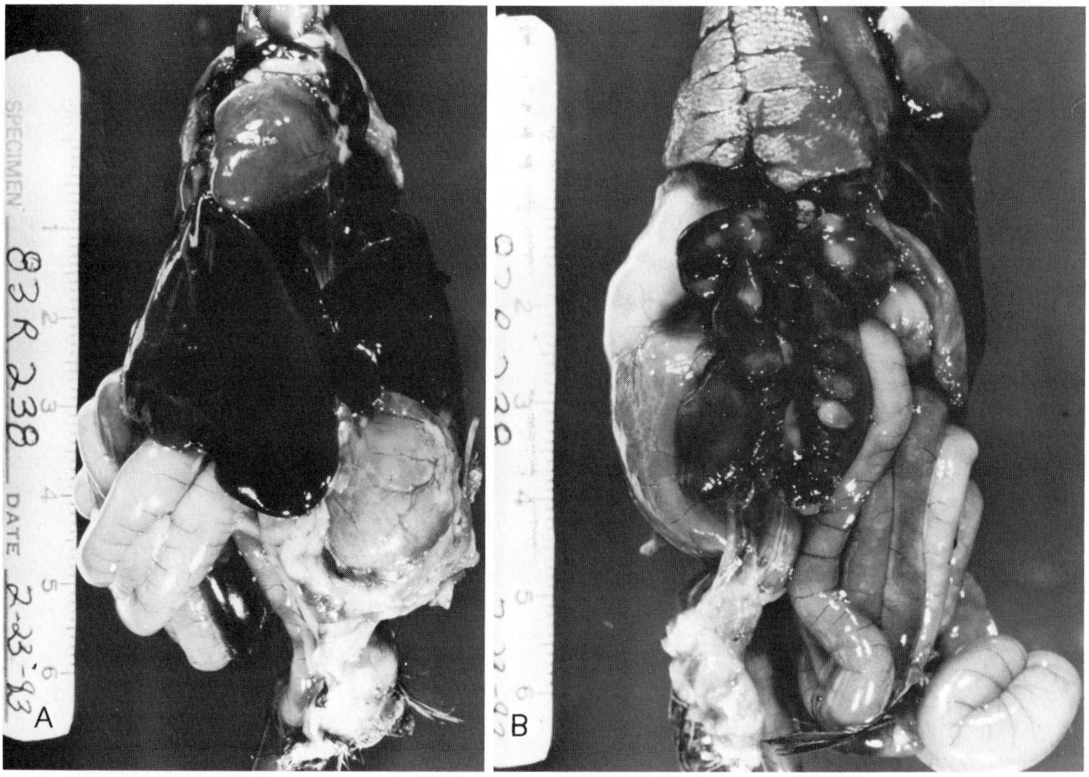


**Figure 23–15.** Viscera may be removed for further examination. Ventral view *(A)* is seen after removal. Dorsal view *(B)* reveals kidney which, in this Pionus Parrot, is dotted by pale foci of inflammation.

entire viscera from the carcass (Fig. 23–15). Once the viscera are removed, additional cultures may be obtained and samples of parenchymal organs may be taken for impression smears (see Chapter 17, Cytology). In very small birds such as finches it is impossible to remove the kidneys from the deep pelvic fossa. The pelvis may be cut from the carcass with kidneys *in situ*, and the entire block fixed in formalin.

The "thoracic" viscera may be removed from the "abdominal" organs by severing the distal esophagus and great vessels anterior to the liver. The esophagus, crop, and trachea should be opened. The mucosa of the esophagus and crop should be smooth and tan or cream or pink. It should not be leathery white or granular. Lesions of parasitic, fungal, bacterial, viral, and nutritional diseases occur in this area (Fig. 23–16).

In large birds, the heart may be opened to allow examination of the valves. Keep in mind that the right atrioventricular valve is a muscular rather than membranous structure. Instead of opening the heart in small birds, cut across the apex and fix *in toto* with the carotids and thyroids attached. Observe the lateral surfaces of the lungs for lesions (Fig. 23–17). Cut through the lungs at intervals of 0.4 to 0.5 cm (as if slicing bread) to look for lesions such as plugged bronchi which may not be readily visible from the pleural surface.

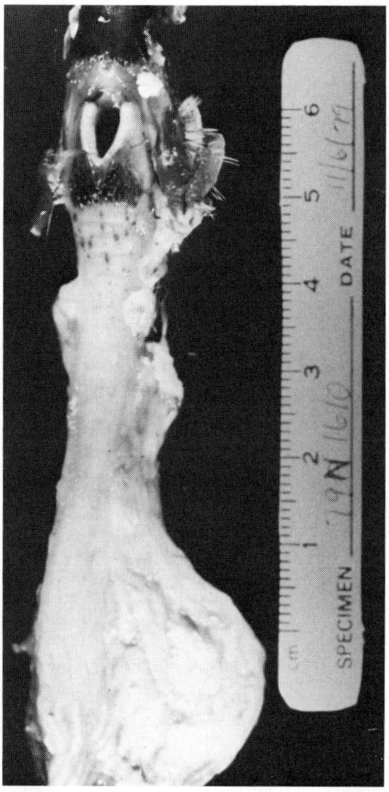

**Figure 23–16.** Esophagus and crop of a Blue-fronted Amazon have been opened to demonstrate nodular pox lesions and rugose thickening from candidiasis.

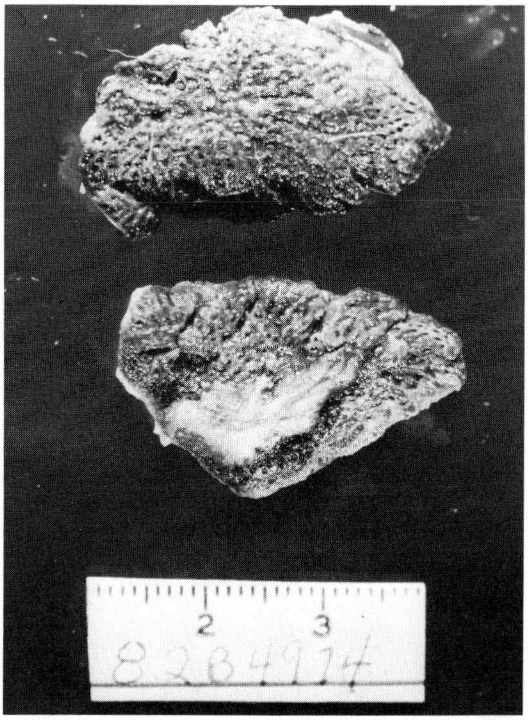

**Figure 23–17.** Bronchopneumonia is evident on lateral surfaces of the lungs from a cockatoo. The grey triangular area was not visible when the lungs were within the body cavity.

Next, examine the "abdominal" viscera. Open the proventriculus and ventriculus. The mucosa of the proventriculus is more opaque than that of the proximal alimentary tract. Pore-like openings of submucosal glands can be appreciated. The ventriculus varies with the natural diet of the bird species. The gizzard of seed-eating and omnivorous birds has a thick muscular wall, and the mucosa is lined by a horney material, called koilin, which is usually bile-stained. In carnivorous and piscivorous birds the gizzard is more fusiform and thinner-walled and blends with the proventriculus.

The bowel may be opened in large birds. In small birds, one should leave the bowel in its coiled configuration and make small cuts at intervals to examine for parasites and to allow penetration of formalin. In very tiny birds the bowel may be fixed *in toto*.

Examine the cloaca and search for the bursa (which resembles a small lymph node) on the dorsum of the cloaca (Fig. 23–18). The exact

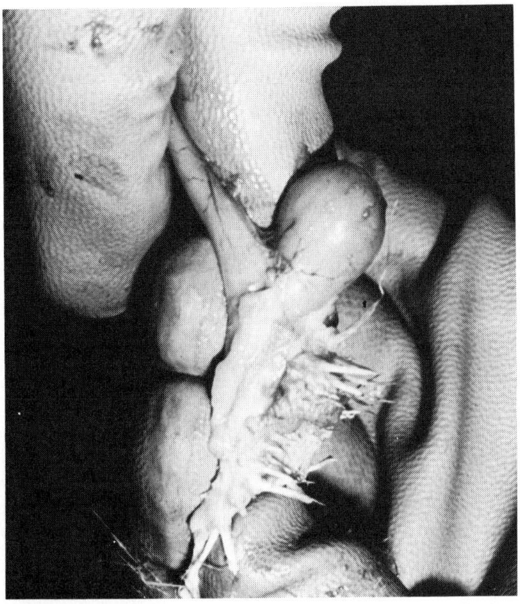

**Figure 23–18.** Vent, cloaca, and the lymph node–like bursa from a chicken are shown. (Courtesy of Dr. A. Bickford.)

times of bursal development and involution are unknown for most of the cage and aviary species.

The kidneys, gonads, adrenals, and oviduct (if a female) can be removed from the rest of the viscera and examined. There is normally a faint reticular pattern appreciated on the kidney surface. The presence of a white slurry within the ureters is common. The presence of stellate or linear white foci scattered throughout the kidney is abnormal, however, and usually indicates renal gout.

In contrast to the kidneys, the avian liver is usually homogeneous, lacking the lobular pattern usually seen in mammalian species. The liver should be sliced and examined in the same manner as the lungs were.

The spleen must be identified and measured. The normal budgerigar has a 1- to 2-mm spleen, a lovebird or cockatiel 3- to 4-mm, and an Amazon 7- to 8-mm. Spleen shape varies with species. It is oval in psittacines and galliforms, elongated to a comma shape in passerines, and very long and narrow in gulls. It varies from pink to red-brown. Tan-colored spleens are usually inflamed or have marked reticuloendothelial hyperplasia (Fig. 23–19). If splenomegaly is present, further diagnostic measures are indicated—impression, culture, histopathology.

The remainder of the carcass consists of musculoskeletal, integumentary, and nervous systems. Separate the head from the neck to remove the brain. Peel the "scalp" forward to explore the calvarium. Ecchymoses within the calvarium are a common agonal change and do not imply head trauma (Fig. 23–20). Cut through the calvarium using scissors, rongeurs, or a vibrating saw, depending upon the size of the bird (Fig. 23–21). The brain can be removed from the calvarium by severing the cranial nerves as one lifts the brain from front to back. This is easy to do in chickens but somewhat more difficult in psittacines because of the bone plate that covers the large optic lobes (optic tecta). In tiny birds do not attempt to remove the brain. Just remove the calvarium and fix the whole head after first removing the eyes.

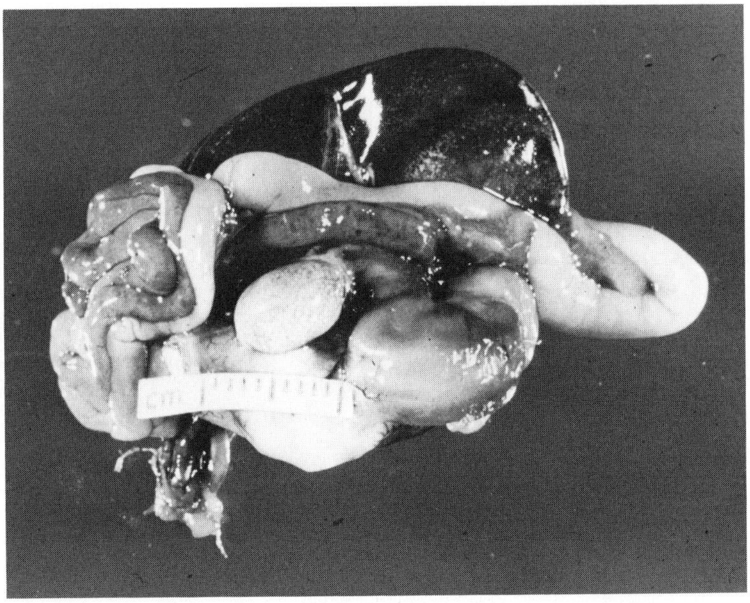

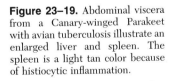

**Figure 23–19.** Abdominal viscera from a Canary-winged Parakeet with avian tuberculosis illustrate an enlarged liver and spleen. The spleen is a light tan color because of histiocytic inflammation.

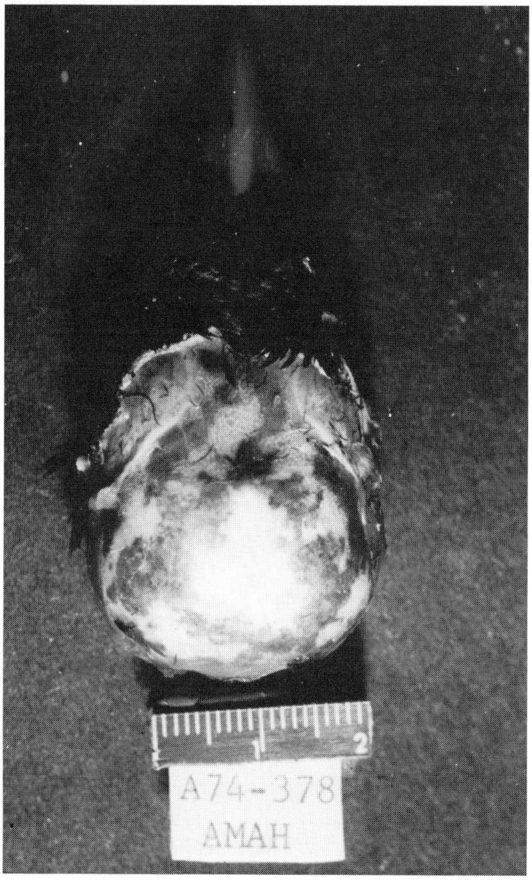


**Figure 23–20.** The ecchymoses, seen in the calvarium of a mynah bird with scalp removed, are agonal pooling of blood within the skull and are not indicative of head trauma. (Courtesy of Angell Memorial Animal Hospital.)

Likewise, removing the spinal cord from the canal can be very difficult in smaller species. Use ronguers to remove the dorsal arches and fix the cord within the column.

The eyes, choanal slit, infraorbital sinuses, ears, and nares should all be examined before the head is fixed.

Joints of wings, legs, and feet should be opened and examined. Surfaces should be smooth and glistening. Bone marrow may be obtained by cracking open a long bone. Since many bird bones are pneumatized, one may have to check a few bones before marrow is found. The femur and tibia are usually marrow-filled. If all else fails, submit a rib. Finally, muscles of the legs and the sciatic nerve running on the posterior surface of the femur should be examined.

## Impression Smears

Impression smears are a useful adjunct to a complete postmortem examination (see Chapter 17, Cytology). In cases suspected of psittacosis, immunofluorescent staining of impression smears for *Chlamydia* is far more reliable than routine histochemistry on fixed paraffin-embedded histologic sections. Tissue phases of parasites such as *Lankesterella,* malarias, or *Toxoplasma* are also more readily identified on impression smears of liver and spleen. Also, the cytology of lymphoreticular and hematopoietic neoplasms is more diagnostic on impressions of liver, spleen, and bone marrow.

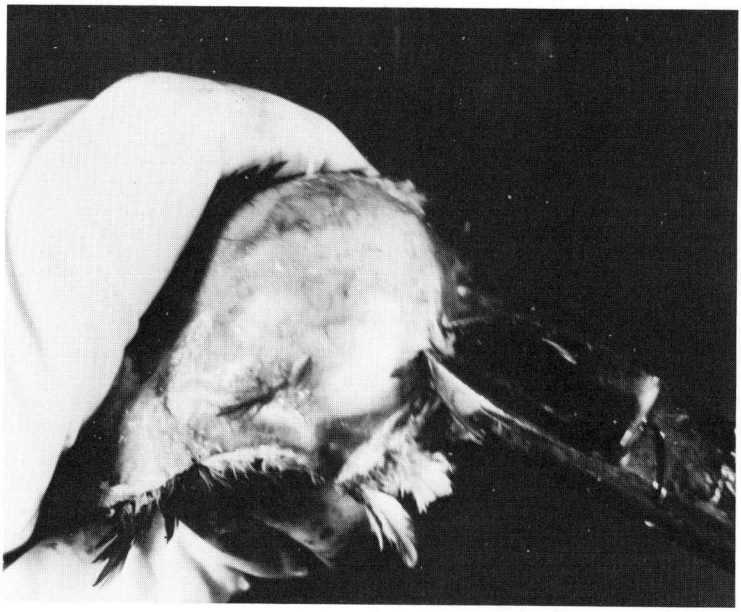

**Figure 23–21.** The calvarium may be removed with rongeurs.

To make a good impression smear (actually a touch preparation), it may be easier to bring the slide to the tissue. Grasp a small piece of the tissue with forceps so that a freshly cut, well-blotted surface faces upward. Lower a clean slide to the tissue, touching it lightly. Retract quickly *without* sliding the slide across the tissue. Make several "touch preps" on each slide. Impressions are generally more useful when air-dried. If other fixation is necessary (e.g., heat fixation for acid-fast stains or acid alcohol for certain Giemsa preparations), it can be done later.

Exudates may be prepared for cytologic evaluation (see Chapter 17, Cytology).

## Fixation

Most tissues are fixed in formalin for optimum histology. Ten per cent buffered formalin penetrates only about 0.2 cm in 24 hours. That means that a 0.1 cm strip on the center of a 0.5 cm piece of tissue remains unfixed. Penetration is slower in very bloody, dense tissue (e.g., congested spleen or liver) and more rapid in relatively porous tissue (e.g., lung). Formalin will not penetrate well into the brain through the unopened calvarium or into the marrow of bone unless the bone has been cracked. The biggest problem with submission of fixed tissues is inadequate fixation due to prior severe autolysis or inadequate volume of fixative, allowing continuing decomposition. Initial fixation is achieved with ten times the volume of formalin as volume of tissue. The amount of formalin may be reduced after 12 to 24 hours of fixation in preparation for mailing. Wet formalin-fixed tissue may be conveniently stored and shipped in plastic heat-sealed bags.

Other fixatives, such as those required for electron microscopic evaluation, are probably not necessary for routine submission. However, if gout is suspected, a small piece of affected tissue should be placed in a vial with absolute alcohol, as urates are water-soluble and are lost in formalin fixation.

**Table 23–1.** TISSUES THAT SHOULD BE SAVED FOR POSTMORTEM ANALYSIS

| | |
|---|---|
| Brain | Cecum and colon |
| Tongue or soft palate | Heart |
| Trachea and syrinx | Thyroids |
| Esophagus | Thymus |
| Crop | Lungs |
| Proventriculus | Adrenals and gonads |
| Ventriculus | Kidneys |
| Duodenum and pancreas | Oviduct |
| | Sciatic nerve and thigh muscle |
| Intestines | Bone marrow |
| Cloaca and bursa | Skin and breast muscle |
| Liver | Spleen |

**Table 23–2.** TISSUES THAT SHOULD BE SUBMITTED TO THE LABORATORY FOR POSTMORTEM ANALYSIS

| | |
|---|---|
| Heart | Kidney, gonad, and adrenal |
| Lung | Piece of intestine |
| Liver | (duodenum and pancreas) |
| Spleen | |

The number of tissues submitted to the histology laboratory may depend on the cost per tissue. If you do not send complete tissues it is wise to save the rest of the viscera in formalin while awaiting a diagnosis (Tables 23–1 and 23–2). If only grossly visible lesions or limited tissues are submitted, a diagnosis may not be possible. Too often, the limited tissues *suggest* a diagnosis, which cannot be confirmed because other tissues have already been discarded. In addition, subsequent submission of supplemental tissues may help generate new information regarding patterns of disease. The retained tissues can be disposed of when the final report is received or can be stored for future reference.

Tissues for histopathology should *not* be frozen. Freezing creates crystals and ruptures cells, making histopathology virtually useless. If a carcass has been frozen, thaw it before fixing the tissues in formalin. If the bird was valuable, it may still be worthwhile to attempt a histologic examination under these conditions.

Tissues for toxicologic analysis *should* be frozen. They may be frozen at −20°C (regular deep-freeze) after being wrapped in aluminum foil. Freezing for virus isolation is best done on Dry Ice, in liquid nitrogen, or in a laboratory freezer at less than −60°C. If this cannot be accomplished, the tissues for viral isolation should be sent (by rapid mail) in sterile containers (Petri dishes or test tubes) on wet ice.

## RECOMMENDED READINGS

Ensley, P. K., et al.: A necropsy procedure for exotic birds. Proceedings of the Annual Meeting of the American Association of Zoo Veterinarians, 1976, pp. 131–144.

Fiennes, R. N. T-W.: *In* Petrak, M. L. (ed.): Diseases of Cage and Aviary Birds. 2nd ed. Philadelphia, Lea & Febiger, 1982, pp. 498–499.

Graham, D. L.: Necropsy procedures in birds. Vet. Clin. North Am. [Small Anim. Med.], 14:173–177, 1984.

Harrison, L. R., and Herron, A.: Submission of diagnostic samples to a laboratory. Vet. Clin. North Am. [Small Anim. Med.], 14:165–172, 1984.

Keymer, I. F.: Postmortem exams of pet birds. Mod. Vet. Pract., 42:35–38, 1961.

van Riper, C., and van Riper, S. G.: A necropsy procedure for sampling disease in wild birds. Condor, 82:85–98, 1980.

*Section Five*

# THERAPY CONSIDERATIONS

# *Chapter 24*

# RELATIONSHIPS OF AVIAN IMMUNE STRUCTURE AND FUNCTION TO INFECTIOUS DISEASES

GEORGE V. KOLLIAS, JR.

Many of the problems frequently encountered in clinical avian practice relate directly or indirectly to infectious diseases. The immune responses elicited to toxins, bacteria, viruses, protozoa, and other parasitic agents are dependent on a number of factors. To a great degree immune responsiveness is genetically controlled; however nutrition (see Chapter 31, Nutritional Diseases), environmental factors, the developmental stage of the host, and products of the endocrine system all play critical roles in the effectiveness of the avian immune response (Fig. 24–1).

With very few exceptions, the research involved with the avian immune system, immune responses, and antimicrobial defense mechanisms has been based on findings utilizing domestic species such as the chicken, turkey, duck, and quail. The material presented in this brief overview of the avian immune system and its relationship to infectious diseases specifically relates to these species; however, the principles outlined are undoubtedly applicable to many of the families of birds commonly encountered in avian clinical practice.

## PHYLOGENY OF THE AVIAN IMMUNE SYSTEM

Phylogenetically, birds are closely allied to mammals with regard to the functional aspects of the immune system. Major differences exist in the organization of lymphoid tissue, in the humoral response, and in the transfer of maternal antibodies (Table 24–1).

## ONTOGENY OF THE AVIAN IMMUNE SYSTEM AND RESPONSE

The thymus of birds develops from the third and fourth pharyngeal pouches at day 5 of 21-day incubation. By the twelfth day of incubation the cellular components of the thymus are predominantly lymphoid. At four months of age the thymus reaches its maximum size and thereafter begins to atrophy. Recrudescence of thymic size and function has been noted in subadult and adult birds following administration of thyroxine.

The cloacal bursa, a globular or spherical

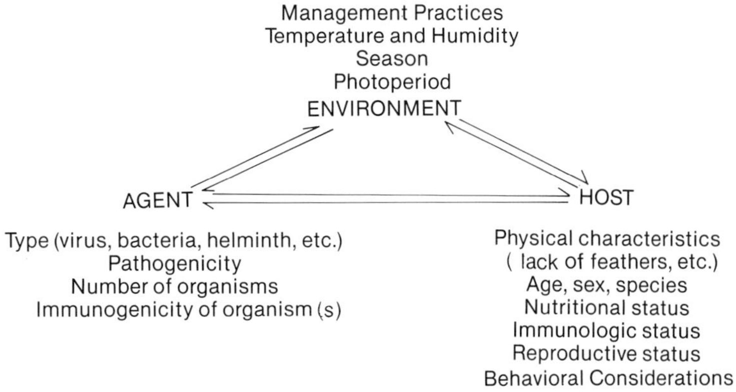


**Figure 24–1.** Factors important in host resistance to infectious diseases. (Adapted from Schwabe, C. W., et al.: Epidemiology in Veterinary Practice. Philadelphia, Lea & Febiger, 1977.)

**Table 24–1.** THREE MAJOR PHYLOGENETIC LEVELS OF IMMUNOEVOLUTION

| Type of Surveillance System | Occurrence | Phylums or Classes Identified | Experimental Evidence |
|---|---|---|---|
| Quasi-immunorecognition | Invertebrates and vertebrates | Coelenterates<br>Tunicates<br>Mammals | Allograft incompatibility<br>Allogeneic incompatibility<br>MLC reactions |
| Primordial cell-mediated immunity | Advanced invertebrates<br>Vertebrates? | Annelids<br>Echinoderms | Allograft incompatibility with specific memory |
| Integrated cell-mediated and humoral antibody immunity | All vertebrates | Fishes<br>Amphibians<br>Reptiles<br>Birds<br>Mammals | Very extensive |

Adapted from Hildemann and Reddy: Fed. Proc., 32:2192, 1973.

lymphoepithelial organ formed by a dorsal diverticulum of the cloacal proctodeum, is unique to birds. In the chicken, this structure begins to develop on day 4 of incubation, attains its maximum size at 10 weeks of age, and becomes atrophic at the time of sexual maturity or at 5 to 6 months of age. Precursor or stem cells migrate from the yolk sac to the bursa at 12 days of incubation, which results in the early formation of antibody-producing cells, or B cells (bursa derived). At day 20 of incubation there is a change, or switch, in the type of immunoglobulin produced, from predominately IgM to more specific and diversified IgG.

Bone marrow, another important source of cellular components central to the immune system, begins development at 8 to 9 days of incubation. Bone marrow stem cells have also been noted to arise in the yolk sac membrane. Bone marrow–derived cells from 17-day embryos have been shown to possess the capability to transform into antibody-producing precursor cells.

Lymphoid aggregates and nodules, discussed below, and the spleen begin development at day 4 of incubation. Lymphocytes in these organs and structures are mature at 19 days of incubation. Full structural differentiation of these cells occurs by one week after hatching.

## NONSPECIFIC FACTORS IN AVIAN IMMUNITY

Innate or nonspecific antimicrobial mechanisms include the action of bactericidal enzymes, phagocytes, and interferon. Phagocytic cells, important as mediators of nonspecific immunity, include thrombocytes, the cells of the mononuclear phagocytic cell system (syn. macrophages), and heterophils. Thrombocytes, actively phagocytic circulating cells, are in many ways analogous to mammalian platelets. They are capable of phagocytizing particles as large as bacteria such as staphylococci. Macrophages are found throughout the body in the respiratory, gastrointestinal, hemopoietic, and nervous systems as well as in the coelomic cavity. Their functional capabilities vary with location. Bactericidal enzymes are present in a variety of fixed and circulating cell types. An example of such a circulating cell is the heterophil, which, unlike its mammalian counterpart, does not possess specific lysozyme activity. The complement system and its activation by the classical and alternate pathways have been described in avian species. Properdin production, present in mammals, has not been described in birds. Interferon, lymphocyte products, plays an important role in antiviral immunity in birds. Whether avian lymphocytes are capable of producing gamma-interferon, a factor responsible for activating macrophages to destroy intracellular parasites in mammals, awaits elucidation.

## SPECIFIC OR ACQUIRED FACTORS IN AVIAN IMMUNITY

Central to the induction of acquired immunity is the acquisition (trapping) and processing of antigenic components of bacteria, viruses, and other parasites. The cloacal bursa is a major antigen-absorbing organ, a function that is closely linked to its anatomic location. Antigenic substances passing through the gastrointestinal and urogenital systems are exposed to bursal cells following absorption through its duct. In addition, vascular networks connecting the bursa and spleen allow for further exposure of splenic T (thymus-derived) and B (bursa-derived) lymphocytes to antigenic substances. The cecal tonsils, another important site for antigen trapping, contain T and B lymphocyte–dependent areas. These tonsils are known also to be an important source of immunoglobulins. The harderian, lacrimal, and nasal (salt) glands located in the head are major antigen-trapping and immunoglobulin-producing sites. Significant production of IgA occurs in the harderian gland. Other structures important in antigen trapping and processing include lymphoid aggregates and nodules found in the walls of hollow organs such as the intestine, trachea, and esophagus. In addition, they are associated with the lymphatic vessels in other organs and in the limbs and neck. Specifically delineated lymph nodes are found in waterfowl, marsh and shore birds, and more recently have been noted in chickens.

The cell types found in the anatomic sites described are important in the recognition, binding, and processing of antigens. Small lymphocytes arise from the yolk sac membrane and intrathymic tissues. Lymphoid stem cells develop in the cloacal bursa and the thymus, neither of which contains pure populations of B or T cells. Pre-B cells are accepted by the bursa at 6 to 18 days of incubation and are under the influence of the hormone bursopoietin. Alternatively, pre-T cells are accepted well into postembryonic life. Birds, unlike mammals, have plasma cell surface immunoglobulins, with a life span of 5 to 6 days. These antibody-producing cells are present in the red pulp of the spleen, harderian glands, ducts of the lateral nasal glands, and intestinal lymphoid aggregates. Direct evidence for the production of immunoglobin classes designated IgG, IgM, and IgA has been documented in birds. There is indirect evidence for the presence of IgE and IgD.

The roles of antigen-sensitive T cells in avian species are very similar to those defined in mammals. Helper functions, such as lymphokine release, mediation of delayed-type hypersensitivity responses, and cell-mediated immune functions have all been identified. Important avian lymphokines identified include interleukin 1 and 2, both cell growth factors, and a soluble factor called thrombocyte migration inhibitory factor (T.M.I.F.). T.M.I.F. prevents cells from migrating away from areas of inflammation and intense antigen stimulation. It is analogous to mammalian macrophage inhibitory factor, yet another lymphokine. T cells are also responsible for mediating delayed-type hypersensitivity reactions in addition to being responsible for immunity to certain bacterial (e.g., mycobacteriosis) and viral (e.g., fowlpox) infections. They also have suppressor functions that are important in graft rejection, an indicator of cellular immunity. Additionally, suppressor T cells are important in antiviral immunity and in antitumor immunity, which has been studied mainly with Marek's disease virus and other related lymphoreticular viruses.

Another form of acquired immunity in birds is passive transfer from the hen to the chick. Passive transfer of immunity in mammals has been extensively studied; however, it is only recently that studies have involved avian species. In the chicken it is clear that the maternal immunoglobulins are passed via the egg yolk fluid in the oviduct. IgM and IgA are acquired in the oviduct with albumin. IgG is absorbed from the yolk and is found primarily in the gut. Once the egg is formed, assuming that there has been no infectious or metabolic process affecting the hen, the embryo should have all the essential immunoglobulins for passive protection against a number of infectious agents. Maternal IgM and IgA are actually swallowed by the embryo via amniotic fluid during development and can be detected at day 17 of incubation. At hatching IgG is expressed in the serum, while only IgM and IgA are found in the intestine. It appears that the avian embryo's swallowing IgA and IgM via amniotic fluid is analogous to a neonatal mammal's consuming colostrum. Passive transfer of IgG via the yolk is analogous to transplacental transfer of antibody in mammals. Egg yolk, an important source of passive humoral immunity, begins to be absorbed by 24 hours after hatching. If the yolk is not totally absorbed, owing to vascular problems or infection, passive immunity gained from IgG can be affected. A protective mechanism operating with avian species to overcome such problems centers on the fact that chick IgG has a half-life twice as long as that in an adult bird. The egg yolk is now known to be an important source of cells for the development of the immune system as well as a source of passive immunity for the chick.

Maturation of immune competence in chicks, in relation to cell-mediated immune responses, occurs at one to three weeks after hatching. One-day-old chicks are capable of antibody production to certain protein antigens; however, a full adult level of immune competence, relative to immunoglobulin production, is not reached until six weeks of age.

## ADDITIONAL FACTORS INFLUENCING THE IMMUNE RESPONSE

Genetic factors are extremely important in the development of effective immune responsiveness. Histocompatibility antigens are a group of genes that are responsible for the control of a diversity of immune functions. In chickens it has been shown that the B histocompatibility locus is responsible for controlling such functions as skin graft rejections, graft-versus-host reactions, complement production, leukocyte antigen production distinct from erythrocyte antigens, resistance to viral diseases such as Marek's disease, tumor regression of lymphoid leukosis, and the regulation of autoimmune reactions. These phenomena have been studied extensively in inbred strains of chickens.

Generally, stress results in adaptive physiologic responses by an animal. One such response is the production of adrenal corticosterone. Chronic high level production of this hormone may result in weight loss and alteration in organ function (e.g., adrenal gland), which may mediate a decrease in the ability to resist disease.

Wilgus[34] reviewed the literature for evidence of practical implications of interactions between disease and vitamins in poultry and found it to be in accord with that reported in other animals and man: (1) dietary vitamin deficiencies can impair body defense mechanisms important in mediating disease resistance, and (2) occurrence of disease can increase vitamin requirements. The major evidence suggesting positive interaction between individual vitamins and immune responsiveness was reported to involve vitamin A and ascorbic acid in bacterial infections and vitamins A and K in parasitic infestations. There is a suggestion of a positive association between ascorbic acid and immune responsiveness to viral infections.

Supplemental vitamin E, at levels greater than that required for normal growth and reproduction, was reported by Nockels[19] to effectively improve humoral immune responses and significantly improve the resistance of chicks challenged with *Escherichia coli*. Other nutritional factors, including caloric exhaustion (undernutrition), amino acids, and zinc, have an effect on immune responsiveness in poultry.

Studies in poultry by Pardue and Thaxton[22, 23] suggest that vitamin C is synthesized in the liver in limited amounts. Supplemental ascorbic acid, in chickens subjected to heat stress, appeared to reduce endogenous stress-related responses (e.g., increased total adrenal weight, plasma corticosterone concentrations, alterations in sodium/potassium ratio, and mortality).

In contrast, McCorkle et al.[17] reported that cell-mediated immunity and antibody response to sheep red blood cells (T cell–dependent) were not influenced by feeding megalevels (1 per cent of the diet) of vitamin C to chickens. However, the antibody response to *Brucella abortus* (T cell–independent) was lower in neonates and higher in adults in the presence of increased levels of vitamin C.

Temperature has an effect on immune response in the developing embryo, the chick, and adult birds. Consistent with clinical observations, lower temperatures, or acclimation of birds to lower temperatures, enhances antibody production and resistance to certain infectious agents. Studies have been carried out which indicate that excessive environmental temperatures have resulted in the suppression of antibody production. These phenomena may have a neuroendocrine basis related to thyroxine and other hormones. Much of the work that has been performed, however, is controversial. In other species, such as reptiles, it is clear that the temperature plays a very important role in relation to certain immune functions such as antibody production.

As in mammals, the endocrine system has profound effects on the immune system. Thyroxine is an important regulator of antibody production and lymphocyte proliferation. Regulation is primarily by $T_3$ and secondarily by $T_4$. Recent studies in domestic fowl indicate corticosterone as the primary adrenal cortical hormone. Corticosterone, unlike other cortisone-like hormones, lacks immunosuppressive properties even at very high levels when administered parenterally. Corticosterone has, in fact, very little effect on lymphocyte function and, in contrast to hydrocortisone, lacks cytotoxicity for lymphocytes. Progesterone and testosterone both have been shown to be important hormones in relation to immune regulation. Early studies with testosterone clearly showed that lymphocyte populations in the cloacal bursa were sensitive to the effects of such agents. As a result of the lymphocytotoxic effects of testosterone, young chickens were shown to be more susceptible to infection by highly encapsulated gram-positive organisms. In addition, a concurrent and marked decrease in antibody production was observed.

An area requiring further study relates to the effects of chemotherapeutic agents on the immune system of birds. It has been shown in domestic chickens that tetracycline, tylosin, and gentamicin can be immunosuppressive. The immunosuppression is manifested by decreased antibody production and by the decreased sensitivity of the test birds to specific antigens. Interestingly enough, immune competence of the birds in these studies could be reinstated through the administration of low doses of immunologic enhancing agents such as levamisole. Studies have shown the paraimmunity inducer PIND ORF to be effective in birds prior to viral exposure (see Chapter 32, Viral Diseases). Future studies with other biologic modifiers, such as bacterial cell wall components, may prove to be important therapeutic alternatives to conventional chemotherapy (see Chapter 27, What to Do Until a Diagnosis Is Made).

## CLINICAL EVALUATION OF THE IMMUNE SYSTEM

Clinical assessment of the immune status of the avian patient is difficult. For example, evaluation of a blood smear may offer some information regarding the absolute number of heterophils and lymphocytes; however, the effects of the stress on a bird from handling or transporting are sufficient to produce some alterations in heterophil and lymphocyte count.

Several tests can be performed to evaluate specific aspects of an individual's immune capabilities. For example, one can assay for (1) response to specific antigens by evaluating T cell functional capabilities, (2) production of antibody to a variety of foreign antigens, (3) concentration of specific immunoglobulins, and (4) mast cell or basophil release of soluble factors (e.g, histamine, serotonin, prostaglandins). There is, however, no absolute correlation between the results of these tests and the ability of the host animal to resist a bacterial or viral infection.

A variety of assays has been developed primarily for avian experimental studies. For example, immunoglobulin concentrations in serum have been determined; however, baseline values are known only for a limited number of nondomestic species. Further basic research needs to be performed to develop useful assays for nondomestic species.

## REFERENCES


1. Abplanalp, H.: The role of genetics in the immune response. Avian Dis., 2:299–314, 1979.
2. Beard, C. W.: Avian immunoprophylaxis. Avian Dis., 2:327–334, 1979.
3. Benedict, A. A. (ed.): Avian Immunology. Adv. Exp. Med. Biol., 88:entire issue, 1977.
4. Caporale, E. P., et al.: Local resistance of chicken respiratory tract to viral infections. Zbl. Vet. Med. B., 25:383–393, 1978.
5. Chang, C. F., and Hamilton, P. B.: The thrombocyte as the primary circulating phagocyte in chickens. J. Reticuloendothel. Soc., 6:585–590, 1979.
6. Glick, B., et al.: Comparison of the phagocytic ability of normal and bursectomized birds. J. Reticuloendothel. Soc., 1:442–449, 1964.
7. Glick, B.: Lymphocyte lifespan in chickens. *In* Wright, R. K., and Cooper, E. R. (eds.): Phylogeny of Thymus and Bone Marrow-Bursa Cells. Amsterdam, Netherlands, Elsevier/North Holland Biomedical Press, 1976, pp. 237–245.
8. Glick, B., and Olah, I.: The morphology of the starling (*Sturnus vulgaris*) bursa of Fabricius: A scanning and light microscope study. Anat. Rec., 204:341–348, 1982.
9. Glick, B.: Bursa of Fabricius. *In* Farner, D. S., et al. (eds.): Avian Biology. New York, Academic Press, 1983, pp. 443–500.
10. Glick, B.: Calorie-protein deficiencies and the immune response of the chicken. I. Humoral immunity. Poultry Sci., 60:2494–2500, 1981.
11. Higgins, D. A.: Physical and chemical properties of fowl immunoglobulins. Vet. Bull., 3:139–154, 1975.
12. Holmes, K. L., and Haar, J. L.: Migration of chicken yolk sac cells to bursa of Fabricius supernatants. Dev. Comp. Immunol., 6:727–736, 1982.
13. Keast, D., and Ayre, D. J.: Antibody regulation in birds by thyroid hormones. Dev. Comp. Immunol., 4:323–330, 1980.
14. Kock, C.: The alternate complement pathway in chickens. Dev. Comp. Immunol., 7:785–786, 1983.
15. Le Douarin, N. M., et al.: The lymphoid stem cells in the avian embryo. *In* Wright, R. K., and Cooper, E. L. (eds.): Phylogeny of Thymus and Bone Marrow-Bursa Cells. Amsterdam, Netherlands, Elsevier/North Holland Biomedical Press, 1976, pp. 217–226.
16. Leslie, G. A., and Martin, L. N.: Studies on the secretory immunologic system of fowl. III. Serum and secretory IgA of the chicken. J. Immunol., 1:1–9, 1973.
17. McCorkle, F., et al.: The effects of a megalevel of vitamin C on the immune response of the chicken. Poultry Sci., 59:1324–1327, 1980.
18. Mueller, A. P., et al.: The chicken lacrimal gland, gland of Harder, caecal tonsil, and accessory spleens as sources of antibody-producing cells. Cell Immunol., 2:140–152, 1971.
19. Nockels, D. F.: Relationship of vitamin E to immune response in animals. Proceedings of the Roche Vitamin Nutrition Update Meeting, Hot Springs, Arkansas, 1978.
20. Olah, I., and Glick, B.: Avian lymph node: Light and microscopic study. Anat. Rec., 205:287–299, 1983.
21. Panigrahy, B., et al.: Antibiotic-induced immunosuppression and levamisole-induced immunopotentiation in turkeys. Avian Dis., 2:401–408, 1979.

22. Pardue, S. L., and Thaxton, J. P.: Interaction of ascorbic acid and cortisol on humoral immunity in broilers. Poultry Sci., 60:7, 1981.
23. Pardue, S. L., and Thaxton, J. P.: Enhanced livability and improved immunologic responsiveness in ascorbic acid supplemented cockerels during acute heat stress. Poultry Sci., 61:7, 1982.
24. Richter, R.: Studies on the enhancement of the immune response by the paraimmunity inducer PIND ORF in birds. University of Munich, Postgraduate Dissertation, 1983.
25. Rose, M. E.: The immune system in birds. J. R. Soc. Med., 72:701–705, 1979.
26. Rose, M. E., et al.: Immunoglobulin classes in the hen's egg: Their segregation in yolk and white. Eur. J. Immunol., 4:521–523, 1974.
27. Schaffner, T., et al.: The bursa of Fabricius: A central organ providing for contact between the lymphoid system and intestinal content. Cell Immunol., 13:304–312, 1974.
28. Schauenstein, K., et al.: Avian lymphokines: 1. Thymic cell growth factor in supernatants of mitogen stimulated chicken spleen cells. Dev. Comp. Immunol., 7:533–540, 1982.
29. Schauenstein, K., and Hayari, Y.: Avian lymphokines. Dev. Comp. Immunol., 7:767–768, 1983.
30. Szenberg, A.: Ontogenesis of the immune system. *In* Marchalonis, J. J. (ed): Comparative Immunology. London, Blackwell Scientific Publications, 1976, pp. 419–431.
31. Thaxton, P.: Influence of temperature on the immune response of birds. Poultry Sci., 57:1430–1440, 1978.
32. Tizard, I.: The immune response of birds. *In* Developments in Biological Standardizations, Vol. 51. Basel, Switzerland, S. Karger, 1982, pp. 3–10.
33. Toivanen, A., et al.: Histocompatibility requirements for cellular cooperation in the chicken. Adv. Exp. Med. Biol., 88:257–265, 1977.
34. Wilgus, H. S.: Disease, nutrition—interaction. Poultry Sci., 59:4, 1980.

## ACKNOWLEDGMENTS

The contributions of Dr. Bruce Glick via personal communication are greatly appreciated. Also, Linda Harrison's editorial assistance is greatly appreciated and has been contributory to the completion of the manuscript.

GEORGE V. KOLLIAS, JR.

# *Chapter 25*

# PHARMACOLOGY OF ANTIBIOTICS

CARL H. CLARK

The literature on the pharmacology of antibiotics in birds is very limited, and many gaps in our knowledge exist. The antibiotics discussed in this chapter include the penicillins, cephalosporins, aminoglycosides, tetracyclines, and chloramphenicol.

The purpose of antibiotic administration to birds is to inhibit or kill bacteria, allowing the bird time to develop resistance and eliminate the disease. It is important to use a relatively nontoxic dose that will produce at least one or two times the MIC (minimum inhibitory concentration) in the vicinity of the bacteria. Bacteriostatic drugs should be dosed frequently enough to maintain continuous inhibitory concentrations, while bactericidal drugs must remain in the vicinity of the bacteria long enough to kill sufficient numbers to reduce the severity of the disease. Many pathogenic organisms are killed with two hours exposure to penicillins or cephalosporins. Aminoglycosides (gentamicin, kanamycin, etc.) are bactericidal in one to three hours.

After oral or parenteral administration time is required to reach peak plasma concentration (intravenous [IV], seconds; intramuscular [IM], 30 to 60 minutes; oral, 60 to 120 minutes). The antibiotic diffuses from plasma through the interstitial spaces to the area of inflammation and then to the bacteria. This diffusion takes considerable time and is related to the concentration of drug in the plasma. The higher the plasma concentration, the quicker the drug reaches the bacteria in adequate concentrations. In mild inflammatory processes the antibiotics may reach the bacteria in adequate concentrations in one to two hours, whereas in severe diseases (anaerobic bacteria, *Clostridium*, *Bacteroides*) or with a great deal of necrotic tissue, edema, and circulatory disruption, the antibiotics may penetrate very slowly and inadequately. The dose in diseases of greater severity must be increased and given parenterally to produce higher plasma concentrations and thus increase diffusion pressure. The second dose of a bactericidal drug should be given soon after the first to assure complete penetration and maintain the concentration, at the bacteria, above one to two MIC for an adequate period of time. This dose interval varies for each drug and should be equal to about four half-lives ($t_{1/2}$) or two to eight hours. Subsequent doses are usually given at eight- to twelve-hour intervals, which will result in times when there is inadequate concentration of drug in the vicinity of the bacteria.

Bacteriostatic drugs are usually given at regular intervals designed to maintain an adequate minimum concentration of drug at the bacteria. The initial dose can be higher than subsequent doses to improve drug penetration. Irregular dosage intervals, resulting in a lack of adequate concentration of drug in the vicinity of the bacteria, should be avoided.

Another consideration in therapy is the toxicity of antibiotics. It is desirable to have a therapeutic index (margin of safety) of five times the therapeutic dose before toxic signs occur. Unfortunately with gentamicin in birds, undesirable side effects (polyuria) occur with the therapeutic dose, indicating that the margin of safety is very narrow. Penicillins, cephalosporins, tetracyclines, and chloramphenicol are much less toxic in birds than are aminoglycosides.

## PENICILLINS

These antibiotics block bacterial cell wall formation in actively growing or multiplying organisms. Resting or inactive organisms are resistant. Penicillins are effective against gram-positive organisms and *Pasteurella* with MIC's of 1 μg/ml or less. The broad-spectrum penicillins—ampicillin, amoxicillin, and hetacillin—are also effective against some gram-negative organisms *(Salmonella, E. Coli)* with MIC's of 1 to 5 μg/ml. Carbenicillin and ticarcillin but not the others are effective against *Pseudomonas* in MIC's of 16 to 60 μg/ml.

### Toxicities

Literature reports of penicillin toxicities are rare. Some practitioners report that procaine penicillin may be toxic in some species (canaries, pigeons). In one study, procaine penicillin was injected I.M. in pigeons in doses of 100 (166,700 IU), 200 (333,400 IU), and 500 (833,500 IU) mg/kg. The birds receiving the 500 mg/kg dose regurgitated food within 5 minutes after injection and began to tremble for 5 to 10 minutes. None of the birds died and the two lower-dose groups exhibited no abnormal signs. All birds remained healthy and ate well during a two-week observation period after the single dose of procaine penicillin. Both procaine penicillin and benzathine penicillin were used in turkeys in doses up to 100 mg/kg with no mention of any toxic side effects.[23]

### Absorption and Metabolism

Ampicillin is absorbed erratically and poorly after oral administration. Plasma concentrations obtained are adequate only for highly susceptible organisms having an MIC of 1 μg/ml or less. In a study involving psittacines, 5 Blue-naped Parrots *(Tanygnathus lucionesis)* were given 150 mg/kg of ampicillin orally. The peak plasma concentrations varied from 0.17 to 0.82 μg/ml[17] and occurred at two hours after dosing in three birds and at 10 hours in two birds. These values are the maximal plasma concentrations reached and are inadequate for all except the most sensitive organs. Four Amazon Parrots were given 175 mg/kg of ampicillin orally. Peak plasma concentrations were 3.2, 3.0, 15.5, and 6.2 μg/ml, which occurred in 2 to 10 hours after the dose. Two of the birds developed plasma concentrations that would be effective against *Salmonella* and *E. coli* while two did not. The inconsistency of absorption after an oral dose makes this route of administration unreliable.

Similarly, chickens given a single oral bolus of 25 mg/kg developed a mean peak serum concentration of 0.6 μg/ml. When chickens were treated for four days with drinking water containing ampicillin 200 mg/L, the peak plasma concentrations ranged from 0.18 to 0.2 μg/ml.[46] The failure of ampicillin to produce higher plasma concentrations is a result of poor absorption (less than 5 per cent) and rapid excretion.

A study in chickens and ducks indicates that amoxicillin produces twice the levels of ampicillin after oral administration.[25] Parenteral injections produce more consistent and much higher plasma concentration in birds. Five Amazon Parrots given an I.M. dose of 100 mg/kg of sodium ampicillin developed mean plasma concentrations of 60 μg/ml in 30 minutes, which decreased to 0.65 μg/ml in four hours.[17] Similar values were found when 100 mg/kg was administered I.M. to chickens.[46]

Ampicillin is very rapidly eliminated from chickens, with a half-life of approximately 30 minutes.[45, 46] Mammals eliminate ampicillin primarily by renal filtration and tubular excretion. The glomerular filtrate volume of birds is only about 1/40 that of mammals, which should result in slower renal elimination. The rapid elimination of ampicillin in birds may mean that it is excreted by other routes. The concentrations of ampicillin in chickens one hour after an I.M. injection of 100 mg/kg were 2.4 μg/ml in plasma, 154 μg/gm in the large intestinal contents, 39.6 μg/gm in the small intestinal contents, 9.2 μg/gm in the large intestinal wall, 3.4 μg/gm in the small intestinal wall, and 5.2 μg/gm in the kidney.[45] The high concentrations in the intestinal wall and contents support the view that a large portion of ampicillin is eliminated via the intestinal tract. Other researchers found that the ampicillin content of bile after oral and parenteral routes was usually higher than that of other tissues.[36]

The available evidence indicates that ampicillin is partially eliminated via the kidney and partially by excretion in bile and directly through the intestinal wall. After oral medication, the drug is absorbed and transported directly to the liver where a large portion is excreted in bile.

### Dosages

Penicillin G is available in three parenteral forms. These are sodium (Na) or potassium (K) penicillin, procaine penicillin, and benzathine penicillin. In turkeys, a dose of K penicillin G of 50 mg/kg I.M. produced a peak plasma concentration of 25 μg/ml in 30 minutes, which declined to 0.15 μg/ml in 8 hours.[23] Two or three doses daily should produce plasma concentrations adequate for most organisms susceptible to penicillin G. Procaine penicillin 100 mg/kg will produce plasma concentrations in turkeys of 2 to 3 μg/ml, decreasing to 0.03 μg/ml in 24 hours. A combination of 50 mg/kg procaine penicillin and 50 mg/kg benzathine penicillin will produce plasma concentrations in turkeys from 2 to 3 μg/ml to 0.06 μg/ml in two days.[23] Potassium penicillin can be given first to obtain high plasma concentrations followed by a combination of procaine and benzathine penicillin. Toxicity studies in various species with procaine and benzathine penicillin will have to be performed to assure safety.

Ampicillin will produce adequate plasma concentrations for about 8 hours after parenteral administration. Orally the drug may be useful in diseases confined to the gastrointestinal tract. Probenecid (dose of 200 mg/kg), which interferes with the excretion of ampicillin, will triple the peak concentrations of ampicillin and double the time that ampicillin remains in plasma.[46]

## CEPHALOSPORINS

Cephalosporins are bactericidal antibiotics with a chemical structure very similar to that of the penicillins. They act by blocking cell wall formation in a manner similar to penicillins. They are effective against actively growing and dividing bacteria but not against resting bacteria. All of the cephalosporins are broad-spectrum antibiotics effective against both gram-positive and gram-negative organisms. There have been three generations of cephalosporins developed. The first generation, exemplified by cephalothin and cephalexin, has a bacterial spectrum similar to that of ampicillin and amoxicillin. Further development of these antibiotics has increased the effectiveness against gram-negative organisms (second generation). Third-generation cephalosporins, cefotaxime and moxalactam, are very effective against *Pseudomonas, Serratia, Citrobacter, Enterobacter, Haemophilus*, and *Bacteroides*.

### Toxicities

Cephalosporins are relatively nontoxic in mammals, and it is expected that they will be relatively nontoxic in birds, although toxicity studies have not been performed to date.

### Absorption and Metabolism

Only a few of the first-generation cephalosporins are absorbed after oral administration. Cephalexin is readily absorbed after oral administration to quail, pigeons, ducks, Sandhill Cranes, and emus. Oral doses of 25 to 50 mg/kg in these species resulted in average plasma concentrations of 20 μg/ml in 0.5 to 1.0 hour and measurable concentrations 1.5 to 5.5 hours later.[8]

### Dosages

The dose of cephalosporin will vary according to the species, and, until sufficient information is available for each species, the dose will have to be extrapolated from those species of birds already studied. There appears to be a correlation between drug half-life and body weight, so that larger birds have a longer drug half-life and thus require longer dosage intervals than smaller birds.

Cephalothin 100 mg/kg every six hours in pigeons, cranes, and emus and every two to three hours in quail and ducks should maintain therapeutic plasma concentrations. Cephalexin orally in doses of 35 to 50 mg/kg every four hours for the larger birds and every two to three hours for the smaller birds should also be effective.[8] Larger doses would be necessary for gram-negative than for gram-positive organisms.

## CHLORAMPHENICOL

Chloramphenicol is a bacteriostatic broad-spectrum antibiotic effective against aerobic gram-positive and gram-negative organisms, anaerobic organisms, *Chlamydia, Rickettsia*, and *Mycoplasma*. Chloramphenicol is effective against *Salmonella*, somewhat less effective against *E. coli*, and mostly ineffective against *Pseudomonas*. In a study of 226 *Salmonella* strains of 84 serotypes isolated from zoo animals, birds, and reptiles, 73 per cent were inhibited by 5 μg, 95 per cent by 10 μg, and 99 per cent by 25 μg/ml of chloramphenicol, indicating that 5 to 10 μg/ml were efficacious concentrations for *Salmonella*.[35]

### Chloramphenicol Safety

A non–dose-related aplastic anemia occurs in man in certain individuals with especially sensitive marrow. Very small doses, such as those absorbed after the ophthalmic use of chloramphenicol, have been incriminated in aplastic anemia in man.[1, 33] It is currently illegal to use chloramphenicol in animals that may be used for food even though there are no detectable residues. Veterinarians should use caution in dispensing the drug to avoid exposing the owners to undue risk.

## Toxicities

Chloramphenicol exhibits similar toxicities in birds and in mammals. A reversible dose-related anemia, central nervous system depression, and loss of appetite are seen in chickens, turkeys, and ducks.[31, 32, 39, 40] The dosage that is necessary to produce these signs is much higher in birds than in mammals. Studies in ducks indicate that 500 mg/kg/day has relatively few toxic effects.[31, 32] An $LD_{50}$ of chloramphenicol in four-week-old chicks was 3100 mg/kg.[39] Although these studies have not been extended to other avian species, chloramphenicol doses of 50 to 200 mg/kg should be relatively nontoxic to birds.

## Absorption and Metabolism

Plasma concentrations following an oral dose of chloramphenicol vary widely, This is caused by the differences in the amounts of drug absorbed and the liver metabolism of the drug after absorption. A mean of 17 per cent of the oral dose given Spot-billed Ducks was absorbed.[14] The peak plasma concentrations in these ducks varied from 2 to 20 μg/ml. Pigeons given 200 mg/kg orally developed inadequate plasma concentrations.[12] These birds metabolize chloramphenicol so rapidly that most of the drug is eliminated by the liver before adequate plasma concentrations can be achieved. Oral doses (50 mg/kg) in chickens produced peak plasma concentrations (6.7 μg/ml) in four hours, whereas turkeys developed peak concentrations (11.7 μg/ml) in one hour.[40] Other investigators found that the oral absorption of chloramphenicol in chickens was good.[22] The plasma concentration of chloramphenicol after a normal dose is inconsistent and varies from species to species.

Intramuscular doses of chloramphenicol yield higher and more consistent plasma concentrations than oral doses. Approximately 90 per cent of the I.M. dose is absorbed and peak concentrations are reached within one hour. Parenteral chloramphenicol is available in propylene glycol formulations or as the succinate. The succinate must be enzymatically removed before chloramphenicol can develop bacteriostatic activity. In those species studied, birds are capable of removing the succinate from chloramphenicol. When identical concentrations of chloramphenicol as either the base or the succinate were injected (I.V.) into turkeys, chloramphenicol plasma concentrations from the succinate ester were 60 per cent of those resulting from injections of chloramphenicol base.[12] This is probably caused by the more rapid excretion rate of the ester.[15] Until data are available in psittacine birds, the dose of chloramphenicol succinate should be increased to 1.5 times the calculated dose of chloramphenicol.

After absorption, chloramphenicol is rapidly distributed to the tissues with a volume of distribution smaller than that found in mammals but greater than one (chickens V'd L/kg—2.04, duck V'd area L/kg—1.97).[12, 14, 22] Chloramphenicol accumulates in tissues of birds as it does in mammals, but to a lesser degree. A study of tissues of chickens 30 minutes after a single I.M. dose of 40 mg/kg revealed that the kidney, liver, spleen, and heart all contained greater concentrations of chloramphenicol than did the plasma.[21] Similar studies in mammals indicate that chloramphenicol tissue concentrations remain higher longer than plasma concentrations. This persistence of chloramphenicol in tissue may improve its therapeutic effectiveness.

Chloramphenicol is metabolized in the liver and excreted through the kidneys. The metabolic products and routes of excretion have not been thoroughly studied in birds.

## Dosages

When Table 25–1 is examined, it appears that closely related birds with similar dietary habits develop similar plasma concentrations. The birds of prey consistently develop higher plasma concentrations that persist longer than in other birds. A dose of 50 mg/kg I.M. at either 8- or 12-hour intervals, depending on the minimum desired plasma concentration, should be adequate. With the 50 mg/kg dose, ducks, geese, and psittacines would require a dosage interval of less than 8 hours.[40]

A pharmacokinetic study in the Spot-billed Duck indicates that the dosage intervals and doses listed in Table 25–2 are necessary to maintain concentration of 5 μg/ml. This same dosage schedule would probably be applicable to other ducks and geese.

**Table 25–1.** AVERAGE PLASMA CHLORAMPHENICOL* CONCENTRATIONS IN DIFFERENT GROUPS OF BIRDS

| | Mean Plasma Concentration (μg/ml) | | | |
|---|---|---|---|---|
| **Type of bird** | $t_{1/2}$ (HR) | 2 HOURS† | 8 HOURS | 12 HOURS |
| Birds of prey | 2.69 | 2.94 | 7 | 2.4 |
| Water fowl | 1.45 | 19.6 | 1.5 | 0.6 |
| Psittacines | 1.03 | 14.2 | 0.7 | — |
| Pigeons | 0.46 | 2.8 | 0 | — |

*Chloramphenicol propylene glycol I.M. 50 mg/kg.[12]
†Post-injection time

## AMINOGLYCOSIDES: GENTAMICIN, SISOMICIN, TOBRAMYCIN, KANAMYCIN, AMIKACIN, NEOMYCIN, STREPTOMYCIN

The aminoglycosides are bactericidal, broad-spectrum antibiotics effective against many varieties of gram-positive and gram-negative aerobic bacteria. They are ineffective against anaerobic organisms *(Clostridium, Bacteroides)*. The aminoglycosides are usually synergistic with penicillins and cephalosporins.

### Toxicities

Toxicities caused by the aminoglycosides are similar in birds and mammals. The primary toxic effects are kidney and vestibular apparatus damage and neuromuscular blockade. Low doses (10 mg/kg/day) of gentamicin in psittacines, chickens, turkeys, and hawks will usually cause a reversible polyuria and polydipsia. When therapy is terminated, the polyuria and polydipsia cease with little permanent renal damage. Higher doses (20 mg/kg/day) may cause renal necrosis and death. Vestibular apparatus injury and loss of balance along with renal necrosis occurred in chickens from a dose of 40 mg/kg/day. Hawks given an I.V. dose of 20 to 40 mg/kg/day developed weakness, apnea, and sudden death after injection. These signs were attributed to neuromuscular blockade. All hawks receiving 40 mg/kg/day died.[6] Giving gentamicin I.V. may have increased the potential for neuromuscular blockade, which is less evident after I.M. injections. Histopathologic lesions were produced in the kidneys with both dosage levels in hawks.[6] Slight renal damage in Lanner Falcons after gentamicin therapy has been reported.[19]

**Table 25–2.** CHLORAMPHENICOL DOSAGE SCHEDULE AND PLASMA CONCENTRATION FOR THE SPOT-BILLED DUCK

| Dosage Interval (hrs) | Dose (mg/kg) | Minimum Plasma Concentration (μg/ml) |
|---|---|---|
| 2.9 | 22 | 5 |
| 6.0 | 102 | 5 |
| 8.0 | 243 | 5 |

From Dein, F. J., et al.: Pharmacokinetics of chloramphenicol in Chinese Spot-billed Ducks. Vet. Pharm. Therap., 3:161–168, 1980.

The available toxicity studies indicate that the maximum daily dose of gentamicin should not exceed 10 mg/kg. Even at this dose polyuria and some renal damage may occur. Tobramycin is less nephrotoxic in dogs, and the renal accumulation is only 50 per cent that of gentamicin. Dogs given a daily dose of 45 mg/kg of gentamicin or tobramycin for 10 days developed renal cortical concentrations of 1877 μg/gm for gentamicin and 994 μg/gm for tobramycin.[30] Unpublished studies in Great-horned Owls, Barred Owls, and Red-tailed Hawks indicate that the accumulation of tobramycin in renal tissue is 40 to 65 per cent less than gentamicin with the same dose.[18]

Gentamicin has been studied more than all other aminoglycosides in birds. An I.M. injection of gentamicin reaches peak plasma concentrations in 30 to 45 minutes. Nearly the total amount of drug is absorbed over the first 12 hours: greater than 95 per cent absorption in hawks, owls, chickens, and turkeys and greater than 70 per cent in eagles.[5, 9, 29] Gentamicin is distributed fairly rapidly to tissue. The flow of gentamicin into the tissues is faster than the elimination; hence the concentrations in the tissues usually exceed the plasma concentrations after a few hours. Renal tissue concentrations in chickens and turkeys reach a peak in 10 to 12 hours, with concentrations of 200 times the plasma.[9, 29] The renal concentration of gentamicin in Sandhill Cranes was 50.2 μg/gm at 6 hours and was still above 30 μg/gm at 18 hours.[7] High renal tissue concentrations were also demonstrated in cranes, various ducks, pigeons,

Bobwhite Quail, emus, and pheasants.[7] Gentamicin leaves the renal tissue very slowly and may require three to four weeks in chickens and turkeys for the kidney to be completely cleared of gentamicin.[9, 21, 38] When a dose of gentamicin is repeated once daily, the plasma concentrations after 24 hours are very low, while the renal tissue concentrations are still high. Each subsequent dose will continue to increase renal tissue concentrations until toxic levels are reached. Plasma concentrations may remain normal and are not a good indication of pending renal damage. The renal accumulation of gentamicin after multiple doses has not been studied in adult birds; however, one study on neomycin renal concentrations in 15-day-old chickens is available.[10] Chickens were injected with 20 mg/kg of Na neomycin daily for four days. The renal tissue concentration of neomycin increased rapidly to a peak in 36 hours (90.4 μg/gm) and then decreased through 96 hours (53.8 μg/gm). This decrease of renal neomycin concentration occurred even though the chicks continued to receive neomycin.

In those species of birds studied, gentamicin also accumulated in the liver, but not as much as the kidney.[7, 9, 29, 38] Liver concentrations were frequently larger than plasma concentrations four to six hours after dosing. Therapeutic concentrations of gentamicin in the livers of chickens and turkeys persist for 12 to 96 hours after a single dose.

### Doses

Table 25–3 lists plasma concentrations of gentamicin obtained after a single I.M. dose. These concentrations should be therapeutically effective for at least 6 and possibly 12 hours, except in the pigeon.

Gentamicin, like chloramphenicol, is eliminated very rapidly in the pigeon. A single dose of 10 mg/kg will decrease to 1 μg/ml in 5.5 hours. The $t_{1/2}$ of gentamicin for pigeons is shorter than for other birds studied. The concentrations of gentamicin in the plasma, kidney, and liver in normal pigeons and in pigeons infected with *Salmonella* revealed that the gentamicin concentrations were lower in the diseased birds.[34] The author questioned the value of gentamicin in this species.

A pharmacokinetic study in birds of prey (hawk, owl, eagle) indicated that an I.M. or I.V. dose of 2.5 mg/kg t.i.d. would produce peak concentrations of 10 to 12 μg/ml and trough concentrations of <2 μg/ml. The total daily dose would be 7.5 mg/kg, which should not be toxic.[5]

### Kanamycin

A study in chickens indicates that a subcutaneous dose of 10 mg/kg of kanamycin produced plasma concentrations of 19.3 μg/ml in 30 minutes, which decreased to 0.24 μg/ml in 12 hours. A dose of 25 mg/kg produced plasma concentrations of 59.0 μg/ml, which decreased to 1.08 μg/ml in 12 hours.[2]

## TETRACYCLINES: CHLORTETRACYCLINE, TETRACYCLINE, OXYTETRACYCLINE, DOXYCYCLINE, AND MINOCYCLINE

The tetracyclines are bacteriostatic and exert their effect by inhibiting protein metabolism. They have a very wide spectrum of antimicrobial activity, including gram-positive and gram-negative aerobic and anaerobic bacteria, *Rickettsia, Ehrlichia, Mycoplasma, Chlamydia, Plasmodium gallinaceum,* and *Spirochaeta.* Many bacteria have developed a resistance to the tetracyclines. In 97 strains of *E. coli* isolated from macaws and parrots imported into Japan, 59 per cent were resistant to oxytetracycline.

**Table 25–3.** PLASMA CONCENTRATION (μg/ml) OF GENTAMICIN IN VARIOUS SPECIES

| Bird | Dose | | Time in Hours | | |
|---|---|---|---|---|---|
| | $t_{1/2}$(HR) | mg/kg | 1 | 6 | 12 |
| Blue and Gold Macaw | 1.6 | 10 | 23 | 2.3 | 0.2 |
| Pigeon | 0.9 | 10 | 27 | 0.7 | 0.0 |
| Pigeon | 1.0 | 20 | 52.1 | 2.1 | 0.04 |
| Raptors† | 2.0 | 10 | 44.1 | 4.7 | 1.0 |
| Ducks, geese | 2.3 | 10 | 35.8 | 6.3 | 1.3 |
| Chickens | 1.8 | 10 | 18.4 | 5.7 | 0.3 |

*Post-injection time
†Hawk, owl, osprey, eagle

For those strains that were sensitive, the MIC varied from 1.56 to 6.25 μg/ml.[27] The average MIC for oxytetracycline for avian strains of *Pasteurella multocida* varied from 0.1 to 0.4 μg/ml.[20] Two μg/ml of doxycycline and tetracycline inhibited 61 and 67 per cent of the strains of *Campylobacter* tested.[41] While the tetracyclines have a wide bacterial spectrum, the MIC varies widely between different species of bacteria. It would require a much larger dose for *E. coli* than for *Pasteurella*.

## Toxicities

Very few toxicology studies have been performed on birds. The acute toxicity ($LD_{50}$) dose of doxycycline given in the crop was 2500 mg/kg in one-week-old chickens.[16]

## Absorption and Metabolism

The tetracyclines are absorbed after oral medication, although the percentage absorbed is very low. Chickens absorbed 0.28 to 1 per cent of the chlortetracycline given orally.[28]

Budgerigars given feed containing chlortetracycline 0.05 per cent for 30 days maintained average plasma concentrations of 1 to 1.6 μg/ml and fecal concentrations of 400 to 800 μg/gm. The large fecal concentrations were a result of the poor absorption of the drug.[26] A sugary liquid containing 0.05 per cent chlortetracycline was effective in producing blood concentrations of 1.7 to 3.6 μg/ml in lories and lorikeets.[4] Yellow-crowned Amazon Parrots fed pelleted feed containing 1, 1.5, 1.75 per cent chlortetracycline produced mean plasma concentrations of 2.6, 3.1, and 3.2 μg/ml, respectively.[24] Parrots and parakeets fed seed containing 0.18 to 0.25 per cent chlortetracycline averaged plasma concentrations of approximately 1 μg/ml.[43] Medicated feed of approximately 0.05 to 1.75 per cent produced plasma concentrations of 1 μg/ml, which is effective for chlamydiosis but may not be effective in salmonellosis or colibacillosis.

A single oral dose of doxycycline (5 mg/kg) produced serum concentrations of 1.15 μg/ml in two hours, which decreased to 0.06 μg/ml by 18 hours.[16]

The tissue concentrations in chickens fed doxycycline (0.02 gm/L) in the drinking water for seven days revealed that all the tissues studied contained higher concentrations than plasma. The tissue concentrations were approximately as follows: kidney, 2 μg/gm; lung, 0.7 to 0.4 μg/gm; serum, 0.3 μg/gm.[16] These values are quite different from those of dogs, which excrete doxycycline through the liver and intestine with very little renal excretion. This may indicate that the excretion and metabolism of doxycycline in birds differ from those in mammals.

Calcium, magnesium, iron, and other cations will interfere with the absorption of an oral dose of the tetracyclines. In mammals, large doses of milk (calcium) given simultaneously with oxytetracycline, chlortetracycline, and tetracycline may reduce the amount of drug absorbed by 70 per cent. Milk will also reduce the amount of doxycycline and minocycline absorbed by 20 per cent. Iron, which might be given birds in oral mineral/vitamin mixtures, reduces the absorption of doxycycline and minocycline by 80 to 90 per cent and the other tetracyclines by 50 per cent.[3] Organic acids like citric acid, when given simultaneously with the tetracyclines, increase the amount of the drug absorbed by three to five times.[3, 4, 13, 24, 26, 28, 43] These acids prevent the cations from combining with the tetracyclines, making more free drug available for absorption. Reducing the calcium content of the diet also helps to increase absorption.[44]

The parenteral route will yield higher plasma concentrations in birds. Studies in Germany with minocycline and doxycycline in psittacines revealed that excellent plasma concentrations could be produced for five to six days with a single I.M. injection of 100 mg/kg (see Chapter 35, Chlamydia). The plasma concentrations with minocycline were 20.68 μg/ml in 12 hours, which decreased to 2.21 μg/ml in five days. Doxycycline produced slightly lower levels that lasted longer (11.85 μg/ml at 12 hours to 3.17 μg/ml at 6 days).[42] The preparations of these drugs available in the United States produce very acid solutions (minocycline pH 2.4), which are irritating after I.M. injection and cause methemoglobin formation after rapid I.V. injection. The preparations of these drugs available in the United States should be used with caution pending toxicity studies. A long-acting oxytetracycline preparation (Pfizer LA 200) has been used successfully to treat *Pasteurella* infection in turkeys (52 mg/kg); however, plasma concentrations of the drug were not determined.[37]

## REFERENCES

1. Abrams, S. A., et al.: Marrow aplasia following topical application of chloramphenicol eye ointment. Arch. Intern. Med., 140:576–577, 1980.

2. Andreini, G., and Pignattelli, P.: Kanamycin blood levels and residues in domestic animals. Veterinaria (Milano), 21:51–72, 1972.
3. Arnson, A. L.: Pharmacotherapeutics of the newer tetracyclines. J. Am. Vet. Med. Assoc., 176:1061–1068, 1980.
4. Arnstein, P., et al.: Chlortetracycline chemotherapy for nectar-feeding psittacines birds. J. Am. Vet. Med. Assoc., 154:190–191, 1969.
5. Bird, J. E., et al.: Pharmacokinetics of gentamicin in birds of prey. Am. J. Vet. Res., 44:1245–1247, 1983.
6. Bird, J. E., et al.: Toxicity of gentamicin in red-tailed hawks. Am. J. Vet. Res., 44:1289–1298, 1983.
7. Bush, M., et al.: Gentamicin tissue concentrations in various avian species following recommended dosage therapy. Am. J. Vet. Res., 42:2114–2116, 1981.
8. Bush, M., et al.: Pharmacokinetics of cephalothin and cephalexin in selected avian species. Am. J. Vet. Res., 42:1014–1017, 1981.
9. Carli, S., et al.: Gentamicin in day old chicks. Clin. Vet., 101:876–880, 1978.
10. Carli, S., et al.: Serum levels, tissue distribution and residues of neomycin following intramuscular administration in chicks. J. Vet. Pharmacol. Therap., 5:203–207, 1982.
11. Clark, C. H.: Unpublished data.
12. Clark, C. H., et al.: Plasma concentrations of chloramphenicol in birds. Am. J. Vet. Res., 43:1249–1253, 1982.
13. Clary, B. D., et al.: The potentiation effect of citric acid on aureomycin in turkeys. Poultry Sci., 60:1209–1212, 1981.
14. Dein, F. V., et al.: Pharmacokinetics of chloramphenicol in Chinese Spot-billed Ducks. J. Vet. Pharmacol. Therap., 3:161–168, 1980.
15. Dorrestein, G. M., et al.: Pharmacokinetic aspects of penicillins, aminoglycosides and chloramphenicol in birds compared to mammals. Vet. Q., 6:216–224, 1984.
16. Drumev, D., et al.: Addition of doxycycline to the drinking water of fowls. Vet. Med. Nauki, 19:100–105, 1982.
17. Ensley, P. K., and Janssen, D. L.: A preliminary study comparing the pharmacokinetics of ampicillin given orally and intramuscularly to psittacines. J. Zoo Anim. Med., 12:42–47, 1981.
18. Evans, R.: Personal communication.
19. Fernandez-Repollet, E., et al.: Renal damage in gentamicin-treated Lanner Falcons. J. Am. Vet. Med. Assoc., 181:1392–1394, 1982.
20. Frost B. M., et al.: Activity of fosfomycin against *Pasteurella*. Avian Dis., 18:578–589, 1974.
21. Gupta, R. C., et al.: Distribution of chloramphenicol in tissues of the white leghorn. Ind. J. Exp. Biol., 18:918–920, 1980.
22. Gupta, R. C., et al.: Pharmacokinetics of chloramphenicol in the white leghorn. Ind. J. Exp. Biol., 18:612–614, 1980.
23. Hirsh, D. C., et al.: Pharmacokinetics of penicillin G in turkeys. Am. J. Vet. Res., 39:1219–1221, 1978.
24. Landgraf, W. W., et al.: Concentration of chlortetracycline in the blood of Yellow-crowned Amazon Parrots. Avian Dis., 26:14–17, 1981.
25. Lashev, L., and Semeralzheiv, U.: Comparative studies of amoxicillin absorption and distribution in poultry. Vet. Med. Nauki, 22:5–6, 1983.
26. Luthgen, Von W., et al.: Relationships between dosage with chlortetracycline, its concentration in the blood and excretion with feces in budgerigars *(Melopsittacus undulatus)*. Berl. Munch. Tierarzth. Wochenschr., 86:454–457, 1973.
27. Nakamura, M., et al.: Drug resistance and R plasmid in *Escherichia coli* strains isolated from imported pet birds. Microbiol. Immunol., 24:1131–1138, 1980.
28. Pollet, R. A., et al.: Pharmacokinetics of chlortetracycline potentiation with citric acid in the chicken. Am. J. Vet. Res., 44:1718–1721, 1983.
29. Pradella, G., et al.: Gentamicin in newborn turkeys. Rev. Zoo. Vet., 6:511–515, 1975.
30. Reiner, N. E., et al.: Nephrotoxicity of gentamicin and tobramycin given once daily or continuously to dogs. J. Antimicrob. Chemother., 4(Suppl):85–101, 1978.
31. Rigdon, R. H., et al.: Anemia produced by chloramphenicol in the duck. Arch. Pathol., 58:85–93, 1954.
32. Rigdon, R. H., et al.: Consideration of the mechanism of anemia produced by chloramphenicol in the duck. Antibiot. Chemother., 5:38–44, 1955.
33. Rosenthal, R. L., and Blackman, A. L.: Bone marrow hypoplasia following use of chloramphenicol eye drops. J.A.M.A., 191:136–137, 1965.
34. Sabrautzki, S.: The course of gentamicin level in serum tissue of the pigeon *(Columba livia* GMEL, 1789 var. domstica). Postgraduate Dissertation, University of Munich, 1983.
35. Schroder, H. D., and Karasck, E.: The distribution of antibiotic sensitivity of *Salmonella* in captive wild animals. Monatsh. Vet. Med., 25:551–554, 1970.
36. Shakaryan, G. A.: Distribution and concentrations of ampicillin in the body of fowl. Veterinariia, 8:105–106, 1976.
37. Skeeles, J. K., et al.: Studies on the use of long-acting oxytetracycline in turkeys. Avian Dis., 27:1126–1130, 1983.
38. Spreat, S. R., and Beckford, S. M.: Pharmacodynamics of gentamicin in day-old chicks. Proceedings of 26th Western Poultry Disease Conference, pp. 101–107.
39. Strakova, J., et al.: Toxicity of chloramphenicol in chickens when administered chronicin. Biol. Chem. Vet. (Praha), 16:255–265, 1980.
40. Strakova, J., et al.: Toxicity of chloramphenicol in young poults treated with chronicin. Biol. Chem. Vet. (Praha), 16:269–276, 1980.
41. Svedhem, A., et al.: Antimicrobial susceptibility of *Campylobacter jejuni* from humans with diarrhea and from healthy chickens. J. Antimicrob. Chemother., 7:30–35, 1981.
42. Teichmann, B. von, and Gerlach, H.: Determination of blood levels after a parenteral dose of minocycline or doxycycline. Prakt. Tierarzt., 57:87–89, 1976.
43. Wachendörfer, G., et al.: Further investigations on the control of psittacosis in parrots and parakeets. Tierarzt. Umschaw., 37:177–194, 1982.
44. Waldroup, P. W., et al.: Comparison of low dietary calcium and sodium sulfate for the potentiation of tetracycline antibiotics in broiler diets. Avian Dis., 25:857–865, 1981.
45. Yesov, B. E.: Pharmacokinetics of ampicillin in the chick. Veterinariia, 56:94–98, 1978.
46. Ziv, G., et al.: Effects of probenecid on blood levels and tissue distribution of ampicillin in fowls and turkeys. Avian Dis., 24:927–939, 1979.

# Chapter 26

# THERAPEUTICS

## INDIVIDUAL AND FLOCK TREATMENT REGIMENS

SUSAN L. CLUBB

Flock treatment regimens for poultry have been extensively researched and are often applicable when treating flocks of pet birds. In contrast, individual therapeutic regimens used in pet bird practice are largely extrapolated from poultry or mammalian research. Owing to the large number of species encountered by the avian practitioner and to the expense of pharmacokinetic studies, it is likely that empirical dosage regimens will continue to be in common usage. In Tables 26–1 to 26–5 some guidelines for therapy will be outlined for use in various situations. Most of these dosages are empirical but have been used successfully in a clinical situation.

Antibiotics are among the most important and highly utilized drugs in avian practice. Bacterial diseases and chlamydiosis are frequently encountered by the avian practitioner. In addition to diseases of primary bacterial etiology, the clinician is faced with secondary bacterial infections when dealing with viral, parasitic, and noninfectious diseases. Because the underlying cause of disease may be very difficult to diagnose, he is in many cases forced to treat for the treatable diseases, the secondary invaders. Interest in pharmacokinetics of antibiotics in birds has increased in recent years, and much new information is now available for treatment of the individual bird (see Chapter 25, Pharmacology of Antibiotics).

*Ideally*, any antibiotic therapeutic regimen should maintain tissue levels at the site of infection equal to or greater than the MIC (minimum inhibitory concentration) of the causative organism for a sufficient time to control the infection. There are, however, many inherent difficulties and variables involved in this approach to antibiotic therapy. First, we must know the pharmacokinetics of each drug in each species; individual, sex, and age variations may also be important. The MIC of the organism being treated is usually not known and is relatively costly to determine. Excretion of many drugs is rapid in birds, and the maintenance of therapeutic blood levels may require frequent dosage and excessive handling. Further studies emphasizing tissue antibiotic levels are needed before accurate dosage regimens can be developed. Tissue levels are often higher and maintained longer than blood levels. Tissue penetration will vary according to the target organ involved and the amount of tissue destruction or vascular compromise.

Drug administration techniques also differ from those commonly used in mammals. For example, it is difficult if not impossible to pill a psittacine (although a pilling instrument is in use in Europe for canaries) (Fig. 26–1). Techniques differ when dealing with the individual tame pet bird, a large flock of imported birds, a breeding flock, and a hand-feeding baby bird.

Physical restraint is stressful to birds, especially wild birds, and must be minimized. All medications and diagnostic tools should be on hand prior to handling the bird. Drugs should be administered quickly and gently, keeping restraint time to a minimum. The value of frequent administration of drugs must be weighed against the stress of frequent handling.

Alteration of gut flora or gut sterilization is a frequent side effect of antibiotic therapy. Supplementation of *Lactobacillus* to restore normal flora after antibiotic therapy is commonly advocated but is controversial. Many natural products, such as some brands of yogurt, do not contain viable *Lactobacillus* cultures. Some researchers feel that species-specific lactobacilli may be required for gut colonization. Clinical improvement does often accompany the use of

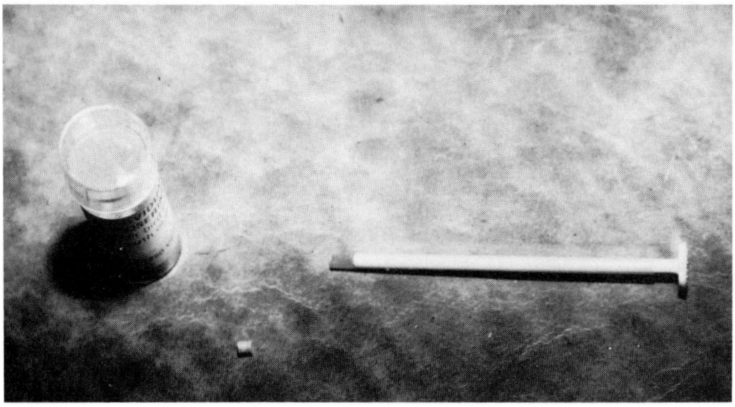

**Figure 26–1.** Pilling instrument used in Europe for administering drugs to canaries. (Courtesy of Greg J. Harrison.)

*Lactobacillus*, possibly resulting from temporary alteration of the gut environment, allowing proliferation of normal strains.

Immunosuppression is another side effect of antibiotic therapy which must be considered. Several antibiotics suppress the immune response in man and animals by interference with protein or immunoglobulin synthesis, elimination of antigen, interference with phagocytosis, or indirect action on the properdin system (see Chapter 24, Relationships of Avian Immune Structure and Function to Infectious Diseases). Levamisole has been shown to restore the immune response of x-irradiated and antibiotic-treated turkeys to normal levels and may be a helpful adjunct to therapy.[23]

Antibiotics are not the only chemicals valuable in the control of pathogens. Chlorhexidine, gentian violet, copper sulfate, chlorine bleach, and food preservatives have all been used with some success. Resistance to these chemical agents is uncommon.

## ROUTES OF ADMINISTRATION

### Parenteral Therapy

Parenteral therapy is the most exact and effective method of administering drugs to birds. Parenteral medications are most commonly administered intramuscularly (I.M.) in the pectoral or leg muscles. Repeated injections in the same side of the breast or the use of extremely irritating drugs I.M. may result in muscle necrosis or atrophy. Drugs administered in the posterior pectoral muscle or legs may pass through the renal portal system prior to entering the general circulation.

Parenteral therapy may be used to deliver high concentrations of antibiotics to the site of a local problem. Air spaces may be effectively reached by intratracheal or air sac injections. Joints and sinuses are sites in which direct instillation of antibiotics may be helpful. Intravenous administration is highly effective for initial therapy in life-threatening diseases such as septicemia and chlamydiosis. Owing to the fragility of avian veins and the uncooperative nature of most birds, continuous intravenous (I.V.) administration is difficult in, if not detrimental to, the avian patient, unless the bird is anesthetized. Subcutaneous injections are an alternative, but owing to minimal amounts of dermis, drugs may be poorly absorbed. Subcutaneous fluids or drug administration should be avoided in the cervical region because of the possibility of deposition into the cervicocephalic air sac system.

### Oral Medications

The addition of medications to drinking water is controversial but is often the only practical means of drug administration. This is the least stressful means of providing medications, especially for those that are palatable. Theoretically the bird will frequently self-dose during the day. While therapeutic blood levels may not be achieved with many antibiotics, levels within the intestine may be sufficient to control enteric infections. Aminoglycosides are frequently used in pet birds owing to their bactericidal activity and high level of efficacy for gram-negative bacterial infections. Aminoglycosides are not absorbed following oral administration if gut mucosa is intact. This may be desirable in treating primary gut infections. Mild systemic bacterial infections often respond to the low, sometimes erratic, blood levels produced by water medications.

One very valuable use of water-borne antibiotics in flocks is in the reduction of the spread of disease by water contamination. A large number of pathogens multiply initially in the oropharynx and are spread by oral contamination of water supplies. Fecal contamination of water supplies is likewise a common means of disease transmission in a flock. Antibiotics in the water supply will decrease the bacterial contamination of water supplies while combating bacteria in their primary port of entry, the pharyngeal area and the intestinal tract.

Antibiotic water mixtures should be prepared fresh daily. For the single pet bird, a volume sufficient to change the solution in the drinking cup several times a day is mixed at one time and the unused portion stored in the refrigerator. Some soluble mixtures (e.g., tetracyclines) should not be stored at all. Other additives to the antibiotic water solution, such as vitamins, should be avoided.

There are many inherent disadvantages to the use of water medications. Ideally water consumption levels should be established and the dose calculated based on body weight. For obvious reasons this is impractical for flock use. Estimations may be based on average weight and consumption and multiplied for the number in the flock, but individual consumption will vary. Medications may be rejected because of color or taste. Zerophilic birds such as budgerigars, Zebra finches, and Australian parakeets may not consume any water with dissolved medications. Environmental temperature may affect water consumption. Anorectic birds may not consume adequate amounts of water; and polydipsic birds may increase water consumption. These factors can result in quite a variation in antibiotic dosage. Overdosage of drugs such as nitrofurazone or dimetridazole may result in toxicity. In breeding flocks the male bird may consume large amounts of water in order to feed the female in the nest, resulting in toxic drug levels in the male and inadequate levels in the female. Some drugs are poorly or erratically absorbed from the gut in birds. Some foods may also interfere with absorption (for example, calcium binding with tetracycline). The presence of concurrent disease, parasitism, or nutritional deficiencies may also alter drug absorption. Some drugs, most notably tetracyclines and vitamin A, quickly lose potency in water mixes. Some medications, especially those containing sugars or vitamins, may promote bacterial growth in water bowls. The increased chance of developing antibiotic-resistant bacterial strains with subtherapeutic administration of antibiotics in drinking water must be considered.

The addition of medications to food stuffs is a more reliable way of medicating birds. The addition of drugs to a favorite food increases the chances of acceptance.

Oral suspensions, ground tablets, or the contents of capsules may be applied to fruits, peanut butter sandwiches, fresh corn, cooked sweet potatoes, monkey biscuits, or any other relished item. Toucans may be easily medicated by the injection of drugs into grapes. A practical method of adding drugs to a seed mixture is by coating a moist food which is then added to the mixture. Freshly cooked whole kernel corn is inexpensive and well-accepted by large and medium-size psittacine species. Canned or fresh corn can be used. Rice may also be readily accepted. Budgerigars, cockatiels, and Australian parakeets may be more reluctant to accept these preparations.

Pelleted feeds are an excellent means for delivering chlortetracycline while providing a balanced diet. Many other drugs can be used in this manner if they can withstand the heat required by the pelleting process. Unfortunately, some species are very reluctant to accept pellets.

Extensive use of pelleted feeds in quarantine stations has revealed some tendencies regarding acceptance among common species. Amazon parrots, Pionus parrots, African parrots, and most conures and cockatoos usually accept pellets readily. Pellets are mixed with corn or rice for the first two to three days, after which only pellets are offered. Cockatiels, lovebirds, and budgerigars will accept pellets but are more easily converted to tetracycline-impregnated millet or a combination of the two. Macaws tend to be stubborn and may refuse a diet of pellets alone. Addition of medicated corn, monkey biscuits, and peanuts will often solve the problem. Birds that are addicted to sunflower seeds are the most difficult to convert (see Chapter 3, Captive Behavior and its Modifications). In these birds tube feeding may be required until the medicated diet is accepted. Addition of sunflower seeds to the medicated diet will tempt the bird to try it, but seeds must be removed after a few days. Many birds will dig through a mix for seeds and will not eat adequate amounts of medicated feeds to maintain their weight.

Hand-feeding babies are easily medicated in their formulas. Most medications that are used in self-feeding birds can be used in hand-feeding babies. It must be kept in mind, however,

*Text continued on page 354*

**Table 26–1.** ANTIBIOTICS FOR USE IN BIRDS

| Generic Name | Form | Route | Species | Dosage | Frequency and Duration | Notes |
|---|---|---|---|---|---|---|
| **Amikacin** (Amiglyde—Bristol) | Injectable (250 and 500 mg/ml) | I.M. | Most | 15–20 mg/kg (0.015 mg/gm) | s.i.d. or b.i.d. | For gram-negative bacterial infections that are resistant to gentamicin, especially strains of *Pseudomonas* and *Klebsiella*. Available products are too concentrated for use in birds and should be diluted. |
| **Amoxicillin** (Amoxi Drops—Beecham Labs) | Suspension (50 mg/kg) | Oral | Most | 150–175 mg/kg | s.i.d. or b.i.d. | Most preparations are palatable and well accepted. Many require refrigeration and expire rapidly. Higher blood levels than with oral ampicillin. |
| **Ampicillin** | Oral suspension | Oral | Amazon parrots | 150–200 mg/kg | b.i.d. or t.i.d. | Ampicillin is poorly absorbed, making oral route unrealiable and rapidly excreted by the kidneys. Poor efficacy against many gram-negative pathogens of pet birds. See Chapter 25, Pharmacology of Antibiotics, for effect of probenecid. |
| | Injectable (Polyflex—Bristol) | I.M. | Amazon parrots | 100 mg/kg (0.1 mg/gm) | Every 4 hours | This dose is required for the maintenance of therapeutic blood levels in septicemia and other serious disorders. Drug of choice for initial therapy in "cat bite" *Pasteurella* septicemias. |
| | Capsule | Water | Chicken | 1.65 gm/L | | See Oral above. |
| | | Water | Pet birds | 250 mg/8 oz | | |
| | | Feed | Most | 250 mg/kg | 5 to 10 days | Sprinkle on favorite food or add to mash or corn mix. |
| **Carbenicillin** | Injectable (Geopen—Roerig) | I.V. or I.M. | Psittacines | 100–200 mg/kg | b.i.d. or t.i.d. | Synergistic with aminoglycosides; excellent in combination with gentamicin, but they must not be mixed in the same syringe. Retains potency for only three days under refrigeration. |

| | | | | | | |
|---|---|---|---|---|---|---|
| | | Intratracheal | Psittacines | 100 mg/kg | s.i.d. or b.i.d. | For use in conjunction with I.M. gentamicin for *Pseudomonas* pneumonias. If used alone, more frequent dosage may be required. |
| | Tablets (382 mg ground) (Geocillin—Roerig) | Water | Most | ½–1 tablet/ 8 oz | 5 to 10 days | Ground tablets have a very objectionable taste and odor that may be masked with sugar, honey, or juice. Not water-soluble and will usually float or settle. |
| | | Feed | Psittacines | 200 mg/kg | 5 to 10 days | Apply to favorite food (cooked sweet potato works well) or mix in mash or hand-feeding formula. |
| **Cefotaxime** (Claforan—Hoechst-Roussel) | Injectable | I.M. | Most | 50–100 mg/kg | t.i.d. | Broad-spectrum drug with low toxicity. May be used in conjunction with aminoglycosides; however, nephrotoxicity may occur. Reconstituted vial good for 13 weeks in freezer. |
| **Cephalexin** (Keflex—Eli Lilly) | Suspension | Oral | Most | 35–50 mg/kg | q.i.d. | Most preparations are well-accepted. |
| **Cephalothin** (Keflin—Eli Lilly) | Injectable | I.M. | Most | 100 mg/kg | q.i.d. | Not absorbed from the intestines. |
| **Chloramphenicol** | Succinate—injectable (100 mg/ml) | I.M. | Most | 80 mg/kg | b.i.d or t.i.d | Lower dose for succinate form can be used if given I.V. |
| | | I.V. | | 50 mg/kg | t.i.d. or q.i.d. | Excretion of I.V. chloramphenicol succinate is extremely rapid but may be required in the initial treatment of bacterial septicemias. |
| | Palmitate—oral suspension (30 mg/ml) (Chloromycetin palmitate—Parke-Davis) | Oral | Turkey<br>Psittacines | 50 mg/kg<br>0.1 ml/30 gm<br>1.0 ml/300 gm | q.i.d.<br>b.i.d., t.i.d., or q.i.d.<br>b.i.d., t.i.d., or q.i.d. | Absorption is erratic, so it should not be used for initial therapy in life-threatening infections. Palatable and very useful for therapy in hand-feeding birds in which food passage has slowed owing to bacterial infection. If crop stasis occurs, parenteral antibiotics must be used. |

*Table continued on following page*

THERAPEUTICS

**Table 26–1.** ANTIBIOTICS FOR USE IN BIRDS *(Continued)*

| Generic Name | Form | Route | Species | Dosage | Frequency and Duration | Notes |
|---|---|---|---|---|---|---|
| | Capsules | Feed | Most | 100–200 mg/kg | 5 to 10 days | Coat corn and add to seed mix or apply to favorite food or mash. Drug of choice for flock treatment of *Salmonella*. |
| **Chlortetracycline** (CTC)* | Soybean meal base (100 gm/lb) (SF 66—Cyanamid) | Feed | Large psittacines | 100 gm/20 lb mash | Only food source for 30 to 45 days | For treatment of chlamydiosis. Prepare mash by cooking rice, corn, and commercial chicken feed in water until soft. Cool completely and add CTC. Brown sugar may be added in equal volumes to CTC. Must be prepared fresh daily. |
| | Capsules or soluble powder (Aureomycin—Cyanamid) | Feed | Lories and lorikeets | 500 mg/L of food and/or nectar (0.5%) | Only food source for 30 to 45 days | Mash may be prepared from overcooked rice with boiled beans, moist commercial chick starter ration, or soaked monkey rations. If nectar is fed, CTC should be added. For treatment of chlamydiosis, must be prepared fresh daily. |
| | | Water | Most | 250 mg/pint | Only source of water | Solution should be mixed fresh b.i.d. or t.i.d. Inadequate for long-term treatment of chlamydiosis but helpful in initial therapy as birds are converted to pellets or mash. |
| | Pelleted feeds (Zeigler Brothers, Lafeber Co., or Bird Life) | Feed | Large psittacines | 1% | Only food source for 30 to 45 days | Excellent for treatment of medium and large psittacines for chlamydiosis. Some birds, especially macaws, are reluctant to accept pelleted feeds. Antifungal ingredients reduce fungal overgrowth. See Nystatin notes in Table 26–2. |
| | Impregnated millet seed (Keet Life—Hartz) | Feed | Small psittacines | 0.5% | Only food source for 30 to 45 days | Excellent for budgerigars and cockatiels. May be inadequate for lovebirds. If constipation occurs greens should be offered. |

*Treatment may be ineffective. See Chapter 35, Chlamydia.

| | | | | | | |
|---|---|---|---|---|---|---|
| **Doxycycline** | Suspension (5 mg/ml) (Vibramycin monohydrate—Pfizer) | Oral | Psittacines | 8–12 mg/lb | b.i.d. | Drug of choice for *Chlamydia*. Less fungal overgrowth and flora disturbance than CTC. |
| | Syrup (10 mg/ml) (Vibramycin calcium syrup—Pfizer) | Oral | Psittacines | 8–12 mg/lb | b.i.d. | Sensitive to iron and calcium in diet. Suspension and injectable have short shelf life unfrozen. |
| | Injectable (10 mg/ml) (Vibramycin hyclate—Pfizer) | I.V. | Psittacines | 10–20 mg/lb | Once or twice | For initial therapy in severe cases of chlamydiosis. Toxicity and muscle necrosis may occur if given I.M. |
| | Injectable (20 mg/ml) (Vibravenös—Pfizer, West Germany) | I.M. | Most<br>Macaws, lovebirds | 1 mg/10 gm<br>0.75 mg/gm | Once every 7 days for 4 weeks; next 2 injections after 6 days each; last injection after 5 days | Unavailable in the U.S. Vibramycin hyclate *cannot* be substituted. |
| | Capsules (100 mg generic) | | | 200 mg/lb food | | Use for 45 days for chlamydiosis |
| | Feed (Vibramycin [140 mg/lb food in soft, moist food] Avi-cake—Lafeber Co.) | Feed | Most | | Only food source for 45 days | Custom-produced by request. |
| **Erythromycin** | Soluble powder | Water | Most | 500 mg/gallon | 10 days on, 5 days off, 10 days on | For chronic respiratory disease, especially if *Mycoplasma* is suspected, air sacculitis, mild sinusitis, and mild enteric infections. |
| | Injectable | Nebulize | Most | 1 ml/10 ml saline | 15 min. t.i.d. | For air sacculitis and chronic respiratory disease. Injectable solution should not be given I.M. owing to severe muscle irritation. |
| | Suspension (40 mg/ml) | Oral | Psittacines | 20–40 mg/lb | b.i.d. 5 to 10 days | Most psittacine gram-negative isolates are resistant. Good efficacy in some cases of sinusitis. |
| **Gentamicin** (Gentocin—Schering) | Injectable (50 mg/ml) | I.M. | Pheasants and cranes | 5 mg/kg | t.i.d. 5 to 10 days | These dosages maintain therapeutic blood levels as |

*Table continued on following page*

**Table 26–1.** ANTIBIOTICS FOR USE IN BIRDS *(Continued)*

| Generic Name | Form | Route | Species | Dosage | Frequency and Duration | Notes |
|---|---|---|---|---|---|---|
| **Gentamicin** (*Continued*) | | | Quail | 10 mg/kg | | required for septicemias and other serious illnesses.* In less serious infections, s.i.d. or b.i.d. dosage is clinically very effective because of the bactericidal effect of gentamicin. May produce a transient polyuria. Overdose may result |
| | | | African Grey Parrot | 10 mg/kg | t.i.d. | |
| | | | Blue and Gold Macaw | 10 mg/kg | b.i.d. | |
| | | | Most species | Probably similar to above | | The above dosages do not exceed blood levels of 12 μg/ml, which is the mammalian nephrotoxic level. This dosage produces blood levels for 12 hours, but the mammalian nephrotoxic level is greatly exceeded initially. The avian nephrotoxic level has not been established. This dosage is clinically effective when used s.i.d. This dosage may be toxic in owls. |
| | | Oral | Most | 40 mg/kg | s.i.d., b.i.d., or t.i.d., 2 to 3 days | For sterilization of the gut or treatment of infections confined to the gut. Gentamicin is not absorbed across intact mucosa. |
| | | Water | Most | 1–5 ml/gallon | 3 days | For infections confined to the gut. |
| | | Nebulize | Most | 1 ml/10 ml saline | 15 min. t.i.d. | For sinusitis or air sacculitis. |
| | | Intratracheal | Most | 5–10 mg/kg | s.i.d. | Useful in treating pneumonia in conjunction with carbenicillin or tylosin administered I.M. |
| | Ophthalmic solution | Intranasal | Most | Several drops in each nostril | s.i.d., b.i.d., or t.i.d | For sinusitis and pharyngitis |
| | Powder (2 gm/30 gm) | Water | Most | ⅛ tsp/2 gallons | 3 days | For infections confined to the gut. |

*Unpublished data.

| | | | | | | |
|---|---|---|---|---|---|---|
| **Kanamycin** (Kantrim—Bristol) | Injectable (50 mg/ml) | I.M. | Most | 10–20 mg/kg | b.i.d. | |
| | | Water | Most, especially finches | 1–5 cc/gallon | 3 to 5 days | For infections confined to the gut. Good for use immediately prior to or after a stressful event such as shipping, reducing the number of potential pathogens. |
| **Lincomycin** (Lincocin—Upjohn) | Suspension (50 mg/ml) | Oral | Budgerigar | 1 drop | b.i.d. 7 to 14 days | May be useful for skin diseases. Poor efficacy for most bacterial infections in psittacines owing to primarily gram-positive spectrum of activity. |
| | | | Amazon Parrots | 0.5 ml/300 gm | b.i.d. 7 to 14 days | |
| | | | | 1 ml/300 gm | s.i.d. 7 to 14 days | |
| | | | Raptors | 100 mg/kg | s.i.d. 7 to 14 days | |
| **Lincomycin and Spectinomycin** (LS-50—Upjohn) | Soluble powder (16.7 gm lincomycin and 33.3 gm spectinomycin/2.55 oz) | Water | Most | ⅛–¼ level tsp/pint | 10 to 14 days | For chronic respiratory disease when *Mycoplasma* is suspected. Therapy may be extended if necessary. May also be effective for mild enteric infections. Sugar may be added to improve acceptance. |
| **Neomycin** (Biosol-M—Upjohn) | Solution with methscopolamine bromide | Water | Most | 1–8 drops/oz | 1 to 3 days | Contains anticholinergic; care must be taken not to overdose. |
| **Nitrofurazone** | Soluble powder (9.3%) | Water | Most psittacines | 1 tsp/gallon | 7 to 10 days | Excellent for the treatment of gram-negative especially *E. coli* enteric infections. Will slow the spread of salmonellosis in a flock. Effective for many strains of coccidia in psittacines but not very effective in mynahs and toucans. Toxic in overdose, resulting in neurologic signs and/or death. If neurologic signs are observed, discontinue use immediately. The lower dose is needed to prevent toxicity in lories, lorikeets, and mynahs. |
| | | | Lories, lorikeets, mynahs | ½ tsp/gallon<br>Do not put in nectar | | |

*Table continued on following page*

**Table 26–1.** ANTIBIOTICS FOR USE IN BIRDS *(Continued)*

| Generic Name | Form | Route | Species | Dosage | Frequency and Duration | Notes |
|---|---|---|---|---|---|---|
| **Oxytetracycline** long-acting (LA 200—Pfizer) | Injectable | I.M. | Most | 200 mg/kg | s.i.d. for 3 to 5 days | Has worked well in treating chlamydiosis in breeding birds to control outbreak and while getting birds to eat form of CTC or doxycycline to finish treatment.* |
| **Procaine penicillin G and benzathine penicillin** | Injectable | I.M. | Turkey | 100 mg/kg of each drug | s.i.d. or every two days | Provides therapeutic blood levels for one to two days. Care must be used in the treatment of small birds because of potential procaine overdose. Used extensively in poultry. |
| **Streptomycin** | Injectable | I.M. | Large birds | 30 mg/kg | b.i.d. or t.i.d. | Should not be used in pet birds owing to potential toxicity. Used extensively in poultry. |
| **Spectinomycin** (Spectam-R—Abbott) | Water-soluble solution | Water | Most | 20 cc/gallon | 5 to 10 days | For gram-negative enteric infections. Good for flock treatment. |
| **Sulfachlorpyrizidine** (Vetasulid—Squibb) | Soluble powder (packets of 5 gm, bottles of 50 gm) | Water | Most | ¼ tsp/L | 5 to 10 days as only water | Effective for many *E. coli* enteric infections. |
| **Tetracycline** | Soluble powder, 10 gm/6.4 oz (Polyotic—American Cyanamid) | Water | Most | 1 tsp/gallon | 5 to 10 days | Tetracycline rapidly loses potency in water and must be changed b.i.d. or t.i.d. |
| | Suspension | Oral or gavage | Most | 200–250 mg/kg | s.i.d. or b.i.d. | For converting regimen to pelleted feeds. Administer by gavage until feeds are accepted. Inadequate for long-term therapy for chlamydiosis. |
| **Ticarcillin** (Ticar—Beecham) | Injectable | I.V. or I.M. | Most | 200 mg/kg | b.i.d., t.i.d., or q.i.d. | More effective against *Pseudomonas* than carbenicillin. Nontoxic and synergistic with aminoglycosides. Stable for 72 hours after reconstitution. |
| **Tobramycin** (Nebcin—Lilly) | Injectable (80 mg/2 ml) | I.M. | Most | Dose as gentamicin | See Gentamicin | For strains of *Pseudomonas* resistant to gentamicin. |

| | | | | | | |
|---|---|---|---|---|---|---|
| **Tylosin** | Injectable (50 mg/ml) (Tylan 50—Elanco) Injectable (200 mg/ml) (Tylan 200-Elanco) | I.M. | Most | 10–40 mg/kg | b.i.d. or t.i.d. | Good in initial therapy of upper respiratory infections and air sacculitis. Nontoxic. |
| | Soluble powder with vitamins (250 gm/8.81 oz) (Tylan plus vitamins—Elanco) | Water | Most | 2 tsp/gallon | 10 days on, 5 days off, and 10 days on | May be poorly accepted because of bitter taste. Dosage may be divided between food and water. For treatment of chronic respiratory disease. |
| | | Eye spray | Cockatiels and others | Mix with water 1:10 | s.i.d., b.i.d., or t.i.d. | For flock or individual treatment of conjunctivitis. Allows frequent treatment without handling. May be used in conjunction with tylosin or preferably lincomycin and spectinomycin in the water for suspected mycoplasmosis. |
| | Nebulize (Tylan 200—Elanco) and DMSO (Domoso—Diamond) | Nebulization | Quail and pigeons | 1 gm tylosin/50 ml DMSO | Nebulize for one hour | Therapeutic tissue concentrations of tylosin are produced after one hour of nebulization and remain high for more than three hours. May be used as an adjunct to systemic therapy or alone. |
| **Trimethoprim and Sulfamethoxazole** | Suspension (40 mg trimethoprim and 200 mg sulfamethoxazole/5 ml) (Bactrim—Roche) | Oral | Psittacines | 0.1 ml/30 gm | b.i.d. or t.i.d. for 5 to 7 days | For respiratory or enteric infections in hand-fed babies. |
| | | | | 1.0 ml/lb | | For respiratory and enteric infections. |
| | | | Toucans and mynahs | 1.0 ml/lb | s.i.d. for 5 days | For coccidiosis. May be added to feed. |
| | Injectable (40 mg trimethoprim and 200 mg sulfamethoxazole/1 ml) (Tribrissen—Coopers) | I.M. | Psittacines | 0.1 ml/lb | s.i.d. or b.i.d. | For respiratory and enteric infections. |

THERAPEUTICS

**Table 26–2.** ANTIFUNGALS FOR USE IN BIRDS

| Generic Name | Form | Route | Species | Dosage | Frequency and Duration | Notes |
|---|---|---|---|---|---|---|
| **Amphotericin B** (Fungizone—Squibb) | Injectable | I.V. | Raptors and psittacines | 1.5 mg/kg | t.i.d. | Use for three days in conjunction with or follow with flucytosine therapy. Potentially nephrotoxic and may cause bone marrow suppression. For aspergillosis. |
| | | Intratracheal | Raptors and psittacines | 1 mg/kg | s.i.d., b.i.d., or t.i.d. for three days | Dilute with sterile water to increase volume for maximum distribution throughout lungs and air sacs. Use in conjunction with flucytosine. |
| | | Nebulize | Raptors and psittacines | 1 mg/ml saline | 15 min b.i.d. | |
| | Lotion, creme, or ointment (3%) | Topical | Psittacines | | s.i.d. | Apply to oral lesions of candidiasis that are refractory to nystatin. Not absorbed from the gut. |
| **Chlorhexidine** (Nolvasan—Fort Dodge; Virosan—Bioceutic) | Solution (2%) | Water | Most | 10–20 ml/gallon | 7 to 14 days | To treat mild flock candidiasis and to slow spread of virus. Not absorbed from gut. Can be used over 30 days. Unpalatable to canaries and in scented form can cause death owing to lack of water consumption. |
| **Flucytosine** (5-fluorocytosine) (Ancobon—Roche) | Capsules (250 and 500 mg) | Gavage | Psittacines | 250 mg/kg | b.i.d. | This dosage can be used safely for extended periods of time for treatment of aspergillosis. Because of bone marrow toxicity, hematologic assessment is recommended during therapy. |
| | | | Raptors | 18–30 mg/kg | every six hours | |
| | | Feed | Psittacines and mynahs | 100–250 mg/lb | | For flock treatment of aspergillosis or severe |

| | | | | | | |
|---|---|---|---|---|---|---|
| | | | | | | candidiasis, especially respiratory candidiasis. Apply to favorite food or mix with mash. Drug is expensive and difficult to obtain. |
| **Gentian Violet** (GV-11—Noremco) | Powder | Feed | Psittacines | 0.5–1.0 gm/kg feed | 7 to 45 days | For treatment of candidiasis or inhibition of *Candida* overgrowth during chlortetracycline therapy. To inhibit fungal overgrowth in feed. |
| **Ketoconazole** (Nizoral—Janssen) | Tablets (200 mg) | Gavage | Psittacines | 5–10 mg/kg | b.i.d. for 14 days | For severe refractory candidiasis. For local effect in the crop, dissolve ¼ tablet (50 mg) in 0.2 ml 1 N hydrochloric acid and 0.8 ml water. Will turn pale pink when dissolved. Mixture is added to food for gavage. |
| | | Water | Most | 200 mg/L | 7 to 14 days | Drug is not water-soluble and will float or sink. Dissolve in acid prior to adding to water. |
| | | Feed | Most | 10–20 mg/kg | 7 to 14 days | Apply to favorite food or add to mash. |
| **Nystatin** | Suspension (100,000 units/ml) (Mycostatin—Squibb) | Oral | Most | 1 ml/300 gm | s.i.d., b.i.d., or t.i.d. for 7 to 14 days | For treatment of candidiasis, after antibiotic therapy or in conjunction with antibiotics. Hand-fed babies should always receive anifungal therapy when being treated with antibiotics. Acts on contact with lesions and is not absorbed from the gut. Should not be given by gavage if treating mouth lesions. |
| | Feed premix in diatomaceous earth base (Myco-20—Squibb) | Feed | Most | 1–2 tbs/5 lb feed | 7 to 14 days | Do not use in conjunction with tetracycline, as calcium content may interfere with absorption. |

**Table 26–3.** ANTHELMINTIC DOSAGES FOR BIRDS

| Generic Name | Form | Route | Species | Dosage | Frequency and Duration | Notes |
|---|---|---|---|---|---|---|
| **Amprolium** (Amprol or Corid—Merck) | Solution (9.6%) | Water | Most | 2 ml/gallon | 5 days or longer | For treatment of coccidiosis. Some strains in mynahs and toucans may be resistant. Cages should be cleaned with live steam to prevent reinfection. Supplement B vitamins. |
| **Chloroquine phosphate** (Aralen phosphate—Winthrop) | Tablets (500 mg) | Oral | Penguin | Initial loading dose of 10 mg/kg followed by three doses of 5 mg/kg | Initial dose followed by lower dose at 6, 18, and 24 hours | For initial therapy of avian malaria caused by *Plasmodium* species. Therapy should include simultaneous dosage with primaquine phosphate. Effective only against circulating forms. |
| **Carbaryl** (Sevin—Southern Agricultural Insecticides, Inc.) | Dust (5%) | Dusting | Most | Cover bird with light dust or 1 tsp to a cockatiel nest box; 2 tbs to large macaw nest box | Once. Repeat as needed. Repeat when cleaning nest box yearly. | For treatment of ectoparasitism. Can be applied with vegetable hand duster. May be added to nest box litter for control of mites and ants. Nontoxic. |
| **Crotamiton** (Eurax—Westwood) | Creme | Topical | Most | Apply to affected areas | Weekly for four weeks | For treatment of *Knemidokoptes* and other cutaneous mite infestations.* Avoid application to feathered areas. |
| **Dimetridazole** (Emtryl—Jensen-Salsbery) | Soluble powder (182 gm/6.42 oz) | Water | Most<br>Lory, mynah<br><br>Pekin Robins | 1 tsp/gallon<br>½tsp/gallon<br><br>Lethal or toxic | 5 days only<br><br><br>0 | May be toxic if this dosage is exceeded. For treatment of trichomoniasis, giardiasis, histomoniasis, and hexamitiasis. Do not treat birds when breeding. A male bird that is feeding the hen on the nest may consume enough of the drug to reach toxic levels. Extended therapy may result in toxicity or overgrowth of *Candida*. Also effective against some infections caused by anaerobic bacteria. Recurrence of clinical giardiasis after therapy may occur. |

| | | | | | | |
|---|---|---|---|---|---|---|
| | | Gavage | Budgerigar | 1.5 gm (1 level tsp)/ pint water. Each bird is given 0.5 ml/30 gm. | 3 doses at 12-hour intervals | For birds that fail to drink adequate amounts of medicated water. |
| **Fenbendazole** (Panacur—Hoechst-Roussel) | Suspension | Oral | Most | 10–50 mg/kg | Once. Repeat in 10 days. | For *Ascaris*. Not to be used during molt (may stunt feathers) or while nesting. |
| | | | | | s.i.d. for 3 days | For microfilaria or flukes. |
| | | | | | s.i.d for 5 days | For *Capillaria*. Not effective against gizzard worms in finches. |
| **Ipronidazole** (Ipropran—Roche) | Soluble powder (61 gm/2.65 oz) | Water | Most | 500 mg/gallon | 7 days | For giardiasis, trichomoniasis, and histomoniasis. |
| **Ivermectin** (Ivomec—Merck) | Injectable (10 mg/ml) | I.M. or oral | Most | 200 μg/kg | Once. Repeat in 10 to 14 days. | Effective for intestinal nematodes, coccidia, *Oxyspiruria*, and gapeworms. Drug of choice for *Knemidokoptes* infestation. Bovine preparation should be diluted 1:4 with propylene glycol and dosed at 0.05 ml/lb. Budgerigar dose, 0.01 ml. Oral dose seems to work as well as I.M. |
| **Levamisole** (Ripercol L—American Cyanamid) | Injectable (13.65%) | Water | Most | 5–15 ml/gallon | 1 to 3 days | Effective for most intestinal nematodes. Repeat in 10 days. For birds that refuse to drink, withhold water prior to treatment. |
| | | | | 0.3 ml/gallon | Several weeks | For immunostimulation in immune-suppressed birds. |
| | | Gavage | Australian Parakeets | 15 mg/kg | Once. Repeat in 10 days. | For treatment of individual birds or in species that fail to drink water (desert species may refuse water for several days). |
| | | I.M. or subcutaneous | Most | 4–8 mg/kg | Once. Repeat in 10 to 14 days. | May be toxic, resulting in vomiting, ataxia, or death. Do not use in debilitated birds. |
| | | | | 2 mg/kg | 3 doses at 14-day intervals | For immunostimulation in immune-suppressed birds. |

*A more effective treatment is available—see Ivermectin.

*Table continued on following page*

**Table 26–3.** ANTHELMINTIC DOSAGES FOR BIRDS *(Continued)*

| | | | | | | |
|---|---|---|---|---|---|---|
| **Mebendazole** (Telmintic—Pitman-Moore) | Powder | Gavage or feed | Raptors and psittacines | 25 mg/kg | b.i.d. for 5 days | Acute toxic hepatitis reported in raptors and some mammals. For *Capillaria*. |
| **Niclosamide** (Yomesan—Bayvet) | Tablets (5-lb dose-357 mg) | Gavage or food | Most | 100 mg/lb | Once. Repeat in 10 to 14 days. | For treatment of tapeworms. Ground tablets are not water-soluble. Add to mash. |
| | | | Finches | 500 mg/kg | Once a week for 4 weeks | Large waterfowl may be pilled. |
| **Piperazine** | Suspension | Oral | Poultry | 100–500 mg/kg | Once. Repeat in 10 to 14 days. | For ascarides in poultry. Not effective in psittacines. |
| **Praziquantel** (Droncit—Bayvet) | Tablets (23 mg) | Oral | Most | ¼ tablet/kg | Once. Repeat in 10 to 14 days. | For tapeworms. Add to feed or administer by gavage. Injectable form should be used with caution owing to potential toxicity. Injectable form toxic to finches. |
| **Primaquine** | | Oral | Penguins | 0.03 mg/kg | s.i.d. for 3 days | For therapy of avian malaria caused by *Plasmodium* sp. Therapy should include simultaneous dosage with chloroquine. |
| **Pyrantel pamoate** (Nemex II—Pfizer) | Suspension (4.5 mg/ml) | Oral | Most | 4.5 mg/kg | Once. Repeat in 10 to 14 days. | Nontoxic and palatable. For intestinal nematodes. |

| | | | | | | |
|---|---|---|---|---|---|---|
| **Pyrethrim** | Spray | Topical | Most | Lightly mist feathers | Repeat as necessary | For external parasites, especially lice, which are resistant to carbaryl. When treating lice spray must be applied in axillary area with wing extended. |
| **Quinacrine** (Atabrine—Winthrop) | Tablets | Gavage | Psittacines | 5–10 mg/kg | s.i.d. for 7 days | Drug is hepatotoxic at high doses (50–150 mg/kg) in cockatoos. Treatment of *Haemoproteus* infection is not recommended. |
| **Rotenone** (Goodwinol—Goodwinol Products) | Creme | Topical | Most | Apply to affected areas | Weekly for four weeks | For treatment of *Knemidokoptes* (see Ivermectin for more effective treatment) and other cutaneous mite infestations. Avoid application to feathered areas. May be toxic if ingested. |
| **Thiabendazole** | | Oral | Most | 250–500 mg/kg | Once. Repeat in 10 to 14 days. | For ascarides. |
| | | | | 100 mg/kg | s.i.d. for 7 to 10 days | For *Syngamus trachea*. |

**Table 26–4.** NUTRITIONAL SUPPLEMENTS FOR BIRDS

| Generic Name | Form | Route | Species | Dosage | Frequency and Duration | Notes |
|---|---|---|---|---|---|---|
| **Calcium** | Syrup (Neo-calglucon—Sandoz) (115 mg calcium/5 ml) | Water | Most | 1 ml/30 ml water | To effect | For calcium deficiency or supplementation during egg laying, growth, bone healing. |
| | Injectable (Calphosan—Carlton Corp) (50 mg Ca gluconate and 50 mg Ca lactate/10 ml) | I.M. | Most | 0.5–1.0 ml/kg | Once; can be repeated weekly | For treatment of hypocalcemia or hypocalcemic tetany. For supplementation during egg laying, rapid growth, bone healing, and tetracycline therapy. Adjunct in treatment of egg binding and soft-shelled eggs. Used in Toucans when on tetracycline, as they are very susceptible to bone deformities. |
| | Calcium gluconate | I.V. | Most | 50–100 mg/kg | To effect; slow | For hypocalcemic tetany. May be diluted and given I.M. if a vein cannot be located. |
| | Powder (Osteoform—Vet-A-Mix) | Feed | Most | ⅛ tsp/kg feed | As needed | For calcium and vitamin $D_3$ supplementation. |
| **Cod liver oil** | | Feed<br>Oral | Most | 10 ml/kg seed<br>1 to 2 drops | Daily | For vitamin A and E supplementation. Should be mixed fresh daily to prevent oxidation. |
| **Iodine** | Strong solution (Lugol's) | Water | Budgerigars | Preparation stock solution 2 ml/30 ml water; 1 drop stock solution added to 250 ml drinking water | Daily | For iodine deficiency or goiter. |
| | | | | | 2–3 times weekly | For preventive supplement in iodine-deficient areas.<br>Store in brown bottle or keep in dark after mixed. Will stain. |
| **Seawater mixture** | 1 liter seawater boiled<br>1 ml Lugol's solution<br>(stock solution) | Water | Most | 5 ml stock solution/250 ml drinking water | Daily | For trace mineral supplementation.<br>Store in refrigerator. |

| | | | | | | |
|---|---|---|---|---|---|---|
| **Vitamin A** (Aquasol A—USV Pharmaceutical) | Injectable | I.M. | Psittacines | 0.1 ml/100 gm | Twice weekly in first week; then weekly as needed. | For hypovitaminosis A. Supplemental therapy for avian pox virus infection, sinusitis, ophthalmic disorders. Long-term therapy may be required to restore liver stores. |
| **Vitamins A, $D_3$, and E** | Injectable (100,000 units vitamin A and 10,000 units vitamin $D_3$/ml) (Injacom 100—Roche) | I.M. | Most | 0.1–0.2 ml/300 gm | Double dosage first treatment; then once weekly as needed. | Company recommends maximum dose of once every 6 weeks; possible vitamin $D_3$ toxicity. Author has treated thousands of Amazons with no side effects. Owing to calcium metabolism problems in African Greys, should not repeat in that species. For hypovitaminosis—seems to help in bone healing, egg binding, and soft eggs and bones. |
| **Vitamins A, D, E, and B complex** | Injectable with vitamin B complex (Injacom 100 + B—Roche) | I.M. | Most | 0.1 ml/300 gm | Once every 7–10 days | Indications as above. Toxicity or anaphylaxis may occur if recommended dosage is exceeded. |
| **Vitamin B complex** | Injectable | I.M. | Most | Dose by thiamine content—1–3 mg/100 gm | Once a week | For muscular weakness, debilitation, and anemia and to stimulate appetite. For supplemental therapy in neurologic disorders, liver, kidney, and intestinal disease, and following long-term antibiotic therapy. Overdose may result in anaphylaxis. |
| **Vitamin B complex** with choline, inositol, and methionine (Methiscol—USV Pharmaceutical) | Capsules | Feed | Most | 1–2 gm/kg feed | Daily | Lipotrophic agent used to facilitate the transport of fats in treatment of liver disease or arteriosclerosis. For vitamin B complex supplementation as indicated above. |
| **Vitamin $B_1$** (thiamine) | Powder | Food | Raptors, penguins, cranes | 1–2 mg/kg | Daily | Prophylactic use when feeding diet of fish containing thiaminase. |

*Table continued on following page*

**Table 26–4.** NUTRITIONAL SUPPLEMENTS FOR BIRDS *(Continued)*

| Generic Name | Form | Route | Species | Dosage | Frequency and Duration | Notes |
|---|---|---|---|---|---|---|
| **Vitamin $B_{12}$** | Injectable | I.M. | Most | 250–500 μg/kg | Once a week | For anemia. May produce pink droppings. |
| **Vitamin C** | Injectable | I.M. | Most | 20–40 mg/kg | s.i.d. to once a week | For supplementation in anorectic or debilitated birds, stress, viral diseases, and liver disease. |
| **Vitamin E and selenium** (Seletoc—Burns Biotec) | Injectable | I.M. | Storks, cranes, flamingos, cockatiels; Most immature psittacines | 0.06 mg/kg selenium (cockatiels—0.01 ml) (macaws—0.1 ml) (Eclectus—0.05 ml) (African grey—0.03 ml) | Prior to or at time of capture or stressful event. (cockatiels—every 3–14 days) | For prevention or treatment of muscular weakness, capture myopathy. Also for paralysis in cockatiels. May assist in early therapy of neonatal leg dysfunction. Overdose may result in selenium toxicity. |
| **Vitamin $K_1$** (Veta-$K_1$—Professional Veterinary Laboratory) (Aquamephyton—Merck, Sharp & Dohme) | Injectable (10 mg/ml) | I.M. | Most | 0.2–2.5 mg/kg | As needed; usually only needs 1–2 injections | For hemorrhagic disorders and to prevent such problems when Amprolium and sulfas are administered. |
| **Vitamins—water soluble** (Plex Sol C—Vet-A-Mix) | Powder—Vitamins A, $D_3$, C, E, etc. | Feed or water | Most | 2 gm/kg feed<br>1 gm/L water | Daily | For vitamin supplementation. Double dose for therapy. Well accepted. Water breaks down vitamin A rapidly. Mix fresh s.i.d. or b.i.d. |

**Table 26–5.** MISCELLANEOUS DRUG DOSAGES FOR BIRDS

| Generic Name | Form | Route | Species | Dosage | Frequency and Duration | Notes |
|---|---|---|---|---|---|---|
| Allopurinol (Zyloprim—Burroughs Wellcome) | Tablets (100 mg) | Oral<br>Water | Parakeet | 1 crushed tablet in 10 ml water; give one drop or 20 drops in 30 ml drinking water. | 4 times daily | For treatment of gout. |
| Aspirin | Tablets (5 grain) | Water | Most | 1 tablet in 250 ml drinking water | Only source of water | Indicated for pain. |
| Atropine | 1/20 gr/ml or 0.5 mg/ml | I.M. or subcutaneous | Most | 0.1–0.2 mg/kg | q.s. | Will not result in pupil dilation. For organophosphate poisoning. |
| | | | | 0.04–0.1 mg/kg | Once preanesthetic | For preanesthetic to strengthen heart. May cause thickening of respiratory secretions resulting in obstruction of endotracheal tube. |
| | | | | | Twice for vomiting | For antispasmodic in birds with GI paralysis. Can make worse. |
| Bromhexine hydrochloride | Injectable (3 mg/ml) | I.M. | Most | 0.05 ml/100 gm | b.i.d. or s.i.d. | Liquefy respiratory mucus |
| Calcium EDTA (Calcium disodium versonate—Riker) | Injectable (200 mg/ml) | I.M. | Most | 10 mg/350 gm | t.i.d. until asymptomatic (5 to 10 days) | Chelating agent for use in lead toxicity. Initial therapy with injectable until patient is asymptomatic, followed by oral therapy. Therapy should not be discontinued until lead has been removed from the gizzard or tissues. |
| | | | | 20 mg/350 gm | t.i.d. | |
| Dexamethasone | Injectable (2 mg/ml) | I.M. or I.V. | Most | 2–4 mg/kg | s.i.d., b.i.d., or t.i.d. | For shock, trauma, gram-negative endotoxemia. Use lower dose for anti-inflammatory. Use decreasing dosage schedule for long-term therapy. |

*Table continued on following page*

**Table 26–5.** MISCELLANEOUS DRUG DOSAGES FOR BIRDS *(Continued)*

| Generic Name | Form | Route | Species | Dosage | Frequency and Duration | Notes |
|---|---|---|---|---|---|---|
| Dextrose—50% | Injectable (50%) | I.V. in slow bolus | Most | 1 gm/kg (2 ml/kg) | As needed | For hypoglycemia. May be diluted and given I.M. if vein cannot be located. |
| Dextrose—50% | Injectable (50%) | I.V. in slow bolus | Most | 1 gm/kg (2 ml/kg) | As needed | For hypoglycemia. May be diluted and given I.M. if vein cannot be located |
| Diazepam (Valium—Roche) | Injectable (5 mg/ml) | I.M. or I.V. | Most | 0.03 mg/30 gm<br>0.6 mg/kg | As needed | Anticonvulsant (see Chapter 45, Anesthesiology). |
| Diethylstilbestrol | 0.25 mg/ml | I.M. | | 0.03–0.10 ml/300 gm | | For reproductive problems in females and plumage problems in both sexes. Overdose may result in anemia. |
| | | Oral | | 1 drop/30 ml water | | |
| Dimethylsulfoxide (Domoso—Diamond) | Liquid or gel | Topical | All | 0.1 cc/100 gm | b.i.d. until pain and/or swelling decrease | Used following trauma. Can be mixed with some liquid antibiotics. Produces garlic smell on breath. |
| Doxapram (Dopram—A.H. Robins) | Injectable (20 mg/ml) | I.M. or I.V. | Most | 5–10 mg/kg (0.007 mg/gm) | Once | To stimulate respiration. To speed recovery from ketamine-xylazine anesthesia. |
| EDTA—TRIS lysozyme solution (Custom-prepared solution)* **Warning:** Oral or parenteral use is contraindicated | Liquid | Topical<br>Intratracheal<br>Intranasal<br>Lavage of wounds | All | As desired | b.i.d. | To increase susceptibility of *Pseudomonas* bacteria to antibiotics and to aid in liquefaction of caseous exudate. Keep refrigerated. |
| Ergonovine maleate | Injectable (0.2 mg/ml) | I.M. | Most | 0.02 mg/300 gm | One dose maximum | To aid in egg expulsion. Use in conjunction with calcium and vitamin A injections. Dilute for use in small birds. |
| Ferric subsulfate | Liquid, powder | Topical | All | q.s. | As needed to stop hemorrhage | Apply with cotton-tip applicator using grinding motion. |
| Flunixin-meglumine (Banamine—Schering) | Injectable (50 mg/ml) | I.M. | Most | 1.0–10 mg/kg | Can be repeated | Analgesic, anti-inflammatory, antipyretic. Helpful in shock and trauma. Nonsteroidal. |
| Furosemide (Lasix—Hoechst-Roussel) | Injectable (50 mg/ml) | I.M. | | 0.05 mg/300 gm | b.i.d. | Diuretic. Lories are very sensitive and easily overdosed. |

| | | | | | | |
|---|---|---|---|---|---|---|
| Iron dextran | Injectable (100 mg/ml) | I.M. | All | 10 mg/kg | Repeat in 7–10 days if PCV fails to return to normal. | For iron deficiency anemia or following hemorrhage. |
| Kaolin and Pectin | Oral suspension | Oral | Most | 2 ml/kg | b.i.d., t.i.d., or q.i.d. | Antidiarrheal. |
| Lactated Ringer's solution | I.V.; fluids in bag or bottle | I.V. | All | Body wt. in gm × 0.10 = fluid deficit (def) (in ml) Maintenance = 50 ml/kg/day. | 50 % def + maintenance in first 24 hrs; 2nd 50 % def + maintenance next 48 hours. Then go to maintenance. Repeat S.P. & PCV to determine if adequately replaced. | Replace fluids prior to gavaging nutrients. Some raptor studies indicate oral tubed fluids do as well as I.V. unless very critical.<br>Adult—⅓ daily dose t.i.d.<br>Weak, debilitated, or baby birds—1/10 to 1/5 every 1½ hours. |
| Lactulose (Cephulac—Merrell Dow) | Liquid | Oral | All | Cockatiel—0.03 cc b.i.d. to t.i.d.<br>Amazon—0.10 cc b.i.d. to t.i.d. | b.i.d. until improvement noted. Can be used for weeks. | To decrease toxins and/or CNS symptoms from liver damage, stimulate appetite, improve intestinal flora. Warning—reduce dosage if diarrhea develops. |
| Levothyroxine (Synthroid—Flint) | Tablets (0.1 mg) | Water | Most | 1 tablet/30 ml water to 1 tablet/4 oz water | Daily | For respiratory clicking, vomiting in budgerigar. Thyroid replacement responsive problems: i.e., slow molt, obesity, lipoma. Overdose; bradycardia, hyperesthesia, weight loss, death. Stir water; offer 15 min and remove. Budgerigar—high dose; water drinkers—low dose. |
| Medroxy-progesterone | Powder (Provera-promone—Upjohn) | Feed | Pigeons | 0.1 % of ration | Continual | To inhibit ovulation. |
| | Injectable (Depo-provera—Upjohn) | I.M. | Most | 0.025–1.0 ml (3 mg/100 gm) | Once ever 4–6 weeks | Antipruritic.<br>To suppress ovulation. Repeat doses tend to cause obesity, polydipsia, polyuria, and lethargy. |

*Table continued on following page*

**Table 26–5.** MISCELLANEOUS DRUG DOSAGES FOR BIRDS *(Continued)*

| Generic Name | Form | Route | Species | Dosage | Frequency and Duration | Notes |
|---|---|---|---|---|---|---|
| Mineral oil | | Oral | Most | 1–3 drops/30 gm<br>5 ml/kg | Once. Repeat as necessary. | Use as laxative and to aid in the elimination of lead from the gizzard. Should be administered by tube or slowly in order to avoid aspiration. |
| Neomycin palmitate–hydrocortisone acetate–trypsin–chymotrypsin concentrate (Kymar ointment—Burns Biotec) | Ointment | Topical | All | As needed | As needed b.i.d. | Apply to debrided necrotic areas. |
| Oxytocin | Injectable | I.M. | Most | 0.01–0.1 ml | Once | For egg expulsion. Should be administered in conjunction with injectable calcium and vitamin A. |
| Pancreatic enzymes: Lipase, protease, and amylase (Viokase V—A. H. Robins Co.) | Powder | Feed | Most | ⅛ tsp/kg | Daily | For pancreatic insufficiency. Mix with moistened feed or administer by gavage. Incubate with food for 15 minutes prior to gavage.<br>Used in birds that are polyphagic (going light), passing whole seeds, and slow in emptying crops. |
| Phosphorated carbohydrate (Emetrol—Rorer) | Solution | Oral | Most | 0.1 ml/30 gm | Repeat 5 times in one hour. | To control vomiting and aid crop emptying. |
| Prednisolone | Tablet (5 mg) | Oral | Most | 0.2 mg/30 gm or 1 tablet in 2.5 ml water—2 drops orally | b.i.d. | Anti-inflammatory—use decreasing dosage schedule in long-term therapy. |
| | Injectable | I.M. or I.V. | Most | 2 mg/kg | | For shock, trauma, endotoxemia. |
| Prednisolone sodium succinate (Solu-Delta-Cortef—Upjohn) | Injectable (10 mg/ml) | I.V.<br>I.M. | All | 0.1–0.2 cc/100 gm | Repeat every 15 minutes to effect. | For shock—can decrease dosage by half in larger birds. |

| | | | | | | |
|---|---|---|---|---|---|---|
| Sodium bicarbonate | 1 mEq/ml | I.V., then subcutaneously | Most | 1 mEq/kg for 15–30 minutes to maximum of 4 mEq/kg | | For treatment of metabolic acidosis. |
| Stanazolol (Winstrol V—Winthrop) | Injectable (50 mg/ml) | I.M. | Most | 0.5 to 1 ml/kg | Once or twice weekly | Anabolic therapy indicated in anemia, anorexia, debility, to promote weight gain and accelerate the recovery from disease. Use with caution in birds with renal disease. |
| Testosterone | Injectable (25 mg/ml) | I.M. | Most | 0.1 ml/300 gm | Once. Repeat weekly as needed for anemia. | To increase male libido (beware of suppression of negative feedback system), for anemia, debilitation, and feather problems. |
| | | | | Canaries 2.5 mg/kg | Weekly for 6 weeks. | See Contraindications to tablets. |
| | Tablet (10 mg) | Water | Canaries | 1 ground tablet in 30 ml water (stock solution); 5 drops stock solution/30 ml drinking water | Daily | To return male canaries to song. If no response in two weeks, double dose. Contraindicated in hepatitis. |

**Table 26–6.** SELECTED PHARMACEUTICAL HOUSES AND PRODUCTS FOR AVIAN USE

**Abbott Laboratories**, Veterinary Division, Abbott Park, North Chicago, IL 60064
Spectam-R—spectinomycin

**American Cyanamid Co.**, P.O. Box 400, Princeton, NJ 08540
Aureomycin—chlortetracycline soluble powder
Polyotic—tetracycline soluble powder
Ripercol-L injectable solution—levamisole
Sodium sulfamethazine soluble powder—sodium sulfamethazine solution 12.5%

**Anaquest**, 2005 W. Beltline Highway, Madison, WS 53713
AErrane—isoflurane anesthesia

**Ayerst Laboratories**, Veterinary Medical Division, 685 Third Ave, New York, NY 10017
Fluothane—halothane

**Bayvet**, Division of Cutter Labs., Inc., Shawnee, KS 66201
Droncit—praziquantel
Yomesan—niclosamide

**Beecham Laboratories**, 501 Fifth Street, Bristol, TN 37620
Amoxi Drops—amoxicillin
Dexamethasone injection
Ticar—ticarcillin

**Bioceutic Laboratories, Inc.**, P.O. Box 999, St Joseph, MO 64502
Piperazine solution
Styptic powder
Virosan solution—chlorhexidine

**Bird Life**, Box 745, Poway, CA 92064
Pelleted parrot diet containing 1 per cent chlortetracycline

**Bristol Laboratories**, Division of Bristol-Myers Co., P.O. Box 657, Syracuse, NY 13201
Amiglyde—amikacin
Ketaset—ketamine
Kantrim—kanamycin
Polyflex—ampicillin

**Burns Biotec Laboratories**, Division Chromalloy Pharmaceutical, Inc., 8530 K Street, Omaha, NE 68127
BVMO—liquid vitamins
Kymar ointment—neomycin palmitate–hydrocortisone acetate–trypsin–chymotrypsin concentrate
Seletoc—selenium and vitamin E injectable

**Burroughs Wellcome Co.**, 3030 Cornwallis Road, Research Triangle Park, NC 27709
Zyloprim—allopurinol

**Carlton Corporation**, 83 North Summit Street, Tenafly, NJ 07670
Calphosan injection—calcium

**Coopers Animals Health**, P.O. Box 167, Kansas City, MO 64141
Tribrissen injectable—trimethoprim-sulfamethoxazole

**Diamond Laboratories**, 2538 SE 43rd Street, Des Moines, IA 50303
Domoso-dimethylsulfoxide gel or liquid

**Elanco Products Company**, P.O. Box 1750, Indianapolis, IN 46206
Tylan plus vitamins—tylosin powder
Tylan 50 injection—tylosin
Tylan 200–tylosin nebulize

**Evsco Pharmaceutical Corp.**, 2285 East Landis Ave., Vineland, NJ 08360
Cardoxin—digoxin elixir

**Flint Laboratories**, Division of Travenol Laboratories, Inc., One Baxter Parkway, Deerfield, IL 60015
Synthroid—levothyroxine

**Fort Dodge Laboratories**, 800 Fifth Street N.W., Fort Dodge, IA 50501
Nolvasan solution—chlorhexidine
Oxytocin injection

**Goodwinol Products Corp.**, E. Northport, NY 11731
Goodwinol ointment—rotenone

**Hartz Mountain Products Corp.**, Harrison, NJ 07029
Keet Life—chlortetracycline-impregnated millet

**Haver-Lockhart Laboratories**, Box 390, Shawnee, KS 66201
Ergonovine maleate solution
Oxytocin injection
Rompun 20 mg/ml injectable—xylazine
Yomesan tablets—niclosamide

**Hoechst-Roussel Pharmaceuticals, Inc.**, Rt. 202-206 North, Somerville, NJ 08876
Claforan—cefotaxime
Lasix—furosemide
Panacur—fenbendazole

**Janssen Pharmaceutica, Inc.**, 40 Kingsbridge Road, Piscataway, NJ 08854
Nizoral—ketoconazole

**Jensen-Salsbery Laboratories**, Division of Richardson Merrell, Inc., 520 W. 21st Street, Kansas City, MO 64141
Dermathycin—thyroid-stimulating hormone
Emtryl—dimetridazole
Hexanthelin—piperazine
Oxytocin

**Lafeber Co.**, RR 2, Odell, IL 60460
Avi-cake—food containing doxycycline (custom)
Pelleted parrot diet containing chlortetracycline and other medications

**Eli Lilly and Company**, 307 E. McCarty St., Indianapolis, IN 46285
Keflex—cephalexin
Keflin—cephalothin
Nebcin—tobramycin

**Med-Tech, Inc.**, P.O. Box 338, Elwood, KS 66024
Dexasone—dexamethasone
Chloramphenicol capsules
Dihydrostreptomycin sulfate injection
Ergonovine maleate injection
Kaopect—kaolin and pectin
Medichol—chloramphenicol oral solution
Lipo B Super—vitamin B complex plus inositol and choline with thiamine 150 mg/ml
Multi B Super—vitamin B complex with thiamine 100 mg/ml
Sodium ascorbate injection
Testosterone aqueous
Triple sulfa solution
Vitamin E injection

**Table 26–6.** SELECTED PHARMACEUTICAL HOUSES AND PRODUCTS FOR AVIAN USE

**Merck Sharp & Dohme AGVET**, Division of Merck & Co., Inc., P.O. Box 2000, Rahway, NJ 07065
- Amprol—amprolium
- Aquamephyton—Vitamin $K_1$
- Corid—amprolium
- Ivomec—ivermectin

**Merrell Dow Pharmaceuticals Inc.**, Subsidiary of the Dow Chemical Company, 2110 E. Galbraith Rd., Cincinnati, OH 45215
- Cephulac syrup—lactulose

**Norden Laboratories**, 601 W. Cornhusker Highway, Lincoln, NE 68521
- Furacin water mix—nitrofurazone

**Noremco**, P.O. Box 1622, Springfield, MO 65805
- GV-11—gentian violet

**Parke-Davis**, Division of Warner-Lambert Company, 201 Tabor Road, Morris Plains, NJ 07950
- Chloromycetin palmitate—chloramphenicol suspension
- Vetalar—ketamine
- Chloromycetin sodium succinate

**Pfizer Laboratories**, A Division of Pfizer, Inc., 235 E. 42nd St., New York, NY 10017
- Dexamethazone
- LA 200—oxytetracycline, long-acting
- Liquamycin R—tetracycline
- Nemex II—pyrantel pamoate
- Vibramycin—doxycycline

**Pfizer, West Geramny**
- Vibravenös

**Pitman-Moore, Inc.**, Washington Crossing, NJ 08560
- Ergonil—ergonavine
- Metofane—methoxyflurane
- Telmintic powder—mebendazole
- Tylocine—tylosin

**Professional Veterinary Laboratory**, 100 Nancy Drive, Belle Plaine, MN 56011
- Veta $K_1$—vitamin K injection

**Rachelle Laboratories**, P.O. Box 2029, 700 Henry Ford Ave, Long Beach, CA 90801
- Chloramphenicol injection
- Chloramphenicol capsules
- Chlororachelle
- Mychel-Vet—chloramphenicol oral solution

**Riker Laboratories, Inc.**, Subsidiary of 3M, 19901 Nordhoff Street, Northridge, CA 91324
- Calcium disodium versonate—calcium EDTA

**A.H. Robins Company**, Pharmaceutical Division, 1407 Cummings Drive, Richmond, VA 23220
- Dopram V injectable—doxapram
- Viokase V—whole pancrease powder

**Roche Laboratories**, Division of Hoffman-La Roche Inc., Roche Park, Nutley, NJ 07110
- Albon oral suspension—sulfadimethoxine
- Ancobon—flucytosine (5-fluorocytosine)
- Bactrim—trimethoprim + sulfamethoxazole
- Injacom 100—vitamins A, $D_3$, and E
- Injacom 100 + B complex—vitamins A, $D_3$, E, and B complex
- Ipropran—ipronidazole
- Valium—diazepam

**Roerig** (A Division of Pfizer Pharmaceuticals), 235 E. 42nd St., New York, NY 10017
- Geocillin—carbenicillin tablets
- Geopen—carbenicillin injectable

**William H. Rorer, Inc.**, 500 Virginia Dr., Fort Washington, PA 19034
- Emetrol solution—phosphorated carbohydrate

**Rugby**, Rockville Center, Long Island, NY 11570
- Nystatin oral suspension (generic)

**Sandoz, Inc.**, Route, 10, East Hanover, NJ 07936
- Neocalglucon syrup—glubionate calcium

**Schering Corporation**, 2000 Galloping Hill Road, Kenilworth, NJ 07033
- Azium solution—dexamethasone
- Banamine—flunixin meglumine
- Gentocin ophthalmic solution—gentamicin
- Gentocin soluble powder—gentamicin
- Gentocin solution—gentamicin

**Southern Agricultural Insecticides, Inc.**, P.O. Box 218, Palmetto, FL 33561
- Sevin dust—carbaril

**E.R. Squibb & Sons Inc.**, P.O. Box 4000, Princeton, NJ 08540
- Fungizone—amphotericin B
- Mycostatin—nystatin oral suspension
- Myco-20—nystastin feed premix
- Vetasulid—sulfachlorpyrizidine

**The Upjohn Company**, 7171 Portage Road, Kalamazoo, MI 49001
- Biosol-M liquid—neomycin
- Depo-provera—medroxyprogesterone
- Kaopectate—kaolin and pectin
- Lincocin Aquadrops—lincomycin
- LS-50—lincomycin and spectinomycin
- Panmycin Aquadrops—tetracycline oral suspension
- Provera-promone—medroxyprogesterone
- Solu-Delta-Cortef—prednisolone sodium succinate
- Veterinary Depo-Medrol

**USV Pharmaceutical Corporation**, 303 South Broadway, Tarrytown, NY 10591
- Aquasol A—vitamin A injection
- Methiscol—vitamin B complex with choline, inositol, and methionine

**Vet-A-Mix**, 604 West Thomas Avenue, Shenandoah, IA, 51601
- Osteoform—powdered mineral supplement
- Plex Sol C—powdered vitamin supplement

**Westwood Pharmaceuticals**, 468 Dewitt St., Buffalo, NY 14213
- Eurax—crotamiton

**Winthrop Laboratories**, 90 Park Ave, New York, NY 10016
- Aralen phosphate—chloroquine
- Aralen with primaquine—chloroquine and primaquine
- Atabrine—quinacrine
- Winstrol V—stanazolol (Veterinary Products Division)

**Zeigler Brothers**, P.O. Box 95, Gardners, PA 17324
- Pelleted parrot diet containing chlortetracycline

that hand-feeders and recently weaned young birds, in many cases, consume twice the volume of food consumed by adult birds, so that most medications added to the mash should be halved. Intestinal stasis is a common complication in most systemic disease processes observed in baby birds. In severe illnesses initial therapy should be parenteral.

Feed medication dosages in the tables are listed by bird weight rather than feed weight. A helpful rule of thumb is that a one-pound bird will consume (or waste) 1/5 to 1/4 pound of feed daily. Food consumption is higher in small birds.

Oral suspensions developed for pediatric use are particularly useful in avian medicine. Most are very palatable, and the concentrations are low enough for treatment of small birds. In larger birds or in cases in which a more concentrated solution is desired, they may be spiked with capsule contents.

Rapid administration of oral suspensions may result in exhalation of the drug through the nares, or in aspiration. Passage of drugs through the choana and nasal passages is usually not dangerous but probably very uncomfortable to the bird. This contributes to fear and stress during subsequent treatments. Oil-based medications, however, may cause foreign body pneumonia if they are aspirated. Oral preparations may be mixed with food and administered by gavage.

### Topical Therapeutics

Judicious use of topical preparations of antibiotics and antifungal agents can be useful in treating skin diseases and wounds. Creams and ointments must be used sparingly on the skin to avoid pasting of the feathers and subsequent loss of insulation. Powdered spray preparations are much safer for topical treatment.

Intranasal application of ophthalmic solutions, especially gentamicin, is helpful in treating rhinitis. Instillation of oily topical preparations in the nares should be avoided, as aspiration may lead to foreign body pneumonia. Intranasal antibiotics are also helpful in the treatment of bacterial pharyngitis. Ophthalmic preparations are too dilute to provide systemic therapy.

Sprays prepared from water-soluble antibiotics are very useful for treating eye diseases, allowing frequent therapy without restraint. Skin diseases and wounds can also be treated in this way if they are accessible.

### Nebulization Therapy

The reader is referred to Chapter 29, Aerosol Therapy, for information on medicating birds by nebulization techniques.

## OUTPATIENT CARE

Depending upon the severity of the disease status of the avian patient, some clients may be willing to treat their birds at home. Client compliance is enhanced if treatment regimens are not stressful for the bird or owner. If the bird is not critically ill, owners can be instructed in injection techniques for home therapy. Oral administration of some medications by the owner may be adequate for some cases requiring long-term therapy.

## REFERENCES


1. Arnstein, P., et al.: Chlortetracycline chemotherapy for nectar-feeding psittacine birds. J. Am. Vet. Med. Assoc., 154:190–191, 1969.
2. Aronson, A. L.: Pharmacotherapeutics of the newer tetracyclines. J. Am. Vet. Med. Assoc., 176:10, 1980.
3. Bauck, L. A., and Haigh, J. C.: Toxicity of gentamicin in Great Horned Owls *(Bubo virginianus)*. J. Zoo Anim. Med., 15:62–66, 1984.
4. Bush, M.: Pharmacokinetics of antibiotics in reptiles and birds. Br. Vet. Zoo. Soc. Newsletter, Sept., 1981.
5. Bush, M., et al.: Pharmacokinetics of cephalothin and cephalexin in selected avian species. Am. J. Vet. Res., 42:1014–1017, 1981.

5a. Campbell, T. W.: EDTA-TRIS buffer lavage for treating psittacine birds with *Pseudomonas aeruginosa* infections. Comp. Cont. Ed., 7(8):598–604, 1985.

6. Clark, C. H., et al.: Plasma concentrations of chloramphenicol in birds. Am. J. Vet. Res., 43:1249–1253, 1982.
7. Clubb, S. L.: Birds. In Johnston, D. E. (ed.): The Bristol Veterinary Handbook of Antimicrobial Therapy. Philadelphia, PA, Veterinary Learning Systems Co., 1982.
8. Clubb, S. L.: Therapeutics in avian medicine—Flock vs. individual bird treatment regimens. Vet. Clin. North Am. [Small Anim. Med.], 14:345–361, 1984.
9. Custer, R. S., et al.: Pharmacokinetics of gentamicin in blood plasma of quail, pheasants, and cranes. Am. J. Vet. Res., 40:892–895, 1979.
10. Dolphin, R. E., and Olsen, D. E.: Antibiotic therapy in cage birds for pathogenic bacteria detected by fecal culture technique. VM/SAC, 72:1504–1507, 1977.
11. Dein, J., et al.: Pharmacokinetics of chloramphenicol in Chinese Spot-billed Ducks. Proceedings of the American Association of Zoo Veterinarians, 1979, pp. 48–50.
12. Ensley, P. K., and Janssen, D. L.: A preliminary study comparing the pharmacokinetics of ampicillin given

orally and intramuscularly to psittacines: Amazon Parrots (*Amazona* sp.) and Blue-naped Parrots (*Tanygnathus luclonensis*). J. Zoo Anim. Med., 12:42–47, 1981.
13. Fudge, A. M.: Avian antimicrobial therapy. Proceedings of the Association of Avian Veterinarians, 1983, pp. 162–183.
14. Fudge, A. M.: Update on chlamydiosis. Vet. Clin. North Am. [Small Anim. Med.], 14:201–221, 1984.
15. Gaskin, J. M., and Jacobson, E. R.: A mycoplasma-associated epornitic in Severe Macaws. Proceedings of the American Association of Zoo Veterinarians, 1979, pp. 56–59.
16. Goren, E., et al.: Some pharmacokinetic aspects of ampicillin and its therapeutic efficacy in experimental *Escherichia coli* infection in poultry. Avian Pathol., 10:43–55, 1981.
17. Hirsh, D. C., et al.: Pharmacokinetics of penicillin G in the turkey. Am. J. Vet. Res., 39:1219–1221, 1978.
18. Kollias, G. V.: The use of antibiotics in birds; a review with clinical emphasis. Proceedings of the American Animal Hospital Association 49th Annual Meeting, 1982.
19. Landgraf, W. W., et al.: Concentration of chlortetracycline in the blood of Yellow-crowned Amazons fed medicated pelleted feeds. Avian Dis., 26:14–17, 1981.
20. Locke, D., et al.: Pharmacokinetics and tissue concentrations of tylosin in selected avian species. Am. J. Vet. Res., 43:1807–1810, 1982.
21. Locke, D., and Bush, M.: Tylosin aerosol therapy in quail and pigeons. J. Zoo Anim. Med., 15:67–72, 1984.
22. Mandel, M.: Lincomycin in treatment of out-patient psittacines. VM/SAC, 73:473–474, 1977.
23. Panigrahy, B., et al.: Antibiotic-induced immunosuppression and levamisole-induced immunopotentiation in turkeys. Avian Dis., 23:401–408, 1979.
24. Pasco, J. W., et al.: Emergency care for pet birds. Avian/Exotic Pract., 1:31–42, 1984.
25. Redig, P.: Methods for the delivery of medical care to trained or captive raptors. Proceedings of the Veterinary Seminar, American Federation of Aviculture, 1981.
26. Spink, R. R.: Nebulization therapy in cage bird medicine. VM/SAC, May, 1980, pp. 791–794.
27. Teare, J. A.: Pharmacokinetics of a long-acting oxytetracycline preparation in three species of birds. A seminar presented to the Faculty of the New York College of Veterinary Medicine of Cornell University, May, 1982.
28. Wachendörfer, G., and Luthgen, W.: Chlortetracycline impregnated food-pellets for the prophylaxis and therapy of psittacosis/ornithosis in psittacines and pigeons. Avian Pathol., 3:105–114, 1974.
29. Williams, J. E., and Whittemore, A. D.: Bacteriostatic effect of five antimicrobial agents on salmonellae in the intestinal tract of chickens. Poultry Sci., 59:44–53, 1980.
30. Woerpel, R. W., and Rosskopf, W. J.: Avian therapeutics. Proceedings of the Veterinary Seminar, American Federation of Aviculture, 1981.

# *Chapter 27*

# WHAT TO DO UNTIL A DIAGNOSIS IS MADE

GREG J. HARRISON

When dealing with a sick bird in a clinical situation, one must consider the complexities of bird diseases, the inter-relationship of multiple etiologies, and some of the difficulties in determining a diagnosis even under ideal conditions. In addition, a number of other factors are involved:

1. The bird may be too small or too weak to be subjected to hematology, radiology, laparoscopy, or other specific diagnostic measures.
2. A particular disease entity may not be considered in the differential diagnosis because the cause may no longer be present or the clinical signs may be masked by secondary problems; thus, inappropriate diagnostic approaches may be selected.
3. Tests for a suspected etiology may not be available (such as multiple passages through embryonated chicken eggs for detection of a psittacine virus).
4. It may take days or weeks for the practitioner to get test results of diagnostic samples submitted to an outside laboratory.
5. The client may not be willing to pay for extensive test procedures.

Much of the "art" of avian practice involves extensive use of information from the history and physical examination as well as the use of good clinical judgment in evaluating the results of the diagnostic tests that are available. In the end, it is the *bird* the practitioner must treat, not the laboratory results. In truth, there are many birds that respond to symptomatic therapy although the etiology of their disorder is never determined.

Once the selected samples have been taken, the practitioner must institute steps for supportive care of the patient until more specific therapeutic measures are indicated.

Chapter 28, Symptomatic Therapy and Emergency Medicine, addresses specific measures for a number of commonly encountered situations. The following are empirical guidelines that may be utilized in the initial care of most sick or injured birds regardless of the cause.

## FLUID THERAPY

In the author's experience fluid therapy is a necessary adjunct to the treatment of most sick birds. Perfusion of the body tissues in a debilitated patient is essential for supplemental therapy to be effective. The most rapid route is intravenous administration.

Based on work reported by Redig,[4] the use of bolus intravenous fluid therapy has transformed the clinical management of sick and injured birds in the author's practice. This therapy is effective for shock and trauma cases, and, in addition, Redig found other avian physiologic deviations caused by diseases and states of malnutrition to respond to this therapy.[3] According to Redig's work with raptors, birds that are emaciated, severely traumatized with blood loss, or afflicted with severe disease are in a modest to severe state of metabolic acidosis. In all instances, intravenous administration of fluids will quickly re-establish the perfusion of the liver and kidney and begin to compensate for the imbalance. The rapid restoration of the circulating volume is considered more important than the type of fluids used. However, lactated Ringer's solution is most often utilized, as it is similar in composition to avian plasma, is economical, and is readily available.[4]

In order to compensate for the assumed body deficit and provide for maintenance of the debilitated patient, a formula for calculating the amount of fluid to administer has been established.[4] For the practitioner, a level of approximately 10 per cent dehydration can be estimated in any bird suffering from trauma or disease. Therefore, the fluid deficit may be calculated: normal body weight (in grams) × 0.1 (10%) = fluid deficit (in ml). The dose for maintenance fluid is estimated at 50 ml/kg per day.

Fifty per cent of the deficit should be replaced in the first 12 hours; the remainder of the deficit (plus the daily maintenance amounts) over the next 48 hours. Some of the maintenance fluids may be administered orally or subcutaneously. The initial treatment, however, should consist of the calculated volume as an intravenous bolus.[4]

In the author's practice, birds the size of cockatiels and smaller are physically restrained and the initial fluid bolus is administered into the jugular vein. Larger birds are tranquilized with isoflurane (see Chapter 45, Anesthesiology) for the treatment. Table 27–1 lists the approximate *maximum* doses that may safely be administered as the initial bolus in a variety of pet birds. If necessary, fluid therapy may be continued every 3 hours for the first 12 hours; every 8 hours for the next 48 hours; then twice daily. The fluids should be kept in an incubator at 96°F so that they are warmed and ready for immediate use.

Several vessels may be approached for multiple injections of fluids that may be required over a several-day period for some critical birds. The reader is referred to Chapter 12, Hematology, for venous sites of blood collection that are also suitable for fluid administration.

Birds with suspected liver dysfunction may be more prone to hematoma formation subsequent to intravenous fluid administration than birds with apparently normal liver function and normal clotting times. Prevention of a hematoma may be aided by sliding the skin toward the body prior to insertion of the needle, then allowing the skin to slide back over the puncture site when the needle is removed. Reduced incidence of hematomas may also relate to the size of the needle used: a 26-gauge, 3/4 inch needle is recommended for large birds, 27-gauge, 3/8 inch for medium-sized birds, and 30-gauge for finches and budgerigars.

**Table 27–1.** APPROXIMATE MAXIMUM INITIAL I.V. BOLUS FLUID DOSES FOR PET BIRDS

| | |
|---|---|
| Finch | 0.5 cc |
| Budgerigar | 1.0 cc |
| Cockatiel | 2.0 cc |
| Conure | 6.0 cc |
| Amazon | 8.0 cc |
| Cockatoo | 12–14 cc |
| Macaw | 12–14 cc |

Because dehydration is usually accompanied by a change in the acid-base balance, the avian practitioner should consider bicarbonate replacement. If no means are available for determining the deficit, a safe rule of thumb is to administer 1 mEq/kg at 15- to 30-minute intervals to a maximum of 4 mEq.[4] With severe states of metabolic acidosis accompanying emaciation, 5 per cent dextrose is contraindicated, as it is a very strong acidifying agent. Glucose may be utilized if calorie replacement is necessary, or 5 per cent dextrose may be diluted with equal amounts of lactated Ringer's solution.[4]

Fluids may be administered subcutaneously for noncritical cases, although absorption is not always complete and cases of pendulous edemas ventral to the site occasionally occur with this method. A 1:1 solution of lactated Ringer's and 2.5 per cent dextrose may be administered in the following doses: budgerigar: 1-2 cc; cockatiel: 3-8 cc; Amazon: 15-20 cc; macaw: 20-35 cc. Several body sites may be used, but the intrascapular region is preferred by the author, although this method is infrequently used (Fig. 27–1).

Although previously reported to have saved several comatose birds in emergency situations,[3] the use of multiple intramuscular injections of 5 per cent dextrose has been abandoned, not only because of the severe damage to the pectoral muscles, but because the ease and effectiveness of intravenous administration are preferred.

## HEAT

A heated environment (85 to 90°F) is indicated for sick birds. However, in cases of critical patients requiring substantial fluid therapy for rehydration, heat should be deferred until some degree of stability has been established. According to Redig, warming of the body before establishment of adequate tissue perfusion may exacerbate the state of acidosis as a result of vascular dilatation.[4]

## PARENTERAL THERAPY

Much of the initial therapy in support of the sick bird involves addressing the body's reaction to stress. Vitamins have not been demonstrated to be pharmacodynamic and do not replace appropriate prophylactic and therapeutic measures. However, because of the stress and po-

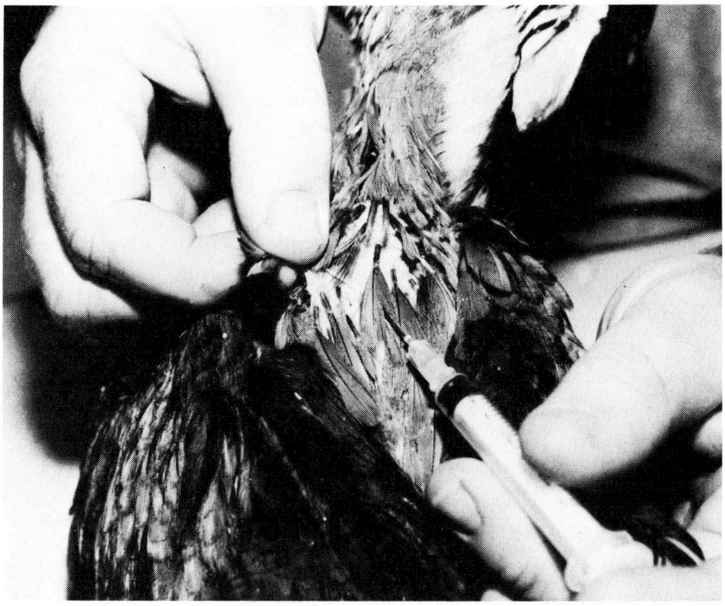

**Figure 27–1.** A secondary choice of fluid administration is subcutaneous injection between the scapulas.

tential nutritional deficiencies and/or liver involvement with most sick birds (see Chapter 31, Nutritional Diseases), initial administration of injectable vitamins A, C, $D_3$, E, and B complex, as well as calcium, is often warranted. Prudent use of parenteral vitamin $D_3$ is advised to prevent potential toxicity of this nutrient. If aggressive vitamin A therapy is indicated for clinical signs of hypovitaminosis A, a separate parenteral product supplies vitamin A alone. Iron dextran injections have been used by Redig to produce dramatic increases in the packed cell volume (PCV) of anemic raptors.[4] A similar increase in PCV was found to occur following iron dextran injections in a group of cockatiels.

Steroids and immunostimulants may be employed as indicated (see Chapter 26, Therapeutics).

Antibiotics may be indicated by Gram's stains of appropriate samples in conjunction with clinical signs. Debilitated patients may also be susceptible to the invasion of enteric organisms through the portal system into the liver.[4] Antibiotics should ideally be selected according to sensitivity results of a bacterial culture and should be administered by injection. The necessity of immediate therapy may dictate the practitioner's choice of an antibiotic pending sensitivity results. Table 27–2 lists some examples of antibiotic choice pending the outcome of sensitivity studies. Indiscriminate use of antibiotics in search of "the one that works" only serves to overtreat the patient, destroy all the normal body flora, and depress the immune system.

## SUPPLEMENTAL THERAPY

Lactulose syrup has been used as an adjunct therapy in avian patients with suspected liver

**Table 27–2.** EXAMPLES OF INITIAL ANTIBIOTIC USE PENDING SENSITIVITY RESULTS

| Clinical Signs | Gram's Stain Results | Antibiotic of Choice |
|---|---|---|
| Depressed, emaciated, polyuric | Normal or slightly low | Ampicillin |
| Depressed, emaciated, polyuric | High percentage gram-negative; low percentage gram-positive | Aminoglycosides + ampicillin (+ bacterin) |
| Respiratory signs, eye involvement in cockatiels | Normal distribution | Lincomycin and spectinomycin soluble powder |
| Respiratory signs | Normal distribution; no yeast, fungus | Erythromycin<br>Ampicillin |
| Chlamydiosis suspected | Low bacterial counts | Vibramycin (+ bacterin) |
| Chlamydiosis suspected | Low percentage gram-positive; high percentage gram-negative | Vibramycin + aminoglycosides (+ bacterin) |

disease. Lactulose apparently reduces the toxic potential of elevated blood ammonia concentration due to liver dysfunction. In addition, the bacterial degradation of lactulose in the intestine acidifies the intestinal contents. This appears to provide a suitable environment for the normal bacterial flora in birds to proliferate and eliminates the need for supplementation with commercial *Lactobacillus* products, which are not species-specific and are often not pure, uncontaminated cultures prepared for sick birds.

## FEEDING THE SICK BIRD

One of the concerns regarding hospitalized birds is to keep them eating during the diagnosis and treatment. For birds that are eating on their own, a wide variety of food can be spread out on a disposable plate or paper towel so that it is all easily available and the bird does not have to pick through it to get what he wants (Fig. 27–2).

Offer familiar and enjoyable foods: seeds, corn on the cob, soaked or dry monkey biscuit, carrot slices, grated cheese, pound cake, grapes, apples, fresh spinach, and peanuts. If the owner says the bird likes chicken, offer chicken. Pelleted diets formulated for exotic birds offer convenience in feeding and cleaning up. However, hospitalized birds will not eat food with which they are not familiar. It is not appropriate to try to change the diet during hospitalization. Tame or imprinted birds will sometimes respond to being offered food by hand. Some bird patients may eat only if the owner is present; therefore, hospital visits are encouraged in these cases. Warming some foods tends to increase palatability.

Contrary to popular opinion, the diet offered in the author's hospital (even to birds with polyuria or diarrhea) includes fruit, especially citrus. Fruit helps keep the food intake up, supplies some vitamin C (which the liver may not be producing), and helps combat dehydration with the high moisture content.

The major weight loss in a sick bird is loss of water; therefore *fresh* water, without contamination by vitamins or medications, must be available, especially for a critically ill bird.

Occasionally a sick bird, especially a baby in the process of being weaned, will overdrink water instead of eating when he is hungry. In these cases, offer the water bowl at selected times during the day rather than leaving it in the cage.

Bland diets may need to be prepared for birds with digestive problems. A soft mixture of baby cereal and water may be offered in a bowl in addition to any necessary forced feeding. Soaked monkey biscuits, pound cake, or strained baby foods can be added as indicated.

Low fat reducing diets may be advised in clinical cases involving obesity or lipomas (see Chapter 42, Miscellaneous Diseases).

### Tube Feeding

If a bird refuses to eat in the hospital, or is losing weight even if it is eating, it must be tube-fed, except under two conditions: vomiting and dehydration. Food is withheld until clinical signs of vomiting have been controlled; rehy-

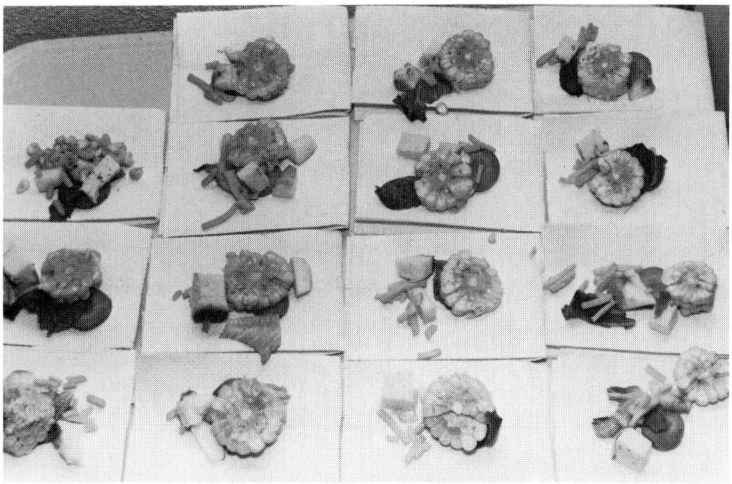

**Figure 27–2.** A wide variety of foods are spread out on disposable paper towels to encourage hospitalized birds to eat. (From Harrison, G. J.: Hospitalization of the avian patient. Vet. Clin. North Am. [Small Anim. Pract.], 14(2):147–164, 1984.)

dration with intravenous fluids is a prerequisite to tube feeding.[4]

The initial formula is a *low*-protein, high-carbohydrate mixture with a concentrated source of vitamins and minerals. The increased stress on the liver for processing high protein diets is contraindicated in sick birds. The protein level may be increased as the bird responds. The formula (e.g., 1 cup baby oatmeal, 1 tablespoon Nutrical, and enough warm water to make a cake batter consistency) is administered into the crop via gavage tube two to four times a day.

With practice, a large psittacine can be tube-fed by a staff member working alone (Fig. 27–3). If the feeder is right-handed, the bird is held in the left hand up against the feeder's body with the fingers controlling the mandible. The tube is placed from the feeder's right side (the bird's left), passed over to the opposite side and gently down the back of the pharynx and esophagus (on the bird's right) into the crop, and the contents of the syringe are emptied. An alternate tubing approach is described in Chapter 17, Cytology. Most birds will not have been tube-fed prior to this; therefore, one would start with small amounts at frequent intervals. As the crop adapts, larger volumes may be administered.

Budgerigars can eventually accommodate approximately 1 to 1.5 cc per feeding; cockatiels, 8 to 9 cc; Amazons, 15 to 20 cc; macaws, 30 to 40 cc. The amount actually given will depend on the bird. If vomiting or crop stasis should occur, a more liquid formula (e.g., lactated Ringer's solution and Nutrical only) should be administered in small amounts until the crop-emptying mechanism returns to normal. Some medications can be mixed with the tube formula and administered orally (see Chapter 26, Therapeutics).

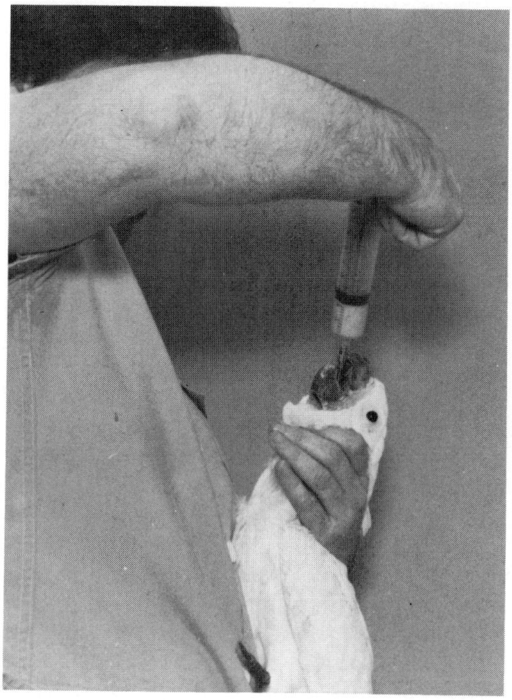

**Figure 27–3.** A technician can become adept at force feeding a large psittacine alone.

## BACTERINS

For most sick birds, the author has found the use of custom-prepared bacterins* to be a successful alternative to antibiotic use in selected cases and for resistant forms of bacteria.

The general use of bacterins in veterinary medicine appears to be most valuable in group medicine, e.g., kennels, herds. This also appears to be true in avian therapeutics. Bacterins are best appreciated when used in birds that are to be, or have recently been, together in groups: in quarantine, brokerage houses, pet shops, zoological institutions, bird shows.

In the author's experience, bacterins have also been apparently successful in the following cases:

1. "Old" birds that are suffering from malnutrition.
2. Hand-raised birds from a low coliform environment in preparation for exposure to outside groups of birds. The outside birds may be carrying various bacterial, low virulence viral, or chlamydial organisms to which the babies may be vulnerable.
3. Birds that show moderate numbers of gram-negative bacteria on Gram's stains of the cloaca and/or choana but do not have clinical signs.
4. Chronic feather picking birds that do not show radiographic lesions (see Chapter 41, Disorders of the Integument).
5. Birds showing chronic and acute upper respiratory signs.

Autogenous vaccines and polyvalent bacterins are developed from and for the birds in the author's practice. A 0.05- to 0.15-cc injection is administered intramuscularly and repeated in two weeks. A minor red swelling may be seen for a few hours at the site of the injection, and some transient lethargy and fluffing may be

*MicroVac, Boca Raton, FL.

evident, usually after the second dose. Any bird showing excessive reaction to the bacterin is not revaccinated, but is cultured and treated accordingly.

The bacterins developed for a specific set of clinical signs may be viable for a number of months. However, owing to the natural mutation of pathogenic organisms, new bacterins are produced frequently from cultures of nonresponsive cases. An estimated 10 per cent of the cases that initially responded to the bacterin appear to require a "booster" after a year.

Commercial bacterins were used by the author in an attempt to decrease cost and make the bacterins more readily available to practitioners, but the porcine and bovine *E. coli* vaccines resulted in several days of lethargy and violent tissue reaction, even in normal birds, so such products cannot be recommended.

## DISEASE OUTBREAK

In cases of a disease outbreak in an aviary or other group of birds, an accurate diagnosis may be achieved more rapidly if one of the birds showing initial signs of the disease is sacrificed for evaluation. As it may not be possible to give injections to each bird, water or food may be the only vehicles available for medicating large numbers of patients (see Chapter 26, Therapeutics).

In the face of a virus outbreak, such as Pacheco's parrot disease, one of the most effective measures is to spread the birds apart, one per cage if possible, so transmission of the organism is stopped. Moving the birds outdoors is advantageous. These steps have also halted the spread of reovirus infection in a group of Eclectus Parrots. Administration of ascorbic acid has been reported to be of some value in the supportive therapy of birds exposed to viral agents (see Chapter 32, Viral Diseases). Chlorhexidine in the drinking water may help deter the spread of the disease, although it may not affect those birds showing clinical signs.[1] Depopulation has been recommended in budgerigar aviaries with papovavirus (budgerigar fledgling disease), followed by stringent disinfection of the premises and repopulation only after 90 days.[2]

A thorough cleaning of the premises is indicated following any disease outbreak (see Chapter 32, Viral Diseases). A number of effective disinfectants are discussed in Chapter 50, Aviculture Medicine. Formaldehyde and potassium permanganate can be used to fumigate incubators or entire rooms. Nest boxes must be replaced.

Vaccines for a number of psittacine viral infections are under investigation but are not commercially available. Poultry vaccines have been used in some cases for Newcastle disease (see Chapter 32, Viral Diseases).

## REFERENCES

1. Clubb, S. L.: Therapeutics in avian medicine. Vet. Clin. North Am. [Small Anim. Pract.], 14:354–361, 1984.
2. Davis, R. B.: Personal communication.
3. Harrison, G. J.: Hospitalization of the avian patient. Vet. Clin. North Am. [Small Anim. Pract.], 14:147–164, 1984.
4. Redig, P. T.: Fluid therapy and acid base balance in the critically ill avian patient. Proceedings, International Conference on Avian Medicine, Sponsored by the Association of Avian Veterinarians, Toronto, Canada, 1984, pp. 59–73.

# *Chapter 28*

# SYMPTOMATIC THERAPY AND EMERGENCY MEDICINE

GREG J. HARRISON
RICHARD W. WOERPEL
WALTER J. ROSSKOPF, JR.
LORRAINE G. KARPINSKI

Evaluation of electrolytes, acid/base balances, and other specialized tests are normally part of emergency medicine in the small animal field. Practical avian emergency care has not evolved to that level. In the emergency treatment of pet birds in life-threatening situations or for those requiring immediate relief of clinical signs, prompt attention may be more important than the choice of therapy.[5]

A brief physical examination is conducted to assess the general condition of the patient and the primary problem. Depending upon the status of the patient, it may be useful to collect some diagnostic samples; however, immediate treatment is usually the priority. The only exception may be in circumstances of flock diseases, in which sacrifice and evaluation of the individual bird may ultimately benefit the flock.

Administration of intravenous fluid therapy, forced feeding techniques, and other measures for general supportive care of avian patients are described in Chapter 27, What To Do Until A Diagnosis Is Made. Those procedures would generally be applied to the "classic sick bird"—the fluffed, lethargic patient showing weight loss, polyuria, and polydipsia, and as adjunct therapy for other ill patients.

This chapter suggests some specific measures that may be incorporated into the care of birds requiring immediate relief of clinical symptoms prior to making a diagnosis. Specific therapeutic agents and doses are listed in Chapter 26, Therapeutics. Nonsurgical management of fractures, symptomatic therapy of disorders of the integument, and those conditions requiring surgical resolution are described in Chapters 30, 41, and 47, respectively. The reader is also referred to Chapter 8, Differential Diagnoses Based on Clinical Signs, and to specific chapters for appropriate diagnostic information.

## SYMPTOMS OF TRAUMA

### Bleeding/Lacerations

The primary therapeutic objectives are to locate the origin of hemorrhage and to provide rapid hemostasis. Some patients with severe lacerations may require blood transfusions or therapy for shock.

In order to prevent subsequent bleeding, fractured and bleeding pin feathers must be entirely removed from the affected follicle. With one hand supporting the bone or tissue surrounding the follicle, one may use a hemostat in the other hand to grasp the base of the feather and pull in the direction of feather growth with a firm, steady motion. Aggressive removal of the feather may injure the germinal epithelium and cause the formation of a feather cyst. Brief finger pressure or cautery may be necessary at the removal site (Fig. 28–1). To prevent wing flapping that may produce further bleeding of affected wing follicles, masking tape can be used to hold the wing against the body in a normal flexed position for a few hours.

Beaks and/or nails that have incurred minor injuries resulting in bleeding may be treated with a cauterizing agent such as ferric subsulfate solution, red hot metal, pen light cautery, electrosurgical electrode, or silver nitrate stick. Chemical cautery must be firmly applied; a superficial layer exposes the tissue to reinjury. A coat of acrylic may be applied as a bandage when the area is dry. Soft foods may need to be offered in cases of beak injuries, although the acrylic appears to reduce the sensitivity of the exposed tissue.

Hemostasis may be difficult to achieve if the injury has occurred to the tongue or elsewhere within the mouth. In these cases, it may be

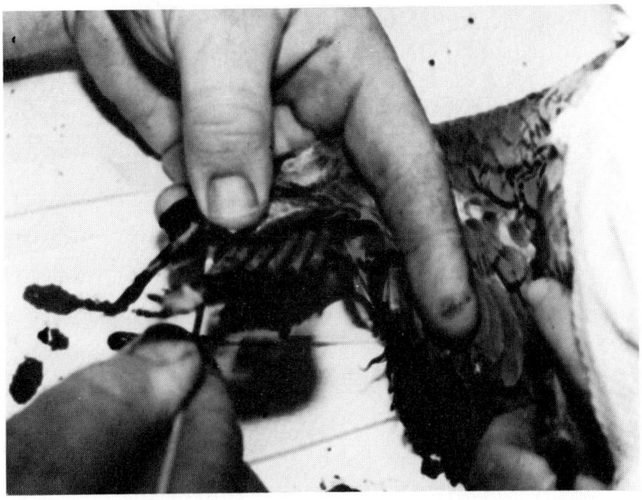

**Figure 28–1.** Although it is infrequently required, chemical cautery is one method of controlling hemorrhage following the removal of a broken, bleeding pin feather.

necessary to anesthetize the patient and suture the wound directly with absorbable suture or employ chemical cautery or electrocautery. Hemostasis sponges (Gelfoam—Upjohn, Kalamazoo, MI) or topical epinephrine is useful for providing hemostasis in relatively large areas where the specific vessel is not visible or readily accessible.

Hemostasis of external wounds (on the head, body, or extremities) may require digital pressure, chemical cautery, or application of a dressing or acrylic bandage. Rubber band tourniquets can be temporarily used on digits. Severely torn nails may require amputation. The decision to suture a wound will depend upon the severity and location of the laceration. Puncture wounds require irrigation and drainage and are usually left open. Although topical or systemic antibiotics are rarely used with avian lacerations, if a puncture wound has resulted from a cat bite, the case is considered critical and antibiotics must be administered immediately to prevent a potential *Pasteurella* bacteremia.

Lacerations over the keelbone usually require surgical resolution. Because they are often caused by impact trauma, the area must be explored for feathers and other debris in the underlying tissue layers prior to suturing. The skin edges are debrided and sutures are closely spaced. In some cases, a protective device may be required to prevent self-trauma to the wounded area (Fig. 28–2).

Administration of vitamin $K_1$ may be necessary in cases of apparent clotting deficits. Whole avian blood transfusions are advised for excessive loss of blood (see Chapter 44, Evaluation and Support of the Surgical Patient). If the patient appears depressed and weak, the clinician should also institute therapy for shock.

The "conure bleeding syndrome" has responded in some cases to parenteral corticosteroids and stress management.

## Burns

The most common burns occur as a result of uncaged birds flying into hot grease or biting

**Figure 28–2.** An orthopedic stockinette is useful as a temporary protective device to prevent self-mutilation of sternal injuries or surgery.

into electrical cords. Administration of steroids and intravenous fluids may be required for shock therapy in either case. Beak and mouth injuries from electrical burns may require localized treatment. If a laceration has simultaneously occurred, debridement and application of an acrylic bandage may be necessary. Use of enzyme products (e.g., neomycin palmitate-hydrocortisone acetate-trypsin-chymotrypsin concentrate ointment or EDTA-TRIS-lysozyme solution[4]) may be necessary if large areas of tissue necrosis are involved. Tube feeding is usually required as well as antibiotics and occasional mycotic therapy.

The oily substances from hot grease burns must be removed from the feathers. This may require several steps of washing/rinsing/drying or repeated applications of corn starch or baby powder on birds that may be inappropriate to bathe (see Chapter 2, Husbandry Practices). Severe skin burns may require debridement of the necrotic tissue and possibly suturing. Application of a thin coat of acrylic bandage helps to protect the epithelium during healing. A hydroactive dressing (Dermaheal—E. R. Squibb & Sons) has been reported to be a successful wound dressing for skin trauma not infected with fungal elements.[3]

### Concussion

The most common cause of head trauma or concussion in small, free-flying birds is impact with windows, mirrors, or household paddle fans. Application of ice bags or other cold material to the bird's head immediately following the trauma is useful in reducing the degree of internal tissue swelling. Oral mannitol, injections of steroids, and very light doses of diuretics may also be of benefit.

### Shock

Intravenous administration of prednisolone sodium succinate or dexamethasone and lactated Ringer's solution (possibly with the addition of sodium bicarbonate)[9] is advised for treatment of shock. Antibiotics are recommended to prevent bacterial invasion. Some practitioners believe that it may be less stressful to a patient in shock to administer a corticosteroid intramuscularly and the fluids subcutaneously, but that is not usually recommended because the therapeutic benefit of these materials is somewhat delayed. Following the initial treatment, the patient should be placed in a warm, quiet environment for recovery. The prognosis for birds in shock is guarded.

## SYMPTOMS OF GASTROINTESTINAL DISTURBANCES

### Anorexia

Birds that have been moved to a new environment, have been overtrained, or are otherwise stressed or fearful may not eat for variable lengths of time, up to 24 to 36 hours in larger species. No therapy is required. Nor is treatment necessary for hens that become anorectic immediately preceding egg-laying.

If the anorexia is a direct result of a gastrointestinal disorder or other illness, administration of intravenous fluids is the initial therapy until a diagnosis is made. Oral supplementation may be offered if the digestive tract is functioning. Use of lactulose to remove toxins appears to increase appetite in many cases. Some practitioners use multiple B vitamins in anorectic birds. If simple food items are not accepted by the birds, formulas can be administered by a crop lavage. The offering of complex food items is delayed until the digestive process has completely resumed.

In those rare instances when oral supplementation may be prevented for an extended period of time, an esophagostomy tube may be necessary. Investigations are currently being conducted on avian duodenal catheterization and total parenteral nutrition. The latter procedure appears to be hindered by the inability to maintain intravenous catheters for the slow drip required of the very hypertonic solutions. Limited studies by Harrison and Ritchie suggest bolus administration of total parenteral nutrition is of no value and may be harmful.

### Crop Disorders

The first step in cases showing crop stasis is removal of food and water. The crop may on occasion be stimulated to empty by the use of phosphorated glycerol (Emetrol). Atropine is contraindicated if proventricular dilatation is suspected or if large amounts of undigested food material are known to be present in the gastrointestinal tract from history or radiography.

If the crop is emptying at a slow rate, a very liquid formula can be tube fed; the consistency can be increased as the crop responds.

"Sour" crop may occur if the stasis has been prolonged. In these cases, the crop contents must be removed. Although many practitioners add sterile water to the crop to suspend the material and express the contents through the mouth by tipping the bird upside down, the procedure is dangerous and can result in irritation to the esophagus and delicate nasal tissues, inhalation pneumonia, upper respiratory infections, or asphyxiation. Emptying the crop with a soft tube is frustrating, time-consuming, and ineffective if a foreign body is present. An ingluviotomy appears to be the fastest and safest method (see Chapter 47, Selected Surgical Procedures). Supportive fluid therapy and injectable antimicrobials as indicated by cytology and culture are necessary. The clinician is advised to check for proventricular atony.

Crop stasis is not uncommon in hand-feeding neonates (see Chapter 53, Pediatric Medicine). In cases in which a simpler food product is required, a purified soybean casein-based formula has been used with success.[14] Crop or esophageal lacerations, scald fistulas, or foreign bodies may present as crop stasis (Fig. 28–3) and can be differentiated by transillumination, insufflation, palpation, and/or endoscopy. Administration of fluids and antibiotics is required in conjunction with surgical correction (see Chapter 47, Selected Surgical Procedures).

## Diarrhea

Pepto Bismol may be helpful in correcting clinical signs of diarrhea until a specific diagnosis is made. Atropine must be administered with caution (see Crop Disorders above). General supportive care, especially intravenous fluids, is recommended. Seeds should be removed from the diet, but fruits and other bland foods can be made available. Commercial canned canine food for enteritis has been used in some cases. If the bird is also anorectic, tube feeding may be necessary. If antibiotics are indicated, ampicillin may be the first choice for mild cases, doxycycline if chlamydiosis is suspected, and newer antibiotics specific for gram-negative or anaerobic bacteria if the patient is very ill with a high number of gram-negative bacteria on the Gram's stain. Chlorhexidine in the water may be effective for cases of suspected viral or minor yeast involvement.

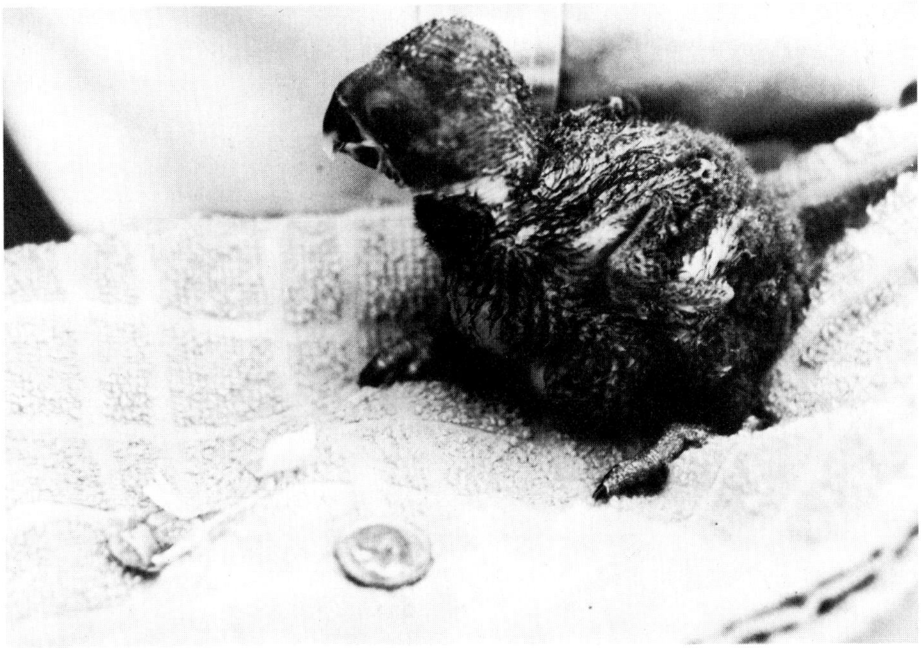

**Figure 28–3.** Crop stasis in this neonatal Eclectus Parrot resulted from ingestion of wood chips, which were quickly and safely removed via an ingluviotomy. The dime in the foreground is for size comparison.

### Emaciation

Until a diagnosis is made in the emaciated avian patient, an important consideration beyond supportive therapy is stress management. All stressful stimuli should be removed from the patient's immediate environment: cage mates, loud noises, excessive handling, rapid changes in ambient temperature.

Because the patient is usually dehydrated, intravenous fluids are essential. Studies of fasting grouse[7] revealed that the birds lost more weight from the pectoralis muscles alone than from adipose tissue, suggesting that protein breakdown may supply 40 per cent of the birds' energy requirements at such times. Perhaps future research may suggest that amino acid supplementation has value in emaciated birds. Parenteral vitamins and antibiotics are usually provided because the patient's weak and debilitated condition often causes an increased susceptibility to infection. Vitamin C, digestive enzymes, lactulose, and anabolic steroids are used when appropriate.

### Passing Whole Seeds

Tube feeding of a soft, easily digested formula may be the first step in therapy while determining the cause of this symptom. If the bird is eating well on its own, provide soft foods that do not require action of the gizzard. Enteric coating products (Kaopectate, Pectolin, Pepto Bismol, Amforal) are useful. Antibiotics as indicated by sensitivity of a cloacal culture may resolve this digestive problem. The appropriate parasiticide may be effective if parasites are the causative agent. This clinical sign frequently accompanies some cases of malnutrition related to suspected malabsorption from giardiasis. Whole seeds can also be passed in response to mineral oil gavage for grit obstruction. Conversely, not having access to any grit may produce these symptoms. If pancreatic disease is diagnosed, the addition of enzyme preparations (Viokase), antibiotics, and antispasmodics is advised, along with a low fat diet.

### Regurgitation

Regurgitation as part of normal courtship or as aberrant sexual behavior may be reduced by extending the daily period of darkness and limiting high fat, moist foods in the diet (see Chapter 3, Captive Behavior and Its Modifications). Medroxyprogesterone injections are used by some practitioners if behavioral modification techniques are not effective. Some clinicians treat this condition with injections of opposite sex hormones, although it is ill-advised to repeat this therapy. Cyproterone acetate has been used experimentally for inhibition of avian male reproductive activity.[2]

Regurgitation as a result of crop infection, poison ingestion, or presence of a foreign body may require emptying the crop by ingluviotomy to properly diagnose and resolve the problem. Minor obstructions respond to mineral oil via stomach tube (4 drops/cockatiel). Surgical intervention may be required for lesions of the ventriculus.

Symptomatic treatment for regurgitation caused by infectious organisms includes the removal of grit, appropriate antimicrobial agents, coating agents (see Passing Whole Seeds), fluid therapy, and withholding of food or administration of soft, easily digestible foods and digestive aids. Knowledge of a full gastrointestinal tract (from history or radiograph) would contraindicate the administration of atropine to prevent an increase in autointoxication from the atony response to atropine. If a food allergy is suspected, an elimination diet can be developed. Regurgitation as a reaction to medication seldom requires therapy. Careful dose calculation and appropriate route of administration are the best preventive measures. Regurgitation from motion sickness may be assisted by gradually increasing the distances the bird travels by car.

### Tenesmus

Tenesmus is frequently noted in female birds with egg-laying problems. In birds not laying eggs, it can be associated with cloacal pathology (e.g., cloacal papillomas), protozoan parasites of the digestive tract, or anaerobic or other bacterial infections. The appropriate therapy would address the offending cause. Mild, nonirritating, lubricating laxatives are sometimes used. No etiology or cure is known for a very acid, foul-smelling stool frequently associated with papillomas in Amazons and, less frequently, macaws.

## CLOACAL TISSUE PROTRUSION

### Cloacal Prolapse

The gross clinical differentiation of cloacal tissue protrusions, and thus the appropriate therapy, lies primarily in the texture of the protruding tissue. A prolapsed cloaca resulting from sphincter problems, nerve interruption, chronic irritation of the rectum, or tenesmus has the appearance of smooth, glistening epithelium. A more complete physical examination differentiates a prolapsed cloaca from a prolapsed uterus or vagina resulting from egg laying problems, or from cloacal papillomatosis as discussed below.

A prolapsed cloaca is potentially serious, as a partial or complete prolapse may cause severe constipation and retention toxemia. In cases of neglected or undetected prolapse, emergency therapy, i.e., manual reduction and aggressive fluid therapy, may be required to stabilize the patient before a surgical repair can be considered. The clinician should be aware that the prolapsed cloaca may contain prolapsed intestinal tissue, the oviduct, and one or both ureters.

Acute avian cloacal prolapses frequently respond to manual reduction (Fig. 28–4) and the placement of two simple interrupted sutures within the vent to hold the cloaca in place until the cause of the straining or intra-abdominal pressure can be determined. After the sutures are placed, if the patient strains excessively or the volume of feces is abnormal, the sutures must be replaced or removed. Chronic or severe cloacal prolapse usually dictates aggressive surgical repair. One method of cloacapexy has been described.[13]

## Cloacal Papillomatosis

A proliferative protrusion of epithelium from the cloaca with a "cauliflower" appearance (Fig. 28–5*A*) may indicate a papilloma or papilloma-like hypertrophy (see Chapter 40, Neoplasia). Specific histopathologic diagnosis of a biopsy sample is indicated in cases of apparent cloacal papillomas. These conditions require excision of the affected tissue. Papillomatous protrusions are highly vascular, and surgical excision by conventional or electrosurgical methods is difficult. Cryosurgery may be utilized (Fig. 28–5*B* and *C*); however, the practitioner is advised to perform the procedure in steps, rather than freezing all the involved tissue to a solid ice stage. The procedure can be repeated after a few days. Some cloacal nerve damage and necrosis of normal tissue may be associated with aggressive freezing.[1] Cloacal obstruction from swelling and scab formation must also be prevented. Papillomas with a high mitotic index may not successfully respond to cryosurgery and may regrow following removal.

Cloacal papillomas are suspected to be of viral origin but recent peroxidase, electron microscopy, and transmission studies have failed to confirm this.[15] Similar-appearing tissues have been observed upon necropsy in the oral cavity, crop, and proventriculus. Topical application of a DMSO/steroid combination has had a degree of effectiveness in reducing the swelling associated with cloacal papillomatous tissue. The possibility of autogenous vaccines has been considered.

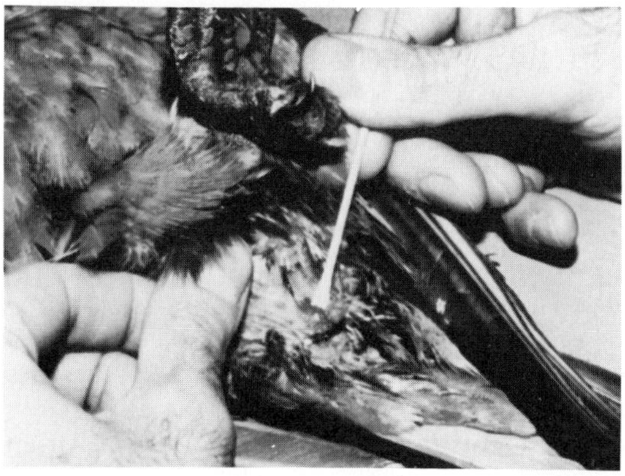

**Figure 28–4.** In some cases a prolapsed uterus or true prolapsed cloaca can be gently replaced manually. (Courtesy of George Kollias.)

## Uterine Prolapse

If the uterus is prolapsed, the tissue may be accompanied by the cloaca, uterers, and rectum. The prognosis depends on how long the uterus has been prolapsed, the degree of sepsis that has occurred, and whether complications have occurred (e.g., ureteral prolapse and uremia, constipation, egg-yolk peritonitis, or mutilation).

A recently prolapsed uterus may be replaced, although stretching of ligamentous attachments may have occurred. The uterus can be gently cleaned of any egg or debris with saline and aqueous carbenicillin. The organ is replaced with a water-soluble (KY) jelly–lubricated thermometer. Appropriate injectable antibiotics are administered, and dexamethasone is given to prevent shock. Administration of oxytocin or ergonil will shrink the uterus and control bleeding. A purse-string suture line may be used until straining stops.

In cases in which the egg is inside the prolapsed uterus, the egg must be collapsed and removed before treatment of the prolapse. If

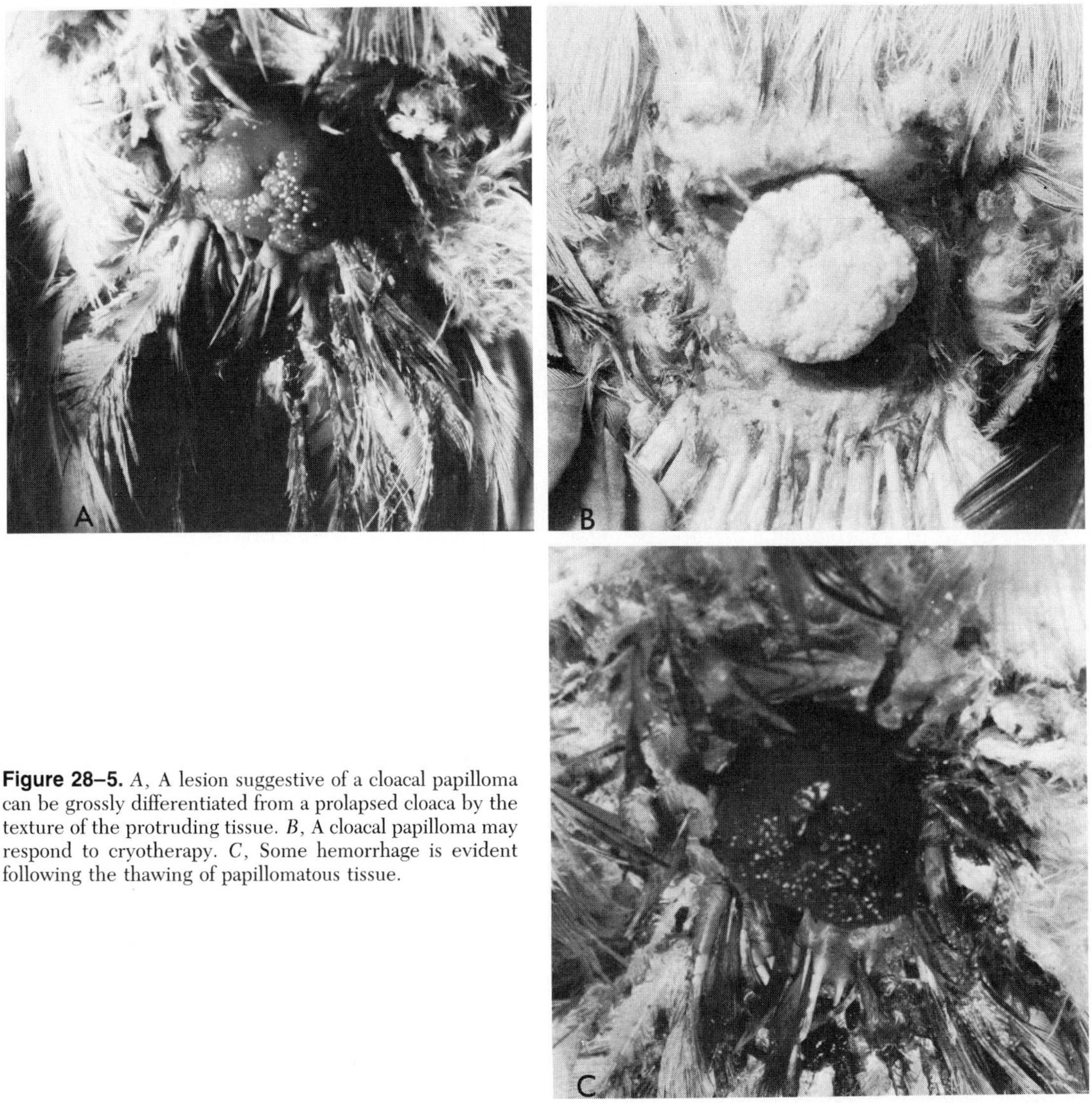


**Figure 28–5.** *A*, A lesion suggestive of a cloacal papilloma can be grossly differentiated from a prolapsed cloaca by the texture of the protruding tissue. *B*, A cloacal papilloma may respond to cryotherapy. *C*, Some hemorrhage is evident following the thawing of papillomatous tissue.

the bird has been in this condition for a prolonged time, the egg may have dried and adhered tightly to the uterine lining. Care must be exercised in gently stripping the egg from the lining to prevent hemorrhage.

If the uterus is damaged beyond repair, the organ must be exteriorized as much as possible, tied off with 4-0 Dexon, and removed, making sure that the ureters are not included in the ligature. The remaining stump is gently replaced with a lubricated thermometer. A therapeutic injection of medroxyprogesterone is given to temporarily stop ovarian activity.[10] A laparotomy may be required to perform the hysterectomy (see Chapter 47, Selected Surgical Procedures).

## SYMPTOMS OF CENTRAL NERVOUS SYSTEM DISTURBANCES

### Coma (See also Hyperthermia below and Emaciation above)

Primary support is provided with intravenous fluids, intravenous administration of prednisolone sodium succinate, antibiotics (cefotaxime and carbenicillin; doxycycline if chlamydiosis is suspected). Administration of heat and oxygen may be indicated. Place the patient in a dark environment to rest. A cerebral ischemia from arteriosclerosis may cause temporary unconsciousness, resulting in falling from the perch, but no therapy has been reported.[8]

## Convulsions

If a convulsion occurs, place the bird under inhalation anesthesia and check the body temperature. If it is increased (over 107°F), refer to the discussion of Hyperthermia (below). Intravenous or intramuscular diazepam may be given as an anticonvulsant. Some practitioners prefer phenobarbital, but intravenous administration of phenobarbital is very dangerous in birds; the drug should be underdosed and repeated only after a 10-minute interval.

If hypoglycemia is the suspected cause of the convulsions, glucose may be given. Parenteral administration of glucose is preferred because oral supplementation may precipitate asphyxiation in a convulsing bird. Calcium gluconate and vitamin $D_3$ supplementation can be given if hypocalcemia is suspected, and calcium EDTA if the etiology is suspected to be lead poisoning (see Chapter 39, Toxicology). Oral anticonvulsants, phenobarbital oral drops, or primadone in fruit or other favorite food may be used in some cases. True epilepsy would require continuous medication. Chlamydiosis or other infectious diseases are treated accordingly (see Chapter 38, Neurologic Disorders).

## Paralysis of the Mouth

Paralysis of the mouth may require tube feeding. If trauma is suspected, a short-acting steroid may be effective. Antimicrobials may be indicated by a Gram's stain of the mouth.

The cockatiel paralysis syndrome, suspected to be related to a malabsorption problem, may affect only the muscles of the jaw or mouth and responds in most cases to injections of vitamin E/selenium (Seltox).

## Paralysis of the Extremities

Symptomatic treatment until the cause is known for "paralysis" of the extremities includes cage rest and an appropriate sling or splint (Fig. 28–6). If taping or application of an acrylic splint to "train" a paralyzed misdirected toe is not successful, amputation is often advised to prevent mutilation by the bird, frequent injuries, abnormal wear to the bird's foot, callus formation, or bumblefoot. Leg paralysis related to calcium deficiency may respond in early cases to injectable calcium supplementation; however, permanent nerve damage is possible. Treatment for shock or swelling in many cases would include antibiotics, intravenous fluids, corticosteroids, and topical DMSO. Body slings and/or femoral head ostectomies may be attempted for dislocated hips; however, these are difficult to treat and the limb paralysis may remain. Nonsurgical and surgical management of fractures are discussed in Chapters 30 and 49, respectively.

Decompression therapy (steroids, intravenous mannitol) may be helpful following trauma that causes whole-body paralysis. Tube feeding and other supportive care is necessary for chronically paralyzed birds. Slings or rolled-up towels may help support the body in a nearly normal position.

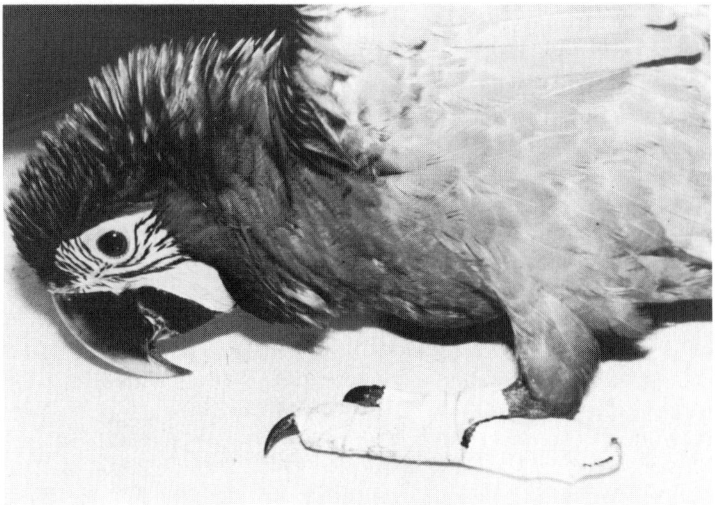

**Figure 28–6.** Splinting is one temporary method of treating curled toe paralysis.

## EXPOSURE

### Dehydration

Rehydration with intravenous lactated Ringer's solution is the primary objective in treating dehydrated patients, after which the bird may be warmed up. Birds that have suffered trauma or have been recaptured after an aviary escape are usually weak, hypothermic, dehydrated, and possibly hypoglycemic. Prednisolone sodium succinate can be added to the fluids if the bird is critical and can be administered every 15 minutes up to 10 times until the bird is aroused. As soon as the bird is strong and rehydrated, tube feedings may begin. Fluids are continued as necessary.

### Oil

"Oiled" feathers may result from indiscriminate use of topical creams and ointments by bird owners and veterinarians. The therapeutic objective in a cage bird with oiled and matted feathers is to retard the rate at which the bird loses heat and to remove the offending substance from the feathers as soon as possible (see Burns).

Prompt removal of oils from the feathers is contraindicated if the patient is seriously ill. Treat the patient for the underlying problem and stabilize it in a warm environment until it is stress-tolerant enough to have the feathers thoroughly cleaned.

### Hyperthermia

If heat stress results in a protracted period of hyperthermia, a bird will become comatose and will die shortly thereafter. Prior to this stage, immediate action taken to reduce the bird's body temperature may result in a reasonably good prognosis. Spray the plumage with alcohol or cold water or hold the bird under a shower or faucet that is running cold water. It is important that the down feathers beneath the contour feathers be sufficiently wet to facilitate the loss of body heat. The bird's feet can be immersed in cold water, and cold water can be infused through the vent and into the cloaca to achieve a rapid decline in the body temperature. Once the patient has stopped panting and its wing posture has returned to normal, it may be necessary to place the bird in a warm (not hot) environment to avoid hypothermia.

If a patient is presented suffering from hyperthermia and is comatose or nearly so, it would be wise to administer dexamethasone and mannitol intravenously to reduce the severity of the cerebral edema that is a likely sequela to hyperthermia. Flunixin meglumine may rapidly reduce hyperthermia but should not be overdosed. One should also take the more conservative steps mentioned above to rapidly reduce the patient's body temperature.

## SYMPTOMS OF THE RESPIRATORY SYSTEM

### Dyspnea

In the severely dyspneic patient, handling should be avoided and direct administration of oxygen may be necessary. When the patient is stabilized, diagnostic measures may be instituted (e.g., check for ascites, radiograph, examine endoscopically) to suggest further therapy.

Acute dyspnea from inhalation of foreign bodies or seeds may require tracheoscopy for recovery attempts (see Foreign Bodies below). Systemic antibiotics and diuretics have been used to a degree for generalized lung disease, as well as oxygen and nebulization therapy in dry environments.

### Foreign Bodies

Seed hulls and necrotic debris from chronic rhinitis are frequently found as foreign bodies in the rhinal cavity. Removal is facilitated by bending the tip of a hypodermic needle to use as a retrieval instrument. Use of a magnifying instrument and sedation of the patient help minimize hemorrhaging. Removal of seed parts can often be accomplished only with physical restraint.

Saline flushes, culture of vital tissue scrapings of the operculum or the conchae, and administration of appropriate antibiotics may be required for resolution of rhinal disorders following removal of necrotic debris.

Access to inhaled seeds or other foreign bodies in the lower trachea occasionally can be gained through a tracheotomy. In some cases involving larger species, the material may be retrieved directly with biopsy forceps; in other cases, a blunt probang can be employed to move the substance up and out of the trachea. Foreign bodies lodged in or below the syrinx are very difficult to remove. In extreme cases, when there is no other choice, emergency relief of respiratory signs has been accomplished by pushing the substance further into the respiratory system. This is a very dangerous procedure that is frequently unsuccessful. Occasionally such a foreign substance has been apparently walled off by the body and the bird survives. Systemic antibiotics are used.

### Respiratory Symptoms in Amazons, Macaws, and Cockatoos

Effective respiratory therapy depends on the etiology. In imported birds, respiratory symptoms may be related to etiologies that are no longer present (e.g., pox) (Fig. 28–7). The respiratory tissue disturbances and secondary invaders may be the only demonstrable pathogens. Nebulization has some potential for therapy (see Chapter 29, Aerosol Therapy).

If a respiratory outbreak has occurred in a flock, the recommendations for a disease outbreak should be followed (see Chapter 27, What To Do Until A Diagnosis Is Made). In the early stages chlorhexidine in the drinking water may help to reduce the morbidity. The birds must be kept eating. In some cases, children's decongestant nosedrops may be used in sneezing birds with rhinorrhea. The nasal sinuses may be flushed with antibiotics (based on sensitivities) (Fig. 28–8). Recent limited experience suggests that in some infections, especially those in which highly resistant *Pseudomonas* is involved, the addition of EDTA-TRIS-lysozyme to flushing solutions enhances antibiotic activity and normal body phagocytosis.[4]

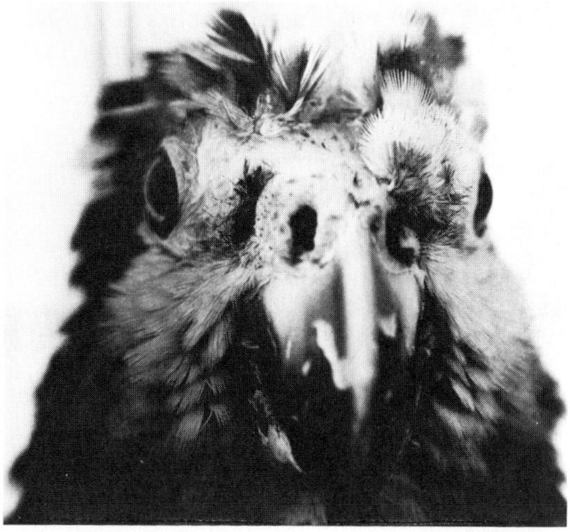

**Figure 28–7.** Chronic respiratory diseases in old Amazons may cause disfiguration of the cere, nares, and beak.

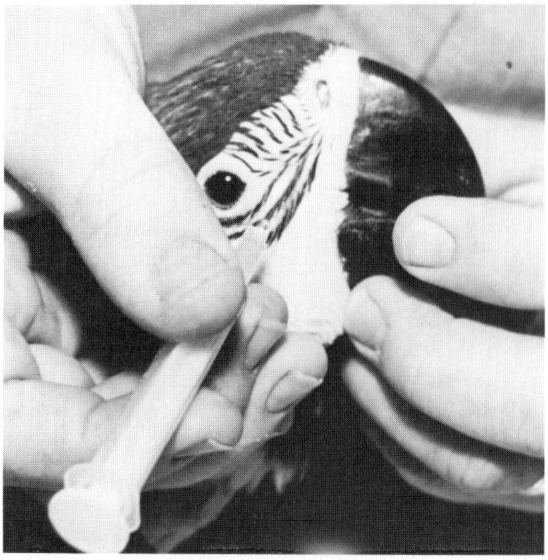

**Figure 28–8.** Injections of antibiotics directly into the sinus may relieve some respiratory symptoms.

### Respiratory Symptoms in Cockatiels

*Mycoplasma* is suspected to be a major cause of respiratory symptoms in cockatiels (see Chapter 34, Mollicutes). Spraying the birds with a tylosine solution and adding lincocin/spectinomycin to the drinking water provides relief from clinical signs in most cases.

### Respiratory Symptoms in Budgerigars

*Mycoplasma* also produces respiratory symptoms in budgerigars; however, other clinical signs may be related to conditions that respond to steroids, L-thyroxine, or injectable and water solutions of iodine supplementation (see Chapter 42, Miscellaneous Diseases).

### Respiratory Symptoms in Finches

The most common cause of respiratory symptoms in finches is mites in the air sacs or trachea. Oral ivermectin is the therapy of choice (see Chapter 37, Parasites).

## SYMPTOMS OF SWELLING

### Ascites

Palpation of the abdomen in an ascitic bird must be done with caution (see Chapter 7, Preliminary Evaluation of a Case). The ascitic fluid may be removed by abdominal tap with an appropriately sized needle and syringe; however, only the amount required to relieve the clinical signs should be removed at one time. A rapid removal of the fluid may put the bird into shock or hypoproteinemia as the body attempts to replace the lost fluid. Diagnostic procedures may be performed on the aspirate (see Chapter 17, Cytology). If egg-related peritonitis is suspected, the aspirated fluid is usually a characteristic yellow color (see Egg-related Peritonitis below).

Depending upon the cause of the ascites, therapeutic measures may include judicious administration of diuretics, intravenous fluids, serum transfusions, and other supportive therapy. Appropriate antimicrobials are determined by culture and sensitivity testing. Contrast radiography may be indicated early in the diagnostic work-up (see Chapter 14, Radiology). Surgery may be required for resolution of reproductive disorders or tumors and other space-occupying lesions.

## Egg Binding

A medical approach to the treatment of egg binding should be instituted prior to any surgical intervention. Increasing the environmental temperature (85 to 90°F), administration of injectable multivitamins, calcium, and oxytocin or ergonovine, and gentle massage of the uterine area should be the first considerations. Intravenous fluids with glucose or dextrose are indicated if the bird is weak. These procedures will deliver the egg 70 to 80 per cent of the time. A gentle, warm enema or immersing the bird up to its neck in warm water has been reported by aviculturists to be effective in some cases.

The smaller the species, the more quickly the practitioner needs to proceed to more aggressive therapy if the above techniques have not been successful. Egg binding in a finch is critical and should be resolved within an hour. Larger species might remain in an egg-bound condition for 8 to 12 hours without danger. A Blue-and-Gold Macaw survived surgery following 72 hours of egg binding, but the situation was risky.

One method of aggressive therapy for a retained egg uses a blunt probang (e.g., a thermometer or rounded bone pin) to mechanically dilate the cloaca and the vaginal orifice (Fig. 28–9). After the egg is visualized, a needle, trocar, or bone pin may be used to penetrate the shell; gentle palpation will then collapse the egg.

Other techniques are aspiration of the contents of the retained egg by abdominal paracentesis or aspiration of the contents through the uterine wall from the mechanically dilated vent. A 22-gauge needle is adequate for smaller species; however, an 18-gauge needle may be required for a macaw owing to the increased viscosity of the egg contents. The centesis procedures often will allow the shell to collapse sufficiently for the hen to pass it on her own. The hen is normally placed back into the warm environment to encourage the effect of the uterine contracting drugs to deliver the remaining portions of the egg. Occasionally it may be necessary to mechanically remove with forceps some portions of the shell that may have adhered to the uterine wall. Although there is no anatomic or physiologic mechanism to prevent it, retrograde movement of the egg contents following careful collapsing of the shell has never been encountered by the authors.

A hysterotomy should be considered instead of the above procedures in cases in which the

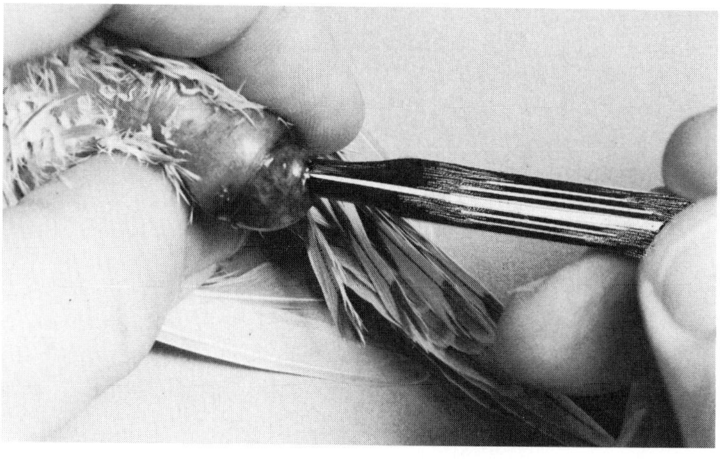

**Figure 28–9.** Aggressive therapy for nonresponsive egg binding includes dilation of the prolapsed vagina so that puncture of the egg can be accomplished with a blunt probang.

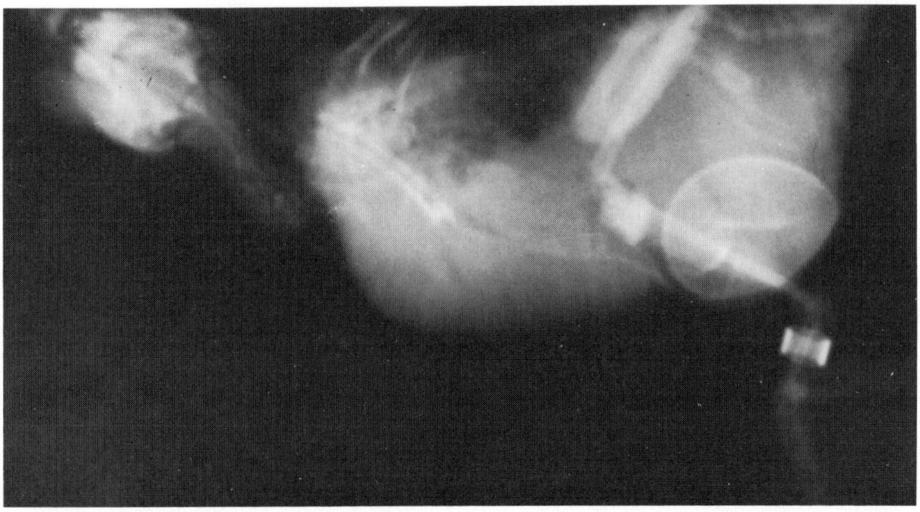

**Figure 28–10.** Radiography confirmed that this retained egg had ruptured the uterus and was located in the peritoneal cavity.

hen is weak, the egg is to be saved, or the exact location of the egg cannot be determined from a radiograph (Fig. 28–10). A laparotomy is required if the egg is known to be outside the uterus (see Chapter 47, Selected Surgical Procedures).

## Egg-related (Egg-yolk) Peritonitis

The abdominal swelling normally associated with egg-related peritonitis needs to be addressed first. A Penrose drain can be installed (Fig. 28–11) and the patient placed on appropriate antibiotics as determined by culture and sensitivity testing prior to surgical intervention.

Egg-related peritonitis in a macaw has been seen by one of the authors (Harrison) only once. The abdominal swelling and fluid accumulation normally associated with this condition in budgerigars, cockatiels, and lovebirds were not present. Gelatinous mucus was present in the crop upon necropsy, and several inspissated yolks in the peritoneal cavity were all that were seen. If larger psittacines fail to exhibit obvious clinical signs of egg-related peritonitis, antemortem diagnosis may be difficult.

## Other Swellings

Air-filled swellings may result from rupture or overinflation of air sacs (Fig. 28–12). Paracentesis or lancing to deflate the air pocket has been a successful procedure in some cases. However, the condition often recurs and must be corrected surgically (see Chapter 47, Selected Surgical Procedures).

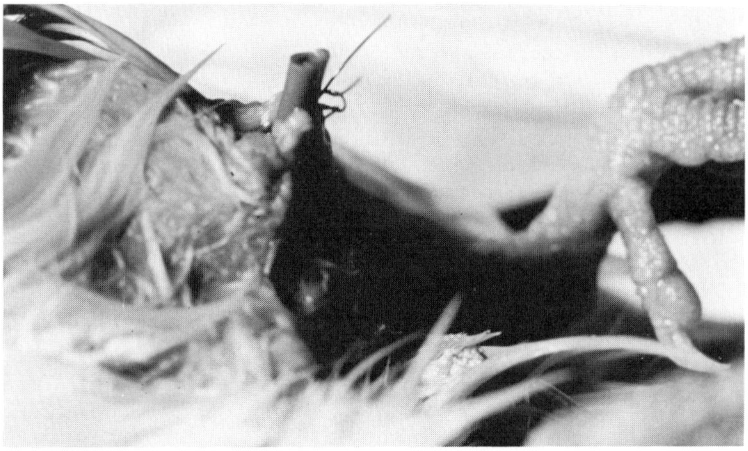

**Figure 28–11.** Insertion of a Penrose drain relieves fluid accumulation from egg-yolk peritonitis. (From Harrison, G. J.: New aspects of avian surgery. Vet. Clin. North Am. [Small Anim. Pract.], 14(2):376, 1984.)

**Figure 28–12.** Rupture or overinflation of air sacs may require surgical correction to prevent recurrence.

Swellings of caseated material may accumulate inside the mouth, under the mandible, or elsewhere. Because the material is not liquid, aspiration is not possible and surgical incision and curettage must be employed to remove the offending substance. Circumferential injections of sterile hypertonic saline around oral lesions reduce hemorrhage associated with subsequent curettage. Flushes with EDTA-TRIS-lysozyme and appropriate antibiotics may help to resolve some conditions. Injections of vitamin A are warranted if the dietary history suggests hypovitaminosis A.

## OPHTHALMOLOGIC LESIONS

If there is periocular swelling, there is often concomitant or previous respiratory disease. Systemic antibiotics are indicated. Vitamin A supplementation may be helpful. A discrete swelling at the medial canthus may be caseated material within the nasolacrimal sac. This material should be disrupted and flushed from the sac. A nasolacrimal cannula or the plastic portion of a feline indwelling intravenous catheter may be used with sterile saline as the flushing solution. Culturing the material will indicate the appropriate antimicrobial to add to the solution.

If the conjunctiva is swollen, examination of a surface scraping, stained with Wright's/Giemsa, Gram's, and/or Macchiavello's stain will be helpful (see Chapter 17, Cytology). If conjunctival cytology is not available, if bacteria are present, or if *Chlamydia* is suspected, a topical antibiotic is indicated. If no bacteria are present, or if the cell type is predominantly heterophilic, lymphocytic, and especially eosinophilic, then a combination antibiotic/steroid preparation should be used topically.

With fluorescein stain retention of the cornea, a topical broad-spectrum antibiotic is indicated. Debridement of the corneal epithelium with a dry cotton-tipped applicator after the instillation of a topical anesthetic will remove devitalized epithelium and aid healing. Parasympatholytic drugs used therapeutically for ulcers in other species are not effective in birds. Collagenase inhibitors such as acetylcysteine (Mycomyst 10 per cent) used topically are indicated for deep corneal ulcers and lacerations. Corneal and corneoscleral lacerations should be sutured using fine diameter nonabsorbable sutures. Perforating wounds, whether ulcerative or traumatic, should be treated with topical solutions, not ointments. The presence of ointments intraocularly may incite uveitis.

Uveitis, with lowered intraocular pressure, cloudy cornea, hyperemic iris, and aqueous turbidity, may be a reflection of some systemic disease. Systemic antibiotics may be indicated and a topical steroid/antibiotic combination used. Ophthalmic preparations should contain dexamethasone or 1 per cent prednisolone to be effective. Mydriatics used in uveitis in other species are not helpful in birds. Hyphema should be treated as a uveitis, but the antibiotics are not necessary.

If non-infectious vitritis or retinitis has been detected, systemic steroids are indicated.

Idoxuridine or trifluridine, antiviral ophthalmic preparations, may have some potential as effective therapy for birds with corneal lesions showing no organisms in corneal scrapings.

## REFERENCES


1. Altman, R. B.: AAV Newsletter, 5(4):101, 1984. From material presented at the 12th Annual Veterinary Surgical Forum, Chicago, IL, 1984.
2. Balthazart, J.: Behavioural and physiological effects of testosterone proprionate and cyproterone acetate in immature male domestic ducks *(Anas platyrhynchos)*. Z. Tierpsychol., 47:410–421, 1978.
3. Cambre, R.: Use of hydroactive dressing in birds. Abstracts from the Annual Meeting of the American Association of Zoo Veterinarians, Louisville, KY, 1984.
4. Campbell, T. W.: EDTA-TRIS buffer lavage for treating psittacine birds with *Pseudomonas aeruginosa* infections—a case report. Comp. Cont. Ed., 7(8):598–606, 1985.
5. Flammer, K.: Emergency procedures for aviary birds. In Proceedings AFA Veterinary Seminar, American Federation of Aviculture, Orlando, FL, 1984.

6. Gallerstein, G. A., and Ridgeway, R. A.: Proventricular dilation in psittacines. In Proceedings of the Annual Meeting of the Association of Avian Veterinarians, San Diego, CA, June, 1983.
7. Grammeltvedt, R.: Atrophy of a breast muscle with a single fibre type (M. pectoralis) in fasting Willow Grouse *(Lagopus lagopus)*. J. Exp. Zool., 205:195–204, 1978.
8. Hasholt, J., and Petrak, M. L.: Disorders of the nervous system. In Petrak, M. L. (ed.): Diseases of Cage and Aviary Birds. 2nd ed. Philadelphia, Lea & Febiger, 1982.
9. Redig, P. T.: Fluid therapy and acid-base balance in the critically ill avian patient. Proceedings of the International Conference on Avian Medicine, sponsored by the Association of Avian Veterinarians, Toronto, Canada, 1984.
9a. Ritchie, B.: Personal Communication, 1985.
10. Rosskopf, W. J., et al.: Egg-binding in an Amboina King Parrot. VM/SAC, Feb., 1982, pp. 231–232.
11. Rosskopf, W. J., et al.: Preservation responses in a dying Military Macaw. VM/SAC, 76(12):1458–1459, 1981.
12. Rosskopf, W. J., et al.: Egg-yolk peritonitis in a cockatiel. Mod. Vet. Pract., 63(5):418–419, 1982.
13. Rosskopf, W. J.: Surgery of the avian digestive system. In Proceedings for Avian Medicine and Surgery, 12th Annual Veterinary Surgical Forum, Chicago, IL, 1984.
14. Roudybush, T.: Material presented at the Annual Convention of the American Federation of Aviculture, San Francisco, CA, 1985.
15. Sundberg, J.: Search for the etiology of avian papillomatosis. AAV Newsletter, 5(4):108, 1984.

*Chapter 29*

# AEROSOL THERAPY

RONALD R. SPINK

Aerosol therapy involves the use of a solution that has been atomized into a fine mist for inhalation therapy. This may be considered primary or supportive treatment for diseases with respiratory signs. Aerosol therapy may be employed alone, as an adjunct to parenteral or other treatment, or in cases that do not respond to conventional therapy.

In mammals, aerosol therapy is used primarily to control or remove secretions, maintain or adjust homeostasis of blood gases, and control respiratory reflexes.[4] Antibiotic aerosol therapy has been reported in some cases to be more effective than intramuscular therapy in treating respiratory tract diseases in mammals, including humans.[3]

The unique respiratory system of birds does not lend itself to the primary goals of nebulization in mammals. In birds the primary use is to humidify and locally treat the respiratory tract.[8] There apparently has been some interest in nebulization for possible therapy of bacterial dermatitis.[1]

Three forms of aerosols used for respiratory therapy include humidification, vaporization, and nebulization. Each has its own degree of effectiveness and may be selected according to the specific indication.

## HUMIDIFICATION

The primary purpose of humidification is to increase the relative humidity of the bird's environment. Humidification of inspired air offers significant advantages in managing upper respiratory disease, especially if the relative humidity of the inspired air is very low.[2] Humidified air may serve to soothe and lubricate the mucous membranes to ease some respiratory symptoms. The role of humidification in treating upper airway disease has not been established in mammals.[2]

Although rarely employed alone in the avian hospital, humidification may be an adjunct or follow-up to concurrent therapy by the client. In the home, the bird may be exposed to humidification in a relatively confined area but a safe distance from the source, such as a household humidifier, steaming tea kettle, or hot running shower.

## VAPORIZATION

Vaporization with a commercial vaporizer will also increase the humidity with a "hot mist" or "cool mist" and has the advantage of dispensing some medications through the mist for superficial contact treatment of the bird's exposed mucous membranes. The relatively large size of mist particles produced by vaporizers prevents effective delivery of medicated vapors to the lower respiratory tract.

Utilized primarily for home treatment, eucalyptus-based products may be used with the vaporizer in a small, closed room. Or a towel may be placed over the cage and the mist directed under the towel. Lacking a vaporizer, one may apply a small amount of these products to the underside of a towel, which is then draped over the top of the cage to assist in decongestion and soothing of the mucous membranes.

## NEBULIZATION

Nebulization is the aerosol therapy of choice in an avian hospital, as it is the most effective method of humidifying, delivering medications, and penetrating to the deeper respiratory tree (Fig. 29–1). Nebulization breaks up the selected solution into extremely small particles, 0.5-6.0 microns in diameter, which penetrate and make direct contact with hard-to-reach, affected respiratory tissues. The natural defense mecha-

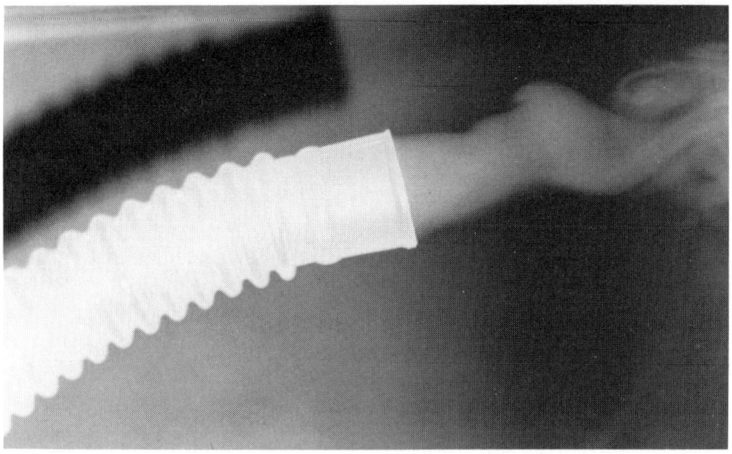

**Figure 29–1.** Nebulized particles are able to penetrate into the lower respiratory tract because of their extremely small size.

nisms of the respiratory epithelium depend upon proper hydration. Nebulization assists in rehydrating this epithelium.

## Indications

Nebulization therapy may be considered in birds with respiratory signs for which humidification and topical treatment of the respiratory tract may be advantageous. Based on clinical judgment, birds with sinusitis, rhinitis, pharyngitis, or air sacculitis may be candidates.

Some clinicians believe nebulization is effective in some cases in which systemic medications alone have failed. Therapeutic agents can be effectively delivered to the tissues alone, in combination, or in sequence.

Nebulization may be considered as alternative therapy for a highly stressed bird, as the degree of handling is minimal.

## Equipment and Procedure

Nebulization in birds is accomplished by using a disposable nebulizer driven by a source of compressed air and delivered into an enclosed chamber* (Fig. 29–2). Commercial medical air compressors are the most practical air source, although compressed air can be used. Oxygen may be used as a delivery vehicle in indicated cases. The nebulizer is easy to use and inexpensive to operate once the initial equipment costs are met. Large birds require a large chamber and therefore proportional volumes of nebulized particles.

Depending upon the clinical signs and response of the patient, birds in the author's practice may be nebulized three or four times

*Therapy Unit Plus Nebulizing Chamber, Corners Limited, Kalamazoo, MI.

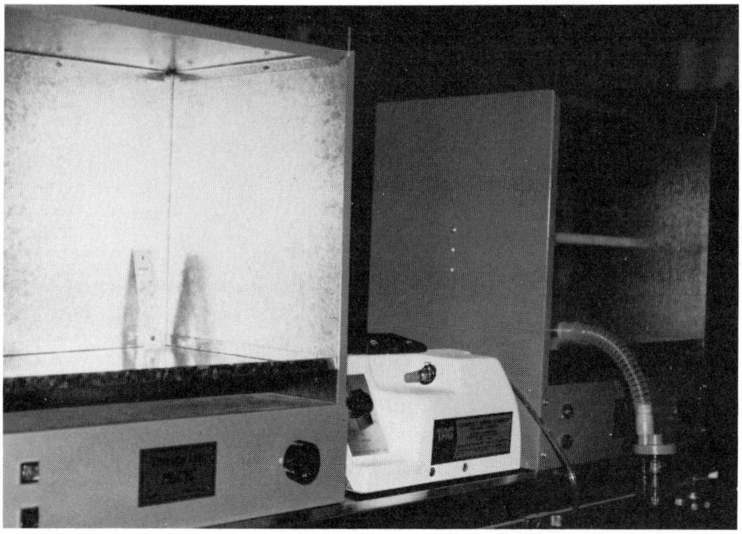

**Figure 29–2.** Varying sizes of nebulization chambers may be set up for avian patients.

per day for 10 to 15 minutes each time. The treatment will extend for several days beyond clinical improvement. Critically ill birds may require 30 minutes three or four times daily.[5]

## Nebulized Medications

The choice of therapeutic agent to be administered by nebulization should be based on sensitivity plates of cultures of the upper respiratory tract, such as of the choana or sinus aspirates, or of the air sacs (see Chapter 10, Clinical Microbiology, and Chapter 15, Endoscopy).

The effectiveness of nebulized medications in mammals depends upon several factors. The amount and concentration of the medication reaching the affected respiratory tissue is governed by the duration and frequency of nebulization; the flow rate, pressure, and character of the pressurized gas that drives the nebulizer; the concentration and viscosity of the medication in the nebulizer; the temperature of the nebulized solution; the toxicity or irritability of drugs used; adjuvants in the nebulized medications; combinations of drugs used via other routes of administration; and the patient's temperament or excitability. Treatment failure is common with short-term therapy.[1]

The differences between birds and mammals in their response to nebulization are the result of the unique anatomy and physiology of the avian lungs and air sacs. However, penetration of nebulized particles into avian lung parenchyma and onto the air sac surfaces has been demonstrated to be effective.[7] After 10 minutes of nebulization into a 1.2 cubic foot chamber, gentamicin had reached significant levels in lung tissues and on air sac epithelia. No systemic absorption was evident.

A recent study with Bobwhite Quail and pigeons[3] showed that tylosine aerosol particles are able to traverse the tertiary bronchi of the lungs to the air sacs. In most cases, tylosine aerosol therapy with DMSO rather than water greatly elevates the tylosine concentrations in or on the respiratory tissues. DMSO enhances the tissue penetration of antibiotics and fungicides by altering the natural selectivity of the cellular membranes. Tylosin aerosol administered in a ratio of 1 gm tylosin to 50 ml DMSO for one hour is required to establish therapeutic concentrations in the lungs and air sacs of quail and pigeons.

Antibiotics are commonly employed as nebulized medications. In mammals, bacterial infections of the upper airways are considered secondary to one or more predisposing conditions, e.g., viral or fungal rhinitis.[2]

One disadvantage of this therapy mode is the difficulty in establishing and accurately delivering appropriate doses of medications. Medications in avian nebulization are used on an empirical basis at this time. Many therapeutic agents that can be toxic when used systemically may be used in aerosol therapy, as there is little or no systemic absorption. Table 29–1 lists some medications and doses in common usage. Those medications with an asterisk are poorly absorbed from the respiratory epithelium and provide primarily topical therapy. The author has also used liquid camphor-containing drugs to help decongest and sooth upper respiratory tissues.

Some avian practitioners recommend the simultaneous use of a mucolytic* agent.[5] Others have reservations about the unique avian anatomy and the inability of birds to cough up or absorb exudates. The author shares these last reservations and uses mucolytic drugs only for upper respiratory diseases in which exudates can be discharged.

Most nebulized solutions contain saline. When added to saline, some medications may alter the toxicity of the nebulized particles, resulting in a reaction or irritation of the respiratory mucosa. Excess secretions with various viscosities may affect the results.

The addition of 13.65 per cent levamisol phosphate (1 ml in 15 ml of saline) to the initial treatment as an immunostimulant seems to hasten recovery in select cases. The use of corticosteroids or other immunosuppressive drugs is contraindicated, however, as these may alter the natural defense mechanisms of the respiratory mucosa.[3]

*Acetylcysteine (Mucomyst), Mead Johnson.

**Table 29–1.** MEDICATIONS COMMONLY USED FOR NEBULIZATION

| Drug | Dosage |
|---|---|
| *Amphotericin B | 100 mg in 15 ml saline |
| Chloramphenicol succinate | 200 mg in 15 ml saline |
| Erythromycin | 200 mg in 10 ml saline |
| *Gentamicin | 50 mg in 10 ml saline |
| *Polymixin B | 333,000 IU in 5 ml saline |
| Spectinomycin | 200 mg in 15 ml saline |
| Sulfadimethoxine | 200 mg in 15 ml saline |
| *Tylosin | 100 mg in 10 ml saline<br>1 gm in 50 ml DMSO |

*Poorly absorbed from the respiratory epithelium; provide primarily topical therapy.

One should not overlook the potential of using nebulization therapy in birds, especially in an environment of low humidity, and in respiratory disorders that are nonresponsive to conventional therapy. Nebulization can be a valuable adjunct in selected cases.

## REFERENCES

1. Ford, R. B.: Management of chronic upper airway disease. In Kirk, R. W. (ed.): Current Veterinary Therapy VIII. Philadelphia, W. B. Saunders Company, 1983.
2. Locke, D., and Bush, M.: Tylosin aerosol therapy in quail and pigeons. J. Zoo Anim. Med., 15:67–72, 1984.
3. McKiernan, B. C.: Principles of respiratory therapy. In Kirk, R. W. (ed.): Current Veterinary Therapy VIII. Philadelphia, W. B. Saunders Company, 1983, pp. 216–221.
4. Miller, T. A.: Nebulization for avian respiratory disease. Mod. Vet. Pract., 65:309–311, 1984.
5. Riviere, J. E., et al.: Aerosol therapy. J. Am. Vet. Med. Assoc., 179:166–168, 1981.
6. Soifer, F., and Spink, R. R.: Material presented at the American Animal Hospital Association Annual Meeting, Las Vegas, NV, 1983.
7. Spink, R. R.: Physical examination and therapeutics. Proceedings of the Association of Avian Veterinarians Annual Meeting, Atlanta, GA, 1982.
8. Spink R. R.: Nebulization therapy in cage bird medicine. VM/SAC, 75:791–795, 1980.

# *Chapter 30*

# EVALUATION AND NONSURGICAL MANAGEMENT OF FRACTURES

PATRICK T. REDIG

Specific treatment recommendations and guidelines for prognostic evaluations of various avian long bone fractures have developed from experiences gained in the management of fractures and other traumatic injuries encountered in over 3000 raptors over the last 10 years, as well as many psittacines and other caged birds presenting with similar orthopedic problems. Although there are limitations to nonsurgical methods in managing some specific fractures, successful clinical techniques for nonsurgical management of the long bone fractures of large and small pet birds have been adapted from those found to provide satisfactory results on a regular basis for birds that have to function at 100 per cent of normal capacity in order to survive in the wild.

Although it has been possible to surgically manage fractures in avian patients as small as 50 grams, the benefits of nonsurgical management where it can be effectively employed (decreased chance of infection, less damage to regional vascularity), outweigh the potential advantages of surgical management in many instances. Birds are ideally suited for external coaptation, especially for wing fractures. The bird's body provides an accurate mold for the shape of the wing.[9] When the wing is properly folded against this mold, it will extend with the proper alignment for flight. The examination of avian skeletons in museums gives evidence for a high rate of spontaneous fracture recovery with nothing more than "cage rest." Therefore, one should always strive first to find an effective, conservative, nonsurgical method of managing fractures.

The utility of coaptation is limited in cases of an open fracture or significant overriding of the fragments. In these circumstances, a decision must be made as to the goal of the repair effort; i.e., is it necessary for the bird to be returned to full flight capability or are cosmetic appearance and general well-being of the bird in a captive environment satisfactory outcomes? Surgical intervention may be the only means of providing a chance for restoration of flight capabilities, but there is an increased risk of failure and perhaps loss of the appendage. If full flight capabilities are not necessary and cost for surgical treatment and postoperative management is an encumbering factor, external coaptation may be an entirely satisfactory mode of treatment even for some serious fractures.

The common sites of fracture occurrence in the avian skeleton are detailed in Figure 30–1. The recommended management methods for each of these fractures along with prognostic information and further literature references are contained in Table 30–1. Further information regarding diagnostic methods, amplified treatment methods, and recovery potential is contained in the following section.

## FRACTURES AND INJURIES TO THE PECTORAL LIMB

### Fractures of the Shoulder Girdle

Fractures and other traumatic injuries affecting the shoulder are usually the outcome of a headlong collision with a solid object such as a wall or a window. Such a patient cannot elevate the affected wing above the horizontal plane when it attempts to fly. Separation of the shoulder joint and/or fracture of the coracoid occurs most frequently. Occasionally there may be some asymmetry to the resting posture of the wings associated with coracoid fractures if the entire shoulder has collapsed. Rarely there may be a fracture of the scapula or furcula (Fig. 30–

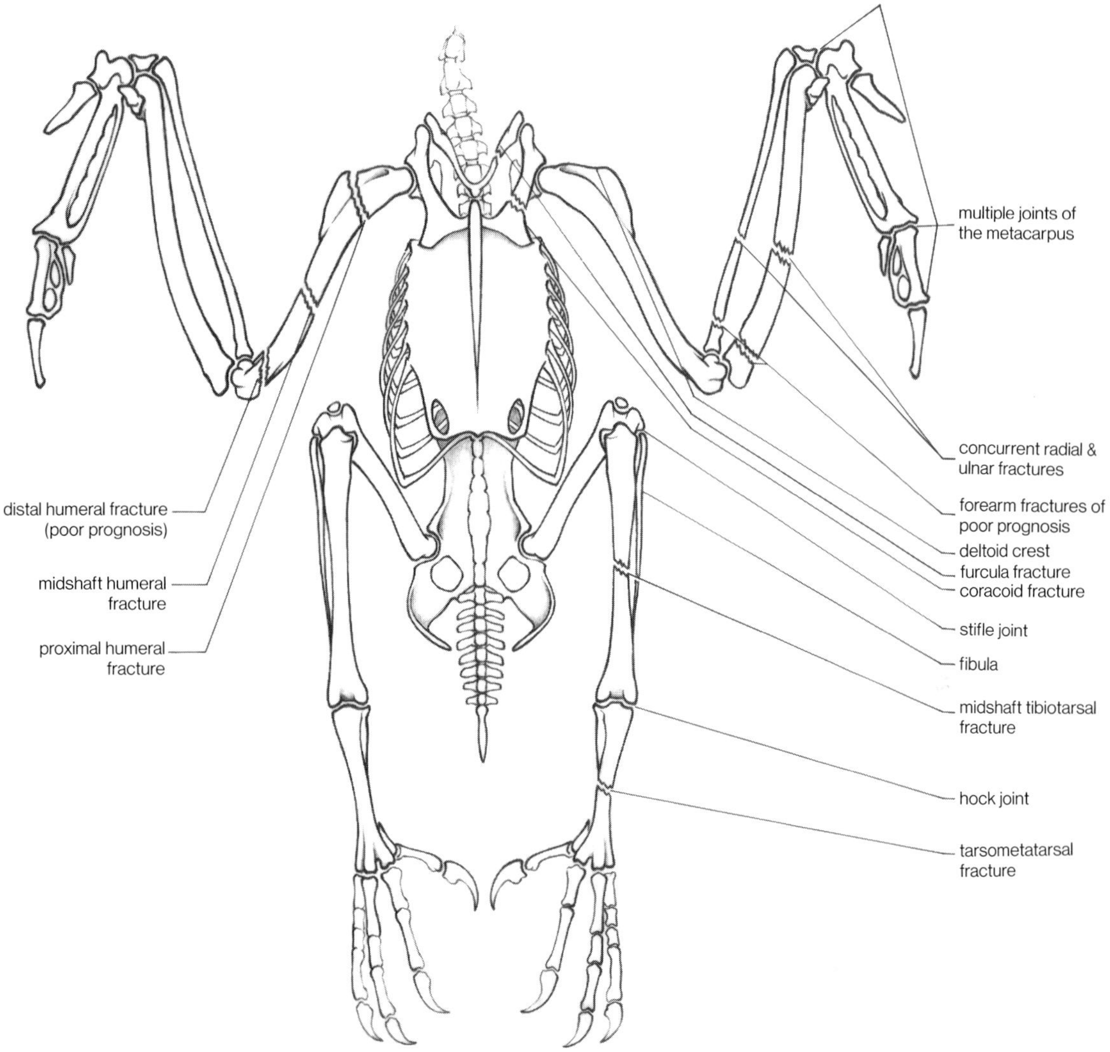


**Figure 30–1.** The avian skeleton, as it would typically be seen in a V-D radiograph, showing sites of regularly occurring fractures.

1). Definitive diagnosis usually requires radiology (Fig. 30–2).

Immobilization of the wing and cage rest for four to six weeks result in a high rate of normal recoveries in cases involving injuries to the shoulder girdle (Table 30–1). Coracoid fractures may require surgery if fragments are displaced and flight of the bird is a desired outcome. Intramedullary pinning has been successful for stabilization of such a fracture.[5] I have encountered two cases in which inadequate immobilization of coracoid fractures that were managed nonsurgically yielded unsatisfactory results. In both of these, large knotty calluses formed on the midshaft of the coracoid and narrowed the passageway for the esophagus. These birds could not swallow large boluses of food properly. In one case involving a Peregrine Falcon that was maintained in captivity, the problem resolved completely in two years. The other case involved a goshawk that was returned to the wild and later was recovered dead and extremely emaciated.

## Fractures of the Humerus

Humeral fractures are less amenable to nonsurgical management than fractures of any of the other bones of the pectoral limb. These fractures are nearly always unstable owing to the displacing action of the powerful flexor muscles that contract following the fracture. The location of the fracture along the length of the humerus has significant effects upon management decisions and prognosis.

**Table 30–1.** SUMMARY OF FRACTURE TYPES, MANAGEMENT RECOMMENDATIONS, AND PROGNOSIS IN AVIAN SPECIES

| Fracture Site | Fracture Condition | Management Recommendation | Prognosis to Heal | Prognosis for Recovery | Reference |
|---|---|---|---|---|---|
| Coracoid | Closed | Bind wing to body | Excellent | Good | Spink, 1978 |
| Coracoid | Closed | I.M.* pinning | Fair | Guarded | Redig, 1983 |
| Furcula, scapula | Closed | Bind wing to body | Excellent | Excellent | Redig, 1983 |
| Proximal humerus | Closed | Bind wing to body | Excellent | Good | Altman, 1977<br>Spink, 1978 |
| Proximal humerus | Open and comminuted | I.M. pinning | Fair | Guarded | Redig, 1983 |
| Midshaft humerus | Closed | Bind wing to body | Good | Good | Altman, 1977<br>Spink, 1978 |
| Midshaft humerus | Open | I.M. pinning | Good | Good | Redig, 1983 |
| Midshaft humerus | Open and comminuted | I.M. pinning | Guarded | Poor | Redig, 1983 |
| Distal humerus | Closed | Bind wing to body | Fair | Guarded | Altman, 1977<br>Spink, 1978 |
| Distal humerus | Open and comminuted | I.M. pinning | Poor | Negligible | Redig, 1981 |
| Distal humerus | Open and comminuted | Modified Kirschner | Fair | Poor | Redig, 1983<br>MacCoy, 1983 |
| Proximal radius | Closed | Bind wing to body | Good | Guarded | Spink, 1978 |
| Midshaft radius | Closed | Figure 8<br>Braille<br>Bind wing to body (small birds) | Excellent | Excellent | Redig, 1983<br>Rousch, 1980<br>Altman, 1977 |
| Distal radius | Closed | Figure 8<br>Braille<br>Bind wing to body (small birds) | Excellent | Excellent | Redig, 1983<br>Rousch, 1980<br>Altman, 1977 |
| Midshaft radius | Open | Figure 8 | Excellent | Excellent | Redig, 1983 |
| Distal radius | Open | Figure 8 | Excellent | Excellent | Redig, 1983 |
| Proximal ulna | Closed | Figure 8 | Excellent | Good | Redig, 1983 |
| Midshaft ulna | Closed | Figure 8<br>Braille | Excellent | Excellent | Redig, 1983<br>Rousch, 1980 |
| Distal ulna | Closed | Figure 8<br>Braille | Excellent | Excellent | Redig, 1983<br>Rousch, 1980 |
| Proximal radius and ulna | Closed | Transarticular<br>Kirschner | Fair<br>Good | Poor<br>Good | Redig, 1981<br>Redig, 1981 |
| Midshaft radius and ulna | Closed | Modified Kirschner with plastic rod | Excellent | Good | Satterfield, 1981<br>Redig, 1983<br>MacCoy, 1983 |
| Distal radius and ulna | Closed | Modified Kirschner with plastic rod | Good | Fair | Satterfield, 1981<br>Redig, 1983<br>MacCoy, 1983 |
| Proximal radius and ulna | Open and comminuted | I.M. pinning | Fair | Negligible | Redig, 1983 |

1. Distally located open fractures are the most difficult to manage nonsurgically.
2. Distal fractures are amenable to nonsurgical treatment if they are closed fractures.
3. Most fractures in the proximal third can be managed nonsurgically.
4. Midshaft fractures are usually extensively overridden and require internal fixation in order to heal. If they are open, the prognosis is poor owing to wound infection, soft tissue damage, and death of the bone fragment projecting outside the wound.

The presence and severity of humeral fractures can be readily ascertained by direct observation of the bird at rest and by physical examination. Not only is the bird unable to fly, as in the case of shoulder injuries, but the wing has the obvious appearance of "hanging from the shoulder" and there is little control over its movements. Fractures in the proximal one fourth tend to be well-stabilized by the heavy musculature and therefore do not exhibit such pronounced clinical signs. Fractures in the middle half tend to be spiral in nature with long, sharp spikes of bone on both the proximal and distal fragments. The contraction of the flexor muscles after the injury typically causes the spike from the proximal fragment to project through the skin on the dorsal aspect of the wing and the spike from the distal fragment to project through the skin on the ventrum. Very distal fractures of the humerus are usually transverse, and the distal fragment is pulled into very tight flexion by the forearm flexor muscles, the tendons of origin of which are located on the distal humerus.[3]

Radiology is not essential as a diagnostic tool

**Table 30–1.** SUMMARY OF FRACTURE TYPES, MANAGEMENT RECOMMENDATIONS, AND PROGNOSIS IN AVIAN SPECIES *Continued*

| Fracture Site | Fracture Condition | Management Recommendation | Prognosis to Heal | Prognosis for Recovery | Reference |
|---|---|---|---|---|---|
| Midshaft radius and ulna | Open and comminuted | Figure 8 with minimal internal fixation | Good | Fair | Redig, unpubl. |
| Distal radius and ulna | Open and comminuted | Figure 8 with stiffener | Fair | Poor | Rousch, 1980 |
| Corpometacarpus | Closed | Figure 8 with stiffener | Good | Good | Rousch, 1980<br>Redig, 1983 |
| Carpometacarpus | Open† | Figure 8 with stiffener | Fair | Fair | Rousch, 1980<br>Redig, 1983 |
| Proximal or mid-femur | Closed and simple | Bind to body<br>Cage rest<br>Pinning (400 gm) | Good | Good | Rousch, 1980<br>Redig, 1983 |
| Distal femur | Closed | Modified transarticular Kirschner | Good | Good | Redig, 1981 |
| Proximal tibiotarsus | Closed | I.M. pinning<br>Transarticular Kirschner | Excellent | Good‡ | Rousch, 1980<br>Redig, 1981<br>Redig, 1983 |
| Midshaft tibiotarsus | Closed§ | I.M. pinning and Schroeder-Thomas finger plate | Excellent | Good‡ | Redig, 1983<br>Rousch, 1980 |
| Distal tibiotarsus | Closed | I.M. pinning and Schroeder-Thomas tape splint (small) | Good | Good‡ | Redig, 1983<br>Altman, 1977 |
| Distal tibiotarsus | Open | I.M. pinning and Schroeder-Thomas tape splint (small) | Fair | Fair‡ | Redig, 1983<br>Altman, 1977 |
| Metatarsus | Closed | Tape splint<br>Schroeder-Thomas<br>Boston | Good | Good | Altman, 1977<br>Redig, 1983<br>Satterfield, 1981 |
| Metatarsus (non-trap) | Open | Tape splint<br>Schroeder-Thomas | Fair | Fair‡ | Altman, 1977<br>Redig, 1983 |

*I.M. = intramedullary.

†Attempts at primary fixation usually result in a nonunion. It is better to allow time for soft tissue repair, and then, after the fibrous connective tissue has resolved, attempt an open reduction and fixation of the bone.

‡These cases are frequently lost owing to delayed healing of the fracture and development of pressure necrosis in the opposite foot.

§If the fracture is comminuted, consider the following: with one butterfly fragment, internal fixation is generally successful; with two or more fragments, internal fixation generally results in a nonunion. External fixation alone is recommended under these circumstances.

in evaluating many fractures of the humerus, but it is most helpful for determining the extent of injury and relative placement of the fragments, particularly in dealing with proximally located fractures.

The specific recommendations for dealing with most humeral fractures are contained in Table 30–1. Note that nonsurgical treatment requires the binding of the wing to the body with a figure-of-8 bandage (Fig. 30–3).[8] Simply binding the wing to itself with a figure-of-8 bandage is insufficient. Most humeral fractures require some form of surgical intervention to bring about adequate alignment of the fragments to ensure flight and to manage the significant amount of soft tissue damage and wound infection that accompany fractures occurring in the distal half of the humerus. I prefer intramedullary pinning for fractures in birds weighing more than 60 to 75 grams. However, most small birds such as finches, canaries, and budgerigars can be adequately managed by simply taping the wing to the body and providing necessary wound treatment. If there is no open wound, cage rest will often provide an adequate mode of treatment for small birds.

Closed humeral fractures occurring at any point along the humerus have an excellent prognosis for complete healing with any reasonable program of immobilization and cage rest. Healing will require three to four weeks. The prognosis for a favorable outcome decreases with the presence of each departure from this condition.

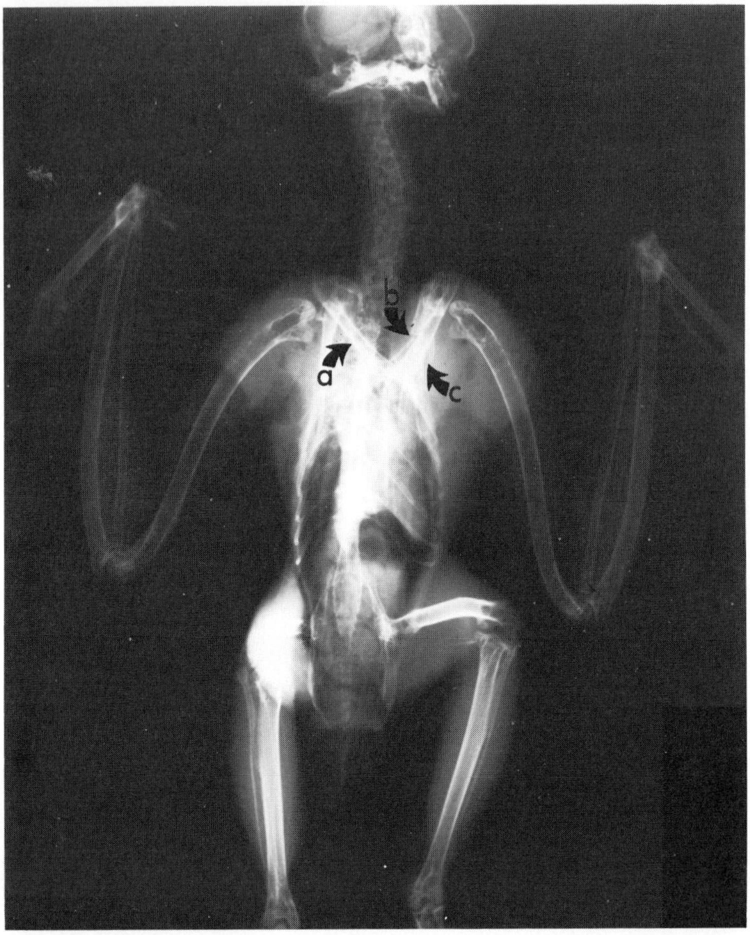


**Figure 30–2.** Radiograph showing massive multiple injuries to the shoulder girdle, including *(a)* displacement of the right coracoid, fracture of the *(b)* left coracoid and *(c)* left scapula, and bilateral fractures of the furcula (Barred Owl).

## Elbow Dislocations

Luxation of the elbow occurs frequently in wild birds. It is not likely to occur in most situations involving caged and aviary birds. Birds with elbow dislocation are unable to fly and are also unable to extend the injured wing at the elbow joint. On physical examination, the joint is usually found to be swollen and there will be a small amount of discoloration under the skin. Minor luxations can be restored to function by a period of immobilization of two to three weeks, with interposed periods of controlled flexion and extension of the joint (usually performed under anesthesia). When there is gross displacement of either the radius or the ulna or both from the humerus, the prognosis is poor. At present, there is no method available that will bring about repositioning and stabilization of the elbow joint.

## Fractures of the Forearm

The majority of forearm fractures involves only the radius or only the ulna. Therefore they are readily managed by external fixation methods. The presence of one intact bone provides an internal splint that effectively stabilizes the wing. Indeed, many birds admitted with an ulnar fracture show no external evidence of a fracture other than the inability to fly. Palpation along the length of both the radius and the ulna will reveal instability and crepitation; however, radiographic examination should be employed to further define the nature and extent of the injury.

There are several effective methods for externally stabilizing forearm fractures. The oldest of these by several hundred years is the braille (Fig. 30–4). We have modified the original design to include a leather pouch that is slipped over the flexed wrist to reduce the chances of a fractious bird's injuring its wrist by traumatizing it against the cage wall. A figure-of-8 bandage (Fig. 30–3) is the most frequently used method in my practice, particularly if there are open wounds associated with the fracture for which it is necessary to maintain a dressing over the injury site. It is not necessary to bind the wing to the body. The prognosis for these fractures is very good in most instances, and average healing times are approximately three weeks to removal of fixation and another three weeks for return to full flight. Proximity of the fracture to either the wrist or the elbow will adversely affect the prognosis.

If both the radius and ulna are fractured, surgical implantation of an intramedullary pin or installation of a modified Kirschner-Ehmer device is almost essential for restoration of flight[2, 5, 7] (Table 30–1; see also Figure 48–1). The Kirschner-Ehmer device is preferred over intramedullary pins, as the latter must exit through the elbow joint, which often leads to interference with the integrity of that joint. I have been successful in a few instances with a minimal internal fixation accompanied by external coaptation (Fig. 30–5). Such an approach should always be considered when there is severe comminution of any of the bones, as surgical intervention is more likely to fail under these circumstances.

A device using a sheath of x-ray film to provide fixation of the wing in extension has been described.[6] I have no experience with this technique and have reservations about its practicality. It is worthy of further consideration for providing traction as a prelude to surgical reduction and fixation in situations in which there is significant overriding of the fragments.

### Fractures of the Metacarpus

Fractures occurring distal to the wrist joint should always be managed nonsurgically. The scarcity of overlying soft tissue renders them very likely to develop into nonunions and massive death of the remaining soft tissue if surgical intervention occurs, especially in the immediate posttrauma period. In addition to physical examination, radiology is very important in evaluating these fractures to ensure differentiation of an actual fracture from the normal points of articulation in the distal portion of the wing (see Fig. 30–1). Wing tip sprains occur frequently with no external sign of damage to the wing, and radiology is necessary to differentiate these from fractures. Coaptation is identical to that used for forearm fractures. Prognosis is guarded in most instances, particularly if there is an open wound in association with the fracture. Healing times are extended, usually requiring three to four months.

When there is an open wound and comminution associated with a metacarpal fracture, the best management mode appears to be external stabilization of the wing for two to three months while the soft tissue damage is repaired, then treatment of the nonunion by surgical stabilization and perhaps bone grafting. Surgical treatment of an open, comminuted metacarpal fracture at the time of injury invariably results in loss of the end of the wing. Successfully employed fixation modes have included intramedullary pinning and modified Kirschner-Ehmer devices.[5]

## FRACTURES AND OTHER INJURIES TO THE PELVIC LIMB

### Hip Dislocations and Femoral Fractures

The femur is covered along most of its length by a web of skin that extends from the lateral body wall to the stifle. Placement of bandages and splints above the stifle is virtually impossible; thus, it is difficult to immobilize the femur or the hip joint by external means. Rousch[6] suggests the use of a modified hip spica cast fashioned out of some malleable material such as Orthoplast or Hexcelite (Fig. 30–6). In this method, the leg is wrapped with the material and a tongue is fashioned which extends laterally up the thigh and curves over the back underneath the wings. This tongue is further affixed to the body by enclosure within a sturdy wrap of conforming gauze and an elastic self-adhesive bandage such as Elastikon. Care must be taken not to restrict respiration or block defecation. I have not personally experimented with this method for treatment of a hip luxation, but I believe it would be an effective means of management. The injured leg should be positioned in a conformation compatible with normal perching posture if the bird is to be expected to accept the fixation for the several weeks necessary for recovery.

Femoral fractures in waterfowl and most small caged birds can be adequately managed by cage rest. On the other hand, femoral fractures of medium to large birds occurring anywhere along the length of the femur will of necessity require surgical reduction and fixation to prevent overriding of the fragments and to provide for rotational stability. A delay in healing brought on by inadequate stabilization on the first attempt results in an excessive extension of the amount of time that the bird is forced to bear weight on the contralateral foot, which invariably begins to undergo pressure necrosis leading to bumblefoot. Hence, it is imperative with fractures of the leg that adequate stabilization be applied at the outset. The exception to this principle again is a situation in which there is significant comminution of the

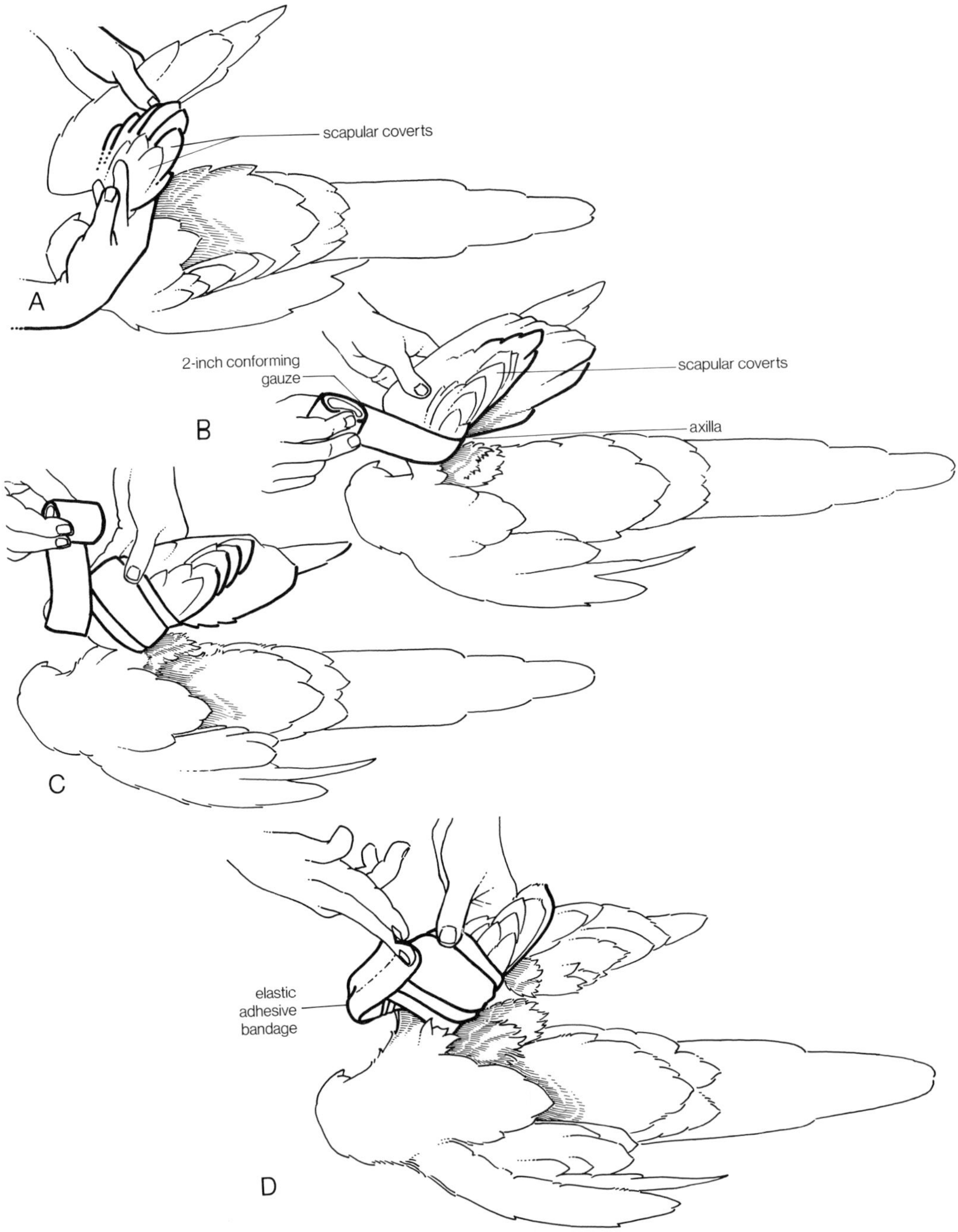


**Figure 30–3.** The proper method for the application of a figure-of-8 bandage. *A*, Positioning of the bird and grouping of feathers for application of the bandage. *B*, Starting the gauze wrap around the back of the humerus. *C*, Extending gauze over the wrist and wrapping in a figure-of-8 configuration. *D*, Applying the single wrap of Elastikon over the wrist joint.

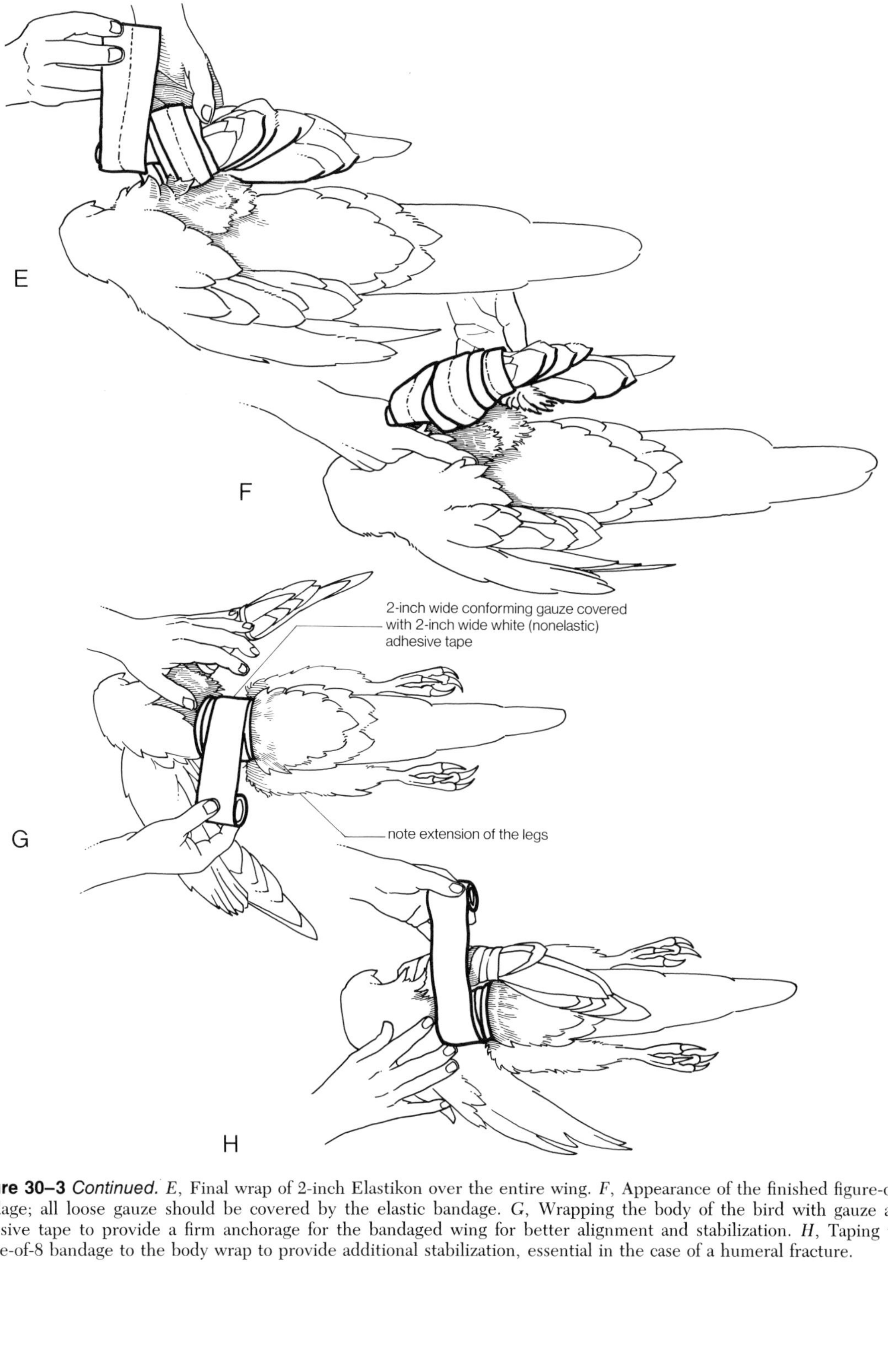


**Figure 30–3** *Continued.* *E*, Final wrap of 2-inch Elastikon over the entire wing. *F*, Appearance of the finished figure-of-8 bandage; all loose gauze should be covered by the elastic bandage. *G*, Wrapping the body of the bird with gauze and adhesive tape to provide a firm anchorage for the bandaged wing for better alignment and stabilization. *H*, Taping the figure-of-8 bandage to the body wrap to provide additional stabilization, essential in the case of a humeral fracture.

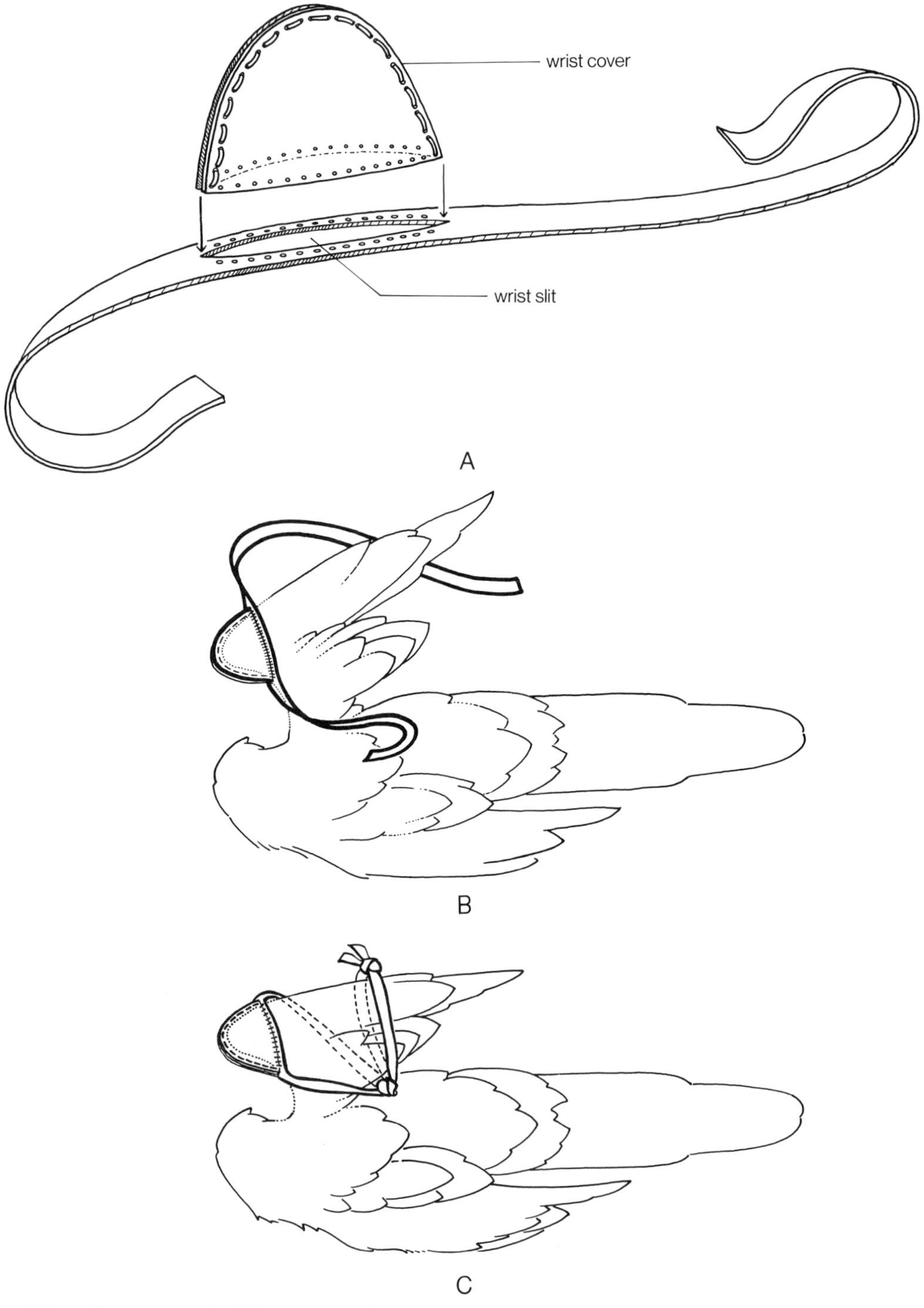


**Figure 30–4.** The proper method for applying a braille to the wing. *A*, The braille is fabricated out of leather. *B*, Applying the braille to the wing. *C*, The braille is affixed to the wing by tying the straps together as shown.

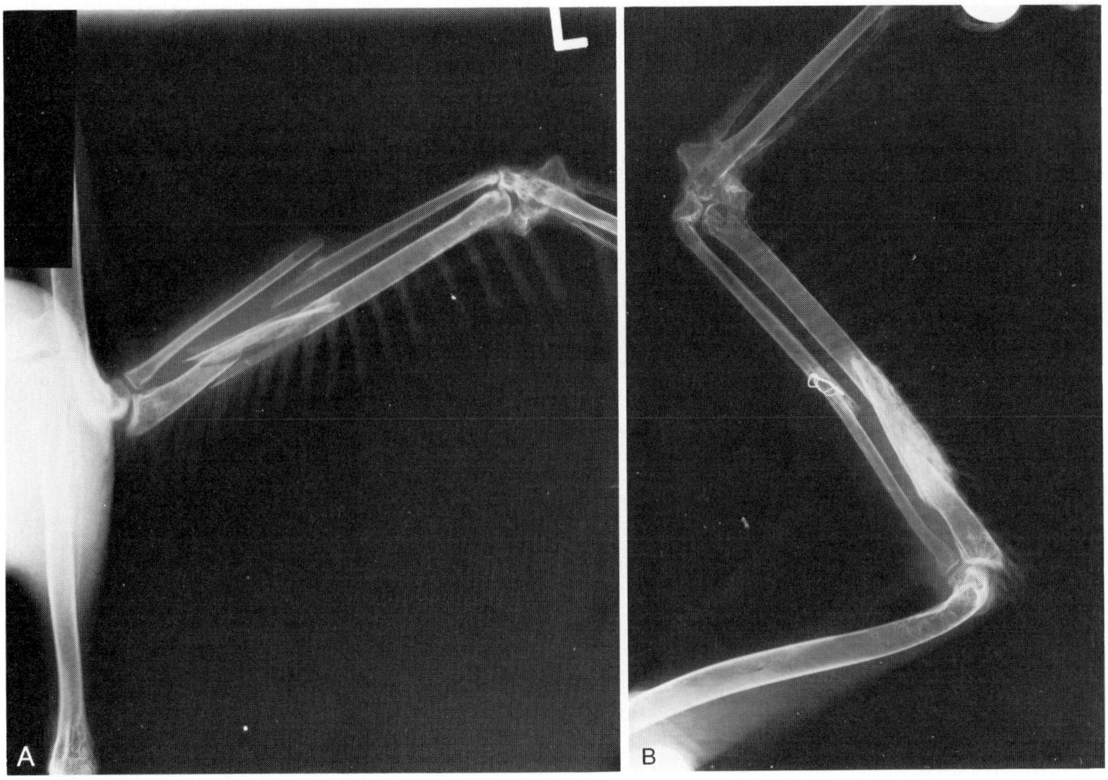


**Figure 30–5.** Radiographs of a goose with a comminuted fracture of the radius and ulna. The radial fragments were opposed with a piece of circlage wire; the ulna was stabilized with a figure-of-8 bandage only. *A*, Radiograph at time of admission. *B*, Radiograph at time the bird could fly.

fracture. Here, coaptation may be the best alternative.

## Tibiotarsal Fractures

Fractures of the tibiotarsus are among those most likely to be encountered among caged birds. Fortunately, they are readily managed and there are a variety of techniques available that may be employed. As with the femur, rapid return to a state of at least partial weight bearing is essential to prevent damage to the other foot. For birds weighing less than 100 to 150 grams, the tape splints described by Altman (Fig. 30–7) work well[1] (see also Chapter 46, Surgical Instrumentation and Special Techniques). A combination of internal and external fixation is recommended for management of most tibiotarsal fractures in larger birds (Table 30–1). Owing to the straightness of this bone and the longitudinal stability afforded by an intramedullary pin, it is the preferred internal device. The pin may exit at either the knee or the hock joint; however, the latter is probably more forgiving to the insult. External fixation is provided by the Schroeder-Thomas splint as I have modified it for use in birds (Fig. 30–8). This device in combination with intramedullary pinning provides excellent stabilization and allows immediate bearing of weight on the affected leg. Severely comminuted fragments tend to die and sequester, leading to nonunions if surgically manipulated; hence, such fractures are better managed by external fixation methods only (Fig. 30–9). Despite their relative bulk, such splints are well tolerated by any bird if comfortably applied (Fig. 30–10). Wire of the size and flexibility of a coat hanger is adequate for Amazons, cockatoos, and macaws. For cockatiels and conures, 14- to 18-gauge copper wire is a convenient material with which to work.

## Tarsometatarsal Fractures

Fractures of the tarsometatarsus occur with less frequency, are more likely to be open fractures, and have a poorer prognosis than fractures of the tibiotarsus. Nevertheless, the approach to management of these is similar to that for the tibiotarsus. There is, however, a higher degree of reliance upon coaptation for

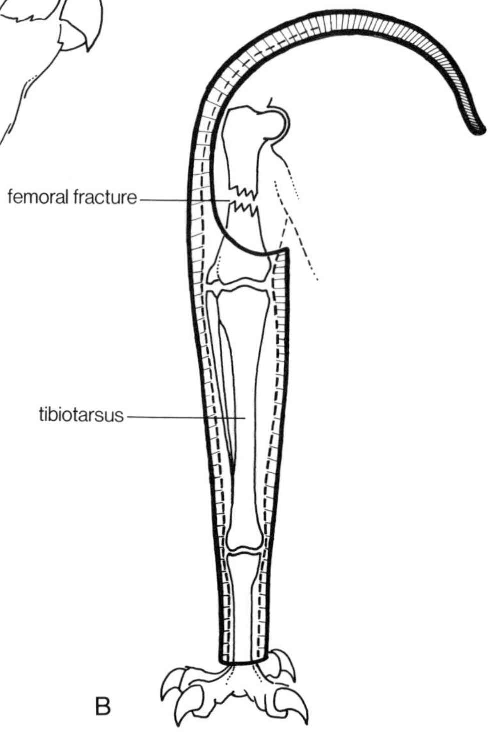

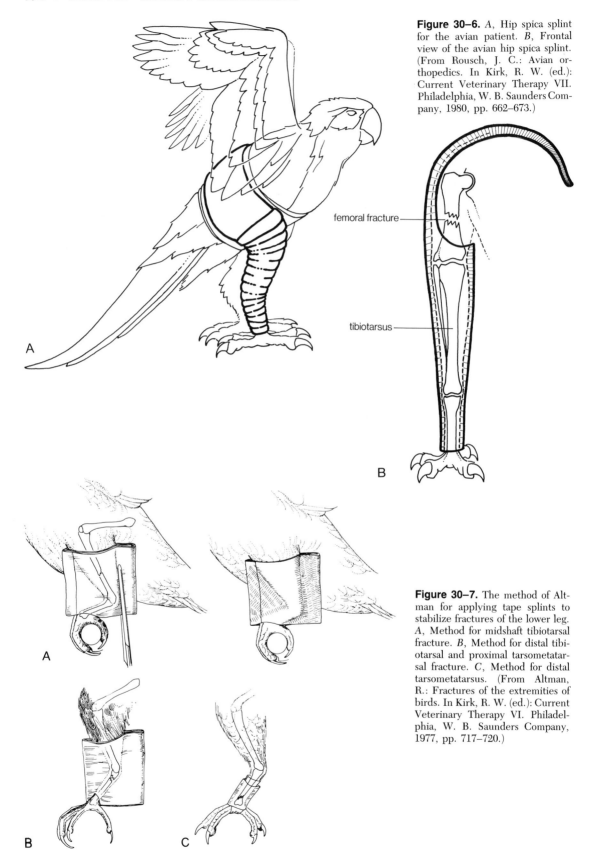

**Figure 30–6.** *A*, Hip spica splint for the avian patient. *B*, Frontal view of the avian hip spica splint. (From Rousch, J. C.: Avian orthopedics. In Kirk, R. W. (ed.): Current Veterinary Therapy VII. Philadelphia, W. B. Saunders Company, 1980, pp. 662–673.)

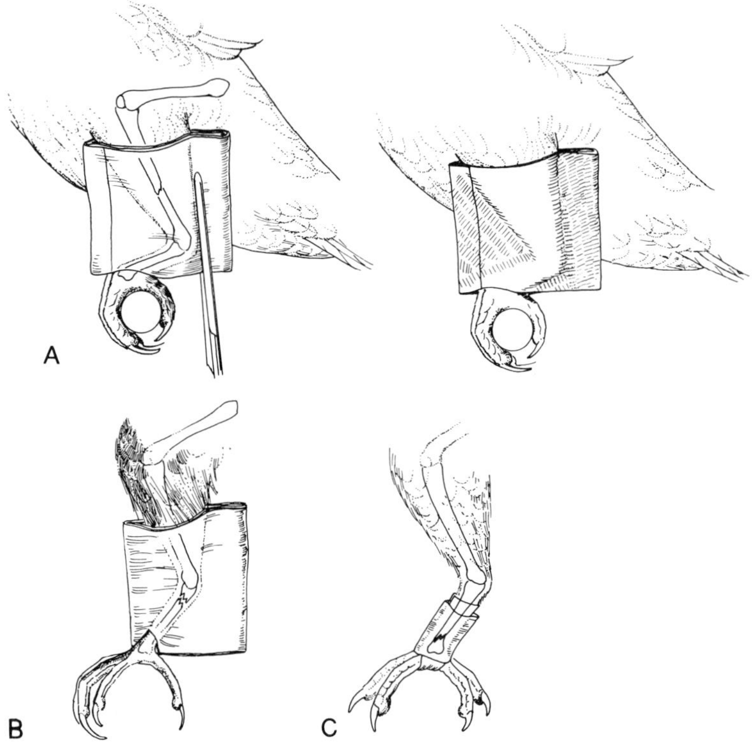


**Figure 30–7.** The method of Altman for applying tape splints to stabilize fractures of the lower leg. *A*, Method for midshaft tibiotarsal fracture. *B*, Method for distal tibiotarsal and proximal tarsometatarsal fracture. *C*, Method for distal tarsometatarsus. (From Altman, R.: Fractures of the extremities of birds. In Kirk, R. W. (ed.): Current Veterinary Therapy VI. Philadelphia, W. B. Saunders Company, 1977, pp. 717–720.)

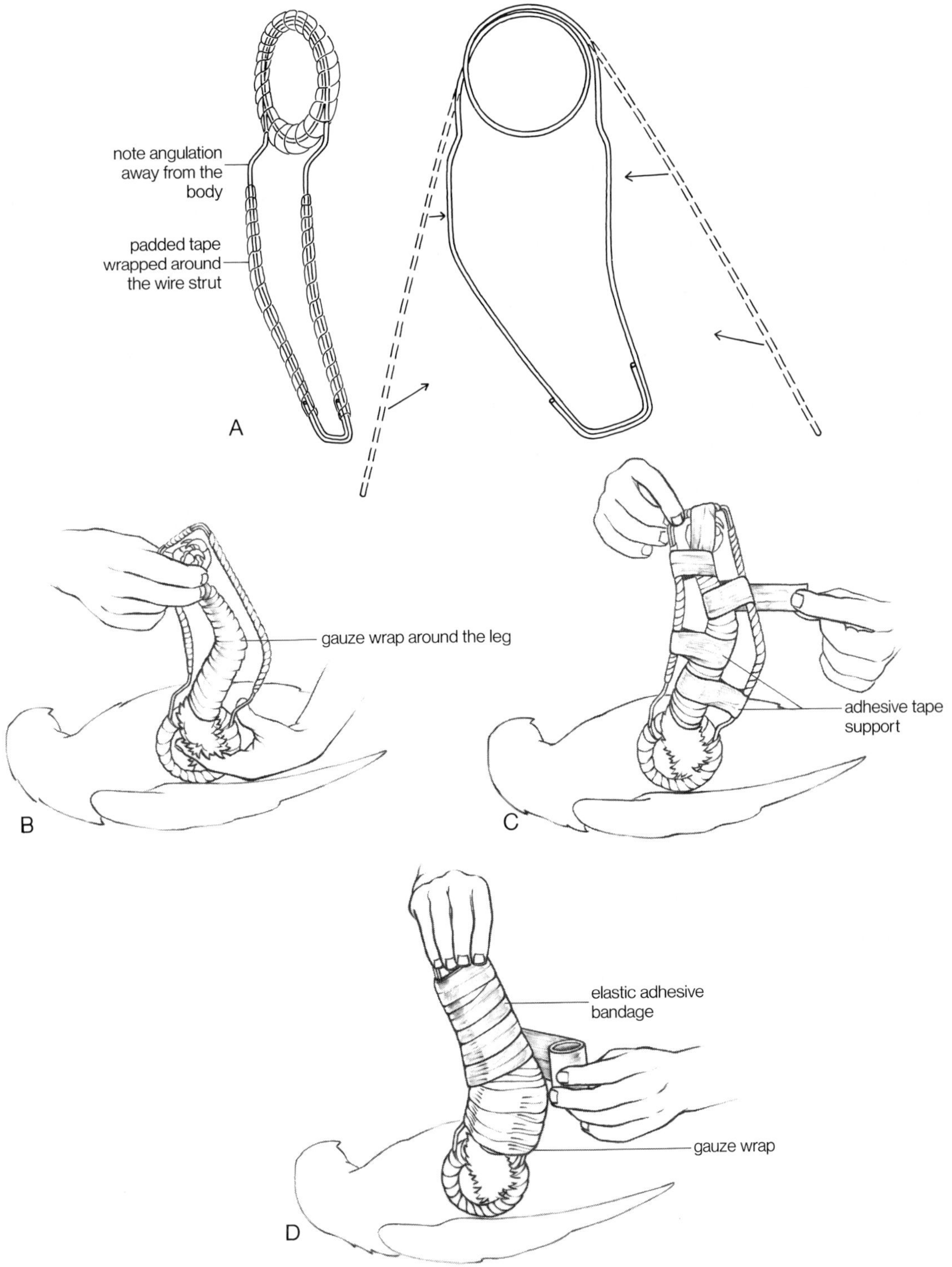


**Figure 30–8.** The method for the construction and application of the Redig modification of the Schroeder-Thomas splint for birds. *A*, Making the splint. *B*, The padded wire splint is fitted to the gauze-wrapped leg and bent to conform to the angulations of the leg. *C*, Adhesive tape cross supports are applied to suspend the leg between the struts of the splint. *D*, The entire leg and splint are wrapped in a heavy layer of gauze and covered with an elastic adhesive bandage.

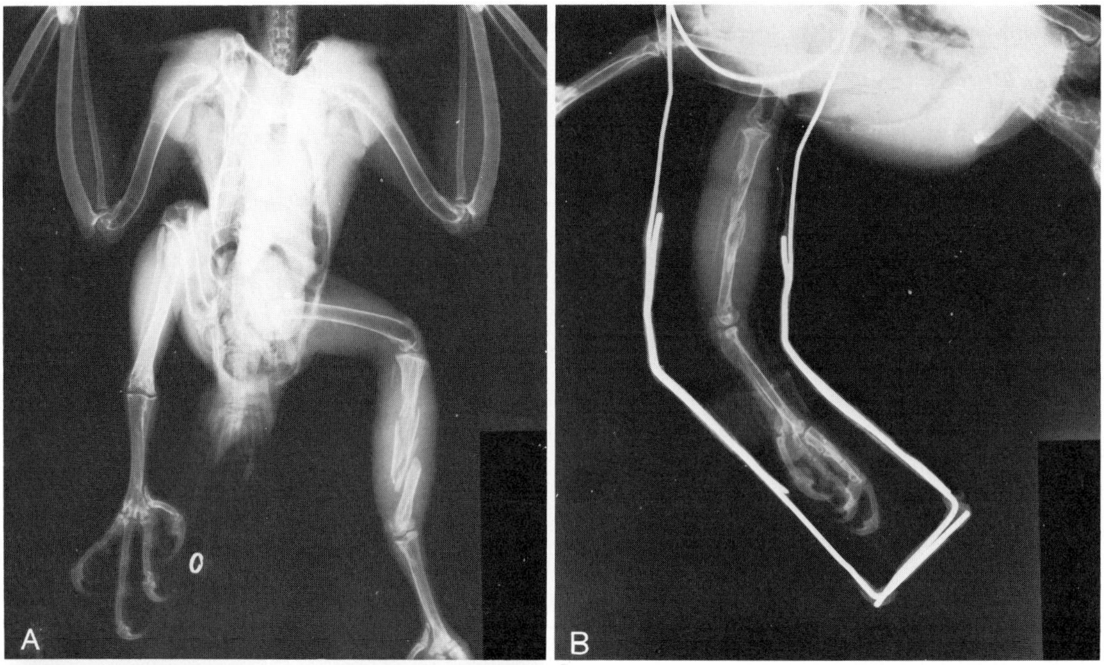


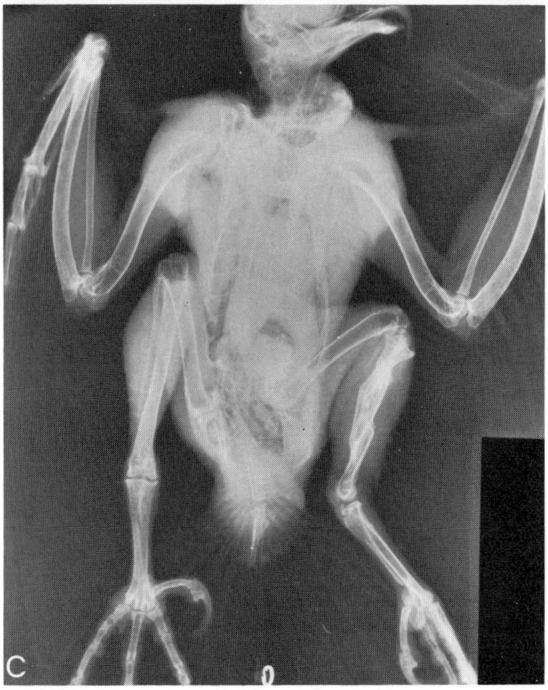

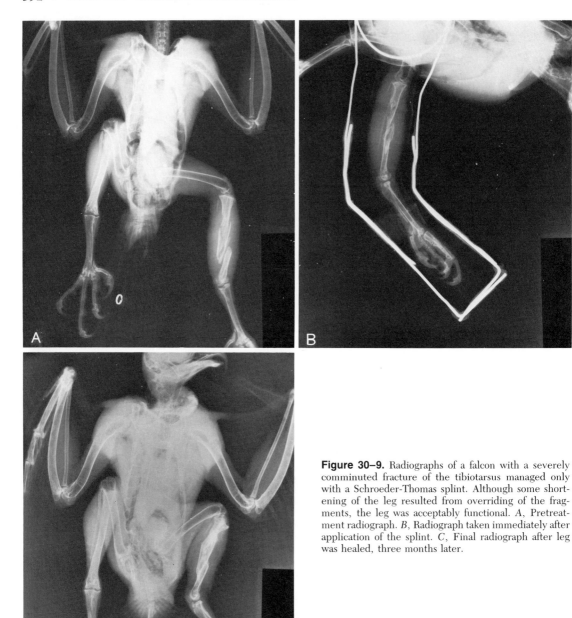

**Figure 30–9.** Radiographs of a falcon with a severely comminuted fracture of the tibiotarsus managed only with a Schroeder-Thomas splint. Although some shortening of the leg resulted from overriding of the fragments, the leg was acceptably functional. *A*, Pretreatment radiograph. *B*, Radiograph taken immediately after application of the splint. *C*, Final radiograph after leg was healed, three months later.

fixation of these fractures. Surgical fixation is generally contraindicated for the same reasons as encountered in managing fractures of the carpometacarpus.

### Fractures of the Toes

These may be readily managed by wrapping the toes around a ball of gauze 2-inch × 2-inch sponges and holding them in place with several layers of a continuous wrap of Kling gauze, then wrapping the entire foot with a single layer of Elastikon for three to four weeks (Fig. 30–11). It is important to stabilize the metatarsophalangeal joint with the bandage to enable the bird to stand on the bandage. A "ball bandage" 6 to 7 cm in diameter is typical of that applied to a macaw. In some cases tongue depressor splints have been used with varying degrees of

**Figure 30–10.** This Blue and Gold Macaw is recovering from a tibiotarsal fracture stabilized in part with a Schroeder-Thomas splint,

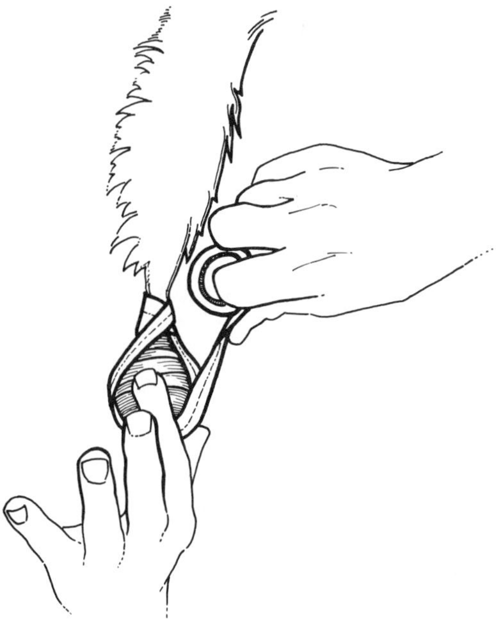

**Figure 30–11.** The digits are stabilized by the application of a ball bandage. Note the longitudinally directed layers of gauze and Elastikon that extend half way up the metatarsus to stabilize the joint.

success. Open fractures of the digits in which there is exposure of the bones and tendons usually result in loss of a portion of the digit below the site of injury.

## GENERAL COMMENTS

Anesthesia is an essential prerequisite to the successful diagnosis, evaluation, and treatment of avian orthopedic problems. If a patient is in shock from injury, only minimal wound cleaning and limb immobilization such as taping the wing against the body with masking tape should be attempted initially. The shock and dehydration may require 24 to 48 hours to stabilize, after which the patient may undergo anesthesia for further proper management of the fracture. To attempt to apply complete fixation initially will result in inadequate application of the coaptation device and possibly the death of the bird from handling stress.

During physical examination of the anesthetized orthopedic patient, the application of 70 per cent alcohol over the joints and along the bones induces a transparency to the skin, making it very easy to visualize some fractures and to observe bruising over sprained joints. I have found birds unable to fly because of a small bruise to an elbow joint which was immediately visible after the application of alcohol over the joint.

Most birds have adapted well to the immobilization of one wing. Although Altman[1] recommends immobilization of both wings to prevent the bird from attempting to fly, I have preferred to bind only the affected wing and have had very good results. Typically, the bird may make one or two attempts, only to discover that flight is impossible. Additionally, the free wing aids them in balancing as they move about their cage.

The temperament of the bird is often a significant factor in selecting a mode of treatment for any fracture. A wild, fractious bird is more likely to require the additional stabilization provided by a combination of internal and external fixation than one that is content to sit quietly in its cage and not react to external stimuli by thrashing about. This behavioral aspect can be an important consideration in selecting a fixation technique.

The use of perioperative and postoperative antibiotics is recommended. My initial choice is amoxicillin or chloramphenicol. These are maintained for the first postoperative week. If these prove inadequate by evidence of postoperative infection, then further choice of antibiotics is predicated upon results of culture and sensitivity studies. When dealing with open wounds, this determination should be made as part of the initial work-up.

## POST-FIXATION CARE

It is generally advantageous to the client and the bird to maintain an orthopedic case in the hospital for at least the first week following application of fixation. Strict immobilization of the fracture is critical during the first week, and this is also the time when one can gauge the acceptance of the device by the bird and react immediately should problems arise. The bird should be housed in a comfortable cage with solid walls and limited visibility to the exterior. It should not be allowed any movement outside the cage. After one week, the device can be removed, radiographs taken to check alignment, and modifications for effectiveness and comfort made as deemed necessary. The bird may then be sent home with instructions for further complete inactivity for another two weeks. It should then be brought back for another physical and radiologic examination.* If all is proceeding well at this point, the home instructions may be altered to allow limited movement, such as climbing around the inside of a cage to exercise a leg. Wing fractures may be sufficiently healed by this point to allow limited wing flapping, but no free flying. Under no conditions should a bird recovering from a broken leg be allowed to fly free inside a house until six weeks after the injury, assuming normal progression of healing. A final examination at about six weeks is usually sufficient to assure satisfactory resolution of the case. Anesthesia is recommended for all postoperative evaluations. Most orthopedic failures in otherwise properly managed fractures of pet birds occur because owners did not adhere to a strict program of restricted activity for the bird.

Birds recovering from broken legs should have their tails taped to protect the feathers and the first digit of the other foot wrapped with tape to prevent pressure sores from developing on the ventral surface. They should be maintained on a thick pad of foam rubber covered with a plastic wrap and overlaid with a turkish towel that is changed at least daily. It is often useful to put a low "step perch" in a cage, which allows the bird to hang its broken leg with the coaptation splint below the level of the other foot. Once *shown* how to use such a perch, birds accept them readily.

Elizabethan collars may be indicated for individual birds that demonstrate a propensity to remove bandages, although I have not personally found them necessary to use. Bandages are most frequently removed by the bird because they are uncomfortable. Aside from a small amount of picking at a bandage when it is first applied, birds accept them readily.

*Editor's Note: Because of the cost factor to the client, the practitioner may need to be selective as to when radiology is necessary.

### List of Products Cited

Elastikon, Johnson and Johnson Products, Inc., New Brunswick, NJ 08903

Hexcelite, Hexcel Medical, 6700 Sierra Lane, Dublin, CA 94566

Kling Elastic Gauze Bandage, Johnson and Johnson Products, Inc., New Brunswick, NJ 08903

Orthoplast, Pitman-Moore, P.O. Box 344, Washington Crossing, NJ 08560

## REFERENCES

1. Altman, R.: Fractures of the extremities of birds. In Kirk, R. W. (ed.): Current Veterinary Therapy VI. Philadelphia, W. B. Saunders Company, 1977.
2. MacCoy, D. M.: High density polymer rods as an intramedullary fixation device in birds. J. Am. Anim. Hosp. Assoc., 19:767–772, 1983.
3. Redig, P. T., and Rousch, J. C.: Surgical approaches to the long bones of raptors. In Fowler, M. (ed.): Zoo and Wild Animal Medicine. Philadelphia, W. B. Saunders Company, 1978.
4. Redig, P. T.: Methods for the delivery of medical care to trained or captive raptors. In Proceedings of the American Federation of Aviculturists, San Diego, 1981.
5. Redig, P. T.: A clinical review of orthopedic techniques used in the rehabilitation of raptors. In Proceedings of the Annual Meeting of the Association of Avian Veterinarians, San Diego, 1983, pp. 34–39.
6. Rousch, J. C.: Avian orthopedics. In Kirk, R. W. (ed.): Current Veterinary Therapy VII. Philadelphia, W. B. Saunders Company, 1980, pp. 662–673.
7. Satterfield, W., and O'Rourke, K. I.: External skeletal fixation in avian orthopedics using a modified through-and-through Kirschner-Ehmer splint technique (the Boston technique). J. Am. Anim. Hosp. Assoc., 17:635–637, 1981.
8. Spink, R. R.: Fracture repair in rehabilitation of raptors. VM/SAC, 73(11):1451–1455, 1978.
9. Ward F. P.: Repair of wing fractures in a red-shouldered hawk. VM/SAC, 65(6):550–552, 1970.

### ACKNOWLEDGMENT

The author would like to acknowledge Rick Volkmar for the artistry he contributed to this chapter.

*Section Six*

# DISEASES

# Chapter 31

# NUTRITIONAL DISEASES

GREG J. HARRISON
LINDA R. HARRISON

With the exception of obesity, the adult single pet bird clinically appears very resistant to gross malforming and well-recognized lesions associated with specific malnutritional syndromes. In the author's opinion, however, chronic generalized malnutrition, whether in terms of deficiencies or excesses, seems to be a primary underlying factor of disease in the majority of birds presented to a clinical practice. Malnutrition becomes particularly evident during periods of high nutritional demand on the body such as reproduction, molting, and other stresses, including exposure to disease organisms and subsequent response to treatment.

Nonbreeding budgerigars normally remain in good health on a diet consisting of little except seeds,[20] even though it has been well-documented that such cereals are deficient in calcium, vitamin A, and the amino acids lysine, methionine, and tryptophan.[21] This would imply that nonbreeding requirements are extremely low.

From a University of California study[13] it appears that vitamin supplementation of a cockatiel seed mixture is not critical for continued maintenance of nonbreeding healthy birds (at least up to 500 days) and that high levels of vitamins (up to 200 times the calculated dosage based on turkey recommendations) for 135 days are apparently not harmful.

Breeding birds and their young, on the other hand, are more likely to suffer from a variety of disease conditions because their requirements for specific nutrients are much higher than those of adult nonbreeding birds.[14]

Clinically, one cannot categorically state that a particular syndrome is the result of a specific nutrient deficiency. Uncomplicated deficiency of a single nutrient seldom, if ever, occurs.[21]

## FACTORS INFLUENCING MALNUTRITION

When one considers the possibility of nutritional deficiency diseases in psittacines, the initial clinical reaction is to look to the diet to see whether the nutrient in question is present. Certainly, the composition of the diet has to be examined (see Chapter 2, Husbandry Practices); an absolute deficiency of a nutrient in an otherwise adequate diet would be an acceptable explanation for a deficiency disease. However, a number of other factors are involved and have to be considered in the pathogenesis of deficiency diseases.[9]

### Individual Bird's Intake of Its Diet

The diet offered to most pet bird species is usually based on a seed mixture, with sunflower seeds forming the bulk of the mixture for species the size of cockatiels and larger. Although precise nutritional requirements for pet bird species are not yet known, educated bird owners tend to offer a wide variety of foods to supplement the basic seed diet. This, however, does not ensure that the bird will eat all that is available. Owners have mentioned that some birds actually become fearful of new foods introduced into the cage; therefore, long-term patience and ingenuity may be required to shift the intake (see Chapter 3, Captive Behavior and Its Modifications).

Clearly, a bird that has some degree of anorexia would not be taking in the usual nutrients and may ultimately manifest the deficiency or absence of one or more of them. More specific causes of anorexia related to lesions of the beak, tongue, oral cavity, or even upper digestive tract, causing problems with prehension and swallowing, can also be related to inadequate intake of nutrients.

## Ability to Absorb the Nutrients

Although such problems as vomiting and diarrhea are obvious, other causes of malabsorption of nutrients are difficult to evaluate. Lesions of the alimentary tract can have a distinct effect upon the ability of the bird to digest and absorb a particular nutrient. Intestinal mucosal lesions can interfere with the bird's ability to convert carotene pigments to functional vitamin A. It is known in the chicken that lesions causing dysfunction of the liver and/or pancreas can affect the assimilation of the fat-soluble vitamins—A, $D_3$, and E. Renal problems may specifically interfere with vitamin $D_3$ absorption.[21] Because vitamin C is normally synthesized in the liver, abnormalities of this organ may increase the need for dietary vitamin C[19] (see Chapter 24, Relationships of the Avian Immune Structure and Function to Infectious Diseases). The presence of gastrointestinal parasites[15, 17] or mycotoxins[10] may also impair intestinal absorption of vitamins.

Other components of the diet may result in the formation of insoluble complexes of essential nutrients so that, although they are present in adequate amounts in the diet, they will not be able to be absorbed from the alimentary tract. High-fat diets, such as those based on sunflower seeds, can cause the formation of insoluble soaps with calcium or iron, rendering them unavailable to the bird, as noted by several authors.[5, 8, 23]

Rations high in lipids may also dilute the fat-soluble vitamins so that a portion of those ingested may be excreted unassimilated. The addition of fatty acids to avian supplements, a recent trend in commercial preparations, is thought to be unnecessary with diets based on high-fat seeds and may only serve to increase the lipid component.

## Patterns of Storage and Utilization of the Nutrients

Chronic liver disease and liver dysfunction can result in a decreased ability of the liver to store vitamin A, for example. As the vitamin A within the liver is depleted, blood levels of the essential vitamin decrease and manifestations of hypovitaminosis A emerge. A marked decrease in the vitamin A content of the liver has also been associated with aflatoxicosis in animals.

## Secretory Functions

Egg production is a classic example of increased secretion. Reproductive function places an increased nutritional demand on the female with regard to calcium for shell material, vitamin A and fats for yolk preparation, and protein and other nutrients. The biological state of the individual needs to be considered in evaluating nutritional needs.

## Changing Requirements

A bird that is actively growing has significantly different nutritional requirements from a bird that has obtained full skeletal, soft tissue, and feather growth. (Some results of nutritional studies with cockatiel chicks are reported in Chapter 53, Pediatric Medicine.) Molt also affects the demands on the body. The development of feathers at a rapid rate in a relatively short period of time during the molting period of the year causes many increased requirements. The need for additional protein is well-recognized. There is no particular increase in mineral requirement during this time, but there is a very significant increase in the need for lipids. One may often overlook the fact that plumage is basically composed of the remnants of cell membranes of the epithelial cells that went into forming the feather, and cytomembranes have a significant lipid component (see Chapter 41, Disorders of the Integument).

Adult birds that are rearing young have higher general nutritional needs, especially for minerals and vitamins that are required in much smaller quantities by solitary adults (see Chapter 52, Reproductive Medicine).

## Other Factors in the Diet That Interfere with Nutrients

The diet may contain components that will inactivate or inhibit the action of certain nutrients. The incorporation into the diet of raw egg white, which contains avidin, makes biotin unavailable and possibly deficient.[6, 11, 14] Analogues may compete with the natural vitamin for sites of action and cause apparent deficiency diseases. Calcium and other minerals in direct contact with vitamins, especially vitamin A, in supplements may be oxidized quickly and rendered useless; vitamins and minerals should be supplied separately unless they are protectively coated within the supplement to maintain their effectiveness.[1] The oxidation of a nutritional supplement is significantly accelerated by its addition to water, and one might assume that the chlorine and fluorine in city tap water contribute further to the degradation.

Obviously, one has to look beyond the mere absence of nutrients in the diet to see clinical syndromes that are suggestive of deficiency disease.

## CLINICAL FINDINGS ASSUMED TO BE RELATED TO DEFICIENT DIETS

Based on the dietary history and clinical observations, the first author has found the "average" pet bird, which has been maintained on basically an all-seed diet (with the possible addition of vitamins in the water and a cuttle bone), has a characteristic appearance. This appearance is suspected to be related to chronic deficiencies of several nutrients which may prevent the proper interrelationships and functions within the body for normal health.

### Physical Examination

The most obvious characteristic that is clinically associated with these birds is a lackluster, "loose" appearance of the feathers (Fig. 31–1). Breakdown of the interlocking barbules that hold the feather together for a smooth, shiny finish, the presence of transverse "stress" lines, presence of debris on microscopic examination, depigmentation of normally blue or green feathers to black or yellow, incomplete molt, chronic heavy pinfeathering, and easily broken wing and tail feathers may all be related to inadequate nutrition.

All epithelial surfaces on these birds, including those of the body skin, eyelids, nares, and ears, may have a dry, flaky texture. The beak and nails may be long and have a rough surface. The beak in particular may appear as if normal cell replacement has occurred but there has been no sloughing of the old dead tissue, resulting in a layered appearance (Fig. 31–2). The skin of the legs and feet often appears thickened and dull; however, the plantar surfaces may be worn smooth and may be devoid of a distinctive scale pattern (see Chapter 7, Preliminary Evaluation of a Case). This smoothness on the bottom of the feet may be related to general inactivity, to which malnutrition may be contributory.[4]

**Figure 31–1.** Lack of a smooth, interlocking, tight appearance to the feathers in an otherwise normal-appearing Amazon may be a sign of chronic malnutrition. Note also the areas of depigmentation.

In birds suspected to be chronically malnourished, the roof of the mouth has been observed to be either very slimy, owing to increased production of mucus, or very dry; the margins of the tongue are often calloused. Various stages of abnormal oral epithelial pigmentation, papillae, or textural variations due to slightly swollen or congested pharyngeal or paralingual glands are common. Some of these may be scars from previous diseases, such as pox in Amazon species.

Obesity from overnutrition is prevalent in pet birds, particularly budgerigars and cockatoos. Aviculturists have found cockatoos especially resistant to eating anything but sunflower seeds. This high-fat diet likely contributes to the frequency of lipomas (Fig. 31–3), which are particularly common in Rose-breasted Cocka-

**Figure 31–2.** The chronically malnourished bird may have rough layers of horny epithelium on the beak.

toos. Placing a bird on a strict low-fat diet has resulted clinically in weight loss and reduction of lipomas prior to surgical attempts (see Chapter 42, Miscellaneous Diseases and Chapter 47, Selected Surgical Procedures).

It is theorized that pica, which may be expressed in a variety of behavioral problems including feather picking, destruction of perch and other environmental items, and cannabalism, may result from vitamin/mineral imbalances. Palpation or radiography may reveal a malformed sternum or spine, suggestive of early malnutrition.

## Abnormal Gram's Stain

A Gram's stain of samples from the gastrointestinal system is considered by the first author to be of some assistance in evaluating possibly inadequate diets in clinically normal birds. A Gram's stain of the droppings or choana reflects the quantitative and qualitative bacterial population in the gastrointestinal tract or oral pharynx (see Chapter 10, Clinical Microbiology, and Chapter 17, Cytology). Significant variations from the accepted norm of more nutritionally sound birds may suggest chronic malnutrition. One can hypothesize that with an incorrect dietary composition, the secretions of the digestive tract change and normal flora cannot tolerate it as well, resulting in a shift in the intestinal bacterial components.

In contrast to the composition of a Gram's stain of a "normal" fecal sample (see Chapter 9, Evaluation of Droppings), a chronically malnourished bird may show very low numbers of *any* bacteria, with abnormal percentages in

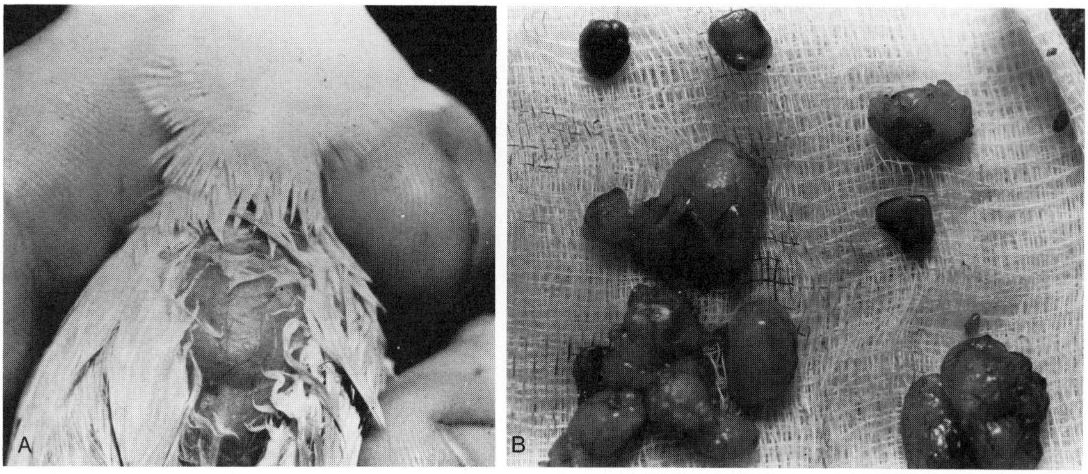


**Figure 31–3.** *A*, Wetting and separating the feathers reveals a developing lipoma in the early stages on the sternum of a budgerigar. *B*, Some lipomas can be surgically removed, such as these from a Rose-breasted Cockatoo.

distribution. For example, a Gram's stain of a fecal sample from a clinically normal pet cockatoo that is maintained on a diet of sunflower seeds only may consist of only 30 organisms per oil field, 94 per cent of which may be gram-positive rods, 4 per cent gram-positive cocci, 1 per cent pleomorphic bacteria, and 1 per cent yeast. By restricting the sunflower seeds and adjusting the diet to include legumes or animal protein, whole grain bread products, fresh vitamin A–containing vegetables, occasional fruits, and a source of calcium, a normal Gram's stain may result with no other treatment. Daily exposure to unfiltered sunlight (not through glass) also helps.

A change in the Gram's stain profile frequently results from simple adjustment of the nutritional intake. This may empirically suggest a cause and effect; however, low bacterial counts on fecal Gram's stains are not pathognomonic for malnutrition. These results are relatively common with young, hand-raised psittacines, and can be attributed to other conditions, such as overtreatment with antibiotics or concurrent virus infections.

## Clinical Chemistries

Clinical chemistries are of little value in diagnosing subclinical malnutrition (see Chapter 13, Clinical Chemistries), although lipemia is obvious when a blood sample from an obese, lethargic bird is centrifuged and a thick layer of fat forms in the upper layer of the serum. Elevated serum alkaline phosphatase levels have been reported to appear in most species in negative calcium balance before radiographic signs are definite.[23] Adjustment of the mineral imbalance is reported to lower the alkaline phosphatase levels to normal in 7 to 15 days.

Serum calcium levels usually remain normal until all other body stores are severely depleted except in some specific disease states, such as those associated with convulsions in African Greys.

In the authors' experience, subclinical malnutrition appears to be extremely common. Thus, it is likely that published reference "normals" of clinical chemistry tests have actually been developed from "abnormal" birds.

## Radiographs

Radiographs may offer little in the way of diagnostic evidence for subclinical nutritional disease, although hepatic enlargement or atrophy may be noted in some birds. Radiographic evidence of bone changes in clinically normal adult psittacines is rare (see Chapter 14, Radiology).

## Increased Susceptibility to Disease

Malnutrition reduces the effectiveness of the disease defense mechanisms, and disease increases nutrient requirements.[26] The optimal health of an animal is partially dependent on the interactions among the various bacterial species of the intestinal tract. Normal bacterial flora not only creates an environment that is nonsupportive of pathogens but is necessary for proper conversion, assimilation, and absorption of other nutrients. A disturbance in the normal flora may be due to antibiotic overuse,[15, 21, 27] starvation, diet change, stress, or vitamin deficiency. Without the appropriate amino acids, vitamins, minerals, and other nutrients available for the body defense system, the normal secretory IgA immunoglobulins would be absent, allowing the transient opportunists and pathogens to attach to the intestinal wall and proliferate[27] (see Chapter 24, Relationships of the Avian Immune Structure and Function to Infectious Diseases).

The earliest lesions of vitamin A deficiency, in particular, may involve a change in the mucous protection over the epithelial membranes. An increase in viscosity reduces the flow rate and epithelial resistance to invasion by pathogens. A laboratory report may indicate "overwhelming *E. coli* infection" or organ malfunction, but one often needs to look to the diet for a possible predisposing factor. It seems safe to assume that if visible epithelium and feathers do not appear healthy, functions of the liver, thyroid, intestine, kidneys, adrenals, and other organs, which have their own epithelium, may also be impaired.

## Decreased Response to Therapy of Disease

Treatment of chronically malnourished birds for a disease may be prolonged and may include long-term supplementation. If the condition is not a severe clinical illness, one might address the nutritional status first and possibly stimulate the immune system. Use of antibiotics alters the intestinal flora[21] and may, in some cases, depress the immune system further, creating a vulnerable environment for fungi and yeasts, requiring judicious use of antibiotics in these birds (see Chapter 27, What To Do Until A Diagnosis Is Made).

Treating the disease condition and correcting the diet will not immediately produce a dramatic change in the appearance of the adult bird. For example, pulling out abnormal feathers to induce new growth is useless in the early stages of the treatment, because until the nutritional status of the animal is improved (a process that may take months to years depending upon the degree of owner compliance), regenerating feathers will appear as unfit as those they replace. Young birds appear to respond more quickly to nutritional therapy.

## Avicultural Manifestations

Many aviculture failures may be attributed to inadequate nutrition. Although pairs of adults may apparently do well during the rest of the year, once the breeding season begins, aviaries with birds on marginal diets may begin to lose females to a multitude of disease organisms. Breeding hens have a significantly increased need for protein, lipids, calcium, vitamins A, B complex, and $D_3$, and trace minerals,[14, 21] which, if not provided, will prevent or inhibit egg laying or result in decreased hatchability of the young (at the very least)[22] or death of the hen due to increased susceptibility to disease. The lack of biotin, for example, may not be evident in the adult breeding birds but may be reflected in dead embryos or lesions in hatchlings.[6]

Generally malnourished parents may produce youngsters with poor immune transfer, and early deaths of hatchlings from a wide variety of causes are not uncommon.

Taylor's work with budgerigars showed that on unsupplemented seed diets, breeding pairs might be able to raise a couple of small clutches of offspring, at the expense of the female's own body tissue, but then reproduction would cease. He found that simply supplementing her diet with milk in a 1:1 ratio with drinking water would permit reproduction to resume.[20]

Until aviculturists are educated as to the nutritional implications of breeding, prize-winning show birds will continue to be the first to hatch in the season. Subsequent clutches often become smaller and less hardy, and "resting" the breeding pair after two clutches is still a common practice. Physiologists have maintained that increasing the amounts of protein, minerals, and vitamins to accommodate the increased needs of the female during this time can result in year-round production of those species not seasonally determined. Adequately nourished laying hens of exotic species are not expected to cease production from senility.[7]

Aviculturists have observed increased coprophagy in birds preparing for egg laying. Whether this is related to an instinctive attempt to deter potential predators or to provide for increased nutritional needs is unknown.

Soft-shelled eggs are relatively common in large psittacine species such as Amazons and macaws for which calcium supplementation has not been provided (Fig. 31–4), although the condition is rare in budgerigars and cockatiels, whose nutritional needs may be more adequately met. The calcium in bone meal, defluorinated phosphate, or gypsum (calcium sulfate) is adequate for growth but inferior to calcium carbonate for egg production[14] (see Chapter 2, Husbandry Practices).

Although genetic tendencies and lack of exercise have been implicated in egg binding, nutritional deficiencies also appear to influence the incidence of this condition. Egg binding is

**Figure 31–4.** Nutritional imbalances, including insufficient calcium in the diet of large psittacines, may result in soft- and thin-shelled eggs.

responsive to intramuscular injections of calcium and uterine contracting drugs and/or surgery; however, adjusting the diet may assist in providing the muscle tone and lubrication necessary for normal delivery of the egg.

## SPECIFIC NUTRITIONAL DISEASE SYNDROMES

### Mineral Malnutrition

More complex than is sometimes appreciated, mineral malnutrition generally involves deficiencies and/or imbalances of calcium and phosphorus, with deficiencies of vitamin D, specifically $D_3$ required by birds.[5, 8, 24] The seed diets provided to most of the granivorous avicultural species are usually adequate in phosphorus but deficient in calcium.[25] The calcium:phosphorus ratio utilized in the body and assumed to be necessary for dietary provision is 1.5:1.0. Ratios of 2.5:1.0 have not been detrimental, and laying hens may need as much as 3:1.[14] The skeletal diseases that are seen as evidence of mineral malnutrition (rickets in a growing bird and osteomalacia in adults) are manifestations of an inequality in these two components.

To maintain normal skeletal homeostasis in either the growing or mature skeleton, the rate of bone deposition should equal the rate of bone resorption. In a growing animal, the essential areas that need to be mineralized for normal skeletal growth are the growth cartilage matrix and any new bone intercellular material being laid down. Should fractures occur, the initial callous material is often cartilage and must be mineralized for ossification to proceed.

In the young bird with a growing skeleton, this lack of mineralization and impaired ossification result in restriction of growth (stunting), curving defects in long bones, the occurrence of pathologic fractures, and, subsequent to that, abnormal or impaired healing of those fractures (Fig. 31–5). With rapid skeletal growth, mineral deficiencies develop rapidly and are not uncommon in avicultural facilities.

Although clinically uncommon, osteomalacia occurs in older birds with mature skeletons. One sees essentially the same manifestations as rickets except for development of curved deformities. Osteomalacia results from more chronic mineral malnutrition, as the skeleton has reached full development.

In both young and mature birds, the parathyroid glands maintain a normal plasma calcium level at the expense of bone tissue, responding to a negative calcium balance by becoming somewhat hyperplastic and, even more significantly, by becoming hyperfunctional and producing increased amounts of parathyroid hormone. The effect of parathyroid hormone is to cause increased bone resorption and to promote the replacement of the resorbed bone by fibrous connective tissue, or "osteitis fibrosa," the classic bone lesion of hyperparathyroidism.

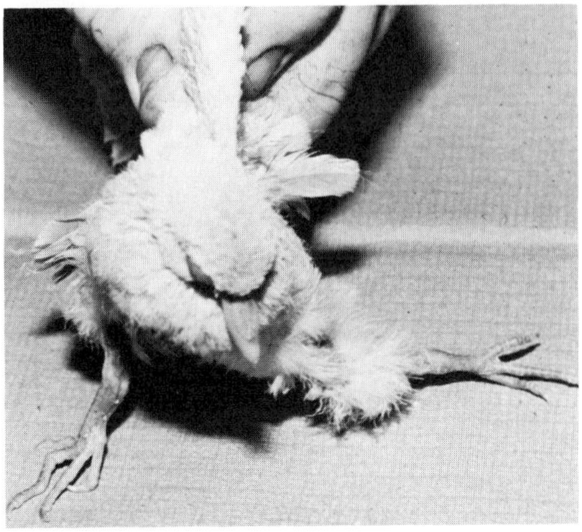

**Figure 31–5.** Hatched from parents maintained solely on white millet, this dove showed evidence of severe nutritional deficiencies and imbalances affecting developing bone.

Thickened cortices found on radiographic evaluation of birds with osteomalacia, although apparently more stout than normal bone, may actually be fragile fibrous connective tissue. One should be aware of the potential for fractures in these birds produced by traditional physical restraint methods.

Another component of the negative calcium balance is hypocalcemia, which can occur in young or mature birds and may present as hypocalcemic tetany (convulsions). These birds respond very well to subcutaneous or intravenous calcium therapy.

One of the common responses to the recognition of mineral/vitamin malnutrition is oversupplementation. Overtreating with calcium does not seem to produce any significant clinical syndromes; however, problems with excessive dosage with vitamin $D_3$ are fairly common. The results of hypervitaminosis D include mineralization of a variety of soft tissues, especially the kidney, and an increase in bone resorption, a paradoxic effect of excessive vitamin D treatment that may possibly also result in a fragile, easily broken skeleton.[23, 24]

## Vitamin A Deficiency

Based upon an evaluation of the diet, many birds would seem to be chronically vitamin A–deficient but may appear relatively normal as long as they are free of stress. These birds are usually also lacking in other vitamins, minerals, and protein. Clinical experience with some pet birds such as the Eclectus Parrot suggests that some species may have a particularly high requirement for vitamin A (Fig. 31–6).

Sources of vitamin A are plant materials, basically pro-vitamin A carotenoids from plants. The most well-characterized and prevalent is beta-carotene, found in deep green and orange vegetables, such as spinach, parsley, endive, yams, and carrots, and also in egg yolk.

Pro-vitamin A carotenoids are normally converted to one of the functional vitamin A forms—the vitamin A alcohol, aldehyde, or acid—when they reach the intestinal mucosa, if the intestinal mucosa is intact and functional. However, if there is some intestinal mucosal lesion, such as diffuse neoplastic infiltrate or perhaps diffuse granulomatous lesions such as with tuberculosis, threatening a major portion of the small intestine, this conversion will not occur.

The functions of vitamin A are classically characterized as being related to vision, skeletal development, and epithelial maintenance. One tends not to see, or at least not to appreciate, the significant effects of vitamin A deficiency on visual and skeletal development in pet birds.

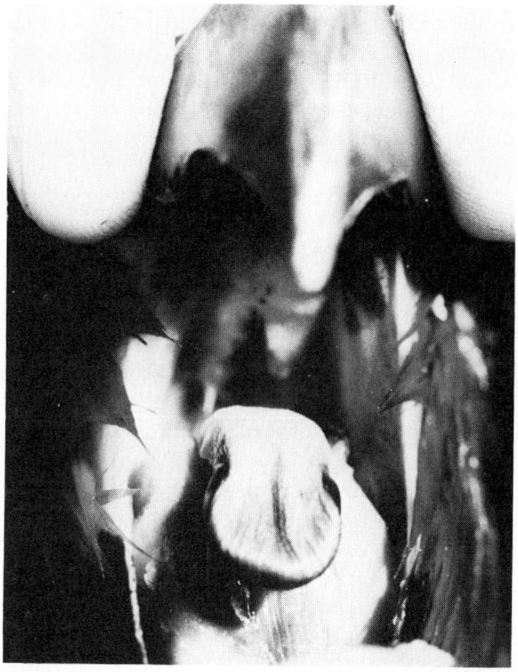

**Figure 31–6.** Oral lesions associated with vitamin A deficiency are frequently observed in Eclectus Parrots.

The epithelial effects of hypovitaminosis A depend on the kind of epithelial surfaces involved. With tissues that are normally squamous epithelium, the major effect is a marked thickening of the keratin layer on the epithelial surface, or hyperkeratosis. Nonsquamous epithelium, such as that of the upper respiratory tract, renal tubules, or collective ducts of the urinary tract, will change *to* squamous epithelium—squamous metaplasia—which is usually followed by hyperkeratotic changes, or increased build-up of keratin from this newly formed squamous epithelial surface.

Relatively early cases of vitamin A deficiency may manifest as oral mucous membranes that are beginning to undergo squamous metaplasia. Advanced oral lesions may involve glands that are entirely converted to squamous epithelium with keratin material. Although these are often called abscesses or pustules, if they are not secondarily infected (as evaluated by Gram's stain, cytology, culture), these are essentially keratin cysts. These also must be differentiated from pox lesions, and less commonly, candidiasis and trichomoniasis, which would not respond to vitamin A therapy.

Although these lesions of vitamin A deficiency are seen in psittacine birds, the absence of lesions in the oral cavity does not rule out vitamin A deficiency. Clinically, oral lesions are most commonly associated with "new" birds that have had recent exposure to other birds (see Chapter 7, Preliminary Evaluation of a Case). The primary author tends not to see these lesions in "old" birds that have had no exposure to other birds, even though these birds may have been maintained on a straight seed diet and are also presumed to be vitamin A–deficient.

Vitamin A deficiency is variable in its manifestations and can present as diverse lesions in birds. In addition to the oral cavity, the conjunctiva, nasal lacrimal duct, upper alimentary tract, and respiratory tract, including the nasal passage, sinuses, trachea, and syrinx, can be affected. Presenting signs may include severe dyspnea or respiratory embarrassment due to severe squamous metaplasia in the syrinx.

Swellings of the sinus that are nonresponsive to treatment with sinus washings or antibiotics, including *Mycoplasma* therapy, should suggest the possibility of vitamin A deficiency. The distended sinus may be entirely filled with laminated keratin.

Chronic cases may present with renal constipation or renal gout from squamous metaplasia and/or hyperkeratosis of tubules in the kidney, resulting in back-up of uric acid and urates.

Although vitamin A deficiency is certainly not the only cause of corns (local plantar hyperkeratosis on the feet), it should be considered in the differential diagnosis of the development of corns. Especially the large, heavy-bodied psittacines such as cockatoos and macaws, which have a greater amount of weight load for the surface area of the feet, may manifest this lesion of inadequate vitamin A nutrition. If the condition is severe and secondarily infected, bumblefoot may result (Fig. 31–7).

Reproductive failure is yet another result of the deprivation of this essential nutrient.

Once uncomplicated vitamin A deficiency is recognized, if the bird is not severely debilitated, it *should* respond to supplementation with vitamin A, often dramatically, within a period of three to seven days, restoring the liver concentration and normal blood levels of the nutrient. Clinically, this seldom occurs, however, as bacterial or fungal agents have often invaded the immunocompromised host; thus, additional diagnostic and treatment regimens are usually necessary for full recovery. Candidiasis is a common secondary infection.

Specific nutritional requirements of vitamin A for exotic bird species have not been determined. The diagnosis for deficiency is based primarily on the nutritional history, clinical signs, and response to therapy.

A successful treatment for a well-characterized vitamin A deficiency has been cod liver oil administered orally once daily for four or five days (1 drop for a budgerigar up to 8 to 10 drops for a macaw).[9] From Graham's point of view, cod liver oil is preferred over beta-carotene supplements because it does not require conversion; it does, however, require absorption. There have been reports of birds that needed to be treated with injectable vitamin A, and one would have to recognize that there are a few cases in which this is necessary. Treatment by injection is based on the suggestion that because vitamin A is necessary for epithelial integrity, its absence would prevent absorption. Graham has yet to see lesions of the intestinal mucosa impair the absorption of vitamin A to the extent that clinical manifestations of hypovitaminosis A were produced.[9] However, from the clinical standpoint of the time required to retrain the bird and owner, injections seem to be the fastest and most efficient route of initial treatment, followed by subsequent adjustment of the diet. Research has shown that heated fish meal contains a substance identified as gizzerosine,[14a, 18] which may be a contributing factor to gizzard erosion ulcerations in poultry. Gizzard erosion may also result from the effects of highly polyunsaturated fatty acids, such as those present in cod liver oil, if these fatty acids are not protected by an adequate dietary level of vitamin E. Similar side effects might be expected in pet birds from cod or other fish liver oils.

Known in mammals, hypervitaminosis A has not been well-characterized in pet and aviary birds; however, one possible significant effect of hypernutrition of vitamin A may be an anti–

**Figure 31–7.** Lack of sufficient vitamin A may predispose birds to corns and subsequent bumblefoot.

vitamin D effect on the growing and mature skeleton.

## Vitamin E Deficiency

Another fat-soluble vitamin, deficiencies of which are seen in cage and aviary birds, is vitamin E, alpha-tocopherol. Common sources of the nutrient are vegetable and seed oils, green leafy vegetables, and eggs or egg products. Wheat germ oil is probably the richest natural source of Vitamin E.[25] Vitamin E basically is an antioxidant, the major effect of which is to prevent the formation of peroxides through oxidation in cells and organelle membranes. Selenium, often functionally linked with tocopherol, has a synergistic effect in the process. It has been reported that very large doses of vitamin E are antagonistic to vitamin A instead of showing an oxidation-preventing and thus vitamin A–sparing effect.[11]

The effects of vitamin E deficiency or hypovitaminosis E are several, basically drawn on information from domestic poultry. Encephalomalacia, "crazy chick disease," is seen in cage and aviary birds from alpha-tocopherol deficiency. White muscle disease and muscular dystrophy have also been reported in these species,[21, 23] and the relationship between selenium and the sulfur-containing amino acids, particularly cystine, has been well-documented in domestic poultry, although less so in cage and aviary birds. Elevated levels of serum creatine phosphokinase may suggest nutritional myopathy from vitamin E/selenium deficiency.[8]

Cockatiels and other species with varying degrees of paralysis have responded clinically to selenium/vitamin E supplementation (Fig. 31–8). In one case in which an attempt at oral supplementation failed, the bird continues to respond to periodic selenium/vitamin E injections. The response to the injection is so dramatic in this case that improvement is evident in the bird before the client leaves the office. It has been reported that these clinical signs may be primarily related to giardiasis,[17] which causes secondary malabsorption of vitamins.

The presence of rancid fats destroys vitamin E, and heavy supplementation of the diet with oily substances interferes with its absorption.[23]

**Figure 31–8.** Some cases of localized paralysis in cockatiels, seen affecting the wings in this bird, have been clinically responsive to vitamin E/selenium therapy. Giardiasis may be a predisposing factor to vitamin malabsorption.

## Vitamin K Deficiency

Vitamin K is normally synthesized by bacterial action in the intestinal tract in adult birds[14] but may need to be provided in the form of fresh greens to growing birds. Changes in the intestinal flora resulting from indiscriminate use of antibiotics[21] or inappropriate amounts of vitamin A[11] may interfere with the availability of vitamin K and lead to a deficiency. Excessive hemorrhage may be seen in viral and other diseases affecting the liver. Not pathognomonic for vitamin K deficiency, prolonged clotting time may sometimes respond clinically to injections of vitamin $K_1$. Some evaluation of clotting time is recommended prior to surgical procedures (see Chapter 44, Evaluation and Support of the Surgical Patient).

## Iodine Deficiency

"Thyroid dysplasia" implies that the basic defect is abnormal formation of the thyroid gland; however, the lesion is in fact hyperplasia of the thyroid gland as a result of iodine deficiency. With iodine deficiency, there is decreased synthesis and secretion of functional thyroid hormone, a corresponding increase in TSH secretion by the adenohypophysis, and TSH stimulation of the gland, which then becomes hyperplastic.

The physical effect of the increased size of the thyroid glands, located just within the thoracic inlet, has been reported to cause some pressure and displacement effects on the trachea and syrinx, resulting in some degree of dyspnea or characteristic change in respiratory sounds. Although there may be an audible "click" or rasp on either inhalation, exhalation, or both, there may be some question as to the cause. Recent surgical removal of an extremely large thyroid from a budgerigar did not change the characteristic respiratory sounds on recovery.

The iodine deficiency, beyond simply causing hyperplastic goiter, produces some degree of hypothyroidism. The reader is referred to Chapter 42, miscellaneous Diseases.

## PREVENTION OF NUTRITIONAL DISEASES

Until such time as definitive nutritional requirements for the variety of species commonly kept as pets are known, prevention of nutritional disease appears to be accomplished by the addition of some supplemental protein, fresh dark green or dark yellow vegetables, fruits, whole grain products, minerals, and salt to the basic fresh seed diet or through the use of a high-quality formulated pelleted diet in addition to fresh vegetables and fruits.

## REFERENCES

1. Adams, C. R.: The effect of environmental conditions on the stability of vitamins in feeds. In The Effect of Processing on the Nutrition of Feeds. Washington, DC, National Academy of Sciences, 1973.
1a. Austic, R. E., and Scott, M. L.: Nutritional deficiency diseases. In Hofstad, M. S., et al. (eds.): Diseases of Poultry, 8th ed. Ames, IA, Iowa State University Press, 1984, pp. 38–64.
2. Best, C. H., and Taylor, N. B.: The Physiological Basis of Medical Practice. 7th ed. Baltimore, Williams & Wilkins Co., 1961.
3. Braunlich, K.: Leg Lesions in Poultry. Roche Information Service, F. Hoffman-LaRoche & Co., AG, Basel, Switzerland, 1972.
4. Coffin, D. L.: Angell Memorial Parakeet and Parrot Book. Angell Memorial Animal Hospital. Boston, 1953.
5. Fowler, M. E.: Metabolic bone disease. In Fowler, M. E. (ed.): Zoo and Wild Animal Medicine. Philadelphia, W. B. Saunders Company, 1978, pp. 53–76.
6. Frye, T. M.: Biotin. Poultry Tribune, April, 1981.
7. Gee, G. (Avian physiologist, Patuxent Wildlife Research Institute): Personal communication.
8. Graham, D. L., and Halliwell, W. H.: Malnutrition in birds of prey. In Fowler, M. E. (ed.): Zoo and Wild Animal Medicine. Philadelphia, W. B. Saunders Company, 1978, pp. 236–242.
9. Graham, D. L.: Pathophysiology of nutritional diseases in cage and aviary birds. Material presented at Annual Meeting of Association of Avian Veterinarians, San Diego, 1983.
10. Hamilton, P. B.: In Schlessinger, D. (ed.): Microbiology. Washington, DC, American Society for Microbiology, 1975.
11. Klaui, H.: Inactivation of vitamins. Proc. Nutr. Soc., 38:135, 1979.
12. Kurnick, A. A., Eoff, H. J., et al.: Vitamin dynamics. Parts I–III. National Feed Ingredients Association, 1978.
13. McDaniel, L. D., and Roudybush, T. E.: Cockatiel experiments. In Proceedings of the 31st Western Poultry Disease Conference, University of California, Davis, 1982.
14. Morrison, F. B.: Feeds and Feeding. Clinton, IA, The Morrison Publishing Co., 1959.
14a. Okazaki, T., et al.: Gizzerosine, a new toxic substance in fish meal, causes severe gizzard erosion in chicks. Agric. Biol. Chem., 47(12):29–49, 1983.
15. Perry, S. C., and Zimmerman, C. R.: Optimum vitamin nutrition. An. Nutr. Health, April-May, 1979.
16. Perry, S. C.: Vitamin allowances for poultry and swine. In Proceedings of the Roche Vitamin Nutrition Update Meeting, Hot Springs, AR, 1978.
17. Roertgen, K.: Avian giardiasis. In Proceedings of the California 95th Annual Science Seminar, Oakland, 1983.
18. Scott, M. L.: Gizzard erosion. Animal Health and Nutrition, Sept., 1985, pp. 22–29.
19. Tagwerker, F.: Die Rolle des Vitamins C in der Physiologie und Ernahrung des Geflugels (The role of vitamin C in the physiology and feeding of poultry). Arch. Geflugelkunde, 24:160, 1960.
20. Taylor, T. G.: The nutrient requirements of budgerigars. Small Anim. Pract., 6:11–13.
21. Tollefson, C. I.: Nutrition. In Petrak, M. L. (ed): Diseases of Cage and Aviary Birds. 2nd ed. Philadelphia, Lea & Febiger, 1982.
22. Toone, C. K.: Causes of embryonic malformations and mortality. Proceedings of the American Association of Zoo Veterinarians, Tampa, 1983.
23. Wallach, J. D.: Nutritional problems in zoos. Proc. Cornell Nutrition Conference for Feed Manufacturers, Buffalo, NY, 1971.
24. Wallach, J. D., and Flieg, G. M.: Nutritional secondary hyperparathyroidism in captive psittacine birds. J. Am. Vet. Med. Assoc., 151:7, 1967.
25. Watt, B. K., and Merrill, A. L. (eds.): Composition of Foods. Agriculture Handbook No. 8, Washington, DC, U.S. Department of Agriculture, 1963.
26. Wilgus, H. S.: Disease, nutrition—interaction. Poultry Sci., 59:4, 1980.
27. Woolcock, J. B.: Bacterial Infection and Immunity in Domestic Animals. Amsterdam, Elsevier Scientific Publishing Company, 1979.

### ACKNOWLEDGMENTS

The authors wish to acknowledge David L. Graham, D.V.M., Ph.D., for providing much of the information for this chapter from his unpublished presentation at the 1983 Annual Meeting of the Association of Avian Veterinarians, San Diego, California. Appreciation is also extended to Ted Frye, Ph.D., from Hoffman-LaRoche.

# *Chapter 32*

# VIRAL DISEASES

HELGA GERLACH

## INTRODUCTION TO INFECTIOUS DISEASES

Conclusions drawn from experiences with poultry and many other bird species have shown that diseases are frequently not induced by a single cause but are produced by the interaction of several supporting factors in the environment. Noninfectious stressors may be as important as microorganisms. This totality approach is also valid if primary pathogenic organisms are involved. Primary pathogens are recognized by their capacity to cause a specific disease that can be reproduced experimentally. This is in contrast to secondary or opportunistic invaders, which do not necessarily fulfill Koch's fourth postulate. The treatment of such agents does not always restore health.

Diseases of multiple etiology are difficult to analyze because primary factors such as nutritional and environmental deficiencies as well as infections with viruses or fastidious microorganisms are frequently not recognizable. Secondary invaders (viruses, bacteria, fungi, parasites) are allowed to penetrate the host's defense system and, according to their pathogenicity, complicate the primary disease process or induce a disease of their own. A disease of multiple etiology is a prolonged ailment with changes of the bacterial component and infections with fungi and various parasites. Fungal infections (with the exception of *Cryptococcus*) are generally secondary infections. Among the parasites, mucosal flagellates are frequently opportunistic agents. Secondary invaders are, as a rule, either ubiquitous or regularly carried in small numbers on mucosal surfaces (frequently in the nasopharynx). Penetration of bacteria and fungi as well as some flagellates through the mucosa and multiplication beyond lead to septicemia and may actually cause the death of the host. However, restoration of health normally requires the correction of the triggering factors involved.

A prerequisite for healthy birds in a reasonably suitable environment is the presence of a normal autochthonous (or indigenous) intestinal and nasal flora. For psittaciforme birds and many other bird groups, Enterobacteriaceae should be absent, except probably for the genus *Enterobacter* and a few other transient organisms, which are always to be expected, especially with ubiquitous bacterial species. Colonization with Enterobacteriaceae indicates in such birds a disturbance of the normal balance between host and environment, which might be the first step for a disease of multiple etiology.

## VIRAL DISEASES

Viral diseases are still a therapeutic problem. Nonspecific treatments include use of vitamin C and inhibition of secondary bacterial or fungal invaders by respective chemotherapeutics. Interferon inducers have not been used in psittaciformes, but paramunity inducers (PIND) like PIND ORF have demonstrated encouraging results.[38] Specific prophylaxis in the form of vaccination for psittaciformes is often not available or not explored at all (see Chapter 27, What To Do Until A Diagnosis Is Made).

The traffic with pet birds, including imports from foreign countries, and the establishment of mixed aviaries or of large breeding farms frequently cause a blending of different populations carrying different microorganisms or antibodies against them. But if one population has a latent virus infection and the other does not, a disease outbreak can occur. The birds with the latent infection produce their own protective antibodies that inhibit the disease, but the virus may be able to propagate at a low level in spite of this antibody and may be excreted via feces, urine, respiratory secretions, and even feather quills. Such virus can then infect members of the other population and, because they have no immunity whatsoever, rapid replication may occur and cause disease. Another mechanism is that the group with the persistent infection has suffered an environmental change that has led to a decrease in resistance. This allows a virus to replicate more rapidly, and both groups of birds may become sick. The use of fertile eggs to mix birds from two distant populations cannot always overcome this disease problem. The female transfers to the eggs not only antibodies for the offspring to better withstand the first weeks of life, but prior to antibody production during viremia, a variety of viruses can be hematogeneously transported into the egg follicle. The contaminated or infected newly hatched chick provides a potent source of infection for its unprotected hatchmates.

In order to prevent disastrous disease outbreaks, knowledge about the occurrence of particular virus infections is necessary, and specific serologic testing or virus culture is recommended as part of a control program. Such knowledge is still greatly insufficient, and the methods of diagnosis (by culture or serologic means) are far from being optimal. International cooperation is mandatory to make further progress. In this chapter important viral diseases in pet birds are discussed.

## POX

Avian pox are diseases caused by a variety of viruses of the genus *Avipoxvirus*, a member of the poxviridae. This DNA virus group has the largest viral particles (250 to 300 nm) so far known. They can be recognized histologically by certain staining procedures. *Avipoxvirus* induces intracytoplasmatic, lipophilic inclusion bodies called Bollinger bodies in the epithelium of the integumentum. Their appearance is pathognomonic. This may be one reason that the disease is recognized in so many species of birds. In a survey, Roehrer[39] cited literature for 57 bird species. Meanwhile, new species have been added to that list. It is interesting that psittaciformes are not included in these early reports. One reason may be that the disease in parrots varies considerably from that in other orders, and frequently a diagnosis cannot be made by histology alone. Isolates from different bird species have been classified into 17 provisional species. From many diseased birds with diagnosed pox infections, isolations have not been attempted. Therefore, since biologic and serologic properties are undetermined, the additional numbers of species in the environment are unknown.

For species differentiation, the following features and tests are used: host spectrum (by intramuscular, intravenous, and intracerebral injection); plaque morphology (of primary, not egg-adapted passages); electron microscopy; thermostability; optimal propagation temperature; and serology.

Generally, *Avipoxvirus* is relatively resistant to factors such as desiccation, humidity, and light, but steam kills the virus easily. Virus surrounded by "crusts" can be viable for years. The virus can survive in infected soil for approximately 1½ years. The chemoresistance of *Avipoxvirus* is also remarkable. However, 1 per cent KOH, 2 per cent NaOH, and 5 per cent phenol are effective against it, particularly at room temperature.

### *Transmission*

The mode of transmission is governed by virus carriers and contaminated environment. In many countries mosquitoes are the main vectors. The disease shows a seasonal incidence in most parts of the world, occurring during late summer and autumn. In some countries this coincides with the peak of the mosquito season. Birds hatched in the spring prior to the vector season are most susceptible. The virus can survive in the mosquitoes' salivary glands (depending on the virus or mosquito species) for two to eight weeks. The spreading of the virus within a given bird population is triggered by close contact between birds and the usual establishment of the "peck order," which can lead to tiny traumatic lesions. During latent infection or after recovery from clinical disease, the virus is intermittently shed via the feces or the skin and feather quills.

### *Pathogenesis*

*Avipoxvirus* is not capable of penetrating intact epithelium. Traumatic lesions, including mosquito bites, are necessary to overcome the epithelial barrier. In the body, the virus spreads via the blood and also locally to neighboring tissues. This mode of spread characterizes avian pox as a generalizing, so-called cyclic virus disease. There are two viremic stages. The first leads from the portal of entry (with little virus propagation) to liver and bone marrow (organs of primary affinity). The quantity of propagation in these organs governs the extent of the secondary viremia. During this secondary viremia the typical lesions develop. Nonpathogenic or slightly pathogenic strains do not replicate enough to produce the secondary viremia. The same is true for vaccine strains, which develop only the primary viremia and almost no virus propagation in liver and bone marrow. These strains cause a local abortive disease confined to the inoculation site.

### Incubation

Regarding the incubation period, there is insufficient information collected for all 17 *Avipoxvirus* species so far identified. However, following natural infection, incubation times are expected to vary between one and two weeks.

### Clinical Disease and Pathology

The following survey gives a general view of avian pox (Table 32–1) and concentrates on the psittaciformes.

The clinical expression of avian pox differs with the virulence of the virus strain, the mode of transmission, and the susceptibility of the host. Different forms of the disease are distinguished:

1. **Cutaneous Form.** This is generally the most common form, but not always in the large parrots. The lesions in the form of papules appear mainly on the unfeathered skin around the eyes, beak, and nares and from the tarsometatarsus downward. Pox lesions on feathered skin are rather rare. The papules, during the course of the disease, change color from yellowish to dark brown, and the lesions are finally (after weeks) spontaneously desquamated, usually without leaving a scar. Bacterial and fungal superinfection can alter the typical lesions (Fig. 32–1).

2. **Diphtheroid Form** (also called "wet pox"). Lesions similar to those of the cutaneous form develop on the mucosa of the beak cavity, tongue, pharynx, and larynx. In accordance with the mucosa being the main reactive tissue,

**Table 32–1.** AVIAN POX AND SOME OF ITS INTERCONNECTIONS

| Virus Species | Serologic Cross-Reactions | Peculiarities of Pathology or Diagnosis | Vaccine |
|---|---|---|---|
| Pigeonpox | → Fowlpox<br>← Falconpox | Cutaneous and diphtheroid forms; occasionally cutaneous tumors; frequently brown to black in color (melanin) | Yes, homologue |
| Fowlpox | → Turkeypox<br>⇌ Waterfowlpox*<br>→ Quailpox<br>⇌ Falconpox<br>← Pigeonpox | Cutaneous and diphtheroid forms | Yes, homologue and heterologue = Pigeonpox |
| Turkeypox | ← Fowlpox<br>⇌ Falconpox† | Cutaneous and diphtheroid forms | Yes, homologue and heterologue = Fowlpox |
| Canarypox | None | Septicemic form predominant, cutaneous form rare; approximately 10% of survivors or latently infected birds develop "adenoma-like" tumorous tissues within the lungs (only histologically demonstrated; clinically severe dyspnea); cutaneous tumors rare (as a rule, no melanin) | Yes, homologue |
| Quailpox (*Coturnix japonica*) | ← Fowlpox | Cutaneous and diphtheroid forms | Yes, heterologue = Fowlpox |
| Waterfowlpox | ⇌ Fowlpox* | Cutaneous form predominant, occasionally diphtheroid enteritis described | Yes, heterologue = Fowlpox |
| Falconpox (true Falcons, Kestrels, Buzzards) | ⇌ Fowlpox<br>⇌ Turkeypox† | Only cutaneous form; wide host range with cutaneous infection in chicken, turkey, and pigeons (nonsymptomatic in pigeons) | Yes, heterologue = Turkeypox |
| Psittaciforme birds (three independent, specific taxons) | | | |
| a. Agapornispox | None | a. cutaneous and diphtheroid forms; cutaneous form regularly without typical skin eruptions; skin looks dehydrated and brownish discolored | a. No |
| b. Psittacinepox (Amazonapox) | None | b. Amazons: clinically main symptom is coryza. General: Severe cutaneous and diphtheroid | b. No |

**Figure 32–1.** Bacterial and fungal elements may further complicate pox lesions. (From Gerlach, H.: Virus diseases in pet birds. Vet. Clin. North Am. [Small Anim. Pract.], 14:299–316, 1984.)

fibrinous exudate is excreted that later becomes grey to brown and caseous. Confluence of multiple foci can form large areas of altered mucosal surface. These exudates cling closely to the surface and cannot be removed without subsequent bleeding or tissue loss, because the whole mucosa is involved in the disease process. In severe cases, the birds cannot swallow feed. They may suffer from dyspnea or even asphyxia if the larynx is closed by exudates.

Cutaneous and diphtheroid forms can be mixed in one bird as well as mixed in a population. Both forms can also be seen in connection with the septicemic form.

3. SEPTICEMIC FORM. This form is characterized by an acute onset of general symptoms, including ruffled plumage, somnolence, cyanosis, and loss of appetite. Some hours to three days later the affected birds are dead. Cutaneous lesions normally are rare or not present at all. Diagnosis by pathology or histopathology might be difficult. The classic exam-

**Table 32–1.** AVIAN POX AND SOME OF ITS INTERCONNECTIONS *Continued*

| Virus Species | Serologic Cross-Reactions | Peculiarities of Pathology or Diagnosis | Vaccine |
|---|---|---|---|
| Host spectrum b: South American parrots and parakeets | | forms; some develop septicemic form; necrosis of myocardium and diphtheroid enteritis frequent, particularly in macaws. Diagnosis: difficult in chicken embryos, adaptation by several passages, no plaque development on CAM, but necrosis of myocardium; better replication in budgerigar embryofibroblast cells. Best virus source: feather quills | |
| c. Budgerigarpox | None | c. No cutaneous lesions nor mortality; experimentally 3-4 DPI dyspnea | c. No |
| Woodpeckerpox (Picidae) | None | Cutaneous form predominant; no replication in chicken embryos; diagnosis to be established only by histology (Bollinger bodies) | No |
| Cranepox (Gruidae) | None | No specific details known } All show cutaneous form | No |
| Shorebirdpox (Charadriiformes) | None | No specific details known } All show cutaneous form | No |
| Pelicanpox (Pelecaniformes) | None | No specific details known } All show cutaneous form | No |
| Silvereyepox (Genus *Zosterops*) | None | Cutaneous form | No |
| House-Sparrowpox *(Passer domesticus)* | None | Cutaneous form | No |
| Juncopox *(J. hyemalis)* | Not completely tested | Intracytoplasmatic and intranuclear inclusion bodies, cutaneous form | No |
| Penguinpox (Spheniscidae) | Probably no specific taxon | No specific details known | No |
| Starlingpox *(Sturnus vulgaris)* | Probably no specific taxon | Cutaneous form, no other details known | No |

*Full protection (probably same virus)
†Full protection

ple is the lung form in canaries. In the psittaciformes, in particular in the genera *Amazona* and *Ara*, *Avipoxvirus* can cause diphtheroid enteritis and/or necrosis of the myocardium, with or without cutaneous or diphtheroid forms. From the clinical point of view, the same rather nonspecific symptoms are seen as in the septicemic form in the canary.

**4. CORYZA.** Infection with *Avipoxvirus* can cause coryza, particularly in the genus *Amazona*. Serous discharge is seen initially, but mucoid or fibrinoid exudates develop later on, mainly because of bacterial or fungal secondary infection. In severe cases conjunctivitis with closed eyelids occurs.

**5. TUMORS.** *Avipoxvirus* has, like other members of the poxviridae, oncogenic properties, and survivors can develop tumors. Most tumors are located within the skin, but in the canary the gas-exchanging tissue is replaced by an adenoma-like neoplasm. Columbiformes seem to be more susceptible to skin tumors than are other orders. Tumors of the pharyngeal mucosa have been seen in Bobwhite Quail (*Colinus virginialis*).

In Table 32–1, details of the host range, the common clinical picture, and the immunologic relations of 17 different types of *Avipoxvirus* are given. In addition to information given in Table 32–1, the Quailpox virus, as defined in the table, comprises only strains from the Japanese Quail (*Coturnix japonica*). These strains are of low virulence and are serologically related to Fowlpox virus. It is likely that the disease seen in Bobwhite Quail is caused by a different type of virus. The disease in these birds is frequently severe, showing diphtheroid forms.

Three distinct virus types have been isolated from psittaciformes so far, i.e., Agapornispox virus,[24] Psittacinepox virus (= Amazonapox virus),[49] and Budgerigarpox virus. Agapornispox virus has been isolated only from *Agapornis;* however, it is not quite clear whether or not *Agapornis* may become infected with the Psittacinepox virus.* In experiments with Agapornispox virus, it was not possible to cause disease in a variety of other psittaciformes.

The clinical picture in *Agapornis* is dominated by the cutaneous form, although "wet pox" can occur (Fig. 32–2). The skin lesions frequently do not develop the typical eruptions, and the altered skin looks dehydrated and be-

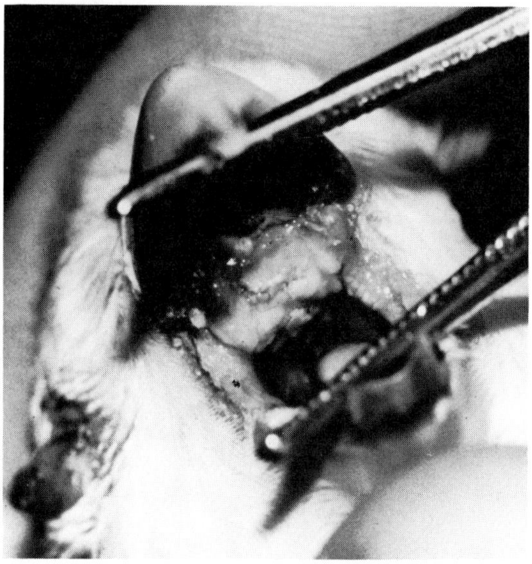

**Figure 32–2.** Substantial amounts of fibrinous lesions are seen in the mouth of a lovebird following infection with pox virus. (Courtesy of Lorraine Karpinski.)

comes brownish in color. Eye lesions develop frequently. They begin as a conjunctivitis with watery exudate and heavy vascularization. The skin of the lower lid and of the facial angular palpebra displays yellow-brown discoloration and palpable induration. Later the exudate becomes mucous or fibrinous (secondary infections with bacteria or fungi). Blepharosynechia can occur, caused by drying crusty exudates. Morbidity and mortality can be high (up to 75 per cent) depending on the virulence of the strain.

The Psittacinepox virus has a wide host range among South American parrots and parakeets. Most susceptible to severe disease with septicemic forms, diphtheroid enteritis, and myocardial necrosis are the genera *Amazona* and *Ara*. Cutaneous and diphtheroid forms are rare. In the genus *Amazona*, coryza is frequently the dominating clinical symptom, but postmortem examination reveals enteric and heart lesions as well. In addition, eye lesions similar to the ones described in *Agapornis* can develop. However, the dry and discolored skin around the eyes and on the eyelids is rare in Amazons.* Dry scurfiness at the marginal eyelid is often the first sign. The lid cleft is closed later by exudates or crusts and sometimes can be opened only by blepharotomy. In such cases secondary infection can cause keratitis and following ulceration and perforation may lead to panophthalmia and finally ophthalmophthisis. So far, there is no

*Editors' Note: Pox virus isolated from psittacine birds originating in Argentina (presumably Amazons) was used as a successful vaccine and challenge in trials with lovebirds.[46a]

*Editors' Note: This condition is frequently seen clinically in Blue-fronted Amazons infected with pox.

reason to believe that a different virus is involved; only the virulence is at variance.

The Budgerigarpox virus appears to be apathogenic. The virus has been found, so far, only in the budgerigar. Experimentally, dyspnea occurring three to four days after infection is the only clinical feature. Neither cutaneous lesions nor death occurs. Interestingly, this virus has been isolated from cases of French molt, and a connection with this disease has not yet been excluded.

### *Diagnosis*

Diagnosis is normally made by histology, using altered parts of the skin and pharyngeal mucosa. Biopsies can be taken from live birds for this purpose. The occurrence of Bollinger bodies is pathognomonic.

The septicemic form as well as the coryzal form can, as a rule, be diagnosed only by culture. Whole carcasses or feather quills (Table 32–1) should be sent for laboratory diagnosis. The cutaneous tumors can be diagnosed histologically as pox tumors by their palisade-like arrangement of the epithelial cords and the presence of Bollinger bodies. The cultural examination of feces, particularly in psittaciforme birds, might help to detect clinically healthy carriers; however, since the virus is excreted irregularly, several cultures might be necessary.

For culture, embryonated chicken eggs are especially suitable because of the typical chorioallantoic membrane (CAM) lesions that display Bollinger bodies histologically. Several passages for adaptation of strains from psittaciformes may be necessary. With strains from this group of birds, CAM lesions may not be prominent. The embryo develops liver necrosis and, from the fourth or fifth passage onward, necrosis of the myocardium.

### *Differential Diagnosis*

As differential diagnosis for the cutaneous form, multiple encrusted traumata and infections with *Trichophyton* spp. or *Knemidocoptes* spp. (rare) should be considered.

For the diphtheroid form the rule-out list includes:

1. Trichomoniasis (the organisms are frequently on the mucosa in cases of pox).
2. Fungal infections such as candidiasis and aspergillosis (they can also be secondary infections).
3. Vitamin A deficiency.
4. In pigeons only, infection with Pigeonherpes virus.
5. In Amazons only, infection with the Amazon-tracheitis virus (*Herpesvirus*).

### *Treatment*

There is no specific therapy against avian pox. But some nonspecific treatments can be attempted and have proved to be beneficial on occasion.

1. Keep the birds under conditions as optimal as possible.
2. Administer vitamin A (2000 IU/kg body weight) and vitamin C (50 mg/kg body weight) via feed or drinking water or intramuscularly.
3. Antibiotics can be given against secondary bacterial infections according to sensitivity testing. Do not use antibiotics empirically because one of the common side effects is immunosuppression. Local application of chemotherapeutics can be beneficial, especially in cases of eye lesions.
4. For single birds and small groups that are easily caught, a mixture of tincture of iodine and glycerol, 3:2, can be applied to the cutaneous or mucosal pox lesions once or twice a day. Bacterial and fungal infections of the lesions are supposed to be inhibited by this treatment. Do not apply directly to the eye. Use a good eye ointment to help keep the eyelids lubricated (see also Chapter 19, Ophthalmology).

### *Control*

Recovery from a clinical disease usually produces immunity that lasts about eight months. This immunity is dependent on cellular or local mechanisms and produces only small amounts of humoral antibodies. Routine serologic methods are, therefore, not suitable for the detection of birds that are carriers or have had contact with the virus. Whether or not there is an age resistance in psittaciformes as in poultry needs to be examined.

Besides hygienic methods, vaccination is the best tool for control. But no specific vaccines for either Agapornispox or Psittacinepox are available. Akrae[1] demonstrated distinct differences between the Psittacinepox, Agapornispox, and Canarypox viruses as well as isolates from poultry, pigeons, and falcons. Kraft and Teufel[24] proved the independence of the Agapornispox virus. Therefore, there is no reason to try heterologous vaccines, which is not en-

**Table 32–2.** SAMPLE SIZES FOR DETERMINATION OF 90 PER CENT ± 5 PER CENT IMMUNITY FOLLOWING CUTANEOUS VACCINATION*

| Negative Reactors | Total Number of Observations Required |
|---|---|
| 0 | 36 |
| 1 | 54 |
| 2 | 70 |
| 3 | 84 |
| 4 | 98 |
| 5 | 111 |
| 6 | 125 |

If more than 6 negative reactors are found within the first 125 observations, the flock is not satisfactorily protected.

*Modified from Siegmann, O.: Kompendium der Gefluegelkrankheiten. 4th ed. Hannover, Verlag M. H. Schaper, 1983.

couraged because of the high capacity of *Avipoxvirus* for recombination.

Canaries and closely related birds of the genus *Serinus* can be immunized, but only healthy birds should be given the vaccine. In a flock already in the incubation period, field and vaccine virus may recombine to cause disease of the whole flock. Immunity following vaccination lasts only three to six months.

Following cutaneous vaccination (wing-web, feather follicle), typical pox lesions develop locally. They are correlated with immunity, and vaccinated birds should be inspected at nine or ten days following vaccination. For statistical reasons in flocks of more than 200, the sample sizes given in Table 32–2 can be used to determine 90 per cent ± 5 per cent immunity.

## HERPESVIRUS

Herpesviridae contain double-stranded DNA and display diameters between 120 and 220 nm according to variations in the envelope. Replication takes place in the nucleus of susceptible cells, but the envelope is added during the cytoplasmal passage.

Avian strains known so far commonly develop intranuclear inclusion bodies of Cowdry type A. Some stimulate lymphocytic or plasma cell proliferations within the host organism, and all can induce long-persistent infections (weeks, months, perhaps lifetime) with periods of recrudescence. They differ in their affinity for the target cell, some being excreted cell-free and some remaining cell-associated. Regarding lesions of the organs, lytic/neoplastic, hemorrhagic, and necrotizing reactions can be recognized.

The *Herpesvirus* group is considered a phylogenetically old group. As with other "old" microorganisms, good adaptation to the natural host (narrow spectrum) is displayed, resulting in low pathogenicity and latent infections, often for life. However, high virulence can occur with infections in other than the natural hosts and in birds suffering from other diseases, environmental stressors, or hormonal changes. In addition, the virulence of virus strains varies. Thirteen different avian *Herpesvirus* types are known (Table 32–3). Host spectrum and pathogenicity of these types are still not satisfactorily documented. Details for decontamination of all these avian types are also not available. However, as typical members of the herpesviridae and therefore having a lipid-containing envelope, they are supposed to be sensitive to most disinfectants in common use (i.e., phenol, cresol, pH under 3 and over 11). Heat resistance is not high (10 to 15 minutes at 56°C).

### *Transmission*

The mode of transmission of strains from many pet bird species is not well-studied. Nasal and oral routes are possible. Vertical transmission seems to be rare (except with Budgerigar-herpes virus). In a small study, York and York[50] could show that susceptible budgerigars infected experimentally with Pacheco's virus shed the virus within the feces 48 hours after infection. Birds used as contacts in the same cages shed the virus in the feces 48 hours later at the same time as "aerogene" controls across the corridor. In birds becoming clinically sick, the fecal virus concentration reached up to $10^6$ to $10^7$ per gm feces. The same was found shortly before death. Birds without clinical symptoms have shed the virus in the feces for approximately three weeks. In dead birds virus concentration was up to $10^6$/gm in lung tissue, whereas in the liver the virus titer was $10^7$/gm. The high lung titers may indicate that the respiratory tract is an important portal of entry.

### *Pathogenesis*

Virulent *Herpesvirus* causes a viremia and, in most of the birds discussed here, necrotizing lesions in the large parenchymatous organs. The distribution of the virus by cell-to-cell transfer can be accomplished by constriction of a "pocket" of cytoplasm containing infective particles and by uptake of these "pockets" into cytoplasmal pseudopods of noninfected cells. This process allows for virus particles to be transferred without contact with the blood (specific and nonspecific defense mechanisms) and helps explain why *Herpesvirus* infections can persist for long periods of time even in the presence of humoral antibodies.

### *Incubation*

The incubation periods following natural infections are only partially known. Once a disease outbreak has started, the high concentration of virus in the environment causes a rapid spread and acute disease three to seven days after the onset. These observations are comparable to data gained with experimental exposure, which repeatedly leads to the first clinical signs within 48 hours.

## Herpesvirus in the Psittaciforme Host

The psittaciformes harbor three distinct, serologically unrelated *Herpesvirus* types (i.e., Pacheco's virus, Amazon-tracheitis virus, and Budgerigarherpes virus). Birds of all age groups can become clinically sick, and no seasonal peaks of disease are known. The most important is Pacheco's virus and a variety of strains causing similar diseases. However, only a few of these isolates have been tested serologically. Lowenstine[26] pointed out that *Herpesvirus* from a variety of psittaciformes can show morphologic differences. She discusses the possibility of the presence of several serotypes specific for an individual host range. To press her point, the author describes three cases:

1. A proliferative tracheobronchitis in *Neophema* species, observed clinically with respiratory signs, sometimes in conjunction with CNS signs such as stargazing, torticollis, and rolling.
2. A Double Yellow-headed Amazon (*Amazona ochrocephala*) with an acute fatal illness, which displayed a thickened parakeratotic lining of the crop, multiple foci of pancreatic acinar cell degeneration, and chronic hepatitis at necropsy.
3. A Moluccan and a Lesser Sulfur-crested Cockatoo (*Cacatua moluccensis* and *C. sulphurea*) with wartlike growths on their feet and legs, primarily on the plantar surface, which appeared histologically to be squamous papillomata (Fig. 32–3).

Inclusion bodies consistent with *Herpesvirus* infections have been found intranuclearly in some cells in all three cases, and electronmicroscopy revealed *Herpesvirus* particles. However, in the *Neophema* case the typical particles failed to grow *in vitro*. The particles from the Amazon and cockatoo cases consisted of virions with spiked outer membranes (unlike Pacheco's virus or ILT virus), suggesting differences of the virus. No attempts were made to culture a virus from the Amazon or the cockatoos.

## Pacheco's Disease

The disease was first described by Pacheco and Bier[36] in parrots from Brazil that died in the course of eight days without having shown clinical signs other than somnolence and ruffled plumage. Since then, many papers have been published on Pacheco's disease and *Herpesvirus* related to it. Kitzing[23] presented a survey. Naturally susceptible hosts of Pacheco's virus "sensu strictu" include at least the following bird groups: macaws (*Ara*), Amazons (*Amazona*), conures (*Aratinga* and *Cyanoliseus*), African Grey Parrots (*Psittacus*), members of the genus *Poicephalus*, lovebirds (*Agapornis*), lories of the genus *Eos*, parakeets of the genus *Psittacula*, cockatoos (*Cacatua*), budgerigars (*Melopsittacus*), king parrots (*Alisterus*), and cockatiels (*Nymphycus*). Some reports of virus isolation do not mention the bird species involved.

### *Clinical Disease and Pathology*

The clinical disease is, as a rule, highly acute and the symptoms are nonspecific (somnolence, lethargy, anorexia, ruffled plumage, and irregular diarrhea). A yellowish coloration of the liquefied feces and sometimes also of the urine may indicate an involvement of the liver in the disease process. Kidneys and lungs are also frequently altered by secondary infections; therefore dyspnea and/or polyuria as well as uricemia may be recognized. Sinusitis and nasal excretions have been observed rarely, as have convulsions or tremors in the neck region and in the musculature of wings and legs. Hemorrhagic diarrhea and conjunctivitis occur only occasionally. A common factor in all reports is distinct change in the environment (import, quarantine station, pet shop, etc.). Under those conditions, recrudescence takes place and virus excretion becomes so heavy that even contact birds become sick and an epidemic is started.

**Table 32–3.** AVIAN HERPESVIRUS AND SOME OF ITS INTERCONNECTIONS

| Virus Species | Serotype | Serologic Cross-Reactions | Peculiarities of Pathology or Diagnosis | Vaccine |
|---|---|---|---|---|
| Marek's disease (only in chickens) | 1 | ← Turkeyherpes | a. Neural form<br>b. Tumorous form<br>c. Ocular form<br>d. Transient paralysis | Yes, heterologue = Turkeyherpes, homologue |
| Infectious laryngotracheitis (chicken, pheasant, peafowl) | 3 | ← SMON*<br>← Amazon-tracheitis | Inclusion bodies in early stages; lesions and mortality vary according to virulence; predominantly in adults | Yes, homologue |
| Turkeyherpes | 2 | → Marek's disease | No clinical disease in turkeys or chickens | Yes, homologue |
| Duck plague (ducks, geese, swans) | 2 | None | Reservoir: wild waterfowl; lesions and mortality vary according to virulence of virus strain and to susceptibility of host species | Yes, homologue |
| Pigeonherpes† | 5 | ⇌ Falconherpes<br>⇌ Owlherpes | Frequently no lesions at all; liver lesions with inclusion bodies, mainly in young or "stressed" birds; central nervous signs are observable only in young birds when they try to fly. Pathogen for budgerigars, at least experimentally; development of specific antibodies. | No, but in preparation in Belgium |
| Falconherpes | 5 | ⇌ Pigeonherpes<br>⇌ Owlherpes | Mainly liver lesions (inclusion bodies) | No |
| Owlherpes (only in "eared" owls) | 5 | ⇌ Pigeonherpes<br>⇌ Falconherpes | Pharynx with lesions resembling those caused by *Trichomonas* (which may be present in high numbers as secondary invaders), liver and spleen resembling avian tuberculosis (Ziehl-Neelsen staining obligatory) | No |

| | | | | |
|---|---|---|---|---|
| Pacheco's virus (wide host spectrum, only Psittaciformes) | 4 | None | Affinity for liver; highly lethal experimentally (chicken refractory). Infection can cause lifelong carriers. Diagnosis: cell culture (chicken or Muscovy Duck); chicken embryo; few CAM lesions, but hepatitis of the embryo. | No |
| Amazon-tracheitis | Not classified | →Infectious laryngotracheitis | Pseudomembranous tracheitis predominantly of the distal segments; about nine months' persistence of clinical symptoms. Juvenile chicks show tracheitis after experimental injection. Diagnosis: chicken embryos; mainly CAM lesions; chicken or duck embryo fibroblasts develop CPE with brownish pigments; virus source: pharynx swabs better than feces. | No |
| Budgerigarherpes | Not classified | Not completely tested | Latent course, no specific; virus predominantly in feathers, rare in organs, infrequent in blood or feces; egg-transmissible, causes reduced hatchability, probably dependent on the dose. Diagnosis: chicken embryo only for antigen replication (agar gel precipitation); no lesions in embryos | No |
| Lake Victoria Cormorant virus | 6 | None | Accidental isolation, no details known | No |
| Craneherpes | 7 | Not completely tested | Liver lesions, no details known | No |
| Herpes ciconiae (so far only from various storks) | 8 | None | Necrotic foci in liver, spleen, and bone marrow. Diagnosis: chicken embryo liver or kidney cells (inclusion bodies) | No |

*SMON = Subacute *M*yelo*o*ptico*n*europathia

†Pigeon, Falcon, and Owlherpes virus are closely related and probably only one type.

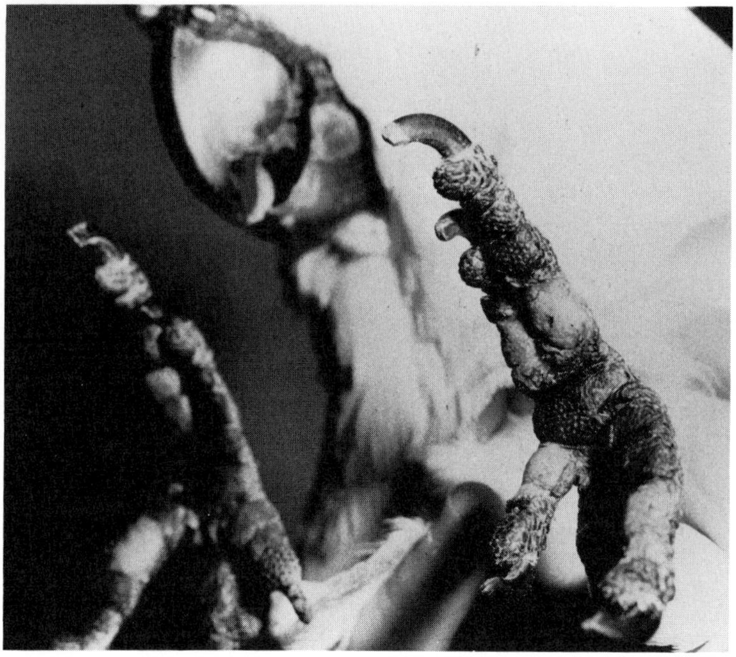

**Figure 32–3.** Wartlike growths on the feet of a cockatoo are suspected to be caused by a herpesvirus. (Courtesy of Greg J. Harrison.)

To be distinguished from Pacheco's virus and its related strains are cases in which a similar disease (course and symptoms) has been caused by a *Herpesvirus* serologically related to the Pigeonherpes virus.[47] This seems to be an example of increased virulence of a *Herpesvirus* in the unnatural host. Vindevogel et al.[45] could demonstrate that budgerigars are susceptible to Pacheco's virus and Pigeonherpes virus, whereas pigeons are susceptible only to their own virus. The host spectrum among the psittaciformes for disease with Pigeonherpes virus–related strains seems to be the same as for Pacheco's disease. The degree of lesions seen at necropsy is most likely due to the course of the disease and includes necrotic foci in liver and kidneys or no significant macroscopically visible alterations at all. The spleen might be swollen. Eosinophilic intranuclear inclusion bodies of Cowdry type A can be found in parenchymal cells surrounding necrotic foci mainly in liver and kidney. The spleen and occasionally the lung harbor cells with inclusion bodies as described.

Gross and histologic lesions of the birds dying from Pacheco's or Pacheco-related and Pigeonherpes-related viruses are without distinct differences. In acute cases caused by these viruses the swollen liver and spleen may display hemorrhages. Conjunctivitis, rhinitis, and sinusitis, as well as hemorrhagic enteritis, can be present.

### *Diagnosis*

The occurrence of typical inclusion bodies is indicative of a *Herpesvirus* infection but does not indicate type. In peracute cases, inclusions may not be found. Thus, for diagnosis, virus isolation is frequently necessary. The virus (all the different strains) propagates in embryonated eggs from chicken and Muscovy Duck or cell cultures (fibroblasts, kidney cells) from those embryos. Compared with other avian *Herpesvirus*, Pacheco's virus and related strains cause only discrete lesions on the CAM, but the embryo develops liver necrosis (with inclusion bodies). The final diagnosis is made by serologic identification. The strains related to the Pigeonherpes virus propagate in the same systems, but the cytopathic effect (CPE) in chicken embryo fibroblasts differs slightly from that of Pacheco's virus.

Humoral antibodies following disease or latent infection are developed inconsistently. Positive titers indicate exposure to the virus and may convey nothing about the immunity of the individual. Negative results do not exclude infection.

### *Differential Diagnosis*

Since liver, kidney, and lung can be involved in the disease process, the rule-out list is large.

The most important causes to rule out are other infections, such as psittacosis, salmonellosis, and Newcastle disease. A more frequent, noninfectious etiology, particularly in single pets, is lead poisoning.

### *Treatment*

No specific treatment is known. Sick birds should be placed in isolation in a warm cage. Vitamin C (50 mg/kg body weight), fluid therapy, and methionine (20 mg/kg body weight) can be tried. Antibiotics (sensitivity-tested) can be administered to suppress secondary bacterial infections, but their hepatotoxicity should be considered as well.

In an exposed flock all clinically healthy birds should be given vitamin C and/or the (expensive) paramunity inducer PIND ORF (2 ml/kg body weight) at intervals of two to three days. PIND ORF has been developed from a *Parapox ovis* strain. Experiences in human medicine, in which pox vaccines have been used for treatment of diseases caused by *Herpesvirus*,[29] have encouraged research for veterinary purposes. PIND ORF seems to be effective in birds, but only if given at least 24 hours before an experimental infection.[38]

### *Control*

Today control is possible only by strict hygiene and quarantine of "new" birds. Since Pacheco virus and related strains can cause carriers with only intermittent shedding of the virus in the feces, quarantine should include sentinel birds as young as possible and those that are unimportant for the breeding program, etc. The quarantine period should last at least six, and preferably twelve, weeks. During this time, virologic, bacteriologic, and parasitologic as well as serologic examinations can be carried out. The extent of testing will depend on the size of the population, its monetary value, and the laboratory services available. Provided that both groups of birds remain clinically healthy during quarantine, the "new" ones can be integrated into the aviary.

For control, a vaccine might be advantageous. Investigations are underway but so far no product is available for use in the field. Vaccines against other species of avian *Herpesvirus* are of no use because Pacheco's virus is serologically unrelated. If conclusions from other avian *Herpesvirus* vaccines are allowed, a vaccine against Pacheco's virus would be expected to reduce primary virus excretion and clinical signs as well as pathological lesions following challenge. But the vaccine would not prevent carriers. For strains related to the Pigeonherpes virus, a vaccine produced by Vindevogel et al.[44] might be tried.

## Amazon-Tracheitis

The Amazon-tracheitis virus is different from Pacheco's virus and related strains. It shows group-specific serologic relations to the infectious laryngotracheitis virus (ILT) and is thought to be a mutant of this virus.[23, 48] So far, only members of the genus *Amazona* have been found to harbor this virus naturally. Whether or not there is a relation between the *Amazona* virus and an isolate from a Bourke's Parrot (*Neophema bourkii*)[18] remains to be investigated. Chicken and Southern Caucasus Pheasant (*Phasianus c. colchicus*) are sensitive experimentally. The Amazon-tracheitis virus is a typical *Herpesvirus*, being sensitive to lipid solvents, temperature of 56°C, and pH 3. Efficacious disinfectants are the same as for other herpesviridae.

### *Transmission*

Transmission takes place horizontally, and contact birds within a group of experimentally infected ones become sick and may die.

### *Pathogenesis*

In contrast to Pacheco's virus, in which necrotizing lesions are predominant, the Amazon-tracheitis virus causes hemorrhagic-necrotic tissue alterations in the trachea.

### *Incubation*

The incubation time in the Green-cheeked Amazon (*Amazona viridigenalis*) following experimental infection is three to four days and results in peracute death. Contact birds died after six days.

### *Clinical Disease and Pathology*

Amazons of a variety of species seem to show the same clinical signs after natural infection. This disease may be peracute to acute or subacute to chronic and has been observed clinically for up to nine months. The lid cleft and the nostrils are frequently thickly encrusted with exudates. The birds display oral breathing and noises like rales, rattles, cough, etc. The beak cavity can contain excess mucus. At postmortem examination, rhinitis, pharyngitis, laryngitis, and tracheitis can be seen. The type of inflammation varies from serous, mucous, or fibrinous to diphtheroid, pseudomembranous, or granulomatous (Fig. 32–4*A* and *B*). Hemorrhagic inflammation also occurs (Fig. 32–4*C*). The lumen of the trachea (and the larynx) can be filled with mucus or fibrinous material, causing death by asphyxia. A bronchopneumonia is commonly seen with the bronchi and bronchioli filled with fibrin. In addition, conjunctivitis, blepharitis, glossitis, ingluveitis, and air sacculitis can be observed irregularly and are probably due to secondary bacterial and fungal infections. Histopathologic lesions resemble classic ILT. Intranuclear inclusion bodies of Cowdry type A are rare and usually are found only in early stages of the disease, when the mucosal epithelium is not yet exfoliated.

### *Diagnosis*

In contrast to other *Herpesvirus* infections, swabs taken from pharynx or larynx are better virus sources than feces.

The virus propagates in embryonated chicken eggs, causing CAM lesions similar to ILT. In cell cultures (fibroblasts or kidney cells from chicken or Muscovy Duck), proliferative foci demonstrate brownish pigment. The final diagnosis is made serologically. Using the immunodiffusion test and the indirect immunofluorescent test, the virus reacts positively with ILT hyperimmune serum, but is negative in the virus neutralization test against ILT serum. Compared with avian herpesviridae other than ILT, no cross-reactions are demonstrable.

Humoral antibodies are developed only irregularly and should be interpreted in the same way as with Pacheco's virus.

### *Differential Diagnosis*

The differential diagnosis includes a long list of respiratory diseases and the diphtheroid form of avian pox. The more important rule-outs, particularly for subacute to acute cases, are Newcastle disease, chlamydiosis, and influenza A virus infections complicated by *Candida*, *Aspergillus*, *Trichomonas*, or a variety of bacterial infections. *Syngamus* invasions are a

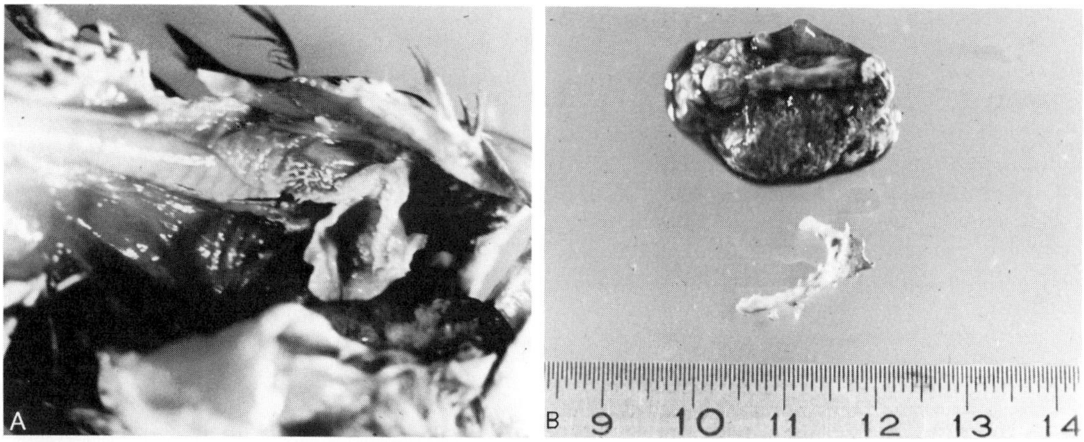


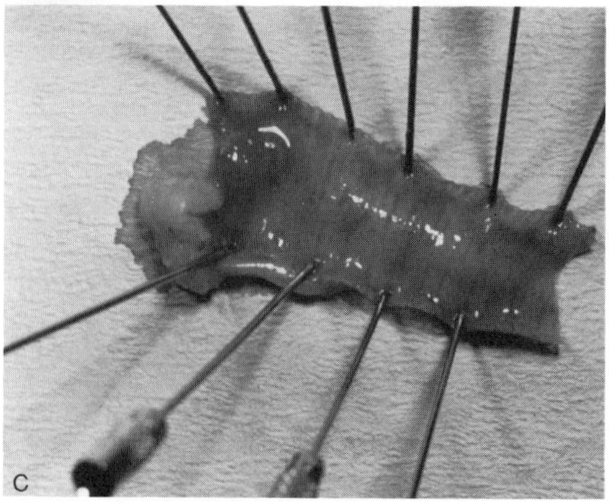


**Figure 32–4.** *A*, A fibrinous cast from the upper end of the trachea as a result of Amazon-tracheitis. *B*, Fibrinous casts from the bronchi with Amazon-tracheitis. *C*, Hemorrhagic exudate in the tracheal lumen caused by Amazon-tracheitis virus.

rather rare possibility. Of noninfectious causes, vitamin A deficiency is the most common.

### *Treatment*

Treatment is nonspecific and limited to the application of the same remedies as for Pacheco's virus. The long duration of clinical disease requires monitoring of secondary invaders (bacteria, fungi) and testing for sensitivity to antibiotics. Since mainly the conjunctivae, beak cavity, nares, etc., are included, local cleaning of exudates and application of suitable topical preparations will help.

### *Control*

The control of Amazon-tracheitis should follow the same principles as for Pacheco's disease. Although chickens vaccinated with Amazon-tracheitis virus and challenged with ILT virus are partially immune, the reverse trial has not yet been conducted and no knowledge about protection by ILT vaccination in Amazons is available. Generally, the vaccination of Amazons against tracheitis is not encouraged, because the virus is not yet widespread, and, most likely, carriers are not prevented.

## Budgerigarherpes Virus

Winterroll[47] isolated a *Herpesvirus* from budgerigars which is demonstrated predominantly within the feathers, and rarely in organs, blood, or feces. The virus is egg transmitted. It causes decreased hatchability of the eggs, probably dependent on the dose.

Since the virus seems to cause no direct lesions, further examination of some characteristics of the agent is still pending. Interestingly, most of the strains were isolated from so-called "feather dusters." Whether or not there is a connection with this disease of uncertain etiology needs further clarification.

The virus causes no alteration in cell cultures or in embryonated chicken eggs. Diagnosis can be made by demonstration of antigen replication in the embryonated chicken egg via an immunodiffusion test.

There is no justification for any treatment, since there is no disease related to the agent, and the "feather duster" contains a lethal genetic factor. "Feather dusters" occur only in the English show budgerigar, and flocks infected with the Budgerigarherpes virus should not be mixed with flocks of common budgerigars without a quarantine.

## PARAMYXOVIRUS

Avian paramyxovirus (PMV) has recently gained considerable attention not only because of its well-known member, Newcastle disease (ND) virus, but also because many new isolates serologically unrelated to ND virus have been reported.

The virion of PMV is pleomorphic, which might be the cause of the varying sizes (120 to 350 nm) reported. PMV is an RNA virus with a single-stranded RNA molecule. The genus PMV contains hemagglutinin (HA) and neuraminidase as glycoproteins. Another glycoprotein (F) is important in the induction of virus penetration, cell fusion, and hemolysis. The genus PMV contains the following avian strains: ND virus (group 1), avian PMV groups 2 to 9, some nonclassified isolates, and parainfluenzae 2 and 3.

Obviously PMV has a broad host spectrum, particularly ND virus. As far as psittaciformes are concerned, strains belonging to groups 1, 2, 3, and 5 have been isolated, and the discussion will, therefore, concentrate on these groups. Groups 4, 6, 8, and 9 are mainly duck and goose strains, and group 7 comes from columbiformes. Parainfluenzae 2 and 3 are probably human and murine types that can infect chicken and turkey incidentally.

Sufficient knowledge about some virus characteristics such as resistance, disinfection, transmission, pathogenesis, etc., is available only for ND virus, but analogous conclusions for the other groups might be acceptable.

## PMV Group 1

PMV group 1 is synonymous with ND virus. The virus is spread worldwide. Birds of all age groups can be susceptible. Although heat may be a triggering factor, no real seasonal peaks are recognizable. It seems as if epizootiological waves occur in 10- to 12-year intervals.

ND virus is sensitive to pH 3, pH 11, temperature (56°C), and daylight. While the virus is rather resistant on eggs, meat, etc., at refrigerator temperatures, in ambient temperatures of empty chicken houses it loses its infectivity during a period of four weeks depending on the access of daylight to the house. Although ND virus has a lipid-containing envelope, disinfection is generally not good with any products in actual use. Products producing pH 3 and pH 11 in the final concentration are corrosive and dangerous for the people who work with them. Lysol, cresol, phenol, and related substances as well as 2 per cent formalin are useful. The efficacy of formalin is dependent on the ambient temperature. At 0°C it is totally ineffective. Increasing temperature increasingly restores efficacy. At about 30°C, it is excellent.

### *Transmission*

Transmission takes place mainly horizontally, with the respiratory and oral routes being equally important. The embryo may become infected by penetration of virus from egg shells contaminated by fecal material in the cloaca. True egg transmission is rare, owing to the rapid cessation of egg production in viremic layers. Carrier status and constant virus shedding are possible, although these birds are immune. In contrast to herpesvirus, persistent infections are limited to weeks or months. Poultry are less prone to become carriers as compared with wild-living waterfowl, pittidae, psittaciformes, some passeriformes, and strigiformes, as well as a variety of other migrating birds. All these groups are considered reservoirs.[14, 21] The virus is excreted mainly in feces.

### *Pathogenesis*

Pathogenesis is governed by the high affinity of ND virus for erythrocytes by which the virus is spread instantly throughout the host's entire body. The pathomorphologic basis for the frequently observed clinical dyspnea is lung congestion following circulatory disturbances and the disturbance of the respiratory center as a result of encephalitis. Virus adherent to the vascular endothelium produces local damage and is considered the cause of the frequently observed petechiae.

### *Incubation*

The incubation period following natural and experimental infection is usually four to seven days, with a maximum of 25 days, depending on the host species, the virus dose, and pathotype of the virus.

### *Clinical Disease and Pathology*

The clinical disease is probably best known in the domesticated chicken, where lentogenic, mesogenic, and velogenic strains are differentiated. Disease caused by the latter is frequently called velogenic-viscerotropic ND (VVND). The host spectrum is very broad, but the clinical expression of the infection is rather different among the various groups of birds, even between two species. The terms lentogenic, mesogenic, and velogenic comprising certain biologic characteristics refer only to the capacity of the strain in question to induce disease in the domesticated chicken and have nothing to do with the type of disease in any other bird species. Several forms of clinical disease are differentiated which may vary decisively in degree:

1. Acute death without prior clinical symptoms or only a few hours of nonspecific ailment associated with viremia.
2. Acute visceral form with diarrhea as the main symptom in connection with anorexia, lethargy, and cyanosis.
3. Acute respiratory form clinically showing nasal exudate, rales, and dyspnea, as well as the general symptoms of anorexia, lethargy, and cyanosis as described in the visceral form.
4. Acute mixed form, in which visceral and respiratory signs may be present.
5. Chronic form with central nervous system (CNS) symptoms, including opisthotonos, torticollis, tremor, clonic-tonic paralysis of the limbs, backward movement, etc. The CNS disease can develop following an acute or subclinical disease. As a rule, with the onset of CNS symptoms, humoral antibodies against ND virus are demonstrable.
6. Persistent infection with no clinical signs. The virus propagates and is shed for periods of time, depending on the bird species, ranging from days up to several months. The latter should be regarded as carriers and natural reservoirs for the virus.

A survey of the better-known bird orders regarding the variability of the ND in different bird species is given. It is emphasized that only examples are cited and that there are reasons to expect exceptions, especially in species not mentioned.

**Phasianiformes** (Gallinaceous Birds). They are generally highly susceptible. Visceral, respiratory, and CNS disease occurs. Persistent infections are known only with lentogenic strains, and carriers seem to be rather rare.

**Anatiformes** (Waterfowl). There is only a low susceptibility. As a rule only velogenic strains cause disease and normally without respiratory symptoms. In geese spontaneous drowning is typical. Ducks may develop CNS signs.

**Columbiformes** (Pigeons). Until recently,

pigeons have been considered rather resistant to ND, only velogenic strains causing CNS symptoms. Since 1982 in Europe a PMV with strong serologic relations to ND virus and some to PMV group 3 has been isolated from domesticated pigeons (homing pigeons and fancy breeds).[37] Clinically, the birds display polydipsia and diarrhea followed by CNS signs. The natural disease seems to be limited to pigeons, although experimental infections induce typical disease in chickens. The pathotyping of one strain demonstrated a mean death time of 96 hours (lentogenic) and a Hansen index of 1.7 (velogenic).[37]

**PSITTACIFORMES** (Parrots and Parakeets). The susceptibility is extremely variable. For instance, the genus *Cacatua* is very susceptible, and surviving birds, as a rule, develop CNS symptoms (so-called head disease). The genus *Lory* seems to be refractory, but they are not all tested. The budgerigar is rather resistant to natural infection and shows only CNS signs but frequently does not produce humoral antibodies in significant titers.

Experimental infections were used in order to study the variability of the clinical disease and the pathologic lesions. Erickson et al.[14] infected six different avian species with a velogenic isolate (budgerigar = *Melopsittacus undulatus;* canary = *Serinus canarius;* Yellow-crowned Amazon = *Amazona ochrocephala;* Black-headed Manniken = *Lonchura malacca atricapilla;* Lesser Indian Hill Mynah = *Gracula religiosa indica;* Orange-fronted Conure = *Aratinga canicularis*). The reaction to the infection varied distinctly. Canaries were rather resistant, and mynah birds died at a rate of 20 per cent without prior clinical symptoms. The three psittaciforme species demonstrated distinct clinical signs, including CNS symptoms. Pathologically, lesions of the digestive tract, like those seen in chickens with the same strain, were not prominent. Luethgen[27] infected several psittaciformes with a VVND strain. The Black-winged Lovebird (*Agapornis taranta*) showed a mortality of 30 per cent, whereas the Alexandrine Parakeet (*Psittacula eupatria nipalensis*) was resistant. Members of the same genus, *Psittacula krameri* and *Psittacula cyanocephala*, as well as the budgerigar, displayed clinical disease, and all sick birds died suddenly. CNS symptoms were frequent, and nonpurulent encephalitis was demonstrated in birds with or without CNS signs.

**PASSERIFORMES** (Finches = Fringillidae; Weaver Finches = Ploceidae, etc.). The susceptibility is extremely variable. Canaries and the Mexican members of the genus *Spinus* do not show disease, but the virus propagates and is excreted. Many weaver finches (Gouldian Finch = *Erythrura gouldiae*, Red-cheeked Cordon-bleu = *Uraeginthus bengalus*) are highly susceptible (asphyxia by pseudomembrane in larynx, conjunctivitis, rarely CNS symptoms). In mixed aviaries only a few species may die, so that an infectious etiology is frequently not considered.

**CORVIDAE** (Ravens and Crows). The susceptibility is moderate. Occasionally death can occur.

**PITTIDAE** (Pittas). The susceptibility is low. Some are known for long-persistent virus shedding. Strains excreted by pittas have been typed as VVND.

**FALCONIFORMES** (Falcons, Hawks, Eagles, etc.). The susceptibility is low. CNS symptoms are predominant. Convulsions of the talon musculature and inability to coordinate flight are characteristic. The latter lesion causes "accidents."

**STRIGIFORMES** (Owls). The susceptibility is low. ND virus has been isolated from birds suffering from "accidents." Shedding of virus via feces has been demonstrated for over four months.

**SPHENISCIDAE** (Penguins). Their susceptibility is high. Infected birds die acutely following viremia.

**STRUTHIONIFORMES** (Ostriches). The susceptibility is low. Clinical disease is dominated by respiratory signs.

**PICIFORMES** (Woodpeckers, Toucans). The susceptibility is low.

**PELICANIFORMES** (Pelicans). The susceptibility is moderate. Acute death occurs occasionally.

Clinical disease and pathologic lesions may be altered by partial immunity. At necropsy, septicemic lesions in the form of petechiae on serosal tissue, fatty tissue, mucosa of larynx, trachea, and proventriculus, as well as hemorrhages into the egg follicles, may be seen depending on the acuteness of the disease. Virulent strains may cause a hemorrhagic necrotizing enteritis, mainly within the jejunum. These lesions, called "boutons," are in a local relation to lymphatic tissue. Boutons are pathognomonic in phasianiformes and suggestive of ND in other birds.

Grossly CNS lesions may be just hyperemia. Histologically a nonpurulent encephalitis with vascular and perivascular infiltrates of mononuclear cells is seen. Glial cells are frequently increased and pseudoneuronophagia may be present. Degeneration of ganglial cells is rare. There seems to be no correlation between clinical signs and the extent of the histologic lesions.

### Diagnosis

Diagnosis is made directly by isolating the virus from organ material (including brain in birds clinically sick longer than four days) or from feces of birds still alive. Chicken embryo fibroblasts or embryonated chicken eggs are used for culture. In contrast to poultry strains, one to three adaptation passages may be necessary even for velogenic strains. The amnioallantoic fluid (AAF) of dead embryos and the supernatent of cultures showing CPE hemagglutinate chicken erythrocytes (and erythrocytes of many other animals). The inhibition of this hemagglutination by specific serum (HI test) ensures the diagnosis. Cross-reaction with other PMV groups is known only for PMV group 3 (significant lower titer). The aqueous humor of birds that died acutely occasionally contains enough virus to hemagglutinate on a slide. Indirect diagnosis can be made by serology using primarily the hemagglutination inhibition (HI) test. HI antibodies appear about eight days after infection and are, therefore, valuable for the diagnosis of more chronic cases, especially those with CNS symptoms and for detection of carriers. The HI test is also used for monitoring vaccination procedures. Exceptions are seen in various bird species. HI antibodies following natural or experimental infections do not develop as expected in budgerigars and some columbiformes.

### Differential Diagnosis

For acute death with intestinal or respiratory lesions, any agent causing septicemia, enteritis, or respiratory disease has to be considered. The main candidates are chlamydiosis, Pacheco's disease, salmonellosis, and other PMV infections. The rule-out list for CNS lesions includes chamydiosis, salmonellosis (for pigeons also Pigeonherpesvirus), and adenovirus. In addition, nutritional deficiencies should be considered, particularly in the genus *Lory*.

### Treatment

There is no specific therapy. The use of hyperimmune serum seems not to harm the birds (about 2 ml/kg body weight intramuscularly once) but is of no benefit if given after the onset of clinical signs. It is understandable that hyperimmune serum is ineffective for treating CNS signs, since they develop while humoral antibodies are already present. The treatment of the nonpurulent encephalitis has been tried with many kinds of drugs, including the various B vitamins, anticonvulsants, etc. The results are discouraging. The disease followed the same course in both untreated controls and treated birds. Several months after onset the CNS signs became less violent but never ceased. Some birds have been observed for more than a year. Violent onset of convulsions, tremors, etc., can be provoked by any disturbance, and unless the bird is tame euthanasia should be considered.

### Control

Since the virus is distributed worldwide and wild bird species may be carriers, vaccination is recommended for aviaries, breeding farms, zoo collections, etc. In contrast to what occurs in chickens, most nondomesticated birds of orders other than phasianiformes have to be vaccinated parenterally preferably by oil emulsion vaccine, or via eye or nasal drops. The dose of live vaccines is generally five times the chicken dose. The use of vaccines produced for chickens is helpful, although variations in the antigenic structure are known. This is also true for the recently described pigeon strains of ND. A whole series of different vaccines is produced for chickens. Only the Hitchner $B_1$ and the truly apathogenic LaSota are recommended as live virus vaccines. The preparation of the suspension is more important than the virus strain for inactivated vaccines. The suspension to be injected should be stable and inert for the host. Many suspensions can cause abscesses surrounding the injected depot. Subcutaneous injections are, therefore, sometimes better because abscesses can be surgically removed more easily.

A combination of live vaccine followed by inactivated vaccine some weeks later will prolong the duration of the protection. The live virus vaccination once repeated after three weeks provides immunity for three to four months, as a rule. The inactivated vaccines normally provide protection for five to seven months. The combination might protect for 9 to 12 months. These figures are from limited experience with a few parrot species in Germany and are considered only guidelines.

Vaccination (with some exceptions referred to above) results in an increase of HI titers. These titers are not directly correlated with immunity but indicate a response of the host following vaccination. In an active epornitic, emergency vaccinations should be conducted only with live virus vaccines and only via eye or nasal drops in order to put the antigen in contact with the area of the harderian gland and the lacrimal gland duct. A local protection that is not measurable by HI antibodies can be built up in these structures within a few days.

## PMV Group 2

The first isolates of PMV Group 2 were made from domesticated chicken and turkey. Wild-living, clinically healthy birds of the order passeriformes frequently harbor PMV Group 2. The Senegal seems to be an area where this virus group is endemic within many passeriformes, particularly in weaver finches (*Zonotrichia, Zosterops, Estrilda*).[15] According to Alexander et al.,[2] psittaciformes are only rarely infected. However, a large group of the African Grey Parrot (*Psittacus erythacus*) suffered disease and death.[8] Clinically, the birds demonstrated only progressive emaciation. At postmortem examination, pneumonia, with excess mucus in the trachea, was the main finding.

One of the finch isolates (Finch/N. Ireland/Bangor 73) also caused the death of the host, a Blue-breasted Waxbill (*Uraeginthus angolensis*). Pathogenicity tests with isolates of PMV Group 2 need to be conducted in order to evaluate the capacity of provoking disease in many bird species.

The diagnosis can be made the same way as with ND; however, most smaller laboratories will have to send hemagglutinating isolates other than ND to reference laboratories for diagnosis. Since respiratory signs appear to be predominant, the rule-out list has to include all respiratory diseases.

Only hygienic measures, as already described, are available for treatment and control.

## PMV Group 3

As with PMV Group 2, the first isolates of PMV Group 3 were made from domesticated birds, mainly turkey. Most of the isolates from nondomesticated species originated from several psittaciformes (*Agapornis roseicollis, Psittacula* spp., *Amazona* spp., *Nymphicus hollandicus, Melopsittacus undulatus, Ara ararauna*). PMV Group 3 has also been found in a passeriforme bird.[3]

The pathogenicity of this group is variable, and more experimental investigations are necessary. As with ND, there are decisive species differences and CNS involvement (torticollis, tremor, paralysis of the limbs) occurs. The *Neophema* isolate[43] was experimentally used to infect a variety of bird species. In *Neophema* spp. and *Platycercus* spp. the CNS signs dominated the clinical picture. Some passeriformes like the canary, Gouldian Finch (*Erythrura gouldiae*), and other weaver finches became sick as well, but the budgerigar, cockatiel, and Japanese Quail proved to be refractory. Ducklings (domesticated Pekin Duck) and pigeons (domesticated) did not succumb to the disease but produced antibodies.

An *Agapornis* isolate described by Hitchner and Hirai[20] was experimentally tested for pathogenicity by Hirai et al.[19] Infected lovebirds suffered from greenish watery diarrhea prior to death. Chicken and budgerigar did not display clinical symptoms but developed antibody titers. Budgerigar isolates, as far as tested, did not provoke disease in the budgerigar.[13]

Isolation methods are not different from ND virus, but adaptation passages may be necessary. PMV Group 3 has a closer serologic relationship to ND virus than the other groups. This serologic relation seems to go together with pathogenicity for day-old chicks, the intracerebral pathogenicity index being 1.3 for the *Neophema* isolate. The differential diagnosis includes ND and other agents causing CNS signs and enteritis. Neither specific treatment nor vaccines are available.

## PMV Group 5

So far all PMV Group 5 isolates have originated from budgerigars, the best known being the Kunitachi virus. Following natural and experimental infections, all budgerigars suffered an acute, frequently lethal disease. Clinical signs include diarrhea, dyspnea, and more rarely, torticollis. Histologically, multiple necrotic foci are present in the liver and kidney with development of giant cells. Pathogenicity for other bird species is not yet known.[34, 35]

In contrast to ND virus, PMV Group 5 has been isolated in embryonated chicken eggs following inoculation only into the amnionic cavity. Hemagglutination of erythrocytes from the chicken, duck, goose, guinea pig, and man occurs regularly only after purification of the AAF.

The rule-out list is the same as for other PMV. The lesions are similar to what can be seen with *Salmonella, Escherichia coli,* and some other bacterial infections. Neither specific treatment nor vaccination is available.

## PMV Unclassified

An unclassified PMV isolated from the budgerigar was described by Mustaffa-Babjee et al.[32] Approximately 50 per cent of the infected budgerigars died, suffering from severe diarrhea. At postmortem examination, splenomegaly and extensive hemorrhages within the mucosa of the proventriculus and ventriculus were seen, as were edemata within the intestinal wall. A number of Rainbow Lory (*Trichoglossus haematodus*) died acutely at the same location and, it was assumed, of the same etiology.[33] The lesion in the lory was described as hemorrhagic to necrotic enteritis. This PMV does not propagate in the embryonated chicken egg following inoculation into the allantoic cavity. It grows in chicken kidney cells and embryo fibroblasts as well as in monkey kidney cells. The cytopathic effect consists of giant cells and numerous nuclear inclusion bodies. A relation with PMV Group 5 is possible.

The differential diagnosis includes all causes for enteritis, i.e., Enterobacteriaceae, in particular *Salmonella* spp., to which lories are extremely susceptible. Noninfectious causes are intoxications and nutritional deficiencies.

Neither specific treatment nor vaccination is available.

## PAPOVAVIRUS

Papovaviridae are small icosehedral particles (45 to 55 nm) containing double-stranded DNA and no envelope. Three members of the group are described in birds.

1. A papilloma-like virus has been isolated from some European finches (genus *Carduelis* and *Fringilla*) with papillomas on the legs. Cases are rare, and therefore this virus is not discussed here.

2. A papilloma-like virus infection in an African Grey Parrot, subspecies timneh, has been described.[22] Proliferative skin lesions lasting for more than one year were distributed over the head, including the palpebrae. Biopsies were taken for histology, electron microscopy, and immunodiagnosis. Histologically, the lesions consisted of long, thin folds of hyperplastic epidermis, which were moderately acanthotic and parakeratotic. Nuclei retained in the stratum corneum stained positively for papillomavirus structural antigens by the peroxidase-antiperoxidase technique with rabbit hyperimmune serum against disrupted bovine Papilloma virus type I. By electron microscopy viral particles of uniform size (47.5 nm) were seen, particularly within retained nuclei in the stratum corneum.

In addition papillomas have been described in Amazons, macaws and budgerigars.[9, 17] The majority of the cases display cloacal papillomas, but skin, eyelids, oropharyngeal mucosa, crop or esophagus, and proventriculus can be affected as well. No virus has been demonstrated from these cases. The history of cloacal papillomas is as a rule that of a recurrent "cloacal prolapse" over several months. The clinical examination under anesthesia reveals a mass protruding from the vent which is attached to the cloacal wall as well as around the vaginal or rectal opening.

Histologic examination of a biopsy is necessary to diagnose a papilloma. Treatment consists of surgical removal of the mass. Since recurrences may be expected and other internal papillomas may be present, the preparation of an autogenous vaccine to be given subcutaneously into the wing web twice at approximately 14-day intervals should be considered.[9, 17]

3. Budgerigar fledgling disease (BFD) (generalized parrot papovavirus infection) has been found by Davis et al.[11] and Bozeman et al.[4] to be caused by a member of the papovavirus family. The virions of the BFD virus are of the same size and symmetry as members of the genus *Polyoma* of the papovaviridae. In the United States and Canada this virus seems to be an important agent of fledgling mortality in commercial budgerigar aviaries. Although the virus alone can be the cause of death, secondary microorganisms such as Enterobacteriaceae and, in particular, *Giardia* spp. increase the losses.

Papovaviridae are stable against treatment with chloroform and resistant to heat (two hours, 56°C) and freeze-thawing. As a small particle without an envelope, it is expected to be resistant to a variety of disinfectants. No details are known about this virus (BFD), but generally for mammalian papovaviridae iodophores are considered useful (two hours of action), as are certain combinations of aldehydes. The agent of BFD does not hemagglutinate erythrocytes of chicken, turkey, budgerigar, guinea pig, or man (type 0).

### *Transmission*

According to Dykstra and Bozeman,[12] several routes are possible for transmission. The breeder birds feeding the offspring by regurgitation can transmit the virus via exfoliated epithelial cells of the crop. The virus is known to propagate in epidermal cells, particularly in feather follicles, and can be in "feather dust." The respiratory route is, therefore, strongly suspected. The presence of virus in the lung tissue further supports respiratory transmission. Cellular debris containing virus in renal tubules points to the possibility that excreta are infectious.

### *Pathogenesis*

The BFD virus causes an acute disease, which is an unusual feature for papovaviridae. The virus probably replicates at the portal of entrance, is released into the blood stream, and is transported through the entire body, where it can replicate in a variety of different target cells. Tissue damage is severe and directly correlated with disease or death. The virus has, in addition, the capacity to destroy or inhibit the normal development of lymphoid tissue. Since papovaviridae are potentially highly oncogenic, the question has been asked whether or not latent infections that occur are related to the rather high tumor rate in budgerigars. So far, no proof could be given that this is the case, but Davis[10] stated that "the virus has possible oncogenic implication."

### *Incubation*

The incubation time in budgerigar fledglings is 10 to 21 days from the day of hatching. Adults are carriers.

### *Clinical Disease and Pathology*

The clinical picture is rather uniform. Sick fledglings show a distended abdomen, lack or malformation of down feathers, and retarded growth of tail and contour feathers. Severe dehydration occurs regularly, and the vent area is soiled by urinary material rather than feces. The skin might be of a rather red color. The crop is frequently still filled, confirming acute disease.

The gross lesions are dominated by hydropericardium, enlarged heart, liver necrosis, and swollen, hyperemic kidneys. Secondary infections eventually alter the lesions. Histologic examination reveals karyomegalic clear or opaque inclusion bodies, composed of virus particles, in a variety of tissues such as kidneys, feather follicles, liver, heart, bone marrow, spleen, and brain. A vesicular degeneration of the crop epithelium with inclusion body formation is frequently observed.[17] Atrophy of lymphoid tissue and inflammatory reactions in the kidneys and occasionally in the liver can be present.

The disease seems to occur in a variety of psittaciforme birds other than the budgerigar such as Hyacinth Macaw (*Anordorhynchus hyacinthinus*), Blue and Gold Macaw (*Ara ararauna*), Military Macaw (*Ara militaris*), Buffon Macaw (*Ara ambigua*), Scarlet Macaw (*Ara macao*), Green-winged Macaw (*Ara chloroptera*), Yellow-crowned Amazon (*Amazona ochrocephala*), Hispanolian Amazon (*Amazona ventralis*), Sun Conure (*Aratinga solstitialis*), Jenday Conure (*Aratinga jendaya*), Gold-capped Conure (*Aratinga auricapilla*), Gold-crowned Conure (*Aratinga aurea*), Queen of Bavaria Conure (*Aratinga guaruba*), Red-masked Conure (*Aratinga erythrogenys*), White-eyed Conure (*Aratinga leucophthalmus*), Slender-bill Conure (*Enicognathus leptorhynchus*), White-crowned Pionus (*Pionus senilis*), African Grey Parrot (*Psittacus erythacus*), Sulfur-crested Cockatoo (*Cacatua galerita*), and Eclectus Parrot (*Eclectus roratus*).[7, 17]

The majority of the bird species mentioned above contract the disease as hand-fed nestlings, but a few parent-reared nestlings are affected as well. The age at the time of death ranged from 20 to 56 days, the older birds belonging to the larger species. Peracute death from chronic debilitation and renal failure can be seen clinically. First signs are reduction in daily weight gain, prolonged crop emptying time, vomiting, and reverse peristaltic waves from the crop followed by depression, glassy eyes, anorexia, and dehydration.[7] The pulling of a feather leads to excessive bleeding, as may an intramuscular injection. The first sign is characteristic and can be used for a tentative diagnosis.[7] Posterior paresis and paralysis have been observed prior to death in a Double Yellow-headed Amazon.[7]

The chronic course of the disease is described by subnormal weights, slow gut transit time, polyuria and renal failure, abnormal feathering, failure to be weaned at the normal age, and secondary candidiasis.[7]

The pathologic lesions vary from those in budgerigars as follows[17]: Epidermis and feather follicles are only rarely producing inclusion bodies. Multiple hemorrhages are common, together with the typical inclusion bodies in the vascular endothelium. The crop epithelium seems to be only rarely involved in the disease process. Hepatic necrosis is extensive, as a rule, and viable hepatocytes may form only a small rim about the portal veins. In conures and caiques a glomerulitis suggestive of an antigen-antibody reaction at the basement membrane can be found occasionally.

### *Diagnosis*

Originally, Bozeman et al.[4] isolated the BFD virus in budgerigar fibroblasts, but this virus adapts to chicken embryo fibroblasts during several passages. However, virus from psittaciforme birds other than the budgerigar cannot always be isolated via chicken tissues, and budgerigar embryo fibroblast cultures may be necessary. A serologic procedure has been developed by Davis[10] for identification of infected birds, including carriers. Clear serum, 200 μl per bird, is necessary for the test. Impression smears of liver and spleen (other affected organs) can be stained for inclusion bodies or for viral antigen by the immunofluorescence technique.

### *Differential Diagnosis*

Chlamydiosis, infection with Pacheco's virus or Pigeonherpes virus, as well as salmonellosis and other bacterial infections, are rule-outs. However, death at 10 to 21 days of age should be considered highly suspicious of BFD in budgerigars.

### *Treatment*

There is no specific treatment. Since the fledglings are fed by their parents, nonspecific treatment will normally cause so much unrest in the breeding facility that no benefit can be expected.

Efforts to develop a vaccine have been unsuccessful so far. More knowledge of the virus, its enzootical behavior, and its pathogenicity is necessary.

### *Control*

According to Davis,[10] depopulation and restocking with clean birds (each to be tested during quarantine) is one means of control. The interruption of budgerigar breeding for several months in order to minimize the number of highly susceptible birds and therefore reduce the viral cycling is another possible method for control.[7] For nonbudgerigar parrots, more intensive epizootiological studies are necessary. Seropositive birds can raise healthy progeny, but probably only when they transfer maternal antibodies into the egg.[7] Since the virus is highly infectious, closed breeding flocks and avoidance of visits between aviaries are recommended.

## ADENOVIRUS

Adenovirus contains double-stranded DNA, has no envelope, and has an average diameter of approximately 75 nm. Surface structures in the form of fibers display an endpiece. Adenovirus is resistant to chloroform and tolerates heat as well as pH changes. Since it is a small virus without an envelope, many commonly used disinfectants are ineffective. Iodophores and aldehydes are useful provided that the component can react with the virus for at least one hour.

Adenovirus is distributed worldwide, and many birds species are known to be susceptible. The virus can produce carriers in which periods of high humoral antibodies and low virus titers are followed by low humoral antibodies and high virus titers (viremia). Transmission takes place primarily by the oral route, possibly by respiratory means, and egg transmission is epornitically important. According to the individual status of the breeder hen, either virus or antibodies are transmitted into the egg. Infected eggs frequently show lower hatchability than noninfected ones.

Birds of all age groups can be susceptible and, as pointed out above, virus and antibodies can be present in the same bird. Infection does not necessarily mean disease, although defined diseases are known and variability in virulence is documented. Avian adenovirus usually acts as a complicating factor under unfavorable conditions for the host.

Avian adenovirus is not uniform. Obviously, there are species-related groups and within these groups several serotypes:

**Group I.** Fowl adenovirus (FAV) — 12 serotypes (1–12). Several serotypes together or different serotypes alone can produce similar clinical disease and pathologic lesions.

Turkey adenovirus — 2 or 3 serotypes
Goose adenovirus — 3 serotypes
A variety of isolates from other avian species

**Group II.** Virus of the hemorrhagic enteritis

of the turkey (HET) and the closely related virus of the marble spleen disease in the pheasant (MSD).

**Group III.** Virus of the egg-drop syndrome (EDS 76) isolated from chickens but originating from ducks (several serologically identical strains).

There is some evidence that adenovirus in an unnatural host might be more virulent than in the natural host. The group III strains causing EDS 76 in chickens are nonvirulent in ducks. FAV1 (= CELO virus) is only rarely the causative agent of a respiratory disease in chickens, but in Bobwhite Quail (*Colinus virginianus*) the same virus as adeno-associated parvovirus causes a highly lethal disease.

The basis for avian adenovirus as a complicating factor is its capacity to induce mild histopathologic lesions that produce no clinical signs. The histologic alterations regularly found are degeneration of liver cells and of cells in the respiratory tract. These lesions are the basis for secondary invasions, mainly by bacteria, but also by fungi and protozoa.

Routine examinations of many samples from psittaciformes have proven, as expected, that adenovirus is not rare in these birds. Most strains have not been characterized or serotyped. In many instances the adenovirus could not easily be linked to the disease in question. Two distinct diseases in which adenovirus is likely to be the primary causative agent are described as follows:

## Inclusion Body Pancreatitis

The disease was first described by Wallner-Pendleton et al.[46] in a Peach-faced Lovebird (*Agapornis roseicollis*). Prior to death, only nonspecific signs such as depression or light olive-green droppings had been observed. At the postmortem examination, an enlargement of the duodenal loop and the proventriculus was striking. Histopathology revealed an acute necrotizing pancreatitis with large basophilic intranuclear inclusion bodies in exocrine pancreatic cells. The mucosal epithelium of the duodenum contained similar inclusion bodies. In another case the same kind of inclusions was detected in the tubular epithelium of the kidney. A third case involving Nyasa Lovebirds (*Agapornis lilianae*) displayed inclusions in liver and spleen cells.

Electron dense and uniformly round particles approximately 70 to 80 nm in size could be demonstrated in the affected pancreatic cells by electron microscopy. In addition to the round-shaped particles, a few hexagonal ones were present. The particles are considered typical for adenovirus. Similar inclusions (histology and electron microscopy) were found in non-necrotic intestinal epithelium of an Amazon.

Graham[17] isolated an adenovirus from Nyasa Lovebirds and demonstrated an enteric adenovirus infection in a cockatoo as well as hepatitis and enteritis in a Moluccan Cockatoo (*Cacatua moluccensis*) by electron microscopy. Interestingly, Graham[17] could isolate an adenovirus from a lovebird with a feather loss syndrome similar in appearance to the feather loss syndrome in cockatoos.

## Budgerigar Encephalitis

The disease was fairly widespread in the Federal Republic of Germany for about one year and has since almost vanished. Adult budgerigars, most one to three years of age, developed a so-called acute onset of CNS signs such as opisthotonus, torticollis, tumor, convulsions, etc. The more subtle signs, such as decrease of optical orientation and failure to perch where aiming, have been frequently overlooked. Some birds did not die naturally but were euthanized because the CNS disease did not improve with or without treatment. At necropsy, no significant gross lesions were found, but histology revealed a nonpurulent encephalitis and a proliferation of glial cells as well as degeneration and lysis of ganglial cells. A focal degeneration and lysis of hepatocytes and some infiltrates containing mainly mononuclear cells were present in the liver. Gassmann et al.[16] isolated adenovirus (two to three passages on chicken embryo liver cells) from two birds. These adenovirus strains could be neutralized by FAV 2 and FAV 11 antisera in one case and by FAV 4 antiserum in the other case. Examination of the genome structure by the restriction enzyme ECO-RI revealed not only typical DNA fragments of the FAV but other DNA fragments that indicate differences in the genomes of the budgerigar isolates.

Experiments to reproduce the clinical disease with the isolates were only partially successful. The budgerigars showed similar histopathologic lesions but no clinical signs following intranasal or oral infection.

### *Diagnosis*

Virus isolation from dead birds can be attempted using chicken or duck embryo kidney or liver cells. Typical inclusion bodies are sometimes produced in these cells. Several adaptation passages might be necessary for material from psittaciformes. Serologic identification may be difficult because new serotypes and even groups are expected from pet birds. Only strains related to poultry and strains containing a large amount of group-specific antigen might be recognized by the immunodiffusion test. Although infected birds, as a rule, seroconvert, the same difficulties as with the identification of isolates apply.

### *Differential Diagnosis*

The rule-out list for inclusion body pancreatitis includes psittacosis, ND, influenza A infection, salmonellosis, and other bacterial infections causing diarrhea and liver disease. Budgerigar encephalitis must be separated from psittacosis, ND, bacterial or fungal encephalitis, and lead intoxication.

### *Treatment*

As with other virus diseases, there is no specific treatment and symptomatic measures should be taken to ease clinical signs. The encephalitis of the budgerigar is resistant to treatment, as already mentioned. Treatment with exocrine enzymes could be tried for pancreatitis.

Vaccines available for poultry should not be tried unless the same serotype is present. But even then, extreme care is necessary because of the possibility of an increase of virulence in unnatural hosts as closely related as the species. Another problem with vaccines is the existence of virus in the presence of humoral antibodies.

## REOVIRUS

Reoviridae are cubic in shape, have a diameter of 70 to 80 nm, lack an envelope but possess a double capsid, and contain 10 segments of double-stranded RNA. Avian reovirus seems to possess no common antigens with human or mammalian reovirus. Avian reovirus is widespread, and 11 serotypes have been distinguished. Avian strains do not hemagglutinate.

Reovirus is resistant to organic solvents, to pH 3, temperatures of 60°C, and most disinfectants in common use. Reovirus is inactivated by 70 per cent ethanol and 0.5 per cent organic iodine. Some aldehydes in combination with alcohols have proved efficacious when allowed to act for two hours.

### *Transmission*

Horizontal and vertical transmission has been proven for poultry. Details for psittaciformes are not available.

### *Pathogenesis*

Although reovirus is prevalent in many bird species, including the psittaciformes,[17, 41] only a few confirmed disease conditions are known, i.e., virus arthritis in chickens, infectious myocarditis in goslings. The influence of reovirus in a variety of enteric conditions in poultry has been previously discussed in the literature. Aside from the fact that there are definite differences between the strains regarding virulence, the pathogenesis of reovirus is not yet clarified. The strains isolated by Meulemans et al.[30] are serologically (virus neutralization) different from the chicken 1133 (virus arthritis) isolate. Rosenberger[40] discusses the possibility of interference with the antibody production against other disease agents. Maslin[28] could show in chickens that reovirus is correlated with lower glucose levels in the blood than in noninfected birds. Maslin found histologic lesions that might suggest the pancreas as a target organ for reovirus.

The role of humoral antibodies is also not fully investigated. While there is protection correlated with humoral antibodies, fecal shedding of the virus by persistently infected birds is seen in the presence of those antibodies.

### *Incubation*

No details are known for psittaciformes and other pet birds.

### *Clinical Disease and Pathology*

The clinical signs reported by Mohan[31] are nonspecific and include emaciation, incoordination, and labored breathing in a cockatoo. In a Grey-cheeked Parakeet (*Brotogeris pyrrhopterus*), loose droppings and a paleness of beak and toes were the main signs. The pathology varies from no lesions to lesions described[30] as hepatitis, enteritis, or pneumonia (Fig. 32–5). Mohan's report[31] is more detailed and describes a necrotic hepatitis and nephritis in the cockatoo and a catarrhal enteritis in the Grey-cheeked Parakeet as the main findings.

Clubb[6] and Graham[17] describe a disease in the African Grey and Jardin Parrot (*Poicephalus gulielmi*) in which a reovirus was isolated among

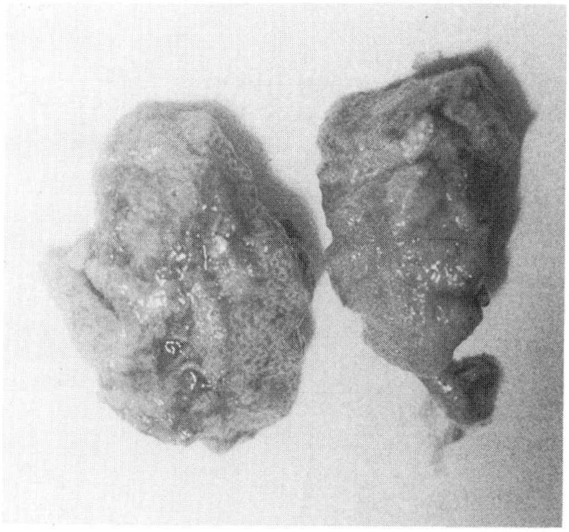

**Figure 32–5.** Edematous lungs from Eclectus parrot; the acute death was associated with reovirus. (Courtesy of Greg J. Harrison.)

a variety of other microorganisms (Paramyxovirus—Group 3, *Salmonella, Aspergillus, Mucor*). An experimental infection of African Greys with a reovirus strain from this species reproduced the significant clinical signs and pathologic lesions.

The first clinical signs are usually abnormal appearance of the eye, and within four days depression, weakness, loss of body weight, and death are observed. The urates are frequently yellow and the abdomen swollen. Prior to death dyspnea, paresis, and paralysis of the legs accompanied by skin edema of the head as well as of the legs occurs. A bloody to dark brown nasal discharge is not rare.

The ophthalmic lesions consist of fixed, dilated pupils, reticular hemorrhages followed by uveitis, hypopyon, and fibrous exudates in the anterior eye chamber. Posterior and anterior synechiae are present in birds that recover from the acute disease.[6]

At necropsy, hepatomegaly and splenomegaly are consistent findings.[17] Histologically, a multifocal coagulation necrosis of the liver is seen. The spleen usually displays congestion and sometimes varying degrees of necrosis of the reticulum sheaths of blood vessels. A depletion of lymphocytes in the white pulp can be present as well, probably depending on the course of the disease. The common presence of thrombi and microthrombi is suggestive of a disseminated intravascular coagulopathy.[17]

### *Diagnosis*

Feces, enteric material, liver, heart, and lung can be taken for culture (embryonated chicken egg or cell cultures from those embryos, i.e., kidney or liver cells). Group-specific antibodies can be demonstrated, at least in some of the cases, by using the immunodiffusion test in the agar gel.

### *Differential Diagnosis*

Since the pathogenicity of reovirus in psittaciforme birds is convincingly established only in the African Grey, other etiologies should be considered.

### *Treatment*

There is no specific treatment available. Symptomatic therapy can be tried according to the clinical signs.

### *Control*

The usual hygienic measures should be instituted. Vaccines available for chickens are not promising in psittaciformes, since the strains described by Meulemans[30] are serologically different from at least one of the common chicken strains.

## OTHER VIRAL GROUPS

The presence of other viral groups in psittaciforme birds is expected (Fig. 32–6).

1. Psittacine beak and feather disease syndrome (see Chapter 41, Disorders of the Integument).

Lester and Gerlach[25] isolated a virus from the feather pulp of an affected Salmon-crested Cockatoo (*Cacatua moluccensis*) by means of an explant followed by a co-culture with primary chick embryo fibroblasts. A constant CPE was observed by the seventh to eighth passage within 48 to 96 hours with titers approximately $10^8$ $TCID_{50}$/ml. The CPE is rather nonspecific, with rapid cell degeneration yielding floating cellular debris. Growth occurs best in rapidly dividing cells, i.e., inoculation of freshly seeded cells. The agent is also highly cell-associated and resistant to organic solvents as well as pH changes. Heat resistance is moderate. By filtration trial the size is less than 100 nm. Electron microscopic examinations of the isolate are still pending. The tentative characterization of this virus points toward papovavirus, but a parvovirus is possible as well.

Electron microscopic studies and indirect histopathologic evidence allow the suggestion that

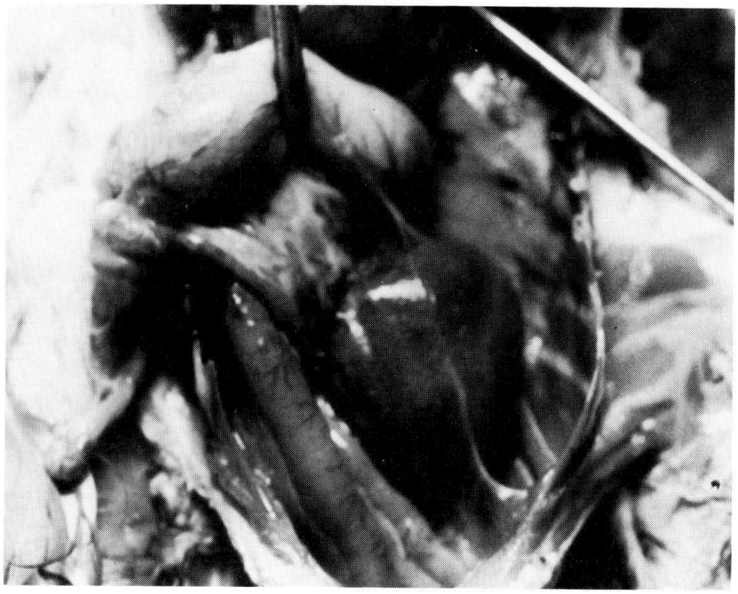

**Figure 32–6.** Lymphoproliferative lesions characterized by extreme splenomegaly and a very high serum protein level in a macaw is one of the diseases of unknown etiology, suspected to be a virus. (Courtesy of Greg J. Harrison.)

a virus is the causative agent of other diseases mentioned elsewhere in this book.

2. Proventricular dilatation of macaws and other species (see Chapter 38, Neurologic Disorders).

3. Bleeding disorders of conures and possibly other psittaciformes (see Chapter 8, Differential Diagnoses Based on Clinical Signs).

At press time, no virus isolation has been reported in the latter two conditions and until it is the etiology is open to speculation.*

*Editor's Note: Personal experience with a recent case of chronic nonresponsive malodorous diarrhea in Cape Parrots (*Poicephalus robustus*) was identified by electron microscopy as a coronavirus. Cornell workers[19] report that such a virus produces diarrhea and hemorrhagic necrosis of the liver and spleen.

## REFERENCES


1. Akrae, M.: Die Identifizierung und Charakterisierung eines Falkenpocken-und eines Psittacidenpockenvirus aufgrund ihres Verhaltens in der Eikultur, sowie serologischer Reaktionen und ihres Wirtsspektrums in Tierversuch. Diss. Med. Vet., Muenchen, 1980.
2. Alexander, D. J., Allan, W. H., Parsons, G., and Collins, M. S.: Identification of paramyxoviruses isolated from birds dying in quarantine during 1980 to 1981. Vet. Rec., 108:571–574, 1982.
3. Ashton, W. L. G., and Alexander, D. J.: A two-year survey on the control of the importation of captive birds into Great Britian. Vet. Rec., 106:80–83, 1980.
4. Bozeman, L. H., Davis, R. B., Gaudry, D., Lukert, P. D., Fletcher, O. J., and Dykstra, M. J.: Characterization of a papovavirus isolated from fledgling budgerigars. Avian Dis., 25:972–980, 1981.
5. Brugh, M., and Beard, C. W.: Atypical disease produced in chickens by Newcastle disease virus isolated from exotic birds. Avian Dis., 28:482–488, 1984.
6. Clubb, S. L.: A multifactoral disease syndrome in African Grey Parrots (*Psittacus erythacus*) imported from Ghana. Proceedings of the International Conference on Avian Medicine, sponsored by the Association of Avian Veterinarians, 1984, pp. 135–149.
7. Clubb, S. L., and Davis, R. B.: Outbreak of a papova-like viral infection in a psittacine nursery—a retrospective view. Proceedings of the International Conference on Avian Medicine, sponsored by the Association of Avian Veterinarians, 1984, pp. 121–129.
8. Collings, D. F., Fitton, J., Alexander, D. J., Harkness, J. W., and Pattison, M.: Preliminary characterization of a paramyxovirus isolated from a parrot. Res. Vet. Sci., 19:219–221, 1975.
9. Cribb, P. H.: Cloacal papilloma in an Amazon Parrot. Proceedings of the International Conference on Avian Medicine, sponsored by the Association of Avian Veterinarians, 1984, pp. 35–37.
10. Davis, R. B.: Budgerigar fledgling disease. Proceedings of the 32nd Western Poultry Disease Conference, Davis, CA, 1983, p. 104.
11. Davis, R. B., Bozeman, L. H., Gaudry D., Fletcher, O. J., Lukert, P. D., and Dykstra, M. J.: A viral disease of fledgling budgerigars. Avian Dis., 25:179–183, 1981.
12. Dykstra, M. J., and Bozeman, L. H.: A light and electron microscopic examination of budgerigar fledgling disease virus in tissue and in cell culture. Avian Pathol., 11:11–28, 1982.
13. Engelhard, E.: Charakterisierung eines aus einem Wellensittich (*Melopsittacus undulatus*) isolierten Paramyxovirus. Diss. Med. Vet., Muenchen, 1982.
14. Erickson, G. A., Gustafson, G. A., Pearson, J. E., Miller, L. D., Proctor, S. J., and Carbrey, E. A.: Velogenic viscerotropic Newcastle disease in selected captive avian species. American Association of Zoo Veterinarians Annual Proceedings, 1975, pp. 133–136.
15. Fleury, H. J. A., and Alexander, D. J.: Paramyxovirus Yucaipa. Bull. Intl. Inst. Pasteur, 76:175–186, 1978.
16. Gassmann, R., Monreal, G., and Bayer, G.: Isolierung von Adenoviren bei Wellenstittichen mit zentralner-

voesen Ausfallserscheinungen. Vol. II. Tagung Krankheiten der Voegel. Muenchen Dtsch. Vet. Med. Gesellschaft, 1981, pp. 44–47.

17. Graham, D. L.: An update on selected pet bird virus infections. Proceedings of the International Conference on Avian Medicine, sponsored by the Association of Avian Veterinarians, 1984, pp. 267–280.
18. Helfer, D. H., Schmitz, J., Seefeldt, S. C., and Lowenstine, L.: A new viral respiratory infection in parakeets. Avian Dis., 24:781–783, 1980.
19. Hirai, K., Hitchner, S. B., and Calnek, B. W.: Characterization of paramyxo-, herpes-, and orbiviruses isolated from psittacine birds. Avian Dis., 23:148–163, 1979.
20. Hitchner, S. B., and Hirai, K.: Isolation and growth characteristics of psittacine viruses in chicken embryos. Avian Dis., 23:139–147, 1979.
21. Ibrahaim, A. L.: Proceedings of the 31st Western Poultry Disease Conference, Davis, CA, 1982, p. 165.
22. Jacobson, E. R., Mladinich, C. R., Clubb, S., Sundberg, J. P., and Lancaster, W. D.: Papilloma-like virus infection in an African Gray Parrot. J. Am. Vet. Med. Assoc., 183:1037–1038, 1983.
23. Kitzing, D.: Zur Charakterisierung eines mit dem ILT-Virus serologisch verwandten Herpesvirus aus Amazonen. Diss. Med. Vet., Muenchen, 1981.
24. Kraft, V., and Teufel, P.: Nachweis eines Pockenvirus bei Zwergpapageien (*Agapornis personata* und *Agapornis roseicollis*). Berl. Muench. Tieraerztl. Wochenschr., 84:83–87, 1971.
25. Lester, T. L., and Gerlach, H.: Isolation of a virus from feathers of a cockatoo with feather-loss syndrome. Proceedings of the Eighth International Congress of the World Veterinary Poultry Association, Jerusalem, Israel, 1985. In press.
26. Lowenstine, L. J.: Diseases of psittacines differing morphologically from Pacheco's disease, but associated with herpesvirus-like particles. Proceedings of the 31st Western Poultry Disease Conference, Davis, CA, 1982, pp. 141–142.
27. Luethgen, W.: Die Newcastle Krankheit bei Papageien und Sittichen. Verlag Paul Parey. Berlin, Fortschr. Veterinaermedizin, Vol. 31, 1981.
28. Maslin, W. R., and Fletcher, O. J.: Effect of a reovirus isolated from broilers with pale bird syndrome on SPF chicks: Proceedings of the 33rd Western Poultry Disease Conference, Davis, CA, 1984, pp. 67–68.
29. Mayr, A.: Praemunitaet, Praemunisierung und paraspezifische Wirkung von Schutzimpfungen. Muench. Med. Wochenschr., 120:239, 1978.
30. Meulemans, G., Dekegel, D., Charlier, G., Froyman, R., Van Tilburg, J., and Halen, P.: Isolation of orthoreoviruses from psittacine birds. J. Comp. Pathol., 93:127–134, 1983.
31. Mohan, R.: Clinical and laboratory observations of reovirus infection in a cockatoo and a Grey-cheeked Parrot. Proceedings of the International Conference on Avian Medicine, sponsored by the Association of Avian Veterinarians, 1984, pp. 29–34.
32. Mustaffa-Babjee, A., Spradborrow, P. B., and Samuel, J. L.: A pathogenic paramyxovirus from a budgerigar. Avian Dis., 18:226–230, 1974.
33. Mustaffa-Babjee, A., and Spradborrow, P. B.: Acute enteritis in Rainbow Lorikeets. Kajian Vet., 5:16–19, 1973.
34. Nakayama, M., Nerome, K., Ishida, M., Fukumi, H., and Morita, A.: Characterization of virus isolated from budgerigars. Med. Biol., 93:449–454, 1976.
35. Nerome, K., Nakayama, M., Ishida, M., Fukumi, H., and Morita, A.: Isolation of a new avian paramyxovirus from budgerigars. J. Gen. Virol., 38:293–301, 1978.
36. Pacheco, G., and Bier, O.: Epizootie chez les perroquets du Brésil. Relations avec la psittacose. C. R. Soc. Biol. [Paris], 105:109–111, 1930.
37. Richter, R.: Paramyxoinfektion bei Tauben. Vol. III. Tagung ueber Vogelkrankheiten, Muenchen. Dtsch. Vet. Med. Gesellschaft, 86–95, 1983.
38. Richter, R.: Untersuchungen zur Steigerung der Immunantwort durch den Paramunitaetsinducer PIND ORF beim Gefluegel. Diss. Med. Vet., Muenchen, 1983.
39. Roehrer, H.: Handbuch der Virusinfektion bei Tieren. Bd II, Jena, Gustav Fischer Verlag, 1967, p. 605.
40. Rosenberger, J. K.: Reovirus interference with Marek's disease vaccination. Proceedings of the 32nd Western Poultry Disease Conference, Davis, CA, 1983, pp. 50–51.
41. Senne, D. A., Pearson, J. E., Miller, L. D., and Gustafson, G. A.: Virus isolation from pet birds submitted for importation into the United States. Avian Dis., 27:731–744, 1983.
42. Siegmann, O.: Kompendium der Gefluegelkrankheiten. 4th ed. Hannover, Verlag M. H. Schaper, 1983.
43. Smit, T., and Rondhuis, P. R.: Studies on a virus isolated from the brain of a parakeet. Avian Pathol., 5:21–30, 1976.
44. Vindevogel, H., Pastoret, P. P., and Leory, P.: Vaccination trials against pigeon Herpesvirus infection (Pigeon Herpesvirus 1). J. Comp. Pathol. 92:483–496, 1982.
45. Vindevogel, H., Pastoret, P. P., Leroy, P., and Coignul, F.: Comparison de trois souches de Virus Herpetique isolees de psittacides avec le Virus Herpes du pigeon. Avian Pathol., 9:385–394, 1980.
46. Wallner-Pendleton, E., Helfer, D. H., Schmitz, J. A., and Lowenstine, L.: An inclusion-body pancreatitis in Agapornis. Proceedings of the 32nd Western Poultry Disease Conference, Davis, CA, 1983, p. 99.

46a. Winterfield, R. W., et al.: Immunization against psittacine pox. Avian Dis., 29(3):886–890, 1985.

47. Winterroll, G.: Herpesvirusinfektionen bei Psittaciden. Prakt. Tierarzt., 5:321–322, 1977.
48. Winterroll, G., and Gylstorff, I.: Schwere durch Herpesvirus verursachte Erkrankung des Respirationsapparates bei Amazonen. Berl. Muench. Tieraerztl. Wochenschr., 92:277–280, 1979.
49. Winterroll, G., Mousa, S., and Akrae, M.: Pockenisolate aus Psittaciden and Falken — Naehere Charakterisierung. Deutsche Vet. Med. Ges., II. Tagung — Krankheiten der Voegel, Muenchen 7. and 8. Maerz, 1979.
50. York, S. M., and York, C. J.: Pacheco virus vaccine studies. Proceedings of the 32nd Western Poultry Disease Conference, Davis, CA, 1983, pp. 101–103.

## ACKNOWLEDGMENTS

The author highly appreciates the assistance of Thomas L. Lester, D.V.M., Ph.D., in preparing this manuscript. The author also thanks Richard J. Hidalgo, D.V.M., Ph.D., for his valuable suggestions.

HELGA GERLACH

# *Chapter 33*

# BACTERIAL DISEASES

HELGA GERLACH

Bacterial organisms from birds frequently cause difficulties taxonomically and in their evaluation as pathogens. A larger proportion of isolates from birds than from mammals is a biochemical variant of an established species. During the more intensive examination of many bird species in recent years, varieties of bacteria have emerged that are not yet taxonomically described and have no valid names. Some of them have erroneously been put in known taxons, e.g., *Moraxella anatipestifer, Bordetella bronchiseptica, Pasteurella haemolytica.* This is not to say that these species do not exist; they do. However, the avian strains listed under these taxons are usually different. Although some of these unclassified strains are definitely pathogens with the capacity for primary infection, many others and the biochemical variants of established species have not been tested for pathogenicity. The difficulties are obvious; the psittaciformes, which are the primary hosts, are not readily available for experimental infections. Chickens or other poultry are suitable for preliminary tests, but the psittaciforme hosts in question might react differently. This scenario is also true for many well-known bacteria. However, analogic conclusions can be drawn by asking the following questions:

1. Has the organism been isolated in large numbers and in almost pure culture? If the answer is yes, the agent should be considered further.
2. Has the organism caused a bacteremia; i.e., was it found in high numbers and almost pure culture in the heart tissue? If the answer is yes, the agent is considered to be part of the disease process.
3. Is this organism known as a component of the autochthonous flora? If the answer is yes, the agent is likely to be only a secondary invader.
4. Does the organism have pathogenicity markers? If the answer is yes, the agent is likely to have played a role in the disease process, although no conclusions can be drawn about primary or secondary pathogenicity.
5. Do the parenchymata, from which the culture in question was isolated, display pathologic and/or histopathologic lesions consistent with a bacterial infection? If the answer is yes, the agent has at least influenced the disease process at a significant level.
6. Are microorganisms, other than the bacterium in question, present (virus, chlamydia, other bacteria, fungi, protozoa)? If the answer is no, the agent may be a primary pathogen. If the answer is virus or chlamydia, the agent is likely to be a secondary invader. If the answer is fungi or protozoa, the agent may be a primary pathogen.

## STAPHYLOCOCCUS (S.)

Staphylococci are gram-positive, nonmotile spherical bacteria (0.8 to 1.0 μm in diameter) that grow readily on common media. Recent literature[1, 3, 4] has shown that 13 *Staphylococcus* spp. can be found in birds. In a study with 463 strains isolated from 10 different bird orders, including many psittaciformes, the groups listed in Table 33–1 were established for easier identification.

Pathogenicity markers include clumping factor, production of hemolysins, DNase, phosphatase, protein A, leucozidine, etc. For the avian strains, the clumping factor seems to have the closest correlation to pathogenicity. However, not much is known about the influence of the other pathogenicity markers on disease processes in pet birds. The isolation of strains with pathogenicity markers in pure cultures and with high numbers of organisms from the heart tissue of dead birds allows the presumption of at least facultative pathogenicity. High frequency and broad host spectrum are described for *S. xylosus*, *S. sciuri*, and *S. aureus*. All three species have been divided into several biotypes. These biotypes might be an adapta-

**Table 33–1.** GROUP DIFFERENTIATION OF STAPHYLOCOCCI

| Group | | Oxidase | Lysostaphin | Lysozyme | Xylose | Maltose | Nitrate reduction | Cellubiose | Pathogenicity markers |
|---|---|---|---|---|---|---|---|---|---|
| I | *S. aureus* | – | s | r | – | v | + | – | + |
| | *S. intermedius* | – | s | r | – | v | + | – | + (–) |
| | *S. hyicus* | – | s | r | – | v | + | – | + |
| | *S. simulans* | – | s | r | – | v | + | – | + (–) |
| II | *S. xylosus* | – | s(r) | r(s) | + | + | + | – | + (–) |
| III | *S. cohnii* | – | s | r | – | + | – | – | + (–) |
| | *S. saprophyticus* | – | s | r | – | + | – | – | – |
| IV | *S. haemolyticus* | – | r | r | – | + | v | – | + (–) |
| | *S. warneri* | – | r | r | – | + | v | – | + (–) |
| | *S. hominis* | – | r | r | – | + | v | – | + (–) |
| | *S. epidermidis* | – | r | r | – | + | v | – | + (–) |
| V | *S. capitis* | – | r | r | – | – | + | – | – |
| VI | *S. sciuri* | + | s(r) | r(s) | – | v | + | + | + (–) |

\+ = more than 90% positive
– = more than 90% negative
v = variable
(–) = clumping factor negative
r = resistant
s = sensitive

tion to specific hosts, since some biotypes have been found only in a limited number of bird species. The genus *Staphylococcus* is considered to be ubiquitous, and many species (except *S. aureus*) are part of the autochthonous flora. There is no doubt, however, that *S. aureus* contains more virulent strains than the other species, and the following discussion applies mainly to this species.

## S. aureus

Virulent *S. aureus* strains from birds form a specific group within the species and are only rarely transmitted to humans or mammals. The organism is found mainly in air and dust. Clinically healthy birds may harbor the organism on the skin and on the mucosa of the respiratory or digestive tract. *S. aureus*, like many other *Staphylococcus* species, displays a high resistance, especially against dryness, and can survive outside the host for long periods. Propagation outside the host under favorable conditions is possible. Resistance to common disinfectants is frequent, and a change of compounds is recommended in case of problems with *S. aureus*.

### *Transmission*

Transmission can take place vertically, horizontally, directly, or indirectly via vectors. Vertical transmission can be a problem with many *Staphylococcus* species.

### *Pathogenesis*

Endogenous infection causes septicemia, as a rule, and if the birds survive the acute stage, local disease processes develop. Local lesions are governed by the occurrence of thrombi in the arterioles, the results of which are necrosis of the affected areas. This is best seen at the tips of the extremities. But the parenchymata, particularly the liver and kidneys, can contain infarcts. The central nervous system (CNS) is occasionally involved as well. Endogenous infections can be primary but frequently are secondary.

Exogenous infection leads to local disease processes of the skin that can spread and finally become septicemic. Since birds are extremely resistant to wound infections, as compared with mammals, triggering factors—damage of the skin, other infections (particularly with *Clostridium*), immunosuppression, environmental stressors, prolonged application of an antibiotic—are essential to establish exogenous infections.

### *Incubation*

Data containing incubation periods following natural infection for psittaciformes are not available. However, since carriers develop, any time period from hours to several months is possible.

### Clinical Disease and Pathology

Staphylococcosis can manifest itself in different clinical and pathological forms, such as the following:

1. High embryonic mortality, which can be caused by egg transmission or lack of hygiene during the breeding process.
2. Navel and/or yolk sac inflammation, which is seen in newly hatched chicks (up to 10 days of age). The navel region is either dry and brownish or smudgy, reddish, and edematous. The yolk sac in the body cavity is not absorbed normally, and its contents may be deteriorated.
3. Septicemia, which occurs with nonspecific clinical signs such as lethargy, anorexia, humpback posture, ruffled plumage, and sudden death. Sudden necrosis of the end parts of the toes or the adnexa of the head and neck is rather specific. The color of the tissue changes to dark brown or black and becomes dehydrated. In the beginning, the involved toes are very painful and the bird shows "lameness." CNS symptoms can appear rather suddenly and consist of tremor, opisthotonus, torticollis, etc. At necropsy the large parenchymata display petechiae, ecchymoses, and infarct-like necrosis. Depending on the course of the disease process, histology shows infiltrates with mainly heterophils as well as granulomata.
4. Arthritis-synovitis can develop with serofibrinous or fibrinous inflammation of the synovial membrane of the joints, the tendon sheaths, and the bursae. Although all these tissues may be involved, preference seems to be for the tarsal and metatarsal joints. In chronic cases and following antibiotic therapy, *Staphylococcus* in the synovial membrane is occasionally present in the L-form or as a protoplast.
5. Osteomyelitis, which may be followed by chronic lesions of the skeleton. The proximal ends of femur, tibiotarsus, and tarsometatarsus, as well as the fifth to seventh thoracic vertebrae, are affected. The latter can lead to the so-called kinky back, in which distention and colliquation resulting in deformation of the sponglosa vertebrae may cause a narrowing of the spinal cord channel and compression of the spinal cord.
6. Exogenous infections, which can show various forms of dermatitis:

a. A vesicular dermatitis, which shows vesicles that contain yellowish exudate. Following rupture, brownish to blackish crusts develop. Infection with avipox virus can be an underlying factor, and histology is necessary for a differential diagnosis.

b. A gangrenous dermatitis, which is seen following edematous/hemorrhagic inflammation of the skin. The affected parts of the skin are blackish, smudgy, and without feathers. In many instances, gangrenous dermatitis is secondarily infected by *Clostridium perfringens* or another *Clostridium* species. Both the *Staphylococcus* and the *Clostridium* need a triggering factor to enter the tissue. Gangrenous dermatitis is rare in psittaciformes.

c. Bumblefoot is a condition frequently seen in raptors, but it can occur in any bird species. It is a serofibrinous or fibrinous inflammation, and even pus may be present. Pus is not produced normally in birds. The disease is manmade, i.e., seen only in captive birds. Although the pathogenesis of bumblefoot is still a matter of controversy, the tissue is frequently secondarily infected by *S. aureus* or other *Staphylococcus* species, as well as many other bacterial groups. Virulent *S. aureus* strains can become generalized from bumblefoot and cause other lesions or sudden death.

### Diagnosis

Apart from the necrosis of the toes, which is highly characteristic, diagnosis is made by culture and differentiation. For quick orientation, pathogenicity markers are determined, such as clumping factor or alpha-hemolysins. Alpha-hemolysins seem to be more toxic for birds than beta-hemolysins. As already mentioned, clumping factor–positive strains are likely to be virulent. These strains should be regarded as potential pathogens, and treatment should be considered for all birds still alive. Strains with neither of these pathogenicity markers are likely to be secondary invaders, and treatment should be considered for the clinical condition. However, emphasis should be placed on finding the primary disease agent.

### Differential Diagnosis

The differential diagnosis includes almost any septicemic disease, and only histology may narrow it down to a bacterial etiology. A special rule-out for the genus *Amazona* is an infestation with *Filaria* in the tendon sheath of the gastrocnemius muscle, causing a serofibrinous inflammation. For exogenous infections, allergies and ectoparasites should be ruled out. Frostbite is a noninfectious rule-out.

### *Treatment and Control*

For treatment, suitable antibiotics are given, preferably by injection, because the bird is likely to be anorectic. Since S. *aureus* is commonly resistant to a variety of antibiotics, sensitivity testing is mandatory. However, in an emergency an antibiotic is administered according to experience in a given area. Carbenicillin or gentamicin is often suitable. For control, hygienic measures are indicated.

## STREPTOCOCCUS (Str.)

Streptococci are gram-positive, coccoid-ovoid bacteria arranged in pairs or in chains, with a diameter of less than 2 μm. Growth is seen on most of the commonly used media. The genus *Streptococcus* comprises many species, which are differentiated by morphologic, biochemical, and serologic means. Streptococci are ubiquitous and found mainly in air and dust. The organisms survive outside the host for long periods. Disinfection with commonly used products is normally no problem.

In birds, *Str. zooepidemicus* (serogroup C) and members of serogroup D (*Str. faecalis, Str. faecium, Str. avium, Str. durans,* and *Str. gallinarum* sp. *nov.*) can be found. The literature shows conflicting evidence as to the pathogenicity of all *Streptococcus* species mentioned. This is possibly due to their opportunistic behavior under conditions in which virus or *Chlamydia* is the primary pathogen, the lack of experimental infections, and confusion in taxonomy or nomenclature.

### *Transmission*

*Str. faecalis, Str. faecium,* and *Str. avium* are components of the autochthonous flora and, as a rule, are nonpathogenic. However, transmitted vertically into the eggs or horizontally by deficient hygiene, streptococci can kill an embryo or cause septicemic death (omphalogenic) in newly hatched chicks. Other routes of transmission are aerogenic or percutaneous (if the skin is previously damaged). Not enough is known of the transmission of "nonavian" *Streptococcus* species, such as *Str. zooepidemicus.*

### *Pathogenesis*

On rare occasions, a *Streptococcus* species is the cause of septicemia in birds older than 10 days, and if the birds survive, a consecutive allergic arthritis, tendovaginitis, or endocarditis can occur.

### *Incubation*

The incubation period is short, from 24 hours to a few days. Details for psittaciformes are not known.

### *Clinical Disease*

Clinically characteristic for a peracute to acute course is an apoplectiform death or a high degree of depression and somnolence that leads to death within two to three days. Diarrhea, dyspnea, paresis, and conjunctivitis may also be seen.

The more chronic disease shows inflammation in the range of joints and tendon sheaths as well as adnexa of the head. Endocarditis can lead to insufficiency of the cardiac valves, which may be difficult to diagnose in live birds, since it is frequently expressed as dyspnea.

### *Pathology*

At necropsy, birds that have died suddenly display an exudative serositis as well as a swollen liver, spleen, and kidney. A prolonged course leads from focal to coalescent necrosis of the myocardium and other large parenchymata, as well as to thromboendocarditis verrucosa or ulcerosa. Lesions of the synovial membranes are fibrinous to necrotizing.

### *Diagnosis*

Diagnosis is made by culture and identification of the organism. Preference is given to liver, heart, and brain as sources for culture in acute cases. Isolation from the brain indicates that *Streptococcus* is a significant part of the disease process; however, a triggering factor might have been necessary to allow the organism to become septicemic.

### *Differential Diagnosis*

The differential diagnosis again includes almost any other septicemic disease that needs to be narrowed to a bacterial etiology by histology.

### *Treatment and Control*

For treatment, administration of an antibiotic, preferably intramuscularly, is recommended as soon as possible. Sensitivity testing is often necessary. The chronic lesions in joints and tendon sheaths merit treatment only in pets and valuable birds. Injections are given into the synovial cavities, and even surgical measures may be necessary. In contrast to mammals, these methods are more likely to succeed in birds, provided that techniques are clean, since the danger of wound infection is much less. Frequently streptococci within synovial membranes are in the L-form or in the form of a protoplast. Treatment consists of a combination of ampicillin and erythromycin. For control only hygienic measures are suitable.

## MYCOBACTERIUM (M.)

Pathogenic myocobacteria are gram-positive, acid-fast, granulated rods, $0.5 \times 3$ μm in size. *M. avium* is the species that commonly causes tuberculosis in birds. So far, all bird species examined have been found susceptible. In contrast to *M. tuberculosis,* large numbers of the causative agent are within the lesions and are excreted with the feces and urine. *M. avium* is pathogenic for birds, pigs, guinea pigs, rabbits, and humans (rare, but frequently resistant to therapy). Recently several (7) serotypes have been distinguished; serotype 2 is the common one in Europe.

*M. avium* has a high resistance and survives outside the host for approximately five months in cages and aviaries. Infected soil of aviaries, zoos, and similar environments is considered infectious for two years. For disinfection, only compounds tested against *Mycobacterium* are recommended.

### *Transmission*

In contrast to mammals, avian tuberculosis is considered an alimentary infection, although the aerogenic route cannot be completely excluded. Arthropods can serve as mechanical vectors. Egg transmission is possible, but is epornitically unimportant, because *M. avium* bacteremia causes an immediate cessation of egg production.

### *Pathogenesis*

Unlike mammals, birds suffer from the visceral form of tuberculosis. The main portal of entrance in birds is the intestinal tract, and the infection spreads into the intestinal wall as well as by means of the portal circulation. Penetration into the general circulation takes place early and in small numbers, usually without clinical signs. From within the blood, *M. avium* is removed by cells of the reticuloendothelial system (RES), mainly in the liver, spleen, and bone marrow. The lack of lymph nodes is instrumental in the rather unlimited hematogenous spread within the host. Locally, *M. avium* causes cellular reactions governed by the cell-mediated immune system. A secondary generalization out of the tubercles within parenchymata is possible, and during these processes the lungs can become involved. Exceptions are pigeons and anserinae, in which tuberculosis of the lungs is frequently the only gross lesion. The tubercles in the intestinal wall commonly have an opening into the intestinal lumen. Avian tuberculosis is therefore *never closed,* but is always an *open tuberculosis.*

### *Incubation*

As in mammals, tuberculosis in birds is a chronic disease, and incubation periods can range from weeks to months depending on the frequency and dose of exposure and the condition of the host.

### *Clinical Disease*

The clinical signs are not typical and consist in adult birds of chronic wasting in spite of a good appetite, recurrent diarrhea, polyuria, anemia, and dull plumage. Immature individuals rarely exhibit clinical signs, although they may already be infected. An intermittent favoring of one or the other leg or even lameness without visible lesions is caused by pain originating from bone marrow tuberculosis. Arthritis, mainly of the carpometacarpal joints and the elbow joint, can be seen occasionally. In these cases, the skin over the diseased joint often shows thickening or even ulceration. In pigeons, a granuloma may develop within the conjunctival sac.

### *Pathology*

The pathology is governed by the presence of miliary to greater than pea-sized nodules in the wall of the intestinal tract (covered by serosa), and in the liver, spleen, and bone marrow (Fig. 33–1). Involvement of lungs, serosa, and trachea (particularly in the peacock), as well as of the reproductive tract and the kidneys, is seen. Virtually all organs can be involved. The nodules are frequently necrotic in the center and, in advanced cases, display calcification. Histology shows accumulations of epitheliloid cells and, in more advanced cases, central necrosis, surrounded by a zone of epi-

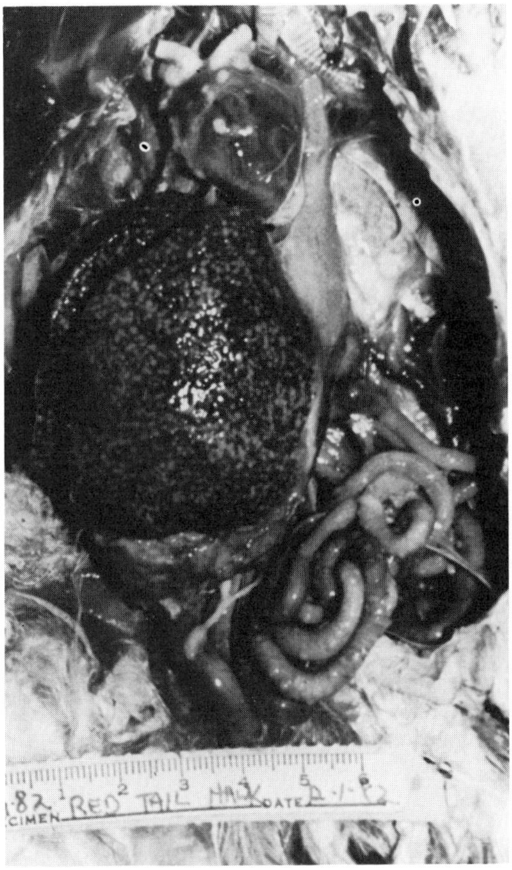

**Figure 33–1.** Characteristic of avian tuberculosis on necropsy are the nodules present in the liver, intestine, and other organs in a Red-tailed Hawk. (Courtesy of Richard Evans.)

thelioid cells interspersed with giant cells (of the Langerhans type) and marginated by connective tissue and lymphocytes. The number of giant cells is often considerably lower than in mammals.

### *Diagnosis*

The diagnosis can be made directly by demonstrating *M. avium* from biopsies or postmortem materials on slide preparations that are acid-fast stained. Care should be taken to get material from inside the nodules onto the slide. High numbers of red stained organisms confirm the diagnosis. Feces are useful sources of diagnostic material for birds with no real clinical signs but should be treated with $H_2SO_4$ or one of the sputum solvents used in human medicine. After centrifugation, the sediment is spread on a slide and acid-fast stained. Culture of *M. avium* is, as a rule, desirable, especially if more information on the strain is required or if it is doubtful that *M. avium* is present at all. Nonpathogenic mycobacteria, particularly in the feces, are wider than *M. avium* and show no granularity.

Indirect diagnosis is possible with the tuberculin test, which is an allergenic test, or with slide agglutination, which is a serologic method. For the former test, avian tuberculin is injected intradermally, preferably into naturally featherless skin, (e.g., the eyelid or cloacal lips); and the reaction is read after 48 hours (swelling, reddening, pain). The test often has false-negative results in early infections and in the late stages of the disease. The slide agglutination test with fresh plasma and *M. avium* antigen gives better results; nevertheless, approximately 10 per cent of the birds tested may have false-negative results. This is in view of the fact that only serotype 2 is available as antigen.

### *Differential Diagnosis*

The differential diagnosis includes infections by other bacteria that induce granulomata, such as *E. coli* granulomatosis, yersiniosis, salmonellosis, fungal granulomata (particularly by *Aspergillus* spp.), and neoplasia (such as leukosis, reticuloendotheliosis, and sarcomatosis).

### *Treatment and Control*

No treatment should be initiated for the following reasons: (a) There are no tested drugs for the visceral form of tuberculosis in birds. (b) A tremendous number of organisms are being shed continuously. (c) The potential danger for man and the absence of an appropriate method of treatment.

Birds that are positive either directly or indirectly for *M. avium* should be euthanized. Negative contact birds should be euthanized as well. If they are too valuable to be euthanized, they should be removed from the contaminated area, placed in cages, and kept singly if possible. Retesting should take place at six-week intervals for approximately one year. Birds still negative *and* in good body condition can then be considered free of the disease. There is almost no absolute means of control. Being alert in cases of chronic wasting and being careful in acquiring birds are all that can be done.

## M. tuberculosis

In contrast to some literature, infections of psittaciformes with *M. tuberculosis* are extremely rare and, as a rule, cause a benign, localized disease of the dermis around the ceres or nares as well as the retro-orbital tissue.[6] The bird in question is usually a pet with very close contact to humans and, as such, can serve as a sentinel for human patent tuberculosis. The affected dermis looks granulomatous and may even ulcerate. The swelling of retrobulbar tissue causes a protrusion of the eye (exophthalmos). Because it is usually unilateral, hyperthyroidism can be excluded. Diagnosis can be made by histology and/or culture of biopsies, the taking of which can cause considerable bleeding. Common sense dictates that infected birds be euthanized.

## ERYSIPELOTHRIX (E.)

The causative agent is *Erysipelothrix rhusiopathiae,* a gram-positive, nonmotile rod, 0.2 to 0.4 × 1 to 2.5 μm in size, and comprising 11 distinct serotypes. *E. rhusiopathiae* is widespread and can propagate outside the host. The organism survives particularly in moist soil and in the water of shallow lakes and ponds. The disease in birds has a seasonal peak during the late autumn, winter, and early spring. Reservoirs are identified as rodents and pigs. Humans are susceptible, and special care should be taken when handling infected birds, particularly in the presence of small skin trauma on the hands.

Disinfection with commonly used compounds seems to be efficacious.

### *Transmission*

The oral route is the main means of transmission. Egg transmission has been proven for the white stork *(Ciconia ciconia).*

### *Pathogenesis*

From the sporadic reports of erysipelas in the literature, it can be concluded that many bird species, including parrots, are susceptible. Waterfowl and birds living on fish or in swamps have a higher exposure and disease rate, particularly during the cold season, when feed is scarce and energy requirements are high. Other diseases and/or poor management seem to be prerequisites for natural infections in exotic birds. The disease has an acute septicemic course, during which most of the birds die. Survival of the acute disease can lead to allergic reactions in the dermis and in the joints.

### *Incubation*

The incubation period for turkeys and ducks is one to eight days. Details for psittaciformes and other bird groups are not reported.

### *Clinical Disease and Pathology*

The clinical picture is dominated by sudden death. Systemic disturbances may be absent or are observed as somnolence, weakness (sitting on the floor of the cage), anorexia, dark red to violet coloration of the featherless and nonpigmented or slightly pigmented skin.

At necropsy, petechiae are observed in the subcutis, musculature, and intestinal mucosa. The friable liver and spleen, with or without necrotic foci, are discolored dark red to black. The rare chronic course (mainly in turkeys and geese) is characterized by thickened, leather-like dermal areas *(E. rhusiopathiae* is found only in these parts) or serofibrinous arthritis or endocarditis valvularis. The parenchymata display degenerations, necrosis, and infarcts. Histology reveals degeneration of the wall of the blood vessels and thrombi.

### *Diagnosis*

The diagnosis is confirmed by isolation of E. rhusiopathiae. In case of progressive putrefaction, a culture of the bone marrow may be successful.

### *Differential Diagnosis*

The differential diagnosis includes pasteurellosis, enteritis of various etiologies, Newcastle disease, psittacosis, and listeriosis.

### *Treatment and Control*

*E. rhusiopathiae* is sensitive to penicillin, and parenteral injections are recommended as soon as possible. The tetracyclines are also suitable and may be preferred for psittaciformes. Hyperimmune serum can be given to valuable birds together with the antibiotic (two different syringes). For control, an adjuvant bacterin containing hemagglutinating strains is available. Vaccination may just sensitize the birds and provoke more cases of chronic disease than without vaccination. Hygienic measures are necessary to keep the birds from infected soil.

## LISTERIA (L.)

*L. monocytogenes* is a gram-positive, motile rod, 0.4 × 0.5 to 2 μm in size, that causes beta-hemolysis on blood agar. The organism shows high resistance in the environment and can propagate outside the host. Disinfection with commonly used compounds is effective.

### *Transmission*

*L. monocytogenes* is ubiquitous, and the oral route is the primary route of infection.

### *Pathogenesis*

There is some controversy about the capacity of *L. monocytogenes* to cause primary infections, but there is no doubt that oral transmission in birds usually leads to a latent or abortive infection. The occurrence of a clinically apparent disease depends on conditioning factors of different origin. In birds, the acute disease is a septicemia, with death occurring within one to two days. The subacute to chronic course of the disease is characterized by the reactions of the cell-mediated immune system, as in other bacterial infections in which the agent is found intracellularly.

*L. monocytogenes* has a broad host spectrum among birds, including psittaciformes. Man is susceptible, although the agent is rarely reported as having originated from birds. Canaries seem to be more susceptible than other bird species.

### *Incubation*

The incubation period following natural infection is not known.

### *Clinical Disease*

The clinical picture is governed by sporadic acute deaths. It is rarely a flock problem (except in canaries and related birds that are kept in dense populations). During a prolonged course, organ involvement occurs. Preferred parenchymata are heart, liver, and rarely brain. Clinically, the only signs regularly observed are torticollis, tremors, stupor, and paresis or paralysis. The white blood count in birds with subacute to chronic disease shows monocytosis (10 to 12 times normal), but this is a rather nonspecific sign for a variety of infectious diseases.

### *Pathology*

At necropsy, the simultaneous presence of a fresh serofibrinous pericarditis and myocardial necrosis is considered indicative. In the brain, no lesions are observable. Birds that die during the acute phase may show no lesions at all or only a few petechiae.

Histologic examination of the heart and liver shows mainly degenerative lesions, and no productive reactions are present. No brain lesions have been demonstrated, although the organism has been isolated from the cerebrum of birds with clinically manifested CNS signs.

### *Diagnosis*

A definitive diagnosis depends on the isolation of *L. monocytogenes*. Serologic results (slide agglutination) have to be evaluated critically because of the regular presence of latent infections.

### *Differential Diagnosis*

The differential diagnosis for the acute disease includes consideration of almost any other septicemia, but in particular erysipelas, pasteurellosis, salmonellosis, and yersiniosis; for the more chronic disease with CNS symptoms, paramyxovirus infections, psittacosis, and *Herpesvirus* infections should be considered.

### *Treatment and Control*

The tetracyclines are recommended for therapy. They are efficacious only in the acute to subacute cases. The number of tissue lesions at the beginning of treatment is decisive to the outcome. Birds with CNS signs do not respond to any treatment.

For control, hygienic measures are used, especially the reduction of the numbers of *L. monocytogenes* in the environment.

## CLOSTRIDIUM (Cl.)

Since the psittaciformes have no caeca, the presence of clostridia in the lower part of the intestinal tract is rare and probably only transitory. This is in contrast to other bird orders, such as phasianiformes, anatiformes, and in particular accipitriformes, in which clostridia are plentiful and are members of the autochthonous flora.

The clostridial species known to be pathogenic in birds are considered to be ubiquitous, and the oral intake of a large number of spores or exotoxins seems to be necessary for the onset of the clinical disease. The spores survive in soil and in feces for several years. Chemical disinfection is of questionable value, and, in most instances, not efficacious. Autoclaving is the method of choice for treating water and feed containers, nest boxes, perches, etc. A blowtorch might be used (with care) in aviaries.

### *Transmission*

The usual routes of transmission are oral and cutaneous.

### *Pathogenesis*

The pathogenesis is not yet fully recognized. Except for *Cl. botulinum,* which causes mainly an alimentary toxicosis, it seems that habitation and propagation of clostridial species are dependent on a decrease in intestinal motility. Reduced peristaltic action can be seen in certain forms of enteritis, following the administration of some drugs, and after infections with certain reovirus strains.

Experimental reproduction of the disease with a culture alone is usually not possible. Once *Clostridium* has colonized and produces exotoxins, lesions develop and death by toxemia is possible.

### *Incubation*

The incubation periods are not known for natural infections. Following experimental infections, a few hours up to three days are reported.

Since the same agent can cause different diseases and similar disease can be caused by different clostridial species, the following two sections are headed by clinical signs.

**Necrotic Enteritis or Ulcerative Enteritis.** The latter form is typical in the Bobwhite Quail *(Colinus virginianus)* and is usually caused by *Cl. colinum* (species incertae sedis). In most other bird species, particularly in game birds of the subfamily tetraoninae, *Cl. perfringens* type A, more rarely type B, C, or D, or most rarely type E, is the etiologic agent.

The clinical pattern is nonspecific and is usually observed in young birds from the second week on. Adult birds are more resistant. In the acute course of the disease, diarrhea, with or without blood in the feces, and an increased intake of water are seen. General signs develop, followed by death within a few hours. The more chronic course is characterized by discontinuation of growth and loss of body weight. The emaciated birds finally die.

At necropsy, a diffuse or focal reddening of the mucosa, usually of the upper jejunum, is seen, which develops into necrotic areas or into ulcers. These lesions start as pinpoint foci and later display a necrotic center with a wall and a reddish halo. Ulcers may coalesce and perforate the intestinal wall. Liver, spleen, and kidneys may be swollen and can contain necrotic foci surrounded by a greyish muddy zone.

The diagnosis is made by pathology and isolation of the agent. For *Cl. perfringens,* the toxins can be demonstrated in serum, intestinal contents, or liver homogenates (sterilized by filtration) by injection into mice or guinea pigs, some of which are protected by antitoxins.

The differential diagnosis includes certain forms of salmonellosis and the local enterotoxic form of *E. coli* infections. Ulcers called "boutons" can also be seen with Newcastle disease. A necrotic enteritis develops following applications of mebendazole in penguins (spheniscidae), cormorants (phalacrocoracidae), and pelicans (pelicanidae).

Treatment for acute cases usually begins too late. Lincomycin or spiramycin can be administered via drinking water. For control, vaccination with a *Cl. perfringens* toxoid is successful but expensive. In Bobwhite Quail, zinc bacitracin as a feed additive (200 parts per metric ton) can be tried.

**Gangrenous Dermatitis.** Most cases have multiple etiologies. The clostridial agents found are *Cl. perfringens* type A, *Cl. septicum,* and *Cl. novyi.*

The portal of entry is the skin. Predisposing factors include necrotic lesions (not always visible macroscopically). Avipoxvirus, staphylococcosis, or a combination of both is also a common triggering factor. Immunosuppression is another factor.

Clinically, a sudden onset of general signs and a loss of feathers in the affected part of the skin are observed. The areas involved are blue-red or almost black in color, swollen because of gas in the tissue (emphysema), and painful. A toxemia causes death, as a rule within 24 hours.

At necropsy, emphysema, edema, and hemorrhages with or without necrosis are seen in the subcutis, skeletal musculature, and myocardium. The diagnosis can be confirmed only by isolation of clostridia. For differential diagnosis, septicemia by *Aeromonas hydrophila*, dermatitis by staphylococci, and a primary infection by avipoxvirus should be ruled out.

Because of the rapid course of the disease, treatment cannot alter the situation. Lincomycin, spiramycin, and local application of $H_2O_2$ (only on small areas) can be tried. It is essential, however, to find the predisposing causes for the clostridial infection and to correct them.

**BOTULISM (SYN. LIMBERNECK).** *Cl. botulinum* produces neurotoxins that are the cause of the signs. Usually it is an alimentary intoxication; only rarely is the organism found producing its toxin within the intestinal tract. Exotoxin types A to F have been identified, and types A and C are predominant in diseased birds. Sources of the toxins are cadaverous proteinaceous feed, including plants and fly maggots, which are resistant to the toxins and which have been feeding on cadavers. The toxins have a specific affinity for the nervous system and also cause damage to the walls of the blood vessels, thus producing edema and small hemorrhages.

The clinical picture is characterized by a flaccid paralysis of skeletal musculature, including the tongue and the muscles of deglutition. Bulbar paralysis and loss of feathers, as well as diarrhea, are additional signs. Pathology and histopathology frequently do not yield characteristic lesions.

Diagnosis centers on various methods of toxin demonstration. Serum and filtered extracts from liver or kidneys, as well as from feed or water (sewage, mud), should be used for diagnosis (preserved by deep-freezing). The toxins are sensitive to temperature and are inactivated slowly even at room temperature. For differential diagnosis, intoxication by certain algae has to be considered in birds that have access to muddy, drying bodies of water.

Treatment can be tried in early cases by parenteral application of antitoxin and by oral administration of a laxative, such as sodium sulfate. Guanidine (15 to 30 mg/kg body weight) counteracts the effect of the toxins at the motor endplates.

For control, hygiene measures are necessary. A vaccination with a toxoid is possible but expensive.

## ENTERIC ORGANISMS

The Enterobacteriaceae are a family of related gram-negative bacteria, motile or nonmotile, that grow easily on commonly used media. For identification, biochemical and serologic patterns are established. Many species have developed a series of biotypes and serotypes. For complete identification, the O, K, and, in motile species, the H antigens are determined. The presence of group-specific antigens can result in cross-reactions not only between species but even between genera.

Given favorable conditions, Enterobacteriaceae can propagate outside the host. Except for encapsulated strains, i.e., many Klebsiella strains, a variety of disinfectants are efficacious. Enterobacteriaceae are ubiquitous and received their name from the fact that many members of this family are typically enteric organisms and are found as part of the autochthonous flora in the intestinal tract of man and many domesticated animals, including poultry. A majority of the clinically healthy psittaciformes, fringillidae (finches), ploceidae (weaver finches), and astrildae (waxbills) do not carry Enterobacteriaceae in their digestive tracts. As far as is known, the respiratory and the reproductive tracts of healthy avian specimens are free of Enterobacteriaceae. However, this group of bacteria frequently acts as secondary invaders as well as primary pathogens at times. There are considerable differences in the virulence of the bacterial strains and in the reactions of various avian hosts. The genera *Shigella* and *Edwardsiella* are not normally cultured from pet birds. The genera *Enterobacter, Serratia, and Proteus*, are generally of a low pathogenicity. The presence of *Enterobacter agglomerans** (a plant pathogen) in feces frequently is indicative of ingestion of seeds that contain more than $10^6$ bacteria/gm. Such seeds are considered to be deteriorated and toxic because of their high content of endotoxins. *Serratia marcescens* is increasingly found in large parrots with chronic debilitating diseases, prior antibiotic treatment, and immunosuppression.[5] As with other opportunistic organisms, once *Serratia* is established beyond a mucosal barrier it can sustain a disease process of its own.

The genera *Escherichia, Salmonella* (including subgenus III), *Klebsiella, Citrobacter,* and *Yersinia* are discussed in more detail below:

### Escherichia (E.)

*E. coli* is the important member of this genus, and in pet birds it may be even more significant

*Nomenclature according to Bergey's Manual of Systematic Bacteriology, Vol. 1. Baltimore, Williams and Williams, 1984.

than *Salmonella.* The size is approximately 0.5 × 3.0 μm. Motile and nonmotile strains occur. Attempts to classify the numerous avian *E. coli* strains by virulence have not been successful. The serologic classification used in humans and mammals does not identify the virulent strains of birds. Virulent and avirulent strains are found in many serogroups; however, within the serogroups 01, 02, and 078 there seems to be a slightly higher frequency of virulent strains. Experimental infections, in connection with biochemical studies, have showed that the lysine decarboxylase–negative strains examined are all virulent. Unfortunately, the lysine decarboxylase–positive *E. coli* strains also include many virulent strains. Experimental studies are desirable in many cases, but often are not possible with pet bird species.

### *Transmission*

Although the oral route is the natural one, aerogenic infections caused by dust containing the agent are not rare. *E. coli* can be egg-transmitted.

### *Pathogenesis*

The pathogenesis of the primary disease in birds is only poorly understood. In contrast to mammals, in birds typical exotoxins are rare except for the enterotoxins. At least three different endotoxins are considered to induce a hypersensitivity angiitis followed by septicemia and lethal shock.

### *Incubation*

As with other ubiquitous organisms, incubation times following natural infections are vague. Freshly hatched chicks or birds after experimental infection show incubation times of 24 to 48 hours.

### *Clinical Disease and Pathology*

The clinical signs of the primary infection are often governed by the mode of infection, i.e., digestive, respiratory, genital, etc. The following differentiation is made:

**COLISEPTICEMIA WITH OR WITHOUT ORGANOPATHY.** A sudden onset of disease with somnolence, ruffled plumage, anorexia, and need for higher environmental temperature is characteristic. Diarrhea and polyuria can occur. The latter is clinically not prominent, although the kidneys are rather regularly involved in the disease process. A penetration of *E. coli* into the CNS during septicemia is rare, except for penetration into the eye, where exudation of fibrin into the anterior eye chamber and/or uveitis occurs. The occurrence of *E. coli* in the synovial membranes of the joints is also uncommon, but serofibrinous arthritis is possible. According to the course of the disease, varying degrees of a fibrinous polyserositis are present at necropsy. Serofibrinous inflammation with infiltrates of mainly plasma cells is seen in liver and kidneys.

**LOCALIZED DISEASE OF THE INTESTINAL TRACT**

**Acute Enteritis** (caused by enterotoxins that are true exotoxins). *E. coli* propagates within the intestinal lumen, and the toxin induces an increased secretion of fluids. The birds show diarrhea and they lose electrolytes.

**Local Invasive Disease.** Certain *E. coli* strains are capable of infection and destruction of the intestinal epithelium, which results in a pseudomembranous or ulcerative enteritis. Clinically, either sudden death occurs or a nonspecific disease develops over a few hours. Usually this type of enteritis is diagnosed only at postmortem examination.

**RESPIRATORY DISEASE CAUSING RHINITIS AND AIR/SACCULITIS.** The latter condition can extend to the peritoneum, and a characteristic fibrinous polyserositis develops. Owing to the special anatomy of the avian lung, pneumonia is not present as frequently as expected. Young chicks that have inhaled a high number of virulent *E. coli* with dust (in unclean hatchers, etc.) are the primary victims of pneumonia (those chicks may display navel inflammations as well).

**ASCENDING INFECTION OF THE GENITAL TRACT.** This condition in females causes a salpingitis and even an oophoritis. The final state is frequently a lethal salpingoperitonitis. The same process in males resulting in orchitis and permanent sterility is rare.

**COLIGRANULOMATOSIS (HJAERRE'S DISEASE).** This is a chronic disease of mature birds caused by the mucoid form of the *E. coli* serotypes 08, 09, and 016. Lesions of the enteric mucosa caused by other agents, particularly parasites, are supposed to provide the portal of entry for this *E. coli* type. It is assumed that the galactans of the mucoid capsule induce the production of granulomata.

Clinically the birds become emaciated and show diarrhea and polyuria. Sometimes granulomata appear in the skin. At necropsy greyish foci of varying sizes are seen in the liver, under the serosa of the intestine, and, more rarely, in the spleen or kidney. Usually there is no opening into the intestinal lumen, which is in contrast to avian tuberculosis. The center of the foci can be mineralized. By histology, a wide demarcation around the central necrosis is seen, comprising numerous multinucleated giant cells and only a few heterophils. Acid-fast staining is necessary to rule out *Mycobacterium avium.*

The secondary *E. coli* infections usually produce the same lesions as the primary ones except for the localized disease producers.

### *Diagnosis*

Septicemia with polyserositis and granulomatosis can be indicative of *E. coli* infection. However, the isolation of *E. coli* is necessary for the diagnosis. Serotyping is of academic interest. The estimation of the virulence and of the capacity to be a secondary or primary pathogen is of more importance for the practitioner.

### *Differential Diagnosis*

For the septicemic disease, a variety of bacteria and virus, most of which are clinically too similar to allow differentiation, have to be ruled out.

For the enteritis acuta, other Enterobacteriaceae including *Salmonella* and the *Pseudomonas-Aeromonas* group have to be considered. The local invasive disease should be compared with the lesions caused by *Clostridium.* For CNS signs, infections with paramyxovirus, *Chlamydia*, and bacteria such as *Salmonella*, *Klebsiella*, *Staphylococcus*, and *Listeria* must be ruled out.

### *Treatment*

Antibiotics are chosen for treatment according to the results of sensitivity tests and to the tissue involvement. In psittaciformes, intramuscular application is frequently the only successful method. The administration of antibiotics is only one part of the treatment. In order to install a proper autochthonous flora, the pH of the intestinal tract should be lowered using lactobacilli, either as commercial cultures or as sour milk products. Both are nonindigenous *Lactobacillus* spp. that do not colonize but if given in high numbers over two to three weeks will usually change the intestinal environment enough for the indigenous *Lactobacillus* spp. to take over.

### *Control*

Because of the variety of pathogenic serotypes, vaccines have not been satisfactory. Since the majority of the strains are secondary invaders, the main effort should be to analyze the situation and monitor the primary disease factors.

## Salmonella (S.)

This includes subgenus III (*S. arizonae, Arizona hinshawii*). The genus *Salmonella* includes approximately 1500 species. The gram-negative rods are 0.3 × 2 to 3 μm in size. Most strains are motile. They grow on most of the routine media, but selective media for easier recognition are advantageous. The identification of the subgenera is made by biochemistry tests, and the species are recognized serologically (O, K [Vi], H antigens). Lysotyping is used for subspecies identification and epornitic research.

*Salmonella* survives outside the host in dry feces for eight months to two years and in water for three weeks. Propagation in the environment is possible depending upon ambient temperatures and available nutrition. The commonly used disinfectants are normally effective.

Almost all *Salmonella* spp. can infect pet birds. Host reactions are quite variable. However, adaptations of *Salmonella* strains to certain host species have occurred, e.g., *S. typhimurium* var. *copenhagen* in pigeons and some European finches.

### *Transmission*

As enteric bacteria, salmonellae are transmitted primarily by the oral route. Aerogenic infection can occur by infective dust from feces or feathers. Egg transmission can be seen regularly with host-adapted strains such as *S. typhimurium* var. *copenhagen* in pigeons and finches. From experimental studies, it is estimated that less than 10 bacteria are transmitted per egg and that the developing embryo can survive and hatch with *Salmonella* as the first organism to be harbored in the intestine. The germ is then spread horizontally. By this means, an infective chain is completed, independent of any other vector. Another route of vertical transmission is the feeding of the offspring via crop contents.

On the other hand, at least the large parrots (African Grey, Amazon, cockatoo, macaw) have been proven to pick up infections from human carriers.

The subgenus III (*S. arizonae*) is indigenous in reptiles, and birds frequently acquire it by close contact with snakes, turtles, lizards, etc.

### *Pathogenesis*

As typical members of the Enterobacteriaceae, salmonellae contain endotoxins that can cause an intoxication, particularly a food intoxication. In birds, however, infection is the main cause of disease. The capacity to penetrate the mucosal barrier seems to be an important factor. It distinguishes between strains that are primary pathogens (penetrate and propagate beyond the mucosa) and secondary invaders that need mucosal lesions of different origin in order to penetrate. Many nonpenetrating strains propagate well and are often capable of inhabiting the intestinal tract for varying periods of time. Birds infected with these strains can shed the organisms without having clinical symptoms (carriers).

Following penetration, bacteremia allows *Salmonella* to enter the parenchymata, where propagation leads to lesions. If a secondary bacteremia occurs, it coincides with the onset of clinical disease. The outcome can be death, elimination of the *Salmonella* and immunity, or persistent infection, sometimes together with localized chronic disease processes.

### *Incubation*

Incubation periods of three to five days are reported for the acute outbreak. Since carriers occur, incubation times can be vague.

### *Clinical Disease and Pathology*

The acute disease starts with general signs like lethargy, anorexia, polydipsia (followed sometimes by polyuria), and diarrhea. The more subacute course displays conjunctivitis, iridocyclitis or panophthalmia, CNS signs (particularly in pigeons), and arthritis (again, frequent in pigeons).

At necropsy, dehydration, degeneration, or necrosis of skeletal musculature, gastroenteritis (occasionally with ulcers), enlargement of the liver and spleen with or without small whitish foci disseminated throughout the tissue, bile congestion, or nephritis is present. The protracted course can lead to pericarditis or epicarditis fibrinosa, development of granulomata in the large parenchymata, and degeneration or inflammation of the egg follicles or testes.

The lories (loriidae) are extremely susceptible and develop a peracute disease with high mortality even when infected with strains that are considered nonpenetrating in other host species. The African Grey Parrot (*Psittacus erythacus*) is very susceptible as well, but here the disease is more subacute to chronic, with granulomatous organic lesions being predominant. Phlegmon and skin granulomata, as well as arthritis and tendovaginitis, are not uncommon. The finches can show a characteristic granulomatous ingluveitis caused by subacute salmonellosis, which can be confused with crop candidiasis or crop capillariasis.

Histology is rather nonspecific, showing an intestinal inflammation and focal necrosis in the heart, lung, liver, spleen, and kidney. More chronic cases reveal granulomata.

### *Diagnosis*

Although some of the lesions are indicative, diagnosis is made by isolation and identification of the organism. A serologic survey of a flock is useful only if the *Salmonella* species is known. Chronic carriers are frequently serologically negative.

### *Differential Diagnosis*

The rule-out list includes infections by other Enterobacteriaceae, *Pasteurella* spp., *Erysipelothrix, Listeria monocytogenes*, as well as *Chlamydia*, paramyxovirus, herpesvirus, and adenovirus. For the ingluveitis, *Candida* and *Capillaria* have to be excluded.

### *Treatment*

Because of the possible public health hazard, pet birds carrying *Salmonella* should be treated, including administration of *Lactobacillus* spp. (see *E. coli*), until the organisms are eliminated. Generally, any treatment regimen should consider antibiotic sensitivity and the tissues involved in the disease process. CNS signs and chronic organic lesions normally do not respond to any treatment. In flocks in which *Salmonella* is egg-transmitted, treatment and possibly serologic monitoring should be concentrated on the breeder stock. The author has found from experience that it is almost impossible to eliminate *Salmonella* from birds infected by egg transmission. The breeder stock is treated and the feces cultured about 2 to 3 weeks after therapy, and again 10 to 15 weeks later. *Salmonella* infections, particularly the better host-adapted ones, seem to cycle in periods of approximately three months. The collection of hatching eggs should be timed to approximately four weeks after therapy. The newly hatched chicks should be cultured (fecal swabs) at once and treated if infected.

### *Control*

Hygienic measures are the only methods available for control and should include control of flies, rodents, and other pests as well as pelleting of feed whenever possible, regular control of feed components, and regular cleaning and disinfection of the premises and the hatchers. Experiments with vaccines, particularly in pigeons, have not been satisfactory so far.

## Citrobacter

This gram-negative rod is of the same size as *E. coli* and *Salmonella. Citrobacter* is not as widely distributed as other Enterobacteriaceae. The tenacity seems to be comparable to that of other members of the group, but detailed studies are not available.

*Citrobacter* is found frequently in Weaver Finches and Waxbills, where it is apparently a secondary invader. Following penetration of the intestinal mucosa, bacteremia occurs and the small birds succumb to disease or die acutely. Clinically, at best, general signs and diarrhea can be observed. The postmortem examination reveals either nothing or petechiae in the heart musculature and parenchymata which is consistent with a bacteremia. Surviving birds frequently become carriers.

The diagnosis is made by isolation and identification. The differential diagnosis includes consideration of other Enterobacteriaceae and paramyxovirus. For treatment, antibiotics as determined by sensitivity testing are given. For organisms that are sensitive, neomycin via drinking water is promising to clear carriers that harbor the organisms in the intestine. For permanent success, the primary disease factor has to be eliminated.

## Klebsiella (K.)

*K. pneumoniae* and *K. oxytoca* are the frequently occurring members of the genus in pet birds. The taxonomical position of *K. oxytoca** is uncertain, as is its capacity for primary infections in pet birds. *K. pneumoniae* is the causative agent of Friedländer's pneumonia in humans. So far, there are no reports of human disease acquired from birds. Whether or not pet birds can become infected by human carriers is unknown.

*Klebsiella* strains are typical members of the Enterobacteriaceae in size, but the majority of the strains are encapsulated and nonmotile. Because of the capsule, disinfection can be a problem; therefore, heat disinfection should be used wherever possible.

Actual data about transmission, pathogenesis, and incubation time for pet birds are not available. Because of the capsule, *Klebsiella* strains are highly resistant to nonspecific host defense mechanisms, and carriers develop. As is typical for Enterobacteriaceae, endotoxin is produced. Following penetration through the intestinal (or respiratory) mucosa (whether primary or secondary), bacteremia develops with or without organopathy. Nephritis is the most frequent lesion, followed by pneumonia, hepatitis, and occasionally encephalomyelitis.

The diagnosis is made by isolation and identification of the organism. The differential diagnosis includes consideration of other Enterobacteriaceae, pasteurellosis, erysipelas, *Listeria*, paramyxovirus, herpesvirus, and *Chlamydia*.

The genus *Klebsiella* is naturally resistant to ampicillin and carbenicillin. Treatment should be given intramuscularly to psittaciformes according to sensitivity testing. Gentamicin or polymyxin B can be tried for emergency treatment. The control consists of the usual hygienic measures.

## Yersinia (Y.)

*Y. pseudotuberculosis* and *Y. enterocolitica* have been isolated from pet birds. Yersinia are gram-negative coccoid to ovoid rods. 0.4 to 0.8 $\times$ 0.8 to 1.0 μm in size.

### 1. Y. pseudotuberculosis

This organism is motile at 20 to 28°C and grows on most of the common laboratory media.

*Still listed as such in Bergey's Manual of Systematic Bacteriology, Vol. 1, 1984.

Six different serotypes are distinguished; serotype I is the most frequent one from birds. The organism propagates in the environment, provided that there is enough organic nitrogen, at ambient temperatures as low as +4°C. Disinfection poses no special problems.

### *Transmission*

*Y. pseudotuberculosis* is not indigenous to the United States. Imported birds from northern and middle Europe theoretically can carry the organism. The oral route of infection is the natural one, and the organism is shed via feces.

The host spectrum is wide and includes many bird species, various mammals, particularly rodents, and man. Birds and rodents are known as potential carriers. Toucans, aracaris, barbets (piciformes), and turacos (musophagidae) are extremely susceptible.

### *Incubation*

Incubation times range from several days to two to three weeks depending on host susceptibility, strain virulence, and dosage.

### *Clinical Disease and Pathology*

Piciformes and musophagidae die peracutely without clinical disease. The acute form of the disease is characterized by severe general signs such as somnolence, emaciation, dehydration, diarrhea, and dyspnea. Locomotor disturbances (flaccid paresis or paralysis of the hind limbs) can occur with a more protracted course. The chronic disease is characterized by wasting as in tuberculosis. At necropsy, splenomegaly and hepatomegaly and enteritis are seen in peracute cases. The acute form displays submiliary to miliary, sharply demarcated, greyish foci within the large parenchymata, including the lungs. The chronic course reveals granulomata in various organs and in the skeletal musculature with or without ascites and osteomyelitis. Histology reveals coagulation necrosis, possibly surrounded by heterophils in the organs. With the prolonged course, development of granulomata is seen.

### *Diagnosis*

Although pathology and particularly histopathology might be characteristic, isolation and identification of the agent are necessary. Cold enrichment may be helpful for primary isolation.

### *Differential Diagnosis*

The rule-out list includes salmonellosis, coligranulomatosis, pasteurellosis, and avian tuberculosis.

### *Treatment*

Treatment of peracute and acute cases comes too late, and, since the birds have ceased eating and drinking, drugs have to be given parenterally. Handling of toucans and turacos can be life-threatening to them even under normal conditions. Treatment in chronic cases does not reach the *Yersinia* that is surrounded by granulomatous tissue. Flock prophylaxis by treating the birds that are clinically still healthy, in addition to sanitary measures, can be tried.

### *Control*

In countries where the organism is not endemic, bacteriologic examination of the feces during quarantine is recommended. Rodents and wild birds are reservoirs in endemic areas, and these animals should be controlled and kept off the premises.

Vaccines have been tried, particularly in zoos, but have not yet been found to be satisfactory.

## 2. Y. enterocolitica

This organism is considered to be a human pathogen. Primarily, children of elementary school age are affected. The occurrence in birds is rare and so far is limited to individuals exposed to sewage plants. From the scattered data available, the isolation of *Y. enterocolitica* from birds may indicate exposure to human carriers. No distinct lesions have been observed in birds. Cold enrichment is probably the best method for isolating small numbers of the organism from material contaminated with various bacteria.

## Pasteurella (P.)

Pasteurellae are gram-negative, nonmotile, ovoid rods that stain bipolarly from tissue smears or first culture passages after fixation in methanol and staining with methylene blue.

Diseases caused by members of the genus *Pasteurella* occur in many bird species, including phasianiformes, anatiformes, psittaciformes, columbiformes, and passeriformes. The susceptibility of the birds is highly variable, but the acute and chronic course of the disease, as well as several pathomorphologic changes, are similar to that of yersiniosis. In contrast to *Yersinia*,, the capacity to propagate outside the host is low and the survival of *Pasteurella* is dependent on temperature, relative humidity, pH, and the kind of carrying material. Disease outbreaks seem to have a seasonal distribution, with a peak occurring during November and December in the northern part of the world, whereas in tropical zones the disease peak coincides with peak ambient temperature and humidity.

Commonly used disinfectants are efficacious as a rule.

### *Transmission*

The genus *Pasteurella* is primarily an etiologic agent for respiratory disease in birds, and the portal of entry is the respiratory tract. Nasal cavities, sinuses, and choanae are the sites where clinically healthy carriers harbor the organisms. Direct contact or indirect transmission by vectors can cause infections. Rodents and wild birds are particularly important live vectors.

*P. multocida*, *P. pneumotropica*, and *P. gallinarum* have been proven to occur in birds. Avian strains classified as *P. haemolytica* usually are not a *Pasteurella* at all (gram-negative pleomorphic rods), and their taxonomic status is uncertain. Many strains seem to belong to the genus *Actinobacillus*.

## 1. P. multocida

The species is serologically nonuniform, and 16 endotoxin types as well as 4 capsular polysaccharides are distinguishable. In birds, serotypes 1 and 3 and capsule types A and D are predominant. Differences in virulence among the strains are marked.

### *Pathogenesis*

Following propagation in the lung, highly virulent strains produce an acute septicemic disease and death, whereas with less virulent strains an organic manifestation is seen after bacteremia. Weakly virulent strains cause a local chronic respiratory disease. Endotoxic damage in the walls of the blood vessels leads to edema and hemorrhages and coagulation necrosis in the liver. Outbreaks of the disease, particularly by less virulent strains, are triggered by unfavorable environmental factors and other diseases.

### *Incubation*

Variations in incubation periods are expected according to host susceptibility. Generally, incubation times are only a few days, but no data are available for pet birds.

### *Clinical Disease and Pathology*

Acute death or general signs such as cyanosis, dyspnea, and diarrhea are seen in the acute form of the disease. Sometimes excess mucus is present around the nostrils or beak. A prolonged disease reveals respiratory rales, exudates from nares and conjunctivae, and/or swelling of the sinus infraorbitalis. Arthritis and CNS signs are possible. Granulomata in the skin can be seen in raptors, owls, and pigeons. These granulomata contain the agent, but they can be removed surgically followed by antibiotic treatment.

At necropsy, the acute form of the disease displays petechiae or ecchymoses over the large parenchymata at best. Prolonged disease produces exudative serositis (white in contrast to the yellow with *E. coli*) and necrotic foci in several organs. Following chronic disease, catarrhal to fibrinous rhinitis, sinusitis, blepharoconjunctivitis, and tracheitis may be seen. A croupous pneumonia is characteristic of the more chronic course. Localized lesions include arthritis, osteomyelitis, otitis media, and skin granulomata. Histology is rather nonspecific.

## 2. P. pneumotropica

This species is indigenous in rodents and is only occasionally seen in pet birds, except in pigeons. As indicated by the name, pneumonia is the main lesion seen at postmortem examination. Since the clinical diagnosis of pneumonia is difficult in birds, acute or subacute death is the only regular clinical observation. Dyspnea prior to death may develop. Detailed data on infections with *P. pneumotropica* in psittaciformes and other pet birds are not available.

## 3. P. gallinarum

The avian host spectrum seems to be the same as for *P. multocida*. Although sufficient data are not available, from experience it appears that this species is far less pathogenic than *P. multocida.* However, respiratory signs such as coryza, conjunctivitis, rales, and dyspnea are observed once the organism has penetrated the mucosa. In psittaciformes, *P. gallinarum* has been isolated from the choanae and nostrils of live birds suffering from air sac and lung aspergillosis. As far as is known, *P. gallinarum* is a secondary invader that needs one or several triggering factors for penetration. At necropsy, a catarrhal to fibrinous inflammation of the upper respiratory tract as well as a pneumonia can be displayed. The air sac may be involved in the disease process.

### *Diagnosis*

Isolation and identification of the causative agent are necessary for diagnosis.

### *Differential Diagnosis*

Other respiratory disease agents, in particular paramyxovirus, *Chlamydia,* avian influenza, other bacteria, *Aspergillus,* and *Syngamus,* have to be ruled out.

### *Treatment*

Acutely sick birds are rarely saved by treatment. Intravenous (or intramuscular) application of long-lasting sulfonamides or broad-spectrum antibiotics can be tried. The prolonged course of the disease, particularly by *P. multocida,* has a better chance for successful treatment. Initially, the parenteral application of chemotherapeutics is necessary, but as soon as the birds are eating again, administration with food and water is possible except in psittaciformes. For very valuable birds a combination of chemotherapy and hyperimmune serum has proven beneficial. Birds in the chronic stage of the disease cannot be treated successfully because of irreversible damage to the parenchymata.

### *Control*

The control of rodents and wild birds is mandatory to avoid transmission into the aviary.

Vaccination against *P. multocida* is available but not satisfactory. This is due to the many serovariants and to the fact that immunity is developed best against the toxins. Autogenous vaccines may be the partial answer for large aviaries with continuous problems.

## Haemophilus (H.)

The genus consists of a group of small (0.3 to 0.6 × 1 to 3 μm) gram-negative coccoid to pleomorphic rods that may need X- and/or V-factor (see Chapter 10, Clinical Microbiology.) for growth. Strains from pet birds are difficult to classify, and many have no valid name. *H. paragallinarum,* which is the agent of coryza contagiosa gallinarum and a primary pathogen, is rare in pet birds. Columbiformes and anatiformes seem to be refractory. A variety of other strains can be isolated from birds with coryza, but their pathogenicity is doubtful. At best they are secondary invaders sustaining an upper respiratory disease. *H. paragallinarum* is serologically not uniform.

*Haemophilus* strains do not survive in the environment for longer than several days. Disinfection is no problem.

### *Transmission*

Transmission takes place by the respiratory route, and clinically healthy carriers are the source.

### *Pathogenesis*

*H. paragallinarum* produces neuraminidase and some strains also hyaluronidase. Many strains are encapsulated. Details for other *Haemophilus* species, particularly from psittaciforme birds, are insufficiently known.

### *Incubation*

The incubation for *H. paragallinarum* is one to five days. No details are known for other species.

### *Clinical Disease and Pathology*

Nasal exudate, which is primarily serous and becomes mucoid to fibrinous, is characteristic. Conjunctivitis and sinusitis can develop. A catarrhal to fibrinous rhinitis is seen at necropsy. Bronchopneumonia and air sacculitis are usually indications for involvement of other microorganisms (virus, other bacteria, *Candida* sp.). By histology, noncharacteristic lesions are observed.

### *Diagnosis*

The isolation and tentative identification of *Haemophilus* are necessary for diagnosis.

### *Differential Diagnosis*

Other respiratory diseases, in particular mycoplasmosis, have to be ruled out. In psittaciformes, *Chlamydia* should be considered.

### *Treatment*

Sulfonamides are the drugs of choice for *Haemophilus* infections. Chronic fibrinous sinusitis may need surgical opening of the sinus, removal of the caseous content, rinsing of the cavity with a mild disinfectant (quarternary ammonium base), and open healing.

### *Control*

In birds in which *Haemophilus* is a problem, culturing (choanae) is recommended, and carriers are treated according to sensitivity testing. Repeated treatment with sulfonamides frequently leads to intoxication and hypersensitivity, pathologically recognized as "hemorrhagic syndrome."

In locations with *H. paragallinarum* problems, vaccination with a bactrin containing several serotypes can be tried.

## Campylobacter (C.)

Pathogenic avian strains of *Campylobacter* have been recently classified as *C. jejuni*. Erroneously, these strains have been labeled as *Vibrio*, particularly as *V. metchnikovi*. This error is the more deplorable when one considers that *Vibrio cholerae*, serotype 01, and noncholera *Vibrio* can be isolated from healthy birds, especially from water fowl.

*C. jejuni* is a gram-negative, motile rod (0.2 to 0.5 × 1.5 to 5 μm) that appears as a short comma, S-shaped form, or long spiral. Coccoid forms usually indicate degeneration with loss of viability. *C. jejuni* requires a microaerobic atmosphere and blood agar or special selective media. Colony formation takes 72 to 96 hours at 37 to 42°C. Outside the host, survival seems to be short (one week). Common disinfectants are efficacious as a rule.

The host spectrum is large, including humans, cattle, sheep, goat, swine, chicken, turkeys, pheasants, geese, pigeons, Pekin Nightingale *(Leiothrix lutea)*, Nandu *(Rhea americana)*, and Great Bustard *(Otis tarda)*. *C. fetus* var. *jejuni* has recently been reported to be found in passerine birds, particularly tropical finches *(estrildidae)*, and to a lesser degree in canaries *(fringillidae)*.[2] Avian strains of *C. jejuni* are serologically not uniform, and details about possible cross-infections between humans, mammals, and birds have been insufficiently investigated. Campylobacter other than *C. jejuni* are known to occur in birds; however, they are considered nonpathogenic.

### *Transmission*

The natural route of infection seems to be the oral one (*C. jejuni* is shed with the feces).

### *Pathogenesis*

Although experimental infection of susceptible birds with *C. jejuni*, as a rule, produces a characteristic hepatitis, the organism can also be found in the alimentary tract of healthy birds. Which factors are necessary to actuate clinical disease is not known. It is speculated that parasites such as coccidia and nematodes, as well as *E. coli*, are triggering factors.

### *Clinical Disease and Pathology*

As with other diseases of the liver, nonspecific signs such as somnolence, anorexia, diarrhea, and emaciation (with chronic course) may be seen. A high mortality was noted in finches, especially among fledglings.[2] At necropsy the liver is enlarged, pale, or greenish in color, and congested with or without hemorrhages. The lobules are more prominent than usual owing to perivascular infiltrations. Necrotic areas, which may be coalescent, can occur. A catarrhal enteritis is common.

### *Diagnosis*

To verify possible involvement of *C. jejuni* in a characteristic disease process, the organism has to be isolated and differentiated. In birds with gallbladders, bile can be inspected directly under a phase-contrast microscope for tentative diagnosis.

### *Differential Diagnosis*

Other diseases with liver lesions such as salmonellosis, pasteurellosis, and infections with herpesvirus and reovirus in African Grey Parrots should be ruled out.

### *Treatment*

There are discrepancies between the results of the sensitivity testing and clinical improvement. Erythromycin or tetracyclines can be tried. Depending on the bird species (*never* in psittaciformes), streptomycin or dihydrostreptomycin given parenterally can be useful. Recidivism occurs sometimes despite treatment.

### *Control*

Because of the recurrences, it seems necessary to change the environment (clean and disinfect cages, houses, and aviaries) and to try to recognize and treat triggering factors, including nutritional regimens.

## Pseudomonas (Ps.) and Aeromonas (Ae.)

Although these two genera are not very closely related taxonomically, they have many common characteristics that may justify a discussion of them together.

*Ps. aeruginosa* and *Ae. hydrophila* are the important species for birds. Both are typical organisms in moist and wet environments. They propagate in water at ambient temperatures of about 20°C and *Ps. aeruginosa* even lower. Both organisms are gram-negative rods of approximately the same size (1 to 3 × 0.3 to 0.6 μm) and grow on the usual media, including MacConkey plates. Both species cause beta-hemolysis on blood plates. *Ps. aeruginosa* produces a blue-grey, diffusible pigment and has a sweetish aroma. *Ae. hydrophila* has the bad smell of the proteinolytic organisms and can be confused with *E. coli* on Endo or MacConkey plates. Both species can be differentiated into several biotypes and serotypes. Studies to investigate the biotypes and serotypes of avian strains have not been conducted.

*Ps. aeruginosa* is extremely resistant to many disinfectants in common use, whereas *Ae. hydrophila* seems to produce no problems in this respect. Steam and boiling water are efficacious for *Ps. aeruginosa,* as are a few specially tested brands of chemical disinfectants.

The host spectrum of both species appears to be very wide, including humans, many mammals, and the majority of all bird species kept in captivity. In wild birds, water fowl seems to be predisposed to infection, but other wild-living birds are susceptible as well.

### *Transmission*

High numbers of each species seem to be necessary to produce an oral infection (enrichment of the organism in the drinker or in the environment). Infection through skin following trauma or other damage appears possible, although birds are usually resistant to wound infections.

### *Pathogenesis*

*P. aeruginosa* and *Ae. hydrophila* are typical secondary invaders. Treatment with antibiotics, immunosuppression, and damage of mucosal surfaces are predisposing factors. Pathogenic strains of *Ps. aeruginosa* produce lecithinase and protease, causing edema, hemorrhages, and necrosis of host tissues. *Ae. hydrophila* strains are variable in their capacity to produce toxins. Several hemolysins, a so-called lethal toxin, and a necrotizing toxin are distinguished. Since *Ae. hydrophila* grows optimally only between pH 6.7 and 8.0, an increase of the pH in the upper alimentary tract seems to be necessary. In a few cases, it could be proved that the number of lactobacilli in this area was significantly reduced.

Because of the capacity for toxin production, an infection with either species is considered life-threatening.

### *Incubation*

No details for pet birds are known. However, incubation periods of several hours for *Ps. aeruginosa* and one to three days for *Ae. hydrophila* can be estimated.

### *Clinical Disease and Pathology*

Bacteremia with pathogenic strains of both species causes severe general signs combined with diarrhea, dyspnea, or dehydration. An edematous or necrotizing dermatitis can develop following skin lesions. Acute death is not rare. At necropsy, hemorrhages and coalescent necrosis of the parenchymata are characteristic. Catarrhal to hemorrhagic enteritis is frequently present. Edema or fibrinous inflammation of serosal membranes can occur as well.

### *Diagnosis*

The isolation and identification of the causative agent are necessary for diagnosis.

### *Differential Diagnosis*

Septicemia and acute death by other agents have to be considered. The rule-outs for the dermatitis are staphylococcosis and avian pox, as well as clostridia.

### *Treatment*

Severely sick birds with extensive necrosis of the large parenchymata normally cannot be saved. *Ps. aeruginosa* is highly resistant to many antibiotics, and sensitivity testing is generally recommended. However, in an acute disease immediate application of an antibiotic is necessary and polymyxin B, gentamicin, or carbenicillin may be tried.

*Ae. hydrophila* is, in many instances, sensitive to the antibiotics in common use; however, exceptions do occur.

### *Control*

It is imperative to avoid enrichment of these germs in the environment. Regular cleaning and boiling (disinfecting) of the drinking cups or other sources of drinking water are mandatory. The same is true for the moisturizing system of incubators. Under certain circumstances waterfowl have to be kept off a natural water body for some critical weeks of the year when *Ae. hydrophila* is heavily propagating.

Generally, for control, the predisposing factors have to be recognized and remedied.

## REFERENCES

1. Brunner, U.: Eigenschaften verschiedener Micrococcen bei Voegeln. Erstisolierung von *Micrococcus lylae, Micrococcus nishinomyaensis* und *Staphylococcus sciuri* aus Vogelmaterial. Diss. Med. Vet., Muenchen, 1981.
2. Dorrestein, G. M., et al.: *Campylobacter* infections in cage birds—clinical, pathological and bacteriological aspects. Proceedings of the Joint Meeting of Veterinary Pathologists, Utrecht, The Netherlands, September, 1984.
3. Filgis, R.: Zur Differenzierung aviaerer Staphylococcen. Diss. Med. Vet., Muenchen, 1981.
4. Gerlach, H.: Biochemische Eigenschaften aviaerer Staphylococcen. Adv. Vet. Med., 35:190–194, 1981.
5. Quesenberry, K. E., and Short, B. G.: *Serratia marcescens* infections in a Blue and Golden Macaw. J. Am. Vet. Med. Assoc., 183:1302–1303, 1983.
6. Woerpel, R. W., and Rosskopf, W. J., Jr.: Retro-orbital *Mycobacterium tuberculosis* infection in a Yellow-naped Amazon Parrot *(Amazona ochrocephala auropalliata)*. Proceedings of the Annual Meeting of the Association of Avian Veterinarians, San Diego, CA, 1983.

**ACKNOWLEDGMENTS**

The author highly appreciates the assistance of Thomas L. Lester, D.V.M., Ph.D., in preparing this manuscript. The author also thanks Richard J. Hidalgo, D.V.M., Ph.D., for his valuable suggestions.

HELGA GERLACH

# *Chapter 34*

# MOLLICUTES

## (MYCOPLASMA, ACHOLEPLASMA, UREAPLASMA)

HELGA GERLACH

Because of their peculiarities, *Mycoplasma* and related organisms have been separated from the class of bacteria and put into the class of mollicutes. Mollicutes are the smallest organisms capable of propagation in cell-free media. They have no cell wall like bacteria, only a membrane around their cytoplasm, and are therefore naturally resistant to antibiotics acting on the cell wall. Because of their small size mollicutes are fastidious. Specialized laboratories are generally needed for isolation and identification. Special transport media (for example, heart infusion broth + 5 per cent horse serum + 1000 units penicillin G per ml) are necessary for shipping tissue or strains. Mollicutes are supposed to survive only for hours outside the host (except in the drinking water, where they survive for two to four days, depending on the strain). Owing to the lack of a cell wall, mollicutes are sensitive to all disinfectants commonly used in veterinary medicine. The covering and protection of the organisms by host excretions can be a problem that is solved only by meticulous cleaning before disinfecting.

The genus *Mycoplasma* has a rather narrow host spectrum; i.e., each species has only one or a few closely related hosts, whereas *Ureaplasma* and *Acholeplasma* have a much wider range. In contrast to mammals, *Ureaplasma* from avian species has been isolated only from the respiratory tract.

### Transmission

Because of the low survivability of the mollicutes in nature, close contact between individuals is necessary for transmission. Respiratory and venereal routes of infection can occur. Densely populated aviaries and breeding farms provide the close contact necessary for outbreaks of disease. Epornitically, transovarian transmission can become important, as can the feeding of offspring via infected crop regurgitation or "crop milk."

### Pathogenesis

Mycoplasmosis is a typical, chronic multifactorial disease. Avian mollicutes colonize on the mucosa of the respiratory and/or urogenital tract and enter into a kind of commensalism with their host. To gain entrance into the mucosa, most mollicutes need cellular lesions caused by other noxae except those that are toxin producers. However, cellular damage may be caused at the site of colonization depending on the virulence of the strain in question. The host answers with an inflammatory response that is usually serofibrinous in nature. In addition, the cell-mediated defense system is activated, and the pathology is governed by these reactions. *Mycoplasma* species are especially known to cause a tranformation of the host lymphoblasts (mainly B cells) by a mutagen. These cells fail to mature and cannot function as expected. This causes the host to produce more and more of the altered cells in huge lymph follicles, with invasion of these lymphoid cells into the infected mucosa. Other mechanisms to inactivate the defense system of the host are the production of cytotoxins (e.g., exotoxins, $H_2O_2$), and of polysaccharides. Mollicutes that have entered the mucosa can be carried hematogenously to other organs. Frequent locations are joints, egg follicles, and brain. Tissue so impaired offers possibilities for secondary invaders, thus altering clinical and pathologic signs.

Isolates of mollicutes that are not yet fully classified have been reported for a variety of psittaciformes. Proof of their being the cause of disease is, in most instances, not yet conclusive. Well-known *Mycoplasma* or *Acholeplasma* species that occur in the poultry industry are *not* the cause of disease in psittaciformes. The presence of typical histologic lesions and/or significant increase in antibody titers (penmates) should be evaluated as proof for pathogenicity with a given isolate. One of the few papers meeting these requirements is Adler's[1] report on a *Mycoplasma* infection in a budgerigar (*Melopsittacus undulatus*). A review of avian mollicutes including pet, game, and zoo birds is given by Gerlach.[2]

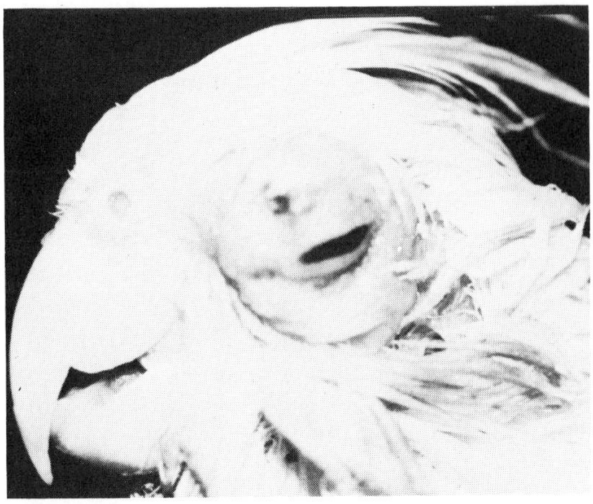

**Figure 34–1.** Mycoplasmosis may be considered in the differential diagnosis of conjunctivitis in cockatiels, although the *Mycoplasma* organism is often difficult to culture. (Courtesy of Lorraine Karpinski.)

## Incubation

Incubation times are uncertain because of long latent periods, egg transmission, and the interaction of environmental factors.

## Clinical Disease and Pathology

Clinical signs indicative of infections with mollicute species in psittaciformes are chronic conjunctivitis (Fig. 34–1), rhinitis, sinusitis, respiratory rales, and arthritis.

At postmortem examination, tissues of the upper respiratory tract, as a rule, contain fibrin. The trachea and air sacs may be mildly inflamed. Only histopathology reveals the typical proliferation of lymphoid cells. With regard to the reproductive system, drops in fertility and hatchability and an increase in early chick mortality are the main signs to be expected. No reports on disease of the reproductive system in psittaciformes are available.

## Diagnosis

A tentative diagnosis can be made histologically. Isolation of the organism is necessary for further characterization and serologic analysis. Swabs from the upper respiratory tract and from the phallus in cocks can be taken for diagnosis in live birds if deemed necessary. Serology as a tool for diagnosis should be regarded with care (see Chapter 18, Serology). The presence of a mollicutes species on a mucosal surface is frequently not followed by the production of humoral antibodies. Only after the organism has penetrated the mucosa can humoral antibodies be detected, and then often only in low titers. Mollicutes, as a rule, can propagate in the presence of humoral antibodies, indicating that these are not correlated with immunity. Sufficient immunity to protect from clinical signs is probably not developed, because the disease is governed by the reactions of the cell-mediated immune system.

## Differential Diagnosis

The rule-out list includes many diseases that are caused by a virus, as well as secondary bacterial and fungal infections. Special attention should be paid to psittaciformes with the clinical signs described above, because these signs coincide with those of psittacosis. The rule-outs for the reproductive tract include other egg-transmissible or venereal diseases, nutritional deficiencies, and failure during the hatching process.

## Treatment

To treat clinical infections, tylosin, erythromycin, spectinomycin and its combination with lincomycin, or pleuromutilin is recommended for poultry. However, the sensitivity may vary with strains from nonpoultry populations. Gerlach[3] has shown that for strains from columbiformes only pleuromutilin has the ability to inhibit approximately 90 per cent (57 of 65) of the isolates tested. The efficacy of the tetracyclines against avian mollicutes has yet to be proven, but for psittaciformes this treatment is encouraged because a number of suspected cases of *Mycoplasma* infection proved to be chlamydiosis.

## Control

As with poultry, problems are to be expected primarily in densely populated aviaries and breeding farms. Because of the low infectivity, control can be accomplished by the methods that should be considered standard operational procedures (e.g., quarantine, disinfection). Since the disease is of the hypersensitivity type, vaccination is theoretically not likely to solve the problem. However, vaccines that interfere with the attachment of the mollicutes to the surface of the mucosal cell might be promising.

## REFERENCES

1. Adler, H. E.: Isolation of a pleuropneumonia-like organism from the airsac of a parakeet. J. Am. Vet. Med. Assoc., 130:408–409, 1957.
2. Gerlach, H.: Infektionen durch Mollicutes bei Voegeln. In Gylstorff I. (ed.): Infektionen durch Mycoplasmatales. Jena, VEBGústav Fischer Verlag, 1985.
3. Gerlach, H.: Untersuchungen ueber Mollicutes bei Tauben. Hab.—Schrift, Muenchen, 1978.

# Chapter 35

# CHLAMYDIA

HELGA GERLACH

*Chlamydia (Chl.) psittaci* is the causative agent of a disease called chlamydiosis (syn.: psittacosis or parrot fever in psittaciformes, and ornithosis in other birds). Infections have been demonstrated in approximately 130 nonpsittaciforme species and in several domesticated mammals. *Chl. psittaci* is highly contagious and can cause disease in man.

*Chl. psittaci* is an obligate intracellular parasite that contains DNA and RNA and is characterized by specific morphogenesis and autonomic synthesis of its specific enzyme systems. The infectious-toxic elementary bodies (approximately 0.3 μm) possess a gram-negative staining cell wall. These bodies gain entrance into the target cell (mononuclear phagocytes) by endocytosis. This is the first step of the propagation cycle. The elementary bodies, which have little metabolic activity, develop into large (0.5 to 1.5 μm), fragile, low-density forms, called initial bodies (syn.: reticular bodies) by using pre-existing intermediate macromolecular products of the host cell. Initial bodies propagate by binary fission. They enzymatically synthesize amino acids, purine bases, pyrimidine bases, etc.[11] Initial bodies are in some respects similar to cell wall defect L-form bacteria and, like them, can persist in the presence of antibodies and antibiotics active against cell wall components.[6] The next step in the cycle is the reorganization of the initial bodies into elementary bodies. During this process, the toxic surface antigens (see later) reappear, and a so-called chlamydial factor becomes demonstrable (chlamydial glycolysis system).[10] The latter destroys the integrity of the host cell and provides hydrolytic enzymes that are in part responsible for cytopathology and lysis. The presence of the chlamydial factor apparently stops the synthesis of membranes and cell organelles. The pre-existing macromolecules in the cytoplasm of the host cell are also disorganized. The result is reversible or irreversible cellular damage, which can cause clinical disease. Rupture of the host cell releases large numbers of infectious elementary bodies.

The dose of elementary bodies, particularly the ratio of elementary bodies to macrophages, is decisive in the outcome of the infection. High doses of virulent *Chlamydia* cause a lethal lytic reaction of the phagocytes. Low doses of virulent *Chlamydia* induce an adequate lytic reaction of the mononuclear and polymorphonuclear phagocytes during which the agent is quickly inactivated. Damage to the macrophages reduces the probability of survival for the *Chlamydia*. On the other hand, low doses of a nonvirulent chlamydial strain induce only an inadequate lytic reaction of the macrophages and transform them into long-lived epithelioid cells, in which the *Chlamydia* survives, converting the host into a carrier of an asymptomatic infection.

*Chl. psittaci*, protected by cells and proteinaceous material, survives outside the host for approximately one month. Bacterial decomposition of the tissue and of fecal material kills the organisms quickly (within hours). Precautions for shipments have to be made accordingly. Formalin is still one of the best compounds for disinfection. Some modern disinfectants, although only sporadically tested, have repeatedly failed to be effective. It must be remembered that the formalin loses efficacy with falling temperatures and is inert at 0°C. Room temperatures (+20°C) are necessary to have a reasonable efficacy; however, temperatures between +30°C and +40°C are even better.

## Transmission and Host Spectrum

*Chl. psittaci* is well-adapted to the avian host. The organism is endemic in all countries where psittaciformes are indigenous. It is estimated that 1 per cent of the wild population are infected and act as carriers. There seems to be no serious disease among the wild populations in those parts of the world. The clinical disease is precipitated by man-made conditions and procedures, such as taking chicks from the nests, transport, drastic changes in feed or environment, and other infections.

During latent infection the organism persists in macrophages and epithelioid cells followed by an incomplete autosterilization. This mechanism favors the selection of strains of low virulence for the host species. The recrudescence of this type of infection results in shedding and distribution of high numbers of infectious, highly virulent *Chlamydia* for other species.

Transmission takes place mainly by the aerogenic route, with feather dust and dust from fecal material being the main sources. Egg transmission has been proven in ducks. The shedding of *Chlamydia* with the feces begins, as a rule, 10 days prior to the onset of clinical disease. Carriers can excrete the organisms intermittently over a period of at least several months. Young birds are, as a rule, more susceptible than mature ones. There are also considerable differences among the various bird species. For instance, the large South American parrots (macaws, Amazons) are more susceptible than the ones from South Asia, Australia, and related islands (cockatoos, lories, king parrots). The latter are again more susceptible than the African parrots. This is, of course, only a generalization. The actual condition of an individual is perhaps more important than a species-specific susceptibility. Dog, cat, horse, and human are susceptible as well, but they are at the end of the infectious chain, and transmission between these individuals seems not to be possible. This is in contrast to cattle, sheep, and goats, in which transmission occurs easily.

## Pathogenesis

*Chl. psittaci* can cause a completely asymptomatic infection in mature primary hosts, or acute systemic, often fatal disease in young birds or secondary hosts. A persistent infection as described earlier is correlated with an adaptation of the chlamydial strain to a host species. If elementary bodies of such a strain infect another susceptible host, virulence and toxicity are normally fully regained. Strains from psittaciformes, turkey, and ducks are known to be particularly toxic for humans, in contrast to strains from pigeons, which rarely cause a clinical disease.

There is no doubt that chlamydial strains vary with regard to virulence, antigenicity, and biologic properties. Virulent strains have an increased growth rate and a wide host spectrum. The cause of the toxicity is a hepatotoxic and nephrotoxic toxin that is inseparably connected to the surface membrane of the elementary bodies and seems to vanish after penetrating the host cell. The toxin has not been isolated purely, so far.[6] This important virulence factor is variable between strains and stimulates the production of antitoxic antibodies, which neutralize not only toxicity but also infectivity.[6, 7] The toxin is related to a few of the labile, specific cell-wall antigens (suggested to be proteins) of the intact elementary bodies, which stimulate none or only low antibody titers following primary infection, in contrast to uniform group-specific, heat-stabile antigens (glycolipids), which induce the production of complement-fixing antibodies. The latter are not correlated with immunity.

The growth of *Chlamydia* in susceptible host cells induces structural and metabolic changes. Changes in pathogenicity and immunogenicity of the chlamydial strains in question are likewise induced.[6] Frequent passages through mammalian or avian hosts (or in different cell cultures) induce new antigens on the surface of their elementary bodies.[1] These antigens are specific to the chlamydial strains in question and to the host species in which they are produced. The interspecies transfer, which undoubtedly occurs in import stations, breeding farms, multispecies aviaries, pet shops, etc., changes physiochemical properties and therefore the antigenicity as well as the toxic component and the host spectrum. However, these newly gained characteristics are not really stable.

The susceptibility of the host species is decided upon at the level of the mononuclear phagocytes. Phagocytosed chlamydiae not coated with opsonins are not killed and can even replicate within the macrophage.[7] Lymphokines from activated lymphocytes cause a microbistasis of the organism in activated phagocytes. This inhibition is only short-lived unless fresh lymphokines are produced daily.[2] Clinically, the results of the described mechanisms are the cause of variable incubation time, signs, pathology, and recovery. For instance, one chlamydial strain produces in different hosts an infection that differs with regard to number and kind of infected tissues, length of the lag phase, propagation rate, and time period for the appearance of the new generation of elementary bodies. Similar conditions are seen in different chlamydial strains infecting a single host species (Table 35–1).

A primary infection is assumed to lead to propagation of the organism within the intes-

**Table 35–1.** EXPERIMENTAL INFECTION OF AMAZONS *(AMAZONA VIRIDIGENALIS)* WITH DIFFERENT CHLAMYDIA STRAINS*

| Chlamydia Strain | Route of Infection | Clinical Signs (Per Cent) | Pathological Lesions (Per Cent) | Exitus (Per Cent) |
|---|---|---|---|---|
| A | intramuscular given once | 50 | 50 | 50 after 20 days |
| | air sac given once | 100 (slight) after 33 days | 100 | 0 |
| B | intranasal given once | 100 after 94 days | 100 | 0 |
| | intranasal given 3 times | 0 | 100 | 100 after 22 days |
| C | intranasal given once | 100 after 27 days | 100 | 0 |
| | air sac given once | 0 | 100 | 0 |
| D | intranasal given 3 times | 100 after 14 days | 100 | 20 |

*From Jacoby, J. R.: Verlauf einer experimentellen Infektion mit Chlamydia psittaci bei Amazonen. II. Tagung Krankheiten der Voegel. Muenchen, Dtsch. Vet. Med. Gersellschaft, 1981, pp. 81–88.

tinal tract, frequently without clinical signs in the host. As already pointed out, the production of antitoxin antibodies is at best low. The survival of an infection, even with clinical disease, can leave a fully susceptible host. However, a latent infection causes a premunity, inhibiting an infection with another strain, even experimentally.

## Incubation Time

The incubation time varies with the chlamydial strain and the bird species. With long-time carriers, the situation becomes even more complicated. Following natural infection the shortest incubation for psittaciformes is 42 days (relevant for forensic cases). Incubation times up to 1.5 years have been observed. Opinions in the older literature stating that after four months an infection within a closed flock will be self-eliminated could not always be confirmed by recent data.

## Clinical Disease

The clinical disease is generally highly variable, and emphasis is given to the psittaciformes. An early description of "pneumo-enteritis" in psittaciformes was made. The acute disease following natural infection shows respiratory signs such as dyspnea, rales, coryza, sinusitis (budgerigar), and conjunctivitis (frequently unilateral), as well as diarrhea and polyuria. During the acute phase the birds are hypothermic, lethargic, and anorectic and have a ruffled plumage. Signs of dehydration may be present as well, but even severe kidney malfunctions are frequently hard to detect clinically. The droppings may appear yellowish, suggesting involvement of the liver in the disease process, but are normally greenish, greyish, watery, and frequently without crystalline uric acid. The subacute to chronic phase of the disease is often characterized by the appearance of CNS signs of the clonic-tonic type. Opisthotonus, torticollis, tremor, convulsive movements, etc., are the main symptoms. The acute phase of the disease frequently may be subclinical or overlooked. Therefore, CNS signs may be the only clinical symptoms observed. Although this can occur in many psittaciforme birds, classic examples are the African Grey Parrot *(Psittacus erythacus)*, several *Cacatua* spp. and *Amazona* spp. An exception occurs in the cockatiel *(Nymphicus hollandicus)*, in which flaccid paresis and paralysis of the hind limbs have been observed. Those birds are seen lying almost helplessly on the floor of the cage. Survival from the acute disease can leave residual disturbances of the feathering, particularly in young birds, supposedly because of damage in the thyroid and/or adrenal gland. A locally circumscribed infection in the form of a keratoconjunctivitis has been described in small Australian parakeets, particularly in the genus *Neophema*. The signs are recurrent, frequently without general symptoms, and cannot be treated successfully. Consequently, the *Neophema* species involved have to be euthanized. Conjunctivitis with or without coryza is a typical sign of chlamydiosis in columbiformes and in a variety of European finches.

Persistent infection may be recognized only by the transmission of the agent to other animals or man or by sudden death of nestlings from an apparently healthy breeder pair.

## Pathology

At necropsy, the lesions are as variable as the clinical picture. Depending on the course of the disease, more acute lesions such as fibrinous peritoneal exudate, air sacculitis, perihepatitis, pericarditis, myocarditis, bronchopneumonia, catarrhal (rarely hemorrhagic) enteritis, and nephrosis can be seen. Tiny disseminated necrotic foci may be present in internal parenchymata. Splenomegaly is somewhat overemphasized in the literature. It certainly is frequently present, but in psittaciformes the occurrence of a mildly fibrinous air sacculitis is more indicative of psittacosis. Splenomegaly can be seen in a variety of other diseases and may be absent in psittacosis owing to the stage of the disease at time of death. *Chl. psittaci* can cause orchitis and epididymitis, particularly in sexually active males, resulting in infertility. The female equivalent, oophoritis, is rather rare.

The more chronic lesions include liver and kidney cirrhosis as well as pancreatic necrosis (budgerigar and pigeons). CNS lesions can be recognized only by histology and consist of a nonpurulent meningitis, which can spread to the cerebrum. Other histologic findings are mostly nonspecific, except for the presence of Levinthal-Coles-Lillie (LCL) bodies, which may be formed in any organ (Fig. 35–1). Bacterial, fungal, and viral secondary infections can alter the lesions and obscure psittacosis.

## Diagnosis

Clinical history is an important tool, since the majority of the psittaciformes kept as pets become sick three to four months after being purchased or after a "new" bird has come in contact with the patient. Imprints or smears can be taken from conjunctivae or nasal exudate. Epithelial cells are necessary for evaluation on these preparations. The slides are heat-fixed and stained (procedures: Stamp, Giemsa, Giménez, Macchiavello's, Castaneda) for the presence of LCL bodies, which are pathognomonic. Negative results do not exclude psitta-

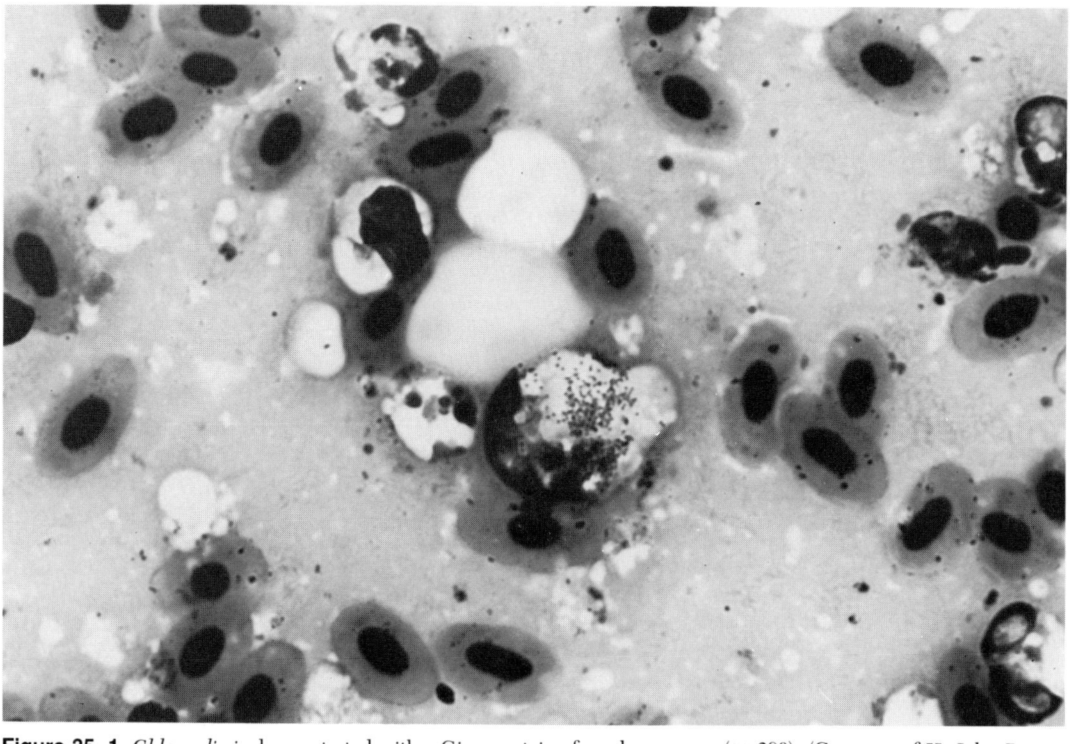

**Figure 35–1.** *Chlamydia* is demonstrated with a Giemsa stain of a spleen smear (× 280). (Courtesy of H. John Barnes.)

cosis, because only the elementary bodies can be stained. Fecal samples can be cultured via cell culture (McCoy cell line, chicken embryo fibroblast) or by injecting the material into mice.

A sucrose-albumin-phosphate solution,[5] pH 7.2, has been used successfully for the past several years to preserve chlamydial stock suspensions and to serve as a transport medium for the shipment of feces or cloacal swabs for isolation of chlamydiae. For use with feces, it must contain 500 μg each of streptomycin, kanamycin, and vancomycin and 5 μg of fungizone per ml to retard or prevent growth of conventional bacteria or fungi. Other antibiotics not effective against *Chlamydia* may be substituted but should not be toxic to cell cultures used for isolation attempts.

Imprints from the air sacs, pericardial sac, and/or liver capsule are taken for staining from dead birds. Spleen, liver, and lung are the favored tissues for culture. Positive cultures contain LCL bodies. Fluorescent antibody staining of the imprints is possible provided that nonspecific reactions, which occur rather frequently, can be neutralized. Whereas the simple staining procedures can be conducted in any veterinary clinic (the reading of the slides can be learned), all other methods are dependent on a specialized laboratory.

Humoral antibodies can be detected via the complement fixation (CF) test (limited use in birds because the $C_1$ of the guinea pig complement is incompatible with the serum of many birds species), hemagglutination inhibition (HI) test with mouse erythrocytes, immunodiffusion, or ELISA. A fourfold rise in titer with paired samples is considered conclusive; however, under treatment with tetracyclines the production of humoral antibodies can be severely inhibited, and no increase in titer may occur.

A recently developed latex agglutination serologic method is described in Chapter 18, Serology.

## Differential Diagnosis

The rule-out list depends on the state of the disease and includes infections with herpesvirus, particularly Pacheco's virus, paramyxovirus, influenza A virus, and Enterobacteriaceae.

## Treatment

The following considerations about treatment are written without regard to any governmental regulation. Only scientific points of view are given.

For treatment, so far, only the tetracyclines have been generally efficacious.[9] Tetracyclines interfere early and reversibly with the propagation cycle by inhibiting the synthesis of chlamydial proteins, growth and fission of the initial bodies, and probably also the reorganization to elementary bodies. Variant initial and probably elementary bodies are produced under therapy and convert to normal states following cessation of treatment in 5.5 days. These reversible variants have to be removed by host defense mechanisms[10] and are one reason for long-term treatment. Other disadvantages to the use of tetracyclines are (1) low intracellular concentration of the drug, (2) immunosuppression, (3) production of chelates with calcium and other alkaline earths, and (4) side effects that include inhibition of the autochthonous intestinal flora and triggering of fungal infections.

This summary of unfavorable effects of the tetracyclines explains the schematic for the treatment of psittacosis or ornithosis. Birds with an intact defense system on a diet low in calcium are given tetracyclines at a dosage that ensures the establishment of a minimum blood concentration of 1 μg/ml over a period of approximately six weeks. Birds suffering from the active disease usually improve rapidly and the *Chlamydia* strain can be eliminated rather easily if it is sensitive to tetracyclines. Resistant strains are still rather rare, but those with reduced sensitivity are emerging in increasing numbers.[4] Problems are arising with chronically or latently infected birds in which the *Chlamydia* is not actively propagating. In those cases, the time period of treatment is particularly important because it appears that the elimination of *Chlamydia* is completed only by the natural replacement of host cells and prevention of the establishment of newly infected cells during this period.

The following procedures are available for treatment:

1. Chlortetracycline (CTC) for oral application can be administered at 5000 ppm in soft mixed feed (cooked grain with or without egg) or in commercial parrot pellets. Commercially dehulled seeds covered with CTC (500 ppm) are recommended for budgerigars and small finches only. CTC is excreted mainly by the kidneys and can cause failure of a predamaged kidney.
2. Oxytetracycline (OTC) can be used for intramuscular injection at a dosage level of 100 mg/kg body weight. Recently an OTC LA for-

mulation has been released that produces a longer-lasting blood level at a dose of 75 to 100 mg/kg body weight than the normal OTC. Details of this treatment have to be worked out. OTC can cause the same damage as CTC.

3. Doxycycline was developed for human intravenous administration. Doxycycline produced in the United States leads to heavy local necrosis in a variety of psittaciformes when given intramuscularly (Fig. 35–2). A similar product manufactured in West Germany can be given intramuscularly safely and gives minimum blood levels of 1 μg/ml for about 7 days at the dose of 100 mg/kg body weight. Repeated injections are necessary for long-term treatment. An increased rate of disappearance from the blood occurs so that for a 45-day treatment the recommended intervals are 7, 7, 7, 7, 6, 5, and 5 days.

Apart from the inconvenience, there is no reason not to try intravenous injections with doxycycline produced in the United States. However, doses and blood concentrations have to be studied before general usage is instituted. Doxycycline is excreted extrarenally mainly (feces, bile, etc.); therefore, side effects to the kidneys are rare, but those to the liver may occur.

A disadvantage of injection is the necessity of catching the birds. However, effective blood levels are established within a few hours, and the shedding of *Chlamydia*, in most instances, ceases within 24 hours. The birds can eat their accustomed feed and need no special diet, except for some multivitamin supplementation. In addition, the birds do not lose body weight by refusing untasty feed.

Treatment via feed does not necessitate catching the birds. But many psittaciformes refuse to eat the treated feed and 10 days are required to achieve effective blood levels. The intestinal flora is influenced far more by treated feed than by injected tetracyclines, and diarrhea can occur as a side effect. The nectar-eating birds are given tetracyclines with their liquid food at a concentration of 500 ppm.

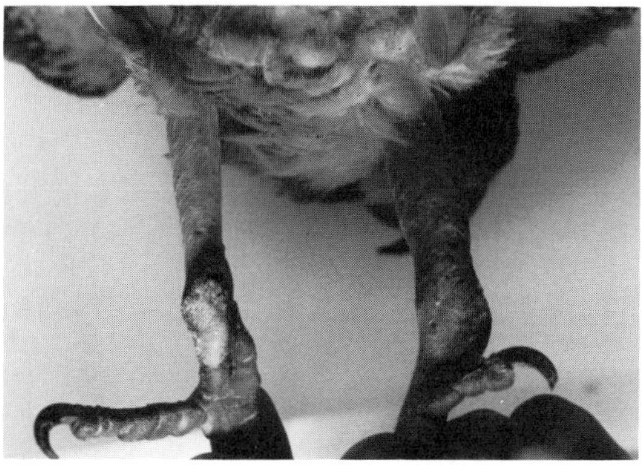

**Figure 35–2.** Unilateral dark tissue necrosis and obvious swelling were produced by inadvertent perivascular contact with I.V. doxycycline. (Courtesy of Greg J. Harrison.)

Application of tetracyclines via drinking water has been unsuccessful, mainly because the birds neither drink enough nor drink at correct intervals to produce minimum blood concentrations (1 μg/ml). The taste of the medicated water plays a role with many psittaciformes as well.

Another problem concerning treatment is that while the psittaciformes are frequently regarded as a uniform group of birds, considerable differences are demonstrable. The compatibility of drugs and their metabolism are variable features among many others.

Birds that do not need the high dosages of tetracyclines are genus *Agapornis*, the large macaws, cockatiels, Bourke's Parrot *(Neophema bourkii)*, Turquoise Parrot *(Neophema pulchella)*, Scarlet-crested Parrot *(Neophema splendida)*, Red-winged Parrot *(Aprosmictus erythropterus)*, Eastern Rosella *(Platycercus eximius)*, Pale-headed Rosella *(Platycercus adscitus)*, Western Rosella *(Platycercus icterotis)*, Mulga Parrot *(Psephotus varius)*, Red-fronted Parakeet *(Cyanoramphus novaezelandiae)*, and in particular, Canary-winged Parakeet *(Brotogeris versicolorus)*, as well as Grey-cheeked Parakeet *(Brotogeris pyrrhopterus)*. CTC concentrations of 2000 to 2500 ppm in the feed are sufficient to produce blood levels higher than 1 mg/ml in the birds mentioned above. It is known that they eat more of the low-medicated feed as well. For injections, 75 mg/kg body weight are given as compared to 100 mg/kg body weight for the less susceptible species.

The success of treatment depends not only on long-term therapy, but, because the survival of infection frequently leads to a fully susceptible host, also on the effective prevention of reinfection.

Editor's Note: In my experience, the severely ill psittacine with chlamydiosis is very toxic and requires aggressive supportive therapy such as intravenous fluids, lactulose, and heat (see Chapter 27, What To Do Until A Diagnosis Is Made). For these cases, a single intravenous dose of injectable doxycycline is administered initially, followed by oral doxycycline for 45 days. Flocks may be treated by the incorporation into the food of the contents of doxycycline capsules for the 45-day time period, or custom-made soft-moist food containing doxycycline (140 mg/lb of food) can be ordered (LaFeber Co., Odell, IL).

## Control

Since the survival of the clinical disease does not necessarily induce immunity, vaccination as means of control is not promising. However, the majority of the commercially available psittaciformes are intended to become pets and as such should be free of *Chl. psittaci*. Quarantine of recently imported birds and treatment for psittacosis are obligatory in many countries, but breeder farms and pet shops should voluntarily submit to control examinations. Serologic and cultural methods probably have to be combined. *Chl. psittaci* is very well-adapted to the avian host, and today's methods of detection of the agent are not good enough for complete elimination. But constant monitoring of breeding flocks and birds in quarantine, as well as in pet shops, will lead finally to success.

## REFERENCES


1. Allan, I., and Pearce, J. H.: J. Gen. Microbiol., 112:61–66, 1979.
2. Byrue, G., and Faukion, C. L.: Lymphokine-mediated microbistasic mechanisms restrict *Chlamydia psittaci* growth in macrophages. J. Immunol., 128:469–473, 1982.
3. Gerbermann, H.: Der Einfluss des Immunsystems auf die Abwehr einer Psittacoseinfektion. II. Tagung: Krankheiten der Voegel. Muenchen, Dtsch. Vet. Med. Gesellschaft, 1981, pp. 89–93.
4. Gerbermann, H.: Die Wirksamkeit von Doxycyclin gegen Chlamydia psittaci bei der Maus. I. Tagung: Krankheiten der Voegel. Muenchen, Dtsch. Vet. Med. Gesellschaft, 1979, pp. 31–36.
5. Grimes, J. E.: Transmission of chlamydiae from grackles to turkeys. Avian Dis., 22:308–312, 1978.
6. Gylstorff, I., Jakoby, J. R., and Gerbermann, H.: Vergleichende Untersuchungen zur Psittakosebekaempfung auf medikamenteller Basis. Berl. Muench. Tieraerztl. Wochenschr., 97:91–99, 1984.
7. Idtse, F. S.: *Chlamydia* and chlamydial diseases of cattle: A review of the literature. Vet. Med., April, 1984, pp. 543–550.
8. Jacoby, J. R.: Verlauf einer experimentellen Infektion mit Chlamydia psittaci bei Amazonen. II. Tagung Krankheiten der Voegel. Muenchen, Dtsch. Vet. Med. Gesellschaft, 1981, pp. 81–88.
9. Meyer, K. F., Eddie, B., Richardson, J. H., Schipkowitz, N. L., and Muir, R. J.: In Beaudette, F. R. (ed.): Progress in Psittacosis Research and Control. New Brunswick, NJ, Rutgers University Press, 1958.
10. Stokes, G. V.: Infect. Immunity, 9:497–499, 1974.
11. Storz, I.: Chlamydia and Chlamydia-induced Diseases. Springfield, IL, Charles C Thomas, Publisher, 1971.

# *Chapter 36*

# MYCOTIC DISEASES

TERRY W. CAMPBELL

Avian mycotic diseases can be divided into two classifications, mycoses and mycotoxicoses. Mycoses result from fungal growth within the host tissues and can be divided into two types, superficial and deep. Superficial mycoses result from fungal growth within the skin, whereas deep mycoses occur when fungal elements invade deeper body tissues or become systemic. Mycotoxicoses occur when toxic fungal metabolites gain entry into the host.

## ASPERGILLOSIS

Aspergillosis is the most common of the avian mycoses, with the majority of cases caused by *Aspergillus fumigatus*. *Aspergillus* is a ubiquitous organism, and birds are continuously exposed to environmental spores. The fungal agents become pathogenic only under certain conditions. Predisposing factors include immunosuppression, malnutrition, and unhygienic environmental conditions. Immunosuppression appears to play a major role in the chronic forms of aspergillosis and often results from stress, malnutrition, pre-existing disease, or prolonged antibiotic or corticosteroid therapy.[35, 39] Stress-induced aspergillosis is frequently seen in birds subjected to oil contamination, capture and confinement (especially penguins, water fowl, and raptors), surgery, and other debilitating conditions or diseases.[20, 30] Malnutrition can cause a primary thiamine deficiency that has been considered by some to predispose birds to aspergillosis.[55] *Aspergillus* and other fungi grow readily in decaying organic matter. Conditions that favor mold growth in the bird's environment include damp feed, feed utensils, or litter, dry feces, and dark, damp conditions with poor ventilation.[30, 55] An inverse relationship occurs between the relative environmental humidity and the concentration of airborne spores and pulmonary lesions.[46] Low humidity and excessive dust interfere with the normal mucociliary activity of the respiratory ciliated epithelium and predispose the bird to respiratory mycoses.[59] Damp material or soil used as nesting material can cause mold growth and contamination of incubating eggs or mycoses in the nestlings.[19] Moldy incubators can result in contaminated eggs that become a potential source of infection to other eggs or hatching chicks, resulting in high mortality. Dusty, deteriorated food or live food cultures (e.g., mealworm cultures) can contain numerous fungal spores. Inhalation of an overwhelming number of spores can result in an acute aspergillosis.[39]

Aspergillosis is predominantly a disease of the lower respiratory tract in birds and can occur as an acute or chronic disease. Acute aspergillosis is a fatal respiratory disease.[40] Gross lesions of the acute form can appear as white mucoid exudation in the respiratory tract, marked congestion of the lungs, thickening of the air sacs, or acute pneumonic nodules surrounded by areas of hyperemia.[30, 55] Histopathology of the acute pneumonic nodular form reveals a central core of cell debris and fungal hyphae surrounded by heterophils and macrophages.[1] Chronic aspergillosis is the more common form with typical granulomatous lesions. Multiple plaques or nodules may be seen disseminated throughout the lungs and air sacs. The nodules vary in size and are most prevalent in the periphery of the lungs and caudal thoracic and abdominal air sacs.[39, 59] Chronic aspergillosis lesions appear as white, yellow, or green caseous granulomatous nodules that often exhibit sporulating fungal colonies. Histopathology of the chronic lesions reveals mycotic nodules surrounded by multinucleated giant cells, macrophages, heterophils, lymphocytes, plasma cells, and a band of fibrous tissue.[1] In severe cases, extensive adhesions between the air sac lesions and the abdominal viscera may occur. Rarely, large tumor-like aspergillomas may occur.[52]

Although the lungs and air sacs are the primary organs affected by *Aspergillus* sp., the trachea, syrinx, and bronchi are frequently involved. Localized obstruction due to mycelial masses in these areas may be the only lesions present in some birds.[40] Extension of the fungal growth from the respiratory tract to pneumatized bone or the peritoneal cavity may occur. Isolated cases of aspergillosis involving the mouth and upper gastrointestinal tract, central nervous system, eye, kidney, adrenal glands, aorta, and bone have been reported.[5, 30, 45]

Clinical signs of avian aspergillosis are not diagnostic and resemble other chronic debilitating diseases. Acute aspergillosis has a rapid onset, unlike the insidious onset of the chronic form. The clinical signs of the acute form may include anorexia, dyspnea, or sudden death without signs.[39] Clinical signs of chronic aspergillosis are variable but commonly include dyspnea, lethargy or depression, and emaciation. Often the disease reaches extensive involvement of the respiratory system before clinical detection. Lesions involving the upper respiratory tract (trachea, syrinx, and primary bronchi) are frequently associated with voice changes (affected birds may exhibit a reduction in the normal voice volume, change in tone, or a reluctance to talk or vocalize), respiratory clicking or gurgling noise, and dyspnea.[20, 30, 39] Avian aspergillosis involving the central nervous system is often associated with ataxia or paralysis. Frequently the first noticeable change is a prolonged recovery of normal breathing after moderate exertion.

Antemortem diagnosis of avian aspergillosis can be difficult. A tentative diagnosis can be made with presenting signs of dyspnea and a history of environmental conditions suitable for fungal growth and recent exposure to stress. A stronger tentative diagnosis can be made if the bird's respiratory condition is unresponsive to antibiotics and radiographs reveal increased density or nodules involving the lungs or air sacs. A history of dyspnea, anorexia, lethargy, and weight loss is helpful. A hemogram showing leukocytosis, heterophilia, monocytosis, lymphopenia, and an increase in serum total protein and globulins is also supportive of the diagnosis. Occasionally a nonregenerative anemia (anemia of chronic disorders) may be seen.[21] A presumptive diagnosis of avian aspergillosis can be made with the additional information of a positive culture of *Aspergillus* sp. from the trachea or pharynx and positive serodiagnostics for aspergillosis. A definitive diagnosis is made by demonstration of typical lesions by laparoscopy or exploratory surgery biopsy, identification of the causative agent by cytologic or histopathologic examination of the lesion, and culture of the organism from the lesion.

Aspergilli grow well on Sabourand dextrose agar or blood agar at room temperature. Initially the colonies appear white but later become green with sporulation. *Aspergillus* sp. are identified microscopically by their characteristic conidiophores, which expand into large vesicles that contain the sterigmata bearing long chains of spores.[47] Smears made by scraping or crushing the mycotic nodules and examined using 10 per cent sodium hydroxide, lactophenol cotton blue stain, new methylene blue stain, or Diff-Quik stain (Harleco, Gibbstown, NJ 08027) can aid in the diagnosis. Aspergilli have branching septate hyphae, and occasionally spores and conidiophores can be seen. Histopathologic specimens can be examined using hematoxylin-eosin stain, Grocott's stain for fungi, or other suitable stains. Typical colonies showing conidial heads are present only in the lungs or air sacs where oxygen is available.

A variety of therapeutic regimens has been used for avian aspergillosis with variable results. In general, this disease is given a poor to grave prognosis with or without therapy. Systemic amphotericin B (Fungizone, E. R. Squibb & Sons, Inc.) has been frequently used in the treatment of aspergillosis.[36, 49] In addition to the intravenous injections, intratracheal injections with amphotericin B with or without an antibiotic (e.g., chloramphenicol) has been suggested.[30, 39, 41, 49, 55] The injections are given directly into the glottis during inspiration, with the bird held upright and tipped first to the right and then left to assure equal distribution of the drug into both bronchi.[41] Nebulization using amphotericin B with Alvaire (Breon), Mucomist (Mead Johnson and Co.), or saline for four hours a day has also been recommended.[41, 55] Oral 5-fluorocytosine (Ancobon, Roche Products, Inc.) has also been suggested as a treatment for aspergillosis.[39] In a recent study,[42a] amphotericin B (Fungazone—Squibb) and 5-fluorocytosine were found to reach therapeutically useful levels in the plasma of raptors and turkeys; although thiabendazole inhibited most isolates in vitro, there were no detectable inhibitory concentrations in the plasma. Rifampicin (Rifadin, Dow Chemical) has also been given as a treatment for systemic fungi.[39] Drugs not available in the United States have been given with varying success.[36] Ketoconazole and miconazole may not be effective in the treatment of aspergillosis.

Supportive care such as forced feeding and fluid therapy, along with the specific antifungal therapy, may be required. Studies with birds of prey suggest that intravenous fluid therapy to prevent dehydration may also reduce the incidence of aspergillosis.[42] Surgical removal of the mycotic lesions, flushing of the affected abdominal cavity with an amphotericin B solution, and the use of immunostimulants (e.g., levamisole HCl [Levasole, Pitman-Moore, Inc.]) has been suggested.

Prevention of avian aspergillosis is centered on the limitation of fungal growth in the environment and reduction of stressful conditions. Good hygienic practices should be applied to the daily care of captive birds. Incubators and brooders require frequent cleaning and disinfecting to prevent fungal contamination. A germling vaccine (germinating spores) has potential for use in birds and may become available in the future.[43] Oral 5-fluorocytosine given twice a day for 10 to 14 days has been suggested for use in birds subjected to stressful situations as a precautionary measure against aspergillosis.[39]

## CANDIDIASIS

Avian candidiasis is a mycotic disease primarily involving the upper alimentary tract (mouth, esophagus, and crop). The most common organism isolated is *Candida albicans*. *Candida* species can be isolated from the digestive tract in normal birds where it resides in low numbers.[29, 31, 32] However, certain conditions favor pathogenic overgrowth of *Candida*. Predisposing factors include prolonged antibiotic therapy (especially tetracyclines), malnutrition (i.e., essential fatty acid deficiencies and hypovitaminosis A), coexisting diseases, and poor sanitation.[11, 29, 54]

Young birds are more susceptible to candidiasis, and the disease is a frequent cause of crop impaction and death in baby psittacines.[11] Candidiasis is usually confined to the crop, causing delayed crop emptying and malnutrition. The organism can invade the mouth, infraorbital sinuses, esophagus, proventriculus, gizzard, and intestinal tract.[28, 29] Venereal, cutaneous, and ocular candidiasis has been reported in birds.[2, 13, 33] Systemic infections can occur, especially in young birds.[6] Crop lesions are characterized by a rough, thickened mucosa. The degree of mucosal involvement varies from mild white streaking to severe diphtheritic membrane formation.[9] Frequently, the crop mucosal surface is coated with a catarrhal to mucoid exudate.[29]

Clinical signs of avian candidiasis vary with the predisposing causes and degree of pathologic involvement. The common signs associated with crop candidiasis include general malaise (frequently the only presenting sign), weight loss or reduced growth rate, delayed emptying with thickening or dilatation of the crop, and frequent regurgitation. Crop candidiasis in raptors often causes impaired food ingestion due to pseudomembrane formation on the crop mucosa.[23]

A diagnosis of candidiasis cannot be made on laboratory culture alone, since the organisms can be normal inhabitants of the avian alimentary tract. The diagnosis is based on clinical signs, a history of predisposing factors, and demonstration of the lesions.[6] Insufflation and illumination of the crop for endoscopic examination of the crop mucosa will reveal the characteristic lesions. Cytologic examination of the crop content can provide a presumptive diagnosis of candidiasis. The presence of *Candida* blastospores and pseudohyphae (or true mycelium) provide strong evidence for candidiasis, since these forms represent *in vivo* tissue invasion.[29] The yeasts can be identified on dry smears using Diff-Quik, Gram's, or new methylene blue stains or a wet mount stained with lactophenol cotton blue. *Candida* yeast measure 3 to 6 μ in diameter and are oval and thin-walled, with broad-based budding. Cytologic specimens can be obtained from swabs of oral or esophageal lesions or crop aspiration. Positive samples reveal typical *Candida* yeasts or pseudohyphae but little inflammatory response (heterophils and macrophages). The organism grows well on blood or Sabourand's glucose agar, producing smooth creamy colonies with a yeastlike odor in one to three days.[3] Cornmeal agar is used to identify the characteristic chlamydospores of *C. albicans*.[8, 29] Differential diagnosis for avian candidiasis includes pox, trichomoniasis, histomoniasis, and hypovitaminosis A, which can be differentiated by history and microscopic evaluation of the lesions.

Treatment for avian candidiasis involves correction of predisposing factors and specific anti-*Candida* medication. Nystatin (Mycostatin, E. R. Squibb & Sons, Inc.), a common treatment for candidiasis, can be given orally.[11, 56] A 3 per cent amphotericin B lotion applied to oral lesions or 5-fluorocytosine has been suggested for cases refractory to nystatin.[11, 41] Ketoconazole (Nizoral, Janssen Pharmaceutica, Inc.) in the drinking water has also been effective in the treatment of avian candidiasis.[21] Ketoconazole acidified with hydrochloric acid or dilute chlorhexidine diacetate (Nolvasan, Fort Dodge Laboratories) can be administered by tube feeding.[25] Flock or aviary outbreaks can be treated using ketoconazole or 2 per cent chlorhexidine solution in the drinking water or 15 per cent formic acid solution sprayed on the food.[11, 57] Ocular lesions can be treated with a 3 per cent amphotericin B ointment or subconjunctival injections of a 25 mg amphotericin B/ml sterile water solution along with oral 5-fluorocytosine for three weeks.[13] Severely affected birds require forced feeding and supplementation with A and B complex vitamins.

## UNCOMMON MYCOSES

Although aspergillosis and candidiasis represent the majority of avian mycoses, other fungal agents can produce disease in birds. Many soil saprophytes can be opportunistic and become pathogenic under certain conditions. Overtreatment with antibiotics, immunosuppressive conditions, unsanitary environments, and coexisting diseases are a few examples. A review of the literature reveals a wide variety of mycotic agents capable of producing lesions in birds.[53]

*Aspergillus fumigatus* is responsible for the majority of the mycotic infections involving the avian respiratory tract; however, other *Aspergillus* species and fungi can be pathogenic. Mucormycosis (primarily caused by *Absidia* sp.) and nocardiosis (caused by *Nocardia asteroides*) involving the lungs and air sacs have been reported.[4, 15, 34] Mycotic rhinitis and sinusitis can be caused by *Candida* sp., *Cryptococcus neoformans, Rhinosporidium* sp., and other saprophytic fungi.[10, 12, 17, 22] Rhinosporidiosis appears to be associated with aquatic habitats and affects primarily water fowl. *Cryptococcus neoformans* is a common saprophyte associated with avian excrement (especially pigeons) but is rarely a cause of disease in birds. Natural immunity of birds to *C. neoformans* involves the high avian body temperature (*C. neoformans* does not grow in temperatures greater than 40°C) and antifungal activity of the normal bacterial flora.[10, 17] Respiratory mycoses can involve abdominal organs by direct extension from air sac lesions.

Superficial mycoses or dermatomycoses are either rare or poorly understood in most avian species. *Trichophyton gallinae* is the primary etiologic agent for avian ringworm (favus) affecting canaries, ducks, chickens, turkeys, and pigeons.[3] *Microsporum gypseum, Trichophyton simii*, and *Candida* sp. can occasionally cause dermatomycoses in birds.[3] Clinical evidence of superficial mycoses includes alopecia (especially on the head and neck), scaliness, and formation of concentric ring-shaped skin lesions.[29] Clinical signs in caged birds include pruritus, automutilation, and feather plucking.[53] Lesions are limited to the skin and feather follicles with minimal tissue reaction. Treatment of avian superficial mycoses can be difficult. Biweekly application of a salicylic and tannic acid solution (3 gm salicylic acid and 3 gm tannic acid q.s. to 100 ml with ethyl alcohol) or copper sulfate solution has been recommended.[53] Nystatin cream can be used for *Candida* infections as a topical treatment.

Detection of avian mycotic diseases can be difficult, and often therapeutic success depends on early diagnosis before advanced lesions can develop. Diagnosis and treatment of deep mycoses are the same as outlined for aspergillosis. Identification of the causative agent requires microscopic examination of swabs or scrapings of lesions using 10 per cent sodium hydroxide, Diff-Quik stain, lactophenol cotton blue stain, Gram's stain, or acid-fast stain. Skin scrapings or crushed feather follicles can be examined in cases of superficial mycoses. The organism can be isolated and identified by culture on blood, Sabouraud's dextrose, or other suitable agar for fungi. It should be emphasized that many fungi are common environmental contaminants, and the isolation of a fungal organism does not provide a diagnosis without demonstrating its involvement in the lesion. Histopathologic evaluation of biopsied lesions or necropsy specimens will aid in the diagnosis. Reproduction of the disease in sentinel birds may be helpful. However, many saprophytic fungi become pathogenic only under certain conditions and tend not to be contagious (except for the dermatophytes). Typically, lesions of avian mycoses have become extensive by the time clinical signs are apparent. Therefore, prevention is the best approach to veterinary management of avian mycotic disease.

## MYCOTOXICOSIS

Avian mycotoxicosis occurs when a bird is exposed to toxic secondary fungal metabolites produced during fungal growth. Only certain fungal strains are capable of mycotoxin production and only under optimal environmental conditions. Birds are exposed to mycotoxins by ingestion, direct contact, or inhalation.

Fungi are ubiquitous in nature and commonly associated with agricultural crops. Fungi associated with seeds can be grouped as field or storage fungi.[48] Field fungi (i.e., *Claviceps* and

*Fusarium*) have high moisture requirements and grow on seed prior to harvest. Storage fungi (e.g., *Aspergillus*, the major mycotoxin producing fungi) have low moisture requirements and tend to grown on seed after harvest. Small areas in stored feedstuffs may have a microenvironment suitable for fungal growth and mycotoxin production, resulting in extremely high amounts of toxin.[50]

Mycotoxins can be grouped according to the organ system most frequently involved, although most produce lesions in more than one organ.[7] The lesions produced vary with the dose of the toxin, length of exposure time, and species susceptibility. Hepatotoxins include the aflatoxins. Nephrotoxins include citrinin and ochratoxin A. Neurotoxicosis is produced by citreovirdin and tremorgens. The trichothecenes (e.g., T-2 toxin) are dermatotoxins and alimentary tract toxins. Of the mycotoxins, aflatoxins (primarily aflatoxin $B_1$), trichothenes (primarily T-2 toxin), and ochratoxin A have significance to avian disease.[7, 37]

Avian aflatoxicosis occurs when a bird ingests the toxic metabolites from certain *Aspergillus* fungi. Aflatoxins are aromatic and heterocyclic metabolites of toxin-producing strains of *Aspergillus flavus* or *A. parasiticus* when allowed to grow in a suitable environment. The toxin-producing strains are widely distributed in nature and can grow on virtually any food or feedstuff.[7] Foods that are potentially at risk include peanuts (usually poor grade), peanut products, nuts, and cereals (usually the incidence is low with proper storage). Foods that provide adequate substrate for aflatoxin production at room temperature include bread, cheese, beans, fruit juices, and meats.[27] Environmental factors that influence aflatoxin production in food are aeration, temperature, moisture, and light.[27] Low oxygen and high carbon dioxide concentrations will decrease aflatoxin production. The optimum temperature for aflatoxin production is between 25 and 30°C; therefore, refrigeration (0.5 to 10°C) should decrease toxin production. Moisture is a critical factor, with the optimum humidity for fungal growth and toxin production being 85 per cent or greater. Aflatoxin production occurs best in darkness, since light has a deleterious effect on aflatoxin formation.

The general effects of aflatoxicosis include reduced reproduction, poor growth, anticoagulant activity, alteration of immunity, and hepatotoxicity.[7, 18, 27] The anticoagulant activity is reflected by prolonged whole blood clotting time and prothrombin time, decreased fibrinogen, and gastrointestinal hemorrhage.[27, 37] The immune system alterations are indicated by a decreased resistance to infection, vaccine failures, a decrease in serum IgG and $\alpha$ and $\beta$ globulins, and an alteration of lymphocyte stimulation and interferon production.[27, 37] Hepatotoxicity is the most frequently recognized effect of avian aflatoxicosis. The liver is swollen and yellow with serosal petechiation in acute aflatoxicosis. Chronic hepatic lesions show nodular formations and fibrosis. Histopathologic evaluation shows bile duct hyperplasia, severe generalized fatty change, large hyperchromatic periportal parenchymal cells, and portal fibrosis.[7, 27] Acute hepatic lesions show massive hepatocyte necrosis and hepatic hemorrhages, whereas more chronic lesions show biliary hyperplasia and cirrhosis. Aflatoxins can also be carcinogenic.

The clinical signs of avian aflatoxicosis vary with the type of toxin, amount of toxin ingested, duration of consumption, and species of bird. Aflatoxin $B_1$ is the most toxic of the aflatoxins. The usual clinical signs of avian aflatoxicosis include anorexia, weight loss, depression, and some mortality. Dietary aflatoxin $B_1$ at a concentration of 10 ppm fed to chickens results in a severe reduction of feed consumption, inhibition of hepatic microsomal drug metabolism, inhibition of protein and nucleic acid synthesis, suppression of mitosis, and an elevation of serum aspartate aminotransferase levels.[14] Susceptibility to aflatoxicosis appears to be greater in males than females and tends to decrease with age.[27]

Avian fusariotoxicosis occurs with the exposure of birds to toxic metabolites of *Fusarium* fungi. *Fusarium* spp. produce metabolites with estrogenic activity and trichothecenes. Trichothecenes appear to be more significant in birds, especially the T-2 toxin. T-2 toxin causes altered feathering and growth and necrosis of the oral mucosa. Oral lesions are characterized by raised yellow-white caseous lesions that may interfere with feed consumption.[58] Similar lesions may occur on the feet and shank, where the T-2 toxin in the litter can cause a contact dermatitis. A bioassay using Muscovy Ducklings can be used to detect trichothecene mycotoxins in grain. Necrotic oral lesions appear within 48 hours after the ducklings ingest 0.5 to 1.0 μg T-2 toxin or diacetoxyscirpenol per gram of contaminated feed.[50] Histologically the lesions appear as areas of focal necrosis with inflammatory exudative infiltration. T-2 toxin can also cause nervous lesions that are indicated by seizure disorders or impaired righting reflexes.

Trichothecenes can cause vasoconstrictive lesions, with the distal extremities (i.e., toes) showing well-defined dry necrosis.[44]

Avian ochratoxicosis is primarily caused by the potent ochratoxin A produced by *Aspergillus ochraceous*. Ochratoxicosis is often associated with kidney disease, but hepatotoxicity, bone marrow depression, and altered lymphoid activity may also occur. Ochratoxin has been associated with field outbreaks of air sacculitis in broiler chickens, possibly as a result of immunosuppression (i.e., lymphoid hypoplasia and decreased serum immunoglobulins).[16] Acute ochratoxicosis may show nervous disorders. Gross pathologic lesions reveal emaciation, visceral gout, and enlarged pale kidneys with urate plugs in the ureters. Histopathology shows tubular epithelial cell swelling, renal tubule dilation with proteinaceous material, hepatocyte degeneration, bone marrow suppression, and lymphoid depletion of the spleen and bursa.[7]

The primary acute mycotoxicoses are often easier to identify than the chronic conditions because the clinical signs are more dramatic, the toxin source is often identifiable, and a sample is available for mycotoxin assay. In chronic cases the clinical signs are less obvious with regard to the etiology (i.e., weight loss, poor growth, reduced reproduction, immune system alterations). Mycotoxins can reduce the birds' resistance to infectious agents (i.e., *Salmonella*, *Candida*, and coccidia), and impair digestive processes (i.e., alter vitamin D and riboflavin metabolism).[24] Therefore, the secondary disease conditions may mask a primary mycotoxicosis. It is often difficult to isolate the mycotoxin from the environment or the bird because of low toxin concentrations, and frequently the toxin source has disappeared from the environment before the disease becomes apparent. Mycotoxin outbreaks have certain characteristics that may aid in the diagnosis. These include a disease outbreak (1) associated with a specific food source that often shows signs of fungal contamination, (2) that is not transmissible directly from one bird to another and is often limited to a certain group of birds, (3) showing seasonal variation (mold growth is climate-dependent) or associated with grains from a particular harvest or storage season, (4) having no apparent infectious cause, and (5) with affected birds showing no response to chemotherapy.[7]

A conclusive diagnosis of mycotoxicosis is subject to the following limitations: (1) The isolation of a potentially toxigenic fungus is not diagnostic, because it is common to have moldy feed without mycotoxin formation. Only certain strains will produce mycotoxins and only with optimal environmental conditions. (2) The toxigenic mold does not need to be present for mycotoxicosis to occur, since the toxin may be present in the feed after the mold has disappeared. (3) The positive identification of a mycotoxin does not provide a definitive diagnosis without considering the amount present and host species susceptibility. (4) Obtaining representative feed samples may be difficult, since one sample may contain no toxin and another may have lethal levels. (5) Animal toxicity trials are time-consuming and expensive. (6) Physical, chemical, and biologic tests have been developed for only a few mycotoxins; therefore, undetected mycotoxins may exist.[7, 38]

The diagnosis of avian mycotoxicosis relies on the clinical history, isolation of the potentially toxic fungus from the feed, necropsy and histopathologic findings, and chemical confirmation of the suspected mycotoxin from the feed, gastrointestinal tract contents, or tissues from necropsied birds. Information concerning sampling and sample handling should be obtained from the diagnostic laboratory performing the mycotoxin assays. A rapid screening test for the detection of aflatoxins in seed involves the examination of the seeds under ultraviolet light. A bright yellow-green fluorescence is suggestive of aflatoxicosis; however, false-positive and false-negative results can occur. Therefore, the use of ultraviolet light to detect aflatoxin in seed has little validity unless supported by more specific chemical chromatographic or spectrophotometric procedures.[51]

No specific therapy is available for mycotoxicosis. Oral antidotes that have been shown to reduce the toxic effects of aflatoxin $B_1$ include activated charcoal, reduced glutathione or cysteine, selenium, and vitamin A or carotenes.[14] The activated charcoal acts as an inabsorbable carrier that adsorbs the toxin and aids in the elimination of the toxin from the intestinal tract. Endogenous thiols (i.e., reduced glutathione) conjugate the absorbed toxins and their metabolites as part of the detoxification process. Adding reduced glutathione or cysteine (a glutathione precursor) to the drinking water of chickens reduced the toxic effects of aflatoxin $B_1$.[14] Oral selenium (sodium selenite in the diet) can also reverse the toxic effects of aflatoxin $B_1$ where selenium acts as a competitive inhibitor to the toxin in the liver. Vitamin A or carotenes (precursors of vitamin A) may also provide protection against aflatoxin $B_1$ toxicity. Supportive treatment should be initiated in suspected cases of mycotoxicosis, and the toxin source should be removed if it can be identified.

The best approach is to minimize the chance for mycotoxin outbreaks by proper control measures. This is achieved by preventing mold formation in the food. Food storage conditions are the most important control measure for bird owners, since most have no control over field conditions, harvesting, and processing. Only good-quality, undamaged seed should be purchased. However, sound whole kernel peanuts without evidence of molding can contain high levels of aflatoxin inside the seed.[27] Thus, contaminated seeds can appear to be of good quality. Moisture is the single most important factor for fungal growth and mycotoxin formation. *Aspergillus flavus* will not invade seeds when the seed moisture is in equilibrium with a relative humidity of 70 per cent or less.[27] Temperature is another important factor, and the limit for aflatoxin production is near 10°C. Refrigeration guidelines to protect perishable foods from molding and mycotoxin formation include the following: (1) Food is promptly chilled to temperatures below 5°C to compensate for fluctuations in refrigerator temperatures. (2) Large items may have air pockets that do not cool sufficiently; therefore, refrigeration of smaller aliquots of food may be required. (3) Air-tight plastic containers seal in moisture, and periodic opening allows airborne spores to enter, creating an environment suitable for mold growth.[27]

## REFERENCES


1. Adrian, W. J., Spraker, T. R., and Davies, R. B.: Epornitics of aspergillosis in mallards *(Anos platyrhynchos)* in north central Colorado. J. Wildl. Dis., 14:212–217, 1978.
2. Beemer, A. M., Kuttin, E. S., and Katz, Z.: Epidemic venereal disease due to *Candida albicans* in geese in Israel. Avian Dis., 17:639–649, 1973.
3. Beneke, E. S.: Mycotic infections. In Hitchner, S. B. (ed.): Isolation and Identification of Avian Pathogens. Ithaca, NY, Arnold Printing Corp., 1975, pp. 133–141.
4. Burr, E. W., Huchzermeyer, F. W., and Made, V. D.: Mucormycosis in a parrot. Mod. Vet. Pract., 63(12):961–962, 1982.
5. Bygrave, A. C.: Leg paralysis in pheasant poults *(Phasianus colchicus)* due to spinal aspergillosis. Vet. Rec., 109:516, 1981.
6. Campbell, T. W.: Disorders of the avian crop. Compend. Contin. Educ. Pract. Vet., 5(10):813–824, 1983.
7. Carlton, W. W.: Pathogenesis and pathology of mycotoxicosis. Western Poultry Disease Conference, 1976, pp. 82–85.
8. Carter, G. R.: Essentials of Veterinary Bacteriology and Mycology. East Lansing, MI, Michigan State University Press, 1976.
9. Chute, H. L.: Fungal infections. Ir Hofstad, M. S. (ed.): Diseases of Poultry. Ames, IA, The Iowa State University Press, 1972, pp. 458–463.
10. Clipsham, R. C., and Britt, J. O.: Disseminated cryptococcosis in a macaw. J. Am. Vet. Med. Assoc., 183(11):1303–1305, 1983.
11. Clubb, S. L.: Infectious diseases. In Proceedings of the 1982 Annual Meeting of the AAV, pp. 91–119.
12. Courtney, C. H., Forrester, D. J., and White, F. H.: Rhinosporidiosis in a wood duck. J. Am. Vet. Med. Assoc., 171(9):989–990, 1977.
13. Crispin, S. M., and Barnett, K. C.: Ocular candidiasis in ornamental ducks. Avian Pathol., 7:49–59, 1978.,
14. Dalvi, R. R., and Ademoyero, A. A.: Toxic effects of aflatoxin $B_1$ in chickens given feed contaminated with *Aspergillus flavus* and reduction of the toxicity by activated charcoal and some chemical agents. Avian Dis., 28(1):61–69, 1984.
15. Dawson, C. O., Wheeldon, E. B., and McNeil, P. E.: Air sac and renal mucormycosis in an African gray parrot *(Psittacus erithacus)*. Avian Dis., 20(3):593–600, 1976.
16. Dwivedi, P., and Burns, R. B.: Ochratoxin in chickens. Vet. Rec., 114(12):298, 1984.
17. Ensley, P. K., Anderson, M. P., and Fletcher, K. C.: Cryptococcosis in a male Becarri's crowned pigeon. J. Am. Vet. Med. Assoc., 175(9):992–994, 1979.
18. Exarchos, C. C., and Gentry, R. F.: Effect of aflatoxin $B_1$ on egg production. Avian Dis., 26(1):191–195, 1981.
19. Flammer, K.: Veterinary care of nestling and hand reared psittacine birds. In Proceedings of the 1983 Annual Meeting of the AAV, pp. 190–206.
20. Fowler, M. E.: Miscellaneous waterbirds. In Fowler, M. E. (ed.): Zoo and Wild Animal Medicine. Philadelphia, W. B. Saunders Company, 1978, pp. 213–217.
21. Fudge, A. M.: Avian antimicrobial therapy. In Proceedings of the 1983 Annual Meeting of the AAV, pp. 162–183.
22. Grinder, L. A., and Walch, H. A.: Cryptococcosis in columbiformes at the San Diego zoo. J. Wildl. Dis., 14:389–394, 1978,
23. Halliwell, W. H.: Diseases of birds of prey. Vet. Clin. North Am., 9(3):541–568, 1979.
24. Hamilton, P. B.: Economic losses and control of mycotoxins in poultry. Western Poultry Disease Conference, 1976, pp. 86–88.
25. Harrison, G. J.: Guidelines for treatment of neonatal psittacines. In Proceedings of the 1983 Annual Meeting of the American Association of Zoo Veterinarians, pp. 176–182.
26. Harrison, G. J.: Personal communication.
27. Heathcote, J. G., and Hibbert, J. R.: Aflatoxins: Chemical and Biological Aspects. Amsterdam, Elsevier Scientific Publishing Co., 1978.
28. Humphreys, P. N.: Debilitating syndrome in budgerigars. Vet. Rec., 101:248–249, 1977.
29. Jungerman, P. F., and Schwartzman, R. M.: Veterinary Medical Mycology. Philadelphia, Lea and Febiger, 1972, pp. 61–74.
30. Keymer, I. F.: Mycoses. In Petrak, M. L. (ed.): Diseases of Cage and Aviary Birds. Philadelphia, Lea and Febiger, 1982, pp. 599–605.
31. Kocan, R. M., and Hasenclever, H. F.: Normal yeast flora of the upper digestive tract of some wild columbids. J. Wildl. Dis., 8:365–368, 1972.

32. Kocan, R. M., and Hasenclever, H. F.: Seasonal variation of the upper digestive tract yeast flora of feral pigeons. J. Wildl. Dis., 10:263–266, 1974.
33. Kutlin, E. S., Beemer, A. M., and Meroz, M.: Chicken dermatitis and loss of feathers from *Candida albicans*. Avian Dis., 20(1):216–218, 1975.
34. Long, P., Choi, G., and Silberman, M.: Nocardiosis in two Pesquet's Parrots *(Psittrichas fulgidus)*. Avian Dis., 27(3):855–859, 1983.
35. Migaki, G.: Mycotic diseases in captive animals—mycopathologic overview. In Montali, R. J., and Migaki, G. (eds.): The Comparative Pathology of Zoo Animals. Washington, DC, Smithsonian Institution Press, 1980, pp. 267–275.
36. Nakeeb, S. M., Babus, B., and Clifton, A. Y.: Aspergillosis in the Peruvian Penguin *(Spheniscus humboldti)*. J. Zoo Anim. Med., 12:51–54, 1981.
37. Pier, A. C.: Biological effects and diagnostic problems of mycotoxicosis in poultry. Western Poultry Disease Conference, 1976, pp. 76–79.
38. Pollock, G. A.: Mycotoxins in animal feeds. Mod. Vet. Pract., 64(4):285–287, 1983.
39. Redig, P. T.: Aspergillosis. In Kirk, R. W. (ed.): Current Veterinary Therapy VIII. Small Animal Practice. Philadelphia, W. B. Saunders Company, 1983, pp. 611–613.
40. Redig, P. T.: Methods for the delivery of medical care to trained or captive raptors. In 1981 Annual Meeting of the AFA, Veterinary Medical Seminar.
41. Redig, P. T.: Mycotic infections of birds of prey. In Fowler, M. E. (ed.): Zoo and Wild Animal Medicine. Philadelphia, W. B. Saunders Company, 1978, pp. 273–290
42. Redig, P. T.: Personal communication.
42a. Redig, P. T., and Duke, G. E.: Comparative pharmacokinetics of antifungal drugs in domestic turkeys, Red-tailed hawks, Broad-winged hawks, and Great-horned owls. Avian Dis., 29:649–661, 1985.
43. Richard, J. L., Thurston, J. R., Cutlip, R. C., and Pier, A. C.: Vaccination studies of aspergillosis in turkeys: Subcutaneous inoculation with several vaccine preparations followed by aerosol challenge exposure. Am. J. Vet. Res., 43(3):488–492, 1982.
44. Robb, J., Kirkpatrick, K. S., Marshall, D. B., and Norval, M.: Association of toxin-producing fungi with disease in broilers. Vet. Rec., 111:389–390, 1982.
45. Rosskopf, W. J., Woerpel, R. W., Howard, E. B., and Walder, E.: Pacheco's disease and aspergillosis in a parrot. Mod. Vet. Pract., 63(4):300–301, 1982.
46. Saif, Y. M., and Moorhead, P. D.: Effect of relative humidity on the pathogenesis of *Aspergillus fumigatus* infection in turkey poults. J. Am. Vet. Med. Assoc., 165(8):745, 1974.
47. Schneierson, S. S.: Atlas of Diagnostic Microbiology. North Chicago, Abbott Laboratories, 1971, p. 44.
48. Schroeder, H. W.: Occurrence and prevention of some mycotoxins in field crops. Western Poultry Disease Conference, 1976, p. 82.
49. Schultz, D. J.: Respiratory disease: Aetiology, diagnosis, therapy. In Hungerford, T. G. (ed.): Proceedings No. 55 Refresher Course on Aviary and Caged Birds. Sydney, Australia, The Post-graduate Committee in Veterinary Science, 1981, pp. 479–495.
50. Shlosberg, A., Klinger, Y., and Malkinson, M.: Mycotoxin detection using a Muscovy Duckling bioassay. Vet. Rec., 114(5):387, 1984.
51. Slanker, M., and Braselton, E.: Aflatoxin and fluorescence under ultraviolet light ("blacklight"). Vet. Diag. News [Michigan State University], 1(3):4, 1984.
52. Stroud, R. K., and Duncan, R. M.: Aspergillosis in a red-crowned crane. J. Am. Vet. Med. Assoc., 183(11):297–298, 1983.
53. Tudor, D. C.: Mycotic infections of feathers as the cause of feather-pulling in pigeons and psittacine birds. VM/SAC, 78(2):249–253, 1983.
54. Wagstaff, R. K., Jensen, L. S., Tripathy, S. B., and Kenzy, S. G.: Essential fatty acid deficiency and *Candida albicans* infection in the chick. Avian Dis., 12:186–190, 1968.
55. Wallach, J. D., and Boever, W. J.: Diseases of Exotic Animals. Philadelphia, W. B. Saunders Company, 1983, pp. 856, 906, 945–946.
56. Woerpel, R. W., and Rosskopf, W. J.: Avian therapeutics. Mod. Vet. Pract., 62:947–949, 1981.
57. Wood, N. A.: Treatment of established infections of candidiasis in partridges with formic acid. Vet. Rec., 87:656–658, 1970.
58. Wyatt, R. D., Harris, J. R., Hamilton, P. B., and Burmeister, H. R.: Possible outbreaks of fusariotoxicosis in avians. Avian Dis., 16(4):1123–1130, 1972.
59. Zinkl, J. G., Hyland, J. M., and Hurt, J. J.: Aspergillosis in common crows in Nebraska, 1974. J. Wildl. Dis., 13:191–193, 1977.

# *Chapter 37*

# PARASITES

H. JOHN BARNES

Considering the increasing variety of avian species being maintained as companion animals and the diversity of parasites found in birds, one might feel overwhelmed by the possible infections (internal parasitisms) and infestations (external parasitisms) that might be encountered in a clinical practice. In actual fact, parasitism is not a common problem but should always be kept in mind when dealing with the clinically ill bird or high-density populations of avian species. Parasites of companion birds are discussed here; for parasites of free-living and zoo birds, the reader is referred to the appropriate references listed at the end of this chapter.[4, 8, 13, 18]

There are several basic points which need to be kept in mind in the clinical approach to avian parasitism:

1. A variety of diagnostic procedures can be performed to determine if parasitic infestations and infections are involved, including scrapings, smears (wet and stained), fecal sedimentation, and flotation. No single technique can be relied upon for a definitive answer.
2. Because of the external changes they cause, infestations are more likely than infections to be noticed by the client.
3. Treatment may effect a clinical cure, but ultimate control of the problem rests with management. Consultation with the client on management changes designed to interrupt the parasite's cycle will be of greatest benefit to the bird, the client, and the practitioner, and will ensure that repeated episodes of parasitic disease do not occur.
4. Unusual parasitic infections and infestations are more likely in imported than in domestically reared birds. In the latter group, the parasites typically have direct life cycles, whereas parasites with complex indirect cycles are much less likely to occur since there is less opportunity for intermediate hosts to be present in the more controlled environments of domestically reared birds.
5. Many avian parasites are not host-specific. Parasites of indigenous, domestic, or free-living birds can spread to companion birds if care is not taken to segregate them from exposure to their parasites (Fig. 37–1).

## EXTERNAL PARASITES

Because of improved management, improved environments, awareness of their effect, and more effective and available treatments, external parasites are no longer common, especially on the individual pet bird. However, problems with external parasites may still occur in larger aviaries.

Of the external parasites, mites constitute the most important and most commonly encountered external parasites (Fig. 37–2). They are small, generally less than 1 mm, have eight legs as adults and nymphs and six legs as larvae. Diagnosis of mite infestations is made by direct visualization of the parasites with the aid of a magnifying lens or by finding them in scrapings from affected tissues (Fig. 37–3). Table 37–1 lists some of the features of mite infestations of

**Figure 37–1.** Engorged hard tick on a recently caught yellow-fronted canary.

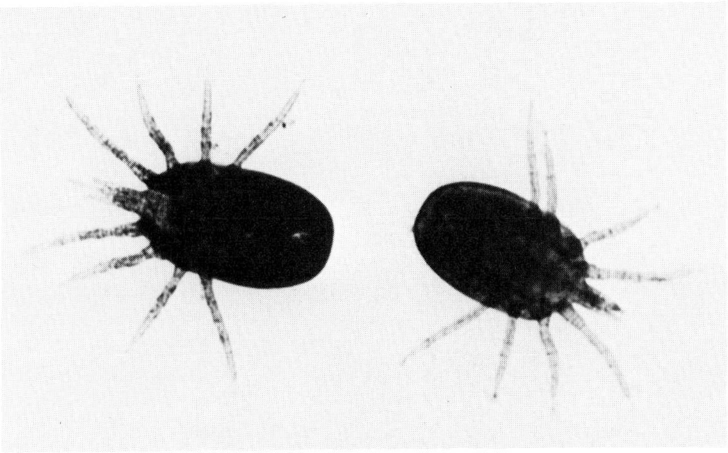

**Figure 37–2.** Northern fowl mites from pigeon nest.

companion birds. Other external parasites are less commonly encountered and are summarized in Table 37–2.

Treatments for external parasites change rapidly; new ones are constantly being developed. Ivermectin, for example, is effective against *Knemidokoptes* mites (Fig. 37–4). One should not attempt to extrapolate treatments from domestic poultry to caged birds. The latter tend to be more susceptible to toxicity. Also, one should not overtreat. Problems have been encountered when a product, generally recognized as safe, was overused and caused intoxication. One should use compounds that have a known wide safety margin such as rotenone and pyrethrin, especially when the latter is potentiated with piperonyl. Malathion and carbaryl are also fairly safe in low concentrations (see Chapter 26, Therapeutics); higher concentrations can be used to treat the environment. One should keep insecticides out of the bird's mouth and eyes. Insecticides that act similarly should never be combined; problems have arisen when carbaryl was used to dust birds, which were then put in an aviary with DDVP-emitting pest strips.

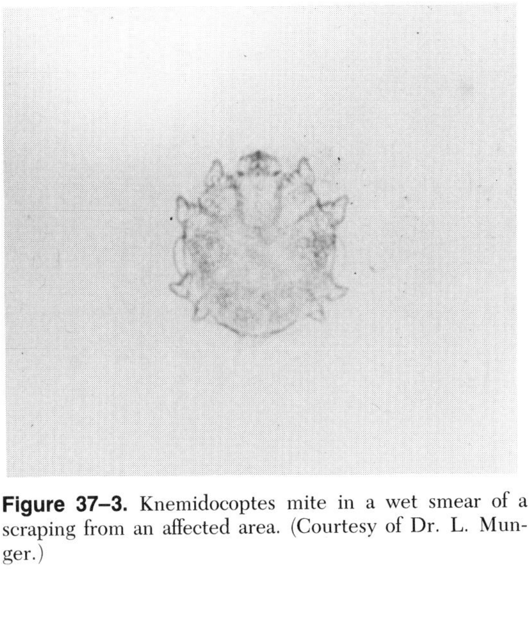

**Figure 37–3.** Knemidocoptes mite in a wet smear of a scraping from an affected area. (Courtesy of Dr. L. Munger.)

Boiling water is good for treating movable objects in the environment. One should burn debris, used nest material, and trash that may have accumulated in the aviary. Flight pens should be constructed without cracks and crevices where parasites can reside off the host.

### Summary

Control of external parasites is based on prevention. Domestically reared birds should be obtained from sources not infested with external parasites. Birds should be maintained in facilities that will exclude biting insects. Caged birds should be segregated from free-living birds, mammals, and their habitations to prevent spread of non–host-specific external parasites.

## GASTROINTESTINAL PARASITES

*Giardia* is reported by California practitioners to be one of the most common gastrointestinal parasites of pet birds, affecting up to 50 per cent of the budgerigars and cockatiels in that state. Diagnosis is often difficult. Trophozoites or cysts can be found intermittently in fresh fecal direct smears. Fudge and McEntee[9] recommend collecting very fresh feces in polyvinyl alcohol and using trichrome stain for identification of the flagellate (Fig. 37–5). Clinical signs associated with giardiasis include weight loss, soft green stools, and dry, flaky skin. This

*Text continued on page 478*

**Table 37–1.** MITES INFESTING COMPANION BIRDS

| Parasite | Clinical Findings | Diagnosis | Treatment/Control |
|---|---|---|---|
| *Dermanyssus* spp. (red mites) | Weakness, anemia, poor appearance, excess preening, pruritus | Demonstration of mites on birds at night or in protected areas in environment during daytime. Dark brown to red "spots" that move quickly. | Cleaning and treatment of environment as well as host. Segregate from wild birds, their roosts and nests. |
| *Ornithonyssus* spp. (fowl mites) | As above | Remain on birds. Mites crawl on hands and arms of handler holding heavily infested birds. | Dust birds with pyrethrins, carbaryl, or rotenone or spray with malathion. Segregate from wild birds. |
| *Knemidokoptes* spp. *Procnemido-koptes* spp. (scaley leg and scaley face mites) | Slowly developing, pitted or honeycombed, proliferative, scaley, crusty, often disfiguring lesions of face, beak, vent, or legs. Young adults most commonly affected. Canaries develop tassle-like growth on digits ("tassle-foot"). | Demonstration of small, round mites with stumpy legs in scraping from lesions. | Soften lesions with mineral oil and use a topical acaricide containing orthophenylphenol, rotenone, crotamiton, etc. Do not overtreat. Loosen and remove crusts as possible. Treat all sites of infestation. Invermectin given I.M. at 200 μg/kg has been found to be safe and effective. |
| *K. laevis* (deplumming mite) | Feather loss or breakage at skin. Focal lesion spreading to legs, neck, and head. Occurs in summer. Psittacines affected. | As above. | Bathe affected skin. Treat topically with acaricide. Control secondary bacterial infections. |
| Epidermoptid mites[12] | Pruritic dermatitis, feather loss, esp. of head and neck. Hyperkeratosis and erythema of defeathered skin. | Demonstration of mites from scrapings. | No effective treatment. Disease often fatal. Life cycle poorly known but may involve lice or hippoboscid flies. |
| *Syringophilus* spp., *Dermatoglyphus* spp., *Pterolichus* spp., *Analges* spp. (quill mites) | Feather loss. New feathers often affected. Part of quill remains in follicles and often contains debris and mites. | Demonstration of mites in quill. Mites can be squeezed out and seen in wet mounts. | None completely effective. Dusting or spraying with acaricides helpful. Isolate infested birds. |

**Figure 37–4.** Knemidocoptes mite infestation, which can produce severe damage to the beak of a budgerigar, can be treated with ivermectin. (Courtesy of Greg J. Harrison.)

**Table 37–2.** EXTERNAL PARASITES, OTHER THAN MITES, THAT INFEST COMPANION BIRDS

| Parasite | Clinical Findings | Diagnosis | Treatment/Control |
|---|---|---|---|
| Lice (biting or chewing lice of many genera) | Unthriftiness, irritability, damaged feathering, scratching, preening, decreased productivity and vitality. Eggs ("nits") attached to feathers. | Demonstration of parasites on birds or nits attached to feathers. Around vent, over back, on neck, and along wing are common locations. Fast-moving, pale brown, elongated, flattened parasites. | Dusting powders or sprays containing pyrethrin, rotenone, carbaryl, or malathion. Repeat in 2-3 weeks. Parasites are host-specific and survive only a few days off the bird. |
| Fleas (avian and mammalian) | Anemia, scratching, unthriftiness, irritated behavior, weakness. Embedded fleas in skin of face, head, neck. | Fleas are uncommon on birds. Nestlings in aviaries most affected. They are difficult to find on the bird but may be located in the environment, esp. nest material. | Clean up environment and treat birds as for lice. The ability of fleas to survive off the host and use various hosts makes control difficult. Segregate aviary from free-living mammals, birds, and their nests. |
| Ticks (soft and hard) | Anemia, unthriftiness, poor growth, and mortality. Tick paralysis and spirochete infections can occur as sequelae. | Most often in outdoor aviary or on recently caught birds. | Soft ticks are nocturnal, intermittent feeders; only larvae attach for any period of time. Other stages live in environment. Remove ticks, treat with acaricide, clean up environment. Segregate aviary from wild birds and their nests. |
| Hippoboscid flies | Anemia with death of severely infested young birds. Transmit *Haemoproteus* spp. | Flat, ked-like flies between feathers. When disturbed, they quickly return to the bird and rapidly crawl beneath feathers. | Treat as for lice. Keep wild birds away. Clean up environment. |
| Other biting insects (mosquitos, gnats, midges, blackflies, and bugs) | Anemia, irritation, poor performance. Dermatitis involving unfeathered areas of body in heavy infestations. Transmit blood protozoa, poxviruses, and arboviruses. | Observation of feeding especially at night. Collection in traps. Pox and blood protozoan infections. | Construct aviaries to prevent insects; remove potential breeding sites for insects; use of "bug zappers" helpful; establish insect control through integrated pest management programs (contact local state university for information on IPM programs). |

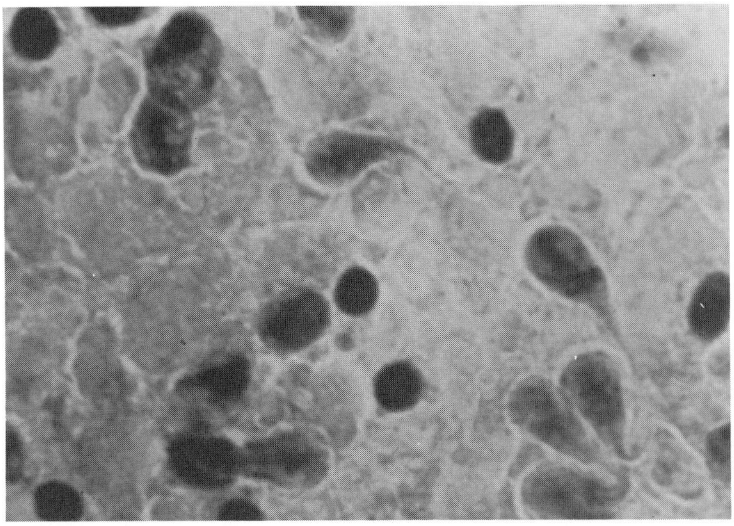

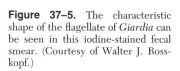

**Figure 37–5.** The characteristic shape of the flagellate of *Giardia* can be seen in this iodine-stained fecal smear. (Courtesy of Walter J. Rosskopf.)

**Table 37–3.** INTESTINAL PARASITES IN CAPTIVE BIRDS*[16]

| Parasite | Host | Transmission/Life Cycle |
|---|---|---|
| **I. PROTOZOA** | | |
| Trichomoniasis<br>*Trichomonas gallinae* | Raptors<br>Pigeons (also canaries, finches, budgerigar) | Contamination of water sources<br>Predation by raptors on infected birds<br>Through pigeon crop milk to young |
| Giardiasis<br>*Giardia lamblia* (same species as infects humans so zoonotic potential) | Budgerigar<br>Cockatiel<br>(Toucan) | Cysts passed in feces (viable for 3 weeks)<br>Fecal-oral transmission<br>Trophozoites liberated from cysts in small intestine, multiply and encyst in lumen. |
| Coccidiosis<br>*Eimeria* sp.<br>*Isospora* sp.<br>(See Fig. 37–6) | Galliformes<br>Pigeons (also budgerigar, finches) | Oocysts passed in feces<br>Fecal-oral transmission<br>Sexual and asexual states in intestinal epithelium, resulting in oocyst formation |
| Histomoniasis<br>*Histomonas meleagridis* | Galliformes<br>(turkeys, peafowl)<br>(also rheas) | The protozoa are transmitted by the ascarid, *Heterakis gallinarum*, which can be transported by earthworms. |
| **II. HELMINTHS** | | |
| *Capillaria* spp. (threadworms)<br>(See Fig. 37–7) | All species (esp. raptors, canaries) | Direct life cycle<br>Egg with infective larvae is ingested; larvae develop into adults in epithelium of GI tract. |
| Ascarids<br>*Ascaridia* spp.<br>(See Fig. 37–8) | Psittacines (esp. Australian Parakeets) | |
| *Porrocaecum* spp. | Passerines, raptors (also other species) | Direct life cycle |
| *Heterakis gallinarum* | Galliformes | |
| Spirurids<br>*Spiroptera incesta* | Psittacines (esp. Australian Parakeets) | Indirect life cycle<br>Arthropod intermediate host (ex: pillbug, cockroach) |
| *Dispharynx nasuta* | Passerines, galliformes | |
| *Tetrameres* sp.<br>(others) | Waterfowl (also psittacines, raptors, pigeons) | |
| Strongyle<br>*Amidostomum* sp. | Waterfowl | Direct cycle—eggs larvate in water; infective larvae are swallowed or penetrate skin—penetrate mucosa of ventriculus. |
| Cestodes<br>Many species | Psittacines (esp. Australian Finches)<br>Passerines<br>Waterfowl | Indirect life cycle (hosts usually arthropods or annelid worms) |

**Table 37–3.** INTESTINAL PARASITES IN CAPTIVE BIRDS* *Continued*

| Clinical Signs | Pathology | Diagnosis | |
|---|---|---|---|
| Anorexia<br>Weight loss | Caseous plaques in oral cavity, esophagus, crop (and upper respiratory system) | Scrape lesion or do crop washing<br>Look for flagellate<br>—characterized by 4 anterior flagellae<br>—progressively motile |  approx.<br>8 × 18 μm |
| Wasting<br>Chronic mucoid diarrhea<br>Up to 20–50% mortality | Usually none<br>Occ. distal small intestine distended with yellow creamy material<br>Organisms lie on surface GI epithelium and interfere with absorption. | Cysts in feces<br>Trophozoites in intestinal scaping (flagellate) |  approx.<br>9-14 μm 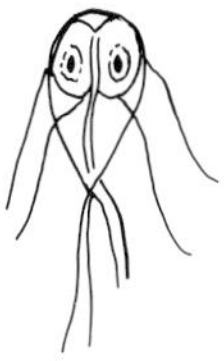 |
| Wasting<br>Mucoid/hemorrhagic diarrhea | Distention, congestion<br>Hemorrhage of small intestine | Fecal smear or fecal flotation<br>Look for double-walled cysts |  approx.<br>20 × 20 μm |
| Acute death in young<br>Wasting | Focal (bull's eye) necrosis of liver<br>Necrotic typhlitis | Characteristic pathology, organisms on histo or cecal scrapings |  approx.<br>6-20 μm |
| Dysphagia due to obstruction<br>Diarrhea when intestine involved<br>Anemia<br>Weakness | Often none grossly<br>Swelling of GI tract due to increased thickness of epithelium<br>(Parasite can be found in oral cavity, esophagus, and small intestine) | Eggs in feces or scrapings<br>—flattened bipolar plugs<br>—striated wall<br>—oval |  approx.<br>25 × 50 μm |
| Not specific | Adults: obstruction of lumen, competition for nutrients, perforation<br>Larvae: destruction due to migration in wall (parasites in intestines) | Eggs in feces<br>—not embryonated<br>—thick, mamillated shell<br>—round to oval | 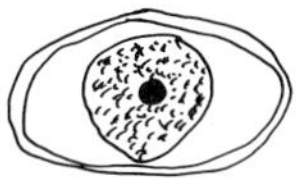 approx.<br>50 × 90 μm<br>(*Heterakis* is smaller—40 × 65 μm) |
| Acute death<br>Wasting | Swelling or nodules on mucosa of the proventriculus or ventriculus | Eggs in feces<br>—thick-shelled<br>—embryonated<br>—oval | 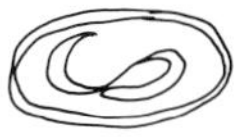 approx.<br>30 × 20 μm |
| Wasting | Ulceration and necrosis of the ventricular mucosa<br>Leads to exudative gastritis | Eggs in feces<br>—thin-shelled<br>—not embryonated (will embryonate when exposed 5–10 hrs.) | 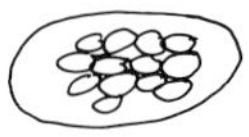 approx.<br>60 × 100 μm |
| Wasting | Catarrhal or hemorrhagic enteritis intestinal obstruction (parasites in small intestine) | Proglottids in feces<br>Egg packets or eggs in feces<br>—thick-shelled<br>—hexaconth oncosphere with 3 pairs of hooks |  approx.<br>75 × 90 μm |

*By Julia Langenberg, V.M.D.[16]

Acantocephalan and Trematode infections are much more rare in caged birds. These parasites require intermediate hosts that usually are not present in the captive environment.

can progress to episodes of feather pulling, alopecia, and pruritus. Fudge has speculated that this is a malabsorption problem and allergic reaction and reports that changes in the accompanying blood picture include hypoproteinemia, esoinophilia, and an occasional increase in monocytes as well as elevations in LDH and SGOT.

Giardiasis has been found to increase mortality in birds with budgerigar fledgling disease.[5]

Table 37–3 describes the intestinal parasites most commonly encountered in captive birds. The reader is referred to Chapter 26, Therapeutics, for information on the use of anthelmintics in caged birds.

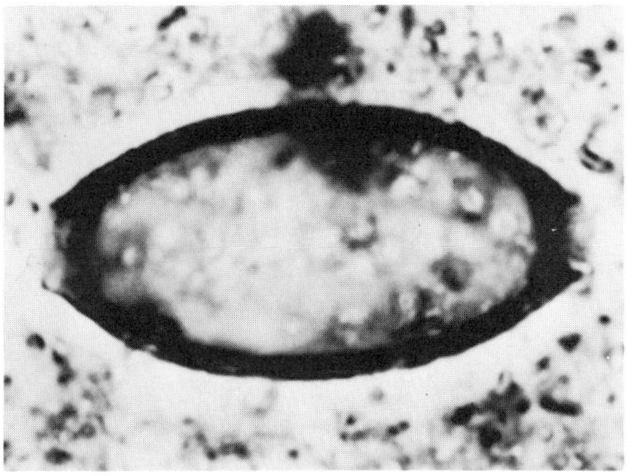

**Figure 37–7.** *Capillaria* ova. © Zoological Society of San Diego 1982.

## OTHER INTERNAL PARASITES

**Respiratory Tract.** Two parasites infecting the respiratory tract are of clinical importance: tracheal mites (*Sternastoma tracheacolum* [Fig. 37–9]), and gapeworms (*Syngamus tracheae* [Fig. 37–10]). A number of other parasites including nasal mites, air sac mites, filaria, trematodes, nematodes, and protozoa (cryptosporidia) can be found in the respiratory tract. They either do not produce clinical disease or are rare in companion birds (Fig. 37–11).

The clinical signs of both tracheal mite and gapeworm infection include dyspnea and voice changes or loss. A characteristic sucking or clicking noise may be heard with mite infection. Gapeworm-infected birds gasp, cough, and may have dried bloody mucus at the corners of the mouth. In both infections, the respiratory problems lead to a generalized debilitation. Sudden death from suffocation can occur, sometimes concurrently with exertion by the bird or handling by the client or veterinarian. Young birds are more severely affected. Canaries and Gouldian Finches have the greatest problems with mites, whereas larger birds, especially gallinaceous birds, are most commonly affected by gapeworms.

Diagnosis can be confirmed by demonstrating the mite or characteristic double-operculated gapeworm eggs in coughed-up mucus. The latter are also present in feces and must be distinguished from *Capillaria* eggs. Extending the neck and transilluminating the trachea while looking through the glottis can allow direct visualization of the parasites. The parasites can

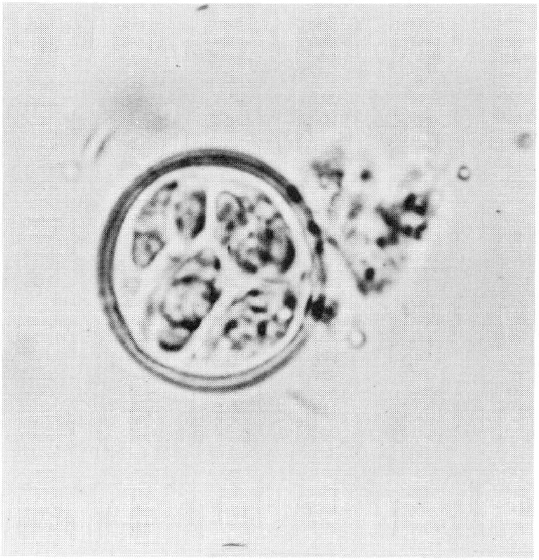

**Figure 37–6.** *Coccidia* oocyst. © Zoological Society of San Diego 1982.

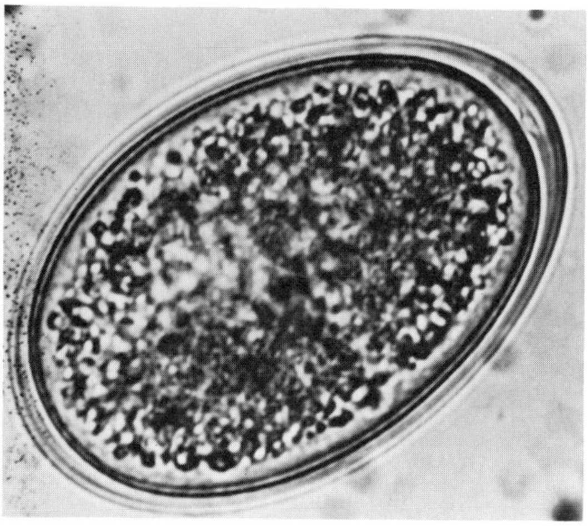

**Figure 37–8.** Ascarid ova. © Zoological Society of San Diego 1982.

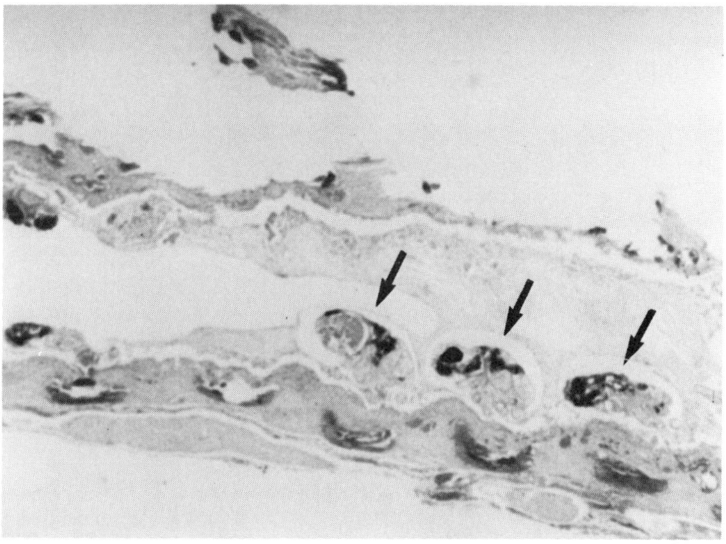

**Figure 37–9.** A photomicrograph of a histologic section from a canary that was presented for respiratory difficulty and died during the examination. Several tracheal mites can be seen causing chronic tracheitis and obstruction of the lumen.

be seen at necropsy. Tracheal mites are small black spots in the trachea and bronchi. Large numbers are often seen in air sacs around the base of the heart. Typically, there is excess mucus in the trachea which can vary from pale to dark brown. Gapeworms are large, rather robust, bright red helminths that are in a Y configuration, with the small male attached to the larger female. Excess mucus is also present in the trachea. Granulomas and even perforation of the trachea may occur where the female is attached to the tracheal mucosa.

The life cycle of both parasites is known (gapeworms) or presumed (tracheal mites) to be direct. Gapeworms can use various transport hosts, notably the earthworm. Tracheal mites are believed to be transferred from infected parents to offspring through parental feeding of the young. Control is achieved by preventing exposure to gapeworm ova or larvae through sanitation and management practices and parenting of offspring by mite-free parents. Flammer has used noninfected Society Finches as foster parents for Gouldian Finches to establish mite-free colonies of the latter.[7]

Treatment of both parasitic infections can be accomplished by the use of ivermectin at a dose of 200 μg/kg (see Chapter 26, Therapeutics). Although intramuscular administration is safe in most species, some toxic reactions have occurred; oral administration of the same dose is safe and effective. The placing of mite- or gapeworm-infected birds in a container with an insecticide powder and shaking them both together can no longer be recommended.

Although there are a number of parasites that may affect other body systems, they are not commonly encountered as clinical entities by

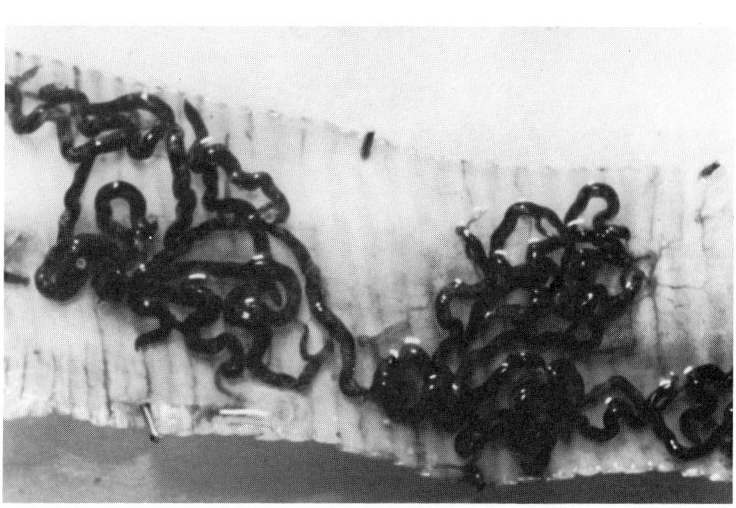

**Figure 37–10.** Gapeworms in the trachea of a pheasant.

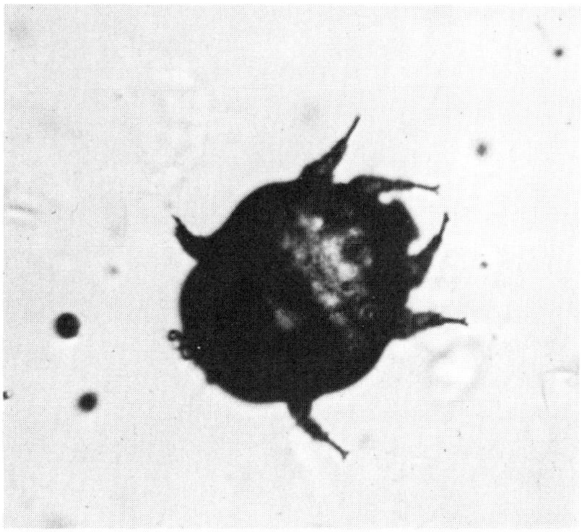

**Figure 37–11.** This air sac mite was one of many retrieved with biopsy forceps from an asymptomatic Sun Conure during a routine laparoscopy for sex determination. (Courtesy of Greg J. Harrison.)

the practicing veterinarian. Various helminths (trematodes and nematodes) may be found in the conjunctival sac.[3] Coccidia and cryptosporidia[10] can infect the kidney. Finding of coccidial oocysts in droppings does not automatically mean that the infection is in the gut, although that is still most likely to be the case. Schistosomes and their ova can affect the liver, kidney, and genital tract. Other trematodes may infect urinary and genital tracts. Migrating ascarid larvae, especially those of *Baylisascaris* from the raccoon, will produce neurologic signs. Toxoplasmosis is uncommonly seen but can also cause nervous system disease. Both of the latter diseases can be prevented by not allowing exposure to cat or raccoon feces. Filarids occur in the lateral ventricles of some avian species. *Sarcocystis* is common in the musculature of imported psittacines, especially cockatoos. Recently, developing schizonts of *Sarcocystis* have been found in avian lung, suggesting the possibility that acute sarcocystosis, as described in mammals, may also occur in birds. Migrating nematode larvae can be found in muscle. Filarids *(Ornithofilaria* spp.), nematodes, and trematodes are possible causes of cutaneous swellings (Fig. 37–12).

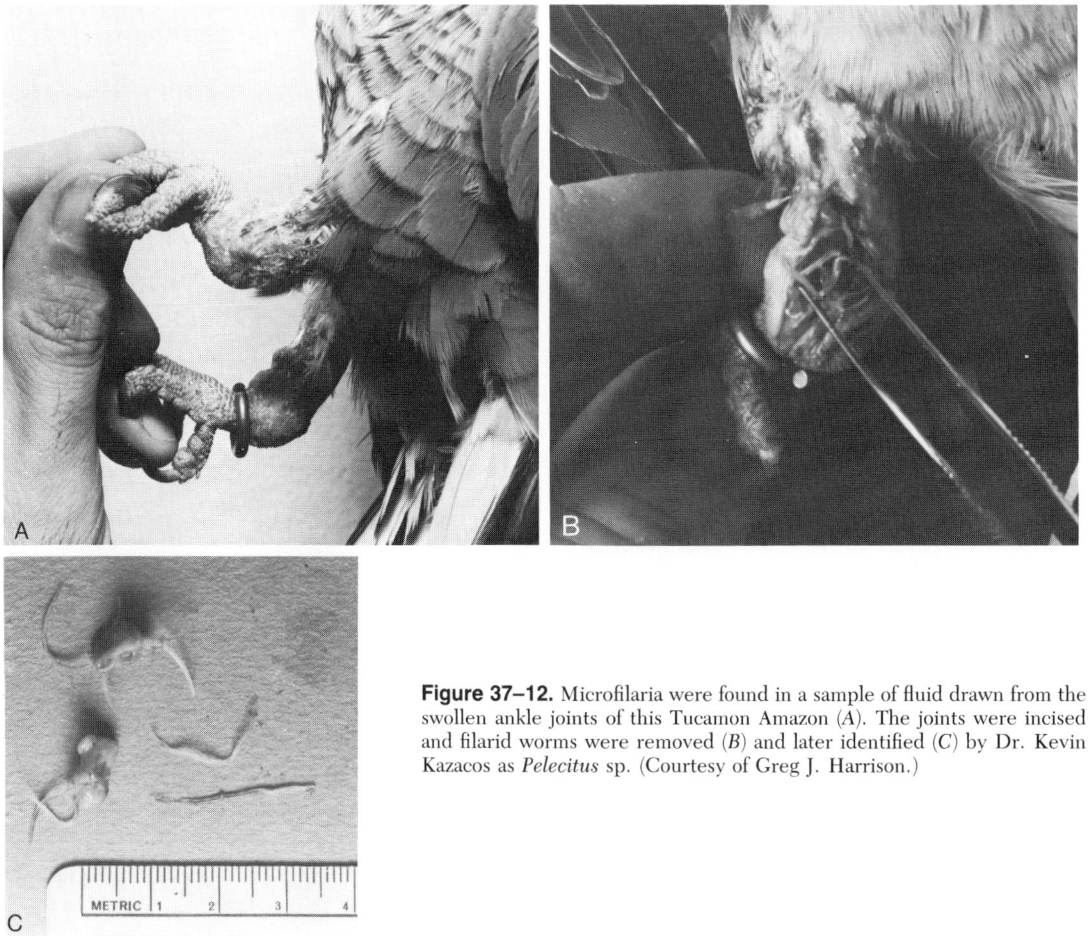


**Figure 37–12.** Microfilaria were found in a sample of fluid drawn from the swollen ankle joints of this Tucamon Amazon (*A*). The joints were incised and filarid worms were removed (*B*) and later identified (*C*) by Dr. Kevin Kazacos as *Pelecitus* sp. (Courtesy of Greg J. Harrison.)

## Summary

Internal parasitism outside of the gastrointestinal tract is uncommon clinically except for parasites in the respiratory tract. Since these have direct life cycles, prevention is based on preventing contact with infected birds, embryonated ova, or invertebrate transport hosts.

## IDENTIFICATION OF BLOOD PARASITES

Parasites and other microorganisms that may be found in the blood of avian species and their possible clinical significance are presented in Table 37–4 and illustrated in Figure 37–13. Although identifying the parasite species may be complicated, genus identification is straightforward and can be accomplished by answering a few questions. For the clinician, genus identification is usually adequate to provide an insight into the potential effect the parasite may have on the bird.

**QUESTION 1: IS AN ORGANISM OR PARASITE PRESENT?** Familiarity with normal blood elements, blood dyscrasias such as binucleate erythrocytes, and artifacts, along with a well-prepared, well-stained smear, is a necessary prerequisite to answering question one (Fig. 37–14). Parasites tend to require overstaining. Avian blood and parasites within it are also better differentiated by using a slightly basic buffer (pH 7.2) compared to the acidic buffer normally used for mammalian blood. Certain organisms in blood can be concentrated in the buffy coat. These include microfilaria, trypanosomes, and septicemic bacteria. *Leucocytozoon* and microfilaria, because of their large size, tend to be found in greatest numbers around the periphery of the smear and can be readily spotted at 100 × magnification by scanning the

**Table 37–4.** PARASITES AND OTHER MICROORGANISMS THAT MAY BE FOUND IN AVIAN BLOOD AND THEIR POSSIBLE CLINICAL SIGNIFICANCE[1, 2, 15]

| Organism | Occurrence | Clinical Significance/Comments |
|---|---|---|
| **Helminths** | | |
| Microfilaria | Common | ±, not known to cause any clinical disease, increased numbers associated with stress. |
| **Protozoa** | | |
| *Haemoproteus*/ *Parahaemoproteus**[17] | Common, esp. in cockatoos | +, high numbers associated with stress and feather picking, reduced stamina in pigeons. Invertebrate vectors: hippoboscid flies (*Haemoproteus*), *Culicoides* spp. (*Parahaemoproteus*). Treatment generally not recommended. |
| *Plasmodium*[11] | Occasional | + + +, anemia and mortality in many species, exoerythrocytic schizonts occlude CNS capillaries. Fatal infections have occurred in Moluccan Cockatoos and passerines. Invertebrate vector: Mosquito. |
| *Leucocytozoon*/ *Akiba**[6] | Occasional | + + +, highly pathogenic for anseriformes, galliformes and some passeriformes. Uncommon in psittacines but fatal infections in parakeets have occurred. Invertebrate vectors: *Simulium* spp. (*Leucocytozoon*), *Culicoides* spp. (*Akiba*). |
| *Atoxoplasma* (Syn. *Lankesterella*, *Toxoplasma avium*) | Occasional | + + +, highly pathogenic for canaries, starlings, and mynahs, rare in other species. Direct life cycle, *Isospora*-type oocysts. |
| *Babesia* (Syn. *Nuttalia*) | Rare | Unknown |
| *Haemogregarines* | Rare | Unknown |
| *Trypanosoma* | Occasional | ±, more common in bone marrow. Increased numbers associated with stress. |
| *Trichomonas* | Rare | + + +, systemic spread can occur with parasites in blood. |
| Other sporozoa[14] | ? | Unknown. *Toxoplasma*, *Sarcocystis*, and certain species of *Eimeria* may be found in blood during systemic involvement. Systemic infections of lories with *Sarcocystis* and cranes with *Eimeria* have been found. |
| **Bacteria** | | |
| *Aegyptianella* | Rare | Unknown. Anaplasma-like organism. Fatal infections in lovebirds reported. |
| *Borrelia* | Rare | + + +, spirochete causing septicemia. Often fatal in infected birds. |
| Other bacteria | ? | + + +, bacteria causing septicemia can often be found in blood films, especially buffy coat smears. |

*Parasites in these two genera are indistinguishable in the bird.

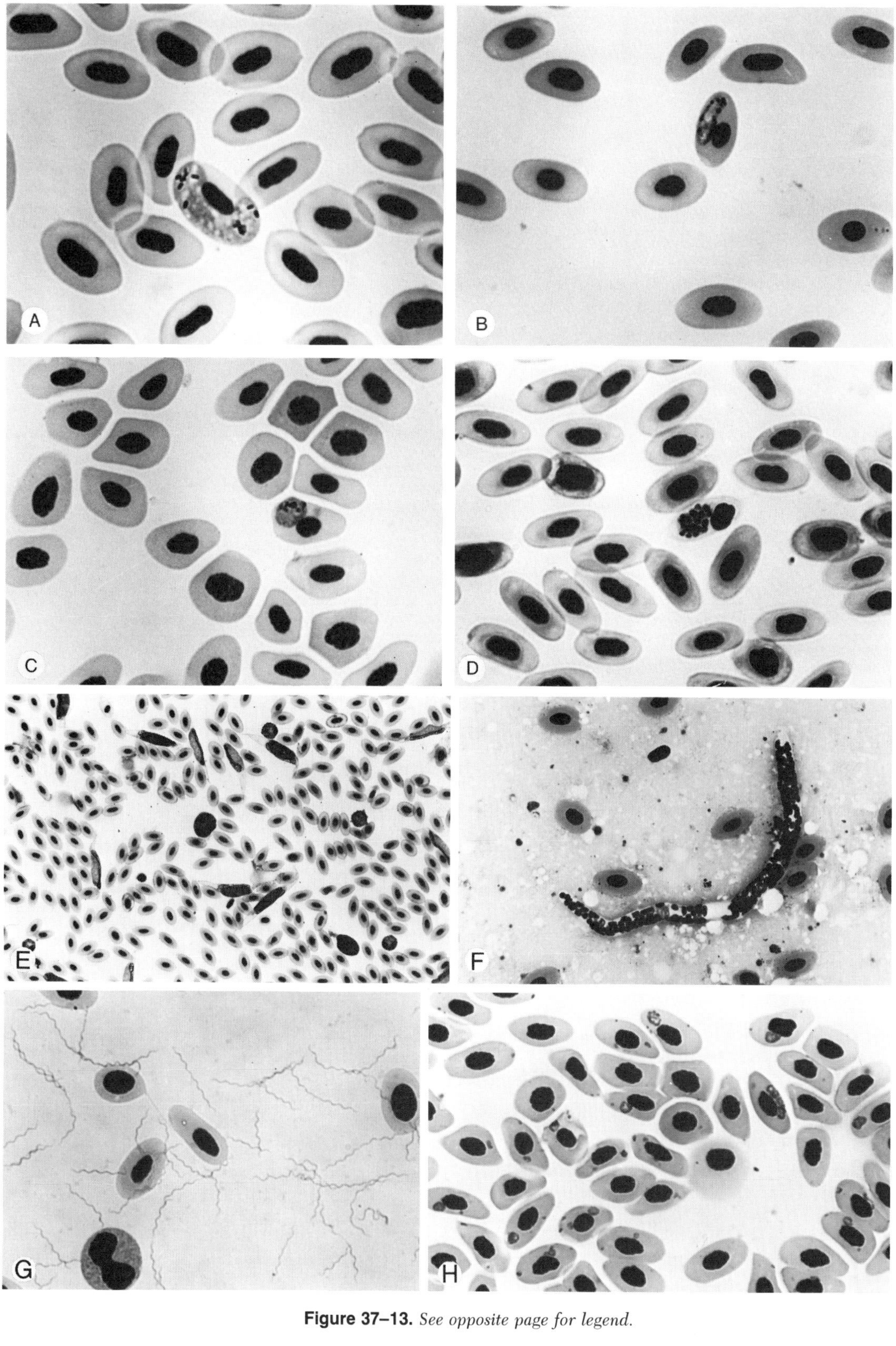


**Figure 37–13.** *See opposite page for legend.*

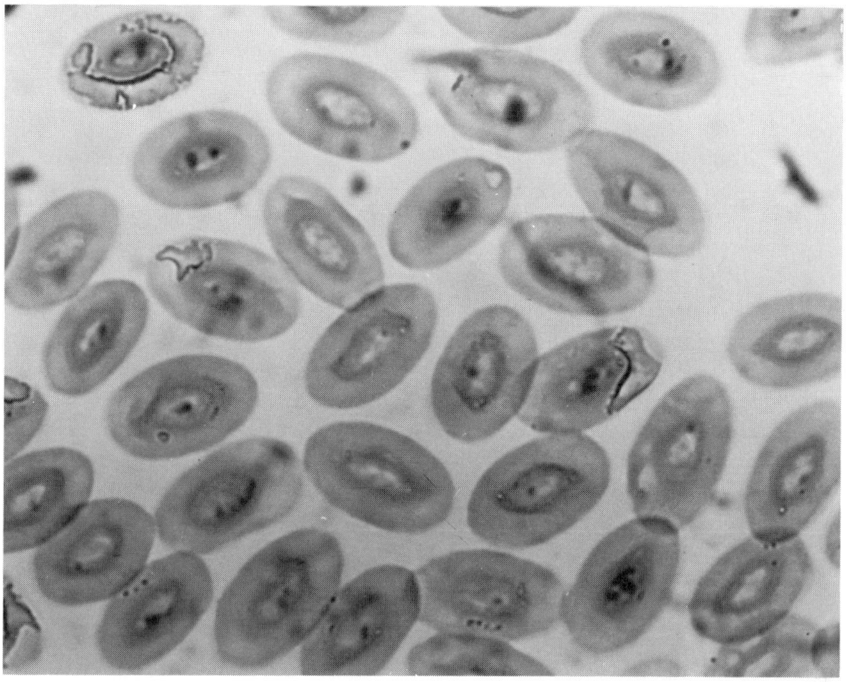

**Figure 37–14.** Artifacts from improper slide preparation must not be confused with intracellular blood parasites.

edges and tail of the blood smear. The other parasites are best located using oil immersion objectives and eyepieces giving magnifications of 600 to 1000 times.

For surveillance or incidence studies, it is necessary to examine a considerable number of microscopic fields before being relatively sure whether parasites are present or not. In general, the smear is completely scanned at low power and at least 200 oil immersion fields are examined for single samples, while 100 fields are checked if sequential smears are going to be examined. Such an extensive examination may be necessary for the clinician trying to establish a causative role for a blood parasite. Often they are numerous when signs are present but may occur only in low numbers or even be absent from the blood in birds with malarial infections (*Plasmodium* sp.), atoxoplasmosis, or aberrant leucocytozoonosis. Affected birds can become acutely ill and die suddenly as a result of numerous, immature schizonts or other multiplying forms in vital tissues, including myocardium, liver, and brain. If one is looking for carriers in a preventive medicine program, a complete examination would also be required.

QUESTION 2: IS THE PARASITE INTRA- OR EXTRACELLUAR? Microfilaria, trypanosomes, trichomonads, and septicemic bacteria, including the spirochete *Borrelia anserina*, are extracellular. All of these can be recognized by their characteristic appearance. The remaining parasites are intracellular.

QUESTION 3: WHAT TYPE OF CELL IS THE PARASITE IN? Those parasites most likely to be found in recognizable erythrocytes are *Plasmodium* and *Haemoproteus; Atoxoplasma* occurs in mononuclear cells, and *Leucocytozoon* in unrecognizable cells (these may have been either erythrocytes or leukocytes initially but

---

**Figure 37–13.** *A*, *Haemoproteus* gametocyte. Note large size and course, elongated granules. *B*, *Plasmodium* gametocyte. Elongated gametocytes occur in the subgenera *Giovannolaia*, *Novyella*, and *Huffia*. They are typically smaller than *Haemoproteus* gametocytes and have fine granules. *C*, Round gametocyte typical of *Plasmodium* species in the subgenus *Haemamoeba*. *D*, *Plasmodium* schizont. Multinucleated schizonts are diagnostic of *Plasmodium* infection. They may contain from as few as 4 to 20 more merozoites, may occur anywhere in the cytoplasm of either mature or immature erythrocytes, and may or may not displace the host cell nucleus. Some species have diffuse, round, or elongated schizonts while others have condensed, fan-shaped ones. *E*, *Leucocytozoon* gametocytes. This species has the elongated type. Some species produce round gametocytes. Regardless of morphologic type, the gametocytes so distort the host cell that it is no longer recognizable. Pale gametocytes are microgametocytes (male); dark ones are macrogametocytes (female). *F*, Microfilariae. *G*, *Borrelia*. *H*, *Aegyptianella*. Note the characteristic "ring-forms" in this heavily infected bird. (Magnifications: *A*, ×1000; *B*, ×800; *C*, ×900; *D*, ×900; *E*, ×225; *F*, ×540; *G*, ×900; *H*, ×800.)

**Table 37–5.** MORPHOLOGIC CHARACTERISTICS OF AVIAN *PLASMODIUM* SUBGENERA[11]

| Subgenus | Erythrocytic Type Parasitized | Gametocyte Shape | Schizont: Nucleus (No. Merozoites) |
|---|---|---|---|
| *Haemamoeba* | Mature | Round | Larger (>8) |
| *Huffia* | Immature | Elongate | Larger (>8) |
| *Giovannolaia* | Mature | Elongate | Larger (>8) |
| *Novyella* | Mature | Elongate | Smaller (≤8) |

have been so distorted by the time the parasite develops that their identity can no longer be ascertained). There remains only the problem of distinguishing between *Plasmodium* and *Haemoproteus* in the erythrocyte.

**QUESTION 4: ARE SCHIZONTS (ASEXUAL, MULTINUCLEATED STAGES) PRESENT OR ABSENT IN THE ERYTHROCYTES?** They are present for *Plasmodium* but absent for *Haemoproteus*. Erythrocytes containing only gametocytes or trophozoites differentiating into gametocytes are found in *Haemoproteus* infections.

There are also other morphologic features that help distinguish between the protozoan parasites in blood. Some *Plasmodium* species develop in immature erythrocytes (subgenus *Huffia*) or have round gametocytes (subgenus *Haemamoeba*). Even the elongated gametocytes that characterize species in the *Plasmodium* subgenera *Giovannolaia* and *Novyella* are smaller, finer, more delicate, and have small pigment granules compared to the large, robust gametocytes of *Haemoproteus* with their numerous, large, coarse pigment granules. *Haemoproteus* and *Plasmodium* are the only blood parasites that contain the characteristic dark golden-brown malaria pigment. *Leucocytozoon* and *Atoxoplasma* lack pigment.

Unpigmented organisms in erythrocytes could be *Babesia*, haemogregarines, or *Aegyptianella*, but these are rare and unlikely to be encountered. *Aegyptianella* is an *Anaplasma*-like organism in birds and is circular or ring-shaped, small, and basophilic and tends to be located along the erythrocyte margin. The morphologic features characterizing the four avian *Plasmodium* subgenera are presented in Table 37–5.

Table 37–6 summarizes the identifying characteristics of the common intracellular blood parasites. Transmission of blood parasites is a potential hazard with blood transfusions, even between dissimilar hosts, since most of the blood parasites are not very host specific. The potential of spreading blood parasites by using the same needles or instruments without sterilization between individuals also needs to be recognized.

**Table 37–6.** IDENTIFYING CHARACTERISTICS OF COMMON INTRACELLULAR AVIAN BLOOD PARASITES

**Haemoproteus/Parahaemoproteus**—large gametocytes and trophozoites developing into gametocytes in mature erythrocytes. Parasites contain malaria pigment. No asexual division in peripheral blood. Infection not readily established or maintained by inoculation of blood into susceptible host.*

**Plasmodium**—Round or elongated gametocytes in mature or immature erythrocytes. Asexual division in peripheral blood (schizonts—multinucleated stages). Parasites contain malaria pigment. Infection readily established and maintained in susceptible host by blood transfer.

**Leucocytozoon/Akiba**—Very large, round, oval, elongated, or spindle-shaped gametocytes in unrecognizable host blood cell. No pigment present. Infection not readily established or maintained by blood transfer to susceptible host.*

**Lankesterella**—Nonpigmented, oval (approx. 3 × 5 μ) parasites in mononuclear cells. Typically lies in an indentation in the host cell nucleus. Usually one or two parasites per cell except terminally, when large numbers may be seen.

*Technically, infection with these parasites should not occur following subinoculation of blood. However, schizonts of these parasites develop in endothelial cells, which can be ruptured during venipuncture. releasing infective merozoites. Rarely, infection can be transmitted through one or two passages but not maintained indefinitely as with *Plasmodium.*

## REFERENCES

1. Bennett, G. F., et al.: Bibliography of the Avian Blood-Inhabiting Protozoa: Supplement I. St. John's, Newfoundland, Memorial University Newfoundland, 1981.
2. Bennett, G. F., et al.: A Host-Parasite Catalogue of the Avian Haematozoa. St. John's, Newfoundland, Memorial University Newfoundland, 1982.
3. Brooks, D. E., et al.: Conjunctivitis caused by *Thelazia* sp. in a Senegal Parrot. J. Am. Vet. Med. Assoc., 183:1305–1306, 1983.
4. Davis, J. W., et al. (eds.): Infectious and Parasitic Diseases of Wild Birds. Ames, IA, Iowa State University Press, 1971, pp. 175–316.
5. Davis, R. B.: Budgerigar fledgling disease (BFD).

Proceedings of the 32nd Western Poultry Disease Conference, 1983, p. 104.
6. Fallis, A. M., and Desser, S. S.: On species of *Leucocytozoon*. Adv. Parasitol., 12:1–67, 1974.
7. Flammer, K.: Preliminary experiments for control of internal parasites in Australian finches. Proceedings of the 31st Western Poultry Disease Conference, 1982, pp. 155–157.
8. Fowler, M. E. (ed.): Zoo and Wild Animal Medicine. Philadelphia, W. B. Saunders Company, 1978, pp. 374–384.
9. Fudge, A., and McEntee, L.: A feather syndrome in cockatiels, associated with giardiasis. Unpublished data.
10. Gardiner, C. H., and Imes, G. D.: *Cryptosporidium* sp. in the kidneys of a black-throated finch. J. Am. Vet. Med. Assoc., 11:1401–1402, 1984.
11. Garnham, P. C. C.: Malaria Parasites and Other Haemosporidia. Oxford, England, Blackwell Scientific Publications, 1966.
12. Greve, J. H., and Uphoff, C. S.: Mange caused by *Myialges (Metamicrolichus) nudus* in a Gray-cheeked Parakeet. J. Am. Vet. Med. Assoc., 185:101–102, 1984.
13. Harrigan, K. E.: Parasitic diseases of birds. Proceedings No. 55 Refresher Course in Aviary and Cage Birds. Sydney, Australia, The Postgraduate Committee in Veterinary Science, 1981, pp. 337–396.
14. Helman, R. G., et al.: Systemic protozoal disease in zebra finches. J. Am. Vet. Med. Assoc., 11:1400–1401, 1984.
15. Herman, C. M., et al.: Bibliography of the Avian Blood-Inhabiting Protozoa. St. John's, Newfoundland, Memorial University Newfoundland, 1976.
16. Langenberg, J.: Intestinal parasites in captive birds. AAV Newletter 4(3), from material presented at the Mid-Atlantic States Avian Veterinary Seminar, Atlantic City, NJ, 1983.
17. Levine, N. D., and Campbell, G. R.: A check-list of the species of the genus *Haemoproteus*. J. Protozool., 18:475–484, 1971.
18. Wallach, J. D., and Boever, W. J.: Diseases of Exotic Animals. Philadelphia, W. B. Saunders Company, 1983, pp. 946–951.

# *Chapter 38*

# NEUROLOGIC DISORDERS

RONALD LYMAN

Clinical signs of neurologic disorders may be associated with specific neuroanatomic lesions, multifocal lesions, or idiopathic neurologic disease. Many cases with multifocal signs occur without proven etiology in the avian species, as is the case in mammalian medicine. Nutritional deficiencies, inflammatory disease, toxicities, spontaneous neuromuscular disorders, or mild metabolic disorders may account for these cases. Until further experience with clinical and experimental avian neurology develops, the clinician should make use of available diagnostic, therapeutic, and pathologic techniques in an attempt to accurately define the disease processs when presented with an avian neurologic disorder (see Chapter 20, Neurologic Examination).

## DISORDERS ASSOCIATED WITH HEAD SIGNS

### Epilepsy

Seizures may occur as the result of metabolic (e.g., hypoglycemia, hypocalcemia, liver failure), toxic (e.g., lead, organophosphates), neoplastic, or inflammatory central nervous system disorders. Seizures may also be due to underlying cardiac arrhythmia or to conduction disturbances (see Chapter 21, Electrocardiography). In addition, there are idiopathic, or "true," avian epileptics.

Careful attention should be paid to environmental history, concurrent symptomatology, and symmetry of the head signs. A whole-body roentgenogram may demonstrate radiopaque material other than grit (e.g., lead). An avian blood profile may reveal hypoglycemia. Evidence of liver disease may be obtained from serum chemistries or organ biopsy. An asymmetrical electroencephalogram (EEG) may indicate a structural cerebral lesion suggesting a neoplasia. An idiopathic epileptic will have a normal diagnostic work-up.

Control of seizures is attempted with diazepam parenterally, phenobarbital elixir, or primidone administered orally. Although canine hepatic cirrhosis from long-term anticonvulsant therapy has been reported, no histologic evidence of liver damage was found in one Amazon that had been maintained on primidone 125 mg/day for five years. A calendar of events should be maintained to facilitate adjustment of dosage in the idiopathic epileptic. Additional specific therapy is necessary when an underlying cause is discovered.

### Head Trauma

History and neurologic examination are usually diagnostic in cases of head trauma. Windows and paddle fans are common instruments of blunt trauma. Unconsciousness associated with loss of the doll's eye response indicates a brain stem lesion with poor prognosis. Management includes maintenance of body heat by incubation, artificial tear replacement, and tube feeding. As long as neurologic improvement occurs, the potential for recovery exists.

### Isolated Cranial Nerve Dysfunction

Isolated unilateral cranial nerve dysfunction occurs rarely in the avian species. It may be associated with trauma or may be idiopathic. Confirmation of neurologic dysfunction may be supported and followed by electromyography. Traumatic neuropathies should be treated with a two-week course of corticosteroids to minimize edema. Idiopathic neuropathies may precede more generalized neurologic dysfunction. Tear replacement solutions must be used when cranial nerve V sensory or VII motor (eye blink) are affected. Hypothyroidism should be considered when idiopathic cranial nerve dysfunction is diagnosed, although it remains to be documented in avian medicine at this time. Replacement therapy may be considered if hypothyroidism is diagnosed with appropriate testing procedures (see Chapter 22, Miscellaneous Diagnostic Tests).

## Lead Toxicity

Birds exhibiting seizures, other head signs, or multifocal neurologic signs may be victims of lead poisoning, especially if gastrointestinal signs are a concurrent problem.

History may include a source of exposure such as ingestion of products containing lead. Heavy exposure to automobile exhaust may contribute to the acquired lead concentrations in body tissues.

In addition to appropriate therapy (see Chapter 39, Toxicology), supportive alimentation may be necessary, and dexamethasone parenterally is indicated to suppress the central nervous system edema accompanying severe lead toxicity. Phenobarbital should be added if seizures are occurring.

## Polytetrafluoroethylene Toxicity

Respiratory distress, incoordination, and seizures followed by rapid death have been observed in birds exposed to toxic fumes produced by excessive heat applied to cooking utensils or range drip pans lined with polytetrafluoroethylene (commercial nonstick surfaces such as Teflon). Budgerigars, parrots, cockatiels, and finches are reported to be sensitive to these fumes. Client education will prevent this toxicity.

## Organophosphate Toxicity

Acute or chronic exposure to environmental insect and parasite control chemicals may result in incoordination, generalized weakness, and seizures accompanied by gastrointestinal signs. History and clinical signs are usually diagnostic.

Removal from the offending exposure and treatment with atropine, diphenhydramine, and pralidoxime chloride may be indicated in severe cases of this toxicity. Supportive fluids and incubation are often required.

Numerous other toxicities with possible neurologic signs (marijuana, hexachlorophene, alcohol, etc.) will be encountered sporadically in avian medicine. The clinician must extrapolate therapy from other species in these cases.

## Cerebral Neoplasia

Any primary or metastatic tumor may result in neurologic head signs. Of special interest are chromophobe pituitary adenomas reported in a series of budgerigars. Polyuria, polydipsia, somnolence, and seizures are common in this syndrome.

EEG, skull roentgenography, and computed tomography (C.T.) scanning may be useful in diagnosis of cerebral tumors (Fig. 38–1). Therapy is not often considered, although empirical adrenolytic therapy with o,p'-DDD might be considered in pituitary tumors, and superficial tumors may rarely be excised. Dexamethasone and phenobarbital may slow the progression of signs in specific cases.

## Vascular Diseases

Acute ischemic infarction resulting in head signs and death has been discovered at necropsy in a budgerigar. In most cases this would be a postmortem diagnosis; however, EEG and C.T. scanning could be useful diagnostic aids in valuable birds. Therapy is directed at the ac-

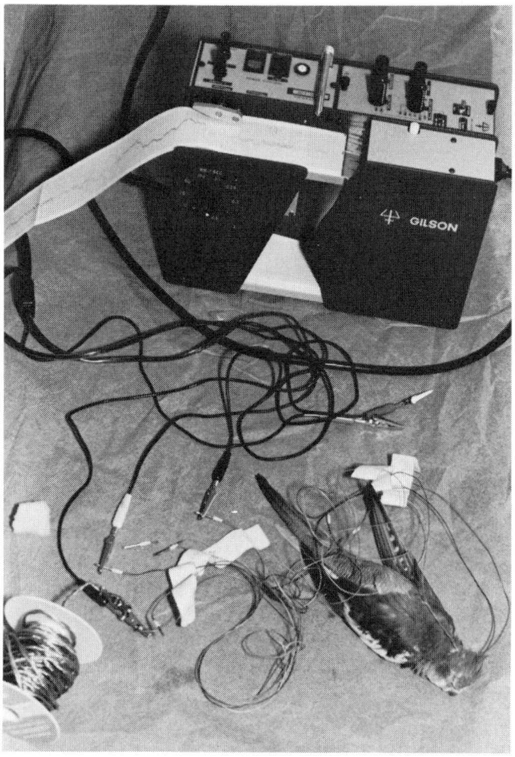


**Figure 38–1.** Additional information for the diagnosis of neurologic disorders in birds may be provided by an electroencephalogram.

companying edema with parenteral dexamethasone, and phenobarbital if seizures are associated with the disease.

### Congenital Anomalies

Hydrocephalus has been reported in caged birds with head signs. Again, diagnosis was made at necropsy. EEG and C.T. scanning could identify mildly affected birds. Therapy is aimed at suppressing the formation of cerebrospinal fluid by chronic administration of dexamethasone. As in other species, numerous other anomalies are possible.

## DISORDERS ASSOCIATED WITH CERVICAL AND THORACOLUMBAR LESIONS

### Spinal Trauma

In avian medicine, traumatic fractures or luxations of the vertebral bodies cause the overwhelming majority of cervical spinal cord lesions. History, physical examination, radiography, and electromyography contribute to the diagnosis of a cervical lesion. Dexamethasone (for at least two weeks) and "tincture of time" appear to be the standard therapy for spinal trauma in birds. As long as gradual neurologic improvement occurs, the potential exists for further improvement. At the present time, myelography and spinal surgery are rarely attempted in birds. The use of naloxone or T.R.H. parenterally immediately after spinal injury has been shown to reduce the magnitude of spinal cord traumatic lesions in some experimental models with animals. This type of neurotransmitter manipulation may soon find a place in avian medicine.

### Renal Carcinoma of Budgerigars

A progressive unilateral paresis of parakeets occurs with some regularity in renal adenocarcinomas. Pressure is actually exerted on the sciatic nerve as it emerges from the spinal nerve roots. The expanding abdominal mass may be palpable or visualized radiographically.

Prognosis is poor, and attempts at surgical removal of the lesion have been unrewarding. Dexamethasone may temporarily improve neurologic function. Although it mimics a thoracolumbar lesion, this disorder is actually a peripheral nerve lesion.

The following disorders, some of which are discussed elsewhere in this book from a broader context, may be expressed by neurologic signs.

## DISORDERS ASSOCIATED WITH MULTIFOCAL LESIONS

### Paramyxoviruses

Viscerotropic velogenic Newcastle disease (VVND), the best-known paramyxovirus, may affect many avian species (see Chapter 32, Viral Diseases). Any bird with respiratory and a variety of neurologic signs must be suspect. Other paramyxoviruses, such as the one most recently described in pigeons, also produce neurologic signs.

### Budgerigar Fledgling Disease (BFD)

Most BFD-affected birds die suddenly at about 10 months of age without previous signs; however, a small percentage show multifocal neurologic signs for 24 to 48 hours before death. These signs are described as generalized tremor, incoordination of head and body, and ataxia. Metabolic studies are not yet reported. Histopathology reveals multifocal viral inclusion bodies in the cerebrum and cerebellum. Most are intranuclear and located in the cerebellar ganglionic layers.

### Chlamydial Infections

Those birds surviving the acute respiratory and/or gastrointestinal phases of chlamydial infections may exhibit central nervous system signs of multifocal lesions (see Chapter 35, Chlamydia). Seizures, tremor, torticollis, and opisthotonus may occur in psittacine species. Some cockatiels with leg signs (weakness or paralysis of the hindlimbs) have been positive for chlamydial infection, but the relationship is unclear.

### Other Inflammatory Agents

Other viruses (e.g., *Herpesvirus*), bacteria (e.g, *Salmonella*), parasites, and fungi are capable of causing multifocal neurologic signs in avian species. Serious difficulty arises in the diagnosis of these sporadic cases. Culture techniques or cytologic evaluation of any concurrently affected organ systems may suggest an etiology. Consultation with clinical laboratories may assist in the determination of appropriate diagnostic tests in persistent obscure cases. Light and electron microscopic evaluation of brain tissue has demonstrated protozoal organisms *(Sarcocystis)* as the causative agent of central nervous system disease in a cockatiel.

## Possible Nutritional Factors

Many clinicians practicing avian medicine believe that various nutritional factors are associated with multifocal neurologic signs. References have implicated deficiencies in vitamins A, C, $D_3$, E, calcium, the B complex, and selenium (see Chapter 31, Nutritional Diseases). In such cases, parenteral supplementation may yield prompt relief from clinical signs. A "cockatiel paralysis syndrome," with localized effects on the jaws, wings, or legs (Fig. 38–2), is responsive in some cases to therapy consisting only of vitamin E/selenium injections. Convulsions in African Grey Parrots have been clinically associated with hypocalcemia. Much work remains to be done in documenting such conditions. However, it is probably wise to use parenteral vitamins in moderate superphysiologic dosages for avians with obscure neurologic diseases, especially in the face of a history of poor diet.

## Botulism Toxicity

Botulism has long been a problem in game farm birds, ducks, and pheasants, and it occurs sporadically in other wild or exotic species. In the syndrome, all cranial nerves and spinal reflexes are *depressed*, and muscle tone is generally poor ("limberneck" in ducks). Death usually follows rapidly. Type "C" is the serotype commonly identified. Electromyographic examination is not reported in this species but should reveal loss of conduction across the neuromuscular junction.

**Figure 38–2.** An etiology has not been established for a localized neuromuscular disorder frequently seen in cockatiels.

Treatment may be attempted with specific antisera, and the response is dramatic in mild cases. The prognosis is poor in birds showing severe neurologic signs (see Chapter 33, Bacterial Diseases).

## Proventricular Dilatation—"Macaw Wasting Disease"

Although this syndrome was first reported in several species of macaws, cockatoos and possibly other species may also be susceptible. Combinations of gastrointestinal and neurologic signs are a key to this syndrome, which is almost uniformly fatal.

Signs include intermittent regurgitation, diarrhea, depression, undigested seeds in the droppings, and variable multifocal neurologic signs.

Radiographs often show an enlarged proventriculus without the radiopaque material suggestive of lead toxicity. Clinical pathology is unremarkable. Electromyographic studies have not yet been reported.

The etiology and pathogenesis are unknown at this time; however, the possibility of a viral etiology is under consideration. Histopathologic changes include multifocal lymphocytic leiomyositis, especially in the proventriculus. Ganglionic plexi cannot be found on section. Lymphocytic poliomyelitis is evident throughout the spinal cord.

Large psittacines with similar clinical and radiographic signs have been reported to respond to surgical removal of proventricular contents and subsequent soft diet maintenance. However, smooth muscle biopsies were not performed in these cases for confirmation of histopathologic changes.

## RECOMMENDED READINGS

Clubb, S.: Viscerotropic velogenic Newcastle disease in pet birds. In Kirk, R. W. (ed.): Current Veterinary Therapy VIII. Philadelphia, W. B. Saunders Company, 1983, pp. 628–630.

Foreyt, W. J., et al.: Maggot-associated type C botulism in game farm pheasants. J. Am. Vet. Med. Assoc., 177:827–828, 1984.

Galvin, C.: Acute hemorrhagic syndrome of birds. In Kirk, R. W. (ed.): Current Veterinary Therapy VIII. Philadelphia, W. B. Saunders Company, 1983, pp. 617–619.
Gerlach, H.: Survivors of macaw wasting disease in Germany. AAV Newsletter, 2:52, 1984.
Graham, D. L.: An update on selected pet bird virus infections. Proceedings of the International Conference on Avian Medicine, sponsored by the Association of Avian Veterinarians, Toronto, Canada, 1984.
Harrison, G. J.: Clostridium bolutinum type C infection on a game fowl farm. Annual Proceedings of the American Association of Zoo Veterinarians, Atlanta, GA, 1974.
Harrison, G. J.: Long-term primidone use in an Amazon. AAV Newsletter, 4:89, 1982.
Hughes, P. E.: The pathology of myenteric ganglioneuritis, psittacine encephalomyelitis, proventricular dilatation of psittacines, and macaw wasting syndrome. Proceedings of the 33rd Western Poultry Disease Conference, 1984, pp. 85–87.
Jackson, C. A. W., and Cooper, K.: Proceedings No. 55 in Refresher Course in Aviary and Cage Birds. Sydney, Australia, The Post-Graduate Committee in Veterinary Science, 1981, pp. 623–640.
Jacobson, E. R., et al.: *Sarcocystis* encephalitis in a cockatiel. J. Am. Vet. Med. Assoc., 8:904–905, 1984.
Janssen, D. C., et al.: Lead toxicosis in three captive avian species. Annual Proceedings of the American Association of Zoo Veterinarians, 1979, pp. 40–41.
Mathey, W. J., and Cho, B. R.: Tremors of nestling budgerigars with BFD. Proceedings of the 33rd Western Poultry Disease Conference, 1984, p. 102.
Schlumberger, H. G.: Neoplasms in the parakeet, spontaneous chromophobe pituitary tumors. Cancer Res., 14:237, 1954.
Petrak, M. (ed.): Diseases of Cage and Aviary Birds. Philadelphia, Lea & Febiger, 1982, pp. 468–476.
Wells, R. E., et al.: Acute toxicosis of budgerigars caused by pyrolysis products from heated polytetrafluoroethylene: Clinical study. Am. J. Vet. Res., 43:1238–1242, 1984.

# Chapter 39

# TOXICOLOGY

GREG J. HARRISON

Compared to infectious and nutritional diseases, the level of mortality in cage birds from toxic causes is very low. According to Peckham in *Diseases of Poultry*, "Most poisons are occult and insidious and masquerade under the guise of chemotherapeutic agents, fungicides, and insecticides. Drugs and chemicals used for control of infectious and parasitic diseases may have therapeutic levels that impinge on toxic levels."[22] He also states the careless use of disinfectants is a common cause of toxic accidents.

The variable ways that cage birds appear to react to potential toxins may be dependent upon the age, body weight, sex, breeding status, and idiosyncrasies of the species. The potential for detoxification (i.e., the health status of the liver and composition of the gastrointestinal bacterial flora) may help determine the toxicologic effects. Compromised avian patients require lower doses of pharmacologic agents than those prescribed for healthy birds.

## AUTOINTOXICATION

Although it is not often considered in a discussion of toxic conditions, autointoxication could well be the most frequently encountered clinical manifestation of toxicity in cage birds. Autointoxication occurs as a result of metabolic waste product absorption from the intestine[22] and may occur in cases of pasting and occlusion of the vent, impaction, obstruction, or other causes of reduction of gastrointestinal motility (e.g., pansystemic illnesses, crop stasis, macaw wasting disease, or presence of foreign bodies in the gastrointestinal tract). Clinical signs include anorexia, polydipsia, polyuria, vomiting, depression, weakness, prostration, and symptoms of central nervous system disturbance. Intravenous fluid therapy, lactulose, broad-spectrum antibiotics, and liquid, easily digestible foods are indicated (see Chapter 27, What To Do Until A Diagnosis Is Made). Surgical removal of crop or proventricular contents may be required.

## LEAD POISONING

Lead is the most widely reported and clinically described ingested poison in cage and wild birds. Table 39–1 suggests some common environmental sources that provide birds with accessibility to lead. Surprisingly, even bone meal, which may be used in large quantities as a calcium supplement by aviculturists, was found to contain lead, but the effect on birds at that level of feeding is not known.

### Clinical Signs

The clinical signs associated with lead poisoning may be nonspecific and are dependent upon the total amount ingested, the surface area of the particles, and the length of time the lead is in the body. Lethargy, depression, weakness, abnormally colored diarrhea (green or black), excessive thirst, vomiting, wing droop, head tilt, convulsions, apparent hallucinogenic activity, and other symptoms of central nervous system disturbances (see Chapter 38, Neurologic Disorders) have been described. Similar clinical signs may be observed as a result of ingestion of zinc, salt, fertilizer containing nitrates, nitrofurazone in susceptible species, and other toxins (Table 39–2).

**Table 39–1.** SOURCES OF LEAD ASSOCIATED WITH CAGE BIRD TOXICITIES

| |
|---|
| Antiques (collapsible tubes, other lead items) |
| Lead frames of stained-glass windows, doors, ornaments, or Tiffany lamps |
| Weighted items (plastic toy penguins, ash trays) |
| Weights for draperies, fishing, scuba diving |
| Boat supplies requiring weights (window covers, lamps) |
| Batteries, solder, bullets, air rifle pellets |
| Old paint, sheetrock, galvanized chicken wire, hardware cloth |
| Some foil from champagne or wine bottles |
| Mirror backing, linoleum, some zippers, costume jewelry, base of light bulbs |
| Dolomite and bone meal products |
| Chronic exposure to leaded gasoline fumes |

## Diagnosis

Radiographs may reveal metal densities in the gastrointestinal tract or elsewhere, although the lack of radiodense material does not rule out lead poisoning, nor does the presence of densities confirm the condition. It was found during one study of waterfowl that 23 per cent of shotgun pellets in the gizzard went undetected on radiograph, primarily because of the size of the well-eroded pellets.[20] Although basophilic stippling and anemia have been reported to be associated with some cases of lead poisoning in avian patients, the relationship to the disease is controversial. Neither is a consistent finding nor is believed to be pathognomonic. Depressed free fatty acid plasma levels and elevated uric acid levels have been reported to suggest increased protein catabolism and perhaps renal dysfunction in lead-poisoned waterfowl.[18]

The inhibition of delta-aminolevulinic acid dehydratase (ALAD) enzyme activity is a sensitive and reliable test for detecting exposure of ducks to lead.[8] The test is specific for lead, is sensitive to a wide range of lead concentrations, and can detect lead intoxication up to three months after lead ingestion. ALAD levels of less than 86 units were found to be highly significant and indicative of lead exposure.[19] In a German study, ALAD was used with clinically normal pigeons in conjunction with evaluation of lead levels in the femur as a bioassay for environmental pollution.[29] Hematocrit values and hemoglobin levels in this study did not show any indication of the existence of a toxic effect of lead.

In a study in which Mallard Ducks were experimentally fed lead shot pellets,[9] lead concentrations of 0.5 ppm in the brain were found to correspond to a 75 per cent reduction of ALAD activity. In a separate study with Canvasback Ducks,[8] concentrations of 200 ppm in the blood also reduced ALAD activity by 75 per cent. The authors believe that the determination of inhibited ALAD activity to detect the presence of lead contamination has tremendous diagnostic potential because biochemical lesions in the brain precede clinical signs, which emerge only in the later stages of plumbism (e.g., wing droop and vent staining) in ducks.

Blood lead determination is currently used for substantiation of a diagnosis of lead toxicosis. Heparinized blood levels greater than 20 μg/dl are suspect and greater than 60 μg/dl are suggestive of lead toxicity in the presence of appropriate signs. Necropsy lead concentrations in the kidney and liver of greater than 6 ppm (undried) have been reported to verify this. However, because wet weights for heavy metal toxicity determinations in duck livers were found to generate sizeable errors in contamination level estimates, it was recommended that samples first be oven dried to constant weight and then analyzed for heavy metal concentrations.

The hepatofluorometer is reported to be extremely useful for diagnosing lead poisoning in waterfowl because it requires only one drop of blood to measure protoporphyrin levels and can process these samples for less than $0.10 each.[26] Blood protoporphyrin levels exceeding 40 ppm indicate lead ingestion, and at 500 ppm motor function impairment is observed.

Wildlife agencies are well aware of sublethal levels of lead poisoning. Nonregenerating neuronal cells in the cerebral portion of the brain appeared to have been destroyed by lead in one Mallard Duck study. This area of the brain integrates functions that are critical to a bird's survival, including its visual, auditory, and motor reflexes; therefore, even partial loss of the cortex from lead poisoning can be severely debilitating.

This author has encountered two cases of chronic poisoning from lead outside the gastrointestinal tract, which were diagnosed by radiography. A Severe Macaw, which had suffered intermittent stupified depression for over two years, was found to have a lead pellet lodged in the marrow of the mandible (Fig. 39–1*A*). Surgical removal of the pellet and calcium EDTA therapy arrested the episodes (Fig. 39–1*B*). Necrosis of the marrow adjacent to the pellet was noted during the surgery. In another case, a Yellow-naped Amazon was presented with hemorrhagic feces. A 22-caliber lead pellet, which had obviously been chronically embedded in the cloacal musculature, was subsequently removed. Both birds showed a marked change in the pigmentation of the feathers from green to black. No hematologic changes were evident in these cases.

## Therapy

If lead toxicosis is suspected from evidence of clinical signs and radiography, chelation therapy with intramuscular injections of calcium EDTA can begin immediately. Chelation therapy should continue until radiographic evidence fails to reveal lead and/or symptoms fail to return following cessation. Rapid response to therapy is further clinical evidence for lead poisoning prior to confirmation by blood lead

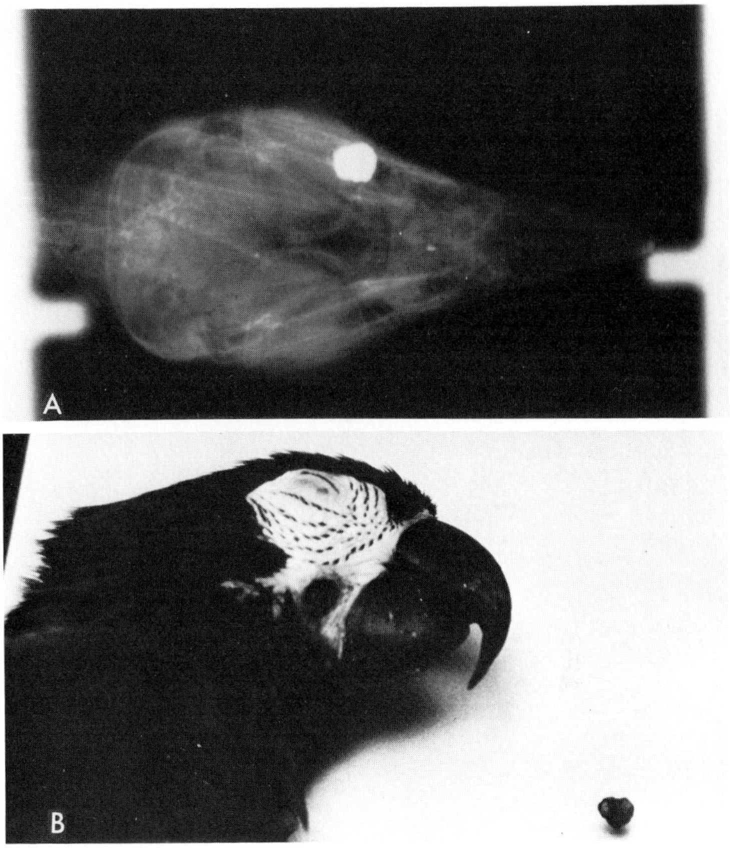

**Figure 39–1.** *A*, An uncommon site for a source of lead toxicosis was revealed in a radiograph. *B*, The clinical signs of intermittent depressive episodes were relieved by surgical removal of the lead and intramuscular calcium EDTA injections.

levels. In cases of gastrointestinal particulate ingestion, oral lubricants such as diluted peanut butter may be gavaged to assist in passing the particles. Supportive therapy is advised (see Chapter 27, What To Do Until A Diagnosis Is Made), and anticonvulsants may be indicated. Depending upon the amount of lead ingested, surgical removal of remaining particles may be necessary. The prognosis is guarded, especially in chronic cases in which permanent neuronal cell degeneration may have occurred.

A study with Pekin Ducklings indicated that while either lead or selenium treatment independently exerts a selective toxic effect on thyroid function, selenium supplementation of lead-treated Pekin Ducklings does protect against the toxic effects of lead on thyroid function.[11]

## PHARMACOLOGIC AGENTS

In the early years of cage bird practice and few pharmacologic guidelines, the toxic effects of many inappropriately prescribed or dosed medications were made known from clinicians' experiences of adverse reactions or sudden death in birds. The publication of informal comments from avian practitioners* revealed that of the significant number of poisonings reported, 62 per cent were directly related to iatrogenic pharmaceutical toxicity, 16 per cent to pesticides, and 16 per cent to household hazards. Careful attention to the weight of the patient (with a gram scale) and precise measuring of medication (a microliter syringe is useful) helps to eliminate some potential danger. Lories and lorikeets appear to be particularly susceptible to some drug overdoses. The dose of nitrofurazone must be halved for safety, and doxycycline is administered to lories (as well as macaws) at 3/4 the recommended dose listed for other species. Table 39–2 lists products and/or doses of medications that have been responsible for producing toxic reactions in pet bird patients.

*Association of Avian Veterinarians Newsletter, Lake Worth, FL.

**Table 39–2.** REPORTED EXPERIENCES OF TOXICITIES IN BIRDS

| Generic Name | Brand Name | Manufacturer | Species Affected | Toxic Dose | Route of Administration | Clinical Signs | Comments |
|---|---|---|---|---|---|---|---|
| | | | **Parasiticides** | | | | |
| Cythioate | Proban | Haver-Lockhart | Gouldian Finches | 35 mg/kg | Oral | Death in 4 hours | |
| Dimetridazole | Emtryl | Salsbury | Cockatiels<br>Budgerigars (breeding males)<br>Pigeons | 1 tsp/gal | Drinking water | Incoordination; unsteadiness; acute seizures; death<br>Acute hepatitis in cockatiel fledglings | Usually occurs in breeding birds.<br>May respond to injection of vitamin $B_1$ and $B_{12}$ |
| | | | Pekin Robins | Uniformly fatal all strengths | | | |
| Fenbendazole | Panacur | Hoechst | Canaries | 20 ml/2.5 liter | Drinking water | Ataxia; depression; mydriasis | |
| | | | Finches<br>California Quail<br>Pigeons<br>Birds in pin-feather | 10 ml/liter (3–5 days) | | Death at higher doses<br><br>Feather stunting, malformation, loss | |
| Ivermectin | Eqvalan<br>Ivomec | Merck | Orange-cheeked Waxbill Finches<br>Budgerigars | 200 μg/kg | I.M. | Death<br>Anorexia<br>Lethargy | Equine formula deaths reportedly from stabilizer |
| Levamisole | Tramisol<br>Ripercol<br>Levasol<br>Nilverm | American Cyanamid<br>Ethnor<br>Pitman-Moore<br>ICI Australia | Budgerigars<br>Mynah bird<br>Cockatoo | 40 mg/kg | Subcutaneous | Depression; ataxia; regurgitation; mydriasis; leg and wing paralysis | Most severe symptoms with I.M. |
| | | | Budgerigars | 25 mg/kg | I.M. | Hepatotoxicity | |
| | | | Pigeons | 35 mg/kg | I.M. | Death | |
| | | | Peach-faced Lovebirds | 66 mg/kg | | Death | |
| | | | White Ibis | 22 mg/kg | | Death | |
| Mebendazole | Telmin | Pitman-Moore | Finches<br>Psittacines | All doses | | Death<br>Intestinal obstruction by nematodes | Toxicity is controversial; not commonly used |
| | | | Columbiformes | 12 mg/kg | | Death | |
| Metronidazole | Flagyl | Searle | Finches | | Oral | Contraindicated | |
| Niclosamide | Yomesan | Bayvet | Pigeons<br>Geese | All doses | Oral | Death | |
| Nitrothiazole | Enheptin | American Cyanamid | Finches | All doses | Water medication | Death | |

| | | | | | | | |
|---|---|---|---|---|---|---|---|
| Quinacrine | Atabrine | Winthrop | Cockatoos | 50–150 mg/kg | Oral | Hepatotoxicity | |
| Praziquantel | Droncit | Bayvet | Society Finches | 100–250 mg/kg | I.M. | Depression; collapse<br>Death | |
| Rotenone | Goodwinol | Goodwinol Products | Avian | Excessive or careless use | Ingestion of topical preparation | Vomiting<br>CNS signs<br>Death | |
| **Pharmacologic Agents** | | | | | | | |
| Cephaloridine | Loridine | Corvel | Amazon | 0.05 cc | Subcutaneous<br>Periorbital | Blindness | |
| Chloramphenicol | | | Avian | 1000 mg/kg | I.M. | Death | |
| Dihydrostreptomycin | | | Avian species | In excess of 1 mg/20 mg | I.M. | Paralysis; curare effect<br>Death | |
| Doxycycline | Vibramycin (I.V. form) | Pfizer | Avian species | Any dose given I.M. | I.M. | Tissue necrosis | |
| Gentamicin | Gentocin | Schering | All, esp. lories | 20 mg/kg | I.M.<br>Subcutaneous<br>I.V. | Collapse; respiratory arrest; nephritis; death | |
| Lincomycin | Lincocin | Upjohn | Avian | | I.V. | Death | |
| Nitrofurazone | | | Lories<br>Lorikeets<br>Mynahs | 1 tsp/gallon drinking water | Water and/or nectar medication | Screaming; falling off perch; convulsions; death | |
| Polymyxin B | | | Amazons | 5–10 mg/kg | | Weakness; shakiness; falling off perch; vomiting; death at high dose | |
| Procaine penicillin | | | All | 1 mg/kg | | Paralysis; death | |
| Ticarcillin | Ticar | Beecham | Rose-breasted Cockatoos | Combined with tobramycin | I.M. | Hepatotoxicity | |
| **Pesticides** | | | | | | | |
| Diazanon 4% | Baygon | | 250 gm +/− avian | | Adjacent spraying | Death | |
| | | | | | Topical application | Weakness; tremors; death | Pulmonary edema |
| Coumarin | Warfarin | | | | Brought into aviary by rodents | Death due to hemorrhage | |
| Lead arsenate | | | Avian species | | Inadvertant feeding of laced seed | Death | |
| Brodifacoum | Talon | | Raptors | | Secondary poisoning by rodenticide | Death | |

**Table 39–2.** REPORTED EXPERIENCES OF TOXICITIES IN BIRDS *Continued*

| Generic Name | Brand Name | Manufacturer | Species Affected | Toxic Dose | Route of Administration | Clinical Signs | Comments |
|---|---|---|---|---|---|---|---|
| **Miscellaneous Potential Toxins** | | | | | | | |
| Alcohol | | | All | | Access to beverages | Lethargy; "drunkenness"; regurgitation | |
| Atropine | | | All | Inappropriate or excessive use | Subcutaneous I.M. | Autointoxication from gut stasis | |
| Chlorine | | | Avian species | | Exposure to fumes | Photophobia, epiphora; coughing; sneezing; hyperventilation | |
| Formaldehyde | Generic | | Canaries | | Exposure to fumes | Eye irritation; respiratory distress; collapse; death | |
| Medroxyprogesterone | Depo-provera Provera | Pitman-Moore Upjohn | All | Chronic use | I.M. | Iatrogenic polydipsia; lethargy; obesity; fatty liver | |
| Megestrol acetate | Ovarid 20 mg | Glaxo-vet, Australia | Sulfur-crested Cockatoos | 5 mg/50 ml (10 days) | Oral | Same as medroxyprogesterone<br>Weight gain | |
| Nicotine | Cigarettes Pesticides | | | | Ingestion | Depression; labored respiration; cyanosis; coma; death | |
| Nitrates | Component of fertilizers | | All | | | Thirst; anorexia; vomiting; diarrhea; hypothermia; cyanosis; CNS signs; death | |
| Polytetrafluoroethylene | Teflon | | All | | Inhalation of fumes | Respiratory distress; death within 15–30 minutes | |
| Selenium sulfide | Selenium shampoo | | Budgeriars | Topical application | Topical | Death | |
| Sodium chloride | Salt | | All | 3.5% in food<br>0.5% in water | | Depression; PU/PD; dyspnea; CNS signs; anemia; death | |
| Surgical acrylics | Generic | | All | As supplied | Inhalation of fumes | Increased effects of inhalation anesthesia; death | |
| Vitamin $D_3$ | Injectable | | All | As supplied | I.M. | Increased calcium in kidneys, other soft tissues | |
| Zinc | "New wire disease" | | | Ingestion | Same as lead poisoning | Treat with Ca EDTA | |

### Anesthetics

Safe anesthetic agents and doses have been established for avian patients (see Chapter 45, Anesthesiology). Toxic effects from anesthesia may be seen in critical patients or in deviations from accepted protocol for anesthesia administration. Appropriate preanesthetic evaluation and surgical support assist in preventing anesthetic accidents.

There are no instances in the author's practice in which local anesthetics are indicated for avian use. Reports of adverse effects of topical analgesic products are believed to be dose-related.

### Parasiticides

Parasiticides appear to account for the majority of iatrogenic poisonings; reported incidences in cage birds are listed in Table 39–2. The dose of medication administered as a water treatment is dependent upon the total water consumed. Male breeding budgerigars and pigeons, which transport large quantities of water to incubating hens, have succumbed to a dose of 1 teaspoon dimetridazole per gallon of drinking water, whereas the same dose did not have any adverse effect on nonbreeding individuals.

The administration of fenbendazole is reported to be contraindicated in molting or nesting birds with young in the pin feather state owing to its effect on feather formation.[6]

The toxic potential of levamisole has been widely publicized. Intramuscular injection increases absorption but may result in toxicity, paresis, or death. Side effects of intramuscular levamisole administration may also include regurgitation, especially in parrots and pigeons. There have also been reported cases of toxicity with intramuscular injections of ivermectin. Because both parasiticides are effective when administered orally, that is the preferred method for this author. The practitioner is advised to observe the patient closely for signs of intestinal obstruction which may result from treatment for parasites. Intestinal lubricants may be used for mild cases of obstruction; severe cases may require treatment of shock (see Chapter 28, Symptomatic Therapy and Emergency Medicine).

## PESTICIDES

Some groups of birds appear to be very resistant to pesticides. One practitioner observed no toxic effects in ducks and geese when D-Con rat and mouse poison was fed as the sole food for 24 hours. Close proximity to a dog with a spot application flea control chemical was suggested in avicultural literature to precipitate vomiting, anorexia, and lethargy in a macaw. However, this author maintained six budgerigars in confinement for 36 hours with a paper towel soaked with a 20-pound dose of fenthion (Spotton—Haver-Lockhart) with no adverse effects over a several-month period following the exposure. Reported incidents of pesticide toxicity are included in Table 39–1. The reader is referred to Chapter 38, Neurologic Disorders, for clinical manifestations and therapy of organophosphate poisoning.

## AFLATOXICOSIS

Although the ingestion of aflatoxins from improperly stored food products is more likely to occur in poultry, aflatoxicosis can occur in cage birds (see Chapter 36, Mycotic Diseases). The presence of aflatoxin $B_1$ in mold-contaminated corn was found to produce a greenish-yellow fluorescence in scientific filtered blacklight; however, approximately 50 per cent of samples positive to blacklight are negative when tested by official laboratory procedures. The blacklight can be used for screening food products prior to submission of subsequent positives for appropriate testing. Other aflatoxins apparently do not fluoresce. The clinician is advised to be aware of the potential for mold growth on sprouts prepared for feeding (see Chapter 50, Aviculture Medicine).

## POISONOUS PLANTS

Many plants have been reported to be poisonous to dogs, cats, humans, and other animals. Table 39–3 lists plants that have historically been listed as poisonous to avians; however, the toxicity to birds is controversial in many instances, and few documented cases of plant poisoning have occurred.

Birds with a degree of freedom in the home frequently chew on split leaf philodendron, diffenbachia, and other house plants that are considered to be poisonous. A macaw was observed to chew on oleander with no adverse reaction, whereas this plant was found to be lethal to geese and ducks fed 6 to 30 grams of the plant. In these cases there may be some question as to the quantity that is actually ingested. The potential for poisoning also depends on the location of the toxic component.

**Table 39–3.** PLANTS REPORTED IN THE LITERATURE AS POISONOUS TO CAGE AND AVIARY BIRDS*

| |
|---|
| Amaryllis—*Amarylidaceae* |
| Azalea—*Rhododendron occidentale* |
| Bird of paradise—*Poinciana gilliesii* |
| Black locust—*Robinia pseudoacacia* |
| Boxwood—*Buxus sempervirens* |
| Buttercup—*Ranunculaceae* |
| Caladium—*Caladium* spp. |
| Castor bean—*Ricinus communis* |
| Cherry—*Prunus* spp. |
| Clematis—*Clematis* spp. |
| Cowslip—*Caltha polustris* |
| Daphne—*Daphne* spp. |
| Datura—*Datura* spp. |
| Diffenbachia—*Dieffenbachia* spp. |
| Elephant's ear—*Colocasia* spp. |
| English ivy—*Hedera helix* |
| Foxglove—*Digitalis purpurea* |
| Hemlock—*Conium maculatum* |
| Horse-chestnut—*Aesculus hippocastanum* |
| Hyacinth—*Hyacinthus orientalis* |
| Hydrangea—*Hydrangea* spp. |
| Iris—*Iris* spp. |
| Jack-in-the-pulpit—*Arisaema* spp. |
| Jerusalem cherry—*Solanum pseudocapsicum* |
| Jimsonwood—*Datura stramonium* |
| Juniper—*Juniperus virginiana* |
| Larkspur—*Delphinium* spp. |
| Lily-of-the-valley—*Convallaria majalis* |
| Lobelia—*Lobelia* spp. |
| Marijuana—*Cannabis sativa* |
| Mistletoe—*Phoradendron villosum* |
| Mock orange—*Prunus caroliniana* |
| Monkshood—*Aconitum* spp. |
| Morning glory—*Ipomoea* spp. |
| Mountain laurel—*Kalmia latifolia* |
| Narcissus—*Narcissus* spp. |
| Nightshade—*Solanum* spp. |
| Oleander—*Nerium oleander* |
| Philodendron—*Philodendron* spp. |
| Pokeweed—*Phytolacca americana* |
| Poinsettia—*Euphorbia pulcherrima* |
| Poison hemlock—*Conium maculatum* |
| Potato (shoots)—*Solanum tuberosum* |
| Privet—*Ligustrum vulgare* |
| Rhododendron—*Rhododendron* spp. |
| Rhubarb—*Rheum rhaponticum* |
| Rosary pea—*Abrus precatorius* |
| Shunk cabbage—*Symplocarpus foetidus* |
| Snowdrop—*Ornithogalum umbellatum* |
| Tobacco—*Nicotiana* spp. |
| Virginia creeper—*Parthenocissus quinquefolia* |
| Wisteria—*Wisteria* spp. |
| Yew—*Taxus cuspidata* |

*Few documented cases.

For example, the root of the pokeweed is highly poisonous. While the ripe berries are widely regarded as edible for birds, *overripe* berries have been observed to induce "drunkenness" and stupor in wild birds. Hemp seed (*Cannabis sativa*, or marijuana) is a common food source for captive birds.

The wood of the Brazilian pepper tree is commonly used by aviculturists for perches, and the fruit is widely eaten by local and migratory birds in Florida. A newspaper article that blamed these berries for high mortality in migratory robins later retracted the story when open bags of pesticides were found in a nearby field. This author believes that a similar predisposing factor may have been involved in the deaths of Cedar Waxwings in which the Brazilian pepper tree was incriminated.[4] Incidental cyanide poisonings have reportedly occurred in cage birds from consumption of large quantities of apple seeds, cherry pits, and immature almonds. It is unclear what conditions may predispose cage birds to toxic reactions to plants and plant parts.

Plants known or highly suspected to be poisonous which one should avoid planting near aviaries include oleander (toxic foliage, wood, and flowers); castor bean (toxic seeds and branches); spurges, which include croton, crown of thorns, pencil tree, poinsettia (toxic, irritant, and co-carcinogenic properties); and rosary pea. However, in a limited test this author chopped rosary peas and gavaged them into the crops of budgerigars without incident.

Some suggestions for plants that can be safely used within planted outdoor aviaries are presented in Appendix 7.

## INHALED TOXINS

Chemicals that are normally irritating to humans and other animals can be acutely toxic to avians. The inhalation of ammonia, naphthalene, or fluoropolymer from spray starch has resulted in mortality. Birds may succumb to carbon monoxide exhaust in unventilated areas. The depolymerization of polytetrafluoroethylene (Teflon), which may occur when a coated pan is accidentally left to overheat on a range, produces fumes that are highly toxic to birds (see Chapter 38, Neurologic Disorders). Hair spray has also been suggested as a possible toxin.

## THERAPY

Toxicoses from ingestion of products other than lead are difficult to substantiate; however, symptomatic procedures including fluid therapy are effective in some cases (see Chapter 27, What To Do Until A Diagnosis Is Made, and Chapter 28, Symptomatic Therapy and Emergency Medicine). A study of emetics in wild birds suggested that the most promising drug was a 0.5 per cent solution of tartar emetic but concluded that emetics are extremely difficult to use safely in birds. Crushed activated charcoal may be gavaged to absorb toxins, followed by mineral oil to encourage its passage through the digestive tract. A commercial antitoxin product (Toxi Ban—Vet-A-Mix) combines activated charcoal and kaolin. Other combinations with activated charcoal to produce antidotes include magnesium sulfate and tannic acid in a 2:1:1 ratio and 1 teaspoon powdered activated aquarium charcoal in an oral lavage with 10 to 12 cc milk of magnesia. An emergency ingluviotomy is effective if recent ingestion of a highly poisonous plant, particulate chemical, or other poisonous substance is known to have occurred. Inhaled toxins may be treated by access to oxygen or fresh air and supportive care. Coughing may respond to administration of steroids and antibiotics. Irritant chemicals coming in contact with the bird's skin or feathers should be washed off immediately.

## REFERENCES


1. Adrian, W. J., and Stevens, M. L.: Wet versus dry weights for heavy metal toxicity determinations in duck liver. J. Wild. Dis., 15:125–126, 1979.
2. Altman, R. B.: Common toxic material in avian medicine. CVMA Speakers' Syllabi, California Veterinary Medical Association Annual Scientific Seminar, 1982.
3. Blackburn, K.: Planting for bluebirds and other wildlife: a weed worth cultivating. Sialia, 6(4):132, 1984.
4. Britt, J. O., and Howard, E. B.: Pepper tree berry poisoning in a flock of wild cedar waxwings. Avian/Exotic Pract., 1(3):11, 1984.
5. Consumer Report: Calcium supplements with a touch of lead. Sept., 1982, p. 478.
6. Cooper, H.: Material presented at Australian Veterinary Poultry Association Cage and Aviary Bird Medicine Seminar, Melbourne, Australia, 1985.
7. Dalvi, R. R., and Adenoyero, A. A.: Toxic effects of aflatoxin B1 in chickens given feed contaminated with *Aspergillus flavus* and reduction of toxicity by activated charcoal and some chemical agents. Avian Dis., 28(1):61–69, 1984.
8. Dieter, M. P.: Blood delta-aminolevulinic acid dehydratase (ALAD) to monitor lead contamination in Canvasback Ducks *(Aythya valisineria)*. In Animals as Monitors of Environmental Pollutants. Washington, DC, National Academy of Sciences, 1979, page 177–191.
9. Dieter, M. P., and Finley, M. T.: Delta-aminolevulinic acid dehydratase enzyme activity in blood, brain, and liver of lead-dosed ducks. Environ. Res., 19:127–135, 1979.
10. Feldman, B. F., and Kruckenberg, S. M.: Clinical toxicities of domestic and wild caged birds. Vet. Clin. North Am., 5(4):653–673, 1975.
11. Fowler, M. E.: Plant Poisoning in Small Companion Animals. Ralston Purina Co., 1981.
12. Fowler, M. E.: Poisoning. In Speakers' Syllabi, California Veterinary Medical Association Scientific Seminar, 1982.
13. Goldman, M., and Dillon, R. D.: Interaction of selenium and lead on several aspects of thyroid function in Pekin Ducklings. Res. Comm. Chem. Pathol. Pharm., 37(3):487–489, 1982.
14. Hunter, B., and Wobeser, G.: Encephalopathy and peripheral neuropathy in lead-poisoned Mallard Ducks. Avian Dis., 24(1):169–178, 1980.
15. Janssen, D. L., et al.: Lead toxicosis in three captive avian species. Annual Proceedings of the American Association of Zoo Veterinarians, Denver, CO, 1979.
16. Kingsbury, J. M.: Poisonous Plants of the United States and Canada. Englewood Cliffs, NJ, Prentice-Hall, Inc., 1964.
17. Lederer, R. J., and Crane, R.: The effects of emetics on wild birds. North Am. Bird Bander, 3:3–5, 1978.
18. March, G. L., et al.: The effects of lead poisoning on various plasma constituents in the Canada Goose. J. Wild. Dis., 12:14–19, 1976.
19. McDonald, S. E.: Lead toxicosis in psittacine birds. Proceedings of the Annual Meeting of the Association of Avian Veterinarians, San Diego, CA, 1983.
20. Montalbano, F., and Hines, T. C.: An improved x-ray technique for investigating ingestion of lead by waterfowl. Proceedings of the 32nd Annual Conference, SE Association of Fish and Wildlife Agencies, 32:364–368, 1978.
21. Panigrahy, B., et al.: Insecticide poisoning in peafowls and lead poisoning in a cockatoo. Avian Dis., 23:760–762, 1979.
22. Peckham, M. C.: Poisons and toxins. In Hofstad, M. S., et al. (eds.): Diseases of Poultry. 8th ed. Ames, IA, Iowa State University Press, 1984.
23. Perry, R. A.: Diseases of Birds; Avian Therapy and Disease Control. The T. G. Hungerford Vade Mecum Series for Domestic Animals. Sydney, Australia, The University of Sydney Post Graduate Foundation in Veterinary Science, 1983.
24. Petrak, M. L.: Poisoning and other casualties. In Petrak, M. L. (ed.): Diseases of Cage and Aviary Birds. 2nd ed. Philadelphia, Lea & Febiger, 1982.
25. Robinson, P. T., and Richter, A. G.: A preliminary report on the toxicity and efficacy of levamisole phosphate in zoo birds. J. Zoo Anim. Med., 8(1):23–26, 1977
26. Roscoe, D. E., et al.: A simple, quantitative test for erythrocytic protoporphyrin in lead-poisoned ducks. J. Wild. Dis., 15:127–136, 1979.
27. Rosskopf, W. J., and Woerpel, R. W.: Lead poisoning in a cockatiel. Mod. Vet. Pract., Jan., 1982.
28. Stowe, C. M., et al.: Ethylene glycol intoxication in ducks. Avian Dis., 25(2):538–541, 1981.
29. Wasler, D.: The pigeon as a bioindicator of environmental lead pollution. Postgraduate Dissertation, University of Munich, 1984.

# Chapter 40

# NEOPLASIA

TERRY W. CAMPBELL

Captive birds have a higher incidence of neoplasia than feral birds. One reason may be the increased lifespan of captive birds, since older birds tend to suffer from neoplastic diseases more frequently than younger birds. Cage birds may have an increased exposure to carcinogens and infectious agents and are affected by factors that influence neoplasia (i.e., genetic selection, nutritional imbalances, and hormonal influences). Also, the availability for study of wild birds suffering from neoplastic diseases is limited.

The primary carcinogens affecting animals are irradiation, chemical agents, and viruses. Other factors also have a role in promoting neoplastic lesions. Irradiation is probably an insignificant cause of neoplasia in cage birds because of their limited exposure to this type of carcinogen. Chemical and viral carcinogens may have a more important role in avian oncogenesis. The best-known chemical carcinogens are aromatic amines, alkylating agents, azo dyes, and the polycyclic aromatic hydrocarbons.[39] Aflatoxins are potential carcinogens, especially with poor sanitary practices and improper food storage. Viral oncogenesis has been demonstrated in poultry with the discovery of the leukosis/sarcoma viruses (oncornavirus type C) that induce a variety of transmissible neoplastic disorders. Also, the Marek's disease virus, a *Herpesvirus*, has been shown to be an oncogenic virus of poultry. Other neoplasms found in birds may have viral etiologies, such as some papillomas and renal adenocarcinomas.[16, 21]

Other factors may influence a bird's response to carcinogens. Hormonal factors may have a role in the development of gonadal and endocrine gland tumors. Genetic predisposition for tumor formation has been demonstrated in mammals (i.e., certain strains of laboratory mice) and may have a role in avian oncogenesis (i.e., lipomas in budgerigars and galahs). Physical factors, such as trauma, may also influence tumor development. Dietary and environmental factors also influence the development of avian neoplasia. It is most likely that a combination of co-carcinogens and the true carcinogens work together in the development of avian neoplasms.

The clinical signs of neoplastic diseases vary with the size and location of the tumor and the degree of pressure or invasion exerted on adjacent tissues. Abdominal tumors frequently show abdominal enlargement, dyspnea, lethargy, body weight loss (indicated by a marked reduction in the breast muscle mass), loose droppings, and intermittent constipation. Ascites may result from intrahepatic neoplastic lesions (either primary or secondary) or infiltration of neoplastic tissue into the serosa of abdominal viscera or mesenteric tissue. High-protein ascites can result from a neoplasm compressing blood flow draining the hepatic sinusoids (e.g., vena cava), whereas low-protein ascites may occur if the obstruction is presinusoidal (e.g., abdominal neoplasms causing obstruction of the portal vein). Low-protein ascites also occur with hypoproteinemias associated with abdominal neoplasms. Abdominal enlargement may result from a hemoperitoneum. Spontaneous hemorrhage into the abdominal cavity in an older bird with no history of trauma should include an ulcerated or ruptured abdominal neoplasm, resulting in hemorrhage as part of the differential diagnosis. The bird may have had signs of illness several weeks or months prior to abdominal enlargement. Metabolic disturbances may result from abdominal neoplasms. Pancreatic neoplasias may induce secondary diabetes mellitus, causing the clinical signs of polydipsia and polyuria. Neoplastic involvement of the gonads may be reflected as a color change in the cere of budgerigars due to hormonal changes. Lameness resulting from pressure on the nerve plexus supplying the legs is a common clinical sign associated with renal and ovarian neoplasms. Ovarian tumors may cause unilateral lameness involving the left leg, with progression toward bilateral lameness as the tumor increases in size.[8] Neoplasms of the extremities may cause lameness or loss of flight owing to pressure on nerve supplies, mechanical obstruction of tendons and joints, or pathologic fractures of bones. Bone tumors may show a sudden onset of lameness or inability to fly, along with palpable swellings in the involved bones. Skin tumors frequently ulcerate owing to self-inflicted trauma or picking by cage mates. Neoplasms of the beak, oral cavity, and alimentary tract may lead to emaciation owing to an inability to gather or properly digest food.

The diagnosis of abdominal neoplasms requires a complete history and physical examination to rule out non-neoplastic disorders that may create similar signs. Traumatic, nutritional, and infectious disorders may also create abdominal distention, dyspnea, lethargy, weight loss, and gastrointestinal disturbances. Abdominal radiographs are often beneficial in revealing abdominal masses. The hemogram may also be helpful. Neoplastic disorders may show an associated poorly regenerative or nonregenerative anemia typical of anemias of chronic disorders. Some abdominal tumors that cause hemorrhage into the abdominal cavity may reveal a regenerative anemia. The examination of the peripheral blood may provide evidence of a leukosis-like disorder (i.e., lymphoid leukosis, erythroblastosis, myeloblastosis, and myelocytomatosis). Blood chemistries may also be helpful, especially when metabolic disturbances are present, such as diabetes mellitus due to a pancreatic neoplasm (the blood glucose would be greater than 800 mg/dl). Neoplastic infiltration of the liver may result in an elevation of aspartate aminotransferase (AST) and lactate dehydrogenase (LDH) enzymes. An elevated uric acid value may indicate renal involvement. Therefore, radiographic evidence of a hepatic mass and elevated AST and/or LDH values would provide a strong presumptive diagnosis for a hepatic neoplasm. Likewise, radiographic evidence of a renal mass with an elevated uric acid level would suggest a renal neoplasm. Cytologic evaluation of the abdominal fluid may also provide a presumptive diagnosis of abdominal neoplasia. Cytologic features of the exfoliated cells may suggest the type of neoplasm involved. Aspiration or excisional biopsy of exposed lesions may be examined cytologically for evidence of neoplasia. Often the presence of a neoplasm is determined only at necropsy, and the histopathologic evaluation of the lesions provides the definitive diagnosis.

Treatment of discrete neoplastic lesions involves surgical excision, if possible, or amputation of an involved limb. Supportive therapy is given for secondary disorders (i.e., bacterial or fungal infections and metabolic disorders). Treatment of disseminated neoplasias is usually unrewarding. Chemotherapy using anticancer agents commonly used in mammalian medicine (i.e., vincristine, vinblastine, 5-fluorouracil, or alkylating agents) can be tried. The author has used adrenocortical hormones (prednisolone) in the treatment of lymphoid leukosis in an Amazon Parrot. After two weeks on therapy, the solid lymphoid masses in the subcutis of the bird diminished to half their original size and the marked peripheral lymphocytosis disappeared. Birds with neoplasias of suspected viral etiologies should not be kept in close contact with other birds owing to the potential for horizontal transmission. Treatment of these birds may create carrier birds and increase the shedding of the virus. These birds should not be used for breeding, since vertical transmission of the oncogenic virus will perpetuate the disease in the flock or aviary. However, no avian leukemia virus has been confirmed in any species other than the domestic fowl.

Neoplasms affecting birds can be classified into three main groups; ectodermal, endodermal, and mesodermal origin. A fourth group includes rare neoplasms such as teratomas and dermoid cysts. Embryonal mesoderm gives rise to mesenchyme, which differentiates into fibroblasts (connective tissue cells), endothelial cells, mesothelial cells, macrophages, histiocytes, mast cells, lymphoid cells, blood cells, muscle cells, and adipose cells.[5] The epidermis develops from embryonal ectoderm, whereas the alimentary tract is derived largely from endoderm.[5] Nervous tissue is derived from ectoderm and supported by mesodermal derivatives.

## NEOPLASIA OF NONHEMATOPOIETIC MESENCHYMAL TISSUE

### Fibroma/Fibrosarcoma

Fibrosarcoma is a malignancy of fibrous connective tissue and is a common neoplasm of birds. Fibroma is the benign form. These neoplasms can be found anywhere on the body but are commonly located on the skin and subcutis, wings or legs, beak, long bones, syrinx, liver, and small intestine[2, 3, 5, 14, 17, 23] (Figs. 40–1 and 40–2). These tumors vary in size and shape (often ovoid or globular) and are typically firm on palpation. The cut surface is slightly granular and often contains a central cavity of necrosis. Fibrosarcomas are moderately vascularized and often lack a capsule, allowing infiltration of the adjacent tissue. They tend to be locally invasive but rarely metastasize.

**Figure 40–1.** A fibroma involving the beak of an eight-year-old female budgerigar *(Melopsittacus undulatus)*.

## Neurofibroma/Neurofibrosarcoma

These neoplasms originate from nerve sheath cells (Schwann cells) and resemble fibrosarcomas. They often appear as firm nodular lesions within the skin or musculature of the breast, neck, legs, and abdominal wall.[19, 36]

## Lipoma/Liposarcoma

Neoplasms of adipose tissue occur frequently in some species of birds (e.g., budgerigars), suggesting a possible genetic predisposition. Lipomas commonly occur on the sternum. However, they can be found in the subcutaneous tissue nearly anywhere on the body.[14, 20, 23, 36] Lipomas are considered benign but can be multiple. They are typically soft, pale yellow, encapsulated, lobulated masses found in the subcutis. The cut section reveals a fatty mass surrounded by a thin capsule. Often there is a necrotic center (Fig. 40–3). Surgical excision is considered the treatment of choice; however, recurrence is common. Lipomas are the most common tumor seen in pet birds, and nutritional management should be a part of the treatment schedule[13] (see Chapter 42, Miscellaneous Diseases). Liposarcomas occur in the same locations as lipomas. They tend to be firm masses resembling fibrosarcomas and can be highly vascularized.[23] Liposarcomas have metastatic potential. It should be emphasized that many masses presumed to be lipomas are circumscribed lipogranulomas.[11]

## Chondroma/Chondrosarcoma and Osteoma/Osteogenic Sarcoma

Primary neoplasms of cartilage and bone are often located in areas of active cartilaginous growth (i.e., epiphyses of long bones) and have the potential of becoming metastatic. Elevated serum alkaline phosphatase in the blood has been reported to be a good indication of osteoblastic activity in birds.[15] Therefore, an elevation of this enzyme in the serum and evidence of a mass involving bone on radiographic and/or

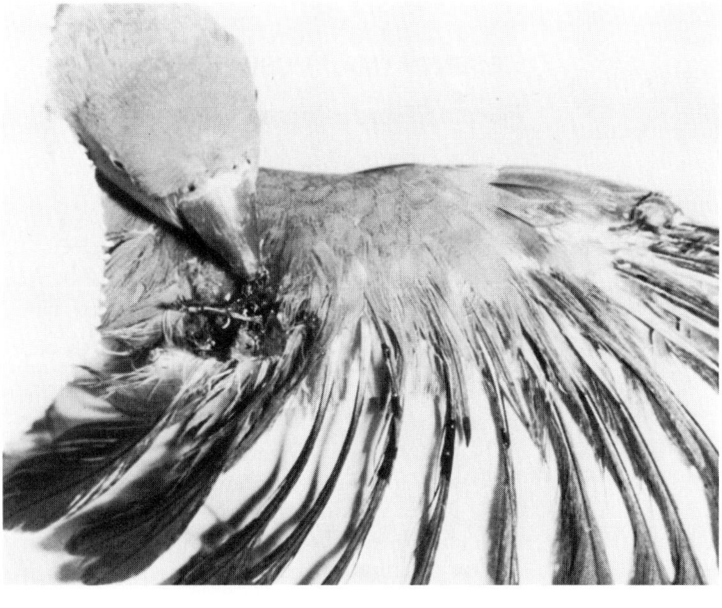

**Figure 40–2.** Multiple fibrosarcomas on the wing of a Rose-ringed Parakeet *(Psittacula krameri manillensis)*. Note the automutilation involving the proximal lesion.

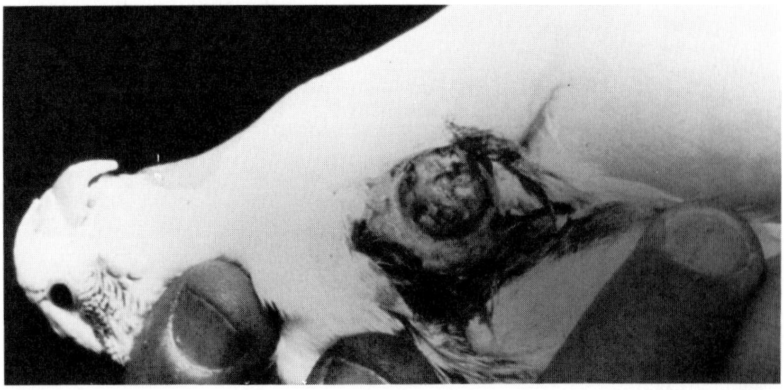

**Figure 40–3.** A large lipoma on the sternum of a three-year-old female budgerigar *(Melopsittacus undulatus)*. Note that the neoplastic mass has ulcerated through the overlying skin.

physical examination provides a presumptive diagnosis of a bone neoplasm. Chondrosarcomas and osteogenic sarcomas often involve the long bones of the legs and wings.[34, 36] Osteogenic sarcomas involving the beak have been reported in budgerigars.[34] These neoplasms tend to be firm; and the cut surface often reveals erosion or invasion of the involved bone.[36] The clinical history may indicate an insidious onset of lameness or inability to fly. Firm swellings involving the long bones of the affected limbs can often be palpated. The treatment usually involves total or partial amputation of the involved limb.

## Rhabdomyoma/Rhabdomyosarcoma and Leiomyoma/Leiomyosarcoma

Primary neoplasia of skeletal muscle—rhabdomyoma and rhabdomyosarcoma—has been reported in the wing or shoulder muscle groups, sartorius-gracilus group, and dorsal lumbar muscles of birds.[5, 23, 27] These tumors tend to be moderately firm and lobulated. The primary neoplasms of smooth muscle are leiomyomas and leiomyosarcomas. These have been reported in the crop, alimentary tract, trachea, spleen, and oviduct.[5, 23] They tend to be firm masses that show diffuse infiltration of the surrounding tissues. They can be highly vascular and cystic.

## Hemangioma/Hemangiosarcoma

These types of neoplasms are derived from vascular endothelium. They can occur as soft reddish to black swellings under the skin and often involve feather follicles.[5] Locations reported for these neoplasms include the dorsum of the neck, wing, legs, and abdominal viscera.[14, 19, 23, 36] Hemangiosarcomas can be locally invasive, metastatic, or multicentric in origin.[5] A spontaneous hemoperitoneum in an older bird with no history of trauma should always include a hemorrhagic abdominal neoplasm as part of the differential diagnosis. Frequently, abdominal hemangiosarcomas result in a spontaneous hemoperitoneum. Cytologic examination of the abdominal fluid may be helpful in detecting the presence of the neoplasm, but hemangiosarcomas exfoliate poorly and may not reveal themselves cytologically.

## Mesothelioma

Mesotheliomas have been reported in the alimentary tracts, mesenteries, and lungs of birds.[5, 19, 36] Abdominal mesotheliomas can be associated with ascites. These tumors are firm and can surround the intestines and cause obstructive lesions.[5]

## Xanthoma and Xanthomatosis

These conditions are not true neoplasms but can appear as discrete masses or diffuse thickening of the skin. They are typically yellow in color. Xanthomas are lobular masses or plaques in the skin. They occur more frequently in psittacine and gallinaceous birds. Xanthomatosis is a diffuse involvement of the skin that causes it to become thickened, friable, and highly vascularized. Xanthomas and xanthomatosis are characterized by the presence of many highly vacuolated macrophages and multinucleated giant cells. The macrophage vacuoles are uniformly round and discrete, suggestive of phagocytized lipid. Cholesterol clefts are present on histopathology (Fig. 40–4). Xanthomatosis can be associated with other skin lesions. The author has observed this reaction to occur several months following the surgical excision of a feather cyst from the wing of a Green-wing Macaw *(Ara chloroptera)*. Treatment consists

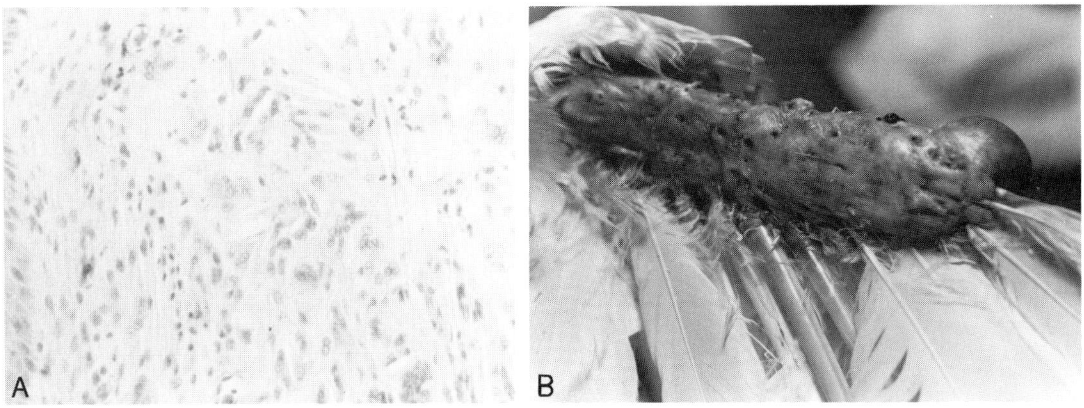

**Figure 40–4.** *A*, Histopathologic section of a xanthomatous reaction in the skin of a Green-wing Macaw *(Ara chloroptera)* (H & E stain, ×200). Note the multinucleated giant cells and cholesterol clefts. *B*, Xanthomatosis on the upper surface of the wing of a Greater Sulfur-crested Cockatoo *(Cacatua galerita galerita)*.

of the excision of the involved skin, if possible (see Chapter 46, Surgical Instrumentation and Special Techniques).

## NEOPLASIA OF EPITHELIAL TISSUE

### Papilloma

Papillomas are neoplasms derived from nonglandular surface epithelium. The appearance of papillomas in birds can vary from the typical finger-like projections to crusty, ulcerated lesions. They tend to bleed profusely when excised or traumatized. Papillomas often occur on the skin of the neck, wings, feet, legs, palpebra, and uropygial area.[12, 16, 23] Papillomas have also been found in the oral cavity, glottis, esophagus, crop, and proventriculus[6, 11, 12] (Fig. 40–5). Many avian papillomas may have a viral etiology.[16, 18] They are usually benign and can become large. Surgical excision is the treatment of choice, but recurrence is common. Harrison[13] has made the following observations concerning cloacal papillomas:

> Cloacal papillomas occur more commonly in macaws (primarily Blue and Gold, Scarlet, Yellow-collared, and Severe), cockatoos, and Amazon Parrots (primarily Orange-wing, Yellow-naped, and Blue-front). Cloacal papillomas can occur as protruding cloacal masses and are often misdiagnosed as cloacal prolapses. Excisional biopsy for histopathologic evaluation is indicated for the differentiation of cloacal prolapses, granulomas, and papillomas or other neoplasms. True papillomas respond favorably to cryosurgery, but those with a high mitotic index often regrow rapidly after surgical removal. Clinical signs include tenesmus, hemorrhage in the vent area, and chronic pasting of the vent area with droppings, resulting in scalding and a foul odor.

Although avian papillomas have not been proven to have a viral etiology, the use of an autogenous vaccine made from a removed papillomatous lesion and administered subcutane-

**Figure 40–5.** *A*, Multiple papillomas removed from the esophagus and crop of a Green-wing Macaw *(Ara chloroptera)*. *B*, The diagnosis of cloacal papilloma is determined only by histologic evaluation of cloacal protrusions.

ously in the wing web may be a consideration in the treatment plan.[9, 12]

## Squamous Cell Carcinoma

Squamous cell carcinomas occur rarely in birds. This neoplasm can occur in the oral cavity, esophagus, crop, and anywhere on the skin.[3, 5, 14, 23] Squamous cell carcinomas have the potential for metastasis or local invasiveness. Surgical excision should be attempted, if possible.

## Keratoma

The normal scales found on the feet and legs of birds are primarily keratin. Keratomas can result from abnormal keratin proliferation in multiple areas on the feet and legs with the lack of normal squamous epithelium and inflammation.[37] These horny growths can resemble papillomas or hyperkeratoses due to mites, inflammation, nutritional imbalances, or senility. This type of growth is considered benign, but the excessive keratin proliferation may cause mechanical interference with the movement of the digits or traumatic ulceration with secondary infections.

## Adenoma/Adenocarcinoma

These neoplasms, which are derived from glandular epithelium, occur as primary tumors or metastatic lesions nearly anywhere in the body. Adenomas or adenocarcinomas are frequently renal or ovarian in origin in birds but have been reported in adrenal, intestinal, lung, splenic, hepatic, and pancreatic tissues as well.[5, 8, 10, 14, 19, 21, 23, 29, 41, 42]

Renal adenomas or adenocarcinomas often show the same nonspecific signs as other abdominal neoplasias. However, they are frequently associated with unilateral (sometimes bilateral) leg paresis or paralysis due to pressure created by the neoplastic kidney on the sciatic (ischiatic) plexus supplying the leg. These neoplasms can appear as oval to lobulated, firm kidney masses, or the kidney may appear pallid and swollen. Renal adenocarcinomas show little tendency for metastasis, but metastasis can occur usually by peritoneal implantation.[10]

Ovarian adenomas or adenocarcinomas also show the same nonspecific signs as abdominal neoplasms. Likewise, a unilateral leg paresis or paralysis frequently occurs, but only the left leg is involved, since the functional ovary in birds is located on the left side. These neoplasms appear as firm, cream-colored, lobulated masses that frequently contain watery cysts. Ovarian cystadenocarcinomas occur as large yellow masses with multiple cystic spaces.[8] These ovarian tumors cannot be differentiated from granulosa cell tumors on gross examination.[23] The adenocarcinomas tend to be locally invasive or to metastasize by peritoneal implantation. Metastasis by infiltration of the intestinal, hepatic, or pancreatic serosa can occur. Surgical excision should be attempted; however, most cases are so advanced by the time the neoplasm is detected that surgery is unsuccessful.

## Hepatoma/Hepatocarcinoma and Bile Duct Carcinoma

Hepatic parenchyma and bile duct epithelium can undergo hyperplasia as a result of neoplastic or non-neoplastic (i.e., aflatoxins) conditions. It may be difficult to differentiate the two conditions by gross appearance. Primary neoplasms of the liver can be seen as an enlarged, mottled, or nodular liver.[38] Abdominal radiographs may reveal a hepatomegaly, and ascites may be present. The clinical signs of primary hepatic neoplasms are often the same as those of other abdominal neoplasias rather than of hepatic failure.[35] Bile duct carcinomas can be nodular and cystic, or the affected liver may appear pale and firm but normal in size.[23, 24] Carcinomas of the bile duct often metastasize by implantation to the serosal surfaces of the abdominal viscera.[44]

## Primary Neoplasia of the Pancreas

The neoplastic changes affecting the pancreas are usually due to metastatic implantation from other abdominal neoplasms (i.e., ovarian, renal, or bile duct carcinomas). Primary pancreatic islet cell tumors have been reported in birds and can cause a secondary diabetes mellitus.[5, 30] The affected pancreas may appear normal on gross examination; therefore, histologic inspection may be required to determine the presence of the neoplasm.

## Nephroblastoma/Malignant Nephroblastoma

These neoplasms arise from embryonal tissue of the kidney. They usually appear unilateral, with a preference for the right kidney in the fowl.[5, 11] They consist of two cell types, mesenchymal and epithelial.[21] In the fowl the involved kidney often has a firm, lobulated, often cystic mass with a vascular capsule.[5] The clinical signs are the same as described for renal adenomas and adenocarcinomas.

### Seminoma

A seminoma is a rare neoplasm of the testis. The involved testis may appear as an irregular, globular mass or maintain its normal shape in the fowl.[5] The mass is well encapsulated and the cut surface is soft, spongy, and hemorrhagic.[23, 40] Seminomas rarely metastasize.

### Sertoli Cell Tumor

This is an uncommon neoplasm of the testis. It can appear as an ovoid, lobulated mass.[23] Because this type of neoplasm may cause hyperestrogenism, male birds could develop female characteristics (e.g., brown cere in male budgerigars).

### Granulosa Cell Tumor

Granulosa cell tumors affecting female birds are typically yellow, irregular, and lobulated[8] (Fig. 40–6). Grossly they resemble ovarian adenomas or adenocarcinomas. The clinical signs associated with this type of tumor are the same as those of other ovarian tumors. Granulosa cell tumors are often locally invasive and can metastasize by implantation to the serosal surfaces of other abdominal organs.

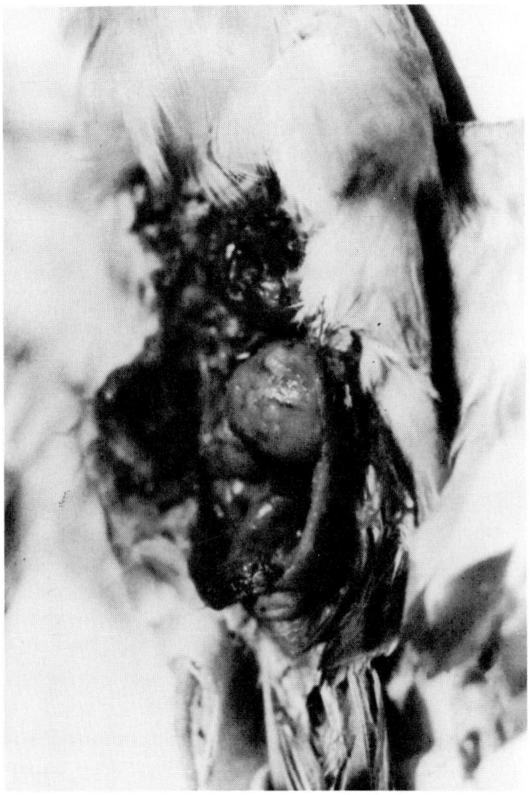

**Figure 40–6.** Granulosa cell tumor in a budgerigar *(Melopsittacus undulatus).*

### Thymoma

In domestic fowl, thymic neoplasms usually occur in the cervical region near the thoracic inlet.[5] Thymomas have two components, lymphoreticular and epithelial.[11] They are slow-growing masses that create pressure on the trachea and esophagus, causing the clinical signs of dyspnea and dysphagia.[45] Thymomas are usually benign.

### Chromophobe Pituitary Tumors

Schlumberger[31–33] has described spontaneous chromophobe pituitary tumors in budgerigars and believes these tumors are among the most frequent seen in that species. He states that the incidence is equally distributed between the sexes and that the tumors occur as diffuse pituitary enlargements, locally invasive growths, and metastasizing neoplasms.[31] The clinical signs of budgerigar pituitary neoplasms include exophthalmos (unilateral or bilateral), blindness, somnolence, polyuria, polydipsia, and obesity.[2, 33] Pituitary tumors occur most frequently in young birds (average age is 2.5 years) and may have associated renal adenocarcinomas or goiter formation of the thyroid glands.[31]

## NEOPLASIA OF NERVOUS TISSUE

Primary neoplasia of nervous tissue is rare in birds except for the neurofibrosarcomas that occur occasionally in broiler chickens.[5] A glioblastoma was diagnosed in a budgerigar with a central nervous system disorder.[28] The tumor involved the diencephalon, mesencephalon, and optic nerves.

## TERATOMA/MALIGNANT TERATOMA

Teratomas are embryonal neoplasms that occasionally occur in the gonads, adrenal glands, and kidneys of chickens.[5] Teratomas can appear as spherical masses disseminated throughout

**Figure 40–7.** Periorbital lymphoid mass in an Amazon (*Amazona* spp.) parrot with lymphoid leukosis.

the body.[1] Teratomas have cystic cavities containing fully developed feathers.

## NEOPLASIA OF HEMATOPOIETIC TISSUE

Neoplasia of hematopoietic tissue is usually detected by the examination of the peripheral blood smear. Lymphoid leukosis is the most common neoplastic disorder affecting this type of tissue. Lymphoid leukosis of poultry has a viral etiology, the leukosis/sarcoma group viruses. These viruses are oncornavirus type C and are responsible for the production of many types of transmissible neoplasms in poultry (i.e., erythroblastosis, myeloblastosis, myelocytomatosis, endothelioma, nephroblastosis, hepatocarcinoma, fibrosarcoma, and osteopetrosis).[25] The target cells for the virus causing lymphoid leukosis in poultry are cells of the bursa-dependent lymphoid system. The result is the abnormal proliferation of B lymphocytes. The virus responsible for Marek's disease, a *Herpesvirus*, in poultry also causes lymphoid neoplasia.

Lymphoid neoplasia has been reported in a number of cage and wild birds.[4, 7, 12, 22, 23, 26, 43] The lymphoid masses can appear anywhere in the body (cutaneous or visceral). The liver and spleen are common abdominal organs affected.[22, 43] A frequent site for cutaneous lymphoid masses is the periorbital area (Fig. 40–7). Cutaneous lesions are often associated with areas of feather loss, probably due to pressure on the feather follicles as the mass enlarges. The masses consist of lymphoid tissue and can be nodular, miliary, or diffuse. The hemogram may reveal a marked leukocytosis with a lymphocytosis. The lymphocytes in the blood may appear immature but more commonly have scalloped cytoplasmic borders (pseudopodia) (Fig. 40–8). Surgical excision of the discrete nodules can be attempted, and chemotherapy with anticancer drugs can be tried. It is important to realize that lymphoid neoplasms in any bird may have a viral etiology (as demonstrated in poultry) and that the affected bird may be a potential threat to other birds. The client should be informed of the possibility of horizontal and vertical transmission of a possible oncogenic virus.

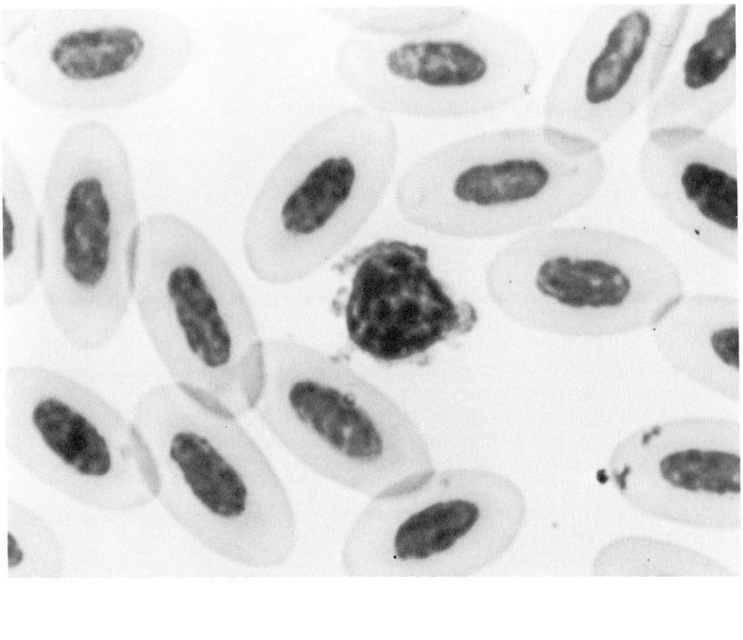

**Figure 40–8.** A small mature lymphocyte in the peripheral blood of the parrot in Figure 41–7 (Wright's stain, ×1000). This bird had an absolute peripheral lymphocyte count of 49,528 lymphocytes/cu mm with the majority of the lymphocytes showing pseudopodia.

## REFERENCES


1. Baker, J. R., and Chandler, D. J.: Suspected teratoma in a black-headed gull *(Larus ridibundus)*. Vet. Rec., 109:60, 1981.
2. Beach, J. G.: Diseases of budgerigars and other cage birds. A survey of postmortem findings. Vet. Rec., 74:10–15, 63–68, 134–140, 1962.
3. Blackmore, D. K.: The clinical approach to tumors in cage birds. 1. The pathology and incidence of neoplasia in cage birds. J. Small Anim. Pract., 7:217–223, 1966.
4. Bowen, S.: What is your diagnosis? J. Am. Vet. Med. Assoc., 173(9):1257–1258, 1978.
5. Campbell, J. G.: Tumours of the Fowl. London, William Heinemann Medical Books Ltd., 1969.
6. Campbell, T. W.: Disorders of the avian crop. Compend. Contin. Educ. Pract. Vet., 5(10):813–824, 1983.
7. Campbell, T. W.: Lymphoid leukosis in an Amazon parrot—a case report. In Proceedings of the Association of Avian Veterinarians, 1984, pp. 229–234.
8. Campbell, T. W., and Stuart, L. D.: Ovarian neoplasia in the budgerigar *(Melopsittacus undulatus)*. VM/SAC, 79(2):215–218, 1984.
9. Cribb, P. H.: Cloacal papilloma in an Amazon parrot. In Proceedings of the Association of Avian Veterinarians, 1984, pp. 35–37.
10. Decker, R. A., and Hruska, J. C.: Renal adenocarcinoma in a sarus crane *(Grus antigone)*. J. Zoo Anim. Med., 9(1):15–16, 1978.
11. Graham, D.: Personal communication, 1984.
12. Graham, D. L.: An update on selected pet bird virus infections. In Proceedings of the Association of Avian Veterinarians, 1984, pp. 267–280.
13. Harrison, G.: Personal communication, 1984.
14. Hubbard, G. B., Schmidt, R. E., and Fletcher, K. C.: Neoplasia in zoo animals. J. Zoo Anim. Med., 14:33–40, 1983.
15. Ivins, G. K., Weddle, G. D., and Halliwell, W. H.: Hematology and serum chemistries in birds of prey. In Fowler, M. E. (ed.): Zoo and Wild Animal Medicine. Philadelphia, W. B. Saunders Company, 1978, pp. 286–290.
16. Jacobson, E. R., Mladinich, C. R., Clubb, S., Sundberg, J. P., and Lancaster, W. D.: Papilloma-like virus infection in an African Gray Parrot. J. Am. Vet. Med. Assoc., 183:1307–1308, 1983.
17. Jessup, D. A.: Fibrosarcoma in a burrowing owl *(Speotyto cunicularia)*. J. Zoo Anim. Med., 10:51–52, 1979.
18. Lina, P. H. C., van Noord, M. J., and de Groot, F. G.: Detection of virus in squamous papillomas of the wild bird species *Fringilla coelebs* (chaffinch). J. Natl. Cancer Inst., 50:567–571, 1973.
19. Montali, R. J.: An overview of tumors in zoo animals. In Montali, R. J., and Migaki, G. (eds.): The Comparative Pathology of Zoo Animals. Washington, DC, Smithsonian Institution Press, 1980, pp. 531–542.
20. Moore, M.: Generalized lipomas in a Hispaniolan Amazon. VM/SAC, 79(5):666–669, 1984.
21. Neumann, U., and Kummerfeld, N.: Neoplasms in budgerigars *(Melopsittacus undulatus)*: Clinical, pathomorphological and serological findings with special consideration of kidney tumours. Avian Pathol., 12(3):353–362, 1983.
22. Palmer, G. H., and Stauber, E.: Visceral lymphoblastic leukosis in an African Gray Parrot. VM/SAC, 76:1355–1356, 1981.
23. Petrak, M. L., and Gilmore, C. E.: Neoplasms. In Petrak, M. L. (ed.): Diseases of Cage and Aviary Birds. Philadelphia, Lea and Febiger, 1982, pp. 606–637.
24. Potter, K., Connor, T., and Gallina, A. M.: Cholangiocarcinoma in a Yellow-faced Amazon parrot *(Amazona xanthops)*. Avian Dis., 27(2):556–558, 1983.
25. Purchase, G. H., and Burnmester, B. D.: Leukosis/sarcoma group. In Hofstad, M. S. (ed.): Diseases of Poultry. Ames, IA, Iowa State University Press, 1978, pp. 418–468.
26. Rambow, V. J., Murphy, J. C., and Fox, J. G.: Malignant lymphoma in a pigeon. J. Am. Vet. Med. Assoc., 179(11):1266–1268, 1981.
27. Raphael, B. L., and Nguyen, H. T.: Metastasizing rhabdomyosarcoma in a budgerigar. J. Am. Vet. Med. Assoc., 177(9):925–926, 1980.
28. Raphael, B. L., Clemmons, R. M., and Nguyen, H. T.: Glioblastoma multiforme in a budgerigar. J. Am. Vet. Med. Assoc., 177(9):923–925, 1980.
29. Rosskopf, W. J., Woerpel, R. W., Howard, E. B., and Britt, J. O.: Pancreatic adenocarcinoma in a mynah bird. Mod. Vet. Pract., 63(7):573–574, 1982.
30. Ryan, C. P., Walder, E. J., and Howard, E. B.: Diabetes mellitus and islet cell carcinoma in a parakeet. J. Am. Anim. Hosp. Assoc., 18:139–142, 1982.
31. Schlumberger, H. G.: Neoplasia in the parakeet. I. Spontaneous chromophobe pituitary tumors. Cancer Res., 14(3):237–245, 1954.
32. Schlumberger, H. G.: Neoplasia in the parakeet. II. Transplantation of the pituitary tumor. Cancer Res., 16(2):149–153, 1956.
33. Schlumberger, H. G.: Tumors characteristic for certain animal species. Cancer Res., 17(9):823–832, 1957.
34. Schultz, D. J.: Disorders of the musculoskeletal system. In Hungerford, T. G. (ed.): Proceedings No. 55 Refresher Course in Aviary and Cage Birds. Sydney, Australia, The Post-graduate Committee in Veterinary Science, 1981, pp. 519–533.
35. Schultz, D. J.: Liver diseases. In Hungerford, T. G. (ed.): Proceedings No. 55 Refresher Course in Aviary and Cage Birds. Sydney, Australia, The Post-graduate Committee in Veterinary Science, 1981, pp. 597–607.
36. Siegfried, L. M.: Neoplasms identified in free-flying birds. Avian Dis., 27(1):86–99, 1983.
37. Speckmann, G., Bundza, A., and Patenaude, R.: Multiple keratomas of a captive pelican. J. Zoo Anim. Med., 8(4):32–35, 1976.
38. Spira, A.: Hepatoma in a mynah. Mod. Vet. Pract., 60(11):925–926, 1979.
39. Thomson, R. G.: General Veterinary Pathology. Philadelphia, W. B. Saunders Company, 1978, pp. 299–388.
40. Turk, J. R., Kim, J., and Gallina, A. M.: Seminoma in a pigeon. Avian Dis., 25(3):752–755, 1981.
41. Van Toor, A. J., Zwart, P., and Kaal, G. Th. F.: Adenocarcinoma of the kidney in two budgerigars. Avian Pathol., 13(2):145–150, 1984.
42. Wadsworth, P. F., and Jones, D. M.: An ovarian adenocarcinoma in a greater flamingo *(Phoenicopterus ruber roseus)*. Avian Pathol., 10(1):95–99, 1981.
43. Wadsworth, P. F., Jones, D. M., and Pugsley, S. L.: Some cases of lymphoid leukosis in captive wild birds. Avian Pathol., 10(4):499–504,1981.
44. Wadsworth, P. F., Majeed, S. K., Brancker, W. M., and Jones, D. M.: Some hepatic neoplasms in non-domestic birds. Avian Pathol., 7(4):551–555, 1978.
45. Zubaidy, A. J.: An epithelial thymoma in a budgerigar *(Melopsittacus undulatus)*. Avian Pathol., 9(4):575–581, 1980.

# *Chapter 41*

# DISORDERS OF THE INTEGUMENT

GREG J. HARRISON

Etiologic agents affecting the integument of birds are discussed in the respective chapters. The information here describes the types of clinical cases most frequently encountered by the author and suggests guidelines for the formulation of a differential diagnosis and direction of therapy.

## FEATHERS

Among the problems of the integument, disorders of the feathers cause the greatest concern to the avian practitioner. Because the condition of the feathers reflects the total physiologic and psychologic health of the bird, one must consider the possibility that more than one factor is contributing to feather ailments. For example, significant emphasis has been placed in the past on psychologic components of "feather picking." Clinical evidence suggests that many cases of feather picking have underlying nutritional or disease problems. Although primary psychologic causes can be responsible for disturbed plumage, the incidence may be far less than suggested by early cage bird literature.

Some diagnoses of abnormal feather conditions can be made directly from the history and physical examination; some cases may require extensive investigation and/or prolonged therapy; and some others may never be resolved. Even when appropriate therapy has been instituted, it may take weeks or months for a change to be reflected in the feathers. Table 41–1 lists some frequent causes of feather disturbances. Clients should be discouraged from buying birds that are not in good plumage at the time of purchase.

### Diagnoses to Consider from the History and Physical Exam

A thorough investigation of the history and a complete physical examination may provide sufficient information for the diagnostic and therapeutic approach to some feather disorders. Categorizing the patient as an "old" or "new" bird (see Chapter 7, Preliminary Evaluation of a Case) may be the first step. Feather abnormalities encountered in an "old" bird from a stable environment may likely relate to chronic dietary or endocrine imbalances or psychologic or reproductive factors. "New" bird feather problems often result from trauma, stress, or infectious causes, or suspected nutritional or immunological compromise in young birds. There appear to be species' tendencies in the incidence and type of feather problems (see Chapter 1, Choosing a Bird). For example, feather loss in mynah birds is usually believed to be related to nutrition rather than to self-mutilation.

Poor feather appearance and a history of dietary deficiency may indicate that generalized malnutrition is a contributing factor in a bird with a feather disorder. These avian patients often benefit from an initial intramuscular injection of multiple vitamins (A, $D_3$, E, B). The owners should be advised that a brief period of increased preening and pin feather growth may follow nutritional injections. Spraying the bird with water appears to offer temporary relief during this time. The client is advised to balance the diet with increased amounts of whole grain bread, dark green vegetables, carrots, protein, and calcium and other minerals, and to provide weekly access to unfiltered sunlight.

#### *Poor Feather Appearance*

Broken, ragged feathers may be the result of improper housing, incompatibility of cage mates, trauma, or delayed molt. Brittle, frizzled feathers (with accompanying scaly skin and dermatitis) have been blamed on deficiencies of nutrients and trace elements such as calcium, zinc, selenium, manganese, magnesium, biotin, pantothenic acid, and salt (see Chapter 31, Nutritional Diseases). "Stress" marks (Fig. 41–1)

**Table 41–1. FACTORS CONTRIBUTING TO FEATHER DISORDERS**

| | |
|---|---|
| **Management** | |
| Malnutrition | |
| Insufficient nutrients | |
| Improper storage of food | |
| Improper housing | |
| Environment too confined | |
| Lack of privacy | |
| Bullying by cage mates | |
| Incompatibility of mates | |
| Humidity too low | |
| Insufficient exposure to unfiltered sunlight | |
| Environmental changes | |
| **Trauma** | |
| Including self-trauma to injuries, pain sites | |
| **Physiologic Factors** | |
| Reproductive frustration | |
| Exaggeration of normal preening | |
| Chronic egg-laying on deficient diet | |
| Self-mutilation from internal disease states | |
| **Psychological Factors** | |
| Defending of territory; displaced aggression | |
| Inappropriate pair bonding | |
| Attention-getting device | |
| New bird, pet, owner, food, toy, cage location | |
| Temporary or permanent loss of mate | |
| Fear, nervousness, jealousy | |
| Boredom, lack of routine or purpose | |
| Habit | |
| **Genetic** | |
| **Metabolic** | |
| **Endocrine Imbalances** | |
| **Internal Pathology** | |
| Pansystemic diseases | |
| Air sacculitis | |
| Abscesses | |
| **Parasites** | |
| **Species' Tendencies** | |
| **Allergies** | |
| **Viral Diseases** and diseases of suspected viral origin | |
| **Mycotic Organisms** | |
| **Bacteria** | |
| **Mycobacteria** | |
| **Mycoplasma** | |

are believed to result from brief episodes of dysfunction of the epidermal collar associated with release of corticosteroid hormones during some stress (e.g., disruption of the feeding schedule).[10] Humidity apparently plays an important role in the initial development of normal feathers in neonates. Retarded feather growth and retention of pin feather sheaths in neonates raised in dry environments (e.g., nursery containers heated with incandescent lights) has been noted to respond to an increase in humidity.

**Figure 41–1.** A disruption at the growth center may result in feather cross lines known as "stress" marks.

### *Abnormal Coloration of Feathers*

Lack of normal pigmentation has been produced experimentally in cockatiels by feeding a purified diet deficient in choline and riboflavin.[30] Although lysine deficiency results in color irregularities in the feathers of chickens, the lack of this nutrient in the diet of cockatiels did not produce the same results in this species.[30] Until a species is studied on a purified diet, one cannot state what effects specific nutritional deficiencies will have.

More advanced diagnostic tests may be required to determine the cause of abnormal feather coloring (see Chapter 8, Differential Diagnoses Based on Clinical Signs). A subtle dark depigmentation appears to occur primarily in feathers of the wing and breast and over the back, and may suggest liver disease. This color change has also been clinically associated with chronic lead poisoning in psittacines and can be reversed to normal following therapy (see Chapter 39, Toxicology).

### *Abnormal Molt*

Dark areas or frayed feathers may suggest that the normal molt is past due and the feathers are showing signs of wear. Chronically ill, malnourished, or recently imported birds commonly show evidence of an interrupted molt and feather replacement cycle. The bird may remain in an extended pin feather stage over large localized or generalized portions of the body. Abnormal molt may be accompanied by pruritus and has been reported to be induced by temperature extremes, severe shock or fright, hypothyroidism, or internal parasites.[2]

### *Disturbances in Feather Formation*

Trauma, malnutrition, and infectious diseases may result in abnormal growth of the developng feathers.

#### PSITTACINE BEAK AND FEATHER DISEASE SYNDROME

One feather disorder that can usually be initially diagnosed solely from the physical examination (and confirmed by biopsy of affected follicles) is psittacine beak and feather disease syndrome (PBFDS).[19, 21–25] It is advantageous to note the identifying characteristics of this disease in the feathers and/or beak as early as possible to prevent suffering of the bird and limit the expenses of the owner for further diagnostic and therapeutic measures.

Many interesting speculations regarding psittacine beak and feather disease syndrome have been made over recent years. At various times the disease has been described as cockatoo beak and feather syndrome, beak rot, nonresponsive fungal dermatitis, cockatoo feather picking or molt disease, feather maturation syndrome, and adrenal gland insufficiency. Recent studies have shown no evidence of adrenal or other endocrine gland insufficiency in affected birds studied[22]; ACTH stimulation of cockatoos with feather loss has shown no lack of adrenal response.[33] The list of affected species has also expanded to include other psittacines in addition to cockatoos. Table 41–2 lists Australian captive psittacine species identified by Perry and Pass[24] to have demonstrated histopathologic evidence of PBFDS.

**Table 41–2.** SOME SPECIES WITH HISTOPATHOLOGIC EVIDENCE OF PSITTACINE BEAK AND FEATHER DYSTROPHY SYNDROME IN AUSTRALIA*

| |
|---|
| Sulfur-crested Cockatoo *(Cacatua galerita)* |
| Major Mitchell Cockatoo *(Cacatua leadbeateri)* |
| Galah *(Cacatua roseicapilla)* |
| Little Corella *(Cacatua sanguinea)* |
| Long-billed Corella *(Cacatua tenuirostris)* |
| Budgerigar *(Melopsittacus undulatus)* |
| Cockatiel *(Nymphicus hollandicus)* |
| Rainbow Lorikeet *(Trichoglossus haematodus)* |
| Eastern Rosella *(Platycerus eximius)* |
| Western Rosella *(Platycerus icterotis)* |
| Hooded parrot *(Psephotus dissimilis)* |
| Red-rumped Grass Parrot *(Psephotus haematonotus)* |
| Mallee Ringneck *(Barnardius barnardi)* |
| Port Lincoln or 28 Parrot *(Barnardius zonarius)* |
| Eclectus Parrot *(Eclectus roratus)* |
| Princess Parrot *(Polytelis alexandrae)* |
| Bourke Parrot *(Neophema bourkii)* |
| Peach-faced Lovebird *(Agapornis roseicollis)* |
| Nyasa Lovebird *(Agapornis lilianae)* |
| Fischer's Lovebird *(Agapornis fischeri)* |
| Masked Lovebird *(Agapornis personata)* |
| King Parrot *(Alisterus scapularis)* |

*Perry, R. A., and Pass, D. A.: Proceedings of the Cage and Aviary Bird Medicine Seminar, Australian Veterinary Poultry Association, Melbourne, Australia, 1985.

According to Australian investigators,[19, 21–25] PBFDS occurs frequently in wild flocks of Sulfur-crested Cockatoos in Eastern Australia, with variable incidence of up to 20 per cent of birds in some flocks reported to be affected. Wild galahs, Rainbow Lorikeets, and budgerigars are similarly affected, although subtle differences may occur among species in the pattern of the disease. PBFDS has not been recognized in nonpsittacine species.

**DIAGNOSIS.** Birds affected with PBFDS exhibit a progressive deterioration in the quantity and quality of normal plumage and eventually lose contour feathers over most areas of the body (Fig. 41–2*A*). Replacement feathers following a molt feature one or more of the following characteristics: retained feather sheath, blood within the shaft of the feather, short clubbed feathers, curled and deformed feathers, feathers with a circumferential constriction, and stress lines in the vane (Fig. 41–2*B*).[21–25] Most birds with this condition are presented in advanced stages of feather loss. Early clinical signs of the disease may be identified by examination of the powder down feathers. These feathers are often the first to be affected by hemorrhage in the follicles and dystrophy of the new growth and are most easily located by wetting down the feathers overlying the hip joints and extending back toward the preen gland and vent.[24] To verify a PBFDS diagnosis from the clinical examination, this author confirms the presence of several short, clubbed feathers with the "hour glass" circumferential constrictions. Magnification aids in the identification. Because of the increased frequency of these characteristics very early in otherwise clinically normal-appearing birds, it is unclear if all cases progress to the full expression of the disease.

Beak lesions associated with the disease include progressive changes in color (from a semigloss to gloss color) and growth, with progressive elongation, development of fault lines, breakage, and underrunning of the outer rim[22] (Fig. 41–3). Bacteria and fungi become secondary invaders. Usually the beak is affected after considerable feather loss, but the reverse may also occur (Fig. 41–4). The most prominent early clinical sign in an affected beak is a brownish tan necrosis of the palatine tissue under the premaxilla. Other beak changes may

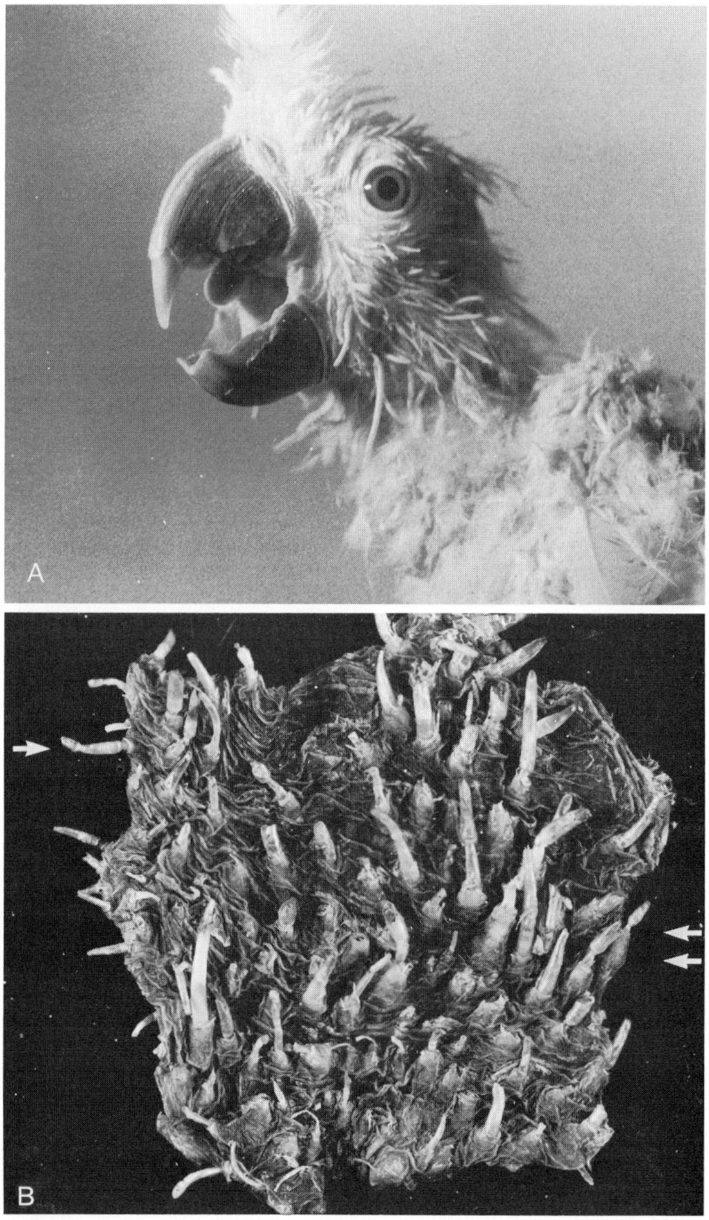


**Figure 41–2.** *A*, Progressive reduction in the quantity and quality of feathers is a characteristic of psittacine beak and feather disease syndrome (PBFDS). *B*, Magnification of a skin section from a PBFDS-affected bird illustrates the "hour glass" constrictions on the developing feathers. (Courtesy of David Pass and Ross Perry.) (From Harrison, G. J.: Feather disorders. Vet. Clin. North Am. [Small Anim. Pract.], 14(2):196, 1984.)

not occur, and not all cases develop beak lesions.

With the unique opportunity to observe captive and wild afflicted birds in Australia, Perry and Pass report:[24]

> Acute, sub-acute and chronic forms of PBFDs occur. The acute form is usually seen in fledgling or immature birds in the process of a normal moult and is characterized by sudden necrosis, fracture, bending, bleeding and/or shedding of what were normally growing blood quills over a period of a few days.
>
> The acute form of PBFD is often accompanied by malaise, depression, green bubbly diarrhea, weight loss and some mortalities over 1 to 2 weeks. Acute PBFD is seen seasonally in late spring and summer in immature trapped Sulfur-crested Cockatoos and immature non-captive Rainbow Lorikeets and galahs in Sydney.
>
> Some birds with acute PBFD slide into the sub-acute or chronic forms of the disease. These differ mostly in the rate of progression of bilaterally symmetrical replacement of normal feathers by deformed dystrophic quills which fail to grow to maturity and are repeatedly moulted and replaced by more severely dystrophic quills. Many birds with sub-acute and early chronic PBFD are presented for veterinary examination because of the more dramatic signs of concurrent diseases such as septicemia, pneumonia, hepatopathy and enteritis to which they seem predisposed (possibly due to immunosuppression). In these birds, a diagnosis of PBFD is easily missed

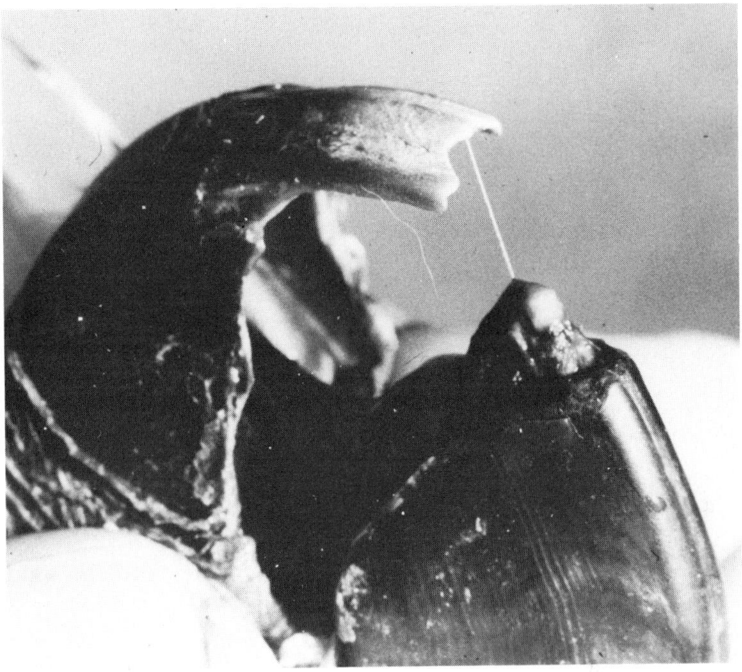

**Figure 41–3.** Characteristic beak lesions in birds affected with PBFDS include changes in color and growth patterns.

unless the plumage is searched for dystrophic feathers.

PBFD is said by some breeders to increase in incidence as the breeding season progresses and the incidence is greater in nests used for several clutches. The disease does occur in young birds produced by birds with no clinical evidence of the disease, but it is neither clear whether affected birds always produce affected young, nor whether they are latent carriers of the virus particles. Strict nest hygiene and sanitation (including elimination of lice and mites), have been said to reduce the incidence of PBFD, but supportive data are not available.

A number of institutions are currently addressing the disease condition. At press time, the disease had been experimentally transmitted,[22a] and a virus had been characterized (see Chapter 32, Viral Diseases). Independent researchers have differing opinions about the identity of the suspected causative organism. One investigator (Gerlach) suggests the possibility that distinctly different viruses may be producing similar clinical and histopathologic signs in various species.

Perry and Pass confirm the clinical diagnosis

**Figure 41–4.** This Umbrella Cockatoo with PBFDS beak lesions (illustrated in Fig. 41–3) does not show obvious feather involvement.

**Figure 41–5.** Feathers from the crest of the Umbrella Cockatoo in Figure 41–4 are pulled for closer examination and exhibit PBFDS characteristics.

of PBFDS by demonstration of characteristic microscopic lesions in the growing portion of affected feathers.[24] In the live bird, this is best achieved by plucking growing (pin) feathers (the larger the quill, the better) and fixing them in 10 per cent buffered formalin (Fig. 41–5). The specimens are submitted for histopathologic examination for the presence of necrosis of feather (and beak) epidermal cells, which results in abnormal growth (dystrophy) of these organs. Virus particles similar to those found in affected feathers have been noted to occur in association with degeneration and atrophy of the thymus and bursa of Fabricius of affected birds. This is consistent with the observation that many birds with the disease appear to suffer from a degree of immunosuppression. Graham demonstrated that an affected cockatoo produced humoral antibodies directed against an antigen contained in tissues of its own affected follicles and in follicles from other affected cockatoos.[11] A fluorescent antibody test for early screening of the disease has been produced by his investigation. While there is no question with a positive response to this test, the negative results may be questionable.

**Treatment.** Clinical attempts at therapy have included support hormones, direct vaccination with feather pulp, parenteral and topical antibiotics and antifungals, immunostimulants, nutritional supplementation, and beak prostheses. None of these therapies has succeeded in curing the condition. One clinician has noted clinical improvement with a series of multivalent autogenous vaccines.[15] In the author's experience, euthanasia is recommended when the bird's eating habits suggest evidence of pain or when the bird is continually under therapy for multiple secondary invaders. The use of beak prostheses or treatment attempts is believed to be inappropriate in these advanced cases.

In the author's experience, a few observations have been clinically associated with some cases of PBFDS, but the relationship to the disease is unknown. Some owners of affected birds have remarked about an apparent change in personality of the bird from aggressive to timid (or vice versa) as the disease progresses. A single case of nonunion in a long bone fracture in a cockatoo was subsequently diagnosed as PBFDS. Perhaps the possibility of microvascular damage in other parts of the body should be investigated. One clinician reported inclusion bodies in the spinal cord similar to those found in affected feathers of a PBFDS bird.[15]

### "FRENCH MOLT"

Apparently "French molt" has not been well-characterized, and the term has been loosely used to describe a number of abnormal molts. As researchers around the world identify the etiology for the condition described as French molt in their country, there is some confusion over the specific disease being described. As an example, the acute form of psittacine beak and feather disease syndrome is often termed "French Moult" by aviculturists in Australia, but it differs from some abnormal molts of the same name described in other countries. In Australia, the pathology of French molt and PBFDS is identical and the terms are synonymous. Lesions described recently by Japanese,

Canadian, and German[16] workers as French molt in budgerigars were attributed to chronic papovavirus infection or budgerigar fledgling disease (BFD). Perry states that the pathology of papovavirus is quite distinct from that recorded for PBFDS in Australia.[24] Although BFD has not been specifically identified in Australia, clinical signs suggestive of generalized parrot papovavirus infection in other species were reported to the author by Australian aviculturists during a 1985 visit.

In the United States, the term "French molt" is generally applied to a condition seen in budgerigars, cockatiels, lovebirds, and occasionally conures in which feather replacement is significantly delayed following a sudden extensive loss of primary feathers. Birds chronically affected are referred to as "creepers" because they frequently remain in a miniaturized or pin feather stage and are never capable of flying. In addition to papovavirus, other viruses have been speculated as the etiology for French molt (see Chapter 32, Viral Diseases). Environmental, hereditary, parasitic, and nutritional influences have been suggested as contributing factors.

#### OTHER CAUSES OF MALFORMED FEATHERS

Australian practitioners have noted a nonfatal condition in budgerigars (and to a lesser extent in Peach-faced Lovebirds), colloquially called "budgerigar short tail disease," which resembles acute PBFDS in the early stages.[24] It differs from PBFDS in that affected feather follicles subsequently develop more than one quill. Short and twisted tail feathers may result in chronic cases. Because feathers tend to grow under the skin, excessive preening and self-mutilation are involved.

*Chlamydia*, *Mycoplasma*, mycobacteria, papovavirus, and other infectious agents[10, 11] are also potential causes of malformed feathers. Although Graham[10] has noted histologic evidence of infected follicles and feather collar epithelium by parrot herpesvirus and adenovirus, the death of the bird precludes distorted feather production.

### *Feather Cysts*

A feather cyst appears to be an ingrown feather that accumulates necrotic debris to form a subcutaneous mass associated with the follicle (Fig. 41–6). Cysts arising from primary or secondary wing feathers (the most frequent sites in the author's experience) appear to relate to prior trauma to the feather follicle. The trauma may have resulted from overpreening by the bird, injury from cage confinement, improper removal of a bleeding feather stub, or wing trims that were too short or included the coverts (see Chapter 6, Management Procedures). Feathers cysts may be removed surgically but frequently recur (see Chapter 46, Surgical Instrumentation and Special Techniques, and Chapter 47, Selected Surgical Procedures).

A genetic factor may be related to feather cysts in specific canary hybrids. The dorsal thoracic area is a frequent site of multiple follicle cysts in affected canaries (Fig. 41–7). In these cases the surgeon is advised to excise the general area of affected skin and suture rather than remove the affected follicles individually.

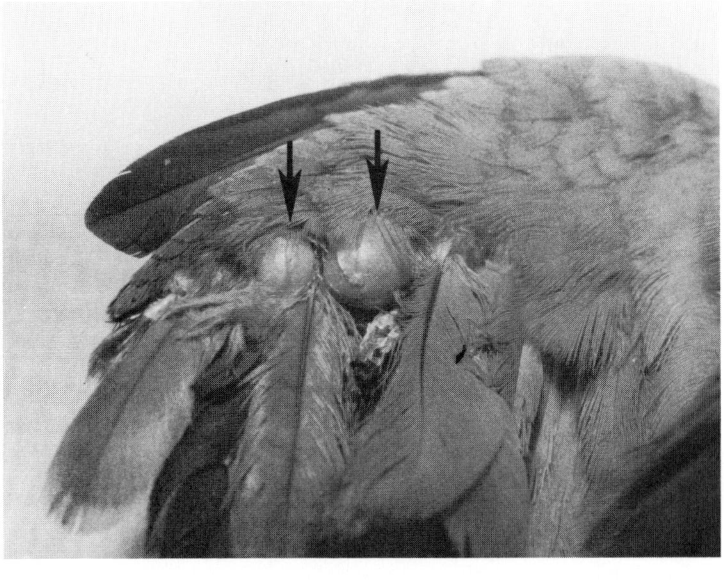

**Figure 41–6.** Feather cysts in psittacines occur most commonly on the wings. (From Harrison, G. J.: Feather disorders. Vet. Clin. North Am. [Small Anim. Pract.], 14(2):195, 1984.)

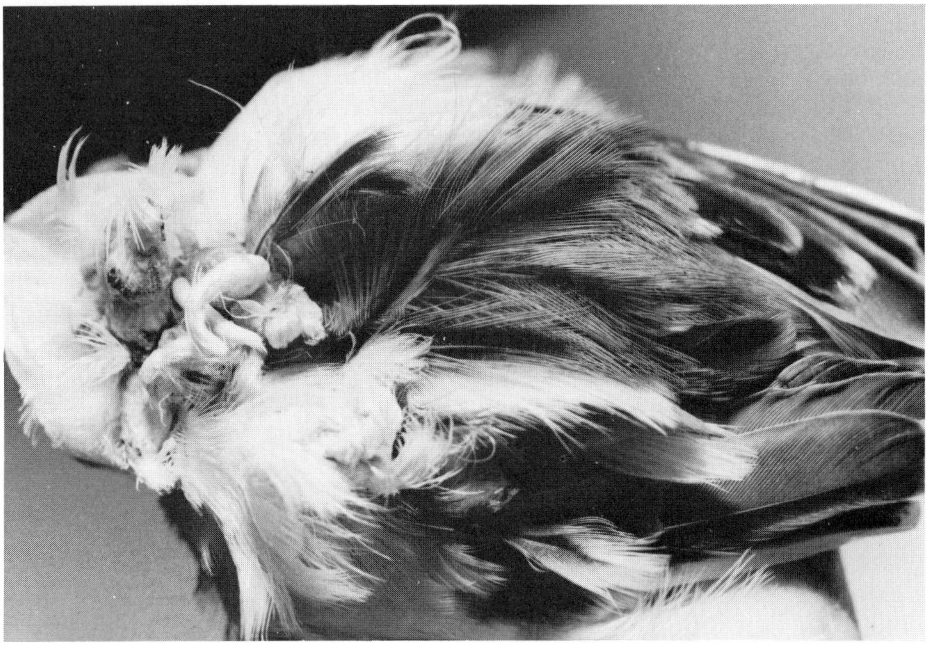

**Figure 41–7.** The back of the neck is a frequent site of multiple feather cysts in affected canaries.

### *Inability to Preen*

Failure of the bird to remove the keratin sheath from growing feathers may be attributed to malnutrition, spinal deformities (Fig. 41–8), injuries such as fractures and subsequent malformation of the beak, or lack of interest from being sick. Maintaining an Elizabethan collar for a prolonged period will prevent normal preening by the bird. Some assistance by the owner in removal of the sheaths may be necessary. Young birds learn to preen with assistance from the parents; thus, hand-raised babies may need some initial "educational" help from the owner.

**Figure 41–8.** Spinal deformities from suspected nutritional causes may prevent normal preening and predispose the bird to further feather damage.

### *Baldness*

Some species (e.g., lutino cockatiels) frequently have genetic-related baldness, especially on the crown of the head. Other feather deficiencies on the head or elsewhere may be the result of aggression by another bird (Fig. 41–9).

Reduction in testosterone levels may disrupt the molt and cause baldness in up to 60 per cent of some species of male canaries. Gerlach states that some of these birds have responded for one molting season to intramuscular injections or oral doses of testosterone, once a week for six weeks. The effectiveness of this therapy decreases with each subsequent treatment. Additionally, this is not a safe treatment in birds with latent liver disease, as unexpected death may occur.

### *Feather Picking in the Solitary Bird*

The "feather picking" bird is identified primarily by the presence of healthy, well-formed

**Figure 41–9.** Feather loss on the back of the head may be due to intraspecies aggression.

feathers on the head and random feather loss in body areas accessible to the bird's beak. "Feather picking" generally applies to all mutilation of the feathers by the beak and includes chewing or plucking. Identifying the cause for the activity requires a process of elimination.

Feather picking due to behavioral causes does exist; acute feather destruction particularly may be psychologically induced. The initial history interview may elicit clues regarding the environment, housing arrangements, family structure, or other factors that may have precipitated the self-abuse as a manifestation of fear, boredom, aggression, jealousy, or other emotions. Behavior modification appears to be the most successful preventive and treatment technique for these cases (see Chapter 3, Captive Behavior and Its Modification).

Cyclic feather picking by the otherwise healthy bird may coincide with reproductive cycles and may in fact be exaggerations of normal courtship preening behavior. With the possible exception of cockatiels and some conures, feather picking single pet birds seldom benefit from the addition of a cage mate. Very rarely is picking associated with external parasite infestation.

## Possible Diagnoses of Feather Disorders from Intermediate Diagnostic Steps

If the cause of a feather disorder is not evident from communication with the owner and the physical examination, some preliminary evaluations are in order. A Gram's stain of the feces and parasite examination may be the initial steps. A culture and sensitivity study would be advised if the Gram's stain shows abnormal distribution and types of bacteria. A skin scraping may identify the presence of *Knemidokoptes* mites or cytologic deviations. Serum chemistry panels and complete hematology evaluations may suggest underlying disease conditions. Radiology is often the most reliable diagnostic measure used by the author for many cases of feather picking, especially those with chronic skin lesions (Fig. 41–10).

Although in some instances of localized feather loss and severe self-mutilation it may be valuable to apply an Elizabethan collar for a temporary period to determine if the follicles are functioning normally, the author rarely uses collars. Failure to grow pin feathers following such a collaring might suggest a disorder of metabolic origin. Some conditions are responsive to thyroid hormones and nutritional changes. Cessation of follicular activity in some chronically malnourished birds may not respond to therapy (Fig. 41–11). One theory is that continuous efforts to grow new feathers may lead to exhaustion of the thyroid gland so that even if a behavioral problem is overcome, new feathers do not grow properly.[12] Nonfunctional follicles can also result from abscesses (internal or feather follicle), scarring, or fibrosis around the follicles' outer edges.

If both the Gram's stain and radiographs are normal, symptomatic therapy of a feather picking syndrome may include administration of levamisole phosphate as a general immunostimulant. Additionally, in the author's practice, two injections of polyvalent avian bacterin are given two weeks apart. If there is obvious pin feather growth at the time of the second bacterin injection and the feather picking continues, a radical grinding of the beak may inhibit self-mutilation. The owners need to be advised in this case that soft foods should be available to the bird, handfed if necessary, to keep him eating while the

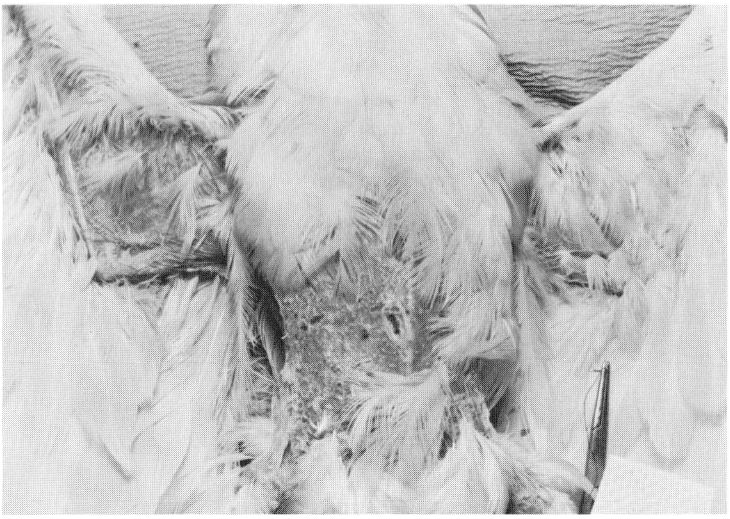

**Figure 41–10.** Self-mutilation and chronic skin irritation frequently suggest the presence of internal lesions, which are detected by radiography. (From Harrison, G. J.: Feather disorders. Vet. Clin. North Am. [Small Anim. Pract.], 14(2):191, 1984.)

beak heals. Instituting behavior modification practices is a component of all feather picking cases, as the habit must be overcome regardless of the cause. The use of tranquilizers (diazepam in the water) has been successful in some cases of feather abuse.

### *Folliculitis*

A Gram's stain of the feather pulp or follicles may suggest the advisability of culture and sensitivity testing for diagnosing and treating folliculitis. Bacterial or mycotic organisms may be responsible.[29, 32] Some chronic conditions may not totally respond to therapy, and affected feathers may not regrow.

**Figure 41–11.** Feather follicles may become nonfunctional in chronic cases. (From Harrison, G. J.: Feather disorders. Vet. Clin. North Am. [Small Anim. Pract.], 14(2):192, 1984.)

## Possible Diagnoses of Feather Disorders from Advanced Diagnostic Tests

Most chronic feather pickers with secondary dermatitis have internal organ pathology and can be otherwise asymptomatic for years. Radiographic evaluation of birds with feather disorders commonly reveals lesions in the air sacs or elsewhere that require more complex techniques for a diagnosis. Endoscopic observation and biopsies of air sacs, lung, kidney, liver, and spleen with subsequent cytology/histopathology/microbiology of the biopsy sample may be required.

If a significant quantity of caseous material is observed on endoscopy, treatment may include surgical intervention for removal of the bulk of the offending material, including portions of affected air sac tissue, if necessary. Based on the differential stains at the time of the surgery, appropriate topical therapy could be applied prior to closure followed by systemic therapy.

### *Hormonal Disorders*

Bilateral feather loss without subsequent replacement could alert the practitioner to evaluate endocrine function (see Chapter 22, Miscellaneous Diagnostic Tests, and Chapter 42, Miscellaneous Diseases) because the adrenals, thyroid, and gonads are associated with normal molt and feathering.

## SKIN

Dermatologic disorders of avian skin that is normally covered with feathers are usually associated with clinical signs reflected in the overlying feathers. However, unfeathered skin may be involved in independent disease processes unassociated with feather disorders. In very young, featherless birds, a general assessment of the overall health can be observed through the translucent skin; congestion of blood vessels due to toxemia or wrinkling of the epithelium from dehydration are obvious (see Chapter 53, Pediatric Medicine).

### Dermatology

Inflammation of the dermis of birds can be caused by fungal, bacterial, parasitic, or traumatic reactions. Viral agents affecting the skin are described in Chapter 32, Viral Diseases. The author has also encountered cases of suspected allergic skin responses. One tends to see skin irritation associated with infection of the periorbital or paranasal area or with upper respiratory disease. Diarrhea and diseases of the gastrointestinal, urinary, and/or reproductive system may induce cloacal skin irritation. A dermatitis specific to Rose-breasted Cockatoos and lovebirds is believed to be associated with stress.

The propatagium (wing web) appears to be particularly susceptible to bacterial infections, which often progress to ulceration, hyperemia, edema, pruritus, and self-mutilation (Fig. 41–12). Although debridement and aggressive local therapy may produce temporary improvement, subsequent infection by other agents often occurs because the loss of elasticity leads to frequent tearing of the skin. Radical therapy may include surgical removal of large areas of the fragile skin tissue. The remaining tissue will eventually stretch to a relatively normal appearance.

#### *Conditions Affecting the Facial Skin*

Facial skin appears to be easily bruised and vulnerable to lesions. Bruises of birds tend to have a dark discoloration for the first 24 hours, then change to green as biliverdin infuses the subcutaneous tissues.

The differential diagnosis for facial lesions may include pox or other viruses (Fig. 41–13), *Knemidokoptes* or other mites, trauma, bacterial or mycotic infections, or insect bites (Fig. 41–14). *Knemidokoptes* mites may be particularly disfiguring to the skin, with epithelial proliferation around the eyes, cere, and legs. One theory on the transmission of these mites suggests that they can be transmitted only in the nest to the featherless offspring. Another theory suggests that susceptibility is a genetically linked, immune-related condition, similar to the pathogenesis of demodectic mange in dogs. In the author's experience, a highly infected and untreated budgerigar did not transmit the disease to a normal cage mate in a two-year period. In another case, a budgerigar that had been maintained as a single pet for five years suddenly developed the characteristic skin lesions.

The cere is subject to change in response to a number of conditions. With chronic upper or lower respiratory diseases, the cere may atrophy and lose its plumpness; the nares may become malpositioned and distorted (see Chapter 28, Symptomatic Therapy and Emergency Medicine). The color may become darker, especially related to aging in conjunction with chronic respiratory problems and malnutrition. Brown hypertrophy of the cere as a normal characteristic of older breeding hen budgerigars requires no treatment. The same cere condition

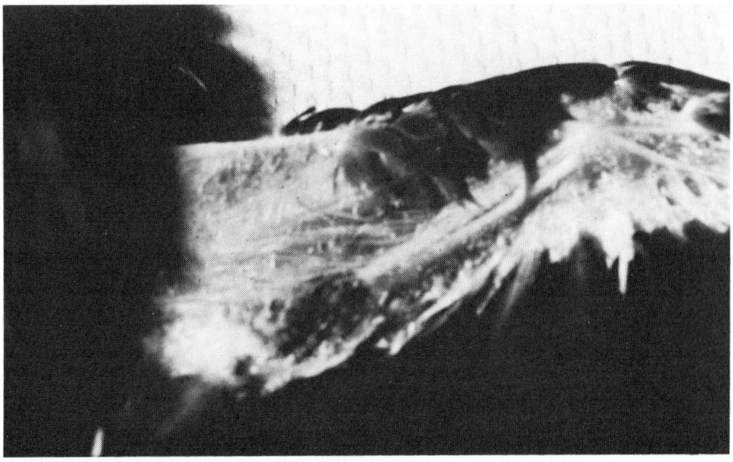

**Figure 41–12.** The wing web is a frequent site of dermatitis and self-mutilation.

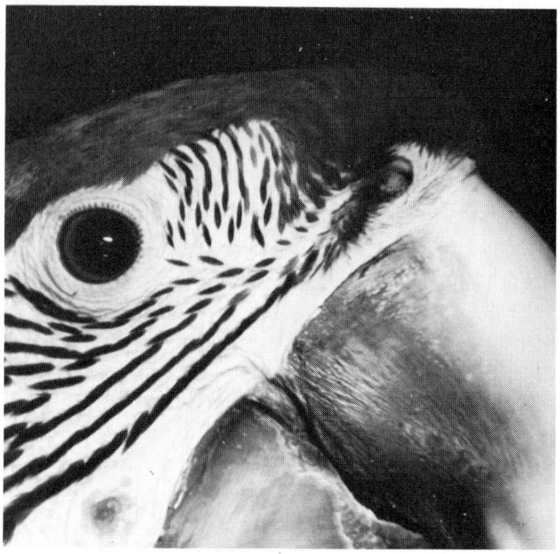

**Figure 41–13.** Pox may be expressed as lesions on the unfeathered facial skin of a macaw. (Courtesy of Lorraine Karpinski.)

in males, believed to be related to the presence of gonadal feminizing hormones from a gonadal tumor, may be temporarily treated by the use of moisturizers, gentle peeling, or light sanding.

### *Skin Conditions of the Legs and Feet*

Plantar pododermatitis (bumblefoot) in psittacines usually includes a thinning, ulceration, and general devitalization of the epithelium of the plantar foot tissue. This condition may also occur on the caudal surface of the tarsometatarsus in cases in which weakening of ligaments from suspected mineral malnutrition may alter the weight-bearing surfaces (Fig. 41–15). Although bumblefoot in raptors may progress to a septic condition that can affect the tendons and joints, this severity is rarely seen in psittacines. Malnutrition, particularly vitamin A deficiency, is believed to be the predisposing factor in pododermatitis (see Chapter 31, Nutritional Diseases). Chronic impact with highly abrasive surfaces or traumatic injuries may compound the condition. Heavy-bodied or obese birds may be particularly susceptible to bumblefoot.

Contrary to common belief, the characteristics of the perch (size, shape, texture) do not appear to be singularly responsible for producing the disease in an otherwise healthy bird, although improper perches may aggravate an already existing condition. Redig could not cause bumblefoot in well-nourished raptors when only narrow edges of snow fence were used as perches, nor could he prevent the disease in malnourished birds by using appropriate perches.[27]

Suggested therapies for psittacine pododermatitis include injections of vitamin A, topical use of povidone-iodine scrub, application of chlorhexidine cream, availability of soft perches, possible injections of polyvalent bacterins, and dietary management. Remarkable healing has occurred in cases of mild to moderate degrees of pododermatitis with the use of a chlorhexidine-soaked padded perch. This treatment is discontinued if the bird shows signs of allergy (redness, swelling) to the chlorhexidine. The prolonged therapy (weeks to months) that is required in some cases raises a suspicion of immune-related response.

Incarcerated sepsis may develop as a possible sequela to rapid healing of the skin; thus, subsequent swelling of a digit or foot may require lancing on the dorsal or lateral surface for draining and resolution. The foot may be band-

**Figure 41–14.** The facial lesions in this macaw are attributed to a skin reaction to wasp stings.

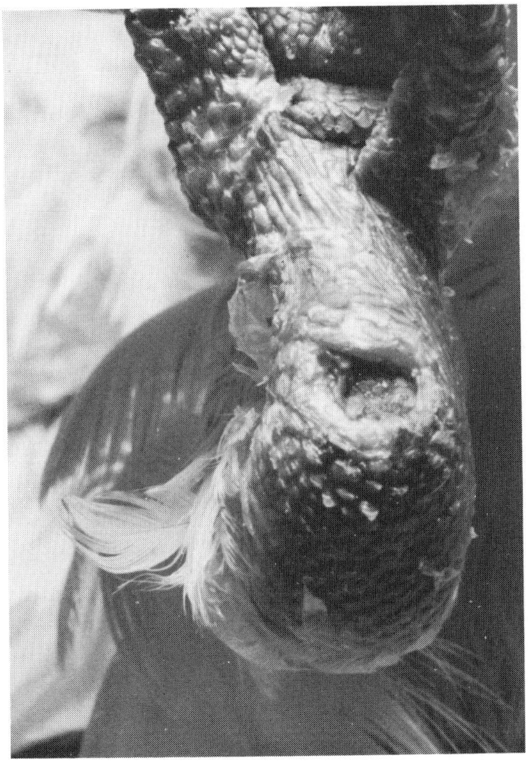

**Figure 41–15.** In cases in which the weight-bearing surfaces of the feet have been altered, pododermatitis-type lesions may be produced.

aged around a ball of folded gauze for padding. Severe cases may require a water bed-type body support during convalescence. If the skin of the foot has been severely damaged, re-epithelialization will not reproduce the normal scale pattern.[10]

In canaries the thickened red scale condition known as "scaley leg" is usually caused by *Knemidokoptes* mite infestation. These birds are frequently deficient in vitamin A and other nutrients. Treatment is oral and topical application of ivermectin, dietary management, and use of moisturizers. Similar clinical signs have been observed by the author in a case of *Staphylococcus* infection of mosquito bites; this was resolved with appropriate antibiotics and relocating the birds indoors.

Projections of proliferated keratin on the foot have been noted primarily in canaries and mynah birds. Severe keratomas resemble additional digits (Fig. 41–16). Sanding may reduce the size, and nutritional therapy may be advised.

A specific condition with lesions suggestive of staphylococcal ulcerative dermatitis (see Chapter 33, Bacterial Diseases) has been noted primarily in Amazons, particularly the Yellow-naped. The birds present with extensive, acute, necrotic-appearing lesions of the leg and digital skin, which linger as dark brown or black scabs (Fig. 41–17). In extreme cases, the birds lose their ability to grasp, so perches have to be removed. Bacteria other than *Staphylococcus* have been isolated from the lesions (e.g., *Pseudomonas, Klebsiella*); however, affected birds in a limited number of cases have responded to early treatment with only parenteral (dexamethasone) and topical corticosteroids. The possibility of some kind of viral etiology or immune-mediated reaction has been considered. If extensive necrosis and self-traumatization have occurred, therapy often requires long-term nursing care of several months' duration (e.g., scrubbing with povidone-iodine, applying chlorhexidine salve t.i.d., keeping the cage extremely clean). Standard immunostimulant

**Figure 41–16.** Severe keratin projections in canaries resemble extra digits.

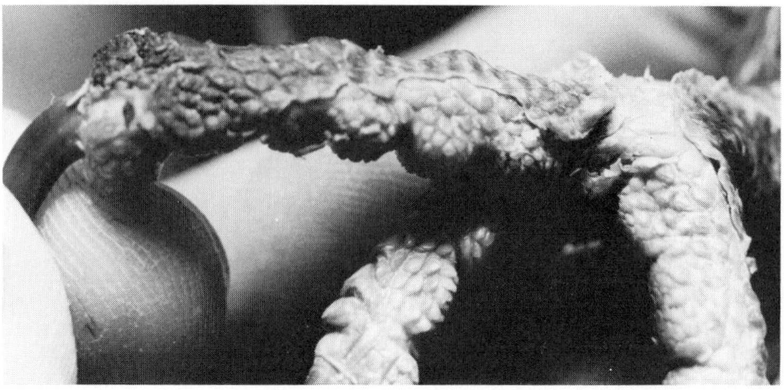

**Figure 41–17.** Amazons are particularly affected by a necrosis of the skin of the feet and legs.

procedures (levamisole, bacterins, injectable vitamins) are used. Injectable antibiotics are often used, but this author questions their efficacy. Application of a collar or blunting of the beak may reduce self-mutilation of the scabs. Circumferential scabs may produce edema from constriction and should be removed. The lancing described for premature healing of the skin with bumblefoot lesions may also be necessary. Recurrence has been noted by the author several months after apparent recovery.

Necrosis of the digits has been reported in finches and other species (Fig. 41–18). Suspected causes may include ergot poisoning, staphylococcosis, frost bite, constriction of the toe by human hair or by string that is sold commercially as nesting material. In the latter case, the fibers can often be seen with magnification deep in the circumferential indentation. Although the tip of a small hypodermic needle can be bent into a hook to pick out the fibers, this is tedious. Amputation of the digit is often the preferred alternative, especially if the constriction is deep or if scar tissue has formed to exacerbate the constriction. Healing of the exposed tissue is rapid after amputation.

Pressure necrosis from leg bands is common, and circumferential constriction of the vessels to the feet can result from the build-up of keratin. Some longitudinal lancing may be required to reduce edema in chronic cases after the band is removed. If the necrosis involves tissue adjacent to the bone, amputation may be necessary.

**Figure 41–18.** Circumferential constriction of a digit may require amputation for resolution.

### Miscellaneous Skin Conditions

Minor skin traumas and burn wounds are seldom treated beyond cleaning and debriding (see Chapter 28, Symptomatic Therapy and Emergency Medicine). One condition that requires suturing is an acute ulceration on the ventral surface of the pygostyle, which appears to occur relatively commonly in chronically malnourished cockatiels (Fig. 41–19). The tear is speculated to be caused by the lack of normal skin elasticity when wing-trimmed birds fall in flying attempts.

Tumors involving the skin (lipomas, other neoplasms) are usually excised in a routine fashion (see Chapter 47, Selected Surgical Procedures). Only one case of uropygial gland pathology has been seen in the author's experience—a ductile tumor. The necessity for expressing the contents from a uropygial obstruction has not been encountered by this author.

Subcutaneous white nodules on the joints may be tophi of gout infections (see Chapter 42, Miscellaneous Diseases) and need to be differentiated from abscesses.

## BEAK

Malformed beaks may occur as a result of malnutrition, improper incubation, *Knemidokoptes* mite infestation, trauma, bacterial or minor viral infection, or liver disease. Any of these may cause germinal tissue damage leading to a permanent beak deformity (Fig. 41–20). An asymmetric upper beak condition in neonates was first believed to result from consistently hand feeding from a single side of the mouth, but this occurs only occasionally in

**Figure 41–19.** Malnourished cockatiels appear to be particularly susceptible to acute skin tears.

otherwise short-term ill birds. Severe nutritional secondary hyperparathyroidism may produce a pliable "rubber" beak (observed most often in pigeons, doves, and cockatiels), which may respond to injectable calcium and vitamins. Corrective grinding appears to offer only temporary improvement for a twisted upper beak. A permanent groove may be worn in the beak as a result of chronic rhinorrhea and should alert the clinician to investigate for other clinical signs relating to chronic respiratory disease.

Overgrowth of the upper beak (and nails) is commonly seen in apparently malnourished birds with suspected liver disease (see Chapter 31, Nutritional Diseases). Although dietary management and liver therapeutics may reduce the frequency of trimming, the condition does not appear to be reversible. Brownish discolorations in the upper beak have been associated with this condition, but intralaminal hemorrhage from impact trauma has also been suggested as a cause.[25]

Puncture wounds of the upper beak or other type of localized beak injuries that penetrate the horny layer may be repaired with medical acrylics (see Chapter 46, Surgical Instrumentation and Special Techniques). Avulsions occurring in the distal third of the beak have potential for beak regeneration. A prosthetic device is usually not considered for any of these cases by the author. A temporary repair with acrylics will protect the exposed tissue until the bird can adapt its eating habits and the beak regrows. Severe injuries to the proximal two thirds of the beak usually result in scarring and failure to regenerate (Fig. 41–21).

Treatment of beak infections with topical or systemic antibiotics often fails because the close adhesion of the layers of the beak and lack of subcutaneous tissue confine swelling and the blood flow cannot reach the affected areas.

**Figure 41–20.** Permanent beak damage may result from infestation with *Knemidokoptes* mites.

**Figure 41–21.** Severe upper beak avulsions will not regenerate, but many birds can adapt their eating habits.

Debridement, drainage, and intralesional therapy are more effective.[10]

Injuries such as old fractures or those caused by obsessive biting on the wire of the cage[25] appear to be primary causes of unilateral lower beak elongation and uneven wearing. Severe fractures of the lower beak seldom heal; however, birds can adapt eating habits to such a fracture. One Scarlet Macaw in the author's practice has successfully reared nine offspring even with a one-inch gap in her lower beak from a previous fracture.

## NAILS

Abnormalities of the nails are most often expressed as elongation or lack of normal curve (exaggerated curling or straightness). *Knemidokoptes* mites, trauma, malnutrition, and liver disease are most often implicated in these disorders.

## REFERENCES


1. Altman, R. B.: Conditions involving the integumentary system. In Petrak, M. (ed.): Diseases of Cage and Aviary Birds. 2nd ed. Philadelphia, Lea & Febiger, 1982, pp. 368–381.
2. Fowler, M. E.: Diseases of the integument. Speakers' Syllabi, California Veterinary Medical Association Annual Scientific Seminar, 1982, pp. 441–444.
3. Franklin, T. E., et al.: Feather follicle cysts—a case report. In Proceedings of the 32nd Western Poultry Disease Conference, Davis, CA, 1983, pp. 115–117.
4. Fudge, A. M.: Approaching the feather picker. AAV Newsletter, 6(2):45, 1985.
5. Fudge, A. M.: Feather picking in birds—what can be done to solve the problem? Avian Med Center of Sacramento Newsletter, 3(4):1, 1985.
6. Galvin, C.: The feather picking bird. In Kirk, R. W. (ed.): Current Veterinary Therapy VIII. Philadelphia, W. B. Saunders Company, 1983.
7. Gerlach, H.: Viral diseases that can interfere with efforts for conservation of birds. Proc Jean Delacour/IFCB Symposium on Breeding Birds in Captivity, Universal City, CA, 1983.
8. Gerlach, H.: Virus diseases in pet birds. Vet. Clin. North Am. [Small Anim. Pract.], 14(2):299–315, 1984.
9. Graham, D. L.: An update on selected pet bird virus infections. Proceedings of the International Conference on Avian Medicine, Association of Avian Veterinarians, Toronto, Canada, 1984.
10. Graham, D. L.: The avian integument: Its structure and selected diseases. Proceedings of the Annual Meeting of the Association of Avian Veterinarians, Boulder, CO, 1985.
11. Graham, D. L.: Parrot reovirus and papovavirus infections and feather and beak syndrome. Proceedings of the Western Poultry Disease Conference, Davis, CA, 1985, pp. 118–120.
12. Greenwood, A.: Disease and medicine. In Stoodley, J., and Stoodley, P.: Parrot Production. Portsmouth, England, Bezels Publications, 1983.
13. Harrison, G. J.: Feather disorders. Vet. Clin. North Am., 14(2):179–199, 1984.
14. Hungerford, T. G.: Diseases of Poultry Including Cage Birds and Pigeons. Sydney, Australia, Angus and Robertson, 1939.
15. Johnson, C.: Clinical improvement of psittacine beak and feather loss birds with autogenous vaccines. AAV Newsletter, 6(3): 1985.
16. Krautwald, M., and Kaleta, E. F.: Untersuchungen zur akuten (Nestling-hepatitis) und chronischen (Franzosische Mauser) Form der Papovavirusinfektion der Wellensittiche. [Studies on the acute (fledgling hepatitis) and chronic (French moult) form of papovavirus infection in budgerigars.] Proceedings of IV Bird Diseases Seminar with Emphasis on Parrots and Gallinaceous Birds, University of Munich, 1985, pp. 58–63.
17. Kray, R. A.: Feather problems. Bird World Magazine, Dec. 1980/Jan. 1981.
18. Lothrop, C. D., et al.: Endocrine diagnosis of feathering problems in psittacine birds. Proceedings of the Annual Meeting of the American Association of Zoo Veterinarians, Tampa, FL, 1983.
19. McOrist, S., et al.: Beak and feather dystrophy in wild Sulphur-crested Cockatoos *(Cacatua galerita).* J. Wild. Dis., 20(2):120–124, 1984.
20. Panigrahy, B., et al.: Mycobacteriosis in psittacine birds. Avian Dis., 27:4, 1983.
21. Pass, D. A., and Perry, R. A.: The pathogenesis of psittacine beak and feather disease. Proceedings of the International Conference on Avian Medicine, Association of Avian Veterinarians, Toronto, Canada, 1984, pp. 113–119.
22. Pass, D. A., and Perry, R. A.: The pathology of psittacine beak and feather disease. Aust. Vet. J., 61:69–74, 1984.

22a. Pass, D. A., and Wylie, S. L.: Psittacine beak and feather disease: Progress report to A.A.V. Research Committee, 1985.

23. Perry, R. A.: Psittacine beak and feather disease syndrome. Proceedings No. 55, Refresher Course on Aviary and Caged Birds, University of Sydney, Australia, 1981.
24. Perry, R. A., and Pass, D. A.: Proceedings of the Cage and Aviary Veterinary Medicine Seminar, Australian Veterinary Poultry Association, Melbourne, Australia, 1985.
25. Perry, R. A.: The avian integument in disease. In Burr, E. (ed.): Companion Bird Medicine. Ames. IA, Iowa State University Press, in press.
26. Redig, P. T.: Basic surgical procedures for avian species. Proceedings of the Annual Meeting of the Association of Avian Veterinarians, Atlanta, GA, 1982.
27. Redig, P. T.: Personal communication.
28. Roertgen, K.: Avian giardiasis. Speakers' Syllabi, California Veterinary Medical Association Scientific Seminar, Oakland, CA, 1983, pp. 631–641.
29. Rosskopf, W. J., et al.: Treatment of feather folliculitis in a lovebird. Mod. Vet. Pract., Nov., 1983, pp. 923–924.
30. Roudybush, T.: Personal communication.
31. Tollefson, C. I.: Nutrition. In Petrak, M. (ed.): Diseases of Cage and Aviary Birds. 2nd ed. Philadelphia, Lea & Febiger, 1982, pp. 220–249.
32. Tudor, D. C.: Mycotic infection of feathers as the cause of feather-pulling in pigeons and psittacine birds. VM/SAC, Feb., 1983, pp. 249–254.
33. Walsh, M.: Endocrine evaluation of normal and feather loss cockatoos. Proceedings of the International Conference on Avian Medicine, Association of Avian Veterinarians, Toronto, Canada, 1984, pp. 101–102.
34. Woerpel, R. W., and Rosskopf, W. J.: Surgical repair of beak trauma in caged birds. VM/SAC, July, 1982, pp. 1068–1072.

# Chapter 42

# MISCELLANEOUS DISEASES

CLINTON LOTHROP
GREG J. HARRISON
DAVID SCHULTZ
TAMMY UTTERIDGE

A number of avian diseases that have not been specifically categorized elsewhere may occur with varying frequency in pet species.

## OBESITY

Obesity is commonly encountered in pet birds, especially in Amazons, budgerigars, Rose-breasted Cockatoos, and other aged psittacines fed free-choice diets consisting primarily of seeds. The evaluation of obesity is based on the individual bird rather than predetermined standards and is enhanced by a practitioner's experience in palpating birds of normal weight. Obese birds appear to have bald areas owing to separation of feather tracts by the fat-induced swelling. Although the species' tendency may be a predisposing factor, obesity is usually the result of excessive food consumption and inadequate exercise and is frequently accompanied by fatty infiltration of the liver. Obesity may be secondary to diabetes mellitus, malnutrition, or suspected hypothyroidism. Clinicians have noted that multiple injections of testosterone or medroxyprogesterone to continually suppress ovulation frequently lead to obesity in the treated birds.

Birds that are fed a wide variety of low-fat food items as youngsters appear to be less likely to develop obesity when exposed at a later date to high-fat items than are birds that had access to high-fat items (e.g., sunflower seeds, oats, peanuts) as neonates.

Obesity may be treated by reduction of caloric intake and an increase in exercise. Table 42–1 suggests an alternate cage bird diet used in Harrison's practice that has produced satisfactory weight loss. The dietary change should take place over a period of two to three weeks and the actual weight loss over several months. The severely obese bird should be weighed twice weekly by the owner and monitored by weekly and then monthly office visits. Birds on reducing diets may need to be supplemented with vitamins A, C, and E. Methionine, choline, inositol, and lactulose are useful supplements for the fatty liver condition (see Chapter 27, What To Do Until A Diagnosis Is Made).

Obese birds usually require the addition of iodine and trace minerals to their diet. Because sea water contains iodine and as many as 44 trace minerals, one method of providing this supplementation to birds is the daily addition of a fresh "sea water mixture" to the bird's drinking water. To make the "sea water mixture," one-fourth teaspoon of Lugol's iodine solution or strong tincture of iodine is added to one quart of clear, high tide ocean water or reconstituted products designed for salt water

**Table 42–1.** EXAMPLE OF AN AVIAN LOW-FAT DIET (Harrison)

Mix the following ingredients together with sufficient water to hold together and form into 15 balls:

- 4 lbs. sweet potatoes, boiled and mashed
- 2 hard-cooked eggs
- 6 drops vegetable oil
- 1 skinless chicken breast, boiled and chopped
- 3 cups of high-fiber, low-fat (e.g., Less) bread crumbs
- ½ cup Clovite or other guaranteed analysis vitamin supplement
- 10 calcium carbonate tablets, crushed

One ball per day will supply nutritional supplementation for a 400-gm Amazon.

Additional items may be selected and fed as desired from:

- Unleavened bread
- Sprouts
- Fresh vegetables
- Plain yogurt
- Fresh fruit (apple, apricot, or cantaloupe)
- White millet (essential for cockatiels and budgerigars; optional in larger species). If it appears that weight loss is occurring too rapidly, additional protein and/or fat sources may need to be added in small amounts.

aquariums* (with separately packaged trace elements). Directions for use are to add one teaspoon "sea water mixture" to one measuring cup of fresh drinking water daily. The supplemental mixture is stored in the refrigerator.

Increased exercise for the overweight bird may be encouraged by providing a larger cage, supervised free flight, or simulated flight for clipped birds 15 minutes twice a day.

Lipomas are a common sequela to obesity. Lipomas will often regress following correction of the diet. The clinical use of thyroid hormones to stimulate metabolism has apparent successful results.

## PSITTACINE LIVER DISEASE

Many psittacine diseases involve the liver. Clinical signs that may be directly or indirectly correlated with liver disease are listed in Table 42–2. Table 42–3 lists numerous conditions that have been reported to affect the avian liver.

A presumed diagnosis of liver disease may be made from the history (primarily diet and exposure to infectious causes) and physical examination, with particular inspection of the urine, urates, and feces for evidence of biliverdin.

There are no clinical diagnostic tests specific for measuring liver function, although some avian practitioners use a number of serum chemistry tests in conjunction with other clinical signs (see Chapter 13, Clinical Chemistries). The correlation of some serum chemistry results with organ pathology is controversial. In one study, pigeons with subclinical cases of chlamydiosis, salmonellosis, and possibly other disease entities failed to show significant elevations of SGOT (AST), LDH, AP, or cholesterol levels. Elevated values of the same tests were noted post-surgically in cockatiels, but no correlation could be made with organ disease symptoms, nor was therapy instituted.

Radiography may be used as the initial diagnostic evaluation of the liver. One method of determining the relative size of the liver is by measuring the space caudal to the heart and anterior to the ventriculus on the lateral view. Additionally, the angle of the proventriculus, suggestive of a 45-degree drop in the normal parrot, appears steeper with liver atrophy and flatter with hepatomegaly. Table 42–4 illustrates how the liver size may vary according to the disease state and therapy.

*Instant Ocean

**Table 42–2.** CLINICAL SIGNS SUSPECTED TO BE ASSOCIATED WITH LIVER DISEASE (Harrison)

| |
|---|
| Anorexia |
| Inactivity |
| Personality change, cessation of talking, irritability |
| Feather disorders, including pigmentary changes, frequent or incomplete molt, feather picking, other integumentary disorders |
| Chronic weight loss |
| Anemia |
| Softening, flaking, overgrowth of beak and nails |
| Obesity |
| Polyuria, polydipsia |
| Biliverdinuria |
| Regurgitation |
| Respiratory embarrassment: lung and air sac fluids, especially in viral hepatic diseases |
| Coagulopathies |
| Diarrhea and pasting of vent feathers |
| Seizures from hepatic encephalopathy |
| Ambulatory difficulties due to hepatomegaly and accompanying ascites; paresis |
| Asymptomatic |
| (Icterus is very rare.) |

Symptomatic therapy for birds with suspected liver disease is discussed in Chapter 27, What To Do Until A Diagnosis Is Made. Response to therapy is suggested by increased appetite and activity, weight gain, and decreased presence of liver pigments in the urine and/or urates. Lack of clinical improvement indicates the necessity for further investigation with endoscopy, biopsy, or other appropriate studies (e.g., cytology, microbiology, histopathology).

An interesting point is brought out by Ratcliffe in discussing the etiologic role of the Philadelphia Zoo hepatitis virus and the varying clinical picture. In contrast to experimental birds, zoo exhibition display birds had a higher incidence of "post-hepatitis" lesions (nodular hyperplasia, necrosis, regeneration, and associated inflammation of slowly progressive hepatitis). "Present evidence suggests . . . that social factors (rather than diet), i.e., intraspecific and interspecific conflicts, have operated through endocrine mechanisms to increase the severity of disease, i.e., to delay both the immune response and the process of healing." His observations of an increased frequency of tuberculosis in captive wild birds with increased population density was used to support his opinion.

**Table 42–3.** REPORTED ETIOLOGIES OF LIVER DISEASE IN PET BIRDS (Harrison)

**Metabolic**
- Fatty liver degeneration due to chronic malnutrition
- Drug-induced (e.g., long-term corticosteroid or tetracycline use)
- Hereditary hepatic lipidosis in budgerigars
- Amyloidosis
- Diabetes mellitus
- Toxins
- Allergies
- Visceral gout
- Hemochromatosis of mynah birds
- Congestive heart failure in mynah birds
- Hypothyroidism
- Starvation

**Chlamydial**

**Bacterial**
- *Escherichia coli*
- *Pasteurella*
- *Pseudomonas*
- *Yersinia*
- *Salmonella*
- *Mycobacterium*
- *Erysipelothrix*

**Mycotic**
- *Aspergillus*
- *Nocardia*
- *Absidia*

**Neoplastic**
- Bile duct adenoma/carcinoma
- Fibrosarcoma
- Hemangioendothelioma
- Hemangiosarcoma
- Rhabdomyosarcoma
- Lymphoid leukosis–like disease in psittacines

**Parasitic**
- Ascarid
- Protozoa (in canaries)
- Histomoniasis (in gallinaceous birds)
- *Plasmodium*
- *Leucocytozoon*
- Microsporidiosis
- Hepatic distomiasis
- *Trichomonas* (in budgerigars)

**Toxic**
- Rapeseed poisoning
- Aflatoxicosis

**Viral**
- Herpesvirus (Pacheco's parrot disease)
- Generalized psittacine papovavirus disease (budgerigar fledgling disease)
- Philadelphia Zoo hepatitis virus
- Pox
- Adenovirus
- Reovirus
- Acute psittacine beak and feather disease

## PET BIRD RENAL DISEASE

Primary renal disease is difficult to diagnose in clinical cases because the kidney, like the liver, is frequently a component of pansystemic diseases in pet birds. No clinical chemistry test is uniformly diagnostic of renal damage or disease, and information on evaluation of urine and urates is insufficient (see Chapter 9, Evaluation of Droppings). Clinical presumption of kidney disease may be made from the presence of polyuria, proteinuria, hyperuremia, and radiographic evaluation (see Chapter 14, Radiology). Chronic cases may require endoscopy and kidney biopsy for diagnostic information (see Chapter 15, Endoscopy, and Chapter 16, Biopsy Techniques). Chronic renal disease generally shows a nonregenerative anemia and stress white blood cell response (increased white blood cell count, lymphopenia). Hematuric cases show a regenerative anemia.

**Table 42–4.** RADIOGRAPHIC MEASUREMENTS OF DOUBLE YELLOW-HEADED AMAZON LIVER DURING AND AFTER CHLAMYDIOSIS (Harrison)

| | Lateral View | Ventrodorsal View |
|---|---|---|
| Day 1 | 2.0 cm | 3.6 cm |
| Day 30 | 2.5 cm | 4.0 cm |
| Day 150 | 0.5 cm | 2.8 cm |

Specific serum chemistry tests that may provide varying levels of assistance in diagnosing renal disease are discussed in Chapter 13, Clinical Chemistries. In a study of gentamicin sulfate toxicity in budgerigars, birds given clinical doses of gentamicin for four days developed proximal tubular necrosis. The uric acid levels were so variable that uric acid was believed to be an unreliable indicator of gentamicin nephrotoxicity in the 20 budgerigars studied. Schultz disagrees with the conclusions of this study for several reasons: it is not known if proximal tubular necrosis significantly affects production of uric acid; a four-day toxic exposure may not be sufficient; and because the renal system of budgerigars is designed for low availability of water, their kidneys may be more resistant to damage. Schultz believes that this may explain why the synovial form of gout is the more common syndrome seen in budgerigars, whereas the serosal form is more frequently noted in other species.

In Harrison's practice, renal tumors are relatively rare; however, Schultz commonly encounters them in budgerigars in Australia. A Canadian veterinary teaching hospital compiled data from a study of the 736 budgerigars that had been admitted over 15 years. Seventy-six per cent of the birds had died; of these, 5 per cent had renal tumors, and of the birds with renal tumors, 78 per cent were males. An aviary of budgerigars that are bred for health and longevity has not produced a single case of renal or testicular tumor in 20 years in Harrison's experience.

The treatment of pet bird renal disease is currently confined to supportive care (see Chapter 27, What To Do Until A Diagnosis Is Made). Some attempts have been made to replenish the renal loss of protein with plasma extenders and serum transfusions, but no conclusion has been drawn on the effectiveness of this therapy.

## Avian Renal Pathology

Histopathology becomes important in the clinical diagnosis of kidney biopsy samples and in the postmortem diagnosis of cases with inconclusive antemortem evidence and expected recommendations regarding treatment of other birds. Necropsy samples from very small birds may be obtained after the part of the synsacrum containing the kidneys is fixed in formalin; the kidneys can then be dissected out.

Single or multiple renal cysts occur with or without renal neoplasia. The kidney has a rich vascular supply and is affected by generalized disease processes such as amyloidosis. Bacterial emboli may lodge in the kidney, sometimes with the production of micro- or macroabscesses. *Yersinia pseudotuberculosis, Pasteurella, Salmonella, Escherichia coli, Streptococcus, Erysipelothrix, Pseudomonas pseudomallei,* and mycobacteria may be isolated. Chronic interstitial nephritis, small calcified foci, and proliferative glomerulonephritis are common findings (particularly in birds with gout) and may reflect previous bacteremia or toxic insults.

Air sac and renal fungal granulomas of *Absidia corymbifera* have been associated with clinical disease, whereas a solitary granuloma of *Aspergillus flavus*—orzyae group has been regarded as an incidental finding.

The role of nutrition as a primary or predisposing factor in renal disease of cage birds is unclear at present, although it is thought to be involved in the pathogenesis of urolithiasis.

The papovavirus isolated from budgerigar fledgling disease results in focal nephrosis, and intranuclear inclusions in renal tubular epithelium occur in both sick and healthy in-contact birds. Eosinophilic intranuclear inclusions have also been seen in Pacheco's disease in Masked Lovebirds (*Agapornis personata*). Kunitachi virus (paramyxovirus group 5) in budgerigars causes nephrosis with development of giant cells. Intranuclear basophilic inclusions in renal tubular epithelium occur in lovebirds, and, although virus has not yet been isolated, an adenovirus is suspected. Although avian leukosis virus GS antigen has been detected by ELISA in budgerigars, no correlation with tumor development has been observed.

Renal coccidiosis may cause chronic active inflammation with schizonts in the epithelium of distal tubules, collecting ducts, and ureters. Toxoplasma may be found in necrotic foci. Microsporidia occur in the kidney but rarely incite an inflammatory response. Similarly, megaloschizonts of *Leucocytozoon* occur, often as an incidental finding, as *Leucocytozoon* is widespread among clinically healthy birds. Schistosomes can occur in renal blood vessels but cause debilitation and death only when present in large numbers. Eucotylids and Renicola compress adjacent renal parenchyma, and large numbers may cause granulomatous inflammation, whereas fibrosis may be seen in fulmars (*Fulmarus* sp.).

Renal neoplasia is particularly common in budgerigars; the tumor is usually a nephroblastoma similar to those described in chickens. Concurrent gonadal neoplasia may be present.

Salt poisoning of canaries results in marked congestion and hemorrhage. Ingestion of *Oxalis* sp. may result in distention of tubules by numerous oxalate crystals. Intranuclear inclusions are occasionally seen in renal tubular epithelium in lead poisoning, although incidence may vary with species. Acute toxic tubular nephrosis (especially affecting the proximal convoluted tubules) is seen following chemical toxin exposure or endogenous toxemia.

## Gout

Gout results when uric acid and urates fail to be excreted by the kidneys and are deposited within the body. In budgerigars, urate deposits (tophi) are often evident in the ankle, tarsal phalangeal joints, and the toes and must be differentiated from abscesses, infectious arthritis and sprains. The bird may exhibit shifting leg lameness from the pain and inability to bend toes. Other clinical signs are nonspecific. The presence of uric acid in the nodules can be confirmed by evidence of crystals on microscopic examination of an aspirate (Fig. 42–1). The incidence of the disease is reported by Australian and California practitioners to be relatively common but is believed not to be a significant clinical entity in some other parts of the United States and Canada. Visceral gout is more commonly seen in Australia than the synovial form. Schultz's histopathologic evaluation of gout supports infectious, toxic, or neoplastic causes rather than nutritional.

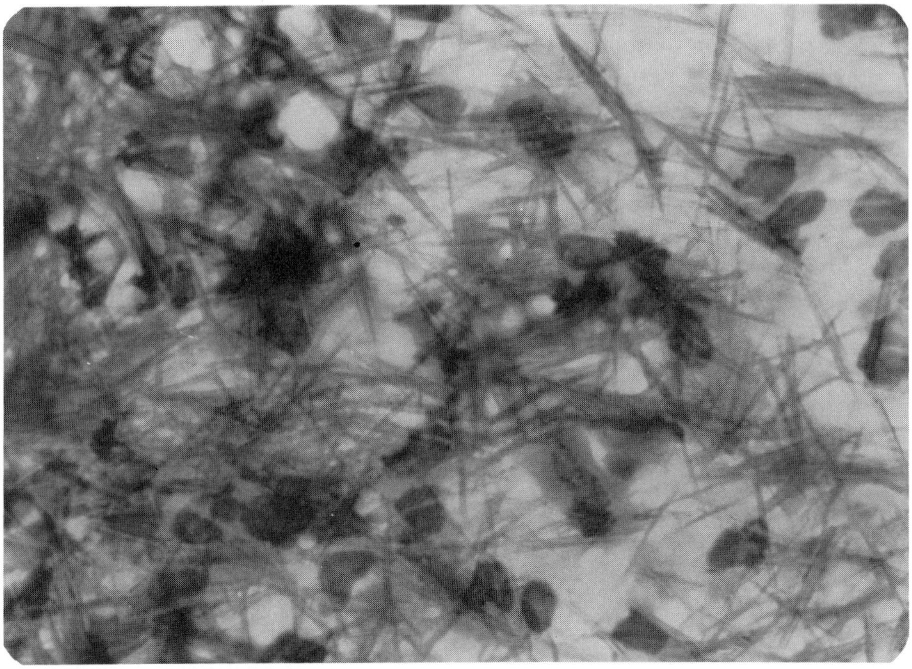

**Figure 42–1.** The uric acid crystals associated with gout in a budgerigar are evident on this impression smear (× 1000).

Profuse hemorrhage results from surgical attempts to remove the deposits, and Schultz believes the patient does not benefit from this therapy. Dietary management consisting of reduced protein and increased dietary sources of vitamin A and moisture (e.g., fresh greens and other vegetables) may offer some relief and alllow the polyuric patient to survive for up to 18 months. Alll of these birds should receive antibiotics, the success depending upon the stage at which they are given in the course of the disease. Allopurinol has been used with varying results.

## DISEASES OF THE THYROID GLAND

Early avian clinical literature often described what was believed to be the most common disease condition of avian pets at that time: thyroid dysplasia in budgerigars. Blackmore reported that 34 out of 120 budgerigars (23.8 per cent) died or were destroyed as a direct result of thyroid dysplasia, second only to neoplasia as a cause of death in budgerigars. The greatest number of deaths were reported to be in the five- to six-year-old age group. This condition is also known as goiter, thyroid hyperplasia, and hypothyroidism.

### Goiter

Some nutritional aspects of thyroid dysplasia are discussed in Chapter 31, Nutritional Diseases. Dietary iodine is necessary for thyroid hormone biosynthesis. Dietary deficiency of iodine inhibits thyroid hormone biosynthesis, decreasing blood concentrations of thyroxine ($T_4$) and triiodothyronine ($T_3$). Decreased thyroid hormone concentrations release the pituitary thyrotrophs from negative feedback control, and the constant production of thyrotropin (TSH) may subsequently result in hyperplasia and thyroid dysplasia in severe cases. Thus, thyroid enlargement with decreased production of thyroid hormones is seen in iodine-deficiency goiter. Biochemically, these birds would be hypothyroid if a thyroid function test were performed.

Budgerigars with thyroid dysplasia have been reported to have thyroid glands weighing up to 1000 mg, compared to the 1.5 mg considered normal for budgerigars, although most affected birds have thyroids weighing between 150 and 300 mg. Goiter has also been seen in canaries and pigeons, and although not reported in other popular cage birds, goiter could occur in any bird on a diet severely deficient in iodine.

Although it has been commonly believed that an enlarged thyroid can impinge on the trachea, causing inspiratory dyspnea and respiratory distress, crop emptying disorders, and direct cardiac pressure, the capability of thyroids to enlarge to such a degree is controversial. Ringer has chemically produced severe thyroid enlargement, but impingement on the trachea was not noted on necropsy evaluation. From clinical cases, Harrison suspects that it is excessive fluid secretions in the crop and lower digestive system that are responsible for the increased swallowing motions and regurgitation and that increased fluids in the lungs produce the dyspnea and subsequent death in advanced cases.

### *Treatment of Goiter*

Although birds with thyroid dysplasia or goiter are functionally hypothyroid, thyroxine replacement is not indicated, since iodine supplementation alone will correct the biochemical defects and re-establish the euthyroid state. Emergency treatment for the severely dyspneic bird may include supplemental oxygen, and intramuscular administration of dexamethasone followed by iodine supplementation (see Chapter 22, Miscellaneous Diagnostic Tests). Injectable 20 per cent sodium iodine is indicated in life-threatening situations only. In less severe cases without respiratory embarrassment oral supplementation is sufficient. Iodine can be added to the diet in the form of iodized seeds or to the drinking water. Although cod liver oil contains 10 mg/kg of iodine, the potential for oil rancidity may result in hypovitaminosis E (see Chapter 31, Nutritional Diseases). A low-fat diet is also used for birds with moderate weight problems.

## Hypothyroidism

Clinical manifestations of hypothyroidism are reduced metabolic rate, fat deposition, and alteration in feather structure. Although not proven, decreased fertility, normochromic normocytic anemia, hypercholesterolemia and triglyceridemia, cardiac embarrassment from carotid artery pressure, nonpruritic feather loss, and predisposition to bacterial and mycotic cutaneous infections are also possible in hypothyroid pet birds. Decreased food intake, lethargy, depression of mental alertness, extreme sensitivity to cold temperature, and heat seeking may also be suggestive of hypothyroidism. Recurrent subcutaneous lipomas have been reported to be associated with decreased resting thyroxine concentration and respond to thyroid replacement.

Empirical doses of thyroid supplementation have been useful in treating some birds with varying degrees of feather loss, although budgerigars diagnosed as hypothyroid from iodine deficiency seldom show feather problems. Lothrop has not correlated abnormal thyroid levels with feather picking. Side effects from overdosing thyroid supplementation may range from hyperactivity and nervousness to vomiting and diarrhea.

## Hyperthyroidism

Spontaneous hyperthyroidism has not been reported in cage birds but should be considered in a cage bird with suggestive symptoms (weight loss, voracious appetite, voluminous stools, cardiac arrhythmias, and hyperactivity). Intestinal parasitism and malabsorption are differential diagnoses. A suspicion of hyperthyroidism can be confirmed by thyroid function testing (see Chapter 22, Miscellaneous Diagnostic Tests). Surgical thyroidectomy or medical ablation with $^{131}$Iodine or propylthiouracil are all possible therapeutic modes for treatment of hyperthyroidism.

# ADRENAL DISEASE

A routine survey of necropsied psittacine birds revealed an unusually high incidence (27 per cent) of adrenal pathology. Although the adrenal pathology was not correlated with antemortem adrenal function or a complete history of clinical symptoms, one may suspect a significant incidence of adrenal disease in pet birds. Diagnosis of adrenal disorders in pet birds should be based on quantification of corticosterone (not cortisol) with the use of ACTH stimulation test (see Chapter 22, Miscellaneous Diagnostic Tests).

Adrenal insufficiency should be suspected in any bird with a decrease in the Na/K ratio ($<27$), episodic weakness and chronic debilitation, low fasting blood glucose concentration, hypercalcemia, decreased urine specific gravity, feather loss, vague gastrointestinal signs such as periodic diarrhea, generalized abdominal tenderness, and nonregenerative anemia.

A decrease in the sodium/potassium (Na/K) ratio implies a deficiency in mineralocorticoid production and may or may not be seen with concurrent glucocorticoid deficiency. A decrease in the Na/K ratio implies a decrease in serum $Na^+$ concentration, increase in serum $K^+$ concentration, or both. Serum $Na^+$ concentration decreases in the absence of adrenal insufficiency with psychogenic polydipsia, inappropriate ADH secretion, or chronic fluid loss due to diarrhea or polyuria. An increase in the serum $K^+$ concentration is seen with hemolysis of the blood sample. Hemolysis should always be considered in the differential diagnosis of hyperkalemia. Hyperkalemia may be seen with acidosis due to the intracellular shift in hydrogen ion and associated extracellular shift in $K^+$. Renal disease should also be considered in the differential diagnosis of hyperkalemia. Both $Na^+$ and $K^+$ concentration can be increased in dehydration.

Episodic weakness, nonregenerative anemia, low fasting glucose concentration, hypercalcemia, low urine specific gravity, and gastrointestinal symptomology are signs of glucocorticoid deficiency.

The signs of adrenal insufficiency are nonspecific and may be seen with renal, gastrointestinal, and/or hepatic disease as well. Not all signs of adrenal insufficiency are necessarily seen at the same time in one bird. Thus, the diagnosis of adrenal insufficiency is difficult to make without performing an adrenal function test.

### Cushing's Syndrome

Cushing's syndrome has not been reported in cage birds but could be diagnosed on the basis of clinical impression and adrenal function testing. Signs suggestive of Cushing's syndrome include polydipsia, polyuria, hepatopathy, muscular weakness and catabolism, and hyperglycemia. It is not known if feather loss is associated with excessive corticosterone production. Administration of exogenous glucocorticoids in excessive doses can cause iatrogenic Cushing's syndrome.

#### *Therapy in Avian Adrenal Disease*

Complete adrenal insufficiency, such as with adrenalectomy, is life-threatening without replacement therapy. Adrenalectomized birds can be maintained with 4 mg/kg deoxycorticosterone acetate daily or 10 mg/kg cortisone acetate daily.

Replacement with dexamethasone and fludrocortisone acetate has also been suggested for treating adrenal insufficiency. It is important not to underestimate the replacement dose. Insufficient replacement could be lethal, whereas it is doubtful that over-replacement could cause serious problems. It is better to overestimate and subsequently decrease the replacement dose, after judging response to therapy.

Cushing's syndrome due to a pituitary tumor or pituitary hyperplasia is most often treated medically with o′p′-DDD in dogs. Cushing's syndrome due to an adrenal tumor is best treated by surgical removal of the affected gland. Medical treatment of suspected Cushing's syndrome in pet birds is currently experimental, since the efficacy of o′p′-DDD therapy is unknown. Response to treatment, if attempted, could be best evaluated by effective remission of clinical signs.

## DIABETES MELLITUS

The pathophysiology of diabetes mellitus in granivorous avian species has only recently begun to be understood and differs considerably from diabetes in mammals and carnivorous birds. In mammals, diabetes is considered to be a bihormonal disease associated with insulin deficiency or ineffectiveness and glucagon excess. The distribution of pancreatic islet cells, pancreatic hormone concentrations, serum glucagon and insulin concentrations, and in general, glucose regulatory mechanisms, are similar in mammals and in carnivorous birds. However, pancreatic islet composition, pancreatic and serum hormone concentrations, and glucose homeostatic mechanisms in granivorous birds are markedly different from those in mammals and carnivorous birds.

In mammals, the pancreatic islets are composed of approximately 70 per cent beta cells (insulin-secreting), 20 per cent alpha cells (glucagon-secreting), 9 per cent delta cells (somatostatin-secreting), and 1 per cent PP cells (pancreatic polypeptide–secreting). In carnivorous birds, the distribution of islet cells is similar to that in mammals. However, in granivorous birds, the distribution of pancreatic islets is markedly different: 50 per cent alpha cells, 37 per cent beta cells, and approximately 12 per cent delta cells. The intraislet cellular composition (alpha/beta ratio) varies in the avian pancreas. Histologically, islets can be classified as light (predominantly beta cells) and dark islets (predominantly alpha cells). Pancreatic and serum glucagon concentrations are approximately 5 to 10 times those found in mammals, and the molar ratio of insulin to glucagon in pancreatic extracts and serum from granivorous birds reflects the islet cell composition and the predominance of glucagon. The insulin/glucagon (I/G) ratio in fasted mammals is 3.0 to 3.5, whereas in granivorous birds it is only 1.3 to 2.0. Based on observations such as these and other experimental studies, it has been suggested that glucagon is the major glucose-regulating hormone in birds. This, of course, includes most of the popular pet bird species.

The predominance of glucagon in avian species demonstrates that these animals are primarily in a catabolic state (I/G ratio). This is analogous to type 2 diabetes (insulin ineffectiveness) in man. However, a number of similarities in the metabolic actions of insulin and glucagon at the cellular level exist between birds and mammals. Insulin is anabolic in nature, stimulating glucose uptake in skeletal muscle, adipocytes, and hepatocytes. Insulin stimulates glycogenesis and glycogen deposition and is antigluconeogenic. However, insulin is not antilipolytic in birds as it is in mammals. Glucagon is strongly catabolic and stimulates gluconeogenesis, lipolysis, and glycogenolysis. The net effect of glucagon secretion is a marked and rapid hyperglycemia.

Pancreatic glucagon secretion is stimulated by hypoglycemia, amino acids, and acetylcholine. Normally glucagon secretion is inhibited by glucose, insulin, and somatostatin. Insulin sensitizes the alpha cells to glucose so that an appropriate amount of glucagon is secreted for the current glucose concentration. The abnormality in diabetes is that glucagon is continually released in the face of an increased glucose concentration. However, the inappropriate release of glucagon in diabetes is not necessarily due to an insulin deficiency (e.g., glucagonoma) but will be compounded by an insulin deficiency.

Insulin secretion is stimulated by glucose, amino acids, and glucagon, and its release is inhibited by somatostatin, hypoglycemia, and hypoproteinemia.

Experimental pancreatectomies have yielded varying results depending on extent of pancreatic resection (complete vs. partial) and the type of bird (granivorous vs. carnivorous). In carnivorous birds, complete pancreatectomy results in diabetes, as it does in mammals. This suggests the predominance of insulin in glucose homeostasis in carnivorous birds and reflects pancreatic histology with beta cell predominance. Incomplete pancreatectomy in granivorous birds results in permanent diabetes if only alpha cells are left in the remaining pancreatic tissue. Complete pancreatectomy with total removal of alpha and beta cells causes a transient diabetes with a later return to normal glucose concentrations.

Serum glucagon and insulin concentrations have been determined in a limited number of granivorous birds with spontaneous diabetes (Table 42–5). Serum glucagon concentrations were markedly elevated in a diabetic toucan and an Amazon parrot, with approximately normal insulin concentrations in both birds. Histologic examination of the pancreas from the diabetic toucan demonstrated a diffuse hyperplasia of alpha cells throughout the pancreas (not shown). It is interesting that the serum glucagon concentration was two to three-fold lower in the diabetic cockatiel than in serum pools from normal cockatiels. An insulin deficiency was not noted in the diabetic cockatiel.

**Table 42–5.** GLUCAGON AND INSULIN CONCENTRATIONS IN AVIAN DIABETES

| | Glucagon (pg/ml) | Insulin (μU/ml) |
|---|---|---|
| **Toco Toucan** | | |
| Diabetic | 5,255 | 2.8 ; 3.8 |
| Normal | 1,368 | — |
| Normal | 955 | — |
| Normal | 625 | — |
| Normal | 577 | 2.4 |
| **Amazon** | | |
| Diabetic | 14,222 | 7.4 |
| Normal | 604 | 7.7 |
| **Cockatiel** | | |
| Diabetic | 235 | 8.4 |
| Normal | 780 | 8.6 |
| Normal | 964 | 7.5 |
| Normal | — | 7.6 |
| Normal | — | 5.8 |

There are several other reports of spontaneous avian diabetes but none with serum hormone measurements. Diabetes and an islet cell carcinoma were documented in a parakeet. The cellular origin of the tumor was not identified in that case, but it was suggested that it could be an alpha cell tumor. Altman and Kirmayer documented diabetes in four parakeets and reported normal pancreatic histology in one bird necropsied. There are sporadic reports of diabetes in association with renal tumors in birds. Normal pancreatic histology *per se* does not preclude abnormal glucagon secretion unless concurrent glucagon and insulin concentrations are determined. The defect could be biochemical in nature, with no abnormal cellular histology. Endocrine, especially pancreatic, tumors are noted for being very active but benign histologically. The association between diabetes and renal tumors is interesting, and a similar correlation has been noted in humans. In rare cases, the renal tumors have been found to elaborate glucagon-like molecules.

The experimental and clinical observations summarized above suggest that avian diabetes, excluding that in carnivorous birds, results from excessive glucagon production rather than from an insulin deficiency. The biochemical basis for diabetes in the aforementioned cockatiel is not clear, but it is possible that glucagon-like peptides or large glucagon molecules that are biologically active but not detected in the glucagon assay are being produced, causing the glucose intolerance.

Clinical symptoms of diabetes in birds are similar to those observed in mammals. The most frequent finding is a remarkable polydipsia and polyuria. Birds will spend almost all day drinking from their water bowl. Frequently, polyphagia is present, but even with the increased appetite there is a significant weight loss in untreated birds. Weight loss is due to the strongly catabolic effects of excessive glucagon secretion. Birds may be depressed and found sitting on the bottom of the cage. Untreated diabetes predisposes birds to secondary infections, and in fact diabetes may be diagnosed incidentally while treating a bird for another problem.

The hallmark of avian diabetes, like diabetes in mammals, is hyperglycemia and glucosuria. Urine glucose is negative in most normal birds, while diabetics may be strongly glucosuric. Normal blood glucose levels range from 250 to 500 mg/dl in birds, which are significantly higher than normal levels in mammals. Blood glucose levels in diabetic birds may range from high normal to almost 1800 mg/dl.

Diagnosis of diabetes can often be suspected on the basis of history and measurement of serum and urine glucose concentrations. However, since PD/PU and glucose intolerance are not pathognomonic for diabetes, other disorders should be ruled out with appropriate laboratory tests at the time the diagnosis of diabetes is considered. Definitive diagnosis of diabetes can be made with serum insulin/glucagon measurements in conjunction with determining the serum glucose concentration.

Currently, treatment of diabetes is daily injections of insulin, as with mammals. As the etiology is probably not an insulin deficiency, this treatment is only palliative in nature. Insulin is not always effective in lowering the blood glucose in diabetic birds but seems consistently to prevent the severe weight loss often seen in diabetic birds.

The rule of thumb when starting insulin therapy in a diabetic bird is to start low and increase as needed to prevent insulin overdose and hypoglycemic shock. An insulin preparation of intermediate duration such as NPH U40 (40 units/cc), after appropriate dilution, is best to begin treatment. However, length of duration after dilution is unknown, since the vehicle is diluted too. In general, smaller birds need more insulin per gram of body weight than do larger birds. The treatment protocol should be tailored for each individual bird to achieve optimum response. Any bird being treated with insulin should be allowed free access to food to decrease the chances of hypoglycemic shock. The primary author suggests that diabetic birds be hospitalized during the initial insulin therapy until the correct dosage scheme is obtained.

## Clinical Experiences with Diabetes Mellitus

From Harrison's clinical standpoint, the necessity for further investigation into a potential diagnosis of diabetes mellitus is indicated by the results of a urinalysis with dip test strips or tablets (Clinitest—Miles Laboratories, Inc.). If the urine glucose level exceeds 0.5 per cent, serum glucose level should be assayed. When the serum glucose level exceeds 1000 mg/dl, insulin therapy is usually instituted on an empiric basis. If the serum glucose levels are between 600 and 1000 mg/dl, the assay should be repeated in 24 hours. Insulin therapy is then initiated if the second test also shows a glucose level that is the same or elevated. However if the glucose level has dropped by the second test, insulin is not given. A transient hyperglycemia is not diabetes; affected birds may become asymptomatic without therapy.

The dose of insulin is highly variable for an individual bird (Table 42–6). Most insulin dosages in the literature are based on the affected species, but it is more accurate to administer the insulin based on units per gram of body weight. Based on average weights for the species, conversions for reported dosages range from 0.000067 units/gm to 0.00333 units/gm in the budgerigar. Based on glucose test results, clinical signs, and long-term clinical survival, Harrison recommends administering insulin twice a day.

## Preparation of Insulin

NPH U40 insulin is diluted with lactated Ringer's solution or sterile water. A preparation is made with 0.3 cc insulin mixed with 2.7 cc lactated Ringer's solution, which results in 4 units/cc insulin for larger birds. A further dilu-

**Table 42–6.** DOSAGE FLUCTUATIONS OF INSULIN FOR A SINGLE INDIVIDUAL (Harrison)

| | |
|---|---|
| Cockatiel: | 0.0014 U/gm to 0.0025 U/gm |
| Toco Toucan: | 0.00001 U/gm to 0.00011 U/gm |
| Budgerigar: | 0.00013 U/gm to 0.0033 U/gm |

tion using 0.1 cc of this solution (0.4 units) in 0.9 cc lactated Ringer's results in 0.04 units/0.1 cc for smaller birds. Diluted insulin can be stored under refrigeration for three to four months.

An overdose of insulin in a five-year-old cockatiel that had been receiving the medication twice a day for three years was reflected in clinical signs of wing droop and sleepiness. Hypoglycemic birds may fall off the perch, shiver, or convulse. These signs respond to oral sugar supplementation. Owners should be advised to monitor the droppings to maintain a slight positive glucose level and adjust the insulin dose accordingly.

Stogdale has prepared a useful client education publication designed for owners of diabetic dogs and cats. Many of the principles apply to the avian diabetic patient.

## EXCESSIVE IRON STORAGE DISEASE

(Hemochromatosis)

In a clinical situation hemochromatosis is seen primarily in mynah birds, although it is reported to occur in Birds of Paradise and Quetzals. The clinical signs include dyspnea, cachexia, and severe abdominal distention from ascites. Severe cases must be handled with care to avoid precipitating the death of the bird. Radiography often reveals severe hepatomegaly. Clinical chemistries may indicate low total protein, high bilirubin, and increased SGOT, SGPT, LDH, and ALP.

Abdominocentesis may offer temporary relief of the dyspnea, but hypoproteinemic patients may suffer from further loss of protein contained in the ascitic fluid. A characteristic yellow fluid that may be recovered is described as a modified transudate with macrophages and mononuclear cells present and a specific gravity of 1.013 to 1.018. Abdominocentesis may be followed by oral administration of a diuretic (furosemide 1 mg t.i.d.) in severely dyspneic birds. Birds with symptoms of a lesser degree can be given dexamethasone (0.01 mg t.i.d. initially, decreased over a 14-day period to 0.01 mg every 48 hours). Antibiotics and diet change (multivitamins and high-quality protein) have been included in some cases (Harrison), but birds do not recover. Weekly phlebotomies are the treatment of choice in human hemochromatosis and were effective in the temporary management of one Greater Indian Hill Mynah.*

The cause has been speculated to be dietary, but from his investigations Gosselin "demonstrated that excessive iron storage in mynah birds was not diet induced. . . ."

## ALLERGIES

Several cases in Harrison's experience suggest that avian patients suffer from a variety of allergies, although these cases have not been confirmed.

A number of birds with clinical signs of

*Dr. S. Avgeris, Personal communication.

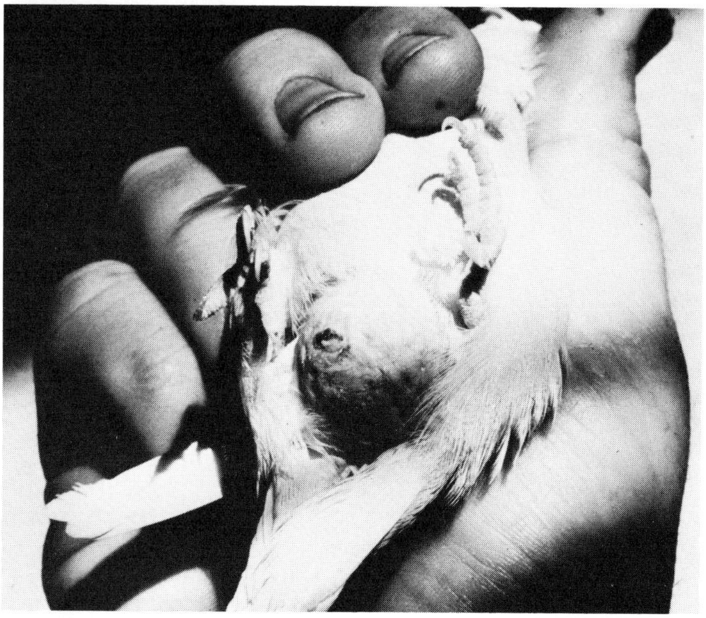

**Figure 42–2.** An endocrine imbalance has been implicated as a possible cause of abdominal herniation in female budgerigars.

tenesmus, flatulence, and inflammation of the cloaca have responded solely to the removal of sunflower seeds from the diet. This same therapy was successful in treating an African Grey for the passing of whole seeds. Sneezing in several species has ultimately been associated with and responded to cessation of the cage being covered with a horsehair blanket; cessation of the use of a heavily perfumed liquid dish detergent; and cessation of the use of fresh avocado branches in the cage as a "treat." In the latter case, a 10 per cent eosinophil count occurred in the differential during the sneezing episodes. In another case, a bird was reported to sneeze when housed with other birds. The sneezing stopped when the bird was housed alone. Tracheal washing cytology during the sneezing episodes showed a high percentage of eosinophils, which disappeared when the housing was changed.

## ABDOMINAL HERNIA XANTHOMATOSIS

Abdominal hernia xanthomatosis is a relatively uncommon condition of female budgerigars which is apparently characterized by a weakening of the abdominal musculature and herniation of the abdominal organs. It is suspected to be related to ovarian or oviductal abnormalities. The use of testosterone has resulted in some shrinkage of the hernial sac. Surgical correction of the condition is reported to be only temporary or not practical. In Harrison's experience, the tissues most often affected by this particular form of xanthoma are diffuse and unsuturable. One attempt at surgical removal of affected tissue resulted in massive abdominal organ exposure precluding closure, and euthanasia was advised.

## RECOMMENDED READINGS


Altman, R. B.: Heterologous blood transfusions in avian species. Proceedings of the Annual Meeting of the Association of Avian Veterinarians, San Diego, 1983, pp. 28–32.

Altman, R. B.: Noninfectious diseases. In Fowler, M. E. (ed.): Zoo and Wild Animal Medicine. 2nd ed. Philadelphia, W. B. Saunders Company, 1985, pp. 497–512.

Altman, R. B.: Diseases of the avian urinary system. In Kirk, R. W. (ed.): Current Vet Therapy VI. Philadelphia, W. B. Saunders Company, 1977.

Altman, R. B., and Kirmayer, A. H.: Diabetes mellitus in the avian species. J. Am. Anim. Hosp. Assoc., 12:531–537, 1976.

Astier, H.: Thyroid gland in birds: Structure and function. In Epple, A., and Stetson, M. H. (eds.): Avian Endocrinology. New York, Academic Press, 1980, pp. 167–189.

Bauck, L.: Renal disease in the budgerigar—a review of cases seen at the Western College of Veterinary Medicine. Proceedings of the International Conference on Avian Medicine, Association of Avian Veterinarians, Toronto, Canada, 1984.

Blackmore, D. K.: The incidence and etiology of thyroid dysplasia in budgerigars (*Melopsittacus undulatus*). Vet. Rec., 75:1068–1072, 1963.

Blackmore, D. K., and Cooper, J. E.: Diseases of the endocrine system. In Petrak, M. L. (ed.): Diseases of Cage and Aviary Birds. 2nd ed. Philadelphia, Lea & Febiger, 1982, pp. 478–490.

Bozeman, L. H., et al.: Characterization of a papovavirus isolated from fledgling budgerigars. Proceedings of the 52nd Northeastern Conference on Avian Diseases, Cornell University, Ithaca, NY, 1981.

Dawson, C. O., et al.: Air sac and renal mucormycosis in an African Grey Parrot (*Psittacus erithacus*). Avian Dis., 20:593, 1972.

Douglass, M.: Diabetes mellitus in a Toco Toucan. Mod. Vet. Pract., 293–295, 1981.

Ensley, P. K., et al.: Congestive heart failure in a Greater Hill Mynah. J. Am. Vet. Med. Assoc., 175(9):1010–1013, 1979.

Epple, A., and Stetson, M. H. (eds.): Avian Endocrinology. New York, Academic Press, 1980.

Gosselin, S. J.: Pathophysiology of excessive iron storage in mynah birds. J. Am. Vet. Med. Assoc., 183(11):1238–1240, 1983.

Harrigan, K. E.: Aviary and Caged Birds. Proceedings No. 55, Postgraduate Committee in Veterinary Science, University of Sydney, Australia, 1981.

Harrison, G. J., et al.: A clinical comparison of anesthetics in domestic pigeons and cockatiels. Proceedings of the Annual Meeting of the Association of Avian Veterinarians, Boulder, CO, 1985, pp. 7–22.

Hazelwood, R. L.: Pancreatic hormones, insulin/glucagon molar ratios and somatostatin as determinants of avian carbohydrate metabolism. J. Expt. Zool., 232:647–652, 1984.

Holmes, W. N., and Cranshaw, S.: Adrenal cortex: Structure and function. In Epple, A., and Stetson, M. H. (eds.): Avian Endocrinology. New York, Academic Press, 1980, pp. 271–300.

Kollias, G. V.: Liver biopsy techniques in avian clinical practice. Vet. Clin. North Am. [Sm. Anim. Pract.], 14(2):287–298, 1984.

Lothrop, C. D.: Diagnosis of adrenal diseases in caged birds. Proceedings of the International Conference on Avian Medicine, Association of Avian Veterinarians, Toronto, Canada, 1984.

Lothrop, C. D.: Diseases of the thyroid gland in caged birds. Proceedings of the International Conference on Avian Medicine, Association of Avian Veterinarians, Toronto, Canada, 1984.

Lothrop, C. D., et al.: Endocrine diagnosis of feathering problems in psittacine birds. Proceedings of the American Association of Zoo Veterinarians, Tampa, FL, 1983, pp. 144–147.

Minsky, L., and Petrak, M.: Metabolic and miscellaneous conditions. In Petrak, M. L. (ed.): Diseases of Cage and Aviary Birds. 2nd ed. Philadelphia, Lea & Febiger, 1982.

Neumann, U., and Kummerfield, N.: Neoplasms in budgerigars (*Melopsittacus undulatus*): Clinical, pathomorphological and serological findings with special consideration of kidney tumors. Avian Pathol., 12:353, 1983.

Panigrahy, B., et al.: Hemorrhagic disease in canaries (*Serinus canarius*). Avian Dis., 28(2):536–541, 1984.

Petrak, M. L.: Liver disease in birds. Proceedings in Aviculture Veterinary Medicine, American Federation of Aviculture, Las Vegas, NV, 1980.
Randell, M. G., et al.: Hepatopathy associated with excessive iron storage in mynah birds. J. Am. Vet. Med. Assoc., 179(11):1214–1217, 1981.
Ratcliffe, H. L.: Hepatitis, cirrhosis, and hepatoma in birds. Cancer Res., 21:26, 1961.
Ratcliffe, H. L.: Tuberculosis in captive wild birds. Increased frequency with increased population density. J. Albert Einstein Med. Ctr., 8:138–142, 1960.
Richkind, M.: Hormonal influences on normal and abnormal feathering and moult. Bird World, Aug.-Sept., 42–56, 1982.
Ringer, R. K.: Thyroids. In Sturkie, P. H. (ed.): Avian Physiology. 3rd ed. New York, Springer-Verlag, 1975, pp. 348–358.
Rosskopf, W. J., and Woerpel, R. W.: Remission of lipomatous growths in a hypothyroid budgerigar in response to L-thyroxine therapy. VM/SAC, 78:1415–1418, 1983.
Rosskopf, W. J., et al.: Chronic endocrine disorder associated with inclusion body hepatitis in a Sulphur-crested Cockatoo. J. Am. Vet. Med. Assoc., 179:1273–1276, 1981.
Rosskopf, W. J., et al.: Normal thyroid values for common pet birds. VM/SAC, 77:409–412, 1982.
Rosskopf, W. J., et al.: Pathogenesis, diagnosis and treatment of adrenal insufficiency in psittacine birds. Cal. Vet., 5:26–29, 1982.
Ryan, C. P., et al.: Diabetes mellitus and islet cell carcinoma in a parakeet. J. Am. Anim. Hosp. Assoc., 18:139–142, 1982.
Sitbon, G., et al.: Diabetes in birds. Horm. Metab. Res., 12:109, 1980.
Skadhauge, E.: Osmoregulation in Birds. New York, Springer-Verlag, 1981.
Spira, A.: Clinical aspects of diabetes mellitus in budgerigars. Proceedings of the Annual Meeting of the American Animal Hospital Association, 1981.
Stacpoole, P. W.: The glucagonoma syndrome: Clinical features, diagnosis and treatment. Endo. Rev., 2:347–360, 1981.
Stogdale, L.: Diabetes mellitus: An owner information guide. Kal Kan Forum, Kal Kan Foods, Winter, 1984.
Sturkie, P. D.: Avian Physiology. 3rd ed. New York, Springer-Verlag, 1975.
Voikevic, A. A.: The Feathers and Plumage of Birds. New York, October House, 1966, pp. 91–215.
Waller-Pendleton, E., et al.: An inclusion body pancreatitis in *Agapornis*. Proceedings of the 32nd Western Poultry Disease Conference, University of California, Davis, 1983.

## ACKNOWLEDGMENTS

I wish to acknowledge Drs. Rosskopf, Woerpel, Reynolds, and Helman, and especially Dr. K. Flammer, for cooperation in studying the hormonal causes of avian diabetes.

CLINTON LOTHROP

# Chapter 43

# ZOONOTIC DISEASES

RICHARD H. EVANS
DANIEL P. CAREY

Pet bird species may be involved in zoonotic diseases, with chlamydiosis being the most common of these. The practicing veterinarian may be required to report these incidents to the appropriate authorities within the United States Department of Agriculture, Animal and Plant Health Inspection Service (USDA/APHIS).

As reportable animal diseases vary from state to state, it is recommended that the clinician contact his state veterinarian for information on those diseases reportable in that state and the proper methods of reporting the disease. The federal authorities would be interested in any avian infectious disease suggestive of an exotic disease for poultry.

Human reportable diseases of interest to the avian veterinarian include chlamydiosis and tuberculosis. The local state health department should be contacted; it in turn is required to report them to the Centers for Disease Control (CDC) in Atlanta, Georgia.

The reader is referred to the appropriate sections for discussion on how the following zoonotic diseases manifest in birds.

### CHLAMYDIOSIS [PSITTACOSIS, PARROT FEVER, ORNITHOSIS] *(Chlamydia Psittaci)*

The host range for chlamydiosis includes birds, humans, and other mammals. Humans contract this disease by ingestion or inhalation of chlamydial organisms from birds. Incubation may be as short as 48 hours but is usually five days to two weeks. Stress, concurrent disease, and age (very old or very young) affect the severity, which can range from inapparent to fatal. It can be acquired from "apparently" healthy birds carrying a latent infection (see Chapter 35, Chlamydia).

Typical symptoms in humans include fever and chills, pneumonia, headache, weakness, and fatigue. Other symptoms seen are myalgia, chest pain, anorexia, nausea and vomiting, dyspnea, and diaphoresis (profuse perspiration). It is an atypical pneumonia and therefore symptomatically typical of "flu" lasting 7 to 10 days. Antibodies are produced in humans but are transitory. Human diagnosis relies upon recognizing the clinical signs and testing for complement fixation antibody on acute and convalescent serum samples. A titer of 1:32 or greater is considered presumptive evidence, although there is cross-reaction between *Chlamydia psittaci* and *Chlamydia trachomatis* (human venereal organism). A new commercial ELISA test for *C. trachomatis* antigen will cross-react with *C. psittaci*, which would be valuable in birds but nonspecific in humans. Some humans have background titers without prior exposure to either organism. Culture is definitive but often difficult on human cases owing to improper sample choice or submission. Sputum, frozen immediately to −70°F, is needed, whereas saliva is often submitted.

### GIARDIASIS (*Giardia* sp.)

Many species of birds can be infected with *Giardia*, including several heron species, house sparrows, turkey vultures, meadowlarks, and psittacine birds. Giardia are not nearly as species-specific (if at all) as once thought, and there is little evidence for labeling each new host isolate as a new species without animal transmission studies.

Infection is acquired by direct fecal to oral transmission of as little as 10 *Giardia* cysts. Food-borne infection is suspected but as yet unproven. There is no intermediate host. Epidemiologic investigations in Colorado indicate that human hikers contaminated the water in mountain lakes, and beavers living nearby became infected, acting as a source of natural infection to man. Diarrhea and malabsorption are the primary effects observed in clinical giardiasis.

Diagnosis relies on microscopic demonstration of cysts or trophozoites in the stool. Either direct-stained (Lugol's iodine) or unstained wet mounts or concentration techniques followed by wet mount examination are employed. Sugar or salt flotations will commonly shrink cysts, making them unrecognizable. Sedimentation techniques utilizing zinc sulfate centrifugation or formalin-ether techniques appear to be best. Fixation solutions such as polyvinyl alcohol or 2 to 5 per cent formalin are recommended; 10 per cent formalin may disrupt cysts. Cysts may be intermittently excreted in the stools; thus serial stool examinations are recommended to completely rule out giardiasis. Recent immunologic and serologic tests have been developed, including immunofluorescence, ELISA, and countercurrent electrophoresis for *Giardia* fecal antigen. These appear to be specific and reproducible indicators of the disease.

Hygiene must be good to avoid the dirty, overcrowded environments that probably promote transmission. Every effort should be made to improve food and water quality. Contaminated water must be boiled to kill the chlorine-resistant cysts. Rats and other vectors need to be controlled in aviaries.

## TUBERCULOSIS *(Mycobacterium avium)*

Transmission of *M. avium* from caged pet birds to humans has been reported and presumably is the result of ingestion of either feces or fecal aerosols.

*M. avium* maintenance hosts include domestic poultry and a myriad of other avian species, both domestic and wild. In wild species, the incidence has been reported to correlate with the proximity to captive birds.

In humans, *M. avium* infection is very rare, as humans are considered highly resistant. In less than 100 confirmed cases worldwide, concurrent invasive or immunosuppressive disease appears to be common and is thought to be the main predisposing cause of *M. avium* infection in humans.

Today *M. avium* infection in commercial poultry operations is rare but can be seen in birds, usually adults, in backyard flocks kept in less than optimal hygiene condition. There is still much controversy over whether wild birds serve as a primary source of infection for poultry. Most sources consider carrier birds or sick, shedding members of the flock to be the most important sources of infection. However, wild birds can spread infection between poultry farms and other food animal operations by themselves becoming infected from poultry. The same controversy exists in zoos, where it appears that any species of bird may become infected, and such birds serve as a source of infection for the entire zoo population of both birds and mammals.

The postmortem diagnosis of tuberculosis is seldom difficult because of the rather characteristic lesions. Cytologic or histologic staining usually demonstrates acid-fast organisms in the lesions. Antemortem diagnosis is more difficult, but positive results have been accomplished by culture, serology, radiology, endoscopy, and biopsy. *Mycobacterium* can be cultured by specific methods, but this may take several months to complete. Fluorescent antibody techniques, animal inoculations, and bacteriophage typing have also been used. Supporting data include elevated white blood counts and hyperproteinemia.

In humans, a combination of radiology, sputum cytology and culture, tuberculin testing, and immunologic testing such as lymphocyte transformation tests and ELISA serology have been used for definitive diagnosis. Tuberculin testing for *M. avium* was attempted at the National Zoo, but results were not definitive.

## SALMONELLOSIS (*Salmonella* sp.)

*Salmonella* have been isolated from virtually every species of domestic and wild animal, including birds. Human salmonellosis is acquired by ingestion.

Stool culture is diagnostic of salmonella presence in all species, although false-negatives do occur, particularly in asymptomatic carriers. Multiple samples may be necessary. In animals, postmortem culturing of fresh specimens will find salmonellae in enlarged spleens and livers and frequently in mesenteric lymph nodes. Serologic testing is available for poultry, but not for psittacine species.

Proper personal hygiene after handling pets and animals and prompt veterinary attention for animals with diarrhea will reduce the incidence of direct transmission.

## CANDIDIASIS *(Candida albicans)*

This organism is a resident of human oral, intestinal, and urogenital tracts and is found in many avian species as well as other animals.

Transmission is usually from human to human but can be from human to animal or vice versa. Contact with secretions may communicate the disease. The gastrointestinal form in all animals is very similar to the disease in birds, manifestations being dysphagia, anorexia, enteritis, and ulcers. Humans usually have superficial disease such as thrush, intertrigo, vulvovaginitis, or paronychia. Ulcers and/or pseudomembranes may be noted in the oral, gastrointestinal, and urogenital forms. Hematogenous spread may occur with respiratory involvement, septicemia, endocarditis, and meningitis. Diagnosis is by demonstration of large numbers of yeasts in exudate. One may avoid damp, molding organic materials and ensure proper ventilation and disinfection of animal quarters in an attempt at prevention. However, this disease is usually secondary to antibiotic therapy or another disease, since the mere exposure to or presence of *C. albicans* does not normally produce disease.

## NEWCASTLE DISEASE

In man, Newcastle disease has been reported most commonly in laboratory workers or poultry pathologists and is usually manifested by mild to severe conjunctivitis or sinusitis. Recovery is usually uncomplicated, occurring within 7 to 20 days. NDV has been isolated from affected humans, and seroconversion has been documented.

## YERSINIA ENTEROCOLITICA

*Yersinia enterocolitica* has been associated with an increasing number of acute to subacute gastrointestinal disturbances in man and animals since the early 1960's.

World wide, *Y. enterocolitica* has been isolated from Canada geese, Pekin Ducks, robins, canaries, pigeons, herring gulls, other birds, and animals. Only the dog and pig have been implicated as possible sources of human infection by epidemiologic studies. Serotypes of other animals almost invariably have not appeared as human pathogens. At present, wildlife in the United States has not been documented to carry human pathogens.

## CAMPYLOBACTERIOSIS *(Campylobacter jejunii)*

*C. jejunii* has been isolated from chickens, ducks, migratory waterfowl, sparrows, starlings, pigeons, blackbirds, turkeys, and finches, as well as mammals. Some of these isolates are distinct biotypes and may not be associated with human disease.

Some of the modes of transmission include ingestion of contaminated poultry or contaminated water, direct contact with infected animals, and person-to-person spread via the fecal/oral route. Efforts to determine the source of infection find only a small number due to animals. Humans are not considered to be a reservoir, however.

Campylobacteriosis in humans is an invasive and inflammatory disease involving the small and large intestines in an enteritis that may be self-limiting or very severe. Clinically it closely resembles salmonellosis, acute appendicitis, or shigellosis.

Specimens should be plated quickly or the organism may die. Direct microscopic fecal examination with dark-phase illumination will often reveal the motile, curved rods.

Good hygiene when handling birds should prevent most cases.

## RABIES

All warm-blooded animals are reported to be susceptible to rabies virus infection, but susceptibilities vary greatly. Wild birds have been found to contain antibody titers to rabies virus, and fatal disease has been produced experimentally in some birds. In experiments with a Great Horned Owl, which was allowed to feed on a rabid skunk, the owl developed antibody to rabies virus but not clinical disease and shed virus in its oral secretions for several months after infection. Despite this there has been no evidence of rabies transmission from birds to humans.

## RECOMMENDED READINGS


Baer, G. M.: The Natural History of Rabies. New York, Academic Press, 1975.

Bottone, E. J.: *Yersinia enterocolitica:* A panoramic view of a characteristic microorganism. CRC Critical Reviews in Microbiology, 1977, pp. 211–224.

Davison, G. J.: Acute diarrhea and dysentery caused by *Yersinia enterocolitica.* In Burford, G. H., and Conver, D. H. (eds.): Pathology of Tropical and Extraordinary Diseases. Washington, DC, Armed Forces Institute of Pathology, 1976, pp. 162–164.

Emerson, J. K.: Psittacosis. J. Am. Vet. Med. Assoc., 180:612–613, 1982.

Gale, N. B.: Tuberculosis. In Davis, J. W., et al. (eds.): Infectious and Parasitic Diseases of Wild Birds. Ames, IA, Iowa State University Press, 1978, pp. 84–94.

Giardiasis: Parasitic Zoonoses. CRC Handbook Series in Zoonoses. Boca Raton, FL, CRC Press, 1982.

Giles, N., and Carter, M. J.: *Yersinia enterocolitica* in budgerigars. Vet. Rec., October, 1980.

Karlson, A. G.: Avian tuberculosis. In Montali, R. J. (ed.): Mycobacterial Infections of Zoo Animals. Washington, DC, Smithsonian Press, 1978, pp. 21–23.

Kirkpatrick, C. E., and Farrell, J. P.: Giardiasis. Comp. Cont. Ed., 5:367–377, 1982.

Kleeburg, H. H.: Tuberculosis and other mycobacterioses. In Hubbert, W. T., et al. (eds.): Diseases Transmitted from Animals to Man. Springfield, IL, Charles C Thomas, Publisher, 1975, pp. 303–360.

Kulda, J., and Mohynkova, E.: Flagellates of the human intestine and of the intestines of other species. In Krier, J. (ed.): Parasite Protozoa. New York, Academic Press, 1978, pp. 1–138.

Langford, E. V.: *Yersinia enterocolitica* isolated from animals in Fraser Valley of British Columbia. Can. Vet. J., 5:109–113, 1972.

Manson-Bahr, P. E. C.: Tuberculosis; buruli olga. In Manson-Bahr, P. E. C. (ed.): Manson's Tropical Disease. London, Bailliere Tindall, 1982, pp. 323–330.

Manson-Bahr, P. E. C.: Amoebiasis, giardiasis and balantidiasis. In Manson-Bahr, P. E. C. (ed.): Manson's Tropical Disease. London, Bailliere Tindall, 1982, pp. 121–145.

Meyer, K. F.: The ecology of psittacosis and ornithosis. Medica, 21:175–206, 1942.

Newcastle disease. In Torten, M., and Kaplan, W. (eds.): Viral Zoonoses. CRC Handbook Series in Zoonoses. Boca Raton, FL, CRC Press, 1981.

Potter, M. (Center for Disease Control, Atlanta, GA): Personal communication, 1985.

Prescott, J. F., and Monroe, D. L.: *Campylobacter jejunii* enteritis in man and domestic animals. J. Am. Vet. Med. Assoc., 181:1524–1530, 1982.

Rigley, C. E., et al.: The isolation of salmonellae, Newcastle disease virus and other infectious agents from quarantined imported birds in Canada. Can. J. Comp. Med., 45:366–370, 1981.

Risser, A. C., et al. (eds.): Proceedings of the First International Birds in Captivity Symposium. North Hollywood, CA, International Foundation for Conservation of Birds, 1978, pp. 185–195.

Schachter, J.: Psittacosis. In Hubbert, W. T., et al. (eds.): Diseases Transmitted from Animals to Man. Springfield, IL, Charles C Thomas, Publisher, 1975, pp. 369–386.

Scholtens, R. G., et al.: The nature and treatment of giardiasis in parakeets. J. Am. Vet. Med. Assoc., 2:170–173, 1982.

Sikes, R. K.: Rabies. In Hubbert W. T., et al. (eds.): Diseases Transmitted from Animals to Man. Springfield, IL, Charles C Thomas, Publisher, 1975, pp. 871–896.

Storz, J.: Chlamydia and Chlamydia Induced Diseases. Springfield, IL, Charles C Thomas, Publisher, 1971, p. 358.

Tuberculosis. In Stoemer, H., et al. (eds.): CRC Handbook in Zoonoses. Boca Raton, FL, CRC Press, 1979.

*Section Seven*

# SURGERY

# Chapter 44

# EVALUATION AND SUPPORT OF THE SURGICAL PATIENT

GREG J. HARRISON

The clinical signs and results of the physical examination or radiology in some cases may indicate that surgery is required for therapy. It is appropriate that some clinicians approach the prospect of avian surgery with a degree of hesitation. Those practitioners who lack the training and skill would better serve their avian patients by referring surgical cases to more experienced colleagues.

Prior to the development of successful techniques for use in birds, patients frequently died from surgical attempts. The justifications for these initial failures ranged from "birds can't stand such trauma" to "we don't have the right equipment." Although the possibility of successful surgical management of avian diseases, especially those involving intra-abdominal access, was enhanced by the incorporation of new anesthetic protocols and specialized techniques as described in succeeding chapters, significant progress was not made until avian surgeons acknowledged the role of presurgical conditioning in the ability of the avian patient to withstand some extensive procedures.

In general, the practitioner must assess a number of factors to determine the feasibility of a safe and successful surgical procedure: the age and physical condition of the patient, the possibility of a co-existing disease, and the extent and significance of the surgical problem. Some nonelective procedures may have to proceed without the benefit of a complete evaluation of the patient. Similarly, diagnostic procedures such as radiology or relatively simple invasion for surgical sexing may be conducted under light anesthetic immobilization with little more than a preliminary physical examination. In most surgical cases, however, sufficient time is available to appraise a number of parameters in order to make a judgment regarding the patient's potential tolerance of anesthesia and surgical intervention. In some cases, surgery may be the only remaining therapeutic alternative for the chronically ill patient. Specific conditioning and rehabilitation procedures may be instituted for days or even weeks prior to the anesthetic/surgical event to improve the physiologic stability of the patient.

## PRESURGICAL EVALUATION AND CONDITIONING OF THE PATIENT

### Evaluation

A thorough physical examination, clinical pathology, radiology, and other diagnostic procedures are components of presurgical evaluations in birds. Table 44–1 lists specific tests for evaluation of a patient prior to a basic surgical procedure requiring anesthesia. Some particular factors must be taken into account depending upon the anesthetic selected (see Chapter

**Table 44–1.** PARAMETERS TO ASSESS IN THE SURGICAL PATIENT

**BASIC**
- Complete physical examination
  - General condition
  - Weight
  - Respiratory recovery time
- Gram's stain of the feces and choana
- Fecal parasite examination

**OPTIONAL**
- Radiograph
- Hematology
  - Packed cell volume
  - Total protein
  - White blood cell count
  - Differential
  - Clotting time
  - Evaluate thrombocytes, red blood cell morphology
  - Check for hemoparasites
- Culture and sensitivity (if suggested by Gram's stain)
- Serum chemistry panel
  - Glucose
  - Uric acid
- Body temperature
- Electrocardiogram

45, Anesthesiology). The smaller the patient, the more critical the presurgical assessment becomes.

One valuable clinical assessment is respiratory recovery time (see Chapter 7, Preliminary Evaluation of a Case). A return to normal respiration in three to five minutes following capture and at least two minutes of handling would suggest respiratory vigor sufficient for most anesthetic and surgical procedures. Although metabolic acidosis and electrolyte balances are not commonly evaluated in pet birds, these have been addressed in raptors[11] and provide valuable information on patient status.

The merit of serum chemistry tests as a presurgical evaluation method is somewhat controversial unless the test results widely deviate from reference values for the species. In one anesthesia study of cockatiels and pigeons,[10] serum chemistry values did not provide a reliable means of predicting the ability of these birds to withstand serial anesthetic and multiple biopsy procedures. However, of the three birds that died in this study, two had preanesthetic values of >700 cholesterol, >650 SGOT, and >600 LDH. Although other individuals tested showed SGOT and LDH levels of this magnitude and higher, it appears that the combination with the elevated cholesterol was significant. The third bird had an extremely high preoperative uric acid value. Highly variable values were noted in the birds following the surgical stress.

The optional tests in Table 44–1 are indicated prior to more complicated procedures in which bleeding and extensive tissue handling are anticipated.

The liver is a prime target for investigation in presurgical radiographs, as hepatomegaly is frequently accompanied by coagulopathies. Serial radiographs may illustrate progressive decreases in organ size from presurgical therapy (Fig. 44–1).

The dehydrating characteristics of Gastrografin (see Chapter 14, Radiology) appear to alter organ size and clinical symptoms when administered presurgically following diagnostic scout films. A repeat radiograph in 24 to 48 hours has in several cases shown the serendipity of reduction in size of a swollen liver, intestines, or kidneys, thereby obviating the need for diagnostic surgery in those cases.

It is valuable to consider some assessment of clotting time. Rabbit thromboplastin routinely used in evaluating prothrombin times in mammals is not effective for avian determinations.[6] Because avian thromboplastin may not be available, the practitioner may have to depend upon

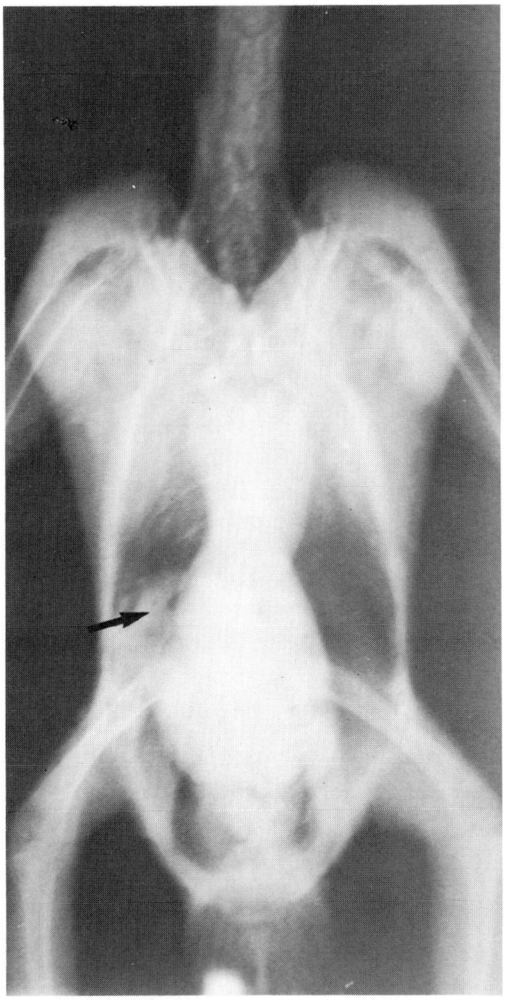

**Figure 44–1.** Radiographs provide a general organ assessment and identification of possible lesions, such as air sac granulomas, prior to surgery.

simple clinical methods of assessing clotting response. Penetration of the ulnaris wing vein with a 30-gauge needle for presurgical blood samples and application of direct pressure for one minute should result in normal clotting at the site. More prolonged times may suggest a bleeding disorder, and surgery should be postponed until the condition is corrected. Bleeding from the skin site following the plucking of a feather or following administration of intramuscular injectable agents may also suggest an abnormality. Alternatively, drawing a needle through a drop of blood on a slide at periodic intervals and recording the clotting time is another simple technique that can be used with avians. Administration of vitamin $K_1$ prior to a procedure to assist clotting may be required in some cases, and blood should be available for possible transfusion if necessary.

Some practitioners recommend recording a

presurgical body temperature and electrocardiogram (ECG) to provide baselines for monitoring during the procedure.[4] Halothane and ketamine anesthesia were found to produce a rapid decrease in heart rate when electrocardiograms were recorded during surgery.[5] The rate rose immediately upon cessation of anesthesia. In a separate study, the heart rate increased in pigeons and cockatiels with methoxyflurane anesthesia and increased slightly in cockatiels with halothane.[10] The ECG, which establishes the heart rate and demonstrates any cardiac abnormalities, may be particularly indicated in ascitic patients.

Further investigation into the preoperative status of the clinically ill patient may include culture and sensitivity studies of specific lesions (e.g., oral, respiratory system), localized radiographic survey (e.g., ventrodorsal and lateral of head, thorax, abdomen, and/or extremities if warranted), diagnostic endoscopy, biopsy and histopathologic/cytologic evaluation of samples, and, in some cases, endocrine function tests.

## Preconditioning the Surgical Patient

The most common pet avian surgeries in the experience of the author are performed for resolution of abdominal disorders, followed in decreasing frequency by oral, skin, crop, and respiratory lesions (e.g., air sacs, trachea). Of the abdominal disorders, removal of acutely ingested foreign bodies usually requires minimal patient preconditioning, whereas tumor removal, egg-related peritonitis, biopsies, and exploratory endoscopy may require some preliminary rehabilitative steps. Chronic disease states may have exhausted organ reserves; therefore, it behooves the practitioner to revitalize the organs as much as possible and eliminate any complicating factors prior to surgical attempts.

Injectable multivitamins, fluid therapy, hematinics, anabolic agents, lactulose, digestive aids, and antimicrobials may be advised (see Chapter 27, What To Do Until A Diagnosis Is Made). Dietary management is often indicated.

If the packed cell volume (PCV) is below 25 to 30, blood transfusions and other supportive therapy are required; if the PCV is above 55 to 60, the patient requires supportive fluids. An estimated total protein below 3.0 gm/dl requires therapy prior to any surgical attempts. A glucose value below 200 mg/dl suggests hypoglycemia and 5 per cent glucose should be provided in the fluid therapy.

### *Antibiotics*

Administration of antibiotics in conjunction with surgery is not necessary for prevention of infection at the surgical site but may prevent postsurgical septicemia from gram-negative bacteria present in the gastrointestinal tract. If antibiotics are employed, the administration should be timed for peak blood levels to coincide with the surgical assault. High potency antibiotics are not usually used for this preventive measure, unless clinical signs indicate a culture and sensitivity for specific therapy. If general preventive antibiotics are used, this author generally prefers ampicillin. Severely compromised birds may be maintained on antibiotics for 5 to 15 days postoperatively.

One practitioner attempts gut sterilization by medicating the drinking water with injectable gentamicin for several days prior to elective gastrointestinal tract procedures.[13]

### *Other Preanesthetic Medications*

No other preanesthetic medications are used by this author. Intramuscular injections of atropine as a preanesthetic in conjunction with parenteral anesthetics (ketamine/xylazine) or isoflurane are avoided because the respiratory secretions thicken into a viscous mucus and tend to interfere with a procedure. This atropine protocol may be indicated, however, in patients with abnormal cardiac rhythms or to increase the stength of the heartbeat. Intramuscular diazepam is another preanesthetic that is reported to reduce stress-related cardiovascular changes of some inhalation agents. Dexamethasone has been administered by some practitioners prior to particularly risky procedures.[13]

### *Fasting*

The varying opinions in the literature regarding presurgical fasting of the avian patient are presumably based on experiences with birds under various anesthetic agents. One opinion is that unless crop surgery is performed, birds should not be fasted because poor glycogen storage in the liver will predispose the bird to hypoglycemia. Small birds are suspected to be susceptible to glycogen depletion in 12 to 24 hours, whereas larger parrots may be affected in 24 to 48 hours. On the other hand, vomiting, regurgitation, and aspiration pneumonia have been reported in unfasted avian surgical patients.[1] As prevention, some practitioners intubate all anesthetic patients and support the tongue with a bent paper clip to prevent occlusion of the glottis. This author believes that in some cases endotracheal tubes may cause respiratory irritation and may easily occlude because of their small lumen size.

Very limited fasting (0 to 3 hours) is the current protocol in use by this and other authors[13] in conjunction with intravenous parenteral (ketamine/xylazine combination) or inhalation anesthetics (isoflurane) (see Chapter 45, Anesthesiology). One hour is usually sufficient for emptying the crop in small birds. Food and water are routinely removed two to three hours prior to anesthesia for elective procedures in large psittacines. Birds requiring emergency surgical repair of traumatic injuries are anesthetized immediately. The crops of birds with crop-emptying disorders are flushed, if possible, and the contents of crops with fistulas can often be expressed prior to administration of anesthesia. One method of preventing the hazard of vomiting and aspiration is to pass a tube into the esophagus with or without an endotracheal tube. Positioning the patient with its head elevated is another method of preventing passive regurgitation. In extreme emergency situations, the entrance to the esophagus can be temporarily blocked with cotton to prevent regurgitation.

### *Special Rehabilitation*

Birds that have egg-related peritonitis complicated by fluid in the abdominal cavity require presurgical treatment of at least 24 hours' duration. A tap with a 20-gauge needle may produce as much as 3 to 4 ml of straw-colored fluid in a budgerigar. If necessary, a Seaton or Penrose drain may be installed in the abdomen to facilitate complete drainage. Supportive care during this time may include injectable antibiotics, multivitamins, lipotrophic factors, and intravenous fluids. Steroids should be used cautiously with rapidly metabolized low doses at first (prenisolone sodium succinate), then longer-acting forms if necessary. Some patients may require a serum transfusion if high-protein fluid loss causes the serum protein level to drop. Failure to drain the fluids preoperatively in any apparently ascitic patient may result in increased stress and possible obstruction of the air sacs when the abdominal incision is made. A midventral approach is essential in these cases (see Chapter 47, Selected Surgical Procedures).

Preconditioning of up to three weeks' time may be necessary for safe removal of uterine tissue during an elective hysterectomy. Suppression of follicular activity and subsequent decreased vascularization have been found to respond to decreased daily photoperiods and low fat diet. Some tumors may respond to direct administration of steroid (dexamethasone).

Because obese avians requiring abdominal surgery often have fatty livers, a prolonged period of low-fat dietary intake and the metabolic stimulation of thyroid supplementation may support these patients (see Chapter 42, Miscellaneous Diseases).

## PATIENT SUPPORT DURING THE SURGICAL PROCEDURE

### Patient Monitoring

The degree of monitoring required in some procedures may depend in part on the choice of anesthetic (see Chapter 45, Anesthesiology). Recommendations in the literature have ranged from a team of two in addition to the surgeon (Fig. 44–2) with multiple monitoring devices (Doppler transducer over brachial artery, oscilloscope or ECG, esophageal stethoscope, cloacal temperature probe, indirect blood pressure pediatric cuff over various arteries, blood gas analyzer) to the other extreme of a solo practitioner's taping a feather over the nares to monitor respiratory rate. Critical suppression of body temperature and heart rate may not be noticed without proper monitoring.[4]

Of the commonly used inhalation anesthetics delivered by precision vaporizer, halothane appears to require the most monitoring. Erection of the neck feathers in birds under halothane anesthesia appears to be a grave sign. Birds anesthetized with parenteral ketamine/xylazine at low intravenous doses may require the least close observation. Gallinaceous birds must be observed closely under any anesthetic.

Mucous secretions may occasionally collect in the pharynx during surgery, especially in those birds suffering from gastrointestinal disease or those that have been improperly fasted; these secretions may be periodically swabbed and removed with cotton-tipped applicators. The secretions appear to occur less frequently with ketamine/xylazine and with isoflurane anesthetics.

### Supplemental Support

One study reported a temperature loss averaging 11.28°F, with variations of 2.7 to 19.3°F in birds under 30 minutes of halothane; temperature loss under ketamine averaged 4.2°F, with a range from 2.6 to 5.3°F.[5] The temperature was found to increase gradually during recovery. In another study,[10] isoflurane was found to depress body temperature an average

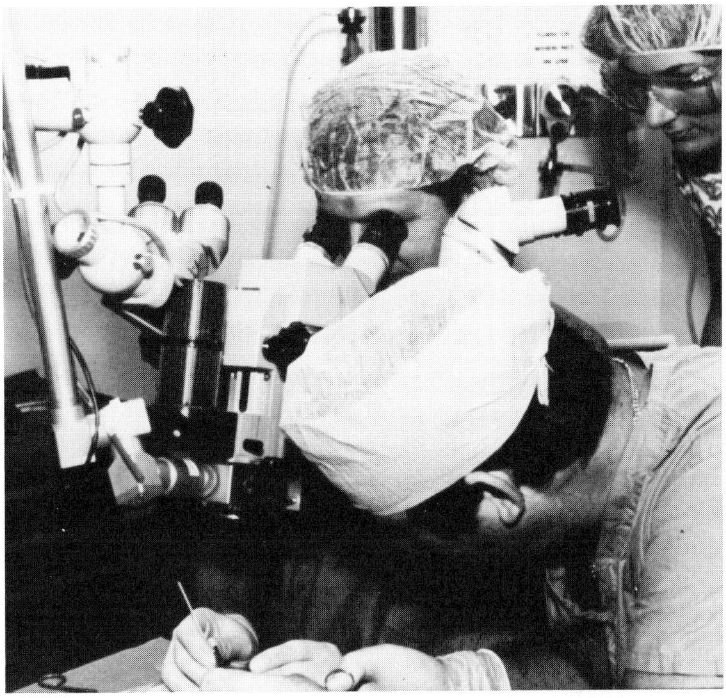

**Figure 44–2.** A full surgical team may be necessary for some avian procedures, such as this experimental laser surgery for the removal of a renal tumor in a budgerigar.

of 4.7°F in cockatiels and 2.8°F in pigeons. To compensate for this loss of body temperature from the effects of anesthesia, supplemental heat is recommended during surgery, especially in anticipation of prolonged procedures (see Chapter 46, Surgical Instrumentation and Special Techniques). To further limit heat loss, a quick swabbing with povodine-iodine scrub solution is preferred by this author for surgical preparation rather than excessive water or alcohol.

Other support measures that may be necessary to have on hand in case of potential hemorrhage are blood for transfusions, lactated Ringer's solution for intravenous fluids, steroids for intravenous use, sodium bicarbonate, an electrosurgical device, vascular clamps, and hemostatic sponges. These are especially important in cases of large abdominal masses that are ultimately diagnosed as tumors or substantial cysts that may have an extensive blood supply.

### *Blood Transfusions*

Blood for transfusions should be available for administration during the surgical procedure when necessary. For small avian surgical patients, the anticoagulant of choice is heparin rather than EDTA to avoid excessive binding of calcium.

Even in birds with normal preoperative PCV values, if excessive hemorrhage is anticipated, such as with large tumor removals, a presurgical blood transfusion is advised, especially in budgerigars (Fig. 44–3). The first choice, if available, is blood collected from the same species; second choice is from a psittacine donor to a psittacine patient; third choice is from a chicken or pigeon to a psittacine.

Work with avian heterologous blood transfu-

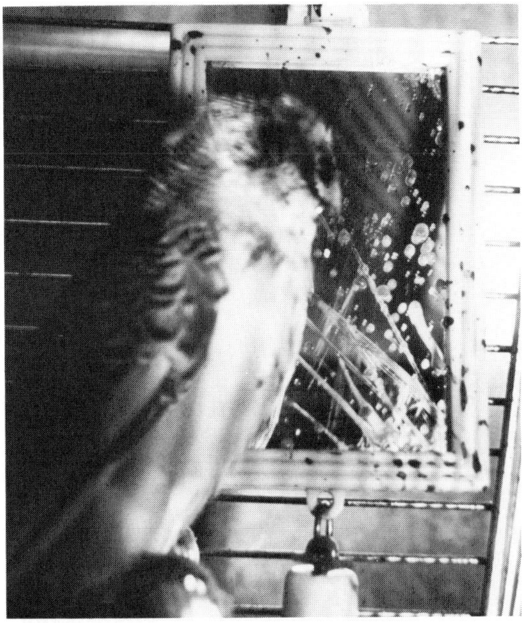

**Figure 44–3.** Although hemostasis is extremely important during avian surgical procedures and the need for blood transfusions may exist, a bird can survive with the loss of more than four or five drops of blood, as evidenced by this budgerigar that had a significant amount of blood splattered in all corners of its cage from a beak injury.

sions has shown that a single transfusion from chickens[9] or pigeons[2, 3] to psittacines appears to be a safe and efficacious means of whole blood replacement. However, serial or multiple heterologous blood transfusions can cause fatal transfusion reactions when administered less than 10 days after the first transfusion. It appears from these studies that antibodies develop within the first several days after a transfusion and can persist for three weeks or longer. After three weeks the risk of a second transfusion decreases significantly. Cross-matching was done in one study by mixing the recipients' sera with the donor's red cells after the red cells were washed with phosphate-buffered saline and incubated at 40°C in a water bath for 30 minutes. Incompatible cross-matches showed hemolysis or agglutination. In another study,[2] when the serum of the donor was mixed with the *unwashed* red cells of the recipient, no incompatibility was evident even though a bird died from the transfusion reaction. Therefore cross-matching with unwashed red cells and sera is not an accurate means of determining compatibility.

### *Bolus Intravenous Fluid Therapy*

Rapid intravenous infusion of lactated Ringer's solution quickly rehydrates a debilitated or stressed surgical patient and has been credited with reversing emergency situations[11] (see Chapter 27, What To Do Until A Diagnosis Is Made). Preventive fluid therapy can be administered prior to, during, and following surgery. On small birds a bolus dose of fluids can be given periodically, whereas larger birds may have continuous drips through intravenous catheters* during prolonged procedures. During surgery the surgeon may notice the increased blood pressure and tendency toward hemorrhage from the bolus fluid therapy.

## POSTSURGICAL CARE OF THE AVIAN PATIENT

Support of the patient does not end when the surgical procedure is complete. The quality of postoperative care has a significant effect on the outcome of many procedures. Recovery from prolonged anesthetic (i.e., injectable) is hastened in some cases by rolling the patient over every 15 minutes until the bird is ambulatory. Fluids increase kidney perfusion to eliminate injectable anesthetics more rapidly.[13] Depending upon the extent of the invasion, the postsurgical patient may be hospitalized and observed for as little as an hour (for surgical sexing) to several days for recovery from more extensive procedures. Serial hematologic tests, especially PCV, are valuable in postoperative monitoring of birds that have had blood transfusions. The white blood cell counts may assist in evaluating the appropriateness of antibiotic therapy following abscess removal. The convalescing patients are weighed daily and tube fed if necessary (see Chapter 6, Management Procedures). Blood transfusions postoperatively are not common, but there may be isolated cases in which cautery was incomplete and pooling of blood may result. Postsurgical radiographs may be indicated in cases of gastrointestinal, liver, or air sac disease.

*Angio-Set Shorty—Deseret, Sandy, UT 84070.

## REFERENCES


1. Altman, R. B.: Avian anesthesia. Comp. Cont. Ed., 11(1):38–42, 1980.
2. Altman, R. B.: Heterologous blood transfusions in avian species. Proceedings of the Annual Meeting of the Association of Avian Veterinarians, San Diego, CA, 1983, pp. 28–32.
3. Altman, R. B.: Supplement to heterologous blood transfusions in avian species from 1983 AAV Proceedings. AAV Newsletter, 4(2):48, 1983.
4. Altman, R. B.: Preoperative evaluation. Proceedings for Avian Medicine and Surgery, 12th Annual Veterinary Surgical Forum, Chicago, IL, 1984, pp. 14–16.
5. Altman, R. B., and Miller, M. S.: Effects of anesthesia on the temperature and electrocardiogram of birds. Proceedings of the Annual Meeting of the American Association of Zoo Veterinarians, Denver, CO, 1979, pp. 61–62.
6. Campbell, T. W.: Avian brain thromboplastin needed for prothrombin time. AAV Newsletter, 6(2):54, 1985.
7. Harrison, G. J.: Recent advances in avian surgery. Proceedings of the Annual Meeting of the American Animal Hospital Association, Orlando, FL, 1985, pp. 30–34.
8. Harrison, G. J.: New aspects of avian surgery. Vet. Clin. North Am. [Small Anim. Pract.], 14(2):363–380, 1984.
9. Harrison, G. J.: Experimental interspecies avian blood transfusions. J. Zoo Anim. Med., Vol. 8, 1977.
10. Harrison, G. J., et al.: A clinical comparison of anesthetics in domestic pigeons and cockatiels. Proceedings of the Annual Meeting of the Association of Avian Veterinarians, Boulder, CO, 1985, pp. 7–22.
11. Redig, P. T.: Fluid therapy and acid-base balance in the critically ill avian patient. Proceedings of the International Conference on Avian Medicine, Association of Avian Veterinarians, Toronto, Canada, 1984, pp. 59–73.
12. Rosskopf, W. J.: Avian preoperative evaluation: Clinical pathology. Proceedings for Avian Medicine and Surgery, 12th Annual Veterinary Surgical Forum, Chicago, IL, 1984, pp. 7–12.
13. Rosskopf, W. J.: Avian preoperative care: Antibiotics and fluid therapy. Proceedings for Avian Medicine and Surgery, 12th Annual Veterinary Surgical Forum, Chicago, IL, 1984, pp. 17–18.

# Chapter 45

# ANESTHESIOLOGY

GREG J. HARRISON

One of the most controversial areas of pet bird practice is anesthesia. Although there is agreement that general anesthesia is essential for major surgical procedures, opinions vary as to other indications, choice of anesthetic agent, route of administration, and dose. Conflicting reports in the literature confuse the issue.

A number of publications on avian anesthesia are given under Recommended Readings. Rather than providing an extensive review of the literature, the author will approach the subject from the viewpoint of an avian practice in which the primary objectives of the use of anesthetic agents in birds are (1) to accomplish diagnostic, therapeutic, and surgical procedures as safely and quickly as possible; (2) to minimize the level of pain and stress and the potential side effects in the animal; and (3) to minimize recovery time and degree of postanesthetic monitoring.

## INDICATIONS

Use of an anesthetic is indicated for situations in which immobilization or analgesia of the patient is necessary and the patient tolerance has been evaluated (see Chapter 44, Evaluation and Support of the Surgical Patient). Anesthesia safety and efficacy are controlled by the choice of agent and dose in relation to the patient and procedure.

For many avian procedures, deep surgical planes of anesthesia are unnecessary and, in the author's opinion, not recommended. Light immobilization levels of anesthesia may be desired for procedures such as surgical sexing, diagnostic endoscopy, biopsy, radiology, fluid therapy, blood collection, blood transfusions, bandaging, and splinting. For some particularly fractious birds, a level of immobilization may be indicated for even the most routine procedures (e.g., nail or wing trims) to minimize the level of stress.

From the literature and the author's experience with a wide variety of anesthetic agents, characteristics of the most common agents have been summarized in Table 45–1.

## CONTRAINDICATIONS

In most cases, general contraindications for the use of anesthesia in birds include clinical signs of shock, ascites, severe anemia, respiratory distress or impairment, fluid-filled or semi–fluid-filled crops, severe emaciation, dehydration, and acidosis. However, many of these very conditions may require a degree of evaluation and treatment that can be adequately accomplished only under some level of immobilization.

Many of the previously published contraindications to the use of anesthesia in birds were specific to the anesthetic agents available and/or the dosages recommended at the time of publication. Consideration should be given to these characteristics when selecting an appropriate anesthetic.

## CHOICE OF ANESTHETIC

General considerations of the choice of anesthetic in a given situation include immediate and long-term safety to the patient, ease of administration, control of dosage, speed of recovery, degree of anesthetic and postanesthetic monitoring required, the specific procedure involved, and safety of surgical support personnel. The most important consideration for the practitioner, however, is familiarity with the agent so that predictable results may be obtained. Rapid recovery from anesthesia may be a significant factor for an avian practice in which auxiliary personnel are not available for prolonged monitoring of the patient.

The most common anesthetic agents reported in recent literature for use in birds are (1) ketamine alone, (2) ketamine/xylazine combination, (3) ketamine/diazepam combination, (4) halothane, (5) methoxyflurane, (6) the previous

**Table 45–1.** CHARACTERISTICS OF COMMON ANESTHETIC AND IMMOBILIZATION AGENTS*

| | Advantages | Disadvantages |
|---|---|---|
| **ISOFLURANE**<br>Induction:<br>up to 5%<br>(2–3 L/min oxygen)<br>Maintenance:<br>2–3% (adjust to bird)<br>(0.5–1 L/min oxygen) | Safe in critical patients<br>Apparent extensive interval between apnea and cardiac arrest<br>Can induce at 5%<br>Only 0.3% metabolized by body; no hangover<br>No organ toxicity<br>Rapid changes in level of anesthesia<br>Less monitoring required<br>Provides excellent muscle relaxation<br>No oral secretions after induction<br>Rapid body temperature recovery<br>Can be administered with face mask<br>Not affected by UV light, soda lime, alkali<br>Insoluble in blood, therefore rapid recovery<br>No CNS excitement<br>Nonexplosive<br>Rapid return to normal eating, drinking | Pungent odor which may increase induction<br>Higher cost/cc<br>May get decreased blood pressure due to muscle relaxation<br>Respiratory depressant (although less than halothane)<br>Depressed body temperature<br>Veterinary approval only for equines<br>Some birds vomit on recovery, esp. if improperly fasted. |
| **HALOTHANE**<br>Induction:<br>3–4% conc.<br>(2–3 L/min oxygen)<br>Maintenance:<br>1.5–2% conc.<br>(0.5–1 L/min oxygen) | Readily available<br>Nonexplosive<br>Limited changes in anesthesia level<br>Moderate muscle relaxant<br>Relatively inexpensive<br>Rapid consciousness recovery<br>Rapid recovery of body temperature<br>Inoffensive odor | Heart arrhythmias<br>Potential for renal and hepatic damage<br>15–20% metabolized<br>Prolonged hangover<br>Some CNS excitement upon recovery<br>Moderately prolonged depressed body temperature<br>Risk to operating room personnel if waste gas escapes<br>Extensive monitoring required<br>Apnea and cardiac arrest usually simultaneous<br>Respiratory depressant<br>Endotracheal intubation recommended<br>Mucous secretions often accumulate in endotracheal tube<br>Unexplained deaths in "healthy" birds |

two in combination with nitrous oxide, and (7) isoflurane. Increased knowledge of the pathogenesis of diseases in cage birds, and the recent availability of information (including manufacturers' drug inserts) on adverse effects of some of these anesthetic agents appear to contraindicate their use in the clinically ill avian patient. In general veterinary medicine, for example, organ toxicities have been reported in various mammals from the use of halothane or methoxyflurane; methoxyflurane is contraindicated by the manufacturer for use in animals with liver disease and toxemia; ketamine is not recommended for use in cats or subhuman primates suffering from renal or hepatic insufficiency; halothane is contraindicated in excited or debilitated birds, dogs, and cats owing to the risk of cardiac arrest or toxicity; and xylazine has been implicated in the potentiation of epinephrine-induced arrhythmias. Specific anesthetic incidences in birds have been confined to halothane or ketamine, with reports of decreased cardiac and respiratory rates, vomiting, slow recovery, and depressed body temperature. Although bizarre ECG tracings were found in raptors given a combination of ketamine and xylazine, these signs disappeared at low intravenous doses. In human and animal studies to date, isoflurane appears to exhibit the least potential for patient toxicity.

In the author's experience, the use of isoflurane inhalation anesthetic provides the opportunity for aggressive diagnostic and therapeutic measures under immobilization and surgical planes of anesthesia in critical patients. A degree of risk is still inherent in the procedure, but with the use of isoflurane, there is less likely to be direct incrimination of the anesthetic agent itself if a severely debilitated patient succumbs to the procedure. Although a

**Table 45–1.** CHARACTERISTICS OF COMMON ANESTHETIC AND IMMOBILIZATION AGENTS *Continued**

| | Advantages | Disadvantages |
|---|---|---|
| **METHOXYFLURANE** | | |
| Induction:<br>3–4% conc.<br>(2–3 L/min oxygen) | Least expensive of gas anesthetics<br>Excellent muscle relaxant<br>Nonexplosive<br>Low vapor pressure; safe in vaporizer | 50% metabolized: recovery hangover<br>Potential for waste gas damage to personnel<br>Respiratory depressant<br>Slow induction, recovery |
| Maintenance:<br>1.5–2%<br>(0.5–1 L/min oxygen) | | Slow to change level of anesthesia<br>Apnea and cardiac arrest may be simultaneous<br>Prolonged depressed body temperature<br>Unexplained deaths<br>Unsafe in critical patients |
| **KETAMINE HYDROCHLORIDE** | | |
| (15–20 mg/kg) | Wide margin of safety | Variable effects on species, individuals<br>Inability to attain analgesia<br>Inadequate muscle relaxation<br>Prolonged recovery at high doses<br>Violent recovery<br>Contraindicated in patients with kidney/liver disease |
| **KETAMINE/XYLAZINE COMBINATION** | | |
| (10–30 mg/kg ketamine) See Table 45–4. | Analgesia provided<br>Rapid induction with I.V. administration<br>Rapid recovery from I.V. administration<br>Safe, even with repeated doses<br>Preferred parenteral combination<br>Readily available<br>Excellent for use in field | Prolonged recovery with IM administration<br>Violent recovery in some birds<br>Higher doses (3×) lead to ECG and respiratory disturbances<br>Hypothermia<br>Occasional unexplained death<br>Decreased respiration, temperature |

*Inhalation agents in precision vaporizers only.

number of other anesthetic agents and routes of administration are safe and effective for the relatively *healthy* bird, because of the numbers of debilitated and critical avian patients that would benefit from chemical immobilization, it behooves the practitioner to select the safest agent for all cases.

Based on experience with critical care cases, the author recommends the anesthetic agents listed in Table 45–2, in order of decreasing safety. Anesthetic agents and/or methods of administration that are no longer believed to be appropriate for avian species are listed in Table 45–3.

The reader is referred to Chapter 44, Evaluation and Support of the Surgical Patient, for additional preanesthetic considerations.

**Table 45–2.** RECOMMENDED ANESTHETICS FOR CRITICAL AVIAN PATIENTS*

1. Isoflurane (inhalation)
2. Ketamine and xylazine combination intravenously
3. Ketamine and xylazine combination intramuscularly
4. Halothane (inhalation)
5. Methoxyflurane (used only with appropriate vaporizer)

*Anesthetic agents listed in order of decreasing safety.

## INHALATION AGENTS

The primary advantages in the use of inhalation anesthetic agents are the speed of induction and total recovery. In addition, inhalation anesthetics delivered from a precision vaporizer with oxygen may be administered directly into the air sacs by trocarization in cases of tracheal

**Table 45–3.** TECHNIQUES AND PRODUCTS NOT RECOMMENDED FOR USE IN PET BIRDS

1. Methoxyflurane (and other volatile anesthetics) when administered by open drop or nose/mouth
2. Barbiturates (use recommended only for convulsions in otherwise healthy bird)
3. Chlorohydrate combinations (extremely long reaction and recovery times)
4. Thiamylal sodium; sodium pentobarbital
5. Intraperitoneal administration
6. Local anesthetics (safety is dose-related; however, errors are common)
7. Xylazine alone
8. Ether (due to explosiveness only)

obstruction. Until the advent of isoflurane, the two most common inhalation anesthetics were halothane and methoxyflurane, both of which required intensive monitoring of the patient. In addition, there has been increased concern among operating room personnel regarding the potential effects of waste gas exposure to veterinarians and veterinary technicians. It appears that isoflurane offers some wider margins of safety for the patient and auxiliary surgical personnel, although the standards of waste gas exhaust must be maintained.

Despite the advantages possessed by halothane, reports of liver damage, cardiorespiratory depression, and sensitivity of the heart to the effects of exogenous or endogenous epinephrine following halothane anesthesia point out obvious deficiencies. Hepatic necrosis, reported to be associated with use of halothane in a dog and a snow leopard, has also been associated with the use of methoxyflurane.

Isoflurane and methoxyflurane are methyl ethyl ethers, whereas halothane is an alkane. Alkanes in general predispose the heart to arrhythmias, whereas ethers do not. In contrast to halothane and methoxyflurane, which require preservatives to maintain their stability, isoflurane is a more stable molecule and preservatives are not required. Preservatives tend to accumulate and become sticky in the vaporizer, thus reducing the accuracy of the calibration.

Isoflurane shares with halothane the feature of low blood solubility, which means that recovery is rapid and that the alveolar concentration of the agent can be controlled with greater precision than with methoxyflurane.

All modern inhaled anesthetics, including isoflurane, depress ventilation in a dose-related fashion. Although this depression incurs the risk of respiratory acidosis and hypoxia, it limits the delivery of the anesthetic and the potential for anesthetic overdose. The depression is partially antagonized by surgical stimulation.

## Isoflurane

Although isoflurane is less known in veterinary medicine, it is the anesthetic of choice in human medicine. At printing time, veterinary approval was limited to equines, althouth there have been reports in the literature of isoflurane use in dogs, cats, and birds.

The most significant feature of isoflurane which contributes to its safe use in avian patients as well as in humans is its invulnerability to biodegradation and metabolization. This limits its potential to produce organ toxicity. Human and animal studies suggest little or no effect on liver or renal function. This feature is vital for critical avian patients, who are usually suffering from disease-related liver and kidney damage.

A clinical trial was conducted to compare aspects of inhalation anesthetics in pigeons and cockatiels. Isoflurane was delivered safely under extreme conditions of patient dehydration, deliberate multiple overdoses, collection of up to 10 per cent of blood volume, and multiple organ biopsies. This margin of safety is especially important in critical cases in which immobilization of the patient may be required several times a day for bolus fluid therapy or other aggressive treatment regimens.

Isoflurane anesthesia may be induced and maintained with a face mask (Fig. 45–1). A concentration as high as 5 per cent with 2 to 3 liters per minute of oxygen may safely be used for induction. As soon as there is noticeable relaxation of the patient (within one to four

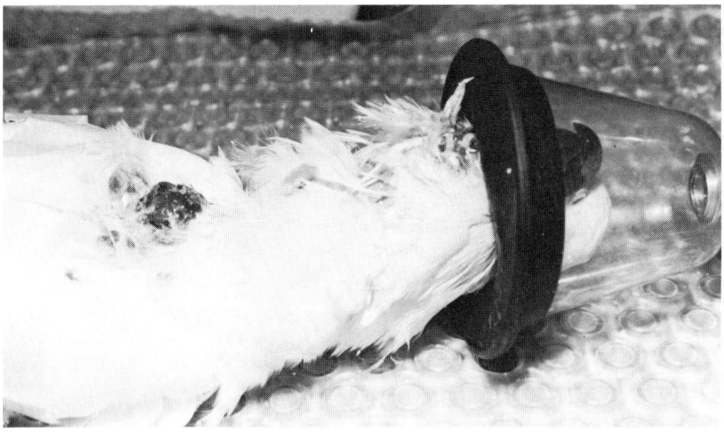

**Figure 45–1.** Isoflurane anesthesia is induced and maintained with a face mask in this cockatoo in preparation for crop surgery.

minutes, depending upon the ambient temperature), the concentration may be maintained at 2 to 3 per cent with 0.5 to 1.0 liter per minute oxygen flow. A slightly higher flow may be required for some larger birds. Some birds, in particular Blue-and-Gold Macaws, occasionally exhibit some respiratory depression at the recommended concentrations and should be maintained at 0.5 to 2 per cent following induction. Following induction, isoflurane anesthesia can be maintained in small birds or in cases requiring access to the head and neck by urinary or intravenous catheters that have been glued to endotracheal tube adapters and placed in the bird's pharynx.

The author attempts to maintain avian patients at the lowest level required for the procedure. Because birds respond so rapidly to changes in concentration of isoflurane, this may entail changing the concentration often in response to effects on the patient. This technique is not tolerated with other inhalation anesthetics.

### Halothane

Halothane is currently the most widely used inhalation agent in avian anesthesia. Induction requires a 3 to 4 per cent concentration with 2 to 3 liters per minute oxygen flow. Maintenance is 1.5 to 2 per cent concentration with 0.5 to 1 liter per minute oxygen flow. Some authors suggest a preanesthetic dose of ketamine/xylazine to assist induction with halothane. Combining halothane 1:1 with nitrous oxide and reducing the halothane concentration to 0.5 to 1 per cent is reported to provide adequate relaxation.

Halothane is most often reported to be administered by endotracheal tube with an Ayres T-piece, nonrebreathing system. Canine urinary/feeding tubes cut to the appropriate length and smoothed at the cut end are softer, more flexible, and less traumatic than standard endotracheal tubes. Smaller birds can be masked.

The primary disadvantages of halothane anesthesia are the potential for liver damage and the cardiorespiratory sensitivities, requiring intensive monitoring of the patient due to the brief interval between apnea and cardiac arrest. One author suggests a three-person surgical team when using halothane: an anesthesiologist, the surgeon, and a third person to position the bird. This may not be possible in all avian practices. The literature for halothane specifically states that ". . . the concentration of vapor should not be suddenly increased."

### Methoxyflurane

In the author's opinion, the state of the art of avian anesthetics dictates that the only acceptable way to administer this agent is with a precision vaporizer. Because 50 per cent of methoxyflurane is metabolized in the body, a high potential for organ toxicity exists in the patient and in operating room personnel who may be exposed to waste gas. Patient recovery is slow and sluggish, and a prolonged hangover is common.

Although some practitioners, and even the manufacturer, recommend use of the open-drop or closed container with wet pledget method of administering methoxyflurane, the dose is extremely difficult to control and is potentially very dangerous to the patient, as well as to the practitioner in close proximity. The manufacturer's literature states: "The concentration of vapor necessary to maintain surgical anesthesia is much less than that required to induce it." It may be difficult to determine the point at which to reduce or otherwise exercise control over the concentration administered to the bird with these methods. Direct oral or nasal administration of any liquid anesthetic is totally unacceptable. The manufacturer further cautions against the use of methoxyflurane in animals with liver disease, providing a further contraindication to its use with most avian intensive care programs.

### Ether

Ether was among the inhalation anesthetics evaluated in pigeons and cockatiels and interestingly, when administered with a precision vaporizer, was used successfully for immobilization and establishing surgical planes of anesthesia in these birds. Ether cannot be recommended because of its extreme flammability and the subsequent cautions required for use, including avoiding the use of spark-generating devices. However, contrary to previous reports, ether is not toxic to birds if accurately delivered, and one may expect responses of the patient to be similar to those with isoflurane use. Ether cannot be expected to perform in this way, however, if it is administered by the open-drop or cone method.

## USE AND CARE OF PRECISION VAPORIZERS

Because the vapor pressures are very similar, isoflurane may be used in a clinical situation in a "tec-type" vaporizer calibrated for halothane, with some restrictions. Initially, accumulated preservatives from halothane use must be completely removed with several eight-hour soaks and flushes with ether prior to instillation of isoflurane. Once the vaporizer has been prepared for isoflurane use, it is contraindicated to alternately use halothane in the machine. Not only is the machine once again contaminated with the preservatives from halothane which interfere with isoflurane calibrations, but the techniques that a practitioner may develop in adjusting the concentration of isoflurane to attain the desired effect are too dangerous to use with halothane. For these reasons, the manufacturers recommend a vaporizer specifically for isoflurane, thus eliminating potential confusion about the contents of the machine. An anesthetic vaporizer designed for methoxyflurane is not suitable for isoflurane.

Components of the anesthetic machine which come in contact with avian patients (e.g, face masks and delivery hoses) must be disinfected following each use to prevent contamination by disease organisms. In the author's practice, face masks are disinfected in a solution of chlorhexidine or quaternary ammonia. Spray disinfectants may be used on the exterior of the machine and inside the soda lime container.

Because of the satisfactory use of isoflurane to facilitate production of diagnostic quality radiographs, and because of the very rapid recovery time when this agent is used, the author has chosen to locate the anesthetic machine adjacent to the radiographic equipment. Previous attempts to remove the mask and rush an isoflurane-anesthetized bird to the radiograph room resulted in evidence of movement on the films from recovery already in progress, or worse, apnea and a need for resuscitation from a deliberate overdose.

## PARENTERAL AGENTS

Parenteral agents may be indicated for situations in which inhalation anesthesia is not available or for birds requiring surgical procedures around the head or mouth. Some of the parenteral chemicals commonly used for "anesthesia" are reported to produce a dissociative state that inhibits movement but may not establish a level of analgesia. Parenteral agents are contraindicated in avian patients with clinical signs suggestive of kidney dysfunction, such as anuria.

The major criticism of parenteral anesthetics concerns the irreversibility of the administered dose. For this reason, the practitioner is advised to initiate the anesthetic level with the minimal amount given to effect, which may be only one eighth to one fourth of the total recommended dose. The wide margin of safety of popular parenteral agents such as ketamine or ketamine/xylazine combinations permits repeated doses until the desired plane of anesthesia is achieved.

Original published doses for intramuscular injections of parenteral anesthetics were determined primarily according to body weight only; however, several other factors need to be considered: age, species, amount of body fat, and general condition of the bird. Young or baby birds and thin or debilitated birds (including those in respiratory distress) require a lower dose. Obese birds or extremely excitable species may require a higher dose to attain a similar level of immobilization.

For intramuscular administration, the predetermined portion of the dose should be administered into the deepest portion of the pectoral musculature, avoiding the sternal area, which may involve major vessels, or the lateral side, where the lungs are located. The forefinger placed alongside the needle serves as a guide to avoid penetration beyond half the determined depth of the muscle. Care must be exercised in administering intramuscular injections to thin birds with sparse breast tissue. Injections in the leg muscles of pet birds are contraindicated owing to the increased potential for nerve damage. Following the injection, the site should be inspected for hemorrhage. The capture, restraint, injection, and placement back into the cage can be accomplished in less than 30 seconds. Providing a dark, quiet environment may facilitate a smoother induction and obviate the need for repeated doses for a frightened bird.

Intraperitoneal injections of parenteral anesthetics are contraindicated, in the author's opinion, because the absorption rate cannot be controlled.

### Ketamine and Xylazine

The cyclohexylamine, ketamine, is the most frequently used component of parenteral anesthesia in birds in the United States. Owing to the synergistic effect produced by combining this product with tranquilizers, ketamine is never used alone by this author. When parenteral agents are indicated, the author has found consistently good results using ketamine (100 mg/cc) with an equal volume of xylazine (20 mg/cc). This combination improves muscle relaxation and analgesia and reduces the incidence of stormy recoveries observed in some individuals or species with ketamine alone. Diazepam is less commonly used in combination with ketamine by some practitioners.

From experience, the author and others have formulated safe and effective doses of injectable ketamine in combination with xylazine (Table 45–4) for any procedures, in all pet species (including pigeons), and in all but the most critically ill birds (for which any anesthetic other than isoflurane would be contraindicated).

The dose and route of administration of these parenteral agents depend upon the degree of immobilization and speed of recovery desired. Ketamine and xylazine may be combined in the same syringe for administration. Xylazine is drawn up first, especially for small birds, to avoid adverse effects from the pure solution remaining in the needle hub. The use of 10 to 30 mg/kg ketamine combined with an equal volume of xylazine and administered intramuscularly induces an anesthetic level adequate for minor surgical or diagnostic procedures. The dose per kilogram of body weight is lower for birds weighing over 250 grams than for birds of less than 250 grams body weight. For brief procedures and rapid recovery times, intravenous administration is recommended. The intravenous dose would be 30 to 50 per cent less volume than the intramuscular dose.

The calculated volume for the initial injection may be one eighth to one fourth of the full dose. This may be repeated if the plane of immobilization is not sufficient. One may recognize within 5 to 10 minutes of an intramuscular injection or within 1 minute of an intravenous injection if a dose needs to be repeated. Using the small increments of the ketamine/xylazine combination rather than the full

**Table 45–4.** MAXIMUM INTRAMUSCULAR DOSES OF KETAMINE[1] FOR USE WITH XYLAZINE[2]*

| Species | Intramuscular | Intravenous |
|---|---|---|
| Budgerigars | 0.01 | 0.005 |
| Cockatiels | 0.02 | 0.01 |
| Conures | 0.05 | 0.025 |
| Lories, rosellas | 0.07 | 0.035 |
| Amazons, miniature macaws | 0.05–0.1 | 0.025–0.05 |
| African Greys | 0.08–0.1 | 0.04–0.05 |
| Cockatoos | 0.12–0.15 | 0.06–0.07 |
| Macaws | 0.15–0.20 | 0.075–0.10 |

*Amount given in volume (ml) of ketamine to be mixed with equal *volume* of xylazine before administering.
[1]Ketamine—100 mg/ml
[2]Xylazine—20 mg/ml

dose all at once allows for control and safety in the use of these agents. There is the occasional case that may require from two to four times the dose listed in Table 45–4.

With low doses of ketamine/xylazine combinations for immobilization, the birds may be back on the perch in 15 to 30 minutes. Repeated doses can be given for longer procedures, or the bird may be prepared for inhalation anesthesia from the initial injection. A face mask is attached with oxygen only supplied; when movement occurs, the inhalant agent can be administered.

### Methoxymol

Methoxymol is a short-acting hypnotic that is available in Canada and Europe. Its effects and safety are reported to be similar to ketamine, and it apparently produces some fluttering in birds during recovery. It is often used as a preanesthetic to halothane. At a dose of 10 to 40 mg/kg administered intramuscularly, the induction time is 1 to 3 minutes, with a duration of 15 to 30 minutes. The lower dose is suggested for raptors. Repeated doses may be given.

## MONITORING THE ANESTHETIZED BIRD

The patient must be monitored during the induction, maintenance, and recovery of any anesthesia. This may be accomplished through observation of the rate and depth of respiration, reaction to toe pinch, eye-blink reflexes, use of electrocardiogram machine, or cloacal temperature (see Chapter 44, Evaluation and Support of the Surgical Patient).

### Anesthesia Levels

Levels of narcosis and anesthesia have been reported in the literature from work done by Arnall (Table 45–5). Predicted responses may differ according to the chemical in question.

With ketamine alone, early stages of anesthesia may include rigidity, tremors, opisthotonus, wing flapping, and occasionally convulsions. With the smoother induction provided by the addition of xylazine, one usually observes easy respirations and body relaxation in the early stages. Flaccidity of the neck and wings is evident in deeper levels. Swallowing and blinking reflexes are present; however, there is no toe-pinch response.

With the gas anesthetics, especially isoflurane

**Table 45–5.** LEVELS OF NARCOSIS AND ANESTHESIA*

**NARCOSIS**

Light: The bird appears sedated, is lethargic; eyelids tend to droop.

Medium: The feathers are ruffled, the head hangs low; the bird can be aroused but offers little resistance to handling.

Deep: No response to sound; fluttering may be provoked by painful stimuli; rapid respiration, regular and deep.

**ANESTHESIA**

Light: Reflexes (palpebral, corneal, cere, pedal) present but lack voluntary movement; no response to postural changes or vibration.

Medium: Palpebral reflex is lost; pedal and corneal reflexes are sluggish; respiration is slow, deep, and regular; best anesthesia for surgery.

Deep: All reflexes are absent; respirations are very slow but may be regular; any deeper level of anesthesia will further depress respirations until they cease.

*Adapted from Arnall, L.: Anesthesia and surgery in caged and aviary birds. Vet. Rec., 73(7):139–142, 1961.

and ether, one notes in the patient a very slow relaxing slump into an anesthetic plane without an excitement phase. Slow, even respirations with a mild cloacal sphincter tone response or mild deep toe-pinch response can be observed during the desired anesthetic level.

When isoflurane is used as the induction agent, respiration rates have been more variable. Administration into a chamber tends to increase the initial respiration rate due to the difficulty in achieving the 5 per cent concentration in the large volume of air. Induction by face mask may also have initial high rates (up to 50 to 80 respirations per minute [R/M]); however, these are decreased to 25 to 35 R/M within 3 to 5 minutes in larger birds and 5 to 10 minutes in smaller birds. A further slowing of respiration (25 R/M) would require dropping the isoflurane concentration to 0 per cent and allowing the residues in the mask and hose to stabilize the bird. In small birds, the flow of oxygen would be reduced to 0.3 liter per minute.

## ANESTHETIC EMERGENCIES

A decrease in respiration, heart rate, or body temperature may alert the practitioner to an impending anesthetic emergency. A heart rate below 120 beats per minute or respiration rates below 25 to 35 for large birds and 35 to 50 for small birds require a cessation of the procedure and intensive monitoring. Table 45–6 summarizes suggested emergency steps.

In cases of respiratory and cardiac arrest on gas anesthetic, one would immediately take the following steps:

1. Remove the bird from the attachment to the anesthetic machine.
2. Institute pulmonary resuscitation by gentle rhythmic finger pressure (40 cycles/minute) on the sternal carina; at the same time remove residual gas from the mask or endotracheal tube by suction.
3. Turn off the flow of anesthesia.
4. Flush the machine and hose with pure oxygen.
5. Reattach the bird to the hose to administer pure oxygen.
6. Continue resuscitation until unaided respiration reaches a safe rate (30/minute).

Use of antishock and antiacidosis therapy is

**Table 45–6.** ANESTHETIC EMERGENCIES

| Condition | Action | Prognosis |
|---|---|---|
| Depressed respiration | Decrease level of anesthesia (decrease gas; administer I.V. fluids)<br>Heated water blanket<br>Dopram-V | Good |
| Depressed body temperature | Same as above | Good |
| Respiratory arrest | Remove bird from anesthetic<br>Sternal CPR 30/60 per minute<br>Suck gas from mask, hose and/or<br>Flush system with oxygen<br>Re-attach bird to oxygen | Good (isoflurane)<br>Poor (others) |
| Cardiac arrest | Remove bird from anesthetic<br>Direct cardiac massage with saline-soaked cotton applicator via laparotomy<br>Intravenous fluids<br>Norepinephrine | Grave |

indicated if the patient is reviving but at a slow rate.

Cardiac arrest is usually irreversible in birds. However, in one surgical patient under isoflurane, the heart was revived by direct stimulation with a saline-soaked cotton-tipped applicator via the thoracic air sac. There have been a few reports of successful avian response to intracardial norepinephrine and use of doxapram hydrochloride as a respiratory stimulant, but the author has had no experience with these.

## RECOVERY

An unobstructed environment of 85 to 90°F should be provided for recovery from anesthesia. One report suggests an ambient temperature of 100 to 104°F for two to three hours following anesthesia, but this has not been used by this author.

Recovery from halothane and methoxyflurane anesthesia may be somewhat prolonged. Although the birds may attain a level of consciousness within a relatively short period, they usually appear fluffed and sluggish for an extended time. In contrast, birds recovering from isoflurane anesthesia are immediately alert and interested in eating.

For the anesthetic studies mentioned previously, all pigeons had been fasted for several hours prior to the surgery. When they were returned to the cages following anesthesia, those pigeons from halothane trials preferred roosting to eating. Isoflurane-anesthetized birds went straight to the food bowl; some were even interested in breeding activity as soon as they were returned to the cage. This suggests that if a relatively long recovery time from anesthesia precludes eating, latent liver problems may be exacerbated owing to glycogen depletion. The pigeons (and cockatiels) also regained normal body temperatures within two to four minutes of removal of isoflurane.

With the parenteral doses of ketamine/xylazine combinations listed in Table 45–2, the bird is expected to be ambulatory in 15 to 30 minutes. As a species, cockatoos are generally slower to recover from intramuscular injections of ketamine/xylazine. Reports in the literature of very prolonged and extremely violent recoveries following ketamine use may be dose-related, although some macaws, particularly Blue-and-Golds, seem to react more strongly to the ketamine component. These birds may require additional protection to prevent trauma during recovery. A single layer of masking tape or cloth towels may be lightly wrapped around the wings (Fig. 45–2). Place the bird in a padded area with close supervision. A padded recovery may also prevent periocular bruising that may occur due to head tossing. Some birds under anesthesia with a ketamine component may show a unilateral or occasionally bilateral wrinkling of the eye globe or loss of pressure within the eye, but this is transitory (Fig. 45–3).

Recovery from repeated intramuscular injections for long procedures is enhanced by massaging the patient's body, especially over the rib area, and turning the bird from side to side every 15 minutes. Other practitioners have suggested doxapram hydrochloride as an effective antagonist for ketamine/xylazine. Slow recoveries from this combination also respond to intravenous lactated Ringer's solution. In the author's experience, furosemide may be combined with fluids.

## ORAL ANESTHETICS

Avian practitioners may be called upon to assist in capture of wild birds such as waterfowl

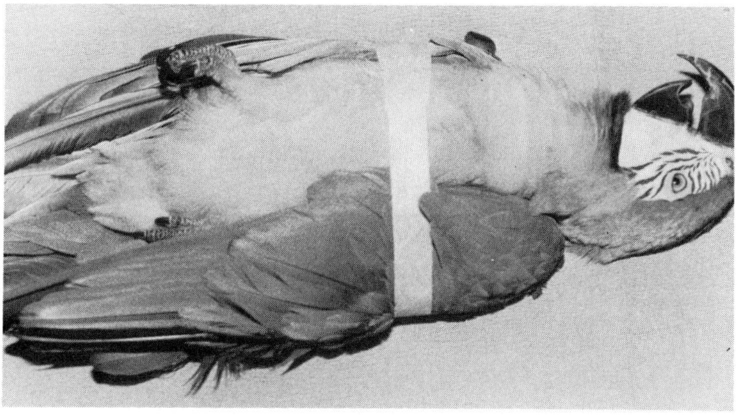

**Figure 45–2.** Masking tape may be lightly wrapped around the wings of Blue-and-Gold Macaws to prevent trauma from ketamine recovery.

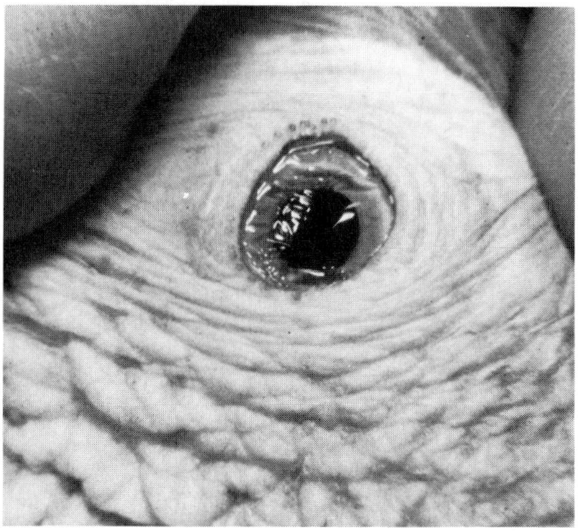

**Figure 45–3.** In a laterally recumbent bird that is under ketamine anesthesia, the anterior chamber of the down eye may collapse. The eye returns to a normal appearance within a few minutes after repositioning. (Courtesy of Lorraine Karpinski.)

for population reduction or transport. Drugged baits are reported to be the most effective method. A combination of alpha-chloralose and diazepam was reported to be faster and safer than either drug alone. The two drugs were dissolved in ethyl alcohol at 5 grams per fluid ounce to form a milky emulsion. The alcohol was allowed to evaporate overnight. Optimal dosage rates for diazepam and alpha-chloralose appear to be 0.30 to 0.40 gram and 0.10 to 0.12 gram per cup of bait, respectively. Whole shelled yellow corn was a successful bait and was used for an extensive prebaiting period prior to administration of the drug.

The use of 1 to 2 grains of pentobarbital mixed with bread has also been reported to immobilize wild ducks sufficiently for capture within 15 to 20 minutes.

The primary drawbacks of the use of drugged baits are the difficulty in controlling the dose and rate of absorption of drugs ingested orally by individuals of various sizes and species under a number of conditions: states of health of the birds, changing weather conditions, and idiosyncrasies of the locations. Sedated uncaptured individuals may be subjected to hypothermia, hyperthermia, overdose, suffocation, aspiration, pneumonia, drowning, predation, and peer-inflicted trauma.

## PRODUCTS MENTIONED IN THE TEXT

Diazepam (Valium—Hoffman LaRoche, Nutley, NJ)
Doxapram hydrochloride (Dopram-V—A.H. Robins Company, Richmond, VA)
Halothane (Fluothane—Ayerst Laboratories, New York, NY)
Isoflurane (AErrane—Anaquest, Madison, WS)
Ketamine Hydrochloride (Vetalar—Parke-Davis, Morris Plains, NJ) (Ketaset—Bristol Laboratories, Syracuse, NY)
Methoxyflurane (Metofane—Pitman Moore, Washington Crossing, NJ)
Methoxymol (Metomidate—Pitman Moore, Washington Crossing, NJ)
Xylazine (Rompun—Haver-Lockhart, Shawnee, KS)

## RECOMMENDED READING


Altman, R. B.: Avian anesthesia. Comp. Cont. Ed., 2:38–42, 1980.

Altman, R. B.: Effects of anesthesia on the temperature and electrocardiogram of birds. In Annual Proceedings of the American Association of Zoo Veterinarians, Denver, CO, 1979.

Amand, W. B.: Avian anesthetic agents and techniques—a review. In Annual Proceedings of the American Association of Zoo Veterinarians, Washington, DC, 1980.

Amand, W. B.: Avian anesthesia. In Kirk, R. (ed.): Current Veterinary Therapy VI. Philadelphia, W. B. Saunders Company, 1977, pp. 705–710.

Arnall, L.: Anesthesia and surgery in caged and aviary birds. Vet. Rec., 73(7):139–142, 1961.

Camburn, M. A., and Stead, A. C.: Anesthesia in wild and aviary birds. J. Small Anim. Pract., 19:395, 1978.

Eger, E. I., II: Pharmacology of isoflurane compared to other general anesthetics. In Annual Meeting Lectures, American Society of Anesthesiologists, St. Louis, MO, 1980.

Fanton, J. W., et al.: Halothane-induced hepatic necrosis in a snow leopard. J. Zoo Anim. Med., 15:108–111, 1984.

Gandal, C. P., and Amand, W. B.: Anesthetic and surgical techniques. In Petrak, M. L. (ed.): Diseases of Cage and Aviary Birds. 2nd ed. Philadelphia, Lea & Febiger, 1982.

Gaunt, P. S., et al.: Hepatic necrosis associated with use of halothane in a dog. J. Am. Vet. Med. Assoc., 184:478–480, 1984.

Haigh, J. C.: Anaesthesia of raptorial birds. In Cooper, J. E., and Greenwood, A. G. (eds.): Recent Advances in the Study of Raptor Diseases. London, Chiron Publications, 1980.

Harrison, G. J.: Recent advances in avian surgery. Proceedings of the Annual Meeting of the American Animal Hospital Association, Orlando, FL, 1985.

Harrison, G. J., et al.: A clinical comparison of anesthetics in domestic pigeons and cockatiels. Proceedings of the Annual Meeting of the Association of Avian Veterinarians, Boulder, CO, 1985, pp. 7–22.

Jessup, D. A.: Chemical capture of upland game birds and waterfowl: Oral anesthetics. In Nielsen, L., et al. (eds.): Chemical Immobilization of North American Wildlife. Milwaukee, WS, The Wisconsin Humane Society, Inc., 1982, pp. 214–226.

Karpinski, L. G., and Clubb, S. L.: Clinical aspects of ophthalmology in caged birds. Proceedings of the Annual Meeting of the Association of Avian Veterinarians, San Diego, CA, 1982.

Kittle, E. L.: Ketamine HCl as an anesthetic for birds. Mod. Vet. Pract., 52:40–41, 1971.
Klide, A. M.: Avian anesthesia. Vet. Clin. North Am., 3(2):175–185, 1973.
Kollias, G. V.: Avian anesthesia—principles, practice and problems. Proceedings of the American Animal Hospital Association Annual Meeting, Las Vegas, NV, 1982.
Mandelker, L.: Ketamine hydrochloride as an anesthetic for parakeets. VM/SAC, 67:55–56, 1972.
Mandelker, L.: A toxicity study of ketamine HCl in parakeets. VM/SAC, 68:487–488, 1973.
Mandelker, L.: Practical technics for administering inhalation anesthetics to birds. VM/SAC, March, 1971.
Maze, M., and Smith, C. M.: Identification of receptor mechanisms mediating epinephrine-induced arrhythmias during halothane anesthesia in the dog. Anesthesiology, 59:322–326, 1983.
Meyer, R. E., et al.: Isoflurane anesthesia as an adjunct to hypothermia for surgery in a dog. J. Am. Vet. Med. Assoc., 184(11):1387–1389, 1984.
Muir, W. W., et al.: Effects of xylazine and acetylpromazine upon induced ventricular fibrillation in dogs anesthetized with thiamylal and halothane. Am. J. Vet. Res., 36:1299, 1975.
Neal, L. A., et al.: Ketamine anesthesia in pigeons *(Columba livia)*: Arterial blood gas and acid-base status. J. Zoo Anim. Med., 12:48–51, 1981.
Pascoe, P. J.: More comments on xylazine. Letters, J. Am. Vet. Med. Assoc., 187(5):456–457, 1985.
Redig, P. T.: An overview of avian anesthesia. In Proceedings of the Annual Meeting of the Association of Avian Veterinarians, Atlanta, GA, 1982.
Steiner, C. V., and Davis, R. B.: Selected Topics in Caged Bird Medicine. Ames, IA, Iowa State University Press, 1981.
Stunkard, J. A.: Diagnosis, Treatment and Husbandry of Pet Birds. 2nd ed. Edgewater, MD, Stunkard Publishing Co., 1984.
U.S. Department of Health, Education and Welfare: Criteria for a recommended standard occupational exposure to waste anesthetic gases and vapors. March, 1977.
Webb, A. I.: Feline anesthesia. Proceedings of the Annual Meeting of the American Animal Hospital Association, Orlando, FL, 1985.
Wingfield, W. E., et al.: Waste anesthetic gas exposures to veterinarians and animal technicians. J. Am. Vet. Med. Assoc., 178:399–402, 1981.

## ACKNOWLEDGMENT

The author acknowledges Phillip Ensley, D.V.M., for his contributions to this chapter.

# Chapter 46

# SURGICAL INSTRUMENTATION AND SPECIAL TECHNIQUES

GREG J. HARRISON

Avian practitioners have found that the familiar surgical instruments, ligature techniques, and unaided 20/20 vision used in small animal surgery are not adequate for surgical resolution of some avian conditions. Therefore, specialized equipment and techniques developed for use in human medical and dental fields have been successfully adapted to avian procedures.

## MAGNIFICATION

The single most significant feature to be incorporated into avian surgery is a form of magnification to enhance the capabilities of the human eye. The ideal equipment for this is an operating microscope. The reader is referred to Chapter 47, Introduction to Microsurgery, for a discussion of the basic principles of working under magnification. Beyond its surgical applications, an operating microscope can be incorporated into clinical use (1) to locate veins of small birds for intravenous therapy; (2) to magnify radiographs; (3) to identify ophthalmic, skin, feather, and other lesions requiring closer scrutiny on physical examination; and (4) to enhance diagnostic capabilities during necropsies. For example, a number of gross postmortem conditions (e.g., bacterial or fungal colonies, parasites in the air sac, evidence of particles that may have produced inhalation pneumonia) are usually visible only by observation under magnification.

One significant benefit of magnification in general avian surgery is that it provides the capacity to identify specific bleeding vessels and, in conjunction with appropriate hemostasis, to minimize hemorrhage associated with the procedure.

The magnification afforded by binocular loupes may have application in some procedures. In limited access areas, a fiberoptic endoscope may provide the level of light required; however, magnification and surgical manipulation of tissues in the viewing field are limited with this instrument.

## ELECTROSURGERY (RADIOSURGERY)

The use of electrosurgery has been incorporated into avian surgical procedures because of the cutting and coagulating features and the capability of the electrode to reach previously inaccessible anatomic areas. Electrosurgery is the use of energy created by a high-frequency alternating current. Modern electrosurgical units* (Fig. 46–1) employ frequencies of 2 and 4 mHz, which include part of the maritime and amateur radio frequency spectrum; therefore, the term "radiosurgery" is used as an alternative term to electrosurgery to distinguish this technique from electrocautery.[2, 4, 8]

*Surgitron FFPF—Ellman International Manufacturing, Inc., Hewlett, NY.

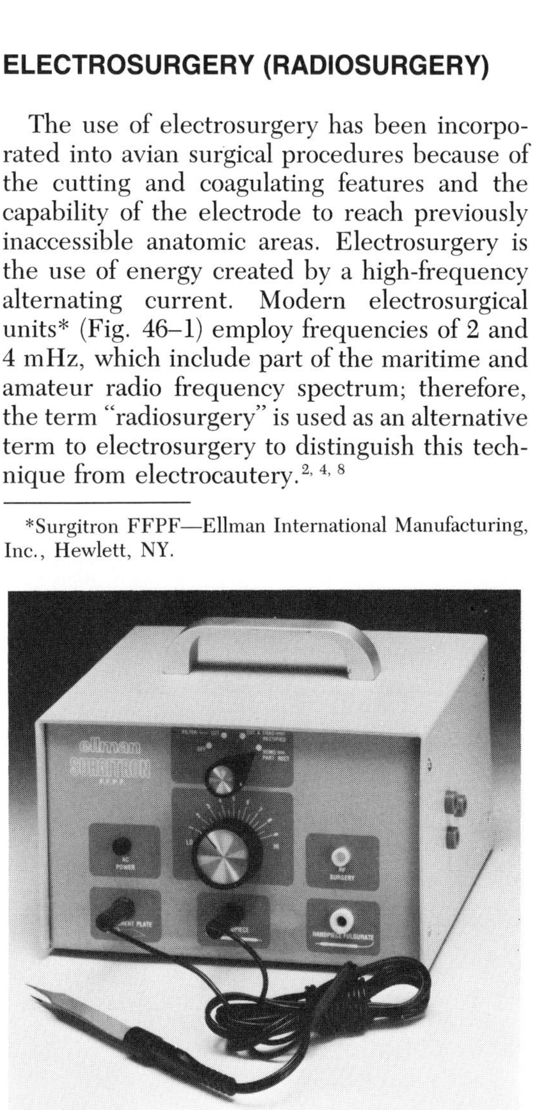


**Figure 46–1.** An electrosurgical unit produces high-frequency alternating current (radio frequencies) for use in avian procedures. (Courtesy of Ellman International Manufacturing, Inc.)

The resistance of tissue to the passage of this high-frequency energy creates internal heat in the tissue while the electrode itself remains cold. By the choice of electrodes and selection and adjustment of the current, the operator controls the effect of this energy on the tissues to achieve the desired results.

## General Principles of Radiosurgery

Cutting is achieved with radiosurgery when the energy emitted from the electrode generates molecular heat in each cell to the point where the fluids in the cell volatilize and the cell explodes. Applying this energy to individual cells in sequence by moving the electrode continuously through the tissue limits the line of destruction and produces the cutting effect. At the same time the capillaries are sealed, resulting in almost bloodless cutting.

When the fine wire electrode is used for cutting, it is suggested that a practice stroke or two be taken before the tip is activated to ensure that the planned stroke can be completed comfortably and correctly. The electrode is activated by a foot pedal or finger switch prior to contacting the tissue to disperse any initial current surge. Using a smooth brushing motion without pressure, the electrode should pass through the tissue without dragging. If it seems to drag or bits of tissue adhere to the tip, the current may not be set high enough; if it sparks or the tissue turns white, the current may be too high. Charred tissue clinging to the electrode tip must be cleaned off immediately with an emory paper or scalpel blade to prevent further tissue damage. It is not necessary to race through the incision, but inordinately slow motion will generate undesirable lateral heat in the tissues, with the possibility of subsequent necrosis and sloughing.

The coagulation of bleeding vessels takes place when the current density is sufficiently concentrated to dehydrate the cells and coagulate their organic contents, but deep penetration into the tissues is avoided. In some cases coagulation may be accomplished by clamping a bleeding vessel with a hemostat and touching the tip of the electrode to the hemostat. In contrast to cutting, the electrode should contact the tissue *before* the switch is activated for coagulation procedures.

A commitment to practicing and refining techniques will improve the surgeon's skills with this equipment. The most common initial discouragement results from a failure to persist in practicing to achieve the balance of touch and current for the desired results.

An extensive choice of electrodes for other purposes is available, such as variable-length fine wires for cutting, and loops, snares, and diamonds for biopsies. A specially adapted handle services a disposable scalpel blade to combine traditional incising with electrosurgical applications. In addition, a standard hypodermic needle may be attached by an adapter to the electrosurgical unit (Fig. 46–2) and used with or without the current for procedures such as dissection of scar tissue or fracture calluses. A flexible cord, which is optionally available as a coiled phone-type cord, attaches the electrode handle to the power source.

Some models of electrosurgical units are capable of producing a spark-gap current, which, when combined with a ball electrode, is useful for desiccation and fulguration for destruction of tumors and cysts.

When using the single hand-held electrode, a ground plate directly under the area of surgical manipulation in the patient serves as the second electrode. This fine tunes the energy frequency and allows lower power settings on the machine.

## Use of Radiosurgery in Birds

The skin and subcutaneous tissues in a bird are not the same as they are in mammals; therefore, some principles of radiosurgery that were previously developed by Greene[6] for use in mammals (paint brush stroking, fine tip cut-

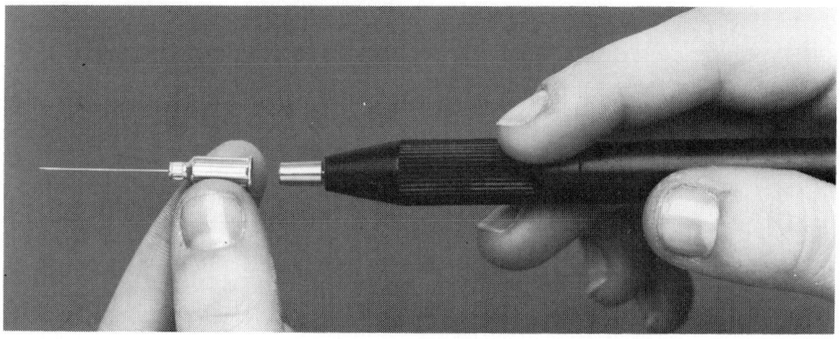

**Figure 46–2.** A variety of electrode tips is available for radiosurgery, including an adaptation for a hypodermic needle. (Courtesy of Ellman International Manufacturing, Inc.)

ting) are not as successfully applied to avian surgery in this author's experience.

Birds have a relatively thin epidermis that is nourished by diffusion of capillaries in the dermis[5]; therefore, bleeding from a skin incision is not as acute as in mammals. The dermis is closely opposed and firmly attached to the muscle fascia with little intervening subcutaneous tissue.[5] Simultaneous cutting/coagulation with the radio current and fine tip electrode is successful in mammals because the tissues surrounding the vessels are heated up by a partial absorption of the cutting energy and help to seal the vessel. Avian blood vessels tend to be less protected by surrounding tissues; therefore, a greater potential for hemorrhage normally exists (in venipuncture as well as surgical procedures).

Cutting through the relatively dry skin tissue in birds with the fine tip electrode requires a high radiosurgical setting (i.e., No. 4). When the wet blood vessel is encountered at this setting, the intense current directed from the fine tip electrode has a tendency to cut rather than coagulate the vessel in avian patients.

Although simply touching the electrode directly to a vessel may stop the bleeding, the use of magnification has revealed the extent of damage that may be created in surrounding tissues when this technique is applied to small avian vessels. Additionally, it has been noted by the author that a vessel that responds to the surface current by retraction may later relax and start oozing again.

Some modifications in radiosurgical techniques have been developed through the use of modified ophthalmic bipolar forceps (Fig. 46–3), which have a broader surface area for dispersement of the current, are used at a lower energy setting, and appear, in this author's opinion, to provide a better method of achieving hemostasis in pet birds than do single fine wire electrodes. Bipolar forceps combine both electrodes in a single instrument, thus eliminating the need for a ground plate.

Coagulation procedures with the bipolar forceps are best performed under magnification so that the surgeon can identify, isolate, and provide hemostasis to the individual bleeding vessel by precise tissue welding procedures. The setting for this coagulation technique with the bipolar forceps is usually No. 1 with fully filtered current. This form of current is normally used only for cutting in mammalian procedures.

It is recommended that gloves be worn when bipolar forceps are used for avian surgery to prevent transfer of the current to the surgeon. In contrast to the firm grip on the forceps that is recommended by the manufacturer for mammalian use, the light touch that is required for these avian techniques may allow current transfer to unprotected hands. Alternately, insulated bipolar forceps may be used.

In addition to facilitating coagulation in birds, bipolar forceps may be employed for radiosurgical incisions. The degree of tissue reaction varies with the contact distance of the instrument to the tissue. Direct firm contact with the tissue by the forceps results in dispersion of the energy away from the point of contact. Elevating the forceps slightly from the tissue creates an arc that localizes and intensifies the current. This latter principle has been used by the author for coagulation prior to traditional incising. The fine-tipped bipolar forceps may be swept across the potential incision line just above the level of the tissue. The resulting color change will indicate where scissors can be used

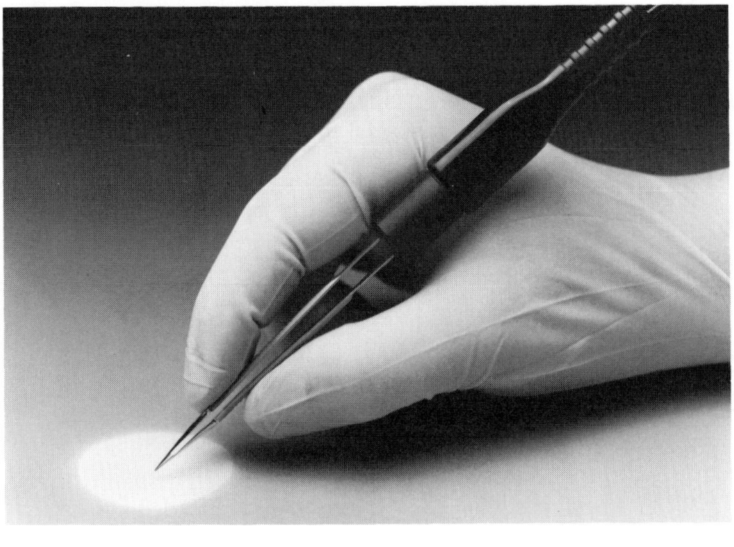

**Figure 46–3.** Bipolar forceps include both electrodes in a single instrument and concentrate the energy transfer to the area between the tips, eliminating the need for a ground plate. (Courtesy of Ellman International Manufacturing, Inc.)

for cutting. This technique is often the fastest, least hemorrhagic means of incising the relatively thick skin of larger species, e.g., Amazons, macaws, cockatoos.

A cutting technique suitable for thin-skinned birds (cockatiels and smaller species) utilizes the forceps in direct contact with the tissue. The incision is made by sequential picking up of bits of tissue with the forceps and activation of the current. This is successful because of the relative lack of connective and fibrous tissue in avians.

The author has altered the ophthalmic bipolar forceps by bending one of the points a few millimeters from the end into a 45-degree angle toward the other tip. The bent tip can be used without current in some instances to bluntly dissect soft tissue. For vessel coagulation the straight tip of the forceps is placed under the vessel to elevate it from the surrounding tissue. Because of the position of the other tip, the transfer of current between the bipolar electrodes occurs at a slight angle, which appears to facilitate adhesion of the vessel walls for hemostasis. This is the primary surgical instrument in the author's practice.

Whereas it was necessary to frequently adjust the settings on the machine in previous techniques to accommodate the varying degrees of vascularization, the current is rarely changed when the latter above techniques are used.

Practicing radiosurgical techniques is ideally done on fresh necropsy specimens or recently euthanized birds in order to appreciate and develop skill in using the coagulation capabilities of the machine.

As an alternative to an electrosurgical unit, the beginning avian surgeon may locate bleeding vessels with magnification and and control hemostasis with a pen light cautery unit. A similar cautery unit is available as an adaptation to the alternating current handle of a portable endoscope system.* (See Chapter 48, Selected Surgical Procedures, for cautions in the use of spark-gap cauteries.)

## CRYOSURGERY

Although practitioners may not have access to commercially available liquid nitrogen cryosurgery units, directed skin refrigerants (Fig. 46–4) such as dichlorodifluoromethane† have made adaptations of cryosurgical principles

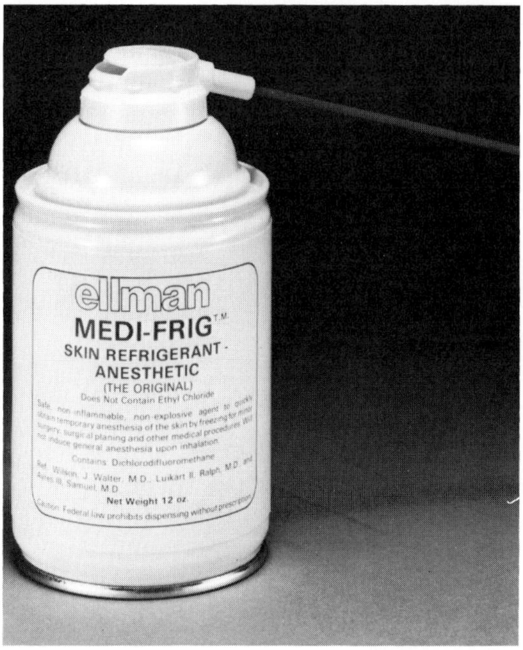


**Figure 46–4.** Pressurized cans of skin refrigerants are useful for cryotherapy of selected avian conditions.

available for some treatment attempts in birds. To confine the effects of the chemical to the targeted lesion, petroleum jelly can be applied around the perimeter of the area.

Selected cases of cloacal papillomas have responded to cryosurgery, although the procedure must be performed in stages because aggressive freezing of cloacal tissue may result in nerve damage.[1] Cryosurgery has been a useful therapy in selected cases of chronic mouth lesions and other granulomas and has been reported as an alternative for enucleation in a cockatiel.[11] Cryotherapy in intra-abdominal procedures is contraindicated.

## USE OF MEDICAL ACRYLICS

Originally developed for the dental field, medical acrylics (Fig. 46–5) have been useful as an adjunct to some avian procedures. Ethyl acrylic hardens to a firm substance and is suitable for splint, bone, or beak work; isobutyl acrylic* has flexible properties and may be used for bandaging or surgical closures. Both forms are tissue compatible. The widely available methyl acrylic (Super Glue) contains tissue-toxic components.

*Focuscope—Medical Diagnostics, Brandon, FL.

†Medi-Frig—Ellman International Manufacturing, Inc., Hewlett, NY.

*Cyanodent or Tissu Glu—Ellman International Manufacturing, Inc., Hewlett, NY.

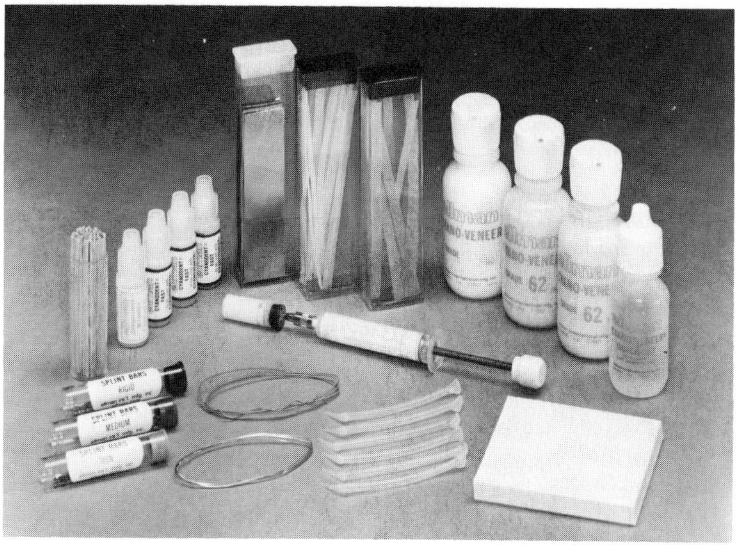


**Figure 46–5.** Commercially available dental acrylic products and accessories can be used for beak repair. (Courtesy of Ellman International Manufacturing, Inc.)

Isobutyl acrylic (tissue glue) physically anchors tissues together until normal cell regeneration occurs. Physical penetration of the acrylic by the cells does not occur; therefore, care must be exercised in the application to prevent its producing a healing barrier. The opposing edges of the tissue are rolled slightly under and the acrylic is applied in a layer over the top of the rolls rather than directly on the exposed tissue margins. A micropipette produced by the company assures accurate application of the product. Application of very thin layers of the substance to thoroughly dry tissue will achieve the best results. More pronounced adhesion will occur in some cases with application of a second coat after the first layer has dried and been peeled off.

Acrylics may also be used as a sealant/bandage until natural healing can take place. For example, after amputation of a finch leg, the acrylic may be coated over the stump following the application of electrosurgical current or ligatures to any bleeding vessels. In one case of xanthomatosis in a Sulfur-crested Cockatoo (Fig. 46–6*A*), in which a significant surface area of the distal quarter of both wings was electrosurgically planed, a coating of acrylic was applied over the exposed surface for protection during healing (Fig. 46–6*B*), as no epithelium was available for traditional closure.

Other applications for the liquid acrylics in the experience of the author include (1) painting over a curled toe to serve as a splint; (2) repositioning an abducted wing on a newly hatched macaw; (3) splinting legs of small birds; (4) holding intravenous and endotracheal catheters in place; (5) reinforcing traditional sutures on friable tissue; (6) covering sutures or wounds to prevent picking.

Liquid acrylic products may be combined with a strengthening powder* and used to seal deep, penetrating cracks, chips, or holes in beaks to protect the underlying tissue from fungal or bacterial contamination. Bone repair (long-bone fractures or skull injuries) has been accomplished with this material.

One beak repair technique begins with a thorough cleaning of the injury with a dental drill, flushing with a dilute povidone-iodine solution, and thorough drying. To reinforce adhesion to the beak, keying, or the drilling of a series of small holes angling in from the top to the bottom, may be necessary on each side of the injury. The strengthening powder is mixed into the acrylic liquid on a glazed mixing pad and applied over the entire area, including the openings of the holes; the holes then serve as dovetails to hold the material in place. A coat of a quick-setting solution immediately hardens the material. To facilitate working with the material in more complicated procedures, such as filling compounds for large depression fractures in beaks, a retarder solution slows the hardening of the product. Tints can be added to match the natural color of the beak.

Some vertical fractures in lower beaks have been repaired with wire reinforcement and application of acrylic material; however, there has been little success with use of these products in beaks fractured beyond the germinal layer. In these cases, several applications of the acrylic can support and splint the injury for the time period required by the bird to adjust its eating habits.

---

*Cyano-Veneer Kit—Ellman International Manufacturing, Inc., Hewlett, NY.

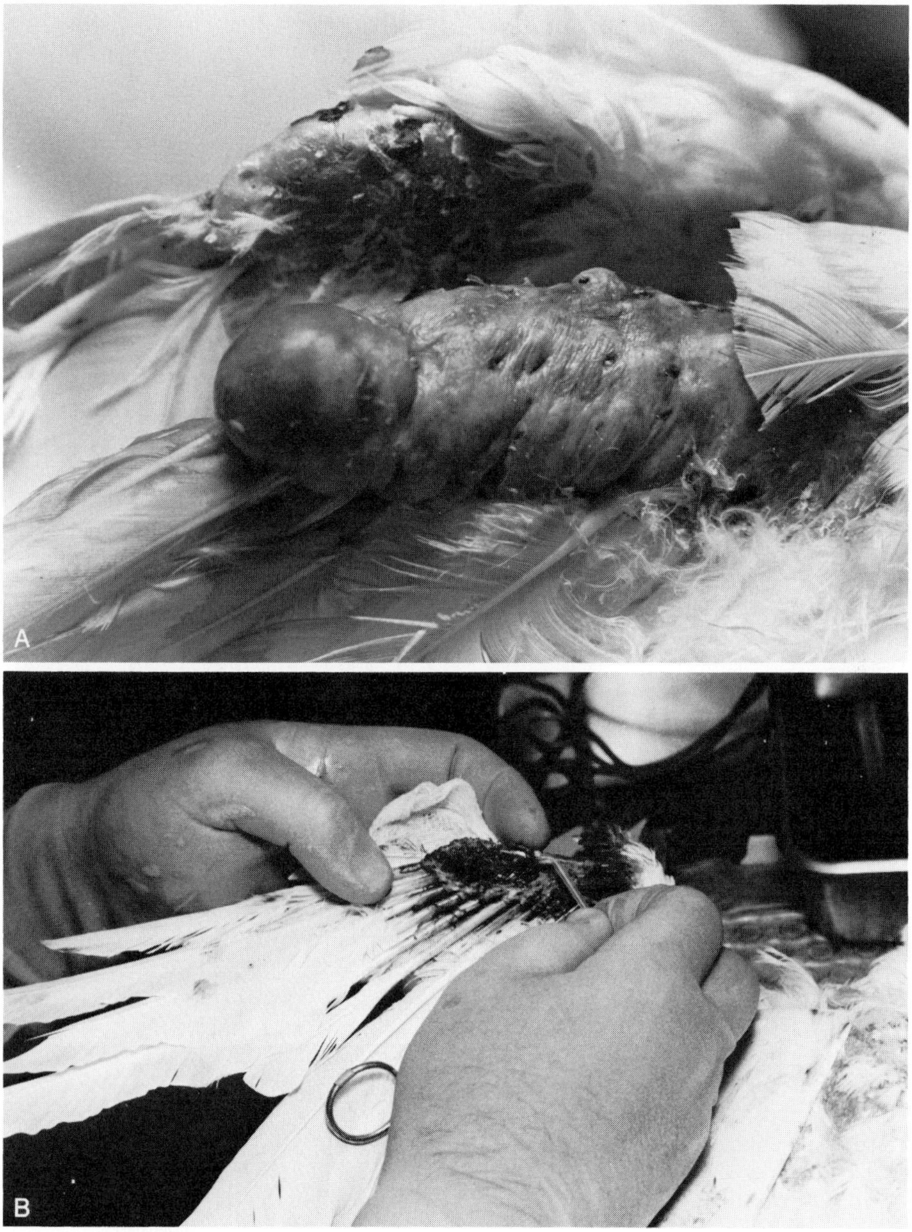


**Figure 46–6.** *A*, Xanthomatosis on the wing of a Sulfur-crested Cockatoo prior to therapy. *B*, Following careful electrosurgical excision of affected tissue, an acrylic bandage is applied with a micropipette.

There are reports of successful attachment of prosthetic beaks with wire implants in a duck,[4] goose,[8] and Moluccan Cockatoo.[3] A malaligned (undershot) beak in a young cockatoo was corrected with a long prosthetic device applied to the filed upper tip, thus forcing the upper beak out over the lower beak.

Some caution must be exercised in the use of acrylics around the nares, air sacs, marrow cavities, or other highly vascular areas of birds under inhalation anesthesia. The fumes have a synergistic effect with the gas and deepen the anesthetic plane. Even in nonanesthetized birds, the fumes tend to be irritating to the eyes and may induce vomiting in birds if used in an unventilated room.

## AUXILIARY SURGICAL INSTRUMENTS AND SUPPLIES

Supportive supplemental heat is recommended for the surgical patient, although a degree of body temperature loss during anes-

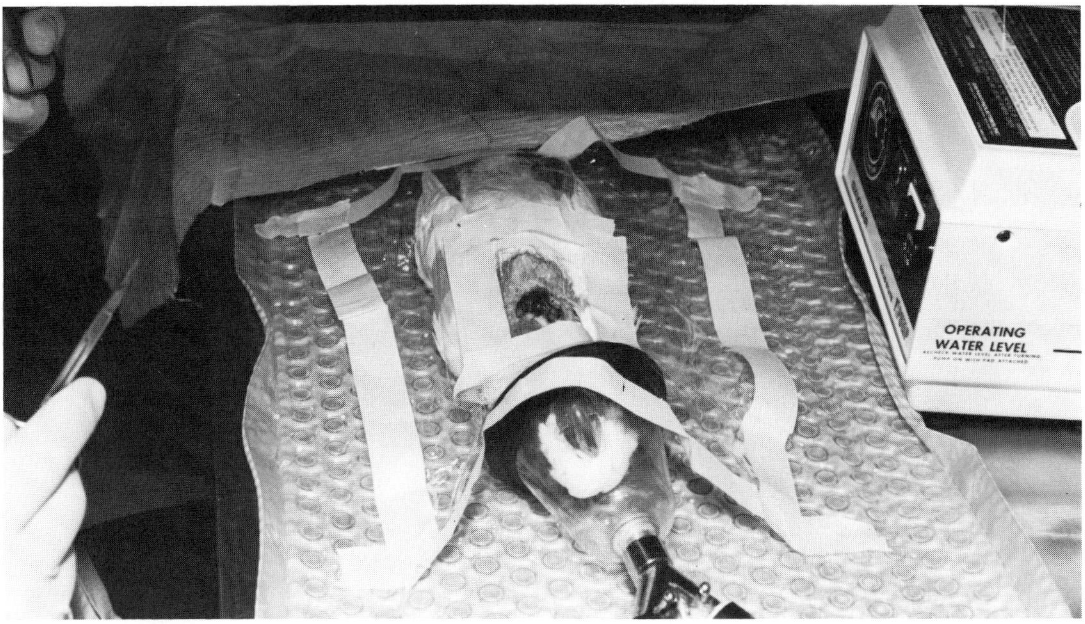


**Figure 46–7.** Inhalation anesthetic is administered to a cockatoo by face mask in preparation for repair of a crop injury. A circulating water blanket provides supplemental heat; clear plastic drape allows direct visualization of the patient.

thesia is unavoidable (see Chapter 45, Anesthesiology). A useful item for supplying heat is a thermostatically controlled, disposable, circulating water blanket (Fig. 46–7), which, in the author's practice, is preset to 102°F. This may be gas sterilized for reuse. Electric heating pads provide poor distribution of heat over the surface area and are not recommended for this purpose unless the patient is well-protected from direct contact with the pad and is monitored carefully. Panting due to hyperthermia may lead to anesthetic overdose.

Ophthalmic instruments (Fig. 46–8), in particular retinal forceps (smooth and toothed-jawed), scissors, iris hooks, and lid retractors, are essential components of the avian surgical pack. The miniaturized handles allow for visualization concurrent with manipulation during intra-abdominal, intratracheal, oral, or intracloacal procedures. The hooks are used for manipulation of delicate tissue, and the lid retractors serve as wound retractors.

Small sections of I.V. tubing may be placed over the ends of fine-tipped instruments to protect them during sterilization and storage. Fine brushes or ultrasonic instrument cleaners and lubricating solutions preserve the finely engineered movements.

Other elements of the avian surgical pack may include rubber bands (useful as tourniquets) and 2-inch × 2-inch gauze pads precut from standard 4-inch × 4-inch pads. Wooden cotton-tipped applicators may be used to absorb blood during procedures. When an applicator is split lengthwise to create a pointed end, it becomes a surgical "instrument" and can be used to position tissues or retrieve material that appears to adhere easily to the wooden surface (e.g., blood, exfoliated cells, exudates).

Suture material for avian procedures should be available from 3-0 to 10-0, the smallest used for vessel repair. Silk or nylon appears to produce the best results. Aneurysm clips show

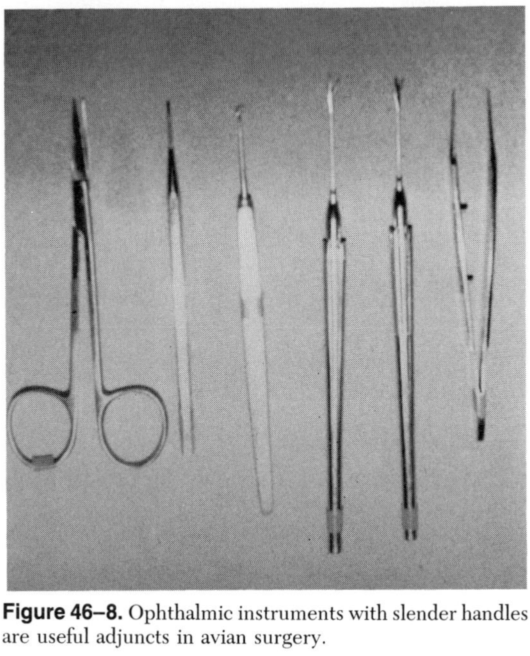

**Figure 46–8.** Ophthalmic instruments with slender handles are useful adjuncts in avian surgery.

promise as ligatures (see Chapter 48, Selected Surgical Procedures).

Disposable transparent surgical drapes can be fashioned from pre-cut pieces of plastic wrap* that are bound with masking tape to prevent tearing. A precut incision area sized appropriately for the species is also bound with tape. These are superior to cloth or paper drapes in that they are inexpensive, disposable, flexible, form fit to the bird, and allow the surgeon complete visualization of the patient during the procedure. These drapes can be folded with paper and gas sterilized. Transparent surgical drapes are also available commercially.†

A small restraint board similar to what may be used for radiography is valuable for restraint of an anesthetized bird. This facilitates any repositioning that may be required during endoscopy, microscopic viewing, or surgical procedures. The author prefers masking tape fasteners to Velcro strips because Velcro is soiled relatively easily by feathers and dust.

Anesthesia face masks can be adapted from those supplied for dog and cat patients. A cat mask can be adjusted to accommodate a cockatiel by stretching a surgical glove over the open cone and cutting a hole of the appropriate size in the glove for insertion of the head. A similar technique may be used to fashion a mask from a plastic 10-cc syringe case for a budgerigar. This is then attached to an endotracheal tube adaptor for connecting to the anesthetic hose.

Soft, flexible intravenous catheters (18-, 20-, or 22-gauge) are suitable as endotracheal tubes for small birds. A tiny drop of surgical glue can be used to adhere the tube to the side of the beak. The glue will easily peel off when the surgery is complete. Exercise the care previously described for use of acrylics with inhalation anesthesia.

Because a number of pieces of equipment may be used simultaneously with avian surgical patients, a small cart on wheels facilitates moving the multiple pieces simultaneously to the surgical area and frees up space on the surgery table.

*Saran—The Dow Chemical Company, Indianapolis, IN.
†General Econopak, Inc., Philadelphia, PA.

## REFERENCES

1. Altman, R. B.: Proceedings of Seminar on Avian Medicine, American College of Veterinary Surgeons 12th Annual Veterinary Surgical Forum, Chicago, IL, 1984.
2. Cresswell, C. C.: Electrosurgery: A treatment for verrucae. The Chiropodist, September, 1984.
3. Frye, F. L.: Prostheses enhance quality of life. Vet. Med., July, 1984, pp. 931–935.
4. Goldstein, A. A.: Radiosurgery in dentistry. Quebec Dental Journal, October, 1977.
5. Graham, D. L.: The avian integument—its structure and selected diseases. In Proceedings of the Annual Meeting of the Association of Avian Veterinarians, Boulder, CO, 1985, pp. 33–52.
6. Greene, J. (ed.): Principles of Electrosurgery. Hewlett, NY, Ellman International Manufacturing, Inc.
7. Harrison, G. J.: New aspects of avian surgery. Vet. Clin. North Am. [Small Anim. Pract.], 14(2):363–380, 1984.
8. Kruase-Hohenstein, U.: Electrosurgery: Fundamental requirements for successful use (I). Oral Surg., 11:1–19, 1983.
9. Paster, M. B., and Isaacs, M.: Repair of damaged beaks in Cherry-headed Conures using cyano-veneer acrylic. Bird World, March, 1983, pp. 52–53, 55.
10. Sleamaker, T. F., and Foster, W. R.: Prosthetic beak for a Salmon-crested Cockatoo. J. Am. Vet. Med. Assoc., 183(11):1300–1301, 1983.
11. Soifer, F. K.: Case report: Cryotherapy as an alternative to avian enucleation. AAV Newsletter, 5(4):105, 1984.
12. Wolf, L.: Prosthetic bill technique for a Canada Goose. Proceedings of the Annual Meeting of the Association of Avian Veterinarians, Boulder, CO, 1985.

# *Chapter 47*

# INTRODUCTION TO MICROSURGERY

JAMES E. DOYLE

There are many procedures in the author's private practice of plastic surgery, specifically hand surgery, that can be adapted to surgical repair of birds. The observations and conclusions described below are drawn from an extensive experience with repair, rehabilitation, and return to the wild of sick and injured birds of prey. The size of the patient and critical nature of many of the repairs make microsurgery and the use of the operating microscope natural adjuncts for avian surgery.

Specific to this area is the use of the operating microscope. The dexterity and manipulation of a surgeon's fingers and hands are far greater than unaided vision will permit. Enhancing these capabilities with the microscope allows the surgeon to use his fingers and hands with a greater degree of precision and care than otherwise attainable. Moreover, magnification in soft tissue procedures affords a closer scrutiny of vessels that may hemorrhage, hemorrhage being one of the major causes of surgical failure in pet birds, particularly with intracavitary procedures (intracranial, intrathoracic, and intra-abdominal).

The impetus for this new work began in the early 1970's with the advent of microvascular surgery, which allowed anastomosis of vessels as small as 0.4 mm to provide a patent antegrade flow. This opened the door to replantation of divided or sectioned digits and limbs and the removal and replacement of large sections of osseous and composite soft tissues from remote distances in a single stage, such as the replacement of a missing thumb with a big toe. All of this is significant in the repair and restoration of wild birds and in the operative therapy of all avian species.

In the past with wild bird rehabilitation, adequate treatment was simply to pin or otherwise stabilize fractures and follow this by the immediate repair of overlying soft tissue. If one considers for a minute the impact required to fracture an avian bone, itself a marvel of light and strong construction, it does not take much imagination to suppose that other more deeply situated, finer, and more delicate structures are similarly injured.

We have reached the point in the surgery, care, and rehabilitation of these injured limbs that simply pinning the fracture and closing the skin will have to give way to more thoughtful investigation, exploration, and repair of injured soft tissues as well, specifically blood vessels, nerves, and musculotendinous units. Repair and restoration of joints will ultimately become a specialized discipline in its own right, but at the present time the engineering tolerances of the joints are so critical that they are relatively intolerant of even the healing process, which itself compromises the ultimate functional range of motion.

Understanding and committing to memory the anatomic relations in a *regional* manner rather than in the usual textbook *systemic* fashion cannot be overemphasized. The author is unaware of any detailed avian surgical text currently available which presents the detailed regional anatomy required by the microsurgeon. The responsibility, then, for acquiring this information lies squarely with the surgeon himself who must do his own dissections and sketches. Careful attention must be paid to fascial and muscular planes and to identifying which vessels and nerve branches lie in each. The surgical principle involved when facing a large hemorrhagic wound that is stained with blood, feathers, and dirt is to proceed proximally and distally, identify the undisturbed structures, and follow them back into the wound. This identification and orientation of structures will prevent the twisting and abnormal rerouting of bones, tendons, vessels, and nerves that commonly occurs.

## MICROSURGICAL TECHNIQUES

In order to accomplish some of the microsurgical procedures, including microvascular repairs, one must become familiar with the technical knowledge and instrumentation and develop some expertise in this area. Essential to microsurgical performance is the avoidance by the surgeon of alcohol, stimulants such as caffeine or nicotine, heavy exercise, and of course, fatigue.

Most macroscopic surgical techniques are applicable to microsurgery; however, a few are replaced or modified. Because the field is magnified, movements that are taking place at a normal pace seem very rapid under the microscope. The velocity increases in direct proportion to the increase in magnification. Normally undetectable hand motions may appear as tremors under the microscope. Therefore, it is necessary that all actions be carried out in a slow, deliberate manner. Particularly when drawing suture through a small vessel, one must progress at a very slow pace in order to avoid tearing the tissue or pulling the suture all the way through the vessel.

Because microscopic fields are so small, it is necessary for the surgeon to keep his instruments, needles, and suture material in an area close around the visible field so that they can be easily retrieved. This may require that the needle be placed in a bit of muscle close to the anastomosis site or on a microscopic clip. When working at some of the higher magnifications, the operator will even find some difficulty in placing his instrument into the microscopic field. This can sometimes be facilitated by sliding the instrument to be placed in the field down the shaft of one that is already located in the field.

It is also essential that the surgeon's wrists be anchored close to the operative site. In contrast to techniques taught in macrosurgery, flicking or snapping the suture through the tissue is not possible in microsurgery because of the great vibration and tissue tearing that may occur during this type of maneuver. Rather, it is necessary to hold the heel of the hand flat against a surface and *roll* the instrument between the fingers. The wrists must rest securely on something near the patient that will provide a sound base: a rolled towel, a special post device, or the edge of the animal board. Resting it on the patient itself is discouraged because the respiration of the patient combined with the surgeon's own respiration will exaggerate the tremor and vibrations that are normally visible through the microscope.

It is necessary for the surgeon to "control" his own respirations by breathing slowly and regularly. Breath holding is almost a reflex for any fine intricate movement but should be avoided to minimize vibrations.

If a hand tremor becomes uncontrollable during the surgery, there are many things that might be suspected, but three are most common. First, the surgeon may find that his wrists are no longer resting on the support. It is very difficult to do microscopic surgery with the wrists in the air. The very large muscles of the shoulder simply cannot hold the hands still enough to operate under the microscope. Additional support is gained with the placement of the feet securely on the floor. Second, again because the field is so small, it is very possible that either the gloves, the suture, or the gown has become entangled in something that is outside the visual field of the microscope and the resulting tether, unrecognized by the surgeon, is preventing him from accomplishing his objective. Finally, when a severe tremor develops, the surgeon may find that he has been holding his breath for the last few minutes and is watching his own hypoxia at work.

It is advantageous in microsurgery to be able to use all instruments in either hand. As the techniques in using these instruments are being developed for the first time, it is worthwhile to acquire ambidextrous skills.

## INSTRUMENTATION

Only the tips and the boxes of microsurgical instruments are required to be microsized. Most of the instruments should have handles of normal length, that is, at least the length of a pencil, so that the handles may extend and rest over the back of the hand (Fig. 47–1). This is to help provide stability to the tips of the instruments and to accommodate the rolling action of the instruments necessary in microsurgery. Because of the required rolling action, nearly all microsurgical instruments, with the exception of scissors, come with round handles. Damage and everyday wear and tear make it useful for the surgeon to acquire some skill in the sharpening and repair of these small instruments. Care is usually done under the microscope.

The needle holder, from an engineering standpoint, needs a given strength and size at the box and proximal part of the point; however, more distally toward the suture and needle-handling end, it is ground down to a dolphin-nosed point.

For vascular repair, an irrigation device is needed to instill heparinized solution into the lumen of the vessel. This step washes out any clots and debris from intravascular blood that remains after clamping. A blunt 30-gauge

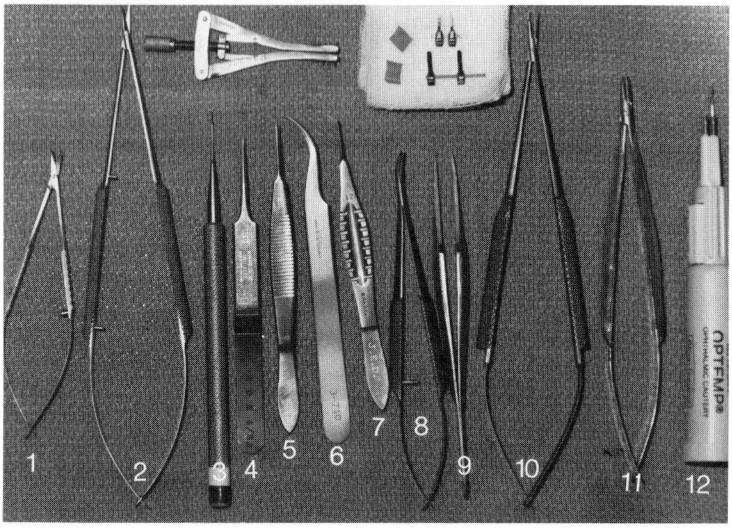


**Figure 47–1.** Microsurgical instruments include (from left) microscissors (1,2), intra-arterial loop (3), micro pickups (4–7), micro needleholders (8–11), disposable ophthalmic cautery (12); (top) self-retaining retractors; microvascular clips with balloon patches.

needle is adequate. A regular 30-gauge hypodermic needle can be used; however, the tip must be amputated and polished under the microscope to prevent snagging the intima of the vessel. The needle is attached to a 30-cc syringe of heparinized lactated Ringer's solution with a plastic connector. This solution may also be used to keep the vessels and surrounding tissues moist throughout the procedure.

The suture material most often used for microvascular repair is a 10–0 nylon with a 75-micron needle. A 6–0 to 9–0 size suture is normally satisfactory for other, less delicate avian repairs.

A microvascular clip is actually two clips on a sliding bar; one clip is placed on the distal and the other on the proximal segment of the vessel when carrying out the repair. A small piece of colored rubber balloon is placed behind the clip and serves as a color backdrop to afford better reference and distinction.

A number of different electrocauteries can be used in microvascular work. The normal bipolar electrocautery is most common. However, for experimental work and for many avian procedures, the ophthalmic battery-operated, nonresterilizable electrocautery is beneficial and has the advantage of having no cords or other connections. (See Chapter 48, Selected Surgical Procedures, for cautions in use.)

The most expensive item, of course, is the microscope itself (Fig. 47–2). Many used models are available through medical schools when the schools update their equipment to include motorized zoom and focus features. These features are not necessary for bird surgery and require additional foot work in order to operate them.

The objective lens of the microscope should be approximately 150 mm for bird surgery; the ocular lens is 12.5 mm. Longer focal lengths may require the surgeon either to sit a great deal higher or to stretch his head and neck in order to reach the oculars, resulting in inability to set his feet flat on the floor with a possible increase in tremor. Nearly all operating microscopes have their own intrinsic light source; however, some may use a reflective light source requiring a mirror.

Some less expensive substitutes for commer-

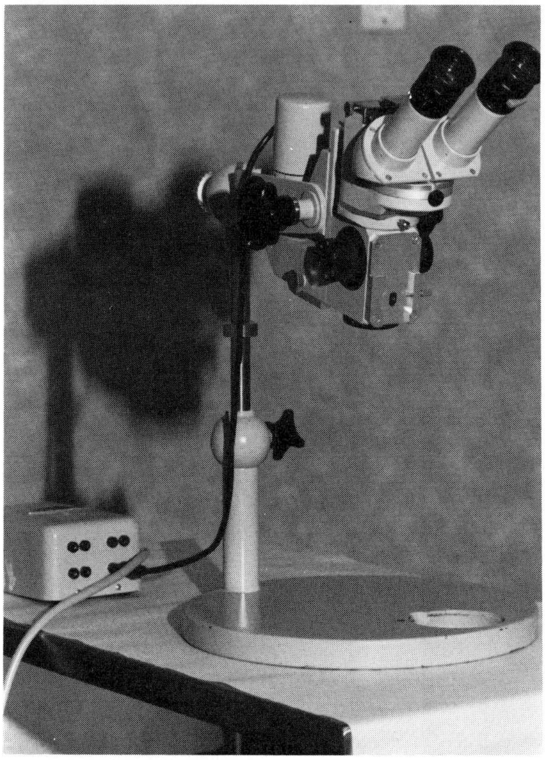

**Figure 47–2.** Tabletop model operating microscope.

cially available microvascular instruments work well in the early stages of experience while the clinician is evaluating the merits of this surgical mode for his practice.

The usual Castroviejo microvascular needle holder may cost from $250 to $450. No. 4 or No. 3 jewelry forceps from a jewelry supply store, with vulcanized silicone (Silastic) applied to the outside of the handles to make them round can replace this instrument. Modification of an ordinary Castroviejo ophthalmology needle holder requires grinding of the tip and application of Silastic to the handles. In addition, the normal clasp or lock on this instrument must be removed because the jar of the click or lock in setting or releasing is enough to tear tissue or fine vessels. These substitute instruments roll well, are strong enough to hold the micro needles and sutures, and are approximately one tenth the cost of an instrument designed specifically for microsurgery.

For general dissection, No. 2 or No. 3 jewelry forceps can be used. The No. 5 jewelry forceps, which is the finest of those available, is used for the very finest dissection, specifically dilating vessels from the inside and stripping the adventitia. Generally two of these are used simultaneously, one in each hand, for the final preparation of the vessel for repair.

A section of latex tubing approximately 1/4- to 5/16-inch in diameter is an adequate replacement for the vascular clip for use in developing initial skills. The top is cut off the tubing, and the tubing itself is then slit in a miter box fashion (Fig. 47–3).

## LABORATORY EXERCISE TO PRACTICE MICROVASCULAR REPAIR

The following is an introductory procedure that can be practiced to develop microsurgical skills. It is advisable to practice microvascular techniques (e.g., suturing of vessels) on laboratory rats to acquire the skill necessary for avian repair. An end-to-end repair is described; however, end-to-side repairs and side-to-side repairs are also possible. Vein grafts may have some potential in replacing arteries lost in gun blast injuries.

In the laboratory specimen rat, the femoral artery and vein are the most accessible and allow refinement of surgical techniques under magnification. Birds will differ in that they have a large network of veins that surrounds most of the major arteries. Accurate repair of an avian artery cannot be accomplished until this network of veins is cleared away.

Four strips of autoclave tape on the extremities and one across the tail are usually sufficient

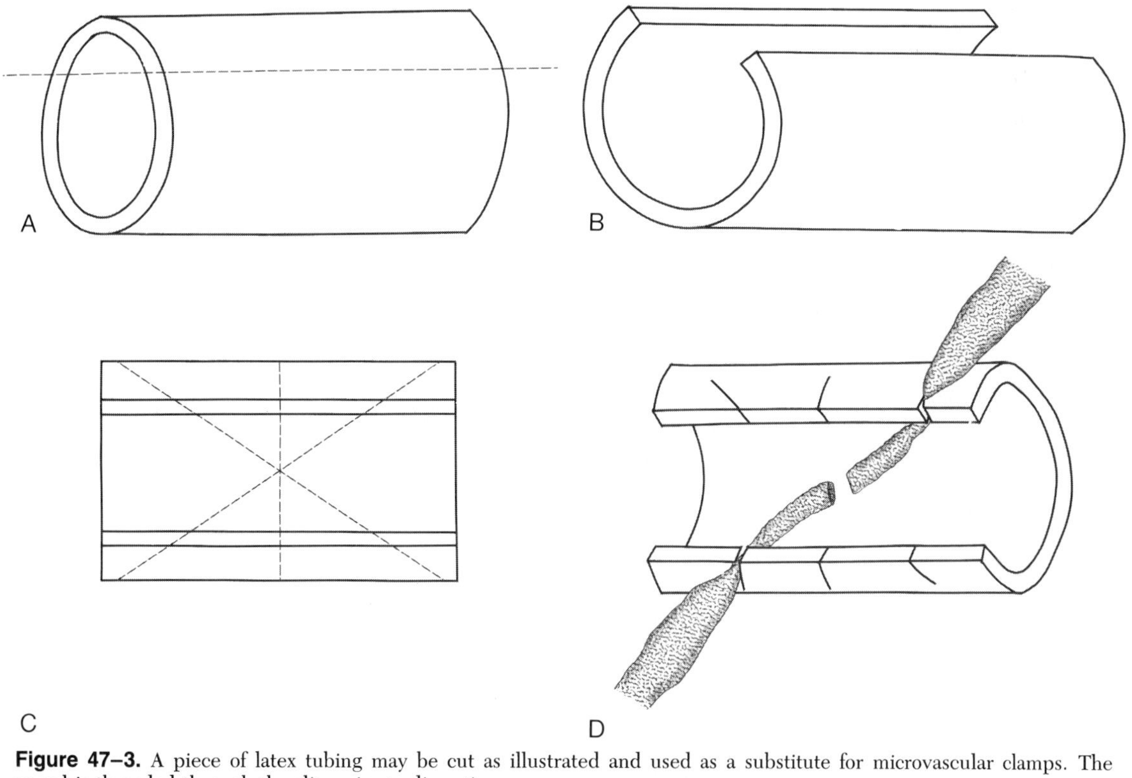


**Figure 47–3.** A piece of latex tubing may be cut as illustrated and used as a substitute for microvascular clamps. The vessel is threaded through the slits prior to dissection.

to position the anesthetized rat on an animal board. The groin area is clipped and surgically prepared. A long incision, usually the entire length of the groin, is recommended, especially in the early stages of practice. This allows the microvascular clip to lie flat instead of in a caudal position, which it will usually assume if a small or keyhole incision is made. The skin and the underlying muscular layer can be opened macroscopically.

At approximately the midsection of the vessel across the groin, a perforating muscular branch will be encountered. This branch must be electrocoagulated and sectioned in order to get adequate length for the anastomosis of the femoral artery. Later, as skill develops, this branch can be anastomosed and rerouted in practicing end-to-side and side-to-side repairs.

As the vessels are isolated, the vein will be found on top and superficial to the artery. Two lymphatic vessels are on the surface of the vein, one superficial and another deep. If these are not stripped free of the vein, subsequent tearing of them may result in lymphorrhea after a vascular repair and provide an avenue for the entrance of infection.

For color contrast, a 1 sq cm section of balloon is then placed beneath the artery, which has been cleaned of the surrounding areolar material (Fig. 47–4). As a comparison, the vessel to be approached for dissection and repair is smaller than the lumen of a 23-gauge needle.

At this point, if it is clear that a considerable amount of vasoconstriction has taken place because of the manipulations, a bath of 0.5 per cent xylocaine without epinephrine will help relax the muscular wall of the artery. A little stroking of the portion of the vessel in the operative field will enhance the vasodilation and the xylocaine effect.

The distal clamp is placed first to get hydrostatic dilation of the vessel before the proximal clamp is situated; the two bar clamps are slid together before sectioning. Try to avoid placing the anastomosis site close to the stump of the muscular branch that was previously electrocoagulated. Transect the vessel between the clamps with sharp scissors.

After sectioning, gentle irrigation of the vessels is critical. The cannula must be inosculated directly into each section of the cut vessel to wash out the blood and fiber. With the No. 5 forceps in one hand and the irrigation device in the other, grasp only the adventitia with the tip of the forceps (Fig. 47–5). Gently flush one side, switch the instruments to the opposite hands, and repeat in the other direction. Any particles of fiber left in the lumen will become the nidus for thrombus formation that will occlude the lumen postsurgically.

After having hydrodilated and irrigated the lumen of the vessel, one of the most delicate maneuvers that is required for this repair is introduction of the No. 5 forceps into the lumen of the vessel and gentle manual dilation of the lumen (Fig. 47–6). This is done both proximally and distally.

The next step involves the stripping back and actual circumcision of the adventitia from the outside of the vessel wall (Fig. 47–7). This is important because even a small bit or fiber of the adventitia projecting into the lumen of the repaired vessel will collect thrombocytes and promote thrombus formation.

The first suture is placed at the superior pole of the interface of the two vessel edges to be

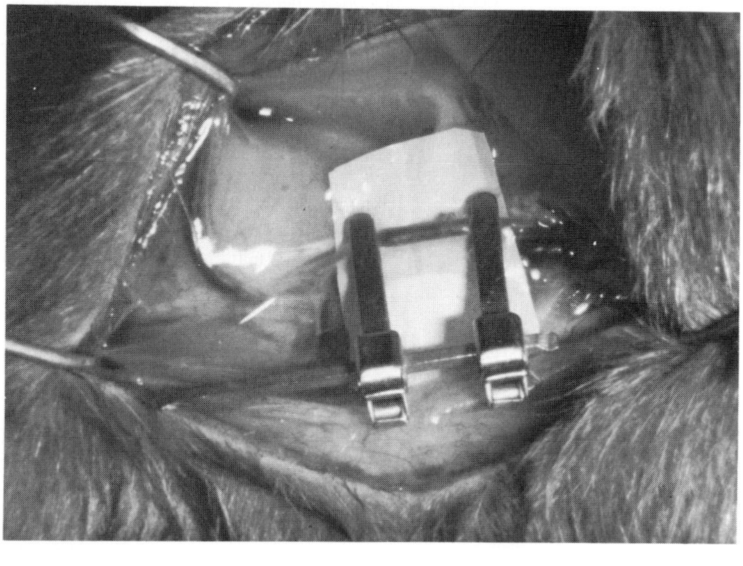

**Figure 47–4.** With low power field, the artery is prepared for sectioning. Microvascular clips and a 1 sq cm piece of colored balloon are in place.

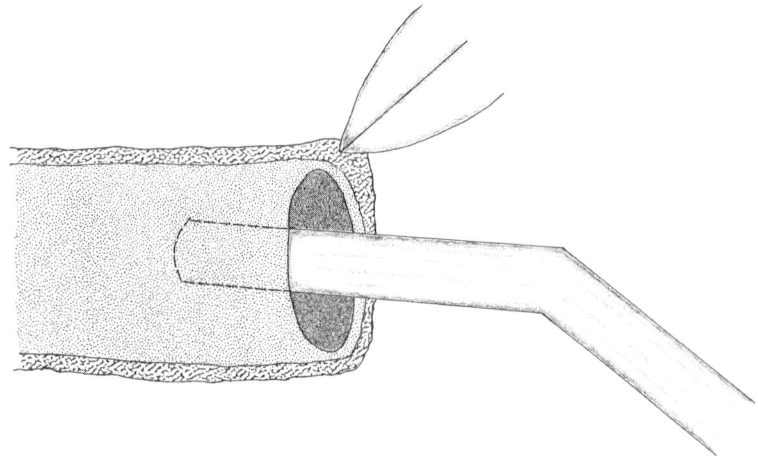

**Figure 47–5.** The tip of the 30-gauge irrigating cannula is placed inside the vessel lumen for relatively forceful irrigation with heparinized lactated Ringer's solution.

anastomosed; the No. 2 suture is placed at the inferior pole and the number three is placed between the two on the first side to be repaired. The whole clip is then rolled over to get to the underside to place sutures 4, 5, and 6 (Fig. 47–8).

The first suture is the most difficult to place because the vessels are floppy and not joined in any manner. Normally the entry is made approximately two wall thicknesses away from the cut margin of the vessel. The most critical suture for the final success of the anastomosis is the second (Fig. 47–9*A*). If the No. 2 suture is not correctly aligned, i.e., not placed exactly the same distance from the No. 1 suture on the distal as it is on the proximal end of the vessel, the resulting deformity in the vessel greatly compromises the cross-sectional area of the repair (Fig. 47–9*B*). This abrupt change in lumen area causes a turbulent pattern in blood flow which will also enhance thrombus formation.

For all sutures, the forceps have to be introduced into the lumen of the vessel to receive the needle (Fig. 47–10*A*). The needle is driven in between the two blades of the forceps, grasped, and retrieved. The forceps protect the needle tip from picking up the rear side of the intima.

On the opposite edge, the forceps will pick up the adventitia, which opens the mouth of the vessel (Fig. 47–10*B*). As the needle is introduced into the mouth, the forceps support the superficial wall of the vessel and hold the vessel steady while the needle is guided out through its exit. The needle is then drawn through with the forceps. It is necessary to use the forceps to steady the distal segment so as to avoid damaging the proximal segment by pulling with the needle.

The suture is then slowly drawn through the two segments of the vessel. If an assistant is available, he can inform the surgeon when the tail of the suture is getting close to the entrance into the vessel. With the limited field it is easy to pull the suture through both sides of the vessel before it is noticed.

Tie the knot square and lay it square. Two throws are enough in 10–0 and 11–0 nylon; a third, unnecessary knot will only increase the

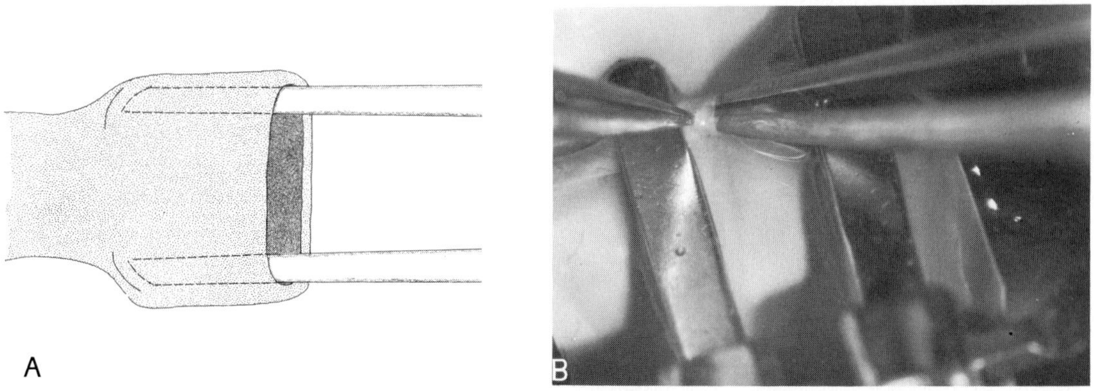


**Figure 47–6.** The vessel is manually dilated with jewelry forceps *(A)*. Compare the magnification of the microvascular clips and balloon piece *(B)* with Figures 47–4 and 47–2.

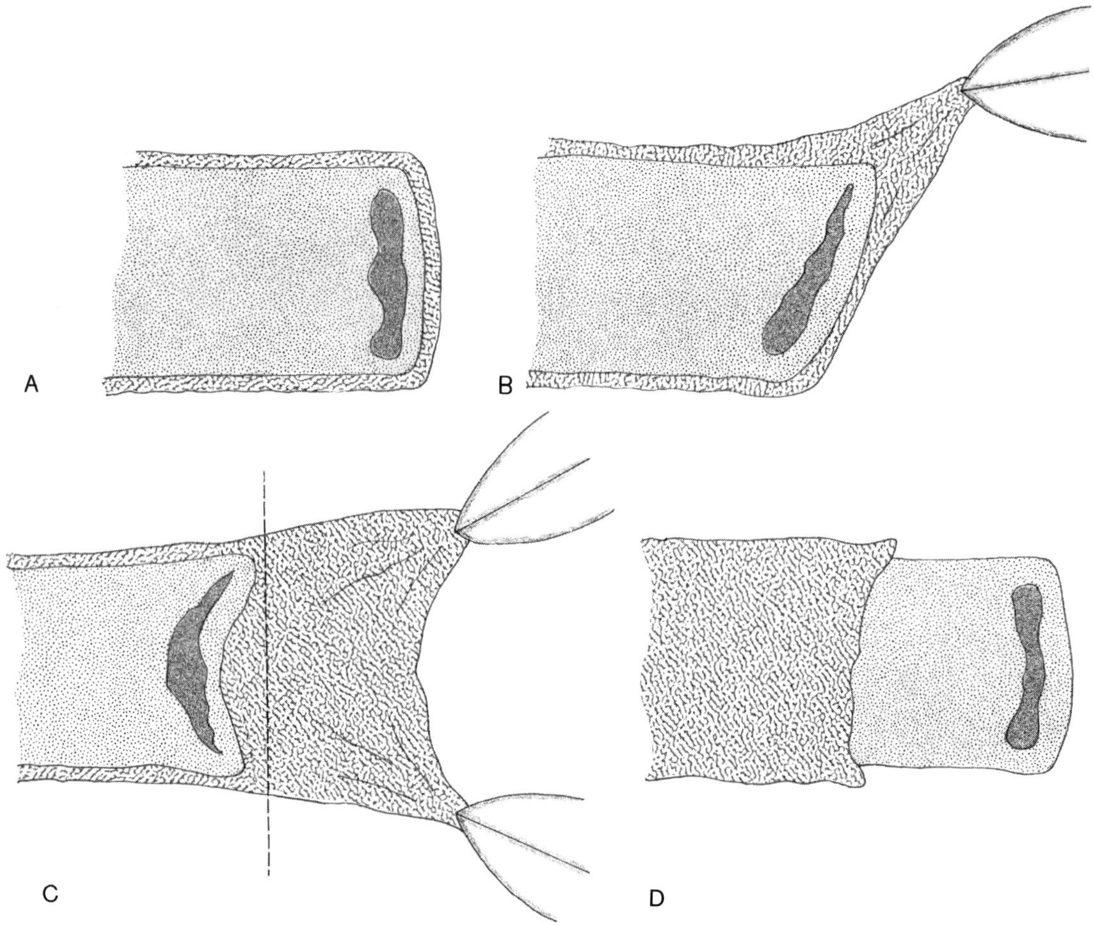


**Figure 47–7.** The adventitia around the vessel *(A)* is first teased over the end *(B)* and prepared for cutting *(C)*. The "circumcision" is complete after retraction of adventitia *(D)*.

risk of tearing the vessel. A surgeon may spend about half his time with a microvascular repair simply in tying the knots.

At this point, the vessel is lying in approximately the same attitude it had when it was sectioned.

In picking up suture material, the surgeon may inadvertently pick up the adventitia and produce a tear. To prevent this, it is helpful to introduce the first blade of the forceps under the distal tail of the suture and either lift it or pick it off the vessel itself before clamping down the second blade.

If a loop occurs while tying, simply tie it down snugly, pull the loop out, and tighten it one more time.

After the third suture is placed, the microvascular clip is turned completely over and again situated flat. Sutures on the reverse side are easier to place, since the first side is already in position. It is also easier to avoid picking up the deep side inadvertently. One may choose to sound both the proximal and distal vessel segments prior to starting this side to make sure none of the first three sutures has violated the deep side of the vessel. Again, tie the knot square and lay it square.

If things are not going smoothly at any

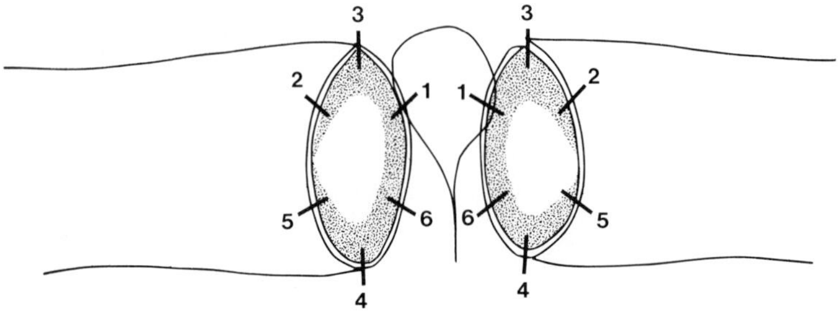


**Figure 47–8.** A diagram of the placement of the six sutures from a side view.

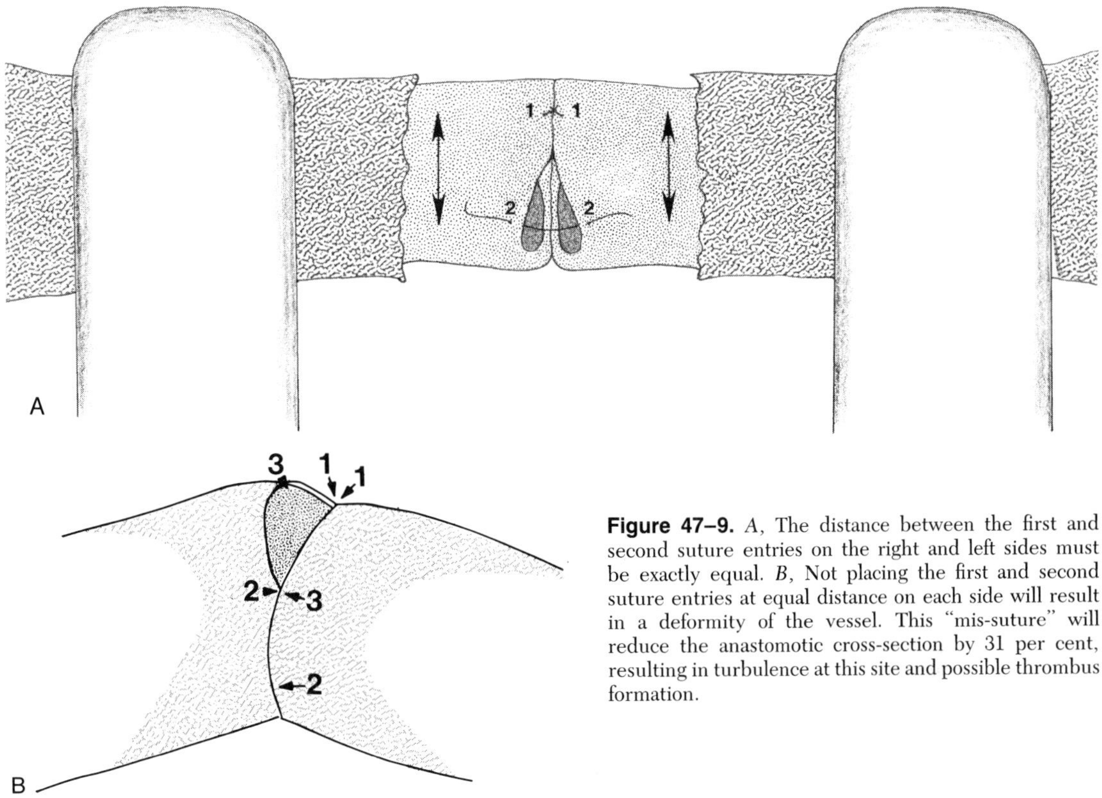


**Figure 47–9.** *A*, The distance between the first and second suture entries on the right and left sides must be exactly equal. *B*, Not placing the first and second suture entries at equal distance on each side will result in a deformity of the vessel. This "mis-suture" will reduce the anastomotic cross-section by 31 per cent, resulting in turbulence at this site and possible thrombus formation.

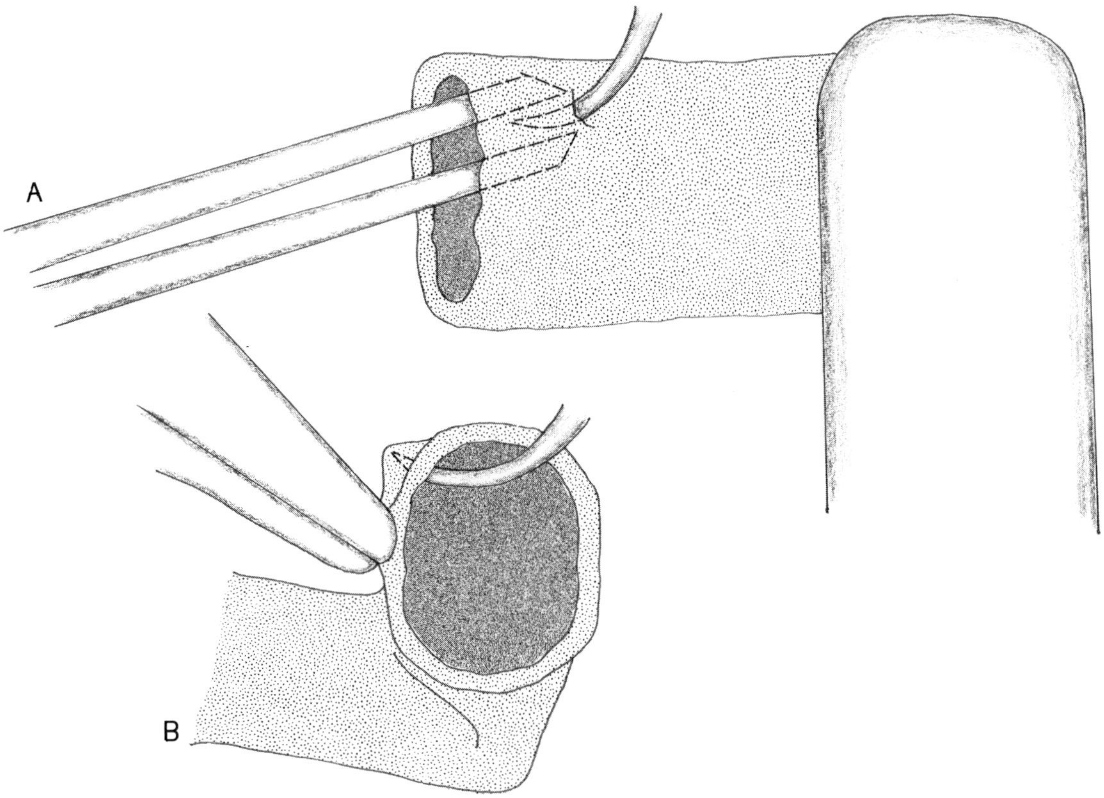


**Figure 47–10.** *A*, In making a suture, the jeweler's forceps must be introduced inside the vessel to prevent picking up the opposite side. *B*, On the second side the mouth of the vessel is turned up so that the inside of the vessel is in full view.

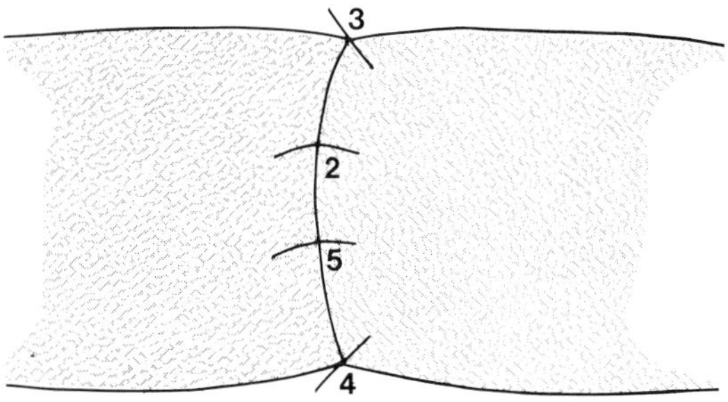


**Figure 47–11.** The completed anastomosis is inspected for possible gaps prior to releasing the clamps.

point—the vessel, needle, or forceps not cooperating—it is better to stop and begin again before the vessel is torn.

Leaving the tails of No. 4 and No. 3 sutures long after tying will help in rotating the vessel for the upper and lower margins. Once the three sutures have been placed on the reverse side, the clip is turned back for the final inspection for holes or gaps that might need to be closed with a final suture before the clamps are removed (Fig. 47–11).

A small piece of Saran or similar plastic wrap may be placed around the anastomosis prior to the release of the clamps. This helps to prevent blood loss while the clotting at the interface is taking place. The distal clamp is removed in order to allow backflow into the repaired site; then the proximal clamp is removed.

A little watchful "negligence" is in order at this point. Avoid getting overly anxious about throwing extra sutures into a bleeding anastomosis. Allow the blood to accumulate around the vessel, snug up the plastic wrap a bit, and wait. Additional xylocaine may be added around the repaired vessel in order to help dilate the lumen and increase the flow. A patent lumen is suggested by expansive pulsations of the vessel in a circumferential manner.

If there is an obstruction of the lumen, the pulsatile motion will be linear, along the long axis of the vessel, rather than circumferential. The linear pulsation indicates that the flow of blood is impacting against the obstruction at the anastomosis site. When the oozing at the anastomosis site has stopped, remove and discard the balloon and the plastic wrap.

A final test of the repair may be performed. The surgeon may use proximal forceps to gently grasp and hold the vessel to prevent flow toward the anastomosis. The blood is then stripped out distally across the anastomosis and into the distal segment of the vessel and held with a second forceps. The proximal forceps is then released, and immediate flow across the anastomosis indicates a patent antegrade flow.

While microvascular repair itself may not be the interest of all avian practitioners, the use of an operating microscope is strongly indicated for avian surgery, and the described procedure offers the clinician the opportunity to develop skills with this equipment. Simple pinning of fractures and tissue closure over damaged limbs or trunk portions is no longer going to be acceptable surgical care. Exploration and identification, if not repair and treatment, of other damaged and diseased areas is vital to current state of the art of surgical therapy.

## ACKNOWLEDGMENT

The author wishes to thank Jody Sayle for the illustrations for this chapter.

# Chapter 48

# SELECTED SURGICAL PROCEDURES

GREG J. HARRISON

The surgical procedures described in this chapter have been developed primarily with the aid of directed intense lighting, magnification of at least 8 to 10 × (Fig. 48–1), and bipolar radiosurgical forceps. While not all avian surgery requires this additional instrumentation, the speed of the procedure, accuracy of tissue handling, hemostatic control, and survivability of the patient are significantly enhanced (see Chapter 46, Surgical Instrumentation and Special Techniques, and Chapter 47, Introduction to Microscopy). Some procedures are not recommended without the auxiliary equipment.

Alternate methods of hemostasis include use of rubber band tourniquets and ligatures. Penlight or other hand-held surgical cauteries are useful. **CAUTION:** When inhalation anesthetics are administered, particularly by face mask, extreme care must be exercised in using cauteries that have a red hot glow or produce spark-gaps. Feathers that have become infiltrated with oxygen and anesthetic agents may be ignited by these types of surgical cautery devices.

One suggestion to help control bleeding of skin vessels is the use of a sharply pointed hemostat to produce a stab wound entry and to crush small sections of skin along the incision line prior to cutting each section with scissors.

## GENERAL CONSIDERATIONS

Preliminary preparation of the surgical site requires the plucking of feathers (small amounts at a time) in the direction of their growth and a brief swabbing of the skin with povidone-iodine (see Chapter 44, Evaluation and Support of the Surgical Patient). Masking tape, lubricating gel, or hair styling products may be used to

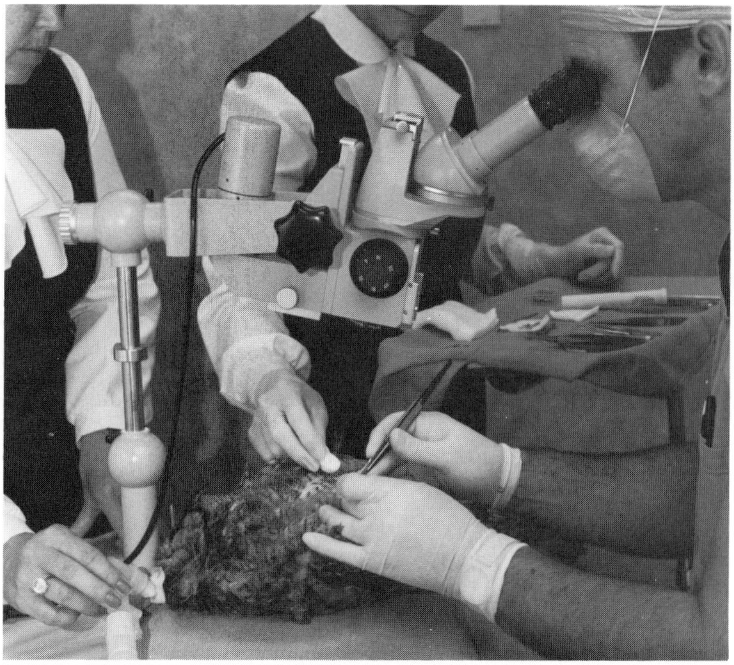

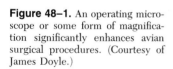

**Figure 48–1.** An operating microscope or some form of magnification significantly enhances avian surgical procedures. (Courtesy of James Doyle.)

retract the peripheral feathers. A clear plastic drape is placed over the site.

The placement of skin incisions in birds requires some consideration of the location of feather tracts so that the post-surgical appearance of the bird and the direction and type of feather growth are not altered. This is particularly important in procedures involving the wing feathers. Initially, a felt-tipped pen may be used to sketch out the incision line.

The development of scar tissue on the skin may result in follicle destruction and subsequent loss of feathers in certain areas. An elliptical incision may be used to subsequently excise the affected tissue and bring adjacent feather tracts into apposition. On some parts of the body where insufficient skin appears to be available for suturing, some attempts at undermining adjacent skin and drawing this toward the wound may be successful, although this procedure is less applicable to the relatively inelastic avian skin than to the skin of mammals. On other nonhealing wounds or areas where skin is insufficiently available for cover, acrylics or hydroactive wound dressings can be used to cover the area to encourage epithelial regeneration (see Chapter 28, Symptomatic Therapy and Emergency Medicine). These can be reevaluated every three to four days. These products also moisturize and protect periosteal tissue that may be exposed following injury over a bony structure.

This author prefers nylon (3-0 to 5-0) suture in a continuous interlocking pattern for most skin closures. Healing is usually uncomplicated.

## PRIMARY SURGERY OF THE SKIN

The most frequent surgical procedures directly involving the skin are management of traumatic wounds, removal of well-encapsulated superficial tumors (lipomas), and treatment of feather cysts.

### Repair of Traumatic Wounds

If surgical management is required of skin trauma, the surgeon is advised to remove any feathers, foreign material, or devitalized tissue from the area, freshen the wound edge by debriding, and place sutures of the surgeon's choice (see Chapter 41, Disorders of the Integument).

Most traumatic wounds are left open for drainage and healing. If they fail to re-epithelialize sufficiently within a week, the lesions may be freshened at the epithelial margins and sutured.

### Removal of Superficial Tumors

Surgery is indicated for lipomas that fail to respond medically to a three- to five-month regimen of dietary and supportive therapy (see Chapter 42, Miscellaneous Diseases), show evidence of necrotic ulcers, or are of sufficient size to mechanically interfere with the bird's activity. Uncomplicated lipoma and other superficial tumor removal is usually accomplished by blunt dissection following surgical incision over the area. During the excision, the mass can be manipulated by an assistant for maximum exposure of attached vessels. Although subcutaneous stitches may appear to be required to reduce the potential for pocketing, they are usually unnecessary; however, excessive skin remaining from removal of exceptionally large tumors may be trimmed prior to suturing. Extensive excision may be required for ulcerated lipomas with mineralized cholesterol clefts in the center (Fig. 48–2).

Surgical treatment of xanthomatosis is described in Chaper 46, Surgical Instrumentation and Special Techniques.

### Feather Cysts

A relatively common avian skin surgery is feather cyst removal. One procedure for treat-

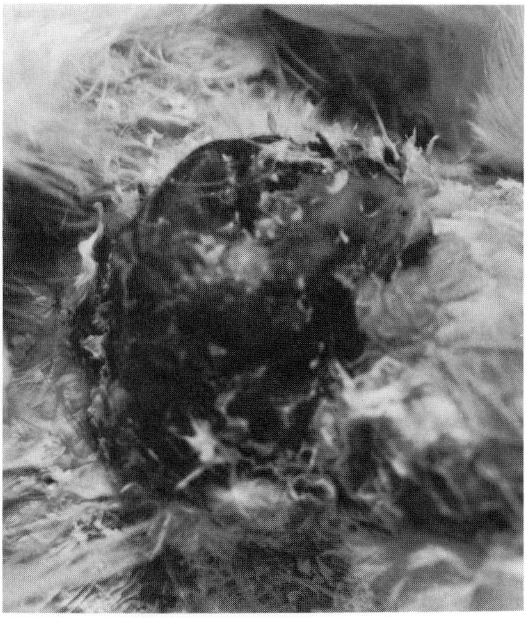

**Figure 48–2.** Lipomas with cholesterol clefts frequently occur in obese birds and can often be removed surgically.

ment involves incision of the cyst for complete exposure, removal of necrotic feather debris, and fulguration of the entire follicle with the ball electrode of the radiosurgical unit (see Chapter 46, Surgical Instrumentation and Special Techniques). An alternative approach is complete removal of the intact cyst by dissection. The prognosis against recurrence is guarded with either technique. It is believed that any disturbance to one follicle may result in damage to adjacent follicles and the potential for subsequent cyst development.

To remove an intact cyst from a primary wing feather, a rubber band tourniquet is applied proximal to the affected follicle and the initial incision is made over the ventral surface following surgical preparation of the site. Because these cysts usually originate on the dorsal surface of the major digit, the bird must be rolled over so that the incision can be extended to this point. The entire feather follicle, including the germinal tissue at the attachment site to the bone, is excised. Hemorrhage is minimal with the use of the tourniquet. Sutures may or may not be applied.

### Uropygial Gland

Tumors of the uropygial gland or ducts, although rarely encountered in this author's clinical practice, can be successfully removed by blunt dissection and bipolar radiosurgery. The ventral fibrinous attachments to the caudal vertebral area are difficult to incise with traditional instruments. Postoperative complications are not expected following a traditional closure.

## SURGERY OF THE HEAD REGION

The most common procedures performed in the region of the head are removal of oral and periophthalmic abscesses or tumors. Beak repair is discussed in Chapter 46, Surgical Instrumentation and Special Techniques, and in Chapter 41, Disorders of the Integument. Enucleation procedures are presented in Chapter 19, Ophthalmology. Cryotherapy has been suggested as an alternative to enucleation in small species.

### Oral Cavity

Surgical intervention may be required for resolution of multiple abscesses of the oral cavity (see Chapter 31, Nutritional Diseases, and Chapter 28, Symptomatic Therapy and Emergency Medicine). The oral lesions caused by *Candida*, pox, or papillomas may respond to standard debridement. Oral surgery for the removal of sublingual foreign bodies, granulomas, or tumors can cause extensive hemorrhage in birds; therefore, radiosurgery and magnification should be employed for these procedures. Ulcerative necrosis, which has been found to accompany some perilingual lesions from foreign body intrusion, may require extensive removal of tissue from the tongue and floor of the mouth. Cryosurgery has been successful in resolving selected cases of oral lesions nonresponsive to traditional surgery, such as chronic ulcers with secondary yeast infections or oral papillomas.

### Infraorbital Sinus

Because of the complex anatomy of the infraorbital sinus of pet birds, sinusitis and other upper respiratory infections that result in rhinorrhea and epiphora are often difficult to treat medically. The extensive collection of exudate in periorbital abscesses resulting from chronic infection may require surgical excision and debridement for resolution. Although irrigation may dislodge portions of the exudate, a more extensive investigation of the sinuses and curettage may be required.

Surgical access to the sinus is not recommended without the aid of magnification and an electrocautery device because of the potential for hemorrhage. Traditional surgical entry with scalpel and scissors results in excessive bleeding in this author's experience. At a level of 8 to 12 × magnification, the vessels can be identified and coagulated with the bipolar forceps.

Swellings ventral to the eye or those responsible for exophthalmia are approached with a horizontal skin incision over the suborbital arch. Figure 48–3 illustrates the complex arrangement of diverticula, any of which may contain molded exudate. The globe may be gently elevated with a flat blunt instrument such as a scalpel handle or cotton-tipped applicator to aid exposure of the infraorbital diverticulum.

The preorbital diverticulum must be investigated in cases of swellings anterior to the eye. A slightly curved perpendicular incision is made over the prefrontal bone. The potential for vascular encounter may be greater than in the previous approach. The anterior dorsal aspect of this sinus is frequently filled with a molded

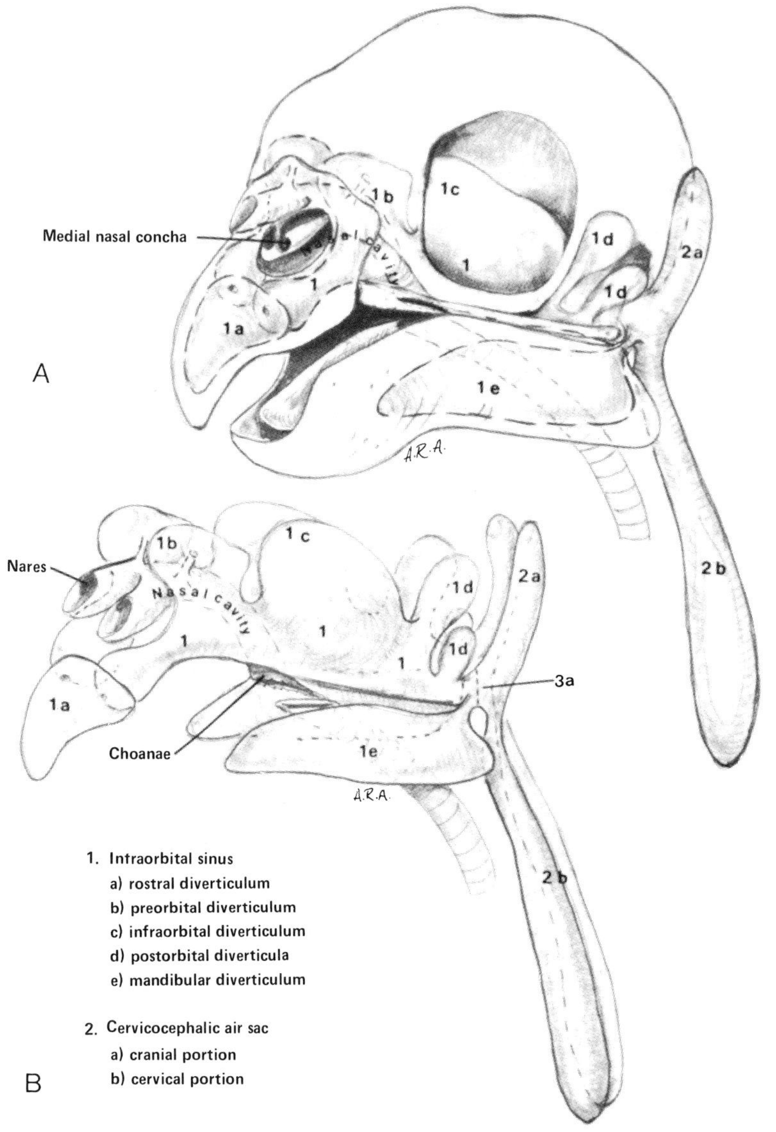


**Figure 48–3.** One can appreciate the multiple potential sites of exudate accumulation in respiratory diseases when the anatomy of the infraorbital sinus is reviewed. (Drawn by A. Allen.)

segment of exudate, which can be gently removed with an iris hook, curved microsurgical thumb forceps, or a standard hypodermic needle bent to form a hook on the end.

## Ear

Surgery of the ear has been limited in this author's experience to a single procedure involving a biopsy and subsequent removal of a nonmalignant tumor that originated from the deep cranial wall of the ear canal in an Amazon. The bird was referred for chronic bleeding from the ear. Examination under high-power magnification revealed that the entire canal was filled with moist, fragile tumor tissue (Fig. 48–4). A lateral ear resection was performed with the bipolar radiosurgical instrument (Fig. 48–5), and the obstructing tissue was removed in pieces until the base of the tumor was uncovered. The base of the canal was fulgurated with the ball electrode, dried, and filled with cyanoacrylic. The bleeding was minimal compared to similar procedures in mammals. When the acrylic plug was removed in 10 days, further areas of necrotic tissue were removed, and subsequent healing was uneventful (Fig. 48–6).

## Skull

One case of repair of a skull fracture in the author's experience illustrates that a bird can have a portion of its brain exposed without immediate clinical signs, that exposed brain

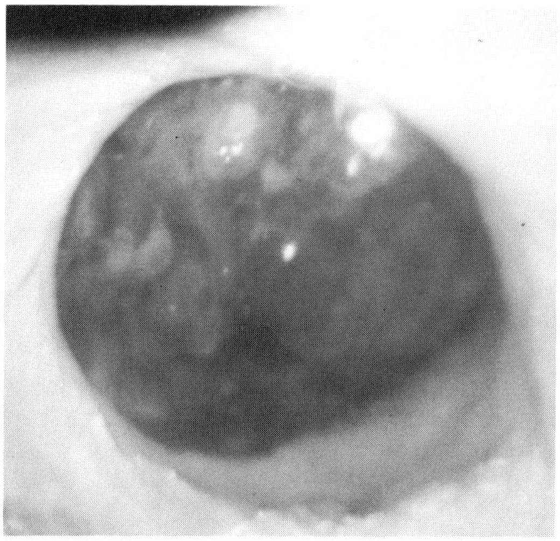

**Figure 48–4.** An Amazon's ear canal is filled with neoplastic tissue.

tissue can be debrided, and that a skull cap can be fashioned from acrylics. A Peruvian Grey-cheeked Parrot that had suffered trauma to the skull and chronic scabbing over a section of the head was presented 10 months later when the bird began to show partial paralysis of the right leg and opisthotonos. Pieces of parietal bone (0.5 cm × 0.75 cm) on both sides of the crest of the skull were missing. A protrusion consisted of vascular tissue and what appeared to be brain material. Histopathologic evaluation of the tissue biopsy confirmed that it was brain (grey matter) covered with inflamed epithelial tissue. With bipolar forceps to control hemorrhage, the tissue was freed from the calvarium and debrided. An artificial skull cap was prepared with alternating layers of cyanoacrylic liquid and powder (see Chapter 46, Surgical

**Figure 48–5.** A resection of the ear was required to provide access for removal of the tumor.

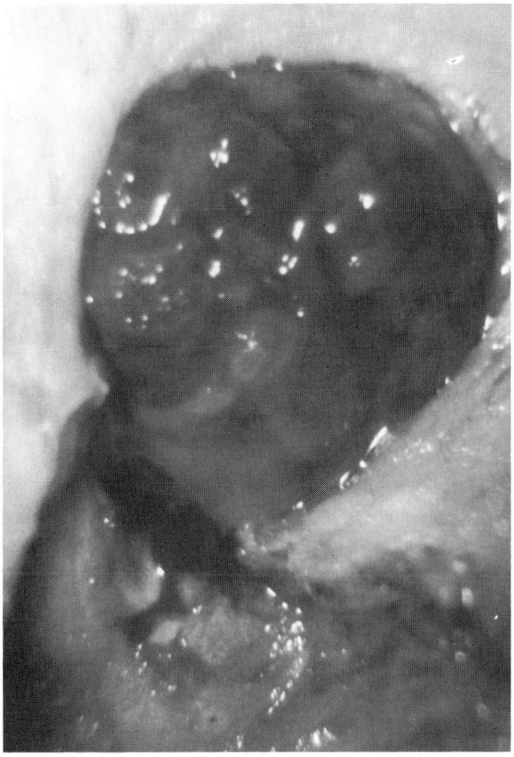

**Figure 48–6.** In a postsurgical visit, the acrylic plug and further necrotic debris are removed from the ear. Subsequent healing was uneventful.

Instrumentation and Special Techniques) directly on top of the brain tissue. The skin was undermined down the neck area to provide sufficient skin for closing over the wound. Chloramphenicol antibiotics were used systemically for two weeks postoperatively. Although the skin eventually retracted a few millimeters away from the acrylic cap, the cap remained in place and the clinical signs disappeared.

## SURGERY OF THE NECK AREA

### Tracheal Surgery

Surgery of the trachea may be necessary for removal of tumors, tracheal granulomas, and inhaled hand-fed or tube-fed formula, and in cases in which foreign bodies cannot be retrieved with biopsy forceps (Fig. 48–7). During the procedure, air can be supplied by trocarization of the abdominal air sacs by an open-ended syringe case or other suitable hollow instrument. For the surgical approach, the author prefers a transection of the trachea immediately cranial or caudal to the surgical area. Longitudinal bisection of the trachea is not

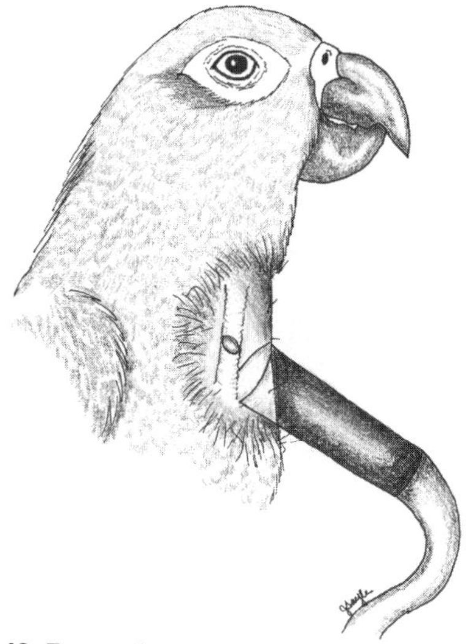

**Figure 48–7.** A tracheotomy may be necessary for retrieval of foreign bodies lodged in the trachea at this location. More cranially located foreign bodies may be retrieved by mouth with biopsy forceps.

recommended. Longitudinal bisection results in limited surgical exposure because of the rigid tracheal rings and may result in inadvertent tearing of the tissues. In addition, subsequent anastomosis of the trachea following longitudinal bisection is more difficult, and scar tissue may reduce the size of the lumen and impair breathing. Tracheal surgery is contraindicated in this author's experience with current technology in species smaller than cockatiels and where access is required caudal to the thoracic inlet (see Chapter 15, Endoscopy).

Foreign bodies in the caudal portion of the trachea just cranial to the syrinx or those lodged in the syrinx have been unapproachable in this author's experience, especially in small birds. Attempts to split these last four to six caudal rings to increase exposure have failed. Even an emergency procedure involving splitting the sternum and the clavicle also failed because of the heavy syringeal musculature and the close apposition of the bronchi and cardiovascular structures.

Investigation by Ritchie with microsurgical approaches on cadavers suggests that a possible approach to this site in the surgical patient might be a transection of the anterior ribs (and possibly the coracoid), freeing of the ventral anterior portion of the appropriate lung lobe to allow reflection of the lung, and entering the syrinx through a bisected bronchus.

## Devocalization

Surgical attempts to approach the syrinx in psittacines for devocalization techniques described for poultry (e.g., electrocautery, placement of wire or plastic structures) have uniformly failed, although adequate reduction of vocalization has been achieved in gallinaceous birds. One new technique that shows promise for gallinaceous birds involves the swabbing of phenol on the syrinx. The syrinx is approached from the left thoracic inlet by blunt dissection and is isolated by transection of the bilateral sternotracheal musculature (Fig. 48–8). Standard ophthalmic phenol solution used for corneal cautery is painted over the syrinx in one or two applications with a cotton-tipped applicator until a uniform blanched appearance is achieved. The author has found that this procedure is facilitated by having the bird in dorsal recumbency, with the head toward the surgeon, the neck slightly elevated, and a focal point light that can be focused down the thoracic inlet. The skin is closed in a continuous suture pattern.

The surgeon is advised to carefully monitor a gallinaceous bird under inhalation anesthesia, as these birds appear to be induced very quickly (often developing apnea) and to recover just as quickly. Intubation is preferred to masking. Because it is difficult to visualize the larynx, Allis tissue forceps can be clamped approximately 1 cm behind the tongue to elevate the larynx into view.

## Correcting Cervicocephalic Air Sac Inflation

The anatomy and disorders involving the cervicocephalic air sacs (Fig. 48–9) have been

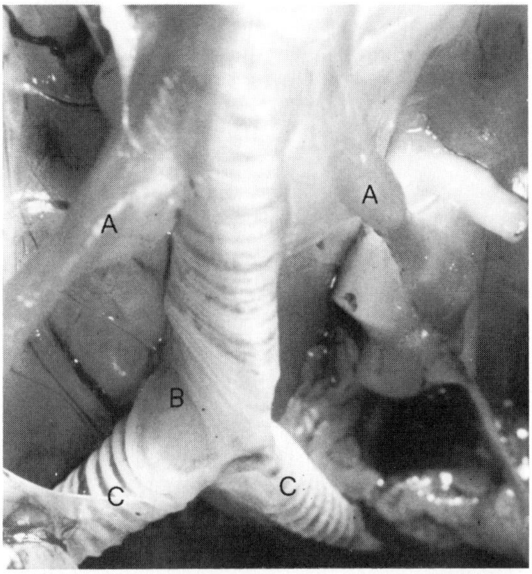


**Figure 48–8.** This surgical approach to the syrinx in a gallinaceous bird illustrates *(A)* the sternotracheal musculature, *(B)* the syrinx, and *(C)* the primary bronchi. (Courtesy of Scott McDonald.)

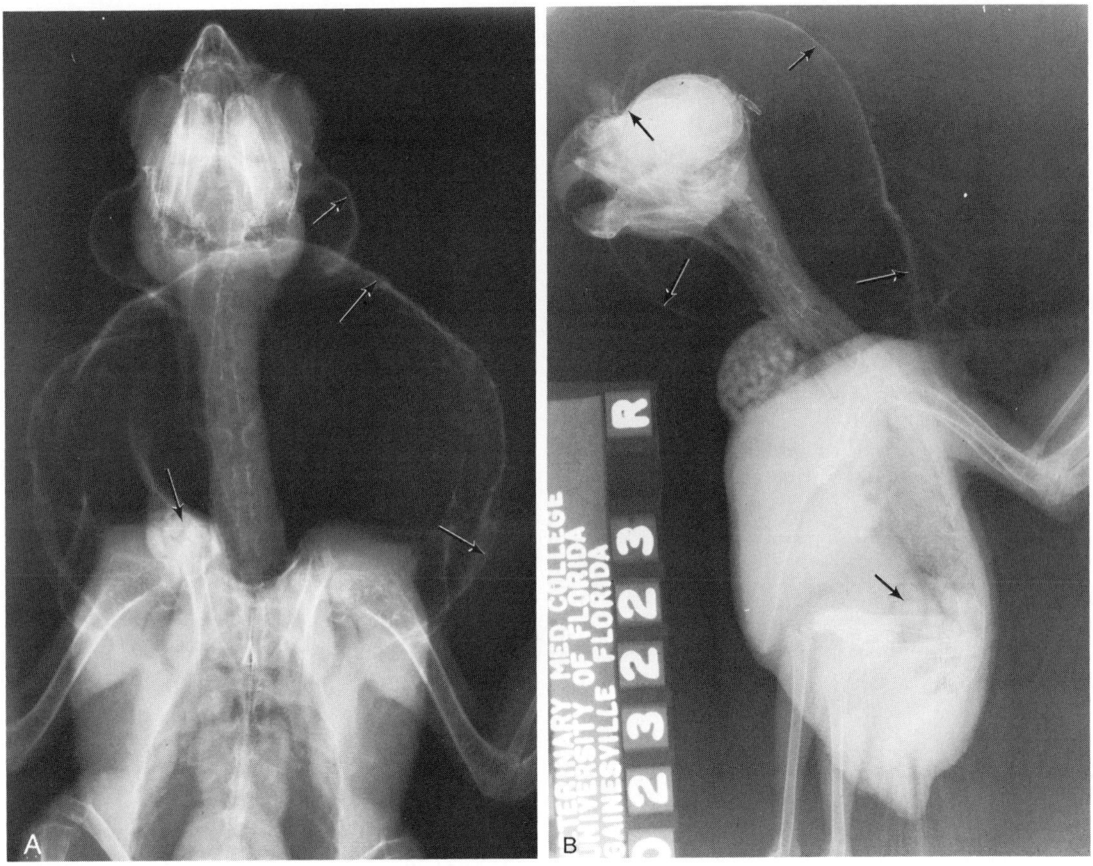


**Figure 48–9.** The degree of distention of overinflated cervicocephalic air sacs is illustrated in these radiographs. (Courtesy of Michael Walsh.)

described (see Chapter 4, Clinical Anatomy, and Chapter 14, Radiology). The following surgical procedure has been successful in preventing recurrence of overinflation of the cervicocephalic air sacs. High power magnification, pin point lighting, microsurgical instruments, and suture needles are critical to the success of this procedure.

With the bird in lateral recumbency, a surgical area encompassing the base of the skull, cervical vertebrae, and the caudal border of the mandible is prepared. At a point approximately 0.5 to 1.0 cm caudal to the mandible and dorsal to a point even with the extended ventral margin of the mandible, a ventral-to-dorsal skin incision is made. The skin is bluntly reflected from the bulging air sac. The air sac is incised and the margins are spread apart. To provide maximum exposure between the cervical muscles, the medial border of the mandible, and the jugular furrow, the skull must be placed in an oblique, acutely flexed angle to the body. During the initial approach a small speculum aids in the exposure needed to place the first two or three stitches. The reader is referred back to Figure 48–3*B*, where the location of the suture placement is illustrated by 3a.

At a point inside the cervicocephalic air sac chamber at the exact junction with the infraorbital sinus (with 8 to 10 × magnification), a single interrupted suture is placed through the air sac wall into the surrounding tissues, where adequate fascia-like tissue exists and back into the dorsal aspect of the air sac. Proceeding ventrally as subsequent sutures are placed, the very thin, closely knit connective tissue surrounding the jugular furrow is picked up and sutured to the dorsolateral muscle fascia of the mandibular area. The space being closed is reduced from approximately 1.5 cm to 0.2 cm; then the suturing direction is reversed. One must choose a very short needle (0.5 cm or less) to allow placement of the last two stitches and raise the magnification up to 25 × to observe the passage of the needle to what appears to be within tenths of a millimeter of the vessels. These last two ventral stitches being applied to ensure final closure of the communication are first placed in the fascia of the mandible. The cervical side of the suture is placed last to enable maximum visualization of the previous needle placement. A total of 7 to 10 stitches is placed, and the neck is manipulated to check for potential openings that would allow reinflation.

The air sac is left as is and the skin is closed with a standard pattern. The procedure is repeated on the opposite side. After the final closure, the air sac is emptied of any remaining air with a 20-gauge needle and syringe. One case recurred and required a second procedure. One case in which the interclavicular air sac herniated into the space previously occupied by the inflated cervicocephalic air sac was corrected by the application of a single hemoclip (see Intra-abdominal Procedures) at that location. The author suspects that the use of hemoclips may obviate the need for sutures in the original procedure. Inflation of the periophthalmic tissue ventral to the orbit, which may accompany the inflation of the cervicocephalic air sac, is also usually corrected by this technique, even though from an anatomic basis this would not be expected.

## SURGERY OF THE CROP AND ESOPHAGUS

Surgery of the cranial aspects of the esophagus is required for repair of traumatic injuries and fistulas (Fig. 48–10*A*) and for removal of foreign bodies. Anesthesia is not always required for emergency crop repair in very young neonates (Fig. 48–10*B*). Injuries are most frequently caused by improper passing of a crop needle or tube when force feeding sick or neonatal birds. Such cases may be presented as critically toxic birds with some swelling or accumulation of food in the cervical area (Fig. 48–11). This may be confused with crop emptying problems. Diagnosis can be made by palpation, crop inflation and transillumination, and/or endoscopy. Massive supportive detoxification procedures (see Chapter 27, What To Do Until A Diagnosis Is Made) need to be instituted and the surgery performed on an emergency basis.

Following anesthesia, the bird is intubated via the trachea and esophagus (to help in identifying the margins). A skin incision is made over the lateral point of swelling. Using blunt-tipped thumb forceps or hemostats and scissors, the skin is incised and reflected from what may be a highly inflamed and necrotic area. Once the cranial and dorsal margins can be identified, the tedious process of debriding the entire area begins (Fig. 48–12). One must exercise caution to avoid the esophageal, tracheal, and cervical vessels and nerves. The site of the traumatic injury in the esophagus is identified, debrided of all devitalized tissue, and sutured with 4-0 to 6-0 silk in an inverting single interrupted suture pattern (Fig. 48–13). Catgut at this strength absorbs too quickly and must be avoided for this procedure. The crop can be insufflated to check for potential leaks. A Penrose drain is sutured in place and anchored to the skin to prevent removal by the bird. Topical

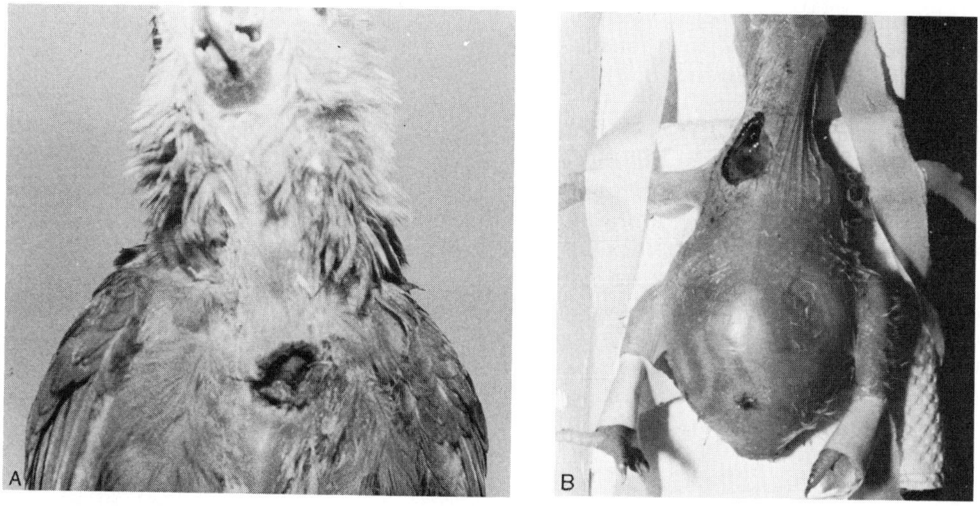


**Figure 48–10.** *A*, A draining fistula of the crop of this 3-month-old conure resulted from its being hand fed a formula that was too hot. *B*, Thermal burns in a 10-day-old macaw were sutured quickly and safely without anesthesia.

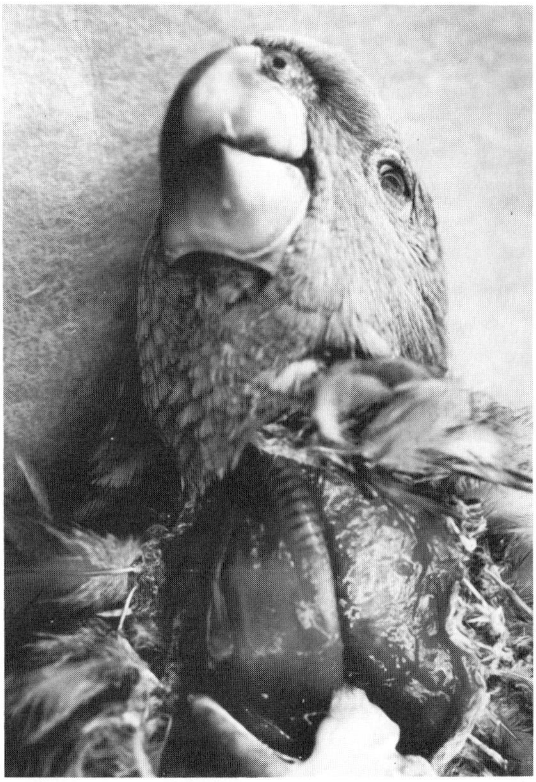

**Figure 48–11.** A penetration of the upper esophagus resulted in a mass of food that remained under the skin for three weeks before the bird died of toxemia.

antibiotics are applied prior to skin closure. An antibiotic with a dilute chlorhexidine solution or EDTA-TRIS lysozyme solution is flushed daily into the surgical site for three to five days, at which time the drain is removed. Feedings for the first 48 hours consist of very small,

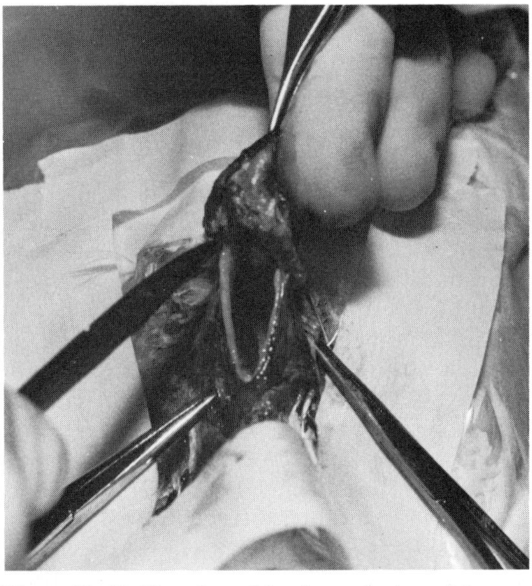

**Figure 48–12.** The edges of the skin and edges of the crop must be debrided and sutured separately.

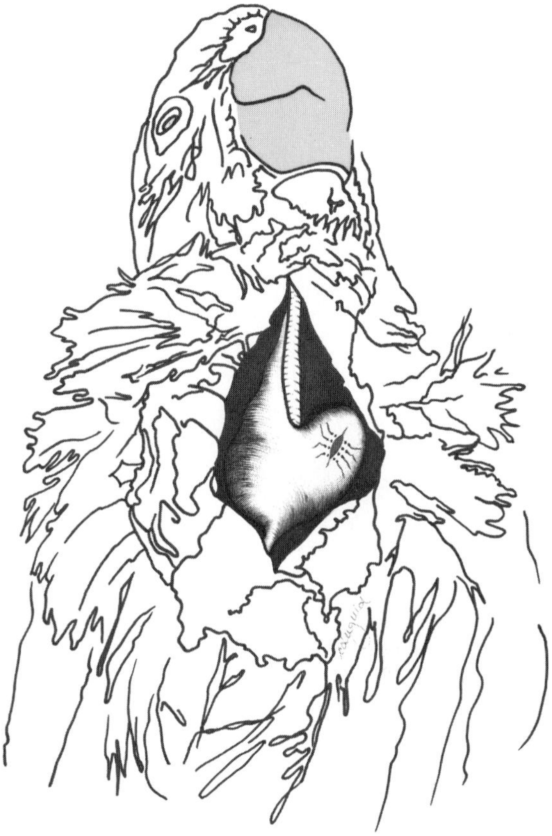

**Figure 48–13.** A simple interrupted inverting suture pattern can be used on crop and proventricular incisions.

liquid meals, with subsequent gradual return to a normal diet.

Foreign bodies ingested by neonates may be removed by prompt retrieval without anesthesia through a crop incision (Fig. 48–14).

Pendulous crops are occasionally encountered and can be surgically reduced if necessary.

## INTRA-ABDOMINAL PROCEDURES

### General Considerations

Intra-abdominal procedures are among those that require the greatest care in control of hemorrhage. For surgical procedures in which ligation may be difficult or contraindicated, a system of surgical stainless steel hemostatic clips* is available for vessel closure. Although the individual forceps for applying each of the various sized clips is relatively expensive, the

---

*Hemoclip—Edward Weck & Co., Inc., Research Triangle Park, NC 27709

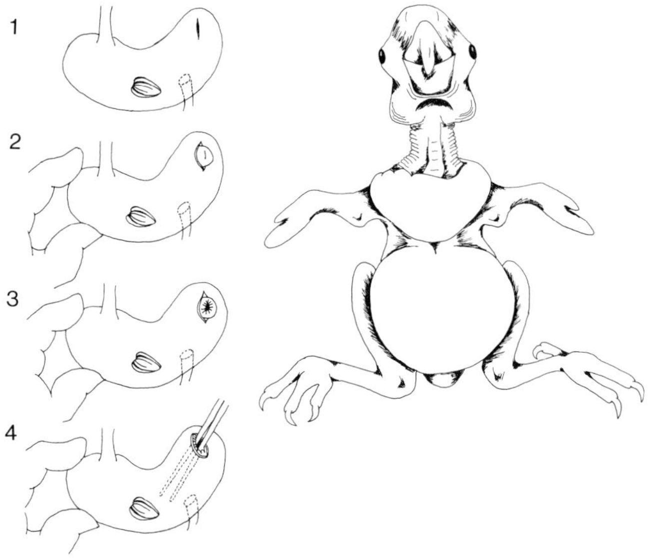


**Figure 48–14.** An ingluviotomy in a neonate for foreign body removal can be quickly performed without anesthesia: (1) A skin incision is made; (2) the crop will bulge out the skin incision when gently squeezed and a second incision can be made in the crop; (3) the foreign body can be gently manipulated toward the incision site; (4) the item is grasped with blunt forceps and removed. The crop and skin are closed in two layers. (Used with permission of The American Association of Zoo Veterinarians.)

clips are inexpensive and require minimal practice to master their application. The No. 10 clips are the most useful for general avian work.

## Exploratory Laparotomy

The need for a complete laparotomy has been significantly reduced by the application of laparoscopy and biopsy techniques to birds, as the latter are less time-consuming and involve less trauma to the patient. However, there are clinical cases in which exploratory surgery may be indicated. Abnormal findings on palpation, radiograph, or laparoscopy, such as the presence of foreign bodies in the ventriculus, abdominal masses, tumors, air sacculitis, or granulomas commonly warrant further investigation. The practitioner must weigh the advantages of the stress of surgery against the potential outcome. Foreign bodies must be removed as soon as possible, especially if they are contributing to vomiting or neurologic or bleeding disorders. In some cases, the risk of death may be great, such as with some abdominal masses, severe diarrhea, emaciation, or overconsumption of grit.

One may use exploratory surgery to confirm a preliminary diagnosis of a suspected terminal case, or to identify the location and degree of involvement of masses evidenced on radiograph. In other cases, diagnostic procedures such as cultures, impression smears of the organs, special stains of ascitic fluids, and biopsies may be performed concurrently.

Although exploratory surgery would be contraindicated in cases of diseases that can be treated medically, it may be elected even in debilitation in which there is no alternative but the demise of the patient. Exploratory laparotomy may be accomplished with either the ventral or lateral surgical approach.

## Surgical Approaches

In the development and refinement of avian surgical techniques, a number of exposures have evolved in an attempt to avail the surgeon of increased access to organs, visibility of vessels, and speed in completing the procedure (Fig. 48–15). Although the traditional ventral midline incision on an avian patient may be employed successfully in some procedures, it severely limits these factors. When the ventral approach is used, a horizontal abdominal incision, which can be extended as necessary, appears to have more applications than a simple midline vertical incision. "Flap" incisions (Fig. 48–15*A* through *D*) offer a greater degree of abdominal cavity exposure for visualization and manipulation.

All midventral incisions must be made with care to avoid the inadvertent sectioning of organs (usually the duodenum or the supraduodenal loop of the ileum) that are located in close proximity to the abdominal musculature. Obese birds also have abdominal fat surrounding these gut segments which may further complicate this approach. The midventral approach is essential, however, for birds with evidence of fluid accumulation in the abdomen to prevent fluid drainage following an incision from obstructing the air sacs.

Lateral recumbency is the author's preference for laparoscopy or laparotomy procedures

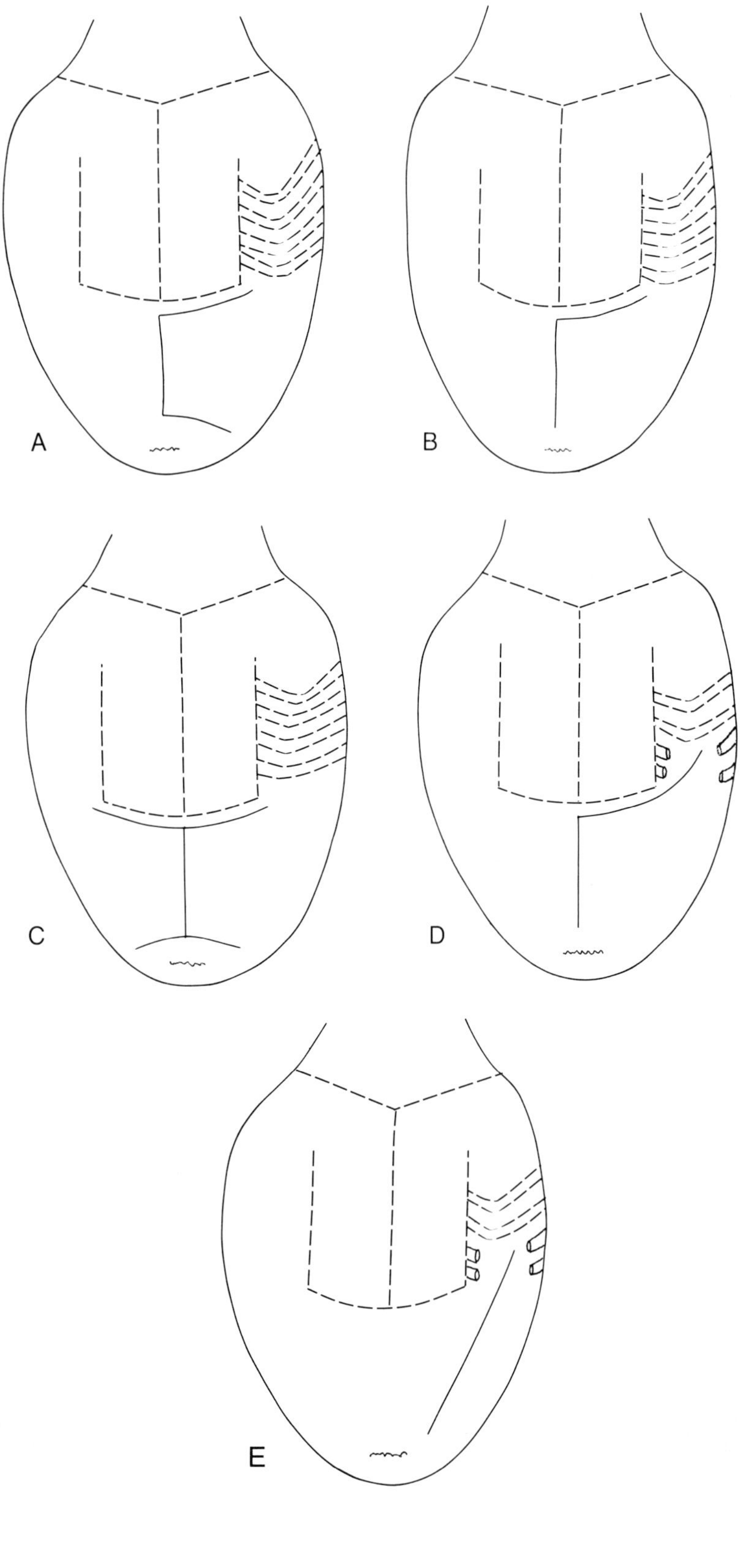


**Figure 48–15.** A number of surgical exposures have evolved in an effort to increase accessibility of the organs and hemostasis in the avian patient.

involving sex determination or evaluation of reproductive potential; examining for evidence of pathology in the air sacs or caudal/medial surfaces of the lung, ovary, anterior kidney or testicle; biopsy or other surgery of the liver, kidney, spleen; proventriculotomy and subsequent exploration of the ventriculus; and most hysterectomies (Fig. 48–15*E*).

## PROCEDURES FROM THE VENTRAL APPROACH

### Ventral Laparotomy

An area along the midline that extends laterally across the ventral two thirds of the abdomen should be plucked and surgically prepared. Redig describes the ventral laparotomy approach in a raptor:

> The skin incision is made along the ventral midline from the tip of the xyphoid to the pubic bones and undermined a sufficient distance on either side. The abdominal musculature is "tented" with a forceps and a stab incision made with a scalpel in the midportion of this ventral midline. A forceps is inserted and used as a groove director to extend the incision cranially and caudally. A large abdominal fat pad will usually be encountered, portions of which may be gently teased away and removed to facilitate entry. If care in the approach has been taken thus far, the abdominal air sacs should be still intact and appear as clear membranous structures that billow gently inward and outward as the patient breathes. Using a small curved Metzenbaum scissors it is possible to work around the lateral sides of the air sacs, freeing them from the abdominal wall. This will allow exposure of more surface area of the air sac so that a clean incision line may be made in it. An incision may now be made in the left abdominal air sac and extended cranially and caudally. The stomach, left side of the intestinal mass, left kidney, adrenal, and gonad may now be observed. Cranially, the posterior wall of the posterior thoracic air sac must be incised to gain access to the caudal border of the lung and to allow visualization of the heart. The same procedure is repeated on the right side, which allows visualization of the duodenal loop, pancreas, gall bladder and contralateral kidney, adrenal and gonad. A thin ligament attaches the liver to the surface of the sternum; cutting it will allow the liver to drop down and permit visualization of the ventral surface of this organ.

This approach must be modified to avoid air sac damage in the presence of abdominal fluid. The hyperventilation associated with the opening of the air sacs in raptors is less noticeable with psittacine species. The most obvious patient reaction is in lightening of the inhalation anesthetic level. A continuous suture is used to close first the musculature and then the skin.

### Tumor Removal

The ventral surgical exposure previously described can be altered for removal of a large abdominal tumor (Fig. 48–16) by connecting the ventral midline incision to two lateral incisions: one just below the sternum and the other above the cloacal area (with care in approaching the femoral vessels). The resulting "flap" is reflected laterally for unilateral abdominal exposure (see Fig. 48–15*A*). A similar flap can be made on the opposite side for bilateral exposure (see Fig. 48–15*C*).

After the air sacs have been incised, the mass is bluntly dissected away from any lateral attachments and lifted up from the abdominal cavity for better visualization of the primary vessels. The vessels are ligated and coagulated with the radiosurgical "welding" technique or clipped with hemostatic clips prior to dissection.

Many abdominal masses are inoperable, especially those attached to the kidney, testes, or other perirenal tissues.

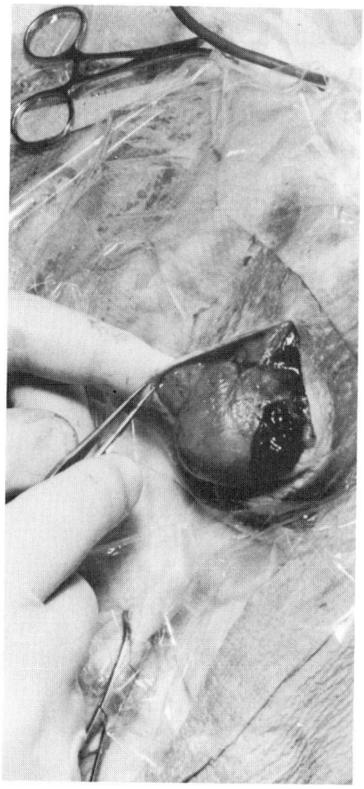

**Figure 48–16.** A relatively superficial tumor can be bluntly dissected for removal, with ligatures, electrocautery, or hemostatic clips applied to the primary vessels.

### Enterotomy and Intestinal Repair

The need for enterotomy and intestinal repair is infrequent but may exist in cases of a transected ileum from trauma or poor surgical technique. The highest degree of success is achieved with the principles described for end-to-end vessel anastomosis (see Chapter 47, Introduction to Microsurgery). Traumatized intestinal ends are squared off and, because the mucosal lining tends to evert, mattress sutures can be relatively easy to place under high magnification. A branch of the cranial mesenteric artery parallels the intestine and retracts when sectioned; therefore, some dissection may be required for hemostasis of this vessel. Standard closure procedures are followed and postoperative care prescribed for a proventriculotomy is required.

### Pancreatic Biopsy

Rosskopf described exploratory laparotomy in a Scarlet Macaw with chronic high values of amylase tests in which a wedge biopsy of the pancreas was removed. The end of the lobe away from a major duct was clamped and surgically excised. Two blood vessels required ligation. The exposed end of the pancreas was oversewn with fine suture to prevent enzyme leaks. The biopsy sample yielded a histologic diagnosis of chronic active pancreatitis.

### Female Reproductive Disorders

Although the lateral approach is preferred by this author for hysterotomy, hysterectomy, or surgical correction of egg binding, these procedures can be performed from a midventral incision in cases in which the distal uterus must be approached to remove an egg or to culture for soft shell or other reproductive disorders. Hysterectomy in a hen with a lacerated uterus is often easier from the midventral approach. Egg-laying disorders accompanied by fluid accumulation must be resolved from this approach (Fig. 48–17).

## SURGICAL PROCEDURES FROM THE LATERAL APPROACH

### Lateral Laparotomy

Unless specifically indicated otherwise, the bird is positioned in right lateral recumbency. The area of the left side from mid-dorsum to mid-sternum, from the cloaca caudally to the third rib cranially and both sides of the left leg are plucked and surgically prepared. With the left leg abducted, an incision line is sketched and the skin is cut from a point at the proximal end of the middle third of the pubic bone cranial to an area just dorsal to the uncinate process of the sixth rib. The left leg can then be taped in a more radically reflected position out of the surgical field.

An incision is made through the mid-lateral abdominal musculature starting just caudal to the last rib and continuing caudally toward the pubic bone at the same level as the skin incision. A major vessel that will need to be ligated or coagulated when the abdominal muscles are transected is located approximately perpendicular and ventral to the acetabulum. This same vessel may be encountered in the initial skin incision, just medial to the femur when the leg is in a normal relaxed position. With thumb forceps elevating the musculature from the abdominal structures, the muscular incision is continued cranially, transecting the intracostal

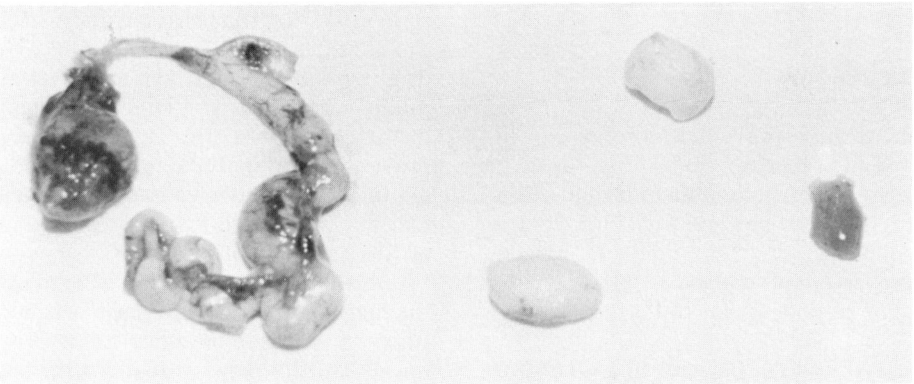

**Figure 48–17.** Abdominal fluids, inspissated egg yolks, and the uterus were removed from a cockatiel presenting with egg-related peritonitis. Because of the fluid accumulation, the ventral surgical approach was used.

muscles, bones, and vessels of the seventh and eighth vertebral ribs. This step may not be required in some species such as pigeons. Ophthalmic lid retractors or cosmetic surgical wound retractors can be used to open the laparotomy incision for improved visualization.

The ribs interfere with adequate exposure, especially in patients requiring proventriculotomies, hysterectomies, or surgical manipulation of the kidney, adrenal, or gonad. In a budgerigar or cockatiel the ribs are relatively flexible and can be bent into a dorsal position to achieve sufficient exposure. However, in Amazons, cockatoos, and macaws the ribs are more rigid, and portions often need to be removed for the speed and safety of the procedure.

To remove the vertebral rib section, an incision is made in a dorsal direction, perpendicular to the previous incision and parallel to the caudal edge of the sixth rib to just ventral to the vertebral column. The seventh and eighth ribs are again transected near the dorsal ends. Another incision behind the eighth rib rejoins the mid-lateral abdominal incision. The seventh and eighth rib segments and corresponding muscles can then be removed.

The surgeon must be aware of potential hemorrhage from intercostal vessels of larger diameter which are located near the dorsal vertebral point of transection of the ribs. These require clamping and electrosurgical sealing. Reflecting the rib dorsally prior to dorsal incisions and visually inspecting the area with magnification will help prevent laceration of lung tissue.

Closure of this large exposure is facilitated with 4-0 nylon by surrounding the sixth rib and the muscle anterior to the pubis and ventral to the pelvis. This is usually ample anchorage to close the gap, especially if the leg is released and allowed to adduct into a more normal position. The pubis can be incorporated independently if necessary. Final closure is routine.

## Proventriculotomy

Indications for a proventriculotomy may include presence of heavy metal (e.g., lead) or other foreign bodies in the proventriculus, vomiting, or symptoms associated with proventricular dilatation (wasting syndrome). If sufficient time is available, intramuscular administration of ampicillin 24 hours prior to the procedure is advised.

Steps described for lateral laparotomy are used for the initial approach to the proventriculus. The intermediate goal is to incise and bluntly dissect the suspensory membranes to enable the proventriculus to be elevated into the surgical site (Fig. 48–18*A* to *E*). A source of intense supplemental light is necessary to detect vascular supply during these steps. The vessels offering the greatest risk of hemorrhage during the blunt dissection of the proventricular membranes are those serving the proventriculus and the liver, especially near the greater curvature of the ventriculus. The left hepatic and ventral gastric arteries and the left branch of the celiac artery must particularly be avoided (see Fig. 4–19). The proventriculus must be handled carefully with smooth, wide-jawed forceps to avoid tearing of the organ. Time spent in properly freeing the proventriculus will facilitate the ease of exploration and evacuation of the contents.

Once the proventriculus can be elevated into the laparotomy incision site, small gauze squares may be placed around the organ to isolate it. The proventriculus can be secured in this position by temporary sutures or Allis tissue forceps connected to the ribs (Fig. 48–18*E*). This step is essential for the control of the proventricular contents when the entry incision is made. The temporary placement stitches are more secure if the suture size is large enough (3-0 or larger) and the penetration is deep enough into the tissues to support the weight of the organ without tearing.

Prior to incising the ventral surface of the proventriculus, the surgeon must note and avoid the location of the dorsal caudal branching of the dorsal proventricular artery and vein. A stab incision opens the proventriculus while the surgeon supports the proventricular wall with forceps. Scissors may be used to enlarge the site. There is some evidence of bleeding, but this often stops spontaneously.

Once the proventriculus is incised, control of the acidic contents is essential. This can be accomplished with the use of a mechanical suction device that also facilitates cleansing of the contents of the proventriculus. The author uses a relatively high volume surgical vacuum pump with a large diameter (1 cm internal diameter) suction tip. This is placed near the proventricular incision to collect the overflow of fluids used to cleanse the proventriculus. One must exercise extreme caution in the use of the pump, as it is powerful enough to eviscerate the bird if placed in direct contact with the incision or internal organs. Warmed lactated Ringer's solution or other fluids are administered into the proventriculus with intravenous drip tubing to further flush out the contents. The needle attachment tip is removed from the tubing and the volume control is completely

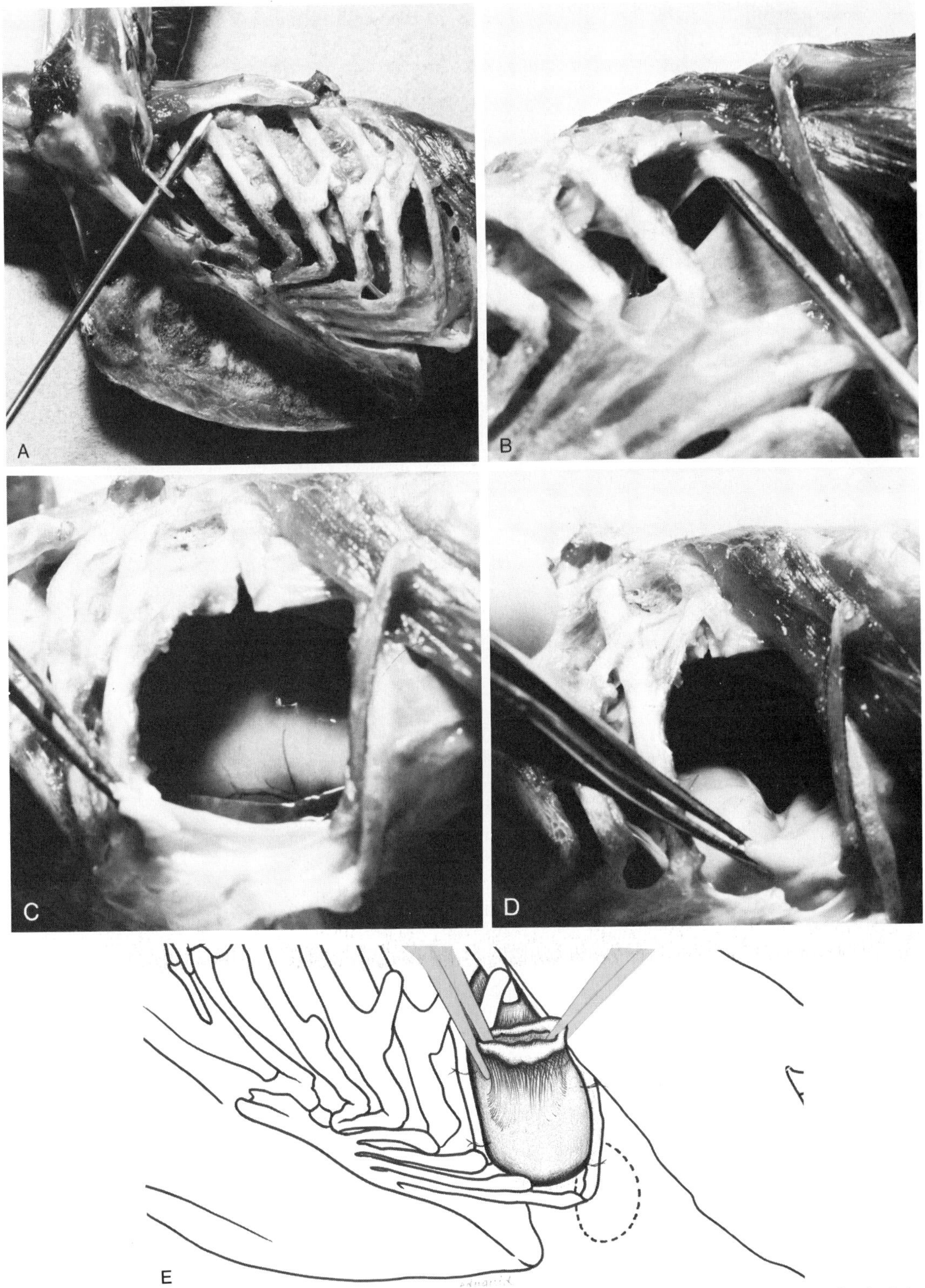


**Figure 48–18.** *A*, The surgical approach to the proventriculus is demonstrated on the cadaver of an Amazon that had died from chronic respiratory disease. *B*, Note the opacity of the air sacs being penetrated. *C*, Following removal of a rib section, the proventriculus is visible inside the cavity; note the margin of the liver. *D*, In order to clean out the contents, the surgeon must elevate the proventriculus into the surgical site. *E*, Sutures hold the proventriculus in place. In small birds, suspending the proventriculus with sutures may produce lacerations; therefore, the organ is elevated with a flat instrument and entry is made at the exact proventricular/ventricular junction. The dotted line indicates the location of the ventriculus. In a surgical procedure, gauze sponges would be packed around the organ prior to the incision to prevent spillage of the acid contents into the abdomen.

opened for maximum flow of the fluids into the proventricular cavity. The flow of fluids is directed against all surfaces of the thoracic esophagus and the proventriculus. Thumb forceps are used to dilate the muscular band that separates the proventriculus from the ventriculus, and the tube is inserted into the ventriculus. All overflow is concurrently suctioned into a standard surgical bottle.

For the procedure to be effective, all contents must be removed, especially in those cases with gastrointestinal stasis or gaseous dilatation. When the organs are apparently cleaned, the fluids are stopped, the negative pressure on the surgical pump is reduced, and the interior of the proventriculus and ventriculus is gently suctioned of remaining fluids and debris. An endoscope is useful for examining the inside of the proventriculus and thoracic esophagus for pockets of food or foreign bodies that may have been missed. Any remaining debris can be scooped out with a bone curette, or the flushing can be resumed. This procedure has been the most successful for thorough cleaning and is much safer and more accurate when performed with sufficient magnification. Previous attempts to remove proventricular or ventricular contents with spatulas only were very time-consuming and often incomplete.

The author uses an atraumatic needle, 5-0 nylon suture in an inverting simple interrupted or Lembert continuous suture pattern to close the proventriculus. There has been no attempt to repair the suspensory membranes except in macaws, and there are no known side effects of failing to do so. The gauze, blood clots, proventricular contents spillage, and stitches to the ribs are removed, the organs are arranged, and the laparotomy is closed using nylon interlocking continuous sutures in a vertical and then horizontal pattern.

Small amounts of easily digested foods such as Nutrical mixed with baby cereal and water may be administered approximately 12 hours postoperatively. The incorporation of grape juice in the initial feeding may serve as a dye to confirm satisfactory passage through the digestive system.

## Ventriculotomy

Direct surgical access to the ventriculus is avoided by this author for several reasons: (1) the ventriculus is more highly vascularized than the proventriculus (see Fig. 4–19); (2) the membranous attachment of the liver to the ventriculus (see Fig. 4–20) is much stronger and more profusely vascularized than it is to the proventriculus; (3) the ventriculus is cranial to the caudal sternal edge and surgical access is limited; (4) some suturing complications arise from the inability to roll the edges of the ventricular incision. Especially in smaller species, some difficulty may be encountered with attempts to free the ventriculus to allow retraction into the surgery site without tearing or major vessel hemorrhage. Therefore the lateral proventricular approach to the ventriculus is believed to be faster and more easily sutured with significantly less chance of hemorrhage. The proventriculus is approached and suspended as previously described, and access to the ventriculus is gained through the muscular band dividing the two organs.

From the ventral approach, the ventriculus is bound by the ventral ligament of the ventriculus, the falciform ligament of the liver, and the hepatoduodenal ligament in a rather fixed position. If necessary, the ventriculus can be incised from this approach on its cranial lateral aspect, which is relatively free of muscle and blood supply. However, considerable dissection and extreme caution are necessary to avoid the numerous and large branches of the celiac vessels. Subsequent closures from this incision appear crudely repaired, and a coating of tissue-compatible acrylic glue is recommended as a sealant. The viscera are replaced in a natural position and standard abdominal closure is executed.

## Hysterectomy

Surgery is indicated in the female bird for treatment of medically nonresponsive egg binding or egg peritonitis (see Chapter 28, Symptomatic Therapy and Emergency Medicine, and Chapter 44, Evaluation and Support of the Surgical Patient), removal of retroperitoneal eggs, metritis, or hysterectomy. A hysterectomy provides resolution of chronic egg-related peritonitis and chronic egg laying. Hysterectomized birds continue to cycle and show signs of impending egg laying such as nest building, broodiness, or posing for copulatory stroking; however, the mechanism that prevents recurrence of egg peritonitis is unclear (see Chapter 52, Reproductive Medicine).

There are no current indications to justify a hysterectomy in the normal preovulatory bird in this author's experience. Although the uterus may be as large as 6 or 7 inches long and a centimeter in diameter in a postovulatory cockatiel, it is extremely small in the preovulatory cockatiel (3 cm × 2 mm). The danger of mistaking a tiny uterus for another anatomic entity, particularly the rectum, is far too great at present to warrant preovulatory surgery.

The procedure currently employed by this author involves removal of the uterus only, the ovaries remaining intact.

The lateral laparotomy approach used for the proventriculotomy provides maximum exposure to perform a hysterectomy that is uncomplicated by fluid accumulation, although the ventral proventricular suspensory membranes are not incised. The ovary is visualized medial and dorsal to the proventriculus after incision of the dorsal proventricular membranes (see Fig. 4–14).

The infundibulum, normally a fanlike, semitransparent membrane located at the caudal lateral aspect of the ovary (see Figs. 4–14 and 4–20), is held with smooth-jawed forceps to avoid laceration and is bluntly dissected away from the ovary with bipolar radiosurgical forceps under a source of magnification. A branch of the ovarian artery requiring ligation passes caudad from the ovary to the dorsal medial surface of the uterus. A ligature or series of three No. 10 hemostatic clips is placed on the uterus near the vagina and the suspensory ligaments of the uterus are bluntly or electrosurgically dissected. In the reproductively active female there is a distinct junction of the uterus with the bulbous section of the vagina. It is at this junction that the final ligatures are placed and then the uterus is transected. The minor suspensory vessels in a cockatiel can be controlled with the radiosurgical instrument. Larger species with larger vessels may require further ligation or application of appropriately sized hemostatic clips.

After the uterus has been removed for the treatment of egg-related peritonitis and prior to closure, the surgeon is advised to investigate and retrieve any egg yolk material (of varying appearances) that may be present in the abdomen.

Although Redig has successfully removed the ovary in raptors by ligation of the ovarian vessels, this author has been unable to achieve that in psittacines owing to hemorrhage. Perhaps some environmental manipulation will make that procedure possible.

Ovariectomies, hysterectomies, and castrations on finches, quail, and canaries have been performed by Dr. Ascenzi of Cornell University as part of a research project on the influence of hormones on the psychology of birds. These procedures are undertaken in relatively young birds following a period of environmental manipulation for the purpose of regressing the reproductive status and reducing vascular supply to the follicles. For three weeks preceding the surgical attempt, access to light is limited to eight hours daily and the dietary moisture (available from fresh greens) is reduced. Ascenzi uses a bipolar electrosurgical unit to remove the ovary by bits and pieces rather than attempting an intact removal. If any hemorrhage occurs, the procedure is stopped, Gel-foam is applied, the incision is sutured, and a further attempt is made within a few days. Ascenzi has not been able to successfully remove the ovary if a mature follicle is present. She believes ligature placement increases the potential for hemorrhage.

### Removal of Lung and Air Sac Lesions

Surgical approaches are made through the air sacs during laparotomies and for the specific purpose of diagnosing or removing air sac granulomas (Fig. 48–19*A*). The lateral approach offers sufficient exposure for removal of large inspissated masses (Fig. 48–19*B*), including the thickened and discolored adjacent tissues that are often in a state of decomposition. A bone curetting tool is satisfactory for these procedures. Some limited experience by this author in removal of lung lesions with radiosurgical and/or biopsy forceps has resulted in favorable response of the bird.

## SURGERY OF THE CLOACAL AREA

A cloacapexy procedure for repair of true cloacal prolapses has been described by Rosskopf et al.:

> A midline incision is made and the cloaca is outlined by having an assistant wearing a finger cot push the cloaca into the incision site. From here, it is carefully sutured to the abdominal wall using 3-0 surgical wire and 3-0 Dexon in a crossing pattern . . . . Care must be taken not to penetrate the cloacal wall.

Those authors concur that this technique is not uniformly successful.

If improperly placed purse-string sutures, which may paralyze the sphincter muscles, or other methods of cloacal prolapse treatment have failed, or in primary cases, this author prefers a cloacotomy, which involves a superficial epithelial debridement on the surface of the cloacal orifice and placement of simple interrupted sutures to permanently reduce the cloacal circumference by 25 to 50 per cent.

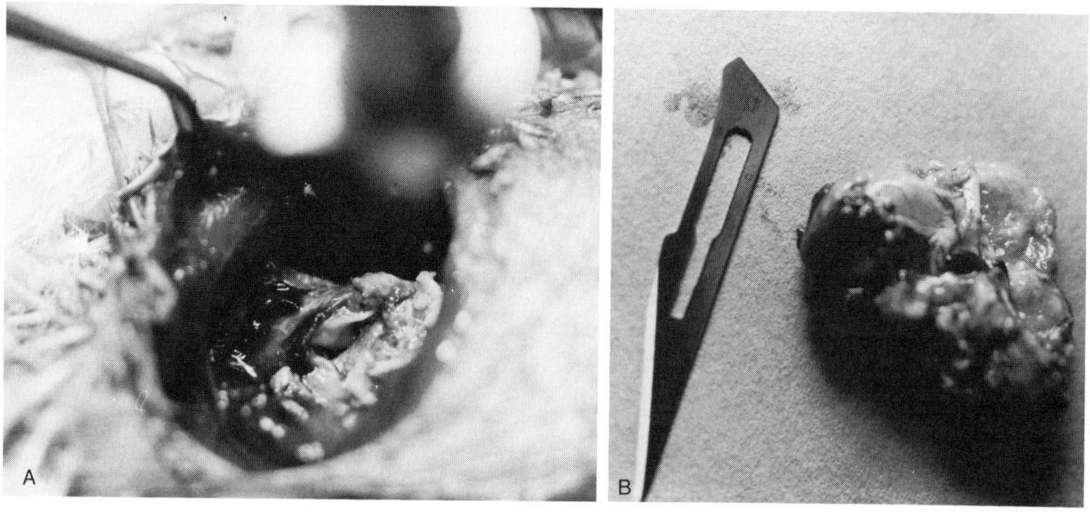


**Figure 48–19.** *A*, The evaluation, diagnosis, and removal of air sac lesions are facilitated by lateral surgical exposure with rib sections removed. *B*, This granuloma was subsequently removed from the air sacs. (*A* from Harrison, G. J.: New aspects of avian surgery. Vet. Clin. North Am. [Small Anim. Pract.], 14(2):376, 1984.)

## SURGERY OF THE EXTREMITIES

### Amputation of Wing Tips

The most common surgery of the psittacine wing tip is amputation, which is often the only alternative to chronic damage from repeated trauma to this area. Amputation may also be required for correction of "angel wing," a condition most commonly observed in waterfowl but also noted in budgerigars, macaws, and conures. An overadduction of the wing extends the phalanges so that the primary feathers on the tip of the wing always protrude and resemble an "angel's wing."

Improperly amputated or pinioned wing tips are susceptible to trauma, therefore, the surgeon is advised to form a protective pad of tissue at the distal end of the bone. To control the hemorrhage of this highly vascular area, a tourniquet is applied to the area proximal to the elbow.

A procedure for pinioning of adult zoo birds involves amputation through the carpometacarpal area just distal to the alular segment. Following surgical preparation, an incision is made completely around the proximal third of the major and minor carpometacarpal area down to the bone. The tissue can be scraped proximally toward the alula with a scalpel blade. The primary artery of concern is located in the interosseous space. A double strand of suture is passed through the interosseous space near the major carpometacarpal bone. One strand is securely tied cranially and the other tied caudally against the bones. The bones are cut, the excess skin is brought around to cover the end of the bone, and absorbable sutures are used in a continuous pattern for closure.

Pinioning at a few days of age, most frequently performed on waterfowl, may have applications in other aviculture birds. At this age, the amputation may be accomplished with scissors and minimal bleeding. The primary mistake is failure to leave the alular segment intact, which often exposes the amputated wing tip to frequent trauma.

A relatively crude procedure that has nonetheless been successful for the amputation of tumored wings in some very small birds (e.g., Zebra Finches) for which the client may wish minimum financial investment is the placing of a wire ligature around the wing to cut off the blood supply. The wing can be amputated following necrosis.

## RECOMMENDED READINGS

Allen, J.: Case report: Use of Duoderm[R] following mycotic therapy. AAV Newsletter, 5(4):106, 1984.

Altman, R. B.: General principles of avian surgery. Comp. Cont. Ed., 3(2)177–183, 1981.

Altman, R. B. Conditions involving the integumentary system. In Petrak, M. L. (ed.): Diseases of Cage and Aviary Birds. 2nd ed. Philadelphia, Lea & Febiger, 1982, pp. 368–381.

Altman, R. B.: General surgical considerations and techniques. Proceedings of the Seminar on Avian Medicine and Surgery, American College of Veterinary Surgeons 12th Annual Veterinary Surgical Forum, Chicago, IL, 1984, pp. 27–33.

Amand, W.: General techniques for avian surgery. Current Veterinary Therapy VI. Philadelphia, W. B. Saunders Company, 1977, pp. 711–716.
Ascenzi, M.: Personal communication.
Harrison, G. J.: Guidelines for treatment of neonatal psittacines. Proceedings of the Annual Meeting of the American Association of Zoo Veterinarians, Tampa, FL, 1983.
Harrison, G. J.: New aspects of avian surgery. Vet. Clin. North Am. [Small Anim. Pract.], 14(2):363–380, 1984.
Madill, D.: Rapid and effective devocalization technique. AAV Newsletter, 6(2):57, 1985.
Madsen, D. E.: A surgical procedure for devocalizing the rooster. VM/SAC, Feb., 1967, pp. 114–118.
McDonald, S.: Cosmetic technique for pinioning of psittacines. AAV Newsletter, 5(4):103–104, 1984.
Redig, P. T.: Basic surgical procedures for avian species. Proceedings of the Annual Meeting of the Association of Avian Veterinarians, Atlanta, GA, 1982.
Ritchie, B.: Personal communication, 1985.
Rosskopf, W. J., et al.: Cropotomy in a parrot. Mod. Vet. Pract., March, 1982, p. 219.
Rosskopf, W. J.: Selected soft tissue pet avian surgical procedures. AFA Watchbird, 12(2):48–53, 1985.
Rosskopf, W. J.: Developing a differential diagnosis. Material presented at Eastern States Veterinary Conference, Orlando, FL, 1985.
Rosskopf, W. J.: Surgery of the avian digestive system. Proceedings of the Seminar on Avian Medicine and Surgery, American College of Veterinary Surgeons 12th Annual Veterinary Surgical Forum, Chicago, IL, 1984, pp. 57–63.
Rosskopf, W. J., et al.: Surgical repair of a chronic cloacal prolapse in a Greater Sulphur-crested Cockatoo *(Cacatua galerita)*. VM/SAC, May, 1983, pp. 719–724.
Rosskopf, W. J., and Woerpel, R. W.: Removal of a lead pellet by gizzardotomy (ventriculotomy) in a Green-winged Macaw. VM/SAC, June, 1982, pp. 969–974.
Smith, R. E.: Hysterectomy to relieve reproductive disorders in birds. Avian/Exotic Pract., 2(1):40–43, 1985.
Soifer, F.: Case report: Cryotherapy as an alternative to avian enucleation. AAV Newsletter, 5(4):104,1984.
Tudor, D. C., and Woodward, H.: A method for devoicing fowl. J. Am. Vet. Med. Assoc., 151(5):616–617, 1967.
Wallach, J.: Surgical techniques for cage birds. Vet. Clin. North Am., 3(2):229–236, 1973.
Wise, R. D.: Surgical removal of uropygial gland tumor in budgerigars. VM/SAC, Oct., 1980, pp. 1601–1604.

*Chapter 49*

# BASIC ORTHOPEDIC SURGICAL TECHNIQUES

PATRICK T. REDIG

A large number of methods and devices for the management of fractures in mammals have been borrowed and modified for repairing avian fractures. Several articles have been published describing the application of intramedullary pins and variants of the Kirschner-Ehmer device.[1–7] Although bone plating has been advocated,[7] it has not been a widely utilized technique.[3] A comprehensive review of the consistent utilization of prescribed orthopedic techniques, detailing rates of or reasons for success or failure and the rate of return to normal function following fracture repair, has recently been published.[5] The following material is an overview of the general principles for avian orthopedic management that have been derived from a consistent approach to the management of fractures occurring in a variety of avian species, including waterfowl, psittacines, and raptors over the last 10 years.

The general guidelines for deciding whether to manage a fracture by coaptation or by surgical means have been discussed in Chapter 30, Evaluation and Nonsurgical Management of Fractures. In my practice there has been an even split between those cases managed surgically (28 per cent) and those managed nonsurgically (28 per cent). The remaining 44 per cent were not dealt with by either means, since the severity of the injury or the debilitated condition of the patient precluded any possibility of recovery.

## SURGICAL SITE PREPARATION

After the induction of anesthesia, feathers are plucked from the surgical site for a distance of 2 to 3 cm on all sides. Where there are rents in the skin at the surgical site, the feathers should be cut close to the surface rather than plucked. Where intramedullary pins are to be inserted in retrograde fashion, the exit site of the pin should also be prepared. Body stockinettes, towels, masking tape, and K-Y jelly are used to prevent adjacent contour feathers from becoming displaced over the surgical site. Open fractures should be preoperatively treated by thorough debridement of necrotic soft tissue and dead bone. The wound is irrigated with a dilute povidone iodine (Xenodine—Squibb) solution and rinsed with sterile saline. The adjoining areas of the site are then washed three times with dilute iodine scrub solution and rinsed with 70 per cent ethyl alcohol.

## SURGICAL APPLIANCES AND APPROACHES

Where open reduction is required, the approaches described by Redig and Roush[6] are broadly applicable to many avian species. Fractures of the long bones are generally managed as described below.

### Humeral Fractures

The implantation of a threaded trochar-tipped intramedullary pin provides the most consistent means of stabilizing humeral fractures. A pin of sufficient size to fill about two thirds of the marrow cavity may be introduced at the fracture site and passed in retrograde fashion out the shoulder, penetrating the midsection of the deltoid crest. A 5/32-inch pin is appropriate for macaws; a 1/8-inch pin is used in Amazons and African Greys. After alignment of the fragments, the pin is driven into the distal fragment so that the tip of the pin and the first two or three threads are projecting through the medial cortex. Cerclage wires may be employed as needed to hold long spiral fragments together. Failure of fracture repair does not appear to be related to the presence of cerclage wire. Postoperatively, the wing is bandaged in a figure-of-8 bandage; then the whole wing is further bound to the body for the first two weeks (see Chapter 30, Evaluation and Nonsurgical Management of Fractures). High humeral fractures may often be sufficiently stabilized by the surrounding muscle mass so as not to require surgical intervention.

## Radial and Ulnar Fractures

As detailed in Chapter 30, internal fixation need not be applied to forearm fractures in most instances if only one of the two bones is broken. Where both are fractured, the radius may be stabilized with a shuttle pin or end-to-end hemicerclage. Shuttle pins should be further stabilized with a small wedge pin to prevent migration. Three variations of the Kirschner-Ehmer (K-E) apparatus may be considered for application to the ulna. The original form of the K-E apparatus, consisting of steel bars and clamps, has proved to be heavy and difficult to align. A modification of this device in which the external bars are replaced with several layered wraps of orthopedic tape[a] provides a very effective means of stabilizing ulnar fractures. Typically, 0.045-inch K-wires are driven through the bone on either side of the fracture at the conventional angles prescribed for the K-E apparatus, but without regard to alignment along the longitudinal axis of the bone. Then the wing is gently folded against the body to provide the proper alignment of the fragments, and orthopedic tape is wrapped and firmly pressed around the four K-wires and held until hardened (Fig. 48–1*A* and *B*). In the third variation of this device, a plastic polypropylene shuttle pin[b] is placed inside the marrow cavity prior to the placement of the 0.045-inch K-wires to provide a more solid substrate for anchoring the pins. This latter is the preferred method for treating fractures of the forearm, where exceptional stability of the fracture is required to allow early return to function to ensure full flight capabilities. The wing should be bandaged in a figure-of-8 bandage for the first 10 days after the application of this device.

[a]Hexcelite—Hexcel Medical, 6700 Sierra Lane, Dublin, CA 94566.

[b]Polypropylene welding rods—Commercial Plastics, Inc., 601 N. Prior Ave., St. Paul, MN 55108.

## Femur

A stack of two to four small-diameter intramedullary pins is the most reliable means of providing rotational and longitudinal stabilization to the femur, since it is not possible to apply any external form of immobilization to this bone. The pins are introduced individually into the proximal fragment and driven in retrograde fashion toward the the hip; following alignment of the fragments, the pins are passed into the distal fragment. Absolute cage rest is required for the first three postoperative weeks to prevent destruction of this fixation.

## Tibiotarsus

Single trochar-tipped, threaded intramedullary pins may be implanted in the tibiotarsus by introduction into the proximal fragment and retrograde passage out the stifle. Rousch[7] suggests having the pin exit at the hock joint rather than the stifle. A 5/32-inch diameter pin is recommended for a macaw. Additional external support to prevent rotation must be provided

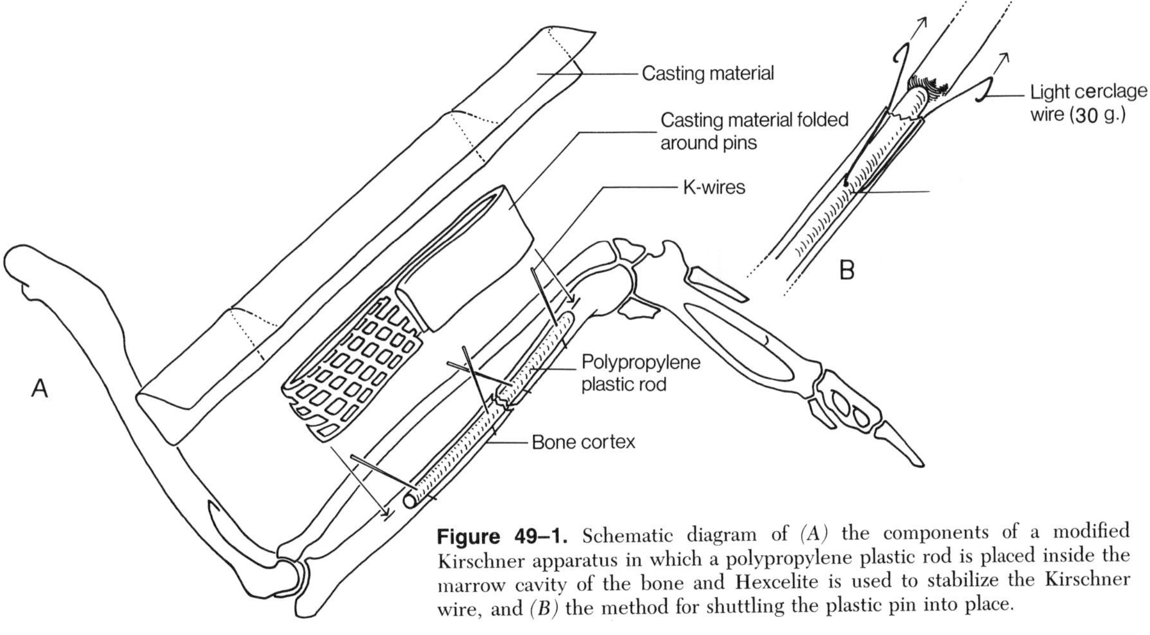


**Figure 49–1.** Schematic diagram of *(A)* the components of a modified Kirschner apparatus in which a polypropylene plastic rod is placed inside the marrow cavity of the bone and Hexcelite is used to stabilize the Kirschner wire, and *(B)* the method for shuttling the plastic pin into place.

by means of a Schroeder-Thomas splint (see Chapter 30, Evaluation and Nonsurgical Management of Fractures).

### Coracoid

The approach for surgical management of coracoid fractures is as follows: The skin is incised over the anterior tip of the coracoid bone, which is readily palpable at the anterior aspect of the shoulder joint. This incision is extended caudally to the junction of the furcula. The muscles are elevated from their attachment to the anterior end of the coracoid and along the length of the furcula. Separation readily occurs along fascial planes, and the entire shoulder joint as well as the coracoid itself is readily exposed. The distal coracoid fragment is located and an appropriately sized intramedullary pin is passed out the point of the shoulder and then driven back into the proximal fragment. The pectoral muscle is sutured to the periosteum of the furcula and the skin closed with 3-0 surgical gut. The entire wing on the affected side should be bound to the body for the first two postoperative weeks to provide additional stability (see Chapter 30, Evaluation and Nonsurgical Management of Fractures).

## MISCELLANEOUS ADJUNCT PROCEDURES

Following the application of fixation, displaced muscles and fascia are tacked together with 4-0 or 6-0 surgical gut. Skin closure is accomplished in most cases with a simple continuous suture pattern of 3-0 surgical gut with an atraumatic needle. Surgical wounds are dressed with a sterile nonadhering pad and bandaged with Kling gauze[c] and a self-adhesive elastic bandage such as Elastikon.[d] Postoperative antibiotics such as amoxicillin[e] should be routinely employed. Whenever possible, antibiotics are initiated prior to surgery. Evaluation of healing is accomplished by palpation at two-week intervals (usually performed under anesthesia), and clinical union, usually achieved in three to six weeks, is verified radiographically. Limited use of limbs may be permitted and passively and actively encouraged when there is sufficient fibrocallus formation to partially stabilize the fracture. This partial return to function often precedes removal of the fixation device, especially in K-E applications, in which there is no inherent danger of joint damage accompanying such movement.

## CONCLUSION

Avian species provide unique challenges in fracture repair by virtue of their hollow bones and their bipedal locomotion. For most birds of Amazon size or larger, the weight of the appliance is not a significant factor in choosing the means for fixation. Intramedullary pins, although the longest used means of internal fixation and one that has been the subject of considerable question, are the method of choice in many instances, particularly when long-term stability and weight-bearing are factors. The development of the many variants of the K-E device, on the other hand, has provided inexpensive, light-weight, and flexible means of stabilizing fractures of the wing, particularly those of the forearm. Since there is no joint invasion, return to function of the wing can be encouraged at an early stage in the healing process, thereby circumventing much of the arthritis in the elbow and the loss of muscle tone that inherently accompanies other modes of fixation of the wing. By choosing appropriately sized hardware (e.g., small hypodermic needles), it is possible to provide surgical treatment of fractures even for very small avian patients.

## REFERENCES

1. Bush, M.: Avian fracture repair using external fixation. In Cooper, J.E., and Greenwood, A. G. (eds.): Recent Advances in the Study of Raptor Diseases. Keighley, England, Chiron Publications, 1981, pp. 176–185.
2. Bush, M., and James, A.E.: Some considerations of practice of orthopedics in exotic animals. J. Anim. Hosp. Assoc., 11(5):587–594, 1975.
3. Gandal, C. P.: Anesthetic and surgical techniques. In Petrak, M. (ed.): Diseases of Cage and Aviary Birds. 2nd ed. Philadelphia, Lea and Febiger, 1982, pp. 304–328.
4. MacCoy, D.M.: Modified Kirschner splints for application in small birds. VM/SAC, 76(6):953–955, 1981.
5. Redig, P. T.: A clinical review of the orthopedic techniques used in the rehabilitation of raptors. In Fowler, M. (ed.): Zoo and Wild Animal Medicine. 2nd ed. Philadelphia, W. B. Saunders Company, 1986, p. 388.
6. Redig, P. T., and Rousch, J. C.: Surgical approaches to the long bones of raptors. In Fowler, M. (ed.): Zoo and Wild Animal Medicine. Philadelphia, W. B. Saunders Company, 1978, pp. 246–253.
7. Rousch, J.C.: Avian orthopedics. In Kirk, R. W. (ed.): Current Veterinary Therapy VII. Philadelphia, W. B. Saunders Company, 1980, pp. 662–673.
8. Satterfield, W., and O'Rourke, K. I.: External skeletal fixation in avian orthopedics using a modified through-and-through Kirschner-Ehmer splint technique (the Boston technique). J. Am. Anim. Hosp. Assoc., 17:635–637, 1981.

---

[c]Kling Elastic Gauze Bandage—Johnson & Johnson Products, Inc., New Brunswick, NJ 08903.

[d]Elastikon—Johnson & Johnson Products, Inc., New Brunswick, NJ 08903.

[e]Amoxi-tabs—Beecham, Inc., Bristol, TN 37620.

*Section Eight*

# AVICULTURE

# Chapter 50

# AVICULTURE MANAGEMENT

KEVEN FLAMMER

Many of the common disease problems encountered in exotic bird collections are related to improper husbandry. The conditions under which exotic birds are kept have a profound influence on their health. The poultry industry learned many years ago what aviculturists are finding today: proper quarantine procedures, aviary design, pest control, and cleanliness are essential to maintaining a healthy collection.

## THE IMPORTANCE OF STRESS IN CAPTIVE BIRDS

Stress is a term that describes the physiologic responses of an animal when adverse conditions are encountered. Adverse conditions include such diverse influences as overcrowding, unfavorable temperature and humidity, and inability to hide from real or imagined threats. Under normal circumstances stress responses are beneficial and permit the animal to react more competently to an emergency. However, when the stressful challenge is prolonged, the responses become maladaptive, leading to decreased resistance to disease and abnormal behavior.[6] A review of the most common causes of death in avian species in one zoological collection showed that most of the disease agents were of relatively low virulence, indicating that predisposing factors such as stress added to the circumstances that allowed these opportunists to cause death.[4]

Stress in exotic birds can be reduced by recognizing some of their basic behavioral needs. Cages should be designed with visual barriers that allow the birds to hide from real and perceived threats. Noisy species, such as Hyacinth Macaws, may upset shyer birds, such as Moluccan Cockatoos, who perceive that the macaws' frequent vocalizations indicate the constant presence of danger. Housing these species in separate areas of the aviary may increase breeding. In species in which intraspecific aggression is common (e.g., Sulfur-crested Cockatoos), cages should have retreats where the submissive birds can hide. Boredom is another cause of stress, particularly for some of the active and playful psittacine birds. Psittacines may chew vegetation in a planted aviary, but nontoxic plants can be grown outside of the cage and will provide a more natural atmosphere than sterile wire aviaries (see Appendix 7). Providing perches that the birds can chew will also keep them occupied.

## QUARANTINE AND ACCLIMATION

New additions to the collection should be carefully investigated, and only healthy stock from reputable sources should be considered for purchase. Clients should be cautioned against buying birds at a bargain price; unreasonably inexpensive birds may be smuggled, diseased, or both. All newly purchased birds should be quarantined for at least 30 to 60 days before introducing them to the rest of the flock. The quarantine facility should be separate from the main collection, and care should be taken to prevent transfer of disease out of the isolation area (Fig. 50–1). The quarantined birds should be fed last, and the caretaker should wear separate clothing or coveralls and wash thoroughly before entering the permanent collection. Maintaining this division is essential to prevent the spread of disease.

During quarantine the birds should be examined and weighed (Table 50–1). Some diseases are common in certain groups of birds (see Chapter 1, Choosing a Bird, and Appendix 2) and can be detected by a careful physical exam (see Chapter 7, Preliminary Evaluation of a Case). It is useful to examine each bird at least twice: immediately after it is purchased so that the sale can be terminated if problems are found, and again near the end of quarantine, since the stress of the quarantine period may reveal disease signs too subtle to note on the initial physical exam.

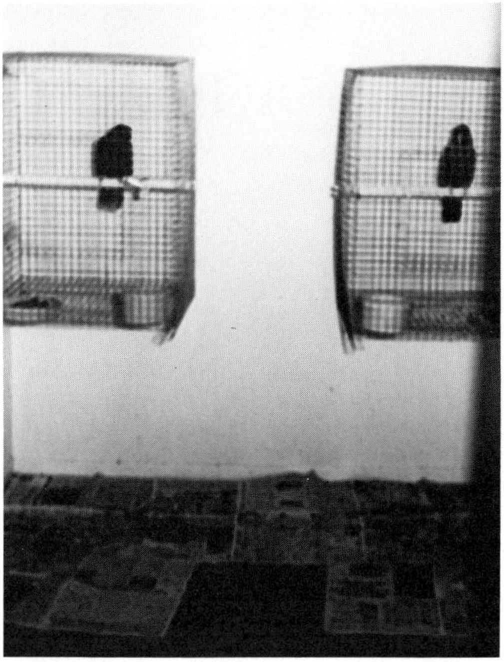

**Figure 50–1.** An isolated room should be available for quarantine of newly acquired birds. (Courtesy of the Aviculture Institute.)

## Disease Survey

All birds should be tested for chlamydiosis and Newcastle disease at the start of the quarantine period. A number of laboratories are equipped to make isolation attempts of these organisms from fecal material. In order to identify intermittent shedders, several fecal samples or cloacal swabs should be collected from each bird and pooled together for the test. A serologic test for chlamydiosis is also available. A Gram's stain of feces can be checked for the presence of large numbers of gram-negative bacteria or yeast. Fecal Gram's stains are not a reliable means of determining the proper course of treatment, so suspect specimens should be backed up by a culture. The presence of internal parasites can be determined by a direct and flotation fecal examination. Even if parasite ova are not seen, prophylactic deworming with levamisole, ivermectin, or pyrantel pamoate is recommended, since one or even several negative fecal examinations do not rule out the presence of parasites. Dusting the feathers with 5 per cent carbaryl to kill external parasites is also advised.

**Table 50–1.** PROCEDURES FOR EVALUATION OF EACH QUARANTINED BIRD

1. Physical exam and weight
2. Test for *Chlamydia* and VVND
3. Fecal Gram's stain
4. Fecal flotation and direct fecal exam
5. Deworm with levamisole, ivermectin, or pyrantel pamoate
6. Dust feathers with 5 per cent carbaryl powder

**Optional Recommended Procedures**

1. Cloacal culture
2. Complete blood count
3. Clinical chemistry panel
4. Radiology/serology
5. Bioassay (use of sentinel birds)

If finances permit, further diagnostic tests will help identify subclinical infections that may pose a threat to other birds in the collection. Culture of cloacal swabs may reveal microbial pathogens such as *Pseudomonas, Salmonella,* and *Candida* (Fig. 50–2). A complete blood count and clinical chemistry tests aid in identifying sick birds and provide background data that are valuable for comparison if the bird later becomes ill (see Chapter 13, Clinical Chemistries). Normal values have not been established for many avian species, and even in species in which large amounts of data have been accumulated, the wide range of individual variation makes it difficult to determine when the values from a particular bird are abnormal. Rather than comparing the test results to a chart of normal values, it is more valuable to see how a particular test value has changed in that individual bird. Obtaining baseline CBC and clinical chemistry results allows such comparisons to be made. Radiology also supplies useful diagnostic information.

Asymptomatic carriers of a number of viral diseases will not be identified by the diagnostic tests listed above (see Chapter 32, Viral Diseases). A significant viral disease in psittacine birds is Pacheco's parrot disease, caused by a

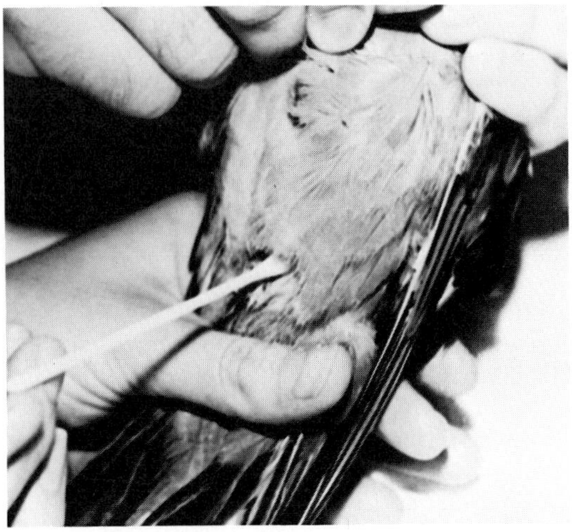

**Figure 50–2.** Swabs cultured from the cloaca can be used to evaluate the microbial status of the gut.

herpesvirus. This herpesvirus may be carried asymptomatically by Patagonian, Nanday, and other conure species. The most practical method for identifying carriers is to expose a susceptible sentinel bird, such as a budgerigar or cockatiel, to quarantined conures (see Bioassay in Chapter 22, Miscellaneous Diagnostic Tests). This is accomplished by mixing a small amount of the suspect bird's feces in the drinking water of the sentinel bird throughout the quarantine period. Unfortunately, carrier birds may shed the virus erratically, so even a 30- to 60-day exposure of the sentinel bird is no guarantee that the suspect bird is free of disease.

Papovavirus infection is a disease of emerging significance in many species of psittacine birds. Adult birds are asymptomatic carriers, but mortality may be high in exposed nestlings at the age when they are forming pin feathers. Serologic and other methods of screening for papovavirus are becoming available. It is useful to review past breeding records, if available, and avoid purchasing birds whose offspring died at the pin feather stage.

Gallinaceous birds, such as pheasants, quail, and exotic fowl, suffer many of the viral diseases of poultry, which are fully discussed in poultry medicine texts.[8]

In addition to viral infections, a number of subclinical microbial infections, such as mycoplasmosis and pulmonary infections by *Aspergillus* and *Haemophilus*, may also be missed with routine testing. It is important to advise the aviculturist that quarantining and testing will identify many, but not all, infectious diseases.

If the laboratory tests are normal and the birds appear healthy, they can then be moved to the permanent collection. As a final precaution they should be placed in an isolated area or housed near the least important birds. That way, if they happen to break out with disease they will not expose the entire collection. The possibility of the birds in the permanent collection harboring asymptomatic infections that might harm the new birds must also be considered.

### Acclimation

The quarantine period can be used to acclimate new birds to local conditions and adapt them to the aviculturist's preferred diet. The birds should be exposed to gradually cooler conditions until the current ambient temperatures are reached. It may take quite a while for birds to fully adapt to a new climate. For example, in the tropics birds enjoy bathing in the warm afternoon rains. Macaws and cockatoos imported to a more temperate climate in California insisted on continuing this habit, even when temperatures dropped to less than 40°F. Although cockatoos have repellent feathers that shed water, the macaws rapidly became soaked to the skin and had to be moved to an indoor shelter to dry out. It took a full winter season for the macaws to learn to take shelter during rains when the temperatures were cool.

## DESIGNING CAGES AND AVIARIES TO AID IN DISEASE CONTROL

Aviaries have traditionally been constructed with large, walk-in flight cages. Most aviculturists now favor cages constructed completely of welded wire that are suspended three to four feet from a concrete floor (Fig. 50–3). Suspended cages offer a number of advantages over flight cages. The elevated wire floors allow droppings to fall through where the birds can't reach them, thus breaking parasite life cycles and preventing reinfection from contaminated feces. Spilled food also falls through, reducing the chance that the birds will eat it after it spoils. Since the cage is not entered, there is less chance of transmitting disease from one cage to another and the birds are disturbed less. Wire floors are easily cleaned by scraping or hosing with hot water.

Suspended cages are not suitable for highly excitable species, such as toucans, or for aggressive species, such as Sulphur-crested and Leadbeater Cockatoos. Larger flight cages that give these species ample room to hide and escape their aggressive mates are more suitable. When territorial species, such as Australian Parakeets, are housed side by side, a double row of wire should separate the cages or the birds will chew each others' toes when fighting. Regardless of the style, the cage should have a sheltered area where birds can get out of the wind, rain, and direct sun.

Welded wire should be used to construct cages for birds that chew readily (e.g., all larger psittacines). Soldered wire (hardware cloth) is easily chewed and contains lead and zinc, which may cause poisoning if enough of the solder is eaten. The galvanized coating on welded wire also contains lead, but usually in quantities too small to cause toxicity, although use of a mild acid wash on new wire is recommended. Macaws and the larger cockatoos require 10- to 14-gauge wire, while smaller psittacine species,

**Figure 50–3.** An L-shaped suspended cage provides visual barriers that allow the birds to hide. The nest box should be placed in the sheltered portion of the cage. (Courtesy of the Aviculture Institute.)

passerines, and other birds can be housed in cages of 16- to 22-gauge wire. Large cages constructed from wire lighter than 14 gauge require a frame for support. Tubular steel makes the best framing material, as it is easily disinfected. Wood is the least suitable material for cage construction (Fig. 50–4), as it is difficult to disinfect and is readily chewed unless protected by wire on the inside surface.

## Flooring

Concrete is an excellent flooring material because it is easy to clean, discourages the completion of parasite life cycles, keeps the birds from digging out of the cage, and keeps rodent pests from digging their way in. An epoxy or other suitable coating may be used to maintain impermeability of the surface. Recently fledged birds in cages with direct access to the floor should be provided with wooden platforms to protect them from the cold, damp surface. Sand floors are popular in some aviaries, especially in those with suspended cages. When birds have access to the floor, sand may not be appropriate, as many psittacine birds, especially cockatiels and lovebirds, may overeat sand and suffer gizzard impactions. Dirt floors are extremely difficult to disinfect and may encourage the entrance of pests.

## Perches

Natural branches of fruit wood, eucalyptus, Australian pine, nut trees, or other nontoxic

**Figure 50–4.** Wooden flight cages are difficult to disinfect and will harbor disease organisms. The placement of the perches in these overcrowded flights allows fecal contamination of food and water bowls below.

plant species make the best perches. Easy replacement of perches is important for psittacine birds that tend to chew them. Commercial stud hangers make excellent perch holders and allow rapid perch replacement. If a number of birds are kept in the same cage, the perches should be placed at approximately the same height to decrease aggression. Most birds prefer to perch at the highest level and will fight for that position.

### Feeding Utensils

Food and water bowls should be constructed of an indestructible, easy-to-clean material such as stainless steel, plastic, or crockery. Metal bowls, particularly aluminum, will corrode and must be cleaned more frequently. The bowls should be placed away from perches and other areas where the birds may sit and defecate in them. If used in flight cages, feeding bowls should be placed up on a platform rather than on the cage floor, where they are more easily soiled and accessible to pests. If free-flowing seed dispensers are used, they should be checked frequently to make sure that they do not plug and prevent the birds from reaching the seed. Well-designed feeders force the bird to perch a slight distance from the seed trough so that the bird will not defecate in the food. Birds will remain calmer if the aviary is designed so that food bowls and feeders can be serviced from the outside of the cage.

### Nest Boxes

Most birds kept by aviculturists, excluding some finches and other passerines, nest in enclosed boxes. Nest boxes must be watertight or protected from rain. Deep nest boxes should have a ladder on the inside that leads to the entrance so the birds can climb out. Otherwise, sick or injured birds might be trapped in the nest box and starve to death before their absence is noted. It is useful to provide a sliding trap door on the back of the nest box for ease in inspecting the contents and removing any birds for medical treatment. The nest box should be placed in the most sheltered portion of the cage. Some psittacine and passerine birds become very defensive of their nests, so the nest area should be visually isolated from other birds.

### Aviary Arrangement

In moderate to large collections the birds may be grouped in a number of ways. From a preventive medicine standpoint, birds may be separated according to their continent of origin or susceptibility to a potential pathogen. For example, it would not be wise to house susceptible Australian cockatoos directly next to South American conures that may be asymptomatic carriers of Pacheco's parrot disease. *Chlamydia* is commonly carried by pigeons, and caution must be extended when mixing psittacines and pigeons in the same aviary. *Mycoplasma* is suspected to be easily transmitted between cockatiels and budgerigars. Another consideration in housing is the effect neighboring birds will have on each other's breeding behavior. If the same species of Australian parakeets are housed side by side, the males may spend so much time defending their territory during the breeding season that the hens are ignored. In contrast, housing macaws, Amazons, and cockatoos next to their own species may actually stimulate breeding if the nest area is sheltered from view.

The aviary should also have a sick room for birds to be isolated and treated. The requirements for this room are the same as for the bird room in a veterinary hospital. It should be well-lit, capable of being heated to 85 to 90°F, and isolated from the main collection.

## HYGIENE AND PREVENTIVE MEDICINE

Maintaining a clean aviary and uncontaminated food and water supplies is especially important in exotic bird collections. Potentially pathogenic bacteria, protozoa, and fungal organisms can proliferate in environmental sources and cause disease in stressed or compromised birds. Many seed-eating and frugivorous birds, including most psittacine and passerine species, have predominantly gram-positive alimentary tract flora. Gram-negative organisms found in the environment can cause infections in these birds under conditions that would not cause infection in other animals. Environmental sources of microbial disease should be investigated when evidence of poor hygiene practices is found or when birds repeatedly become reinfected after antimicrobial treatment.

### Potential Contaminants of Food

Exotic bird dietary rations frequently contain fresh fruits, vegetables, and seeds. Fresh fruits and vegetables are not considered to be a significant source of disease for people if stored in a refrigerator, rinsed before eating, and consumed when fresh. The same should hold true for avian species except that birds do not always eat food immediately after it is offered. Food that is left in the cage or falls to the floor may be eaten some days later. The author found that whole fresh fruits and vegetables contained relatively few bacterial contaminants, but cut fruit and vegetables grew relatively large numbers of gram-negative organisms (especially *E. coli*, *Enterobacter*, and *Pseudomonas*) when held at room temperature for 24 to 48 hours.[3] This indicates that fruits and vegetables should not be left in the cage for more than one day and probably less in hot weather.

Seeds and grains are not a significant source of microbial contamination if kept dry and rodent-free. Seeds may become contaminated at growing but usually do not carry large numbers of bacteria because of their low moisture content. However, if seed becomes damp from rain or humidity, it can support the growth of a number of contaminants, including yeasts and molds. Contamination from rodents is discussed later.

Seeds that are soaked in water until a small sprout is visible (usually 24 to 48 hours) are relished by psittacine birds and finches. It is believed that the sprouting process may convert some of the fats present in the endosperm into more readily assimilated carbohydrates. In spite of these advantages, a number of problems may be encountered during the sprouting procedure. The soaking process provides an excellent environment for the growth of potentially harmful bacteria and fungi. Even when fungistats (e.g., calcium propionate) are added to the soaking solution, the risk of fungal growth and release of mycotoxins persists. This potential may be reduced by performing the sprouting in an air-conditioned environment. The author has seen outbreaks of hemorrhagic enteritis in psittacine birds associated with feeding sprouted sunflower seed. The causative agent could not be isolated from either the food or the affected birds' digestive tracts, but the problem stopped as soon as sprouted seed was deleted from the diet. Many aviculturists feed sprouted seed with success, but the potential for disease originating from this source should be considered if a flock outbreak of alimentary tract disease occurs.

Commercially available foodstuffs vary widely in the bacterial flora that can be isolated from them. In an investigation at one avicultural collection, several brands of dog food and monkey biscuits grew only scant numbers of nonpathogenic gram-positive bacteria. However, commercial poultry feeds, including chick starter, turkey starter, and chicken scratch, grew relatively large numbers of gram-negative organisms, including *Pseudomonas*, *E. coli*, and *Klebsiella*. Tabid et al.[16] examined the microbiologic quality of poultry feed from 11 different sources and found markedly different coliform counts. These results may indicate that poultry products should be tested before feeding to exotic birds. Some premixes for lory nectar and some milk-based bacterial replacement products have also contained high levels of bacterial contaminants.

The utensils used to prepare food must be carefully cleaned to avoid introducing bacteria into the food. *Enterobacter*, *Pseudomonas*, and *E. coli* were the most common gram-negative organisms isolated from various appliances, utensils, and cutting boards at one breeding facility. In one instance, an outbreak of *Pseudomonas* enteritis was traced to a contaminated blender used to prepare a special nesting food for Australian finches.[3] Stricter hygiene practices significantly reduced the number of bacteria isolated. Utensils and appliances should be scrubbed in detergent and then disinfected after each use. The only utensil to be introduced into a common food source should be a clean scoop; food dishes should never be dipped into a container for filling. Nor should feed dishes be collected for filling and returned randomly to various cages.

## Potential Contaminants of Water

*Pseudomonas* and *Alcaligenes* are common water contaminants. *E. coli* and other gram-negative organisms may be important in local areas where well water is used. Very large numbers of *Pseudomonas* can be isolated from garden hoses and water systems using PVC pipe. Drinking water should be collected directly from the tap instead of from a hose. Allowing water to flow through the tap for a few minutes will flush it out and reduce the numbers of bacteria that can be isolated. In one nursery, flushing the tap for two minutes reduced the *Pseudomonas* contamination from greater than 100 to less than 25 colonies per milliliter of water. *Pseudomonas* infections were also traced to a bottled water dispenser that harbored the bacteria in its holding tanks. New dispensers ordered from the water company quickly developed the same problem, so the use of bottled water to prepare nursery formula was discontinued.

Various chemicals can be added to water to decrease microbial growth. Although none of these disinfectants has been specifically tested in exotic birds, their use in other species may provide useful information. Hulan and Proudfoot[11] tested the effects of chlorine at different concentrations on the growth and mortality rates of baby and juvenile broiler chickens. Concentrations less than 37 ppm of available chlorine had no effect on mortality, weight gains, or water consumption. At greater concentrations, water consumption decreased, and at 1200 ppm increased mortality occurred. Most water control districts maintain available chlorine at less than 5 ppm, so there is a considerable safety range when using commercial water supplies. Iodine can be used in much the same manner as chlorine. The addition of 0.25 ml Lugol's iodine per gallon of water used in tube-type waterers controlled the growth of slime and bacteria and had no adverse affects on Australian finches after two years of use in one facility. Copper water pipes are preferred over PVC.

## Other Preventive Practices

The choice of nest material is important in controlling nest-borne disease. The nest material must be kept dry and should be changed frequently to prevent the overgrowth of potentially pathogenic organisms. Aspergillosis, mucormycosis, and sporotricosis in nestling birds have been traced to contaminated nest material. There is increasing evidence that the incidence of some viral diseases may be reduced by rigorous attention to clean nesting material. Peat moss and plant potting soils support the growth of fungi and should not be used. Wood shavings, if kept dry and changed at least once yearly, make a safe and inexpensive nesting material (Fig. 50–5). There is some evidence that pine may have carcinogenic properties in laboratory mammals. Although there is no documentation of this in birds, hardwood shavings may be preferred. Following any disease the nest material should be changed and the box disinfected or discarded to prevent recurrence. Applying a blow torch to the interior of a wooden nest box may be necessary in some disease outbreaks.

Footbaths can be used to disinfect shoes when walking from one area to another. In order to be effective, a broad-spectrum nontoxic disinfectant that is not affected by organic matter should be used. Phenolic products, such as Environ One Stroke (Ceva Laboratories), are ideal. In investigations at an aviculture facility, footbaths filled with chlorhexidine or quaternary ammonium disinfectants were less effective and rapidly became contaminated with *Pseudomonas*. The disinfectant must be changed frequently to maintain effective action against the disease organisms.

Exact recommendations as to how often cages, water bowls, and feeders should be cleaned varies from aviary to aviary. In general, water should be changed daily and the bowl scrubbed and disinfected at least every five to seven days. If vitamins are added to the water, the bowls must be washed more often because the vitamins encourage microbial growth. Seed bowls and feeders need to be cleaned less frequently; they should be scraped whenever fecal matter builds up and washed two to three times monthly. Bowls used for fruit and soft foods should be emptied, washed daily, and returned to the same cage. Fecal matter should be scraped from the cage floors and perches weekly, or more often if a number of birds are housed in the same cage. Some aviculturists believe that it is necessary to completely clean and disinfect the cage and everything in it every few weeks. This practice unnecessarily exposes the birds to disinfectants and disturbance. If spilled food and gross fecal matter are removed promptly, there is no need to routinely disinfect

**Figure 50–5.** Pine shavings are a suitable nesting material. Note that the aviculturist's access door is lined with wire to prevent destruction by the birds. (Courtesy of Greg J. Harrison.)

the cage unless disease problems are encountered.

## Disinfectants

Disinfectants vary in the spectrum of microorganisms they kill and also in the physical properties that make them suitable for particular environmental conditions. No single disinfectant is active against all microorganisms or useful in all situations, so the chemical must be carefully chosen and used according to directions (see Chapter 6, Management Procedures). In order to work properly, disinfectants must be in solution or in a gaseous form and must come in contact with the disease agent for a certain period of time, often several minutes. Organic matter and dirt will prevent the disinfectant from contacting the disease agent and will deplete the action of bleach and quaternary ammonium products. For this reason heavily soiled areas such as cages and perches should be cleaned before disinfecting. Lightly spraying a dirty surface or quickly dipping an item in a disinfectant is not sufficient. Some commercial products combine a detergent with disinfectant and advise using these products for one-step cleaning and disinfection. If products are used in this manner, the surfaces should be thoroughly scrubbed to make sure the disinfectant contacts the pathogen. In addition, higher concentrations of the product should be used, according to label directions, to prevent organic matter from depleting the disinfectant's action. Listed below are some of the commonly used and readily available disinfectants. Mention of a particular product name is included to familiarize the reader with the class of disinfectant and does not imply that one product is better than another not listed.

**CHLORINE (BLEACH).** The most readily available form of chlorine is sodium hypochlorite, which is 70 per cent chlorine. Household bleach is diluted to 5 per cent chlorine. When used properly, chlorine is active against most bacteria and many viruses. Spore formers and mycobacteria are resistant to bleach. Organic matter significantly inhibits the killing action of chlorine, so surfaces must be cleaned prior to disinfection. Temperature and pH also influence the germicidal activity of chlorine. A 10°C rise in temperature decreases the time needed to kill bacteria by 50 to 60 per cent. Decreasing the pH also increases the killing activity. Recommended concentrations for drinking water are 1 to 3 ppm, for disinfecting the surface of fruits and vegetables—4 to 5 ppm, and for disinfecting cages, counters, and hard surfaces—50 to 100 ppm.[8] A recommended dilution for household bleach used as a general disinfectant is 200 ml/4 liters of water.[5] Of 35 disinfectants tested against viruses of feline origin, household bleach diluted at one ounce per quart of water (0.175 per cent solution) was the most effective.[15] Chlorine is corrosive to most metals except stainless steel. Granulated chlorine products used in swimming pools should not be used around birds. Dangerous levels of chlorine gas are produced within 20 to 30 minutes after the granules become damp.

**QUATERNARY AMMONIUM COMPOUNDS (ROCCAL; MICRO QUAT).** Quaternary ammonium compounds (quats) have broad antibacterial activity and are viricidal for lipophilic viruses at moderate concentrations. Some compounds are particularly effective against *Chlamydia* and are recommended for disinfection after psittacosis outbreaks. They are not effective against spore-forming bacteria, *Mycobacteria,* or hydrophilic viruses. Like chlorine, quats are rapidly inactivated by organic matter, so surfaces must be cleaned prior to disinfection. Quats are cationic surfactants and have a cleaning action of their own. They are not compatible with detergents (which are anionic surfactants), so surfaces cleaned with detergents must be thoroughly rinsed before applying the quat. In diluted form quats are neither corrosive nor odoriferous and are suitable for disinfecting metal surfaces and food dishes.

**PHENOL TYPE DISINFECTANTS.** The most commonly used phenolic type disinfectants are sodium orthophenylphenate and orthochlorophenol (e.g., Environ One Stroke). These agents, in combination with other coal tar derivatives, offer a wider range of germicidal activity than most other disinfectants. They are effective against most bacteria, including *Pseudomonas* and *Mycobacteria, Candida, Tricophyton* fungi, and lipophilic viruses (including Newcastle disease). Phenols are less affected by organic matter than other disinfectants, and a detergent is usually added to commercial products to enhance cleaning action. Phenols are particularly suitable for use in footbaths and in cleaning wooden cages and similar areas where other disinfectants would be inactivated by the organic matter.

**CHLORHEXIDINE** (Nolvasan; Virusan). Chlorhexidine is a disinfectant of low toxicity that has been used extensively as a viricide but also has other antimicrobial properties.[12] It is not effective against some gram-positive bacteria, *Pseudomonas,* or *Mycobacteria.* Chlorhexidine is less affected by organic matter than other disinfectants.

The disinfectants listed above are not especially toxic to most mammals and poultry when used as directed and kept out of food and water. However, until more is known about the specific reactions of exotic birds, extra caution should be taken when using disinfectants around them. Birds continually groom themselves, so any overspray that gets on their feathers will rapidly be ingested. Birds should be removed from the aviary when disinfectants are being applied, and the aviary and its contents should be thoroughly rinsed before birds are returned.

### Aviary Visitors

As another step in a preventive medicine program, visitors to the aviary should be kept to a minimum. Not only do strangers upset the birds, but they may bring diseases into the collection on their shoes and clothing. Visitors should wear clothing and shoes different from what they wear when servicing their own collection. Upon arrival they should dip their shoes in a footbath containing a phenol type disinfectant. Disinfectants for clothes, hair, and hands are recommended (see Chapter 6, Management Procedures). If Newcastle disease has been diagnosed within the state, no visitors should be allowed and traffic is kept to a minimum.

## PEST CONTROL

Pests in the aviary can cause a number of problems. Rodents consume food, urinate on stored seed, and can spread disease organisms such as *Salmonella, E. coli,* and other bacterial agents. Mice and rats move into seeders and nest boxes, disrupting the birds and decreasing breeding. Rats may chew the toes or kill and eat baby birds and can catch adult birds the size of a cockatiel. Wild birds may carry a number of viral, bacterial, protozoan, and parasitic diseases and will consume food. Insects, such as ants, mosquitoes, and cockroaches act as mechanical vectors of disease.

Rodents are difficult to control, and the advice of a pest control specialist is helpful. The biology and control of mice and rats have been reviewed in two useful publications.[9, 10] Signs of rodent infestation include sightings, droppings, urine odor, and dark grease-like smudges on heavily traveled routes. Rodents will multiply in direct proportion to the abundance of food and the availability of cover and nesting sites. Reducing the attractiveness of the aviary to the rodents will help decrease the population. Spilled food should be cleaned up and seed should be stored in metal cans with tight-fitting lids. Larger amounts of seed can be stored in metal-lined rooms with tight-fitting doors and concrete floors (Fig. 50–6). Reduce plants and weeds and clean up debris piles around the aviary. An apron of gravel at least 6 inches deep and extending 3 to 10 feet around aviary buildings will provide permanent weed control. It will also reduce the influx of rodents from surrounding areas, since they find it difficult to burrow through and are reluctant to cross the open space. Plug all holes larger than one-fourth inch and surround the base of buildings with metal flashing that extends 8 to 12 inches above and below the ground. Drain pipes are favorite runways for mice and should be covered with screen.

Concrete will reduce, but not prevent, ro-

**Figure 50–6.** Dry food for birds may need to be stored in a metal-lined, concrete-floored room to prevent access by rodents.

dents from burrowing through the building floor. Mice and rats can gnaw through thin cement floors. At the time of construction, laying a metal screen beneath the concrete will prevent this. Rodent holes in concrete should be promptly plugged; place steel wool or a rodenticide in the hole and fill with patching compound.

Once rodents are established in an aviary, traps and poisons are necessary to get rid of them. The control program will be most successful if started at the end of winter before the animals breed and increase in numbers. Trapping is the safest way to eliminate pests inside the aviary. Saturation trapping, in which many traps are placed in a small area, is usually more successful than dispersing traps over a larger area. Repeating traps that catch 10 to 15 animals without resetting* are particularly effective against mice and can be placed inside cages of birds that are larger than the size of a cockatiel. Smaller birds, such as lovebirds, may be accidentally caught in the trap. The traditional snap trap is best for rats but must not be placed inside the bird's cage. Locate traps along runways where the rodents travel and in areas where food is available. The trigger or door of the trap should be placed so that the animals will contact it in the course of normal movement. Mice are curious about new items in their environment, so the traps should be placed in a different arrangement weekly. Rats have the opposite attitude and will shy away from changes. Rat traps should be left in the same place for several weeks to catch shy animals. Favorite baits for both rats and mice are peanut butter and oat seeds, ground chicken, beef, or sardines. Traps can be disguised by covering them lightly with sand to catch particularly smart rats. Glue boards that catch rodents in a sticky adhesive are less effective than traps but useful in areas where traps cannot be placed. A pregnant cat in the neighborhood can significantly reduce the mouse population.

A number of different poisons are recommended for controlling rodents, but not all are suitable for use in an aviary. Acute rodenticides, such as arsenic, zinc phosphide, and sodium fluoroacetate are too dangerous for use around birds, as rats often carry baits into the feeding dishes. Poisonous tracking powders are easily spread into cages and should not be used in aviaries. The safest rodenticides are anticoagulants that cause death by interfering with vitamin K utilization. These are chronic poisons and must be eaten several times before causing death. If a bird is accidentally poisoned, the toxic effects can be counteracted by multiple doses of vitamin K (Table 50–2).

**Table 50–2.** EMERGENCY PROTOCOL FOR ANTICOAGULANT RODENTICIDE INTOXICATION

| |
|---|
| **Clinical Signs** |
| Any signs of abnormal or prolonged bleeding, especially around the nostrils. Suspect poisoning in any birds showing sudden signs of depression and labored or noisy breathing. Poisoned birds will frequently bleed into the lungs. |
| **Treatment** |
| It is extremely important to minimize stress, which will increase hemorrhage into the lungs. Administer 2 mg/kg water-soluble vitamin $K_3$ I.M. and 20 mg/kg prednisolone succinate I.M. Immediately place the bird in a dark environment heated to approximately 80 to 85°F, and do not check or disturb for at least two hours. Repeat treatment every four to eight hours until the bird is stable. Once the bird is stable, measure the time it takes for the blood to clot in a nonheparinized capillary tube. Continue supplementation by giving a multivitamin preparation that contains vitamin K in the drinking water. |
| **Notes** |
| In the very few cases that I have treated, the birds were too stressed to attempt intravenous therapy. In less compromised individuals, whole blood transfusions and administration of the drugs intravenously may be more beneficial. |

Anticoagulants are available in a number of bait forms that are attractive to rodents: water soluble powders, coated seeds, and pellets. The most appropriate bait form for aviary use is pellets that can be readily distinguished from the birds' normal food. Some studies indicate that poisoned seeds dyed a special color will not be eaten by wild birds, but there has been limited testing in exotic species.[1] Poisoned water stations are effective only if no other convenient source of water is available. The bait should be placed inside an enclosed box large enough for the rodents to eat it inside and not spread it around. Bait boxes should be placed 10 to 15 feet apart in the aviary aisles, outside along runways, and wherever signs of rodents are found. Under no circumstances should poison bait be placed in any location where birds may reach it. Rats will carry baits back to their burrows for later consumption, so the aviary should be checked daily for the presence of spilled poisons. It may be necessary to "prebait" rats, that is, to provide a bait identical to the poison bait except for absence of the rodenticide, until the rats accept it as a source of food. Prebaiting can greatly enhance the elimination of shy pests. In most cases, it is best to combine saturation trapping with careful use of rodenticides to clear an aviary of rodents.

Ultrasonic pest control devices may not be

*Ketch-All, Kness Mfg. Co., Albia, IA 52531

appropriate for use around birds. A cessation of reproduction was reported to occur when ultrasonic pest control devices were installed in a mixed species breeding aviary. When the devices were removed, reproductive activity resumed, starting with smaller species (e.g., budgerigars) in the first week, followed months later by the larger species (e.g., macaws, Amazons).

Wild birds can be discouraged from visiting the aviary by reducing spilled food and screening buildings with narrow mesh screen. Predatory mammals, such as racoons, opposums, cats, and skunks, may kill birds and eat eggs. Cats in particular enjoy walking on the tops of the aviaries, and birds may die when they panic and fly into the cage walls. Fences and traps may be necessary to keep these animals away. Snakes may also take babies and eggs from the nest. Elevating the nestbox and minimizing places where the snakes can climb will prevent this. Having taken these precautions, snakes are welcome in many aviaries because they do much to control the mice population.

Ants are a particular problem where fruit, cooked food, and certain types of dog food are fed. They are more than a nuisance, since birds will not eat food contaminated by ants. It is best to contact a professional exterminator, since different ant species require different strategies for control. Even with professional help ants can be surprisingly difficult to exterminate, especially in warm weather. Usually repeated doses of pesticides must be applied. Chlorinated hydrocarbons (such as the DDT derivatives aldrin, dieldrin, and heptachlor) have many toxic effects, including eggshell thinning,[8] and should not be used around birds. Amdro, a pesticide designed to be taken by fire ant workers back to kill the queen, is effective. Carbaryl (5 per cent) dust incorporated into nest box material (1 teaspoon per cockatiel nest) helps control pests in the nest box.

The author has safely used bendiocarb (Ficam, Fisons, Inc., Bedford, MA) around the outside of aviaries and in the aisles but has not used any insecticide where the birds or their food might come in contact with it. Care must also be taken to make sure that applied pesticides will not run off into cages when it rains or the grounds are watered. Bendiocarb is a methylcarbamate pesticide, so if accidental toxicity occurs, atropine can be tried as an antidote. Organophosphates (e.g., diazionon and malathion) may also be used if properly applied. Respiratory distress has been reported as the major sign of toxicity in accidentally poisoned birds.[13]

The veterinarian can help the aviculturist and pest control operator by reviewing the pesticides and rodenticides to be used. In case accidental toxicity occurs, the aviculturist should be advised to look for appropriate signs and take appropriate action. It might be wise to supply capable aviculturists with injectable antidotes, such as vitamin K, along with directions on how to use them. If accidental toxicity occurs, there might not be time to transport the bird to a veterinarian.

## DISEASE MANAGEMENT IN AN EXOTIC BIRD COLLECTION

It is a common misconception among bird owners that a fair number of birds will die each year and that there is nothing anyone can do about it. Some cockatiel breeders, for example, will accept nestling mortality rates of 25 to 30 per cent as the normal attrition of breeding captive birds. The fact is that much of the accepted mortality can be reduced by implementing good management techniques and recognizing the signs of disease early so that control measures can be started before the disease spreads through the entire flock.

Aviary birds are essentially captive wild animals and will make every attempt to hide their signs of disease. Often problems become well-advanced before signs are recognized. For this reason, the aviculturist should check each bird in the collection daily. This will enable him to watch trends in individual birds and spot sick birds at the earliest possible time. The cages should be approached silently, so that the bird is seen before it sees the spotter. When birds know they are being observed they will assume alert postures, and subtle signs of disease will be hidden. The birds are checked first, looking for fluffed feathers, dull eyes, abnormal posture, and individuals that separate themselves from the group. Next the cage should be scanned, looking at the droppings, noting food consumption, and checking for blood and other abnormalities.

Sick birds are immediately quarantined, and the exposure between these birds and the rest of the flock is reduced. Diagnostic procedures employed for the single pet bird are instituted, and the degree of spread through the aviary is determined. If it is necessary to know the status of each individual, as with expensive species, separate samples should be collected from each bird. Often it is more important to survey a group of birds for a particular pathogen than to do an extensive work-up on an individual. In this case, specimens should be collected from a number of birds and pooled together into one culture. If it later becomes necessary to sample individuals, only the groups with pathogens need be retested. This method can quickly identify how far a particular pathogen has spread in a collection and save the time and costs of culturing each bird.

A major concern for the aviculturist with an established collection is the development of resistant strains of bacteria. In many cases the only practical means of treating large groups of birds is to mix drugs in the food or water supply, but this should be approached with caution. In a limited study, several groups of birds with *E. coli* present in their cloaca received drugs to which the bacteria was sensitive. In all cases, five days after the end of treatment, *E. coli* could still be recovered from the cloaca, and the bacteria was resistant to related antibiotics as well as to the antibiotic used.

Since few drugs can be administered through the drinking water, their use should be reserved for the situations in which they are absolutely necessary. The bird owner should be warned against periodic, prophylactic administration of antibiotics unless a specific disease agent has been identified.

Following the treatment of any disease the aviary should be cleaned and disinfected. Bacteria and *Chlamydia* can survive for weeks or months in dry fecal material and reinfect the birds once treatment stops. The cage, perches, food and water containers, nest boxes, and aviary aisles should be included in the clean-up program. In addition, seed exposed to the sick birds should be discarded.

## THE ECONOMICS OF PREVENTIVE MEDICINE SERVICES

Serious aviculturists are aware that preventive medicine procedures and good hygiene are essential to the success of their operation and that competent veterinary advice and evaluation of their facilities are worthwhile. With proper management, mortality can be reduced and the production of breeding birds can be increased to more than offset the cost of professional services. For example, consider a client who owns 25 breeding pairs of cockatiels with a yearly production of 150 salable offspring (at the current wholesale price of approximately 25 dollars). If the veterinarian can increase production by 15 per cent, several hundred dollars in veterinary services can be justified.

## REFERENCES


1. Brunner, H., and Coman, B. J.: The ingestion of artifically coloured grain by birds, and its relevance to vertebrate pest control. Aust. Wildlife Res., 10:303–310, 1983.
2. Dychdala, G. R.: Chlorine and chlorine compounds. In Lawrence, C. A., and Block, S.: Disinfection, Sterilization and Preservation. Philadelphia, Lea & Febiger, 1968.
3. Flammer, K., and Drewes, L.: Environmental sources of gram-negative bacteria in an exotic bird farm. Proceedings of the Jean Delacour/IFCB Symposium on Breeding Birds in Captivity, Universal City, CA 1983, pp. 83–93.
4. Foster, J. W.: Animal behavior and its application to aviculture. Veterinary seminar notes from the American Federation of Aviculturists Annual Meeting, Las Vegas, NV, 1980.
5. Fowler, M. E.: Sanitation and disinfection. In Fowler, M. E. (ed.): Zoo and Wild Animal Medicine. Philadelphia, W. B. Saunders Company, 1978, pp. 31–34.
6. Fowler, M. E.: Stress. In Fowler, M. E. (ed.): Zoo and Wild Animal Medicine. Philadelphia, W. B. Saunders Company, 1978, pp. 31–34.
7. Gilsleider, E., and Oehme, F. W.: Some common toxicoses in raptors. Vet. Hum. Toxicol., 24:169–170, 1982.
8. Hofstadt, M. S., et al.: Diseases of Poultry. 8th ed. Ames, IA, Iowa State University Press, 1984.
9. Howard, W. E., and Marsh, R. E.: The house mouse: Its biology and control. Leaflet 2945, Division of Agricultural Sciences, University of California, 1981.
10. Howard, W. E., and Marsh, R. E.: The rat: Its biology and control. Leaflet 2896, Division of Agricultural Sciences, University of California, 1981.
11. Hulan, H. W., and Proudfoot, F. G.: Effect of sodium hypochlorite (Javex) on the performance of broiler chickens. Am. J. Vet. Res., 43:1804, 1982.
12. Lawrence, C. A.: Antimicrobial activity, in vitro, of chlorhexidine. Ft. Dodge Biochem. Rev., 30:3, 1961.
13. Reece, R. L., and Handson, P.: Observations on the accidental poisoning of birds by organophosphate insecticides and other toxic substances. Vet. Rec., 111:453–455, 1982.
14. Reuber, H. W., Rude, T. A., and Jorgenson, T. A.: Safety evaluation of a quaternary ammonium sanitizer for turkey drinking water. Avian Dis., 14:211–218, 1970.
15. Scott, F. W.: Virucidal disinfectants and feline viruses. Am. J. Vet. Med., 41:410–414, 1980.
16. Tabid, Z., Jones, F. T., and Hamilton, P. B.: Microbiological quality poultry feed and ingredients. Poultry Sci., 60:1392–1397, 1981.

# Chapter 51

# SEX DETERMINATION TECHNIQUES

SUSAN L. CLUBB

Accurate sex determination techniques have allowed aviculturists in recent years greater opportunities for successful propagation of monomorphic avian species. The first requirement for a successful captive breeding program is a true pair. Many species of psittacine birds and other species important in aviculture show no obvious sexual dimorphism.

## SEXUAL DIMORPHISM

Eclectus Parrots (*Eclectus* sp.) are an example of striking sexual dimorphism in which the male bird is a brilliant emerald green and the female bird is deep red and purple. For many years the male and female were considered to be different species, leading to obvious difficulties in captive propagation. Unfortunately sexual dimorphism is not obvious in many species of psittacines.

There are a few characteristics that are helpful in visually sexing monomorphic birds. These are, however, only indications of sex and cannot be depended upon for accurate sexing of pairs for breeding. Size and shape of the beak and head often give an indication of sex in psittacines. The male bird usually has a larger head and a broader, heavier beak than the female. Male cockatoos are usually significantly larger than females.

Behavioral differences are helpful in sexing some species. The male will often exhibit a courting song and dance while the female observes. This is more obvious in passerines than psittacines and is the most practical means of sexing Society Finches. Likewise, a parrot that spends an excessive amount of time rooting around on the cage floor is likely to be a female that is interested in nest building. Male birds of many species tend to be more aggressive and less fearful than female birds. It is also apparent that the female bird is more likely to bite and will protest more loudly when restrained.

Copulation is observed frequently in improperly paired birds (two males or two females) and is no indication of a true pair. Homosexual pairs are uncommon if birds are allowed to choose a mate by placing birds in a group. Groups must be watched carefully for the formation of pair bonds. If one pair decides to breed, they may become very aggressive toward other birds in the group. Several nest boxes should be provided to reduce fighting over nesting sites.

Palpation of the pelvic bones is a commonly reported sexing method. The pelvic bones are located slightly cranial to the vent and can be palpated with the bird in dorsal recumbency in the palm of the hand with the head restrained. Some people prefer to palpate the bird while it is in a standing position and grasping a perch or finger. In the adult female the pelvic bones are reportedly farther apart than the corresponding distance in the adult male. In addition, the pelvic bones of the female are supposedly more pointed and in the male they are more rounded and directed medially. When used on birds of known sex, however, this test is highly inaccurate.

Sexual dimorphism in psittacines is summarized in Table 51–1. Species that are uncommon in aviculture have been omitted. Australian and Asian species are often dimorphic, whereas South American and African species are usually monomorphic. Birds that inhabit arid climates show a higher incidence of sexual dimorphism than jungle species. This may indicate more dependence on sight identification among species in arid climates.[9, 13]

With intensive observation of a species, slight differences may be observed which will be helpful in visual sex determination. For example, African Grey Parrots (*Psittacus erithacus erithacus*) can be sexed with some accuracy by the shade of grey of the plumage. The female has a pale powder grey breast and back, whereas the male is a darker charcoal grey.

**Table 51–1.** SEXUAL DIMORPHISM IN PSITTACINES*[9, 11, 13]

**PARROTS OF AUSTRALIAN AND PACIFIC DISTRIBUTION**

**LORIIDAE**—Most species are monomorphic. The head is usually larger in male birds.

Dusky Lory—Genus *Pseudeos.* Monomorphic.

Black Lory, Duvienbode—Genus *Chalcopsitta.* Monomorphic.

Red Lories—Genus *Eos.* Monomorphic.

Chattering Lory—Genus *Lorius.* Monomorphic.

Rainbow Lory—Genus *Trichoglossus.* Meyer's Lorikeet (*T. flavoviridis meyeri*)—the male has a larger and brighter yellow ear patch than the female.

Stella Lory—Genus *Charmosyna.* Stella Lory is dimorphic. The female has a yellow patch on the rump and lower back.

Blue Lories—Genus *Vini.* Monomorphic.

**CACATUIDAE**

**CACATUINAE**

Palm Cockatoo—Genus *Probosciger.* Monomorphic but male is usually larger and has a larger beak. The size difference varies geographically and according to subspecies.

Black Cockatoo—Genus *Calyptorhynchus.* Sexual dimorphism is striking in some species and barely evident in others. In the Banksian Cockatoo (*C. magnificus*), the male plumage is black except for bands of red in the tail, whereas the female plumage is dotted and barred with yellow-orange.

Gang Gang Cockatoo—Genus *Callocephalon.* The male is slate grey with a red head and crest. The female has a grey head and crest and plumage that is barred with greyish-white.

White and Pink Cockatoos—Genus *Cacatua* and *Eolophus.* Adult birds except for the Bare-eyed Cockatoo (*C. sanguinea sanguinea*) can be sexed by eye color. The female has a red iris, whereas the iris of the male is dark brown to black. A bright light may be needed to determine eye color in some species. The female of most species is smaller than the male. Red-eyed males and dark-eyed adult females have been reported but are rare. The iris is brown in immature birds of both sexes.

**NYMPHICINAE**

Cockatiel—Genus *Nymphicus.* In the wild type (grey) the sexes are easily distinguished by the bright yellow facial markings in the male which are absent in the female. Cinnamon, white, or albino cockatiels can be sexed by the faint diagonal bars on the ventral surface of the primary and secondary flight feathers of the female, which are absent in the male. Pied cockatiels are difficult to sex and if heavily pied require surgical sexing. Cinnamon cockatiels are sexed by faint yellow facial coloration in the male. Pearl cockatiel males lose the pearl coloration upon maturity and resemble the normal grey coloration. Only female and immature birds will exhibit the pearl coloration. Most birds exhibit mature plumage coloration at six to nine months of age.

**PSITTACIDAE**

**PSITTACINAE**

Budgerigar—Genus *Melopsittacus.* In the normal green variety, the cere of the male is blue, whereas the cere of the female is pinkish-brown. This is not dependable in hybrid color variations, e.g., yellow, blue, or white birds.

Rosellas—Genus *Platycercus.* The male of most rosella species is slightly brighter than the female or immature. Female and young of several species have a row of white spots on the ventral surface of seven or eight primary and secondary flight feathers. These are lost by the male upon reaching sexual maturity. The wing spots are retained in adult female Yellow Rosella (*P. flaveolus*), Golden-mantled Rosella (*P. eximius*), Mealy Rosella (*P. adscitus*), and Stanley Rosella (*P. icterotis*).

Red-rumped Parakeet—Genus *Psephotus.* Most species in the genus exhibit sexual dimorphism. The Red-rumped Parakeet exhibits pronounced sexual dimorphism. The male has a red patch on the rump. The female is more drab. Other species are uncommon in aviculture.

Neophemas—Genus *Neophema.* Sexual dimorphism occurs in this genus and varies from slight difference in coloration in the Bourke's Parakeet to obvious dimorphism in the Scarlet-chested Parakeet. The Scarlet-chested male has a scarlet chest, whereas the chest of the female is green. In most species the female is duller in color than the male.

Barabands, Rock Pebblers, etc.—Genus *Polytelis.* Members of this genus show some degree of sexual dimorphism. The male in this genus is often smaller than the female and the female and young are usually duller in color. The female Barabands Parakeet (*P. swainsonii*) lacks the yellow feathers of the male. In the Rock Pebbler the ventral surface of the male's tail feathers are black, whereas they are margined and tipped in pink in the female. The female Princess of Wales (*P. alexandrae*) is duller in color and the bill is paler red than that of the male.

Kakarikis—Genus *Cyanoramphus.* Monomorphic.

Crimson Wings—Genus *Aprosmictus.* The male Crimsom-winged Parakeet is easily distinguished from the female by his black mantle.

King Parrots—Genus *Alisterus.* Sexual dimorphism is present in plumage and beak coloration in some but not all species of King Parrots. Some, but not all, subspecies of Green-winged King Parrots are dimorphic.

Eclectus Parrots—Genus *Eclectus.* These exhibit striking sexual dimorphism in which the male is a brilliant green and the female is red and maroon. This color difference is evident at the time of eruption of the first tail and contour feathers in the baby bird. Both sexes have black down.

Greatbills, Blue Napes, and Muller's Parrots—Genus *Tanygnathus.* Only the Muller's Parrot (*T. mulleri*) is dimorphic, with the beak being red in the male and white in the female. The male Greatbill (*T. megalorynchos*) has a much larger beak than the female.

Fig Parrots—Genus *Psittaculirostris.* Most are dimorphic in plumage coloration.

Pesquet's Parrot—Genus *Psittrichas.* Male has a red line behind the eye which is absent in the female.

**Table 51–1.** SEXUAL DIMORPHISM IN PSITTACINES*[9, 11, 13] *Continued*

**PARROTS OF AFROASIAN DISTRIBUTION**
**PSITTACIDAE**
**PSITTACINAE**

Ring-necks—Genus *Psittacula*. All male birds in this genus have a ring encircling the neck or a wide black mustache ring. In some species this is lacking in the female and immature. Common members of the genus include the Ring-neck Parakeet and the Alexandrine Parakeet. Adult male plumage may not be evident until two and a half years in some individuals. In some species the bill color is different in males and females. For example, the female Mustache Parakeet (*P. alexandri fasciata*) and Derbyan Parakeet (*P. derbyana*) have black upper beaks, whereas those of the males are red.

Hanging Parrots—Genus *Loriculus*. Adult birds are sexually dimorphic in plumage and, in some species, in eye color.

Vasa Parrots—Genus *Coracopsis*. Hypertrophy of tissues of the vent is evident in the male and is especially pronounced during the breeding season.

Lovebirds—Genus *Agapornis*. The commonly available species are monomorphic. Some of the uncommon species of lovebirds, however, show striking dimorphism. For example, the male Madagascar Lovebird (*A. cana*) has a grey head, whereas that of the female is green. The male Abyssinian Lovebird (*A. taranta*) has a red patch on the forehead and lores which is absent in the female.

African Grey Parrots—Genus *Psittacus*. Slight sexual dimorphism is evident in African Greys in coloration. The breast and back of the female is a pale chalky grey, whereas the male is a darker grey. This difference is helpful only when comparing birds from the same region, as color varies geographically.

Senegals and Related Species—Genus *Poicephalus*. Some members of this genus show marked sexual dimorphism, whereas others are monomorphic. The male Red-bellied Parrot (*P. rufiventris*) has a deep red-orange breast and abdomen, whereas the female's breast is greyish-brown. The female Ruppells Parrot (*P. rueppellii*) is more brightly colored than the male, having a bright blue rump patch that is absent in the male.

**PARROTS OF SOUTH AMERICAN DISTRIBUTION**
**PSITTACIDAE**
**PSITTACINAE**

Macaws—Genus *Ara*— Blue & Gold, Scarlet, etc. Monomorphic.
Genus *Andorhynchus*—Hyacinth. Monomorphic.

Conures—Genus *Aratinga*—Sun Conure, etc. Monomorphic.
Genus *Pyrrhura*—Painted Conure, etc. Monomorphic.
Genus *Nandayus*—Nanday. Monomorphic.
Genus *Enicognathus*—Austral and Slender-billed. Monomorphic.
Genus *Cyanoliseus*—Patagonian. Monomorphic.

Hawkheads—Genus *Deroptyus*. Monomorphic.

Thick-billed Parrots—Genus *Rhynchopsitta*. Monomorphic.

Quaker Parakeet—Genus *Myiopsitta*. Monomorphic.

Mountain Parakeets—Genus *Bolborhynchus*. Only one species, the Golden-fronted Mountain Parakeet (*B. aurifrons*), is dimorphic. The male has yellow markings on the lores, forehead, throat, and part of the cheek, whereas the female is predominantly green.

Bee Bee Parakeets—Genus *Brotogeris*. Monomorphic.

Parrotlets—Genus *Forpus*. All are dimorphic. In most species the male will have blue markings on the rump and/or wings whereas the female is predominantly green.

Amazon Parrots—Genus *Amazona*. All species are monomorphic with two exceptions: in the Spectacled Amazon (*A. albifrons*), the male has red markings on the small upper wing coverts and the edge of the carpus; the female is usually green in this area but may have a limited amount of red. The female Yellow-lored Amazon (*A. xantholora*) lacks the white on the head and the red markings of the male and is in general more drab-appearing than the male.

Pionus Parrots—Genus *Pionus*. Monomorphic.

Pileated Parrot—Genus *Pionopsitta*. The male has a red head and the female has a green head. This dimorphic coloration is evident in immature plumage. Other members of the genus are monomorphic.

Caiques—Genus *Pionites*. Monomorphic.

Taxonomic system from Forshaw, J. M.: Parrots of the World. New York, Doubleday Press, 1973

Color differences have also been observed from one geographical region to another; therefore, this test is helpful only if the birds are from the same region. In the Double Yellow-headed Amazon (*Amazona ochrocephala oratrix*) the mature male will usually have a faint red tipping on the feathers of the nape. In Blue-and-Gold Macaws (*Ara arauana*) the female tends to have heavier feathering on feather lines that cross the bare facial skin patches.

Many birds molt into a prenuptial plumage prior to the breeding season. Slight differences in plumage coloration may be easier to detect at this time of year.[11]

In most species that exhibit color dimorphism the immature bird resembles the female. Some species do not exhibit adult dimorphic plumage until 1.5 to 2.5 years of age. Sexual maturity may precede the development of adult male plumage in Ring-neck Parakeets (genus *Psitta-*

*cula*). Aviculturists often state that the male must earn his ring. Surgical sexing is the logical alternative to waiting an excessive period of time to distinguish the sexes by plumage.[13]

## VENT SEXING

Palpation or observation of the vent is a rapid and accurate method for sexing some avian species and is widely used for sexing newly hatched poultry and waterfowl. A rudimentary male copulatory organ is present in domesticated fowl, other gallinaceous birds, waterfowl, and ratites. In sexually mature female psittacines the cervix may be observed on the left cloacal wall.[1, 11, 14]

In canaries the sex may be determined by examination of the vent during breeding season. The female's vent is more rounded and the abdomen is slightly distended, whereas the male's vent is conical in shape and protruding.

## SURGICAL SEXING

Caponization techniques were originally performed by early poultrymen to increase the rate of weight gain and decrease male aggressiveness. Ornithologists utilized laparotomy techniques in the 1950's and 1960's for monitoring seasonal differences in gonadal size and for sex determination. In the past 10 years these techniques have been refined by endoscopy. With the development of rigid endoscopes for arthroscopy, the method has become safer and applicable to smaller species.[6, 10, 17–19]

Surgical sexing by laparoscopy has become widely accepted among aviculturists and is the most common method of sex determination available today. It is rapid, accurate, and not dependent on sexual maturity. The risk factor is low if the veterinarian is experienced.

Advantages of laparoscopy include the opportunity to examine other organs and systems as well as direct observation of the gonads (see Chapter 15, Endoscopy). Size and development of the gonads may help to determine readiness for a breeding program. This is especially true of the female, as the presence of large follicles, and in some cases corpus albicans, may indicate that the female is actively cycling. In the male, testicular size is not so obviously an indication of age or sexual maturity. Many birds examined immediately before or after fledging will have testicles comparable in size to those of an adult bird.

### Technique

Physical examination should precede restraint and induction of anesthesia for endoscopy. The use of anesthetics for laparoscopy is controversial. Surgical sexing can be accomplished rapidly without anesthesia and the bird can be returned immediately to its cage. Small species may be easily restrained. In larger species excessive movement may make the procedure difficult. Stress due to fear is reduced when birds are anesthetized; however, the slight risk of anesthetic death is present. The question of whether or not the bird experiences excessive pain or fear has not been resolved but should be considered by the clinician.

Fasting is not necessary unless excessive amounts of food or water are palpated in the crop. In obese birds fasting is helpful in reducing the size of the proventriculus, which otherwise makes visualization of the gonads difficult (see Chapter 15, Endoscopy).

The bird is positioned in right lateral recumbency, since most species have only one ovary on the left side. An assistant is needed in most cases to hold the wings together over the back. The wings should always be held by the humerus so that any sudden flapping will not result in a fractured elbow. The head and left, or both, legs may be restrained by cords, rubber bands, tape, or Velcro fastening strips on a restraint board. The left leg must be extended caudally and securely fastened to prevent injury. This procedure is accomplished so quickly that supplemental heat is rarely needed.

The site of entry is behind the last rib in a palpable depression located approximately at the midshaft and cranial to the femur. The feathers in this area are plucked and the site prepared with a suitable surgical preparation. Draping is ideal but not necessary.

A skin incision is not necessary when using a sharp trocar, and bleeding is minimized. The trocar, with cannula in place, is punched through the skin and muscular layer, taking care to direct the trocar craniodorsally toward the lung. This allows rapid entrance and a wide margin of safety. The distance between the entry site into the thoracic air sac and any organ is great. If the lung is penetrated, bleeding is minimal and adverse effects are rarely noticed. Direction of the trocar cranially also decreases the chance of penetration of the internal iliac vein. The trocar is removed and the endoscope inserted into the cannula. The air sac wall between the thoracic and cranial abdominal air sacs is visualized and followed caudally until the endoscope is pointed toward the middle or the caudal divisions of the kidney. An area in the air sac that is clear and devoid of vessels is located and the scope is pushed through the air sac. This again allows maximum distance between the site of penetration of the air sacs and the organs beneath. The kidney is followed cranially to the triad of the adrenal gland, gonad, and kidney.

Testicular tissue appears elliptical or cylindrical with a smooth vascular surface (Fig. 51–1). The developing follicles of ovarian tissue in the adult female resemble a cluster of grapes (Fig. 51–2). Both testes and ovaries may be pigmented, especially in white cockatoos. The immature ovary appears as a white blanket with an undulating to smooth surface that spreads over the surface of the kidney and adrenal gland. In order to definitively distinguish an immature ovary from a testicle, the cranial and caudal poles should be observed. A testicle will have rounded poles, whereas the poles of the ovary will be flattened. Sex determination is safe and accurate at a very young age even prior to fledging. The testicle is in many cases larger at the age of fledging than at four to eight months of age. In some birds a small fat pad may cover the gonads, obscuring visualization.

Suture of the entry site is not necessary if the puncture method is utilized. When the leg returns to its normal position, the sliding layers of muscle and skin effectively seal the wound.

Bleeding may be encountered by rupture of a vessel or organ. In most cases the bleeding ceases quickly. If bleeding is excessive the bird should be held in an upright position. This allows collection of blood in the air sacs and prevents asphyxiation. Subcutaneous emphysema occurs rarely and in most cases is self-limiting. Post-surgical infections are very rare, even when only cold sterilization procedures are used. Obstruction of vision is the most common reason for failure to determine sex (see Chapter 15, Endoscopy). This method can be

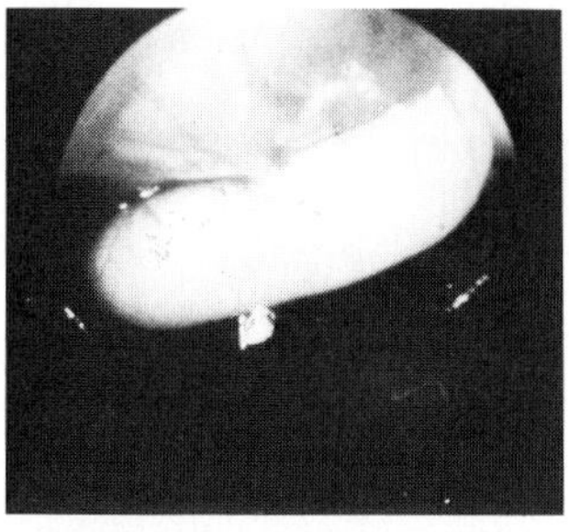

**Figure 51–1.** Through the endoscope, the mature testicle appears elongated and smooth with evidence of vascularization. (Courtesy of W. Satterfield.)

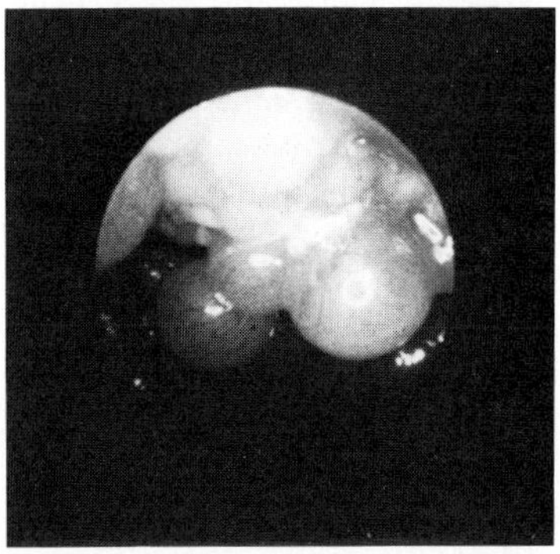

**Figure 51–2.** The developing follicles on a mature ovary give an appearance of a cluster of grapes. (Courtesy of W Satterfield.)

modified for surgical sexing utilizing the otoscope.[8, 12]

Safety and accuracy of surgical sexing vary significantly with the experience of the endoscopist. Mortality rates for surgical sexing by an experienced endoscopist are very low, ranging from 0.2 to 1.0 per cent. As with diagnostic laparoscopy, the veterinarian should first practice with dead birds to study the anatomy and learn to handle the equipment. Practice with birds such as pigeons should precede attempted sexing of clients' birds.[1]

Surgical sexing is widely applied to psittacine collections with a high degree of success and few complications. Fertile eggs have been laid on many occasions within a month following surgical sexing.

## GENETIC SEXING METHODS

Avian chromosome analysis techniques have evolved slowly over the last 15 years. While highly accurate and applicable to any bird regardless of age or condition, these methods are expensive and of limited availability.*[1, 5, 15, 20, 21]

The diploid (2n) chromosome number in birds ranges from 52 to 92. The female bird is homogametic (ZZ), whereas the male is heterogametic and is designated (ZW). Avian W chromosomes are smaller than the Z chromosome. Karyotypic evaluation is accomplished by matching chromosomes based on gross morphologic characteristics such as length, shape, centromeric position, and specific staining qualities.

Karyograms are prepared by three basic methods: (1) squash preparations of feather pulp cells from pin feathers, (2) culture of peripheral

*Marc Valentine, Avian Genetic Sexing Laboratory, Memphis, TN. Call ahead for test kit. (901) 323-4045.

blood leukocytes, or (3) culture of feather pulp cells. Mitotic cells are arrested in late prophase or metaphase, stained, and photographed, allowing pairing of chromosomes and identification of sex chromosomes.

The squash method is simple; however, the cells are often traumatized, resulting in loss of chromosomes. Staining is also variable, and artifacts from debris may interfere with the test. For leukocyte culture, approximately 2 ml of heparinized blood is required. The blood is cultured for 72 hours and the cells are arrested with colcemid. Staining of these preparations is more consistent, and loss of chromosomes is less frequent. The difficulty of culturing avian blood is the primary disadvantage in this procedure. Tissue culture of feather pulp markedly improves the morphology of the metaphase cells and the staining properties of the preparations. This method allows very accurate sex determination at any age and is currently available commercially.* The need for close proximity to the laboratory or rapid shipping time coupled with the expense and time involved makes this method impractical for routine sex determination.

Specific banding and staining techniques are under active investigation and may prove to be a viable technique for sex determination in the future. Chromosome maps of unique fluorescent surface patterns can be produced, evaluated, and documented by special staining and fluorescent techniques. The W chromosome of birds is composed almost entirely of constitutive heterochromatin similar to the centromeres of all chromosomes. The uniform staining of the W chromosome enhances its identity from other macrochromosomes with large nonfluorescing areas.

## HORMONAL SEXING METHODS

### Fecal Steroid Analysis

Sex determination through measurement of sex steroid hormones provides a viable alternative to invasive techniques. These techniques have been adapted from human pregnancy testing methods.[1, 2, 16]

Total excretory estrogen (E) and testosterone (T) can readily be measured from the mixed fecal droppings of unrestrained birds. Only the urate portion is needed, but both feces and urates are collected for a two- to four-hour period each morning for two or three days. Samples are frozen and shipped to the lab for analysis. Estrogen and testosterone levels are determined by radioimmunoassay (RIA) methods that are sensitive to picogram (10 to 12) changes in concentration. The relative amounts of estrogen and testosterone are expressed in a simple E/T ratio. At the San Diego Zoo mature female parrots are identified by a relatively high ratio (2.78 ± 0.58), and mature males are denoted by a low ratio (0.64 ± 0.13). The values differ for each laboratory. Fecal steroid analysis is not currently commercially available.

The obvious advantage of this method is in the fact that the birds require no restraint and readily provide the necessary test material. The disadvantages include the inability to sex immature or inactive birds. Seasonal inactivity as well as poor health affect the accuracy of this method. Accuracy is 95 to 99 per cent in mature active birds; however, 20 to 40 per cent of birds tested at the San Diego Zoo could not be sexed. The Avian Hormone Laboratory at Michigan State University discontinued offering the test because 33 per cent of the samples were unsuitable for accurate sex determination.[1, 16]

### Egg Waste Estrogen Analysis

A method developed at the San Diego Zoo in 1983 allows sexing of hatchlings by steroid analysis of fecal material present in the egg at hatching. Active steroidogenesis and sex hormone production occur in embryonic gonads, and these hormones are present in the urates excreted immediately prior to or during hatching. Mixed urates and feces are collected after hatching, allowing noninvasive sex determination of hatchlings. Estradiol fractions are isolated and quantified by high performance liquid chromatography. This procedure is costly and is not currently available commercially.[1, 3]

### Plasma Hormone Analysis

Plasma hormone ratios can also be used for sexing birds. Testosterone, free estriol, and estriol levels are determined by RIA analysis of plasma. The ratio of these hormones is used to determine sex in the same way as fecal steroids. Blood may be collected in two to six heparinized hematocrit tubes and spun down, and the tubes are snapped and capped with clay. Serum may be submitted in serum separator tubes. The

---

*Marc Valentine, Avian Genetic Sexing Laboratory, Memphis, TN. Call ahead for test kit. (901) 323–4045.

samples are frozen and shipped with cool packs to the laboratory. The samples are stable for at least six days in transit. Results are reported in approximately one to two weeks.*[7]

Advantages of plasma hormone analysis over fecal steroid analysis include reduced chance of contamination and less dependence on light cycles and seasonal activity of the bird. While hormone levels are lower in immature birds, the ratio of hormones may still be an accurate indication of sex; however, there is some controversy over the accuracy of this method. Stress and surgical and anesthetic risks are obviously reduced.

Genetic and endocrine sexing methods are specialized, expensive, and still only experimentally applied. Extensive training and special equipment are required, and some methods are very time consuming. Genetic determinations have the distinct advantage of being highly accurate in sexing birds of any age or condition. Endocrine evaluations offer the most direct assessment of functional activity of reproductive organs. Surgical sexing remains the most practical and readily available method of sex determination in monomorphic avian species.

---

*National Development & Research, 4850 156 Ave. N.E., #45, Redmond, Washington 98052

## REFERENCES


1. Bercovitz, A. B.: Annotated Review of Avian Sex Identification Techniques. *In* Burr, E. (ed.): Companion Bird Medicine. Ames, IA, Iowa State University Press, in press.
2. Bercovitz, A. B.: Fecal steroid analysis: A non-invasive approach to bird sexing. Proceedings of the American Federation of Aviculture, Veterinary Seminar, San Diego, CA, 1981.
3. Bercovitz, A. B.: Endocrine fecology of immature birds. Watchbird Mag., Volume 11, No. 2, 1984.
4. Biederman, B. M., and Lin, C. C.: A leukocyte culture and chromosome preparation technique for avian species. In Vitro, 18:415–418, 1982.
5. Bloom, S. E.: Current knowledge about the avian W chromosome. Bio. Sci., 24:340–344, 1974.
6. Bush, M., et al.: Sexing birds by laparoscopy. Int. Zoo Yearbook, 18:11, 1978.
7. Davis, S.: Personal communication, 1984.
8. Fletcher, K.: Surgical Sexing of Birds of Prey. AAZPA Regional Conference Proceedings, 1981, pp. 432–437.
9. Forshaw, J. M.: Parrots of the World. New York, Doubleday Press, 1973.
10. Harrison, G. J.: Endoscopic examination of avian gonadal tissue. VM/SAC, 73:479–484, 1978.
11. Harrison, G. J.: Personal communication, 1984.
12. Ingram, K. A.: Otoscopic technique for sexing birds. In Kirk, R. (ed.): Current Veterinary Therapy VII. Philadelphia, W. B. Saunders Company, 1980.
13. Low, R.: Parrots, Their Care and Breeding. Poole, Dorset, England, Blandford Press, 1980.
14. Masui, K.: The rudimentary copulatory organ of the male domesticated fowl with reference to the sexual differentiation of chickens. In Masui, K. (ed.): Sex Determination and Sexual Differentiation in the Fowl. Ames, IA, Iowa State University Press, 1967, pp. 3–15.
15. Mengden, G. A., and Stock, A. D.: A preliminary report on the application of current cytological techniques to sexing birds. Int. Zoo Yearbook, 16:138–141, 1976.
16. Nachreiner, R.: Personal communication, 1984.
17. Satterfield, W. C.: Diagnostic laparoscopy in birds. In Kirk, R. (ed.): Current Veterinary Therapy VII. Philadelphia, W. B. Saunders Company, 1980.
18. Satterfield, W. C., and Altman, R. B.: Avian sex determination by endoscopy. Proceedings of the American Association of Zoo Veterinarians, 1977, pp. 45–48.
19. Thomas, B.: Otoscopic sexing of birds: Emphasis on identification of the gonads. Auburn Vet., 39(1):23–26, 1983.
20. Toone, W. D.: Improvements in cytogenetic techniques for gross karyotypic and morphological studies of avian chromosomes. Proceedings of the American Federation of Aviculture Veterinary Seminar, San Diego, CA, 1981.
21. Van Tuinen, P., and Valentine, M.: A non-invasive technique of avian tissue culture (feather pulp) for banded chromosome preparations. Mammalian Chromo. Newsletter, 23(4):182–186, 1982.

# *Chapter 52*

# REPRODUCTIVE MEDICINE

GREG J. HARRISON

Considering the large number of serious aviculturists in the world and the relatively small number of offspring produced in aviaries each year, one would be inclined to suggest that the most common aviculture problem is simply lack of reproduction. Pet bird medicine is still in its infancy with regard to solving why a pair of apparently healthy, sexually mature, obviously compatible birds fails to lay eggs, incubate to hatching, feed to weaning, or is otherwise unsuccessful in producing viable young. Aviculturists have expected and accepted low production as normal and only recently have begun to consult with avian veterinarians about preventive medicine, psychologic factors in breeding, and reproductive health of their birds.

Avian veterinarians most often see emergency aviculture situations such as those associated with egg binding, egg peritonitis, mutilated young, and battered mates, all of which may be obvious, exaggerated expressions of the same underlying causes of lack of reproduction. Etiologic agents affecting the urogenital system are described in other chapters.

## DEVELOPING BREEDING STOCK

A number of factors may have a significant effect on the reproductive potential of a bird and should be considered in the development of breeding stock. The reader is referred to Chapter 1, Choosing a Bird, for some general characteristics desired in a breeding bird.

One element that appears to be considered with less frequency is the relationship of a species' native climate to the captive climate. Not all species can be expected to adapt well to a new environment and reproduce successfully. Hot, humid climates, for example, may be ideal for birds naturally found in subtropical regions of the world. On the other hand, Australian species such as Rose-breasted Cockatoos and other desert-dwelling birds may have greater reproductive success in drier climates. One would advise potential aviculturists to acquire species most physiologically suited to the captive environment.

In the poultry industry, which may serve as a model for cage bird aviculture, information is gained from reviewing three generations of records of pure-bred birds for potential inherited characteristics: age at sexual maturity, egg size, per cent hatchability, per cent survival, rate at which eggs are laid, size of clutch, and interval between clutches. Lack of broodiness is an important poultry trait because egg production increases in birds that quickly stop brooding behavior when eggs or young are removed. Hens showing persistence in production (i.e., laying during the molt and/or longer intervals of laying between rest periods) produce more eggs per year. Some of these tendencies are beginning to be evident in some flocks of genetically managed cockatiels.

Many personality traits are also believed to be genetically related. As an example, aggressive behavior that is frequently associated with breeding cockatoos has been suggested to be influenced by parentage. Aviculturists may choose those individuals with care.

Breeding birds should not be selected solely on the basis of color, shape, and size. In an effort to attain show quality standards, budgerigar, cockatiel, and canary breeders are genetically selecting only for large robust bodies, which may result in anomalies that render the birds incapable of natural copulation, such as has occurred in the turkey industry.

Progressive avicultural facilities are evaluating adult birds for disease or reproductive problems prior to establishing pairs (see Chapter 50, Aviculture Management). It is the opinion of the author that there are sufficient challenges in the breeding of some psittacines that beginning a program with an "abnormal" bird with the hope of clearing up a problem is not a wise move. Starting the program with the youngest, healthiest birds offers the best possibility for success. Selecting breeding stock from domestically bred and raised birds cannot be overemphasized. If one heeds the principles developed in the poultry field, aviculture success cannot usually be expected from constant introduction of wild imported stock.

## FACTORS ASSOCIATED WITH NORMAL REPRODUCTION

The reproductive endocrine system of birds in the wild responds to increased hours of daylight, renewed abundance of succulent foods, and other stimuli (see Chapter 5, Selected Physiology for the Avian Practitioner). Privacy and lack of stress play a role. Aviculturists have attempted to induce captive reproductive behavior with simulation of some of these factors. Control of light cycles and use of flush diets (see Chapter 2, Husbandry Practices) have had some degree of success. Work at the University of California in Davis further suggests that environmental manipulation may improve reproductive results.[16] Although information on psittacines is limited, aviculturists and avian veterinarians alike must avail themselves of what is known about an individual species if captive propagation is to be successful. The reader is referred to van Tienhoven's review of the literature in regard to reproductive influences.[25]

### Reproductive Stimulants

In studies of birds native to the Northern hemisphere, the role of light is the primary factor for inducing the sexual response.[25] These birds responded to a photoperiod in excess of 11 hours per day up to 16 hours per day after the onset of a natural or artificial dawn. Most species required an initial period of less than 11 hours of light in order to be preconditioned to respond with sexual activity to stimulatory periods of light.

The quality of light is significant. Birds have been found to be sensitive only to wavelengths within the visible light spectrum (400 to 750 nanometers). Birds have been stimulated sexually by incandescent, fluorescent, mercury, metal halide, and high pressure sodium lights. The longer red rays emitted from plant stimulatory lamps are believed to be the most stimulatory to birds. Although the light intensity has not been determined for cage birds, 1 to 2 foot-candles (10 to 20 lux) maintain optimum laying in chickens. Fluorescent lamps produce eight times the illumination of incandescent lamps of the same wattage.

Conures, macaws, budgerigars, and cockatiels appear to be less dependent on photoperiods than are most cockatoos, Amazons, African Grey Parrots, Indian Ringnecks, and other parrots and parakeets, as the former genera breed year round. However, 14 to 16 hours of daylight per day was found to be more favorable for egg production in budgerigars. Conversely, cockatiels laid more eggs and fledged more chicks under eight hours of light and 16 hours of darkness. The aviculturist may be able to use a change in day lengths as a factor to artificially induce reproduction.

Some authors[1] believe that the use of artificial lights to induce breeding during the normal off season increases the incidence of weak youngsters, French molt, death in shell, and nestling deaths. This author has not noted these characteristics in Amazons, cockatoos, macaws, and African Grey Parrots that have reproduced for years exclusively under artificial illumination.

The mere presence of a nest box is important as a reproductive stimulus; however, the size, shape, variety of material, location in the aviary, and amount of darkness in the nest box all appear also to play a role (Fig. 52–1). Budgerigars were found to respond to a nest box with an opening at least 10 cm above the perch.[25] While this seems to be a subtle factor, it was found to be of crucial importance. Nest box size can determine the number of eggs laid in some species. The survivability of the young is en-

**Figure 52–1.** A 40-gallon garbage can will suffice as a nest box for some large psittacine species if it meets other requirements, such as depth and degree of darkness.

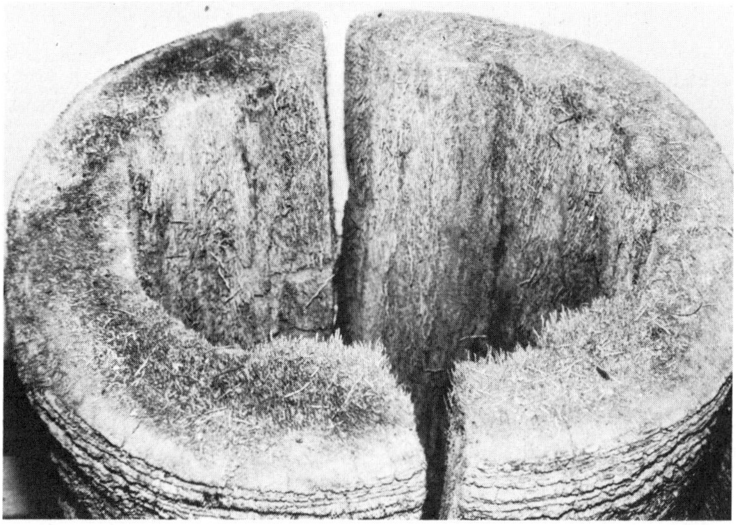

**Figure 52–2.** Some species require a nest box that more closely approximates what they would encounter in the wild.

hanced by a larger space, although some species require a particularly small box for breeding. South African aviculturists provide nest boxes in a variety of sizes, shapes, materials, and placement in the same flight, and are sometimes surprised by which of these nests the pair may choose. Birds whose natural instinct is to chew their way into a tree or log to establish a nest will respond to this type of nest box in captivity (Fig. 52–2).

The quality and quantity of food and presence of specific items used in nest building or courtship (e.g., palm fronds for lovebirds or eucalyptus buds and leaves for some cockatoos) may stimulate pairs. Wild-caught sparrows were shown to develop larger ovaries than controls when fed green wheat sprouts.[5]

The color of leg bands apparently makes a difference in the pairing of Zebra finches (Fig. 52–3). In one study, males preferred black- and pink-banded females; females preferred red-banded males over non-banded males; both sexes avoided the opposite sex with light blue or light green bands.[25]

The presence of other mating pairs may stimulate reproduction. Particularly with Amazons and cockatoos, aviculturists believe that reproductive activity is initiated by neighboring pairs.[21] Specific vocalizations of male budgerigars have been found to stimulate ovarian and oviductal development. Female canaries were found to lay more eggs in response to males with a large vocal repertoire with more syllables.[25] The presence of older birds may stimulate youngsters to breed early. Domestically raised Amazons and macaws apparently reach sexual maturity at an earlier age than their wild-caught counterparts.

**Figure 52–3.** Zebra finches were found to prefer specific characteristics in prospective mates, including the leg band color. (From Harrison, G. J.: Feather disorders. Vet. Clin. North Am. [Small Anim. Pract.], 14(2):181, 1984.)

Stennett[21] reports that a hen cockatoo can be brought into reproductive readiness by stroking the vent and putting hand pressure comparable to the male's weight on her back. An inexperienced male or hen can be taught to mate by a sexually aggressive bisexual surrogate hen. Aviculturists often give up on such hens because they won't lay eggs or are poor parents, and they don't recognize their value as surrogates.

### Behavioral and Physical Changes

A number of behavioral and physical changes accompany the reproductive season in most avian species. The male will begin to defend its territory (cage) against intruders. Many species display their most brilliant plumage during this time, as they have often just completed a molt. The male will call, sing, perform ritual dances, or display his feathers in an attempt to attract the female. Nest preparation and copulation will begin. Coprophagia has been observed by some aviculturists prior to egg laying; fecal matter is not allowed to accumulate in cages of some birds preparing to nest, whereas nonbreeding birds allow stool accumulation.

Impending breeding and laying are reportedly predictable by monitoring the testosterone/estrogen ratio.[2] Females develop a brood patch, and swelling and dilation of the cloacal area may be obvious, especially in conures and canaries. The owner may notice a smaller number, yet more voluminous stools. The female's appetite will increase, then decrease immediately prior to egg laying. Bleeding may be observed in females several hours before, during, or after egg laying, and all symptoms of potential egg binding or egg peritonitis should be attended to (see Chapter 28, Symptomatic Therapy and Emergency Medicine).

Many of the observations of behavioral characteristics of reproductive activity have been developed from field studies of birds in the wild. One must exercise caution in extrapolating information from wild bird species directly to captive situations. Table 52–1 illustrates some morphologic and behavioral differences that have been recorded between feral and captive Zebra finches.

## LACK OF REPRODUCTIVE ACTIVITY

Without the appropriate stimuli captive reproduction may not occur. Other factors may also be responsible for absence of reproductive activity. Aviary disturbances have a significant effect and are discussed in Chapter 50, Aviculture Management.

### Lack of Synchronized Breeding Cycles

In the wild, the male of the species tends to come into reproductive season first. If a natural phenomenon such as drought forces the abortion of the breeding season at this point, the female will not have wasted her energy and bodily components on egg formation. In the interest of energy conservation, it takes less energy for the male to come into reproductive readiness than for the female.

Physiologically, the *captive* male of most species is also ready to reproduce before the female. Lacking the natural outlets of nest selection and defense, this increased energy may be expressed as aggression toward the perches, cage, feeder, or even the female. Confining the birds in neighboring cages until the female is receptive may avoid any danger of brutalization of the hen.

In some species (e.g., cockatoos,[21] Eclectus) or in some individuals of other species, the female may achieve reproductive readiness prior to the male and begin to express dominance. In the author's practice, one pair of Amboina King Parrots was presented for feather picking by the male and aggressive behavior by the female. Laparoscopic examinations revealed fully developed follicles in the female and an abscess on the male's testes, which prevented him from sexually responding.

### Incompatability

More is involved in setting up for successful breeding than putting a male and a female of a species in a cage with a nest box. One responsibility that the aviculturist has assumed—choosing mates—has not always achieved positive results. The trend with successful aviculturists appears to be to allow the birds to form their own pair bonds, if enough individuals of the species are available. This has been a frustrating aspect of working with limited populations of rare birds. Randomly selected and paired wild-caught adults may achieve tolerance only.

One method is to allow a female to be placed in a flight cage with several males and allow her to choose her own mate. Another technique suggests putting 10 to 12 pairs together in a

**Table 52–1.** SOME MORPHOLOGIC AND BEHAVIORAL DIFFERENCES BETWEEN FERAL AND DOMESTICATED ZEBRA FINCHES

| Character | Feral | Domesticated |
|---|---|---|
| Variability | Narrow range | Wide range |
| Body weight | | |
| a. males | Aug. 11.8 gm<br>Nov. 11.1 gm | Aug. 12.7 gm<br>Nov. 11.9 gm |
| b. females | Aug. 11.7–11.8 gm<br>Nov. 10.7–10.9 gm | Aug. 13.1 gm<br>Nov. 12.0–12.1 gm |
| Wing length of females | 54.3 mm | 54.8–56.2 mm |
| Body length | 52–60 mm | 48–65 mm |
| Bill width of females | 6.22–6.24 mm | 6.34–6.48 mm |
| Eye color | Red-brown | Chestnut brown, rust brown, dark brown |
| Plumage colors | | |
| a. Face markings | Black and white | Dilute gray |
| b. dorsal body | Gray with shade of brown | Brown |
| c. ventral body | Female: white | Light brown or light gray |
| d. color variety | Little variation in population | Many varieties: cinnamon, silver, etc. |
| Plumage density | Dense | Less dense |
| Juvenile molt (age) | 2–3 months | More variation |
| Change of beak color (age) | 6–9 weeks | More variation |
| Behaviors | | |
| a. general | Behaviors less easily induced by releasers | Behaviors easily induced by releasers |
| b. gregariousness | Very gregarious | Less frequent |
| c. courtship by males | Less frequent; stops after copulation | More frequent; continues after copulation |
| | More sexual activity | Less sexual activity |
| d. pair bond | Strong; little promiscuity | Weak; promiscuity common |
| e. reciprocal preening | Limited to pair | Random |
| f. nest building | Well-developed | Hyper; nest building interferes with incubation |
| g. incubation behavior | Male and female alternate | Male and female incubate together; sometimes abandon nest in favor of gregarious behavior with conspecifics |
| h. care of young | Attracts fledglings back to nest by feeding in nest | Feed fledglings outside nest |
| i. locomotor activity | Greater | Reduced |
| j. food intake activity | Greater | Smaller |
| Song | Little variation among individuals; song not sexualized | Greater variability among individuals; song sexualized (many attraction calls) |
| Gonadal development of young males | More rapid | Slower |
| Response to favorable breeding conditions (nestbox and nesting materials) | Faster | Slower |
| Egg size | Smaller | Larger |
| Effect photoperiod on clutch size | Increase with photoperiod | No effect of photoperiod |

*From van Tienhoven, A.: Environment and reproduction of pet birds. Proceedings of the Annual Meeting of the Association of Avian Veterinarians, San Diego, CA, 1983, pp. 110–161. Based on work by Immelmann, K.: Vergleichende Beobachtungen uber das Verhalten domestizierter Zebrafinken in Europa und ihrer wilden Stammform in Australien. Z. Tierzucht. Auchtungsbiol., 77:198–216, 1962, and Sossinka, R.: Ovarian development in an opportunistic breeder, the zebra finch *Poephila guttata castanotis*. J. Exp. Zool., 211:225–230, 1980.

colony; the first two or three pairs to bond are kept for breeding; the rest may be sold to the pet market. Pair bonding is evident when the birds prefer to sit close together on the perch, are seen preening each other, and prevent other birds from approaching them.

Some mutation progeny may have difficulty finding a compatible mate. White hybrid Zebra finch mutants were unable to recognize members of their own species including the opposite sex. In producing mutation lovebirds from one parent that normally carried palm fronds under the tail feathers as a species characteristic of normal nesting behavior, and another parent that instinctively used another body location to carry the courtship fronds, the hybrid offspring acted confused about frond carrying and did not breed.

Homosexual pairs are found in the wild and may be inadvertently established in captivity if the aviculturist assumes pair bonding activity is the differentiation of gender. One theory is that homosexuality may correlate with pathologic lesions such as ovarian neoplasia.[1] It has also been suggested to be the result of hormonal treatment (thyroid and estrogen) for plumage disorders and was apparently reversed by testosterone therapy.[1] This author has observed frequent homosexual activity among various species in captivity (e.g., a dominant male copulating with a subordinate male; or strong pair bonds between same sexed individuals that are reared together); there has been no evidence of pathologic or pharmacologic relationships in these cases.

Psittacines have been reported to be monogamous. Although strong pair bonding apparently does occur between individuals, Wiley has reported in his field observations of the Puerto Rican Amazon Parrot that a dominant male will take over another's territory, nest, and female. Highly domestic species such as budgerigars and cockatiels are not at all monogamous; nonbreeding Amazons and macaws have resumed reproductive behavior when new mates were introduced in a captive situation by this author.

### Malnutrition

As suggested in Chapter 31, Nutritional Diseases, malnourished parents cannot be expected to produce healthy, robust young for very long. Lack of an adequate diet may totally prevent egg laying because the calcium, vitamin A, and other nutrients are not available for egg formation. Chronic malnutrition may be implicated in laying of soft-shelled eggs, egg binding, and other situations requiring complex interrelationships of nutrients for normal function.

### Age

In the early days of surgical sexing, veterinarians tended to suggest pairing of birds based on maturity of gonads. Generally, this was found to be unsuccessful. Recently imported, mature birds may have successfully reproduced in the wild with a specific set of physiologic and psychologic stimuli. Despite an adequate diet, privacy, and a compatible mate, the stimuli, if known, would be difficult to duplicate in an artificial captive setting. The trend now, especially in larger species, is to select young birds, preferably domestically bred, allow them to form pair bonds of their choice, provide an appropriate environment, and wait. Waiting a few years for sexual maturity for more assured breeding results is preferred to a situation without the potential for reproduction.

Attempting to pair a mature with an immature bird may lead to displaced aggression.

### Physical Interference

Loose perches, lameness, or missing toes or feet have been incriminated in infertilities, as the birds cannot stabilize for proper copulation. Some birds may seek their own flat surface, or one may be provided as a perch.

Pigeon breeders have reported that heavy growth of feathers around the vent will interfere with deposition of semen; therefore these feathers are clipped prior to the breeding season.

## DISORDERS OF THE HEN

### Egg Binding

The tendency for egg binding may be genetic in part, and it seems to be more prevalent in domesticated species such as finches, budgerigars, and cockatiels. Although there appears to be a relationship between malnutrition and the incidence of egg binding, the condition is not uniformly found in a group of birds fed a similar diet. Overproduction is often believed by aviculturists to be the only factor involved; but nutritional authorities disagree. They believe that the nutritionally sound hen is capable of continuing egg laying.[8] Lack of normal muscle tone and lubrication of the oviduct often cause the egg to adhere to the uterine surface and may contribute to a prolapsed oviduct (see Chapter 28, Symptomatic Therapy and Emergency Medicine). Egg binding appears to be more common after a stress such as cold weather. Other possible factors suspected to be related to egg binding include lack of exercise, first clutch, spasm of cervix, atony of uterus, eggs too large, hen too old, forced breeding out of season, tumor of oviduct, and presence of soft-shelled egg. Gonadal trophic hormones (pregnant mare's serum, or PMSG) that were used with success to induce egg laying in canaries led to a high incidence of soft-shelled eggs, eggs laid out of the nest, and egg binding, in which 13 of 58 birds died as a result.[19, 20]

Clinical signs of egg binding are primarily related to swelling of the abdomen with palpable or radiographic evidence of the egg(s). Usually the history includes nesting activity or previous egg laying. Hens are often weak and may exhibit signs of periodic contractions. Advanced cases resemble the "classic sick bird" syndrome. Egg binding due to soft-shelled eggs is difficult to diagnose, especially if a single egg has been removed and more are still in the reproductive tract.

Egg binding is treated medically (see Chapter 28, Symptomatic Therapy and Emergency Medicine) or surgically, if necessary (see Chapter 48, Selected Surgical Procedures).

## Ectopic Eggs

Eggs may occur outside the uterus as a result of uterine rupture. Affected birds present with clinical signs similar to egg binding, but the abdominal swelling and radiographic position of the egg are more ventral and the bird fails to respond to egg-binding therapy. Treatment consists of hysterotomy from the ventral laparotomy approach, removal of the egg material, and repair of the uterus. A hysterectomy may be indicated (see Chapter 48, Selected Surgical Procedures).

## Egg-related Peritonitis

Although the condition is commonly known as egg *yolk* peritonitis, the infection that occurs has not been proven to be related to the mere presence of the extrauterine yolk, but presumably is a bacterial infection of the yolk material present in the abdomen. In response to a description by the author of the behavioral cycling of hysterectomized cockatiels, Ringer[18] postulates that ovulation probably continues to occur in these birds and the benign yolks that are deposited in the abdominal cavity are absorbed. In earlier studies with chickens, Ringer surgically implanted egg yolks from separate birds into an individual hen's abdomen with no adverse effect. A further theory is that it is the albumin component that may contribute to the peritonitis reaction followed by infection.

## Chronic Egg-laying

Methods of manipulating the environment for cessation of chronic egg laying are described in Chapter 3, Captive Behavior and Its Modification. Hormones and other physiologic alterations have failed to permanently stop egg laying in pet birds or those in which egg laying may be contraindicated. The author's therapy of choice is hysterectomy. Temporary cessation of ovulation is dicussed in Chapter 48, Selected Surgical Procedures.

## Abnormal Eggs

Soft-shelled eggs are believed to occur primarily as a result of nutrition deficiencies of the hen, specifically calcium as well as insufficient vitamins A and $D_3$ and trace minerals. Metritis may also be a cause. These rubber-appearing eggs are the hardest for the hen to deliver normally because of the difficulty in propelling the soft egg down the oviduct. In addition, the lack of calcium may contribute to a weakening of the uterine muscles. Treatment may involve multiple vitamins and calcium injections, uterine culture (via dilated cervix immediately after egg laying), uterine curettage, and/or injections of appropriate antibiotics. Dietary management is instituted. Hysterectomy may be appropriate for birds in which maintenance of reproduction is not a factor.

Yolkless eggs, which are small and sterile, may be caused by metritis, deposition of yolk into the peritoneal cavity, or abnormalities of the ovaries, including fibrous, cancerous, or degenerative tissue.

Some hybrids or conditions causing slow passage of the egg through the albumin production portion of oviduct may result in extra large eggs. Double yolks are also postulated to be related to abnormal egg passage.

Eggs with rough-textured surfaces have been attributed in poultry to excess chlorine in the drinking water when calcium chloride was used to supply supplemental calcium. Treatment involves injectable calcium and multivitamins, appropriate antibiotics, and dietary management.

## Infectious Metritis

Distorted eggs may reflect infections of the uterus if eggs are laid at all with this condition. An increase in estrogen has been found in poultry to coincide with uterine coliform population increase. Coliform metritis complicated by poor diet has been encountered by this author. From personal observations, death rates appear to be highest in hens during ovulation and egg laying.

Salpingitis may be an extension of air sacculitis or retrograde infection of the uterus from the vagina. Extensions of liver and/or respiratory infections may include the reproductive system and should be suspected. Etiologies responsible for liver diseases are listed in Chapter 42, Miscellaneous Diseases.

## DISORDERS OF THE MALE

Infertility of the male may be due to age, inbreeding, excessive environmental heat, testicular neoplasia, testiculitis from septicemia, and malnutrition. Etiologies responsible for infectious metritis may also affect the male. In one case noted by the author, semen samples collected from Amazons that had been maintained solely on frozen vegetables, fresh fruits, and boxed parrot treat showed few sperm present; administration of fat-soluble vitamins produced an immediate increase in sperm quantity and quality.

Gonadal dysfunction may be demonstrated by laparoscopy (see Chapter 15, Endoscopy). Serial laparoscopic examinations may be required to determine a change in testicle size from reproductive response.

## PAIR OR NEST PROBLEMS

Observations of birds in the wild suggest that young, newly mated pairs lay more eggs in the first few clutches, but rear fewer young than more experienced parents who lay fewer eggs but have much greater reproductive success. This suggests that parenting is a learned behavior. Therefore the concern of the aviculturist does not end when eggs are laid. Abandonment of the eggs is relatively common in inexperienced larger psittacines. Small nests apparently are abandoned more readily than larger nests.

Cockatoos are among the several species known to eat or destroy their own eggs. Aviculturists committed to raising these species develop techniques for determining pending egg laying so that the egg can be removed promptly for incubation. These techniques may include keeping accurate records of laying patterns, palpation of pubic bones, and installing "trap door" nests used in poultry operations.

Cold abandoned eggs may be salvaged if they are reheated to normal incubation temperature before any attempt is made to replace them with the original parents. If the original parents are unreliable, the services of a foster parent may need to be enlisted. Experimental work is being done with cockatiels to remove the eggs as they are laid, keep them cool (55°F) until a normal-sized clutch is reached, then return them all to the nest to begin incubation at the same time.[3] In this way, all the clutch mates are the same age and the competitive edge of size is reduced at feeding time.

**Table 52–2.** EGG PRODUCTION AND HATCHABILITY, IN COCKATIELS KEPT UNDER DIFFERENT TEMPERATURES

| Chamber Temperature °C | (°F) | Mating Pairs | Eggs/ Hen (3 mo.) | Hatch (%) |
|---|---|---|---|---|
| 15.6 | (60) | 3 | 6.0 | 11.1 |
| 23.9 | (75) | 3 | 8.3 | 52.0 |
| 32.2 | (90) | 3 | 6.0 | 47.3 |

From Woodward, A. E.: Environmental factors affecting the breeding environment. In Grau, C. R., et al. (eds.): Breeding and Raising Cage Birds. Proceedings of the University of California Extension Class, 1981.

Some control must be exercised over the environmental temperature in the aviary and nest. Table 52–2 illustrates the effect of three temperatures that were recorded for cockatiels; of these, the optimum for egg production was 75°F.[9]

### Foster Parents

It may be advisable in some cases (e.g., a first time hatching of a rare species) to remove the neonate immediately for hand feeding or to place it with experienced foster parents.

Society finches have long been appreciated as capable of raising other finch species. Budgerigars, Monk Parakeets, and conures have also been used as foster parents in the author's aviaries. These species have readily accepted the young of other genera, including Amazons and macaws, for the first few critical days. The neonates can then be removed for hand-feeding. Timing is the most significant factor in foster parents' acceptance of other nestlings. The foster parents must be at the same reproductive stage, i.e., feeding young of their own of a similar age/size. The true offspring of the foster parents are then moved to yet another active nest for rearing.

In working with conures and other small psittacines, attempts by the author to foster another species' *eggs* worked only if the foster parents had never heard their own kind pipping. Apparently the pipping sounds are distinctly different for each species and any variance in cadence from previous experience with their own offspring may be noticed by the parents and the new eggs will be destroyed.

**Table 52–3.** SUMMARY OF POSSIBLE CAUSES OF INFERTILE EGGS

| |
|---|
| Sexual inexperience |
| Normal occurrence as part of clutch |
| Loose perches |
| Lameness, missing toes or feet |
| Heavy cloacal feathering |
| Aviary disturbances |
| Lack of visual barriers |
| Lack of flock stimulants |
| Lack of environmental stimulants |
| Environmental temperature too hot |
| Inappropriate nest size, perch placement |
| Male too young or too old |
| Malnutrition |
| Poor quality semen |
| Urogenital or systemic infections |
| Inbreeding |
| Cloacal growths or abnormalities |
| Insufficient number of males in colony breeding |
| Possible thyroid deficiency |

## HATCHABILITY PROBLEMS

Table 52–3 suggests some frequent causes of infertile eggs. One resource for the aviculture practitioner is a wall chart "How to Identify Infertile Eggs and Early Dead Embryos."* Eggs selected for examination can be cracked open over a Petri dish, inspected under magnification, and compared to the reference chart. Microbiologic procedures can be conducted on the contents.

Nonhatchability of eggs from actively breeding and laying pairs may be attributed to infertility of the male, chronic infection in the female, endocrine imbalances, inbreeding, damaged oviduct, broken or cracked egg, disruption in incubation, other dead eggs that have not been removed from the nest, or dietary deficiencies. The last eggs of the clutch can become dirty from environmental contaminants such as *Salmonella*. Eggs removed from the nest within hours of being laid are less likely to be damaged by surface contaminants. Other causes of early hatch, late hatch, and weak hatch are discussed in Chapter 53, Pediatric Medicine. Eggs that are normal in size for the species hatch best. If the eggs are too large or too small for the species, poor hatching or early death of the hatchling may result. Poultrymen routinely cull eggs of abnormal size prior to hatching.

*Order No. 4032 from Agriculture Sciences Publications, University of California, 6701 San Pablo Ave., Oakland, CA 94608.

## ARTIFICIAL INSEMINATION

Artificial insemination of avian species is not a new practice but is in its infancy with regard to psittacines. A massage technique to induce ejaculation in poultry, cranes, and raptors has failed to be effective in all but the most domesticated psittacine species, the budgerigar and cockatiel. The first psittacine to be produced by artificial insemination was a cockatiel.[11] The massage technique was used to collect the semen. An electroejaculator designed for use in felines was modified by Wasmund[11] for semen collection attempts in larger psittacine species (Fig. 52–4). A unipolar probe was placed into the cloaca and a flat return electrode was placed over the plucked sacrum. The most successful procedure that evolved in using the machine was (1) prestimulation manual expression to remove any feces and/or urates prior to the insertion of the probe (this also evaluated the bird's potential response to manual stimulation); (2) electrostimulation with the rectal probe inserted and the saddle ground plate in place; (3) post-stimulation manual expression.

In most cases, there was no ejaculate at the time of the electrostimulation. However, the ease of manually expressing the semen sample was significantly increased following the elec-

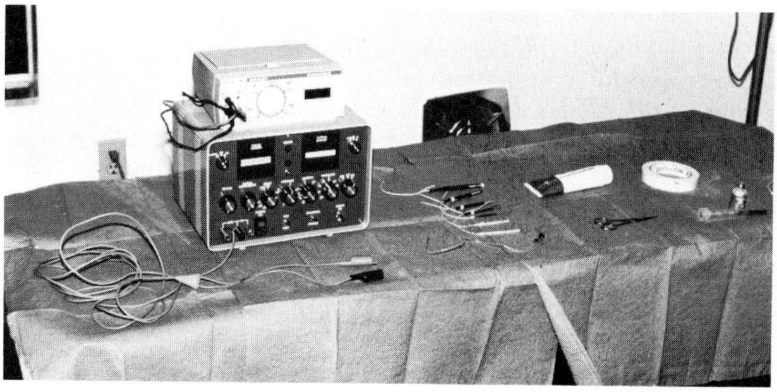

**Figure 52–4.** An electronic device can provide electrostimulation to facilitate manual semen collection in psittacines.

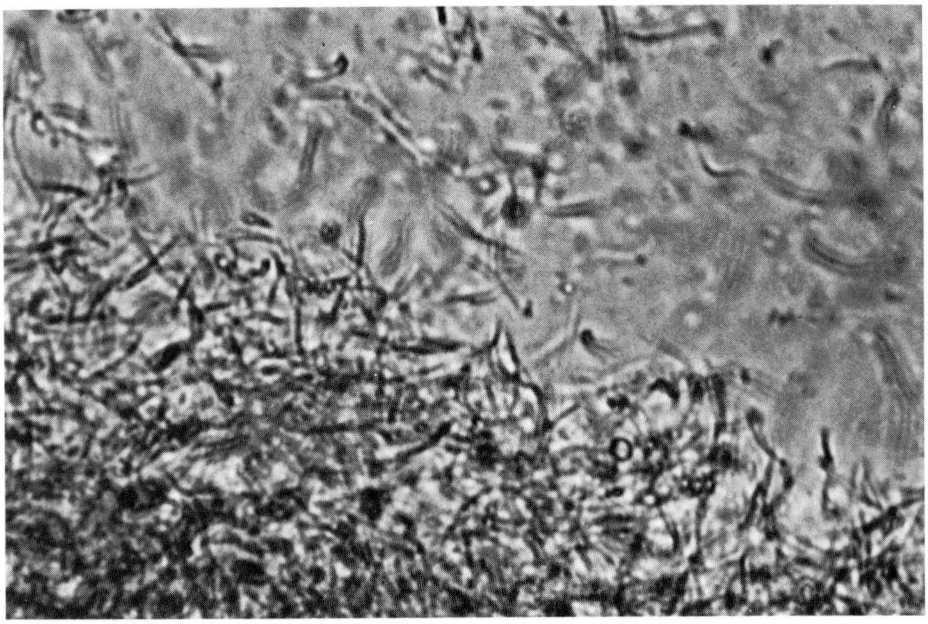

**Figure 52–5.** A microscopic view of budgerigar semen illustrates the high density of the sperm.

trostimulation. Although the mechanism for ejaculation is not known in psittacines, it was speculated during this study that the electrostimulation encouraged the propulsion of spermatozoa into a storage area (perhaps the receptacle of the ductus deferens?) where the semen was concentrated until it was manually expressed.

Some male budgerigars that were conditioned over a period of time to ejaculate with manual collection techniques produced 10 to 15 times the volume of semen over nonconditioned birds, even those aided by electrostimulation (Fig. 52–5).

The effectiveness of cryopreservation of budgerigar semen is being studied. Preliminary results suggest that the motility and viability of frozen/thawed and lyophilized (freeze-dried) budgerigar semen may be adequate for fertilization.[10] Lyophilization does appear to have an adverse effect on the morphology of the sperm cells, but the effect of this on fertilization is unknown.

## ARTIFICIAL INCUBATION

Artificial incubation of eggs may be chosen in cases of rare and valuable species, inexperienced or unreliable parents, injury or death of one of the parents, for production of multiple clutches or in any circumstance in which the risk involved in artificial incubation may be preferred to that of the natural process. New genetic strains of a species may be introduced into the aviary by the acquisition of fertile eggs from a reliable outside source. The primary disadvantage, in the author's opinion, is that artificial incubation and hand-rearing may in some cases perpetuate the genetically weak of the species.

Successful artificial incubation of eggs from psittacines requires total commitment of attention to the details of record keeping, care and monitoring of the equipment, egg handling, and survival of the hatchling.

### Equipment

The three major factors that need to be simulated in an artificial environment for incubation are temperature, humidity, and turning of the eggs. Vital to these is the incubator itself. Several models are available on the market; however, it behooves the serious aviculturist to acquire the most refined, highest quality model for maximum control over the results.

The best incubators available on the market today are of furniture-quality wood (Fig. 52–6). Manufacturers suggest stabilizing the temperature and humidity control by activating the equipment one to two weeks prior to introduction of eggs. This allows the natural swelling of the wood to occur for maximum maintenance of the interior environment. A periodic treatment of the wood exterior with a lemon oil product is advised.

Styrofoam incubators may be useful for prac-

**Figure 52–6.** The serious aviculturist may choose to invest in a high-quality incubator for more control of the artificial incubation environment.

ticing some initial techniques, but they are unreliable and risky in working with eggs of valuable species (Fig. 52–7).

## Incubator Monitoring Devices

Optimum temperature and humidity settings for the stages of embryo development have evolved from aviculturists' experience with a wide variety of species. Table 52–4 suggests incubator parameters for some psittacine species. Multiple monitoring devices are necessary to maintain these variables at a controlled level. Refined mercuroid thermostats are guaranteed to be accurate to within 0.25°F. In the author's work with cockatiels, a variation in temperature of as little as 0.25°F over the 18-day incubation period made the difference in whether the yolk sac was absorbed or not prior to hatching. The accuracy of gas-filled wafer thermostats, most often found in inexpensive styrofoam or plastic incubators, is dependent upon the barometric pressure and thus is unreliable. In addition, when the lid of this type of incubator is removed to gain access to the eggs, an immediate drop in the interior temperature occurs. In contrast, the mass of wood incubators retains the heat and limits the temperature drop when it is opened. Forced-draft incubators have an electric fan to circulate the heated air and maintain a more uniform temperature throughout the incubator than do still-air incubators. Still-air incubators must have an interior temperature 2 to 4°F higher than forced-air incubators.

The ventilation and control of the interior incubator temperature are enhanced by an exterior room temperature of 65 to 70°F.

Monitoring of the humidity level is crucial to successful hatching. More than a single monitoring device must be included in the incubator

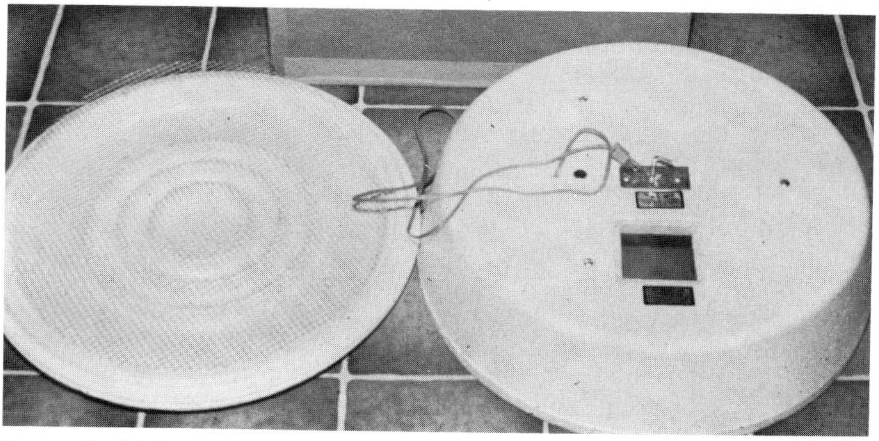

**Figure 52–7.** Unpredictable hatching results are found with styrofoam incubators because of the significant variations in temperature and humidity levels.

**Table 52–4.** SUGGESTED ENVIRONMENTAL INCUBATION CONDITIONS FOR SOME PSITTACINES* (FORCED AIR)

| | Incubation | Hatching |
|---|---|---|
| Dry bulb temperature | 99.5°F | 98.5°F |
| Wet bulb temperature | 86–87°F | 90°F |
| Egg rotation (automatic turning) | 90 degrees per hour | None |

*Cockatiels,[3] Amazons, Macaws, Eclectus Parrots.

for standardization. A whirling hygrometer contains wet and dry bulb thermometers. A wet bulb thermometer is dependent upon the cleanliness of the wick for accuracy, as the accumulation of dust and other debris will change the rate of evaporation of the distilled water, thus altering the reading. Comparison of the readings of the wet bulb and dry bulb thermometers on an appropriate slide rule determines the relative humidity in per cent. A hair hygrometer reads out the per cent relative humidity but must be rechecked often against the whirling hygrometer.

The marking of eggs with two arrows in opposition to each other will assist in determining the next appropriate direction for hand turning of the eggs. The egg is rotated 180 degrees, then counter-rotated at the next interval. Continually turning the egg in the same direction may cause tearing of the chalazae or suspensory ligaments and result in the death of the embryo. Automatic turning devices on some incubators rotate the eggs 90 degrees at a time.

## Care and Handling of Eggs

In retrieving eggs from the nest, precautions are necessary to avoid transmission of disease within the aviary or into the incubator. The greatest numbers of contaminating organisms in carrier birds are often found at the time of egg laying, and these may be found on the surface of the egg. Personnel should be required to wash hands and apply a hand disinfectant between nests. A foaming alcohol (Alcare—Vestal Laboratories) is preferred because it does not contain ingredients that may be toxic to the embryo; the evaporating alcohol does not result in a slippery surface to the hands which may contribute to egg droppage.

To safely handle eggs, place both hands into the nest or incubator and place the egg in the palm of the upturned hand before removing. Foam padded boxes with wells in which to place the retrieved eggs may be used as egg transfer containers.

The egg should be candled to check for cracks, inclusion bodies such as blood clots, and floating air cells, with careful attention to avoid overheating during candling. Small amounts of fecal material can be removed with fine sandpaper, but excessively dirty eggs should not be set. Any cracked eggs can be sealed with hot dripped candle wax. This is best performed during the candling procedure so the full extent of the crack can be adequately visualized and the wax seal extended beyond the edges. The species, date, egg number, aviary number, and any other identifying information can be recorded directly on the egg shell with a No. 2 pencil or felt-tipped pen on thin-shelled eggs. If aviculturists followed the poultry procedure of culling all eggs of abnormal size or appearance at this point, the total hatchability rate would significantly increase.

Although there has not been enough work in this area with psittacines to recommend dipping as a standard procedure, if the eggs are known to be from a contaminated source, the procedure is advised. The egg is passed into a room-temperature solution of one part povidone-iodine solution and nine parts water until completely submerged, then removed and placed in the incubator.

Eggs should be removed from the nests at the same time every day so that the eggs do not start to incubate in the nest. If the eggs are not cool, allow them to cool to room temperature before transferring them to the incubator. Physiologically, the embryo does not develop beyond the first few cell divisions in the uterus, and the hen allows for a cooling period after the egg is laid before initiating the actual incubation process. Eggs intended for artificial incubation can also be safely stored at 55°F for up to six days prior to incubating. Placing the eggs in sealed plastic bags allows carbon dioxide accumulation and curtails the fluid loss.[3] Stoodley recommends incubation under the hen for the most critical first three days, then removing for artificial incubation.[22] Although not a standard procedure, many aviculturists incubate eggs with the large end up, as they believe it is easier for the bird to pip in this position.

## Fumigation

Fumigation is useful for decontamination of newly gathered eggs, egg handling equipment, and incubator. This procedure may be employed to decontaminate entire rooms. Fumigation of incubators is performed weekly in the poultry industry, but such a practice is considered risky and possibly unnecessary in exotic birds. Fumigation of the equipment is best accomplished without the presence of eggs. However, if a disease outbreak can be traced back to organisms in the incubator or on the surface of the eggs, fumigation is essential. Roudybush has enountered no embryo mortality in freshly laid cockatiel eggs from fumigation, but eggs with embryos younger than three days or those starting to pip should be removed for the procedure.[3] For immediate use following the fumigation, the incubator must be fitted with an exhaust system to the outside and an air exit valve for the fumigant.

Potassium permanganate ($KMNO_3$) and formalin are combined inside the incubator. For each 100 cubic feet of incubator space, 1 1/3 ounces of potassium permanganate crystals are used with 2 1/2 fluid ounces 40 per cent commercial grade formalin. The $KMNO_3$ crystals are placed first in a small container that is surrounded by a larger open container to prevent staining with the purple residue. The formalin is added at the last minute and the door immediately shut. The chemical reaction is almost explosive, as foaming and sputtering immediately occur, releasing the toxic gas. The incubator is operated at 90°F and 88 to 90°F wet bulb. The recommended length of fumigation time varies from 10 minutes to two hours, depending upon the degree of contamination of the egg shells. In most cases, shorter times are sufficient. The fumigant is then exhausted.[9]

The formaldehyde odor will remain after the exhaust, but it is safe to return any young eggs that have been removed. Rapid dissipation of the odor can be accomplished with the use of full-strength ammonia hydroxide (26 per cent) sprayed on the incubator walls or with a container of sodium bisulfite liquid or crytals in close proximity.

## Monitoring Incubation

Eggs in the incubator need to be candled once a week to determine fertility or early deaths of the embryos. Small psittacines will begin to show evidence of fertility (blood development) in three days; however, this is not evident in larger species such as macaws until seven days. A lack of normal vascular architecture is the first sign of an early dead embryo. During the candling procedure, a soft "flop" appearance during the turning of the egg suggests an intact yolk or embryo, whereas a dead egg would appear to have a watery consistency. Psittacine eggs often exhibit a greyish-black cast once the embryo has died, and the eggs should be immediately removed. Suspicious eggs can be maintained in the incubator until the decision is final; however, they should be placed below other healthy eggs, so if they explode, the contents will not get on other eggs.

Necropsy procedures can be performed on dead embryos when the eggs are sterilely opened. The contents can be cultured and/or submitted for histologic evaluation to determine the cause of death. One can consult the previously mentioned reference chart for diagnostic suggestions.

Three days prior to hatching (see Appendix 5 for species' incubation times), the incubator temperature is reduced approximately 1°F, the humidity is increased, and rotation of the egg is stopped. Serious aviculturists often weigh the egg during the incubation stages and expect 16 per cent weight loss at hatching.[22] If the egg is losing weight too fast, the humidity is further increased, and vice versa. Additional incubators are necessary to accomplish this if multiple clutches are being set.

## Hatching

The first sign of pipping can be heard when the chick's head is in the air cell. The rotating is avoided so that any liquid in the air cell will not drown the embryo. It may be one or two days after pipping starts before the chick actually hatches. One aviculturist[22] listens to the egg and monitors the movement of the chick. Stoodley can determine by the sounds if the chick needs help in breaking through the shell. In some cases he has removed the end of the shell over the air cell and observed the chick. If it appears that the membrane has dried around the chick's head, he uses a camel's-hair brush to apply distilled water several times a day to moisten the membrane. He then caps the top with a shell portion from a previously hatched egg of the same species. Plastic wrap has worked as well for this purpose in this author's experience.

As soon as the blood vessels are gone from the surface of the membrane, Stoodley believes it is safe to assist the chick out. Premature assistance may cause sufficient hemorrhage to bleed the chick to death. Another aviculturist has gone so far as to actually feed day-one formula to macaws in the egg if he sees they are having difficulty emerging or the sounds suggest that the bird is becoming weak.

When hatching is complete, the bird is transferred to a brooder to dry out.

## REFERENCES


1. Arnall, L., and Keymer, I. F.: Bird Diseases. Neptune City, NJ, TFH Publications, 1975.
2. Bercowitz, A. B.: Annotated review of avian sex determination techniques. In Burr, E. (ed.): Companion Bird Medicine. Ames, IA, Iowa State University Press. In press.
3. Cutler, B., et al.: Viability of cockatiel *(Nymphicus hollandicus)* eggs stored up to 10 days under several conditions. Proceedings of the 34th Western Poultry Disease Conference, University of California, Davis, 1985, pp. 104–106.
4. Dolensek, E., et al.: Hatching problems in exotic birds. Proceedings of the Annual Meeting of the American Association of Zoo Veterinarians, Denver, CO, 1979.
5. Ettinger, A. O., and King, J. R.: Consumption of green wheat enhances photostimulated ovarian growth in white-crowned sparrows. Auk, 98:832–833, 1981.
6. Flammer, K.: Emergency procedures for aviary birds. In Proceedings of the American Federation of Agriculture Veterinary Seminar, Orlando, FL, 1984.
7. Flammer, K.: Aviculture medicine of psittacine birds. Vet. Clin. North Am. [Small Anim. Pract.], 14(2):381–401, 1984.
8. Gee, G.: Personal communication, 1982.
9. Grau, C. R., et al. (eds.): Breeding and Raising Cage Birds. Proceedings of University of California Extension Class, 1981.
10. Hargrove, T.: Preliminary results of cryopreservation of budgerigar semen. AAV Newsletter, 5(4):108, 1984.
11. Harrison, G. J., and Wasmund, D.: Preliminary studies of electrostimulation to facilitate semen collection in psittacines. Proceedings of the Annual Meeting of the Association of Avian Veterinarians, San Diego, CA, 1983.
12. Hinde, R. A., and Putman, R. J.: Why budgerigars breed in continuous darkness. J. Zool. (London), 170:485–491, 1973.
13. Hren, B. J., and Greenwell, G.: Factors influencing reproductive behavior in the CRC crane collection, especially *G. a. antigone*, and *G. vipio*. Bird World, January, 1984, p. 6.
14. Immelmann, K.: Experimentelle Untersuchungen Uber die biologische Bedeutung artspezifischer Merkmale beim Zebrafinken *(Taenoipygia castanotis* GOULD). Zoo. Jb. (Syst.), 86:437–592, 1959.
15. Low, R.: Parrots: Their Care and Breeding. Dorset, United Kingdom, Blandford Press, 1980.
16. Millam, J. R., et al.: Improving reproduction in captive cockatiels via environmental manipulation. Proceedings of the 34th Western Poultry Disease Conference, University of California, Davis, 1985.
17. Morris, D.: The reproductive behavior of the zebra finch *Poephila guttata* with a special reference to pseudofemale behavior and displacement activities. Behavior, 6:271–322, 1954.
18. Ringer, R. K.: Personal communication, 1985.
19. Steel, E., and Hinde, R. A.: Effect of exogenous serum gonadotrophin (PMS) on aspects of reproductive development in female domesticated canaries. J. Zool. (London), 149:12–30, 1966.
20. Steel, E., and Hinde, R. A.: Influence of photoperiod on PMSG-induced nestbuilding in canaries. J. Reprod. Fert., 31:425–431, 1972.
21. Stennett, D.: Short cuts to pairing of cockatoos—total immersion techniques. AAV Newsletter, 6(2):58–59, 1985.
22. Stoodley, J., and Stoodley, P.: Parrot Production. Porstsmouth, England, Bezels Publications, 1983.
23. Thompson, D.: Strategies for captive reproduction of psittacine birds. Proceedings of the Jean Delacour/IFCB Symposium on Breeding Birds in Captivity, Universal City, CA, 1983.
24. Toone, C. K.: Causes of embryonic malformations and mortality. Proceedings of the Annual Meeting of the American Association of Zoo Veterinarians, Tampa, FL, 1983.
25. van Tienhoven, A.: Environment and reproduction of pet birds. Proceedings of the Annual Meeting of the Association of Avian Veterinarians, San Diego, CA, 1983, pp. 110–161.
26. Wiley, J. W.: The role of captive propagation in the conservation of the Puerto Rican Parrot. Proceedings of the Jean Delacour/IFCB Symposium on Breeding Birds in Captivity, Universal City, CA, 1983.
27. Witman, P.: Techniques and problems in a brooder facility. Proceedings of the Annual Meeting of the American Association of Zoo Veterinarians, Tampa, FL, 1983.


## ACKNOWLEDGMENT

The author acknowledges Dr. George Gee for his contributions in developing incubation procedures.

# Chapter 53

# PEDIATRIC MEDICINE

KEVEN FLAMMER

Birds are broadly classified according to their state of maturity at hatching. Precocial birds, such as poultry, pheasants, and waterfowl, are covered with down and able to see, walk, and feed themselves at hatching. Altricial species, such as parrots, song birds, and pigeons, are often born naked with eyes closed and depend totally on their parents for food and warmth. This chapter will discuss specific problems of birds prior to weaning. Veterinary advice is most often sought for psittacine birds, so their care will be emphasized.

At hatching psittacine birds lack a fully competent immune system and are more susceptible to disease than older birds. Because they are helpless, the conditions under which they are kept, the diet they are fed, and the amount of parental care they receive all have a profound influence on their health.

## MEDICAL PROBLEMS OF THE NEONATE: THE FIRST FIVE DAYS

The nutritional, health, and genetic status of the hen at the time the egg is formed can influence the health of the newly hatched chick. If the diet of the hen is marginal, the first egg laid in the clutch may be normal, but subsequent eggs may be affected (see Chapter 31, Nutritional Diseases). Inbreeding, particularly evident with rare or endangered species, can cause lethal and life-threatening genetic mutations, resulting in a number of lesions that affect viability of both the egg and the hatchling. For these reasons it is important to know the pedigree and nutritional status of the parents when investigating problems in their hatchlings.

Any anomaly in the incubation conditions of the egg can cause small, weak chicks that show poor weight gains even if reared under good conditions. The most important parameters are temperature and humidity (see Chapter 52, Reproductive Medicine). The amount of damage done is proportional to the degree of the error and the stage of embryo development when the anomaly occurs. Chicks hatched under low incubation temperatures may fail to retract their yolk sac, fail to absorb all of the albumin in the egg (resulting in "sticky chicks"), and may have bent necks.[19] Elevated incubation temperatures speed development and accelerate hatching, potentially resulting in scruffy down and neurologic problems. Skeletal abnormalities may result if the humidity is too low and the shell membranes dry out. The membranes stick to the shell and prevent mobilization of calcium, which is needed for mineralization of the bones. If the humidity is too high, the yolk sac may not be completely retracted and the chicks are described as "soft and blobby."[19] Chicks hatched under less than ideal conditions may suffer a poor start but are by no means doomed. They will require more diligent care and must be closely watched to make sure that they do not develop secondary microbial infections.

The yolk sac is a diverticulum of the intestines which contains the remnant of the egg yolk. It should be drawn into the abdominal cavity through the umbilicus prior to hatching (Fig. 53–1). Once withdrawn, the yolk is gradually absorbed and provides the chick with nutrition and maternal immunoglobulins during the first days of life (see Chapter 24, Relationships of the Avian Immune Structures and Functions to Infectious Diseases). Like the umbilicus in mammals, the point of entrance of the yolk sac is prone to infection. Gram-negative organisms, especially *Escherichia coli*, may cause infection of the umbilical area or the yolk sac itself (omphalitis). To prevent omphalitis, the chick should be placed in a clean environment and the umbilicus swabbed with a dilute solution of povidone-iodine. If the yolk sac is not withdrawn, packing the chick in tissue paper will restrict it from stepping on the sac and tearing it. If the sac is in danger of rupture, it should be ligated and removed.

Following retraction, problems with the yolk

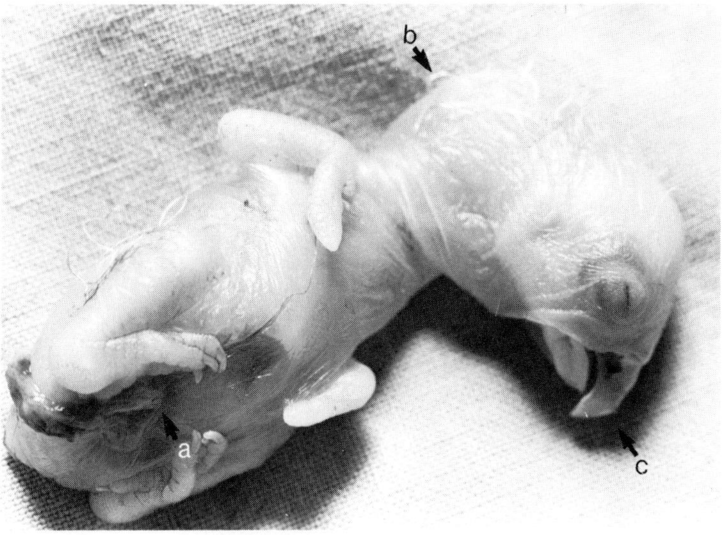


**Figure 53–1.** The yolk sac has been absorbed into the abdomen at the point of the umbilicus (a) in this normal newly hatched Green-winged Macaw. Note the enlarged complexus (b) and egg tooth (c), which aid the bird in hatching. These will diminish in size during the first week. (Courtesy of Greg J. Harrison.)

sac are less frequent but do occur.[13] Trauma or infection can cause rupture inside the chick, which is usually fatal. Failure to absorb the yolk can also compromise the bird. The chick not only loses the nutritional value of the stored yolk but also the maternal antibodies, placing it in a situation similar to that of a mammal that fails to receive colostrum. Absorption times are not known for psittacine birds, but the yolk should probably be absorbed within 10 days of hatching. The cause of yolk retention is not known, and no specific treatment is available.

All species of animals establish a population of resident bacteria in their alimentary tract. This normal flora aids in maintaining the structure and function of the gut, synthesizes certain nutrients such as vitamin K, and protects the gut from colonization by pathogenic organisms.[21] At hatching the alimentary tract of psittacine birds is sterile, but a resident population of gram-positive bacteria, primarily anaerobes of unknown species, is rapidly established. The species of aerobic bacteria most frequently isolated are *Lactobacillus, Corynebacterium, Bacillus, Streptococcus,* and *Micrococcus.*[9] In parent-reared birds this bacterial flora comes from the regurgitated crop contents of the parents. Chicks hatched from artificially incubated eggs do not receive the parental flora but apparently establish a resident population of gram-positive organisms anyway. It is not known if exotic bird chicks fed flora of parental origin grow better than those that establish flora from other sources, but species-specific flora inoculations in poultry have a definite beneficial effect. After oral inoculation with multiple species of bacteria isolated from the cecal contents of adult birds, day-old chicken and turkey poults were protected from colonization by *Salmonella* and *E. coli.*[18] To gain the most positive effect, the entire complement of bacteria present in the cecum (predominantly anaerobic organisms numbering more than 40 species) had to be given. Trials with a single species of bacteria, such as *Lactobacillus,* were much less successful.[1] The bacteria with the most protective effect are species-specific. Chickens were not protected by flora isolated from horse, cow, or mice donors.[15, 20] The benefits of using commercial flora replacers containing a few species of bacteria isolated from poultry to establish normal flora in exotic birds are not known. It is possible that even if the foreign bacterial species does not colonize the gut, it might alter local conditions and favor the proliferation of what normal flora is present.

In the nursery for hand-raised babies of Behavioral Study of Birds (Newhall, CA), an attempt was made to introduce species-specific normal flora into incubator-hatched chicks, and the author believes this practice reduced the incidence of opportunistic bacterial infections. This was accomplished by feeding the neonate samples collected from the crop of an adult bird of the same species, preferably a parent. A tube was passed into the crop of the donor bird, the crop washed with 5 to 10 ml of sterile saline, and the fluid withdrawn. A Gram's stain of the sample was made to rule out gross contamination by gram-negative bacteria and yeast and to assess the amount of gram-positive bacteria present. If the sample was suitable, it was centrifuged to concentrate the bacteria, and 0.2 to 1.0 ml was fed to the chick as its first feeding.

Arbitrarily, additional portions of the crop wash were fed 30 minutes prior to the chick's next five meals. The sample was also cultured to identify any pathogens missed by the Gram's stain.

## COMMON PROBLEMS OF PARENT-REARED BIRDS IN THE NEST

Nestlings are most vulnerable during the first week of life, when fledging (leaving the nest), and again at weaning. It is at these times that peak disease morbidity will occur. Most larger psittacines will not permit frequent handling of their young, so during a disease outbreak the nestlings must be removed from the nest for treatment and hand reared, or medicated by adding drugs to the food. Fortunately, medicated food is more readily accepted when parents are feeding nestlings; they eat more and are less particular about taste. Parents preferentially feed the babies moist, soft foods, with the largest meals fed in the morning and late evening. Medicated food should be available at these times. Semi-domesticated birds, such as budgerigars, cockatiels, and lovebirds, will tolerate handling of their young, and if necessary their offspring can be medicated directly with injectable and oral antibiotics. Abandonment, adverse weather, and poor parental care are the most common problems of parent-reared nestlings. Parents may neglect the chicks owing to inexperience with the rearing process, cold or hot weather that makes it uncomfortable to brood, and disturbances around the nest that make the birds feel insecure. Raising young is a learned behavior, so first-time parents often provide poorer care than more experienced birds. The chick must interact with the parents in order to elicit a feeding response. If the chick is sick or hypothermic, it will not stimulate the parents to feed and will be abandoned. Also if there is a difference in size between nestlings, as is common in Eclectus Parrots, the larger, stronger chick will be fed in preference to the smaller, weaker one.

Chicks that feel cool to the touch, have empty crops, or act listless and unresponsive are sick or receiving poor parental care. Abandoned nestlings are frequently hypoglycemic, hypothermic, and dehydrated and often suffer secondary bacterial infections. Treatment of these disorders is described in Table 53–1. Chicks that are panting with wings outstretched are hyperthermic and should be moved to a cooler environment. Shading the nestbox, providing better ventilation, and cooling the aviary with water spray will help prevent hyperthermia. To identify compromised nestlings, nest boxes should be checked by the aviculturist daily to make sure the chicks are alive, warm, and fed. If nest observations are made carefully throughout the year, even tempermental birds will become accustomed to it and the disturbance will be minimal.[12] If a particular pair of birds repeatedly fails to care for their offspring, their eggs should be artificially incubated, or the nestlings should be fostered to other parents or hand reared.

**Table 53–1.** SYMPTOMATIC TREATMENT OF CLINICAL SIGNS IN NESTLING BIRDS

| | |
|---|---|
| **Hypothermia** | Warm in a 95 to 100°F incubator or with a heat lamp. |
| **Dehydration** | Give lactated Ringer's solution subcutaneously (1 to 3 ml/100 gm) in the wing web or back of neck, or dose orally (3 to 10 ml/100 gm). |
| **Hypoglycemia** | Add 5 per cent glucose to parenteral fluids or 20 per cent glucose to oral fluids. |
| **Shotgun Antimicrobial Treatment** | 1. *Bird in serious condition* Gentamicin (2.5 mg/kg I.M. b.i.d.) + Cefotaxime (100 mg/kg I.M. b.i.d.) + Nystatin (300,000 IU/300 gm P.O. t.i.d.). 2. *Bird ill but not in grave condition* Chloromycetin palmitate (75 mg/kg P.O. t.i.d.) or Bactrim (1 ml/300 gm P.O. b.i.d.) + Nystatin (300,000 IU/300 gm P.O. t.i.d.). |
| **Shock** | Dexamethasone (5 mg/kg I.M.) or prednisolone succinate (20–30 mg/kg I.M.). Add to I.V. fluids if possible. |

Traumatic injuries are common but often preventable. Large psittacines (e.g., Amazons, macaws, and cockatoos) may bite or trample their nestlings while defending the nest. To prevent this, the nestbox should be equipped with a sliding door that can exclude the birds from the entrance. Then if the parents must be caught or the baby examined, the parents can be prevented from entering the nest and trampling the chicks. When nestlings are close to weaning, some parents become anxious to breed again and raise a new clutch. At this time, the hen may pick feathers from the chicks (especially about the head and back) or aggressively drive them out of the nest. If the feather picking is not severe, it can be ignored and the baby will feather normally at the next molt. If the picking is more aggressive, the baby should be removed from the nest and hand raised.

Two conditions may be confused with simple trauma. When a nestling dies, many parents will eat the body or throw it from the nest and may traumatize the body in the process. This should not be confused with parental aggression. In addition, calcium and vitamin $D_3$ deficiencies can cause metabolic bone disease with folding fractures, mimicking traumatic lesions (see Chapter 31, Nutritional Diseases).

Nutritional and alimentary tract problems can occur if an improper diet is fed to brooding parents. Psittacine birds prefer to feed their nestlings large quantities of seed and moist, soft foods. If these are not provided in adequate amounts, they will stuff anything they can find into the baby, including pieces of the perch, nest material, and other dangerous items. Plenty of easily digestible food should be available at all times.

## HAND RAISING OF PSITTACINE BIRDS

Hand raising birds is a time-consuming, demanding process, but aviculturists may undertake the task for the following reasons: (1) to raise offspring hatched from artificially incubated eggs; (2) to save sick or abandoned offspring; (3) to reduce the burden of parental care on a compromised parent; (4) to encourage the parents to produce another clutch of eggs; (5) to prevent transmission of diseases (other than transovarian) from the parents; (6) to tame and socialize the bird to people, so that the bird will be better adapted to captivity, making both a better breeder and a better pet.

The age of the baby at the time it is brought into the nursery influences how easy it is to raise. Photographs of psittacine birds at different stages of development are shown in Figures 53–2 to 53–4. Newly hatched birds can be difficult to feed. Ideally, chicks should be allowed to remain with the parents through early development and brought into the nursery at approximately two to three weeks of age. This gives the birds a good start, and parental immunity has been transferred. Birds removed from the nest after three weeks may be fearful of people and are more difficult to adapt to hand feeding.

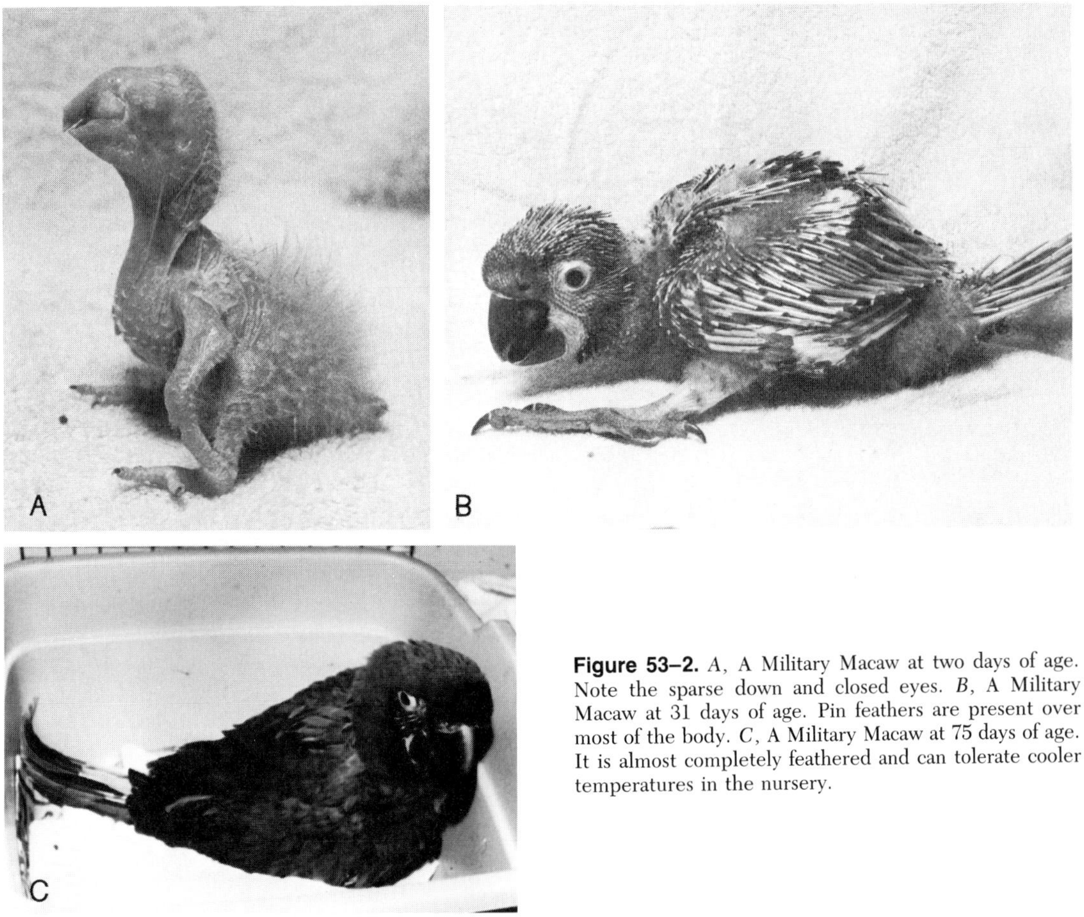


**Figure 53–2.** *A*, A Military Macaw at two days of age. Note the sparse down and closed eyes. *B*, A Military Macaw at 31 days of age. Pin feathers are present over most of the body. *C*, A Military Macaw at 75 days of age. It is almost completely feathered and can tolerate cooler temperatures in the nursery.

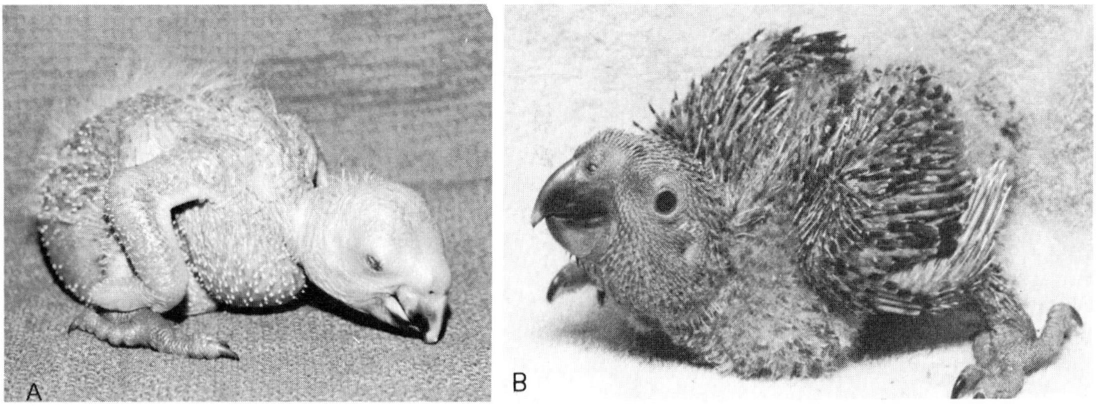


**Figure 53–3.** *A*, A Yellow-naped Amazon Parrot at six days of age. *B*, A Yellow-naped Amazon Parrot at 27 days of age. (Courtesy of the Aviculture Institute.)

## ENVIRONMENTAL CONDITIONS AND PREVENTIVE MEDICINE

The nestling's age and amount of feathering determine the environmental conditions under which it should be kept in the nursery. During the first 7 to 14 days of life, psittacine birds should be kept in a brooder with precise and constant temperature maintenance (Fig. 53–5). Ideally, the air surrounding the chick should be heated to 90 to 95°F and the chick placed on a warm surface (92 to 94°F) such as a heating

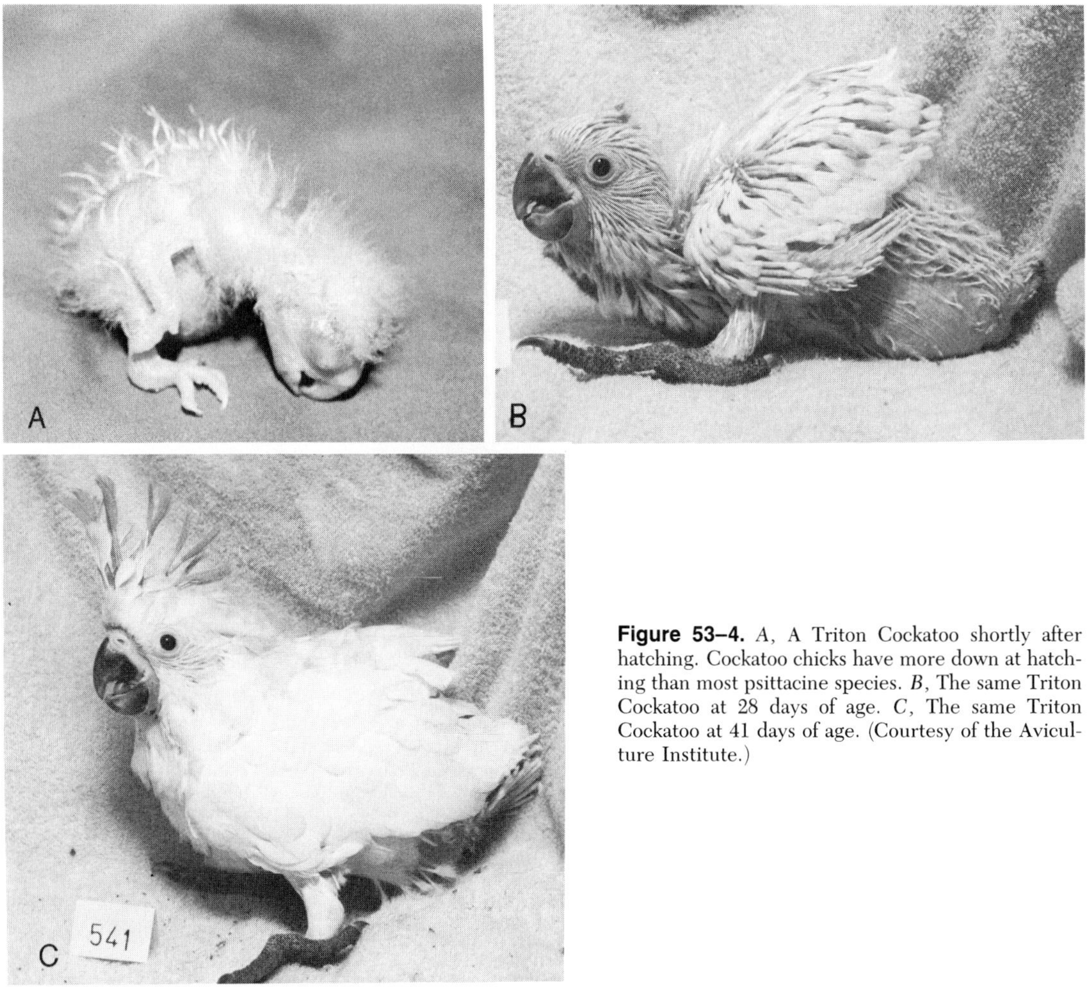


**Figure 53–4.** *A*, A Triton Cockatoo shortly after hatching. Cockatoo chicks have more down at hatching than most psittacine species. *B*, The same Triton Cockatoo at 28 days of age. *C*, The same Triton Cockatoo at 41 days of age. (Courtesy of the Aviculture Institute.)

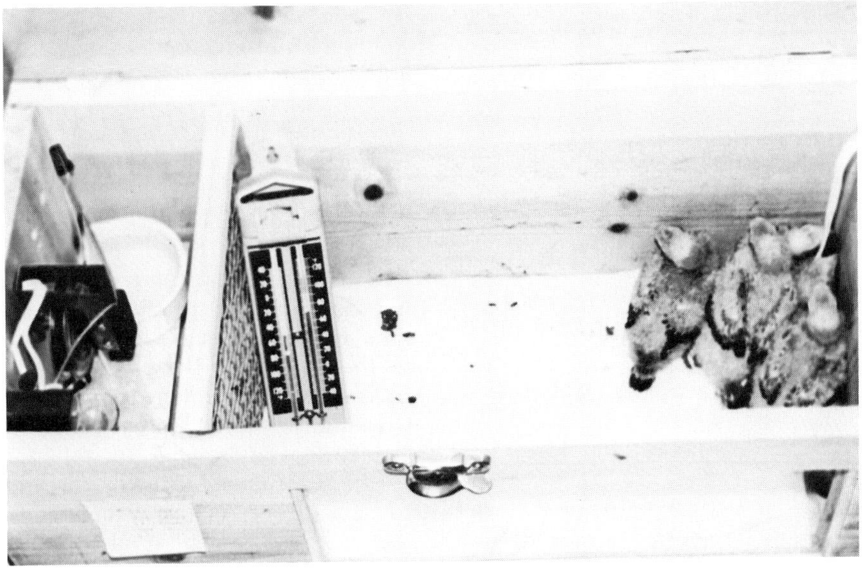

**Figure 53–5.** A temperature-controlled brooder for housing recent hatchlings. (Courtesy of the Aviculture Institute.)

pad. As the bird grows, the temperature can be gradually reduced to 85 to 90°F. Young birds do best if the humidity is kept above 50 per cent relative humidity (RH); below 30 per cent RH the birds suffer dry skin and dehydration. Birds this age should also be placed in a small container to provide support and prevent them from running around the brooder and exhausting themselves. After two weeks, when the chick is starting to feather, it can be moved to a cardboard box or open plastic pan. The temperature should be maintained at approximately 80 to 85°F by heating the room or placing a heating pad under the box. If an enclosed box is used, the top of the box should be opened as the bird becomes more curious. When the bird is nearly feathered, it can be moved to an open cage in a room heated to 70 to 80°F. To reduce the stress of moving the bird from a dark box to the open spaces, portions of the cage should be covered with white sheeting and gradually removed over a week's time.

The substrate inside the brooder must provide sufficient traction so the bird can stand without slipping, should absorb excess moisture, must be easy to change or clean, and should not entangle the bird's feet or cause digestive problems if chewed. Hemmed cloth diapers or indoor-outdoor carpeting have worked well in successful nurseries. While some aviculturists prefer ground dry corn cobs, these have been responsible in one case for cloacal impaction. Sawdust should not be used as the birds may eat or inhale it, resulting in impacted crops and respiratory problems. Wood chips should be avoided until the bird has enough tongue dexterity to crack seeds. Tissue paper is too slippery unless the bird is kept in a small container.

Healthy birds of similar size and taken from the same nest can be housed together in the same brooder but should be separated if problems occur. Occasionally, nestlings will "feed" on one another, locking beaks and pumping their necks in a typical feeding response. Trauma to the beak and other areas of the body may result. Also, if there is a large difference in size between the birds, the larger will trample the smaller in the excitement at feeding time. If a nursery is experiencing repeated microbial infections, each bird should be placed in a separate box or brooder to slow the spread of disease and make it easier to examine each bird's droppings.

## DIETS AND FEEDING TECHNIQUES

There are many diets used to hand raise baby psittacine birds, reflecting the lack of knowledge of exotic bird nutritional requirements. Most of the diets in current use are composed of a cereal base (usually a commercial product such as monkey biscuit or dry baby cereal), to which is added a variety of items: baby food, fruit, vegetables, nuts, vitamins, and other nutritional supplements. The complexity of the diet seems to be more related to the amount of time the feeder has to devote to his birds rather than to differences in feeding success. Clubb and Clubb[5] have described a successful, easily

**Table 53–2.** DIET USED FOR SUCCESSFULLY HAND RAISING PSITTACINE BIRDS*

**Ingredients**
- 1 quart monkey biscuit
- 1 quart water
- 3 heaping tablespoons peanut butter
- ½ jar creamed corn baby food (4.5 oz jar)
- ½ jar oatmeal with applesauce baby food (4.5 oz jar)
- 5 ice cubes

**Directions**
1. Soak the monkey biscuit in the water for 30 minutes
2. Simmer on stove or cook in microwave oven for several minutes
3. Stir in the baby food and peanut butter
4. Cool by adding the ice cubes
5. Mix thoroughly
6. Test temperature before feeding (95 to 105°F)

*From Clubb, S. L., and Clubb, K. J.: Psittacine pediatrics. Proceedings of the American Federation of Aviculture Veterinary Seminar, 1984, pp. 17–33.

prepared diet (Table 53–2). Several commercially prepared diets are also available.

Although specific nutritional requirements are not known, following the guidelines for avian species for which more information is available is a good place to start. The diet can be based on estimations from the nutritional requirements of poultry, from research being conducted at the University of California with purified diets in cockatiels, and from the success of other aviculturists. If the feeder chooses to prepare his own hand feeding ration, monkey biscuit is a logical choice for the major component of the diet. It is formulated for fruit-eating omnivores, provides balanced nutrients, and is available in several protein concentrations. Milk products containing lactose should not be included in the diet, since birds are unable to digest lactose. Data from nutritional experiments in cockatiels[17] and the opinions of successful aviculturists indicate that the total protein content of the diet should be approximately 20 per cent. The diet should contain a complete vitamin mix. Roudybush[16] lists suggested levels for cockatiels that are probably appropriate for other avian species as well. Calcium and phosphorus should be balanced in approximately a 2:1 ratio. Dietary levels of 1 to 2 per cent calcium and 0.7 to 1.2 per cent phosphorus have been suggested. Other nutrients are under investigation. For example, contrary to the effects of lysine deficiency in some other avian species, cockatiels fed purified diets deficient in lysine developed normal feather pigment but the feather growth was poor and body size was also limited.[17a]

Water content, consistency, and temperature of the diet are also important. Cockatiels, hand fed from hatching, grew best when fed a diet containing 7 per cent solids for the first 48 hours, then 30 per cent solids until weaning.[14, 17] The solids content of the diet should be determined by weighing the solid and liquid portions separately and calculating the percentage of each. Thickness, or viscosity, of the diet is a poor indication of the solids content, since cooking time will greatly influence diet consistency.[16] It is also important that the solid and liquid portions of the diet do not readily separate. If such a diet is fed, the solids will settle to the bottom of the crop, leaving a doughy mass that will prolong crop emptying time. Cooking the diet will often gel the starches and reduce separation. Once the total solids content has been formulated, the diet should be cooked to a consistency of soft ice cream. Too fibrous a diet will also prolong crop emptying time and reduce the number of feedings that can be given. The temperature of the food is critical to its acceptance. Young birds may reject food that

**Figure 53–6.** A fistula in the crop of this conure was caused by hand feeding overly heated formula and must be surgically repaired. Note the crooked beak, a phenomenon observed in several species of hand-raised psittacines, of unknown etiology. (Courtesy of Greg J. Harrison.)

is less than 98 to 100°F, and food hotter than 105°F can scald the crop, leading to sloughs and external fistulas (Fig. 53–6).

Three utensils are used by aviculturists to deliver food: spoons, crop tubes, and syringes. Each of these utensils has its own advantages and disadvantages. Spoon-feeding allows a coarser diet to be fed, more closely mimicking what the parents would feed, but it is slower than the other methods. Some aviculturists believe that birds are easier to wean when spoon-fed. Passing a tube into the crop, either forcibly or by training the bird to swallow it, allows large quantities of food to be delivered quickly, but this method is dangerous in the hands of inexperienced feeders, as the tube may be accidentally placed into the trachea. Catheter-tipped syringes are the most popular feeding utensil, are available in a variety of sizes, and are useful when the food must be accurately measured and quickly fed.

During feeding, baby birds display a feeding response that consists of rapid, thrusting head movements and bobbing up and down. These movements can be stimulated by touching the commissures of the beak or pressing lightly under the mandible. While the bird is displaying this behavior, the glottis is closed and large amounts of food can be delivered quickly without fear of passing food down the trachea (Fig. 53–7). If the bird resists feeding or a feeding response is not displayed, the chance of tracheal aspiration is greater. As some birds get older, they display less of a feeding response and are more difficult to feed.

Younger birds should be fed more often than older, larger birds. The total amount of food fed, the weight gained, and the morphologic development of the bird are more important indicators of adequate nutrition than the number of feedings. As a guide, birds one to five days old should be fed five or six times daily. It is not necessary to feed them through the night. At one to two weeks they should receive four or five feedings each day. By the time the birds are three weeks old, three feedings each day should be adequate. The crop should be filled to capacity and then empty completely before the next meal. It is very important to feed young birds large amounts of food early to stimulate growth and increase crop capacity.

## Weaning

Weaning is a stressful time for both the bird and the hand feeder. Some birds wean themselves by refusing to be hand fed, but most must be strongly encouraged to wean. The amount of food fed should be reduced at the time the parents would ordinarily wean their young or when the birds start to eat solid food on their own. The midday feeding should be eliminated first, followed by the morning and then the evening meal. At the same time, the birds should be placed in cages with low

**Figure 53–7.** Pressing lightly under the mandible elicits the feeding response for safe delivery of hand-feeding formula. Pine shavings should not be used as nursery substrate until the bird is fully feathered and has some tongue dexterity, as in this Sun Conure. (Courtesy of Greg J. Harrison.)

**Figure 53–8.** At weaning birds should be surrounded by food cups placed at perch level to encourage them to eat on their own. (Courtesy of the Aviculture Institute.)

perches and surrounded by food (Fig. 53–8). Hulled sunflower seed, corn-on-the-cob, spray millet, and peanut butter sandwiches are often the first foods they will eat. Ideally, they will start eating solid food at the same time hand feeding is reduced. Weanlings ordinarily lose weight during this period, and losses up to 10 per cent of their peak body weight are normal. Usually, in the month prior to weaning, the birds gain extra weight, and this loss represents normal streamlining as the birds assume adult conformation and exercise more. Sometimes losses up to 15 to 20 per cent of the peak weight must be tolerated with stubborn birds. If the birds lose more than this or act depressed, feeding should be resumed and weaning attempted at a later date. The birds should be watched carefully during this time to make sure they remain healthy. This is not a good time for the feeder to sell the bird or radically change its environment.

Some birds will resist hand feeding before they are capable of maintaining body weight on their own. This is especially true of Eclectus Parrots and Rose-breasted Cockatoos. Birds that have been malnourished during development may also resist hand feeding before weaning is appropriate. It seems that behaviorally they mature to the point that they want to wean, but morphologically they are underdeveloped. The feeder is then faced with the dilemma of forcing the bird to hand feed and risking aspiration of food, or stopping the hand feeding altogether and risking considerable weight loss. In most cases it is best to take a balanced approach and attempt to feed the bird once daily, encouraging it to eat on its own. Such birds may also be tube fed. In contrast to the birds that try to wean themselves too soon, some species and individuals, especially macaws and larger cockatoos, must be forced to wean or they will continue to hand feed indefinitely. Sometimes an inexperienced owner will not realize when a bird can be weaned. Suggested weaning ages are listed in Table 53–3.

## HYGIENE AND PREVENTIVE MEDICINE

Hygiene and preventive medicine are very important in the nursery. Psittacine birds are valuable enough from both an economic and an ecologic standpoint to justify maximizing their care. The most important sources of potentially

**Table 53–3.** APPROXIMATE WEANING AGES FOR HAND-RAISED BIRDS

| | |
|---|---|
| Conures | 50–60 days |
| Amazon Parrots | 75–90 days |
| Small macaws | 75–90 days |
| Large macaws | 95–120 days |
| Medium cockatoos | 75–90 days |
| Large cockatoos | 95–120 days |
| African Grey Parrot | 75–90 days |
| Eclectus Parrot | 100–110 days |

pathogenic bacteria are the food, water supply, other birds in the nursery, and contamination by the nursery feeder.

Bacterial levels in the food and water that would have little effect on adult birds may cause infections in baby birds. The components of the diet should therefore be carefully selected. Poultry feeds, such as chicken scratch and turkey starter, should be cultured before including in hand-feeding diets for psittacine birds (see Chapter 50, Aviculture Management). Some perishable foods (e.g., hulled sunflower seed and shelled peanuts) purchased from open bins at a health food store may contain unacceptable levels of *E. coli*. It is best to purchase seeds and nuts in sealed bags intended for human use or directly from a food distributor.

The mixed diet must be prepared fresh for each feeding or kept refrigerated to prevent spoilage. Yeasts such as *Candida* are common contaminants of spoiled food. Standards for cleanliness should be higher than the feeder would maintain for himself. A good motto for evaluating stored food is "when in doubt, throw it out." Cleanliness of the utensils used to prepare the food is also important (see Chapter 50, Aviculture Management).

New birds, brought in from the nest at various ages, are the greatest potential source of contamination in the nursery. Whenever possible, it is valuable to take cloacal cultures from birds before they actually enter the nursery. Then, if the bird is infected, treatment can be started and the problem eliminated before it spreads to other birds. This also allows the babies to act as sentinels for bacterial problems in the breeding flock.

Bird-to-bird transmission can be reduced if the birds are kept in separate boxes and a few simple practices followed. Each bird should be fed from a separate feeding utensil and the feeding area lightly sprayed with a disinfectant. Hands can be disinfected between handling of each bird by washing or using a disinfectant foam. Sick birds should be isolated, fed last, and special care taken to disinfect their area after use.

If possible, different people should be used to take care of the nursery and the breeding flock. If this is not possible, different clothing or coveralls should be used when servicing the adult birds, and the feeder should wash thoroughly before feeding the babies.

## MEDICAL PROBLEMS OF NESTLING BIRDS WITH AN EMPHASIS ON HAND-RAISED BIRDS

The best way to assess the health and growth of developing birds is to measure their body weight at the same time daily (Fig. 53–9). They should gain weight every day until they are fully feathered and start to wean. Often, the only sign of a disease problem is a lack of growth and weight gain. If a bird loses weight two days in a row, the cause should be investigated. Stress, changes in the environment,

**Figure 53–9.** Baby birds should be weighed daily to assess their growth. (Courtesy of the Aviculture Institute.)

and changes in diet may cause temporary weight loss.

Baby psittacine birds can be examined in the same manner as adults but share a number of characteristics that differ from their parents. These characteristics are listed in Table 53–4.

**Table 53–4.** PHYSICAL CHARACTERISTICS OF NESTLING PSITTACINE BIRDS

**Pectoral Muscles**
Early in life these are thinner and much less developed; as the bird completes weaning and exercises more, its body will assume more adult proportions.

**Abdomen**
Ordinarily nestlings have a protuberant gizzard, perhaps reflecting their major goal in life—eating. This characteristic is very pronounced in parrots, giving the bird a tripod effect that helps hold them upright.

**Crop**
The crop of baby birds is more distensible, and on a weight-proportional basis holds approximately twice as much food as that of an adult.

**Skin Texture and Color**
The color and texture of nestling skin are important indicators of general health and state of hydration. Nestlings should be a healthy, yellowish-pink color with soft, translucent skin. Birds that are reddish in color with dry, easily wrinkled skin are dehydrated. A bird that is very pale, especially "china white," is in grave condition.

**Droppings**
Droppings are more watery and the fecal portion is less formed than in adult birds, reflecting the more fluid diet.

**Sitting Position**
Baby psittacine birds, especially cockatoos, sit on their hocks rather than on their feet. This can lead to calluses on the hocks if the birds sit on an unpadded surface. This position is normal; as the bird grows older he will place his feet underneath him.

**Sleeping Position**
Baby birds may sleep in almost any position, since they fall asleep quickly. These odd positions could be mistaken for illness. Cockatoos in the first two weeks of life can barely hold their heads up. They sleep in a crouched position with their heads touching their feet. Most parrots and macaws sprawl out on their sides. Conures often sleep on their backs in a food cup or corner of the cage.

**Feathering**
Most psittacine nestlings have sparse down. The first feathers usually appear on the wings, head, and in some species the tail. The last areas to feather are the dorsal and ventral body surfaces. At the time wing and tail feathers are mostly formed, body pin feathers should also be present. Gross discrepancies, such as a fully feathered head and wings but no feathers elsewhere, are an indication of stunted development.

## Malnutrition and the Stunting Syndrome

Deficiency syndromes of a specific nutrient are rare if commercial products, such as monkey biscuit, are the major component of the diet. Most aviculturists supplement the diet with a multivitamin and mineral preparation, and, in fact, oversupplementation may be more of a problem (see Chapter 31, Nutritional Diseases). In contrast, underfeeding of total calories with resulting malnutrition is one of the most common underlying causes of disease and general debilitation in hand-raised birds. Currently, even in the best nurseries, early weight gains of hand-raised baby birds lag behind those of parent-raised babies. If the difference is not too great, the birds will grow to normal or nearly normal size and make up the weight difference prior to weaning. However, if the difference is marked, then stunted birds, with characteristic morphologic changes, will result.[5] These undernourished birds are more susceptible to disease and often suffer chronic microbial infections that are refractory to treatment.

Stunted birds can be identified by recognizing the clinical signs of the "stunting syndrome" as described by Clubb[5] (Figs. 53–10 and 53–11) and by comparing the weight gains of the

**Figure 53–10.** Misdirected feathers at the back of the head may be an early sign of stunted development.

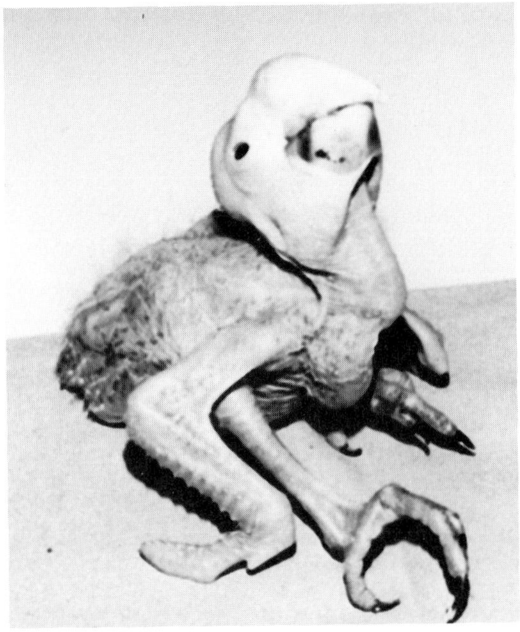

**Figure 53–11.** A severely stunted Green-Winged Macaw. Note the thin appendages and disproportionate head size.

suspect bird with gains of successfully reared birds. Sample weight gains for selected psittacine birds hand-fed at one nursery are listed in Appendix 6. As hand-feeding diets and techniques improve, there is no doubt that these weight gains will be exceeded.

### *Clinical Signs of the Stunting Syndrome*

1. Failure to feather at the normal time; feathering in an uneven manner; misdirected feathers at the back of the head
2. Thin chest, feet, toes, and wing tips; narrow back
3. Head large in proportion to the body
4. Pale skin color
5. Sluggish behavior
6. Chronic microbial infections
7. History of being underfed

There are several reasons for the high incidence of malnutrition. First of all, hand rearing is a relatively new practice, and aviculturists accept undernourished birds as normal. As information regarding the nutritional requirements of these birds becomes available, success in hand rearing will increase. To treat malnutrition, the bird's crop should be fed to capacity each time it empties throughout the day. The water content and nutritional composition of the diet should be evaluated and microbial infections controlled.

## ORTHOPEDIC PROBLEMS

A number of orthopedic problems are seen in nestling birds. Deviation of the tarsometatarsus occurs if birds are raised in confining containers or containers with rounded bottoms. If parents are disturbed in the nest, they may bite the nestling in frustration or attempt to move the baby. Most commonly, they grab it by the leg and fracture the tibiotarsus. Calcium and vitamin $D_3$ deficiencies can cause folding fractures and deformed bones. Leg deviation from the hip (spraddle leg) can occur spontaneously or when the babies are raised on a hard, slippery surface.

Treat orthopedic problems by stabilizing the lesion and eliminating its cause. Very young birds with fractures are best treated by simply propping them up in a container with folded towels. Often they will heal in adequate alignment. Larger birds are more tolerant of splints, such as those made with acrylic (see Chapter 46, Surgical Instrumentation and Special Techniques) or casts, but if the bird is actively growing, the cast should be removed every four or five days to prevent compression of growing bones. The bones of young birds heal rapidly and support can often be removed within one to two weeks. Calcium-deficient young birds with multiple fractures respond well to oral calcium supplementation, and the fractures often heal within a few weeks. Spraddle leg is difficult to correct, but one can try taping the legs together for support (Fig. 53–12). Harrison[12] suggests this condition sometimes responds to injectable vitamin E and selenium if the problem is recognized and treated early.

A lesion that occurs in Grand Eclectus Parrots and less commonly in macaws is a circumscribed constriction around one or more toes, usually on the distal digit (Fig. 53–13). It looks similar to what would happen if a string wrapped around the toe. The etiology of this problem is unknown—no microorganisms have been recovered and histologic examination reveals only marked edema. If the condition is detected early, the toe may be salvaged by repeated massage and soaking in warm water. If the distal segment is markedly swollen, it should be surgically removed.

## ASPIRATION PNEUMONIA

Aspiration pneumonia occurs when food is inhaled during feeding. Birds that are difficult

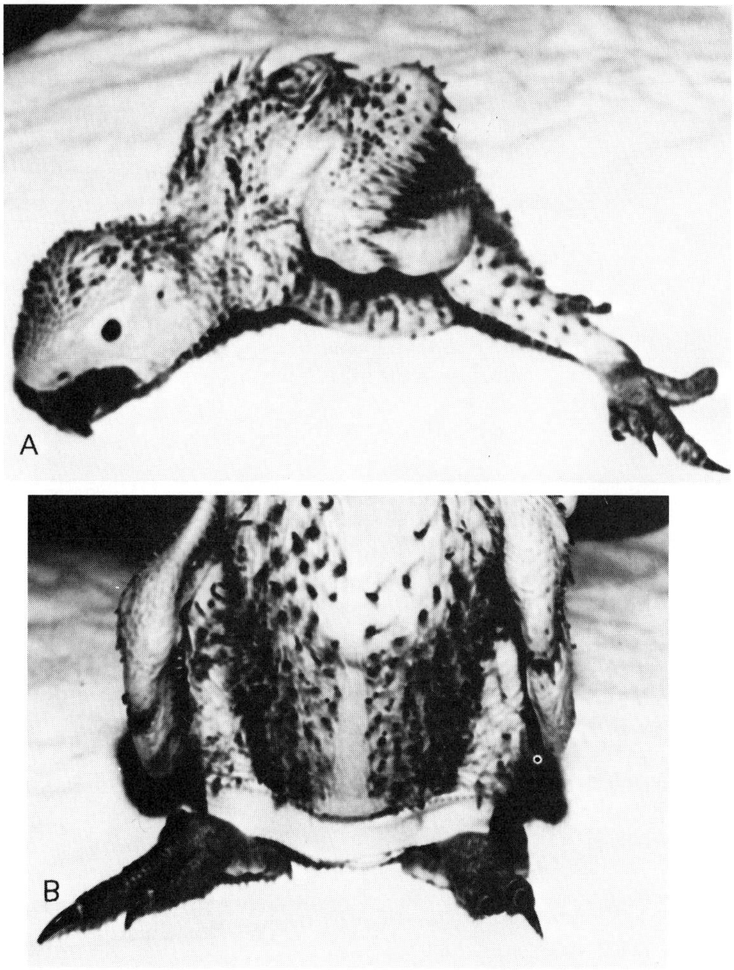

**Figure 53–12.** *A*, A Blue and Gold Macaw with a splayed left leg. *B*, This was successfully treated by taping both legs together with a tape splint for two weeks.

**Figure 53–13.** Swelling of the toe of a Grand Eclectus Parrot caused by an idiopathic constricting lesion.

to feed, such as recently hatched birds, those removed from the nest after three weeks of age, and birds starting to wean are most likely to experience this problem. If a large amount of food is aspirated, the bird will quickly die from asphyxiation. Small amounts of food may be repeatedly aspirated over a long period of time and not noticed by the feeder. Affected birds develop a pneumonia, which may or may not be clinically apparent, and have poor weight gains, sluggish behavior, and persistent, elevated WBC counts that are minimally responsive to antibiotic therapy. It may be difficult to identify the cause of these signs, as food particles may be visible only on histologic sections of lung tissue. Aspiration pneumonia should be included in the differential diagnosis of debilitating diseases of hand-raised birds. Long-term antimicrobial treatment may support the bird long enough for it to encapsulate the lesions.

## CROP STASIS IN NEONATES

Failure of the crop to empty normally is cause for concern (see Chapter 28, Symptomatic Therapy and Emergency Medicine). Some causes of crop stasis are listed in Table 53–5.

If stasis is suspected, the crop should be gently palpated to identify foreign objects or food masses. A Gram's stain of the crop contents may show microbial pathogens; especially important are intracellular gram-negative organisms and yeast. Cultures of the crop contents should be interpreted with scrutiny, since transient gram-negative organisms may be confused with pathogens, and the real cause of stasis missed. If more apparent causes cannot be found, a cloacal culture may identify enteric problems responsible for the stasis. Plain and contrast radiographs will further identify crop foreign bodies and problems with the distal alimentary tract.

Regardless of the cause, a bird with crop stasis should not be fed until the condition is corrected. If the bird appears to be otherwise normal, the problem can often be solved by giving 2 to 15 ml of warm water and gently massaging the crop to break down impacted food. If food does not pass within four or five hours, the crop should be emptied and flushed. A feeding tube with an open end is lightly lubricated with K-Y jelly and passed into the crop. A small amount of saline is then gently flushed back and forth to loosen the food and draw material into the syringe. The crop should be flushed until all food is removed. This procedure should be done in stages if the bird becomes overly stressed. If the crop contents are sour, a teaspoon of baking soda can be added to 25 ml of water and used as the final flush. Following an episode of crop stasis, less volume and a more liquid consistency of food should be fed until the crop empties normally. If bacteria or yeast are suspected as a cause or sequelae, appropriate antimicrobial therapy should be instituted. Foreign objects can be removed by passing forceps into the crop or opening the crop surgically (see Chapter 48, Selected Surgical Procedures).

**Table 53–5.** CAUSES OF CROP STASIS

**Primary Causes**

1. Foreign bodies (wood chips and seed hulls are most common)
2. Infection of the crop lining
3. Atony caused by overstretching of the crop
4. Burn of the crop caused by hot food
5. Fibrous food impaction

**Secondary Causes**

1. Dehydration or inadequate liquid in the food
2. Change in the amount of food fed
3. Environmental temperature too low
4. Blockage of the alimentary tract distal to the crop
5. Malnutrition that causes overall slowing
6. Generalized infection that causes overall slowing
7. Feeding food that is too watery
8. Feeding food that separates into solid and liquid portions
9. Overfeeding at the time of weaning

## BACTERIAL DISEASE

Bacterial infections are the most common diseases diagnosed in baby psittacine birds (see Chapter 33, Bacterial Diseases). Following a stressful episode birds may become infected from the ubiquitous opportunistic bacteria present in the environment and the bird's own alimentary tract. It is also important to note that bacteria are the easiest disease agents to diagnose and treat. Many environmental, nutritional, chlamydial, mycoplasmal, and viral problems may be missed because potentially pathogenic bacteria are isolated and held responsible for the entire disease syndrome.

General signs of bacterial disease include weight loss, reduced feeding response, depression, reduced activity, separation from nest mates, and crop stasis. If the bird is dehydrated, the skin will be reddened and dry (Fig. 53–14). Pale skin color is a grave clinical sign. Baby birds are prone to septicemia, and it is not unusual for bacteria to start in the alimentary tract and spread rapidly to many organs. Highly virulent organisms may spread so rapidly that the bird dies without clinical signs, and little pathology is noted at gross necropsy examination. Less virulent organisms produce clinical signs that depend on the organs they infect.

Respiratory tract infections are relatively uncommon and are usually secondary to food aspiration or lung involvement from a septicemia. Diagnosis is made from the clinical signs and radiographs. Rhinitis and sinusitis are commonly caused by food that is inhaled into the nasal cavity through the choanal slit. The food should be flushed out of the nares with a small amount of saline or picked out with a blunt needle. Infections of internal organs are usually the result of a septicemia rather than a specific organ disease.

The same drugs used to treat adult birds are used for nestlings (see Chapter 28, Symptomatic Therapy and Emergency Medicine) with the

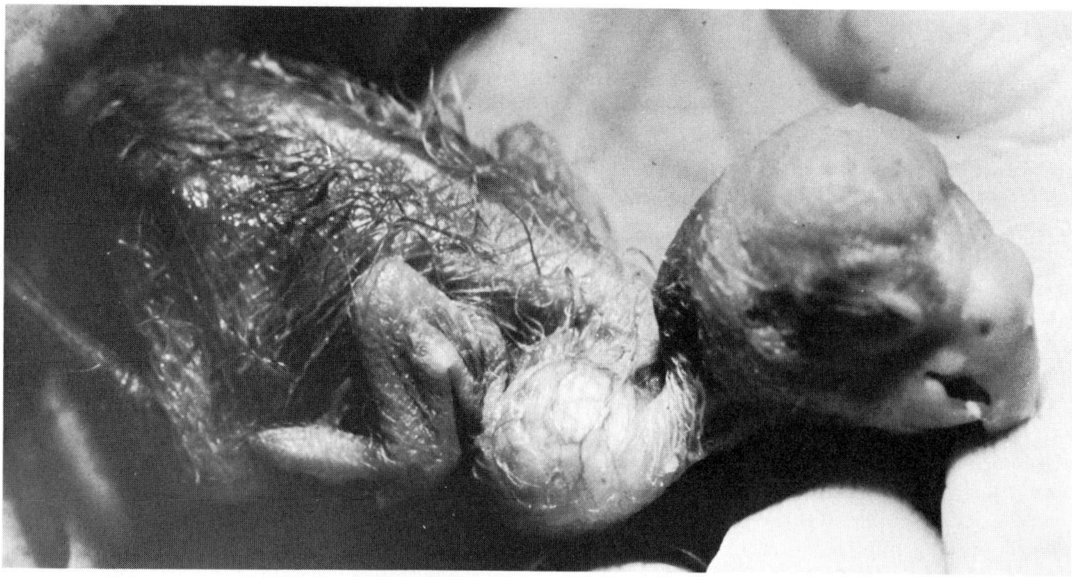

**Figure 53–14.** Signs of dehydration, as noted in this neonatal Rose-breasted Cockatoo, include dark congestion of the vessels of the crop, a "sunburned" look to the body, wrinkling of the skin, crop stasis, and weakness. (Courtesy of Greg J. Harrison.)

following recommendations. Aminoglycoside antibiotics cause severe polyuria in young birds and should be used only for short periods of time and only if there is no alternative drug. Since baby birds are accustomed to being hand fed, oral drugs can easily be delivered as often as necessary in the feeding formula. Chloramphenicol palmitate and trimethoprim-sulfamethoxazole (Bactrim) will successfully eliminate most enteric infections caused by susceptible gram-negative bacteria. The latter drug will occasionally cause regurgitation and should therefore be fed with a small amount of food. *Pseudomonas* is best treated with a combination of an aminoglycoside and a penicillin or cephalosporin derivative. Combinations such as gentamicin or tobramycin and carbenicillin or cefotaxime have synergistic activity against *Pseudomonas* in people and have proved efficacious in birds.

## MYCOTIC DISEASES

Yeasts most commonly infect the alimentary tract of neonates, where they cause malnutrition by interfering with motility and the absorption of nutrients. In debilitated birds, yeasts may spread to internal organs such as the liver and lung. *Candida* is the genus most frequently isolated and is particularly common in warm, humid environments. Signs of candidiasis are nonspecific and include poor weight gain, crop stasis, general debilitation, and diarrhea. In severe cases, whitish or brownish plaques are seen in the mouth, esophagus, or crop. Brownish deposits at the corners of the mouth may also be due to *Candida*. Yeasts commonly reside in the gut as opportunistic organisms and cause disease following a stressful episode or treatment with a broad-spectrum antibiotic that destroys competing bacterial flora. Birds may also become infected when fed spoiled food or contaminated sprouted seed.

Yeast infections are diagnosed by Gram's staining or culturing material from the crop and cloaca. The same agents used to treat adult birds (nystatin, fluorocytosine, ketoconazole, and topical amphotericin B) can be safely used in baby birds. The incidence of yeast infections can be reduced if good hygiene is practiced and the ingredients of the hand-feeding diet are selected with care. If a nursery has a history of candidiasis, prophylactic nystatin therapy should accompany the use of broad-spectrum antibiotics.

Young birds are susceptible to mold infections of the respiratory system, which usually occur when spores build up in nest material. Spores can also build up in an unsanitary brooder, leading to the classic brooder pneumonia of poultry. *Aspergillus fumigatus* and *Aspergillus flavus* are the most common isolates, but even normally innocuous fungi such as *Mucor*, *Rhizopus*, and *Penicillium* species, can cause disease if spores supermultiply in damp nest material, or if the chick is compromised by another disease. Diagnosis and treatment of mycotic infections are described in Chapter 36, Mycotic Diseases. Disease prevention is important, as mycotic therapy in young birds is seldom successful.

## VIRAL DISEASES

Avian pox and papovavirus are the viral diseases of greatest importance in young birds. Pox virus most frequently affects Amazon and Pionus Parrots recently imported from South and Central America.[4] Papovavirus is the cause of budgerigar fledgling disease (BFD) and a similar condition that occurs in other psittacines. Signs of disease include abdominal distention, lack of down feathers on the back and abdomen, hydropericardium, enlarged heart, enlarged liver with multifocal white or yellow spots, pale or congested kidneys, congested lungs, and ascites.[2, 7] Lateral spread in a nursery may cause high mortality[3] (see Chapter 32, Viral Diseases).

## CARE OF JUVENILE WILD BIRDS

Veterinarians are often asked to care for wild baby birds found by clients. In most cases it is best to refer these cases to volunteer wildlife rehabilitation centers found in most cities or to call the State Fish and Game Department. Some simple techniques may be necessary for maintaining birds until the proper help can be obtained.

If at all possible, baby birds that are naked or partially feathered should be returned to their nest. The parents will not be daunted by the touch of human hands. Instruct clients not to pick up young, fully feathered birds found on the ground or in low bushes. Most likely these are recent fledglings that are being cared for on the ground by their parents. If the bird is weak, it should be fed and then put back where it was found so that the parents can reclaim it.[6] If necessary, it can be housed at night and put out again in the morning.

If it becomes necessary to hand raise a wild bird, it must first be identified and its feeding requirements determined. Most altricial birds can be fed the same diet given to neonatal psittacines. The mixture can be waved in front of the baby's beak. Most passerines will gape and the food can be forced down their throat. The birds will stop gaping when full. Small babies should be fed every one to two hours from dawn to dusk or a similar period of time. The same environmental conditions recommended for parrots are suitable for most passerines. If an incubator is not available, warmth can be provided with a 60-watt light bulb, heat lamp, or heating pad. As the chick weans, feeding will become more difficult and a shallow pan with the feeding formula and solid food should be offered.

Precocial birds such as poultry and pheasants are easier to care for, as they will eat on their own.[8] Place the bird in a box and provide heat with a 60-watt light bulb protected by a wire screen. Commercial poultry starter food moistened with a small amount of water makes a good diet. Sometimes scratching the food in front of the chicks or placing a few meal worms on top is necessary to get the birds started. A shallow dish of water is also necessary.

Common waterfowl such as ducks and geese can be reared in the same manner as poultry. These birds should not be allowed to swim, as they are not waterproof until fully feathered. Coots and grebes are normally fed by their parents and are especially difficult to hand raise. Their care is best left to experienced rehabilitators, as is the care of raptors, which are protected by law in all states. One of the most common problems encountered when raptor chicks are tended by inexperienced people is nutritional secondary hyperparathyroidism as a result of a calcium-deficient, all-meat diet. Raptor chicks should be fed whole chicks or mice (chopped if necessary), dusted with a calcium and vitamin supplement.

Successfully returning birds to the wild is often more difficult than rearing them and is best left to experienced rehabilitators.

## REFERENCES


1. Adler, H. E., and DaMassa, A. J.: Effect of ingested lactobacilli on *Salmonella infantitis*, and *Escherichia coli* on intestinal flora, pasted vents, and chick growth. Avian Dis., 22:868–878, 1975.
2. Bernier, G., et al.: A generalized inclusion body disease in the budgerigar (*Melopsittacus undulatus*) caused by a papova-like agent. Avian Dis., 25:1083–1093, 1981.
3. Clubb, S. L.: Outbreak of a papova-like virus infection in a psittacine nursery. Proceedings of the International Conference on Avian Medicine, sponsored by the Association of Avian Veterinarians, 1984, pp. 212–230.
4. Clubb, S. L.: Recent trends in the diseases of imported birds. Proc. of the Jean Delacour/IFCB Symposium on Breeding Birds in Captivity, Universal City, CA, pp. 63–72, 1983.

5. Clubb, S. L., and Clubb, K. J.: Psittacine pediatrics. Proceedings of the American Federation of Aviculture Veterinary Seminar, 1984, pp. 17–33.
6. Cooper, J. E., and Eley, J. T.: First Aid and Care of Wild Birds. North Pomfret, VT, David and Charles, Inc., 1979.
7. Davis, R. B., et al.: A viral disease of fledgling budgerigars. Avian Dis., 25:179–183, 1981.
8. Detrick, J. F., and Raff, M. I.: Husbandry of captive wild birds. In Kirk, R. W. (ed.): Current Veterinary Therapy, Philadelphia, W. B. Saunders Company, 1977.
9. Drewes, L., and Flammer, K.: Preliminary data on aerobic microflora of baby psittacine birds. Proceedings of the Jean Delacour/IFCB Symposium on Breeding Birds in Captivity. Universal City, CA, 1983, pp. 73–81.
10. Flammer, K., and Drewes, L.: Environmental sources of gram-negative bacteria in an exotic bird farm. Proceedings of the Jean Delacour/IFCB Symposium on Breeding Birds in Captivity. Universal City, CA, 1983, pp. 83–93.
11. Grau, C. R., and Roudybush, T. E.: Protein requirements of growing cockatiels. Proceedings of the 34th Western Poultry Disease Conference, University of California, Davis, 1985.
12. Harrison, G. J.: Guidelines for the treatment of neonatal psittacines. Proceedings of the Annual Meeting of the American Association of Zoo Veterinarians, Tampa, FL, 1983.
13. Langenberg, J., and Montali, R. T.: Avian neonatal pathology. Proceedings of the Annual Meeting of the American Association of Zoo Veterinarians, Tampa, FL, 1983.
14. Nearenberg, D. S., et al.: Hand-fed vs. parent-fed cockatiels: A comparison of chick growth. Proceedings of the 34th Western Poultry Disease Conference, University of California, Davis, 1985.
15. Rantala, M., and Nurmi, E.: Prevention of *Salmonella infantis* in chicks by the flora of the alimentary tract of chickens. Br. Poultry Sci., 14:627–630, 1973.
16. Roudybush, T. E.: Early detection of hand-feeding problems. Proceedings of the American Federation of Aviculture Veterinary Seminar, 1984, pp. 8–16.
17. Roudybush, T. E., and Grau, C. R.: Solids in the diets for hand raising cockatiels. Proceedings of the 32nd Western Poultry Disease Conference, University of California, Davis, 1983, pp. 94–95.
17a. Roudybush, T. E., and Grau, C. R.: Lysine requirement of cockatiel chicks. Proceedings of the 34th Western Poultry Disease Conference, Davis, CA, 1985, pp. 113–115.
18. Snoyenbos, G. H., et al.: Protecting chicks and poults from *Salmonella* by oral administration of "normal" gut microflora. Avian Dis., 25:696–705, 1978.
19. Toone, C. K.: Causes of embryonic malformations and mortality. Proceedings of the Annual Meeting of the American Association of Zoo Veterinarians, Tampa, FL, 1983.
20. Weinack, O. M., et al.: Reciprocal competitive exclusion of *Salmonella* and *Escherichia coli* by native intestinal microflora of the chicken and turkey. Avian Dis., 26:585–595, 1982.
21. Woolcott, J. B.: Bacterial Infections and Immunity in Domestic Animals. New York, Elsevier Scientific Publishing Company, 1979.

# *Appendices*

# *Appendix 1.*

# COMMON AND SCIENTIFIC NAMES OF MOST FREQUENTLY KEPT BIRDS

Depending on the country, the common usage names of cage and aviary birds may vary; thus, some knowledge of the scientific names is useful. The following list is presented with the common name in the United States given first, followed by alternate names, and the scientific name.

## PSITTACINES

### Amazon Parrot

Blue-fronted [*Amazona aestiva aestiva*]
Double Yellow-headed (Yellow-headed) [*Amazona ochrocephala oratrix*]
Festive [*Amazona festiva festiva*]
Hispanolian [*Amazona ventralis*]
Lilac-crowned (Finsch's) [*Amazona finschi*]
Mealy [*Amazona farinosa farinosa*]
Mexican Red-headed (Green-cheeked) [*Amazona viridigenalis*]
Orange-winged [*Amazona amazonica amazonica*]
Panama [*Amazona ochrocephala panamensis*]
Red-lored (Orange-cheeked, Yellow-cheeked) [*Amazona autumnalis autumnalis*]
Spectacled [*Amazona albifrons albifrons*]
Yellow-crowned (Yellow-fronted) [*Amazona ochrocephala ochrocephala*]
Yellow-naped [*Amazona ochrocephala auropalliata*]

### Budgerigar (Parakeet, Budgie) [*Melopsittacus undulatus*]

### Caique Parrot

Black-headed [*Pionites melanocephala melanocephala*]
White-bellied [*Pionites leucogaster leucogaster*]

### Cockatiel [*Nymphicus hollandicus*]

### Cockatoo

Bare-eyed (Little Corella) [*Cacatua sanguinea sanguinea*]
Citron-crested [*Cacatua sulphurea cintrinocristata*]
Eleonora [*Cacatua galerita eleonora*]
Goffin's [*Cacatua goffini*]
Lesser Sulphur-crested (Medium Sulphur-crested) [*Cacatua sulphurea sulphurea*]
Major Mitchell's (Leadbeater) [*Cacatua leadbeateri*]
Moluccan (Salmon-crested) [*Cacatua moluccensis*]
Red-vented (Philippine) [*Cacatua haematuropygia haematuropygius*]
Rose-breasted (Galah, Roseate) [*Cacatua roseicapilla roseicapilla*]
Sulphur-crested [*Cacatua galerita galerita*]
Triton [*Cacatua galerita triton*]
Umbrella (White, White-crested) [*Cacatua alba*]

### Conure

Blue-crowned [*Aratinga acuticaudata*]
Cherry-headed (Red-headed) [*Aratinga erythrogenys*]
Crimson-bellied [*Pyrrhura rhodogaster*]
Dusky (Dusky-headed) [*Aratinga weddellii*]
Finsch's [*Aratinga finschi*]
Gold-capped (Golden-capped) [*Aratinga auricapilla*]
Gold-crowned [*Aratinga aurea*]
Green-cheeked [*Pyrrhura molinae*]
Halfmoon [*Aratinga canicularis eburniostrum*]
Jenday (Jandaya) [*Aratinga jandaya*]
Maroon-bellied (Red-bellied) [*Pyrrhura frontalis*]
Mitred [*Aratinga mitrata mitrata*]
Nanday [*Nandayus nenday*]
Orange-fronted (Peach-fronted) [*Aratinga canicularis*]
Painted [*Pyrrhura picta*]
Patagonian (Lesser Patagonian) [*Cyanoliseus patagonus patagonus*]

Queen of Bavaria (Golden) [*Aratinga guarouba*]
Red-fronted (Wagler's) [*Aratinga wagleri wagleri*]
Slender-billed [*Enicognathus leptorhynchus*]
Sun (Yellow) [*Aratinga solstitialis*]
White-eyed [*Aratinga leucophthalmus*]

### Lory

Black-capped [*Domicella lory*]
Chattering [*Domicella garrula garrula*]
Dusky [*Pseudeos fuscata*]
Ornate [*Trichoglossus ornatus*]
Rainbow [*Trichoglossus haematodus*]
Red (Moluccan) [*Eos bornea*]
Violet-naped [*Eos squamata*]
Yellow-streaked [*Chalcopsitta sintillata*]

### Lovebird

Black-winged [*Agapornis taranta*]
Fischer's [*Agapornis fischeri*]
Masked [*Agapornis personata*]
Nyasa [*Agapornis lilianae*]
Peach-faced [*Agapornis roseicollis*]

### Macaw

Blue and Gold (Blue and Yellow) [*Ara ararauna*]
Buffon [*Ara ambigua*]
Green-winged [*Ara chloroptera*]
Hahns (Red-shouldered) [*Ara nobilis nobilis*]
Hyacinth [*Anodorhynchus hyacinthinus*]
Illiger's [*Ara maracana*]
Military [*Ara militaris*]
Noble [*Ara nobilis cumanensis*]
Red-fronted [*Ara rubrogenys*]
Red-bellied [*Ara manilata*]
Scarlet [*Ara macao*]
Severe (Chestnut-fronted) [*Ara severa*]
Yellow-collared (Yellow-naped) [*Ara auricollis*]

### Parakeet

African Ring-necked [*Psittacula krameri krameri*]
Alexander Ring-necked (Alexandrine) [*Psittacula eupatria nipalensis*]
Canary-winged (Bee Bee) [*Brotogeris versicolorus*]
Grey-cheeked [*Brotogeris pyrrhopterus*]
Indian Ring-necked (Rose-ringed) [*Psittacula krameri manillensis*]
Monk (Quaker) [*Myiopsitta monachus*]
Moustache (Moustached) [*Psittacula alexandri fasciata*]
Plum-headed [*Psittacula cyanocephala rosa*]

### Parrot

African Grey (Grey, Congo, Guana) [*Psittacus erithacus erithacus*]
Amboina King [*Alisterus amboinensis amboinensis*]
Australian King (King) [*Alisterus scapularis scapularis*]
Barraband's [*Polytelis swainsonii*]
Blue-headed (Red-vented) [*Pionus menstruus*]
Blue-winged [*Neophema chrysostoma*]
Bourke's [*Neophema bourkii*]
Dusky (Violet) [*Pionus fuscus*]
Eclectus [*Eclectus roratus*]
Elegant [*Neophema elegans*]
Grand Eclectus (Sacred Temple) [*Lorius roratus roratus*]
Hawk-headed (Guiana) [*Deroptyus accipitrinus accipitrinus*]
Many color [*Psephotus varius*]
Maximilian's [*Pionus maximiliani maximiliani*]
Meyer's (Brown) [*Poicephalus meyeri meyeri*]
Princess (Princess of Wales, Princess Alexandra) [*Polytelis alexandrae*]
Red-rumped [*Psephotus haematonotus*]
Red-winged [*Aprosmictus erythropterus*]
Scarlet-chested [*Neophema splendida*]
Senegal (Yellow-bellied) [*Piocephalus senegalus senegalus*]
Timneh (African Grey) [*Psittacus erithacus timneh*]
Turquoisine (Turquoise) [*Neophema pulchella*]
White-crowned (White-capped) [*Pionus senilis*]

### Rosella

Adelaide [*Platycercus elegans adelaidae*]
Crimson [*Platycercus elegans elegans*]
Golden Mantle (Eastern) [*Platycercus eximius cecilae*]
Stanley (Western) [*Platycercus icterotis icterotis*]
Pale-headed [*Platycercus adscitus*]

## PASSERIFORMES

### Canary [*Serinus canarius*]

### Finch

Cordon bleu [*Uraeginthus bengalus*]
Lady Gouldian [*Poephila gouldiae gouldiae*]
Society (Bengalese) [*Lonchura domestica*]
Zebra [*Poephila castanotis*]

### Mynah

Common Mynah [*Acridotheres tristis*]
Greater Indian Hill Mynah [*Gracula religiosa*]
Lesser Indian Hill Mynah [*Gracula religiosa indica*]

## PICIFORMES

### Toucans

Sulfur-breasted Toucan [*Ramphastos sulfuratus sulfuratus*]
Toco Toucan [*Ramphastos toco toco*]

### REFERENCES

Bates, H., and Busenbark, R.: Parrots and Related Birds. Neptune City, NJ, T.F.H. Publications, Inc., 1969.
Forshaw, J.M., and Cooper, W.T.: Parrots of the World. Garden City, NY, Doubleday & Company, Inc., 1973.
Low, R.: Parrots: Their Care and Breeding. Poole, United Kingdom, Blandford Books Ltd., 1980.

*Appendix 2.*

# DISEASES OF IMPORTED BIRDS AS RELATED TO COUNTRY OF ORIGIN AND SPECIES

Susan L. Clubb

| | Countries of Origin | Highly Susceptible Species or Conditions | Occasional Susceptibility |
|---|---|---|---|
| Chlamydiosis | Mexico<br>Central America<br>South Africa (captive-raised)<br>Argentina/Paraguay<br>Indonesia | Amazons<br>Young birds<br>Cockatiels | Cockatoos<br>Macaws |
| Newcastle Disease (VVND) | Mexico/Central America<br>Indonesia<br>Bolivia<br>Peru<br>Argentina/Paraguay | Cockatoos<br>Cockatiels<br>Amazons<br>Budgerigars | Macaws<br>Conures |
| Parrot Pox | Bolivia<br>Guyana/Suriname<br>Argentina/Paraguay<br>Central America/Mexico | Blue-fronted Amazons<br>Blue-headed Pionus<br>Maximilians<br>Dusky Pionus | Other Amazon and Pionus species<br>Lovebirds<br>Rosellas<br>Australian parakeets<br>Conures<br>Young macaws |
| Canary Pox | Europe | Canaries | |
| Pacheco's Parrot Disease | Argentina/Paraguay<br>Bolivia<br>Guyana | Amazons<br>Macaws<br>Cockatoos<br>African Parrots<br>(Almost all psittacines) | (Carriers: Patagonian, Nanday, and other conures suspected) |
| Salmonellosis | Ghana/Tanzania<br>Guyana/Suriname<br>Taiwan (captive)<br>Bolivia<br>Indonesia<br>Peru | African Greys<br>Amazons<br>Conures<br>Finches<br>Eclectus | Macaws<br>Cockatoos |
| *Escherichia coli* and Other Enteric Bacteria | Countries where water quality and sanitation are poor<br>Bolivia<br>Taiwan (captive)<br>Peru<br>Mexico/Central America<br>South Africa (captive) | All species<br>Strains esp. pathogenic in cockatiels, finches, Amazons, budgerigars | Carriers, esp. macaws |
| Reovirus | Indonesia/Malaysia<br>Tanzania/Ghana | Cockatoos<br>African Greys | |
| Adenovirus | South Africa (captive-raised) | Lovebirds | |
| Candidiasis | Bolivia<br>Paraguay/Argentina<br>Taiwan (captive)<br>South Africa (captive) | Amazons<br>Cockatiels<br>Lovebirds<br>Young macaws | Conures<br>Australian parakeets<br>Eclectus |
| Aspergillosis | Ghana/Tanzania<br>Bolivia<br>Paraguay<br>Guyana | African Greys<br>Mynahs<br>Post-pox birds | |

| | Countries of Origin | Highly Susceptible Species or Conditions | Occasional Susceptibility |
|---|---|---|---|
| Suspected Papovavirus | Guyana/Suriname | Conures may be carriers | |
| Suspected Mycoplasma | South Africa (captive)<br>Central America/Mexico | Cockatiels<br>Amazons | |
| Nonspecific Rhinitis and Sinusitis of Unknown Etiology | Guyana<br>Suriname<br>Central America<br>South Africa | Amazons | Budgerigars<br>Cockatoos<br>Macaws<br>Cockatiels |
| Tracheitis of Unknown Etiology | Central America<br>Mexico<br>Argentina | Amazons | |
| Pneumonia of Unknown Etiology | Guyana/Suriname<br>Indonesia/Malaysia | | |
| Collapsed Sinusitis of Unknown Etiology | Bolivia | Macaws | |
| Keratitis of Unknown Etiology | Central America/Mexico<br>Guyana/Suriname | Amazons<br>Conures | |
| Proventricular Hypertrophy of Unknown Etiology | Bolivia | Macaws, esp. Blue and Golds<br>Cockatoos | |
| Psittacine Beak and Feather Disease Syndrome | Indonesia/Malaysia<br>Phillipines | Cockatoos | |

**Parasites**

| | | | |
|---|---|---|---|
| Coccidia | Central America<br>Mexico<br>Argentina<br>Peru<br>Taiwan (captive) | Toucans<br>Mynahs<br>Budgerigars<br>Finches<br>Lories<br>Pigeons<br>Amazons | |
| Ascarids | South Africa<br>Guyana<br>Suriname | Captive-raised<br>Cockatiels and conures<br>Amazons | |
| Capillaria | | Macaws<br>Pigeons<br>Gallinaceous birds | |
| Tape Worms | Tanzania/Ghana<br>Indonesia/Malaysia | Australian parakeets (captive-raised)<br>Lories<br>Cockatoos<br>Eclectus<br>African Greys | |
| Lice | Bolivia<br>Guyana/Suriname | Macaws (may have resistant species) | |
| Air Sac Mites | Guyana/Suriname<br>Indonesia/Malaysia<br>Taiwan (captive) | Conures<br>Lories<br>Finches<br>Canaries | |
| Giardia | South Africa (captive-raised) | Cockatiels | |
| Haemoproteus and Microfilaria | | Cockatoos (asymptomatic) | Lories<br>Fig Parrots<br>Other Asian parrots<br>African Greys |
| Trypanosomes | Bolivia | Macaws | |

*Appendix 3.*

# REFERENCE VALUES OF CLINICAL CHEMISTRY TESTS

The tables in this appendix are reference values for avian species that have been reported by a number of authors. There may be some variation in the techniques used to obtain these values.

## 1. TABLES OF HEMATOLOGIC AND BLOOD CHEMISTRY NORMAL VALUES*

| | Table A | Table B | Table C | Table D | Table E |
|---|---|---|---|---|---|
| WBC (× 10³/cu mm) | 5–11 | 6–11 | 4–11 | 3–8 | 4.5–9.5 |
| Differential | | | | | |
| Heterophils (%) | 45–75 | 30–75 | 40–70 | 45–70 | 40–75 |
| Lymphocytes (%) | 20–50 | 20–65 | 20–50 | 20–45 | 20–60 |
| Monocytes (%) | 0–3 | 0–3 | 0–2 | 0–5 | 0–3 |
| Eosinophils (%) | 0–2 | 0–1 | 0–1 | 0–1 | 0–1 |
| Basophils (%) | 0–5 | 0–5 | 0–5 | 0–5 | 0–5 |
| PCV (%) | 43–55 | 45–55 | 44–60 | 45–57 | 46–58 |
| RBC (× 10⁶/cu mm) | 2.4–4.5 | 2.5–4.5 | 2.4–4.1 | 2.5–4.5 | |
| Total protein (gm/dl) | 3.0–5.0 | 3.0–5.0 | 2.6–5.0 | 2.5–4.5 | 2.5–4.5 |
| Glucose (mg/dl) | 190–350 | 220–350 | 180–300 | 200–400 | 200–350 |
| Calcium (mg/dl) | 8.0–13.0 | 8.0–13.0 | 10.0–15.0 | | |
| SGOT (IU/L) | 100–350 | 130–350 | 150–350 | 150–350 | 150–400 |
| LDH (IU/L) | 150–450 | 160–420 | 200–550 | 150–450 | 150–450 |
| Creatinine (mg/dl) | 0.1–0.4 | 0.1–0.4 | 0.1–0.3 | 0.1–0.4 | 0.1–0.4 |
| Uric acid (mg/dl) | 4.0–10.0 | 2.0–10.0 | 4.0–12.0 | 4.0–14.0 | 4.0–12.0 |
| Potassium (mEq/L) | 2.6–4.2 | 3.0–4.5 | 3.0–4.5 | | |
| Sodium (mEq/L) | 134–152 | 136–152 | 130–150 | | |
| Thyroxine (μg/dl) | 0.3–2.0 | 0.05–1.0 | 0.20–1.1 | 2.5–4.4 | 0.2–2.4 |

| | Table F | Table G | Table H | Table I | Table J |
|---|---|---|---|---|---|
| WBC (× 10³/cu mm) | 4–9 | 5–10 | 5–11 | 4–11 | 4.5–13.0 |
| Differential | | | | | |
| Heterophils (%) | 20–50 | 40–70 | 45–75 | 40–75 | 30–70 |
| Lymphocytes (%) | 40–75 | 25–55 | 20–50 | 20–50 | 20–65 |
| Monocytes (%) | 0–1 | 0–2 | 0–4 | 0–3 | 0–3 |
| Eosinophils (%) | 0–1 | 0–2 | 0–2 | 0–3 | 0–4 |
| Basophils (%) | 0–5 | 0–6 | 0–5 | 0–5 | 0–5 |
| PCV (%) | 45–60 | 45–57 | 40–55 | 42–55 | 30–43 |
| RBC (× 10⁶/cu mm) | 2.5–4.5 | 2.5–4.7 | 2.2–4.5 | 2.5–4.5 | 2.3–3.5 |
| Total protein (gm/dl) | 3.0–5.0 | 2.2–5.0 | 2.5–5.0 | 2.5–4.5 | 2.5–6.0 |
| Glucose (mg/dl) | 200–450 | 200–450 | 190–350 | 200–350 | 150–300 |
| Calcium (mg/dl) | | 8.5–13.0 | 8.0–13.0 | 8.0–15.0 | 10.0–18.0 |
| SGOT (IU/L) | 150–350 | 100–350 | 150–350 | 125–350 | 5–100 |
| LDH (IU/L) | | 125–450 | 225–650 | 125–420 | 150–800 |
| Creatinine (mg/dl) | | 0.1–0.4 | 0.1–0.4 | 0.1–0.5 | 0.1–0.5 |
| Uric acid (mg/dl) | 4.0–12.0 | 3.5–11.0 | 3.5–11.0 | 2.5–10.5 | 2.0–12.0 |
| Potassium (mEq/L) | | 2.5–4.5 | 2.5–4.5 | 3.4–5.0 | 3.0–4.5 |
| Sodium (mEq/L) | | 132–150 | 131–157 | 134–148 | 130–155 |
| Thyroxine (μg/dl) | 0.7–3.4 | 0.7–2.4 | 0.8–4.4 | 0.25–0.9 | 0.8–3.3 |

Table A: African Grey Parrots *(Psittacus erithacus erithacus* and *Psittacus erithacus timneh)* (Sample size: 108)
Table B: Amazon Parrot *(Amazona)* (640)
Table C: Blue-headed Parrots *(Pionus menstruus)* (16)
Table D: Budgerigars *(Melopsittacus undulatus)* (251)
Table E: Grey-cheeked Parakeets *(Brotogeris pyrrhopterus)* (32) and Canary-winged Parakeets *(Brotogeris versicolorus)* (11)
Table F: Canaries *(Serinus canaria)* (62)
Table G: Cockatiels *(Nymphicus hollandicus)* (364)
Table H: Cockatoos (242)
Table I: Conures *(Aratinga* sp.) (85)
Table J: Domestic ducks (31)

*From Woerpel, R. W., and Rosskopf, W. J.: Clinical experience with avian laboratory diagnostics. Vet. Clin. North Am., 14(2): 254, 1984.

## 2. BLOOD CHEMISTRIES OF BLUE AND GOLD MACAWS*

| | | Range | Mean |
|---|---|---|---|
| Sodium | mEq/L | 138–153 | 145.7 |
| Potassium | mEq/L | 5.0–10.4 | 7.4 |
| Chloride | mEq/L | 103–110 | 106.5 |
| Calcium | mg/dl | 8.8–12.3 | 11.1 |
| Ionized Ca | mg/dl | 4.6–6.2 | 5.4 |
| Phosphorus | mg/dl | 1.9–2.6 | 2.3 |
| Total Proteins | g/dl | 4.3–5.6 | 4.95 |
| Albumin | g/dl | 1.9–2.6 | 2.25 |
| Globulin | g/dl | 2.1–3 | 2.55 |
| A/G | | 0.6–0.9 | 0.75 |
| Iron | μg/dl | 79–135 | 107 |
| Glucose | mg/dl | 286–332 | 309 |
| BUN | mg/dl | 1–5 | 3 |
| Creatinine | mg/dl | 0.3–0.5 | 0.4 |
| BUN/Creatinine | | 3–13 | 8 |
| Uric Acid | mg/dl | 4–10.1 | 7 |
| Cholesterol | mg/dl | 139–202 | 170 |
| Triglycerides | mg/dl | 33–170 | 101 |
| Total Bilirubin | mg/dl | 0.1–0.2 | 0.15 |
| Alk. Phos. | U/L | 162–580 | 371 |
| LDH | U/L | 183–664 | 423.5 |
| SGOT | U/L | 197–297 | 247 |
| SGPT | U/L | 99–263 | 186 |

*From Raphael, B. L.: Hematology and blood chemistries of macaws. Annual Proceedings of the American Association of Zoo Veterinarians, Washington, DC, 1980.

## 3. HEMATOLOGIC, SALT, AND SALP VALUES IN SIX CLINICALLY HEALTHY QUAKER PARROTS*

| Test | Mean | Standard Deviation | Range |
|---|---|---|---|
| RBC ($10^6$)/μl | 3.35 | 0.27 | 2.81–3.89 |
| PCV (%) | 50 | 10 | 30–70 |
| Hb (gm/dl) | 14.7 | 1.3 | 12.1–17.3 |
| MCV (fl) | 150 | 32 | 86–214 |
| MCH (pg) | 44 | 5 | 34–54 |
| MCHC (gm/dl) | 30 | 4 | 22–38 |
| TP (gm/dl) | 4.4 | 0.3 | 3.8–5.0 |
| WBC/μl | 5958 | 2367 | 1224–10,692 |
| Heterophils % | 11.5 | 6.4 | 0–24.3 |
| /μl | 808 | 192 | 424–1192 |
| Lymphocytes % | 82 | 4 | 74–90 |
| /μl | 6410 | 2474 | 1462–11,358 |
| Monocytes % | 2.5 | 0.7 | 1.0–4.0 |
| /μl | 203 | 120 | 0–443 |
| Eosinophils % | 0.5 | 0.7 | 0–1.9 |
| /μl | 30 | 42 | 0–114 |
| Basophils % | 1.5 | 2.1 | 0–5.7 |
| /μl | 144 | 203 | 0–550 |
| Azurophils % | 2 | 1 | 0–4 |
| /μl | 155 | 52 | 51–259 |
| ALT (IU/L) | 7 | 7 | 0–21 |
| ALP (IU/L) | 521 | 151 | 219–823 |

*From Goodwin, J. S., Jacobson, E. R., and Gaskin, J. M.: Effects of Pacheco's Parrot disease virus on hematologic and blood chemistry values of Quaker Parrots (*Myopsitta monachus*). J. Zoo Anim. Med., 13:127–132, 1982.

## 4. DIAGNOSES TO CONSIDER FROM HEMATOLOGIC RESULTS

Alan M. Fudge

| | Psittacosis | | Bacterial | |
|---|---|---|---|---|
| | **Subclinical** | **Active** | **Pharyngitis** | **Sepsis** |
| **White Blood Cells** | | | | |
| Normal | Yes | With prev. drug TX | Usually | No |
| Decreased | No | No | No | When Severe |
| Leukopenia | No | No | No | Critically Ill |
| Increased | Sometimes | Often | Sometimes | Early Stages |
| >25,000 | No | Often in Lg. Psitt. | Pseudomonas | Salmonella |
| >40,000 | No | | Rare | |
| **Packed Cell Volume** | | | | |
| Increased (dehydration) | No | Sometimes | Sometimes | Sometimes |
| Decreased (dep. anemia) | No | Often in Lg. Psitt. | Rare | Often |
| Decreased (reg. anemia) | Sometimes | With TX | Rare | With TX |
| Protein Increased | No | Chronic Infl. Dis. | Rare | Chronic Infl. Dis. |
| Protein Decreased | No | Wasting & Liver Dis. | Rare | Sometimes |
| **Heterophilia** | Sometimes | Common | Common | Common |
| **Left Shift** | Sometimes | Common | Rare | Common |
| **Lymphocytosis (relative)** | No | Rare | Rare | Rare |
| **Reac. Lymphos.** | No | Sometimes | No | Sometimes |
| **Monocytosis (rel. or abs.)** | Yes | Often | Sometimes | Sometimes |
| **Eosinophilia (relative)** | No | No | No | No |
| **Basophilia (relative)** | Often | Often | Sometimes | Sometimes |
| **Reac. Thromb.** | No | Sometimes | No | Sometimes |
| **Aniso/Poly** | No | Sometimes | No | Sometimes |

## 5. THE DETERMINATION OF SEVERAL ENZYMES OF BLOOD PLASMA-SERUM IN DIFFERENT BIRD SPECIES (DATA IN IU/L)*

| Species | GOT | GPT | LDH | AP | SP |
|---|---|---|---|---|---|
| Green-cheeked Amazon | 107.96<br>(±22.66) | 9.15<br>(±2.34) | 266.54<br>(±75.03) | 122.30<br>(±51.68) | 4.68<br>(±1.76) |
| Blue-fronted Amazon | 130.48<br>(±19.75) | 15.93<br>(±4.30) | 244.42<br>(±76.32) | 129.48<br>(±24.48) | 8.75<br>(±3.33) |
| African Grey | 63.23<br>(±14.46) | 9.77<br>(±3.33) | 209.19<br>(±33.66) | 47.60<br>(±14.12) | 3.51<br>(±1.26) |
| Budgerigar | 101.89<br>(±15.20) | 14.96<br>(±4.48) | 104.70<br>(±28.43) | 194.12<br>(±62.92) | |
| Homing Pigeon (Male) | 46.01<br>(±10.94) | 18.12<br>(±3.51) | 65.51<br>(±27.92) | 196.48<br>(±56.35) | 32.24<br>(±11.94) |
| Homing Pigeon (Female) | 33.29<br>(±11.44) | 16.95<br>(±2.37) | 105.30<br>(±60.32) | 225.07<br>(±98.96) | 24.42<br>(±3.87) |

*From Baron, H. W.: Die Aktivitätsmessung einiger Enzyme in Blutplasma bzw.–serum verschiedener Vogelspezies. University of München, 1980.

| **Granulomatous (T.B., *Aspergillus*)** | **Parasitism (*Giardia,* Tapes, RW)** | **Hemoparasites (Of RBC or WBC)** | **Lead Toxicosis** | **Viral: Active** |
|---|---|---|---|---|
| No | Usually | Usually, except *Leucocytozoon* | Yes, or Sl. Elevation | No |
| When Severe | Sometimes | No | No | Often |
| No | With Sepsis | No | No | Often |
| Yes | With Concurrent Infections | *Leucocytozoon* | Sometimes | Not Common |
| Yes | With Concurrent Infections | *Leucocytozoon* | No | Rare |
| Yes | Rare | *Leucocytozoon* | No | No |
| Sometimes | Sometimes | No | No | Sometimes |
| Often | Sometimes | Sometimes | Early | Sometimes |
| With TX | With TX | With TX | Common with RBC Ballooning and Chromatin Clump | With Recovery |
| Common | Rare | Rare | Dehydration | Dehydration |
| Sometimes | Common | Rare | Rare | Rare |
| Common | Rare | With *Leucocytozoon* | Moderate | Rare |
| Common | Rare | With *Leucocytozoon* | Rare | No |
| Rare | Rare | Fairly Common | Rare | Common |
| No | Rare | Common | No | Common |
| Often Elevated | Sometimes | Sometimes | No | Rare |
| No | Often | Often | No | No |
| Sometimes | Sometimes | Sometimes | Sometimes | No |
| Sometimes | No | Sometimes | No | Sometimes |
| Often | No | Often | Often | Sometimes |

From: California Avian Laboratory Newsletter No. 2, 1985.

*Appendix 4.*

# SURVEY WEIGHTS OF COMMON AVICULTURE BIRDS*

Keven Flammer

| Species (No. Weighed) | Average Weight (SD) | Range |
|---|---|---|
| Scarlet Macaw (57) | 1001 (151) | 800–1659 |
| Blue and Gold Macaw (43) | 1039 (102) | 889–1313 |
| Green-winged Macaw (28) | 1194 (91) | 1035–1576 |
| Military Macaw (24) | 925 (57) | 807–1030 |
| Hyacinth Macaw (18) | 1355 (95) | 1197–1466 |
| Great Green Macaw (14) | 1290 (109) | 1186–1594 |
| Red-fronted Macaw (12) | 490 (42) | 410–556 |
| Caninde Macaw (2) | 752 (42) | 741–762 |
| Rose-breasted Cockatoo (29) | 403 (57) | 317–515 |
| Umbrella Cockatoo (21) | 577 (71) | 481–713 |
| Leadbeater Cockatoo (19) | 423 (41) | 381–474 |
| Moluccan Cockatoo (24) | 853 (97) | 658–1003 |
| Bare-eyed Cockatoo (13) | 375 (79) | 294–513 |
| Triton Cockatoo (17) | 643 (78) | 515–781 |
| Citron Cockatoo (16) | 357 (34) | 297–423 |
| Medium Sulphur-crested Cockatoo (4) | 465 (39) | 438–510 |
| Greater Sulphur-crested Cockatoo (7) | 843 (147) | 655–1105 |
| Red-vented Cockatoo (7) | 298 (25) | 263–329 |
| Gang Gang Cockatoo (4) | 293 (43) | 247–350 |
| Slender-billed Cockatoo (3) | 593 (88) | 513–687 |
| Queen of Bavaria Conure (8) | 262 (8) | 252–276 |
| Grand Eclectus Parrot (26) | 432 (40) | 347–512 |
| African Grey Parrot (25) | 554 (66) | 370–534 |
| Blue-crowned Amazon Parrot (8) | 740 (123) | 618–998 |
| Blue-fronted Amazon Parrot (9) | 432 (51) | 361–485 |
| Mexican Red-headed Parrot (2) | 360 (24) | 343–377 |
| Yellow-naped Amazon Parrot (20) | 596 (85) | 476–795 |
| Double Yellow-headed Amazon (24) | 568 (61) | 463–694 |
| Princess of Wales Parakeet (10) | 108 (8) | 102–129 |
| Red-capped Parakeet (4) | 111 (8) | 104–120 |
| Kakariki Parakeet (6) | 56 (10) | 35–43 |
| Red-rumped Parakeet (6) | 65 (4) | 62–69 |
| Bourkes Parakeet (9) | 40 (4) | 35–43 |
| Turquoisine Parakeet (6) | 35 (2) | 31–37 |
| Scarlet-chested Parakeet (9) | 38 (5) | 31–42 |
| Keel-bill Toucan (7) | 386 (34) | 317–420 |
| Ariel Toucan (2) | 423 (36) | 405–446 |
| Toco Toucan (2) | 746 (56) | 706–785 |

*All weights were taken from morphologically normal adult birds housed in an avicultural setting.

## AVERAGE WEIGHTS OF ADDITIONAL COMMON AVIAN SPECIES

Susan L. Clubb

| Species | Average Weight |
|---|---|
| Zebra or Society Finch | 10–16 gm |
| Canary | 12–30 gm |
| Budgerigar | 30 gm |
| Lovebird | 42–48 gm |
| Cockatiel | 75–85 gm |
| Conure | 80–100 gm |

# *Appendix 5.*

# AVERAGE BREEDING CHARACTERISTICS OF SOME COMMON PSITTACINE SPECIES

Keven Flammer

| | Breeding Season | Eggs per Clutch/ Offspring | Incubation Period (days) | Fledging Age (days)* | Weaning Age Parent-Raised (days)† | Weaning Age Hand-Reared (days)‡ | Breeding Age§ |
|---|---|---|---|---|---|---|---|
| Budgerigar | All year | 4–9/4–8 | 16 | 22–26 | 30–40 | 30 | 6 mo. |
| Cockatiel | All year | 3–7/4–5 | 18 | 32–38 | 47–52 | 42–49 | 6 mo. |
| Australian Parakeet | Spring | 3–5/3–5 | 18–19 | 30–45 | 50–65 | NA | 1–3 yr. |
| Princess Parakeet | Spring/late spring/summer | 3–5/3–5 | 18 | 32–38 | 50–55 | NA | 1 yr. |
| Ring-neck Parakeet | Early summer | 3–5/4 | 23–24 | 40–45 | 55–65 | NA | 3 yr. |
| Lovebirds | All year | 2–6/3–5 | 18 | 30–35 | 45–55 | 40–45 | 6 mo. |
| Lories/Lorikeets | Early spring/spring | 2–4/2 | 21 | 42–50 | 62–70 | 50–60 | 2 yr. |
| Conures | Spring/late spring/summer | 2–8/2–6 | 21–23 | 35–40 | 45–70 | 60 | 2–3 yr. |
| Amazon Parrots | Spring/late spring/summer | 2–4/2 | 23–24 | 45–60 | 90–120 | 75–90 | 4–6 yr. |
| Small-sized macaws | Spring/late spring/summer | 2–4/2 | 23–24 | 45–60 | 90–120 | 75–90 | 4–6 yr. |
| Large-sized macaws | Early spring/spring/late spring/summer | 2–4/2–3 | 26–28 | 70–80 | 120–150 | 95–120 | 5–7 yr. |
| African Grey | Spring/late spring/summer | 2–4/2–4 | 24–26 | 50–65 | 100–120 | 75–90 | 4–6 yr. |
| Medium-sized cockatoos | Early spring/spring/late spring/summer | 2–4/2 | 23–25 | 45–60 | 90–120 | 75–100 | 3–4 yr. |
| Galah Cockatoo | Early spring/spring/late spring/summer | 2–4/2–3 | 24 | 45–55 | 90–120 | 80–90 | 1 yr. |
| Large-sized cockatoos | Early spring/spring/late spring/summer | 2–4/2 | 24–26 | 60–80 | 120–150 | 95–120 | 5–6 yr. |
| Eclectus Parrot | All year | 1–2/1–2 | 26 | 72–80 | 120–150 | 100–110 | 4 yr. |

Note: These data are provided as an approximate guide only. Some species may vary from these values, depending on individual characteristics and environmental conditions. In general, larger birds have older fledging, weaning and breeding ages, and longer egg incubation periods. For more information on individual species, see Parrots: Their Care and Breeding, by R. Low.

*Time when the young leave the nest.

†Time when the young no longer require parental care in captivity.

‡NA—not applicable; rarely hand-reared.

§Age of sexual maturity—corresponds to the best age to encourage breeding except for budgerigars, cockatiels, and lovebirds. These birds mature earlier, but should not be set up for breeding until 9 months of age (budgerigars), or 12 months (cockatiels and lovebirds).

*Appendix 6.*

# SAMPLE WEIGHT GAINS OF SELECTED HAND-RAISED PSITTACINES

Keven Flammer

These data were derived from the weight gain records of morphologically normal birds hand raised from hatching to weaning. These weights are less than the gains of parent-raised birds, and there is no doubt that these weight gains will be increased by future improvements in hand-feeding diets and techniques. These data represent what can be expected with today's techniques and should therefore be used for comparison—as growth rates that are suboptimal but will produce morphologically normal birds at weaning.

SAMPLE WEIGHT GAINS OF SELECTED HAND-RAISED AMAZON PARROTS

| | Body Weight (gm) | | | | |
|---|---|---|---|---|---|
| Age (days) | Double Yellow-headed Amazon | Yellow-naped Amazon | Mexican Red-headed Amazon | Panama Yellow-crowned Amazon | Blue-fronted Amazon |
| 0 | 14–17 | 14–17 | 11–16 | 14–15 | 14–15 |
| 1 | 16–19 | 15–22 | 14–18 | 20–22 | 15–18 |
| 2 | 17–21 | 18–25 | 16–21 | 22–26 | 18–22 |
| 3 | 23–25 | 20–30 | 20–24 | 27–32 | 22–26 |
| 4 | 25–35 | 30–40 | 22–28 | 35–40 | 25–35 |
| 5 | 30–45 | 35–50 | 25–35 | 40–45 | 35–45 |
| 6 | 35–55 | 45–60 | 30–40 | 50–55 | 45–55 |
| 7 | 45–65 | 60–70 | 40–45 | 55–60 | 55–65 |
| 8 | 55–70 | 70–85 | 60–65 | 60–65 | 65–75 |
| 9 | 60–80 | 90–100 | 65–70 | 70–80 | 75–80 |
| 10 | 85–100 | 100–115 | 70–75 | 80–95 | 80–95 |
| 11 | 100–115 | 115–125 | 80–95 | 105–115 | 95–105 |
| 12 | 125–135 | 130–155 | 100–110 | 115–125 | 105–115 |
| 13 | 135–145 | 160–175 | 110–125 | 125–135 | 120–130 |
| 14 | 140–160 | 170–180 | 120–135 | 130–155 | 140–155 |
| 15 | 160–195 | 190–210 | 140–165 | 155–175 | 150–170 |
| 16 | 200–220 | 215–230 | 155–185 | 170–180 | 160–180 |
| 17 | 230–255 | 240–255 | 190–200 | 190–200 | 180–200 |
| 18 | 235–275 | 260–280 | 195–210 | 205–220 | 200–210 |
| 19 | 250–290 | 270–300 | 200–220 | 220–235 | 210–225 |
| 20 | 260–300 | 300–320 | 210–230 | 230–255 | 225–245 |
| 21 | 275–315 | 310–330 | 215–245 | 245–265 | 245–260 |
| 22 | 285–330 | 330–350 | 220–250 | 270–280 | 260–280 |
| 23 | 295–370 | 350–390 | 230–255 | 285–295 | 265–290 |
| 24 | 330–385 | 370–410 | 240–265 | 300–315 | 280–310 |
| 25 | 360–405 | 390–430 | 270–280 | 315–330 | 310–325 |
| 26 | 380–430 | 410–470 | 290–300 | 320–335 | 325–335 |
| 27 | 400–440 | 440–490 | 305–310 | 335–350 | 335–355 |
| 28 | 410–460 | 450–500 | 310–320 | 345–360 | 350–365 |
| 29 | 425–475 | 470–520 | 320–330 | 355–380 | 355–370 |
| 30 | 440–480 | 490–540 | 330–345 | 365–395 | 360–380 |
| 35 | 450–520 | 500–550 | 350–360 | 385–410 | 370–390 |
| 40 | 470–550 | 540–600 | 355–380 | 420–450 | 380–420 |
| 45 | 500–570 | 570–625 | 340–360 | 450–480 | 400–440 |
| 50 | 510–550 | 580–600 | 330–350 | 460–470 | 400–435 |
| 55 | 490–520 | * | 325–340 | 450–460 | 400–420 |
| 60 | 480–510 | * | 310–330 | 440–455 | 390–410 |
| 65 | 460–500 | * | * | 420–445 | 370–390 |

*Data not available.

SAMPLE WEIGHT GAINS OF SELECTED HAND-RAISED COCKATOOS AND PARROTS

| Age (days) | Body Weight (gm) | | | | |
|---|---|---|---|---|---|
| | Umbrella Cockatoo | Triton Cockatoo | Rose-breasted Cockatoo | African Grey Parrot | Grand Eclectus Parrot |
| 0 | 15–16 | 17–20 | 9–11 | 13–16 | 14–16 |
| 1 | 17–18 | 22–25 | 10–15 | 14–17 | 15–17 |
| 2 | 20–25 | 25–29 | 15–20 | 15–18 | 16–18 |
| 3 | 25–30 | 30–35 | 20–25 | 18–20 | 20–24 |
| 4 | 30–35 | 35–40 | 25–30 | 21–25 | 25–30 |
| 5 | 35–40 | 40–45 | 30–35 | 25–35 | 35–40 |
| 6 | 45–50 | 50–55 | 35–40 | 30–35 | 45–55 |
| 7 | 55–60 | 55–65 | 40–45 | 35–40 | 55–60 |
| 8 | 65–70 | 65–70 | 45–50 | 45–50 | 60–65 |
| 9 | 75–80 | 80–85 | 50–55 | 50–55 | 65–70 |
| 10 | 85–90 | 90–105 | 55–60 | 60–65 | 70–75 |
| 11 | 100–110 | 100–115 | 65–75 | 65–70 | 75–95 |
| 12 | 120–125 | 120–135 | 80–90 | 80–90 | 80–110 |
| 13 | 125–130 | 150–160 | 95–100 | 85–95 | 90–120 |
| 14 | 140–155 | 165–175 | 100–115 | 115–120 | 100–135 |
| 15 | 170–180 | 190–210 | 115–125 | 120–135 | 120–150 |
| 16 | 185–195 | 220–230 | 130–140 | 145–155 | 125–165 |
| 17 | 210–220 | 250–265 | 150–155 | 155–165 | 135–180 |
| 18 | 215–230 | 275–290 | 165–170 | 165–185 | 150–195 |
| 19 | 225–240 | 280–310 | 175–190 | 170–200 | 175–205 |
| 20 | 240–265 | 300–340 | 190–205 | 180–210 | 180–215 |
| 21 | 270–280 | 320–350 | 200–225 | 210–230 | 200–230 |
| 22 | 290–300 | 360–370 | 210–240 | 225–245 | 215–250 |
| 23 | 300–320 | 375–395 | 225–260 | 240–260 | 225–265 |
| 24 | 325–340 | 390–420 | 240–260 | 250–270 | 235–280 |
| 25 | 340–360 | 410–440 | 250–275 | 280–295 | 250–295 |
| 26 | 360–380 | 430–470 | 280–295 | 285–300 | 265–295 |
| 27 | 375–400 | 460–495 | 285–305 | 295–315 | 280–310 |
| 28 | 390–440 | 475–510 | 305–320 | 305–325 | 300–320 |
| 29 | 410–460 | 490–535 | 310–345 | 320–340 | 315–335 |
| 30 | 440–500 | 520–560 | 320–355 | 320–360 | 320–350 |
| 35 | 510–600 | 560–625 | 350–375 | 340–380 | 340–380 |
| 40 | 540–630 | 615–660 | 355–385 | 380–410 | 390–420 |
| 45 | 590–680 | 665–700 | 365–400 | 430–480 | 410–450 |
| 50 | 610–730 | 710–760 | 375–410 | 450–500 | 420–440 |
| 55 | 670–750 | 750–780 | 360–390 | 460–510 | 430–410 |
| 60 | 700–780 | 780–840 | 360–375 | 480–520 | 390–400 |
| 65 | 720–760 | 810–880 | 350–360 | 460–510 | 380–395 |

## SAMPLE WEIGHT GAINS OF SELECTED HAND-RAISED MACAWS

| Age (days) | Body Weight (gm) | | |
|---|---|---|---|
| | Blue and Gold Macaw | Scarlet Macaw | Green-winged Macaw |
| 0 | 19–25 | 18–20 | 20–21 |
| 1 | 20–30 | 20–25 | 25–30 |
| 2 | 30–40 | 25–35 | 35–45 |
| 3 | 35–45 | 30–40 | 40–55 |
| 4 | 40–50 | 35–45 | 50–60 |
| 5 | 45–60 | 50–70 | 55–70 |
| 6 | 55–75 | 65–80 | 65–90 |
| 7 | 65–90 | 85–100 | 80–110 |
| 8 | 90–110 | 100–140 | 110–140 |
| 9 | 100–145 | 115–160 | 130–170 |
| 10 | 120–160 | 135–175 | 135–180 |
| 11 | 135–175 | 170–200 | 165–200 |
| 12 | 185–210 | 200–220 | 200–225 |
| 13 | 200–230 | 225–250 | 220–245 |
| 14 | 210–250 | 250–275 | 230–275 |
| 15 | 240–280 | 280–300 | 280–300 |
| 16 | 265–300 | 325–345 | 300–350 |
| 17 | 280–330 | 360–375 | 345–380 |
| 19 | 310–360 | 370–390 | 385–410 |
| 20 | 365–400 | 380–420 | 410–440 |
| 21 | 400–430 | 410–450 | 440–475 |
| 22 | 415–455 | 460–510 | 500–530 |
| 23 | 465–500 | 485–540 | 550–570 |
| 24 | 495–530 | 500–560 | 560–585 |
| 25 | 510–550 | 540–590 | 600–630 |
| 26 | 540–580 | 580–625 | 640–660 |
| 27 | 560–600 | 600–640 | 670–690 |
| 28 | 590–615 | 620–650 | 685–710 |
| 29 | 620–650 | 640–670 | 720–750 |
| 30 | 645–685 | 680–700 | 760–780 |
| 35 | 750–800 | 800–880 | 840–920 |
| 40 | 820–930 | 900–1000 | 990–1100 |
| 45 | 950–1100 | 980–1080 | 1080–1200 |
| 50 | 1000–1150 | 1000–1150 | 1200–1300 |
| 55 | 1060–1150 | 1010–1100 | * |
| 60 | 950–1050 | 1020–1050 | * |
| 65 | 850–1000 | 950–1050 | * |

*Data not available.

# *Appendix 7.*

# PLANTS SUITABLE FOR USE IN AVIARIES

Developed for South Florida Aviaries by George Staples

## SUGGESTED LANDSCAPE PLANTS

| Common Name | Scientific Name | Comments |
|---|---|---|
| Acacia | *Acacia* sp. | Prickly shrub, nesting sites |
| Areca palm | *Chrysalidocarpus lutescens* | Young plants good foliage, background |
| Bouganvillea | *Bouganvillea* sp. | Prickly plants for shelter, nesting; colorful blooms in full sun |
| Ceriman, monstera | *Monstera deliciosa* | Large split-leaf philodendron |
| Cycads | *Cycas* sp., *Zamia* sp. | Foliage, ground cover |
| Fan palm | *Livistona chinensis* | Foliage, shade tolerant, small size |
| Fig, creeping | *Ficus pumila* | Covers walls, stone, wood, as background |
| Fig, fiddle-leaf | *Ficus lyrata* | Foliage plant, small size for a fig |
| Fig, laurel-leaf | *Ficus microcarpa* | Dense foliage, requires trimming |
| Fig, weeping | *Ficus benjamina* | Trailing foliage, requires pruning |
| Firethorn | *Pyracantha* sp. | Dense foliage, edible berries, nest sites; likes full sun, requires trimming |
| Lady palm | *Rhapis excelsa* | Forms clumps, small size, shade tolerant |
| Natal plum | *Carissa* sp. | Forms natural hedge, **thorns**, edible fruit |
| Philodendron, self-heading | *Philodendron selloum* | Tropical foliage, likes sun |
| Philodendron, split-leaf | *Philodendron pertusum* | Tropical foliage, shade |
| Pittosporum | *Pittosporum tobira* | Evergreen shrub, likes sun, dense growth |
| Shrubby yew | *Podocarpus* sp. | Evergreen, trim for compact growth |
| Silk oak | *Grevillea robusta* | Tree, requires pruning to maintain size |
| Strawberry guava | *Psidium cattleianum* | Evergreen shrub, edible fruit, sun |
| Umbrella tree, schefflera | *Brassaia actinophylla* | Evergreen, requires pruning, edible fruit |

## TREES PROVIDING SUITABLE PERCHES

| Common Name | Scientific Name | Comments |
|---|---|---|
| Australian pine | *Casuarina* sp. | Brittle when green, dries like iron |
| Guava | *Psidium guava* | |
| Florida holly, Brazilian pepper | *Schinus terebinthifolius* | Berries edible; beware of herbicides |
| Hibiscus tree, mahoe | *Hibiscus tiliaceus* | Flowers and seeds edible |
| Melaleuca, paper bark tree | *Melaleuca quinquenervia* | Soft wood, spongy bark, edible flowers |
| Oak | *Quercus* sp. | Hard wood |
| Seagrape | *Coccoloba uvifera* | Wood not readily chewed, edible fruit |

## FLOWERING PLANTS THAT PRODUCE NECTAR OR ATTRACT INSECTS

| Common Name | Scientific Name | Comments |
|---|---|---|
| Flame vine | *Pyrostegia ignea* | Abundant nectar, large vine |
| Geiger tree | *Cordia sebestena* | Small, slow growing tree, sun |
| Ixora | *Ixora coccinea* | Abundant nectar, berries, good hedge |
| Lantana | *Lantana* sp. | Nectar, insects drawn to plant, edible berries |
| Orange jessamine | *Murraya paniculata* | Flowers attract insects, berries edible; good hedge or specimen plant |
| Orchid tree | *Bauhinia* sp. | Deciduous, nectar, attracts hummingbirds |
| Melaleuca | *Melaleuca quinquenervia* | Edible blossoms, insects drawn to tree |
| Hibiscus tree, mahoe | *Hibiscus tiliaceus* | Edible flowers and seeds, insects |
| Bromeliads | *Aechmea* sp., *Guzmania* sp., *Neoregelia* sp. | "Tank" in center of plant attracts insects |

SOUTH FLORIDA PLANTS USEFUL FOR EDIBLE PRODUCTS

| Common Name | Scientific Name | Comments |
|---|---|---|
| Asparagus fern, "sprengeri" | *Asparagus densiflorus* | Red berries, also foliage eaten by birds |
| Barbados cherry, acerola | *Malpighia glabra* | Red berries, high in vitamin C |
| Singapore holly | *Malpighia coccigera* | As above |
| Elderberry | *Sambucus simpsoni* | Edible berries, wild, attracts insects |
| Firethorn | *Pyracantha coccinea* | Edible waxy berries |
| Florida holly | *Schinus terebinthifolius* | Wild plant, edible berries and twigs |
| Mulberry | *Morus nigra, M. rubra* | Excellent berries in spring |
| Orange jessamine | *Murraya paniculata* | Berries, insects, bears fruit several times each year |
| Strawberry tree | *Muntingia calabura* | Excellent berries year round, insects |
| Surinam cherry | *Eugenia uniflora* | Fruit, good hedge plant |
| West Indies almond | *Terminalia catappa* | Corky fruit with edible kernel, drift seeds from beach make good parrot toy |

# *Appendix 8.*

# SELECTED SOURCES OF AVIAN LITERATURE

## MEDICAL INFORMATION

**Association of Avian Veterinarians (AAV):** Publishes quarterly A.A.V. Newsletter and Proceedings of Annual Meetings. For membership information, contact Association of Avian Veterinarians, P.O. Box 299, East Northport, NY 11731.

**American Association of Avian Pathologists (AAAP):** Publishes quarterly journal, Avian Diseases, regional meeting proceedings and other publications. For membership information, contact Robert J. Eckroade, University of Pennsylvania, New Bolton Center, Kennett Square, PA 19348-1692.

Arnall, L., and Keymer, I. F.: Bird Diseases. Neptune, NJ, TFD Publications, 1975.

Baumel, J. J., et al. (eds.): Nomina Anatomica Avium. New York, Academic Press, 1979.

Cooper, J. E., and Eley, J. T. (eds.): First Aid and Care of Wild Birds. Newtown Abbot, England, David & Charles, 1979.

Cooper, J. E., and Greenwood, A. G. (eds.): Recent Advances in the Study of Raptor Diseases. Proceedings of the International Symposium on Diseases of Birds of Prey, London, 1980.

Cottral, G. E. (ed.): Manual of Standardized Methods for Veterinary Microbiology. Ithaca, NY, Comstock Publishing Associates, 1978.

Dein, F. J.: Avian Hematology Handbook. Association of Avian Veterinarians, 1985.

Fowler, M. E. (ed.): Zoo and Wild Animal Medicine. 2nd ed. Philadelphia, W. B. Saunders Company, 1986.

Harrison, G. J. (Guest ed.): Caged Bird Medicine. Vet. Clin. North Am. [Small Anim. Pract.], Vol. 14, No. 2, 1984.

Hoff, G. L., and Davis, J. S: Noninfectious Diseases of Wildlife. Ames, IA, Iowa State University Press, 1982.

Hofstad, M. S., et al. (eds.): Diseases of Poultry. 8th ed. Ames, IA, Iowa State University Press, 1984

Hungerford, T. G. (ed.): Proceedings No. 55, Refresher Course on Aviary and Caged Birds. University of Sydney, Australia, 1981.

King, A. S., and McLelland, J.: Birds: Their Structure and Function. London, Bailliere Tindall (distributed by W. B. Saunders Company), 1984.

Kirk, R. W. (ed.): Current Veterinary Therapy VI, VII, VIII. Philadelphia, W. B. Saunders Company, 1977, 1980, 1983.

Perry, R. A.: Vade Mecum No. 2: Diseases of Birds. Sydney, Australia, Post Graduate Committee in Veterinary Science, 1983.

Petrak, M. L. (ed.): Diseases of Cage and Aviary Birds. 2nd ed. Philadelphia, Lea & Febiger, 1982.

Skadhauge, E.: Osmoregulation in Birds. Berlin, Springer-Verlag, 1981.

Steiner, C. V., and Davis, R. B.: Caged Bird Medicine, Selected Topics.Ames, IA, Iowa State University Press, 1981.

Stunkard, J. A.: A Guide to Diagnosis, Treatment and Husbandry of Caged Birds. Rev. ed. Edgewater, MD, Stunkard Publishing Company, 1984.

Whiteman, C. E., and Bickford, A. A. (eds.): Avian Disease Manual. 2nd ed. American Association of Avian Pathologists, 1983.

## CLIENT EDUCATION

Axelson, R. D.: Caring for Your Pet Bird. Willowdale, Ontario, Canada, Canaviax Publications, 1981.

Gallerstein, G. A.: Bird Owner's Home Health and Care Handbook. New York, Howell Book House, 1984.

Gerstenfeld, S.: The Bird Care Book. Reading, MA, Addison-Wesley Publishing Co., 1981.

Lafeber, T. J.: Tender Loving Care. Niles, IL, Lafeber Products, 1977.

## AVICULTURE/ORNITHOLOGY

Forshaw, J. M., and Cooper, W. T.: Parrots of the World. New York, Doubleday & Co., Inc., 1973.

LaRosa, D.: How to Build Everything You Need for Your Birds. Simi, CA, LaRosa Publications, 1973.

Low, R.: Parrots, Their Care and Breeding. Poole, England, Blandford Books Ltd., 1980.

Proceedings of the Jean Delacour/IFCB Symposium on Breeding Birds in Captivity. North Hollywood, CA, International Foundation for the Conservation of Birds, 1983.

Risser, A., et al. (eds.): Proceedings First International Birds in Captivity Symposium. Seattle, WA, 1978.

American Ornithologists' Union: Publishes quarterly journal of ornithology, The Auk. National Museum of Natural History, Smithsonian Institution, Washington, DC 20560

## PERIODICALS FOR THE BIRD-OWNING PUBLIC

AFA Watchbird, American Federation of Aviculture, P.O. Box 1568, Redondo Beach, CA 90278.

American Cockatiel Society, 9812 Bois D'Arc Ct., Ft. Worth, TX 76126.

Bird World Magazine, Box 70, No. Hollywood, CA 91603.

# *Index*

Page numbers in *italics* refer to illustrations; a "t" following a page number indicates tabular material.